Biology of the red algae

Biology of
the red algae

Edited by

Kathleen M. Cole
Robert G. Sheath

CAMBRIDGE
UNIVERSITY PRESS

CAMBRIDGE UNIVERSITY PRESS
Cambridge, New York, Melbourne, Madrid, Cape Town, Singapore,
São Paulo, Delhi, Dubai, Tokyo, Mexico City

Cambridge University Press
The Edinburgh Building, Cambridge CB2 8RU, UK

Published in the United States of America by Cambridge University Press, New York

www.cambridge.org
Information on this title: www.cambridge.org/9780521202466

© Cambridge University Press 1990

This publication is in copyright. Subject to statutory exception
and to the provisions of relevant collective licensing agreements,
no reproduction of any part may take place without the written
permission of Cambridge University Press.

First published 1990
Reprinted 1995
First paperback edition 2010

A catalogue record for this publication is available from the British Library

ISBN 978-0-521-20246-6 Paperback

Cambridge University Press has no responsibility for the persistence or
accuracy of URLs for external or third-party internet websites referred to in
this publication, and does not guarantee that any content on such websites is,
or will remain, accurate or appropriate.

Contents

List of contributors		vii
Preface		ix
CHAPTER 1	An introduction *Wm. J. Woelkerling*	1
CHAPTER 2	Cell structure *Curt M. Pueschel*	7
CHAPTER 3	DNA: Microspectrofluorometric studies *Lynda J. Goff* and *Annette W. Coleman*	43
CHAPTER 4	Chromosomes *Kathleen M. Cole*	73
CHAPTER 5	Genetics *John P. van der Meer*	103
CHAPTER 6	Cell division *Joe Scott* and *Sharon Broadwater*	123
CHAPTER 7	Solute accumulation and osmotic adjustment *Robert H. Reed*	147
CHAPTER 8	Carbon metabolism *John A. Raven, Andrew M. Johnston* and *Jeffrey J. MacFarlane*	171
CHAPTER 9	Pigmentation and photoacclimation *Elisabeth Gantt*	203
CHAPTER 10	Cell walls *James S. Craigie*	221
CHAPTER 11	Development *Susan D. Waaland*	259
CHAPTER 12	Vegetative growth and organization *Roy J. Coomans* and *Max H. Hommersand*	275

CHAPTER 13	**Sexual reproduction and cystocarp development** *Max H. Hommersand* and *Suzanne Fredericq*	305
CHAPTER 14	**Sporangia and spores** *Michael D. Guiry*	347
CHAPTER 15	**Marine ecology** *Joanna M. Kain* and *Trevor A. Norton*	377
CHAPTER 16	**Freshwater ecology** *Robert G. Sheath* and *Julie A. Hambrook*	423
CHAPTER 17	**Reproductive strategies** *Michael W. Hawkes*	455
CHAPTER 18	**Taxonomy and evolution** *David J. Garbary* and *Paul W. Gabrielson*	477
	Taxonomic Index	499
	Subject Index	508

Contributors

Sharon Broadwater
Department of Biology, College of William and Mary, Williamsburg, Virginia 23185

Kathleen M. Cole
Department of Botany, University of British Columbia, Vancouver, British Columbia V6T 2B1, Canada

Annette W. Coleman
Biology Department, Brown University, Providence, Rhode Island 02912

Roy J. Coomans
Biology Department, North Carolina Agricultural and Technical State University, Greensboro, North Carolina 27411

James S. Craigie
Atlantic Research Laboratory, 1411 Oxford Street, Halifax, Nova Scotia B3H 3Z1, Canada

Suzanne Fredericq
Department of Botany, National Museum of Natural History, Smithsonian Institution, Washington, D.C. 20560

Paul W. Gabrielson
Department of Botany, University of British Columbia, Vancouver, British Columbia V6T 2B1, Canada

Elisabeth Gantt
Department of Botany, University of Maryland, College Park, Maryland 20742

David J. Garbary
Department of Biology, Saint Francis Xavier University, Antigonish, Nova Scotia B2G ICO, Canada

Lynda J. Goff
Center for Marine Studies, University of California, Santa Cruz, California 95064

Michael D. Guiry
Department of Botany, University College, Galway, Ireland

Julie A. Hambrook
Byrd Polar Research Center, Ohio State University, Columbus, Ohio 43212

Michael W. Hawkes
Department of Botany, University of British Columbia, Vancouver, British Columbia V6T 2B1, Canada

Max H. Hommersand
Department of Biology, University of North Carolina, Chapel Hill, North Carolina 27514

Andrew M. Johnston
Department of Biological Sciences, University of Dundee, Dundee DD1 4HN, United Kingdom

Joanna M. Kain
Port Erin Marine Laboratory, University of Liverpool, Port Erin, Isle of Man, United Kingdom

Jeffrey J. MacFarlane
Department of Botany, University of Adelaide, South Australia 5001, Australia

Trevor A. Norton
Port Erin Marine Laboratory, University of Liverpool, Port Erin, Isle of Man, United Kingdom

Curt M. Pueschel
Department of Biology, State University of New York, Binghamton, New York 13901

John A. Raven
Department of Biological Sciences, University of Dundee, Dundee DD1 4HN, United Kingdom

Robert H. Reed
Department of Chemical and Life Sciences, Newcastle Upon Tyne Polytechnic, Newcastle Upon Tyne, NE1 8ST, United Kingdom

Joe Scott
Department of Biology, College of William and Mary, Williamsburg, Virginia 23185

Robert G. Sheath
Department of Botany, University of Rhode Island, Kingston, Rhode Island 02881

John P. van der Meer
Atlantic Research Laboratory, 1411 Oxford Street, Halifax, Nova Scotia B3H 3Z1, Canada

Susan D. Waaland
Department of Biology, University of Puget Sound, Tacoma, Washington 98416

Wm. J. Woelkerling
Department of Botany, LaTrobe University, Bundoora, Victoria 3083, Australia

Preface

Since the publication of "Biology of the Rhodophyta" by Peter Dixon in 1973, red algal research has greatly advanced and diversified. Recent findings in many areas have been accomplished using new technologies which have provided insights beyond the realm of possibility seventeen years ago. This book is designed to summarize current knowledge of the field and provide a much needed reference for researchers and teachers concerned with phycology, aquatic biology, limnology, oceanography and plant evolution.

The presentation is by levels of organization, starting with cells and constituents, followed by organisms and their growth and reproduction, ecology, systematics and evolution. This theme complements that of recently published books on other algal divisions. It was not possible to cover every aspect of red algal biology, and this book does not contain applied topics. The scope and length of coverage among the chapters varies so that both traditional and newly emerging areas could be included.

We are extremely grateful to the contributors of this volume who took much time and effort away from their busy schedules to provide chapters and follow through with our editorial changes. We extend our thanks to the reviewers whose helpful suggestions enhanced the quality of the final presentation. The reviewers were as follows: Y.M. Chamberlain, A.R.O. Chapman, James Craigie, Robert DeWreede, Mathew Dring, Leonard Evans, Paul Gabrielson, David Garbary, Sarah Gibbs, Lynda Goff, Michael Gretz, Michael Guiry, Louis Hanic, David Hart, Michael Hawkes, Johan Hellebust, Max Hommersand, Donald Kapraun, Gerald Kraft, Shigeru Kumano, Arthur Mathieson, Akio Miura, Elizabeth Percival, Curt Pueschel, Joseph Ramus, John Raven, J. Rueness, Joe Scott, Richard Searles, John van der Meer, Maret Vesk, John West, Brian Whitton, Wm. J. Woelkerling and Hiroshi Yabu. We extend our thanks to the many colleagues who generously permitted reproduction of published and unpublished material. Help in manuscript preparation from Marion Jenkins, Kathy St. Germain and Livia Tyler is greatly appreciated. This project was partially supported by N.S.E.R.C. grant 0645 to K.M.C; N.S.F. grant BSR 860792 to R.G.S.; and the Departments of Botany at the University of British Columbia and University of Rhode Island. Lastly, we wish to acknowledge professional inspiration from Elsie Conway and Johan Hellebust and personal support from Phyllis Taylor and Mary Koske.

Chapter 1

An introduction

WM. J. WOELKERLING

CONTENTS

I. Biological significance and characteristics / 1
II. Phylogeny and evolutionary relationships / 2
III. Species numbers and classification / 2
IV. Utilization and future research / 4
V. References / 5

I. BIOLOGICAL SIGNIFICANCE AND CHARACTERISTICS

The biological significance of red algae is only beginning to be appreciated. Even among professional biologists, knowledge of these organisms is often minimal and based on the cursory information contained in general botany textbooks. Nevertheless, there is a growing body of scientific information about red algae that has revealed, among other things, that they predominate over extensive areas of the continental shelves (e.g., see Adey 1971; Lüning 1985; Round 1981; Steneck 1986), that coral reefs could not survive without the cementing action of certain calcified red algae (e.g., see Littler & Littler 1984), that various species of red algae provide a base for an agronomic industry worth hundreds of millions of dollars ($US) annually (Doty 1979; Tseng 1981), and that some calcareous forms show promise for use in human bone implants (Dechau 1987; Kasperk & Ewers 1986). The red algae are indeed a unique, intriguing, and intrinsically interesting group of organisms about which much remains to be learned.

Red algae are presumed to share the combination of characteristics listed in Table 1-1. Although this presumption has consistently been supported by newly provided evidence, it has yet to be fully verified in a substantial majority of included species.

In over 90% of red algal species, chlorophyll *a* and several carotenoids occur along with phycoerythrin and phycocyanin (Ragan 1981); varying proportions of these pigments can result in red, pink, violet, blue, brown, black, yellowish, or greenish thalli. Most marine species, however, normally are red or pink, whereas most freshwater species are bluish to black. In plants of a few genera, pigmentation and chloroplasts are scarce or apparently absent. Plants in such genera often look whitish or somewhat creamy, are small in size and grow only on certain red algal hosts, and are considered to be partly or entirely parasitic (see Goff 1982). Despite the absence of pigments and chloroplasts, plants in these genera possess all other characteristics of red algae, and their sexual cycles and reproductive structures are also distinctly red algal.

The particular combination of six characteristics listed in Table 1-1 does not occur in any other group of organisms. Consequently, red algae are considered to constitute either a separate division in the Kingdom Plantae [variously named the Rhodophyta (e.g., Bold & Wynne 1985; Dixon 1973; Kraft & Woelkerling 1990; Kylin 1956), the Rhodophycota (e.g., Parker 1982; South & Whittick 1987), or the Rhodophycophyta (e.g., Bold & Wynne 1978; Papenfuss 1955)] or a separate phylum in the Kingdom Protoctista [named the Rhodophyta (e.g., Margulis & Schwartz 1982)].

Table 1-1. Attributes presumed to occur in species of Rhodophyta

Eukaryotic cells
Flagella absent
Food reserves stored principally as floridean starch (an α1–4, α1–6 linked glucan resembling amylopectin); storage in cytoplasm, not in chloroplasts
Phycoerythrin, phycocyanin, and allophycocyanins present as accessory pigments
Chloroplasts with nonaggregated (unstacked) thylakoids
Chloroplasts lacking external endoplasmic reticulum

II. PHYLOGENY AND EVOLUTIONARY RELATIONSHIPS

Current debate over whether red algae should be treated as plants or as protists does not affect the common view that they are a very distinct group of organisms whose relationships to other groups are by no means certain, and Hori and Osawa (1987, pp. 456–7) state that "the emergence of red algae is the most ancient event so far detected in the evolution of eukaryotes...." As noted by Gabrielson et al. (1985, p. 343), possible evolutionary links between red algae and cyanobacteria (blue-green algae) have been suggested because both groups have phycoerythrin and phycocyanin as accessory pigments, both groups lack flagella, and both have nonaggregated thylakoids. Cyanobacteria, however, are prokaryotic (and thus lack chloroplasts), and they store food reserves as cyanophycean starch (an α1–4, α1–6 linked glucan resembling glycogen and floridean starch). Evolutionary links to other groups of eukaryotic algae (see Bold & Wynne 1985, Table 1-1; South & Whittick 1987, Table 2-2) such as the Cryptophyceae and the Chlorophyceae (see Gabrielson et al. 1985) are also tenuous; none contains floridean starch (but amylopectin similar to floridean starch) and all contain aggregated thylakoids. Moreover, flagella are present in most or all species of other divisions/classes of algae, and phycoerythrin and phycocyanin do not occur except in some species of Cryptophyta (cryptomonads). Obviously, a clearer picture of the origins and evolutionary relationships of the Rhodophyta cannot emerge until new, more definitive evidence (perhaps of a biochemical or molecular nature) becomes available. As Gabrielson et al. (1985, p. 344) state, the ancestry of the Rhodophyta remains an enigma.

Clues on red algal phylogeny and evolution are more likely to emerge from new studies of living species (e.g., see Hori & Osawa 1987) than from the fossil record. Tappan (1980, p. 138) has suggested that red algae probably have existed for 2 billion years, while Hori and Osawa (1987, p. 456) have estimated 1.3 to 1.4 billion years. Unfortunately, the fossil record for red algae is extremely uneven, and only a small minority of taxa are represented (see Tappan 1980, Table 2.1). The most commonly recorded red algal fossils are of taxa that have been placed in the Corallinaceae. These include over 60 exclusively fossil genera, as well as at least 13 living genera to which fossil species have been assigned (Johansen 1981; Woelkerling 1988; Wray 1977). In an analysis of the 57 exclusively fossil nongeniculate genera, however, Woelkerling (1988) concluded that virtually all were of doubtful taxonomic status. Many are based on specimens in which characteristics essential for unequivocal family or genus placement have not been preserved, and thus a full phylogenetic interpretation is not possible. Similarly, most fossil specimens assigned to genera with living species do not show the diagnostic features of those genera, and the status of most of these species is therefore also in doubt. Most other groups of fossil red algae are attended by parallel problems, thus making detailed evolutionary interpretations difficult.

III. SPECIES NUMBERS AND CLASSIFICATION

There are over 10,000 described species of red algae. How many real species there are, however, remains an unanswered question. Estimates vary by over 100%; these include 2,500 (Dring 1982), 3,700 (Peterfi & Ionescu 1977), 3,900 (Bold et al. 1987; Dixon 1973), 4,000 (Kraft & Woelkerling 1990), 3,500 to 4,500 (van den Hoek & Jahns 1978), 5,100 (Tappan 1980), 5,200 (Dixon 1973), and 6,000 (Pritchard & Bradt 1984); additional estimates are provided by Altman and Dittmer (1972). Although partly based on tabulations of names, such estimates are largely subjective and thus must be considered notional. It is extremely difficult to make accurate estimates of species numbers when reliable species concepts scarcely exist within many genera (e.g., see Garbary 1987; Woelkerling 1983, 1988), when the status of substantial numbers of inadequately described species remains unresolved (e.g., see Woelkerling 1984, pp. 7–18; Womersley 1979, p. 459), when there are virtually no modern, worldwide monographs of genera containing more than a few species, and when many parts of the world (e.g., northern Australia, much of Central and South America, much of Africa) remain largely unexplored phycologically.

In the absence of modern monographic studies,

many checklists, catalogues, and floristic compilations assume an unwarranted mantle of authority in spite of any cautions offered by authors, and it becomes deceptively easy to presume that there are accurate, trustworthy accounts of the red algal floras for many parts of the world, when the opposite is more nearly correct. Such checklists, catalogues, and floristic compilations (e.g., see references in South & Tittley 1986 and in Woelkerling 1988) should be considered more as starting points than as end points of research into species concepts, species delimitation, and the production of broad-based monographic accounts. Eventually, more accurate estimates of species numbers may emerge from these types of research.

One value of generating notional estimates of species numbers, in spite of all their current limitations, is to gain some insight into both the apparent diversity of red algae relative to other algal groups and the relative proportion of red algae occurring in different geographic areas (Table 1-2). Dring (1982, p. 3), for example, has calculated that 27.1% of all known species of marine plants are red algae. His calculations also show that in benthic marine en-

Table 1-2. Selected estimates of numbers of species of marine Rhodophyta, Phaeophyta, and Chlorophyta occurring in particular regions or on a world scale

Region	Rhodophyta	Phaeophyta	Chlorophyta	Reference
Worldwide	2,540	997	900	Dring 1982
North Atlantic Ocean	589	324	258	South 1987
Southern Australia	800	231	123	Womersley 1981, 1984, 1987

Table 1-3. Orders of Rhodophyta recognized by various authors since 1980. Circumscriptions of particular orders vary from account to account.

Order	Lee 1980	Dawes 1981	Dixon 1982	Pritchard & Bradt 1984	Bold & Wynne 1985	Gabrielson et al. 1985	Gabrielson & Garbary 1986	Gabrielson & Garbary 1987	South & Whittick 1987	Kraft & Woelkerling 1990	Chapters 2, 13 & 18 in this book
Acrochaetiales	*	N	N	N	N	*	*	*	N	*	*
Ahnfeltiales	N	N	N	N	N	N	N	N	N	N	*
Bangiales	*	*	*	*	*	*	*	*	*	*	*
Batrachospermales	N	N	N	N	*	*	*	*	N	*	*
Bonnemaisoniales	N	N	N	N	*	*	*	*	N	*	*
Ceramiales	*	*	*	*	*	*	*	*	*	*	*
Compsopogonales	N	N	*	*	*	*	N	*	*	*	*
Corallinales	N	N	N	N	*	*	*	*	N	*	*
Cryptonemiales	*	*	*	*	*	N	N	N	*	N	N
Erythropeltidales	*	N	N	N	N	*	*	*	N	N	N
Gelidiales	*	*	N	N	*	*	*	*	N	*	*
Gigartinales	*	*	*	*	*	*	*	*	*	*	*
Goniotrichales	N	*	N	*	N	N	N	N	N	N	N
Gracilariales	N	N	N	N	N	N	N	N	N	N	*
Hildenbrandiales	N	N	N	N	N	*	*	*	N	*	*
Lemaneales	*	N	N	N	N	N	N	N	N	N	N
Nemaliales	*	*	*	*	*	*	*	*	*	*	*
Palmariales	N	N	*	N	*	*	*	*	*	*	*
Porphyridiales	*	*	*	*	*	*	*	N	*	*	*[a]
Rhodochaetales	*	N	*	N	N	*	*	*	*	*	*[a]
Rhodymeniales	*	*	*	*	*	*	*	*	*	*	*

[a] Porphyridiales appears to be a polyphyletic order, and the monotypic Rhodochaetales may belong in Compsopogonales (see Chapter 18).
* = order recognized; N = order not recognized.

vironments on world scale, there appear to be more species of red algae than of brown algae (Phaeophyta) and green algae (Chlorophyta) combined. Regional data compiled by Womersley (1981, p. 297; 1984, p. 56; 1987, p. 19) generally show similar patterns, and so do data of South (1987) for the North Atlantic Ocean.

Sheath (1984, pp. 89, 92, Table 1) has enumerated 159 species of truly freshwater red algae and estimated that this represents about 3.0% of the total number of red algal species. He also estimated that red algae account for only 0.1 to 1.7% of the total number of algal species present in inland waters of North America, Asia, and Europe in contrast to the presently perceived picture in marine environments. The broad patterns that emerge from these analyses, however, should never be allowed to obscure the large, fundamental gaps in our knowledge of species and their delimitation and the ongoing need for sound monographic studies.

The classification and the taxonomic distribution of species within orders and families of Rhodophyta are unresolved, and there is no overall consensus as to which or how many orders (or families) should be recognized (Table 1-3). Dixon (1982) and South and Wittick (1987), for example, recognize 10 orders of red algae. Lee (1980), in contrast, recognizes 12 orders, Dawes (1981) and Pritchard and Bradt (1984) list 9 orders, Bold and Wynne (1985) recognize 13 orders, Gabrielson and Garbary (1986, 1987) and Kraft and Woelkerling (1990) recognize 15 orders, and Gabrielson et al. (1985) recognize 16 orders. Recently, two additional orders have been proposed, the Ahnfeltiales (Maggs & Pueschel 1989) and the Gracilariales (Fredericq & Hommersand 1989). The criteria used to delimit orders vary to some extent, and thus the circumscription of orders and their constituent families and genera differs from scheme to scheme. Of the 21 orders listed in Table 1-3, only 5 (Bangiales, Ceramiales, Gigartinales, Nemaliales, Rhodymeniales) are recognized by all authors concerned, but circumscriptions among these authors differ substantially for all but the Ceramiales (see Chapter 18 for further discussion).

Data provided by Kraft and Woelkerling (1990) on the estimated numbers of families, genera, and species in one suite of orders/subclasses are summarized in Table 1-4. Kraft and Woelkerling (1990) emphasize the composite nature of these figures and stress that they are only approximations that may not coincide with those of other authors.

Another problem underlying red algal classification is that some characteristics that have been used to delimit orders and families relate to rather

Table 1-4. Estimates of numbers of families, genera and species within orders/subclasses of Rhodophyta mentioned by Kraft & Woelkerling (1990)

Order/Subclass	Estimates of numbers of included taxa		
	Families	Genera	Species
Acrochaetiales	1	1–7	100
Bangiophycidae[a]	Not given	30	110
Batrachospermales	3	8	80
Bonnemaisoniales	2	8	25
Ceramiales	4	325	1,500
Corallinales	1	37–40	Uncertain
Gelidiales	2	12	130
Gigartinales	42	170	1,120
Hildenbrandiales	1	2	11
Nemaliales	2	18	350
Palmariales	1	6	20
Rhodymeniales	3	42	280

[a] Includes: Bangiales, Compsopogonales, Porphyridiales, Rhodochaetales.

obscure and ephemeral events in the sexual cycle (e.g., the time of formation, location, and nature of the generative auxiliary cell) that are difficult to observe and indeed have not been confirmed to occur in a majority of included species. In many species, moreover, male and/or female plants have never been described, and for some, virtually no information on sexual reproduction is available. The extent to which new data on the reproductive biology of red algae will affect their classification both at order and family levels is difficult to forecast but could be substantial.

IV. UTILIZATION AND FUTURE RESEARCH

Although red algae have been consumed by humans for at least 2,800 years (Waaland 1981), their full agronomic and biotechnological potential has yet to be realized. Tseng (1981, p. 681) notes that at least 344 species are considered to be of economic value, but only species of *Porphyra*, *Gelidiella*, *Gloiopeltis*, *Eucheuma*, and *Gracilaria* have been cultivated to any significant extent. Papers published in volumes edited by C. J. Bird and Ragan (1984) and K. T. Bird and Benson (1987) provide examples of and insight into research involving macroscopic red algae in relation to polyelectrolytes, pharmaceutical products, human nutrition, antimicrobial activity, polysaccharide production, resource management, etc. Aspects of the biology and commercial potential of the microscopic red alga *Porphyridium* are summarized by Vonshak (1988). In addition to the possible use of *Porphyridium* pigments for food coloring, living cells produce considerable quantities of arachidonic acid, an essential dietary constituent for humans. Red algal utilization by humans is only one aspect of red algal biology requiring further research, however.

In his book on the biology of red algae, Dixon (1973) emphasizes structure, morphology, reproduction and life histories; minor treatments of fossil representatives, economic utilization, and systematics also are provided. In his final chapter, Dixon concludes: "It is not generally appreciated how little is actually known about most marine and freshwater algae and the Rhodophyta is probably the worst known of any algal group" (pp. 242–3). The current book, produced by a collaboration of 26 authors, covers a broad spectrum of cytological, physiological, ecological, morphological, and systematic topics, and it is obvious that considerable progress toward a better understanding of red algae has been made in the past 17 years. This added knowledge, however, also has led to many intriguing new questions and unresolved problems relating to red algal biology, thus guaranteeing numerous exciting and scientifically rewarding lines of red algal research during the next 17 years and beyond.

V. REFERENCES

Adey, W. H. 1971. The sublittoral distribution of crustose corallines on the Norwegian coast. *Sarsia* 46: 41–58.

Altman, P. L. & Dittmer, D. S. 1972. *Biology Data Book*. Vol. 1. 2nd ed. Bethesda, M.: Federation of American Societies for Experimental Biology.

Bird, C. J. & Ragan, M. A. eds. 1984. *Eleventh International Seaweed Symposium. Proceedings of the Eleventh International Seaweed Symposium, held in Qingdao, People's Republic of China, June 19–25, 1983*. Dordrecht, The Netherlands: Dr. W. Junk. (Reprinted from *Hydrobiologia*, Vol. 116/117.)

Bird, K. T. & Benson, P. H., eds. 1987. *Seaweed Cultivation for Renewable Resources*. Amsterdam: Elsevier Science.

Bold, H. C., Alexopoulos, C. J. & Delevoryas, T. 1987. *Morphology of Plants and Fungi*. 5th ed. New York: Harper & Row.

Bold, H. C. & Wynne, M. J. 1978. *Introduction to the Algae*. Englewood Cliffs, N.J.: Prentice-Hall.

Bold, H. C. & Wynne, M. J. 1985. *Introduction to the Algae*. 2nd ed. Englewood Cliffs, N. J.: Prentice-Hall.

Dawes, C. J. 1981. *Marine Botany*. New York: Wiley.

Dechau, C. 1987. Surgeons mend bone with algal skeletons. *New Scient.* 115(1569): 38.

Dixon, P. S. 1973. *Biology of the Rhodophyta*. New York: Hafner.

Dixon, P. S. 1982. Rhodophycota. In *Synopsis and Classification of Living Organisms* Vol. 1, ed. S. P. Parker, pp. 61–9. New York: McGraw-Hill.

Doty, M. S. 1979. Status of marine agronomy with special reference to the tropics. *Proc. Int. Seaweed Symp.* 9: 35–58.

Dring, M. J. 1982. *The Biology of Marine Plants*. London: Edward Arnold.

Fredericq, S. & Hommersand, M. H. 1989. Proposal of the Gracilariales ord. nov. (Rhodophyta) based on an analysis of the reproductive development of *Gracilaria verrucosa*. *J. Phycol.* 25: 213–27.

Gabrielson, P. W. & Garbary, D. J. 1986. Systematics of red algae (Rhodophyta). *CRC Crit. Rev. Plant Sci.* 3: 325–66.

Gabrielson, P. W. & Garbary, D. J. 1987. A cladistic analysis of Rhodophyta: Florideophycidean orders. *Br. Phycol. J.* 22: 125–38.

Gabrielson, P. W., Garbary, D. J. & Scagel, R. F. 1985. The nature of the ancestral red alga: inferences from a cladistic analysis. *BioSystems* 18: 335–46.

Garbary, D. J. 1987. *The Acrochaetiaceae (Rhodophyta): An Annotated Bibliography*. Berlin: J. Cramer. (*Bibliotheca Phycologica* 77).

Goff, L. J. 1982. The biology of parasitic red algae. *Prog. Phycol. Res.* 1: 289–369.

Hori, H. & Osawa, S. 1987. Origin and evolution of organisms as deduced from 5S ribosomal RNA sequences. *Mol. Biol. Evol.* 4: 445–72.

Johansen, H. W. 1981. *Coralline Algae, A First Synthesis*. Boca Raton, Fl.: CRC.

Kasperk, C. & Ewers, R. 1986. Tierexperimentelle Untersuchungen zur Einheilungstende synthetische, koralliner und aus Algen gewonnener (phykogener) Hydroxlapatit materialien. *Dt. Z. Zahnarztl. Implantol.* 2: 242–8.

Kraft, G. T. & Woelkerling, W. J. 1990. Rhodophyta – systematics and biology. In *Biology of Marine Plants*, eds. M. N. Clayton & R. J. King, in press, Melbourne: Longman Cheshire.

Kylin, H. 1956. *Die Gattungen der Rhodophyceen*. Lund: CWK Gleerups.

Lee, R. E. 1980. *Phycology*. Cambridge: Cambridge University Press.

Littler, M. M. & Littler, D. S. 1984. Models of tropical reef biogenesis: the contribution of algae. *Prog. Phycol. Res.* 3: 323–64.

Lüning, K. 1985. *Meeresbotanik*. Stuttgart: G. Thieme.

Maggs, C. A. & Pueschel, C. M. 1989. Morphology and development of *Ahnfeltia plicata* (Rhodophyta): Proposal of Ahnfeltiales ord. nov. *J. Phycol.* 25: 333–51.

Margulis, L. & Schwartz, K. V. 1982. *Five Kingdoms*. San Francisco: W. H. Freeman.

Papenfuss, G. F. 1955. Classification of the algae. In *A Century of Progress in the Natural Sciences*, pp. 115–224. San Francisco: California Academy of Sciences.

Parker, S. P., ed. 1982. *Synopsis and Classification of Living Organisms*. Vol. 1. New York: McGraw-Hill.

Peterfi, S. & Ionescu, A. 1977. *Tratat de Algologie. II. Rhodophyta – Phaeophyta*. Bucuresti: Editura Academiei Republicii Socialiste Romania.

Pritchard, H. N. & Bradt, P. T. 1984. *Biology of Nonvascular Plants*. St. Louis: Times Mirror/Mosby.

Ragan, M. A. 1981. Chemical constituents of seaweeds. In *The Biology of Seaweeds*, eds. C. S. Lobban & M. J. Wynne, pp. 589–626. Oxford: Blackwell.

Round, F. E. 1981. *The Ecology of Algae*. Cambridge: Cambridge University Press.

Sheath, R. G. 1984. The biology of freshwater red algae. *Prog. Phycol. Res.* 3: 89–157.

South, G. R. 1987. Biogeography of the benthic marine algae of the North Atlantic Ocean – an overview. *Helgol. Meeresunters.* 41: 273–82.

South, G. R. & Tittley, I. 1986. *A Checklist and Distributional Index of the Benthic Marine Algae of the North Atlantic Ocean.* London: British Museum (Natural History).

South, G. R. & Whittick, A. 1987. *Introduction to Phycology.* Oxford: Blackwell.

Steneck, R. S. 1986. The ecology of coralline algal crusts: convergent patterns and adaptive strategies. *Annu. Rev. Ecol. Syst.* 17:273–303.

Tappan, H. 1980. *The Paleobiology of Plant Protists.* San Francisco: W. H. Freeman.

Tseng, C. K. 1981. Commercial cultivation. In *The Biology of Seaweeds*, eds. C. S. Lobban & M. J. Wynne, pp. 680–725. Oxford: Blackwell.

van den Hoek, C. & Jahns, H. M. 1978. *Algen. Einfürung in die Phykologie.* Stuttgart, G. Thieme.

Vonshak, A. 1988. *Porphyridium.* In *Micro-algal Biotechnology*, eds. M. A. Borowitzka & L. J. Borowitzka, pp. 122–34. Cambridge: Cambridge University Press.

Waaland, J. R. 1981. Commercial utilization. In *The Biology of Seaweeds*, eds. C. S. Lobban & M. J. Wynne, pp. 726–41. Oxford: Blackwell.

Woelkerling, W. J. 1983. The *Audouinella* (*Acrochaetium – Rhodochorton*) complex (Rhodophyta): present perspectives. *Phycologia* 22: 59–92.

Woelkerling, W. J. 1984. *M. H. Foslie and the Corallinaceae: An Analysis and Indexes.* Vaduz: J. Cramer. (*Bibliotheca Phycologica* 69).

Woelkerling, W. J. 1988. *The Coralline Red Algae: An Analysis of the Genera and Subfamilies of Nongeniculate Corallinaceae* (Rhodophyta). London: British Museum (Natural History) & Oxford: Oxford University Press.

Womersley, H. B. S. 1979. Southern Australian species of *Polysiphonia* Greville (Rhodophyta). *Aust. J. Bot.* 27: 459–528.

Womersley, H. B. S. 1981. Biogeography of Australasian marine macroalgae. In *Marine Botany: An Australasian Perspective*, eds. M. N. Clayton & R. J. King, pp. 292–307. Melbourne: Longman Cheshire.

Womersley, H. B. S. 1984. *The Marine Benthic Flora of Southern Australia. Part I.* Adelaide, S. Australia: Government Printer.

Womersley, H. B. S. 1987. *The Marine Benthic Flora of Southern Australia. Part II.* Adelaide, S. Australia: Government Printer.

Wray, J. L. 1977. *Calcareous Algae.* Amsterdam: Elsevier.

Chapter 2

Cell structure

CURT M. PUESCHEL

CONTENTS

- I. Introduction / 8
- II. Cell membrane / 9
- III. Golgi apparatus / 10
  - A. Structure and function / 10
  - B. Organelle associations / 11
- IV. Endoplasmic reticulum / 12
- V. Nucleus / 13
- VI. Chloroplasts / 15
  - A. Thylakoid disposition / 15
  - B. Chloroplast development / 17
  - C. Thylakoid substructure / 17
  - D. Chloroplast envelope / 17
  - E. Genophore / 18
  - F. Pyrenoids / 18
  - G. Other chloroplast inclusions / 18
- VII. Mitochondria / 19
- VIII. Microbodies / 20
- IX. Vacuoles and lysosomes / 20
- X. Cytoskeleton and motility / 20
  - A. Cytoskeletal components and cytoplasmic streaming / 20
  - B. Cell motility / 22
  - C. Purported flagella / 22
- XI. Reserve materials / 22
  - A. Starch / 22
  - B. Protein crystals / 23
  - C. Lipid and other hydrocarbon inclusion / 24
- XII. Pit connections and pit plugs / 25
  - A. Formation of pit connections / 25
  - B. Structure and formation of pit plugs / 26
  - C. Function / 28
- XIII. Cell coverings / 28
- XIV. Specialized vegetative cells / 29
  - A. Rhizoids / 29
  - B. Hair cells / 29
  - C. Secretory and storage cells / 29
  - D. Transfer cells / 29
- XV. Reproductive cells / 29

A. Female reproductive apparatus / 29
B. Spermatia / 32
C. Spores / 32
XVI. Summary / 34
XVII. References / 35

I. INTRODUCTION

Features of cell structure provide some of the most distinguishing characteristics of the red algae: unstacked photosynthetic membranes, phycobilin pigment granules, pit plugs, cytoplasmic starch grains, absence of centrioles and flagella, unusual associations of the Golgi apparatus with other organelles. Features of cell structure also reflect the diversity of the red algae. Being an ancient lineage, the red algae have undergone a broad range of modifications in cellular organization. Even the spectrum of morphological possibilities, from unicellular forms (Fig. 2-1) to simple filaments, to complex pseudoparenchymatous (Fig. 2-2) and parenchymatous thalli, fails to convey the degree of cellular diversity. In the category of branched filaments, we find species with small, uninucleate cells, species with giant cells each with hundreds of nuclei (Fig. 2-3), species with uninucleate cells that increase ploidy as cells enlarge, and species in which both ploidy and number of nuclei change during vegetative development (Goff & Coleman 1986; Chapter 3). Furthermore, the diversity of developmental processes in red algae must be appreciated. The red algae have at least three common methods by which a vegetative cell may achieve multinuclearity (see Chapter 3). Similarly, there are several means by which a cell can establish pit connections with another cell. As more cytological features are found to have multiple character states, the systematic potential of these features are beginning to be explored.

The cell structure of red algae has been previously reviewed by Duckett and Peel (1978) and Brawley and Wetherbee (1981). This chapter attempts to up-

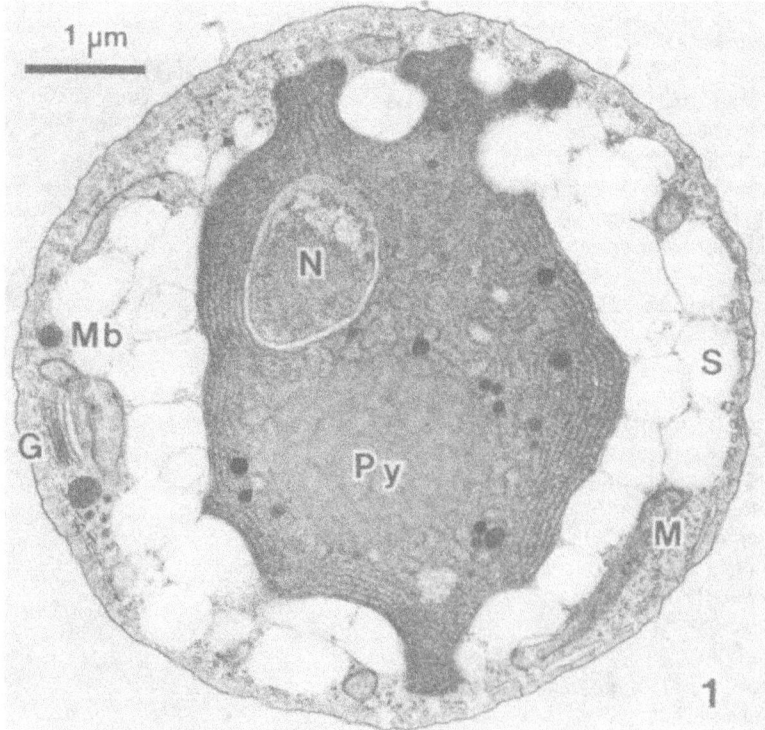

Fig. 2-1. *Porphyridium purpureum*. Nucleus (N) occupies a lobe of stellate chloroplast, which contains a pyrenoid (Py). Starch grains (S) are free in cytoplasm. Mitochondria (M), a layer of ER, Golgi bodies (G), and microbodies (Mb) are present near cell surface. The cell is surrounded by thick layer of electron-transparent mucilage.

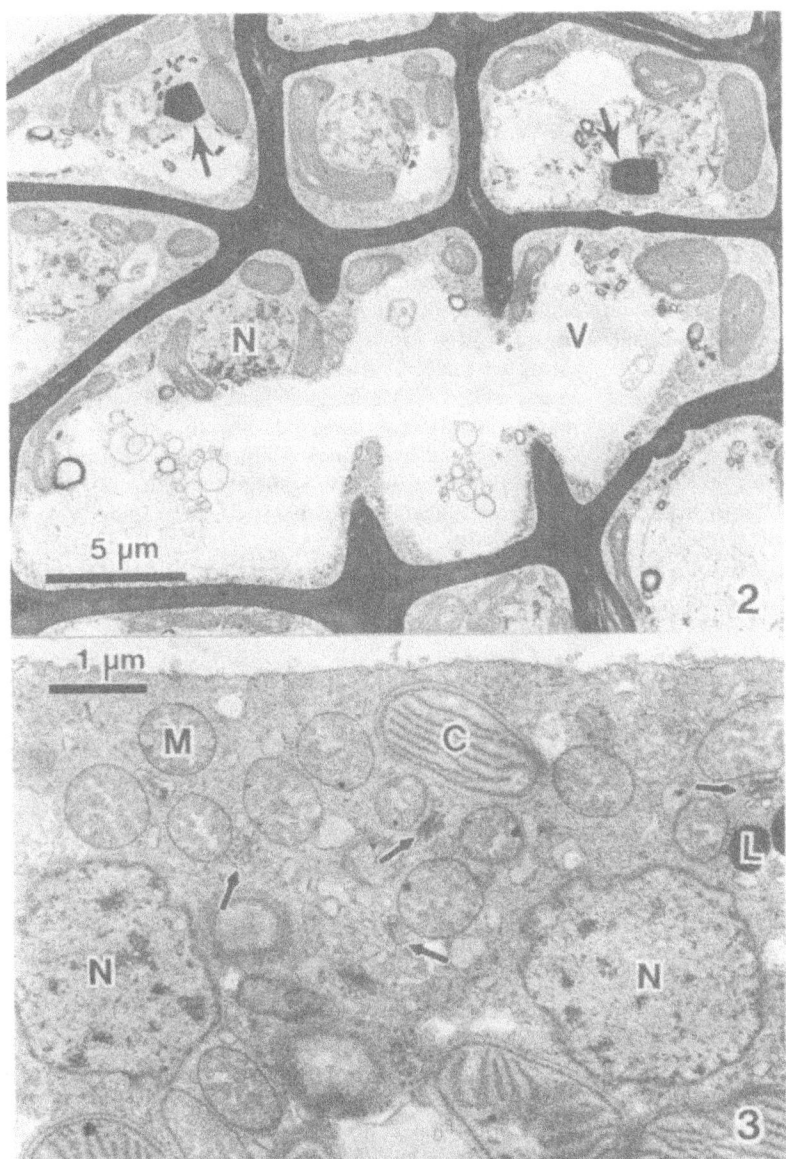

Figs. 2-2, 2-3. 2-2: Inner cortex of *Palmaria mollis*. Lobed outline of large cell stems from its origin by direct fusion of three cells. Only one nucleus (N) of this multinucleate cell is present in section. In contrast to unicellular species, most of the cell volume is vacuole (V). Protein crystals (arrows) are located in cytoplasm. 2-3: Layer of cytoplasm lining central vacuole of *Griffithsia pacifica* contains many nuclei (N), chloroplasts (C), mitochondria (M), and Golgi bodies (arrows), as well as some lipid bodies (L).

date and expand these treatments. It also seeks to provide a perspective on the current state of understanding of red algal cell structure and to identify directions for future work. The components of red algal cells are surveyed and selected specialized cell types, vegetative and reproductive, are described.

II. CELL MEMBRANE

The cell membrane (plasmalemma) of red algal cells is typical in its thickness (8–10 nm) and trilaminate appearance (Figs. 2-6, 2-16). Elaborations of the plasmalemma have been reported, but it is difficult to separate genuine features from artifacts. Plasmalemmavilli were believed to be tubular evaginations of the plasmalemma (Cottler 1971; Duckett & Peel 1978; Young 1978). These structures were hypothesized to be involved in such functions as wall deposition (Cottler 1971) and nutrient absorption by parasites (Goff 1979). However, similar configurations were found to arise as artifacts of specimen preparation (Pueschel 1980a). On the other hand, the authenticity of lomasomes and related vesicular elaborations of the plasmalemma (e.g., Ramus 1969; Tsekos 1981) is supported by freeze-fracture observations (Brown & Weier 1970), and acid phosphatase activity in lomasomes has been reported (Tsekos 1981). The freeze-fracture technique has also been used to demonstrate the response of the plasmalemma to changing osmotic environments (Knoth & Wiencke 1984; Wiencke & Läuchli 1980) and has revealed nonrandom distribution of fracture face particles in the plasmalemma and several other organelles (Tsekos & Reiss 1988; Tsekos et al. 1985).

The plasmalemma of some spores and spermatia have attached tubules, 30–50 nm in diameter and up to 2 μm in length, that project into the cell cortex (Avanzini & Honsell 1984; Cole & Sheath 1980; Kugrens & West 1973a; Mandura et al. 1986; Peyrière 1970; Triemer & Vasconcelos 1977; Vesk & Borowitzka 1984; Wetherbee 1978a). The membranous substructure of the tubule and its continuity with the plasmalemma have been demonstrated (e.g., Avanzini & Honsell 1984; Wetherbee 1978a). The distal end of the tubule appears to be open; therefore, its contents are extracellular (Avanzini & Honsell 1984). Some authors find tubules to be present only during spore formation (e.g. Vesk & Borowitzka 1984; Wetherbee 1978a), whereas others find them only in released spores (Avanzini & Honsell 1984; Mandura et al. 1986). Speculation as to the function of the tubules is likewise variable.

Tubules of similar features were reported to be attached to the cap membrane of pit plugs in gonimoblast cells (Delivopoulos & Kugrens 1984b, 1985b).

III. GOLGI APPARATUS

A. Structure and function

Golgi bodies (dictyosomes) in red algae are generally composed of 4 to 15 stacked cisternae (Figs. 2-4 to 2-6). They are common in active vegetative cells, but they are especially prominent during sporogenesis.

As the spores and their investing mucilages are formed, the number, size, and contents of the dictyosomes and their vesicles change as part of a programmed sequence of Golgi activity (e.g., Kugrens & West 1973a, 1974; Peyrière 1970; Pueschel 1979; Scott & Dixon 1973a; Wetherbee 1978b).

The mechanism of Golgi operation is not completely understood for any cell type. In the animal literature, a model of stationary cisternae among which vesicles exchange products has recently gained ascendancy over the earlier model of cisternal progression (Farquhar 1985). Morphological support for both these models can be found in red algal cells. Evidence for cisternal progression is especially strong; entire cisternae become dilated with product and move from the dictyosome (Fig. 2-4; Alley & Scott 1977; Konrad Hawkins 1974; Scott & Dixon 1973a; Tripodi 1971b; Tsekos 1985; Wetherbee 1978b).

The central and peripheral regions of an individual cisterna may have visibly different contents (Alley & Scott 1977; Kugrens & West 1974; Wetherbee 1978b) and strikingly different morphology. Furthermore, the central regions of adjacent cisternae may become appressed, such that no intercisternal space or material separates them (Fig. 2-4; Alley & Scott 1977). Using the analogy of stacked regions of granal lamellae in plastids, these appressed cisternal membranes may create a special domain enriched with particular membrane proteins.

Ultrastructural autoradiography has been used to demonstrate that sulfation of mucilage occurs in the Golgi apparatus (Evans et al. 1974). Radioactive labels also have been used to study the rate of mucilage production. It took less than 2 h for *Rhodella* to take up labeled sulphate from the medium, to bind it to the mucilage in the dictyosomes, and for the labeled mucilage to be secreted (Evans et al. 1974). In *Porphyridium* (Ramus 1972), a carbon label was incorporated into mucilage that was then released into the medium by dictyosome secretion, all within 30 min.

Ramus and Robins (1975) performed a morphometric analysis of the Golgi apparatus in stationary phase cells versus log phase cells of *Porphyridium*. The number and size of dictyosomes were greater in log phase cells, occupying a total volume three times that of stationary phase cells. Paradoxically, when Golgi volume was high, the amount of secretion was not as great as when there was less Golgi activity. This apparent contradiction between synthesis and secretion rates was resolved by the finding that part of the product synthesized during

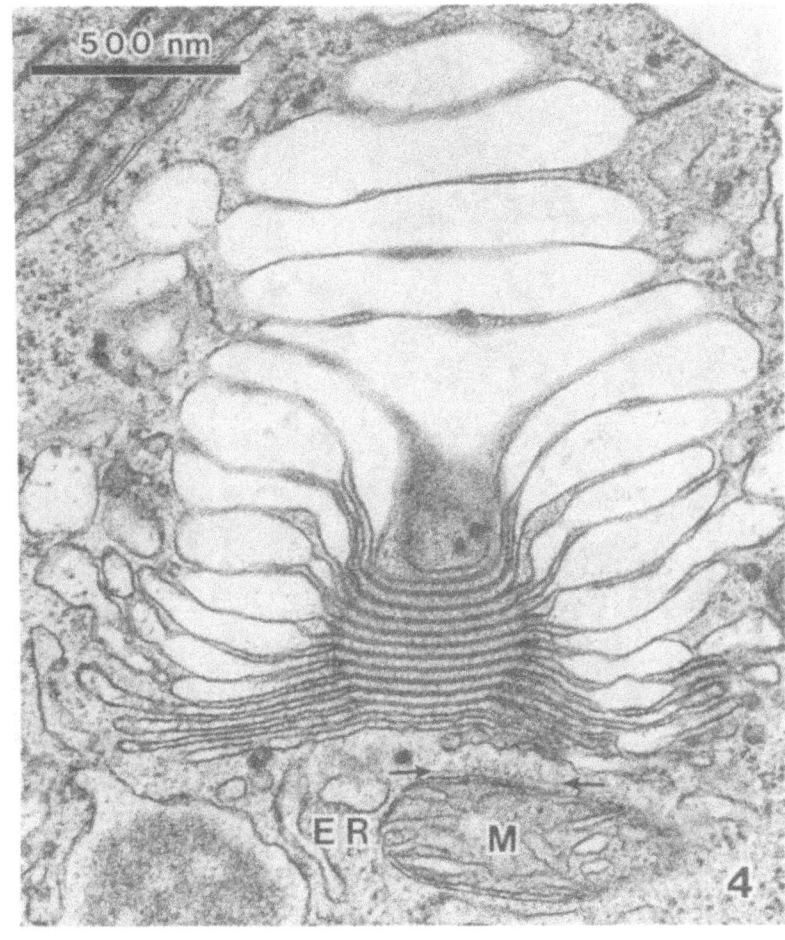

Fig. 2-4. Golgi body in tetrasporangium of *Polysiphonia denudata*. Note apposition of central regions of cisternae, and the fibrillar material (arrows) where a mitochondrion (*M*) is associated with cis face of Golgi body. Endoplasmic reticulum (*ER*) is also in close proximity to Golgi. (Courtesy J. Scott)

log phase accumulated and was not secreted until stationary phase.

Wetherbee and West (1976, 1977) demonstrated the apparent involvement of Golgi bodies in reprocessing proteinaceous inclusions, the striated vesicles, in carposporangia. Another proposed activity that is out of the ordinary for the Golgi apparatus is a role in starch synthesis (Tripodi 1971a,b). The evidence for this suggestion, which has not found support, was the juxtaposition of electron-transparent Golgi vesicles and starch grains.

Ultrastructural localization of acid phosphatase activity within some Golgi cisternae was demonstrated by Tsekos (1981), but no other data of this type are available. Variation in the reactivity of cisternal membranes to PTA-chromic acid staining was observed in *Nemalion* (Pueschel 1980b). Staining increased with increasing size of the cisternae, presumably reflecting differences in membrane-bound constituents across the Golgi stack.

B. Organelle associations

One of the most intriguing features of the red algal Golgi apparatus is the consistent presence of other organelles in proximity to the cis (forming) face. In most eukaryotes, the cis-Golgi is associated with endoplasmic reticulum. However, in studying red algae, many authors have commented on the striking association of mitochondria with Golgi bodies. Recently, it has become clear that this is only the most common and conspicuous association. Scott (1984) recognized three patterns of Golgi associations in the red algae and he considers this feature to have systematic value. All of the florideophytes critically examined (e.g., Figs. 2-3, 2-4), members of the Bangiales (Cole & Sheath 1980), *Porphyridium* (Fig. 2-16; Gantt & Conti 1965), and *Flintiella* (Scott 1986) have dictyosome–mitochondrial associations. Although less conspicuous, cisternae of ER are found nearby, and presumably it is the ER that

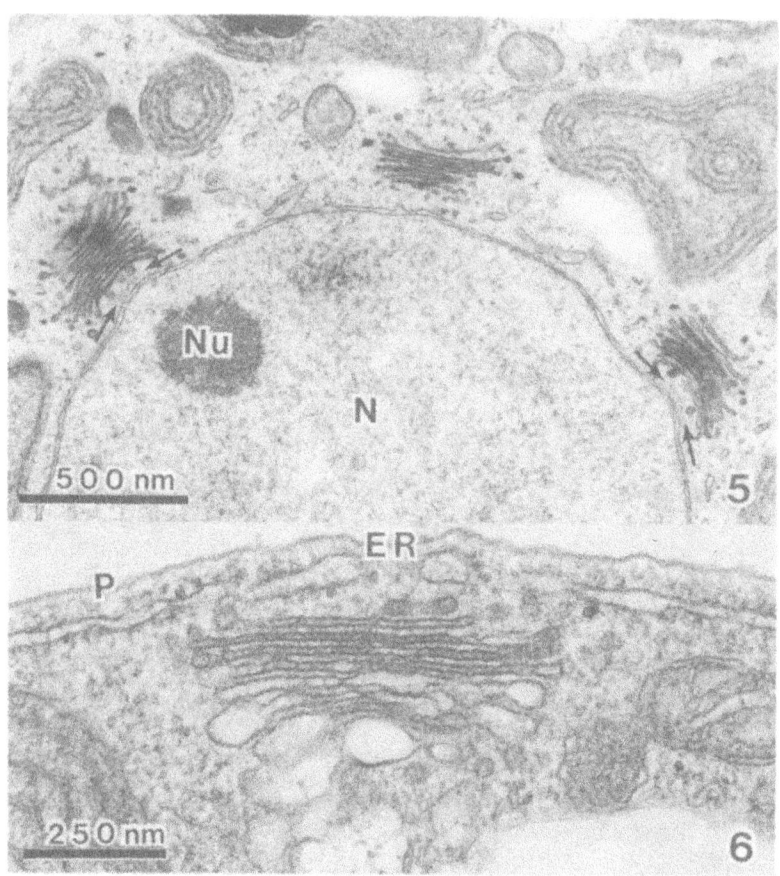

Figs. 2-5, 2-6. *2-5*: Golgi bodies in *Rhodella reticulata* are associated with nucleus (N). Fibrillar material (arrows) is present between nuclear envelope and cis face of Golgi bodies. *Nu* = nucleolus. (Courtesy J. Scott.) *2-6*: Cis face of Golgi body in *Compsopogon coeruleus* is associated with only ER (*ER*). *P* = plasmalemma. (Courtesy J. Scott)

contributes vesicles to the cis-Golgi. In contrast, in *Rhodella cyanea* (Billard & Fresnel 1986; Fresnel & Billard 1987) and *R. reticulata* (Deason et al. 1983), the cis-Golgi is closely associated with the nucleus (Fig. 2-5). There is no ER involvement in this type. In *Rhodella maculata* (Evans 1970), *R. violacea* (Wehrmeyer 1971), *Smithora* (McBride & Cole 1971), *Cyanidium* (Seckbach et al. 1981; Ueda & Yokochi 1981), *Rhodochaete* (Pueschel & Magne 1987), and *Compsopogon* (Fig. 2-6) only ER is associated with the cis face of the Golgi bodies. Because this last configuration is common to most eukaryotes, it may represent the character state of the ancestral red alga.

In some micrographs, a thin line of apparently fibrillar substance is visible between the forming face of the Golgi body and its associated organelles (Figs. 2-4, 2-5; Alley & Scott 1977, Figs. 6-8). These fibrils may represent an element of the cytoskeleton, such as actin, and may be responsible for maintaining these striking associations of dictyosomes with other organelles.

Although not consistent in occurrence, even more complex organelle associations are sometimes observed, such as dictyosomes associated with mitochondria that are, in turn, appressed to chloroplasts (e.g., Pueschel 1979).

IV. ENDOPLASMIC RETICULUM

Rough and smooth endoplasmic reticulum (ER) of typical eukaryotic form is present in red algal cells, especially in the cell periphery (Fig. 2-6) and around the nucleus (Fig. 2-17). In particular cell types, the ER can adopt more distinctive configurations. In some unicells, *Porphyridium* (Figs. 2-1, 2-16; Gantt & Conti 1965; Gantt et al. 1968), *Rhodella* (Deason et al. 1983; Evans 1970; Wehrmeyer 1971) and *Flintiella* (Scott 1986), a cisternal network is found beneath the plasmalemma. Projections of these cisternae form tubules (Fig. 2-16) that may be continuous with the plasmalemma (Gantt & Conti 1965). It is unclear whether these structures bear any functional relationship to the plasmalemmal tubules discussed in Section II, but it is noteworthy that both occur in cells without consolidated walls. In rhizoidal cells of a parasitic red alga, ER forms direct connections with the plasmalemma (Goff 1982). These cisternae are of considerably different morphology and disposition than those in the unicells. The intriguing possibility of open communication of ER cisternae with the cell exterior could be tested with electron-dense tracers or by freeze-fracture electron microscopy. Many unusual configurations of ER appear in fusion cells and other carposporophyte cells, but the most striking are the ER-derived membranous lattices reported by Tripodi and de Masi (1977) and Tsekos and Schnepf (1985).

Several authors suggest that smooth ER functions as a conduit. In parasites, the abundant smooth ER is thought to translocate absorbed nutrients (Goff 1979). Broadwater and Scott (1982) propose that, in the Ceramiales, the signal that fertilization has occurred and an auxiliary cell should be formed is carried through the carpogonial branch by smooth ER.

Perinuclear ER is common in red algae, often becoming temporally abundant in connection with mitosis and meiosis. Annulate lamellae, loose cisternae resembling ER but perforated with annuli like those of the nuclear envelope, are reported to occur in a number of cell types, generally in association with reproduction (e.g., Tripodi 1974; Vesk & Borowitzka 1984; Wetherbee & Wynne 1973; Wetherbee et al. 1974). Most often the annulate lamellae are stacked in a perinuclear position, but single lamellae have been reported within the nucleus, as well (e.g., Vesk & Borowitzka 1984; Wetherbee et al. 1974). In postmeiotic tetrasporangia of some members of the Corallinaceae, a prominent mass of electron-dense material, presumed to be rich in RNA, accumulates in association with the perinuclear ER (Duckett & Peel 1978; Peel et al. 1973; Vesk & Borowitzka 1984).

A variety of other ER activities has been described. Crystalline and amorphous proteinaceous inclusions may be deposited near (Fig. 2-22) or within rough ER. Structures suggested to be autophagic lysosomes (Tsekos 1981, 1985) are believed to originate from ER. Acid phosphatase activity was demonstrated in ER of *Batrachospermum* (Aghajanian & Hommersand 1978), but no indication of the transfer of these enzymes from ER to lysosomes was found. The origin of mucilage sacs from ER was demonstrated in several species (Pueschel 1979; Tsekos 1985; Tsekos & Schnepf 1985). Broadwater and Scott (1983) have demonstrated that the massive vesicles formed by developing spermatia arise from an unusual, highly specialized form of ER (Figs. 2-36, 2-37). They suggest that these mucilage-containing structures are distinctive enough to be considered unique organelles and have dubbed them "fibrous vacuole associated organelles." Such structures may also occur in sporangia (Broadwater & Scott 1983). The direct involvement of both ER and the Golgi apparatus in the production and secretion of carbohydrate-rich products raises the question of how these organelles partition activity.

V. NUCLEUS

Interphase nuclei of red algae are generally small, only about 3 to 5 μm in diameter (Figs. 2-1 to 2-3, 2-5, 2-34, 2-37), but there are some striking exceptions, such as the endopolyploid nuclei found in some members of the Ceramiales (Goff & Coleman 1986; Chapter 3). Multinuclearity is common in multicellular red algae and can arise by cell fusions (Fig. 2-3) or by an uncoupling of karyokinesis and cytokinesis (Fig. 2-2). The ultrastructure of interphase nuclei is typically eukaryotic. The most distinctive feature is the spindle pole body, also called the nucleus-associated organelle (Chapter 6). This discoid or ring-shaped structure is between 0.1 and 0.2 μm in diameter and is found near or even attached to the nuclear envelope during interphase and karyokinesis (e.g., Scott et al. 1981).

In ultrastructural images, the nuclear envelope and its nuclear pores are typical. Annulate lamellae, cisternae perforated by nuclear pores, are present in red algae (see Section IV). A fibrous lamina might not be expected in a group of organisms with persistent nuclear envelopes during karyokinesis and it is not readily apparent in published micrographs, but critical assessment is lacking. Synaptonemal complexes (e.g., Broadwater et al. 1986a; Kugrens & West 1972a; Sheath et al. 1987) and their possible derivatives, polycomplexes (Broadwater et al. 1986b), are present during meiosis.

Only partial characterization of the chromatin organization in red algae has been reported (Barnes et al. 1982). Micrococcal endonuclease digestion of chromatin from *Porphyridium* indicates that nucleosome core particles are present and that, judging from the sizes of the DNA repeat units, the nucleosomes seem to be like those of most other eukaryotes. Ultrastructural visualization of nucleosomes has not yet been reported.

Nucleoli in red algae appear to be typical (Fig. 2-5). Some cells show nucleolar "vacuoles" – spherical, less electron-dense zones within the nucleolus (e.g., Broadwater et al. 1986a). Nucleolar "vacuoles"

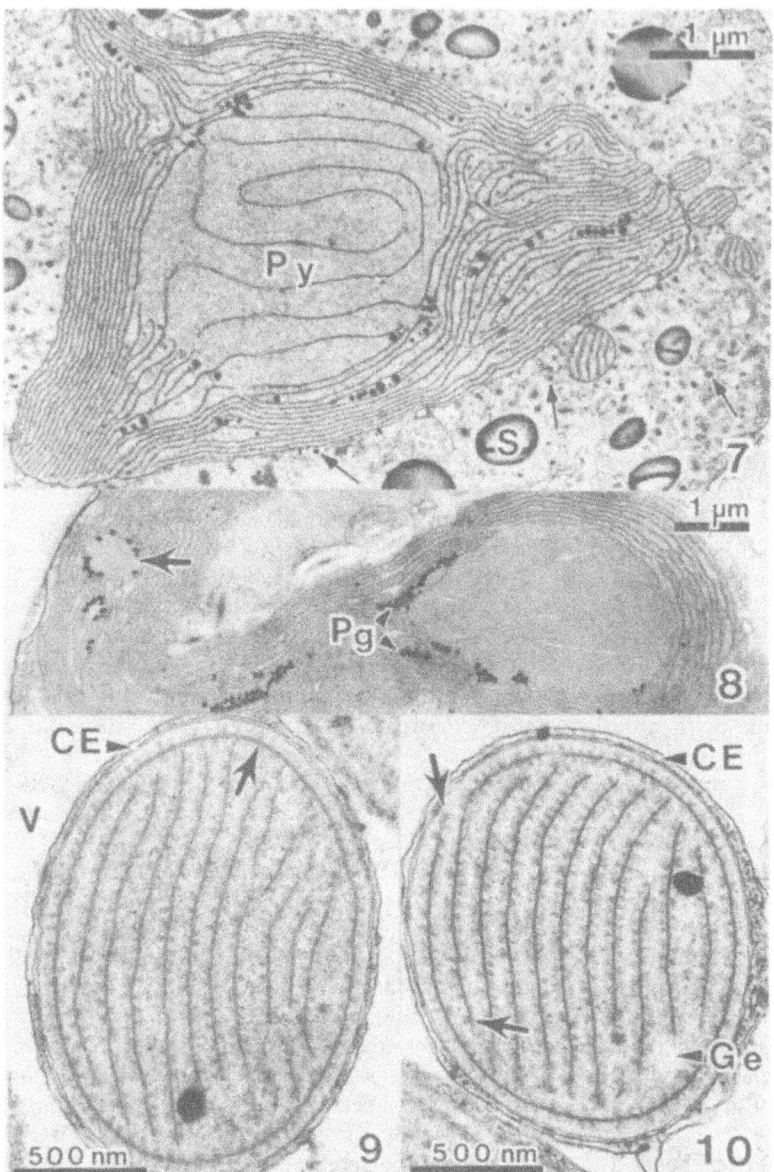

Figs. 2-7 to 2-10. *2-7*: *Porphyra miniata* carpospore with chloroplast containing a large pyrenoid (*Py*) and lacking peripheral encircling thylakoid. Cytoplasm contains many small adhesive vesicles (arrows). *S* = starch. (Courtesy K. Cole.) *2-8*: Germinating *P. variegata* carpospore with a second pyrenoid (arrow) forming de novo in an arm of the originally stellate plastid. Plastoglobuli (*Pg*) are clustered and abundant. *2-9, 2-10*: Chloroplasts of *Griffithsia pacifica* with single peripheral encircling thylakoid that may appear closed (Fig. 2-9, arrow) or have free ends (Fig. 2-10, arrows). Phycobilisomes appear as granules on the stromal surface of thylakoids. *CE* = chloroplast envelope, *Ge* = genophore, *V* = vacuole.

and a marked increase in nucleolus size are common during diffuse diplotene.

In addition to changes in the size and appearance of the nucleolus during particular stages, the nucleus itself enlarges and develops an irregular outline that is assumed to increase surface area for egress of transcripts (e.g., Pueschel 1979, 1980c; Tripodi 1974; Wetherbee & Wynne 1973). In many organisms, the nuclear envelope itself is studded with ribosomes on the cytoplasmic surface, and products of protein synthesis accumulate in the nuclear envelope, similar to rough ER. Accumulation of mucilage, which is largely carbohydrate, within the nuclear envelope has been reported (Delivopoulos & Tsekos 1985b). The direct application of chloroplast envelope to nuclear envelope, even to the exclusion of ribosomes, is found in both unicells (Fig. 2-1) and more complex species (Fig. 2-2).

VI. CHLOROPLASTS

Chloroplasts provide several of the most distinctive features of the red algae, the most obvious of which is thallus color. The water-soluble phycobilin pigments, which are responsible for this unusual coloration, also provide the ultrastructural feature of electron-dense granules, phycobilisomes. If not removed by fixation, the phycobilisomes are found attached to the stromal surface of photosynthetic membranes (Figs. 2-1, 2-9, 2-10, 2-21). Another ultrastructural feature that characterizes red algal plastids is the lack of affiliation of the photosynthetic membranes. The red algae typically have unstacked, evenly spaced thylakoids. The chloroplasts in many algal groups are surrounded by a sheath of ER (e.g., Dodge 1973); such specialized periplastidic ER is not present in the red algae.

The chloroplasts of red algae vary in morphology from being stellate (Figs. 2-1, 2-7) to discoid (Figs. 2-2, 2-3, 2-9 to 2-13) to ramifying, highly lobed forms. Contrasting morphologies in different life history stages are found when cellular morphologies differ between stages (Pueschel & Cole 1985). Plastids are present in all red algae; even the colorless, parasitic forms have plastids, though with few thylakoids (Goff 1982). Specialized cells, such as the vesicle cells of *Antithamnion*, may have few or no plastids due to their exclusion during the formative cell division (Young & West 1979). In some species, the basic description of chloroplast number and morphology is made difficult by extensive lobing of one or a few plastids. This may produce a false impression of many individual plastids. For example, on the basis of ultrastructural study, *Compsopogon* was reported to have multiple plastids (Nichols et al. 1966), but serial section study demonstrated only one highly lobed plastid per cell (Gantt et al. 1986). The presence of lobing in differentiated, nondividing cells (Fig. 2-22) also calls into question the common assumption that constrictions are indicative of plastid division.

A. Thylakoid disposition

The disposition of thylakoids in the red algae usually consists of apparently individual, flat thylakoids surrounded by a single peripheral thylakoid (Fig. 2-9). In some species, several encircling thylakoids appear to be present (Fig. 2-11). Another variation, one that has received considerable attention, is the absence of peripheral encircling thylakoids (Fig. 2-7). This arrangement was once believed to be a characteristic of members of the "lower Rhodophyceae," as opposed to the presumably more advanced florideophytes, which have encircling peripheral thylakoids (Bisalputra 1974). As discussed by Pueschel and Magne (1987), the original appraisal of bangiophytes as generally lacking a peripheral thylakoid was based on a small taxonomic sample and appears to be incorrect. Among the bangiophytes, only the gametophyte-phase of the Bangiales and some members of the Porphyridiales possess this condition. It should be noted that it is easy to distinguish encircling and platelike thylakoids in a discoid plastid, but it can be difficult when a plastid has a stellate morphology (e.g., Fig. 2-1). Nonetheless, if one accepts that the Porphyridiales are reduced rather than primitive, it is clear that the lack of peripheral thylakoids is a specialization (Pueschel & Magne 1987).

Our understanding of thylakoid disposition in red algae is limited because our concepts are often based on randomly selected two-dimensional thin-section images that are then projected into three dimensions. Although Nichols et al. (1966) diagrammed the thylakoids of *Compsopogon* as being concentric, their micrographs show that some thylakoids are coiled. Bergfeld (1970) demonstrated coiled thylakoids in *Batrachospermum*, and DeLuca et al. (1979) found them in *Cyanidium*. This feature also occurs in *Palmaria palmata* (Fig. 2-11). Gantt et al. (1986) reinvestigated *Compsopogon*, demonstrating coiled thylakoids, and they have made the important observation that in some planes of section, a single coiled thylakoid may appear as numerous, independent concentric thylakoids.

The peripheral, encircling thylakoid found in

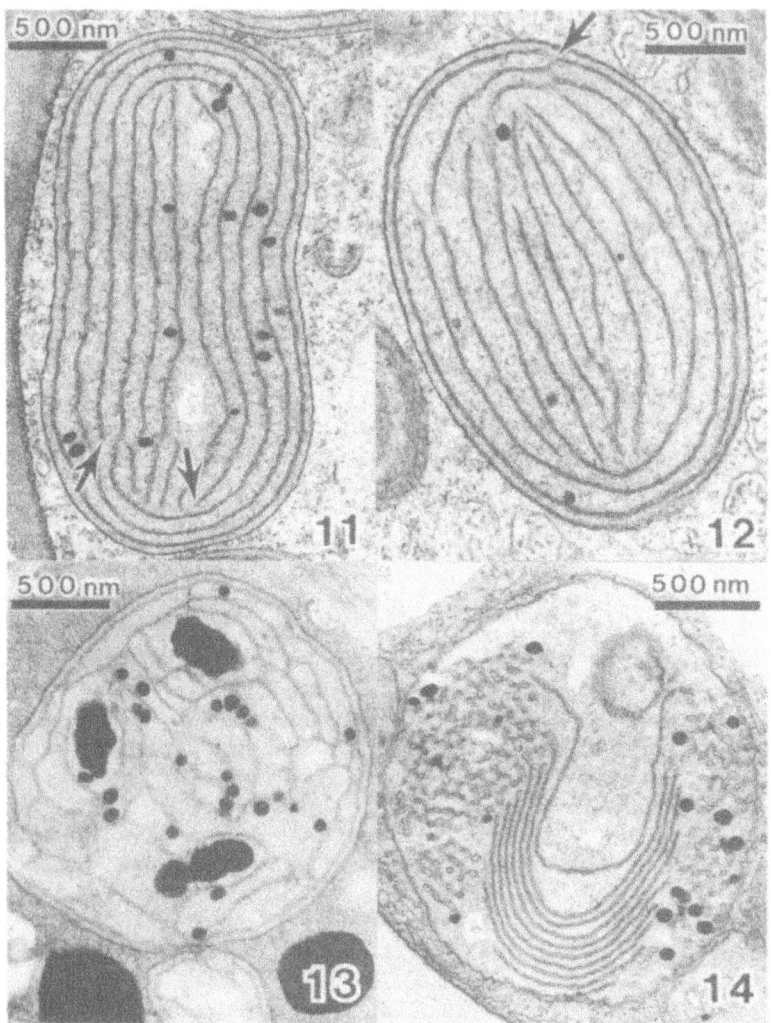

Figs. 2-11 to 2-14. 2-11: Chloroplast of *Palmaria palmata* showing two peripheral thylakoids surrounding a coiled thylakoid (ends marked by arrows) and flat, central thylakoids. Genophore visible between central thylakoids as electron-transparent region containing fibrils. 2-12: *P. palmata*, outermost encircling thylakoid continuous with other thylakoids (arrow). 2-13: Tetrasporocyte chloroplast of *Hildenbrandia rubra*, photosynthetic membrane system forms a continuous reticulum. Large electron-dense inclusions may be lipids or phenolics. 2-14: Plastid of green-pigmented mutant of *P. palmata*. Branching tubular membranes resemble prolamellar body; peripheral encircling thylakoid is lacking.

most red algae is often called the girdle lamella, or girdle band. Implicit in most descriptions of chloroplasts is the view that the peripheral thylakoid is complete and closed (cf. Figs. 2-9, 2-10). Perforations of the thylakoids are considered rare (Dodge 1973). However, if concentric thylakoids are complete, they would define separate stromal compartments, and movement of a molecule from one compartment to another would require transport across a thylakoid. In addition, parts of the stroma would be separated from the genophore by membranes.

Associated with the problem of disposition of thylakoids is the question of the number of thylakoids in a plastid. As Gantt et al. (1986) have demonstrated, we can be misled by images from individual sections. Multiple points of continuity between thylakoids are sometimes apparent (Fig. 2-12). In immature tetraspores of *Hildenbrandia rubra* (Fig.

2-13), the thylakoids form a reticulum, that is, each plastid has only one thylakoid. Similar reticulate lamellae occur in spermatia of *Bangia* (Cole & Sheath 1980) and carpospores of *Gracilaria* (Delivopoulos & Tsekos 1986). No one has determined the extent of interconnection of thylakoids in mature plastids of vegetative cells, and until this is done, we are not in a position to discount even the unlikely possibility that red algal plastids contain a single, continuous thylakoid.

B. Chloroplast development

Chloroplast development has been commented on by numerous authors (Borowitzka 1978; Bouck 1962; Brown & Weier 1968; Delivopoulos & Kugrens 1984b, 1985a; Honsell et al. 1978; Lichtlé & Giraud 1969; Tripodi & Gargiulo 1984; Tsekos 1982). Proplastids, consisting of an envelope and only a partial girdle lamella or none at all (although claims of this were not based on serial sections) are present in vegetative apical cells and in developing sporangia. The most consistent view of the development of the photosynthetic membrane system is that the peripheral thylakoid gives rise to all other thylakoids by localized growth and evagination of folds to one side of the original lamella (Fig. 2-12). When the girdle band is lacking, the inner membrane of the chloroplast envelope gives rise to thylakoids. Less commonly observed developmental features include circular or coiled membrane bodies and tubular membranes. Coiled membrane bodies have been proposed to be the source of thylakoid membrane in developing proplastids (Delivopoulos & Kugrens 1984b, 1985a). Although conspicuous when present, these structures are not universal, a fact that complicates attempts to understand their significance. Hara and Chihara (1974) and Hara (1975) recognized as a distinct type those chloroplasts characterized by membranous tubules, in addition to normal thylakoids. Sheath et al. (1977, 1979) demonstrated that tubular plastid membranes were induced in certain taxa by low levels of illumination, both in the field and in culture. Members of other groups, the "higher florideophytes," generally did not share this feature of darkness-induced changes (Sheath et al. 1979), but see Lüning and Schmitz (1988).

Degeneration of plastids has received less attention than their development. Changes in the thylakoid system and the accumulation of plastoglobuli have been demonstrated (Lichtlé 1973; Pellegrini & Pellegrini 1983). Koslowsky and Waaland (1987) described the destruction of chloroplasts as a result of a cytoplasmic incompatibility reaction induced by the fusion of somatic cells from two different strains of *Griffithsia pacifica*.

Prolamellar bodies, lattices of tubular membranes, are present in plastids of higher plants that develop in the dark. Structures in red algae have been termed prolamellar-like bodies (Borowitzka 1978; Sheath et al. 1979), but only those in a green-pigmented mutant of *Palmaria* (Pueschel & van der Meer 1984) are similar in construction to prolamellar bodies in higher plants (Fig. 2-14). The prolamellar bodies in *Palmaria* are formed in the mutant but not in the wild type, suggesting that the general absence of prolamellar bodies in red algae is due to suppression of their development, rather than to an absence of capacity to form them. Another noteworthy feature of this mutant is the absence of peripheral encircling thylakoids (Fig. 2-14; Pueschel & van der Meer 1984), in contrast to one or several encircling thylakoids in the wild type (Figs. 2-11, 2-12).

C. Thylakoid substructure

Several authors have studied the size and distribution of intramembranous particles seen in freeze-fractured thylakoids of red algae (Bisalputra & Bailey 1973; Guérin-Dumartrait et al. 1970; Lefort-Tran et al. 1973; Neushul 1970; Staehelin et al. 1978; Wiencke 1982). Because grana are absent in red algae, the distinct particle domains associated with thylakoid stacking are absent. The fracture face particles are smaller in red algae than in green plants (Staehelin et al. 1978). Phycobilisomes are attached to the stromal surface of the thylakoids. Immersed within this outer half of the membrane and revealed on the exoplasmic fracture face (EF) are particles believed to be photosystem II complexes; these receive the light energy captured by the phycobilisomes (Staehelin et al. 1978). In some specimens the phycobilisomes are randomly distributed; in others they occur in rows. This variation is probably due to growth under different light intensities (Staehelin 1986). The EF face particles are distributed in the same pattern as the phycobilisomes (Staehelin et al. 1978), and protoplasmic fracture face particles, presumed to represent photosystem I complexes, also appear to be spatially associated with the phycobilisomes (Gantt 1986).

D. Chloroplast envelope

Evolutionary implications have been attached to the chloroplast envelopes of *Bangia* and *Cyanidium*. Textural differences of fracture faces of the inner and outer envelope membranes of *Bangia* were believed by Bisalputra and Bailey (1973) to reflect separate evolutionary origins of the two membranes. Wiencke and Läuchli (1980) found similar textural differences in *Porphyra* as a response to osmotic stress.

Rather than having a normal chloroplast envelope composed of two membranes, *Cyanidium*, a thermophilic alga of debated systematic affinities, was believed to have a single bounding chloroplast membrane (DeLuca et al. 1979; Seckbach & Fredrick 1980; Toyama 1980). It was proposed that this alga is an evolutionary link between red algae and the blue-gree algae (Seckbach & Fredrick 1980; Seckbach et al. 1981) or, alternatively, that this plastid feature was sufficient basis for establishing a new class of red algae (Merola et al., 1981). However, Staehelin (1967) had earlier demonstrated a normal chloroplast envelope by freeze-etching, and this has been confirmed by other methods (e.g., Ford 1984; Ueda & Chida 1987; Ueda & Yokochi 1981). It is now clear that *Cyanidium* has conventional ultrastructure for a red alga.

E. Genophore

The DNA in red algal plastids appears as fibrils in an electron-transparent region within the stroma (Figs. 2-10, 2-11; Bisalputra & Bisalputra 1967). The chloroplast chromosomes appear to be scattered, rather than forming aggregations, as in some algal groups (Bisalputra 1974; Coleman 1985). The genophore is rarely, if ever, found between the most peripheral thylakoid and the chloroplast envelope. A number of authors have commented on membranous tubules or rings in proplastids and mature chloroplasts, to which the plastidic genophores appear to be attached (Borowitzka 1978; Tripodi 1980; Tripodi & Gargiulo 1984; Tripodi et al. 1986).

F. Pyrenoids

Pyrenoids are present in some bangiophytes (Figs. 2-1, 2-7, 2-8) and in some members of the Nemalliales. It is assumed that their main constituent is the same as in other algal groups, the enzyme ribulose bisphosphate carboxylase (RuBPCase) (Griffiths 1980). Total RuBPCase has been isolated from *Porphyridium* and *Cyanidium* as part of a protocol that demonstrated that, unlike the situation in green plants, both enzyme subunits are coded for by the plastid DNA (Steinmüller et al. 1983). The role of RuBPCase in carbon fixation during the dark reactions of photosynthesis is well understood; however, it has yet to be determined why in some species this enzyme aggregates to form the entity we recognize as a pyrenoid. The functional relationship between pyrenoids and starch deposition also needs to be elucidated. Although the starch is deposited in the cytoplasm, if the pyrenoid is close to the chloroplast surface, starch grains occur in close proximity to the pyrenoid (Hara & Chihara 1974; Lee 1974). If the pyrenoid is separated from the plastid surface by several thylakoids, no such association occurs.

Pyrenoids can arise by fission in the equal division of a plastid. However, when a plastid divides unevenly, for instance, by detachment of arms of a stellate plastid, pyrenoids develop de novo (Fig. 2-8) in the smaller plastid. This is seen during the transition of plastid morphologies in germinating carpospores of *Porphyra* (Pueschel & Cole 1985).

The structure of pyrenoids was surveyed as part of a broad ultrastructural analysis of red algal chloroplasts (Hara 1975; Hara & Chihara 1974). Some features were found to be consistent within a species but different between taxa; these include penetration of the pyrenoid by thylakoids, the arrangement of the penetrating thylakoids, and whether the pyrenoid was completely surrounded by thylakoids or was bounded in part by the chloroplast envelope. Pyrenoids clearly have systematic value besides their simple presence or absence. *Rhodella* was originally distinguished from other unicellular genera by the presence of an irregularly shaped pyrenoid not penetrated by thylakoids and not separated from the plastid envelope by thylakoids (Evans 1970).

G. Other chloroplast inclusions

Plastoglobuli are common if not universal constituents of plastids, including those of red algae (Figs. 2-7 to 2-14). In the unicellular species, *Rhodella reticulata*, regular arrays of apparent plastoglobuli in the periphery of the plastid lobes are believed to correspond to orange spots visible in the living cell. The term "eyespot" has been applied to these aggregations (Deason et al. 1983). Gliding motility allows phototaxis in some unicellular species; therefore, there is need for a mechanism that allows orientation to light (e.g., Hill et al. 1980), but the connection with these plastid granules remains to be demonstrated. Large osmiophilic inclusions, possibly phenolics, are found in some plastids of macroalgae, especially those of crustose species (Fig. 2-13).

A variety of inclusions presumed to be protein-

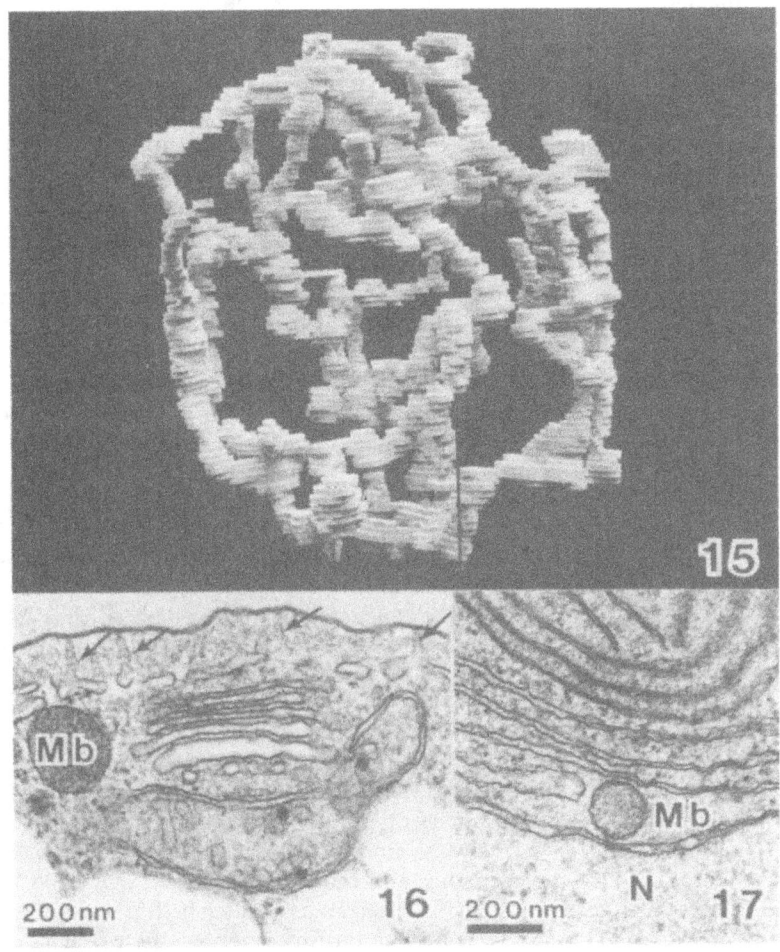

Figs. 2-15 to 2-17. *2-15*: Model of single mitochondrion in cell of *Rhodella reticulata* reconstructed from serial sections. (From Broadwater & Scott 1986, with permission). *2-16*: Microbody (*Mb*) in *Porphyridium purpureum*. Cisternae of ER have tubules (arrows) that approach plasmalemma. *2-17*: Microbody (*Mb*) between perinuclear ER and nucleus (*N*) in *Palmaria palmata*.

aceous are encountered in red algal plastids. The iron-containing protein, phytoferritin, is present in senescent blades of the perennial alga *Constantinea* (Pueschel & Cole 1980; Pueschel & Parthasarathy 1984) and was found in one sample of *Hildenbrandia rubra* (Pueschel 1988a). Amorphous inclusions were present in some chloroplasts of a green-pigmented mutant of *Palmaria palmata* (Pueschel & van der Meer 1984) and spherical inclusions with granular substructure were found in *H. rubra* plastids (Pueschel 1988a). Crystalline inclusions have also been reported to occur in plastids of a few red algae (Goff 1982; Young 1979a).

VII. MITOCHONDRIA

Red algal mitochondria are not at all unusual in their structure (Figs. 2-3, 2-4, 2-16). Although the cristae are sometimes described as being tubular, the folding of the inner mitochondrial membrane to form a series of plates (Fig. 2-4) is much the same as in multicellular animals and vascular plants. Truly tubular cristae, which are found in about half of the protistan groups (Stewart & Mattox 1984; Taylor 1978), give rise to circular images when cut in cross-section, an image rarely observed in red algal mitochondria.

Several groups of eukaryotes have species whose cells contain a single, highly branched mitochondrion. The red algae are one such group. By meticulous thin-section reconstruction, Broadwater and Scott (1986) analyzed the three-dimensional structure of mitochondria in three cells of *Rhodella reticulata*. In one cell, the mitochondrial profiles all represented a single organelle (Fig. 2-15). In other cells, small segments of mitochondria were present in addition to the large organelle, presumably reflecting a dynamic process of budding or fusing (Broadwater & Scott 1986).

During sporogenesis in some red algae, the mitochondrial genophore appears to be attached to a morphologically specialized region of a crista. In section, this region consists of a single or multilamellar ring of membrane (Tripodi 1980). As is true of similar structures in chloroplasts, these are not commonly observed and are of uncertain significance.

Tripodi et al. (1972) reported a twisted structure located in the mitochondrial matrix of *Polysiphonia* cystocarp cells. Both the twisted structure and the typical genophore were reported to be removed by DNase digestion. In contrast, Delivopoulos and Kugrens (1984a) concluded that similar twisted structures in mitochondria of *Faucheocolax* fusion cells were membranous and might represent a specialized cristal morphology.

VIII. MICROBODIES

Microbodies are among the most poorly studied of red algal organelles. Because a microbody has few distinguishing morphological features (Figs. 2-1, 2-16, 2-17) (granular contents, a single bounding membrane, and in some organisms a crystal), demonstration of marker enzymes is often used to verify their identity. Although microbodies typically contain a variety of enzymes, catalase activity is considered to be diagnostic for microbodies of animals and higher plants. Many authors have reported microbodies in red algae (Bronchart & Demoulin 1977; Ford 1984; Oakley & Dodge 1974; Pueschel 1979, 1980c; Schornstein & Scott 1982; Scott 1986; Seckbach et al. 1981), but in only two studies were enzyme localization procedures attempted (Oakley & Dodge 1974; Pueschel 1980c). Both were unsuccessful in demonstrating catalase activity. Peroxidase was believed to be responsible for positive cytochemical staining in *Porphyridium* (Oakley & Dodge 1974). In *Palmaria*, microbodies apparently derived from ER were numerous adjacent to the nucleus (Fig. 2-17), especially during meiotic prophase I (Pueschel 1980c). No evidence of either catalase or peroxidase was found in these microbodies. However, peroxidase activity was found in a different population of structures, and it was suggested that the peroxidase might play a role in halogenation (Pueschel 1980c).

IX. VACUOLES AND LYSOSOMES

The term "vacuole" is often loosely applied to any large, single-membrane bounded structure, including structures that are clearly transient secretory vesicles. In the present discussion, the term "vacuole" is reserved for organelles whose principal, but not necessarily exclusive, role is osmotic regulation. Vacuoles are most prominent in differentiated cells; medullary cells in some species consist of little other than vacuole (Figs. 2-2, 2-18). In contrast, spores (Fig. 2-7), spermatia (Figs. 2-36, 2-37), and unicellular species that lack firm walls (Figs. 2-1, 2-5) are less able to utilize vacuoles to generate turgor. Small vacuoles in such cells may function primarily as lysosomes, although little work has been done to determine whether vacuoles in red algae contain acid hydrolases.

The response of vacuoles to osmotic stresses has been studied by several methods, including electron microscopy (Knoth & Wienke 1984; Wiencke & Läuchli 1980, 1983; Wiencke et al. 1983). Osmotic stress induces dramatic changes in vacuolar volume, changes in the density of intramembranous particles of the tonoplast, the bounding membrane of the vacuole, and changes in the concentration of inorganic ions in the vacuole.

The role of vacuolation in development has not received sufficient study. For instance, the germination of spores commonly involves the formation of large vacuoles (e.g., Pueschel & Cole 1985). Some spores, such as the *Gelidium* type (Dixon 1973), are described as being evacuated during germination. Presumably, the cell maintains its integrity but is filled almost entirely by vacuoles.

Structures with lysosomal activity have been reported in many red algae (e.g., Brown & Weier 1970; Goff 1979; Kazama & Fuller 1970; Tsekos 1981, 1985), but no one has documented the presence of acid hydrolase activity in these presumptive lysosomes. Tsekos (1981) demonstrated acid phosphatase activity in Golgi cisternae but not in structures he identified as lysosomes. Often lysosomal activity is assumed when starch grains appear to be surrounded by membranes or appear to be within the vacuole. In some cases, this appearance can be attributed to disrupted membranes or misleading planes of section. Vacuoles or related structures may function as lysosomes in red algae, but interpretations of lytic activity would be strengthened by use of the usual diagnostic enzyme localization techniques.

X. CYTOSKELETON AND MOTILITY

A. Cytoskeletal components and cytoplasmic streaming

The difficulty of obtaining good preservation of most red algae and the lability of microtubules and

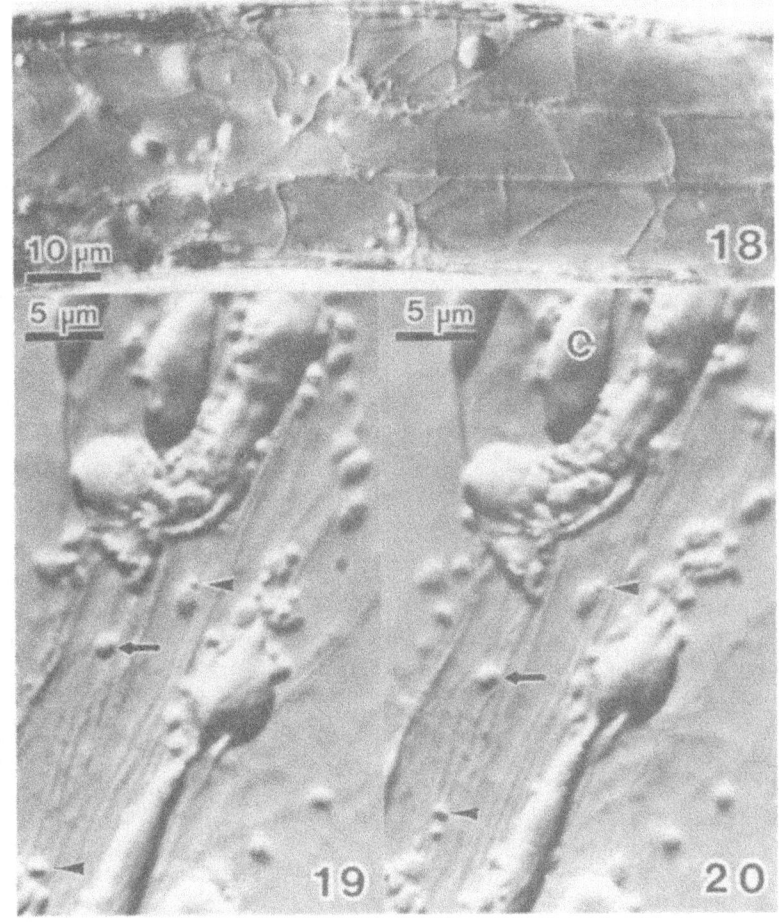

Figs. 2-18 to 2-20. Cytoskeleton, interference contrast microscopy. *2-18*: Interlaced cytoplasmic strands traverse central vacuole of *Antithamnionella spirographidis*. *2-19, 2-20*: Cytoskeletal cables in peripheral cytoplasm of *Griffithsia pacifica*. Cytoplasmic structures (arrows) changed positions along cables during 15-minute interval. C = chloroplast.

microfilaments make it difficult to assess the abundance of these structures. Furthermore, in the early ultrastructural literature, structures with some morphological similarity to tubulin-based microtubules or actin-based microfilaments were often incorrectly categorized as such. There is no question that microtubules are involved in nuclear division (see Chapter 6), and it appears microfilaments play a role in cytokinesis (Schornstein & Scott 1982; Scott 1986); however, the degree to which these two structural elements are integrated to form a cytoskeleton similar to that of higher plants and animals is not known.

Other than during nuclear division, microtubules seem to be sparse in red algal cells. A prominent bundle of parallel microtubules runs the length of the cell that subtends spermatangia in *Corallina* (Peel & Duckett 1975). The unusual morphology of spermatial vesicles (Figs. 2-36, 2-37; Broadwater & Scott 1983) may be maintained by associated microtubules, such as those demonstrated by Tripodi and de Masi (1983).

Filamentous structures bearing some resemblance to actin microfilaments have been found in nuclei (Bronchart & Demoulin 1977; Deason et al. 1983; Delivopoulos & Kugrens 1984a) and the cytoplasm (Delivopoulos & Kugrens 1984a), but diagnostic tests were not performed. More convincing ultrastructural evidence of microfilaments comes from studies of cytokinesis. Cleavage of animal cells by plasmalemmal furrowing involves a contractile ring composed of actin. Cytokinesis in red algae also occurs by furrowing, and a contractile ring is apparent in *Flintiella* and some other unicellular red algae (Schornstein & Scott 1982; Scott 1986).

Cytoplasmic streaming driven by microfilaments is conspicuous in some species of algae, such as *Chara* and *Vaucheria*, in part because the plastids are transported by the stream. Streaming in red algae is less conspicuous and much slower. The cytoplasm in axial cells of *Compsopogon* is said to circulate (Fritsch 1945), and in some ceramialean species cytoplasmic "pseudopodia" and cytoplasmic strands traversing the vacuole (Fig. 2-18) have streaming particles (Phillips 1925). Chloroplasts in red algae apparently do not participate in cytoplasmic streaming to the same extent as in *Chara*. However, plastids in *Griffithsia* can redistribute themselves by movement along cytoskeletal cables (Wilson & Pueschel unpubl.) and such structures also support the streaming of smaller organelles (Figs. 2-19, 2-20).

B. Cell motility

Some unicellular species (Pringsheim 1968; Sommerfeld & Nichols 1970) and reproductive cells (Chemin 1937; Rosenvinge 1927) have long been known to exhibit gliding motility. Gliding in *Porphyridium* is phototactic (Pringsheim 1968; Sommerfeld & Nichols 1970) and is thought to involve posterior secretion of mucilage (Lin et al. 1975). Ameboid movement of spores has also been reported (e.g., Nichols & Lissant 1967). Whether contractile processes involving actin microfilaments might play a role in generating either gliding or ameboid movement has yet to be tested.

C. Purported flagella

One of the most distinctive and phylogenetically significant features of the red algae is the absence of flagella and centrioles. Polar rings and similar structures found at the poles of division spindles have no structural similarity to centrioles. Reports of cryptic internal flagella and basal bodies (Simon-Bichard-Bréaud 1971, 1972a,b; Tripodi & de Masi 1978, 1983) did not demonstrate the diagnostic 9 × 2 + 2 microtubular substructure of flagella shafts or the 9 × 3 + 0 structure of basal bodies and centrioles. More convincing interpretations of those observations have been suggested (Broadwater & Scott 1983; Kugrens & West 1972b; Young 1977). Unusual features of spermatial vesicles are undoubtedly responsible for the original reports of internal flagella-like structures (Figs. 2-36, 2-37; Broadwater & Scott 1983).

If one allows that a vestigial flagellum might not have the full circle of nine doublet microtubules – flagella with only three or six doublets, but capable of active bending, are known (Prensier et al. 1980) – the minimum amount of structure that could distinguish a flagellar remnant from ordinary cytoplasmic microtubules is the presence of microtubular doublets with shared protofilaments. Similarly, shared protofilaments should be an acceptable criterion for recognizing vestigial centrioles. Even using such liberal criteria, no structures described in red algae qualify as flagella or centrioles. At the present time, there is no direct evidence that any extant red alga has flagella, cilia, or centrioles, or that any red alga ever did.

XI. RESERVE MATERIALS

A. Starch

In contrast to higher plant and green algal starch, floridean starch is deposited free in the cytoplasm (Figs. 2-1, 2-7, 2-21), rather than in chloroplasts, and generally is composed solely of amylopectin, rather than a mixture of amylose and amylopectin. However, amylose has been found in several unicellular red algae (McCracken & Cain 1981).

In red algae with pyrenoids, starch is usually appressed to the surface of the plastid close to the pyrenoid. In species without pyrenoids, starch is often in promixity to the nucleus, rough ER, or other structures. The significance of particular sites of starch deposition has been pondered by several authors (Borowitzka 1978; Lee 1974; Pueschel 1979; Tripodi 1971a,b; Tsekos 1982) and ascribed considerable import by some. Tripodi (1971a,b) advocated a dictyosomal origin of starch grains based on juxtaposition of electron-transparent Golgi vesicles and starch grains. In contrast, Aghajanian (1979) reported starch-mitochondrion-dictyosome associations that he believed might be important in rapid utilization of starch to power the activity of the Golgi. As far as is known, the precursor of starch polymerization and the product of starch mobilization – glucose – and the enzymes involved in these activities – ADPG transferase, phosphorylase, amylase (see Sheath et al. 1981a) – are soluble; therefore, the need for spatial associations with organelles is not obvious. Generally, the morphology and substructure of starch grains has not received much attention. Sheath et al. (1981b) demonstrated the value of scanning electron microscopy in examining starch grain morphology. They also showed concentric layering and a radiating fibrillar substructure in thin-sectioned, gelatinized starch grains.

Starch is the visible portion of a cell's carbohydrate reserve, which also includes soluble compounds like floridoside. Wiencke and Läuchli (1980) found that starch abundance decreased dramatically when *Porphyra* cells were subjected to osmotic stress. Presumably, the starch was converted into osmotically active monomers that could contribute

Cell structure

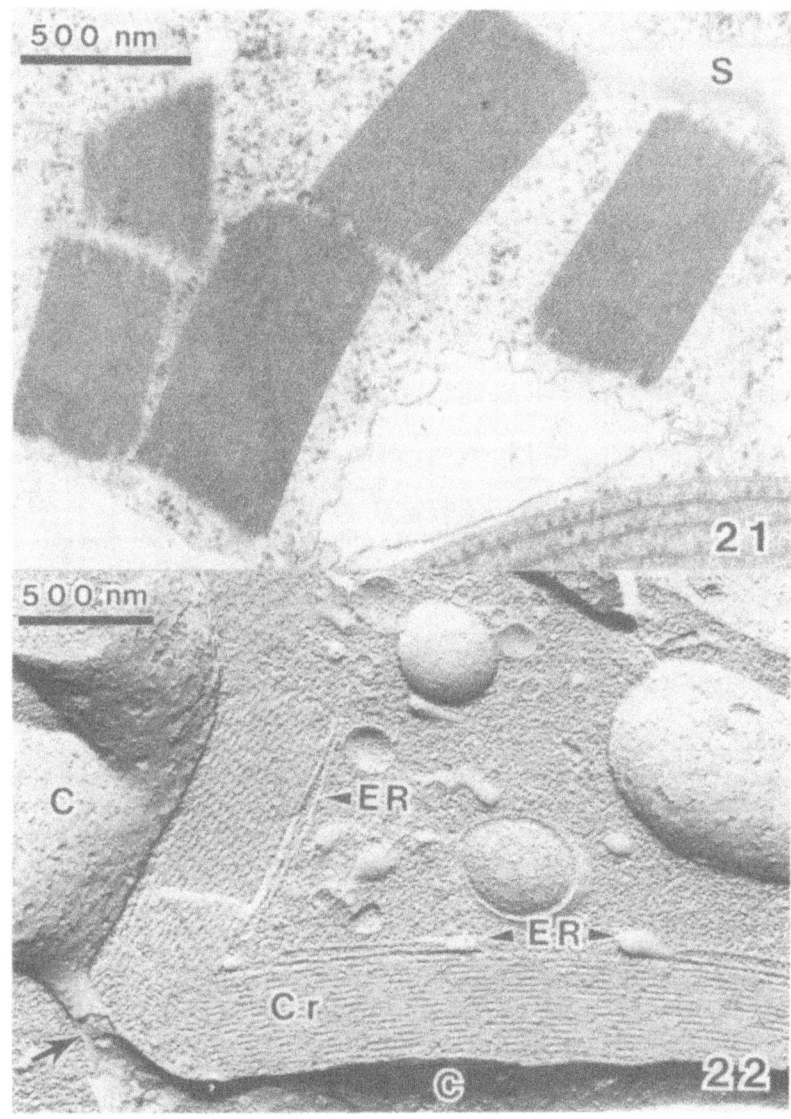

Figs. 2-21, 2-22. 2-21: *Gelidium pusillum*. Protein crystals without bounding membranes are present in cytoplasm. S = starch. 2-22: Protein crystals (Cr) in *Palmaria palmata* form sheets; they lack bounding membranes but develop in association with ER (ER) and are appressed to envelope of lobed (arrow) chloroplast (C). Freeze-fracture.

to osmoregulation. Osmoregulation during formation, discharge, dispersal, and attachment of spores might well involve interconversions of starch and osmotically active carbohydrates. This would aid in accommodating the changes in cell volume and the transitions back and forth between the walled and unwalled conditions.

B. Protein crystals

Albuminous crystals have long been known to occur in the red algae (e.g., Fritsch 1945; Kylin 1956), but their function has not yet been clearly established. They are found in abundance in some specialized cells, as in *Opuntiella* (Young 1979b), but they are also widespread and common in normal vegetative cells (Figs. 2-2, 2-21, 2-22). With the possible exception of the Porphyridiales, it is likely that protein crystals occur in all orders of red algae, though certainly not in all individuals or under all conditions. Structures with a crystalloid morphology typically have an ultrastructurally demonstrable crystalline substructure, but even spherical inclusions sometimes have such substructure (Giraud & Cabioch 1983). Kugrens (1983) reported tubules associated with crystals and he suggested that the two structures represent alternative outcomes of the polymerization of the same proteinaceous subunits. Not all crystalline inclusions are proteinaceous; those in vesicle cells of *Antithamnion* are composed of polysaccharides (Young & West 1979).

The location of the protein crystals is often described as vacuolar (e.g., Fritsch 1945), but this appearance may be the result of relocation. The formation of the proteinaceous subunits occurs in the cytoplasm, often in association with ER (Fig. 2-22), but not necessarily within ER. Bounding membranes may (Kugrens 1983) or may not be present (Fig. 2-21; Wetherbee et al. 1984a; Young 1979a,b).

The function of the protein crystals is assumed to be protein storage, but no one has experimentally tested this possibility. Wetherbee et al. (1984a) described the formation, seasonality, and amino acid composition of crystals in *Wrangelia*, and they suggested that protein crystals may be mobilized to support reproduction. It is equally likely that vegetative growth draws on protein stores if ambient nitrogen supplies are inadequate. In *Palmaria*, for example, protein crystals are abundant in medullary and inner cortex cells of some specimens (Figs. 2-2, 2-22) and may be an important factor in explaining seasonal variations of total thallus protein (reviewed by Morgan et al. 1980). In the temperate North Atlantic, a summertime diminution of nitrate concentration in ambient seawater is paralleled by a depression of nitrogen content in many macroalgae (Chapman & Craigie 1977). Under nitrogen (protein) limitation, photosynthate cannot be fully utilized for growth, so it is sequestered as starch, lipids, or soluble compounds, such as floridoside, known to peak in abundance during the summer months (e.g., Chaumont 1978; Colin & Guéguen 1930). Stored nitrogen in the summer and stored carbon in the winter can probably delay or mitigate some seasonal limitations to growth. Protein inclusions are one possible pool of stored nitrogen. Soluble compounds, such as nitrate and amino acids (Bird et al. 1982), provide another, and the phycobilin pigments may provide a third (Bird et al. 1982; Wyman et al. 1985).

Protein crystals have some potential as systematic characters. For example, *Palmaria palmata* has slender, sheetlike crystals (Fig. 2-22), whereas *P. mollis*, until recently considered conspecific with the *P. palmata*, has crystals that appear as polyhedrons (Fig. 2-2). Presumably, the contrasting crystal morphologies reflect genetically based differences in the protein monomers. Some workers have used the presence or absence of crystals as a systematic character (e.g., Robins & Kraft 1985; Searles & Ballantine 1986), but if crystals are deposited and mobilized in response to environmental conditions, this character may be suspect.

C. Lipid and other hydrocarbon inclusions

Cellular inclusions that are spherical, refractile in the light microscope, and blackened by osmium are assumed to be lipid bodies (Figs. 2-3, 2-13). Such structures occur sporadically in ordinary vegetative cells, and in some groups, especially the Bonnemaisoniales and Ceramiales, they occupy much of the volume of specialized cells. We have just begun to probe the nature of these inclusions, but it is clear they may have both immediate and long-term, large-scale ecological significance (Gschwend et al. 1985; Theiler et al. 1978).

Fat bodies, or spherosomes, are common carbon reserves in higher plants, but there is less evidence of their role in red algae. Nitrate-starved cells of *Porphyridium* store carbon as inclusions rich in triacylglycerols and as floridean starch (Köst et al. 1984). With the replenishment of nitrate, the lipids are utilized to promote the regeneration of the plastid; the stored lipids appear to function as a reservoir for membrane components (Wanner & Köst 1984). A similar process may occur in macroalgae. Osmiophilic inclusions in *Palmaria palmata* (Pueschel 1977b) resemble the lipid bodies in *Porphyridium* cells recovering from nitrogen limitation (Wanner & Köst 1984).

A variety of hydrophobic compounds and membrane-bounded accumulations of hydrophilic compounds, such as phenolics, can also be refractile, osmiophilic, and spherical. Secondary metabolites are common in red algae, and many are halogenated (Fenical 1975; Hewson & Hager 1980). Halogenated hydrophobic compounds may make up 5% of the alga's dry weight (Fenical 1975). The corps en cerise is a refractile inclusion of distinctive morphology found in the cytoplasm of trichoblast cells of some *Laurencia* species (Bodard 1968; Godin 1970a). Young et al. (1980) used x-ray spectroscopy to identify bromine in these inclusions, which appear to be brominated sesquiterpenoid alcohol.

In addition to the discrete localization of bromine in prominent inclusions, bromine has also been found by x-ray spectroscopy to occur in chloroplasts, the hyaloplasm, pit plugs, cell walls, and the cuticle (Hofsten & Pedersén 1980; Hofsten et al. 1977; Pedersén et al. 1979, 1980a,b; Roomans 1979; Westlund et al. 1981). The possibility of artifactual redistribution of compounds during specimen prep-

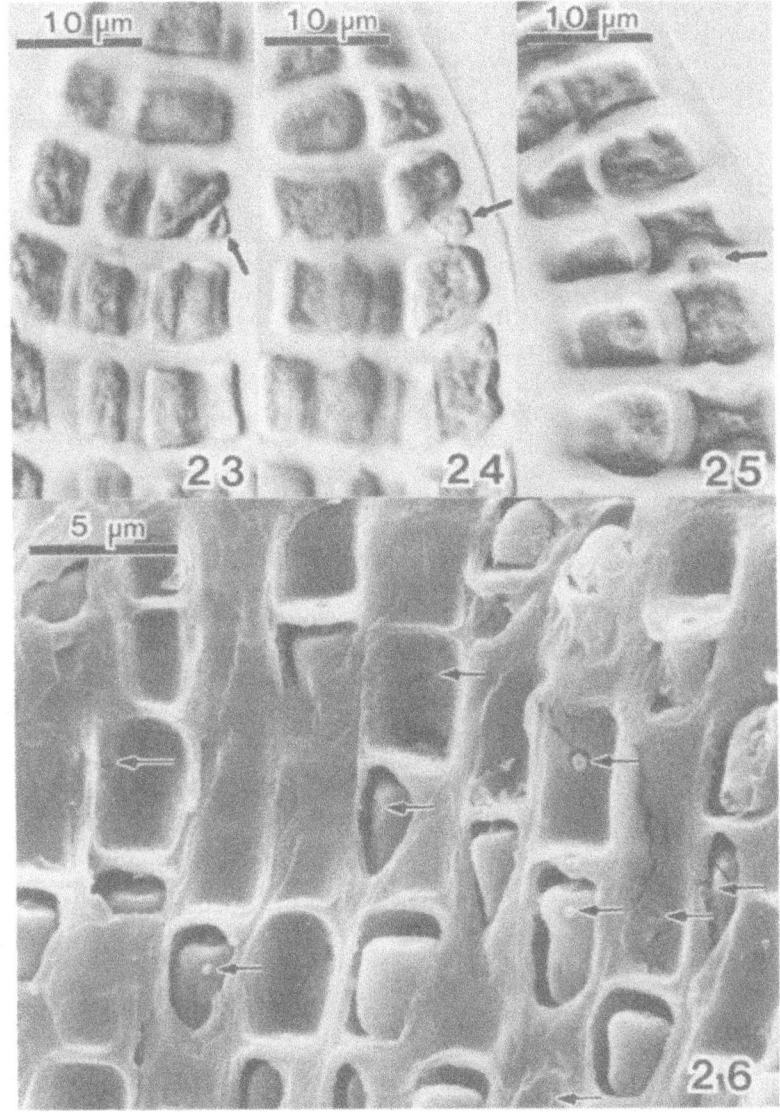

Figs. 2-23 to 2-26. 2-23 to 2-25: Stages in formation of secondary pit connections. *Polysiphonia pacifica*, interference contrast. Pericentral cell divides to produce conjunctor cell (arrows) (2-23) that contacts (2-24) and fuses with nonkindred cell below (2-25). 2-26: Scanning electron microscopy shows abundant secondary pit connections (arrows) penetrating the lateral cell walls of *Hildenbrandia crouanii*.

aration poses a serious problem, but one that can be avoided (Pallaghy et al. 1983).

XII. PIT CONNECTIONS AND PIT PLUGS

A. Formation of pit connections

In florideophytes, the only cell cleavages that continue to completion are those of the tetrasporangia. Some bangiophytes have incomplete septation in all life history stages, some always have complete septation, and some have both, depending on the stage of the life history. When septation is arrested short of completion, a connection, called a pit connection, maintains contact between the daughter cells. Pit connections do not permit cytoplasmic continuity between cells, because shortly after the furrowing ceases, a plug (Figs. 2-27 to 2-30) is deposited in the septal aperture. Pit connections lacking plugs were reported to occur in *Rhodochaete* (Boillot 1978), but plugs have now been demonstrated (Pueschel & Magne 1987).

Primary pit connections join kindred cells, cells formed by the division of a common parent cell. Since growth is apical in most red algae, most cells have only two primary pit connections. However, cells with numerous connections are not uncommon; furthermore, connections may occur between nonkindred cells. These are secondary pit connections. One pathway by which additional pit connections arise has been well known since the work of Rosenvinge (1888) and has been confirmed by modern methods (Goff & Coleman 1985, 1986; Wetherbee & Quirk 1982b; Wetherbee et al. 1984b). An unequal cell division gives rise to a small, nucleated daughter cell (Fig. 2-23) that contacts (Fig. 2-24) and subsequently fuses with (Fig. 2-25) an adjacent, nonkindred cell. As a result, the pit connection formed during the cell division ultimately is shared between nonkindred cells, and a cell is rendered multinucleate by the cell fusion. This pattern is most conspicuous in the Ceramiales. An alternative means of acquiring secondary pit connections is believed to be present in the Hildenbrandiales and in the Corallinales. Despite the presence of numerous pit connections (Fig. 2-26), the cells involved remain uninucleate (Pueschel 1988b). Presumably, the connections arise by perforation of a shared wall, followed by deposition of a pit plug in the aperture (Cabioch 1971). Another process by which cells may acquire numerous pit connections, which could be primary in origin, is by direct cell fusion. Evidences of this process include a pit connection across a partial septum and anatomical configurations in which the outlines of the original unfused cells are still recognizable (Fig. 2-2).

B. Structure and formation of pit plugs

The only structural constituent common to all pit plugs is the plug core (Fig. 2-27). This is also the only component of the plug whose formation is well understood. Within the membrane-lined aperture left by incomplete furrowing, tubular membranes appear (Davis & Scott 1986; Pueschel & Cole 1982; Ramus 1969; Scott et al. 1980) and a homogeneously granular mass of protein (Pueschel 1980b), the plug core, is deposited around them. The tubular membranes eventually disappear. The presence of additional structural components depends on the taxonomic group involved (Pueschel 1987, 1989; Pueschel & Cole 1982), as summarized in Table 2-1. *Ahnfeltia*, *Rhodochaete*, and *Compsopogon* have nothing more than a plug core (Maggs & Pueschel 1989; Pueschel & Magne 1987; Scott et al. 1988). Most orders have an associated membrane, called the cap membrane, at each end of the plug core (Fig. 2-27). The cap membrane appears to be entirely absent from some orders (Fig. 2-28; Pueschel 1987) and absent from particular cells of other orders (Broadwater & Scott 1982; Kugrens & Delivopoulos 1985; Peyrière 1977; Wetherbee 1979; Wetherbee & Kraft 1981). In several orders, one or two layers of material covers each end of the plug core (Figs. 2-28 to 2-30). When two cap layers and a cap membrane are present, the membrane separates the two layers. The outer layer may be a thin layer (Fig. 2-30) or a large dome (Fig. 2-28). This outer layer contains a carbohydrate component (Pueschel 1980b; Pueschel & Cole 1982). The Golgi apparatus and ER are reported to contribute to the deposition of the outer cap layer (Aghajanian & Hommersand 1978; Millson & Moss 1985). In *Palmaria* the outer layer appears to assemble from plaques or vesicles of uncertain origin (Pueschel & Cole 1982). Little is known about the composition or origin of the inner cap layer.

The cap membrane looks, stains, fractures, and extracts like a membrane (Figs. 2-27, 2-29 to 2-31), but its junction with the plasmalemma is so anomalous that one must question whether it is truly a unit membrane (Pueschel 1977a). The plasmalemma is continuous from cell to cell along the sides of the plug core. The cap membrane appears in both thin-section (Fig. 2-29) and freeze-fracture micrographs (Fig. 2-31) to be continuous with the plasmalemma by means of a junction that could be viewed as three phospholipid monolayers associated in paired permutations (Pueschel 1977a). Analogous configurations in other organisms are extremely rare. The origin of the cap membrane is unknown.

Table 2-1. Ordinal distribution of pit plug features

	Number of cap layers	Cap membrane
Rhodochaetales	0	−
Compsopogonales	0	−
Bangiales	1	−
Ahnfeltiales	0	−
Acrochaetiales	2[a]	+
Nemaliales	2[b]	+
Palmariales	2[b]	+
Corallinales	2[c]	−
Batrachospermales	2[c]	?
Hildenbrandiales	1	+
Gelidiales	1	+
Gracilariales	0	+
Bonnemaisoniales	0	+
Gigartinales[d]	0	+
Rhodymeniales	0	+
Ceramiales	0	+

[a] Some species have thin outer cap layer; others have domelike outer cap layer.
[b] Outer cap layer forms a thin plate.
[c] Outer cap layer forms a large dome.
[d] Includes the Cryptonemiales (Kraft & Robins 1985).

Cell structure

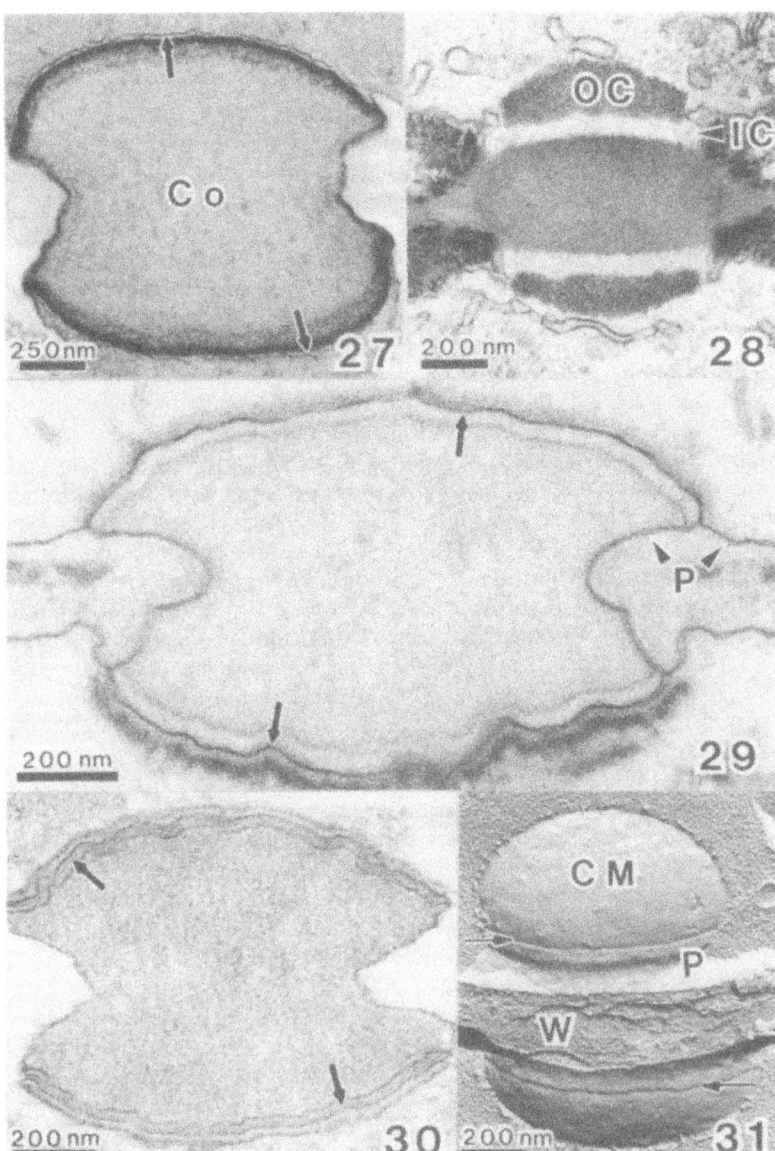

Figs. 2-27 to 2-31. Pit plugs. *2-27*: *Neodilsea borealis*. Plug core (*Co*) is separated from cytoplasm by only cap membrane (arrows), normal fixation. *2-28*: *Corallina officinalis*. Two-layered plug cap is present on either side of plug core, permanganate fixation. Inner cap (*IC*) and outer cap (*OC*) layers are not separated by cap membrane. *2-29 to 2-31*: Demonstration of cap membrane between two layers of plug cap in Palmariales. *2-29*: *Palmaria mollis* fixed with potassium permanganate to make membranes conspicuous. Cap membrane (arrows) appears continuous with plasmalemma (*P*), which goes from cell to cell along sides of pit plug. (From Pueschel 1987, with permission.) *2-30*: *Devaleraea ramentacea*. Lipid extraction causes cap membrane (arrows) to appear in negative image. (From Pueschel & Cole 1982, with permission.) *2-31*: *P. palmata*. Freeze-fracture, cap membrane (*CM*), and plasmalemma (*P*) are continuous along line marked by an arrow (exoplasmic fracture faces). The plasmalemma lining the plug core is hidden by septal wall (*W*).

Degeneration of pit plugs occurs when one of the connected cells dies (Giraud & Cabioch 1981; Pueschel 1988a; Pueschel & van der Meer 1985) or the distal cell differentiates into a spore or spermatium (Pueschel 1979). The plug is then sealed off by deposition of wall material by the living cell. In postfertilization development, pit plugs may be resorbed (Fig. 2-35) or ruptured (Delivopoulos & Kugrens 1984a, 1985b; Kugrens & Arif 1981; Tsekos & Schnepf 1985); this has also been found in some vegetative cells (Aghajanian & Hommersand 1978).

C. Function

The function of pit connections and their plugs has been the subject of much speculation, but proof has been elusive. The pit connection is the only means of symplastic communication in red algae, so a role in transport or cellular communication is a reasonable hypothesis, although the cap membrane would appear to provide a membranous barrier. In some ceramialean species, the plug core enlarges several-fold after the cap membranes are in place (e.g., Konrad Hawkins 1972). If constituents can be transported across the cap membranes to support growth of the core, surely, small molecules can be transported from cell to cell through the plug. Pit plugs lacking cap membranes and with enlarged surface area between particular cells in the carposporophyte were dubbed "transfer connections" by Wetherbee (1979) with the assumption that this configuration enhances intercellular transport. This syndrome has been reported in the reproductive apparatus (Broadwater & Scott 1982; Tsekos & Schnepf 1985) and in some vegetative cells (Scott et al. 1982; Wetherbee & Kraft 1981). However, Ramm-Anderson and Wetherbee (1982) found that plugs in likely positions to carry on extensive transport in *Nemalion* did not exhibit the "transfer connection" syndrome. Broadwater and Scott (1982) noted that the pit plugs lacking cap membranes in the female reproductive apparatus of *Polysiphonia harveyi* were eventually removed; they suggest that the absence of cap membranes may mark plugs for removal.

One of the most intriguing recent findings regarding intercellular communication comes from studies of somatic hydrids of different strains of *Griffithsia* (Koslowsky & Waaland 1984, 1987). A cytoplasmic incompatibility factor moves by diffusion from the fused somatic cell to adjacent cells, but it does not diffuse through lateral cell walls. One must conclude that either the crosswalls are different from the lateral walls and permit diffusion of the factor, or passage from cell to cell is through the pit plug. Bauman and Jones (1986) were unable to demonstrate dye coupling or movement of cobalt ions through pit plugs of *Griffithsia*, but some of their electrophysiological data suggest electrotonic coupling of cells. The structural diversity of pit plugs (Pueschel 1987; Pueschel & Cole 1982) greatly complicates the explication of a universal function of pit plugs.

If there is a difference in the function of primary and secondary pit connections, it is not reflected in fine structure. Despite the differences in their formation, the pit plugs of primary and secondary connections are identical in organization, except in the special case of secondary connections joining cells of two different species (Kugrens & West 1973b; Wetherbee & Quirk 1982a). *Hildenbrandia* provides an extreme example of the development of secondary pit connections (Fig. 2-26; Pueschel 1988b). The most plausible explanation for the abundance of secondary connections in this alga is that they improve cohesiveness of the filaments (Pueschel 1988b). This does not suggest that they are incapable of transporting, but transport does not provide a credible reason for their abundance.

XIII. CELL COVERINGS

There is probably no red algal cell, vegetative or reproductive, that entirely lacks a cell covering. Depending on the chemical constituents present (see Chapter 10), the covering may be a continuously dissolving mucilaginous film (Fig. 2-1), or a thick, firm matrix that contributes to thallus integrity (Figs. 2-2, 2-26); it may even be calcified. Ultrastructural images of the wall are not always accurate representations. The shrinkage and swelling of both the cell contents and the cell covering during different stages of specimen preparation alters dimensions and can leave empty space. In addition, wall constituents that do not bind the normal fixatives and stains may not be retained or, if electron transparent, will be invisible.

A typical vegetative cell wall consists of microfibrils in an electron-transparent, amorphous matrix (Fig. 2-2; see Chapter 10). The microfibrils in florideophytes are often referred to as cellulose by ultrastructuralists (Brawley & Wetherbee 1981; Duckett & Peel 1978). However, this view is not shared by wall chemists (McCandless 1981), and it is not supported by ultrastructural enzyme digestion results (Lichtlé 1975). Microfibrils in *Polysiphonia* cell walls were removed by β-glucuronidase, but not by cellulase, demonstrating that noncellulosic constituents, like cellulose, can form microfibrils.

The surface of most macroalgae is covered by a layer of material called the cuticle that is chemically and structurally distinct from the walls. It is formed from protein-rich cellular secretions that presumably migrate through the cell wall to the surface of the thallus. In some species, the cuticle is multilayered, and this layering is believed to be responsible for the phenomenon of iridescence, as in *Iridaea* (Gerwick & Lang 1977) and *Chondrus* (Pedersén et al. 1980b). In contrast, intracellular structures are believed to cause iridescence in other algae (Feldmann 1970a,b).

XIV. SPECIALIZED VEGETATIVE CELLS

A. Rhizoids

Rhizoids may be either unicellular or multicellular, but their negative phototropism and elongation by tip growth distinguish them from vegetative shoots (e.g., Waaland et al. 1977). Although rhizoids are less pigmented than upright cells, this is due in part to extensive vacuolation. Plastids are present in rhizoids (Rawlence & Taylor 1972), even in the penetrating rhizoids of parasitic red algae (Goff 1979). Most work on rhizoidal cells has been done on host-penetrating rhizoids of epiphytes (Rawlence 1972; Rawlence & Taylor 1972) and parasites (Goff 1979). Demonstration of the cellular processes associated with attachment of rhizoids to inorganic substrate is lacking.

B. Hair cells

Elongate, usually colorless hair cells, are found on the surface of many red algae. Environmental induction of hairs has been demonstrated (West 1971), and it is likely that they have a role in nutrient uptake. These cells may be short-lived and replaced by regeneration from the subtending cell (Duckett et al. 1974). Magruder (1984) proposed that hairs on female gametophytes may provide the initial attachment site for spermatia, which later bind to trichogynes. Some hair cells are apoplastidic as a result of the cell division that forms them; during cytokinesis, the mitotic product that will become a hair does not receive chloroplasts (Duckett et al. 1974; Hymes & Cole 1983a). At maturity, the hairs of *Audouinella hermannii* are thin-walled and highly vacuolated (Hymes & Cole 1983a).

Branching, multicellular filaments called trichoblasts are found in some members of the Ceramiales. The presence of prominent inclusions composed of brominated compounds (Young et al. 1980) has prompted some work on these cells. Plastids are present in trichoblasts of *Laurencia* (Godin 1970b).

C. Secretory and storage cells

Cells clearly distinctive in character but of uncertain function have been given names like gland cells, secretory cells, and vesicle cells. Some sorting of these cells is now possible. For example, the so-called gland cells in the inner cortex of *Botryocladia* have extensive Golgi and secretory activity (Young 1978). Apparently these cells provide much of the mucilage that fills the bladderlike thalli of this alga. Large, colorless cells in the cortex of *Opuntiella* store proteinaceous material that is eventually mobilized (Young 1979b).

Ascertaining the function of many cell types is difficult because distinctive inclusions persist rather than being secreted. Cells that specialize in accumulating halogenated products are especially common in members of the Bonnemaisoniales and Ceramiales. The vesicle cells of *Antithamnion* (Codomier et al. 1983; Feldmann & Guglielmi 1976; Young & West 1979) and of some members of the Bonnemaisoniaceae (Codomier et al. 1983; Wolk 1968) are particularly well studied.

D. Transfer cells

The epithallial cells of many members of the Corallinales possess prominent ingrowths of the distal cell wall. In higher plants, cells with such structural specializations are referred to as transfer cells, and the large plasmalemmal surface associated with the labyrinthine wall is believed to enhance solute transport. Borowitzka and Vesk (1978) first applied the term "transfer cell" to the coralline epithallial cells. They proposed that these cells function in nutrient uptake, perhaps in addition to polysaccharide secretion, as had been earlier suggested (Cabioch & Giraud 1978; Giraud & Cabioch 1976). Bressan et al. (1981) believe that transfer cells engage in the uptake of calcium ions and are important in calcification.

XV. REPRODUCTIVE CELLS

A. Female reproductive apparatus

The female reproductive apparatus, the fertilization process, and the early postfertilization development of the carposporophyte are among the least studied aspects of red algal fine structure. The only ultrastructural study that deals specifically with fertiliza-

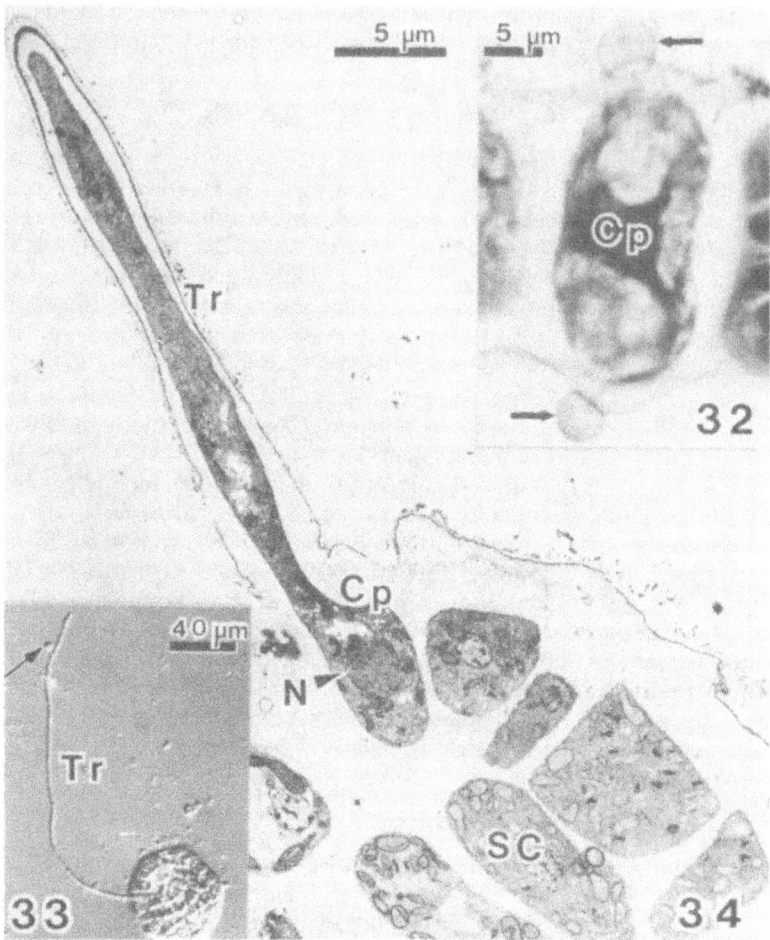

Figs. 2-32 to 2-34. 2-32: *Porphyra gardneri*, spermatia (arrows) attached to the carpogonium (Cp) in two regions; brightfield image of stained section. (From Hawkes 1978, with permission.) 2-33: *Palmaria palmata*, spermatium (arrow) attached to trichogyne (Tr), which is larger than the entire gametophyte; interference contrast. 2-34: Unfertilized carpogonial branch of *Polysiphonia harveyi*. Carpogonium (Cp) is terminal on four-celled branch borne by support cell (SC). Swollen base of carpogonium contains nucleus (N), but otherwise contents of the trichogyne (Tr) are much the same as in swollen portion of carpogonium. Other than nuclei, the most conspicuous structures in carpogonial branch cells are proplastids. (From Broadwater & Scott 1982, with permission)

tion is that of Hawkes (1978) on the bangiophyte, *Porphyra gardneri*. In this monostromatic species, the spermatia have access to the carpogonia from both sides of the blade (Fig. 2-32), and both exposed faces of the carpogonia are receptive. Although the receptive surface may be elaborated somewhat, a distinct trichogyne is not formed (cf. Figs. 2-32, 2-33). A channel in the carpogonial wall develops; then the nucleus and a portion of the spermatial cytoplasm pass into the carpogonium. The rest of the spermatium's cytoplasm and its wall remain behind (Hawkes 1978). Several Feulgen-stained structures, presumably nuclei, were found in the carpogonium; apparently after one spermatial nucleus achieves karyogamy, the others degenerate. It might be expected that changes in the carpogonial wall would be necessary to make it a receptive surface for attachment of spermatia. This changing wall chemistry in fertile thalli of *Bangia* has been examined (Cole et al. 1985).

The ultrastructural literature on development of the carpogonial branch and its comparative fine

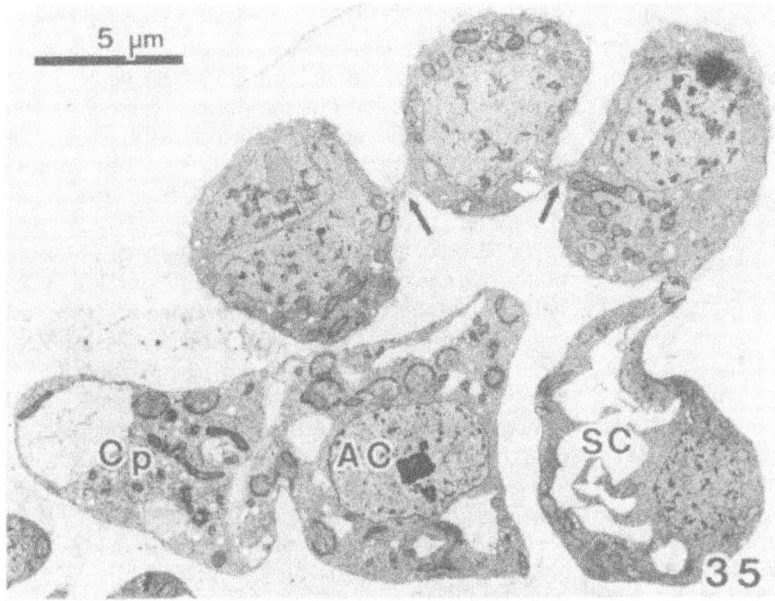

Fig. 2-35. Fertilized carpogonium (Cp) of *Polysiphonia harveyi* has shed its trichogyne and fused with auxiliary cell (AC), which was formed by support cell (SC) after fertilization occurred (see Chapter 13 for alternate interpretation). Pit plugs between cells of carpogonial branch dissolved, leaving open connections (arrows); the connection between carpogonium and its subtending cell is not shown. (From Broadwater & Scott 1982, with permission)

structure after fertilization is sparse. In *Polysiphonia harveyi* the cells of the carpogonial branch usually are binucleate, except for the carpogonium itself (Broadwater & Scott 1982). A long-standing dispute of whether the trichogyne possesses a nucleus in addition to the one at the base of the carpogonium (Dixon 1973; Fritsch 1945) has been resolved for this species, at least; a single nucleus resides in the swollen base (Fig. 2-34). The same appears to be true in *Leachiella* (Kugrens 1982). On the basis of light microscopic study, carpogonial branch cells in most orders were thought to lack chloroplasts (Fritsch 1945), but proplastids have been found throughout the carpogonial branch of *Polysiphonia* (Figs. 2-34, 2-35), even in the trichogyne (Broadwater & Scott 1982). Kugrens (1982) reported plastids in the carpogonium of a parasitic species. The trichogyne is an interesting but poorly studied structure. Broadwater and Scott (1982) found the trichogyne wall to have a weakly consolidated coating of material, perhaps important in receptivity to spermatia. The rapid formation of the short-lived trichogyne, the attachment and penetration of the trichogyne wall by the spermatium, the migration of the spermatial nucleus, and the walling off of the trichogyne after fertilization are processes that require detailed ultrastructural study.

Progress has been made in examining the post-fertilization development (Fig. 2-35) that gives issue to the carposporophyte (Brown 1969; Delivopoulos & Kugrens 1984a,b, 1985b; Delivopoulos & Tsekos 1985a; Kugrens 1982; Kugrens & Arif 1981; Kugrens & Delivopoulos 1985, 1986; Ramm-Anderson & Wetherbee 1982; Tripodi & de Masi 1977; Tsekos & Schnepf 1983, 1985; Wetherbee 1980). Some of the emerging patterns of early carposporophyte development are discussed by Kugrens and Delivopoulos (1986). Particular attention has been given to the fusion cell, which contains both haploid gametophyte nuclei and diploid nuclei derived from zygotic mitoses. One of the most intriguing questions of carposporophyte development is how the nuclei are sorted so that only the diploid nuclei give rise to carposporophyte cells. Spherical, electron-dense structures in multinucleate cells have been interpreted as remnants of haploid nuclei (Kugrens & Delivopoulos 1985; Tripodi & de Masi 1977). However, similar inclusions are present in tetrasporangia of *Dasya* (Broadwater et al. 1986a) and dense, granular materials of various morphologies are commonly found in carposporophyte cells (Delivopoulos & Tsekos 1985a; Wetherbee 1980). One of the more unusual findings concerning carposporophyte development is the subdivision of the fusion cell's cytoplasm in *Asterocolax* by a barrier of electron-dense material (Kugrens & Arif 1981). This compartmentation is believed to separate the haploid and diploid nuclei (Kugrens & Arif 1981).

B. Spermatia

Spermatia are less similar to the sessile female gametes than they are to spores, which share their requirements for liberation, dispersal, and attachment to an appropriate surface. The appropriate surface for spermatium attachment is the carpogonium. This particular substratum differs from most in shape, size, frequency, and position; therefore, one might expect some specializations of the spermatia not shared by spores. Spermatia of some members of the Corallinales have trailing appendages that are almost flagella-like in appearance (see Duckett & Peel 1978). Peel and Duckett (1975) found these appendages to be of the same composition as the spermatial coat. Magruder (1984) found that under very specific methods of preparation, elaborate surface appendages could be demonstrated in *Aglaothamnion*, and these were shown to be the primary means of attachment of the spermatium to the trichogyne. Mucilaginous projections also are found on *Bangia* spermatia (Cole et al. 1985).

Mature spermatia lack some of the cytoplasmic structures that are so abundant in spores. Although the cytoplasm of developing spermatia may contain abundant Golgi vesicles, these vesicles discharge their contents either to the cell exterior, creating an investing layer of mucilage, or to the spermatial vesicles (see below; Figs. 2-36, 2-37). The released spermatia are colorless, and they have few cytoplasmic vesicles and little starch (Fetter & Neushul 1981; Kugrens 1980; Kugrens & West 1972b; Scott & Dixon 1973b). If plastids are present, they are small and have few lamellae (Figs. 2-36, 2-37). In contrast, released spores typically have an abundance of vesicles that presumably aid in attachment; chloroplasts are numerous and highly pigmented, and starch is abundant (Fig. 2-7). These differences in cell contents reflect the differing fates of the cells.

Both spores and spermatia are covered by a thick layer of mucilage, and they have the same need for release and dispersal. One of the striking features of the differentiation of spermatia is the formation of one or two massive, mucilage-containing vesicles (Figs. 2-36, 2-37) called spermatial vesicles (Broadwater & Scott 1983; Fetter & Neushul 1981; Kugrens 1980; Scott & Dixon 1973b). These derive originally from ER (Fig. 2-36) but may also receive contributions from Golgi vesicles. By maturity, these vesicles may occupy the majority of the cell volume. Discharge of the spermatial vesicles occurs as the spermatia are released (e.g., Kugrens 1980; Peyrière 1974; Scott & Dixon 1973b). Although the function of these huge vesicles has been postulated to involve discharge and separation from the subtending pit connection (e.g., Kugrens & West 1972b), spermatial vesicles are present in members of the Bangiaceae (Cole & Sheath 1980; Hawkes 1978), which lack pit plugs in the gametophytes. Another explanation for the function of the spermatial vesicles is that they are the most likely source of the mucilaginous appendages and sheets (Fetter & Neushul 1981; Magruder 1984).

The reader is referred to Duckett and Peel (1978), who reviewed the red algal male gamete in more detail than is possible here.

C. Spores

Studies of the differentiation of spores make up a large fraction of the red algal ultrastructural literature. However, our understanding of the biology of the spores is severely limited because so little work has been done on released spores and on the processes of spore attachment and spore germination.

In monosporogenesis and carposporogenesis the sporocyte becomes the spore. In bisporogenesis, tetrasporogenesis, and polysporogenesis, the sporocyte undergoes mitosis, meiosis, or both, so that two, four, or more spores issue from one sporangium. Although these types differ in the number of karyokinetic and cytokinetic events, the cytoplasmic activities in all types of spore production follow the same basic pattern: modification of the preexisting sporocyte wall, production of mucilage by Golgi (Fig. 2-4) and sometimes ER, multiplication of plastids, accumulation of starch, and formation of adhesive vesicles (e.g., Vesk & Borowitzka 1984). When multiple spores are produced, the nuclear processes can provide a benchmark to help deduce the progress of cytoplasmic activities. For instance, in *Palmaria* (Pueschel 1979), *Hildenbrandia* (Pueschel unpubl.), and *Dasya* (Broadwater et al. 1986a), there is a diffuse diplotene stage during which the tetrasporocyte enlarges considerably. In at least the first two genera, mucilage deposition is initiated at this time. Without such indices, it is difficult to compare sporogenesis in different

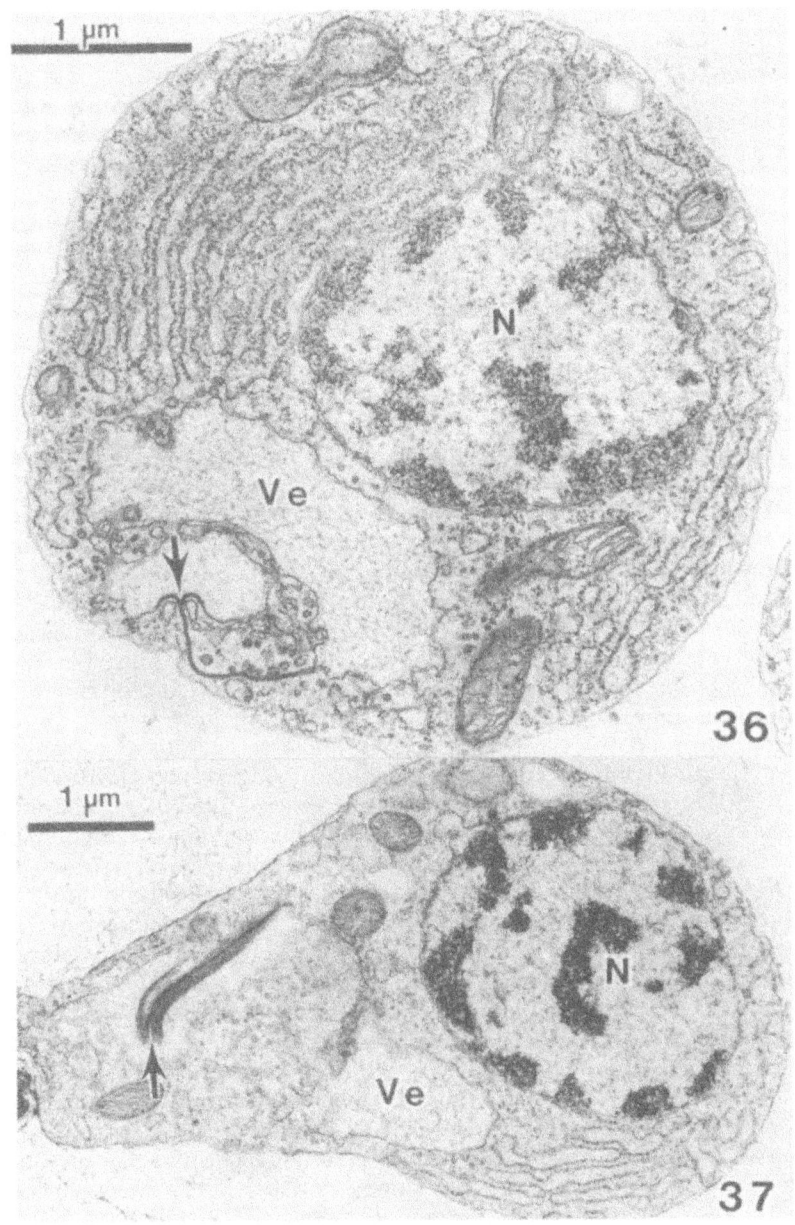

Figs. 2-36, 2-37. Differentiating spermatia of *Polysiphonia harveyi* contain massive, ER-derived vesicles (Ve). The unusual configuration of the vesicle membrane at the site where ER is attached (arrows) may be responsible for some claims of flagella in red algae. N = nucleus. (From Broadwater & Scott 1983, with permission)

species, as is apparent from Vesk and Borowitzka's (1984) tabulation of studies on tetrasporogenesis.

Several different types of vesicles are produced during sporogenesis; their names describe their appearance: striated, striped, dark-cored, fibrous. A nomenclature based entirely on function is not possible at this time; however, the vesicles can be categorized according to time and manner of their formation and disappearance. Striated vesicles are formed from dilated ER. Their proteinaceous contents have a periodic substructure that produces a striated appearance in thin-section images (Kugrens & West 1973a; Wetherbee & West 1976, 1977). These vesicles are formed early in sporogenesis, but they disappear before the spore is fully mature. Their mobilization involves an intimate and unusual interaction with the Golgi apparatus (Wetherbee 1978b; Wetherbee & West 1976, 1977). Striated vesicles have been reported only in members of the Ceramiales. Striped vesicles, also found in a ceramialian alga (Kugrens & West 1974), are formed by ER and, like striated vesicles, they are mobilized before spore maturation. The function of these structures is unknown.

During the early stages of sporogenesis, the Golgi apparatus produces vesicles that contribute to mucilage secretion (Fig. 2-4; Chamberlain & Evans 1973; Kugrens & West 1974; Pueschel 1979). The contents of these vesicles are discharged to the cell surface or may be added to the much larger mucilage sacs for later secretion. The abundance of mucilage vesicles in the cytoplasm diminishes late in sporogenesis, a time at which the Golgi apparatus switches from mucilage production to the formation of a different type of vesicle, the adhesive vesicle (Fig. 2-7). In most species examined, adhesive vesicles have a dense core surrounded by a fibrillar cortex (Kugrens & West 1973a, 1974; Pueschel & Cole 1985; Ramm-Anderson & Wetherbee 1982; Scott & Dixon 1973a), but in some species, the vesicles' contents are evenly electron dense (Pueschel 1979; Vesk & Borowitzka 1984). Vesk and Borowitzka (1984) consider these differences to be real, but a differing view is taken by Kugrens and Delivopoulos (1986). Regardless of their appearance, adhesive vesicles share the features of a dictyosomal origin, of contents rich in protein and carbohydrate (Chamberlain & Evans 1973; Pueschel 1979), and of still being present in the cytoplasm of the releasing spores. The last observation led to the suggestion that upon settling of the spore, these vesicles discharge a substance involved in fixing the spore to the substrate (Chamberlain & Evans 1973). This inference has gained support from study of carpospores of *Porphyra variegata* (Pueschel & Cole 1985). The attached spore no longer contains adhesive vesicles; instead, material resembling the vesicle contents coats the spore and forms a confluent ventral mass. The formation of a true cell wall follows this initial attachment process (Pueschel & Cole 1985). Germination of the spore breaks through the layer of adhesive mucilage.

It is not possible to review the sporogenesis literature in detail here. The reader is referred to Chapter 14 and to recent studies and reviews of monosporogenesis (Hymes & Cole 1983b), carposporogenesis (Kugrens & Delivopoulos 1986), tetrasporogenesis (Vesk & Borowitzka 1984), and polysporogenesis (Sheath et al. 1987).

XVI. SUMMARY

The red algae show a wide diversity of cellular organization, ranging from unicellular species with a single nucleus, chloroplast, and mitochondrion, to species with dozens of nuclei and hundreds of chloroplasts and mitochondria. However, it is the organelles that provide the most distinctive attributes of red algal cytology. The chloroplasts have accessory pigment granules, the phycobilisomes, on the stromal surface of single, unstacked photosynthetic membranes. The three-dimensional organization of the photosynthetic membrane system is not well understood, but it appears that in some species these membranes are coiled. Pyrenoids are present in the chloroplasts of relatively few red algae.

The Golgi apparatus forms regular spatial associations with other organelles. Depending on the species, the cis face of the Golgi body may be in proximity to endoplasmic reticulum, the nuclear envelope, or both endoplasmic reticulum and a mitochondrion. In contrast to animal cells, the Golgi apparatus in red algae appears to function principally by cisternal progression. The ER in unicellular species is interesting in that it forms a peripheral network from which outwardly directed tubules appear to connect to the plasmalemma.

It is evolutionarily significant that centrioles and flagella are apparently absent in red algae. Other types of structures, most commonly ring shaped, occupy the poles of dividing nuclei. Although flagellated motility is absent, gliding and ameboid motion are exhibited by some spores and unicellular species.

Many components of red algal cells are not well characterized. Structures with the appearance of microbodies are common, but catalase localization, which is diagnostic for microbodies in higher plants and animals, has not been demonstrated. Little is known about lysosomes or vacuoles. A cytoskeleton of cables along which organelles move is apparent in some cells; the relative contributions of microtubules and microfilaments to this system is unknown.

A variety of cytoplasmic inclusions that presumably have a storage function are present in vegetative and reproductive cells. Starch occurs free in the cytoplasm. Protein crystals are deposited within or in association with ER. Lipids and other hydrocarbons, some of them halogenated, also form inclusions.

Intercellular connections in red algae are occluded by structures, pit plugs, that are variable in several aspects of their organization. Despite this diversity of form, these connections provide the only possible means of symplastic communication. Cell fusions play an important role in both vegetative and reproductive developmental processes.

Vegetative cells of specialized structure and function are found in red algae; these include rhizoids, hair cells, secretory cells, and storage cells. Few of these approach the complex differentiation exhibited by the spores and spermatia. We have only begun to explore the many complexities of fertilization and postfertilization development at the ultrastructural level.

XVII. REFERENCES

Aghajanian, J. G. 1979. A starch grain-mitochondrion-dictyosome association in *Batrachospermum* (Rhodophyta). *J. Phycol.* 15: 230–2.

Aghajanian, J. G. & Hommersand, M. H. 1978. The fine structure of the pit connections of *Batrachospermum sirodotii* Skuja. *Protoplasma* 96: 247–65.

Alley, C. D. & Scott, J. L. 1977. Unusual dictyosome morphology and vesicle formation in tetrasporangia of the marine red alga *Polysiphonia denudata*. *J. Ultrastruct. Res.* 58: 289–98.

Avanzini, A. & Honsell, G. 1984. Membrane tubules in the tetraspores of a red alga. *Protoplasma* 119: 156–8.

Barnes, K. L., Craigie, R. A., Cattini, P. A. & Cavalier-Smith, T. 1982. Chromatin from the unicellular red alga *Porphyridium* has a nucleosome structure. *J. Cell Sci.* 57: 151–60.

Bauman, R. W. & Jones, B. R. 1986. Electrophysiological investigations of the red alga *Griffithsia pacifica* Kyl. *J. Phycol.* 22: 49–56.

Bergfeld, R. 1970. Die Feinstruktur und Entwicklung der Plastiden einiger Süsswasserrotalgen. *Cytobiologie* 1: 411–19.

Billard, C. & Fresnel, J. 1986. *Rhodella cyanea* nov. sp., une nouvelle Rhodophyceae unicellulaire. *C. R. Acad. Sci.* (Paris) sér. 3, 302: 271–6.

Bird, K. T., Habig, C. & DeBusk, T. 1982. Nitrogen allocation and storage patterns in *Gracilaria tikvahiae* (Rhodophyta). *J. Phycol.* 18: 344–8.

Bisalputra, T. 1974. Plastids. In *Algal Physiology and Biochemistry*, ed. W. D. P. Stewart, pp. 124–60. Berkeley: University of California Press.

Bisalputra, T. & Bailey, A. 1973. The fine structure of the chloroplast envelope of a red alga, *Bangia fusco-purpurea*. *Protoplasma* 76: 443–54.

Bisalputra, T. & Bisalputra, A. 1967. The occurrence of DNA fibrils in chloroplasts of *Laurencia spectabilis*. *J. Ultrastruct. Res.* 17: 14–22.

Bodard, M. 1968. L'infrastructure des "corps en cerise" des *Laurencia* (Rhodomelacées, Ceramiales). *C. R. Acad. Sci.* (Paris) sér. D, 266: 2393–6.

Boillot, A. 1978. Les ponctuations intercellulaires du *Rhodochaete parvula* Thuret (Rhodophycées, Bangiophycidées). *Rev. Algol.* N.S. 13: 251–8.

Borowitzka, M. A. 1978. Plastid development and floridean starch grain formation during carposporogenesis in the coralline red alga *Lithothrix aspergillum* Gray. *Protoplasma* 95: 217–28.

Borowitzka, M. A. & Vesk, M. 1978. Ultrastructure of the Corallinaceae. I. The vegetative cells of *Corallina officinalis* and *C. cuvierii*. *Mar. Biol.* 46: 295–304.

Bouck, G. B. 1962. Chromatophore development, pits, and other fine structure in the red alga, *Lomentaria baileyana* (Harv.) Farlow. *J. Cell Biol.* 12: 553–69.

Brawley, S. H. & Wetherbee, R. 1981. Cytology and ultrastructure. In *The Biology of Seaweeds*, eds. C. S. Lobban & M. J. Wynne, pp. 248–99. Berkeley: University of California Press.

Bressan, G., Ghirardelli, L. A. & Bellemo, A. 1981. Research on the genus *Fosliella* (Rhodophyta, Corallinaceae): structure, ultrastructure and function of transfer cells. *Bot. Mar.* 24: 503–8.

Broadwater, S. & Scott, J. 1982. Ultrastructure of early development in the female reproductive system of *Polysiphonia harveyi* Bailey (Ceramiales, Rhodophyta). *J. Phycol.* 18: 427–41.

Broadwater, S. & Scott, J. 1983. Fibrous vacuole associated organelles (FVAOs) in the Florideophyceae: a new interpretation of the "appareil cinetique." *Phycologia* 22: 225–33.

Broadwater, S. & Scott, J. 1986. Three-dimensional reconstruction of the chondriome of the unicellular red alga *Rhodella reticulata*. *J. Cell Sci.* 84: 213–19.

Broadwater, S., Scott, J. & Pobiner, B. 1986a. Ultrastructure of meiosis in *Dasya baillouviana* (Rhodophyta). I. Prophase I. *J. Phycol.* 22: 490–500.

Broadwater, S., Scott, J. & Pobiner, B. 1986b. Ultrastructure of meiosis in *Dasya baillouviana* (Rhodophyta). II. Prometaphase I – telophase II and post-division nuclear behavior. *J. Phycol.* 22: 501–12.

Bronchart, R. & Demoulin, V. 1977. Unusual mitosis in the red alga *Porphyridium purpureum*. *Nature* (Lond.) 268: 80–1.

Brown, D. L. 1969. Ultrastructure of the freshwater alga *Batrachospermum*. Ph.D. thesis. University of California, Davis.

Brown, D. L. & Weier, T. E. 1968. Chloroplast development and ultrastructure in the freshwater red alga *Batrachospermum*. *J. Phycol.* 4: 199–206.

Brown, D. L. & Weier, T.E. 1970. Ultrastructure of the freshwater alga *Batrachospermum*. I. Thin-section and freeze-etch analysis of juvenile and photosynthetic filament vegetative cells. *Phycologia* 9: 217–35.

Cabioch, J. 1971. Étude sur les Corallinacées. I. Caractères généraux de la cytologie. *Cah. Biol. Mar.* 12: 121–86.

Cabioch, J. & Giraud, G. 1978. Apport de la microscopie électronique à la comparaison de quelque espèces de *Lithothamnium* Philippi. *Phycologia* 17: 369–81.

Chamberlain, A. H. L. & Evans, L. V. 1973. Aspects of spore production in the red alga *Ceramium*. *Protoplasma* 76: 139–59.

Chapman, A. R. O. & Craigie, J. S. 1977. Seasonal growth in *Laminaria longicruris*: relations with dissolved inorganic nutrients and internal reserves of nitrogen. *Mar. Biol.* 40: 197–205.

Chaumont, J. P. 1978. Variations de la teneur en composes azotes du *Rhodymenia palmata* Grev. *Bot. Mar.* 21: 23–9.

Chemin, M. E. 1937. Le développement des spores chez les Rhodophycées. *Rev. Gén. Bot.* 49: 205–34.

Codomier, L., Chevalier, S., Jupin, H., Francisco, C. & Banaigs, B. 1983. Rhodphycées à ioduses et Rhodophycées à bromuques. Étude réalisée à la microsonde électronique. *Biol. Cell.* 48: 75–80.

Cole, K. M., Park, C. M., Reid, P. E. & Sheath, R. G. 1985. Comparative studies on the cell walls of sexual and asexual *Bangia atropurpurea* (Rhodophyta). I. Histochemistry of polysaccharides. *J. Phycol.* 21: 585–92.

Cole, K. & Sheath, R. G. 1980. Ultrastructural changes in major organelles during spermatial differentiation in *Bangia* (Rhodophyta). *Protoplasma* 102: 253–79.

Coleman, A. W. 1985. Diversity of plastid DNA configuration among classes of eukaryote algae. *J. Phycol.* 21: 1–16.

Colin, H. & Guéguen, E. 1930. Variations saisonnières de la teneur en sucre chez les Floridées. *C. R. Acad. Sci.,* (Paris), 190: 884–6.

Cottler, M. H. 1971. Plasmalemmal extensions in *Chondrus crispus* (L.) Stackh. *J. Ultrastruct. Res.* 37: 31–6.

Davis, E. & Scott, J. 1986. Ultrastructure of cell division in the marine red alga *Lomentaria baileyana*. *Protoplasma* 131: 1–10.

Deason, T. R., Butler, G. L. & Rhyne, C. 1983. *Rhodella reticulata* sp. nov., a new coccoid rhodophytan alga (Porphyridiales). *J. Phycol.* 19: 104–11.

Delivopoulos, S. G. & Kugrens, P. 1984a. Ultrastructure of the fusion cell in *Faucheocolax attenuata* Setch. (Rhodophyta, Rhodymeniales). *J. Cell Sci.* 72: 307–19.

Delivopoulos, S. G. & Kugrens, P. 1984b. Ultrastructure of carposporogenesis in the parasitic red alga *Faucheocolax attenuata* Setch. (Rhodymeniales, Rhodymeniaceae). *Am. J. Bot.* 71: 1245–59.

Delivopoulos, S. G. & Kugrens, P. 1985a. Thylakoid formation from coiled lamellar bodies during carposporogenesis in *Faucheocolax attenuata* Setch. (Rhodophyta, Rhodymeniales). *J. Cell Sci.* 75: 215–24.

Delivopoulos, S. G. & Kugrens, P. 1985b. Ultrastructure of carposporophyte development in the red alga *Gloiosiphonia verticillaris* (Cryptonemiales, Gloiosiphoniaceae). *Am. J. Bot.* 72: 1926–38.

Delivopoulos, S. G. & Tsekos, I. 1985a. Ultrastructure of the fusion cell in *Gracilaria verrucosa* (Huds.) Papenfuss (Rhodophyta, Gigartinales). *New Phytol.* 101: 605–12.

Delivopoulos, S. G. & Tsekos, I. 1985b. Nuclear envelope activity in gonimoblast generative cells of the red alga *Gracilaria verrucosa* (Hudson) Papenfuss (Gigartinales). *Flora* 177: 309–15.

Delivopoulos, S. G. & Tsekos I. 1986. Ultrastructure of carposporogenesis in the red alga *Gracilaria verrucosa* (Gigartinales, Gracilariaceae). *Bot. Mar.* 29: 27–35.

DeLuca, P., Gambardella, R. & Merola, A. 1979. Thermoacidophilic algae of North and Central America. *Bot. Gaz.* 140: 418–27.

Dixon, P.S. 1973. *Biology of the Rhodophyta*. New York: Hafner.

Dodge, J. D. 1973. *The Fine Structure of Algal Cells*. London: Academic.

Duckett, J. G., Buchanan, J. S., Peel, M. C. & Martin, M. T. 1974. An ultrastructural study of pit connections and percurrent proliferations in the red alga *Nemalion helminthoides* (Vell. in With.) Batt. *New Phytol.* 73: 497–507.

Duckett, J. G. & Peel, M. C. 1978. The role of transmission electron microscopy in elucidating the taxonomy and phylogeny of the Rhodophyta. In *Modern Approaches to the Taxonomy of Red and Brown Algae*, eds. D. E. G. Irvine & J. H. Price, pp. 157–204. London: Academic.

Evans, L. V. 1970. Electron microscopal observations on a new red algal unicell, *Rhodella maculata* gen. nov., sp. nov. *Br. Phycol. J.* 5: 1–13.

Evans, L. V., Callow, M. E., Percival, E. & Fareed, V. 1974. Studies on the synthesis and composition of extracellular mucilage in the unicellular red alga *Rhodella*. *J. Cell Sci.* 16: 1–21.

Farquhar, M. G. 1985. Progress in unraveling pathways of Golgi traffic. *Annu. Rev. Cell Biol.* 1: 447–88.

Feldmann, G. 1970a. Sur l'ultrastructure des corps irisants des *Chondria* (Rhodophycées). *C. R. Acad. Sci.* (Paris) sér. D, 270: 945–6.

Feldmann, G. 1970b. Sur l'ultrastructure de l'appareil irisant du *Gastroclonium clavatum* (Roth.) Ardissone (Rhodophycae). *C. R. Acad. Sci.* (Paris) sér. D, 270: 1244–6.

Feldmann, G. & Guglielmi, G. 1976. Ultrastructure des cellules sécrétrices des *Antithamnion* (Rhodophycées). *C. R. Acad. Sci.* (Paris) sér. D, 282: 2163–6.

Fenical, W. 1975. Halogenation in the Rhodophyta. *J. Phycol.* 11: 245–59.

Fetter, R. & Neushul, M. 1981. Studies of developing and released spermatia in the red alga, *Tiffaniella snyderae* (Rhodophyta). *J. Phycol.* 17: 141–59.

Ford, T. W. 1984. A comparative ultrastructural study of *Cyanidium caldarium* and the unicellular red alga *Rhodosorus marinus*. *Ann. Bot.* 53: 285–94.

Fresnel, J. & Billard, C. 1987. Contribution à la connaissance du genre *Rhodella* (Porphyridiales, Rhodophyceae). *Cryptogam. Algol.* 8: 49–60.

Fritsch, F. E. 1945. *Structure and Reproduction of the Algae*. Vol. 2. Cambridge: Cambridge University Press.

Gantt, E. 1986. Phycobilisomes. In *Photosynthesis III*, eds. L. A. Staehelin & C. J. Arntzen, *Encyclopedia of Plant Physiology* new ser., 19: 260–8. Berlin: Springer-Verlag.

Gantt, E. & Conti, S.F. 1965. The ultrastructure of *Porphyridum cruentum*. *J. Cell Biol.* 26: 365–81.

Gantt, E., Edwards, M. R. & Conti, S. F. 1968. Ultrastructure of *Porphyridium aerugineum* a blue-green colored rhodophyte. *J. Phycol.* 4: 65–71.

Gantt, E., Scott, J. & Lipschultz, C. 1986. Phycobiliprotein composition and chloroplast structure in the freshwater red alga *Compsopogon coeruleus* (Rhodophyta). *J. Phycol.* 22: 480–4.

Gerwick, W. H. & Lang, N. J. 1977. Structural, chemical and ecological studies on iridescence in *Iridaea* (Rhodophyta). *J. Phycol.* 13: 121–7.

Giraud, G. & Cabioch, J. 1976. Étude ultrastructurale de l'activité des cellules superficielles du thalle des Corallinacées (Rhodophycées). *Phycologia* 15: 405–14.

Giraud, G. & Cabioch, J. 1981. Sur l'existence de cicatrisations cellulaires internes chez les Corallinacées (Rhodophycées, Cryptonémiales) et leur interprétation. *C. R. Acad. Sci.* (Paris) sér. 3, 292: 1037–41.

Giraud, G. & Cabioch, J. 1983. Inclusions cytoplasmiques remarquables chez les Corallinacées (Rhodophytes, Cryptonemiales). *Ann. Sc. Nat., Bot.* (Paris) sér. 13, 5: 29–43.

Godin, J. 1970a. Ultrastructure du pédicelle du corps en cerise chez *Laurencia scoparia*. *C. R. Acad. Sci.* (Paris) sér. D, 271: 1669–71.

Godin, J. 1970b. Ultrastructure des trichoblastes chez *Laurencia scoparia*. *C. R. Acad. Sci.* (Paris) sér. D, 271: 2290–2.

Goff, L. J. 1979. The biology of *Harveyella mirabilis* (Cryptonemiales, Rhodophyceae). VII. Structure and proposed function of host-penetrating cells. *J. Phycol.* 15: 87–100.

Goff, L. J. 1982. The biology of parasitic red algae. *Prog. Phycol. Res.* 1: 289–369.

Goff, L. J. & Coleman, A. W. 1985. The role of secondary pit connections in red algal parasitism. *J. Phycol.* 21: 483–508.

Goff, L. J. & Coleman, A. W. 1986. A novel pattern of apical cell polyploidy, sequential polyploidy reduction and intercellular nuclear transfer in the red alga *Polysiphonia*. *Am. J. Bot.* 73: 1109–30.

Griffiths, D. J. 1980. The pyrenoid and its role in algal metabolism. *Sci. Prog.* (Oxford) 66: 537–53.

Gschwend, P. M., MacFarlane, J. K. & Newman, K. A. 1985. Volatile halogenated organic compounds released to seawater from temperate marine macroalgae. *Science* 227: 1033–5.

Guérin-Dumartrait, E., Sarda, C. & Lacourly, A. 1970. Sur la structure fine du chloroplaste de *Porphyridium*. *C. R. Acad. Sci.* (Paris) sér. D, 270: 1977–9.

Hara, Y. 1975. Studies on the chloroplast ultrastructures and their contributions to the taxonomy and phylogeny of the Rhodophyta II. *Bull. Jap. Soc. Phycol.* 23: 67–78.

Hara, Y. & Chihara, M. 1974. Comparative studies on the chloroplast ultrastructure in the Rhodophyta with special reference to their taxonomic significance. *Sci. Rep. Tokyo Kyoiku Daigaku* Ser. B, 15: 209–35.

Hawkes, M. W. 1978. Sexual reproduction in *Porphyra gardneri* (Smith et Hollenberg) Hawkes (Bangiales, Rhodophyta). *Phycologia* 17: 329–53.

Hewson, W. D. & Hager, L. P. 1980. Bromoperoxidases and halogenated lipids in marine algae. *J. Phycol.* 16: 340–5.

Hill, S. A., Towill, L. R. & Sommerfeld, M. R. 1980. Photomovement responses of *Porphyridium purpureum*. *J. Phycol.* 16: 444–8.

Hofsten, A., Liljesvan, B. & Pedersén, M. 1977. Localization of bromine in the chloroplasts of the red alga *Lenormandia prolifera*. *Bot. Mar.* 20: 267–70.

Hofsten, A. & Pedersén, M. 1980. Bromine localization in the red alga *Odonthalia dentata*. *Z. Pflanzenphysiol.* 96: 103–14.

Honsell, E., Avanzini, A. & Ghirardelli, L. A. 1978. Two ways of chloroplast development in vegetative cells of *Nitophyllum punctatum* (Rhodophyta). *J. Submicr. Cytol.* 10: 227–37.

Hymes, B. J. & Cole, K. M. 1983a. The cytology of *Audouinella hermannii* (Rhodophyta, Florideophyceae). I. Vegetative and hair cells. *Can. J. Bot.* 61:3366–76.

Hymes, B. J. & Cole, K. M. 1983b. The cytology of *Audouinella hermannii* (Rhodophyta, Florideophyceae). II. Monosporogenesis. *Can. J. Bot.* 61: 3377–85.

Kazama, F. & Fuller, M. S. 1970. Ultrastructure of *Porphyra perforata* infected with *Pythium marinum*, a marine fungus. *Can. J. Bot.* 48: 2103–7.

Knoth, A. & Wiencke, C. 1984. Dynamic changes of protoplasmic volume and of fine structure during osmotic adaptation in the intertidal red alga *Porphyra umbilicalis*. *Plant Cell Envir.* 7: 113–19.

Konrad Hawkins, E. 1972. Observations on the developmental morphology and fine structure of pit connections in red algae. *Cytologia* 37: 759–68.

Konrad Hawkins, E. 1974. Golgi vesicles of uncommon morphology and wall formation in the red alga, *Polysiphonia*. *Protoplasma* 80: 1–14.

Koslowsky, D. J. & Waaland, S. D. 1984. Cytoplasmic incompatibility following somatic cell fusion in *Griffithsia pacifica* Kylin, a red alga. *Protoplasma* 123: 8–17.

Koslowsky, D. J. & Waaland, S. D. 1987. Ultrastructure of selective chloroplast destruction after somatic cell fusion in *Griffithsia pacifica* Kylin (Rhodophyta). *J. Phycol.* 23: 638–48.

Köst, H. P., Senser, M. & Wanner, G. 1984. Effect of nitrate and sulphate starvation on *Porphyridium cruentum* cells. *Z. Pflanzenphysiol.* 113: 231–49.

Kraft, G. T. & Robins, P. 1985. Is the order Cryptonemiales defensible? *Phycologia* 24: 67–77.

Kugrens, P. 1980. Electron microscopic observations on the differentiation and release of spermatia in the marine red alga *Polysiphonia hendryi* (Ceramiales, Rhodomelaceae). *Am. J. Bot.* 67: 519–28.

Kugrens, P. 1982. *Leachiella pacifica*, gen. et sp. nov., a new parasitic red alga from Washington and California. *Am. J. Bot.* 69: 306–19.

Kugrens, P. 1983. Electron microscopic observations on tubular structures involved in crystal formation in the red alga *Pleonosporium squarrulosum* (Harv.) Abbott. *J. Phycol.* 19: 507–10.

Kugrens, P. & Arif, I. 1981. Light and electron microscopic studies of the fusion cell in *Asterocolax gardneri* Setch. (Rhodophyta, Ceramiales). *J. Phycol.* 17: 215–23.

Kugrens, P. & Delivopoulos, S. G. 1985. Ultrastructure of auxiliary and gonimoblast cells during carposporophyte development in *Faucheocolax attenuata* (Rhodophyta). *J. Phycol.* 21: 240–9.

Kugrens, P. & Delivopoulos, S. C. 1986. Ultrastructure of the carposporophyte and carposporogenesis in the parasitic red alga *Plocamiocolax pulvinata* Setch. (Gigartinales, Plocamiaceae). *J. Phycol.* 22: 8–21.

Kugrens, P. & West, J. A. 1972a. Synaptonemal complexes in red algae. *J. Phycol.* 8: 187–91.

Kugrens, P. & West, J. A. 1972b. Ultrastructure of spermatial development in the parasitic red algae *Levringiella gardneri* and *Erythrocystis saccata*. *J. Phycol.* 8: 331–43.

Kugrens, P. & West, J. A. 1973a. The ultrastructure of carpospore differentiation in the parasitic red alga *Levringiella gardneri* (Setch.) Kylin. *Phycologia* 12: 163–73.

Kugrens, P. & West, J. A. 1973b. The ultrastructure of an alloparasitic red alga *Choreocolax polysiphoniae*. *Phycologia* 12: 175–86.

Kugrens, P. & West, J. A. 1974. The ultrastructure of carposporogenesis in the marine hemiparasitic red alga *Erythrocystis saccata*. *J. Phycol.* 10: 139–47.

Kylin, H. 1956. *Die Gattungen der Rhodophyceen*. Lund: Gleerup.

Lee, R. E. 1974. Chloroplast structure and starch grain production as phylogenetic indicators in the lower Rhodophyceae. *Br. Phycol. J.* 9: 291–5.

Lefort-Tran, M., Cohen-Bazire, G. & Pouphile, M. 1973. Les membranes photosynthétiques des algues à biliprotéines observées après cryodécapage. *J. Ultrastruct. Res.* 44: 199–209.

Lichtlé, C. 1973. Dégénérescence du plaste du *Rhodochorton purpureum* (Lightft.) Rosenvinge. Rhodophycée. Acrochaétiale. *C. R. Acad. Sci.* (Paris) sér. D, 277: 2341–4.

Lichtlé, C. 1975. Étude ultrastructurale de la paroi du *Polysiphonia elongata* (Harv.) Rhodophycée, Floridée, à l'aide d'actions ménagées d'enzymes. *J. Microscopie Biol. Cell.* 23: 93–104.

Lichtlé, C. & Giraud, G. 1969. Étude ultrastructurale de la zone apicale du thalle du *Polysiphonia elongata* (Harv.) Rhodophycée, Floridée. Évolution des plastes. *J. Microscopie* 8: 867–74.

Lin, H., Sommerfeld, M. R. & Swafford, J. R. 1975. Light and electron microscope observations on motile cells of *Porphyridium purpureum* (Rhodophyta). *J. Phycol.* 11: 452–7.

Lüning, K. & Schmitz, K. 1988. Dark growth of the red alga *Delesseria sanguinea* (Ceramiales): lack of chlorophyll, photosynthetic capability and phycobilisomes. *Phycologia* 27: 72–7.

Maggs, C. A. & Pueschel, C. M. 1989. Morphology and development of *Ahnfeltia plicata* (Rhodophyta): proposal of Ahnfeltiales ord. nov. *J. Phycol.* 25: 333–51.

Magruder, W. 1984. Specialized appendages on spermatia from the red alga *Aglaothamnion neglectum* (Ceramiales, Ceramiaceae) specifically bind with trichogynes. *J. Phycol.* 20: 436–40.

Mandura, A., Boney, A. D. & Bowes, B. G. 1986. Tubular plasmalemmal structures in newly released spores of some red algae. *Nova Hedwigia* 42: 277–82.

McBride, D. L. & Cole, K. 1971. Electron microscopic observations on the differentiation and release of monospores in the marine red alga *Smithora naiadum*. *Phycologia* 10: 49–61.

McCandless, E. L. 1981. Polysaccharides of seaweeds. In *The Biology of Seaweeds*, eds. C. S. Lobban & M. J. Wynne, pp. 559–88. Berkeley: University of California Press.

McCracken, D. A. & Cain, J. R. 1981. Amylose in floridean starch. *New Phytol.* 88: 67–71.

Merola, A., Castaldo, R., DeLuca, P., Gambardella, R., Musacchio, A. & Taddei, R. 1981. Revision of *Cyanidium caldarium*. Three species of acidophilic algae. *Giorn. Bot. Ital.* 115: 189–95.

Millson, C. & Moss, B. L. 1985. Ultrastructure of the vegetative thallus of *Phymatolithon lenormandii* (Aresch. in J. Ag.) Adey. *Bot. Mar.* 28: 123–32.

Morgan, K. C., Wright, J. L. C. & Simpson, F. J. 1980. Review of chemical constituents of the red alga *Palmaria palmata* (dulse). *Econ. Bot.* 34: 27–50.

Neushul, M. 1970. A freeze-etching study of the red alga *Porphyridium*. *Am. J. Bot.* 55: 1231–9.

Nichols, H. W. & Lissant, E. K. 1967. Developmental studies of *Erythrocladia* Rosenvinge in culture. *J. Phycol.* 3: 6–18.

Nichols, H. W., Ridgway, J. E. & Bold, H. C. 1966. A preliminary ultrastructural study of the freshwater red alga *Compsopogon*. *Ann. Missouri Bot. Gard.* 53: 17–27.

Oakley, B. R. & Dodge, J. D. 1974. The ultrastructure and cytochemistry of microbodies in *Porphyridium*. *Protoplasma* 80: 233–44.

Pallaghy, C. K., Minchinton, J., Kraft, G. T. & Wetherbee, R. 1983. Presence and distribution of bromine in *Thysanocladia densa* (Solieriaceae, Gigartinales), a marine red alga from the Great Barrier Reef. *J. Phycol.* 19: 204–8.

Pedersén, M., Roomans, G. M. & Hofsten, A. 1980a. Cell inclusions containing bromine in *Rhodomela confervoides* (Huds.) Lamour. and *Polysiphonia elongata* Harv. (Rhodophyta; Ceramiales). *Phycologia* 19: 153–8.

Pedersén, M., Roomans, G. M. & Hofsten, A. 1980b. Blue iridescence and bromine in the cuticle of the red alga *Chondrus crispus* Stackh. *Bot. Mar.* 23: 193–6.

Pedersén, M., Saenger, P., Rowan, K. S. & Hofsten, A. 1979. Bromine, bromophenols and floridorubin in the red alga *Lenormandia prolifera*. *Physiol. Plant.* 46: 121–6.

Peel, M. C. & Duckett, J. G. 1975. Studies of spermatogenesis in the Rhodophyta. In *The Biology of the Male Gamete*, eds. J. D. Duckett & P. A. Racey. *Biol. J. Linn. Soc.* 7, suppl. 1.: 1–13.

Peel, M. C., Lucas, I. A. N., Duckett, J. G. & Greenwood, A. D. 1973. Studies of sporogenesis in the Rhodophyta. I. An association of the nuclei with endoplasmic reticulum in post-meiotic tetraspore mother cells of *Corallina officinalis* L. *Z. Zellforsch.* 147: 59–74.

Pellegrini, M. & Pellegrini, L. 1983. Particularités ultrastructurales des cellules végétatives de l'*Alsidium helminthocorton* Kützing (Rhodophycées, Rhodomélacées). *Ann. Sci. Nat., Bot.* (Paris) sér. 13, 5: 211–27.

Peyrière, M. 1970. Évolution de l'appareil de Golgi au cours de la tétrasporogenèse de *Griffithsia flosculosa* (Rhodophycée). *C. R. Acad. Sci.* (Paris) sér. D, 270: 2071–4.

Peyrière, M. 1974. Étude infrastructurale des spermatocystes et spermaties de differentes Rhodophycées floridées. *C. R. Acad. Sci.* (Paris) sér. D, 278: 1019–22.

Peyrière, M. 1977. Infrastructure des synapses du *Griffithsia flosculosa* (Ellis) Batters et de quelques autres Rhodophycées Floridées. *Rev. Algol.* N.S. 12: 31–43.

Phillips, R. W. 1925. On vascular pseudopodia in a species of *Callithamnion*. *Rev. Algol.* 2: 14–18.

Prensier, G., Vivier, E., Goldstein, S. & Schrevel, J. 1980. Motile flagellum with "3 + 0" ultrastructure. *Science* 207: 1493–4.

Pringsheim, E. G. 1968. Kleine Mitteilungen über Flagellaten und Algen. XV. Zur Kenntnis der Gattung *Porphyridium* (Rhodophyceae). *Arch. Mikrobiol.* 61: 169–80.

Pueschel, C. M. 1977a. A freeze-etch study of the ultrastructure of red algal pit plugs. *Protoplasma* 91: 15–30.

Pueschel, C. M. 1977b. Unusual lipid bodies in the red alga *Palmaria palmata* (= *Rhodymenia palmata*). *J. Ultrastruct. Res.* 60: 328–34.

Pueschel, C. M. 1979. Ultrastructure of tetrasporogenesis in *Palmaria palmata* (Rhodophyta). *J. Phycol.* 15: 409–24.

Pueschel, C. M. 1980a. On the authenticity of plasmalemmavilli in red algae. *Phycologia* 19: 139–42.

Pueschel, C. M. 1980b. A reappraisal of the cytochemical properties of rhodophycean pit plugs. *Phycologia* 19: 210–17.

Pueschel, C. M. 1980c. Evidence for two classes of microbodies in meiocytes of the red alga *Palmaria palmata*. *Protoplasma* 104: 273–82.

Pueschel, C. M. 1987. Absence of cap membranes as a characteristic of pit plugs of some red algal orders. *J. Phycol.* 23: 150–6.

Pueschel, C. M. 1988a. Cell sloughing and chloroplast inclusions in *Hildenbrandia rubra* (Rhodophyta, Hildenbrandiales). *Br. Phycol. J.* 23: 17–23.

Pueschel, C. M. 1988b. Secondary pit connections in *Hildenbrandia* (Rhodophyta, Hildenbrandiales). *Br. Phycol. J.* 23: 25–32.

Pueschel, C. M. 1989. An expanded survey of the ultrastructure of red algal pit plugs. *J. Phycol.* 25: 625–36.

Pueschel, C. M. & Cole, K. M. 1980. Phytoferritin in the red alga *Constantinea* (Cryptonemiales). *J. Ultrastruct. Res.* 73: 282–7.

Pueschel, C. M. & Cole, K. M. 1982. Rhodophycean pit plugs: an ultrastructural survey with taxonomic implications. *Am. J. Bot.* 69: 703–20.

Pueschel, C. M. & Cole, K. M. 1985. Ultrastructure of germinating carpospores of *Porphyra variegata* (Kjellm.) Hus (Bangiales, Rhodophyta). *J. Phycol.* 21: 146–54.

Pueschel, C. M. & Magne, F. 1987. Pit plugs and other ultrastructural features of systematic value in *Rhodochaete parvula* (Rhodophyta, Rhodochaetales). *Cryptogam. Algol.* 8: 201–9.

Pueschel, C. M. & Parthasarathy, M. V. 1984. X-ray microanalysis of phytoferritin in *Constantinea* (Cryptonemiales, Rhodophyta). *Phycologia* 23: 465–9.

Pueschel, C. M. & van der Meer, J. P. 1984. Ultrastructural characterization of a pigment mutant of the red alga *Palmaria palmata*. *Can. J. Bot.* 62: 1101–7.

Pueschel, C. M. & van der Meer, J. P. 1985. Ultrastructure of the fungus *Petersenia palmariae* (Oomycetes) parasitic on the alga *Palmaria mollis* (Rhodophyceae). *Can. J. Bot.* 63: 409–18.

Ramm-Anderson, S. M. & Wetherbee, R. 1982. Structure and development of the carposporophyte of *Nemalion helminthoides* (Nemalionales, Rhodophyta). *J. Phycol.* 18:133–41.

Ramus, J. 1969. Pit connection formation in the red alga *Pseudogloiophloea*. *J. Phycol.* 5: 57–63.

Ramus, J. 1972. The production of extracellular polysaccharide by the unicellular red alga *Porphyridium aerugineum*. *J. Phycol.* 8: 97–111.

Ramus, J. & Robins, D. M. 1975. The correlation of Golgi activity and polysaccharide secretion in *Porphyridium*. *J. Phycol.* 11: 70–4.

Rawlence, D. J. 1972. An ultrastructural study of the relationship between rhizoids of *Polysiphonia lanosa* (L.) Tandy (Rhodophyceae) and tissue of *Ascophyllum nodosum* (L.) Le Jolis (Phaeophyceae). *Phycologia* 11: 279–90.

Rawlence, D. J. & Taylor, A. R. A. 1972. A light and electron microscopic study of rhizoid development in *Polysiphonia lanosa* (L.) Tandy. *J. Phycol.* 8: 15–24.

Robins, P. A. & Kraft, G. T. 1985. Morphology of the type and Australian species of *Dudresnaya* (Dumontiaceae, Rhodophyta). *Phycologia* 24: 1–34.

Roomans, G. M. 1979. Quantitative x-ray microanalysis of halogen elements in biological specimens. *Histochemistry* 65: 49–58.

Rosenvinge, L. K. 1888. Sur la formation des pores secondaries chez les *Polysiphonia*. *Bot. Tidsskrift* 17: 10–19.

Rosenvinge, L. K. 1927. On mobility in the reproductive cells of the Rhodophyceae. *Bot. Tidsskrift* 40: 72–80.

Schornstein, K. L. & Scott, J. 1982. Ultrastructure of cell division in the unicellular red alga *Porphyridium purpureum*. *Can. J. Bot.* 60: 85–97.

Scott, F. J., Wetherbee, R. & Kraft, G. T. 1982. The morphology and development of some prominently stalked southern Australian Halymeniaceae (Cryptonemiales, Rhodophyta). I. *Cryptonemia kallymenioides* (Harvey) Kraft comb. nov. and *C. undulata* Sonder. *J. Phycol* 18: 245–7.

Scott, J. 1984. Electron microscopic contributions to red algal phylogeny. *J. Phycol.* 20, suppl.: 6.

Scott, J. 1986. Ultrastructure of cell division in the unicellular red alga *Flintiella sanguinaria*. *Can. J. Bot.* 64: 516–24.

Scott, J., Bosco, C., Schornstein, K. & Thomas, J. 1980. Ultrastructure of cell division and reproductive differentiation of male plants in the Florideophyceae (Rhodophyta): cell division in *Polysiphonia*. *J. Phycol.* 16: 507–24.

Scott, J. L. & Dixon, P. S. 1973a. Ultrastructure of tetrasporogenesis in the marine red alga *Ptilota hypnoides*. *J. Phycol.* 9: 29–46.

Scott, J. L. & Dixon, P. S. 1973b. Ultrastructure of spermatium liberation in the marine red alga *Ptilota densa*. *J. Phycol.* 9: 85–91.

Scott, J., Phillips, D. & Thomas, J. 1981. Polar rings are persistent organelles in interphase vegetative cells of *Polysiphonia harveyi* Bailey (Rhodophyta, Ceramiales). *Phycologia* 20: 333–7.

Scott, J., Thomas, J. & Saunders, B. 1988. Primary pit connections in *Compsopogon coeruleus* (Balbis) Montagne (Compsopogonales, Rhodophyta). *Phycologia* 27: 327–33.

Searles, R. B. & Ballantine, D. L. 1986. *Dudresnaya puertoricensis* sp. nov. (Dumontiaceae, Gigartinales, Rhodophyta). *J. Phycol.* 22: 389–94.

Seckbach, J. & Fredrick, J. F. 1980. A primaeval alga bridging the blue-green and the red algae: further biochemical and ultrastructural studies of *Cyanidium caldarium* with special reference to the plastid membranes. *Microbios* 29: 135–47.

Seckbach, J., Hammerman, I. S. & Hanania, J. 1981. Ultrastructural studies of *Cyanidium caldarium*: contribution to phylogenesis. *Ann. N.Y. Acad. Sci.* 361: 409–25.

Sheath, R. G., Cole, K. M. & Hymes, B. 1987. Ultrastructure of polysporogenesis in *Pleonosporium vancouverianum* (Ceramiaceae, Rhodophyta). *Phycologia* 26: 1–8.

Sheath, R. G., Hellebust, J. A. & Sawa, T. 1977. Changes in plastid structure, pigmentation and photosynthesis of the conchocelis stage of *Porphyra leucosticta* (Rhodophyta, Bangiophyceae) in response to low light and darkness. *Phycologia* 16: 265–76.

Sheath, R. G., Hellebust, J. A. & Sawa, T. 1979. Effects of low light and darkness on structural transformations in plastids of the Rhodophyta. *Phycologia* 18: 1–12.

Sheath, R. G., Hellebust, J. A. & Sawa, T. 1981a. Floridean starch metabolism of *Porphyridium purpureum* (Rhodophyta). III. Effects of darkness and metabolic inhibitors. *Phycologia* 20: 22–31.

Sheath, R. G., Hellebust, J. A. & Sawa, T. 1981b. Ultrastructure of the floridean starch grain. *Phycologia* 20: 292–7.

Simon-Bichard-Bréaud, J. 1971. Un appareil cinétique dans les gamétocystes males d'une Rhodophycée: *Bonnemaisonia hamifera* Hariot. *C. R. Acad. Sci.* (Paris) sér. D, 273: 1272–5.

Simon-Bichard-Bréaud, J. 1972a. Formation de la crypte flagellaire et évolution de son contenu au cours de la gamétogenèse mâle chez *Bonnemaisonia hamifera* Hariot (Rhodophycée). *C. R Acad. Sci.* (Paris) sér. D, 274: 1796–9.

Simon-Bichard-Bréaud, J. 1972b. Origine et devenir des vacuoles à polysaccharides des gamétocystes mâles de *Bonnemaisonia hamifera* Hariot (Rhodophycée). *C. R. Acad. Sci.* (Paris) sér. D, 274: 1485–8.

Sommerfeld, M. R. & Nichols, H. W. 1970. Comparative studies in the genus *Porphyridium* Naeg. *J. Phycol.* 6: 67–78.

Staehelin, L. A. 1967. Chloroplast fibrils linking the photosynthetic lamellae. *Nature* (Lond.) 214: 1158.

Staehelin, L. A. 1986. Chloroplast structure and supramolecular organization of photosynthetic membranes. In *Photosynthesis III*, eds. L. A. Staehelin & C. J. Arntzen, *Encyclopedia of Plant Physiology* new ser., 19: 1–84. Berlin: Springer-Verlag.

Staehelin, L. A., Giddings, T. H., Badami, P. & Krzymowski, W. W. 1978. A comparison of the supramolecular architecture of photosynthetic membranes of blue-green, red, and green algae and of higher plants. In *Light Transducing Membranes*, ed. D. W. Deamer, pp. 335–55. New York: Academic.

Steinmüller, K., Kaling, M. & Zetsche, K. 1983. In vitro synthesis of phycobiliproteids and ribulose-1, 5-bisphosphate carboxylase by non-poly-adenylated-RNA of *Cyanidium caldarium* and *Porphyridium aerugineum*. *Planta* (Berl.) 159: 308–13.

Stewart, K. D. & Mattox, K. R. 1984. The case for a polyphyletic origin of mitochondria: morphological and molecular comparisons. *J. Mol. Evol.* 21: 54–7.

Taylor, F. J. R. 1978. Problems in the development of an explicit hypothetical phylogeny of the lower eukaryotes. *BioSystems* 10: 67–89.

Theiler, R., Cook, J. C., Hager, L. P. & Siuda, J. F. 1978. Halohydrocarbon synthesis by bromoperoxidase. *Science* 202: 1094–6.

Toyama, S. 1980. Electron microscope studies on the morphogenesis of plastids VIII. Further studies on the fine structure of *Cyanidium caldarium* with special regard to the photosynthetic apparatus. *Cytologia* 45: 779–90.

Triemer, R. E. & Vasconcelos, A. C. 1977. The ultrastructure of carposporogenesis in *Caloglossa leprieurii* (Delesseriaceae, Ceramiales). *Am. J. Bot.* 64: 825–34.

Tripodi, G. 1971a. Some observations on the ultrastructure of the red alga *Pterocladia capillacea* (Gmel.) Born. et Thur. *J. Submicr. Cytol.* 3: 63–70.

Tripodi, G. 1971b. The fine structure of the cystocarp in the red alga *Polysiphonia sertularioides* (Grat.) J. Ag. *J. Submicr. Cytol.* 3: 71–9.

Tripodi, G. 1974. Ultrastructural changes during carpospore formation in the red alga *Polysiphonia. J. Submicr. Cytol.* 6: 275–86.

Tripodi, G. 1980. A convoluted membranous structure associated to fibrils in the mitochondrial and plastidial matrix during sporogenesis in red algae. *Endocytobiology* 1: 817–23.

Tripodi, G. & de Masi, F. 1977. The post-fertilization stages of red algae: the fine structure of the fusion cell of *Erythrocystis. J. Submicr. Cytol.* 9: 389–401.

Tripodi, G. & de Masi, F. 1978. A possible vestige of a flagellum in the fusion cell of the red alga *Erythrocystis montagnei. J. Submicr. Cytol.* 10: 435–9.

Tripodi, G. & de Masi, F. 1983. Unusual structures in the spermatial vesicles of the red alga *Erythrocystis montagnei. Plant Syst. Evol.* 143: 197–206.

Tripodi, G. & Gargiulo, G. M. 1984. Relationships among membranes in plastids of the red alga *Nitophyllum punctatum* (Stackh.) Grev. *Protoplasma* 119: 55–61.

Tripodi, G., Gargiulo, G. M. & de Masi, F. 1986. Electron microscopy of membranous bodies and genophore in chloroplasts of *Botryocladia botryoides* (Rhodymeniales, Rhodophyta). *J. Phycol.* 22: 560–3.

Tripodi, G., Pizzolongo, P. & Giannattasio, M. 1972. A DNase-sensitive twisted structure in the mitochondrial matrix of *Polysiphonia* (Rhodophyta). *J. Cell Biol.* 55: 530–2.

Tsekos, I. 1981. Growth and differentiation of the Golgi apparatus and wall formation during carposporogenesis in the red alga, *Gigartina teedii* (Roth) Lamour. *J. Cell Sci.* 52: 71–84.

Tsekos, I. 1982. Plastid development and floridean starch grain formation during carposporogenesis in the red alga, *Gigartina teedii. Cryptogam. Algol.* 3: 91–103.

Tsekos, I. 1985. The endomembrane system of differentiating carposporangia in the red alga *Chondria tenuissima*: occurrence and participation in secretion of polysaccharidic and proteinaceous substances. *Protoplasma* 129: 127–36.

Tsekos, I. & Reiss, H. D. 1988. Occurrence and transport of particle "tetrads" in the cell membranes of the unicellular red alga *Porphyridium* visualized by freeze-fracture. *J. Ultrastruct. Mol. Struct. Res.* 99: 156–68.

Tsekos, I., Reiss, H. D. & Schnepf, E. 1985. Occurrence of particle tetrads in the vacuole membrane of the marine red algae *Gigartina teedii* and *Ceramium rubrum. Naturwiss.* 72: 489–90.

Tsekos, I. & Schnepf, E. 1983. The ultrastructure of carposporogenesis in *Gigartina teedii* (Roth) Lamour. (Gigartinales, Rhodophyceae): auxiliary cell, cystocarpic plant. *Flora* 173: 81–96.

Tsekos, I. & Schnepf, E. 1985. Ultrastructure of the early stages of carposporophyte development in the red alga *Chondria tenuissima* (Rhodomelaceae, Ceramiales). *Plant Syst. Evol.* 151: 1–18.

Ueda, K. & Chida, Y. 1987. Chloroplast structure of *Cyanidium caldarium* shown by freeze-substitution. *Br. Phycol. J.* 22: 61–5.

Ueda, K. & Yokochi, J. 1981. Structure of *Cyanidium caldarium. Bot. Mag.* (Tokyo) 94: 159–64.

Vesk, M. & Borowitzka, M. A. 1984. Ultrastructure of tetrasporogenesis in the coralline alga *Haliptilon cuvieri* (Rhodophyta). *J. Phycol.* 20: 501–15.

Waaland, S. D., Nehlsen, W. & Waaland, J. R. 1977. Phototropism in a red alga, *Griffithsia pacifica. Plant Cell Physiol.* 18: 603–12.

Wanner, G. & Köst, H. P. 1984. Membrane storage of the red alga *Porphyridium cruentum* during nitrate- and sulphate starvation. *Z. Pflanzenphysiol.* 113: 251–62.

Wehrmeyer W. 1971. Elektronenmikroskopische Untersuchung zur Feinstruktur von *Porphyridium violaceum* Kornmann mit Bemerkungen über seine taxonomische Stellung. *Arch. Mikrobiol.* 75: 121–39.

West, J. A. 1971. Environmental control of hair and sporangial formation in the marine red alga *Acrochaetium proskaueri* sp. nov. *Proc. Int. Seaweed Symp.* 7: 377–84.

Westlund, P., Roomans, G. M. & Pedersén, M. 1981. Localization and quantification of iodine and bromine in the red alga *Phyllophora truncata* (Pallas) A. D. Zinova by electron microscopy and x-ray microanalysis. *Bot. Mar.* 24: 153–6.

Wetherbee, R. 1978a. The presence of tubular plasmalemmal structures during carposporogenesis in the red alga *Polysiphonia. Protoplasma* 94: 341–5.

Wetherbee, R. 1978b. Differentiation and continuity of the Golgi apparatus during carposporogenesis in *Polysiphonia* (Rhodophyta). *Protoplasma* 95: 347–60.

Wetherbee, R. 1979. "Transfer connections": specialized pathways for nutrient translocation in a red alga? *Science* 204: 858–9.

Wetherbee, R. 1980. Postfertilization development in the red alga *Polysiphonia*. 1. Proliferation of the carposporophyte. *J. Ultrastruct. Res.* 70: 259–74.

Wetherbee, R., Janda, D. M. & Bretherton, G. A. 1984a. The structure, composition and distribution of proteinaceous crystalloids in vegetative cells of the red alga *Wrangelia plumosa. Protoplasma* 119: 135–40.

Wetherbee, R. & Kraft, G. T. 1981. Morphological and fine structural features of pit connections in *Cryptonemia* sp., a highly differentiated marine red alga from Australia. *Protoplasma* 106: 167–172.

Wetherbee, R. & Quirk, H. M. 1982a. The fine structure and cytology of the association between the parasitic red alga *Holmsella australis* and its red algal host *Gracilaria furcellata. Protoplasma* 110: 153–65.

Wetherbee, R. & Quirk, H. M. 1982b. The fine structure of secondary pit connection formation between the red algal alloparasite *Holmsella australis* and its red algal host *Gracilaria furcellata. Protoplasma* 110: 166–76.

Wetherbee, R., Quirk, H. M., Mallett, J. E. & Ricker, R. W. 1984b. The structure and formation of host–parasite pit connections between the red algal alloparasite *Harveyella mirabilis* and its red algal host *Odonthalia floccosa*.

Protoplasma 119: 62–73.
Wetherbee, R. & West, J. A. 1976. Unique Golgi apparatus and vesicle formation in a red alga. *Nature* (Lond.) 259: 566–7.
Wetherbee, R. & West, J. A. 1977. Golgi apparatus of unique morphology during early carposporogenesis in a red alga. *J. Ultrastruct. Res.* 58: 119–33.
Wetherbee, R., West, J. A. & Wynne, M. J. 1974. Annulate lamellae in postfertilization development in *Polysiphonia novae-angliae* (Rhodophyta). *J. Ultrastruct. Res.* 49: 401–4.
Wetherbee, R. & Wynne, M. J. 1973. The fine structure of the nucleus and nuclear associations of developing carposporangia in *Polysiphonia novae-angliae* (Rhodophyta). *J. Phycol.* 9: 402–7.
Wiencke, C. 1982. Effect of osmotic stress on thylakoid fine structure in *Porphyra umbilicalis*. *Protoplasma* 111: 215–20.
Wiencke, C. & Läuchli, A. 1980. Growth, cell volume, and fine structure of *Porphyra umbilicalis* in relation to osmotic tolerance. *Planta* (Berl.) 150: 303–11.
Wiencke, C. & Läuchli, A. 1983. Tonoplast fine structure and osmotic regulation in *Porphyra umbilicalis*. *Planta* (Berl.) 159: 342–6.
Wiencke, C., Stelzer, R. & Läuchli, A. 1983. Ion compartmentation in *Porphyra umbilicalis* determined by electron-probe x-ray microanalysis. *Planta* (Berl.) 159: 336–41.
Wolk, C. P. 1968. Role of bromine in the formation of the refractile inclusions of the vesicle cells of the Bonnemaisoniaceae (Rhodophyta). *Planta* (Berl.) 78: 371–8.
Wyman, M., Gregory, R.P.F. & Carr, N. G. 1985. Novel role for phycoerythrin in a marine cyanobacterium, *Synechococcus* strain DC2. *Science* 230: 818–20.
Young, D. N. 1977. A note on the absence of flagellar structures in spermatia of *Bonnemaisonia*. *Phycologia* 16: 219–22.
Young, D. N. 1978. Ultrastructural evidence for a secretory function in the "gland cells" of the marine red alga *Botryocladia pseudodichotoma* (Rhodymeniaceae). *Protoplasma* 94: 109–26.
Young, D. N. 1979a. Ontogeny, histochemistry and fine structure of cellular inclusions in vegetative cells of *Antithamnion defectum* (Ceramiaceae, Rhodophyta). *J. Phycol.* 15: 42–8.
Young, D. N. 1979b. Fine structure of the "gland cells" of the red alga, *Opuntiella californica* (Solieriaceae, Gigartinales). *Phycologia* 18: 288–95.
Young, D. N., Howard, B. M. & Fenical, W. 1980. Subcellular localization of brominated secondary metabolites in the red alga *Laurencia synderae*. *J. Phycol.* 16: 182–5.
Young, D. N. & West, J. A. 1979. Fine structure and histochemistry of vesicle cells of the red alga *Antithamnion defectum* (Ceramiaceae). *J. Phycol.* 15: 49–57.

Chapter 3

DNA: Microspectrofluorometric studies

LYNDA J. GOFF
ANNETTE W. COLEMAN

CONTENTS

I. Introduction / 43
II. Methodologies / 44
   A. DNA fluorochromes / 44
   B. Staining protocols / 45
   C. Microscopy / 45
   D. Problems and pitfalls / 46
III. Nuclear studies / 46
   A. Nuclear number, size, and positioning; cell cycle / 46
   B. Ploidy of life history stages and site of meiosis / 54
IV. Unusual aspects of red algal nuclei / 60
   A. Nuclear migration / 60
   B. Nuclear DNA in isomorphic forms / 63
   C. Nuclear ploidy levels during morphogenesis and development / 67
   D. Polyploidy and polygenomy as a genetic buffer / 68
V. Summary / 69
VI. Acknowledgments / 70
VII. References / 70

I. INTRODUCTION

Red algae are most remarkable in their complex life histories and morphologies. Only after culture methods for red algae were introduced was it finally verified that there are three phases in the life histories of most Florideophycidae. However, uncertainty about the site, or even the existence, of meiosis frequently remained. Consequently, the appropriate ploidy level of particular life history stages for most red algae could only be inferred. The problem lay chiefly in staining methods that would enable investigators to identify unequivocally the stages of meiosis and the number of chromosomes (see discussions by Godward 1966; Goff & Coleman 1984a; West & Hommersand 1981; Chapter 4). Classical absorption-type stains, such as hematoxylin, brazilin, aceto-carmine, aceto-orcein, and Feulgen, were used with varying success by earlier workers attempting to count chromosomes. Nonspecific binding and the sensitivities of these conventional stains limited their applications.

The problems inherent with using absorption-type stains in visualizing red algal DNA have been overcome by the development and application of fluorescent stains (fluorochromes) that are specific for double-stranded DNA. Nuclei and in some cases DNA of plastid and mitochondrial nucleoids can be readily observed when stained with fluorochromes and examined with epifluorescence microscopy. Some of the available DNA fluorochromes are so sensitive that, under appropriate conditions, genomes as small as that of a T_4 bacteriophage or single plastid genome (ca. 1×10^8 daltons) can be observed, even though these nucleoids are far smaller than the lower limit of resolution of the light microscope (Coleman et al. 1981). Since many of the DNA-specific fluorochromes bind quantitatively (either intercalating into the double helix or binding in one of the grooves), the amount of stain associated with the DNA is directly proportional to the amount of DNA present. Therefore, the ploidy level of any particular nucleus may be determined simply by measuring the amount of light emitted from an excited nucleus stained with a DNA-specific fluorochrome.

In this chapter we will present critical aspects of the methodologies and instrumentation employed in fluorescence microscopy studies of red algal DNA, discuss their application to studies of life histories, and summarize some of the unusual aspects of red algal nuclei and nuclear–cytoplasm interactions revealed by these techniques.

II. METHODOLOGIES

A. DNA fluorochromes

Several water-soluble DNA fluorochromes, e.g., ethidium bromide, acridine orange, Hoechst 33258, 4'-6 diamidino-2-phenolindole (DAPI), and mithramycin, are commercially available and can be used to visualize red algal nuclei. Of these, DAPI and mithramycin (Fig. 3-1, Table 3-1) give the brightest image and suffer least from nonspecific binding, which precludes the use of other fluorochromes in measuring DNA amounts.

DAPI is the preferred fluorochrome for most red algae, even though, at certain concentrations (higher than required for staining nuclei), it binds to polyphosphates (Coleman 1978) and may also bind nonspecifically with wall and cytoplasmic carbohydrates (Table 3-1). However, its blue fluorescence, when bound to DNA, is easily distinguished from the yellow fluorescence of DAPI-polyphosphate and the yellow, orange, and red autofluorescence emission of pigments like the phycobiliproteins and chlorophyll. Excited mithramycin-DNA fluoresces yellow; consequently, in many highly pigmented red algal cells, it is difficult to distinguish from that of the autofluorescing plastids, particularly in tissues fixed in glutaraldehyde.

Mithramycin is less stable than DAPI to excitation irradiation and undergoes rapid photobleaching (fading), thereby limiting its use in quantitative studies. However, in some cases, mithramycin has been employed to stain nuclei that were not well stained with DAPI. The success may be related to differences in the nucleotide requirements of the two dyes: mithramycin binds specifically to GC-rich regions of the genome, whereas DAPI binds to AT-rich regions. Therefore, genomes that are AT rich, such as the plastid DNA of red algae (Goff & Coleman 1988), stain with DAPI but not with mithramycin. In addition, the greater quantum yield of DAPI-complexed DNA, compared to mithramycin-stained DNA, makes DAPI more useful in detecting very

Table 3-1. Fluorochrome characteristics

	DAPI	Mithramycin
λ absorption	340–380 nm	450–500 nm
λ emission	460–480 nm	550–600 nm
Fluorescence color	Blue-white	Yellow
Base requirements	A, T	G, C
pH optima	4.0–7.5	~7.0
Staining requirement	—	Mg^{2+} (1–10 mM)
Sites stained	Nucleus, plastid DNA, and some cell wall and intracellular carbohydrates	Nucleus
Dye concentrations[a]	0.25–2 μg/mL	20–100 μg/mL
Staining time[a]	15–30 min	≥60 min

[a] Higher staining concentrations and longer staining times are required for living cells.

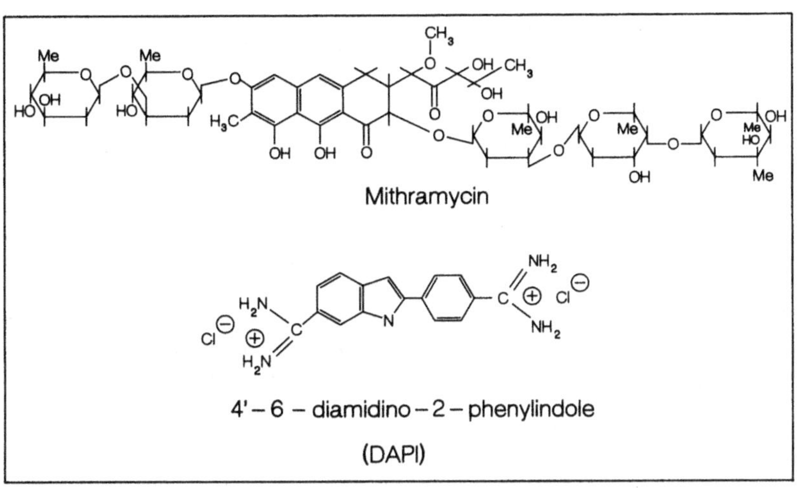

Fig. 3-1. Chemical structure of the two double-stranded DNA fluorochromes, DAPI and mithramycin.

small amounts of DNA, as in plastids and mitochondria. However, this greater sensitivity can also make accurate determination of nuclear DNA difficult in cells with large amounts of plastid DNA because of the neighboring organelle fluorescence. In such cases (Coleman 1982), mithramycin provides more accurate quantification, as the background fluorescence of plastid DNA does not interfere. When using either DAPI or mithramycin, fluorochrome fading can be greatly reduced by locating nuclei with phase optics prior to exposing the nucleus to the excitation light.

B. Staining protocols

Since both DAPI and mithramycin are water soluble dyes, they may be used to stain either living or chemically fixed cells. General staining protocols are summarized in Table 3-1 and discussed in detail by Coleman and colleagues (1981). Generally, aldehyde fixatives, particularly those containing formaldehyde, should be avoided if quantitative measurements are to be made. For most red algae, Carnoy's fixative (3 parts 95% ethanol and 1 part glacial acetic acid) gives optimal results, although for very large, highly vacuolate cells, it may cause extensive cell shrinkage and cytoplasmic rearrangement. In cytological studies of *Griffithsia* (Goff & Coleman 1987a), microwave fixation (using a standard household microwave oven) in 0.5 µg/mL DAPI in seawater gave excellent cellular preservation, while providing brilliant and quantitative staining of nuclear and extrachromosomal DNA. This fixation procedure has proven extremely useful with other red algae as well as with some green and brown algae (McNaughton & Goff unpubl.).

C. Microscopy

An epifluorescence (incident light) microscope is essential for imaging nuclear or organelle DNA stained with either DAPI or mithramycin. These microscopes offer many advantages over the older, transmission fluorescence microscopes. For example, in contrast to transfluorescence microscopes, the intensity of illumination and consequently the amount of excitation energy increases as a function of increasing magnification. In addition, dichroic mirror systems (i.e., beam-splitting filters) employed by epifluorescence microscopes efficiently eliminate the background illumination that reduces contrast in transmission fluorescence microscopes (Ploem & Tanke 1987).

Since DAPI is excited in the wavelength range of near-UV light (340–380 nm) (Table 3-1), a lamp that can provide these wavelengths is essential. The 365 nm line of a high-pressure mercury vapor lamp (50, 100 or 200 watt) is standardly used to visualize DAPI-complexed DNA. This same lamp provides ample excitation energy in the ca. 450–500 nm range required to excite mithramycin. It should be noted that either a quartz halogen or tungsten lamp may be used to image mithramycin-stained DNA, although these lamps generally do not provide enough energy to excite DAPI-DNA sufficiently to be easily detected by eye or photographic film. However, a silicon intensified video camera (SIT) or photon-counting camera can easily image the emission energy from tungsten or quartz halogen-excited DNA stained with DAPI. This is particularly valuable in in vivo studies, where UV excitation would damage living cells.

Different filter combinations (i.e., exciting,

beam-splitting, suppression) are required to image mithramycin and DAPI. For mithramycin-stained nuclei, an FITC type (blue) filter combination (e.g., Leitz I$_2$ cube–513418 or Zeiss 48 77 09) is employed, whereas DAPI requires a UV–short blue filter system (Leitz A cube–513410 or Zeiss 48 77 02). In addition, neofluor-type objectives are essential for imaging DAPI staining since conventional apochromate objectives often contain so many UV-absorbing lens surfaces that the intensity of the excitation and emission energy is reduced (by absorption and reflection) below the limit of detection.

The amount of light emitted from a particular region of fluorochrome-stained DNA can be measured either using a microspectrofluorometer or a video interfaced digital image processor. Since we have previously presented the quantitative methods using Zeiss (Coleman et al. 1981; Goff & Coleman 1984a) and Leitz (Goff & Coleman 1987a) microspectrofluorometers, a detailed discussion is omitted here. The methods for measuring emitted light energy using digital video image processors is discussed in Inoue's recent text (1986) on video microscopy. From our experience with both systems, microspectrofluorometry is preferred, as measurements are obtained much more easily and rapidly, and the data are less variable.

D. Problems and pitfalls

There are a few problems and pitfalls that await the newcomer to these studies. First, it should be anticipated that fixation and staining protocols successful for one particular species or generation may not be so for another. Second, special problems are confronted in examining multicellular algae as the tissues must be sectioned or the cells dissociated and flattened (by squashing) to clearly visualize nuclei and organelle DNA.

As these fluorochromes are water soluble, they readily stain tissues embedded in methyacrylate resins like JB4 and Lowicryl but are unable to penetrate epoxy resins like Spurrs and Epon. Prior to embedding, tissues are fixed in Carnoy's fixative or, if necessary, glutaraldehyde; however, in no case should osmium tetroxide, iodine, or strong oxidizing fixatives like permanganate be employed, as their presence in the tissue strongly quenches fluorochome fluorescence. Successful squashing and flattening of multicellular tissues is accomplished after the cells are "softened". The most useful softening agents are a saturated chloral hydrate solution (3–12 h at room temperature) or EDTA (0.5 M, pH 8–9 at 70°C for 1–3 h); both softening agents act to gelatinize the cells walls, permitting the dissociation of multicellular tissue and the flattening of cells.

Measuring DNA microscopically offers additional problems and challenges. One must always be concerned about the length of the light path, intervening materials, and sources of reflection. For example, in measuring the DNA fluorescence of a stained nucleus that has a plastid situated in the focal plane directly above, the excitation and emission energy reaching and emanating from the nucleus would have passed through the plastid en route. The decrease in fluorescence due to the absorption of the plastid or, conversely, the increase in fluorescence due to the pigment autofluorescence of the plastid would provide a false value for the DNA fluorescence of the underlying nucleus. If this nucleus were stained with DAPI, the contribution to signal due to the pigment autofluorescence could be eliminated using a 490 or 500 nm cut-off filter (Zeiss 46 79 60 KP500) that absorbs effectively 100% of light of wavelength longer than 500 nm. Since the wavelength of DAPI fluorescence is less than 500 nm, it would not be filtered out.

Finally, when undertaking quantitative measurements of DNA, instrument stability and staining repeatability are critical. A prerequisite is a voltage-stabilized lamp that eliminates fluctuations of line voltages. Additional procedures to monitor and compensate for fluorescence loss as a function of lamp aging and internal standards for monitoring the consistency of staining are discussed in Coleman et al. (1981) and Goff and Coleman (1984a, 1987a).

III. NUCLEAR STUDIES

A. Nuclear number, size, and positioning; cell cycle

Epifluorescence microscopy and DNA-fluorochrome studies reveal that red algae are exceedingly diverse with respect to their nuclear cytology. The number, size, and positioning of nuclei in cells, as well as their DNA content (ploidy), vary considerably in different taxa (Table 3-2). In nearly all unicellular, filamentous, and parenchymatous forms previously allied taxonomically with the Bangiophycidae, there is a single nucleus per cell, generally situated adjacent to the cell periphery. DNA measurements reveal that, except in rare cases (e.g., *Rhodella*), these nuclei do not become polygenomic (either polyploid or polytene) and, like nearly all other red algae (one notable exception is *Cyanidium*) and unlike most other eukaryotic cells, the major portion of their cell cycle is spent in G_2, with at least two copies of the nuclear genome present during most of interphase (Goff & Coleman 1984a, 1987a).

In the Florideophycidae, both uninucleate and multinucleate taxa occur; however, the "evolutionary" trend appears to be from cells with a single, medially and peripherally positioned, nonpolygenomic nucleus (Fig. 3-2), as in the Nemaliales, to cells with many nonpolyploid nuclei (Fig. 3-3) or with a single, highly endopolyploid nucleus (Fig. 3-4). The observation of multiple chromosomes sets in nuclei of very high ploidy (>128C) indicates that these nuclei are polyploid rather than polytene (Gonzalez & Goff 1989). The transition from cells with a single, nonpolygenomic nucleus to polynucleate or polyploid cells appears in members previously aligned with the order Cryptonemiales. For example, the somatic cells of *Dudresnaya crassa* and *Cumagloia andersonii* are uninucleate and nonpolyploid, whereas the larger axial cells of *Endocladia muricata* have a single, highly endopolyploid (ca. 8–512C) nucleus (Fig. 3-5).

Growth in all Florideophycidae is primarily due to the division of apical cells and the subsequent elongation and enlargement of their progeny. Therefore, developmentally, the number and ploidy of nuclei in cells derived from the apical cell should be related to the patterns present in the apical cells. The major developmental patterns that give rise to multinucleate or endopolyploid uninucleate cells in the Florideophycidae are summarized in Fig. 3-6. The apical cell may have either a single nonpolyploid nucleus, a single highly endopolyploid nucleus, or numerous nonpolyploid nuclei. In the first case, the uninucleate, nonpolyploid apical cell may give rise to non-polyploid, uninucleate cells, to cells with a single polygenomic (either polyploid or polytene) nucleus, or to cells with many nonpolyploid (or occasionally polyploid) nuclei (Goff & Coleman unpubl.). In the second case, the apical cell with a single highly polygenomic nucleus (8–128C or more) gives rise to cells in which the number of nuclei increases during development, while the ploidy of any single nucleus decreases (Goff & Coleman 1986). And last, some apical cells may have many nonpolyploid nuclei, and the cells derived from these multinucleate apical cells are also multinucleate (Goff & Coleman 1987a). Additional variants of these patterns are possible, though they have not yet been seen in red algal cells.

It also should be noted that some cells may become secondarily multinucleate upon the transfer of a nucleus from one cell to another. This occurs during the process of secondary pit-connection (plug) formation (discussed later), as described in the cases of *Choreocolax* and *Harveyella* (Goff & Coleman 1984b, 1985). It is also quite possible that nuclei, transferred from one uninucleate cell to another, may fuse with the single resident nucleus, thereby contributing to its increase in ploidy. This has been suggested by observations of *Choreocolax* (Goff & Coleman 1985) and in *Plocamium cartilagineum* (Goff & Coleman unpubl.), where the apical cell is uninucleate and nonpolyploid, the periaxial cells contain many nonpolyploid nuclei, and each axial cell contains a single huge endopolyploid nucleus (Fig. 3-7) with which many planetic nuclei, transferred via secondary pit plug formation, appear to fuse (Fig. 3-8).

Whether uninucleate or multinucleate, it is clear from an examination of the nuclear cytology of many of the Florideophycidae that the total nuclear DNA within a cell is very closely correlated to total cytoplasmic volume. Developmentally, this means that, within a filament of cells cut off from a single apical cell, the total number of nuclei (Fig. 3-9) or the ploidy of a single nucleus (Fig. 3-10) increases proportionately with the increase in cell size. Consequently, there remains a reasonably constant ratio between nuclear DNA and cytoplasmic volume (Goff & Coleman 1986, 1987a, unpubl.).

The Florideophycidae appear to lack cyclosis (Goff & Coleman 1987a; 1988b; Koslowsky & Waaland 1984; McNaughton and Goff unpubl.). Thus, organelles are "fixed" in the peripheral cytoplasm that surrounds the large central vacuole. In multinucleate forms, such as *Griffithsia*, the positioning of nuclei within the cytoplasm is highly regular, assuming a distinct, characteristic hexagonal arrangement that is maintained throughout karyokinesis and cell elongation (Goff & Coleman 1987a; Fig. 3-11). The amount of cytoplasm (area and volume) surrounding each nucleus is a function of nuclear ploidy. Consequently, twice as much cytoplasm surrounds a diploid (2–4C) nucleus as surrounds a haploid (1–2C) nucleus, and proportionately greater amounts of cytoplasm surround the occasional polyploid nucleus.

The plastids and mitochondria are arranged in a single layer in the cytoplasmic domain surrounding each nucleus. Approximately twice as many plastids surround a diploid nucleus as a haploid nucleus. And since all plastids have about the same amount of plastid DNA, there is a constant ratio of plastid DNA to nuclear DNA throughout a cell (Goff & Coleman 1987a).

Table 3-2. Nuclear cytology of some red algae

Organism	Source	Number of nuclei per Meristematic cell	Number of nuclei per Derived cell	Apical cell nuclei polyploid	Derived cell nuclei polyploid	G_1 vs. G_2 cell cycle	SPC formation	Nuclear transfer
Cyanidium sp.	Beil	1	1	—	—	G_1	—	—
Porphyridiales								
Stylonema alsidii (as *Goniotrichum elegans*)	UTEX LB1957	1	1	—	—	G_1	—	—
Porphyridium purpureum	UTEX 161	1	1	—	—	G_1	—	—
Rhodella sp.	Deason via J. Scott	1	1	—	$+^a$	G_2	—	—
Rhodella sp.	UTEX	1	1	—	—	G_2	—	—
Compsopogonales								
Erythrocladia sp.	UTEX LB 1637	1	1	—	—	G_2	—	—
Erythrotrichia carnea	UTEX LB 1690	1	1	—	—	G_2	—	—
Bangiales								
Bangia atropurpurea (conchocelis stage)	UTEX 1691	1	1	—	—	G_1	—	—
Batrachospermales								
Batrachospermum sirodotii	UTEX LB1495	1	1	—	NT	G_2	—	—
Palmariales								
Palmaria palmata	Nova Scotia	1	MN	—	—	G_2	+	+
Acrochaetiales								
Acrochaetium pectinatum	UTEX LB 1607	1	1	—	$+^a$	G_2	—	—
Acrochaetium purpureum	UTEX LB1975	1	1	—	—	G_2/G_1^b	—	—
Nemaliales								
Cumagloia andersonii	California	1	1	—	—	G_2	—	—
Gigartinales								
Choreocolax polysiphoniae	California	1	1^c	—	+	G_2	+	+
Harveyella mirabilis	California	1	1^c	—	+	G_2	+	+
Holmsella pachyderma	England	1	1^c	—	+	G_2	+	+
Corallina officinalis	Rhode Island	1	1	—	+	G_2	+	+
Dudresnaya crassa	St. Croix	1	1	—	—	G_2	—	—
Melobesia mediocris	California	1	1	NT	$+^c$	G_2	—	$+^d$
Endocladia muricata	California	1	1	NT	—	G_2	—	—
Sarcodiotheca gaudichaudii	California	1	MN	—	$-^e$	G_2	+	+
Gardneriella tuberifera	California	1	MN	—	NT	G_2	+	+
Hypnea sp.	UTEX	1	MN	—	$-^e$	G_2^f	+	+
Chondrus crispus	Rhode Island	1	MN	—	—	G_2	+	+
Iridaea splendens	California	1	MN	—	—	G_2	+	+

Species	Location	Col3	Col4	Col5	Phase	C1	C2	C3
Gracilariopsis lemaneiformis	California	1	MN	—	G_2^f		+	+
Gracilariophila oryzoides	California	1–4	MN	NT	G_2		+	+
Plocamium cartilagineum	California	1	MN^g	NT	G_2		+	+
Rhodymeniales								
Minium parvum	California	4–8	4–16	—	G_2		+	+
Ceramiales								
CERAMIACEAE								
Antithamnion defectum	UTEX	1	1	—	G_2		—	—
Scagelia pylaisaei	Woods Hole	1	1	—	G_2		—	—
Wrangelia plumosa[h]	S. Australia	1	1	—	G_2		—	—
Ceramium rubrum	Long Island	1	1	—	G_2		—	—
Microcladia coulteri[h]	California	1	1	—	G_2		—	—
Callithamnion baileyi	UTEX 2306	1	1	NT	G_2	NT	—	—
Callithamnion byssoides	UTEX 2296	1	1	NT	G_2	NT	—	—
Callithamnion corymbosum	UTEX 2298	1	1, 2, 4, 8, 16	NT	G_2	NT	—	—
Callithamnion halliae	UTEX 1411	1	1	NT	G_2	NT	—	—
Callithamnion paschale	UTEX 2300	20+	20–200+	—	G_2		—	—
Pleonosporium squarrulosum	California	2+	4, 8, 16+	—	G_2		—	—
Bornetia californica	California	20+	20–200+	—	G_2		—	—
Griffithsia pacifica[h]	California	MN	MN	—	G_2		—	—
Seirospora griffithsiana	UTEX 1510	1	1	+	G_2		—	—
DELESSERIACEAE								
Phycodrys setchelli	California	1	MN	NT	G_2		+	+
Polyneura latissima	California	1	MN	NT	G_2		+	+
Grimmellia americana	Rhode Island	1	MN	NT	G_2		+	+
DASYACEAE								
Dasya baillouviana (as *D. pedicellata*)	UTEX 1513	1	1–18+	+	G_2		—	+
Botryoglossum farlowianum	California	1	MN	NT	G_2		+	+
RHODOMELACEAE								
Bostrychia radicans	Brazil	1	MN	$+^i$	G_2		+	+
Polysiphonia mollis[h]	California	1	2″ or 3×2″	$+^i$	G_2		+	+
Laurencia spectabilis	California	1	MN	$+^i$	G_2		+	+
Janczewskia gardneri	California	1+	MN	$+^i$	G_2		+	+

[a] A few polyploid cells ($8C^+$) are seen in late log and stationary cultures. C-value is defined on p. 54.
[b] Nuclei in cells of the ultimate branches remain in G_1, whereas others remain in G_2.
[c] G_1 planetic nuclei may be transferred into cells via SPC, resulting in a multinucleate cell. The planetic nucleus may fuse with the resident endopolyploid nucleus.
[d] Nuclei are transferred upon cell fusion near apical tip. Nuclei appear to fuse.
[e] Ca. 0.1% of the nuclei in multinucleate cortical cells may be polyploid.
[f] The nuclei of the largest, most highly multinucleate cells appear to be G_1, whereas all other nuclei are G_2.
[g] Axial cell remains uninucleate and becomes highly polyploid.
[h] All species examined have the same pattern.
[i] Eventually in differentiated (multinucleate) pericentral cells, each nucleus is nonpolyploid.
MN = multinucleate; NT = not determined.

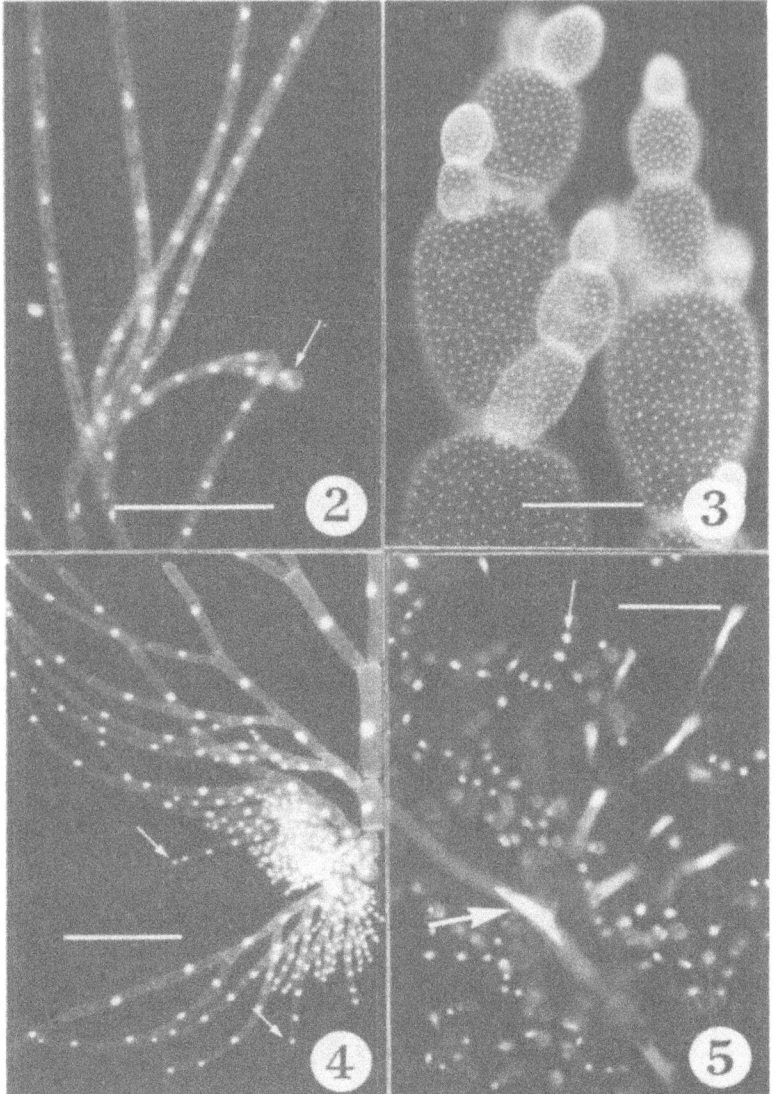

Figs. 3-2 to 3-5. Nuclear patterning in several Florideophycidae. Scale bar = 100 μm. *3-2: Acrochaetium pectinatum* (UTEX), microwave fixed, DAPI stained. Each cell contains a single 1C or 2C nucleus situated peripherally in the midregion of the cell. Note differentiating monospore (arrow). *3-3: Griffithsia globulifera* (2N). Microwave fixed, DAPI stained. Each cell, including apical cell of each branch, is highly multinucleate, with nuclei positioned peripherally and spaced in a very regular lattice. *3-4: Wrangelia plumosa* (2N), fixed in Carnoy's, stained with 20 μg/ml mithramycin. Cells are uninucleate with a small 1–2C nucleus in apical cells (arrows), and larger, polyploid nuclei in larger cells. *3-5: Endocladia muricata* (2N) fixed in Carnoy's, squashed, DAPI stained. Small cortical cells have a single 1–2C nucleus (upper arrow), whereas the larger axial cells (larger lower arrow) have a single polyploid nucleus that may exceed 512C in DNA content.

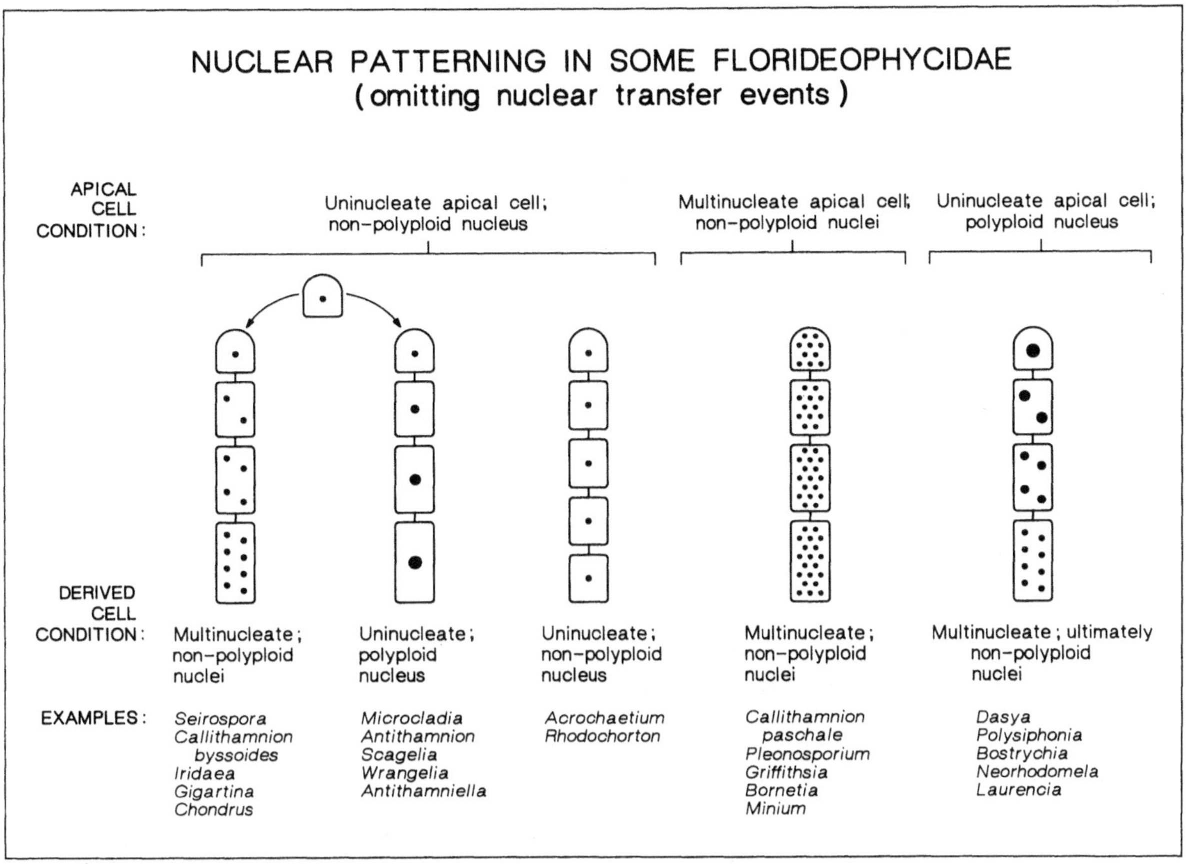

DNA: Microspectrofluorometry 51

NUCLEAR PATTERNING IN SOME FLORIDEOPHYCIDAE
(omitting nuclear transfer events)

APICAL CELL CONDITION:	Uninucleate apical cell; non-polyploid nucleus			Multinucleate apical cell; non-polyploid nuclei	Uninucleate apical cell; polyploid nucleus
DERIVED CELL CONDITION:	Multinucleate; non-polyploid nuclei	Uninucleate; polyploid nucleus	Uninucleate; non-polyploid nucleus	Multinucleate; non-polyploid nuclei	Multinucleate; ultimately non-polyploid nuclei
EXAMPLES:	Seirospora Callithamnion byssoides Iridaea Gigartina Chondrus	Microcladia Antithamnion Scagelia Wrangelia Antithamniella	Acrochaetium Rhodochorton	Callithamnion paschale Pleonosporium Griffithsia Bornetia Minium	Dasya Polysiphonia Bostrychia Neorhodomela Laurencia

Fig. 3-6. Major nuclear patterns observed in some Florideophycidae, omitting transfer of nuclei from one cell to another, which results in additional patterns. (*Antithamniella* should read *Antithamnionella*).

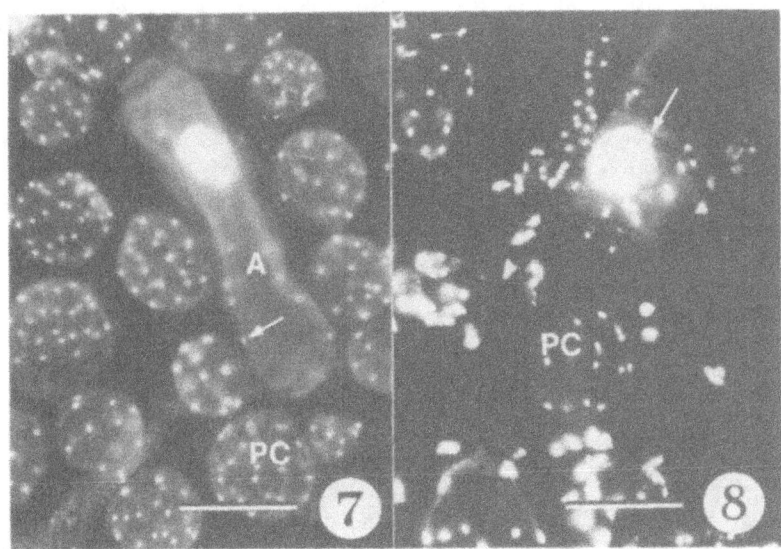

Figs. 3-7, 3-8. *Plocamium cartilagineum*, fixed in Carnoy's, squashed and DAPI stained. Scale bars = 100 μm. *3-7*: Multinucleate pericentral (periaxial) cells (*PC*) contain many 1–2C nuclei; axial cells (*A*) have a single, highly polyploid nucleus. Several small nuclei (arrow) within the axial cell have been transferred to this cell via secondary pit connections with pericentral cells. *3-8*: Planetic nuclei, transferred from adjacent pericentral cells to the axial cell, appear to fuse with the axial cell nucleus (arrow), thereby contributing to its ploidy increase.

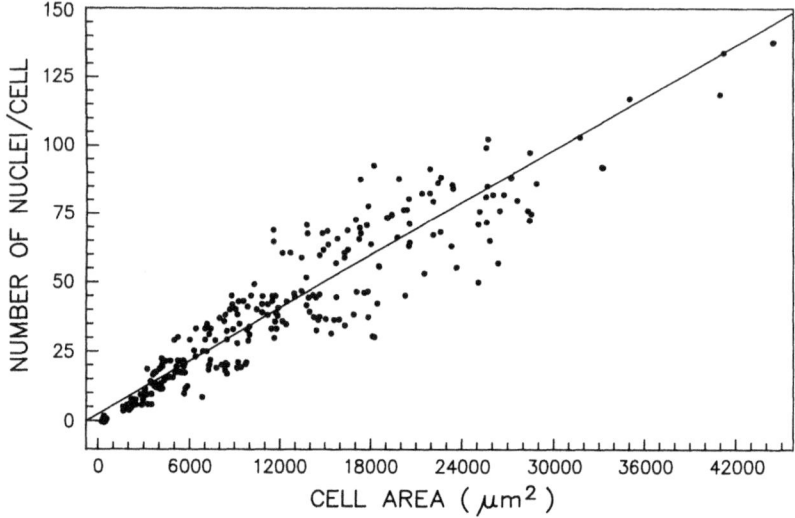

Fig. 3-9. Strict correlation of the number of nuclei per cell with cell size in pericentral cells of *Polysiphonia mollis*. The line represents the calculated regression. At any one cell size there is only a twofold difference in the number of nuclei per cell. The lower number is found in cells that have elongated but have not yet divided, whereas the upper value represents a cell that has not yet elongated but is one in which the nuclei have divided.

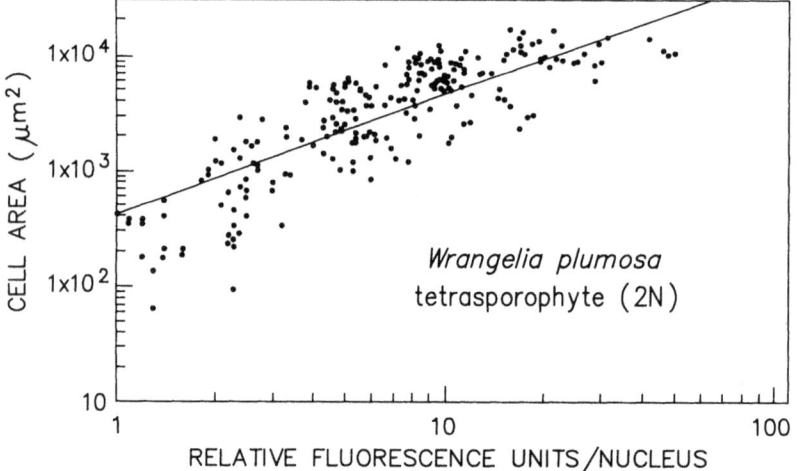

Fig. 3-10. In *Wrangelia plumosa*, the DNA content, measured as fluorescence of the single DAPI-stained nucleus, increases proportionately to cell size increases. Measurements were made of nuclei in vegetative cells.

The regular spacing of nuclei (and plastids) creates cytoplasmic domains that may represent the distance over which diffusion can effectively move gene products in noncyclosing cells (Goff & Coleman 1987a). In uninucleate cells such as those in *Antithamnion*, *Scagelia*, and *Wrangelia*, in which the single nucleus increases in ploidy in proportion to the increase in cell volume, the nucleus rapidly assumes a flattened plate shape as the central vacuole increases in size (Figs. 3-12, 3-13). The area of the nuclear "plate" is directly proportional to the cytoplasmic volume of the cell (Fig. 3-14). The change in shape from a sphere to a plate effectively increases the nuclear–cytoplasm contact area and may thereby facilitate transport of gene products from the nucleus to the cytoplasm. In many taxa (e.g., *Wrangelia*, *Antithamnion*, *Scagelia*, *Microcladia*, and *Platythamnion*), the nucleus undergoes rather dramatic changes in shape in the largest cells. In these cells, the nucleus may assume an elongate band shape (Fig. 3-15), or long extensions of the nucleus can protrude into the far reaches of the cytoplasm. Occasionally, these cells appear to become multinucleate (Figs. 3-16, 3-17), but careful observations reveal that the multiple nuclei within a cell are interconnected by long, thin nuclear extensions (Fig. 3-18). Magne (1964) and Dixon (1973) have also reported the occurrence of long fusiform nuclei in other red algae.

These patterns are quite pertinent to discussions of the "C-value paradox" in biology. The C-value paradox, as discussed by Thomas (1971), Gall (1981), and Cavalier-Smith (1978, 1985), points out that even in eukaryotic cells with the smallest nu-

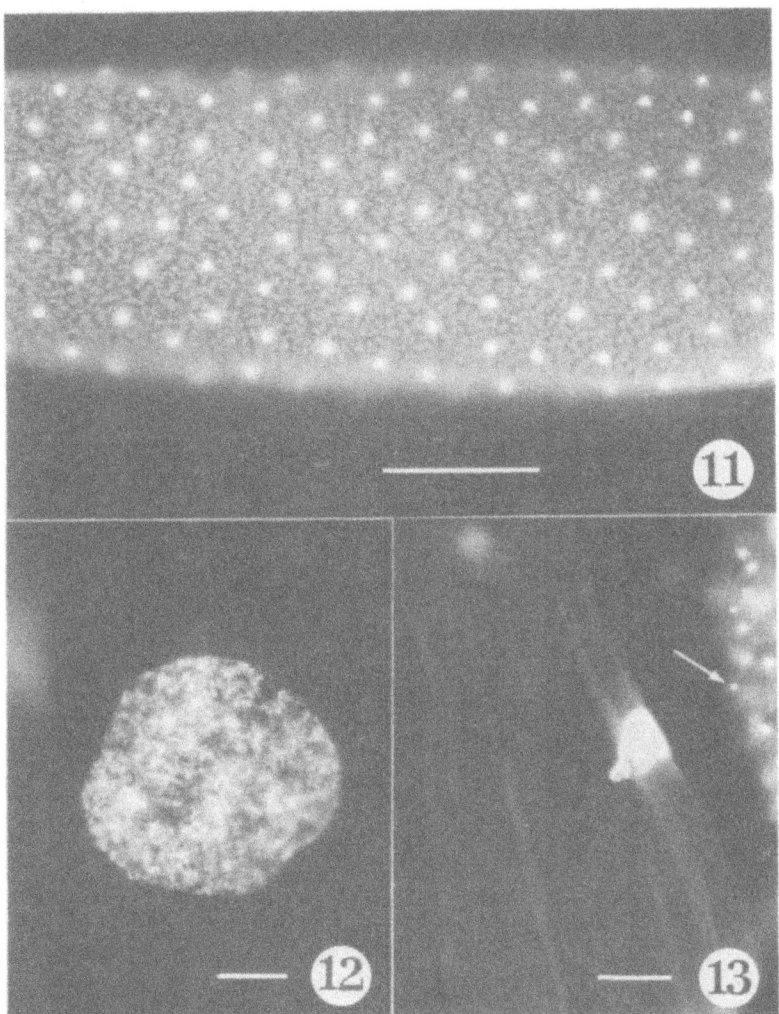

Figs. 3-11 to 3-13. *3-11*: *Griffithsia pacifica* (2N) (fourth cell from apex), microwave fixed, stained with DAPI. Note many 2–4C nuclei positioned in a hexagonal array; each nucleus is surrounded by a cytoplasmic domain, the size of which is related to nuclear ploidy. Scale bar = 100 μm. *3-12, 3-13*: Nuclei in axial cells of the uninucleate taxon *Wrangelia plumosa* (1N), fixed in Carnoy's, DAPI stained; scale bars = 100 μm. *3-12*: Nuclei may be huge with DNA levels exceeding 1024C. They are plate shaped and appear to lack any nucleolar region (a nucleolus would appear as a "hole" in the DAPI-stained nucleus). *3-13*: A plate-shaped polyploid nucleus wrapped around the inner periphery of an axial cell. The arrow indicates the size of 1–2C nucleus in an apical cell.

clear genome size, there exists vastly more DNA than is required to prime transcription of all gene products. Many hypotheses have been proposed to account for this phenomenon (Dawkins 1976, 1983), including one by Cavalier-Smith (1978, 1985) that maintains that DNA may serve both a genetic and structural role in the cell. The nontranscribed nuclear DNA would serve the structural role of determining the size of the nucleus, which in turn would determine the size of the cell. With respect to this idea, Lewis (1985) questioned why, if the main role of much of the DNA is simply to determine the ratio of nuclear size to cytoplasm, flattened or branched nuclei have not evolved to improve the surface-to-volume ratio of the nucleus with regard to the entire cell.

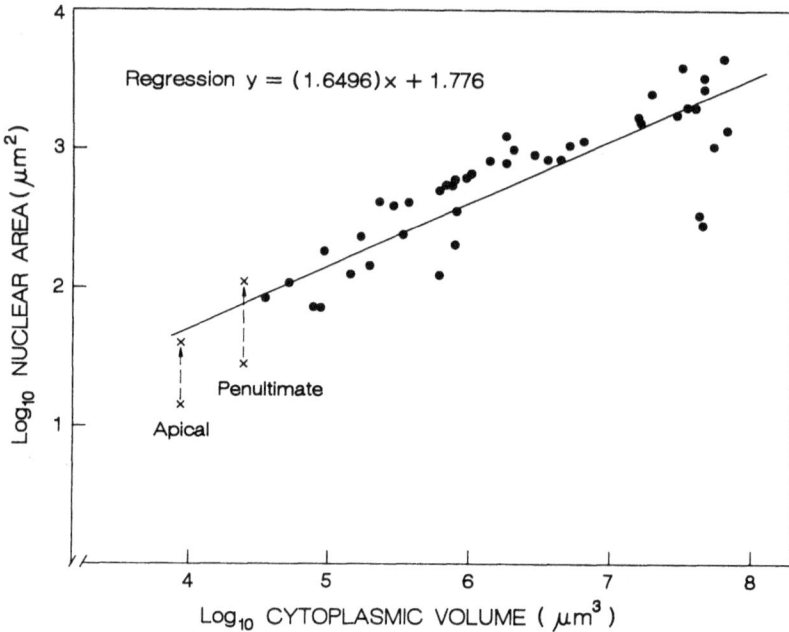

Fig. 3-14. Direct correlation of nuclear size with calculated cytoplasmic volume in *Wrangelia plumosa*. Since most nuclei are plate shaped, this correlation is apparent by measuring nuclear area. In apical and penultimate cells, calculations of nuclear areas fall well below the regression line (lower x's) since these nuclei are not spherical. If their volume is recalculated as a sphere (upper x's), the ratio of nuclear to cytoplasmic volume is consistent with all other larger cells.

B. Ploidy of life history stages and site of meiosis

Since the amount of light emitted from a fluorochrome-stained nucleus is proportional to the amount of DNA present in a nucleus, ploidy values of stages of even the most complex life history can be readily determined. Fluorescence from single nuclei is measured in relative fluorescence units (rfu), which can be directly related to the C-value (i.e., the basic amount of DNA in the nonreplicated haploid chromosome complement) of the species. In organisms with haploid and diploid developmental stages, the lowest DNA value, the 1C level, should occur in the haploid nuclei that have not yet replicated their DNA (i.e., during G_1 of interphase). C-values and N-values need not be the same.

An example of how microspectrofluorometry may be used to investigate a red algal life history is seen in a study of *Wrangelia plumosa*, collected at Kangaroo Island, South Australia. Measurements of the fluorescence of DAPI-stained nuclei from somatic cells reveal a large spread in relative fluorescence values. The lowest levels of DNA (ca. 0.8 rfu) are found in apical and penultimate cells of the female and male gametophyte (Fig. 3-19). Nuclei with two or four times the DNA (ca. 1.6 and 3.2 rfu respectively) are also seen in these most apical cells. The lowest DNA level apparent in the vegetative cells of the tetrasporophyte is twice that (ca. 1.6 rfu) found in the gametophytes. The other two peaks of nuclear values in the apical and penultimate cells of the tetrasporophyte have correspondingly greater amounts of DNA (ca. 3.6 and 7.2 rfu, respectively). Measurements of the DNA level in nonapical vegetative cells show little difference between the gametophytes and sporophytes (Fig. 3-20). Relative fluorescence values ranging from ca. 1 to more than 1000 rfu have been measured in nuclei of these cells.

From these data it might be presumed that the 1C value corresponds to 0.8 rfu, the lowest value measured in the apical cells of the gametophyte. The lowest value measured in the apical cells of the tetrasporophyte is ca. 1.6 rfu, corresponding presumably to the 2C value. In both the gametophyte and sporophyte generations, there are two dominant peaks: 0.8 and 1.6 rfu (gametophyte) and 1.6 and 3.2 rfu (sporophyte). The distribution between

DNA: Microspectrofluorometry

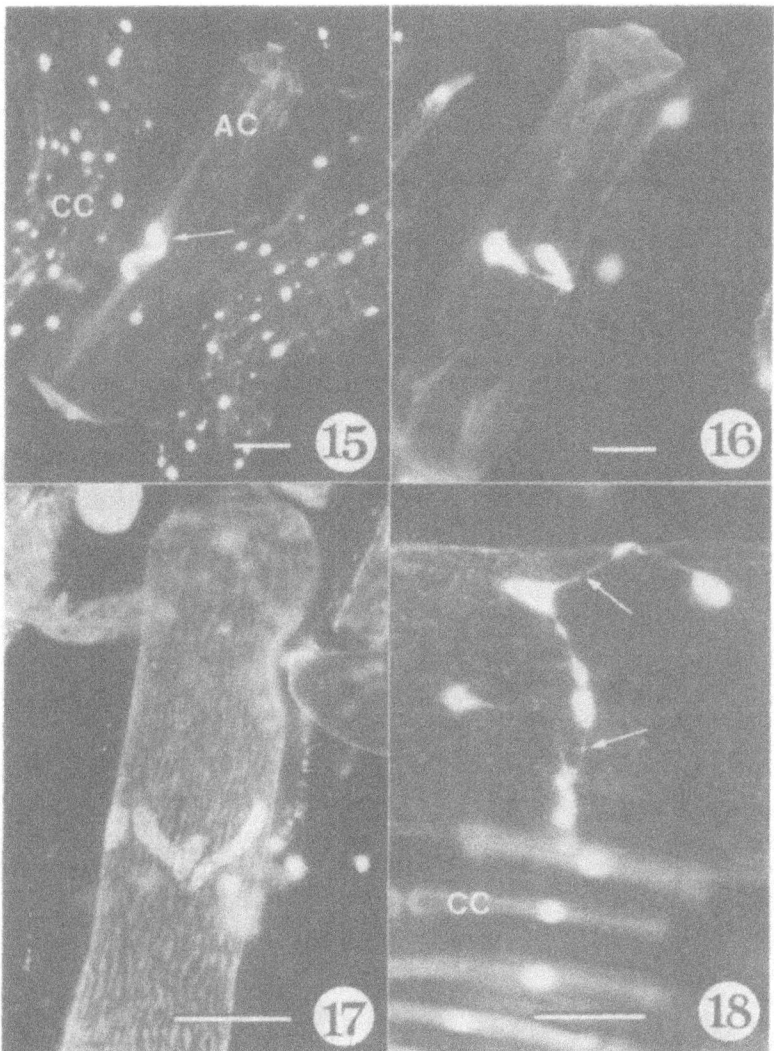

Figs. 3-15 to 3-18. Nuclear shape in *Wrangelia plumosa* and *Scagelia pylaisaei*. Scale bars = 100 μm. *3-15:* Elongate, helical, band shaped nucleus (arrow) in larger axial cell (AC) of *Wrangelia plumosa* (2N) with DNA level exceeding 1000C. The smaller nuclei within the surrounding corticating cells have DNA levels ranging from 8 to 32C+. *3-16:* "Pull apart" nucleus resulting in an apparent multinucleate axial cell of *W. plumosa*. *3-17:* Four nuclei in an apparent multinucleate axial cell of *S. pylaisaei*. *3-18:* Numerous "nuclei" of *W. plumosa* appear to be interconnected by elongate threads that contain DNA (arrows). Nuclei of the corticating cells (CC) are ca. 32C.

the two peaks in each represents the proportion of cells measured in either G_1 or G_2. Since the DNA level doubles between the G_1 and G_2 stages of the cell cycle, a haploid cell in G_2 has the same amount of nuclear DNA as a diploid cell in G_1 (both are 2C).

There are two approaches to determining if the lowest value of 0.8 rfu does represent the 1C value of *Wrangelia plumosa*. In the absence of reproductive cells, one can measure mitotic figures in dividing somatic cells (Fig. 3-21). If the dividing apical cells are not polyploid, the prophase and metaphase figures should have G_2 rfu values corresponding to the 2C or 4C DNA levels in the gametophyte and sporophyte, respectively. Telophase nuclei should

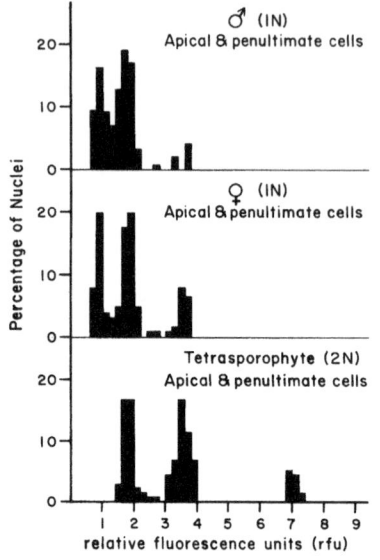
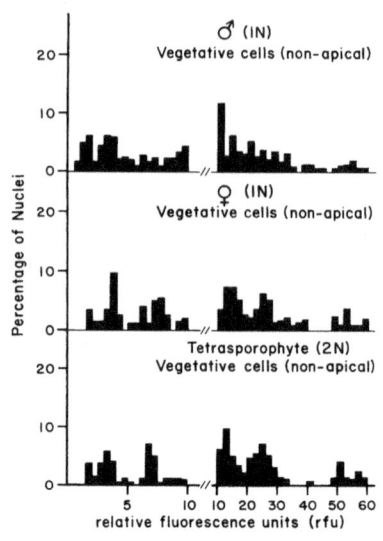

Figs. 3-19 (left), 3-20 (right). Histograms of the DNA levels, measured as relative fluorescence units (rfu), in apical and penultimate cells of gametophyte and tetrasporophyte *Wrangelia plumosa*. For each data set, a minimum of 400 nuclei were measured. See text for interpretations.

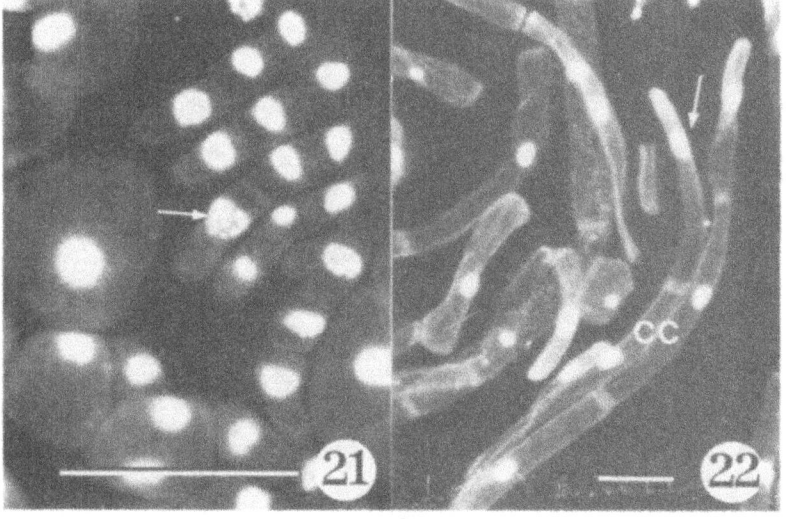

Figs. 3-21, 3-22. Female gametophytes of *Wrangelia plumosa*, fixed in Carnoy's and DAPI stained. Scale bars = 100 μm. *3-21*: Numerous (ca. 27) chromosomes occur in the nucleus of this dividing apical cell (arrow) with a DNA level of 2C. *3-22*: Endopolyploid nuclei of corticating cells, including the apical cells (arrow) have DNA levels ranging from 8 to 32C+.

each have rfu values that correspond to the 1C (gametophyte) or 2C (sporophyte) DNA level if these nuclei have not yet undergone DNA synthesis and are still in G_1. In *W. plumosa*, dividing apical cells with visible prophase chromosomes or metaphase plates have rfu values of 1.6 in the gametophyte and ca. 3.2 in the sporophyte, and in each respective generation, the telophase nuclei have values of 0.8 and 1.6.

Although measurements of division figures in apical cells provide data consistent with the interpretation that the 1C level of DNA in *Wrangelia plumosa* corresponds to the 0.8 rfu readings, it remains possible that the apical cell may be polyploid, as in the case of many of the Rhodomelaceae (Goff & Coleman 1986). Measurements of nuclei in the cells of corticating filaments (Fig. 3-22) that ensheath the axial cells of *W. plumosa* clearly reveal that these are highly polyploid, with DNA values ranging from ca. 6–60 rfu (Fig. 3-23). Dividing nuclei in the apical cells of these filaments have DNA values that correspond to ca. 8–32C, if 0.8 rfu represents the 1C value.

A preferable way of establishing the numerical relationship between rfu and the C-value of a species is to measure the relative fluorescence of

DNA: Microspectrofluorometry

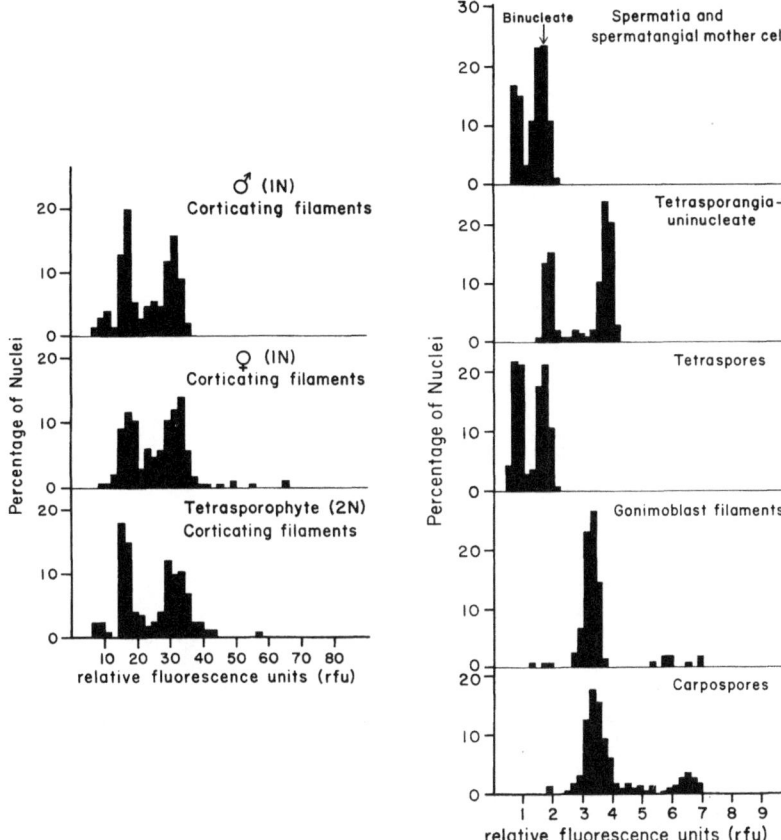

Figs. 3-23 (left), 3-24 (right). Histograms of the DNA level (measured as relative fluorescence units) of corticating filaments of gametophytic and tetrasporophytic cells, and reproductive cells of *Wrangelia plumosa*. For each data set, a minimum of 200 nuclei were measured. 3-23: There is no discernible difference in the amount of DNA in nuclei of the corticating cells from either the 1N (gametophyte) or 2N (tetrasporophyte) generations. 3-24: These data indicate that typical meiosis occurs during the formation of tetraspores. Nuclei of mature spermatia (many of which are binucleate), tetraspores, and carpospores all are in G_2 at the time of release and have DNA levels of 2C, 2C, and 4C, respectively.

nuclei in reproductive cells. Measurements of mature spermatia and spermatangial parental cells reveal two peaks, of 0.8 and 1.6 rfu, respectively (Fig. 3-24). The lowest value is that of each newly formed nucleus found in the spermatangial parental cell just subsequent to the division that forms the spermatium and in the newly formed spermatium. The mature and released spermatia have nuclear DNA values of 1.6 rfu, and many become binucleate at maturity. According to our interpretation, the lowest value of the spermatangial parental cell and spermatangium corresponds to the 1C (G_1) DNA amount, whereas the mature spermatium has a single 2C (G_2) nucleus or two 1C (G_1) nuclei. In the mature 2C spermatium, chromosomes are evident, forming a ring around the inner periphery of the nuclear envelope. This "shell stage," noted by earlier cytologists (Dixon 1973; Grubb 1925), indicates that these nuclei are in prophase at the time of spermatium release.

DNA readings of other reproductive stages are consistent with the premise that the lowest value of DNA seen in the apical cell of the gametophyte is the 1C value of this species. Tetraspores have two DNA levels, with the lower (1C) value found in immature tetraspores just after the completion of meiosis and the higher (2C) value in the mature tetraspores (Fig. 3-24). Thus, like the spermatia, the tetraspores are in G_2 at the time of their release. The uninucleate tetrasporangia have values of 2C

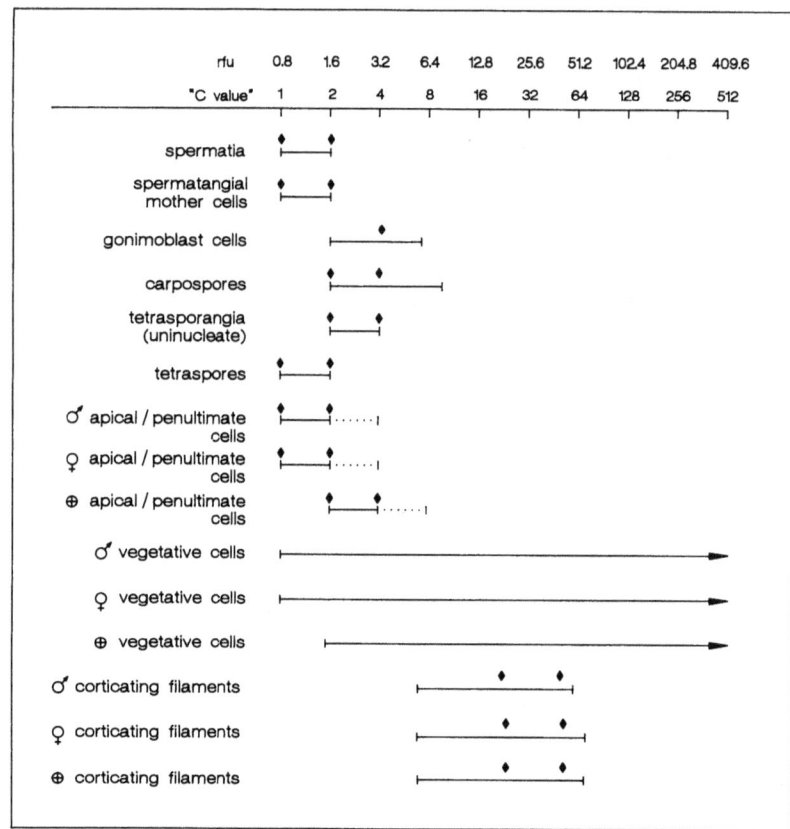

Fig. 3-25. A summary of microspectrofluorometry data from nuclei of somatic and reproductive cells in *Wrangelia plumosa*. The conversion of rfu to C-values are presented on the top scale. ◆ indicates rfu values where mitotic figures were observed.

in newly differentiated sporangia, and 4C in the mature sporangium just prior to meiosis. And last, the gonimoblast cells, which form as a consequence of fertilization, have DNA values corresponding to the 4C value (G_2) of DNA (Fig. 3-24). The carpospores that are produced terminally from the gonimoblast also are 4C, although a small proportion appear to be polyploid (8C) (Fig. 3-24). A summary of the rfu and C-values for the various somatic and reproductive cells of *Wrangelia plumosa* is presented in Fig. 3-25.

If meiosis occurs in the tetrasporangium during the formation of tetraspores, the microspectrofluorometry results depicted in Fig. 3-26 would be expected. Similar results have been seen in studies of tetrasporogenesis in *Wrangelia*, *Antithamnionella pacifica*, and several other red algae. However, several cases have now been documented using microspectrofluorometry in which tetrasporangia may develop apomictically.

Examination of *Scagelia pylaisaei* in culture and from field populations (collected at Woods Hole, Massachussetts), clearly demonstrates that this organism undergoes the classic triphasic isomorphic red algal life history (Goff & Coleman unpubl.). Indeed, our microspectrofluorometric measurements have confirmed that the gametophytes are haploid and the tetrasporophytes are diploid. However, measurements of DNA levels during tetrasporogenesis indicate that no reduction in nuclear DNA occurs during this process (Fig. 3-26).

Measurements of DNA levels during the germination of the tetraspores reveal that meiosis occurs during the first two divisions of the dividing tetraspores (Fig. 3-27). The first cleavage of the tetraspore results in a bipolar germling in which one cell is destined to become the apical cell, whereas the other forms the attaching primary rhizoid. This first cleavage reduces the ploidy of the tetraspore nucleus from 4C to 2C (Fig. 3-27). The following divison of the apical cell nucleus results in an apical and penultimate cell, each with a 1C nucleus. Nuclear and cell division are followed by an immediate round of DNA synthesis in each nucleus, resulting in an increase in nuclear DNA to the G_2 (2C) level.

Apomictic tetrasporangial development has also been shown in three other red algae, using microspectrofluorometry. In the first case, an undescribed

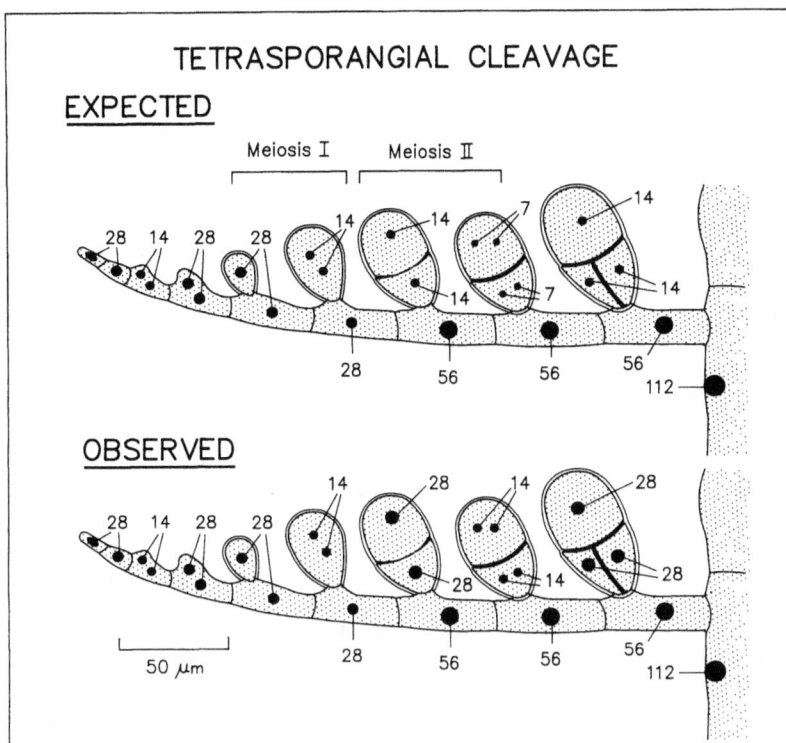

Fig. 3-26. Numbers indicate expected and observed relative fluorescence units in nuclei of *Scagelia pylaisaei* undergoing putative meiosis. If meiosis occurred during tetrasporangial cleavage, DNA levels would be reduced during meiosis I and II, as indicated in the "expected". However, actual measurements of DNA in nuclei during tetrasporogenesis indicate that DNA levels are not reduced; rather, these nuclei divide mitotically.

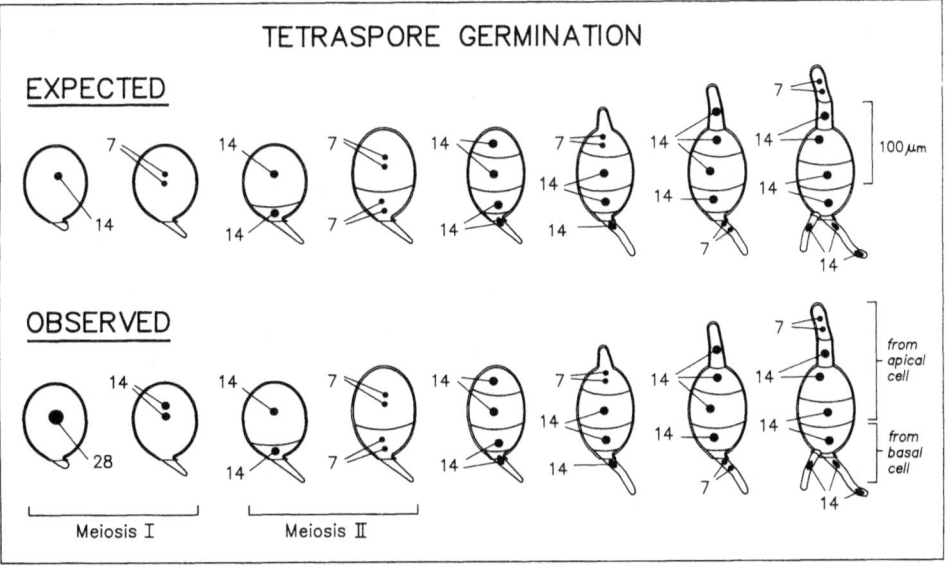

Fig. 3-27. Numbers indicate expected and observed relative fluorescence units in nuclei of germinating *Scagelia pylaisaei* tetraspores. No reduction in DNA is expected if nuclear division is mitotic. However, measurements of nuclear fluorescence indicates that DNA levels are reduced from 28 rfu (4C) to 7 rfu (1C) during the first cell divisions of the germinating tetraspores. This indicates that meiosis occurs during tetraspore germination rather than during tetraspore formation.

red algal parasite of *Palmaria palmata* produced apomictic tetraspores that repeated the same generation (Goff unpubl.). No sexual plants have been observed in this population. In another study, the presumed meiotic polysporangia of *Gonimophyllum skottsbergii* (collected in central California) developed apomictically, and meiosis did not occur during subsequent polyspore germination (Goff unpubl.). This observation is most curious as both male and female gametophytes occur in this population. Last, the bispores of *Gardneriella tuberifera*, a parasite of *Sarcodiotheca gaudichaudii*, may form either meiotically or apomictically (Goff 1981). Each meiotic bispore has two haploid (1–2C) nuclei, whereas the bispores that develop apomictically have a single diploid (2–4C) nucleus. The haploid binucleate bispores give rise in culture to gametophytes, whereas the diploid uninucleate bispores repeat the diploid bispore-producing generation.

A further variant on these patterns may provide one explanation for the many reported observations, particularly in the Ceramiaceae, of the presence of gametes and tetraspores on the same individual (Drew 1934; Hassinger-Huizinga 1952; Knaggs 1969; Lee & West 1980; Rueness & Rueness 1973; Sundene 1965; West & Norris 1966; Whittick & West 1979). In the few cases where this was observed in *Scagelia pylaisaei*, the individual was determined to be genetically haploid (1–2C), as were the tetraspores (Goff & Coleman unpubl.). Upon release from the gametophyte, the haploid tetraspores divide mitotically and give rise to another male gametophyte. Thus, what appears morphologically to be tetrasporogenesis was occurring in haploid plants.

IV. UNUSUAL ASPECTS OF RED ALGAL NUCLEI

A. Nuclear migration

A well-known example of nuclear migration in red algae occurs when a zygote nucleus moves from the site of syngamy (the carpogonium) to another cell (auxiliary cell), where it is replicated. Nuclear transfer may occur simply via the fusion of the carpogonium containing the zygote nucleus to a closely situated cell, or the nucleus may be partitioned into a cell (the ooblastema). This cell elongates and ramifies between vegetative cells of the female gametophyte until it contacts, fuses with, and deposits the zygote nucleus into a distant auxiliary cell. Within this gametophytic cell, the diploid zygote nucleus divides and is finally partitioned into cells of the developing gonimoblast.

There are also many examples of the movement of nuclei between adjacent vegetative cells of red algae. Perhaps the best known is that of nuclear transfer during the formation of secondary pit connections, or pit plugs, in the red alga *Polysiphonia* (Goff & Coleman 1986; Rosenvinge 1888). The processes that result in the formation of a secondary pit connection (SPC) are distinct from those of primary pit connection (PPC) formation (summarized in Figs. 3-28a,b). Unlike PPCs, SPCs form between two cells not derived from a common cell division. During this process, the nucleus of a pericentral cell, when it reaches a position approximately two to four cells behind the apical cell, undergoes division and migration to the outer corner of the cell. The nucleus then moves into a cellular protrusion ("bud") that cleaves off from the parent cell by an incomplete centripetal infurrowing of the plasmalemma. This short-lived uninucleate (polyploid) conjunctor cell fuses completely with the adjacent (more distally positioned) pericentral cell, thereby transferring the nucleus. As a result, the recipient pericentral cell becomes binucleate and gains a pit connection to its more apical neighbor. In all red algae that we have examined, SPC formation results in the transfer of a nucleus from one cell to another.

According to light microscopic observations, nuclei may also be transferred from cell to cell via the direct fusion of vegetative cells (Cabioch 1972; l'Hardy-Halos 1969). In the one case we have examined, adjacent cells (ca. three cells from the tip) of *Corallina officinalis* fuse and consequently two or more cells may become interconnected at the site of cell fusion (Figs. 3-29a,b). A single nucleus is situated at this fusion site and there are no other nuclei in any other region of the interconnected cells. Preliminary microspectrofluorometry data indicate that the nuclei of the newly fused cells may themselves fuse, and consequently, the single nucleus situated at the site of cell fusion is polyploid.

SPC formation and/or cell fusion may also occur between cells of genetically different individuals, establishing a heterokaryotic cell. Waaland (1978) clearly demonstrated that heterokaryons could be produced experimentally by fusing male and female gametophytes via the cell fusion (i.e., wound repair) system that functions in *Griffithsia* and many other large-celled algae of the Ceramiaceae. These heterokaryons were shown to be genetically different from either gametophyte, giving rise to reproductive cells (nonviable tetrasporangia) characteristic of the tetrasporophyte generation.

We have examined "natural" heterokaryon formation during the interaction of genetically dif-

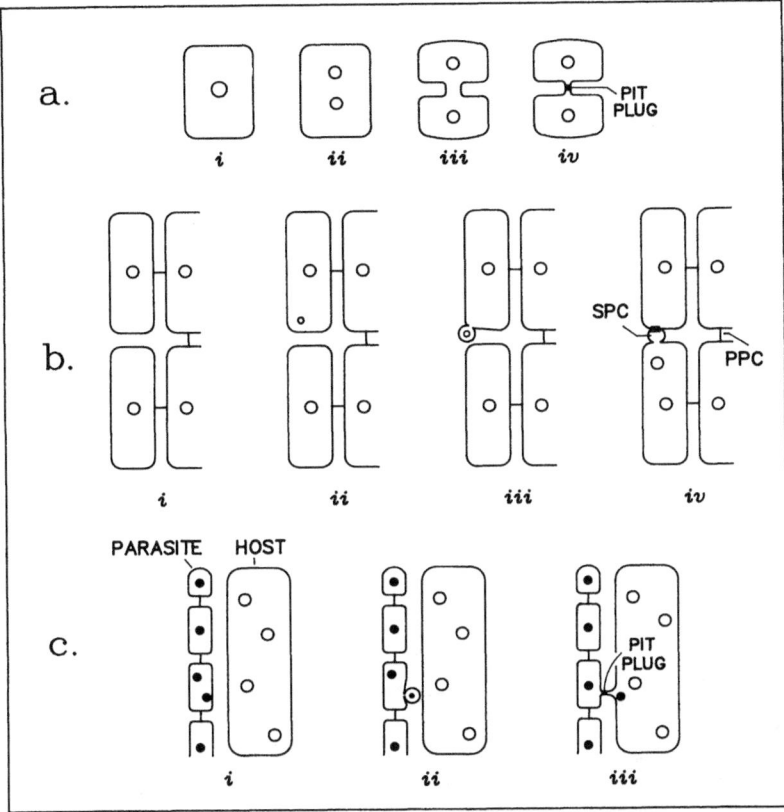

Fig. 3-28. Pit connection formation in Florideophycidae. a. Primary pit connections (pit plugs) occur upon the occlusion of the opening left by incomplete infurrowing of the plasmalemma during cytokinesis. b. Secondary pit connections (SPC) form between two cells not derived from a common cell division. The process results in the interconnection of two cells via the SPC and transfer of one or more nuclei from donor to recipient cell. c. SPC may also form between genetically unrelated cells, as in the case of red algal parasitism. The parasite initiates the process, transferring a nucleus and other cytoplasmic constituents to host cells.

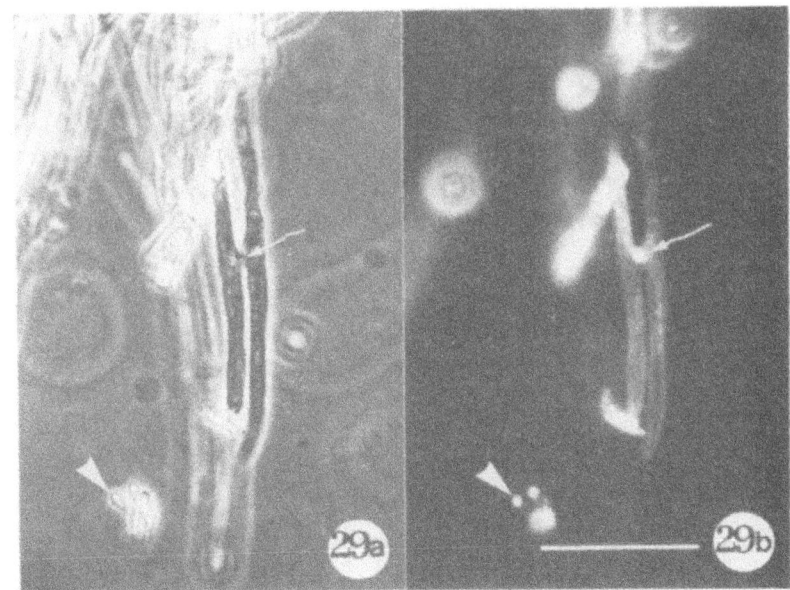

Fig. 3-29. *Corallina officinalis*, fixed in Carnoy's, decalcified in EDTA, squashed, and DAPI stained. Phase contrast (a) and epifluorescence (b) microscopy. Scale bar = 100 μm. Note fusion of two cells near the apical tip (a, arrow). This nucleus appears to be ca. 2–4 times larger and contains proportionately 2–4 times more DNA than nuclei of the spermatangia (arrowheads).

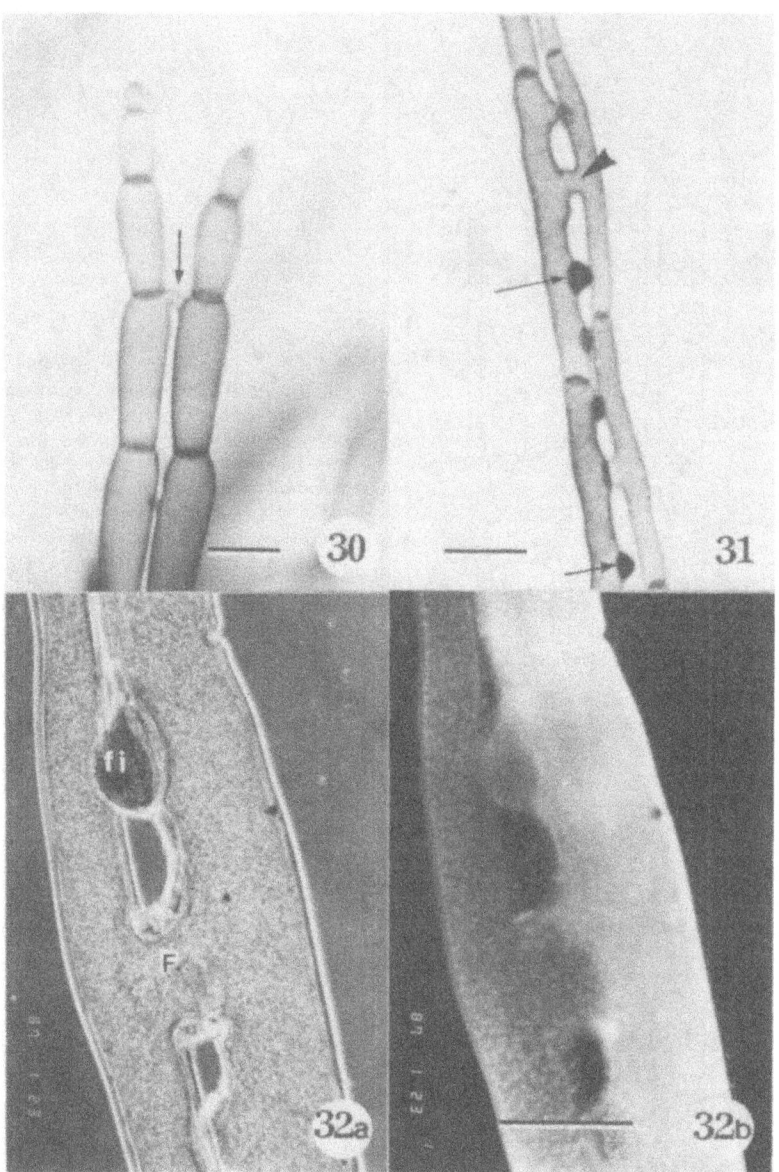

Figs. 3-30, 3-31. Cell fusion in *Griffithsia* and *Bornetia*. Scale bars = 200 μm. *3-30*: In *G. tegis*, a lateral rhizoidlike outgrowth from the cell of the right filament (2N) fusing directly with the cell (1N) on the left. *3-31*: Cells of adjacent filaments fusing in *B. californica*. Nuclei and plastids increase in density at the site of branch initial initiation (arrows). Upon contact with a cell of an adjacent filament, the branch initial fuses (arrowhead), opening a cytoplasmic channel between the cells.

Fig. 3-32. Fusion of male (left) and female (right) cells of *Bornetia*. a: Phase contrast. b: Epifluorescence microscopy. Scale bar = 200 μm. Cells microwave fixed and DAPI stained. A fusion initial (*fi*), dense with plastids and nuclei, is formed from the female cell and will fuse with the male filament. Numerous nuclei occur in the fusion site (*F*) (Fig. 3-32b).

ferent cells in another *Griffithsia* species, *G. tegis*, collected in Victoria, Australia, and the related taxon *Bornetia californica*, collected in central California (unpubl. observ.). Both taxa have a rather extensive prostrate vegetative system of cells which are frequently interconnected by lateral branches (Figs. 3-30, 3-31). The sequence of development of these laterals is the same in both taxa, and connections may form between genetically identical cells (of the same clone), or different cells (i.e., gametophyte-gametophyte or gametophyte–sporophyte). This process appears to be a modification of SPC formation described in *Polysiphonia*.

Two adjacent filaments become interconnected when one or both cells produce a branch initial from the side of the cell adjacent to the juxtaposed filament. Branch formation is first indicated by the migration of nuclei, often hundreds, to a localized cytoplasmic region from which the branch initial develops (Fig. 3-32). The highly multinucleate protuberance is then cut off from the parent cell but remains attached by a pit connection. Developmentally, this initial may grow into a rhizoidlike cell, characterized by tip growth (in *Griffithsia tegis*) or, it may form a cell identical to a normal apical cell of the shoot system (*Bornetia*). In either case, the cell grows toward the adjacent cell, contacts it, and fuses, transferring the numerous nuclei and other organelles of the connecting cell into the cytoplasm of the recipient cell (Figs. 3-32a,b).

In both *Griffithsia tegis* and *Bornetia*, intraspecific fusions of male and female gametophytes have been observed. Responses indicating the cytoplasmic incompatibility of chloroplasts, such as those described by Koslowsky and Waaland (1984, 1987), have not been seen in any case. The biological consequences of fusion and the establishment of heterokaryotic cells remain unknown. However, it is of interest that both taxa in nature form an extensive network of interconnected cells in the basal portion of the plant, and at least in *Bornetia*, numerous laterally interconnected cells have been seen in plants collected from field populations.

Another case of possible heterokaryosis was suggested by observations of sporeling coalescence in *Chondrus crispus* (Tveter-Gallagher & Mathieson 1976, 1980). In this taxon SPCs form between cells of two different crusts when these coalesce. If SPCs do form between different individuals, then nuclear transfer from one crust to another, via SPC formation, may occur, forming heterokaryotic cells at the interface of the coalesced crusts. To investigate this possibility, we examined (Goff & Coleman unpubl.) the interaction of genetically identical (same clone) or different crusts of the red algal *Minium parvum* (Moe 1979). Since this organism is monostromatic at its apical growing edge, the interaction of cells upon crust coalescence can be readily observed. In addition, SPCs form frequently between adjacent cells within individual crusts (near the apical growing region), and this process results in nuclear transfer (Fig. 3-33). Preliminary studies reveal that SPCs do form between different individuals upon contact of their apical cells (Figs. 3-34, 3-35) and that this process results in nuclear transfer from one individual to another (Figs. 3-36a,b, 3-37a,b).

Nuclei may also be transferred between different genera (in some cases from different orders) in the interactions of red algal parasites and their red algal hosts (Goff & Coleman 1984b, 1985, 1987b; Lewin 1984). A nucleus and other cytoplasmic components (proplastids, mitochondria, ribosomes) are transferred from parasite to host upon the formation of an SPC between the two cells (Fig. 3-28c). In all cases examined, the parasite initiates the process; it forms the conjunctor cell (which contains the parasite "planetic" nucleus) that fuses with an adjacent host cell, thereby transferring the nuclear and organelle genomes of the parasite into the host cell (Fig. 3-38). In some associations (i.e., *Gracilariophila* and *Gracilaria*), host and parasite cells may simply fuse with one another, thereby establishing a heterokaryon.

In the alloparasite associations examined (Goff & Coleman 1984b, 1985), parasite nuclei do not undergo DNA synthesis or karyokinesis in the host's cytoplasm. However, in the adelphoparasites examined (Goff & Coleman 1987b), parasite nuclei actively undergo DNA synthesis and karyokinesis within the cytoplasm of the host (Fig. 3-39). These replicated parasite nuclei are transferred from one infected host cell to another, thereby spreading the parasite genome(s) intracellularly throughout the host (Goff & Coleman 1987b).

B. Nuclear DNA in isomorphic forms

The realization that nuclear polygenomy (endopolyploidy, polyteny, or the multinucleate state) seems to be the norm rather than the exception among the Florideophycidae provides a solution to a most curious cytological paradox, that of isomorphy (Goff & Coleman 1987a). In both eukaryotic and prokaryotic cells there is a very strict correlation between cell size and DNA content (Brodsky & Uryvaena 1985; Cavalier-Smith 1978; Commoner 1964; Jacobi 1925; Lewis 1985; Shuter et al. 1983; Watanabe & Tanaka 1982). Therefore, cells of larger

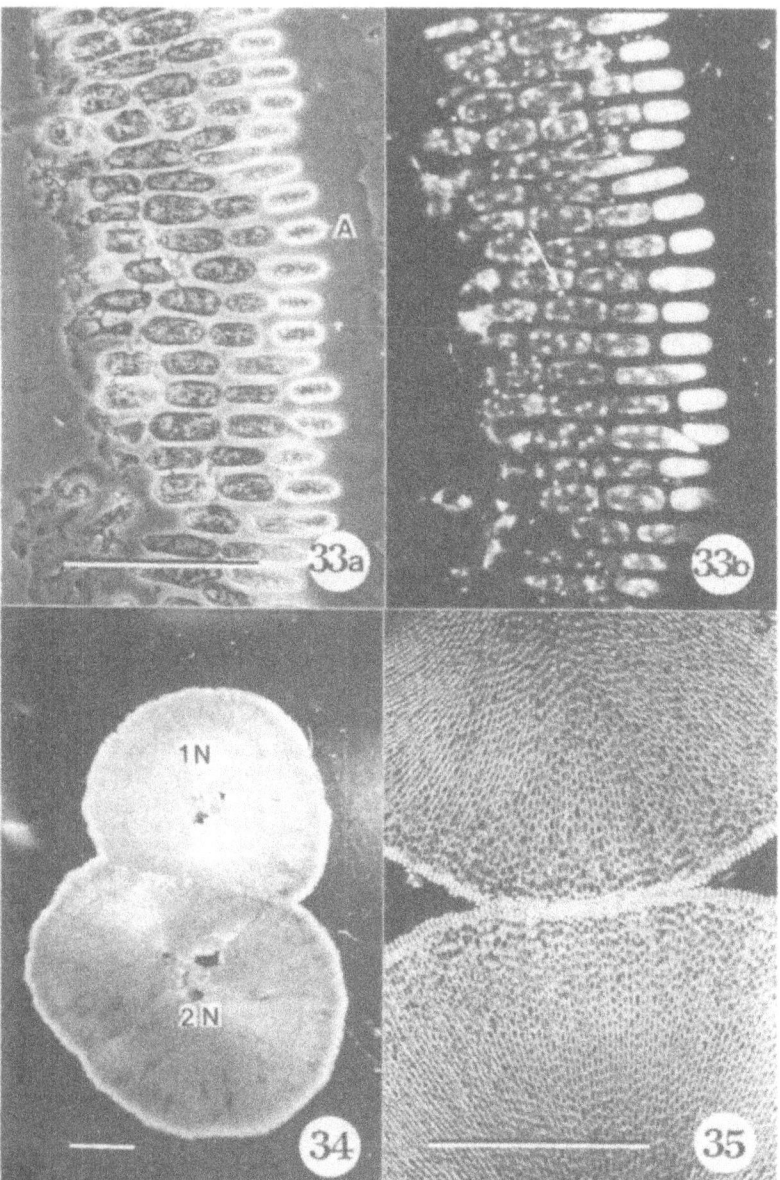

Figs. 3-33 to 3-35. Sporeling coalescence in the crustose alga *Minium parvum* (Rhodymeniales). *3-33*: Numerous apical cells (A) occur along the edge of a rapidly growing crust. Secondary pit connections form between adjacent vegetative cells, two to three cells behind the apical cells. In the process, nuclei are transferred from one cell to another (arrows). Any cell may be simultaneously donating and receiving nuclei from contiguous cells. Phase (a) and epifluorescence (b) microscopy, DAPI stained. Scale bar = 100 μm. *3-34*: A 2N cultured crust coalesced with a smaller 1N crust. Scale bar = 3 cm. *3-35*: Two individual crusts (2N) of the same genetic clone contacting along a margin of apical cells. Darkfield microscopy, scale bar = 0.5 mm.

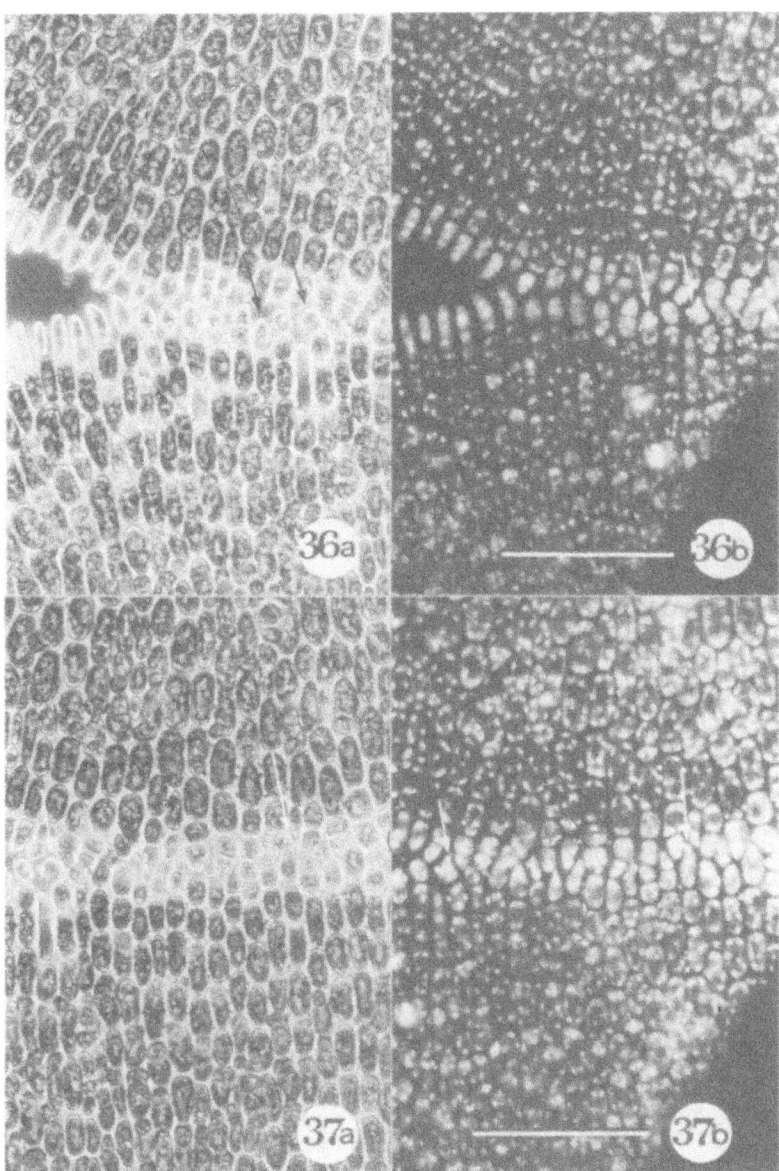

Figs. 3-36, 3-37. Sporeling coalescence in *Minium parvum*. Crusts grown on glass, fixed, DAPI stained. Scale bars = 100 μm. *3-36a, 3-37a*: Phase contrast. *3-36b, 3-37b*: Epifluorescence microscopy. *3-36*: Coalescence of two genetically different individuals (grown from different tetraspores); note apical cells produce nucleated conjunctor cells (arrows), which fuse with apical cells of the adjacent crust. *3-37*. Numerous cells of the two crusts, interconnected by secondary pit connections (arrows) in a region where coalescence has proceeded further.

size have more DNA and, in eukaryotes, larger nuclei. This correlation is clearly seen in organisms that undergo polyploidy (Bennett 1972; Epstein 1967; Epstein & Gatens 1967; Gunge & Nakatomi 1972; Rees 1972; Shuter et al. 1983; Sweeney et al. 1979), and it has also been observed in interspecific comparisons (Holm-Hansen 1969; Oeldorf et al. 1978; Olmo & Morescalchi 1975; Pedersen 1971; Price et al. 1973; Shuter et al. 1983; Soldo et al. 1981).

An apparent paradox emerges upon the consideration of organisms that have isomorphic life histories. In the Rhodophyta, Phaeophyta, and

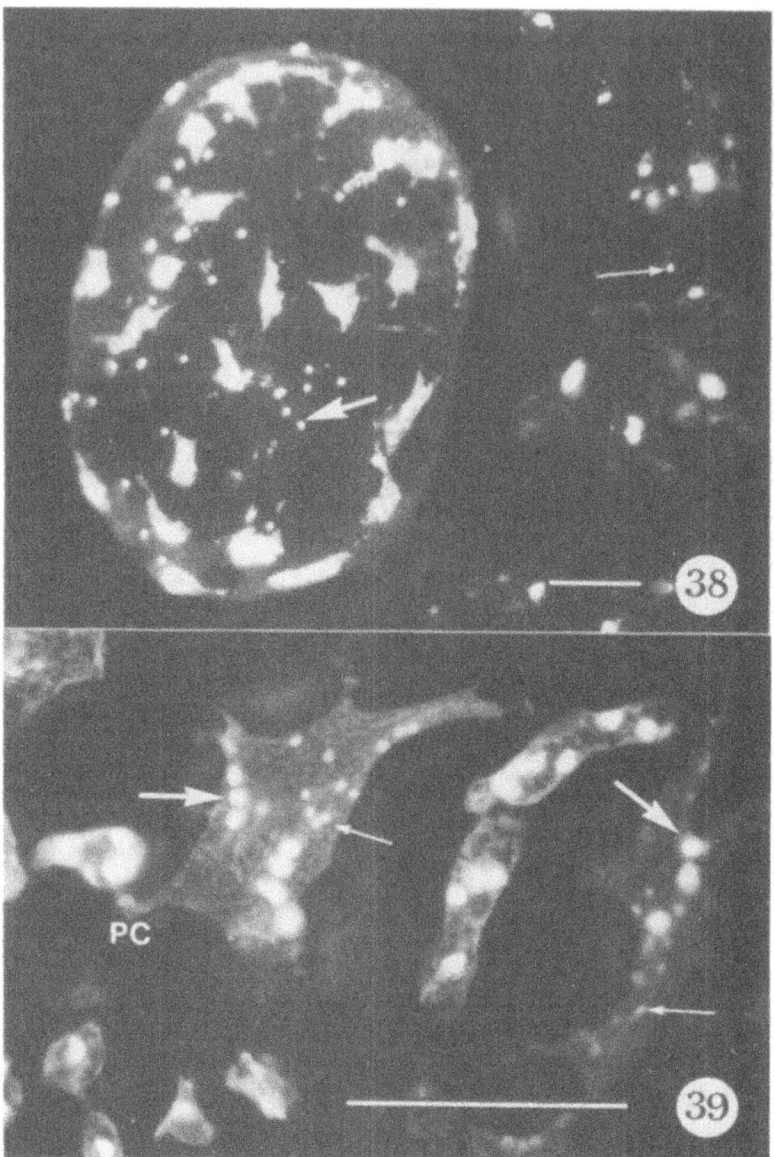

Figs. 3-38, 3-39. Host cells containing host and parasite nuclei. Fixed, squashed, DAPI stained, epifluorescence microscopy. Scale bars = 100 μm. *3-38*: A cell of *Polysiphonia confusa* has been pit connected by several cells of the alloparasite *Choreocolax* (syn = *Leachiella*). The large arrow indicates a G_1 parasite nucleus within a host cell. Larger nuclei within this cell are polyploid host nuclei. The smaller arrow indicates the G_1 nucleus of a conjunctor cell. In this association, parasite nuclei remain in G_1 in the host cytoplasm and do not divide. *3-39*: A heterokaryon pericentral cell of *Laurencia spectabilis* that has received one or more nuclei via a pit connection (PC) from an adjacent cell of the adelphoparasite *Janczewskia gardneri*. Note larger nuclei of the parasite (large arrows) and smaller ones (smaller arrows) of the host *Laurencia*. Within the cytoplasm of the host, parasite nuclei undergo DNA synthesis and divide, eventually outnumbering those of the host.

Chlorophyta, there are taxa that undergo a regular "alternation of generations" between haploid and diploid individuals that, except for reproductive stages, are morphologically indistinguishable. Homologous cells (same age and developmental stage) in the two generations should theoretically differ twofold in their DNA content and consequently their cell volume. Yet no such differences have been reported. The solution, at least in the isomorphic Florideophycidae that we have examined, is that cells are not just haploid or diploid; they are multinucleate (hence, polygenomic), or they have a single polyploid nucleus (Goff & Coleman 1987a, unpubl.). Homologous somatic cells of the two generations have the same amount of DNA and are the same size.

In the case of uninucleate endopolyploid forms like *Wrangelia*, *Antithamnion*, *Scagelia*, *Microcladia*, and *Choreocolax*, only the apical and penultimate cells of the two generations differ in nuclear DNA (Goff & Coleman 1984a, unpubl.; Gonzalez & Goff 1989). Cells derived from the apical cell increase in nuclear DNA proportionately to nuclear and cell size, and independent of the ploidy (i.e., 1N or 2N) of the generation. Consequently, homologous cells of the sporophyte and gametophyte have the same amount of nuclear DNA, contained within a single endopolyploid nucleus (Goff & Coleman 1984a, unpubl.; Gonzalez & Goff 1989).

In multinucleate forms like *Griffithsia* and *Bornetia*, nuclei of the diploid generation contain twice as much DNA as do those of the haploid generation (Goff & Coleman 1987a). The hundreds or thousands of nuclei are arranged in a nearly perfect hexagonal array in the thin cytoplasmic layer just beneath the cell surface. Although the cell size is nearly identical when homologous cells of the two generations are compared, each nucleus of the diploid cell is surrounded by a region of cytoplasm (a domain) nearly twice that surrounding a haploid nucleus. Cell size does not differ between homologous cells of the two generations because total nuclear DNA (sum of the DNA in all nuclei in a cell) per cell does not differ.

C. Nuclear ploidy levels during morphogenesis and development

In some Florideophycidae, sequential changes in the ploidy level of cells during development may play a role in determining cell patterning, branching, and the final morphology of the plant. In

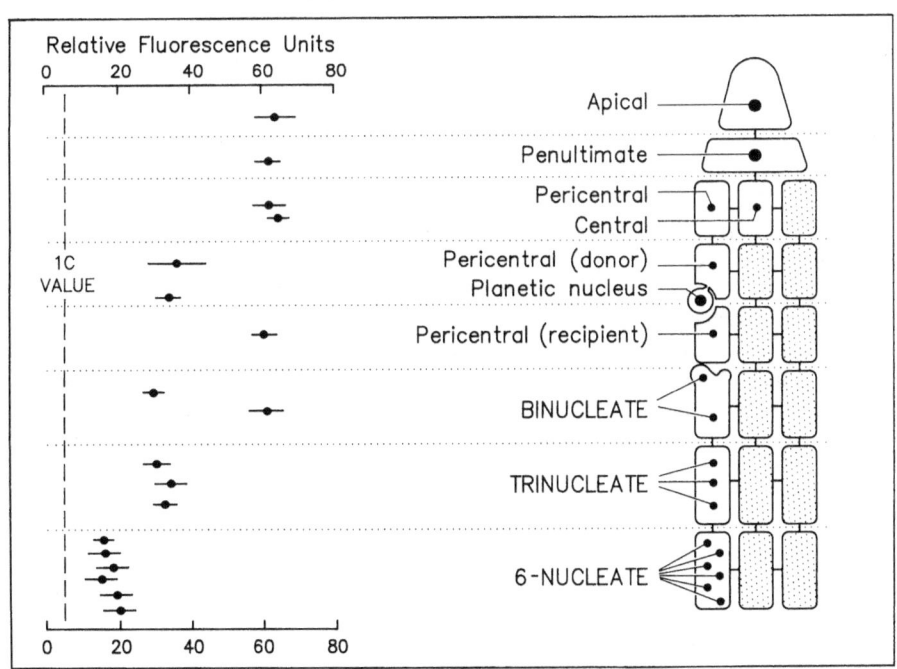

Fig. 3-40. Mean and standard deviation of ploidy level during the development of *Polysiphonia mollis*. The relative fluorescence data are summarized from six haploid individuals after 72 h of development. The 1C value for this species is ca. 4 rfu. Only two of the four pericentral cells of each tier are depicted in this diagram.

several species of Polysiphonia (Goff & Coleman 1986), as well as in other rhodomelacean species, apical cells of the main, "indeterminant" axis and determinant lateral branches are highly polyploid (8–64C and higher). Tiers of elongate pericentral cells are formed from the polyploid apex via a regular developmental process (Fig. 3–40) that involves a stepwise reduction in the degree of polyploidy in nuclei and an increase in nuclear number per cell (both by nuclear division and intercellular nuclear transfer). Like the polyploid cells of the apex, the differentiated pericentral cells contain multiple copies of the nuclear genome. However, rather than having a single endopolyploid nucleus, these cells eventually contain a regular number of G_2 (2C in gametophyte cells and 4C in tetrasporophyte cells), nonpolyploid nuclei.

The ploidy level of the apical cells in Polysiphonia is correlated with and may determine the number of nuclei that reside within a derived pericentral cell and consequently the cell's size. Furthermore, in any one tier the number of nuclei in the maturing pericentral cells, once they reach the G_2 level, is strictly correlated with the ploidy level of the single polyploid nucleus of the central cell (see mathematical formulae in Goff & Coleman 1986).

The level of polyploidy of the apical cell gradually declines with age. Apical cell nuclei of the main indeterminant axis in rapidly growing germlings have a higher ploidy level than that of older, larger plants. This decrease in apical cell endopolyploidy, with the concomitant decrease in nuclear ploidy of its division products, may limit the number of nuclei in mature pericentral cells and accordingly their size. This may account for the apico-basal gradient in cell size and nuclear number evident in pericentral cells of the main axis (Goff & Coleman 1986).

From this point of view, the apex may not then be truly indeterminant in growth. In Polysiphonia, the ultimate size (length) of the plant is a function of the number of polysiphonous tiers formed and the size of each pericentral cell. Whereas cell size reflects the initial ploidy of the apical cell, the number of cell tiers formed reflects the number of times the apical cell divides; this in turn reflects the coordination of DNA replication and cell division (Goff & Coleman 1986).

The relative length of lateral branches in Polysiphonia may also be determined, at least in part, by the ploidy of the apical cell of the indeterminant branch, which in turn establishes the ploidy of the branch apical cells that are formed along this axis. Whenever the rate of apical cell division exceeds that of DNA synthesis in the main axis, thereby reducing its ploidy level, there is a resultant decrease in the ploidy of the branch apices subsequently formed. Consequently, longer branches, composed of more polysiphonous tiers, occur toward the base of the plant, and progressively shorter branches are formed toward the apex (Goff & Coleman 1986).

D. Polyploidy and polygenomy as a genetic buffer

The presence of a multicellular haploid generation in a life cycle of an organism results in the loss of some of the major advantages of diploidy, including complementary gene action and genetic buffering (Lewis 1985; Paquin & Adams 1983; Stebbins 1950). Nearly all the red algae examined have cell cycles in which DNA replication occurs very soon after mitosis, and nuclei predominantly reside in the G_2 stage of the cell cycle. This and the polygenomy of the nuclear genome in both uninucleate endopolyploid and multinucleate cells provide red algal cells with a genetic buffering mechanism that protects them from potentially lethal mutations. Because of the duplication of the nuclear genomes, cells may escape the effects of newly arising recessive mutations in both the haploid and diploid generations and have considerable protection even from newly arisen "dominant" mutations. For example, in the case of uninucleate polyploid cells (e.g., Scagelia, Microcladia, and Wrangelia), the only cells in which a single copy of the genome is present are the developing gametes and tetraspores, and the apical cells of the gametophyte (Goff & Coleman unpubl.). Even in these "germ line" cells, the period of time during which a mutation can occur and subsequently be expressed in all progeny cells is very brief. In the gametes and tetraspores of these organisms, the single genome is replicated during gamete and spore maturation; therefore, the mature spermatia and tetraspores have two copies of each allele.

Even if a mutation occurs in the G_2 nucleus of a haploid reproductive cell (spermatium or tetraspore), it might not appear in subsequent progeny. In the case of mature spermatia, the nucleus undergoes a division prior to fertilization, to produce two sperm nuclei that enter into the egg (carpogonium) (Goff & Coleman 1984a, unpubl.). One of these nuclei would carry the potentially lethal mutation, whereas the other would not. In the case of tetraspores, a lethal mutation in one allele in the germling would be passed with equal probability either to the apical cell of a two-cell bipolar germling or to the rhizoid. Its effect would no doubt be lethal if it occurred in the apical cell as this cell divides to form the entire erect portion of the plant. However, it may not be lethal if expressed in the rhizoid, since in many taxa, upon damage to the rhizoid, the basal cells of the erect filament may produce secondary rhizoids that resume the attachment function of the primary rhizoid (Goff & Coleman unpubl.; Gonzalez & Goff 1989).

Similar genetic buffering potential occurs in multinucleate forms like *Griffithsia* and in forms with polyploid apical cells, such as *Polysiphonia*. In these, as in the previous case, a single allelic mutation would be expressed in the resulting individual only if it occurred during the G_1 stage of spermatial or tetraspore formation. In both cases, the mature sperm and tetraspores have two copies (G_2) of the genome at the time of release, and the spores of these species have the "added protection" of becoming either multinucleate (*Griffithsia*) (Goff & Coleman 1987a) or highly endopolyploid (*Polysiphonia*) (Goff & Coleman 1986) prior to spore germination.

The genetic buffering offered by nuclear polygenomy limits the expression of mutations in germlines (i.e., mutations that affect every cell of the organism). However, as in the case of higher plants, sectorial mutations may arise and be maintained by the normally functional part of the plant. Indeed, most naturally occurring red algal mutations are sectorial (van der Meer 1979, 1986, pers. comm.). If the mutant arose in the apical cell of a haploid plant, even 2C, it may be expressed, either in the apex or in a branch formed later in development. If the mutation occurs in a 2C (G_2) tetraspore, a mutant sector would almost certainly appear (van der Meer pers. comm.). In addition, in diploid plants, no matter how many copies of the diploid genome occur, the diploid "buffer" may help a mutation survive because it is masked by a normal allele on the homologous chromosome. When the diploid individual eventually reaches reproductive maturity, large sectors may be heterozygous for the mutation, resulting in large numbers of mutant-producing spores. The expression of the mutation is delayed to the next generation, but then it can be present in many spores. If this appeared in a haploid nucleus, it might be lethal, selecting out these genotypes from the gene pool (van der Meer pers. comm.).

V. SUMMARY

Using the techniques presented in this chapter, one can readily determine the DNA content of nuclei and thereby determine the ploidy of various life history stages of red algae without having to count chromosomes. The occurrence of meiosis or of apomictic life histories can be established unequivocally and the exact site of meiosis readily ascertained.

The observation and quantification of nuclear DNA explains many of the very unusual characteristics of red algae. For example, we now have demonstrated that the formation of a secondary pit connection, a structure whose function has long been debated, provides a means to transfer one or more nuclei between two neighboring cells of the same individual or between different individuals, as in the case of sporeling and filament coalescence and host–parasite interactions.

Polyploidy, or nuclear polygenomy in the case of multinucleate cells, is of very frequent occurrence among the Florideophycidae, and the red algae have rather ingeniously "employed" this condition. The highly unusual apical cell polyploidy described in *Polysiphonia* and other Rhodomelaceae and the tightly regulated sequence of nuclear and cell divisions and intercellular nuclear transfers may determine, in part, the complex cell patterning, the branching patterns, and the ultimate morphology of these plants.

Polyploidy and polygenomy also provide for the maintenance of very large cells in which the availability of gene transcripts and their distribution may otherwise be limiting. Because red algal cells lack cytoplasmic streaming, the requirement for the distribution of nuclear products to organelles may underlie several prominent cell features, such as the unusual shapes observed in highly polyploid nuclei of large uninucleate cells, the regular distribution of nuclei in highly multinucleate cells, and the rather tight correlation between the number of genomes in a nucleus and the number of plastids and plastid genomes associated with it. In addition, the ability of Florideophycidae to reduplicate their nuclear genome may have permitted the evolution of isomorphic life histories, which characterize many of these organisms. The total DNA content of the cell is not a function of its generation, but of cell and nuclear size.

And last, the evolution of nuclear genomic reduplication provides an extremely effective mutation buffer, functionally comparable to the genetic buffering provided for the plastids and mitochondrial genes by the many duplications of the plastid and mitochondrial genomes. Rarely and briefly is there only a single copy of the nuclear genome during the life history and development of most Florideophycidae; consequently, spontaneous germ-line mutations that affect all cells of the resulting organism are rarely expressed in these organisms.

VI. ACKNOWLEDGMENTS

The authors acknowledge the National Science Foundation (Systematics Biology, Cell Biology, Visiting Professorships for Women, Biological Instrumentation, and the International Programs) and the Fulbright and Guggenheim Foundations, whose generous funding has made this research possible. Our thanks are also extended to Drs. Sarah Gibbs and John van der Meer for their careful reviewing and most useful comments. The authors dedicate this chapter to Dr. Kathleen Cole, co-editor of this volume, on the occasion of her retirement as Professor of Botany at the University of British Columbia.

VII. REFERENCES

Bennett, M. D. 1972. Nuclear DNA content and minimum generation time in herbaceous plants. *Proc. R. Soc. Lond. B. Biol. Sci.* 181: 109–35.

Brodsky, V. Y. & Uryvaena, I. V. 1985. *Genome Multiplication in Growth and Development.* Cambridge: Cambridge University Press.

Cabioch, J. 1972. Étude sur les Corallinacées. *Cah. Biol. Mar.* 12: 121–86.

Cavalier-Smith, T. 1978. Nuclear volume control by nucleoskeletal DNA, selection for cell volume and cell growth rate, and the solution of the DNA C-value paradox. *J. Cell Sci.* 34: 247–78.

Cavalier-Smith, T. 1985. Cell volume and the evolution of eukaryote genome size. In *The Evolution of Genome Size*, ed. T. Cavalier-Smith, pp. 105–84. New York: Wiley.

Coleman, A. W. 1978. Visualization of chloroplast DNA with two fluorochromes. *Exp. Cell Res.* 114: 95–100.

Coleman, A. W. 1982. The nuclear cell cycle in *Chlamydomonas* (Chlorophyceae). *J. Phycol.* 18: 192–5.

Coleman, A. W., Maguire, M. J. & Coleman, J. R. 1981. Mithramycin- and 4'-6, diamidino-2-phenolindole (DAPI)–staining for fluorescence microspectrophotometric measurement of DNA in nuclei, plastids, and virus particles. *J. Histochem. Cytochem.* 29: 959–68.

Commoner, B. 1964. Roles of deoxyribonucleic acid in inheritance. *Nature* (Lond.) 202: 960–8.

Dawkins, R. 1976. *The Selfish Gene.* Oxford: Oxford University Press.

Dawkins, R. 1983. *The Extended Phenotype.* Oxford: Freeman Press.

Dixon, P. S. 1973. *Biology of the Rhodophyta.* Edinburgh: Oliver & Boyd.

Drew, K. M. 1934. Contributions to the cytology of *Spermothamnion turneri* (Mert.) Aresh. I. The diploid generation. *Ann. Bot.* 48: 549–73.

Epstein, C. J. 1967. Cell size, nuclear content, and the development of polyploidy in the mammalian liver. *Proc. Natl. Acad. Sci. USA* 57: 327–34.

Epstein, C. J. & Gatens, E. A. 1967. Nuclear ploidy in mammalian parenchymal liver cells. *Nature* (Lond.) 214: 1050–1.

Gall, J. 1981. Chromosome structure and the C-value paradox. *J. Cell Biol.* 91: 35–145.

Godward, M.B.E. 1966. *The Chromosomes of the Algae.* New York: St. Martin's.

Goff, L. J. 1981. The role of bispores in the life history of the parasitic red alga *Gardneriella tuberifera* (Solieriaeaceae, Gigartinales). *Phycologia* 20: 397–406.

Goff, L. J. & Coleman, A. W. 1984a. Elucidation of fertilization and development in a red alga by quantitative DNA microspectrofluorometry. *Devel. Biol.* 102: 173–94.

Goff, L. J. & Coleman, A. W. 1984b. Transfer of nuclei from a parasite to its host. *Proc. Natl. Acad. Sci. USA* 81: 5420–4.

Goff, L. J. & Coleman, A. W. 1985. The role of secondary pit connections in red algal parasitism. *J. Phycol.* 21: 483–508.

Goff, L. J. & Coleman, A. W. 1986. A novel pattern of apical cell polyploidy, sequential polyploidy reduction and intercellular nuclear transfer in the red alga *Polysiphonia. Am. J. Bot.* 73: 1109–30.

Goff, L. J. & Coleman, A. W. 1987a. The solution to the cytological paradox of isomorphy. *J. Cell Biol.* 104: 739–48.

Goff, L. J. & Coleman, A. W. 1987b. Nuclear transfer from parasite to host: a new regulatory mechanism of parasitism. In *Endocytobiology III*, eds. J. J. Lee & J. F. Fredrick, pp. 402–23. New York: New York Academy of Sciences.

Goff, L. J. & Coleman, A. W. 1988. The use of plastid DNA restriction endonuclease patterns in delineating red algal species and populations. *J. Phycol.* 24: 357–68.

Gonzalez, M. & Goff, L. J. 1989. Studies on the red algal epiphytes *Microcladia coulteri* Harvey and *Microcladia californica* Farlow (Rhodophyceae, Ceramiaceae) I: Taxonomy, life history and phenology. *J. Phycol.* 25: 545–58.

Grubb, V. 1925. The male organs of the Florideae. *J. Linn. Soc. Lond. (Bot.)* 47: 177–255.

Gunge, N. & Nakatomi, Y. 1972. Genetic mechanisms of rare mating of the yeast *Saccharomyces cerevisiae* heterozygous for mating type. *Genetics* 70: 41–58.

l'Hardy-Halos, M. T. 1969. La formation des anastomoses chez *Pleonosporium borreri* (Smith) Naegeli ex Hauck et *Bornetia secundiflora* (J. Ag.) Thuret (Rhodophyceae-Ceramiaceae). *C.R. Acad. Sci.* (Paris) sér. D, 268: 276–8.

Hassinger-Huizinga, H. 1952. Generationswechsel und Geschlechtsbestimmung bei *Callithamnion corymbosum* (Sm.) Lyngb. *Arch. Protist.* 98: 91–124.

Holm-Hansen, O. 1969. Algae: amounts of DNA and organic carbon in single cells. *Science* 163: 87–8.

Inoue, S. 1986. *Video Microscopy.* New York: Plenum.

Jacobi, W. 1925. Über das rhythmische Wachstrum der Zellen durch Verdopplung ihres Volumes. *Wilhelm Roux Arch. Entwicklungs. Org.* 106: 125–92.

Knaggs, F. 1969. A review of florideophycidean life histories and the culture techniques employed in their investigation. *Nova Hedwigia* 18: 293–330.

Koslowsky, D. J. & Waaland, S. D. 1984. Cytoplasmic incompatibility following somatic cell fusion in *Griffithsia pacifica* Kylin, a red alga. *Protoplasma* 123: 8–17.

Koslowsky, D. J. & Waaland, S. D. 1987. Ultrastructure of selective chloroplast destruction in *Griffithsia pacifica* Kyl. *J. Phycol.* 23: 638–48.

Lee, I. K. & West, J. A. 1980. *Antithamnion nipponicum.* Yamada et Inagaki (Rhodophyta, Ceramiales) in culture. *Jap. J. Phycol.* 28: 19–27.

Lewin, R. 1984. A new regulatory mechanism of parasitism. *Science* 226: 427.

Lewis, W. M. 1985. Nutrient scarcity as an evolutionary cause of haploidy. *Am. Nat.* 125: 692–701.

Magne, F. 1964. Recherches caryologiques chez les

Floridées (Rhodophycées) *Cah. Biol. Mar.* 5: 461–671.

Moe, R. L. 1979. *Minium parvum* gen. et sp. nov., a crustose member of the Rhodymeniales (Rhodophyta). *Phycologia* 18: 38–46.

Oeldorf, E. M., Nishioka, M. & Bachmann, K. 1978. Nuclear DNA amounts and developmental rate in holartic anura. *Z. Zool. Syst. Evol.* 16: 216–24.

Olmo, E. & Morescalchi, A. 1975. Evolution of the genome and cell sizes in salamanders. *Experimentia* 31: 804–6.

Paquin, C. & Adams, J. 1983. Frequency of fixation of adaptive mutations is higher in evolving diploid than haploid yeast populations. *Nature* (Lond.) 302: 495–500.

Pedersen, R. A. 1971. DNA content, ribosomal gene multiplicity, and cell size in fish. *J. Exp. Zool.* 177: 65–78.

Ploem, J. S. & Tanke, H. J. 1987. *Introduction to Fluorescence Microscopy.* Oxford: Alden Press.

Price, H. J., Sparrow, A. H. & Nauman, A. F. 1973. Correlations between nuclear volume, cell volume and DNA content in meristematic cells of herbaceous angiosperms. *Experimentia* 29: 1028–9.

Rees, H. 1972. DNA in higher plants. In *Evolution of Genetic Systems*, ed. H. H. Smith, pp. 394–418. New York: Gordon and Breach.

Rosenvinge, L. K. 1888. Sur la formation des pores secondaires chez *Polysiphonia. Bot. Tidskr.* 17: 10–19.

Rueness, J. & Rueness, M. 1973. Life history and nuclear phases of *Antithamnion tenuissimum*, with special reference to plants bearing both tetrasporangia and spermatangia. *Norw. J. Bot.* 20: 205–10.

Shuter, B. J., Thomas, J. E., Taylor, W. D. & Zimmerman, A. M. 1983. Phenotypic correlates of genome DNA content in unicellular eukaryotes and other cells. *Am. Nat.* 122: 26–44.

Soldo, A. T., Brickson, S. A. & Larin, F. 1981. The kinetic and analytical complexities of the DNA genomes of certain marine and freshwater ciliates. *J. Protozool.* 28: 377–83.

Stebbins, G. L., Jr. 1950. *Variation and Evolution in Plants.* New York: Columbia University Press.

Sundene, O. 1965. *Antithamnion tenussimum* (Hauck) Schiffner in culture. *Nytt Mag. Bot.* 11: 5–10.

Sweeney, G. D., Cole, F. M., Freeman, K. B. & Patel, H. V. 1979. Heterogeneity of rat liver parenchymal cells. *J. Clin. Med.* 95: 718–25.

Thomas, C. A. 1971. The genetic organization of chromosomes. *Annu. Rev. Genet.* 5: 237–56.

Tveter-Gallagher, E. & Mathieson, A. C. 1976. Sporeling coalescence in *Chondrus crispus* (Rhodophyceae). *J. Phycol.* 12: 110–18.

Tveter-Gallagher, E. & Mathieson A. C. 1980. An electronmicroscopic study of sporeling coalescence in the red alga *Chondrus crispus. Scan. Elec. Micro.* 3: 571–80.

van der Meer, J. P. 1979. Genetics of *Gracilaria* sp. (Rhodophyceae, Gigartinales). V. Isolation and characterization of mutant strains. *Phycologia* 18: 47–54.

van der Meer, J. P. 1986. Genetic contributions to research on seaweeds. *Prog. Phycol. Res.* 4: 1–38.

Waaland, S. D. 1978. Parasexually produced hybrids between female and male plants of *Griffithsia tenuis* (C. Agardh), a red alga. *Planta* (Berl.) 105: 196–204.

Watanabe, T. & Tanaka, G. 1982. Age-related alterations in the size of human hepatocytes – a study on mononuclear and binuclear cells. *Virchows Arch. B. Cell Pathol.* 39: 9–20.

West, J. A. & Hommersand, M. H. 1981. Rhodophyta: life histories. In *The Biology of Seaweeds*, eds. C. S. Lobban & M. J. Wynne, pp. 133–93. Los Angeles: University of California Press.

West, J. A. & Norris, R. E. 1966. Unusual phenomena in the life history of Florideae in culture. *J. Phycol.* 2: 54–7.

Whittick, A. & West, J. A. 1979. The life history of a monoecious species of *Callithamnion* (Rhodophyta, Ceramiaceae) in culture. *Phycologia* 18: 30–7.

Chapter 4

Chromosomes

KATHLEEN M. COLE

CONTENTS

I. Introduction / 73
II. Karyotypes / 74
III. Chromosome number / 74
  A. Trends / 74
  B. Life histories / 86
  C. Polyploidy / 87
IV. Chromosome morphology / 89
V. Taxonomic aspects / 90
  A. Chromosome number and morphology / 90
  B. Karyotype evolution / 93
VI. Summary / 94
VII. Acknowledgments / 94
VIII. References / 94

I. INTRODUCTION

"... the chromosomes of a species ... are the products of evolutionary pressures, and because no organism is without one or more chromosomes, they must therefore have had a high selective value from the time of their inception." (Swanson et al. 1981 p. 3)

Interest in red algal chromosomes dates from the turn of the present century, generated by the recognition that knowledge of chromosomes is basic to an understanding of life histories and provides valuable corroborative data in systematic studies. Significant contributions in the first five decades, outlined by Bold (1951), Rao (1959) and Magne (1964), include the classic reports by Yamanouchi (1906b) on *Polysiphonia flexicaulis* (as *P. violacea*) and Drew (1934, 1943) on *Spermothamnion repens* (as *S. turneri*), featuring detailed chromosome figures and descriptions of the meiotic process. Yamanouchi (1906 a,b) also determined chromosome numbers for all stages in the life history of *P. flexicaulis*, thus being the first to demonstrate the cytological alternation of isomorphic gametophytic and tetrasporophytic generations in the Rhodophyta.

Prior to the 1950s, many preparative techniques were unsuitable for nuclear studies, especially because of the fixatives and sectioned material used. Early cytologists obtained only scant information on cell division and chromosomes, some of which must be considered prudently, such as the chromosome counts of $n = 10$ and 20 that were so common in the Florideophycidae. Subsequently, it was found possible to obtain more accurate chromosome data by adapting relatively simple squash procedures employing acetic stains carmine, orcein, and hematoxylin, as well as the DNA-specific Feulgen technique and DNA-binding fluorochrome 4'-6, diamidino-2-phenylindole (DAPI) (e.g., Austin 1959; Cole 1962; Goff & Coleman 1985, 1986, 1987; Hanic 1973; Jónsson & Chesnoy 1982; Magne 1964; Mumford & Cole 1977; Rao 1953; Thirb & Benson-Evans 1982; Yabu & Tokida 1966), providing an impetus for further studies.

Chromosome numbers are difficult to determine

in most algae, and discrepancies in cytological data on particular species may arise due to differences in the strain, techniques, and/or investigators. Studies on the Rhodophyta, in particular, have been handicapped by a number of features, including complexities of florideophycidean life histories, smallness of nuclei and chromosomes, and periodicity and rapidity of cell division (e.g., Dixon 1963a, 1966; Kapraun 1977, 1989).

The first comprehensive tables of rhodophytan chromosome numbers were prepared in the 1960s by Altman & Dittmer (1962, pp. 24–33), Magne (1964), and Dixon (1966), although very little was known at that time about chromosomes of algae in the subclass Bangiophycidae. Subsequently, lists were compiled for particular groups of red algae, e.g., red algae in Japan (Yabu 1975), freshwater red algae (Sheath 1984), *Gracilaria* species (Bird & McLachlan 1982), Gelidiales (Akatsuka 1986; Kaneko 1968); *Porphyra* species (Kapraun & Freshwater 1987; Kito 1978; Krishnamurthy 1984; Mumford & Cole 1977; Tseng & Sun 1989; Yabu 1978a), *Bangia vermicularis* (Cole et al. 1983). In addition, the ongoing publication *Index to Plant Chromosome Numbers* (e.g., Cave 1958–1960, 1961–5; Goldblatt 1981, 1984; Moore 1973, 1974, 1977; Ornduff 1967) includes numbers for all algal groups.

II. KARYOTYPES

Karyotypes provide information on chromosome numbers and morphology, as well as on chromosome pairing (homology) and chiasmata when meiotic cells are available; they possess a specific individuality for a group of organisms. Most chromosome analyses of red algae are limited to numbers, usually determined when chromosomes are readily visible in late prophase. A few karyotypes also include chromosome size differences and centromere position, which are most evident during mid-prophase.

A few figures and diagrams of the chromosome complement arranged in order of decreasing length, known as ideograms or karyograms, are available for the Rhodophyta. Relative chromosome sizes within a complement appear to remain consistent during the prophase contraction process and, rather than absolute sizes, are acceptable data for intra- and interspecific comparisons (Cole et al. 1983; Kapraun & Freshwater 1987). Average relative size differences are determined from several prophase stages, except late prophase, but karyograms incorporating different prophase or metaphase stages are not directly comparable. In his examination of the life history and reproduction of *Furcellaria lumbricalis* (as *F. fastigiata*). Austin (1960b) prepared an excellent comparative chart showing photographs of all 34 bivalents in three stages of prophase I meiosis (Fig. 4-1). A series of small mitotic karyograms for six Japanese *Porphyra* species (Fig. 4-2) were presented by Yabu (1969a), and more current publications include ideograms showing mitotic chromosome sizes and centromere positions for several population types in the genus *Bangia* from the northeastern Pacific (Fig. 4-3; Cole et al. 1983), mitotic karyograms of *Porphyra* species from the North Atlantic and Mediterranean (Fig. 4-4; Kapraun & Freshwater 1987) and karyograms of metaphase I bivalents in Atlantic *Polysiphonia* species (Fig. 4-5; D. F. Kapraun pers. comm.).

Karyotype data and their applications in life history and taxonomic studies of the Rhodophyta are surveyed in this chapter, mainly from publications of the past 20 years.

III. CHROMOSOME NUMBER

A. Trends

Currently available chromosome numbers of the Rhodophyta, together with cell types studied, are listed in Table 4-1. Numbers have been published

Fig. 4-1. Comparative chart of 34 bivalents in tetrasporocytes of *Furcellaria lumbricalis* (as *F. fastigiata*). A = early diplotene, B = late diplotene, C = diakinesis. Arrowheads in B point to some median and submedian centromere positions. Acetocarmine. Scale bar = 5 μm. (From Austin 1960b, with permission)

Chromosomes

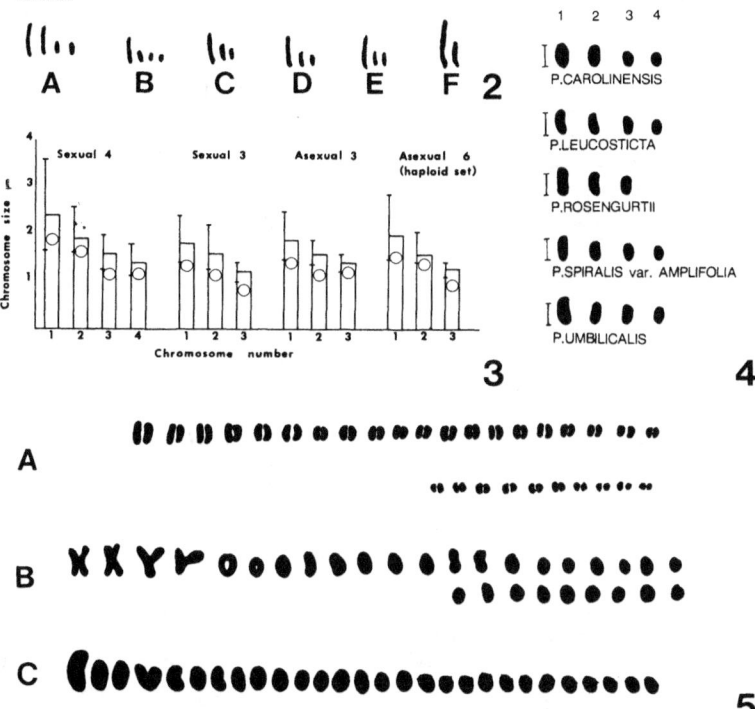

Figs. 4-2 to 4-5. Karyograms of red algae. *4-2*: Mitotic prophase chromosomes in *Porphyra* species from Japan. A = *P. pseudolinearis*, B = *P. okamurae*, C = *P. yezoensis*, D = *P. pseudocrassa*, E = *P. seriata*, F = *P. suborbiculata*. Spermatangial (A, C, D, E, F) and haploid vegetative (B) material: hematoxylin. Scale bar = 4 μm. (From Yabu 1969a, with permission) *4-3*: Haploid karyograms of four chromosome types of *Bangia vermicularis* in British Columbia, Canada, representing average mitotic prophase lengths; range of lengths for each chromosome indicated on the left side of each bar (I); 0 = centromere position. (From Cole et al. 1983, with permission.) *4-4*: Typical mitotic karyograms prepared from haploid vegetative and reproductive material of five *Porphyra* species from the North Atlantic and Mediterranean. Aceto-orcein. Scale bars = 1 μm. (From Kapraun & Freshwater 1987, with permission) *4-5*: Karyograms of meiotic chromosomes in tetrasporangia of three *Polysiphonia* species from the North Atlantic. A = *P. urceolata*, 30 bivalents in late metaphase; symmetric chromosome complement showing gradual decrease in chromosome length from left to right. B = *P. denudata*, 30 bivalents in late metaphase; asymmetric chromosome complement; note four large submetacentric chromosomes at the left end. C = *P. binneyi*, 28 bivalents in late prophase; symmetric chromosome complement; aneuploid. Aceto-orcein. Scale bar = 5 μm. (Courtesy D. F. Kapraun unpubl.)

for about 5% of the recognized rhodophytan species, representing all orders except the Hildenbrandiales. They are approximate for many species, and in a few cases the ploidy state of the material studied is uncertain.

The haploid number of chromosomes in rhodophytan species extends from 2 in the Porphyridiaceae (Porphyridiales) and Bangiaceae (Bangiales) to 68–72 in the Polyideaceae (Gigartinales). Low numbers have been recorded for species in the Acrochaetiales, Bangiales, Batrachospermales (Batrachospermaceae), Compsopogonales, Gelidiales, Nemaliales, Porphyridiales, and Rhodochaetales [n = 2–10 (15)] (Figs. 4-2 to 4-4, 4-6 to 4-8). With the exception of many species in the Ceramiales and Gigartinales that have much higher numbers (n = ca. 25–72) (Figs. 4-5, 4-9), those in the other five orders have haploid chromosome numbers ranging from about 10 to 25 (Fig. 4-10).

In his survey of freshwater red algae, Sheath (1984) observed that haploid chromosome numbers range from 3 to 22; simpler forms like *Bangia*

Table 4-1. Chromosome numbers in the Rhodophyta

(Tissues studied are indicated as follows: C = unfertilized carpogonia, CA = carpospores or carpospore formation, carposporophyte (in Florideophycidae) (includes gonimoblast), CH = chantransia or pseudochantransia, CO = conchocelis, CS = conchosporangia or conchospores, FC = fertilized carpogonia (zygotes), G = gametophyte, M = monospore formation, PS = parasporangia, R = meiosis, S = spermatia or spermatangia, ST = sterile or asexual, T = tetrasporophyte, TS = tetrasporangia, ——— = no information on cell types studied)

Taxon[a]	Chromosome Number[b] n	2n	Cell type[c]	Reference[d]
Acrochaetiales				
ACROCHAETIACEAE				
Audouinella botryocarpa				
(as *Acrochaetium botryocarpum*)	ca. 5	ca. 10	CA, T	Woelkerling 1970**
Audouinella concrescens				
(as *Rhodochorton concrescens*)	—	12–14	T	West 1970*
Audouinella floridula				
(as *Rhodothamniella floridula*)	ca. 20		ST	Magne 1964
(as *Rhodochorton floridulum*)	10 + 1	20 + 2	R, G, T	Knaggs 1964**
Audouinella hermannii	6	—	ST	Hymes & Cole 1983*
Bangiales				
BANGIACEAE[e]				
Bangia atropurpurea	3	—	G	Sheath & Cole 1980*
(as *B. fuscopurpurea*)	2	—	G	Dangeard 1927*
	3, 4	6, 8	G, FC	Yabu 1967
	3	—	G, CO	Sommerfeld & Nichols 1970a*
	4	8	G, CA	F. Magne pers. comm. (Cole 1972)**
Bangia vermicularis	3, 4 sexual	6, 8 sexual	G, C, FC, CA	Cole et al. 1983
	3 asexual	6 asexual	G, M	Cole et al. 1983
(as *B. fuscopurpurea*)		10	G, CO	Richardson & Dixon 1968**
	4	—	S	Cole 1972**
Porphyra abbottae	3	—	G, S, CA	Conway et al. 1975**
	3	6	G, S, CA	Mumford & Cole 1977**
Porphyra amplissima (**1**)[f]	3	6	G, S, FC, CA	Kito et al. 1967*
	3	6	G, S, CA	Yabu 1970
Porphyra atropurpurea	3	—	S	Yabu et al. 1974
Porphyra brumalis	4	—	S	Mumford 1975
	4	—	S, CO	Mumford & Cole 1977
Porphyra carolinensis	4	—	G, S, CA, CO, CS	Freshwater & Kapraun 1986*
	4	—	G, S, CA, CO, CS	Kapraun & Freshwater 1987
Porphyra columbina				
(Chile)	3	—	S	Avila et al. 1986**
(Australia)	4	—	S	J. Coll & E. C. Oliveira pers. comm.
Porphyra crispata	3	—	S	Yabu 1971*
Porphyra dentata	3	6	S, CA	Yabu 1971
	4	8	G, S, CA	Kito 1978
Porphyra fucicola	3(2)	—	S	Mumford & Cole 1977**
	5	—	S, CA	Krishnamurthy 1984**
Porphyra gardneri	4	8	G, S, CA, CO	Hawkes 1977, 1978
Porphyra haitanensis	5	10	G, S, CA	Tseng & Sun 1989
Porphyra irregularis				
(as *P. umbilicalis* f. *linearis*)	4	8	G, S, FC, CA	Kito et al. 1967
	4	8	G, S, CA	Kito 1978
Porphyra kanakaensis	3	—	G	Mumford 1973
	3	6	G, S, CA, CO	Mumford & Cole 1977
Porphyra katadae	4	8	G, S, CA	Kito 1966*
	5	10	S, CA	Yabu 1972a
	5	10	G, S, CA	Kito 1978

Table 4-1. (cont.)

Taxon[a]	Chromosome Number[b]		Cell type[c]	Reference[d]
	n	2n		
Porphyra katadae var. *hemiphylla*	5	10	G, S, CA, CO	Tseng & Sun 1989
Porphyra kinositae	3	6	G, S, CA	Yabu 1972a
Porphyra kuniedae	2	4	S, CA	Yabu 1971
	3	6	G, S, CA	Kito 1978
Porphyra lanceolata	3	—	G	Pringle & Austin 1970
	3	—	S	Mumford & Cole 1977
Porphyra leucosticta	4	—	S, CA	Coll & Oliveira 1977a,b
	3	6	G, S, CA	Yabu 1978a
	4	—	G, S, CA, CO, CS	Kapraun & Freshwater 1987
Porphyra linearis	4	8	G, S, FC, CA	Magne 1952
	5	10	S, CA	Yabu 1978a
(as *P. umbilicalis* f. *linearis*)	2	4	R, G, FC	Dangeard 1927
Porphyra maculosa	3	—	G, S, CO	Conway et al. 1975**
Porphyra marginata	3	6	G, S, CA	Tseng & Sun 1989
Porphyra miniata (**1**)[f]	3	6	G, S, CA, CO	Kito et al. 1971
	3	—	S, CO	Conway et al. 1975**
	3	—	S, CO	Mumford & Cole 1977
	3	6	G, S, CA	Kito 1978
(as *P. cuneiformis*)	5	—	S, CA, CO	Krishnamurthy 1984**
Porphyra moriensis	4	8	G, S, FC, CA	Kito et al. 1967
	4	8	G, S, CA	Kito 1978
Porphyra nereocystis	3	—	S	Yabu 1970
	3	—	S	Mumford & Cole 1977**
	3	—	S	Krishnamurthy 1984**
Porphyra okamurae	4	8	G, S, CA	Yabu 1969a
	3	6	G, S, CA	Kito 1978
Porphyra oligospermatangia	3	6	G, S, CA	Tseng & Sun 1989
Porphyra onoi	3	—	G, S	Yabu & Tokida 1963
	3	6	G, S, CA	Yabu 1972a
	3	6	G, S, CA	Kito 1978*
Porphyra papenfussii	3	—	G, S, CO	Conway & Cole 1973
	3	—	G, S, CO	Mumford & Cole 1977**
Porphyra perforata	2	—	S	Conway et al. 1975**
	2	4	G, S, CA, CO	Mumford & Cole 1977
	3	—	S	J. Coll & E. C. Oliveira pers. comm.
Porphyra pseudocrassa	3	—	G, S	Kito et al. 1967
	3	—	G, S	Yabu 1969a
	3	6	G, S, CA	Kito 1978**
Porphyra pseudolanceolata	5	—	S, CA	Conway et al. 1975**
	5	—	G, S	Mumford & Cole 1977
	5	—	S, CA	Krishnamurthy 1984**
Porphyra pseudolinearis	4	—	G, S	Kito et al. 1967
	4	8	G, S, FC, CA	Yabu 1969a
	4	8	R, G, S, CA, CO, CS	Kito 1974, 1978
Porphyra purpurea	5	10	G, S, CA	Kito et al. 1971
	5	10	G, S, CA	Kito 1978
(as *P. umbilicalis* var. *laciniata*)	5	—	G, S, CA, CO	Krishnamurthy 1959
Porphyra rosengurttii	3	—	G, S, CA	Kapraun & Luster 1980*
	3	—	G, S, CO	Kapraun & Freshwater 1987
Porphyra sanjuanensis (**2**)[f]	3	—	G	Mumford & Cole 1977**
Porphyra schizophylla	2	—	S	Conway et al. 1975**
	2	4	G, S, CA	Mumford 1975
	2	4	G, S, CA, CO	Mumford & Cole 1977
(as *P. norrisii*) (**3**)[f]	5	10	S, CA, CO	Krishnamurthy 1984**
Porphyra seriata	3	6	G, S, CA	Yabu 1969a
Porphyra smithii	3	—	S	Mumford & Cole 1977**
Porphyra spiralis var. *amplifolia* (Venezuela)	4	—	G, S, CO, CS	Kapraun & Freshwater 1987

Table 4-1. (cont.)

Taxon[a]	Chromosome Number[b]		Cell type[c]	Reference[d]
	n	2n		
(Venezuela & Brazil)	3	—	S, CO	J. Coll & E. C. Oliveira pers. comm.
Porphyra suborbiculata	2	—	G, S	Yabu 1969a
	3	—	S	Yabu 1971
	3	6	G, S, CA	Kito 1978
Porphyra subtumens (2)[f]	12–13	—	G	Conway & Wylie 1972**
Porphyra tasa	3	—	S	Yabu 1972a
Porphyra tenera	3	—	G, S	Ishikawa 1921*
	5	10	R, S, FC, CA	Tseng & Chang 1955
	4	8	G, CA, CO	Fujiyama et al. 1955
	4	8	G, CA, CO	Fujiyama 1957
	3	6	R, G, S, FC, CA, CO, CS	Kito 1968, 1974, 1978
	3	6	G, S, CA	Yabu 1969b
	3	6	R, G, S, CA, CO, CS	Tseng & Sun 1989
Porphyra torta	3	—	S, CA	Conway et al. 1975**
	3	—	G, S	Mumford 1975
	3	—	G, S	Mumford & Cole 1977**
	3	—	S, CA, CO	Krishnamurthy 1984**
	3	6	R, S, CA, CC, CO, CS	Burzycki & Waaland 1987
Porphyra umbilicalis	4	8	G, FC	Kito et al. 1971
	4	8	G, S, CA	Kito 1978
	5	10	S, CA, CO	Yabu 1978a
	4	8	S, CA	Kapraun & Freshwater 1987
	3	—	S	J. Coll & E. C. Oliveira pers. comm.
(as *P. umbilicalis* var. *laciniata*)	4	8	R, G, CA, CO, CS	Giraud & Magne 1968
Porphyra variegata	3	—	S, CO	Mumford & Cole 1977**
Porphyra yezoensis	3	6	S, FC, CA	Yabu & Tokida 1963
	3	6	R, G, S, CO, CS	Kito 1967, 1974, 1978
	3	6	R, G, S, FC, CA, CO, CS	Migita 1967
	3	6	G, S	Yabu 1969a
	3	6	R, G, CS	Ma & Miura 1984
	3	6	R, G, CA, CO, CS	Tseng & Sun 1989
Porphyra sp. no. 1 "Ooba-asakusanori"	3	6	G, S, CA	Kito 1978
Porphyra sp. no. 2 "Narawa-susabinori"	3	6	G, S, CA	Kito 1978
Porphyra sp. no. 3 "Muroneamanori"	7	14	S, CA	Yabu 1971
Porphyra sp. no. 1	5	10	G, S, CA	Yabu 1971
Porphyra sp. no. 2	6	—	S	Yabu 1971
Porphyra sp. no. 3	2	—	S	Yabu 1971
Porphyra sp. no. 4	3	—	———	Fujiyama & Yokouchi 1955
Porphyra sp.	3	6	G, S, CA	Tseng & Sun 1989
Porphyra sp. aff. *P. yezoensis*				

Batrachospermales

BATRACHOSPERMACEAE

Batrachospermum mahabaleshwarensis	7	14	R, G, CA, CH	Balakrishnan & Chaugule 1980*
Batrachospermum moniliforme	ca. 10	—	S	Kylin 1917*
	10	20	G, FC, CA	Faridi 1971*
	12	—	CA	Del Grosso & Pogliani 1977
Batrachospermum sp.	ca. 10	—	G, S, CA	Yoshida 1959*
	ca. 22	—	———	Drew unpublished (in Dixon 1966**)
Sirodotia huillensis	14	—	CA	Del Grosso 1981

Table 4-1. (cont.)

Taxon[a]	Chromosome Number[b]		Cell type[c]	Reference[d]
	n	2n		
LEMANEACEAE				
Lemanea australis	ca. 10	—	S, CA	Mullahy 1952*
Lemanea fluviatilis	24	—	G	Del Grosso 1978**
	18–19	36–38	R, G, S, CA, CH	Thirb & Benson-Evans 1982*
Lemanea mamillosa	20 or 21	39–42	R, G, CA	Magne 1967a**
	21	41	R, G, CH	Magne 1967b
Lemanea rigida	15–17	31–34	G, CA	Magne 1961a
Tuomeya americana (as *T. fluviatilis*)	ca. 8	—	G, S, CA	Webster 1958*
Bonnemaisoniales				
BONNEMAISONIACEAE				
Asparagopsis armata	10	20	R, G, S, FC	Svedelius 1933*
	ca. 20	>30	G	Magne 1964
Bonnemaisonia asparagoides	ca. 20	—	G, S	Kylin 1916a*
	ca. 18	—	G, S	Svedelius 1933*
	ca. 18	35–36	G, S, FC	Magne 1960a
	ca. 30	—	T	Rueness & Åsen 1982*
Bonnemaisonia hamifera	20–25	>40	G	Magne 1964
Ceramiales				
CERAMIACEAE				
Antithamnion cruciatum	85–110		T	Whittick & Hooper 1977*
	ca. 24	ca. 48	R, TS	D. F. Kapraun pers. comm.
Antithamnion heterocladum	46 + 4	86 + 8	R, G, T, TS	Athanasiadis 1983*
Antithamnion tenuissimum	ca. 32	ca. 64	G, T	Rueness & Rueness 1973*
Antithamnionella sarniensis	ca. 34	>60	G, T	Magne 1986**
Antithamnionella spirographidis (as *Antithamnion spirographidis*)	32–34	—	---	Rao 1959*
Callithamnion byssoides	28–33	—	---	Harris 1962**
	ca. 30	ca. 60	R, TS	D. F. Kapraun pers. comm.
(as *Aglaothamnion byssoides*)	22–24	44–48	R, TS	Rao 1967a
Callithamnion corymbosum	30	60	R, G, T, TS	Hassinger-Huizinga 1952** (**4**)[f]
	28–33	—	---	Harris 1962**
Callithamnion hookeri (also as *C. brodiaei*)	28–33	—	---	Harris 1962**
Callithamnion roseum	39	—	---	Harris 1962**
Callithamnion sepositum (as *C. purpurascens*)	28–33	—	---	Harris 1962**
Callithamnion tetragonum	28–33	—	---	Harris 1962**
(as *C. brachiatum*)	9–10	18–20	R, G, S, TS	Mathias 1927, 1928* (**5**)[f]
Callithamnion tetricum	ca. 25	ca. 50	G, S, T	Westbrook 1930*
	90–100		---	Harris 1962**
Ceramium cruciatum	21	42	R, TS	Rao et al. 1978
Ceramium deslongchampii	20	40	R, TS	Dammann 1930 (**6**)[f]
Ceramium fastigiatum	28	ca. 56	R, G, CA, T, TS	Yabu et al. 1981
Ceramium fimbriatum	21	42	R, TS	Rao & Sundari 1977
	21	42	R, TS	Rao et al. 1978
Ceramium gracillimum (as *C. gracillimum* var. *byssoideum*)	42	84	R, TS	Sundari & Rao 1977**
	42	84	R, TS	Rao et al. 1978
Ceramium japonicum	ca. 30	—	G, S	Notoya & Yabu 1979a
Ceramium rubrum	6–8	—	G, S	Grubb 1925**
	?8	—	R, TS	Petersen 1928 (**6**)[f]
	34	—	---	Austin 1956, 1959** (**7**)[f]
	—	ca. 64	T	Magne 1964
Griffithsia corallinoides (as *G. corallina*)	20	40	R, S, T, TS	Kylin 1916b*
	>10	ca. 13–14	G, S	Grubb 1925**
Griffithsia globulifera (as *G. bornetiana*)	7	14	R, G, T, TS	Lewis 1909

Table 4-1. (cont.)

Taxon[a]	Chromosome Number[b] n	2n	Cell type[c]	Reference[d]
Griffithsia pacifica	5–8	10–16	G, T	Goff & Coleman 1987*
Halurus equisetifolius	ca. 10	—	G, S	Grubb 1925**
Plumaria elegans	31	62	R, G, T, TS	Drew 1939
		3n = 93	PS	
	ca. 29	ca. 54	R, TS	Magne 1964**
		3n = ca. 90+	PS	Whittick 1977
Pterothamnion crispum	24 + 2	48 + 4	G, T	Athanasiadis 1985**
Pterothamnion plumula	24 + 2	48 + 4	G, T	Athanasiadis 1985**
(as *Antithamnion plumula*)	23	46	R, T, TS	Magne 1964
Ptilota serrata (as *P. pectinata*)	ca. 34	68	R, TS	Yabu 1979
(as *P. pectinata* f. *litoralis*)	ca. 32	40–60	R, G, T, TS	Yabu 1979
Scagelia pylaisaei	—	ca. 50	T	A. Whittick pers. comm.
Spermothamnion repens (as *S. turneri*)	30	60	R, G, CA, T, TS	Drew 1934, 1943
		3n = 90	ST	
(as *S. roseolum*)	20	—	G, S	Schussnig & Odle 1927*
Tiffaniella snyderae (as *Spermothamnion snyderae*)	32	64	R, G, S, T, TS	Drew 1937
Wrangelia argus	26–28	52–56	R, TS	Rao 1970, 1971
Wrangelia penicillata	28	56	R, TS	Magne 1964
Wrangelia plumosa	ca. 27	—	G	Goff & Coleman Chapter 3*
DASYACEAE				
Dasya baillouviana (as *D. elegans*)	ca. 20	ca. 40	G, T	Rosenberg 1933**
Dasya hutchinsiae (as *D. arbuscula*)	—	ca. 40	T	Westbrook 1935
Heterosiphonia plumosa		44	ST	Magne 1964
Heterosiphonia pulchra	ca. 35 (30–36)	ca. 60	R, G, T, TS	Notoya & Yabu 1979b*
DELESSERIACEAE				
Apoglossum ruscifolium	ca. 20	—	G	Kylin 1923**
Caloglossa leprieurii	30	ca. 60	R, G, T, TS	Rao 1967a
Cryptopleura ramosa	30	—	---	Austin 1956, 1959** (7)[f]
(as *Nitophyllum laceratum*)	ca. 8	—	G, S	Grubb 1925**
Delesseria sanguinea	20	40	R, G, S, CA, T, TS	Svedelius 1911, 1912, 1914a
	31	62	R, G, TS	Magne 1964
	30	—	G	Austin 1959
Hypoglossum hypoglossoides	58	—	---	Austin 1959** (7)[f]
Membranoptera alata	32	—	---	Austin 1956, 1959**
Nienburgia andersoniana	8–10	16–20	T	Stewart & Rüdenberg 1978**
Nitophyllum hollenbergii	20–26	40–52	R, TS	Stewart & Rüdenberg 1978*
Nitophyllum punctatum	20	40	R, G, T, TS	Svedelius 1914b, c*
Phrix gregarium	7–8	—	G	Stewart & Rüdenberg 1978**
Phycodrys rubens (as *P. sinuosa*)	20	—	G	Kylin 1923**
Sorella repens	30	ca. 60	R, TS	Yabu et al. 1981
RHODOMELACEAE				
Acanthophora spicifera	32	64	R, TS	Cordeiro-Marino et al. 1974
Chondria armata var. *plumaris*	—	50–55	T	Rao 1967a
	28–29	56–58	R, TS	Pal 1981
Chondria crassicaulis	20	40	R, T, TS	Yabu & Kawamura 1959
Chondria dasyphylla	31	61–62	R, T, TS	Rao & Gujarati 1973
	ca. 25	ca. 50	R, G, T, TS	Westbrook 1928, 1935
Chondria nidifica	—	40	T	Tözün 1974
Chondria tenuissima	20	40	R, T, TS	Tözün 1974
	ca. 25	ca. 50	R, G, T, TS	Westbrook 1935
Chondrus crispus	30	60	R, ST, TS	Magne 1964
	32–35	51–82	G, T	Hanic 1973

Table 4-1. (cont.)

Taxon[a]	Chromosome Number[b]		Cell type[c]	Reference[d]
	n	2n		
Halopitys pinastroides	14	—	G	Celan 1941**
Laurencia arbuscula	29	58	R, TS	Cordeiro-Marino et al. 1983
Laurencia catarinensis	28	56	R, G, TS	Cordeiro-Marino & Fuji 1985
Laurencia hybrida	ca. 20	ca. 40	R, TS	Westbrook 1935
	30	—	---	Austin 1959** (7)[f]
Laurencia caraibica	29?	58	R, TS	Rao & Annapurna 1983
(as *L. nana*)		(52–66)		Annapurna & Rao 1987*
Laurencia nipponica	28	56	R, G, T, TS	Yabu 1978b
Laurencia obtusa				
(as *L. obtusa* var. *majuscula*)	20	40	R, T, TS	Yabu & Kawamura 1959*
Laurencia okamurae	32	64	R, TS	Yabu 1985
Laurencia papillosa	20	40	R, T, TS	Yabu & Kawamura 1959
	26	52	R, TS	Cordeiro-Marino et al. 1974
Laurencia parvula	16	32	R, TS	Pal 1981
Laurencia pinnata	32	64	R, TS	Yabu 1985
Laurencia pinnatifida	ca. 20	—	G, S	Kylin 1923**
	29	58	G, T	Austin 1956, 1959*
	29	—	---	Magne 1964**
	15–16	?ca. 20	G, S	Grubb 1925**
	ca. 20	ca. 40	R, T, TS	Westbrook 1928, 1935
Laurencia undulata	30	60	R, TS	Yabu 1985
Leveillea jungermannioides	23–24	46–48	G, T	Rao 1967b
Odonthalia floccosa	5	10	S, T	Goff & Cole 1973
Polysiphonia binneyi	28	56	R, TS	D. F. Kapraun pers. comm.
Polysiphonia breviarticulata	30	60	R, TS	D. F. Kapraun pers. comm.
Polysiphonia brodiaei	29–31	ca. 60	R, T, TS	Magne 1964
Polysiphonia denudata	30	60	R, TS	D. F. Kapraun pers. comm.
Polysiphonia elongata	36	—	G	Austin 1959**
Polysiphonia ferulacea				
(N. Carolina)	30	60	R, G, TS	Kapraun 1977
(Bermuda)	27	54	R, G, TS	Kapraun 1977, 1978
	14	28	R, T, TS	Annapurna & Rao 1985
Polysiphonia flexicaulis (as *P. violacea*)	20	40	R, G, S, T, TS	Yamanouchi 1906a, b
Polysiphonia harveyi var. *arietina*	32	64	R, TS	Kapraun 1978
Polysiphonia harveyi var. *olneyi*	28	56	R, TS	Kapraun 1978
Polysiphonia japonica	20	40	R, G, S, CA, T, TS	Yabu & Kawamura 1959*
Polysiphonia lanosa	26	—	---	Austin 1959**
Polysiphonia mollis	ca. 16	ca. 32	G, T	Goff & Coleman 1986* (8)[f]
Polysiphonia nigrescens	ca. 20	ca. 40	G, FC, T	Kylin 1923**
	30	60	G, T	Austin 1956, 1959*
Polysiphonia platycarpa	ca. 26	ca.52	R, G, S, T, TS	Iyengar & Balakrishnan 1950
Polysiphonia subtilissima	27	54	R, TS	D. F. Kapraun pers. comm.
Polysiphonia urceolata	30	60	R, TS	D. F. Kapraun pers. comm.
Polysiphonia sp.	30	60	R, TS	D. F. Kapraun pers. comm.
Polysiphonia violacea	16	32	G, T	Ravanko 1987
Rhodomela confervoides	32	—	---	Austin 1956, 1959
	32	—	---	Magne 1964**
(as *R. subfusca*)	ca. 20	ca. 40	R, TS	Westbrook 1935
Rhodomela virgata	20	40	R, S, T, TS	Kylin 1914
Compsopogonales				
BOLDIACEAE				
Boldia erythrosiphon		8 + 1	ST	Nichols 1964a*
COMPSOPOGONACEAE				
Compsopogon coeruleus		7 + 1	ST	Nichols 1964b**
	12	—	ST	Rao 1966*
Compsopogon sp.	7 or 8	—	ST	Shyam & Sarma 1980*

Table 4-1. (cont.)

Taxon[a]	Chromosome Number[b]		Cell type[c]	Reference[d]
	n	2n		
ERYTHROPELTIDACEAE				
Erythropeltis subintegra	7	—	G	D. F. Kapraun & M. Gargiulo pers. comm.
Corallinales				
CORALLINACEAE				
Corallina officinalis	24	—	G, S	Suneson 1937*
	20–25	48	R, T, TS	Magne 1964
Corallina officinalis var. *mediterranea*	24	48	R, G, S, C, T, TS	Yamanouchi 1921** (9)[f]
Jania rubens	—	48	T	Magne 1964**
(as *Corallina rubens*)	24	—	G, S	Suneson 1937*
Lithophyllum corallinae	16	32	R, G, T, TS	Suneson 1950*
Lithophyllum litorale	—	ca. 30	T	Suneson 1950*
Marginisporum aberrans				
(as *Amphiroa aberrans*)	24	—	S	Segawa 1941*
Mastophora lamourouxii	24	—	S	Suneson 1945*
Melobesia farinosa	ca. 24	ca. 48	R, TS	Balakrishnan 1947*
Gelidiales				
GELIDIACEAE				
Acanthopeltis japonica	15	30	R, TS	Kaneko 1968*
Gelidium latifolium	4–5	9–10	G, CA	Dixon 1954**
		ca. 18	G	Boillot 1963**
Gelidium latifolium var. *luxurians*		25–30	ST	Magne 1964
Gelidium sesquipedale (as *G. corneum*)	4–5	9–10	G, CA	Dixon 1954**
Gelidium vagum	7–10	—	G	Kaneko 1966
	14–15	18–30	R, TS	D. Renfrew pers. comm.
GELIDIELLACEAE				
Gelidiella acerosa	4	8	R, T, TS	Rao 1974*
Gigartinales				
CHOREOCOLACEAE				
Harveyella mirabilis	6	12	R, G, S, T, TS	Goff & Cole 1973
CYSTOCLONIACEAE				
Calliblepharis jubata (as *C. lanceolata*)		35–40	ST	Magne 1964
Cystoclonium purpureum	—	50	---	Austin 1956
(as *C. purpurascens*)	ca. 20	—	G, S	Kylin 1923**
Rhodophyllis divaricata (as *R. bifida*)	ca. 20	—	G	Kylin 1923**
DUMONTIACEAE				
Dilsea carnosa	26	ca. 50	R, G, TS	Magne 1964
Dumontia contorta				
(as *D. incrassata*)		22–24	ST	Magne 1964
(as *D. filiformis*)		ca. 7	G	Dunn 1917**
FURCELLARIACEAE				
Furcellaria lumbricalis (as *F. fastigiata*)	ca. 16	—	G, S	Grubb 1925
	34	68	R, G, S, FC, CA, TS	Austin 1955, 1957, 1959, 1960a, b
	ca. 30	—	G	Magne 1964
GIGARTINACEAE				
Gigartina pistillata		32	ST	Magne 1964**
Iridaea cordata	4	8	G, T	Fralick & Cole 1973
Iridaea heterocarpa	4	8	G, T	Fralick & Cole 1973
Mastocarpus stellatus		15–20	T	Drew (in Marshall et al. 1949)*
(as *Gigartina stellata*)	—	ca. 40–50	G	Maggs 1988**
GRACILARIACEAE				
Gracilaria bursa-pastoris	24	48	R, TS	Bird et al. 1982
Gracilaria chilensis	24	48	R, TS	Bird et al. 1986**
Gracilaria coronopifolia	24	48	R, TS	Bird & McLachlan 1982**

Table 4-1. (cont.)

Taxon[a]	Chromosome Number[b] n	2n	Cell type[c]	Reference[d]
Gracilaria foliifera (as *G. multipartita*)	6 or 7	—	CA	Greig-Smith 1954** (**10**)[f]
	24	48	R, TS	McLachlan et al. 1977
Gracilaria pacifica (as *G. verrucosa*)	24	48	R, TS	Bird et al. 1982
Gracilaria sjoestedtii	29–30	58–60	R, TS	Zhang & van der Meer 1988b
Gracilaria tikvahiae	2n = 48	4n = 96	R, TS	Patwary & van der Meer 1984
(as *G.* sp.)	24	48	R, TS	McLachlan et al. 1977
Gracilaria verrucosa	32	—	G	Magne 1964
	32	64	R, TS	Bird et al. 1982
Gracilaria sp. (Italy)	24	—	R, TS	Bird & McLachlan 1982**
PEYSSONNELIACEAE				
Peyssonnelia obscura var. *bombeyensis*	23–24	46–48	R, G, S, TS	Rao 1967a
Peyssonnelia squamaria		40–46	ST	Magne 1964
Rhodophysema elegans	—	30–35	T, TS	South & Whittick 1976*
PHYLLOPHORACEAE				
Ahnfeltia plicata	4, 8	—	G	Gregory 1930** (**11**)[f]
	4	—	G	Rosenvinge 1931a, b*
Gymnogongrus devoniensis	—	ca. 46	G	Maggs 1988**
Gymnogongrus griffithsiae	4	8	G, T	Gregory 1930** (**11**)[f]
Gymnogongrus linearis	ca. 6	—	G	Doubt 1935*
Phyllophora pseudoceranoides	ca. 16–18	28–34	G, T, TS	Newroth 1972
Phyllophora truncata	ca. 15–16	ca. 32	R, G, T, TS	Newroth 1971
(as *P. brodiaei*)	4	8	R, G, CA, TS	Claussen 1929*
Stenogramme interrupta		30		R, TS	Westbrook 1928
POLYIDEACEAE				
Polyides rotundus	ca. 20	—	G	Kylin 1923**
(as *P. caprinus*)	68–72	136–144	R, TS	Rao 1956, 1959
SOLIERIACEAE				
Anatheca montagnei	7–8?	14–16	R, G, CA, T, TS	Bodard 1966**
Turnerella pennyi	18–20	36–40	G, CA, T	South et al. 1972**
Nemaliales				
GALAXAURACEAE				
Galaxaura corymbifera	ca. 10 (9–12)	ca. 20 (15–20)	R, G, TS	Svedelius 1942*
Galaxaura diesingiana	10–12	ca. 17	R, G, S, CA, TS	Svedelius 1942*
Galaxaura tenera	ca. 10	ca. 20	R, G, S, TS	Svedelius 1942*
Nothogenia erinacea				
(as *Chaetangium saccatum*)	8–10	—	G	Martin 1939a*
Scinaia capensis		2n?		
(as *Pseudogloiophloea capensis*)	>12	ca. 15–20	S, FC	Svedelius 1956*
Scinaia confusa				
(as *Pseudogloiophloea confusa*)	4 or 5	8 or 10	G, FC, CA, T	Ramus 1969*
Scinaia forcellata (as *S. furcellata*)	ca. 10	20	R, G, S, FC	Svedelius 1915*
	7–9	15–17	G, CA	Magne 1961a
	6	12–15	T, TS	Boillot 1972
Scinaia turgida	6	ca. 12	R, T, TS	Boillot 1972
HELMINTHOCLADIACEAE				
Helminthocladia calvadosii	ca. 8	16–18	G, S, T	Boillot 1971
Helminthocladia papenfussii	ca. 10	—	S	Martin 1939b*
Helminthocladia senegalensis	8	16	R, G, T, TS	Bodard 1971*
Helminthora divaricata	8–10	16–20	R, FC, CA	Kylin 1928*
	9	18	R, T, TS	Boillot 1972
Nemalion helminthoides	8	16	G, S, CA, T	Magne 1961b
	10	20	G, S, FC, T	Chen et al. 1978*
(as *N. multifidum*)	ca. 8	ca. 16	S, CA	Wolfe 1904
	8	16	R, G, S, FC	Cleland 1919
	ca. 10	ca. 20	R, S, FC	Kylin 1916c*

Table 4-1. (cont.)

Taxon[a]	Chromosome Number[b] n	2n	Cell type[c]	Reference[d]
Palmariales				
PALMARIACEAE				
Devaleraea ramentacea				
(as *Halosaccion ramentaceum*)	ca. 24	ca. 48	R, G, T, TS	van der Meer & Chen 1979
	ca. 24	ca. 48	R, G, FC, TS	van der Meer 1981a*
	8	16	R, G, TS	Jónsson & Chesnoy 1982
Devaleraea yendoi (as *H. saccatum*)	17	ca. 30	R, T, TS	Yabu 1976
Palmaria mollis	21	42	R, T, TS	van der Meer & Bird 1985
(as *Rhodymenia palmata* f. *mollis*)	ca. 14		G	Sparling 1961**
Palmaria palmata	18–23	38–48	R, G, T, TS	van der Meer 1976
	21	—	(R), T	van der Meer & Chen 1979
(as *Rhodymenia palmata*)	>20	—	G, T, TS	Westbrook 1928
	20	—	---	Austin 1959 (**7**)[f]
	26	52	R, T, TS	Yabu 1972b
	21 or 26	ca. 40–50	R, G, TS	Yabu 1976 (**12**)[f]
	14		G, T, TS	Magne 1959**
Porphyridiales				
PORPHYRIDIACEAE				
Porphyridium aerugineum	2		ST	Sommerfeld & Nichols 1970b*
Porphyridium purpureum	8–10		ST	Schornstein & Scott 1982** (**13**)[f]
(as *P. cruentum*)	2		ST	Sommerfeld & Nichols 1970b*
Rhodochaetales				
RHODOCHAETACEAE				
Rhodochaete parvula	4	8	G, S, FC	Magne 1960b, c
	4	8	G, S, T	Boillot 1969, 1975
Rhodymeniales				
CHAMPIACEAE				
Champia parvula	12	24	R, TS	D. F. Kapraun pers. comm.
Chylocladia verticillata (as *C. kaliformis*)	ca. 20	—	G	Kylin 1923**
LOMENTARIACEAE				
Lomentaria articulata	10	20	R, G, T, TS	Magne 1964
Lomentaria baileyana	10	20	R, TS	D. F. Kapraun pers. comm.
Lomentaria clavellosa	10	20	R, G, S, T, TS	Svedelius 1935, 1937*
	22 or 23	44 or 46	R, T, TS	Magne 1964
Lomentaria orcadensis				
(Brest)	10		R, T, TS	Magne 1964
(Morlaix)	20			
(as *L. rosea*)	ca. 20		T, TS	Svedelius 1935, 1937*
RHODYMENIACEAE				
Rhodymenia pertusa	28	30–40	R, G, T, TS	Yabu 1976
Rhodymenia pseudopalmata	10	20	R, TS	D. F. Kapraun pers. comm.

[a] The classification system for the Rhodophyta used in this chapter is in accord with that of Gabrielson et al. (1985) and Gabrielson and Garbary (1986), which recognizes the single class Rhodophyceae, except that the earlier validly published name Compsopogonales replaces Erythropeltidales proposed by Garbary et al. (1980). In this table there is no segregation of the orders into two subclasses Bangiophycidae and Florideophycidae; the orders, families, genera, and species are listed alphabetically. However, the subclasses are occasionally referred to in the text because they are used in the literature.

[b] Chromosome numbers centered between the *n* and 2*n* columns were obtained from material of uncertain *n*/2*n* status. — = no number reported.

[c] In most species, chromosome numbers were obtained from mitotic material. In a few reports, designated R, TS only, mitotic chromosomes were not counted; meiotic chromosomes were observed in tetrasporangia, and the diploid number was extrapolated from the number of bivalents observed in prophase I.

[d] ** No illustration of chromosomes in the report.
 * Illustrations of chromosomes are low-magnification micrographs or sketches, and/or do not show morphological details and/or the complete chromosomal complement.

Chromosomes

Table 4-1. (cont.)

e J. Coll and E. C. de Oliveira (pers. comm.) recently determined haploid (*n*) chromosome numbers from spermatangial material (S) of *Porphyra acanthophora* (3), *P. drewii* (3), *P. lucasii* (3), *P. pujalsii* (4), *P. rizzinii* (4), *P. segregata* (3), and *P. thuretii* (4).

f Notes 1–13 indicated in bold type in the "Taxon" and "Reference" columns:
 (**1**) *Porphyra amplissima* is included in the current checklist of benthic algae of Japan (Yoshida et al. 1985), although Conway et al. (1975) synonymized it in *P. miniata*.
 (**2**) *Porphyra sanjuanensis* and *P. subtumens* were considered to be monophasic by the authors cited, and the chromosome numbers are designated as haploid. The chromosome number for *P. subtumens* is unusually high; chromosome counts were limited to vegetative cells, and no illustrations were provided.
 (**3**) Krishnamurthy (1984) contends that *Porphyra norrisii* should not be synonymized into *P. schizophylla*, according to Conway et al. (1975).
 (**4**) Hassinger-Huizinga (1952) reported chromosomes of $n = 30$ in haploid sexual plants, $2n = 60$ in diploid plants bearing only tetrasporangia, and $2n = 62$ or 64 in diploid plants bearing both tetrasporangia and sexual organs. Considering the statistical evidence and lack of pictures in her paper, Dixon (1966) stated that the latter "aneuploidy" data are suspect, and they are not included in this table.
 (**5**) Westbrook (1930, 1935) was critical of the work of Mathias on *Callithaminion tetragonium*, suggesting that the nucleolus was incorrectly identified as the nucleus.
 (**6**) Westbrook (1930, 1935) and Magne (1964) argued that Dammann (1930) and Petersen (1928) mistook nucleolar fragments as chromosomes.
 (**7**) Austin (1959) indicated that these chromosome numbers were based on few counts.
 (**8**) Goff and Coleman (1986) did not specify the *n* and 2*n* chromosome numbers for *P. mollis*. The values here have been extrapolated from their data.
 (**9**) This paper by Yamanouchi was translated from the Japanese article in *Bot. Mag. Tokyo* 27: 279–85 (1913).
 (**10**) McLachlan et al. (1977) indicated that the chromosome number reported in this paper was probably incorrect.
 (**11**) Dixon (1966) questioned the chromosome numbers by Gregory (1930) because they were not included in a later article (Gregory 1934). Gregory (1930) suggested that the eight-chromosome condition in medullary cells may have been produced by cell fusions.
 (**12**) Yabu (1976) indicated that the $n = 21$ and $n = 26$ chromosome numbers were representative of two different races of *Palmaria palmata* (= *Rhodymenia palmata*).
 (**13**) Schornstein and Scott (1982) equated kinetochore numbers, determined from electron micrographs, with chromosome numbers.

atropurpurea ($n = 3$) and *Audouinella hermannii* ($n = 6$) have lower numbers than the more complex *Lemanea* spp. ($n = 10–22$). He also noted that numbers in some marine ceramialean genera are much higher than those published for freshwater Rhodophyta.

Although chromosome numbers in most red algal orders tend to be uncertain, those in the Bangiales are unequivocal. The genus *Porphyra* has been investigated cytologically more than any other in the Rhodophyta (Mumford & Cole 1977); chromosome numbers have been determined for most of the 65+ species. Bangialean species offer the advantage of having small numbers of chromosomes ($n = 2–7$) contained in uninucleate cells.

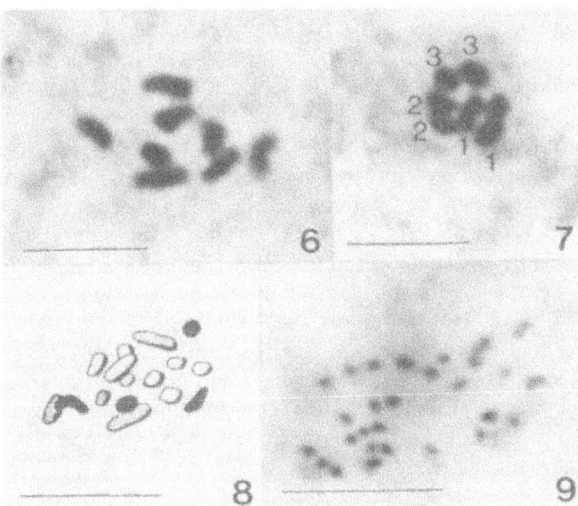

Figs. 4-6 to 4-9. 4-6, 4-7: Micrographs of mitotic prophase chromosomes in *Bangia vermicularis* prepared using Wittmann's (1965) rapid hematoxylin squash technique and showing somatic pairing. Scale bars = 4 µm. (From Cole et al. 1983, with permission) 4-6: Eight chromosomes of fertilized carpogonium in female filament; three closely associated central pairs, fourth pair consisting of single chromosomes at the extreme left and right. 4-7: Three pairs of chromosomes in cells of monospore-producing asexual filament, closely associated in a radial configuration incomplete on the right side because of displaced chromosome number 1. 4-8: Composite drawing of mitotic prophase chromosomes in apical cell of germinating carpospore of *Nemalion helminthoides* ($n = 8$). Feulgen. Scale bar = 5 µm. (From Magne 1961b, with permission) 4-9: Micrograph of midprophase mitotic chromosomes in germinating tetraspore of *Chondrus crispus* ($n = 32–35$). Feulgen-Fe-propionocarmine. Scale bar = 15 µm. (From Hanic 1973, with permission)

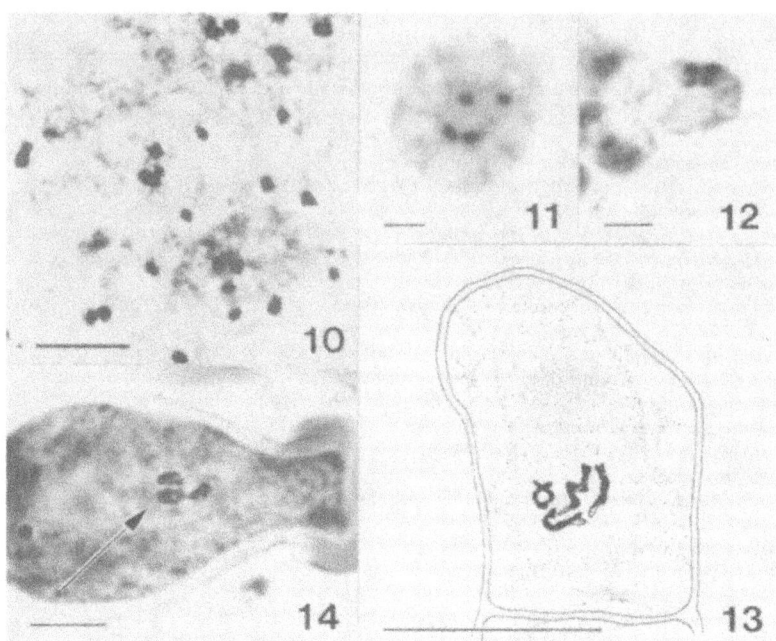

Figs. 4-10 to 4-14. *4-10*: Micrograph of diakinesis in tetrasporangium of *Chondria tenuissima* ($n = 20$), showing 13 bivalents and 14 univalents. Aceto-carmine. Scale bar = 10 μm. (From Tözün 1974, with permission.) *4-11, 4-12*: Micrographs of mitotic chromosomes in spermatangial cell of *Porphyra leucosticta* ($n = 4$). Hematoxylin. Scale bars = 5 μm. (From Coll & Oliveira 1977a, with permission.) *4-11*: Fresh fixed material. *4-12*: Material dried in 1897. *4-13*: Drawing of prophase I meiosis chromosomes in terminal cell of conchosporangial filament of *Porphyra umbilicalis* (as *P. umbilicalis* var. *laciniata*) ($n = 4$). Feulgen. Scale bar = 10 μm. (From Giraud & Magne 1968, with permission.) *4-14*: Micrograph of prophase I meiosis (diakinesis) in germinating conchospore of *Porphyra yezoensis* ($n = 3$); arrow indicates three paired chromosomes. Hematoxylin. Scale bar = 5 μm. (From Ma & Miura 1984, with permission)

Chromosomes in spermatangial material are usually condensed during an extended prophase stage and readily stained and spread, even in specimens that have been pressed on herbarium sheets for a number of years (e.g., Figs. 4-11, 4-12; Cole & Sheath 1980; Cole et al. 1983; Coll & Oliveira 1977a).

B. Life histories

Drew (1955) defined a life history as the recurring sequence of somatic and nuclear phases characteristic of a species. Both morphological and cytological information, including chromosome numbers and the occurrence and position of synapsis and meiosis, are requisite to a thorough understanding of life histories (Dixon 1982). For example, following culture and cytological studies, a new type of rhodophytan life history with extreme sexual dimorphism of the male and female gametophytes was established in three members of the Palmariaceae: *Palmaria palmata* (e.g., van der Meer & Chen 1979; van der Meer & Todd 1980), *Devaleraea ramentacea* (as *Halosaccion ramentaceum*) (e.g., van der Meer 1981a; van der Meer & Chen 1979) and *Devaleraea yendoi* (as *H. saccatum*) (Yabu 1976). However, complete data are available for only about 10% of the red algal species for which chromosome numbers have been published (Table 4-1). Discussions of cytological studies on life histories of some Rhodophyta are included in several review papers (e.g., Dixon 1963b, 1970, 1982; Drew 1955; Sheath 1984; West & Hommersand 1981).

Chromosome numbers of a few florideophycidean species were secured only from meiotic material, because of technical problems in obtaining different morphological phases or in preparing somatic chromosomes (see cell type R,TS in Table 4-1). Compared to mitotic nuclei, which average 3 μm in diameter (e.g., Dixon 1966), nuclei in meiosis I generally are considerably larger, e.g., 7–12 μm in tetrasporocytes of *Furcellaria lumbricalis* (as *F. fastigiata*) (Austin 1960b), providing information on chromosome morphology and pairing during zygotene and pachytene, in addition to more accurate counts when chromosomes are extremely contracted in diakinesis.

According to the current survey, it appears that meiotic chromosomes were counted in very few *Porphyra* species (Fig. 4-13; see R, Table 4-1), reports differing on the position of reduction division in the life histories, even for the same species. For example, meiosis was observed in *P. tenera* during division of the fertilized carpogonium (Tseng & Chang 1955), conchospore formation (Kito 1974, 1978), and conchospore germination (Tseng & Sun 1989), and in *P. yezoensis*, during conchospore formation (Kito 1974, 1978) and conchospore germination (Fig. 4-14; Ma & Miura 1984; Tseng & Sun 1989).

The phenomenon of somatic chromosome pairing during mitosis in *Bangia vermicularis* assisted in the determination of chromosome numbers and homology in diploid vegetative and carpogonial cells (Figs. 4-6, 4-7; Cole et al. 1983). Somatic pairing also may occur in conchosporangial cells of *Porphyra torta* (Burzycki & Waaland 1987) and conchocelis cells of *P. yezoensis* (Tseng & Sun 1989).

Chromosome numbers are available for relatively few life history stages of species in the Acrochaetiales, Batrachospermales, Bonnemaisoniales, Compsopogonales, Gelidiales, and Rhodochaetales, and they tend to be uncertain (Table 4-1). Chromosome studies of some species in the Acrochaetiales and Compsopogonales also appear to be extremely limited because material available was sterile or asexual (e.g., Hymes & Cole 1983; Magne 1964; Nichols 1964a; Rao 1966; Shyam & Sarma 1980).

Cells of the alternating macroscopic gametophyte (vegetative blade and spermatia) and microscopic sporophyte (carpospores, conchocelis) in a number of *Porphyra* species are all haploid (Table 4-1), suggesting that development of the carpospores could be apogametic in some exceptional cases (Krishmamurthy 1984), e.g., *P. brumalis* (Mumford & Cole 1977), *P. carolinensis* (Freshwater & Kapraun 1986; Kapraun & Freshwater 1987), *P. fucicola* (Krishnamurthy 1984), *P. maculosa* (Conway et al. 1975), *P. papenfussii* (Conway & Cole 1973), *P. pseudolanceolata* (Conway et al. 1975; Krishnamurthy 1984), *P. rosengurttii* (Kapraun & Freshwater 1987; Kapraun & Luster 1980), *P. spiralis* var. *amplifolia* (Kapraun & Freshwater 1987), *P. variegata* (Mumford & Cole 1977). Kapraun and Freshwater (1987) noted that this absence of alternating haploid and diploid phases occurred in life histories of the most southerly North Atlantic species in their studies.

C. Polyploidy

Polyploidy is common throughout the plant kingdom and is a significant factor in plant evolution (Jackson 1976). However, it has been studied very little in nonvascular eukaryotes (Nichols 1980). Drew (1934, 1939, 1943) discovered polyploidy in field material of two members of the Ceramiales, *Spermothamnion repens* (as *S. turneri*) and *Plumaria elegans*, collected at different times of the year over a broad distribution area. She documented the occurrence of haploid, diploid, and triploid plants (Figs. 4-15 to 4-17), as well as some tetraploid gonimoblast cells in diploid plants of *S. repens*.

Generally, it is assumed that higher haploid numbers in plants have arisen by polyploidy following natural hybridization and that they are derivatives of basic genomes (x) with much smaller chromosome numbers (Stebbins 1971). This is also probably true for red algae, mainly genera in the Ceramiales and Gigartinales. Establishment of the basic genome for a group of organisms is assisted by information on chromosome pairing and chiasmata, as well as haploid chromosome numbers for as many plants as possible. However, these data are usually incomplete for most rhodophytan species, and the basic number is subject to interpretation. For example, Drew (1934, 1943) used a basic number of 30 to describe ploidy in *Spermothamnion repens* as haploid ($n = 30$), diploid ($2n = 60$), triploid ($3n = 90$) and tetraploid ($4n = 120$). Plants with 45–50 chromosomes were explained as progeny from triploid tetrasporangiate plants. However, Barber (1947) and Feldmann (1952) interpreted her data differently, using the basic number of 15: $n = 15$, $2n = 30$, $3n = 45$, $4n = 60$, $6n = 90$, $8n = 120$.

Considering the range of chromosome numbers in the Ceramiales, $n = 7-42$ (Table 4-1), Rao et al. (1978) and Annapurna and Rao (1985) suggested that a common ceramialean ancestry of $x = 7$ was plausible. They observed meiotic chromosomes in several Indian species: three *Ceramium* spp. (Rao et al. 1978), *Laurencia caraibica* (as *L. nana*) (Rao & Annapurna 1983) and *Polysiphonia ferulacea* (Annapurna & Rao 1985), and noted univalents, bivalents, and polyvalents during prophase I (Figs. 4-18, 4-19). Using these data in conjunction with previously published information on chromosome numbers,

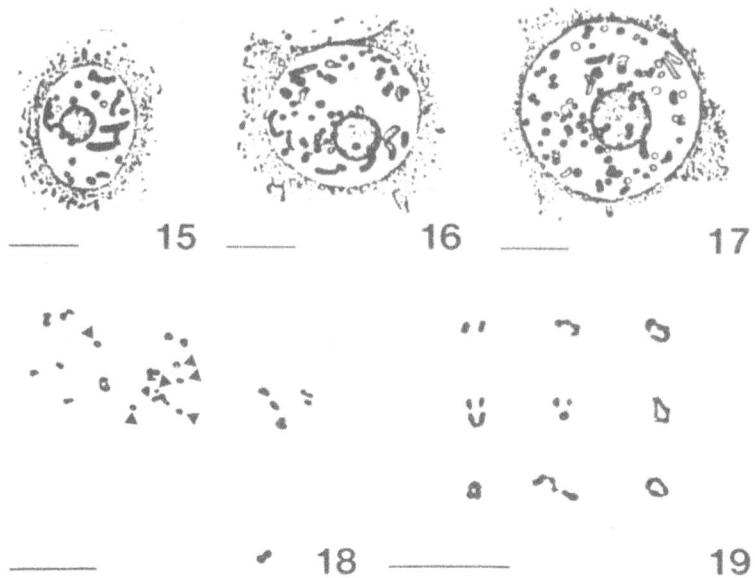

Figs. 4-15 to 4-19. Polyploids. *4-15 to 4-17*: Composite drawings of nuclei of *Plumaria elegans* (n = 31). Outlined chromosomes are in different focal plane. Note three long chromosomes in each haploid complement. Brazilin. Scale bars = 5 μm. (From Drew 1944, with permission) *4-15*: Apical cell of haploid male plant. *4-16*: Apical cell of diploid tetrasporic plant. *4-17*: Triploid parasporangium. *4-18, 4-19*: Camera lucida drawings of diakinesis in *Ceramium fimbriatum* (n = 21). Acetocarmine. Scale bars = 10 μm. (From Rao et al. 1978, with permission) *4-18*: Nucleus containing 18 bivalents and 6 univalents (arrowheads). *4-19*: Common quadrivalent types.

they then concluded that hybridization and polyploidy were the major factors in the evolution of species within the genera *Ceramium, Laurencia*, and *Polysiphonia*. Similar "divergent ploidy levels" are attained by *Callithamnion* species, with chromosome numbers ranging from $n = 9-10$ to $n = 90-100$ (Rao et al. 1978). According to the studies of D. F. Kapraun (pers. comm.), the basic chromosome number ($n = 30$) for the genus *Polysiphonia* is probably a triploid derivative of the basic genome ($x = 10$). Polyploidy also was noted in several species of *Antithamnion* (Athanasiadis 1983; Kapraun pers. comm.; Whittick & Hooper 1977), and in *Plumaria elegans* (Whittick 1977). In other orders, a basic genome number of five was suggested for the Gelidiales (Kaneko 1968) and eight for the genus *Devaleraea* (as *Halosaccion*) (Palmariales) (Jónsson & Chesnoy 1982).

It appears that some insight is being gained into the polyploidy process in red algae. From her field data on *Spermothamnion repens*, in which haploid sexual plants normally alternate with diploid tetrasporic plants, Drew (1934, 1943) reasoned that triploidy ($3n = 90$) and tetraploidy ($4n = 120$) stemmed from the sexual functioning of unreduced as well as reduced gametes, a process common in higher plants (e.g., de Wet 1980). More recently, autopolyploids of *Gracilaria tikvahiae* (Gigartinales) ($3n = 72$ and $4n = 96$) were constructed in culture using tetrasporophytes that produced unreduced gametes following mitotic recombination in vegetative fronds (Patwary & van der Meer 1984; van der Meer 1981b; van der Meer & Patwary 1983; van der Meer & Todd 1977), and repeated self- and crossfertilizations of mutant bisexual plants (no tetrasporophytic stage) (Zhang & van der Meer 1988a). This unique system shows promise for further investigations of polyploidy and its role in rhodophytan evolution.

Endopolyploidy results either from nuclear fusion in multinucleate cells or from the abortion of mitosis before anaphase separation of chromatids, producing a doubling of chromosomes or polyteny if strands are held at the centromere. It is a normal feature of development and differentiation in some plant and animal tissues (e.g., Brown 1972, p. 171), and may complicate determination of basic haploid and diploid chromosome numbers. Endopolyploidy has been reported in specific cells and nuclei of several red algae. For example, in addition to the expected haploid nuclei ($n = 30$) in polysporangia of *Spermothamnion repens*, Drew (1937) observed nuclei containing 60 ($2n$) and 90($3n$), as well as 45–50, chromosomes, attributing these to coalescence of nuclei in young sporangia. Giant triploid nuclei with 30 chromosomes were also produced by nuclear fusion in auxilliary cells of *Lomentaria clavellosa* (Svedelius 1937). Recently, using DNA-binding fluorochrome dyes and quantitative epifluorescence microscopy in conjunction with chromosome counts, Goff and Coleman (1987) noted a small proportion (0.5%) of spontaneously polyploid nuclei in multinucleate haploid and diploid vegetative cells of *Griffithsia pacifica* ($n = 5–8$). A few nuclei with 10–16 chromosomes occurred in haploid cells and with >20 chromosomes in diploid cells. In other florideophycidean species, increased amounts of DNA were identified in certain derived uninucleate cells of *Choreocolax polysiphoniae* and following nuclear transfer (via secondary pit connections) in *Plocamium cartilagineum* and *Polysiphonia confusa* (Goff & Coleman 1985, Chapter 3); chromosomes were not observed.

Goff and Coleman (1986, Chapter 3) reported several examples of endopolyploidy occurring during development of apical cells in some rhodomelacean species (Ceramiales) and enlargement of carpospores and tetraspores prior to germination in *Polysiphonia mollis*. Sequential polyploidy reduction by failure of DNA replication during early nuclear divisions in these cells was documented by microspectrofluorometry. In addition, the occurrence of polyteny, resulting in enlarged chromosomes and chromomeres, is suspected during the germination of carpospores and tetraspores of *Chondrus crispus* (L. A. Hanic pers. comm.) and in polyploid axial cells of an apomeiotic strain of *Scagelia pylaisaei* (Figs. 4-20, 4-21; A. Whittick pers. comm.).

Aneuploidy, originating from an increase or decrease in chromosome number by less than a complete genome (Jackson 1971), is often characteristic of polyploid evolution in plants (e.g., de Wet 1980). It is an important factor in reproductive isolation and has been implicated in the derivation of several ceramialean genera that have a range of haploid numbers similar to aneuploid series in higher plants. Kapraun (1978) suggested that aneuploidy may have been a significant feature of evolution at the species and population level in *Polysiphonia* ($n = 20–36$) (Table 4-1; Fig. 4-5), and that *Callithamnion* species ($n = $ ca. $25–39$) may have had a similar aneuploid origin. In addition, aneuploid derivatives were observed in the polyploid *Ceramium gracillimum* (as *C. gracillimum* var. *byssoideum*) (Rao et al. 1978).

IV. CHROMOSOME MORPHOLOGY

Limited morphological data are available for red algal chromosomes. The shapes are generally referred to as elongate, sausage-shaped, or spherical (e.g., Dixon 1973), and absolute sizes are not recorded routinely. It is very difficult to discern centromere positions, nucleolar organizing regions, and banding patterns, all of which serve as markers for structural changes in chromosomes.

In contrast to mitotic chromosomes in some higher plants and green algae, which average 6 to 10 µm in length and may reach 30 µm (e.g., Sarma 1983; Stebbins 1971), red algal chromosomes are diminutive, ranging in length from the limit of resolution of the light microscope (ca. 0.25 µm) to ca. 3.5 µm (e.g., Cole et al. 1983; Del Grosso 1981; Goff & Cole 1973; Hanic 1973; Kapraun & Fresh-

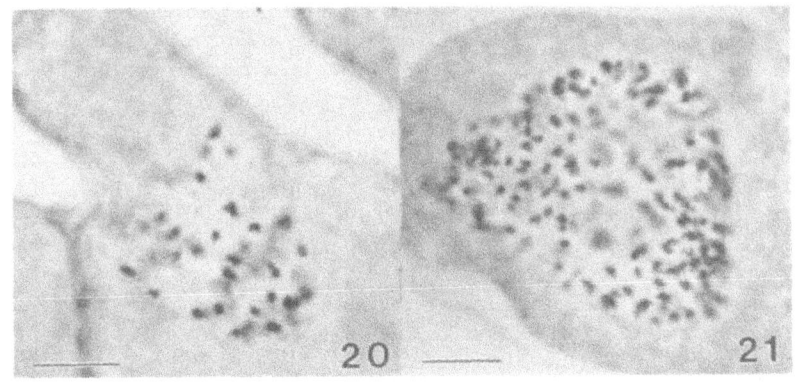

Figs. 4-20, 4-21. Mitotic chromosomes in *Scagelia pylaisaei*. Acetoorcein. Scale bars = 10 µm. (Courtesy A. Whittick unpubl.) *4-20:* Diploid apical cell containing ca. 50 chromosomes. *4-21:* Endopolyploid axial cell in same plants as cell in Fig. 4-20, containing 300+ chromosomes. Note polytene appearance of chromosomes (thick and banded).

water 1987; Rao 1971). Meiotic chromosomes in midprophase I tend to be longer. For example, in *Furcellaria lumbricalis* (as. *F. fastigiata*) the 34 pairs of pachytene chromosomes range from 2.5 to 9.5 µm and are half these lengths in diakinesis (Fig. 4-1; Austin 1960b).

The relative length of chromosomes, described as long, medium, or short, was used as a comparative feature in studies of *Porphyra* species (e.g., Mumford & Cole 1977; Yabu 1969a). In the Florideophycidae, which have larger numbers of small chromosomes, there are a few reports of several distinctively large or small chromosomes in haploid sets; e.g., in the ceramiacean species: four long chromosomes in *Polysiphonia flexicaulis* (as *P. violacea*) (Ravanko 1987), three long chromosomes in *Plumaria elegans* (Figs. 4-15 to 4-17; Drew 1939) and *Wrangelia argus* (Fig. 4-22; Rao 1970, 1971); two large chromosomes in *Gracilaria tikvahiae* (Fig. 4-23; McLachlan et al. 1977); one large chromosome in *Jania rubens* (von Stosch 1969), *Palmaria mollis* (van der Meer & Bird 1985), *P.* (as *Rhodymenia*) *palmata* (van der Meer 1976; Yabu 1972a, 1976), *Devaleraea yendoi* (as *H. saccatum*) (Yabu 1976); *Ptilota serrata* (as *P. pectinata*) (Fig. 4-24; Yabu 1979), and *Laurencia undulata* (Yabu 1985); one very small chromosome in *L. nipponica* and *L. pinnata*; and three in *L. okamurae* (Fig. 4-25; Yabu 1978b, 1985).

Centromeres and their positions have been observed during meiosis or mitosis in several red algal species, chromosomes described as metacentric, submetacentric, or subacrocentric, e.g., *Furcellaria lumbricalis* (Fig. 4-1B; Austin 1960b), *Polysiphonia flexicaulis* (Ravanko 1987), *Wrangelia argus* (Rao 1970), *Harveyella mirabilis* and *Odonthalia floccosa* (Goff & Cole 1973), *Bangia vermicularis* (Figs. 4-3, 4-26, 4-27; Cole et al. 1983), and *Porphyra papenfussii* (Fig. 4-28; Conway & Cole 1973). Based on the orientation of chromosomes to the division poles, telocentrics and metacentrics were recorded in *Rhodothamniella floridula*, *Delesseria sanguinea*, and *Corallina officinalis* (Magne (1964). A distinct nucleolar organizing region, the secondary constriction, was also observed in each of the three chromosomes of *P. papenfussii* (Fig. 4-28; Conway & Cole 1973), and in at least one chromosome of *B. vermicularis* (Fig. 4-26; Cole et al. 1983).

The presence and distribution of eu- and heterochromatic regions provide distinctive light and dark banding patterns when chromosomes are stained selectively following special pretreatments (e.g., Berlyn & Miksche 1976; Swanson et al. 1981), and also without pretreatment in some species, usually during extended early to midprophase. Giemsa banding was observed in the two large midprophase chromosomes in spermatangia of *Porphyra schizophylla*, following application of a technique modified from Arrighi and Hsu (1971) (Figs. 4-29, 4-30; Cole unpubl.).

Using standard Feulgen staining of early and late meiotic stages in the tetrasporangia of *Wrangelia argus*, Rao (1970, 1971) observed heteropycnosis of four prophase I chromosomes that formed a ringlike configuration indicative of a translocation heterozygote. Three of the chromosomes are the largest in the diploid complement, one being particularly long (Fig. 4-22). Each of the four is distinguishable on the basis of length, centromere position, and distribution of light- and dark-staining blocks and knobs in the chromosome arms. Rao (1971) proposed that these may be sex chromosomes X_1, X_2, Y_1 and Y_2, likening the longest to the very large heteropycnotic X chromosome present in some laminarialean (Phaeophyta) gametophytes (e.g., Evans 1965; Yabu & Sanbonsuga 1981). Earlier, Drew (1955) suggested that segregation of sex chromosomes during meiosis is responsible for the production of equal numbers of male and female gametophytes in *Antithamnion spirographidis*. There appear to be no other reports of sex chromosomes in the red algae, although the allelic basis for gametophytic sex determination was demonstrated recently in *Gracilaria tikvahiae* (e.g., van der Meer 1981b, 1986; Chapter 5).

V. TAXONOMIC ASPECTS

"During past geologic periods plant and animal taxa of various kinds have made their appearance, and to judge from those extant today, their acquisition of taxonomic identity has often been paralleled by karyotype uniqueness as well." (Swanson et al. 1981, p. 518)

A. Chromosome number and morphology

Although not included routinely in taxonomic studies, chromosome numbers are valuable in determinations of natural algal species (Bird & McLachlan 1982). In studies of *Gracilaria* species, Bird et al. (1982) noted that the chromosome number of representatives of the type species from England, *G. verrucosa* ($n = 32$), differed from that of a morphologically similar northeastern Pacific form ($n = 24$), also identified as *G. verrucosa*. From hybridization attempts they determined that these two stocks are sexually incompatible and that the latter was misnamed.

The importance of chromosome number as a taxonomic character was emphasized by Krishnamurthy (1984), who questioned the merging of several northeastern Pacific *Porphyra* species that are

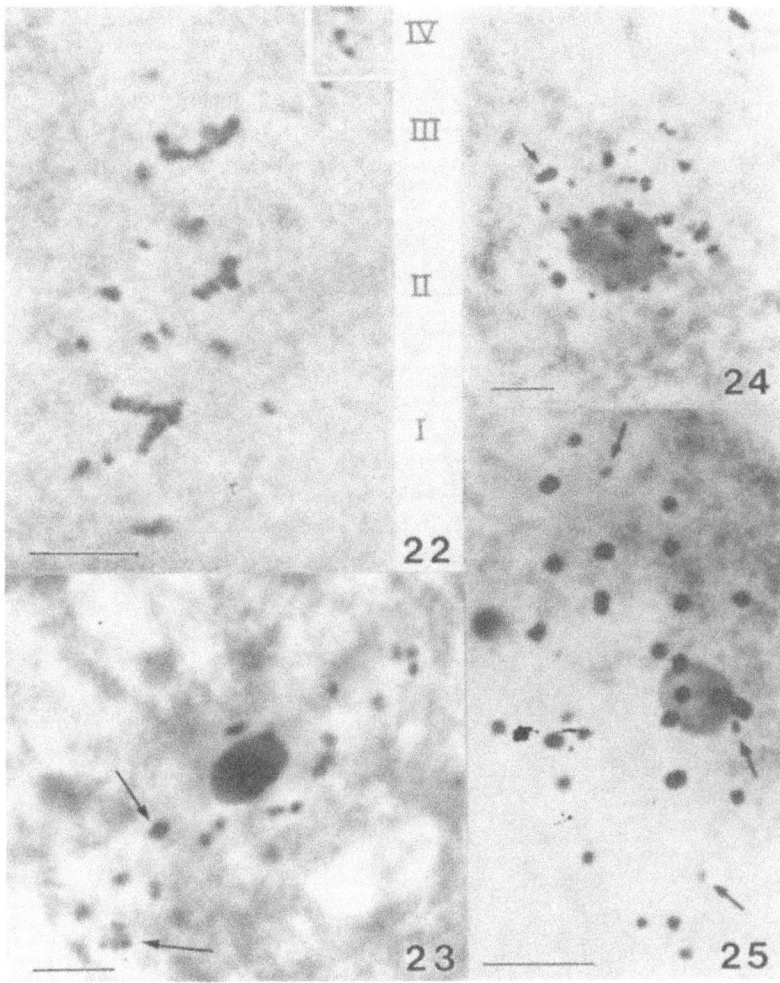

Figs. 4-22 to 4-25. Meiotic chromosomes in red algal tetrasporangia. 4-22: Diplotene in *Wrangelia argus* ($n = 26-28$), showing slightly heteropycnotic sex chromosomes I–IV; note large conspicuous chromosomes I–III and very small chromosome IV (inset), located to the left of chromosome II in the spread. Feulgen. Scale bar = 5 μm. (From Rao 1971, with permission) 4-23: Diakinesis in *Gracilaria foliifera* ($n = 24$); arrows indicate two large bivalents. Aceto-carmine. Scale bar = 5 μm. (From McLachlan et al. 1977, with permission.) 4-24: Mid diakinesis in *Ptilota pectinata* ($n = $ ca. 32); arrow indicates single large bivalent. Hematoxylin. Scale bar = 9 μm. (From Yabu 1979, with permission.) 4-25: Diakinesis in *Laurencia okamurae* ($n = 32$); arrows indicate three small bivalents. Hematoxylin. Scale bar = 10 μm. (From Yabu 1985, with permission)

similar in form but have different chromosome numbers. Earlier, chromosome numbers assisted in discriminating between several nearly indistinguishable species: *P. abbottae* ($n = 3$) and *P. perforata* ($n = 2$) (Mumford & Cole 1977); *P. perforata* ($n = 2$) and *P. torta* ($n = 3$) (Mumford 1975; Mumford & Cole 1977); and *P. lanceolata* ($n = 3$) and *P. pseudolanceolata* ($n = 5$) (Mumford & Cole 1977).

From the results of several studies it is clear that discrepancies in chromosome numbers for certain morphologically equivalent taxa may be explained by the fact that field material was not collected at exactly the same time or place, even within a particular site, and that the karyotype may have changed following some form of isolation. For example, Kapraun (1977) determined different chromosome numbers for isolates of *Polysiphonia ferulacea* from North Carolina ($n = 30$) and Bermuda ($n = 27$). He also showed that they did not hybridize, indicating genetic distinctiveness. Similarly, material of Euro-

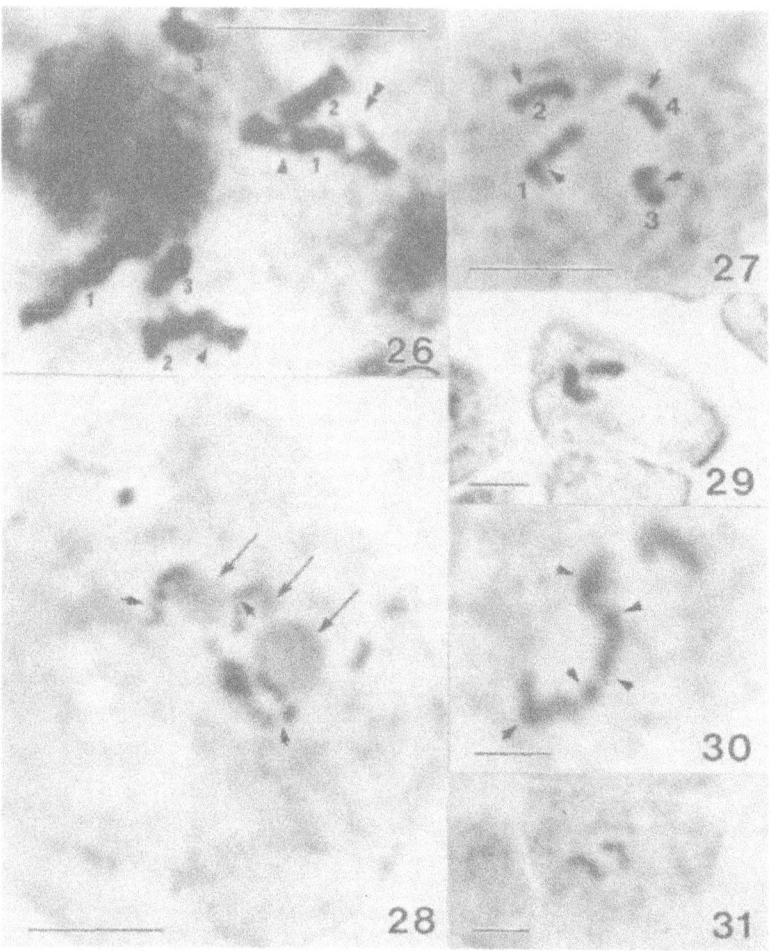

Figs. 4-26 to 4-31. Micrographs showing morphology of mitotic bangiacean chromosomes. Hematoxylin, except Fig. 4-30. *4-26, 4-27*: Subacrocentric positions (single arrowheads) in chromosomes of *Bangia vermicularis*. Scale bars = 4 μm. (From Cole et al. 1983, with permission) *4-26*: Six chromosomes (three pairs) in division preceding monospore production in asexual filament; also note nucleolar organizing region (double arrowhead). *4-27*: Four chromosomes in spermatangial cell of male filament. *4-28*: Three chromosomes in spermatangial cell of *Porphyra papenfussii*: note centromere positions (arrowheads), nucleolus located at the nucleolar organizing region of each chromosome (arrows), two upper subacrocentric chromosomes, and large lower submetacentric chromosome. Scale bar = 6 μm. (From Conway & Cole 1973, with permission) *4-29, 4-30*: Two large chromosomes in spermatangium of *P. schizophylla*. Scale bars = 3 μm. *4-29*: Hematoxylin. *4-30*: Banding technique: Giemsa stain following softening fixed material in 50% chloral hydrate, smearing on slides and then pretreating in trypsin; arrows indicate some of the bands. (Fig. 4-29 from Mumford & Cole 1977, with permission; Fig. 4-30 from Cole unpubl.) *4-31*: Two chromosomes in spermatangium of *P. perforata*: compare size to larger chromosomes in *P. schizophylla* at similar prophase stage shown in 4-29. Scale bar = 3 μm. (From Mumford & Cole 1977, with permission)

pean *Lomentaria orcadensis* collected at Bristol and Morlaix differed in chromosome number, $n = 10$ and 20, respectively (Magne 1964). Yabu (1976) established two different races, represented by haploid chromosome numbers 21 and 26, for *Palmaria* (as *Rhodymenia*) *palmata* collected in various places around Hakodate, Japan. In systematic studies including morphological and cytological aspects of *Bangia* populations in North America, Sheath and Cole (1984) observed a north-to-south trend of increased filament size for *B. vermicularis* on the west coast, coupled with different haploid chromosome numbers, 3 and 4 (Fig. 4-3), and suggested that more than one taxon is present.

Concerning species relationships in the red algae, Rao et al. (1978) indicated that, in addition to numbers, chromosome sizes and meiotic pairing phenomena provide information useful in evaluating cytogenetic similarities and differences between species. However, to date, morphological features of chromosomes have played a limited role in taxonomic investigations; there are relatively few records. In their studies of northeastern Pacific *Porphyra* species, Mumford and Cole (1977) noted that chromosomes of *P. schizophylla* ($n = 2$) are distinctive (Fig. 4-29): they are twice the length and width of those in all others, including *P. perforata*, which has the same chromosome number (Fig. 4-31). Regarding $n = 4$ species from the North Atlantic and Mediterranean, chromosomes of *P. umbilicalis* are larger than those in *P. carolinensis*, *P. leucosticta*, and *P. spiralis* var. *amplifolia* (Fig. 4-4; Kapraun & Freshwater 1987). Within the Ceramiales, chromosomes of *Ceramium gracillimum* are larger than those of *C. fimbriatum* and *C. cruciatum* (Rao et al. 1978); chromosomes of *Sorella repens* are larger than those in *C. fastigiatum* (Yabu et al. 1981); and, as mentioned in Section IV, several species of *Laurencia* can be distinguished on the basis of one large or one to three very small chromosomes (Yabu 1985).

B. Karyotype evolution

Karyotypes evolve as chromosomes undergo translocations, inversions, fusions, deletions, and nondisjunction, often providing valuable signposts conveying some information on evolutionary processes and trends. For example, in flowering plants the predominant trend that accompanies speciation and specialization is toward increased asymmetry of the karyotype, which becomes more heterogeneous through the shift of centromere positions, change in relative sizes of chromosomes in the complement, and/or change in chromosome number (Stebbins 1971). Although investigations of karyotype evolution in the Rhodophyta tend to be handicapped by the presence of many small, spherical chromosomes, several recent studies have utilized comparative karyogram analyses successfully.

Noting haploid chromosome numbers 2, 3, 4, 5, 6, and 7 in worldwide *Porphyra* species, Krishnamurthy (1984) speculated that the basic number for this genus is the most common, 3 (Table 4-1), and that the others evolved from this by loss or gain of chromosomes during irregular separations in mitosis, chromosome doubling, or hybridization. A similar basic chromosome number for Japanese and northeastern Pacific *Porphyra* species had been suggested earlier by Yabu (1975) and Mumford and Cole (1977). More recently, Kapraun and Freshwater (1987) proposed that the basic haploid chromosome number for North Atlantic *Porphyra* species is 4, because it occurs in four of the five species studied (Fig. 4-4) and that the 3-chromosome karyotype for the fifth species arose by deletion of the shortest chromosome, number 4. Chromosomes 1, 2, and 3 in all five species have similar relative sizes.

In an extensive cytological study of *Bangia* over a wide geographical area of the northeastern Pacific, sexual populations with $n = 3$ and $n = 4$ and asexual populations with $n = 3$ and $2n = 6$ were identified (Figs. 4-6, 4-7, 4-26, 4-27; Cole et al. 1983). According to karyotype analysis (Fig. 4-3), it appears that the basic number may be 3; chromosomes 2, 3, and 4 in $n = 4$ sexual populations are similar to 1, 2, and 3 in the other three population types. The longest chromosome in the $n = 4$ sexual populations, number 1, could have originated by breakage and fusion, possibly followed by nondisjunction.

D. F. Kapraun (pers. comm.) observed differences in chromosome size and number in karyograms of *Polysiphonia* species from the North Atlantic. For example, while *P. urceolata* (Fig. 4-5A) has the basic haploid number ($n = 30$) and symmetric chromosome complement characteristic of primitive *Polysiphonia* karyotypes, karyotype repatterning is evident in the asymmetric chromosome complement of *P. denudata* ($n = 30$; Fig. 4-5B) and the aneuploidy of *P. binneyi* ($n = 28$; Fig. 4-5C). He suggests the evolution of increasingly asymmetric karyotypes by centric fission and fusion, inversions and unequal translocations, as well as by meiotic nondisjunction.

In recent transatlantic hybridization experiments on *Palmaria palmata*, van der Meer (1987) observed poor viability of tetraspores, abnormal sporelings, and chromosome rings and chains during meiosis in hybrid tetrasporangia. The latter are indicative of the presence of heterozygous chromosome translocations in the hybrids and the divergence of karyotypes accompanying a splitting of species in the widely separated Irish and Canadian populations.

VI. SUMMARY

The current list of chromosome numbers in the Rhodophyta (Table 4-1) contains more than double the number of species included in previous lists published approximately 25 years ago (Dixon 1966; Magne 1964). However, in the larger orders Ceramiales and Gigartinales, only 18% and 10% of the genera, respectively, are represented. Numbers are still required for approximately 95% of the currently recognized species in this primitive algal division.

Generally, red algal chromosomes are minute (ca. 0.25–3.5 µm), and numbers tend to be variable; data are too fragmentary to establish definite trends within and between genera and orders. Haploid chromosome numbers range widely and appear to fall into three groups: low [2–10(15)], medium (ca. 10–25), and high (ca. 25–72), possibly reflecting the morphological and reproductive complexities of species represented in each group.

Polyploidy and aneuploidy are now documented in several genera, particularly within the Ceramiales and Gigartinales. According to Drew in 1944, there was a primary need at that time to reexamine material collected over a wide geographical range. This still holds more than 45 years later and, in addition, more individuals in more populations should be studied. The construction of polyploids in culture has been achieved in *Gracilaria*, and the role of endopolyploidy during development and differentiation in red algae, documented by light microscopy and microspectrofluorometry, is receiving more attention.

A better understanding of life histories and taxonomy of a number of red algae has been gained from chromosome data obtained in the past 15 years. In conjunction with culture work, these data provided comprehensive accounts of life histories for several species, most notably a new rhodophytan life history type in the Palmariaceae. However, in approximately half the orders, chromosome numbers are available for relatively few life history stages.

Chromosome information was used as a critical taxonomic feature in distinguishing morphologically similar species within several genera, including *Gracilaria* and *Porphyra*, and also in noting that morphologically similar populations of one species may exhibit karyotype changes. Cytogenetic similarities and differences between a few species in several genera in the Bangiales and Ceramiales were estimated from observations of meiotic pairing and relative chromosome sizes. In spite of difficulties dealing with very small chromosomes, recent detailed studies of *Bangia*, *Porphyra*, and *Polysiphonia* provided data leading to speculations on mechanisms of karyotype evolution within these genera, such as chromosome deletion, centric fission and fusion, inversions, and translocations, as well as meiotic nondisjunction.

As red algal methodology advances, further requirements for chromosome data will emerge. To illustrate, protoplasts of several commercially valuable seaweeds are being isolated and cultured with a view to improving strains by somatic hybridization and transformation (e.g., Cheney et al. 1986; Fujita & Migita 1985). Nuclear studies conducted in conjunction with these investigations would determine whether in vitro propagation of algal cells causes abnormal mitosis and chromosome breakage similar to that detected in cultured higher plant protoplasts (Chaleff 1981).

VII. ACKNOWLEDGMENTS

The author is most grateful to A. P. Austin, Y. B. K. Chowdary, J. Coll, L. J. Goff, L. A. Hanic, M. W. Hawkes, F. Magne, A. Miura, E. C. de Oliveira, J. C. Oliveira, J. P. van der Meer, J. A. West, and H. Yabu for their kind cooperation in making available published material and photographs for reproduction. Special thanks to D. F. Kapraun and A. Whittick who provided unpublished data and illustrative material so generously, to P. W. Gabrielson for carefully checking the status of species in Table 4-1, to B. R. Oates for capable assistance with the illustrative material, and to P. M. Taylor for assistance in reference searching and preparing the table and reference list. Appreciation is expressed to colleagues, postdoctorates, graduate students, and research assistants for their cooperation and interest in "Cole's Chromosome World" over the years, and to N.S.E.R.C.C. (operating grant 0645) for continued financial support of the author's research.

VIII. REFERENCES

Akatsuka, I. 1986. Japanese Gelidiales (Rhodophyta), especially *Gelidium*. *Oceanogr. Mar. Biol. Annu. Rev.* 24: 171–263.

Altman, P. L. & Dittmer, D. S. (eds.) 1962. *Growth Including Reproduction and Morphological Development*. Washington, D.C.: Fed. Amer. Soc. Exper. Biol.

Annapurna, B. & Rao, B. G. S. 1985. Cytological studies in Rhodophyceae: meiosis in *Polysiphonia ferulacea*. *Proc. All*

India Mar. Plants 1985: 103–8.
Annapurna, B. & Rao, B. G. S. 1987. Cytological studies in *Laurencia nana* and their significance in the evolutionary system of the red algae. *Ind. J. Bot.* 10: 89–92.
Athanasiadis, A. 1983. The life history of *Antithamnion heterocladum* (Rhodophyta, Ceramiales) in culture. *Bot. Mar.* 26: 153–7.
Athanasiadis, A. 1985. The taxonomic recognition of *Pterothamnion crispum* (Rhodophyta, Ceramiales), with a survey of the carposporophyte position in genera of the Antithamnieae. *Br. Phycol. J.* 20: 381–9.
Arrighi, F. E. & Hsu, T. C. 1971. Localization of heterochromatin in human chromosomes. *Cytogenetics* 10: 81–6.
Austin, A. P. 1955. Meiosis in *Furcellaria fastigiata* (L.). *Nature* (Lond.) 175: 905.
Austin, A. P. 1956. Chromosome counts in the Rhodophyceae. *Nature* (Lond.) 178: 370–1.
Austin, A. P. 1957. Studies in the autecology and cytology of *Furcellaria fastigiata* (L.) Lamour. *Br. Phycol. Bull.* 1(5): 16–17.
Austin, A. P. 1959. Iron-alum aceto-carmine staining for chromosomes and other anatomical features of Rhodophyceae. *Stain. Technol.* 34: 69–75.
Austin, A. P. 1960a. Life history and reproduction of *Furcellaria fastigiata* (L.) Lam. 1. The haploid plants and the development of the carposporophyte. *Ann. Bot.* N.S. 24: 257–74.
Austin, A. P. 1960b. Life history and reproduction of *Furcellaria fastigiata* (L.) Lam. 2. The tetrasporophyte and reduction division in the tetrasporangium. *Ann. Bot.* N.S. 24: 296–310.
Avila, M., Santelices, B. & McLachlan, J. 1986. Photoperiod and temperature regulation of the life history of *Porphyra columbina* (Rhodophyta, Bangiales) from central Chile. *Can. J. Bot.* 64: 1867–72.
Balakrishnan, M. S. 1947. The morphology and cytology of *Melobesia farinosa* Lamour. *J. Ind. Bot. Soc.* (Iyengar Commemoration Volume): 305–19.
Balakrishnan, M. S. & Chaugule, B. B. 1980. Cytology and life history of *Batrachospermum mahabaleshwarensis*. *Cryptogam. Algol.* 1: 83–97.
Barber, H. N. 1947. Genetics and algal life cycles. *Aust. J. Sci.* 9: 217–18.
Berlyn, G. P. & Miksche, J. P. 1976. *Botanical Microtechnique and Cytochemistry*. Ames: Iowa State University Press.
Bird, C. J. & McLachlan, J. 1982. Some underutilized taxonomic criteria in *Gracilaria* (Rhodophyta, Gigartinales). *Bot. Mar.* 25: 557–62.
Bird, C. J., McLachlan, J. & de Oliveira, E. C. 1986. *Gracilaria chilensis* sp. nov. (Rhodophyta, Gigartinales), from Pacific South America. *Can. J. Bot.* 64: 2928–34.
Bird, C. J., van der Meer, J. P. & McLachlan, J. 1982. A comment on *Gracilaria verrucosa* (Huds.) Papenf. (Rhodophyta: Gigartinales). *J. Mar. Biol. Assoc. U.K.* 62: 453–9.
Bodard, M. 1966. Sur le développement des tétrasporocystes d'*Anatheca montagnei* Schmitz (Soliériacees, Gigartinales). *Bull. Inst. Fond. Afrique Noire* 28: 867–94.
Bodard, M. 1971. Étude morphologique et cytologique d'*Helminthociadia senegalensis* (Rhodophycées), Némalionale nouvelle à carpotétraspores et à cycle haplodiplophasique. *Phycologia* 10: 361–74.
Boillot, A. 1963. Recherches sur le mode de développement des spores du genre *Gelidium* (Rhodophycées, Gélidiales). *Rev. Gén. Bot.* 70: 130–7.
Boillot, A. 1969. Sur le cycle de *Rhodochaete parvula* Thuret. *C.R. Acad. Sci.* (Paris) sér, D, 269: 2205–7.
Boillot, A. 1971. Sur le cycle d'*Helminthocladia calvadosii* (Lamouroux) Setchell. *Soc. Phycol. Fr.* 16: 106–10.
Boillot, A. 1972. Cycle biologique de quelques Némalionales. *Botaniste* 55: 207–50.
Boillot, A. 1975. Cycle biologique de *Rhodochaete parvula* (Thuret) (Rhodophycées, Bangiophycidées). *Pubbl. Staz. Napoli* 39 (Suppl.): 67–83.
Bold, H. C. 1951. Cytology of algae. In *Manual of Phycology*, ed. G. M. Smith, pp. 203–27. Waltham, Mass.: Chronica Botanica.
Brown, W. V. 1972. *Textbook of Cytogenetics*. St. Louis, Mo.: Mosby.
Burzycki, G. M. & Waaland, J. R. 1987. On the position of meiosis in the life history of *Porphyra torta* (Rhodophyta). *Bot. Mar.* 30: 5–10.
Cave, M. S. (ed.) 1958–60. *Index to Plant Chromosome Numbers. Vol. I (Nos. 1–4 & Suppl.)*. Chapel Hill: University of North Carolina Press.
Cave, M. S. (ed.) 1961–65. *Index to Plant Chromosome Numbers. Vol. II (Nos. 5–9)*. Chapel Hill: University of North Carolina Press.
Celan, M. 1941. Recherches cytologiques sur les Algues rouges. *Rev. Cytol. Cytophysiol. Végét.* 5: 1–168.
Chaleff, R. S. 1981. *Genetics of Higher Plants. Applications of Cell Culture*. Cambridge: Cambridge University Press.
Chen, L. C-M., Edelstein, T., Bird, C. & Yabu, H. 1978. A culture and cytological study of the life history of *Nemalion helminthoides* (Rhodophyta, Nemaliales). *Proc. N. S. Inst. Sci.* 28: 191–9.
Cheney, D. P., Mar, E., Saga, N. & van der Meer, J. 1986. Protoplast isolation and cell division in the agar-producing seaweed *Gracilaria* (Rhodophyta). *J. Phycol.* 22: 238–43.
Claussen, H. 1929. Zur Entwicklungsgeschichte von *Phyllophora brodiaei*. *Ber. Dtsch. Bot. Ges.* 47: 544–7.
Cleland, R. E. 1919. The cytology and life-history of *Nemalion multifidum*. Ag. *Ann. Bot.* 33: 323–51.
Cole, K. 1962. Aceto-carmine stain in the cytogenetical investigation of some marine algae of the Pacific coast. *Proc. Ninth Pacific Sci. Cong.*, 1957. 4: 313–15.
Cole, K. 1972. Observations on the life history of *Bangia fuscopurpurea*. *Soc. Bot. Fr., Mém.* 1972: 231–6.
Cole, K. M., Hymes, B. J. & Sheath, R. G. 1983. Karyotypes and reproductive seasonality of the genus *Bangia* (Rhodophyta) in British Columbia, Canada. *J. Phycol.* 19: 136–45.
Cole, K. & Sheath, R. G. 1980. Ultrastructural changes in major organelles during spermatial differentiation in *Bangia* (Rhodophyta). *Protoplasma* 102: 253–79.
Coll, J. & Oliveira, E. C. de. 1977a. Chromosome counting on 79-year-old dried seaweed, *Porphyra leucosticta* (Rhodophyta). *Experientia* 33: 102.
Coll, J. & Oliveira, E. C. de. 1977b. The nuclear state of "reproductive" cells of *Porphyra leucosticta* Thuret *in* Le Jolis (Rhodophyta, Bangiales). *Phycologia* 16: 227–9.
Conway, E. & Cole, K. M. 1973. Observations on an unusual form of reproduction in *Porphyra* (Rhodphyceae, Bangiales). *Phycologia* 12: 213–25.
Conway, E. Mumford, T. F., Jr. & Scagel, R. F. 1975. The genus *Porphyra* in British Columbia and Washington. *Syesis* 8: 185–244.
Conway, E. & Wylie, A. P. 1972. Spore organization and reproductive modes in two species of *Porphyra* from New Zealand. *Proc. Int. Seaweed Symp.* 7: 105–7.
Cordeiro-Marino, M. & Fujii, M. T. 1985. *Laurencia*

catarinensis (Rhodomelaceae, Rhodophyta), a new species from Ilha de Santa Catarina, Brazil. *Revta Brasil. Bot.* 8: 47–53.

Cordeiro-Marino, M., Fujii, M. T. & Yamaguishi-Tomita, N. 1983. Morphological and cytological studies on Brazilian *Laurencia* 1: *L. arbuscula* Sonder (Rhodomelaceae, Rhodophyta). *Rickia* 10: 29–39.

Cordeiro-Marino, M., Yamaguishi-Tomita, N. & Yabu, H. 1974. Nuclear divisions in the tetrasporangia of *Acanthophora spicifera* (Vahl) Boergesen and *Laurencia papillosa* (Forsk.) Greville. *Bull. Fac. Fish., Hokkaido Univ.* 25: 79–81.

Dammann, H. 1930. Entwicklungsgeschichtliche und zytologische Untersuchungen an Helgoländer Meeresalgen. *Helgol. Meeresunters.* 18: 1–36.

Dangeard, P. 1927. Recherches sur les *Bangia* et les *Porphyra*. *Botaniste* 18: 183–244.

Del Grosso, F. 1978. Studio cariologico in *Lemanea fluviatilis* Ag. *Inform. Bot. Ital.* 10: 238–9.

Del Grosso, F. 1981. Studio cariologico in *Sirodotia huillensis* (Welw.) Skuja (Rhodophyta). *Inform. Bot. Ital.* 13: 126–7.

Del Grosso, F. & Pogliani, M. 1977. Studio cariologico in *Batrachospermum moniliforme* Roth. *Inform. Bot. Ital.* 9: 36–7.

de Wet, J. M. J. 1980. Origins of polyploids. In *Polyploidy: Biological Relevance*, ed. W. H. Lewis, pp. 3–15. New York: Plenum.

Dixon, P. S. 1954. Nuclear observations of two British species of *Gelidium*. *Br. Phycol. Bull.* 1(2): 4. (Abstract)

Dixon, P. S. 1963a. Variation and speciation in marine Rhodophyta. In *Speciation in the Sea*, Publ. No. 5, eds. J. P. Harding & N. Tebble, pp. 51–62. London: Systematics Association.

Dixon, P. S. 1963b. The Rhodophyta: some aspects of their biology. *Oceanogr. Mar. Biol. Annu. Rev.* 1: 177–96.

Dixon, P. S. 1966. The Rhodophyceae. In *The Chromosomes of the Algae*, ed. M.B.E. Godward, pp. 168–204. New York: St. Martin's.

Dixon, P. S. 1970. The Rhodophyta: some aspects of their biology. II. *Oceanogr. Mar. Biol. Annu. Rev.* 8: 307–52.

Dixon, P. S. 1973. *Biology of the Rhodophyta*. Edinburgh: Oliver and Boyd.

Dixon, P. S. 1982. Life histories in the Florideophyceae with particular reference to the Nemaliales *sensu lato*. *Bot. Mar.* 25: 611–21.

Doubt, D. G. 1935. Notes on two species of *Gymnogongrus*. *Am. J. Bot.* 22: 294–310.

Drew, K. M. 1934. Contributions to the cytology of *Spermothamnion turneri* (Mert.) Aresch. I. The diploid generation. *Ann. Bot.* 48: 549–73.

Drew, K. M. 1937. *Spermothamnion snyderae* Farlow, a floridean alga bearing polysporangia. *Ann. Bot.* N.S. 1: 463–76.

Drew, K. M. 1939. An investigation of *Plumaria elegans* (Bonnem.) Schmitz with special reference to triploid plants bearing parasporangia. *Ann. Bot.* N.S. 3: 347–67.

Drew, K. M. 1943. Contributions to the cytology of *Spermothamnion turneri* (Mert.) Aresch. II. The haploid and triploid generations. *Ann. Bot.* N.S. 7: 23–30.

Drew, K. M. 1944. Nuclear and somatic phases in the Florideae. *Biol. Rev.* 19: 105–20.

Drew, K. M. 1955. Life histories in the algae with special reference to the Chlorophyta, Phaeophyta and Rhodophyta. *Biol. Rev.* 30: 343–90.

Dunn, G. A. 1917. Development of *Dumontia filiformis*. II. Development of sexual plants and general discussion of results. *Bot. Gaz.* 63: 425–67.

Evans, L. V. 1965. Cytological studies in the Laminariales. *Ann. Bot.* N.S. 29: 541–62.

Faridi, M. A. F. 1971. Occurrence of meiosis in *Batrachospermum moniliforme*. *Biologia* 17: 113–14.

Feldmann, J. 1952. Les cycles de reproduction des algues et leurs rapports avec la phylogénie. *Rev. Cytol.* (Paris) 13: 1–49.

Fralick, J. E. & Cole, K. 1973. Cytological observations on two species of *Iridaea* (Rhodophyceae, Gigartinales). *Syesis* 6: 271–2.

Freshwater, D. W. & Kapraun, D. F. 1986. Field, culture and cytological studies of *Porphyra carolinensis* Coll et Cox (Bangiales, Rhodophyta) from North Carolina. *Jap. J. Phycol.* 34: 251–62.

Fujita, Y. & Migita, S. 1985. Isolation and culture of protoplasts from some seaweeds. *Bull. Fac. Fish., Nagasaki Univ.* 57: 39–45 (in Japanese).

Fujiyama, T. 1957. Cytological studies on the crown-gall disease of *Porphyra tenera* Kjellm. *Symposium of Fisheries Studies*: 829–4 (in Japanese). (Tokyo Univ. Press)

Fujiyama, T., Kabutan, J. & Fujiyama, K. 1955. Cytological studies on *Porphyra* I & II. *Report of Annual Meeting of the Japanese Society of Scientific Fisheries, April 1955*: 22 (in Japanese).

Fujiyama, T. & Yokouchi, Y. 1955. Cytological studies on *Porphyra*, III. *Report of Annual Meeting of the Japanese Society of Scientific Fisheries, April 1955*: 22 (in Japanese).

Gabrielson, P. W. & Garbary, D. 1986. Systematics of red algae (Rhodophyta). *C R C Crit. Rev. Plant Sci.* 3: 325–66.

Gabrielson, P. W., Garbary, D. J. & Scagel, R. F. 1985. The nature of the ancestral red alga: inferences from a cladistic analysis. *Bio Systems* 18: 335–46.

Garbary, D. J., Hansen, G. I. & Scagel, R. F. 1980. A revised classification of the Bangiophyceae (Rhodophyta). *Nova Hedwigia* 33: 145–66.

Giraud, A. & Magne, F. 1968. La place de la méiose dans le cycle de développement de *Porphyra umbilicalis*. *C. R. Acad. Sci.* (Paris) sér. D, 267: 586–8.

Goff, L. J. & Cole, K. 1973. The biology of *Harveyella mirabilis* (Cryptonemiales, Rhodophyceae). I. Cytological investigations of *Harveyella mirabilis* and its host, *Odonthalia floccosa*. *Phycologia* 12: 237–45.

Goff, L. J. & Coleman, A. W. 1985. The role of secondary pit connections in red algal parasitism. *J. Phycol.* 21: 483–508.

Goff, L. J. & Coleman, A. W. 1986. A novel pattern of apical cell polyploidy, sequential polyploidy reduction and intercellular nuclear transfer in the red alga *Polysiphonia*. *Am. J. Bot.* 73: 1109–30.

Goff, L. J. & Coleman, A. W. 1987. The solution to the cytological paradox of isomorphy. *J. Cell. Biol.* 104: 739–48.

Goldblatt, P. (ed.) 1981. *Index to Plant Chromosome Numbers 1975–1978*. St. Louis, Mo.: Missouri Bot. Garden Monograph.

Goldblatt, P. (ed.) 1984. *Index to Plant Chromosome Numbers 1979–1981*. St. Louis, Mo.: Missouri Bot. Garden Monograph.

Gregory, B. D. 1930. New light on the so-called parasitism of *Actinococcus aggregatus* Kütz., and *Sterrocolax decipiens* Schmitz. *Ann. Bot.* 44: 767–9.

Gregory, B. D. 1934. On the life-history of *Gymnogongrus griffithsiae* Mart. and *Ahnfeltia plicata* Fries. *Linn. Soc. Lond. Bot.* 49: 531–51.

Grieg-Smith, E. 1954. Cytological observations on *Gracilaria multipartita*. *Br. Phycol. Bull.* 1(2): 4–5.

Grubb, V. M. 1925. The male organs of the Florideae. *J. Linn. Soc. (Bot.)* 47: 177–255.

Hanic, L. A. 1973. Cytology and genetics of *Chondrus crispus* Stackhouse. *Proc. N.S. Inst. Sci.* 27 (Suppl.): 23–52.

Harris, R. E. 1962. Contribution to the taxonomy of *Callithamnion* Lyngbye emend. Naegeli. *Bot. Not.* 115: 18–28.

Hassinger-Huizinga, H. 1952. Generationswechsel und Geschlechtsbestimmung bei *Callithamnion corymbosum* (Sm.) Lyngb. *Arch. Protistenk.* 98: 91–124.

Hawkes. M. W. 1977. A field, culture and cytological study of *Porphyra gardneri* (Smith & Hollenberg) comb. nov., (= *Porphyrella gardneri* Smith & Hollenberg), (Bangiales, Rhodophyta). *Phycologia* 16: 457–69.

Hawkes, M. W. 1978. Sexual reproduction in *Porphyra gardneri* (Smith *et* Hollenberg) Hawkes (Bangiales, Rhodophyta). *Phycologia* 17: 329–53.

Hymes B. J. & Cole, K. M. 1983. The cytology of *Audouinella hermannii* (Rhodophyta, Florideophyceae). I. Vegetative and hair cells. *Can. J. Bot.* 61: 3366–76.

Ishikawa, M. 1921. Cytological studies on *Porphyra tenera* Kjellm. I. *Bot. Mag. Tokyo* 35: 206–18.

Iyengar, M. O. P. & Balakrishnan, M. S. 1950. Morphology and cytology of *Polysiphonia platycarpa* Boergesen. *Proc. Indian Acad. Sci.* (B) 31: 135–61.

Jackson, R. C. 1971. The karyotype in systematics. *Annu. Rev. Ecol. System.* 2: 327–68.

Jackson, R. C. 1976. Evolution and systematic significance of polyploidy. *Annu. Rev. Ecol. System.* 7: 209–34.

Jónsson, S. & Chesnoy, L. 1982. Étude du cycle chromosomique de l'*Halosaccion ramentaceum* (Rhodophyta, Palmariales) d'Islande. *Cryptogam. Algol.* 3: 273–8.

Kaneko, T. 1966. Morphological and developmental studies of Gelidiales. I. Behaviour of the nucleus in early stages of tetraspore germination in *Gelidium vagum* Okamura. *Bull. Jap. Soc. Phycol.* 14: 62–70 (in Japanese).

Kaneko, T. 1968. Morphological and developmental studies of Gelidiales. II. On *Acanthopeltis japonica* Okamura. *Bull. Fac. Fish. Hokkaido Univ.* 19: 165–72.

Kapraun, D. F. 1977. Asexual propagules in the life history of *Polysiphonia ferulacea* (Rhodophyta, Ceramiales). *Phycologia* 16: 417–26.

Kapraun, D. F. 1978. A cytological study of varietal forms in *Polysiphonia harveyi* and *P. ferulacea* (Rhodophyta, Ceramiales). *Phycologia* 17: 152–6.

Kapraun, D. F. 1989. Karyological investigations of chromosome variation patterns associated with speciation in some Rhodophyta. In *Coastal Oceanography: North Carolina Model.* ed. E. Y. R. George, in press. Chapel Hill: University of North Carolina Press.

Kapraun, D. F. & Freshwater, D. W. 1987. Karyological studies of five species of *Porphyra* (Bangiales, Rhodophyta) from the North Atlantic and Mediterranean. *Phycologia* 26: 82–7.

Kapraun, D. F. & Luster, D. G. 1980. Field and culture studies of *Porphyra rosengurtii* Coll et Cox (Rhodophyta, Bangiales) from North Carolina. *Bot. Mar.* 23: 449–57.

Kito, H. 1966. Cytological studies of several species of *Porphyra*. I. Morphological and cytological observations on a species of *Porphyra* epiphytic on *Grateloupia filicina* var. *porracea* (Mert.) Howe. *Bull. Fac. Fish., Hokkaido Univ.* 16: 206–8 (in Japanese).

Kito, H. 1967. Cytological studies of several species of *Porphyra*. II. Mitosis in carpospore-germlings of *Porphyra yezoensis*. *Bull. Fac. Fish., Hokkaido Univ.* 18: 201–2 (in Japanese).

Kito, H. 1968. Cytological studies of several species of *Porphyra*. III. Chromosome number of *Porphyra tenera* Kjellman. *Bull. Fac. Fish., Hokkaido Univ.* 19: 137–40 (in Japanese).

Kito, H. 1974. Cytological observations on the conchocelis-phase in three species of *Porphyra*. *Bull. Tohoku Reg. Fish. Res. Lab.* 33: 101–17 (in Japanese).

Kito, H. 1978. Cytological studies on genus *Porphyra*. *Bull. Tohoku Reg. Fish. Res. Lab.* 39: 29–84 (in Japanese).

Kito, H., Ogata, E. & McLachlan, J. 1971. Cytological observations on three species of *Porphyra* from the Atlantic. *Bot. Mag. Tokyo* 84: 141–8.

Kito, H., Yabu H. & Tokida, J. 1967. The number of chromosomes in some species of *Porphyra*. *Bull. Fac. Fish., Hokkaido Univ.* 18: 59–62.

Knaggs, F. W. 1964. Cytological and life-history studies in the genus *Rhodochorton*. *Br. Phycol. Bull.* 2: 393.

Krishnamurthy, V. 1959. Cytological investigations on *Porphyra umbilicalis* (L.) Kütz. var. *laciniata* (Lightf.) J. Ag. *Ann. Bot.* N.S. 23: 147–76.

Krishnamurthy, V. 1984. Chromosome numbers in *Porphyra* C. Agardh. *Phykos* 23: 185–90.

Kylin, H. 1914. Studien über die Entwicklungsgeschichte von *Rhodomela virgata* Kjellm. *Sven. Bot. Tidskr.* 8: 33–69.

Kylin, H. 1916a. Die Entwicklungsgeschichte und die systematische Stellung von *Bonnemaisonia asparagoides* (Woodw.) Ag. nebst einigen Worten über den Generationswechsel der Algen. *Z. Bot.* 8: 545–86.

Kylin, H. 1916b. Die Entwicklungsgeschichte von *Griffithsia corallina* (Lightf.) Ag. *Z. Bot.* 8: 97–123.

Kylin, H. 1916c. Über die Befruchtung und Reduktionsteilung bei *Nemalion multifidum*. *Ber. Dtsch. Bot. Ges.* 34: 257–71.

Kylin, H. 1917. Über die Entwicklungsgeschichte von *Batrachospermum moniliforme*. *Ber. Dtsch. Bot. Ges.* 35: 155–64.

Kylin, H. 1923. Studien über die Entwicklungsgeschichte der Florideen. *K. Sven. Vetenskapsakad. Handl.* 63: 1–139.

Kylin, H. 1928. Entwicklungsgeschichtliche Florideenstudien. *K. Fysiogr. Sällsk. Handl.*, N.F. 39: 1–127.

Lewis, I. F. 1909. The life history of *Griffithsia bornetiana*. *Ann. Bot.* 23: 639–90.

Ma, J. H. & Miura, A. 1984. Observations of the nuclear division in the conchospores and their germlings in *Porphyra yezoensis* Ueda. *Jap. J. Phycol.* 32: 373–8 (in Japanese).

Maggs, C. A. 1988. A karyological study of life history in *Gymnogongrus* and *Mastocarpus* (Rhodophyta). *Br. Phycol. J.* 23: 293.

Magne, F. 1952. La structure du noyau et le cycle nucléaire chez le *Porphyra linearis* Greville. *C. R. Acad. Sci.* (Paris) sér. D, 234: 986–8.

Magne, F. 1959. Sur le cycle nucléaire du *Rhodymenia palmata* (L.) J. Agardh. *Soc. Phycol. Fr. Bull.* 5: 12–14.

Magne, F. 1960a. Sur le lieu de la méiose chez le *Bonnemaisonia asparagoides* (Woodw.) C. Ag. *C. R. Acad. Sci.* (Paris) sér. D, 250: 2742–4.

Magne, F. 1960b. Sur l'existence d'une reproduction sexuée chez le *Rhodochaete parvula* Thuret. *C. R. Acad. Sci.* (Paris) sér. D, 251: 1554–5.

Magne, F. 1960c. Le *Rhodochaete parvula* Thuret (Bangioidée) et sa reproduction sexuée. *Cah. Biol. Mar.* 1: 407–20.

Magne, F. 1961a. Sur la caryologie de deux Rhodophycées considérées jusquici comme à cycle cytologique entièrement haplophasique. *C. R. Acad. Sci.* (Paris) sér. D, 252: 4023–4.

Magne, F. 1961b Sur le cycle cytologique du *Nemalion helminthoides* (Velley) Batters. *C. R. Acad. Sci.* (Paris) sér. D, 252: 157–9.

Magne, F. 1964. Recherches caryologiques chez les Floridées (Rhodophycées). *Cah. Biol. Mar.* 5: 461–671.

Magne, F. 1967a. Sur l'existence, chez les *Lemanea* (Rhodophycées, Némalionales), d'un type de cycle de développement encore inconnu chez les algues rouges. *C. R. Acad. Sci.* (Paris) sér. D, 264: 2632–3.

Magne, F. 1967b. Sur le déroulement et le lieu de la méiose chez les Lémanéacées (Rhodophycées, Némalionales). *C. R. Acad. Sci.* (Paris) sér. D, 265: 670–3.

Magne, F. 1986. Anomalies du développement chez *Antithamnionella sarniensis* (Rhodophyceae, Ceramiaceae). II: nature des individus issus des tétraspores. *Cryptogam. Algol.* 7: 215–29.

Marshall, S. M., Newton, L. & Orr, A. P. 1949. *A Study of Certain British Seaweeds and Their Utilisation in the Preparation of Agar*. London: His Majesty's Stationery Office.

Martin, M. T. 1939a. The structure and reproduction of *Chaetangium saccatum* (Lamour.) J. Ag–II. Female plants. *Linn. Soc. Lond. Bot.* 52: 115–44.

Martin, M. T. 1939b. Some South African Rhodophyceae. – I. *Helminthocladia papenfussii* Kylin. *J. Bot.* 77: 234–44.

Mathias, W. T. 1927. The cytology of *Callithamnion*. *Brit. Assoc. Rept. 95th, Leeds*: 380.

Mathias, W. T. 1928. The cytology of *Callithamnion brachiatum* Bonnem. *Publ. Hartley Bot. Labs., Liverpool Univ.* 5: 5–27.

McLachlan, J., van der Meer, J. P. & Bird, N. L. 1977. Chromosome numbers of *Gracilaria foliifera* and *Gracilaria* sp. (Rhodophyta) and attempted hybridizations. *J. Mar. Biol. Assoc. U.K.* 57: 1137–41.

Migita, S. 1967. Cytological studies on *Porphyra yezoensis* Ueda. *Bull. Fac. Fish., Nagasaki Univ.* 24: 55–64.

Moore, R. J. (ed.) 1973. *Index to Plant Chromosome Numbers 1961–71*. Utrecht: Int. Bureau Plant Tax. Nomen.

Moore, R. J. (ed.) 1974. *Index to Plant Chromosome Numbers 1972*. Utrecht: Int. Bureau Plant Tax. Nomen.

Moore, R. J. (ed.) 1977. *Index to Plant Chromosome Numbers 1973/74*. Utrecht: Int. Bureau Plant Tax. Nomen.

Mullahy, J. H. 1952. The morphology and cytology of *Lemanea australis* Atk. *Bull. Torrey Bot. Club* 79: 471–84.

Mumford, T. F., Jr. 1973. A new species of *Porphyra* from the west coast of North America. *Syesis* 6: 239–42.

Mumford, T. F., Jr. 1975. Observations on the distribution and seasonal occurrence of *Porphyra schizophylla* Hollenberg, *Porphyra torta* Krishnamurthy, and *Porphyra brumalis* sp. nov. (Rhodophyta, Bangiales). *Syesis* 8: 321–32.

Mumford, T. F., Jr. & Cole, K. 1977. Chromosome numbers for fifteen species in the genus *Porphyra* (Bangiales, Rhodophyta) from the west coast of North America. *Phycologia* 16: 373–7.

Newroth, P. R. 1971. Studies on life histories in the Phyllophoraceae. I. *Phyllophora truncata* (Rhodophyceae, Gigartinales). *Phycologia* 10: 345–54.

Newroth, P. R. 1972. Studies on life histories in the Phyllophoraceae. II. *Phyllophora pseudoceranoides* and notes on *P. crispa* and *P. heredia* (Rhodophyta, Gigartinales). *Phycologia* 11: 99–107.

Nichols, H. W. 1964a. Developmental morphology and cytology of *Boldia erythrosiphon*. *Am. J. Bot.* 51: 653–9.

Nichols, H. W. 1964b. Culture and developmental morphology of *Compsopogon coeruleus*. *Am. J. Bot.* 51: 180–8.

Nichols, H. W. 1980. Polyploidy in Algae. In *Polyploidy: Biological Relevance*, ed. W. H. Lewis, pp. 151–61. New York: Plenum.

Notoya, M. & Yabu, H. 1979a. Culture and cytology of *Ceramium japonicum* Okamura and *C. kondoi* Yendo (Ceramiales, Rhodophyta). *Bull. Fac. Fish., Hokkaido Univ.* 30: 129–32 (in Japanese).

Notoya, M. & Yabu, H. 1979b. *Heterosiphonia pulchra* (Okamura) Falkenberg (Ceramiales, Rhodophyta) in culture. *Bull. Fac. Fish., Hokkaido Univ.* 30: 187–9 (in Japanese).

Ornduff, R. (ed.) 1967. *Index to Plant Chromosome Numbers 1965*. Utrecht: Int. Bureau Plant Tax. Nomen.

Pal, D. 1981. Comparative cytological observations on two genera of Rhodomelaceae. *New Bot.* 8: 105–8.

Patwary, M. V. & van der Meer, J. P. 1984. Growth experiments on autopolyploids of *Gracilaria tikvahiae* (Rhodophyceae). *Phycologia* 23: 21–7.

Petersen, H. E. 1928. Nogle lagttagelser over Cellekernerne hos *Ceramium* (Roth) Lyngbye. *Dansk. Bot. Ark.* 5: 1–5.

Pringle, J. D. & Austin, A. P. 1970. The mitotic index in selected red algae *in situ*. II. A supralittoral species, *Porphyra lanceolata* (Setchell & Hus.) G. M. Smith. *J. Exp. Mar. Biol. Ecol.* 5: 113–37.

Ramus, J. 1969. The developmental sequence of the marine red alga *Pseudogloiophloea* in culture. *Univ. Calif. Pub. Bot.* 52: 1–42.

Rao, B. G. S. & Annapurna, B. 1983. Cytological studies in Rhodophyceae: meiosis in *Laurencia nana*. *Adv. Biol. Res.* 1: 1–7.

Rao, B. G. S., Mantha, S. & Rao, M. U. 1978. Chromosome behaviour at meiosis and its bearing on the cytotaxonomy of *Ceramium* species. *Bot. Mar.* 21: 123–9.

Rao, B. G. S. & Sundari, M. 1977. Some observations on the cytology of *Ceramium fimbriatum*. *Curr. Sci.* 46: 88–9.

Rao, C. S. P. 1953. Acetocarmine as a nuclear stain in Rhodophyceae. *Nature* (Lond.) 172: 1197.

Rao, C. S. P. 1956. The life-history and reproduction of *Polyides caprinus* (Gunn.) Papenf. *Ann. Bot. N.S.* 20: 211–30.

Rao, C. S. P. 1959. Cytology of red algae. *Proc. Symp. Algol.* (UNESCO–ICAR, Delhi), pp. 37–45.

Rao, C. S. P. 1966. Observations on the cytology of *Compsopogon coeruleus* Mont. *Phykos* 5: 91–4.

Rao, C. S. P. 1967a. A contribution to the cytology of some red algae from the west coast of India. *Proc. Sem. Sea, Salt and Plants, Bhavnagar, India. 1965*: 209–16.

Rao, C. S. P. 1967b. On the cytology and life-history of *Leveillea jungermannioides* (Mart. et Her.) Harv. (Preliminary note). *Bot. Mar.* 10: 167–8.

Rao, C. S. P. 1970. Morphology of the chromosomes of red algae. *Indian Biol.* 2: 37–40.

Rao, C. S. P. 1971. Sex chromosomes of *Wrangelia argus* Mont. *Bot. Mar.* 14: 113–15.

Rao, C. S. P. & Gujarati, A. R. 1973. Second meiotic division in the tetrasporangium of *Chondria dasyphylla* (Woodw.) C. Ag. *Curr. Sci.* 42: 361–2.

Rao, C. S. P. 1974. Some observations on the cytology of *Gelidiella acerosa* (Forsskal) Feldmann et Hamel. *Cytologia* 39: 391–5.

Ravanko, O. 1987. Preliminary studies on cells and chromosomes, in *Polysiphonia violacea*. *Memo. Soc. Fauna Flora Fenn.* 63: 45–50.
Richardson, N. & Dixon, P. S. 1968. Life history of *Bangia fuscopurpurea* (Dillw). Lyngb. in culture. *Nature* (Lond.) 218: 496–7.
Rosenberg, T. 1933. Studien über Rhodomelaceen und Dasyaceen. *Akad. Abhandl. Lund.*
Rosenvinge, L. K. 1931a. The marine algae of Denmark. Part IV Rhodophyceae IV. (Gigartinales, Rhodymeniales, Nemastomatales). *D. K. Danske Vidensk. Selsk. Skrifter.* 7 Raekke, 7: 491–630.
Rosenvinge, L. K. 1931b. The reproduction of *Ahnfeltia plicata*. *D. K. Danske Vidensk. Selsk. Biolog. Meddel.* 10: 1–29.
Rueness, J. & Åsen, P. A. 1982. Field and culture observations on the life history of *Bonnemaisonia asparagoides* (Woodw.) C. Ag. (Rhodophyta) from Norway. *Bot. Mar.* 25: 577–87.
Rueness, J. & Rueness, M. 1973. Life history and nuclear phases of *Antithamnion tenuissimum*, with special reference to plants bearing both tetrasporangia and spermatangia. *Norw. J. Bot.* 20: 205–10.
Sarma, Y. S. R. K. 1983. Algal karyology and evolutionary trends. In *Chromosomes in Evolution of Eukaryotic Groups, Vol. I*. eds. A. K. Sharma & A. Sharma, pp. 177–223. Boca Raton, Fla.: CRC Press.
Schornstein, K. L. & Scott, J. 1982. Ultrastructure of cell division in the unicellular red alga *Porphyridium purpureum*. *Can. J. Bot.* 60: 85–97.
Schussnig, B. & Odle, L. 1927. Zur Frage des Generationswechsels bei *Spermothamnion roseolum* (Ag.) Pringsh. *Arch. Protistenk.* 58: 220–52.
Segawa, S. 1941. Systematic anatomy of the articulated Corallines (III). *J. Jap. Bot.* 17: 164–74 (in Japanese).
Sheath, R. G. 1984. The biology of freshwater red algae. *Prog. Phycol. Res.* 3: 89–157.
Sheath, R. G. & Cole, K. M. 1980. Distribution and salinity adaptations of *Bangia atropurpurea* (Rhodophyta), a putative migrant into the Laurentian Great Lakes. *J. Phycol.* 16: 412–20.
Sheath, R. G. & Cole, K. M. 1984. Systematics of *Bangia* (Rhodophyta) in North America. I. Biogeographic trends in morphology. *Phycologia* 23: 383–96.
Shyam, R. & Sarma, Y. S. R. K. 1980. Cultural observations on the morphology, reproduction and cytology of a freshwater red alga *Compsopogon* Mont. from India. *Nova Hedwigia* 32: 745–67.
Sommerfeld, M. R. & Nichols, H. W. 1970a. Developmental and cytological studies of *Bangia fuscopurpurea* in culture. *Am. J. Bot.* 57: 640–8.
Sommerfeld, M. R. & Nichols, H. W. 1970b. Comparative studies in the genus *Porphyridium* Naeg. *J. Phycol.* 6: 67–78.
South, G. R., Hooper, R. G. & Irvine, L. M. 1972. The life history of *Turnerella pennyi* (Harv.) Schmitz. *Br. Phycol. J.* 7: 221–33.
South, G. R. & Whittick, A. 1976. Aspects of the life history of *Rhodophysema elegans* (Rhodophyta, Peyssonneliaceae). *Br. Phycol. J.* 11: 349–54.
Sparling, S. R. 1961. A report on the culture of some species of *Halosaccion*, *Rhodymenia* and *Fauchea*. *Am. J. Bot.* 48: 493–9.
Stebbins, G. L. 1971. *Chromosomal Evolution in Higher Plants*. London: Edward Arnold.
Stewart, J. & Rüdenberg, L. 1978. Nuclear number in cells of some species of Delesseriaceae (Rhodophyta). *Br. Phycol. J.* 13: 391–401.
Sundari, M. & Rao, B. G. S. 1977. A cytological assessment of species delineation of two *Ceramiums* (Rhodophyceae). *Curr. Sci.* 46: 461–2.
Suneson, S. 1937. Studien über die Entwicklungsgeschichte der Corallinaceen. *K. Fysiogr. Sällsk. Handl.* 48: 1–101.
Suneson, S. 1945. On the anatomy, cytology and reproduction of *Mastophora*. With a remark on the nuclear conditions in the spermatangia of the Corallinaceae. *K. Fysiogr. Sällsk. Lund. Forh.* 15: 251–64.
Suneson, S. 1950. The cytology of the bispore formation in two species of *Lithophyllum* and the significance of the bispores in the Corallinaceae. *Bot. Not.* 4: 429–50.
Svedelius. N. 1911. Über den Generationswechsel bei *Delesseria sanguinea*. *Sven. Bot. Tidskr.* 5: 260–324.
Svedelius, N. 1912. Über die Spermatienbildung bei *Delesseria sanguinea*. *Sven. Bot. Tidskr.* 6: 239–65.
Svedelius, N. 1914a. Über die Zystokarpienbildung bie *Delesseria sanguinea*. *Sven. Bot. Tidskr.* 8: 1–32.
Svedelius, N. 1914b. Über die Tetradenteilung in den vielkernigen Tetrasporangiumaniagen bei *Nitophyllum punctatum*. *Ber. Dtsch. Bot. Ges.* 32: 48–57.
Svedelius, N. 1914c. Über Sporen an Geschlechtspflanzen von *Nitophyllum punctatum*: ein Beitrag zur Frage des Generationswechseis der Florideen. *Ber. Dtsch. Bot. Ges.* 32: 106–16.
Svedelius, N. 1915. Zytologisch-entwicklungsgeschichtliche Studien über *Scinaia furcellata*. Ein Beitrag zur Frage der Reduktionsteilung der nicht Tetrasporenbildenden Florideen. *Nova Acta Regiae Soc. Sci. Upsal.* Ser. IV. 4: 1–55.
Svedelius, N. 1933. On the development of *Asparagopsis armata* Harv. and *Bonnemaisonia asparagoides* (Woodw.) Ag. *Nova Acta Reg. Soc. Sci. Upsal.* 9: 1–61.
Svedelius, N. 1935. *Lomentaria rosea*, eine Floridee ohne Generationswechsel, nur mit Tetrasporenbildung ohne Reduktionsteilung. *Ber. Dtsch. Bot. Ges.* 53: (19)–(26).
Svedelius, N. 1937. The apomeiotic tetrad division in *Lomentaria rosea* in comparison with the normal development in *Lomentaria clavellosa*. *Symbol. Bot. Upsal.* 2: 1–54.
Svedelius, N. 1942. Zytologisch-entwicklungsgeschichtliche Studien über *Galaxaura* eine diplobiontische Nemalionales-Gattung. *Nova Acta Reg. Soc. Sci. Upsal.* 13: 1–154.
Svedelius, N. 1956. Are the haplobiontic Florideae to be considered reduced types? *Sven. Bot. Tidskr.* 50: 1–24.
Swanson, C. P., Merz, T. & Young, W. J. 1981. *Cytogenetics: The Chromosome in Division, Inheritance, and Evolution*. Englewood Cliffs, N.J.: Prentice-Hall.
Thirb, H. H. & Benson-Evans, K. 1982. Cytological studies on *Lemanea fluviatilis* L. in the River Usk. *Br. Phycol. J.* 17: 401–9.
Tözün, B. 1974. Nuclear division in the red algae *Chondria nidifica* and *Chondria tenuissima*. *Br. Phycol. J.* 9: 363–70.
Tseng, C. K. & Chang, T. J. 1955. Studies on *Porphyra* III. Sexual reproduction of *Porphyra*. *Acta Bot. Sin.* 4: 153–66 (in Chinese).
Tseng, C. K. & Sun, A. 1989. Studies on the alternation of the nuclear phases and chromosome numbers in the life history of some species of *Porphyra* from China. *Bot. Mar.* 32: 1–8.
van der Meer, J. P. 1976. A contribution towards elucidat-

ing the life history of *Palmaria palmata* (= *Rhodymenia palmata*). *Can. J. Bot.* 54: 2903–6.

van der Meer, J. P. 1981a. The life history of *Halosaccion ramentaceum*. *Can. J. Bot.* 59: 433–6.

van der Meer, J. P. 1981b. Genetics of *Gracilaria tikvahiae* (Rhodophyceae). VII. Further observations on mitotic recombination and the construction of polyploids. *Can. J. Bot.* 59: 787–92.

van der Meer, J. P. 1986. Genetics of *Gracilaria tikvahiae* (Rhodophyceae). XI. Further characterization of a bisexual mutant. *J. Phycol.* 22: 151–8.

van der Meer, J. P. 1987. Experimental hybridization of *Palmaria palmata* (Rhodophyta) from the northeast and northwest Atlantic Ocean. *Can. J. Bot.* 65: 1451–8.

van der Meer, J. P. & Bird, C. J. 1985. *Palmaria mollis* stat. nov.: a newly recognized species of *Palmaria* (Rhodophyceae) from the northeast Pacific Ocean. *Can. J. Bot.* 63: 398–403.

van der Meer, J. P. & Chen, L. C-M. 1979. Evidence for sexual reproduction in the red algae *Palmaria palmata* and *Halosaccion ramentaceum*. *Can. J. Bot.* 57: 2452–9.

van der Meer, J. P. & Patwary, M. V. 1983. Genetic modification of *Gracilaria tikvahiae* (Rhodophyceae). The production and evaluation of polyploids. *Aquaculture* 33: 311–16.

van der Meer, J. P. & Todd, E. R. 1977. Genetics of *Gracilaria* sp. (Rhodophyceae, Gigartinales). IV. Mitotic recombination and its relationship to mixed phases in the life history. *Can. J. Bot.* 55: 2810–17.

van der Meer, J. P. & Todd, E. R. 1980. The life history of *Palmaria palmata* in culture. A new type for the Rhodophyta. *Can. J. Bot.* 58: 1250–6.

von Stosch, H. A. 1969. Observations on *Corallina*, *Jania* and other red algae in culture. *Proc. Int. Seaweed Symp.* 6: 389–99.

Webster, R. N. 1958. The life history of the freshwater red alga *Tuomeya fluviatilis* Harv. *Butler Univ. Bot. Stud.* 13: 141–59.

West, J. A. 1970. The life history of *Rhodochorton concrescens* in culture. *Br. Phycol. J.* 5: 179–86.

West, J. A. & Hommersand, M. H. 1981. Rhodophyta: life histories. In *The Biology of the Seaweeds*, eds. C. S. Lobban & M. J. Wynne, pp. 133–93. Berkeley: University of California Press.

Westbrook, M. A. 1928. Contributions to the cytology of tetrasporic plants of *Rhodymenia palmata* (L.) Grev., and some other Florideae. *Ann. Bot.* 42: 149–72.

Westbrook, M. A. 1930. The structure of the nucleus in *Callithamnion* spp. *Ann. Bot.* 44: 1012–15.

Westbrook, M. A. 1935. Observations on nuclear structure in the Florideae. *Beih. Bot. Centralbl.* 53A: 564–85.

Whittick, A. 1977. The reproductive ecology of *Plumaria elegans* (Bonnem.) Schmitz (Ceramiaceae: Rhodophyta) at its northern limits in the western Atlantic. *Exp. Mar. Biol. Ecol.* 29: 223–30.

Whittick, A. & Hooper, R. G. 1977. The reproduction and phenology of *Antithamnion cruciatum* (Rhodophyta: Ceramiaceae) in insular Newfoundland. *Can. J. Bot.* 55: 520–4.

Wittmann, W. 1965. Aceto-iron-haematoxylin-chloral hydrate for chromosome staining. *Stain Technol.* 40: 161–4.

Woelkerling, W. J. 1970. *Acrochaetium botryocarpum* (Harv.) J. Ag. (Rhodophyta) in southern Australia. *Br. Phycol. J.* 5: 159–71.

Wolfe, J. J. 1904. Cytological studies on *Nemalion*. *Ann. Bot.* 18: 607–30.

Yabu, H. 1967. Nuclear division in *Bangia fuscopurpurea* (Dillwyn) Lyngbye. *Bull. Fac. Fish., Hokkaido Univ.* 17: 163–4.

Yabu, H. 1969a. Observations on chromosomes in some species of *Porphyra*. *Bull. Fac. Fish., Hokkaido Univ.* 19: 239–43.

Yabu, H. 1969b. Mitosis in *Porphyra tenera* Kjellm. *Bull. Fac. Fish., Hokkaido Univ.* 20: 1–3.

Yabu, H. 1970. Cytology in two species of *Porphyra* from the stipes of *Nereocystis luetkeana* (Mert.) Post. et Rupr. *Bull. Fac. Fish., Hokkaido Univ.* 20: 243–51.

Yabu, H. 1971. Observation on chromosomes in some species of *Porphyra* II. *Bull. Fac. Fish., Hokkaido Univ.* 21: 253–8.

Yabu, H. 1972a. Observation on chromosomes in some species of *Porphyra* III. *Bull. Fac. Fish., Hokkaido Univ.* 22: 261–6.

Yabu, H. 1972b. Nuclear division in tetrasporophytes of *Rhodymenia palmata* (L.) Grev. *Proc. Int. Seaweed Symp.* 7: 205–7.

Yabu, H. 1975. Cytological studies of the Rhodophyta and Chlorophyta. In *Advance of Phycology in Japan*, eds. J. Tokida & H. Hirose, pp. 125–35. The Hague, Netherlands: W. Junk.

Yabu, H. 1976. A report on the cytology of *Rhodymenia palmata*, *Rh. pertusa* and *Halosaccion saccatum* (Rhodophyta). *Bull. Fac. Fish., Hokkaido Univ.* 27: 51–62.

Yabu, H. 1978a. Chromosome numbers in species of *Porphyra* from Nova Scotia, Canada. *Jap. J. Phycol.* 26: 97–104.

Yabu, H. 1978b. Nuclear divisions in *Laurencia nipponica* Yamada. *Jap. J. Phycol.* 26: 35–9 (in Japanese).

Yabu, H. 1979. Cytological observations on *Ptilota pectinata* (Gunn.) Kjellm. and *Pt. pectinata* f. *litoralis* Kjellm. (Ceramiales, Rhodophyta). *Jap. J. Phycol.* 27: 17–24.

Yabu, H. 1985. Meiosis in three species of *Laurencia* (Ceramiales, Rhodophyta). *Jap. J. Phycol.* 33: 288–92.

Yabu, H., Cordeiro-Marino, M. & Yamaguishi-Tomita, N. 1974. Mitosis in two Brazilian species of *Porphyra*. *Rickia* 6: 21–5.

Yabu, H. & Kawamura, K. 1959. Cytological study on some Japanese species of Rhodomelaceae. *Mem. Fac. Fish., Hokkaido Univ.* 7: 61–72.

Yabu, H., Notoya, M. & Fukui, K. 1981. Nuclear divisions in *Ceramium fastigiatum* Harvey and *Sorella repens* (Okam.) Hollenberg (Ceramiales, Rhodophyta). *Bull. Fac. Fish., Hokkaido Univ.* 32: 221–4 (in Japanese).

Yabu, H. & Sanbonsuga, Y. 1981. A sex chromosome in *Cymathaere japonica* Miyabe et Nagai. *Jap. J. Phycol.* 29: 79–80.

Yabu, H. & Tokida, J. 1963. Mitosis in *Porphyra*. *Bull. Fac. Fish., Hokkaido Univ.* 14: 131–6.

Yabu, H. & Tokida, J. 1966. Application of aceto-iron-haematoxylin-chloral hydrate method to chromosome staining in marine algae. *Bot. Mag. Tokyo* 79: 381.

Yamanouchi, S. 1906a. The life history of *Polysiphonia violacea*. *Bot. Gaz.* 41: 425–33.

Yamanouchi, S. 1906b. The life history of *Polysiphonia violacea*. *Bot. Gaz.* 42: 401–49.

Yamanouchi, S. 1921. Life history of *Corallina officinalis* var. *mediterranea*. *Bot. Gaz.* 72: 90–6.

Yoshida, T. 1959. Life-cycle of a species of *Batrachospermum* found in northern Kyushu, Japan. *Jap. J. Bot.* 17: 29–42.

Yoshida, T., Nakajima, Y. & Nakata, Y. 1985. Preliminary

checklist of marine benthic algae of Japan–II. Rhodophyceae. *Jap. J. Phycol.* 33: 249–75.
Zhang, X. & van der Meer, J. P. 1988a. Polyploid gametophytes of *Gracilaria tikvahiae* (Gigartinales, Rhodophyta). *Phycologia* 27: 312–18.
Zhang, X. & van der Meer, J. P. 1988b. A genetic study on *Gracilaria sjoestedtii*. *Can. J. Bot.* 66: 2022–6.

Chapter 5

Genetics

JOHN P. VAN DER MEER

CONTENTS

I. Introduction / 103
II. Mutants and mutagenesis / 104
    A. Spontaneous mutations / 104
    B. Chemical mutagenesis / 104
III. Nuclear inheritance / 106
    A. Life history considerations / 106
    B. Classical genetics / 106
IV. Nonnuclear inheritance / 112
    A. Maternal transmission / 112
    B. Phenotypes of nonnuclear mutations / 113
V. Genetic regulation / 113
    A. Sex determination in *Gracilaria tikvahiae* / 113
    B. Phase determination in *Gracilaria tikvahiae* / 114
    C. Sex and phase determination in other red algal species / 114
VI. Molecular and somatic cell genetics / 115
    A. Molecular genetics / 115
    B. Somatic cell genetics / 117
VII. Quantitative genetics and plant breeding / 117
    A. Selection of *Porphyra* cultivars / 117
    B. Strain improvement in other red algal species / 118
VIII. Summary / 118
IX. Acknowledgments / 118
X. Endnote / 118
XI. References / 118

I. INTRODUCTION

The Rhodophyta are an ancient assemblage of algae that diverged from other phylogenetic lines hundreds of millions of years ago (Schopf 1970). The division now encompasses more than 5200 species (Dixon 1973), exhibiting a high degree of variation in morphology and reproduction. In view of this antiquity and diversity, it is reasonable to expect that the Rhodophyta will exhibit rich genetic variation and some unique genetic features; however, the extent to which this expectation will be fulfilled remains to be determined. Genetic information for the Rhodophyta is still very undeveloped in comparison to that available for favored species like the fruit fly *Drosophila melanogaster* or the bacterium *Escherichia coli*.

The late start for genetic studies of red algae derives from the fact that they were poorly known outside phycological circles, difficult to culture, and economically unimportant when pioneering genetic experiments were initiated several decades ago.

Over the years, some of the plants, animals and microorganisms that were selected for those early genetic experiments have become very well characterized. As a consequence, geneticists prefer to work with these organisms, examining new species only when there is a specific scientific or economic reason to do so. There has been no particularly compelling reason to study the Rhodophyta.

In *The Genetics of Algae* (Lewin 1976) the total genetic information for red algae was summarized in just two brief paragraphs (Fjeld & Løvlie 1976). Although genetic studies on red algae are still limited, the length of the present review indicates that there has been progress.

II. MUTANTS AND MUTAGENESIS

Mutations are the principal tools of a geneticist. They range from subtle DNA changes with no visible effect to changes that are lethal for the organism. Between these extremes are mutations that engender specific, visible changes in the phenotype of the affected organism. Mutations that provide distinctive, stable, phenotypic changes (e.g., Figs. 5-1 to 5-4), without seriously weakening the plant, are the most useful for genetic studies. Particularly useful are stable color mutants, which can be used as marker genes to detect sexual fertilization (van der Meer 1981c, 1987a; van der Meer & Todd 1980), to distinguish between sexual and asexual processes (van der Meer et al. 1984), and to distinguish between self- and cross-fertilization in monecious species (van der Meer 1987b). They can also draw attention to unanticipated aspects of algal life histories (e.g., Ohme et al. 1986; van der Meer 1977; van der Meer & Todd 1977).

A. Spontaneous mutations

Among red algae the frequency of mutants appears to vary from species to species. Color mutants of *Gracilaria tikvahiae* arise quite frequently in culture, approximately 1×10^{-4} for haploid sporelings (van der Meer 1979a). Similar mutants of *Chondrus crispus* are common in some intertidal populations, though more frequently as frond sectors than as whole plants (van der Meer 1981d). In contrast, spontaneous color mutants of *Palmaria palmata* and *Devaleraea ramentacea* are infrequent, both in culture and in the field. Spontaneous mutants of *Porphyra umbilicalis* may occur even less frequently; none has been observed among thousands of plants in culture (Mitman & van der Meer unpubl.). They have been found, however, for other *Porphyra* species (e.g., Miura 1977).

The frequency of spontaneous mutants also appears to vary from strain to strain within a species (e.g., Steele et al. 1986). Certain mutations strongly promote secondary mutations during somatic growth of fronds. For example, plants of *G. tikvahiae* that express the *bu* mutation (thin, highly dissected fronds), undergo frequent secondary mutations for color and morphology. Similar mutations arise on a morphological mutant of *G. secundata* (A. Lignell & M. Pedersen pers. comm.). At present it is not clear whether these plants have a higher mutation rate or whether their highly branched morphology simply facilitates the expression of mutant sectors.

Spontaneous mutations usually occur as rare events that yield a single altered plant or plant sector. Occasionally, however, a particular mutation arises repeatedly within a line, e.g., yellow sporelings of *Champia parvula* (Steele et al. 1986), suggesting an underlying mechanism different from those that cause isolated spontaneous mutations.

When spontaneous mutations arise during somatic growth of a plant, they often exhibit one of two sorting-out patterns for the segregation of mutant from normal tissue. In haploid plants, tissue derived from a cell that has undergone a nuclear mutation quickly becomes separated from nonmutant tissue through mitosis, generating a reasonably sharp boundary (van der Meer 1981d). By contrast, cells with nonnuclear mutations separate from normal tissue more slowly, producing a mottled boundary where mutant and normal cells are intermixed (van der Meer 1981d). The gradual sorting-out pattern of nonnuclear mutants is the same as that observed for plastid mutations in higher plants (Kirk & Tilney-Bassett 1978), and reflects the fact that there are usually several copies of plastid DNA in a cell, only one of which is initially mutant. Several rounds of cell division must occur before a purely mutant cell line emerges from the mottled tissue.

B. Chemical mutagenesis

Treatment with a chemical mutagen greatly facilitates the isolation of mutants, especially when a specific type of mutation is sought. Ethyl and methyl methanesulfonate (EMS & MMS), and N-methyl-N'-nitro-N-nitrosoguanidine (NNG) are very effective for red algae. These alkylating agents react with the nitrogenous bases in DNA, primarily guanine, to yield alkylated purines. This alkylation has two

Genetics

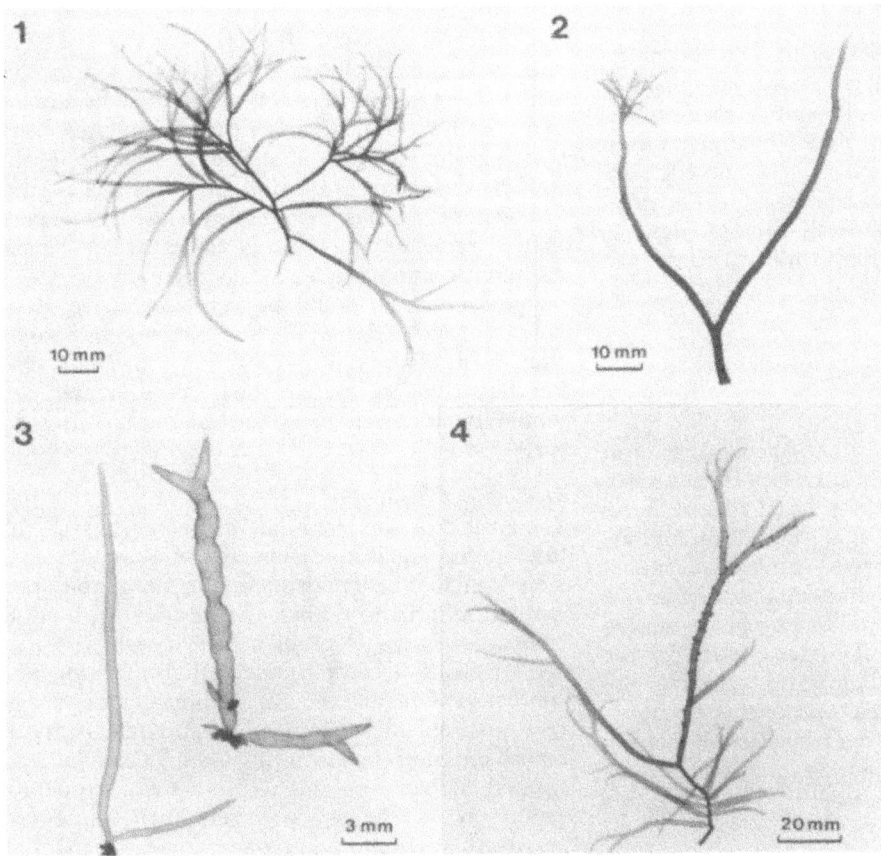

Figs. 5-1 to 5-4. Mutants of *Gracilaria tikvahiae*. 5-1: Spontaneous mutant sector arising from a small lateral branch showing altered, more highly branched morphology. 5-2: Spontaneous, finely subdivided mutant sector arising from an apical tip. This mutant was characterized as an allele of the *bu* mutation. 5-3: A comparison of an EMS-induced puffy mutant and wild-type sporeling (6 weeks old). 5-4: Spontaneous sex mutation (= *bi*), where a bisexual frond appeared on a male plant. (From van der Meer et al. 1984, with permission)

potentially mutagenic effects; first, the alkylation may result in ionization of the purine, which can lead to incorrect base pairing during DNA replication and second, alkylation destabilizes the N-glycosylic bond between the base and the sugar, which can lead to complete loss of the base. If the site is not repaired before replication occurs, any base can be inserted, which leads to base pair substitutions (Drake & Baltz 1976).

Tetraspores of *Gracilaria tikvahiae* that were treated with EMS in situ (0.2 M EMS for 45 min at 20°C) gave a mutant sporeling frequency of 1–5%, about a thousandfold increase over the spontaneous rate. Treatment of small haploid disks was technically easier, and also produced many (0.5–1.0%) mutants (van der Meer 1979a,b). EMS and MMS have been used successfully with a number of other red algae, namely *Callithamnion byssoides* (Spencer et al. 1981), *Gelidium coulteri* (B. Macler per. comm.), and *Champia parvula* (van der Meer & Kehoe unpubl.). Of the three alkylating agents mentioned, NNG appears to be the most mutagenic for red algae. Treatment of small, haploid *Gelidium* sporelings (25 µg NNG.mL^{-1} for 30 min at 20°C) yielded approximately 10–15% visibly mutant plants, a frequency considerably higher than that obtained with EMS (0.2%) in a parallel treatment (van der Meer 1987b). Similarly, NNG (50 µg NNG.mL^{-1} for 30 min at 20°C) produced several mutants of *Porphyra umbilicalis*, whereas no mutants were obtained after treatments with EMS or MMS (Mitman & van der Meer unpubl.).

III. NUCLEAR INHERITANCE

During genetic studies, crosses have to be made between selected individuals in a controlled manner in order to obtain meaningful results. This is essentially impossible in the field or even in large tanks in a greenhouse, because the danger of genetic contamination would be unacceptably high. For practical purposes, genetic research on red algae is limited to sexual species that grow and reproduce well in laboratory cultures.

A. Life history considerations

1. Haploid genetic analysis

Genetic segregation in red algae usually occurs in the haploid gametophyte phase, where recessive mutations are not masked by normal alleles, thereby simplifying the genetic analysis. For diploid organisms it is often necessary to conduct additional test crosses to determine whether a particular individual is completely normal or carries a recessive mutation. The inheritance of characteristics specific to the diploid phase of red algae (for example, the formation of tetrasporangia or conchocelis morphology) have to be analyzed by diploid genetic techniques; however, no such diploid-specific characteristics have been studied.

2. Tetrad analysis

Tetrad analysis is a powerful genetic technique that, under appropriate conditions, can be used to map the position of genes relative to chromosomal centromeres. It also permits detailed analysis of segregation irregularities caused by phenomena like gene conversion, which is not possible with randomly collected spores. Because the meiotic nuclear tetrads of florideophytes are released as spores from individual tetrasporangia, tetrad analysis is possible in many species. Also, for at least some *Porphyra* species, tetrad analysis appears possible using the frond sectors established by meiotic tetrads in germinating conchospores (Ohme et al. 1986).

The isolation of even a few complete tetrads can be extremely useful, because it assures that all of the segregating alleles present in the parent tetrasporophyte are represented among the offspring (for randomly collected spores, one genotype might always be missing due to low viability). For example, genetic segregation in sporeling tetrads (Fig. 5-9) featured prominently in detecting the elusive female gametophytes of *Palmaria palmata* (van der Meer 1981b; van der Meer & Chen 1979; van der Meer & Todd 1980).

3. Zygote amplification

An interesting genetic feature of many red algal species is the amplification of zygotes that occurs when the zygote nucleus is duplicated hundreds, even thousands, of times in the carpospores. Individual cystocarps of most species release genetically uniform carpospores derived from a single zygote. For genetic studies that require extensive experimental replication to distinguish between genetic and environmental effects, this naturally occurring "zygote cloning" could be very useful. In some species, cystocarps may enclose carpospores derived from more than one zygote (e.g., *Gelidium*, van der Meer 1987b), and in these cases greater care would be required to obtain genetically uniform carpospores.

4. Crossing techniques

The basic crossing technique for red algae is to culture mature male and female gametophytes in a single agitated container so that the nonmotile spermatia can find and fertilize the carpogonia. Monoecious species create problems for crosses, because self-fertilization competes with hybridization and confuses the results. For some monoecious species it is possible to separate male and female tissue before crossing, but for others (e.g., *Porphyra yezoensis* and *Gelidium* sp.) the reproductive components are too intimately associated. For such species, recessive pigmentation mutants can be used to identify hybrid offspring (Miura & Kunifuji 1980; Ohme et al. 1986; van der Meer 1987b).

B. Classical genetics

Male and female gametophytes of red algae usually arise from tetraspores in equal frequency, and it has long been recognized that this 1:1 ratio indicates a genetic mechanism for sex determination (Hassinger-Huizinga 1953; von Stosch 1969). It was not until quite recently that the inheritance of a characteristic other than sex was examined. In the first truly genetic reports on red algae, naturally occurring variants for morphology and color were examined in crosses and shown to be inherited according to Mendelian rules (Rueness & Rueness 1975; van der Meer & Bird 1977).

1. Genetics of Gracilaria tikvahiae

Most of what is known about the genetics of red algae comes from a series of studies on *Gracilaria tikvahiae*. Genetic characterization of several mutants established that nuclear genes follow the fundamental genetic rules laid down by Mendel and others. Only details of the life history distinguish red algal genetics from that of other plants and animals. Most classical genetic phenomena were soon observed (van der Meer 1978, 1979a,b; van der Meer & Bird 1977), and these are summarized below with examples from both published and unpublished data. (Abbreviation: wt = wild type.)

a. Segregation. Nuclear mutants were transmitted in a Mendelian fashion, giving clean 1:1 haploid segregation ratios in the gametophytic progeny of single-factor crosses. Reciprocal crosses had identical patterns.

Cross	Tetrasporophytes	Gametophytes
green (*vrt*) × wild type	wild type	95 wt: 97 green
wild type × green (*vrt*)	wild type	96 wt: 88 green

b. Dominant mutants. Among the mutants isolated, dominant as well as recessive phenotypes were found, though much less frequently. Partial dominance was observed for some mutants.

Cross	Tetrasporophytes	Gametophytes
green (*Grs*) × wild type	green	49 wt: 50 green
wild type × pink (*Pur*)	pink	43 wt: 47 pink

c. Multiple alleles. Through complementation and recombination studies, several mutant alleles were discovered for some individual genetic loci. As for multiple alleles in other organisms, phenotypes associated with different mutant alleles of a series varied considerably. Recombination was detected between alleles (i.e., within individual genes) through the appearance of a very low frequency of wild-type plants in crosses.

Cross	Tetrasporophytes	Gametophytes
green × brown-green (*grn*) (*grn³*)	brown-green	121 green: 116 brown-green (+ 0.02% wild type)
green × bright green (*grt*) (*grt²*)	green	36 green: 29 bright green (+ 0.1% wild type)

d. Complementation. When two recessive mutants appeared very similar but actually had defects in different genes, crosses between them produced wild-type tetrasporophytes because the two mutant genomes compensated for each other when present together in diploid cells.

Cross	Tetrasporophytes
green (*sha*) × green (*grn*)	wild type
pink (*lts*) × pink (*dus*)	wild type

e. Independent assortment and recombination. In two factor crosses between nuclear mutations, the two pairs of alleles were usually transmitted in an unskewed, 1:1:1:1 haploid dihybrid ratio, in which parental and recombinant phenotypes were equally represented. Exceptions were caused by reduced viability of one of the phenotypes, epistasis, or linkage.

Cross	Tetrasporophytes	Gametophytes
green (*grn*) × pink (*Pur*)	pink	29 wt: 23 green: 32 pink: 30 gray-green
bushy (*bu*) × green (*jde*)	wild type	77 wt: 81 bushy: 88 green: 78 bushy green

f. Epistasis. In crosses between two similar mutations, the gametophytic progeny did not always exhibit the 1:1:1:1 haploid dihybrid ratio. Occasionally the phenotypes associated with two or three of the genotypes were too similar to distinguish from each other, resulting in 1:2:1 or 1:3 ratios.

Cross	Tetrasporophytes	Gametophytes
green × bright geen (*grt*) (*uai*)	wild type	61 wt: 49 green: 119 bright green
pink (*dus*) × pink (*str*)	wild type	49 wt: 153 pink

g. Linkage. In a small proportion of dihybrid crosses, the two pairs of alleles did not segregate independently, but rather, were transmitted together in the parental combination much more frequently than they recombined. Such genetic linkage has been detected between a number of different genes, and the known associations now identify 11 map fragments (Fig. 5-5).

Cross	Tetrasporophytes	Gametophytes
green (*grn*) × tan (*cha*)	wild type	413 green: 411 tan: 79 wt: 82 light green
wild type × light green (*cha, grn*)	wild type	24 green: 21 tan: 143 wt: 151 light green

h. Nonnuclear mutants. Color mutations showing strict maternal inheritance in reciprocal crosses were frequently encountered among the color variants. These are discussed in more detail in the following section.

Cross	Tetrasporophytes	Gametophytes
green (*NMG-1*) × wild type	green	All green
wild type × green (*NMG-1*)	wild type	All wild type
pink (*NMP-1*) × wild type	pink	All pink
wild type × pink (*NMP-1*)	wild type	All wild type

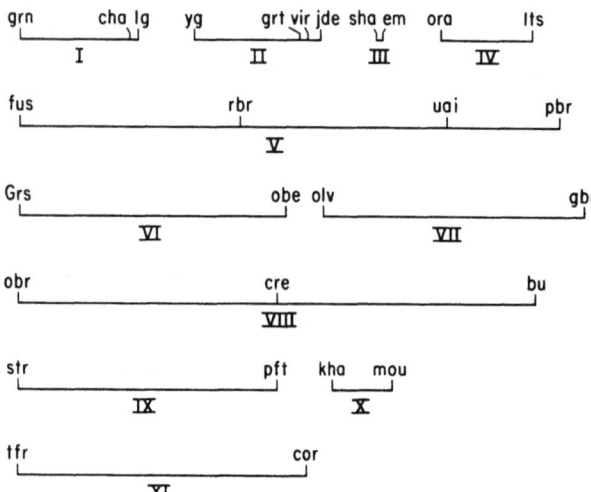

Fig. 5-5. Linkage map for *Gracilaria tikvahiae*. Linkage groups are identified with Roman numerals. Distances between genes are proportional, that between *cha* and *lg* being 1 map unit. (1 map unit = 1% recombination)

i. Unstable nuclear mutants. Most of the color mutants were stable; however, a few repeatedly reverted to wild type during somatic growth, giving the plants a mottled appearance (van der Meer 1979b; van der Meer & Zhang 1988). In crosses, unstable green mutants generally behaved as recessive genes. However, a few (0.5%) of the F_1 tetrasporophytes expressed the unstable green phenotype and appeared to have become homozygous for the unstable mutant allele by some form of gene conversion. The wild-type revertants were stable during somatic growth and in crosses. These unstable mutants are suspected to have mobile genetic elements such as transposons, but no direct evidence is available.

j. Parental effects. Spores released from tetrasporophytes were always the color of the parent plant, regardless of their own genotype. Thus, gametophytes that were genotypically green due to a recessive mutation, were initially red due to the carry-over of pigment from the heterozygous tetrasporophytes. The true phenotypes of the gametophytes could not be scored until there had been sufficient growth to dilute the parental contribution. [This parental effect was especially noticeable with a bright green mutant of another red alga, *Palmaria palmata*, where the spore-borne red pigment from the tetrasporophyte persisted for many weeks in a cluster of cells near the center of an otherwise green gametophytic holdfast disk (van der Meer & Todd 1980)].

k. Mitotic recombination. On mature tetrasporophytes, somatic cells near tetrasporangia underwent a high frequency of mitotic recombination leading to partial homozygosity of the genome. As a consequence, recessive mutations in heterozygous plants became expressed as somatic sectors having mutant color or morphology (Fig. 5-6). These often occurred as "twin spots" of mutant tissue when appropriate markers were present (van der Meer 1981a; van der Meer & Todd 1977). Mitotic recombination was also detected by the production of diploid gametophytic sectors on otherwise normal tetrasporophytic plants (discussed in Section V). A high frequency of mitotic recombination has also been observed for *Gracilaria foliifera* and *G. sjoestedtii*, but it has not yet been observed in other genera of red algae.

2. *Nuclear mutations of* Gracilaria tikvahiae
The rich variety of photosynthetic pigments in red algae permits a broad spectrum of color variation in mutants (van der Meer 1979a,b). The photosynthetic pigments can be altered both quantitatively and qualitatively, yielding plants colored various shades of red, brown, green, yellow, orange, pink, and purple (Fig. 5-7). Large changes in the phycobiliprotein concentrations are possible without loss of plant viability; however, most color mutants are less robust than wild type and grow at slower rates.

Pigmentation mutants of *Gracilaria tikvahiae* have been characterized in some detail (Kursar et al. 1983a,b; van der Meer 1979a). Biochemical changes in these mutants are complex, with concentrations of several pigments altered in each mutant. A few examples of the types of changes that have been described are reviewed below and illustrated in Fig. 5-7. A catalogue of pigmentation mutants of *G. tikvahiae* is included in Table 5-1.

A recessive bright green mutant, vrt^2, contains only trace amounts of phycoerythrin (Kursar et al. 1983a). The phycobilisomes of these plants are composed mostly of phycocyanin and allophycocyanin and are much smaller and more labile than those of wild type (Kursar et al. 1983b). A recessive, greenish-brown mutant, mos^2, contains a large excess of phycocyanin, 90% of which is not bound to the phycobilisomes (Kursar et al. 1983a,b). When irradiated with a moderately bright light, the fronds

Genetics

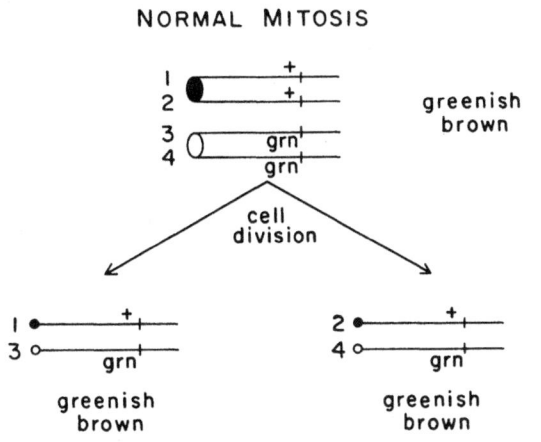
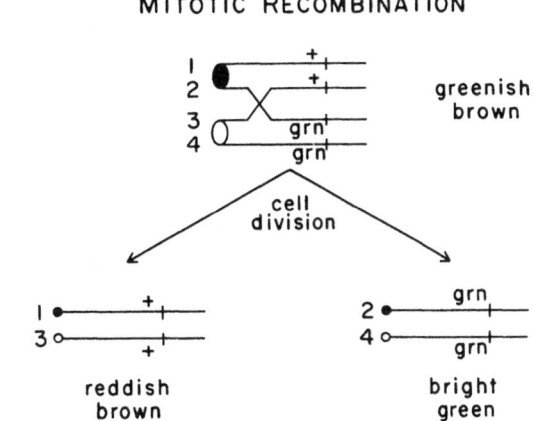

Fig. 5-6. A comparison of the products expected from a normal mitosis with those from a mitosis following mitotic recombination. Note that with appropriate segregation of the chromatides, each of the daughter cells becomes homozygous for genes distal to the crossover point, which allows the expression of recessive alleles. (From van der Meer & Todd 1977, with permission)

fluoresce a reddish color (van der Meer unpubl.) because energy absorbed by the unattached phycocyanin cannot be transferred to the reaction centers for photosynthesis and is released as fluorescence. Another recessive greenish-brown mutant, *obr*, has a spectrally altered phycoerythrin (Kursar et al. 1983a) that lacks the typical absorption peak at 545 nm and has a phycoerythrobilin to phycocyanobilin chromophore ratio of 2.6, in contrast to 4.2 found in wild plants. The mutant Pur^2 is one of a few dominant mutants that have been found (van der Meer 1979b). It has a deep purple color resulting from an excess of all phycobiliproteins and a simultaneous reduction in chlorophyll-*a* concentration. The phycobilisomes are somewhat larger than normal, with more phycoerythrin per particle (Kursar et al. 1983a,b). An orange-colored plant results from the recessive mutation *ora*, which produces only half the phycocyanin and allophycocyanin found in wild type (Kursar et al. 1983a; van der Meer 1979a) and has fewer but larger phycobilisomes than wild plants (Kursar et al. 1983b).

Morphological mutants have not been characterized to the same extent as pigmentation mutants. Because developmental pathways for morphology are completely unknown, it is difficult to target meaningful measurements, making it essentially impossible to identify specific defects. Mutants that have been genetically characterized to some extent are included in Table 5-1.

The most studied – and perhaps the most inter-

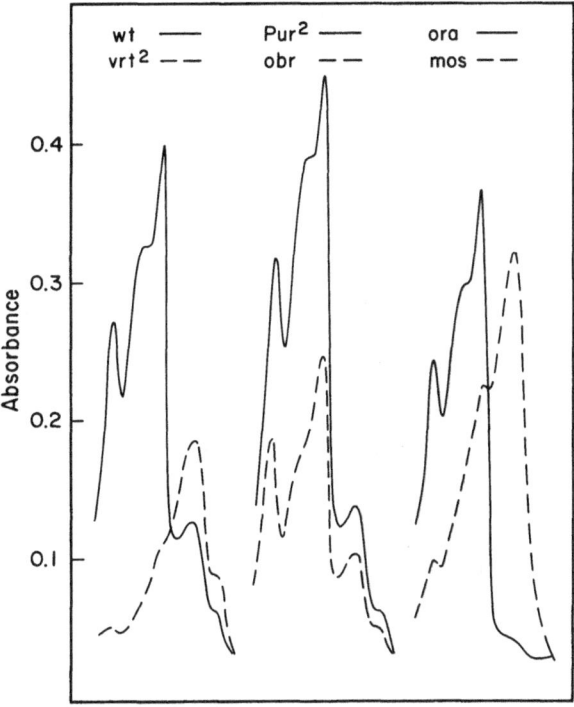

Fig. 5-7. A comparison of water soluble pigments in wild-type and selected pigmentation mutants of *Gracilaria tikvahiae*. Measurements were semiquantitative (0.5 g fresh, laboratory-grown fronds extracted into 10 ml of 25 m M Tris-HCl buffer pH 8.0); thus, relative rather than absolute peak heights should be compared. Spectrophotometer scans were from 475 to 675 nm.

Table 5-1. Identified genes of *Gracilaria tikvahiae*

Code number	Gene symbol	Phenotype of variant[a]	Code number	Gene symbol	Phenotype of variant[a]
A. Mendelian transmission			B-3	stu	dwarf plants, much branched
G-1	grn	green frond color			
G-2	grt	green frond color	B-4	hbr (= MP-44)	profuse apical branching
G-3	vrt	green frond color	B-5	tfr (= MP-46)	thin, much branched fronds
G-4	jde	green frond color	B-6	inc (= MP-48)	more highly branched than related plants
G-5	yg	yellow-green frond color			
G-6	olv	green frond color	B-7	clb	clublike frond tips, becomes multibranched
G-7	mgr	green frond color			
G-8	em	bright green frond color	B-8	fil	uniseriate filaments grow out of thallus
G-9	Grs	green frond color			
G-10	lg	light green frond color	B-9	sdi	disklike tissue subtends branches
G-11	sha	green frond color			
G-12	vir	green frond color	B-10	thn	thin, darkly pigmented fronds
G-13	uai	bright green frond color			
G-14	kha	brownish green frond color			
G-15	obr	brownish green frond color	H-1	hls	lacks hyaline hairs, (female sterile)
G-16	gbr	brownish green frond color			
G-17	mos	brownish green frond color	H-2	bld	lacks hyaline hairs
G-18	mou	brownish green frond color	F-1	obe	puffy fronds
			F-2	plu	puffy fronds
L-1	cha	yellowish brown frond color	F-3	sto	puffy fronds
L-2	ivo	pale brown frond color	F-4	puf (= MP-64)	puffy fronds
L-3	tan	yellowish brown frond color	F-5	thi (= MP-70)	thick fronds
L-4	bgr	greenish brown frond color	F-6	fa (= MP-67)	puffy fronds
L-5	lbr	light brown frond color	F-7	cor (= MP-61)	puffy fronds
L-6	pbr	light brown frond color	S-1	mt	gametophyte sex determination
P-1	Pur	purple frond color	S-2		
P-2	lts	dark pink frond color			
P-3	fus	bright pink frond color		bi	bisexual, constitutive female expression
P-4	pom	dark pink frond color			
P-5	dus	pink frond color	B. Maternal transmission[b]		
P-6	str	pink frond color			
P-7	ora	orange frond color	NMG-1		green frond color
P-8	pft	pink frond color	NMG-2		green frond color
P-9	rbr	pinkish brown frond color	NMG-3		green frond color
P-10	brp	pinkish brown frond color	NMG-4		green frond color
Y-1	amb	yellow frond color	NMG-5		green frond color
Y-2	whe	yellow frond color	NMG-6		green frond color
B-1	bu (= MP-40)	thin, very branched fronds	NMP-1		purple frond color
B-2	cre (= MP-42)	small holdfast, basal branching of fronds	NMP-2		purple frond color
			NMY-1		yellow frond color
			NMY-2		yellow frond color
			NMY-3		yellow frond color

[a] See Patwary & van der Meer 1982; van der Meer 1978, 1979b, 1986a; van der Meer et al. 1984. Some mutants have only been listed in the literature. Only a portion have been phenotypically characterized.

[b] Only a few clones, representing different color types, have been kept. Pigmentation mutants having maternal transmission are all assumed to be in chloroplast DNA, but it is unknown whether they are all in different genes.

esting – morphological variant obtained thus far is the mutant *bu* (Fig. 5-2), a semidominant, spontaneous mutant with such finely dissected fronds that plants superficially resemble *Polysiphonia* (Patwary & van der Meer 1982, 1983a,d). Agar extracted from a strain that carries the *bu* mutation forms much firmer gels than does agar from wild-type plants (Craigie et al. 1984; Patwary & van der Meer 1983c). This agar has an altered chemical composition resulting from quantitative rather than qualitative changes in the constituent sugars (Craigie et al. 1984). Young apical tissue of wild-type plants has a similar agar composition (Craigie & Wen 1984) and a somewhat similar anatomy, which suggests that

Table 5-2. Genetically characterized mutants in species other than *Gracilaria tikvahiae*

Species	Mutant	Reference
A. Mendelian transmission		
Pterothamnion plumula	morphology altered	Rueness & Rueness 1975
Callithamnion byssoides	green (several)	Spencer et al. 1981
Champia parvula	pink (pnk1-1)	Steele et al. 1986
	green (grn1-1)	Steele et al. 1986
	yellow green (yg1-1)	Steele et al. 1986
	yellow (yel1-1)	Steele et al. 1986
	stubby (st1-1)	Steele et al. 1986
	small, stubby (st2-1)	Steele et al. 1986
	inflated (inf1-1)	Steele et al. 1986
	highly branched (hb1-1)	Steele et al. 1986
	green (grn2-1)	van der Meer & Kehoe unpubl.
	green (grn3-1)	van der Meer & Kehoe unpubl.
	yellow-green (yg2-1)	van der Meer & Kehoe unpubl.
	pink (pnk2-1)	van der Meer & Kehoe unpubl.
	pink (pnk3-1)	van der Meer & Kehoe unpubl.
	translucent (net1-1)	van der Meer & Kehoe unpubl.
Chondrus crispus	green	van der Meer 1981d
	green (cold sensitive)	van der Meer 1981d
	bright green	van der Meer 1981d
Devaleraea ramentacea	green	van der Meer 1981c
Gelidium coulteri	green	Macler unpubl.
Gelidium sp.	grn1-1	van der Meer 1987b
	green (several)	van der Meer unpubl.
	dwarf (2)	van der Meer unpubl.
	spore germination defect	van der Meer unpubl.
	no cystocarp ostiole	van der Meer unpubl.
	female sterile	van der Meer unpubl.
	male sterile (2)	van der Meer unpubl.
Gigartina teedii	green	Guiry 1984
Gracilaria foliifera	green	van der Meer unpubl.
	brown	van der Meer unpubl.
	unstable green	van der Meer & Zhang 1988
Gracilaria sjoestedtii	green (3)	Zhang & van der Meer 1988b
	unstable green	Zhang & van der Meer 1988b
	yellow-green	Zhang & van der Meer 1988b
	pink (2 loci)	Zhang & van der Meer 1988b
Palmaria palmata (NS)[a]	green-1	van der Meer & Todd 1980
(NS)	green-2	van der Meer unpubl.
(NS)	pink	van der Meer unpubl.
(Irl)[a]	green	van der Meer unpubl.
Porphyra yezoensis	red	Ohme et al. 1986
	green	Ohme et al. 1986
	light red	Miura 1985
	light green	Miura 1985
B. Maternal transmission		
Champia parvula	pink	van der Meer & Kehoe unpubl.
	pale brown	van der Meer & Kehoe unpubl.
Chondrus crispus	green (several)	van der Meer 1981d
Gelidium sp.	green (several)	van der Meer unpubl.
	yellow-green	van der Meer unpubl.
	brownish-green	van der Meer unpubl.
	pink (several)	van der Meer unpubl.
Gracilaria sjoestedtii	green (6 isolates)	Zhang & van der Meer 1988b
	yellowish (5 isolates)	Zhang & van der Meer 1988b
	pink (3 isolates)	Zhang & van der Meer 1988b
Palmaria palmata (NS)	yellow	van der Meer unpubl.

[a] *Palmaria palmata* from Nova Scotia, Canada (NS), and from Ireland (Irl). These two populations are interfertile, but hydrids leave few viable offspring (van der Meer 1987a).

the agar characteristics of the *bu* mutant are a secondary consequence of altered plant morphology. A number of other morphological mutants appear puffy (e.g., Fig. 5-3). The anatomy of such plants is greatly disturbed due to enlarged cells in the medulla, resulting in weak, sometimes flaccid, fronds (Patwary & van der Meer 1982). In one mutant of this type, *fa*, the medullary cells have nearly twice the normal cell diameter (Patwary & van der Meer 1982). Some puffy mutants literally rip apart as expansion of large medullary cells forces the cortical cells to separate (van der Meer unpubl.). Agar has been extracted from the puffy mutant *cor*. It forms only weak gels (Patwary & van der Meer 1983c), but the sugar composition of this agar has not been determined. Another, rather unusual mutant, *hls*, lacks the hyaline hairs present on normal fronds, especially near the apex (van der Meer 1979a). Fortunately this mutant was discovered as a male plant, because females affected by the *hls* mutation are completely sterile. The sterility of hairless females, together with an inability to separate hairlessness and sterility genetically, has led to speculation that the *hls* mutation is defective in a cell elongation process needed for the formation of both hairs and trichogynes (van der Meer 1979a).

3. Nuclear mutations in other species

Nuclear mutations have been characterized in a few other red algal species (Table 5-2), and most behave the same as mutants of *Gracilaria*. One yellow mutant of *Champia parvula* is an exception. It arises repeatedly from a wild-type tetrasporophyte clone, and all of these isolates appear to be defective in the same gene (Steele et al. 1986). In crosses to wild type, they are inherited as recessive nuclear mutations, with normal 1:1 ratios of yellow to wild-type plants. However, no viable carpospores are obtained when yellow plants are crossed to each other. It seems unlikely that yellow color and carposporophyte infertility could both result from a defect in a single gene; the two phenotypes are more likely caused by a complex mutation, for example, a chromosomal deletion.

Color variants studied in *Porphyra* have pigment changes similar to those in *Gracilaria* mutants as determined by in vivo absorption (Aruga & Miura 1984; Migita & Fujita 1983) and absorption in crude extracts (Fujita & Migita 1984; Merrill et al. 1983). Green mutants have reduced phycoerythrin content, whereas reddish mutants have slightly elevated levels. An orange mutant is low in phycocyanin. Green mutants of *Callithamnion byssoides* (Spencer et al. 1981), *Palmaria palmata*, and *Chondrus crispus* (van der Meer unpubl.) all have much reduced levels of phycoerythrin, the *Palmaria* mutant also having a radically altered chloroplast lamellar organization (Pueschel & van der Meer 1984; Chapter 2). The *Chondrus* mutant is cold sensitive, having a much greener phenotype at 5°C than at 20°C (van der Meer 1981d).

IV. NONNUCLEAR INHERITANCE

A number of stable mutants in various species of the Florideophyceae have a non-Mendelian inheritance pattern. These mutations are transmitted to the tetrasporophyte only by the maternal parent and do not exhibit any segregation in the subsequent gametophytic generation. When such mutants first arise, they may undergo an extended period of sorting out during which mutant cell lines gradually segregate from wild-type and mixed areas (van der Meer 1981d). Plant color variants inherited in this maternal fashion often result from mutations in chloroplast DNA (Kirk & Tilney-Bassett 1978), and it is likely that nonnuclear mutations affecting pigmentation of red algae are also defects in the plastid genome.

A. Maternal transmission

The inheritance of non-Mendelian mutations in *Gracilaria tikvahiae* was examined in reciprocal crosses to wild-type plants (van der Meer 1978). When the female parent carried the mutation, the F_1 tetrasporophyte inherited the mutant phenotype, and all the subsequent gametophytes were mutant. In contrast, when the mutation was in the male parent, the F_1 tetrasporophyte was wild type and the mutation did not reappear in the next gametophytic generation or subsequently. In culture dishes containing many tens of thousands of F_1 tetrasporophytes, no plants or plant sectors with the phenotype of the male parent were ever observed. Thus, the transmission of nonnuclear mutations in *G. tikvahiae* appears to be strictly maternal, with complete elimination of the male phenotype. Exclusively maternal transmission of nonnuclear mutants has also been observed for *Chondrus crispus*, *Champia parvula*, *Palmaria palmata* and *Gelidium* sp. (Table 5-2); biparental inheritance of nonnuclear mutations has not been observed in any red algal species. For terrestrial plants, chloroplast inheritance is also strictly maternal for most species, but there are several exceptions, such as *Oenothera* and *Pelargon-*

Genetics

ium, which have biparental transmission (Kirk & Tilney-Bassett 1978).

The genetic results indicate that only the chloroplast genome from the female parent is transmitted during sexual reproduction in many florideophytes. Plastids from the male parent are excluded in some manner, before or after fertilization. Cytological observations on the development of spermatia (Kugrens & West 1972; Scott & Dixon 1973) have demonstrated that plastids rarely occur in cross-sections of mature male gametes. The situation is somewhat different in the Bangiophyceae, where chloroplasts become very reduced in size and complexity but continue to divide and are found regularly in mature spermatia (Cole & Sheath 1980; Hawkes 1978). Whether the plastids contain functional DNA and can become part of the chloroplast complement of the zygote cannot be determined from the cytological studies. At present there is no genetic information available for non-Mendelian mutants in this group.

B. Phenotypes of nonnuclear mutations

Plant colors of nonnuclear mutants are very similar to those observed for nuclear mutations and include several shades of green, yellow, and pink. The phenotypes of only a few have been examined (Kursar et al. 1983a,b). Phycobilisomes of a yellow non-Mendelian mutant, *NMY-1* (van der Meer 1978, 1986a), are composed of phycoerythrin and allophycocyanin, with almost no phycocyanin present. Electrophoresis of isolated phycobilisome proteins indicates that a 29 kd protein, which may be required to assemble the phycobilisome core (Yamanaka & Glazer 1981), is missing in this mutant (Kursar et al. 1983a,b). The two nonnuclear green mutants that have been examined, *NMG-1* and *NMG-2*, are somewhat less interesting. Both have greatly reduced levels of phycoerythrin in cell extracts [only a trace in *NMG-2* (Kursar et al. 1983a)], similar to some bright green nuclear mutations. The biliprotein composition of isolated phycobilisomes from *NMG-2* reflects the low phycoerythrin levels in cell extracts (Kursar et al. 1983b).

V. GENETIC REGULATION

There is only a little information on genetic regulation in red algae, none of it very specific. Descriptive observations have been made on what appear to be regulatory genes involved in sex and phase determination; however, these serve more to expose our ignorance than to illustrate our understanding.

A. Sex determination in *Gracilaria tikvahiae*

The development of equal numbers of male and female plants from tetraspores demonstrates that the primary control of sex determination is through a single pair of Mendelian factors. The observation that these are capable of undergoing mitotic recombination (van der Meer 1981a; van der Meer & Todd 1977) indicates that they are found in a limited region of a chromosome. We have designated the sex factors as alleles, mt^m for the male and mt^f for the female; however, it is not established that they are simple alternatives of a single regulatory gene. A control mechanism involving a complex locus or a cluster of linked cistrons remains a possibility. Mapping the chromosomal locus of the *mt* gene would be helpful, but thus far no linkage with other mutations has been discovered.

A recessive Mendelian mutation, *bi*, permits constitutive expression of female-specific functions even in the presence of the male mt^m allele, where genes for female functions are normally silent (Fig.

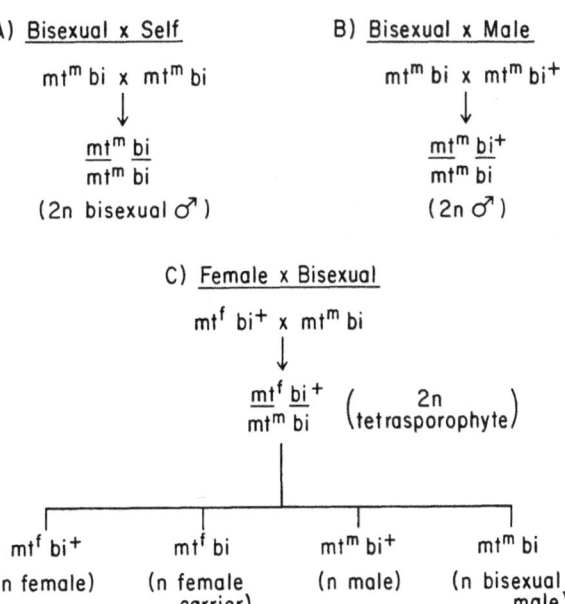

Fig. 5-8. Genetic interpretation of results obtained in crosses involving the mutant *bi* allele of *Gracilaria tikvahiae*. mt^m and mt^f are male and female alleles of the sex-determining locus, *bi* and bi^+ are the mutant and wild-type alleles of the "bisexual" locus. (From van der Meer 1986b, with permission)

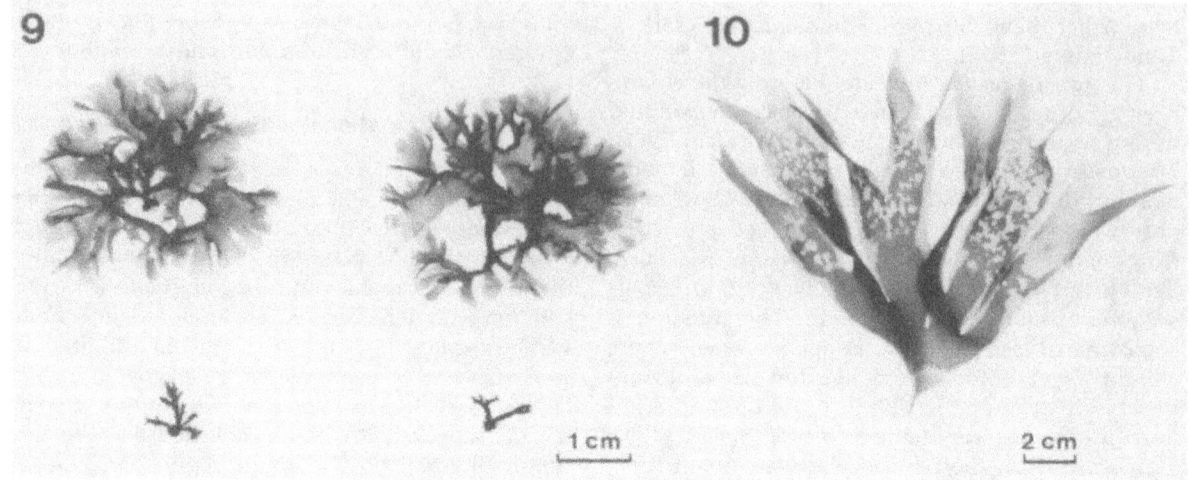

Figs. 5-9, 5-10. *Palmaria palmata*. 5-9: Sporeling tetrad obtained from the spores of a single tetrasporangium, illustrating the slow and stunted development of the two unfertilized female plants as compared to the two male sporelings. 5-10: Haploid tetrasporophyte obtained from an excised female frond. Abundant tetrasporangia are produced on the frond, but these undergo haploid meiosis and the resulting spores abort before release. (From van der Meer & Chen 1979, with permission)

5-4). Haploid male plants with the *bi* mutation form both spermatia and cystocarps (van der Meer 1986b; van der Meer et al. 1984). Haploid female plants can transmit the mutation to their offspring but are themselves unaffected (van der Meer 1986b). In other words, the *bi* mutation renders males bisexual; it does not promote a general bisexuality in all gametophytes (Fig. 5-8). Diploid tetrasporophytes that are homozygous for the mutant *bi* allele form carpogonia in addition to tetrasporangia (van der Meer 1986b). Although the molecular action of the *bi* mutation has not been studied, it appears to be an important gene in the regulation of sexual differentiation. It is speculated that in male and tetrasporophytic plants, the product of the normal bi^+ allele functions as a suppressor of key female-specific regulatory genes.

B. Phase determination in *Gracilaria tikvahiae*

Normal diploid tetrasporophytes are heterozygous for the male- and female-determining *mt* alleles. However, through mitotic recombination some diploid cells become homozygous for one or the other of these alternatives, and under appropriate conditions such cells give rise to visible patches of recombinant tissue. Interestingly, tissue homozygous for an *mt* allele expresses gametophytic rather than tetrasporophytic characteristics, despite its diploid condition and its location on a normal tetrasporophyte (van der Meer 1981a; van der Meer & Todd 1977). This response to the genetic condition of the *mt* alleles, reveals that the heterozygous condition of these alleles, not the diploid state, determines tetrasporophyte development. Further, the observation that very small clusters of gametophytic cells can develop on a normal tetrasporophyte indicates that the action of the *mt* gene is within the cell and not mediated by a diffusible substance, such as a plant hormone.

C. Sex and phase determination in other red algal species

The genetic control of phase development in some red algae appears to be different than that in *Gracilaria*. *Palmaria palmata* and *Devaleraea ramentacea* are dioecious and also have a 1:1 Mendelian segregation of male and female gametophytes (Fig. 5-9; van der Meer 1981c; van der Meer & Todd 1980). Female plants are normally very small, but if unfertilized fronds are grown in the absence of spermatia, they slowly develop into sizable blades resembling males and tetrasporophytes (Fig. 5-10). When these female-derived fronds become reproductive, they resemble tetrasporophytes strikingly and produce abundant tetrasporangial sori; however, the tetrasporangia undergo an abortive haploid meiosis and produce no viable tetraspores. This transformation of haploid female fronds into tetrasporophytes appears to indicate that both sex-determining alleles do not need to be present for sporophyte development in all species.

Haploid tetrasporophytes of *Chondrus crispus* also have been described (van der Meer et al. 1983). These arise from asexual spores produced in gall-like structures on male plants in culture. The production of tetrasporophytes directly from haploid male plants again suggests that the role of sex-determining alleles in phase development may differ from species to species. Unfortunately, neither the *Palmaria* nor the *Chondrus* phenomenon can be studied effectively, because genetic experiments cannot extend beyond the sterile haploid sporophytes.

Champia parvula is also normally dioecious, but a female clone of this species occasionally develops patches of spermatia in old cultures (van der Meer unpubl.). The phenomenon occurs irregularly and unfortunately is not inducible at will. Nevertheless, self-fertilization of the "monoecious" (female) plant and cross-fertilization between this female and other females can be achieved. In both situations the resulting diploid plants develop as tetrasporophytes, not as gametophytes, as would be expected based on results for *Gracilaria*. In crosses between two genetically marked females, the color mutations undergo normal Mendelian segregation and assortment during meiosis in the tetrasporophyte, but there is no segregation for sex within the gametophytic progeny. Only female plants are obtained from the "homozygous female" tetrasporophytes. The life cycle of *C. parvula* clearly can be completed from a single "monoecious" (female) gametophyte, through a normal-looking tetrasporophyte, although there are no males in the next generation. A report by Rueness and Rueness (1973) describing the origin of both male and female plants of *Antithamnion tenuissimum* from asexual spores produced on a presumed male plant is interesting in this connection.

VI. Molecular and somatic cell genetics

With the advent of fast and reliable gene isolation, cloning, and sequencing techniques, molecular genetics quickly has become one of the most productive approaches for biological studies in areas like development, genetic regulation, evolution, and systematics. Considerable progress has also been made in learning how to manipulate the tissues, cells, and protoplasts of multicellular plants, opening additional possibilities for genetic research. Both of these developments will have major impact on genetic studies of the Rhodophyta in the coming decade.

A. Molecular genetics

Although genetic engineering of red algae by molecular genetic techniques has not yet been accomplished, basic studies are underway. The initial research has largely focused on red algal chloroplasts, whose genomes are somewhat easier to study due to their smaller size and greater number of copies per cell.

1. Plastid DNA

A great deal is now known about the gene content and organization of chloroplast DNA in green plants (Palmer 1985, 1986). Indeed, the complete nucleotide sequence for chloroplast DNA of *Marchantia polymorpha* was recently announced (Ohyama et al. 1986). Much less is known about plastids in other groups, including the red algae, but interest in nongreen systems is growing (e.g., Fain et al. 1988; Kuhsel & Kowallik 1987; Reith & Cattolico 1986), principally because they provide independent data on the origins and relationships of plastids and, through them, on the relationships of the plants themselves.

One of the most significant discoveries made thus far involves the enzyme ribulose-1,5-bisphosphate carboxylase (RuBPCase), which fixes carbon dioxide in photosynthesis. The functional RuBPCase enzyme has eight large and eight small subunits. The large subunit is encoded and synthesized in the plastid (Palmer 1985). In vascular plants and green algae, the small subunit protein is encoded by a small gene family in the nucleus (e.g., Goldschmidt-Clermont 1986). The protein is synthesized in the cytoplasm, then imported into the chloroplast. In contrast, both of the RuBPCase subunits are encoded and synthesized in the plastid of various nongreen photosynthetic organisms, including the red algae *Porphyridium aerugineum* and *Cyanidium caldarium* (Steinmüller et al. 1983), the protist *Cyanophora paradoxa* (Heinhorst & Shively 1983), and the chrysophyte *Olisthodiscus luteus* (Reith & Cattolico 1986). The transfer of the small subunit gene from plastid to nucleus, which occurred in the evolution of green plants, apparently did not occur in the red algae and some other lines of photosynthetic eukaryotes.

Genes for the phycobiliproteins also appear to be located in the plastid genome of organisms that have these pigments (Egelhoff & Grossman 1983; Lemaux & Grossman 1984, 1985; Steinmüller et al.

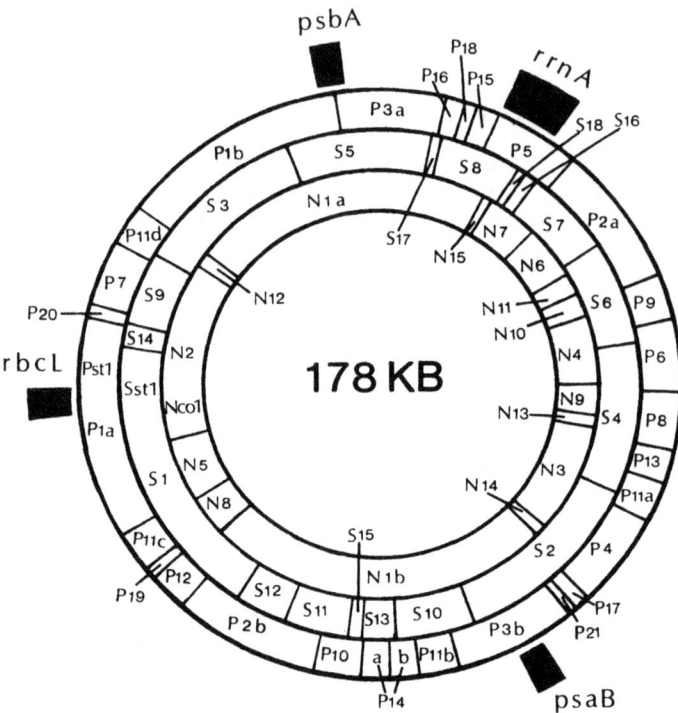

Fig. 5-11. Restriction map for chloroplast DNA of Griffithsia pacifica showing a circular genome of 178 kb, and the approximate position of four chloroplast genes (rbcL, psbA, psaB, and rrnA). Restriction enzymes used to construct the map were Pst1, Sst1 and Nco1. DNA fragments generated by each enzyme are identified with a corresponding letter and a number. KB = kilobases. (From Li & Cattolico 1987, with permission)

1983). Sequence data have revealed that the genes for the two subunits of phycocyanin are located in adjacent coding regions in the plastid DNA of Cyanophora paradoxa (Lambert et al. 1985; Lemaux & Grossman 1985), essentially the same arrangement as that found in the blue-green alga Agmenellum quadruplicatum (de Lorimier et al. 1984; Pilot & Fox 1984). More than 20 different genes have been positioned on the plastid DNA of C. paradoxa (Lambert et al. 1985).

Research on the genomes of larger red algae is also beginning to emerge. The plastid DNA of Griffithsia pacifica has been mapped by restriction analysis (Li & Cattolico 1987). It consists of a circular, 178 kb molecule that contains only a single copy of the genes for ribosomal RNA (Fig. 5-11). This is in sharp contrast to most terrestrial plants and green algae, where chloroplast DNA generally contains two copies of the ribosomal RNA genes in the form of an inverted repeat (Palmer 1985, 1986). Progress is being made in characterizing the plastid DNA of additional taxa such as Porphyra (Shivji & Cattolico unpubl.), Gracilaria (Goff & Coleman 1988), and Palmaria (van der Meer & Douglas unpubl.). Preliminary data for Porphyra yield a plastid DNA molecule of about 185 kb also with a single copy of the ribosomal RNA genes, whereas those for Palmaria palmata indicate a plastid genome of approximately the same size but with an inverted repeat for the ribosomal RNA genes.

2. Other studies

Interesting results are beginning to emerge from nucleic acid sequence studies. Sequence determination for nuclear 5S ribosomal RNA from a limited number of red algal species permits the construction of a rudimentary "tree" (Lim et al. 1986; Takaiwa et al. 1982); however, much more molecular data are needed to establish truly meaningful phylogenetic relationships for the Rhodophyta. The 5S ribosomal RNA has limitations for evolutionary comparisons because of its small size; thus, it will be important to add new information from other molecules, such as 16S ribosomal RNA, and selected protein-encoding genes as these become available. Nevertheless, comparison of red algal 5S RNA sequences with those obtained for a large number of other organisms supports an ancient origin of the Rhodophyta and their early divergence from other algae and from eukaryotes in general (Hori & Osawa 1987).

Mitochondrial DNA of *Griffithsia pacifica*, present as a contaminant in chloroplast DNA preparations, has been partially characterized. The size of the mitochondrial DNA molecules is variable, from 27 kb to 350 kb (Li & Cattolico 1987). Small, circular, plasmidlike DNA molecules appear to be common in nucleic acid preparations isolated from species of *Gracilaria* (Douglas unpubl.; Goff & Coleman 1988). In other work the bacterial plasmid pBR325 was fused with fragments of red algal DNA to obtain vectors capable of autonomous replication in yeast (Loppes et al. 1985). Their ability to replicate in algae cannot be tested at present due to the lack of an appropriate host and transformation system.

B. Somatic cell genetics

The ability to manipulate tissues, cells, and protoplasts of red algae in culture would provide new genetic opportunities and is likely a prerequisite for molecular genetic modification of the living plants. Gene transfer between unrelated red algae could become a reality through genetic transformation of protoplasts that are capable of regenerating into fronds.

1. Tissue and cell culture

Routine tissue culture techniques suitable for red algae are still being developed. Cells isolated from various *Porphyra* species have been cultured in vitro and regenerated into new plants in a number of laboratories (Chen 1986; Liu et al. 1984; Polne-Fuller & Gibor 1984; Polne-Fuller et al. 1984; Zhao & Zhang 1981). Chen and Taylor (1978) regenerated *Chondrus crispus* fronds from colorless axenic cores of medullary tissue. More recently, calli have been obtained for *Agardhiella subulata* (D. Cheney pers. comm.), and a number of other red algal species (Polne-Fuller & Gibor 1987); however, at the present time, calli from only a few species (e.g., *Eucheuma alvarezii* and *Porphyra perforata*) can be maintained indefinitely in culture.

2. Protoplasts

Protoplast formation and plant regeneration from protoplasts have been accomplished with various species of *Porphyra* (*P. linearis*, L. C.-M. Chen pers. comm.; *P. yezoensis*, Fujita & Migita 1987; *P. perforata*, Polne-Fuller & Gibor 1984; Polne-Fuller et al. 1986). Protoplasts with good viability have been isolated from the red alga *Gracilaria tikvahiae* (Cheney et al. 1986), and regeneration of plants from these protoplasts was recently achieved (D. Cheney pers. comm.). Protoplast fusion has been achieved between normal and mutant green cells of *P. yezoensis* (Fujita & Migita 1987). Unfortunately, the report gives no indication whether the mutation was in the nucleus or the plastid, which leaves an ambiguous interpretation of observed color segregation. Very wide crosses have been attempted by fusing protoplasts of *P. perforata* with protoplasts of the green alga *Enteromorpha intestinalis* which yielded inviable hybrid cells (Saga et al. 1986), and by fusing protoplasts of *Porphyridium purpureum* (as *P. cruentum*) with protoplasts of the green algae *Dunaliella bardawil* and *D. salina*, which may have yielded modified *Porphyridium* clones (Lee & Tan 1987).

VII. QUANTITATIVE GENETICS AND PLANT BREEDING

Classical plant breeding, based on quantitative genetic characteristics, has been applied to a limited extent for the improvement of commercial red algal cultivars. Despite the advances of molecular genetics, it is likely that such traditional selective breeding, rather than genetic engineering, will be the most important commercial aspect of red algal genetics in the foreseeable future.

A. Selection of *Porphyra* cultivars

Selective breeding of *Porphyra yezoensis* and *P. tenera* progressed rapidly in Japan after the connection between *Porphyra* blades and shell-boring conchocelis filaments was firmly established (Drew 1949). Complete control of reproduction became possible, and selection of genetically superior strains was greatly facilitated (Miura 1975, 1979).

Serious strain selection has been underway for about 10 to 15 years in Japan; there are now more than 30 recognized cultivars in addition to hundreds of strains selected informally by growers (T. F. Mumford pers. comm.). Significant improvement in the size of fronds has been achieved through strains with delayed fertility (Miura 1975), which allows a longer vegetative growth period. Some color mutants of *Porphyra* are being cultivated on a small scale as specialty products (Aruga & Miura

1984). At the present time, plant breeding objectives for *Porphyra* are directed toward producing strains with increased handling efficiency and quality, for example, those that produce monospores (to compensate for poor spore survival after initial seeding of the cultivation nets), or that are disease resistant and adapted to local conditions. There has been experimentation with interspecific crosses, but such hybrids are unstable (Suto 1963).

B. Strain improvement in other red algal species

Simple phenotypic selection of superior *Eucheuma* strains led to the discovery of an extremely vigorous and nonsporulating clone of *E. alvarezii* known in the Philippines as Tambalang (Doty 1978; Doty & Alvarez 1975). The possibility of selecting fast-growing clones from wild populations also has been demonstrated for *Chondrus crispus* (Cheney et al. 1981; Neish & Fox 1971), *Gigartina exasperata* (Waaland 1979), *Gracilaria tikvahiae* (Ryther et al. 1979), and *G. sjoestedtii* (Hansen 1984).

Attempts to improve the growth of *Gracilaria tikvahiae* by constructing polyploid sporophytes were not successful (Patwary & van der Meer 1984). Triploid tetrasporophytes grow no better than diploids, and most tetraploids grow more slowly. This is true both for polyploids formed within a single genetic line and for hybrid polyploids with genomes from different populations. Gametophytes of *G. tikvahiae* are adversely affected by polyploidy more quickly than tetrasporophytes. Diploid gametophytes resemble normal haploids, but triploid gametophytes are distinctly stunted, and tetraploid gametophytes are grossly abnormal (Zhang & van der Meer 1988a). Although polyploids of *G. tikvahiae* have no commercial potential, they may be useful for basic genetic studies. There are few good experimental systems for studying algal polyploids (Nichols 1980), and the difference between polyploid sporophytes and polyploid gametophytes of *G. tikvahiae* poses an interesting genetic puzzle.

Hybrid vigor, common in terrestrial plants, has not been convincingly demonstrated in red algae. Hybrid tetrasporophytes of *Gracilaria tikvahiae* show no significant heterosis when compared to partially inbred plants (Patwary & van der Meer 1983b). Completely homozygous diploid gametophytes of *G. tikvahiae* [obtained by self-fertilization using the *bi* (bisexual males) mutation] grow as well as hybrid diploid gametophytes (Zhang & van der Meer 1987). These preliminary observations suggest that heterosis may be of minor significance for some species, but more data are needed before a definitive statement is possible.

VIII. SUMMARY

Genetic studies on red algae were initiated late, but are now well underway, and will continue to contribute to our knowledge of this fascinatingly diverse division of the plant kingdom. Over the past decade, studies on mutants have established that inheritance in red algae is not fundamentally different from that of other organisms. Nuclear genes follow well-established Mendelian principles, and nonnuclear genes show strict maternal inheritance. Characterized mutations have been used as marker genes in various ways, both for genetic studies (e.g., to identify hybrid plants in crosses between monoecious or bisexual plants), and for life history studies (e.g., to detect females in the life cycle of *Palmaria palmata* and to determine the genetic basis of mixed reproductive phases in the life cycle of *Gracilaria tikvahiae*). Through molecular genetic studies on 5S ribosomal RNA and the plastid genomes, interesting new data on the systematics, evolution, and origins of the Rhodophyta are just beginning to emerge, and it can be predicted that a flood of molecular data will become available over the next decade. Selective breeding of red algae (e.g., commercial development of *Porphyra* cultivars) will increase in importance as more economically valuable species become cultivated.

IX. ACKNOWLEDGMENTS

I wish to thank the National Research Council of Canada for its continued support of my research and to acknowledge contributions made over the years by colleagues and staff at ARL, particularly Jane Osborne and Edna (Todd) Staples.

X. ENDNOTE

NRCC No. 29084

XI. REFERENCES

Aruga, Y. & Miura, A. 1984. In vivo absorption spectra and pigment contents of the two types of color mutants of *Porphyra*. *Jap. J. Phycol.* 32: 243–50.

Chen, L. C.-M. 1986. Cell development of *Porphyra miniata* (Rhodophyceae) under axenic culture. *Bot. Mar.* 29: 435–9.

Chen, L. C.-M. & Taylor, A. R. 1978. Medullary tissue culture of the red alga *Chondrus crispus*. *Can. J. Bot.* 56: 883–6.

Cheney, D. P., Mar, E., Saga, N. & van der Meer, J. 1986. Protoplast isolation and cell division in the agar-producing seaweed *Gracilaria* (Rhodophyta). *J. Phycol.* 22: 238–43.

Cheney, D., Mathieson, A. & Schubert, D. 1981. The application of genetic improvement techniques to seaweed cultivation: I. Strain selection in the carrageenophyte *Chondrus crispus*. *Proc. Int. Seaweed Symp.* 10: 559–67.

Cole, K. & Sheath, R. G. 1980. Ultrastructural changes in major organelles during spermatial differentiation in *Bangia* (Rhodophyta). *Protoplasma* 102: 253–79.

Craigie, J. S. & Wen, Z. C. 1984. Effects of temperature and tissue age on gel strength and composition of agar from *Gracilaria tikvahiae* (Rhodophyceae). *Can J. Bot.* 62: 1665–70.

Craigie, J. S., Wen, Z. C. & van der Meer, J. P. 1984. Interspecific, intraspecific and nutritionally-determined variations in the composition of agars from *Gracilaria* spp. *Bot. Mar.* 27: 55–61.

de Lorimier, R., Bryant, D. A., Porter, R. D., Liu, W.-Y., Jay, E. & Stevens, S. E., Jr. 1984. Genes for the α and β subunits of phycocyanin. *Proc. Natl. Acad. Sci. USA* 81: 7946–50.

Dixon, P. S. 1973. *Biology of the Rhodophyta*, p. xii. New York: Hafner Press.

Doty, M. S. 1978. Status of marine agronomy with special reference to the tropics. *Proc. Int. Seaweed Symp.* 9: 263–71.

Doty, M. S. & Alvarez, V. B. 1975. Status, problems, advances, and economics of *Eucheuma* farms. *Mar. Technol. Soc. J.* 9: 30–5.

Drake, J. W. & Baltz, R. A. 1976. The biochemistry of mutagenesis. *Annu. Rev. Biochem.* 45: 11–38.

Drew, K. M. 1949. Conchocelis-phase in the life-history of *Porphyra umbilicalis* (L.) Kütz. *Nature* (Lond.) 164: 748–9.

Egelhoff, T. & Grossman, A. 1983. Cytoplasmic and chloroplast synthesis of phycobilisome polypeptides. *Proc. Natl. Acad. Sci. USA* 80: 3339–43.

Fain, S. R., Druehl, L. D. & Baillie, D. L. 1988. Repeat and single copy sequences are differentially conserved in the evolution of kelp chloroplast DNA. *J. Phycol.* 24: 292–302.

Fjeld, A. & Løvlie, A. 1976. Genetics of multicellular marine algae. In *Botanical Monographs* Vol. 12. *The Genetics of Algae*, ed. R. A. Lewin, pp. 219–35. Oxford: Blackwell Scientific.

Fujita, Y. & Migita, S. 1984. Photosynthetic pigments of the different color types of *Porphyra yezoensis* Ueda and *P. tenera* Kjellman. *Bull. Fac. Fish. Nagasaki Univ.* 56: 7–13 (in Japanese with English abstract).

Fujita, Y. & Migita, S. 1987. Fusion of protoplasts from thalli of two different color types in *Porphyra yezoensis* Ueda and development of fusion products. *Jap. J. Phycol.* 35: 201–8.

Goff, L. J. & Coleman, A. W. 1988. The use of plastid DNA restriction endonuclease patterns in delineating red algal species and populations. *J. Phycol.* 24: 357–68.

Goldschmidt-Clermont, M. 1986. The two genes for the small subunit of RuBP carboxylase/oxygenase are closely linked in *Chlamydomonas reinhardtii*. *Plant Mol. Biol.* 6: 13–21.

Guiry, M. D. 1984. Structure, life history and hybridization of Atlantic *Gigartina teedii* (Rhodophyta) in culture. *Br. Phycol. J.* 19: 37–55.

Hansen, J. E. 1984. Strain selection and physiology in the development of *Gracilaria* mariculture. *Proc. Int. Seaweed Symp.* 11: 89–94.

Hassinger-Huizinga, H. 1953. Generationswechsel und Geschechtsbestimmung bei *Callithamnion corybosum* (Sm.) Lyngb. *Arch. Protistenk.* 98: 19–124.

Hawkes, M. W. 1978. Sexual reproduction in *Porphyra gardneri* (Smith et Hollenberg) Hawkes (Bangiales, Rhodophyta). *Phycologia* 17: 329–53.

Heinhorst, S. & Shively, J. M. 1983. Encoding of both subunits of ribulose-1,5-bisphosphate carboxylase by organelle genome of *Cyanophora paradoxa*. *Nature* (Lond.) 304: 373–4.

Hori, H. & Osawa, S. 1987. Origin and evolution of organisms as deduced from 5S ribosomal RNA sequences. *Mol. Biol. Evol.* 4: 445–72.

Kirk, J. T. O. & Tilney-Bassett, R. A. E. 1978. *The Plastids: Their Chemistry, Structure, Growth and Inheritance*, 2nd ed. Amsterdam: Elsevier/North Holland Biomedical.

Kugrens, P. & West, J. A. 1972. Ultrastructure of spermatial development in the parasitic red alga *Levringiella gardneri* and *Erythrocystis saccata*. *J. Phycol.* 8: 331–43.

Kuhsel, M. & Kowallik, K. V. 1987. The plastome of a brown alga, *Dictyota dichotoma* II. Location of structural genes coding for ribosomal RNAs, the large subunit of ribulose-1,5-bisphosphate carboxylase/oxygenase and for polypeptides of photosystem I and II. *Mol. Gen. Genet.* 207: 361–8.

Kursar, T. A., van der Meer, J. P. & Alberte, R. S. 1983a. Light-harvesting system of the red alga *Gracilaria tikvahiae* I. Biochemical analysis of pigment mutations. *Plant Physiol.* 73: 353–60.

Kursar, T. A., van der Meer, J. P. & Alberte, R. S. 1983b. Light-harvesting system of the red alga *Gracilaria tikvahiae* II. Phycobilisome characteristics of pigment mutants. *Plant Physiol.* 73: 361–9.

Lambert, D. H., Bryant, D. A., Stirewalt, V. L., Dubbs, J. M., Stevens, S. E., Jr. & Porter, R. D. 1985. Gene map for the *Cyanophora paradoxa* cyanelle genome. *J. Bact.* 164: 659–64.

Lee, Y. K. & Tan, H. M. 1987. Genetic recombination through protoplast fusion in algae. Abstracts of the 4th International Meeting of the Société pour l'Algologie Appliqué. In *Recent Progress in Algal Biotechnology*, eds. T. Stadler, Y. Karamanos, H. Morvan, M. C. Verdus, J. Mollion & D. Christiaen. Villeneuve d'Ascq, France: Université des Sciences et Techniques de Lille.

Lemaux, P. G. & Grossman, A. R. 1984. Isolation of genes encoding phycobiliprotein subunits. *Plant Physiol.* 75 (Suppl. 1): 64.

Lemaux, P. G. & Grossman, A. R. 1985. Major light harvesting polypeptides encoded in polycistronic transcripts in a eukaryotic alga. *EMBO* 4: 1911–19.

Lewin, R. A. (ed.) 1976. *The Genetics of Algae, Botanical Monographs* Vol. 12. Oxford: Blackwell Scientific.

Li, N. & Cattolico, R. A. 1987. Chloroplast genome characterization in red alga *Griffithsia pacifica*. *Mol. Gen. Genet.* 209: 343–51.

Lim, B.-L., Kawai, H., Hori, H & Osawa, S. 1986. Molecular evolution of 5S ribosomal RNA from red and brown algae. *Jap. J. Genet.* 61: 169–76.

Liu, W. S., Tang, Y. L., Liu, X. W. & Fang, T. C. 1984. Studies on the preparation and properties of sea snail enzymes. *Proc. Int. Seaweed Symp.* 11: 319–20.

Loppes, R., Denis, C. & Bairiot-Delcoux, M. 1985. Isolation of *Cyanidium caldarium* and *Porphyridium purpureum* DNA fragments capable of autonomous replication in yeast. *J. Gen. Microbiol.* 131: 1745–51.

Merrill, J. E., Mimuro, M., Aruga, Y. & Fujita, Y. 1983. Light-harvesting for photosynthesis in four strains of the red alga *Porphyra yezoensis* having different phycobilin contents. *Plant Cell Physiol.* 24: 261–6.

Migita, S. & Fujita, Y. 1983. Studies on the color mutant types of *Porphyra yezoensis* Ueda, and their experimental culture. *Bull. Fac. Fish. Nagasaki Univ.* 54: 55–60 (in Japanese with English abstract).

Miura, A. 1975. Studies on the breeding of cultivated *Porphyra* (Rhodophyceae). *Third Int. Ocean Development Conference*, August 5–8, 1975, Tokyo, Reprint Vol. III, Marine Resources, pp. 81–93.

Miura, A. 1977. Chimeral variegation found in *Porphyra*. *J. Phycol.* 13 (Suppl.): 45.

Miura, A. 1979. Studies on genetic improvement of cultivated *Porphyra* (Laver). In *Proc. 7th Japan–Soviet Joint Symp. Aquaculture*, ed. G. Yamamoto, pp. 161–8. Tokyo: Tokai University Press.

Miura, A. 1985. Genetic analysis of the variant color types of light red, light green and light yellow phenotypes of *Porphyra yezoensis* (Rhodophyceae, Bangiaceae). In *Origin and Evolution of Diversity in Plants and Communities*, ed. H. Hara, pp. 270–84. Tokyo: Academia Scientific Book.

Miura, A. & Kunifuji, Y. 1980. Genetic analysis of the pigmentation types in the seaweed Susabi-Nori (*Porphyra yezoensis*). *Iden* 34: 14–20 (in Japanese).

Neish, A. C. & Fox, C. H. 1971. Greenhouse experiments on the vegetative propagation of *Chondrus crispus* (Irish Moss). *Tech. Rep. Atl. Reg. Lab. Nat. Res. Counc. Can. Halifax* 12: 1–35.

Nichols, H. W. 1980. Polyploidy in algae. In *Polyploidy: Biological Relevance*, ed. W. H. Lewis, pp. 151–62. New York: Plenum.

Ohme, M., Kunifuji, Y. & Miura, A. 1986. Cross experiments of the color mutants in *Porphyra yezoensis* Ueda. *Jap. J. Phycol.* 34: 101–6.

Ohyama, K., Fukuzawa, H., Kohchi, T., Shirai, H., Sano, T., Sano, S., Umesono, K., Shiki, Y., Takeuchi, M. Z., Chang, Z., Aota, S., Inokuchi, H. & Ozeki, H. 1986. Chloroplast gene organization deduced from complete sequence of liverwort *Marchantia polymorpha* chloroplast DNA. *Nature* (Lond.) 322: 572–4.

Palmer, J. D. 1985. Comparative organization of chloroplast genomes. *Annu. Rev. Genet.* 19: 325–54.

Palmer, J. D. 1986. Evolution of chloroplast and mitochondrial DNA in plants and algae. In *Monographs in Evolutionary Biology: Molecular Evolutionary Genetics*, ed. R. J. MacIntyre, pp. 131–240. New York: Plenum.

Patwary, M. U. & van der Meer, J. P. 1982. Genetics of *Gracilaria tikvahiae* (Rhodophyceae). VIII. Phenotypic and genetic characterization of some selected morphological mutants. *Can. J. Bot.* 60: 2556–64.

Patwary, M. U. & van der Meer, J. P. 1983a. Improvement of *Gracilaria tikvahiae* (Rhodophyceae) by genetic modification of thallus morphology. *Aquaculture* 33: 207–14.

Patwary, M. U. & van der Meer, J. P. 1983b. An apparent absence of heterosis in hybrids of *Gracilaria tikvahiae* (Rhodophyceae). *Proc. N. S. Inst. Sci.* 33: 95–9.

Patwary, M. U. & van der Meer, J. P. 1983c. Genetics of *Gracilaria tikvahiae* (Rhodophyceae). IX: Some properties of agars extracted from morphological mutants. *Bot. Mar.* 26: 295–9.

Patwary, M. U. & van der Meer, J. P. 1983d. Growth experiments on morphological mutants of *Gracilaria tikvahiae* (Rhodophyceae). *Can. J. Bot.* 61: 1654–9.

Patwary, M. U. & van der Meer, J. P. 1984. Growth experiments on autopolyploids of *Gracilaria tikvahiae* (Rhodophyceae). *Phycologia* 23: 21–7.

Pilot, T. J. & Fox, J. L. 1984. Cloning and sequencing of the genes encoding the α and β subunits of C-phycocyanin from the cyanobacterium *Agmenellum quadruplicatum*. *Proc. Natl. Acad. Sci. USA* 81: 6983–7.

Polne-Fuller, M. & Gibor, A. 1984. Developmental studies in *Porphyra*. I. Blade differentiation in *Porphyra perforata* as expressed by morphology, enzymatic digestion, and protoplast regeneration. *J. Phycol.* 20: 609–16.

Polne-Fuller, M. & Gibor, A. 1987. Calluses and callus-like growth in seaweeds: Induction and culture. *Hydrobiologia* 151/152: 131–8.

Polne-Fuller, M., Biniaminov, M. & Gibor, A. 1984. Vegetative propagation of *Porphyra perforata*. *Proc. Int. Seaweed Symp.* 11: 308–13.

Polne-Fuller, M., Saga, N. & Gibor, A. 1986. Algal cell, callus, and tissue cultures and selection of algal strains. In *Algal Biomass Technologies: An Interdisciplinary Perspective*, eds. W. R. Barclay & R. P. McIntosh, pp. 30–6. Berlin: J. Cramer.

Pueschel, C. M. & van der Meer, J. P. 1984. Ultrastructural characterization of a pigment mutant of the red alga *Palmaria palmata*. *Can. J. Bot.* 62: 1101–7.

Reith, M. & Cattolico, R. A. 1986. Inverted repeat of *Olisthodiscus luteus* chloroplast DNA contains genes for both subunits of ribulose-1,5-biphosphate carboxylase and the 32,000-dalton Q_B protein: Phylogenetic implications. *Proc. Natl. Acad. Sci. USA* 83: 8599–603.

Rueness, J. & Rueness, M. 1973. Life history and nuclear phases of *Antithamnion tenuissimum*, with special reference to plants bearing both tetrasporangia and spermatangia. *Norw. J. Bot.* 20: 205–10.

Rueness, J. & Rueness, M. 1975. Genetic control of morphogenesis in two varieties of *Antithamnion plumula* (Rhodophyceae, Ceramiales). *Phycologia* 14: 81–5.

Ryther, J. H., Williams, L. D., Hanisak, M. D., Stenberg, R. W. & DeBusk, T. A. 1979. Biomass production by marine and freshwater plants. In *Third Annual Biomass Energy Systems Conference Proceedings*, pp. 13–23. Golden, Colo.: Solar Energy Research Institute.

Saga, N., Polne-Fuller, M. & Gibor, A. 1986. Algal cell, callus, and tissue cultures and selection of algal strains. In *Algal Biomass Technologies: An Interdisciplinary Perspective*, eds. W. R. Barclay & R. P. McIntosh, pp. 37–43. Berlin: J. Cramer.

Schopf, J. W. 1970. Pre-Cambrian micro-organisms and evolutionary events prior to the origin of vascular plants. *Biol. Rev.* 45: 319–52.

Scott, J. L & Dixon, P. S. 1973. Ultrastructure of spermatium liberation in the marine red alga *Ptilota densa*. *J. Phycol.* 9: 85–91.

Spencer, K. G., Yu, M.-H., West, J. A. & Glazer, A. N. 1981. Characterization of green mutants of *Callithamnion byssoides* (Rhodophyceae). *J. Phycol.* 17 (Suppl.): 15.

Steele, R. L., Thursby, G. B. & van der Meer, J. P. 1986. Genetics of *Champia parvula* (Rhodymeniales, Rhodophyta): Mendelian inheritance of spontaneous mutants. *J. Phycol.* 22: 538–42.

Steinmüller, K., Kaling, M. & Zetsche, K. 1983. In-vitro synthesis of phycobiliproteins and ribulose-1,5-bisphosphate carboxylase by non-poly-adenylated-RNA of *Cyanidium caldarium* and *Porphyridium aerugineum*. *Planta* (Berl.) 159: 308–13.

Suto, S. 1963. Intergeneric and interspecific crossings of

the lavers (*Porphyra*). *Bull. Jap. Soc. Sci. Fish.* 29: 739–48.

Takaiwa, F., Kusuda, M., Saga, N. & Sugiura, M. 1982. The nucleotide sequence of 5S rRNA from the red alga, *Porphyra yezoensis*. *Nucleic Acids Res.* 10: 6037–40.

van der Meer, J. P. 1977. Genetics of *Gracilaria* sp. (Rhodophyceae, Gigartinales). II. The life history and genetic implications of cytokinetic failure during tetraspore formation. *Phycologia* 16: 367–71.

van der Meer, J. P. 1978. Genetics of *Gracilaria* sp. (Rhodophyceae, Gigartinales) III. Non-Mendelian gene transmission. *Phycologia* 17: 314–8.

van der Meer, J. P. 1979a. Genetics of *Gracilaria* sp. (Rhodophyceae, Gigartinales). V. Isolation and characterization of mutant strains. *Phycologia* 18: 47–54.

van der Meer, J. P. 1979b. Genetics of *Gracilaria tikvahiae* (Rhodophyceae). VI. Complementation and linkage analysis of pigmentation mutants. *Can. J. Bot.* 57: 64–8.

van der Meer, J. P. 1981a. Genetics of *Gracilaria tikvahiae* (Rhodophyceae). VII. Further observations on mitotic recombination and the construction of polyploids. *Can. J. Bot.* 59: 787–92.

van der Meer, J. P. 1981b. Sexual reproduction in the Palmariaceae. *Proc. Int. Seaweed Symp.* 10: 191–6.

van der Meer, J. P. 1981c. The life history of *Halosaccion ramentaceum*. *Can. J. Bot.* 59: 433–6.

van der Meer, J. P. 1981d. The inheritance of spontaneous pigment mutations in *Chondrus crispus* Stackh. (Rhodophyceae). *Proc. N. S. Inst. Sci.* 31: 187–92.

van der Meer, J. P. 1986a. Genetic contributions to research on seaweeds. *Prog. Phycol. Res.* 4: 1–38.

van der Meer, J. P. 1986b. Genetics of *Gracilaria tikvahiae* (Rhodophyceae). XI. Further characterization of a bisexual mutant. *J. Phycol.* 22: 151–8.

van der Meer, J. P. 1987a. Experimental hybridizations of *Palmaria palmata* (Rhodophyta) from the northeast and northwest Atlantic Ocean. *Can. J. Bot.* 65: 1451–8.

van der Meer, J. P. 1987b. Using genetic markers in phycological research. *Hydrobiologia* 151/152: 49–56.

van der Meer, J. P. & Bird, N. L. 1977. Genetics of *Gracilaria* sp. (Rhodophyceae, Gigartinales). 1. Mendelian inheritance of two spontaneous green variants. *Phycologia* 16: 159–61.

van der Meer, J. P. & Chen, L. C.-M. 1979. Evidence for sexual reproduction in the red algae *Palmaria palmata* and *Halosaccion ramentaceum*. *Can. J. Bot.* 57: 2452–9.

van der Meer, J. P., Guiry M. D. & Bird, C. J. 1983. Sporogenesis in male plants of *Chondrus crispus* (Rhodophyta, Gigartinales). *Can. J. Bot.* 61: 2261–8.

van der Meer, J. P., Patwary, M. U. & Bird, C. J. 1984. Genetics of *Gracilaria tikvahiae* (Rhodophyceae). X. Studies on a bisexual clone. *J. Phycol.* 20: 42–6.

van der Meer, J. P. & Todd, E. R. 1977. Genetics of *Gracilaria* sp. (Rhodophyceae, Gigartinales). IV. Mitotic recombination and its relationship to mixed phases in the life history. *Can. J. Bot.* 55: 2810–17.

van der Meer, J. P. & Todd, E. R. 1980. The life history of *Palmaria palmata* in culture. A new type for the Rhodophyta. *Can. J. Bot.* 58: 1250–6.

van der Meer, J. P. & Zhang, X. C. 1988. Similar unstable mutations in three species of *Gracilaria*. *J. Phycol.* 24: 198–202.

von Stosch, H. A. 1969. Observations on *Corallina*, *Jania*, and other red algae in culture. *Proc. Int. Seaweed Symp.* 6: 389–99.

Waaland, J. R. 1979. Growth and strain selection in *Gigartina exasperata* (Florideophyceae). *Proc. Int. Seaweed Symp.* 9: 241–7.

Yamanaka, G. & Glazer, A. N. 1981. Dynamic aspects of phycobilisome structure : modulation of phycocyanin content of *Synechococcus* phycobilisomes. *Arch. Microbiol.* 130: 23–30.

Zhang, X. C. & van der Meer, J. P. 1987. A study on heterosis in diploid gametophytes of the marine red alga *Gracilaria tikvahiae*. *Bot. Mar.* 30: 309–14.

Zhang, X. C. & van der Meer, J. P. 1988a. Polyploid gametophytes of *Gracilaria tikvahiae* (Gigartinales, Rhodophyta). *Phycologia* 27: 312–18.

Zhang, X. C. & van der Meer, J. P. 1988b. A genetic study on *Gracilaria sjoestedtii*. *Can. J. Bot.* 66: 2022–6.

Zhao, H. D. & Zhang, X. C. 1981. Isolation and cultivation of the vegetative cells of *Porphyra yezoensis* Ueda. *J. Shandong Coll. Oceanol.* 11: 61–6 (in Chinese, with English abstract).

Chapter 6

Cell division

JOE SCOTT
SHARON BROADWATER

CONTENTS

I. Introduction / 123
II. Composite view of mitosis / 124
III. Chromosomes and kinetochores / 125
IV. Nucleus-associated organelles / 126
V. Spindle poles / 133
VI. Spindle / 135
VII. PER and other membranes / 137
VIII. Cytokinesis / 138
IX. Meiosis / 138
X. Taxonomic and evolutionary considerations / 142
XI. Acknowledgments / 144
XII. References / 144

I. INTRODUCTION

Red algae (Rhodophyta) comprise one of the most difficult groups of organisms to place in an evolutionary scheme (Cavalier-Smith 1981; Demoulin 1985; Stewart & Mattox 1980). One of the major problems relates to two common rhodophytan features at the subcellular level, the absence of the flagellar apparatus, and the presence of unique, seemingly primitive chloroplasts. However, the red algal cell in general is not demonstrably more primitive than many other eukaryotic algal cells, a fact not so surprising if one accepts the endosymbiont theory for the evolution of chloroplasts.

Not only is there controversy surrounding the ancestry of this group and its relationship to other organisms, there is also much debate over taxonomic relationships within the red algae. Recently, a great deal of attention has focused on these problems (Gabrielson & Garbary 1986; Gabrielson et al. 1985; Garbary et al. 1980; Kraft & Robins 1985; Pueschel & Cole 1982; Silva & Johansen 1986), but much work yet remains in order to understand fully the evolutionary relationships within this group (see Chapters 1 and 8).

The study of cell division was of incalculable value in revising the taxonomy of the green algae (Pickett-Heaps 1975; Stewart & Mattox 1975) and of major importance in recent modifications of fungal classification (Fuller 1976; Heath 1980a, 1981a,b). It is hoped that a thorough examination of cell division within the Rhodophyta will prove equally useful. This review will be limited primarily to ultrastructural studies since light microscopic work is covered elsewhere (Chapters 3 and 12). Unfortunately, electron microscopic studies of cell division have been published on only a very few genera (Table 6-1). This paucity of information is due to the difficulty of obtaining sufficiently large numbers of dividing cells. Compounding this problem is the fact that relatively few people are active in this area of research.

Table 6-1. Electron microscopic studies of red algal nuclear division

Taxon	Reference
Bangiophycideae	
Porphyridiales	
Porphyridiaceae	
Porphyridium purpureum	Bronchart & Demoulin 1977
	Schornstein & Scott 1980, 1982
Phragmonemataceae	
Flintiella sanguinaria	Scott 1986
Compsopogonales	
Erythropeltidales	
Rhodochaetales	
Bangiales	
Florideophycideae	
Acrochaetiales	
Nemaliales	
Batrachospermales	
Batrachospermum ectocarpum	Scott 1983
Gelidiales	
Hildenbrandiales	
Corallinales	
Palmariales	
Bonnemaisoniales	
Cryptonemiales	
Gigartinales	
Rhodymeniales	
Lomentariaceae	
Lomentaria baileyana	Davis & Scott 1986
Ceramiales	
Ceramiaceae	
Griffithsia flosculosa	Peyrière 1971
Pleonosporium vancouverianum	Sheath et al. 1987*
Delesseriaceae	
Apoglossum ruscifolium	Dave & Godward 1982
Membranoptera platyphylla	McDonald 1972
Dasyaceae	
Dasya baillouviana	Phillips & Scott 1981
	Broadwater & Scott 1986a*, b*
Rhodomelaceae	
Polysiphonia denudata	Scott et al. 1980
Polysiphonia harveyi	Scott et al. 1980

* Asterisk indicates meiosis.

II. COMPOSITE VIEW OF MITOSIS

The first electron microscopic investigations of mitosis treated only two orders, the Ceramiales, which are usually considered to be the most advanced order of red algae, and the Porphyridiales, which are considered to be among the least advanced. Most work concentrated on the Ceramiales (Dave & Godward 1982; McDonald 1972; Peyrière 1971; Phillips & Scott 1981; Scott et al. 1980) and showed a high degree of uniformity, but investigations of *Porphyridium* (Bronchart & Demoulin 1977; Schornstein & Scott 1980, 1982) revealed characteristics significantly different from those in the Ceramiales. These differences emphasized the need for more studies from a broad array of red algae before phylogenetic implications could be made. The next three papers published on cell division broadened the spectrum with studies of *Batrachospermum* (Batrachospermales; Scott 1983), *Flintiella* (Porphyridiales; Scott 1986), and *Lomentaria* (Rhodymeniales; Davis & Scott 1986). All of the studies were conducted on uninucleate cells; however, in several multicellular genera, examination of nuclear division in multinucleate cells showed no apparent differences. In total, studies of mitosis in 11 species representing four orders have been completed. Two of these studies were on unicells; the remainder were on vegetative (three) and reproductive (six) cells of multicellular species.

A compilation of the data suggests there may be five ultrastructural patterns of mitosis among the red algae, although the limited number of studies makes any phylogenetic deductions tentative. Consequently, characteristics of possible importance (Table 6-2) are discussed in detail under separate headings (Sections III to VIII), with the better known ceramialean condition usually presented first. These disparate characters are tied together in Section X to present the different mitotic "types" that seem the best interpretation of current information.

Before discussing observed differences, however, it is important to note those aspects of cell division and cytokinesis that are shared by all genera studied. Dividing nuclei appear to have typical eukaryotic spindles composed of two opposing sets of interdigitating microtubules. The mature spindles are enclosed within the nuclear envelope, which remains intact throughout the process of mitosis except at the polar regions. These areas have interruptions in the form of either variously sized gaps or smaller fenestrations. The nucleolus largely disperses by metaphase; chromatin condenses normally and aggregates into an equatorial arrangement (see also chapter 12). Chromosomes are generally very small and are attached to the spindle by kinetochores. Each division pole contains a "nucleus-associated organelle" (NAO) that, with the exception of *Porphyridium*, has a fairly consistent ring-shaped morphology. All investigated genera undergo the normal sequence of mitosis from prophase to telophase, and all produce an interzonal

Table 6-2. Ultrastructural features of nuclear division in red algae

Feature	Porphyridium	Flintiella	Batrachospermum	Lomentaria	Polysiphonia[a]
Cell organization	Uninucleate unicell	Uninucleate unicell	Uninucleate filament	Multinucleate filament	Multinucleate filament
NAO type	Bipartite: large proximal granule and small distal cylinder	Bipartite: small ring upon a ring	Bipartite: small ring within a larger ring	Bipartite: large ring upon a ring	Bipartite: large ring upon a ring
Late prophase poles	One nuclear pocket at each pole	Two nuclear pockets at each pole	Two nuclear pockets at each pole	Two nuclear pockets at each pole	One NE protrusion at each pole
Prometaphase MT-filled cytoplasmic tunnels traversing nucleus	(−)	(+)	(+)	(−)	(−)
Metaphase NE condition	Intact except for one small gap at each pole	Intact except for one large gap at each pole	Intact except for one intermediate-sized gap at each pole	Intact except for two small gaps and numerous fenestrations at each pole	Intact except for numerous fenestrations at each pole
PER	(−)	(−)	(+)	(++)	(+++)
Kinetochore structure	Small, simple	Small, simple	(?)	Medium-sized, layered	Large, layered
Organelle associated with cytokinesis	Enlarged central chloroplast	Enlarged central chloroplast	Central vacuole	Central vacuole	Central vacuole

[a] *Polysiphonia* is just one of several genera studied in the Ceramiales, all of which share comparable features.
(+), presence and/or relative abundance of feature; (−), absence of feature; (?), not known.

spindle midpiece that persists until two separate daughter nuclei are produced.

Meiosis has been investigated thoroughly in only one species, *Dasya baillouviana* (Broadwater et al. 1986a,b) and the results indicate that characteristics of meiosis, for the most part, are basically similar to those observed in the mitotic studies. Discussion of meiosis has been placed under a separate heading (Section IX).

III. CHROMOSOMES AND KINETOCHORES

In the morphologically advanced red algae, a heterochromatin network is present in interphase–prophase nuclei but is much less pronounced in simpler forms (Fig. 6-1). Prometaphase–anaphase chromosomes in the former group are relatively easy to recognize as discrete structures (Figs. 6-2 to 6-4). At anaphase chromosomes move to the poles attached to microtubules and can still be seen at telophase appressed to the inner membrane of the nuclear envelope. In contrast, clumping of metaphase chromosomes in the simpler forms (Figs. 6-5, 6-26) makes individual chromosomes difficult to distinguish. In *Porphyridium*, Schornstein and Scott (1982) concluded that the chromosomes are very poorly condensed, a factor that could account for the difficulty in differentiating individual chromosomes and the low chromosome counts determined by light microscopy. By counting kinetochore numbers in metaphase nuclei, and assuming a one-to-one relationship between kinetochores and chromosomes, Schornstein and Scott (1982) obtained a chromosome count of 8–10 for *Porphyridium*, rather than the previously determined number of two (Sommerfeld & Nichols 1970).

All red algal chromosomes appear to have kine-

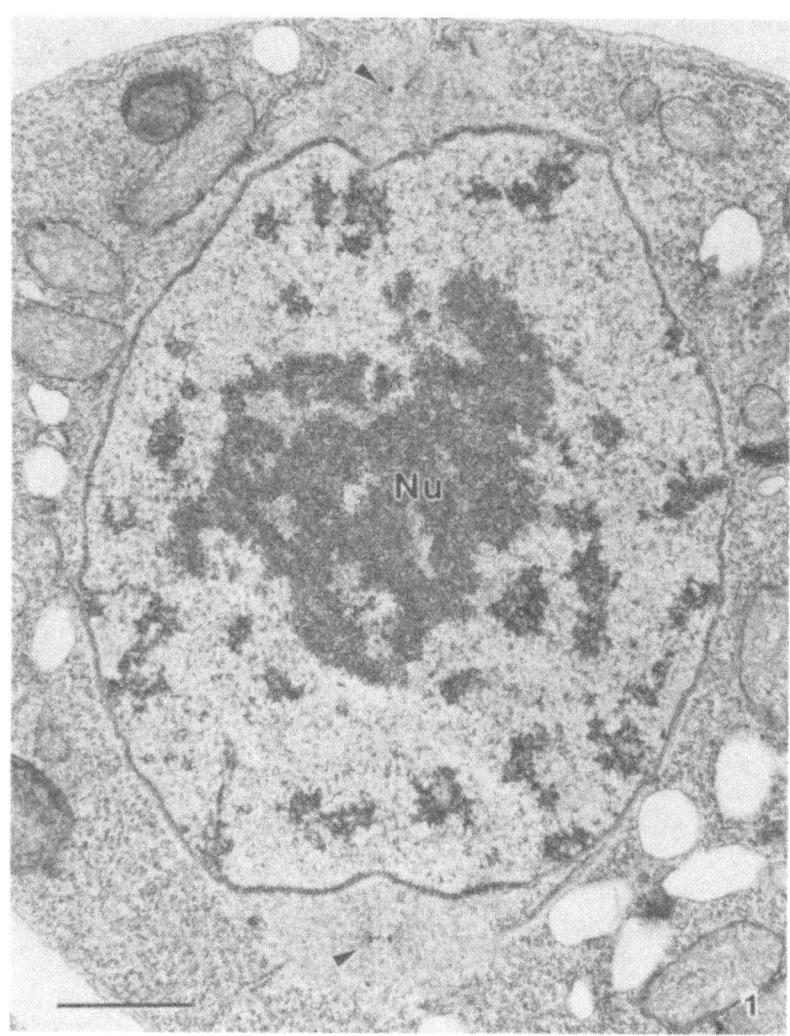

Figs. 6-1 to 6-31. *Mitosis.*
Fig. 6-1. Longitudinally sectioned late prophase nucleus of *Chondria baileyana* showing large polar zones of exclusion. Note an NAO (arrowheads) in each exclusion zone, condensing euchromatin and partially dispersed nucleolus (*Nu*). Scale = 1 μm.

tochores. In some genera – *Lomentaria* (Davis & Scott 1986), *Porphyridium* (Schornstein & Scott 1982), *Batrachospermum* (Scott 1983), *Flintiella* (Scott 1986) – the kinetochores are indistinct and not well defined (Fig. 6-5), whereas in other genera – *Apoglossum* (Dave & Godward 1982), *Membranoptera* (McDonald 1972), *Dasya* (Phillips and Scott 1981), *Polysiphonia* (Scott et al. 1980) – they display a typical three-layered morphology (Fig. 6-3). Only one microtubule per kinetochore was found in *Porphyridium* (Fig. 6-5; Schornstein & Scott 1982) and *Flintiella* (Fig. 6-30; Scott 1986), whereas up to seven were seen in *Polysiphonia* (Scott et al. 1980). The number of microtubules per kinetochore was not determined for the other species examined but appeared to be much lower than 10.

IV. NUCLEUS-ASSOCIATED ORGANELLES

Most lower eukaryotes have either centrioles or some other type of structure at the spindle poles during nuclear division. The noncentriolar structures, most appropriately termed nucleus-associated organelles (NAOs) (Heath 1980a,b), can vary considerably in size and morphology, even within taxonomic groups, although some groups such as the Basidiomycetes and Ascomycetes show a high degree of intragroup homogeneity (Heath 1981b).

The red algae fall into the latter category. All of the species investigated to date, except *Porphyridium*, have a pair of short, hollow, cylindrical NAOs generally referred to as "polar rings" (Figs. 6-1, 6-7 to 6-14, 6-17 to 6-21, 6-23, 6-24; McDonald 1972).

Cell division

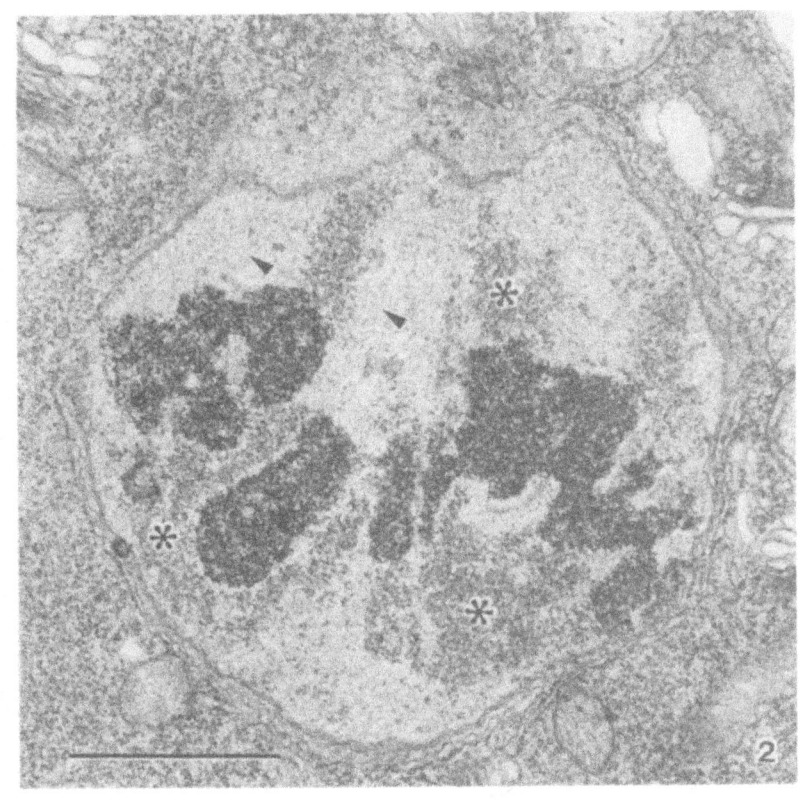

Fig. 6-2. Oblique longitudinal section of a prometaphase nucleus of *Polysiphonia harveyi*. Nucleolar fragments (asterisks) are associated both with the condensing chromosomes and intranuclear spindle microtubules (arrowheads). Perinuclear ER partially surrounds the nucleus. Scale = 1 μm. (From Scott et al. 1980, with permission)

Figure 6-22 is a scale drawing of prophase NAOs, representing their variation in red algae seen to date. *Griffithsia* (Peyrière 1971) was the first alga for which a micrograph of an NAO was published. Although it was referred to as a "polaire et corpuscule dense," it is typical for ceramialean NAOs, which show a high degree of conformity in both structure and size. In all cases the ceramialean NAO at early prophase appears as a double-ringed structure with one doughnut-shaped (distal) portion directly above and attached to a second similarly shaped (proximal) portion. The proximal portion appears to be attached to the nuclear envelope by fine struts. The orientation is such that the longitudinal axis of the NAO is always perpendicular to the nuclear envelope. The diameter of the NAO varies from 120 nm in *Dasya* (Phillips & Scott 1981) to 190 nm in *Membranoptera* (McDonald 1972), and the height varies from 35 nm in *Dasya* to 70 nm in *Membranoptera*.

Lomentaria (Rhodymeniales; Davis & Scott 1986) and *Agardhiella* (Gigartinales; Klepacki & Scott unpubl.) have NAOs whose morphology and size fit within the range found in the Ceramiales. However, in interphase–prophase cells of these algae an extension of the outer membrane of the nuclear envelope runs through the hollow center of each NAO (Figs. 6-23, 6-24). Though not significantly different in shape, NAOs in the unicellular red algae *Flintiella* (Scott 1986), *Rhodella reticulata* (Figs. 6-19 to 6-21; Scott et al. unpubl.), *Rhodella violacea* (Patrone & Scott unpubl.), and *Rhodosorus marinus* (Scott unpubl.), as well as the macroscopic bangiophytes *Compsopogon* (Fig. 6-18; Scott & Broadwater unpubl.) and *Boldia* (Fig. 6-17; Scott unpubl.) are reduced in size. The NAOs of *R. violacea*, for example, have a diameter of 30 nm and a height of 16 nm. Although *Batrachospermum* has an NAO with dimensions comparable to those in the Ceramiales (outside diameter, 120 nm; height, 50 nm), its morphology is quite different (Figs. 6-12, 6-22C; Scott 1983). It appears to have a ring-within-a-ring configuration, with a smaller ring (70 × 30 nm) tucked inside the larger one.

The only alga studied to date whose NAO is not ring-shaped is *Porphyridium* (Schornstein & Scott 1982). The interphase NAO is an asymmetrical bipartite structure (Figs. 6-15, 6-16, 6-22A): Although the distal component is a cylinder 70 nm in diameter and 50 nm high, it appears predominantly solid

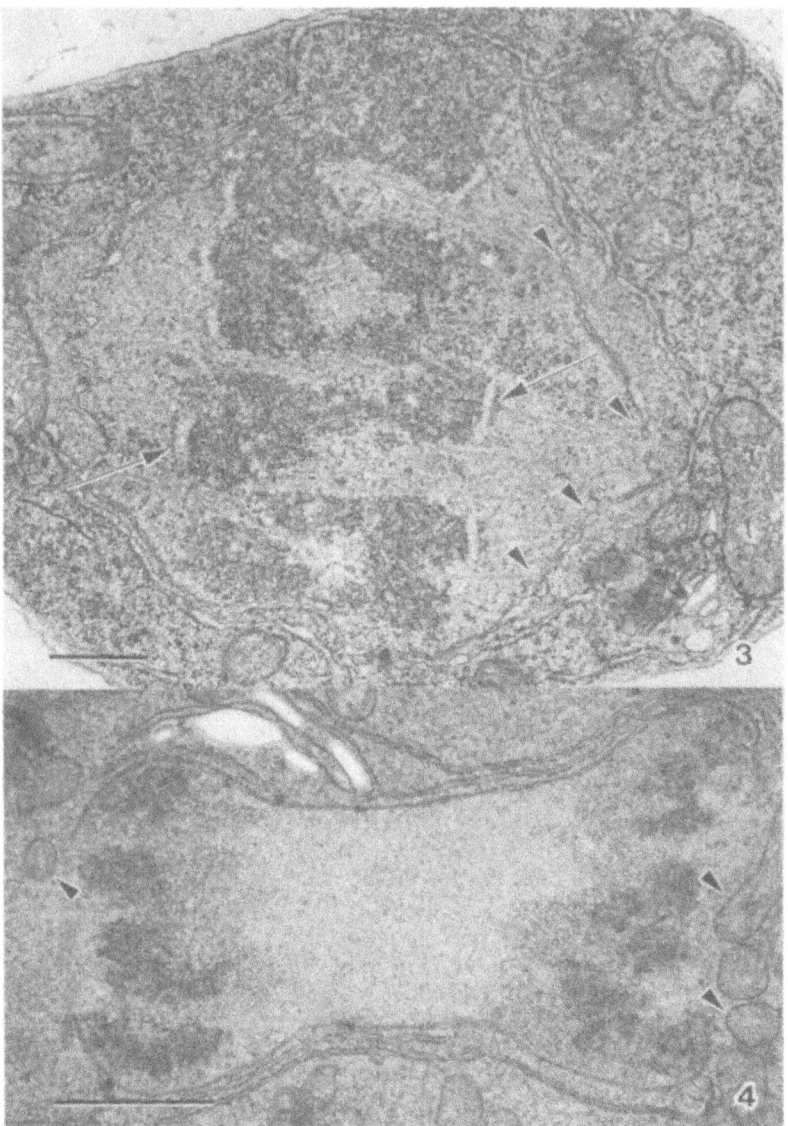

Figs. 6-3, 6-4. 6-3: Early anaphase nucleus of *Polysiphonia harveyi* almost totally surrounded by perinuclear ER. As noted in other planes of sectioning, at least five to seven microtubules are attached to the conspicuous, layered kinetochores (arrows). Microtubules abut the inner membrane of the polar regions of the nuclear envelope along a relatively wide area (arrowheads). Scale = 1 µm. (From Scott et al. 1980, with permission) 6-4: Late anaphase nucleus of *Lomentaria baileyana*. The chromatids have moved to the division poles, separated by an interzonal midpiece containing numerous microtubules (not conspicuous in this plane of sectioning). Mitochondria (arrowheads) are often gathered over the polar regions. Scale = 1 µm. (From Davis & Scott 1986, with permission)

rather than hollow. The cylinder is connected by amorphous material to the proximal portion, which is a poorly defined, electron-dense 140 × 55 nm truncated cone.

Besides morphology, the behavior of the ring-shaped NAOs also differs from that in *Porphyridium*. During preprophase in *Polysiphonia* there are two NAOs attached to the nuclear envelope approxi-

Cell division

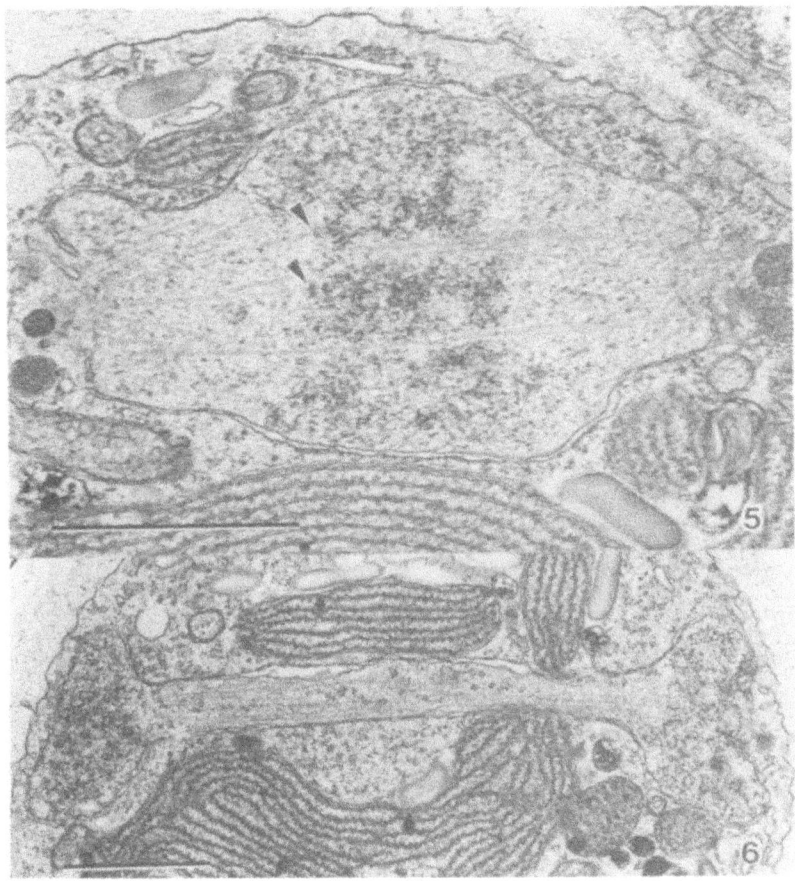

Figs. 6-5, 6-6. 6-5: Metaphase nucleus of *Porphyridium purpureum*, characterized by poorly condensed chromosomes with small kinetochores (arrowheads), each attached to only one microtubule. A single large gap is present at each pole, occupied by an exclusion zone that is the focus of the spindle microtubules. Perinuclear ER is absent. Scale = 1 µm. 6-6: Late anaphase in *Porphyridium* showing an elongated interzonal midpiece containing microtubules separating the nascent daughter nuclei. Scale = 1 µm. (From Schornstein & Scott 1982, with permission)

mately 0.5 µm from each other (Scott et al. 1980). As prophase progresses one migrates 180 degrees along the surface of the nucleus to form the opposing pole. At prometaphase the two halves of each NAO appear to "snap apart," such that the distal portion lies "above" and somewhat lateral to the proximal portion, which is still connected to the nuclear envelope. As evidence of tension prior to the detachment of the proximal portion from the nucleus, the nuclear envelope subjacent to the NAO deforms into an outward bulge called the nuclear envelope protrusion (Fig. 6-7; also seen in *Membranoptera*, Fig. 6-11). The two halves of the NAO remain separated throughout the rest of mitosis, although by late telophase the distal portion moves closer to the nuclear envelope. It then lies alongside the proximal portion, a position similar to the original prophase NAO configuration. This observation led the authors to conjecture that the two NAO halves serve in a semiconservative fashion as templates for NAO biogenesis at either late telophase or early prophase, a time consistent with NAO generation in other organisms (Heath 1981b).

This same behavior occurs in *Dasya* (Phillips & Scott 1981) and *Lomentaria* (Davis & Scott 1986), and is fairly well established for other ceramialean algae studied. There is also some evidence for the same cycle in *Flintiella* (Scott 1986). Very little is known of

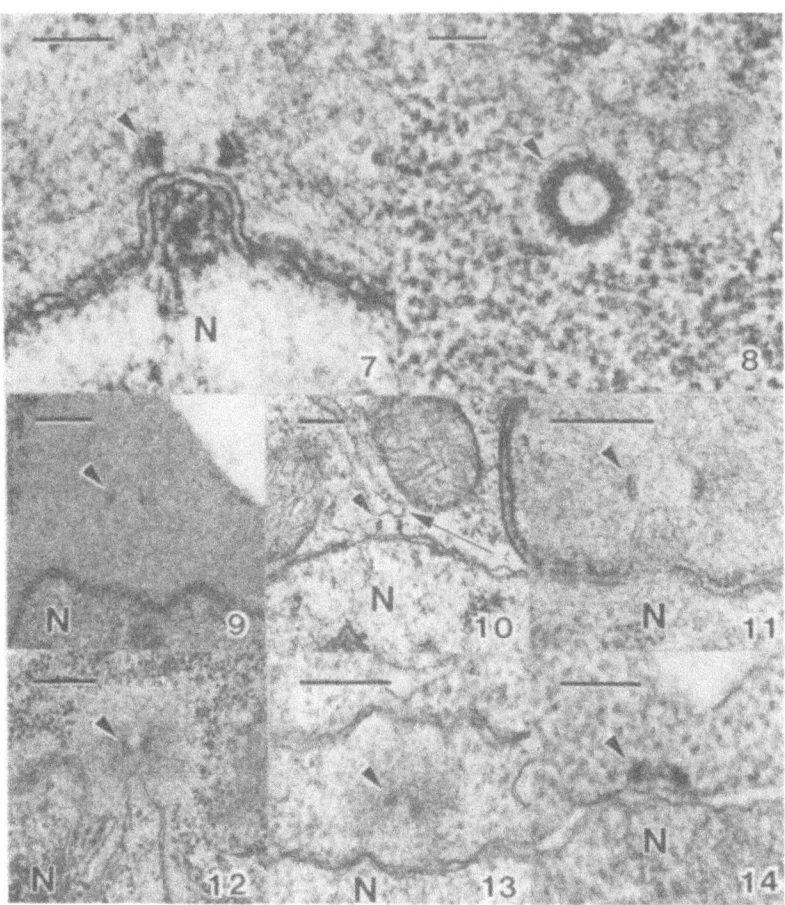

Figs. 6-7 to 6-14. Except for Fig. 6-8, the arrowhead in each figure points to the left side of a longitudinally sectioned NAO. N = nucleus. 6-7: Median, longitudinal section through a polar ring, the most common type of "nucleus-associated organelle" (NAO), and the subtending nuclear envelope protrusion of a late prophase nucleus of *Polysiphonia harveyi*. Scale = 0.1 µm. 6-8: Transverse section through an interphase–early prophase NAO of *P. harveyi*. Note the absence of an exclusion zone at this time. Scale = 0.2 µm. (Figs. 6-7, 6-8 from Scott et al. 1980, with permission) 6-9 to 6-14: NAOs (arrowheads) of several genera of red algae. 6-9: *Antithamnion kylinii* prophase NAO. Scale = 0.2 µm. (Unpubl., courtesy of J. Charleston) 6-10: *Dasya baillouviana*, interphase. Note close association of ER (arrow) with the NAO. Scale = 0.2 µm. 6-11: *Membranoptera platyphylla*. Note the faintly staining NEP associated with the late prophase NAO. Scale = 0.2 µm. (Unpubl., courtesy of K. McDonald) 6-12: *Batrachospermum ectocarpum* early prometaphase NAO. Note conspicuous NAO-associated material here and in next figure. Scale = 0.2 µm. (From Scott 1983, with permission) 6-13: *Palmaria palmata* prophase NAO. Scale = 0.2 µm. (Unpubl., courtesy of C. Pueschel.) 6-14: *Bonnemaisonia nootkana* interphase NAO. Scale = 0.1 µm.

NAO behavior in *Batrachospermum*, but the unconventional NAO design of this genus suggests the pattern will deviate from that of *Polysiphonia*.

The NAO cycle in *Porphyridium* is different from that of any other known eukaryote (Schornstein & Scott 1982). One of the two prophase NAOs appears to move to establish the opposing pole. By late prophase, a conspicuous, pocketlike depression develops beneath each NAO (Fig. 6-27). At the same time, the proximal portion of the NAO becomes diffuse and irregular in shape and soon disperses. Coincident with the disappearance of the proximal portion, the nuclear envelope of the pocket breaks down to form a relatively large gap

Cell division

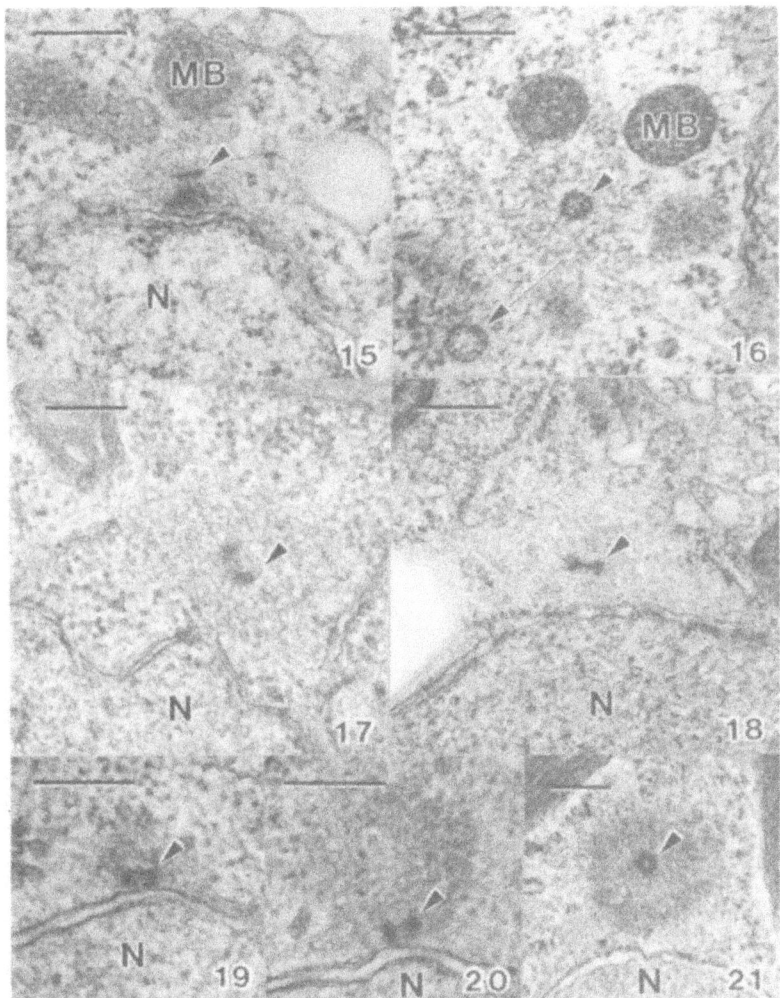

Figs. 6-15 to 6-21. 6-15: Longitudinal section through the bipartite prophase NAO of *Porphyridium purpureum*. The distal NAO portion (arrowhead) sits "above" the relatively larger proximal portion, which is attached to the nucleus (N). Several microbodies (MB) are near the spindle pole. Scale = 0.2 µm. 6-16: Transverse view of the distal portion of a prophase NAO (arrowhead) in *Porphyridium*. Note nuclear pore (arrow). Scale = 0.2 µm. (From Schornstein & Scott 1982, with permission) The arrowhead in Figs. 6-17 to 6-20 points to the right side of an NAO. N = nucleus. Scales = 0.2 µm. 6-17: *Boldia erythrosiphon* prophase NAO. 6-18: *Compsopogon coeruleus* prophase NAO. 6-19 to 6-21: *Rhodella reticulata* NAOs and NAO-associated material. 6-19: Interphase–early prophase; the NAO appears attached to the nucleus, with little or no associated NAO material. 6-20: Late prophase, after establishment of the division poles; note that the conspicuous, large "ball" of NAO-associated material develops "above" the polar ring, which is still attached to the nucleus. 6-21: Late prophase–early metaphase; the NAO, in transverse section (arrowhead), becomes detached from the nucleus and resides in the middle of the NAO-associated material. Scale = 0.2 µm.

(Fig. 6-5). The distal portion of the NAO moves down into a central position in the gap at the nucleoplasm/cytoplasm border. Therefore, only the smaller distal portion of the NAO is present during most of mitosis (Fig. 6-16). It is not known when or how new NAOs are regenerated for the next cell cycle.

Although red algal NAOs are central elements at

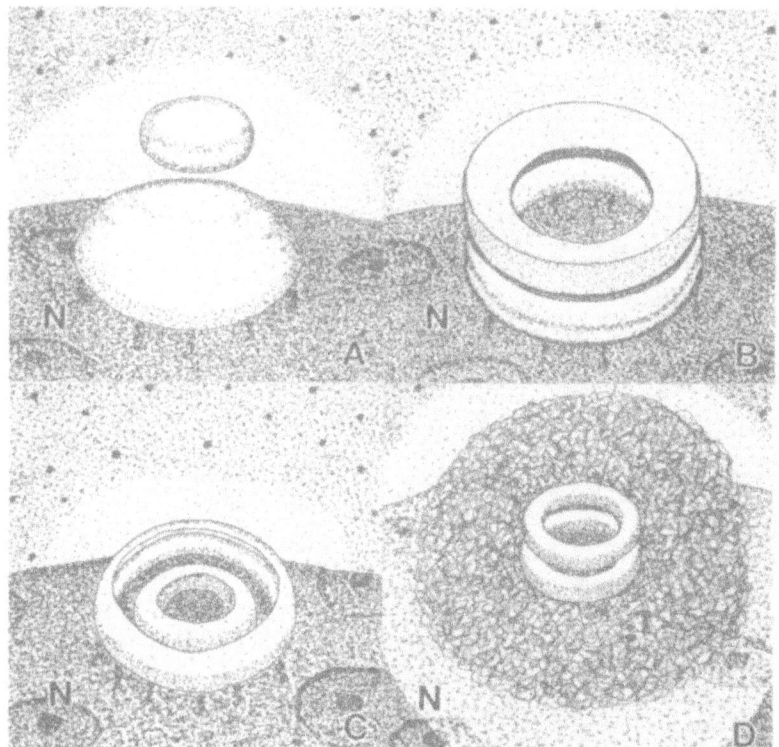

Fig. 6-22. Diagrammatic representation, drawn to scale, of the four basic NAO types discussed in the text, exemplified by *Porphyridium* (A), *Polysiphonia* (B), *Batrachospermum* (C), and *Rhodella* (D). Each NAO is depicted near the nuclear surface (N) during prophase. The conspicuous NAO-associated material shown in 6-22D has been seen only in *R. reticulata* and *Glaucosphaera*, but the NAO size is typical of several other unicellular species.

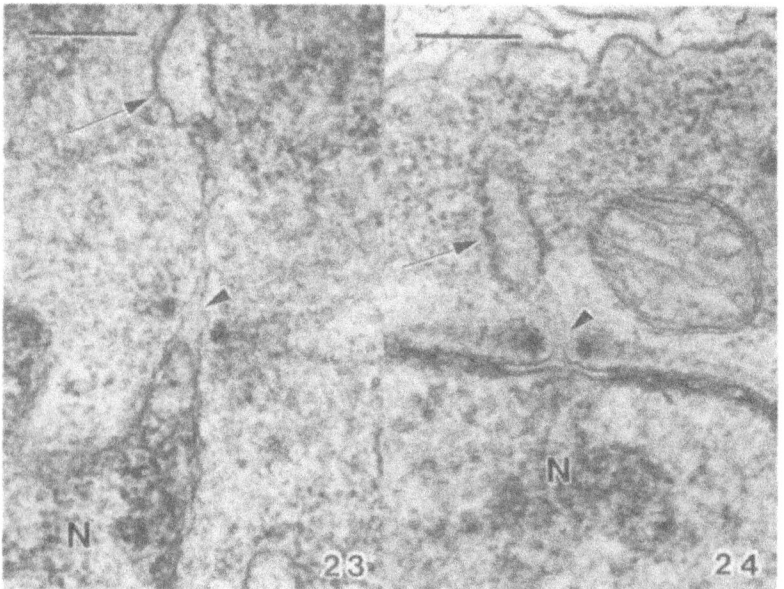

Figs. 6-23, 6-24. *6-23: Agardhiella subulata* late prophase NAO. Note a tubular extension of the outer membrane of the nuclear envelope (arrowhead) running through the center of the NAO. The extension is expanded into a swollen cisternum of rough ER (arrow). N = nucleus. Scale = 0.2 μm. (Unpubl., courtesy of K. Klepacki.) *6-24: Lomentaria baileyana* prophase NAO. Again, note the tubular extension of the nuclear envelope outer membrane (arrowhead) through the center of the NAO and the swollen ER cisternum (arrow). N = nucleus. Scale = 0.2 μm.

the division poles, spindle microtubules do not directly insert upon the NAO, which is relatively small in comparison to the size of the spindle. Instead, spindle microtubules either abut moderately electron-dense "NAO-associated material", which is usually conspicuous in species with large, polar gaps (see Section V) or, instead, terminate on the inner membrane of the nuclear envelope in those species that possess small fenestrations and inconspicuous NAO-associated material (Figs. 6-3, 6-4). Heath (1980b) suspects that red algal NAOs are merely "going along for the ride", using instead of generating the spindle. Since NAOs have been shown to be persistent organelles in postmitotic cells of *Polysiphonia* (Scott et al. 1981), it is possible that they are involved in a function(s) unrelated to nuclear division, even though they still remain closely but randomly associated with the nucleus.

The chemical composition of red algal NAOs is unknown, although in fungi there is some evidence for DNA and protein during mitosis and reports of an RNA component during interphase (Heath 1981b). There is no histochemical or morphological evidence indicating that red algal NAOs are related to centrioles or nuclear pores, as suggested by Dave and Godward (1982). They do not appear to be produced de novo; they persist throughout the cycle (Scott et al. 1981) and reproduce using a preexisting, postdivision portion as a template (Scott et al. 1980).

V. SPINDLE POLES

The nuclear envelope in all genera studied remains intact except at the spindle poles, where it is broken down to some degree. By late prophase each pole has an NAO surrounded by a ribosome-free "zone of exclusion". The NAO-associated zones of exclusion appear and gradually grow in size during the time of NAO migration and are thought to contain spindle precursors (Dave & Godward 1982; Scott et al. 1980). They maintain their maximum size during late prophase and early anaphase, gradually diminishing by telophase. Exclusion zones in *Porphyridium*, however, do not increase greatly in size, possibly because there are fewer precursors needed in this relatively small spindle (Schornstein & Scott 1982). NAOs in several red algae, in addition to being surrounded by exclusion zones, are embedded in "clouds" of moderately electron-dense material referred to as NAO-associated material (Figs. 6-12, 6-13, 6-15, 6-16, 6-20, 6-21). The abundance and morphology of this material may fluctuate appreciably throughout the mitotic cycle in some genera (Figs. 6-19 to 6-21).

Another difference between *Porphyridium* and the other red algae studied is the presence of microbodies at the poles in this genus (Bronchart & Demoulin 1977; Schornstein & Scott 1982). Microbodies have been observed at the poles in nonrhodophytan algae (Barlow & Cattolico 1981; Nilshammer & Walles 1974; Spector & Triemer 1981), but their function is unknown. Most likely, they are merely taking advantage of the spindle for allocation into the daughter cells (Schornstein & Scott 1982).

At least two other organelles, aside from smooth-surfaced membranes (discussed later), are known to consistently associate with division poles of some species. In both the unicell *Flintiella* (Fig. 6-30; Scott 1986) and the macrophyte *Lomentaria* (Fig. 6-4; Davis & Scott 1986) profiles of mitochondria crowd late prophase–early anaphase nuclei at the polar regions. The other organelle is the pyrenoid of the unicell *Rhodella violacea* (Patrone & Scott unpubl.). As in *R. maculata* (Evans 1970), interphase nuclei of *R. violacea* have one or more narrow extensions that can deeply penetrate the pyrenoid (Fig. 6-25). By metaphase, the pyrenoid forms a cup partially enveloping the nucleus; in two-dimensional sections, the pyrenoid appears to intrude into the midregion of the large gap at each pole (Fig. 6-26). The function of this anomalous pyrenoid–nucleus association is as mysterious as the one involving interphase nuclear penetrations.

Major differences in the spindle poles of the various genera of red algae become obvious by prometaphase–metaphase. In the Ceramiales, the nuclear envelope in the polar area becomes fragmented, leaving many small fenestrations in the nuclear envelope, although the area subjacent to the NAO always remains intact. Prior to becoming fenestrated, the nuclear envelope is characterized by numerous, close-packed nuclear pores, which suggests that the fenestrations are a result of nuclear pore dissolution. Supporting this idea is the fact that in those algae with polar areas possessing a single, relatively large gap instead of many, small fenestrations, polar concentrations of nuclear pores are also lacking. In *Porphyridium* (Schornstein & Scott 1982), the nuclear envelope beneath the NAO appears to have only a single nuclear pore at late prophase. This region invaginates to form a "nuclear pocket" by prometaphase (Fig. 6-27). The membranes of the pocket break down so that a large gap exists in the nuclear envelope at each pole. Concommitant with gap production, the relatively large

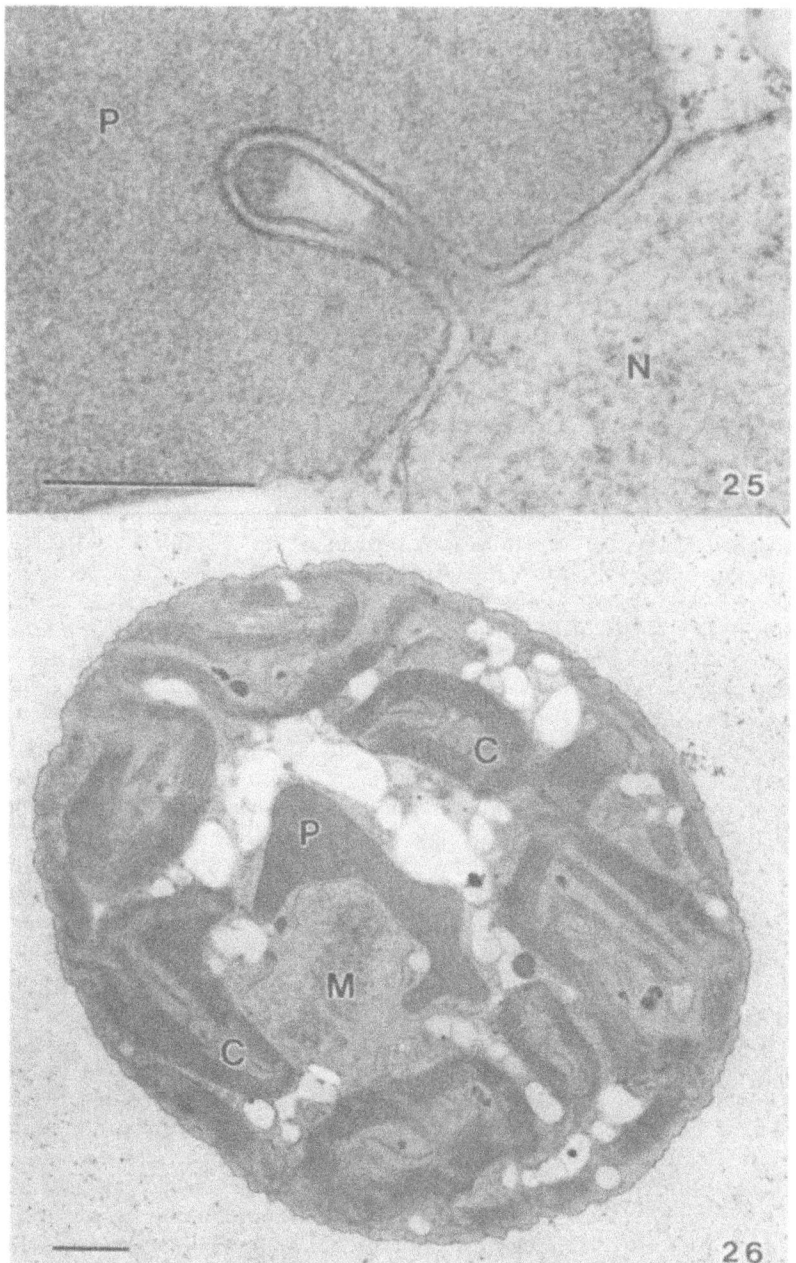

Figs. 6-25, 6-26. *6-25*: Persistent interphase extension of the nucleus (N) into the pyrenoid (P) of *Rhodella violacea*. Scale = 0.5 μm. *6-26*: Metaphase cell of *R. violacea* showing nucleus with a single large gap at each pole. Pyrenoid lobes (P) closely approach each division pole. M = metaphase plate, C = chloroplast lobes. Scale = 1 μm. (Figs. 6-25, 6-26 unpubl., courtesy of L. Patrone)

proximal portion of each NAO disperses, as previously mentioned.

In contrast, *Batrachospermum* (Scott 1983), *Flintiella* (Scott 1986), and *Agardhiella* (Klepacki & Scott unpubl.) usually have two pockets at each pole, with the NAO located in between (Fig. 6-28). These pockets, which contain microtubules, eventually extend completely through the nucleus from pole to pole, forming two to four cytoplasmic channels, or "tunnels" (Fig. 6-29). In these three genera the tunnels break down leaving a large gap at each pole (Fig. 6-30).

The only member of the order Rhodymeniales studied to date, *Lomentaria* (Davis & Scott 1986), appears to have very early prometaphase characteristics similar to those described above, but with

Cell division

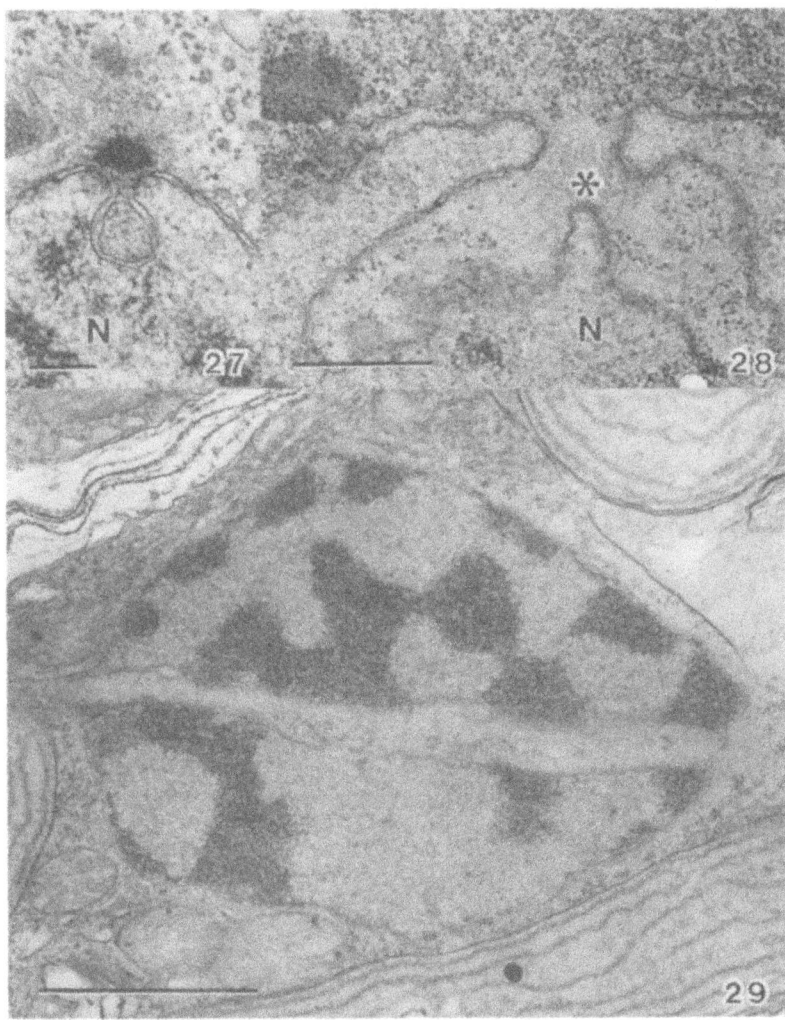

Figs. 6-27 to 6-29. *6-27*: Early prometaphase in *Porphyridium purpureum*. Note a single invagination of the nuclear envelope into the nucleus under the proximal NAO portion. Scale = 0.2 μm. (From Schornstein & Scott 1982, with permission.) *6-28*: Two microtubule-containing invaginations at a prometaphase division pole in *Batrachospermum ectocarpum*. Adjacent sections show the NAO in the position indicated by the asterisk. Scale = 0.5 μm. (From Scott 1983, with permission) *6-29*: *Agardhiella subulata* prometaphase pole-to-pole tunnel. Microtubules present in the channels are not obvious in this micrograph. Scale = 1 μm. (Unpubl., courtesy of K. Klepacki)

metaphase poles resembling those of the ceramialean genera. Two shallow, apparently microtubule-free pockets are produced at each pole during early prometaphase. These break down without deeply penetrating the nucleus and produce two relatively large gaps at each pole, with a central section of nuclear envelope intact beneath each NAO; the remainder of the polar nuclear envelope region appears to be fenestrated. More rhodymenialean genera obviously need examining in order to establish this clearly as a distinct pattern, but the current evidence suggests that prometaphase events in *Lomentaria* may represent an intermediate position between the so-called lower and higher orders of red algae (see Section X).

VI. SPINDLE

Microtubule numbers are low in the cytoplasm at any time in the cell cycle, so it is difficult to judge

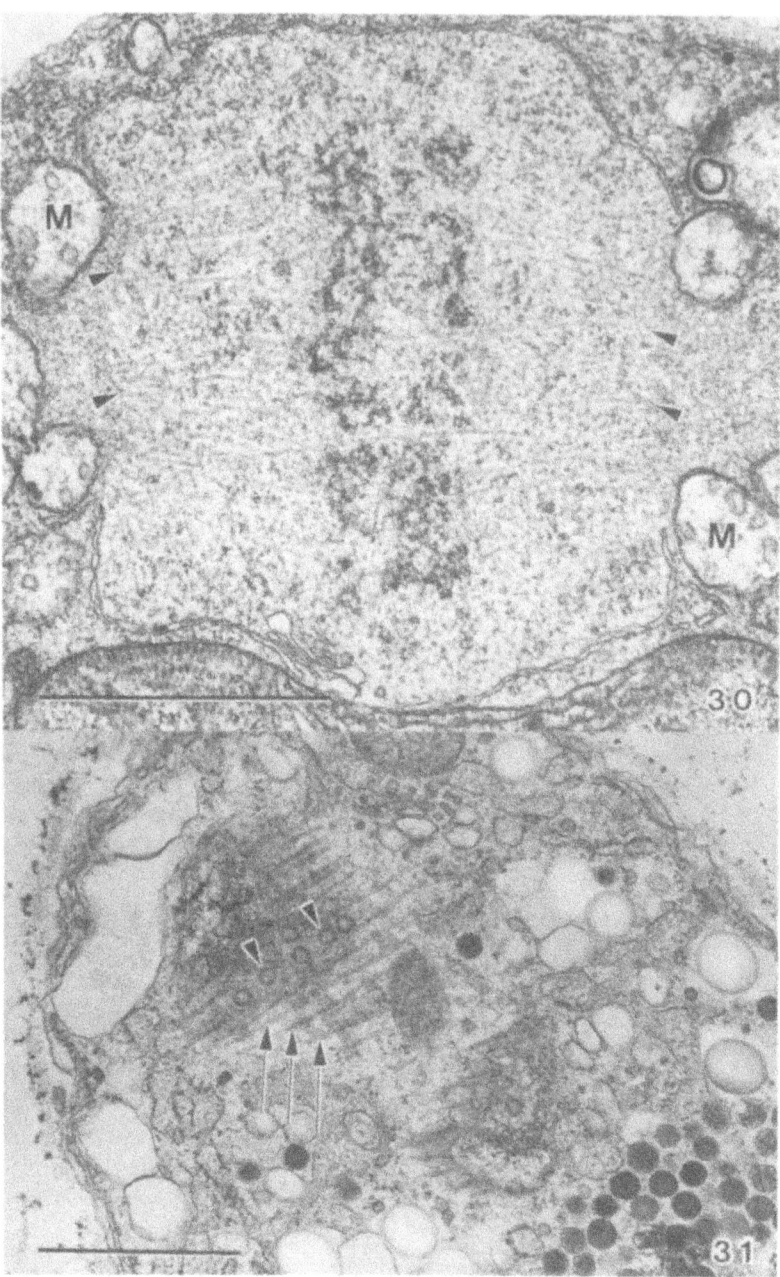

Figs. 6-30, 6-31. 6-30: Early anaphase nucleus in *Flintiella*. A single large gap, always closely associated with mitochondria (M), is present at each division pole. Spindle microtubules end broadly at the cytoplasm/nucleoplasm interface (arrowheads). Scale = 1 μm. 6-31: Glancing section across a prophase spermatangial nucleus of *Dasya baillouviana*. An extranuclear "cage" of microtubules (arrows) closely envelopes the nucleus and appears to be derived from the polar exclusion zones. Note the evenly spaced nuclear pores (arrowheads). Scale = 1 μm.

whether there is any significant change in number during mitosis, as might be expected if cytoplasmic microtubules were utilized for spindle production. Phillips and Scott (1981) did note a perinuclear "cage" of microtubules around prophase nuclei of *Dasya* that disappeared at metaphase (Fig. 6-31). Scott et al. (1981) reported a similar nuclear cage in mature, nondividing vegetative cells of *Polysiphonia*, whereas McDonald (1972) documented perinuclear microtubules in *Membranoptera* at metaphase after production of the spindle. Charleston (1984) noted microtubules radiating from the nucleus in cells of *Antithamnion kylinii* and suggested their possible involvement in orientation of the division plane and in daughter nuclei movement following anaphase. There are also some scattered, short microtubules in the zones of exclusion at all stages of mitosis in all species studied. These microtubules are not numerous, and their organization is not as pronounced as that of astral microtubules.

On first inspection, all red algae appear to have a spindle that develops totally within the nuclear envelope, and in all species, regardless of the type of polar opening, the great majority of spindle microtubules stop at the nucleoplasm/cytoplasm interface. Closer inspection, however, reveals important differences in spindle origin between *Porphyridium*, *Flintiella*, and *Batrachospermum* and all other genera studied. During prometaphase in the latter two genera, microtubule-filled pockets are present. These later develop into cytoplasmic tunnels that traverse the nucleoplasm from pole to pole (Scott 1983, 1986). The nuclear envelope surrounding these tunnels breaks down, and the microtubules presumably enter the nucleoplasm, although they are not sufficient in number to account for the whole spindle. A large gap is present after the breakdown of the nuclear envelope at each division pole. In *Porphyridium* (Schornstein & Scott 1982) there are a number of extranuclear microtubules between the NAOs as one migrates during establishment of the division poles, but there is no evidence that they enter the nucleus as preformed microtubules, since only a single polar pocket devoid of microtubules subtends the NAO at each pole prior to gap formation.

The microtubules in *Porphyridium* and *Batrachospermum* focus sharply upon conspicuous NAO-associated material, whereas in *Flintiella*, where the polar gaps are considerably larger than in the other two genera (Fig. 6-30), the microtubules are not as sharply focused upon each NAO region and instead end abruptly at the cytoplasm/nucleoplasm interface, where a faintly conspicuous fibrillar-granular material is located. In all other species examined, the nuclear envelope at the metaphase–anaphase poles is fenestrated (except for *Lomentaria*, which, as mentioned earlier, has a combination of relatively large gaps and fenestrations), and all spindle microtubules abut either the inner membrane of the nuclear envelope or end at the cytoplasm/ nucleoplasm interface. Direct microtubule connections or bridges to the nuclear envelope were not seen.

These two basic modifications of spindle origin may prove to be significant for understanding phylogenetic relationships within the Rhodophyta. Algae like *Polysiphonia*, *Dasya*, *Membranoptera*, and *Apoglossum* appear to have microtubule-organizing centers (MTOCs) for the bulk of spindle formation just within the nucleus, whereas genera like *Porphyridium*, *Flintiella*, and *Batrachospermum* have MTOCs just outside the nucleus. Stewart and Mattox (1980) hypothesized that there are two basic spindle types, I and II, the latter having subtypes a and b. Type I refers to a totally extranuclear spindle, such as in certain dinoflagellates and hypermastigote flagellates. Type IIa refers to spindles that first develop in the cytoplasm and later enter the nucleoplasm. The nuclear envelope may or may not remain intact. Type IIb includes spindles that develop totally within the nuclear envelope, which may or may not have polar openings. Type IIb spindles are considered slightly more advanced than type IIa. According to this classification, the so-called lower red algae possess a spindle developed in a fashion fairly close to the description of a type IIa spindle, whereas the others are best characterized by spindle type IIb. These observations are evidence for a sequence of spindle evolution within the red algae that is consistent with even the most conservative taxonomic rankings (see Section X). One can readily imagine how MTOC material located outside of nuclei could move inside to coat the inner membrane of the nuclear envelope, in time giving rise to a fundamentally different location of spindle formation.

VII. PER AND OTHER MEMBRANES

Many types of organisms have a system of membranes that totally or partially enclose the nucleus during mitosis (Heath 1980b). These membranes are referred to as perinuclear ER (PER) and are found largely in multinucleate cells. They appear to function in helping separate the milieu of the dividing nucleus from the rest of the cell (Phillips & Scott 1981). All of the multinucleate forms of the Rhodophyta examined by electron microscopy display some degree of PER during nuclear division (Figs. 6-2 to 6-4, 6-38). Perinuclear ER generally forms during prophase, peaks at metaphase and disperses by late anaphase–telophase. It is not found in the unicells *Porphyridium* (Figs. 6-5, 6-6; Schornstein & Scott 1982) and *Flintiella* (Fig. 6-30; Scott 1986). However, it is present in reduced quantities in *Batrachospermum* (Scott 1983) and *Compsopogon* (Scott & Broadwater unpubl.), two macroscopic freshwater genera possessing predominately uninucleate cells.

There are membranes in mitotic cells of many organisms that are believed to be involved in calcium regulation (Hepler & Wolniak 1984). No such membranes are detectable inside actively dividing red algal nuclei, although single cisternae of ER are always associated with the prophase NAOs of *Dasya* (Fig. 6-10; Phillips & Scott 1981) and with the early, prometaphase polar invaginations of *Batrachospermum* (Scott 1983). It is possible that during prometaphase, breakdown of cytoplasmic tunnels, such as in *Flintiella, Batrachospermum,* and *Agardhiella*, may be a means of introducing transient calcium-regulating membranes into the nucleus during spindle formation or that the nuclear envelope and/or PER could be performing a calcium regulatory function. Currently, however, there is no experimental evidence for the presence of calcium-sequestering systems during nuclear division in red algae.

VIII. CYTOKINESIS

Excluding division in multinucleate cells, in which cytokinesis may not be directly associated with karyokinesis, cytokinesis and karyokinesis in all other red algal cells appear to be temporally close together, with incipient cleavage furrows present as early as prophase in some species (Scott et al. 1980).

In the multicellular red algae, cytokinesis does not appear to involve microtubule or microfilament interactions. The ingrowing, centripetally directed cleavage furrow is preceded by a region of cytoplasm unremarkable in appearance. In contrast, cytokinesis in *Porphyridium* (Schornstein & Scott 1982) and *Flintiella* (Scott 1986), based purely on electron microscopic observations, appears to be mediated by a conspicuous contractile ring of microfilaments (Schornstein & Scott 1982; Scott 1986).

During cytokinesis, new cell wall material, including the "mucilaginous" coats of unicells, spermatia, and spores, appears to be somewhat randomly deposited at the periphery of the cell, including the cleavage furrow region. One exception occurs during postmeiotic cytokinesis of tetrasporangia in the parasitic red alga *Harveyella*, where preformed "cleavage channels", probably derived from ER, are formed and delimit the cytoplasmic domains of the four future spores (Kugrens & Koslowsky 1981). The actual deposition of extracellular material occurs later within the preformed cleavage channels, usually proceeding from the sporangial center centrifugally, but can also be random or centripetal.

Plant cells have devised several strategies for maintaining separation of daughter nuclei prior to cytokinesis: the phragmoplast, the phycoplast, a persistent interzonal spindle, the cleavage furrow itself (Barlow & Cattolico 1981), and the intrusion of cell organelles. All red algae appear to employ the last mechanism. Interestingly, *Porphyridium* and *Flintiella*, the only two unicells on which information is published, appear to utilize a single, central chloroplast, which moves between postdivision nuclei following the breakdown of the interzonal spindle (Fig. 6-32). In all other species studied, an interposing vacuole maintains separation of the nuclei (Fig. 6-33). This dichotomy probably relates more to the type of chloroplast present in the dividing cell than to whether or not the alga is unicellular or filamentous, since daughter nuclei separation in the unicell *Rhodosorus* (Scott unpubl.) is effected by a vacuole, and not by the parietal chloroplast.

IX. MEIOSIS

Only one reasonably thorough study of meiosis has been completed (in *Dasya baillouviana*; Broadwater et al. 1986a,b), and the following discussion is largely based on this single account. Spindle organization and most other ultrastructural features of meiosis are essentially the same as those for mitosis. The differences between features of meiosis and mitosis are primarily associated with prophase I, which is much longer and more complex than mitotic prophase. Almost all differences in this stage can be attributed either to special chromosome interactions (e.g., presence of synaptonemal complexes) or to the physiological changes involved in going from a diploid to a haploid cell. During diplotene, most components of the cell increase greatly in size or number. These changes are associated with chromatin decondensation and apparent resumption of RNA synthesis typical at this time in eukaryotic cells. Along with these changes, the cytoplasm becomes more opaque as electron-dense packets of

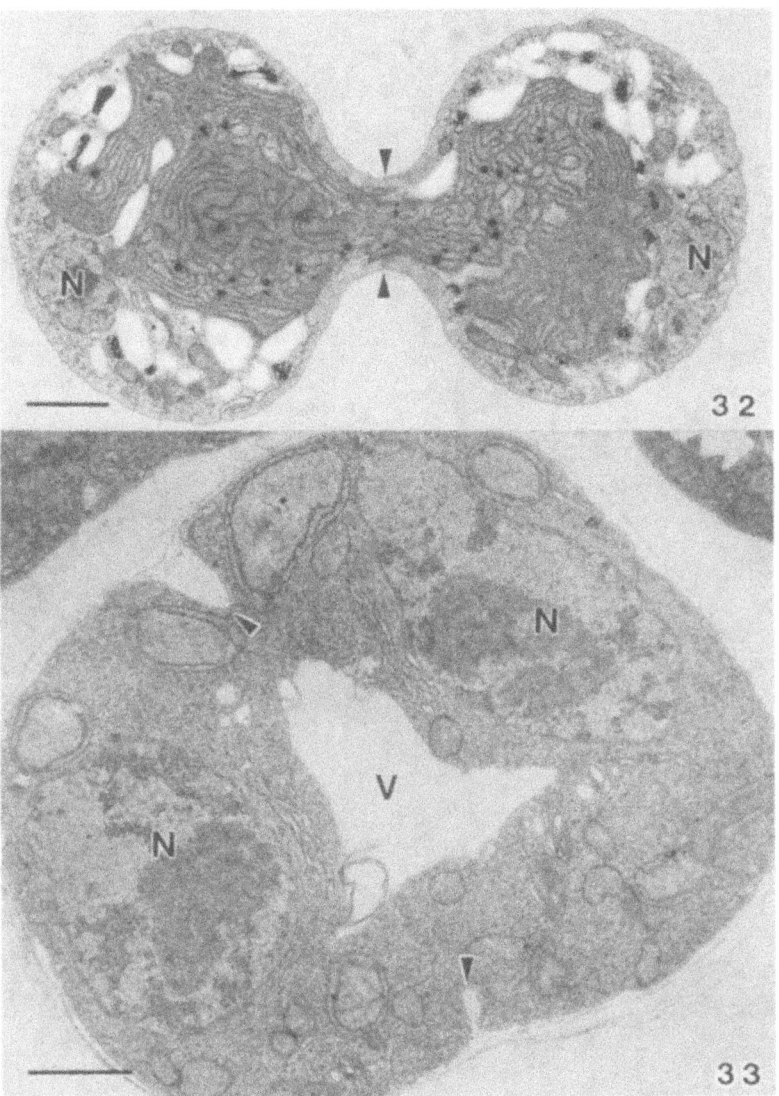

Figs. 6-32, 6-33. 6-32: Late cytokinesis in *Porphyridium purpureum*. The interzonal midpiece is detached, and the daughter nuclei (N) are separated by the large, central chloroplast, soon to be divided by the cleavage furrow (arrowheads). Scale = 1 μm. (From Schornstein & Scott 1982, with permission) 6-33: Early cytokinesis in *Polysiphonia harveyi*. Lacking a large, central chloroplast, a central vacuole (V) separates the daughter nuclei until the cleavage furrow (arrowheads) cuts off two cells. Scale = 1μm. (From Scott et al. 1980, with permission)

material appear. These are believed to be "nuage", which are cytoplasmic accumulations of ribosomal material described for at least eight animal phyla (in Broadwater et al. 1986a). It has been suggested that the lack of general cellular density in early prophase I cells may be due to the process of cytoplasmic clearing of ribosomes and mRNA, and the presence of nuage is related to the repopulation of these components. Both processes are believed to be involved in changing the cell from sporophyte to gametophyte control.

Numerous small vesicles are found at the prophase I pole and to a lesser extent near nuclei in postmeiotic cells. Similar structures have been

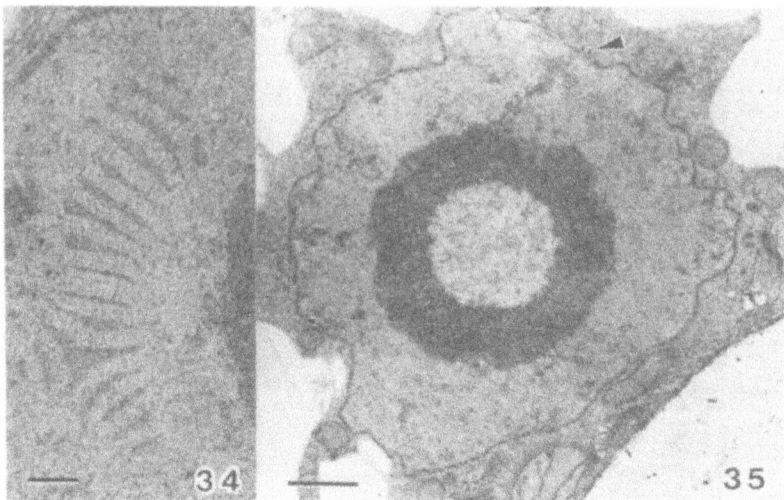

Figs. 6-34 to 6-37. Meiosis.
Figs. 6-34, 6-35. 6-34: Linearly arranged polycomplex in postmeiotic nucleus of *Dasya baillouviana*. Scale = 0.2 μm. 6-35: Diplotene nucleus of *Dasya* with typical "ring-shaped" nucleolus. An NAO is also visible (arrowhead). Scale = 1 μm.

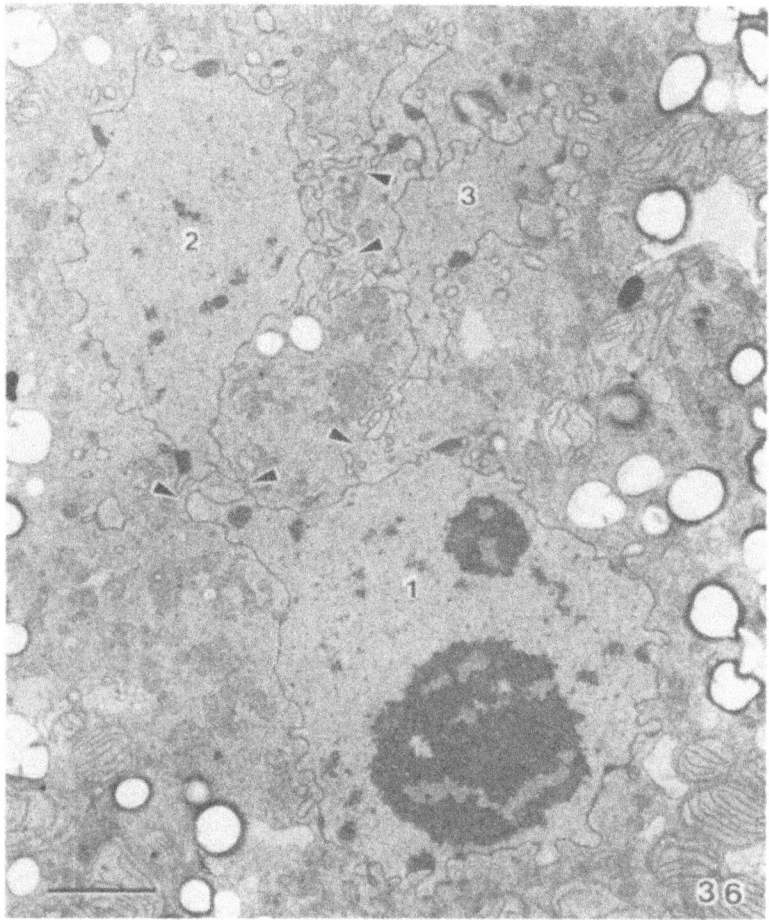

Fig. 6-36. Postmeiotic tetrasporangium in *Chondria baileyana* showing three (1, 2, 3) of the four nuclei. Examination of numerous serial sections reveal that these three nuclei are all interconnected in the areas indicated by the arrowheads. Scale = 1 μm.

Cell division

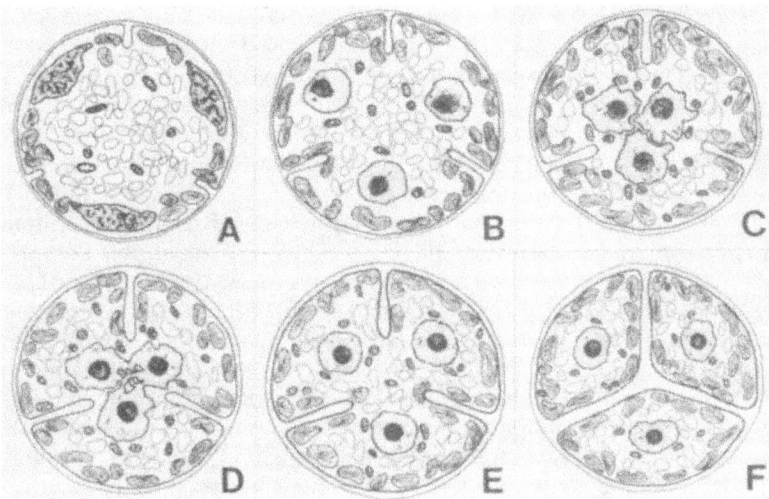

Fig. 6-37. Proposed postmeiotic movement of nuclei in tetrahedrally arranged tetrasporangia in ceramialean algae as described in text. A: Typical late telophase II cell. Note the cleavage furrows, three of the four widely separated nuclei near the plasma membrane, and the central starch deposits. B: The same cell after the nuclei have rounded up and assumed a position in the center of the nascent tetraspores. Note the more advanced cleavage furrows but no other obvious changes. C: The nuclei have moved to the center of the cell, displacing significant accumulations of starch. Note large numbers of mitochondria near the nuclei, which are very irregular in contour and closely appressed to each other. D: Hypothesized stage for *Dasya* that has been observed in other algae. Nuclear envelopes fuse, allowing continuity of the nucleoplasms. E: The cell after the nuclei have disjoined and returned to the centers of the newly forming spores. It is possible that this stage does not exist. F: The cleavage furrows impinge on and separate the nuclei, which then move to the center of the young spores. (From Broadwater et al. 1986b, with permission)

reported in sporocytes of *Palmaria* (Pueschel 1979) but have never been seen during mitosis in any red alga. Massive accumulations of smooth ER are found at the poles of meiotic cells during metaphase–anaphase I in *Dasya* and radiating from postmeiotic nuclei in tetrasporangia of *Haliptilon cuvieri* (Vesk & Borowitzka 1984). With so little information available, it is not known whether the proliferation of smooth ER adjacent to nuclei is typical of meiotic cells, but this conspicuous organelle association has never been found in mitotic cells.

There were three unexpected features of meiosis discovered in *Dasya*. One of these was the presence during diplotene of "ring nucleoli", which are nucleoli characterized by unusually large central "vacuoles" (Fig. 6-35). Ring nucleoli have been observed in many cell types (in Broadwater et al. 1986a), but there is no consensus in the literature regarding their significance. Their extreme hypertrophy in diplotene nuclei indicates an unknown function in red algae.

Polycomplexes (polymerized units of the synaptonemal complex) have been noted in many different organisms during prophase I, so their appearance during meiosis in *Dasya* was not surprising except for the time of their appearance. Polycomplexes were normally found in postmeiotic cells, where they sometimes attained extremely odd morphologies (Fig. 6-34). Their appearance so long after dissolution of the synaptonemal complex implies that they may be more than a breakdown product.

The most unexpected and most puzzling aspect of meiosis, however, was the unusual behavior of postdivision nuclei (Figs. 6-36, 6-37). At telophase II the four flattened haploid nuclei are located far apart from each other. These nuclei round up, and each attains a central position in the nascent spores. Then, simultaneously, the four nuclei move centripetally through massive deposits of starch to a central position in the tetrasporangium. The nuclei become very irregular in shape and are associated with annulate lamellae and large numbers of mitochondria, both of which suggest high metabolic activity. In *Dasya* actual nuclear fusion was not seen, but the nuclear envelopes of all four nuclei were closely appressed; in other species nuclear fusion was very obvious (Fig. 6-35; Scott & Thomas 1975). It is possible that the lack of a central vacuole or large chloroplast, typical of postmitotic cells, allows these nuclear movements to occur.

Although its function is unknown, this post-division movement does satisfyingly explain an erroneous account of "uninuclear meiosis" in red algae, whereby both divisions occur within the same nuclear envelope, with no intervening interkinesis (Scott & Thomas 1975). It seems most likely that red algal meiosis proceeds in the usual two discrete divisions separated by a typical interkinesis since interkinetic nuclei have been reported in earlier light microscopic literature and, more convincingly, in several electron microscopic studies (Kugrens & West 1972; Pueschel 1979; Vesk & Borowitzka 1984).

X. TAXONOMIC AND EVOLUTIONARY CONSIDERATIONS

The greatest problem with assessing the usefulness of cell division characteristics in red algal systematics and phylogeny is simply the current lack of sufficient data, especially at the electron microscopic level. Of the eight genera for which comprehensive ultrastructural information is published, four are in the Ceramiales, one in the Rhodymeniales, another in the Batrachospermales, and two in the Porphyridiales. This means that there are no accounts of mitosis in 12 of the 15 or 16 currently established orders. Although it would be extremely premature to draw many conclusions from so little information, some preliminary analysis is possible.

The electron microscopic characters that appear to be of taxonomic value are the shape, size, and behavior of the NAO, the presence or absence of PER, and the character of the prometaphase-anaphase polar areas (see Chapter 12 for a discussion of light microscopic criteria considered to be useful). Based on these features, there appear to be five basic combinations of ultrastructural features that characterize nuclear division in red algae that, for lack of better terminology, are termed the *Porphyridium*, *Batrachospermum*, *Flintiella*, *Polysiphonia*, and *Lomentaria* types of mitosis (Fig. 6-38).

The *Porphyridium* type has currently been seen in only one alga. It is characterized by an NAO of unique morphology and mode of replication, absence of PER, presence of a single polar pocket without microtubules at each prometaphase pole, an apparent ring of chromatids at metaphase, and a wide polar gap in the nuclear envelope at metaphase-anaphase. Based on Scott (1986), who compared light microscopic details of mitosis in *Porphyridium* with those seen in *Flintiella*, the *Porphyridium* type appears to correspond well to algae belonging to the "Bangiophycidae" of Coomans (1986; Chapter 12).

Two fairly similar groups of ultrastructural mitotic characters have been referred to as the *Batrachospermum* and *Flintiella* types. At prometaphase both of these algae have two microtubule-filled tunnels, arising from each pole, that completely transect the nucleus. The nuclear envelope of the tunnels is short-lived, and the subsequent metaphase-anaphase poles possess a single, large gap. Both have NAOs with polar ring morphologies, although the NAO of *Batrachospermum* has a "ring within a ring" structure differing from the normal red algal "ring upon a ring" construction. However, PER is lacking in *Flintiella*, and at the light microscopic level the *Flintiella* type fits best in Cooman's "Bangiophycidae" group, whereas the *Batrachospermum* type is more representative of his "lower Florideophycidae" group. These differences illustrate the need to examine a variety of both light and electron microscopic characters when using cell division features as taxonomic aids.

The *Polysiphonia* type of mitosis is typical of all ceramialean algae studied. It is characterized by conventional ring-shaped NAOs, the presence of extensive PER, metaphase poles consisting of many small fenestrations, the absence of prometaphase pockets or channels, the presence of a cleavage furrow roughly aligned with the previous metaphase plate and a number of other shared light microscopic features. This type of mitosis would correspond to Cooman's "higher Florideophycidae" group.

The *Lomentaria* type appears to have nearly the same characteristics as the *Polysiphonia* type, except that at prophase there are two shallow invaginations or pockets at each pole and the nuclear envelope at the metaphase-anaphase poles is interrupted by two small gaps as well as numerous fenestrations. *Lomentaria* is the only rhodymenialean alga studied for details of mitosis with the electron microscope. According to current classification schemes, one would expect mitosis in the Rhodymeniales to be very similar to the *Polysiphonia* type of the Ceramiales. Therefore, features, such as the reduced polar invaginations, which could be interpreted as "vestigial" structures when compared wih the well-developed pockets of *Batrachospermum* and *Flintiella*, could easily be viewed as intermediate

Cell division

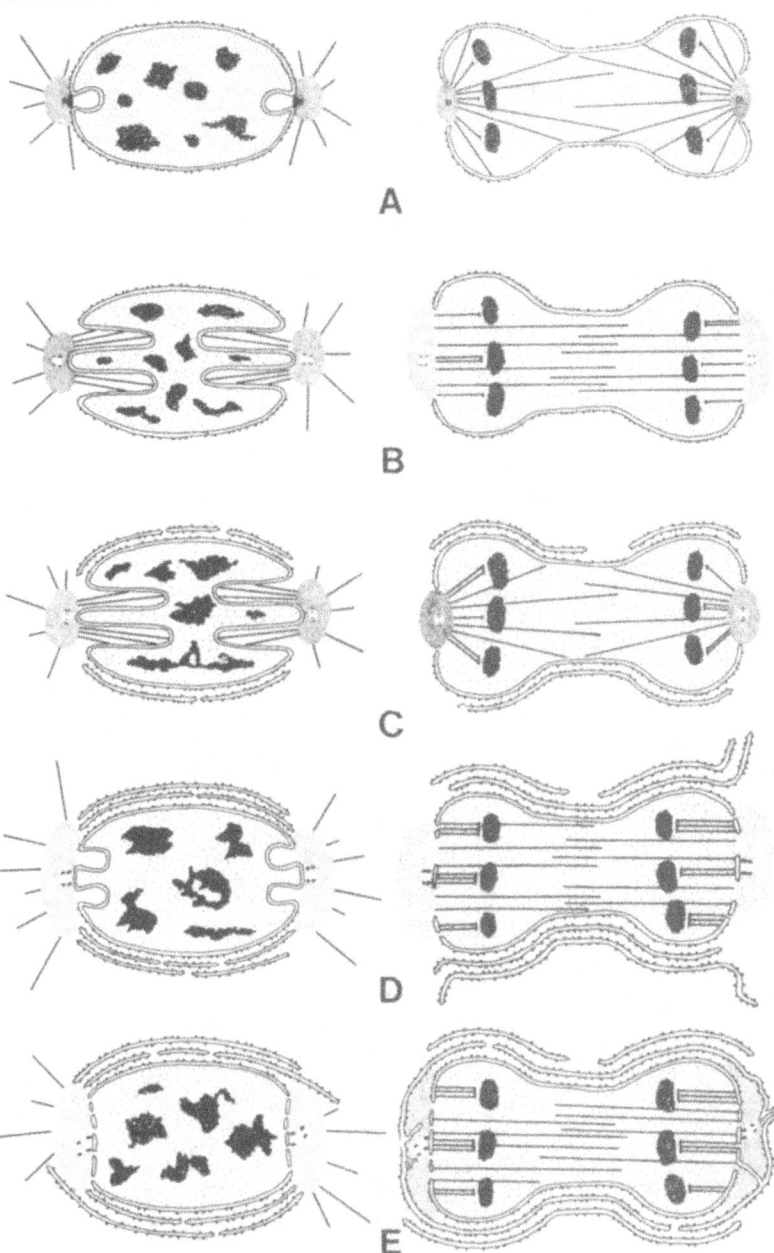

Fig. 6-38. Diagrammatic representation of ultrastructural modifications of early prometaphase nuclei (left column) and the resulting categories of anaphase nuclei (right column) that characterize the five "types" of mitosis discussed in the text, represented by *Porphyridium* (A), *Flintiella* (B), *Batrachospermum* (C), *Lomentaria* (D), *Polysiphonia* (E).

characters in the evolution of the spindle type of the Ceramiales.

The establishment of the five "mitosis types" outlined above should be considered extremely tentative since so very few red algae have been examined for ultrastructural features of dividing cells. However, there is great promise that along with other ultrastructural, biochemical, and light microscopic studies, the sometimes subtle and often dramatic features of dividing cells can be utilized as valuable criteria in the delineation of red algal phylogeny.

XI. ACKNOWLEDGMENTS

We wish to thank Jewel Thomas and Bill Saunders for their technical assistance and Curt Pueschel for a thorough and critical reading of the manuscript. We also thank Leslie Baker for preparation of the drawings in Fig. 6-22. Financial support was provided by NSF grant BSR 8615288.

XII. REFERENCES

Barlow, S. B. & Cattolico, R. A. 1981. Mitosis and cytokinesis in the Prasinophyceae. I. *Mantoniella squamata* (Manton and Parke) Desikachary. *Am. J. Bot.* 68: 606–15.

Broadwater, S., Scott, J. & Pobiner, B. 1986a. Ultrastructure of meiosis in *Dasya baillouviana* (Rhodophyta). I. Prophase I. *J. Phycol.* 22: 490–500.

Broadwater, S., Scott, J. & Pobiner, B. 1986b. Ultrastructure of meiosis in *Dasya baillouviana* (Rhodophyta). II. Prometaphase I–telophase II and post-division nuclear behavior. *J. Phycol.* 22: 501–12.

Bronchart, R. & Demoulin, V. 1977. Unusual mitosis in the red alga *Porphyridium purpureum. Nature* (Lond.) 268: 80–1.

Cavalier-Smith, T. 1981. The origin and early evolution of the eukaryotic cell. *Symp. Soc. Gen. Microbiol.* 32: 33–84.

Charleston, J. S. 1984. Light and electron microscopic investigations of mitosis in the marine red alga *Antithamnion kylinii* (Ceramiales). *J. Phycol.* (Suppl.) 20: 4a.

Coomans, R. J. 1986. A light microscopic study of mitosis and vegetative development in the Rhodophyta. Ph.D. dissertation, University of North Carolina, Chapel Hill.

Dave, A. J. & Godward, M.B.E. 1982. Ultrastructural studies in the Rhodophyta. I. Development of mitotic spindle poles in *Apoglossum ruscifolium*, Kylin. *J. Cell Sci.* 58: 345–62.

Davis, E. & Scott, J. 1986. Ultrastructure of cell division in the marine red alga *Lomentaria baileyana. Protoplasma* 131: 1–10.

Demoulin, V. 1985. The red algal–higher fungi phyletic link: The last ten years. *BioSystems* 18: 347–56.

Evans, L. V. 1970. Electron microscopical observations on a new red algal unicell, *Rhodella maculata* gen. nov., sp. nov. *Br. Phycol. J.* 5: 1–13.

Fuller, M. S. 1976. Mitosis in fungi. *Int. Rev. Cytol.* 45: 113–53.

Gabrielson, P. W. & Garbary, D. 1986. Systematics of red algae (Rhodophyta). *CRC Crit. Rev. Plant Sci.* 3: 325–66.

Gabrielson, P.W., Garbary, D. J. & Scagel, R. F. 1985. The nature of the ancestral red alga: inferences from a cladistic analysis. *BioSystems* 18: 335–46.

Garbary, D. J., Hansen, G. I. & Scagel, R. F. 1980. A revised classification of the Bangiophyceae (Rhodophyta). *Nova Hedwigia* 33: 145–66.

Heath, I. B. 1980a. Fungal mitoses, the significance of variations on a theme. *Mycologia* 72: 229–50.

Heath, I. B. 1980b. Variant mitoses in lower eukaryotes: indicators of the evolution of mitosis? *Int. Rev. Cytol.* 64: 1–80.

Heath, I. B. 1981a. Mechanisms of nuclear division in fungi. In *The Fungal Nucleus*, eds. K. Gull & S. G. Oliver, pp. 85–112. Cambridge: Cambridge University Press.

Heath, I. B. 1981b. Nucleus-associated organelles in fungi. *Int. Rev. Cytol.* 69: 191–221.

Hepler, P. K. & Wolniak, S. M. 1984. Membranes in the mitotic apparatus: their structure and function. *Int. Rev. Cytol.* 90: 169–238.

Kraft, G. T. & Robins, P. 1985. Is the order Cryptonemiales (Rhodophyta) defensible? *Phycologia* 24: 67–77.

Kugrens, P. & Koslowsky, D. J. 1981. Electron microscopic studies on a unique cytokinetic structure in tetrasporocytes of the red alga *Harveyella* sp. (Cryptonemiales, Choreocolaceae). *Protoplasma* 108: 197–209.

Kugrens, P. & West, J. A. 1972. Ultrastructure of tetrasporogenesis in the parasitic red alga *Levringiella gardneri* (Setchell) Kylin. *J. Phycol.* 8: 370–83.

McDonald, K. 1972. The ultrastructure of mitosis in the marine red alga *Membranoptera platyphylla. J. Phycol.* 8: 156–66.

Nilshammer, M. & Walles, B. 1974. Electron microscope studies on cell differentiation in synchronized cultures of the green alga *Scenedesmus. Protoplasma* 79: 3127–32.

Peyrière, M. 1971. Étude infrastructurale des spermatocystes du *Griffithsia flosculosa* (Rhodophycée). *C.R. Acad. Sci.* (Paris) sér. D, 273: 2071–4.

Phillips, D. & Scott, J. 1981. Ultrastructure of cell division and reproductive differentiation of male plants in the Florideophyceae (Rhodophyta). Mitosis in *Dasya baillouviana. Protoplasma* 106: 329–41.

Pickett-Heaps, J. D. 1975. *Green Algae: Structure, Reproduction and Evolution in Selected Genera.* Sunderland, Mass.: Sinauer Associates.

Pueschel, C. M. 1979. Ultrastructure of tetrasporogenesis in *Palmaria palmata* (Rhodophyta). *J. Phycol.* 15: 409–24.

Pueschel, C. M. & Cole, K. 1982. Rhodophycean pit plugs: an ultrastructural study with taxonomic implications. *Am J. Bot.* 69: 703–20.

Schornstein, K. L. & Scott, J. 1980. Reevaluation of mitosis in the red alga *Porphyridium purpureum. Nature* (Lond.) 283: 409–10.

Schornstein, K. L. & Scott, J. 1982. Ultrastructure of cell division in the unicellular red alga *Porphyridium purpureum. Can. J. Bot.* 60: 85–97.

Scott, J. 1983. Mitosis in the freshwater red alga *Batrachospermum ectocarpum. Protoplasma* 118: 56–70.

Scott, J. 1986. Ultrastructure of cell division in the unicellular red alga *Flintiella sanguinaria. Can. J. Bot.* 64: 516–24.

Scott, J., Bosco, C., Schornstein, K. & Thomas, J. 1980. Ultrastructure of cell division and reproductive differentiation of male plants in the Florideophyceae (Rhodophyta): Cell division in *Polysiphonia. J. Phycol.* 16: 507–24.

Scott, J., Phillips, D. & Thomas, J. 1981. Polar rings are persistent organelles in interphase vegetative cells of *Polysiphonia harveyi* Bailey (Rhodophyta, Ceramiales). *Phycologia* 20: 333–7.

Scott, J. L. & Thomas, J. P. 1975. Electron microscope observations of telophase II in the Florideophyceae. *J. Phycol.* 11: 474–6.

Sheath, R. G., Cole, K. M. & Hymes, B. J. 1987. Ultrastructure of polysporogenesis in *Pleonosporium vancouverianum* (Rhodophyta). *Phycologia* 26: 1–8.

Silva, P. C. & Johansen, H. W. 1986. A reappraisal of the order Corallinales (Rhodophyceae). *Br. Phycol. J.* 21: 245–54.

Sommerfeld, M. R. & Nichols, H. W. 1970. Comparative studies on the genus *Porphyridium*. *J. Phycol.* 6: 67–78.

Spector, D. L. & Triemer, R. E. 1981. Chromosome structure and mitosis in the dinoflagellates: An ultrastructural approach to an evolutionary problem. *BioSystems* 14: 289–98.

Stewart, K. D. & Mattox, K. 1975. Comparative cytology, evolution and classification of the green algae with some considerations of the origin of other organisms with chlorophylls a and b. *Bot. Rev.* 41: 104–35.

Stewart, K. D. & Mattox, K. 1980. Phylogeny of phytoflagellates. In *Phytoflagellates*, ed. E. R. Cox, pp. 433–62. New York: North Holland Elsevier.

Vesk, M. & Borowitska, M. A. 1984. Ultrastructure of tetrasporogenesis in the coralline alga *Haliptilon cuvieri* (Rhodophyta). *J. Phycol.* 20: 501–15.

Chapter 7

Solute accumulation and osmotic adjustment

ROBERT H. REED

CONTENTS

I. Introduction / 147
II. Metabolized inorganic ions / 149
    A. Nitrogen / 149
    B. Phosphorus / 152
III. Nonmetabolized inorganic ions / 153
    A. Sodium / 153
    B. Potassium / 154
    C. Chloride / 155
IV. Organic solutes / 156
V. Osmoacclimation / 157
VI. Osmoadaptation / 163
VII. Summary / 164
VIII. Acknowledgments / 165
IX. References / 165

I. INTRODUCTION

Water is an essential intracellular constituent and is responsible for maintaining macromolecular structure. The high permeability of cell membranes to water results in a continuous intracellular liquid phase, with water acting as a solvent for the substrates and products of metabolism. Water is also a reactant in several essential metabolic processes, including photosynthesis and hydrolysis. Furthermore, water is required to maintain cell size and turgidity.

Aquatic plants contain a wide range of intracellular solutes with a diversity of roles, generating an osmotic pressure within the cell (π_i) in accordance with the van't Hoff relationship:

$$\pi = RT\Sigma_j C_j \qquad \text{(equation 1)}$$

where R is the universal gas constant (8.3 J mol^{-1} K^{-1}), T is the absolute temperature (K), and $\Sigma_j C_j$ is the sum of the concentration of all (ideal) solutes (mol m^{-3}). Deviations from ideality are dealt with by the inclusion of an activity coefficient (Nobel 1983). The minimum intracellular osmotic pressure consistent with metabolic requirements is approximately 0.1 MPa (Raven 1976). In the absence of any other forces, immersion in fresh water would lead to water uptake and cell lysis. However, this is prevented in most freshwater algae by the presence of a rigid or semirigid cell wall that allows a positive hydrostatic (turgor) pressure (P_i) to develop, restricting water entry and cell enlargement. These two parameters are the major components of intracellular water potential (ψ_i), a term derived from the chemical potential of water (Dainty 1976) and used to specify the state of water in biological systems (Passioura 1980), since

$$\psi_e = \psi_i = P_i - \pi_i = -\pi_e \quad \text{(equation 2)}$$

where ψ_e and π_e represent external water potential (Pa) and external osmotic pressure (Pa). For most freshwater algae the external osmotic pressure will be minimal and turgor pressure will be equivalent to the intracellular osmotic pressure. In contrast, algae growing in seawater (osmotic pressure $\simeq 2.5$ MPa, equivalent to $\simeq 500$ mol m^{-3} NaCl) must counterbalance the external salinity by the internal accumulation of solutes. Additionally, intracellular solute accumulation must generate a positive turgor pressure, in excess of a critical value (P_c: the yield threshold), to allow cell wall extension and therefore to permit growth and long-term survival (Lockhart 1965):

$$(1/v)\, dV/dt = \varepsilon_c (P_i - P_c) \quad \text{(equation 3)}$$

where $(1/v)\, dV/dt$ is the relative rate of increase of cell volume (s^{-1}) and ε_c is a coefficient describing cell wall extensibility (Pa^{-1} s^{-1}). Thus, turgor pressure is of fundamental significance, acting to coordinate water uptake with wall yielding (growth). Actively growing plant cells regulate turgor pressure to maintain growth under conditions of constant π_e (Cosgrove 1986). P_i is sensitive to changes in π_e (equation 2), decreasing in response to an increase in external osmotic pressure, and vice versa.

Repeated observations using a wide variety of plant cells, including algae, have shown that a change in P_i caused by variation in external osmotic pressure may be followed by adjustment of the number of osmotically active solute molecules per cell, with the restoration of P_i to a value close to the original level (Reed 1984a). The proposal that there is a finely controlled reference value for P_i is fundamental to the concept of turgor regulation, envisaged as a negative feedback system to restore P_i following a change in π_e (Cram 1976; Zimmermann 1978).

Intracellular osmotic pressure is generated principally by inorganic ions and low molecular weight organic solutes. In a gross oversimplification of metabolism, the major inorganic ions can be subdivided into (1) the metabolized solutes (especially nitrogen and phosphorus compounds), which may exist in inorganic form in osmotically significant quantities under conditions of nutient sufficiency (luxury accumulation) but more frequently are assimilated into organic molecules and (2) the nonmetabolized ions (principally K$^+$, Cl$^-$, and Na$^+$).

Organic solutes are formed as a consequence of the photofixation of inorganic carbon, and this chapter considers those features of organic solute accumulation and metabolism that may be involved in osmoacclimation (modifications in cell structure or function resulting from osmotic stress) and/or osmoadaptation (genetically determined changes resulting from natural selection under conditions of osmotic stress: Reed 1984a).

The uptake or exclusion of inorganic ions from the cell interior is affected by the combined physical driving forces of (1) the concentration gradient across the plasmalemma and (2) the transmembrane electrical potential difference (interior-negative), established by the differential permeability of the membrane to ions and by electrogenic (i.e., charge-carrying) ion transport systems (Gutknecht & Dainty 1968). The transplasmalemma equilibrium potential (E_j) for a given ion j, can be calculated using the Nernst equation:

$$E_j = RT/z_j F \times \ln(C_j^e / C_j^i) \quad \text{(equation 4)}$$

where z_j is the valency of the ion, F is the Faraday constant (9.65 $\times$ 10^4 J mol^{-1} v^{-1}), and C_j^e and C_j^i represent external and internal ion concentrations (mol m^{-3}), respectively. As in equation 1, nonideal solutes may be accommodated by including the appropriate activity coefficient (γ_j). Thus, the equilibrium potential of an ion can be calculated and compared with the (measured) plasmalemma potential (E_m) to see whether that ion is in electrochemical equilibrium. Alternatively, the value for the membrane potential can be substituted into equation 4 to predict the intracellular ion concentration at a given external ion concentration (e.g., equation 4 predicts a tenfold accumulation of a passively distributed monovalent cation and a tenfold exclusion of a monovalent anion when $E_m = -58$ mV). If the observed ion distribution is not in agreement with the predicted value from equation 4, then active transport is the most likely explanation, either inwardly (internal concentration greater than predicted) or outwardly directed (internal concentration lower than predicted).

It is necessary to consider the transmembrane rates of ion movement, to distinguish between facilitated diffusion and active transport. Passive diffusion of an ion results in an equilibrium potential equal to the membrane potential, with low rates of transmembrane ion movement (equivalent to those for artificial lipid bilayers; Raven 1980). Facilitated diffusion also leads to electrochemical equilibrium across the membrane ($E_j = E_m$) but with high rates of ion transport, consistent with the involvement of specific carriers (porters). Facilitated diffusion systems may show saturation kinetics and analogue inhibition, in common with active transport but with no direct involvement of energy. This contrasts with active ion transport, which occurs against the electrochemical potential gradient and at a rate greater than that due to passive diffusion. Active transport is energy dependent and is sensitive to metabolic inhibitors. It may be further subdivided into (1) primary active transport, which is coupled to exergonic biochemical reactions (e.g., ATP hydrolysis) and (2) secondary active transport, which utilizes the electrochemical potential gradient of another ion to provide the driving force for transport as cotransport (symport) or countertransport (antiport). Both types of active transport may be electrogenic (i.e., charge carrying); primary active uniport is always electrogenic.

The membrane potential in turn may be governed by the distribution of ions on both sides of the membrane, taking into account any differences in ion permeability, in accordance with the constant field (Goldman) equation (Raven 1984). Since K^+ commonly has the highest permeability, a membrane potential that is more negative than the potassium equilibrium potential may indicate an electrogenic pump (i.e., cation efflux or anion influx). In freshwater algae the available evidence supports the hypothesis that the plasmalemma potential results from a primary active H^+ extrusion system (Raven 1980). Electrogenic ion pumps in marine algae may include Cl^- influx and/or Na^+ efflux, although the evidence for smaller-celled marine red algae is scant. In the giant-celled red alga *Griffithsia monilis* the membrane potential appears to have no electrogenic component (Findlay et al. 1969).

This chapter is concerned with the major intracellular solutes in red algae and the effects of osmotic stress on solute accumulation and metabolism. Experimental data are limited mostly to intertidal (marine) red macroalgae, and this is reflected in the textual emphasis on (red) seaweeds.

II. METABOLIZED INORGANIC IONS

The incorporation of the major macronutrients nitrogen and phosphorus into intracellular organic solutes poses particular problems in assessing the role of active transport. Furthermore, both ammonia and phosphate are weak electrolytes, existing in more than one inorganic molecular form according to the pH and adding further complications to any assessment of the role of active transport based on the prevailing electrochemical gradient.

The transport and fixation of inorganic carbon is dealt with in Chapter 8. Recent general coverage of nutrient assimilation in algae is available in articles by DeBoer (1981), Healey (1973) and Raven (1976, 1980).

A. Nitrogen

Nitrogen is generally considered to be the nutrient that most frequently limits the growth of marine macroalgae (DeBoer & Ryther 1977) and phytoplankton (Syrett 1981). Evidence that nitrogen limitation may restrict the growth of intertidal red algae includes the following observations: (1) increased growth is often recorded following the addition of nitrogenous nutrients in field experiments (Hanisak 1983) and in laboratory culture (Morgan & Simpson 1981; Neish et al. 1977); (2) a comparison of the elemental composition of marine algae with seawater shows that nitrogen is generally in short supply (DeBoer 1981); (3) pronounced seasonal variations in the nitrogen content of natural populations of several marine red algae, e.g., *Chondrus crispus* (Asare & Harlin 1983; Butler 1936) and *Porphyra tenera* (Iwasaki & Matsudaira 1954), may be linked to patterns of growth and nitrogen availability.

The significance of nitrogen limitation is represented by Fig. 7-1. The intercept on the abcissa represents the minimum nitrogen content for growth, $\simeq 0.8\%$ dry weight for *Gracilaria tikvahiae* (Hanisak 1983). Above this point, growth will be proportional to internal nitrogen content until a critical value is reached, $\simeq 2\%$ dry weight for *G. tikvahiae* (Hanisak 1983). Higher internal nitrogen levels indicate a capacity for nutrient storage (luxury uptake: Raven 1976) with the accumulation of organic and/or inorganic nitrogen reserves.

Without the capacity to fix elemental N_2, the two

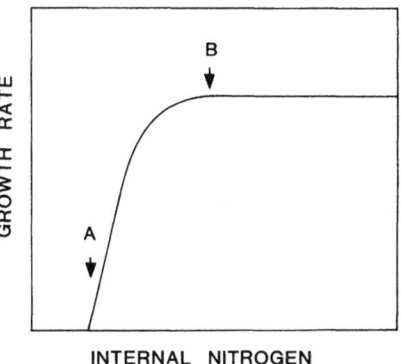

Fig. 7-1. The relationship between growth rate and internal nitrogen content, showing maintenance (A) and nonlimiting (B) levels of nitrogen.

Table 7-1. Uptake kinetics for ammonia

Species	K_s	V_{max}	Reference
Agardhiella subulata	2.3–4.9	94–5000	D'Elia & DeBoer 1978
Gracilaria tikvahiae	1.6	397	D'Elia & DeBoer 1978
Hypnea musciformis	16.6	—	Haines & Wheeler 1978

Mean values of K_s (mmol m^{-3}) and V_{max} (mol g^{-1} h^{-1}, expressed in terms of tissue dry weight) for high-affinity ammonia uptake in selected red algae.

most important forms of inorganic nitrogen for the growth of marine red algae are nitrate (NO_3^-) and ammonia (i.e., NH_3 plus NH_4^+), which occur in oceanic waters at concentrations up to 40 mmol m^{-3} and 4 mmol m^{-3}, respectively (Parsons & Harrison 1983), while coastal seawaters may contain more than ten times these levels (DeBoer 1981). Nitrite (NO_2^-) is a third source of inorganic nitrogen. It is generally present at lower levels in seawater than either nitrate or ammonia and is often regarded as a more toxic form of inorganic nitrogen (Healey 1973), although Fries (1963) has shown that *Stylonema* (as *Goniotrichium*) *alsidii* and *Nemalion helminthoides* (as *N. multifidum*) grew better in axenic culture on nitrite than on nitrate (at 3.5 mmol m^{-3}). High concentration of ammonia can also have toxic effects on algae, due to uncoupling of photophosphorylation (Raven 1980).

When provided with nitrate and ammonia, most algae will assimilate ammonia more rapidly (Stewart 1980; Thomas & Harrison 1987). However, the exact mechanism of ammonia uptake is poorly understood because of the coexistence in solution of NH_3 and NH_4^+, according to the pH, as follows:

$$\log \{NH_3\} / \{NH_4^+\} = pH - pK_a \quad \text{(equation 5)}$$

With a pK_a for NH_3/NH_4^+ of approximately 9.3, the dominant form in seawater is NH_4^+, with <10% of the total ammonia present as NH_3. It is generally accepted that NH_3 has a substantially higher permeability to lipid bilayers than NH_4^+ (Raven 1984) and that passive diffusion of NH_3 is the principal means of ammonia uptake at high external pH, where the lower intracellular pH can lead to NH_4^+ accumulation due to ion trapping (Guarino & Cohen 1979). However, mediated transport of NH_4^+ is also likely in seawater, although the precise mechanism remains unknown (Syrett 1981).

D'Elia and DeBoer (1978) showed, using medium-depletion techniques, that the kinetics of ammonia uptake in *Gracilaria tikvahiae* and *Agardhiella subulata* could not be described by a single Michaelis–Menten relationship but were consistent with the operation of a dual-phase system. At low external concentrations (i.e., below 10 mmol m^{-3}) a saturable, high-affinity system was observed, whereas at higher concentrations (up to 50 mmol m^{-3}) a second, linear component was operative (due to either passive diffusion of NH_3 or a low-affinity transport system). The half-saturation constants (K_s) and maximum rates (V_{max}) calculated for the high-affinity system of each alga are shown in Table 7-1. In contrast, Friedlander and Dawes (1985) reported a single, linear (nonsaturable) component for ammonia uptake in *G. tikvahiae* over a wider concentration range ($\simeq$0.2–600 mmol m^{-3}) and suggested that the absence of saturation kinetics at lower concentrations may have been due to the use of plants that were not nitrogen deficient. However, dual-phase ammonia uptake kinetics were observed by D'Elia and DeBoer (1978) using nitrogen-deplete and nitrogen-replete plants.

The kinetics of ammonia uptake in *Hypnea musciformis* have been described in terms of a truncated hyperbola over the concentration range 1–25 mmol m^{-3} (Haines & Wheeler 1978). Since saturation of ammonia uptake was not achieved at the highest concentration used, Haines & Wheeler (1978) speculated that ammonia uptake by *H. musciformis* does not follow Michaelis–Menten kinetics. However, a half-saturation constant for ammonia uptake was obtained (see Table 7-1).

DeBoer and Whoriskey (1983) proposed a dual-phase model for ammonia uptake in *Ceramium rubrum*, with a high-affinity system operating at concentrations below 10 mmol m^{-3} and a linear (diffusion) component at higher concentrations. The production of hyaline hairs was sensitive to external ammonia concentration, with a stimulation of hair formation below 0.5 mmol m^{-3} and an inhibition above 20 mmol m^{-3}. The enhancement of ammonia uptake rates in plants grown at low external ammonia concentrations was attributed to an increase in membrane surface area due to hair formation (DeBoer & Whoriskey 1983). Hyaline hairs are also known to occur in *Gracilaria tikvahiae*,

Table 7-2. Uptake kinetics for nitrate

Species	K_s	V_{max}	Reference
Agardhiella subulata	2.4	11.7	D'Elia & DeBoer 1978
Cyanidium caldarium	<1.0	—	Fuggi et al. 1984
Gracilaria tikvahiae	2.5	9.7	D'Elia & DeBoer 1978
Hypnea musciformis	4.9	28.5	Haines & Wheeler 1978

Mean values of K_s (mmol m^{-3}) and V_{max} (mol g^{-1} h^{-1}, expressed in terms of tissue dry weight) for high-affinity nitrate uptake in selected red algae.

although their involvement in nitrogen assimilation is unknown.

No attempts have been made to estimate the intracellular pH or NH_3 and NH_4^+ concentrations in any red alga, to assess the role of active (NH_4^+) transport. However, the slowly metabolized analogue methylamine ($CH_3NH_3^+$) was used to study the kinetics of ammonia uptake in the marine brown macroalga Macrocystis pyrifera (Wheeler 1979), and has proved useful in establishing the role of active transport of ammonia in other algae, where intracellular accumulation of $CH_3NH_3^+$ exceeds the concentration that can be explained in terms of CH_3NH_2 permeation and the internal pH (Smith & Walker 1978). Other techniques, including voltage clamping (Raven 1980), are possible only with larger-celled red algae. Under field conditions, where nitrogen is a limiting nutrient, the intracellular ammonia level is likely to be governed to a greater extent by the enzymes of ammonia assimilation than by the (steady-state) electrochemical potential gradient (Falkowski 1983).

The uptake of nitrate generally exhibits saturation kinetics in marine red algae (D'Elia & DeBoer 1978), as in most seaweeds (DeBoer 1981; Harlin & Craigie 1978). A saturable uptake rate is consistent with the hypothesis that NO_3^- uptake is likely to be an active process, owing to the electronegativity of the cell interior. However, as NO_3^- uptake is often strongly correlated with the activity of the enzyme nitrate reductase (Raven 1976), any interpretation in terms of the electrochemical potential gradient for NO_3^- is difficult. It is clear that in cells with an interior-negative membrane potential, together with a significant internal accumulation of NO_3^- (the intracellular concentration can be up to three orders of magnitude greater than in the external medium), an active NO_3^- transport process must be involved. Further evidence includes the stimulatory effects of light (Lapointe & Duke 1984) and the effects of metabolic inhibitors on NO_3^- uptake (Rigano et al. 1979), while desiccation-enhanced uptake of NO_3^- and ammonia is reported in Mastocarpus papillatus (as Gigartina papillata) (Thomas et al. 1987).

Data for the half-saturation constant and maximum uptake rate, calculated from the Michaelis-Menten relationship, are shown in Table 7-2 for several marine red algae and for a strain of the acidothermophilic unicell Cyanidium caldarium. In general, the values of V_{max} are lower for nitrate than for ammonia (cf. Table 7-1). In C. caldarium the high-affinity NO_3^- uptake system appears to operate as a proton cotransport system, with a stoichiometry of $2H^+$ per NO_3^- (Fuggi 1985). In marine red algae Na^+ is a more likely candidate for cotransport than H^+, although the possibility remains untested. However, in Porphyra purpurea the observed fluxes of Na^+ are too low to support NO_3^- cotransport (Raven 1984).

There is a growing amount of evidence that the NO_3^- concentration in the cytosol of eukaryotic algae is regulated at ≤ 1 mol m^{-3} and that the level of NH_4^+ is also likely to be kept low, to prevent uncoupling due to NH_3 permeation (Raven 1980). However, excess NO_3^- may be stored in vacuoles. Annual variations in tissue NO_3^- in subtidal marine brown macroalgae have been interpreted in terms of vacuolar storage during nitrogen-sufficient periods, when vacuolar NO_3^- concentrations may contribute significantly to π_i (Davison & Reed 1985). Certain red algae also possess the capacity to establish internal nitrogen reserves and to utilize these reserves for subsequent growth. Thus, Gracilaria tikvahiae can accumulate sufficient nitrogen in 6 h to provide for 14 d of nitrogen-sufficient growth (Ryther et al. 1981). There is direct evidence of intracellular NO_3^- storage in G. tikvahiae (Bird et al. 1982; Rosenberg & Ramus 1982) and Chondrus crispus (Asare & Harlin 1983), with highest levels in winter and lowest values during the summer period, when growth is minimal. The (vacuolar) accumulation of NO_3^- in these algae must be an active process. However, tissue NO_3^- generally constitutes a small proportion of the total nitrogen reserves

(Lapointe & Duke 1984). Organic nitrogen reserves include amino acids, which may represent a significant proportion of the total nitrogen content of *Porphyra yezoensis* (Oohusa et al. 1977), *Polysiphonia lanosa* (Reed 1983a), and *G. tikvahiae* (Rosenberg & Ramus 1982), decreasing in response to nitrogen deprivation in the manner predicted for a major intracellular nitrogen reserve (Bird et al. 1982). In contrast, the dipeptide citrullinylarginine can account for more than half of the nitrogen reserves of *C. crispus*, exceeding 300 µmol g^{-1} (dry weight) in late winter and reaching an annual minimum in late summer (Laycock & Craigie 1977). Laboratory experiments have shown that citrullinylarginine accumulation may occur as a result of reduced growth (at low temperature) and high external concentrations of inorganic nitrogen (Laycock et al. 1981). While the osmotic effects of dipeptide accumulation in *C. crispus* have yet to be considered, it is likely that intracellular citrullinylarginine may balance up to 10% of the external osmotic pressure. Citrullinylarginine is also present in *G. tikvahiae* (Rosenberg & Ramus 1982).

Other organic nitrogen reserves in marine red algae include the phycobiliproteins (Lapointe 1981; Lapointe & Duke 1984; Ryther et al. 1981) and unspecified cell proteins (Bird et al. 1982).

B. Phosphorus

Although phosphorus-limited growth of freshwater microalgae has been reported (Kuhl 1962), phosphorus is not considered to be a limiting nutrient for most intertidal marine macroalgae, and there have been few studies of phosphorus accumulation and metabolism in seaweeds. Phosphorus in seawater exists as free inorganic orthophosphate (mostly as $H_2PO_4^-$ and HPO_4^{2-}), together with metallophosphate complexes (with Mg^{2+}, Ca^{2+}, and Na^+), polyphosphates, and organic phosphorus compounds; free inorganic orthophosphate may thus represent less than one-third of the total phosphorus in seawater (Cembella et al. 1983). However, the principal form in which algal cells acquire phosphorus is as $H_2PO_4^-/HPO_4^{2-}$ and the ability of algae to utilize organic phosphorus or polyphosphate relies on extracellular (enzymic) hydrolysis with subsequent uptake of inorganic (ortho)phosphate (Raven 1980).

There are few reports for red algae that can be interpreted in terms of the electrochemical potential gradient for phosphate. However, Eppley (1958a) recorded intracellular phosphate in *Porphyra perforata* at 2.5 mol m^{-3}, whereas Reed et al. (1981) give a value of 14.8 mol m^{-3} for *P. purpurea*. These values are more than two orders of magnitude greater than the external inorganic phosphate concentration, strongly supporting an active phosphate uptake mechanism.

Saturation kinetics have been described for phosphate uptake in *Agardhiella subulata*, with a K_s of 0.4 mmol m^{-3} and an average V_{max} of 0.47 µmol g^{-1} h^{-1} – expressed on a dry weight basis (DeBoer 1981). In contrast, Friedlander and Dawes (1985) reported triphasic kinetics for phosphate uptake in *Gracilaria tikvahiae*. Whether the latter result is due, in part at least, to an experimental design based on three external phosphate concentrations remains to be determined. However, the data show linear accumulation of phosphate at high external concentrations (≥ 2 mmol m^{-3}), with two saturable phases at lower concentrations. The high-affinity uptake system, which is likely to be of most ecological significance, showed an apparent K_s under experimental conditions of <0.05 mmol m^{-3} and a V_{max} of approximately 0.02 µmol g^{-1} h^{-1} (expressed with respect to tissue fresh weight).

Although it is an inescapable conclusion that phosphate influx must be an active process, based on the electronegativity of the cell and the internal accumulation of phosphate, it is not clear whether primary active transport is involved or whether uptake is driven by ion gradients (secondary active transport). The results of Ullrich-Eberius and Yingchol (1974) might support the hypothesis that secondary active transport (Na^+ cotransport) is involved, since phosphate uptake in the marine red unicell *Porphyridium purpureum* (as *P. cruentum*) was stimulated up to 100-fold by Na^+, whereas the freshwater species *P. aerugineum* showed a smaller increase in phosphate transport upon addition of Na^+. However, no measurements of the all-important counter-effects of phosphate upon Na^+ influx were carried out to support the cotransport hypothesis. Rigby et al. (1980) showed that the cation requirement for phosphate influx in the blue-green alga *Synechococcus leopoliensis* is not due to cotransport, but is more likely due to cation interactive effects upon the phosphate transporter.

At near-neutral intracellular pH, HPO_4^{2-} will be the dominant form of inorganic phosphate and cytoplasmic phosphate is known to be finely controlled, due to its fundamental role as a metabolic regulator (Walker & Sivak 1986). Excess intracellular phosphate can be stored as vacuolar phosphate (Cembella et al. 1983) or as cytoplasmic polyphosphate granules, e.g., in *Ceramium* sp. (Kuhl 1962). The excess negative charges of intracellular phos-

Solute accumulation and osmotic adjustment

Table 7-3. Intracellular Na^+, K^+, Cl^- concentration and membrane potential of selected marine red algae

Species		Na^+ (mol m^{-3})	K^+ (mol m^{-3})	Cl^- (mol m^{-3})	E_m (mV)	Reference
Centroceras clavulatum	(w)	90	480	380	n.d.	Kirst & Bisson 1979
Ceramium rubrum	(w)	31	515	575	n.d.	Kesseler 1964
Corallina officinalis	(w)	280	520	410	n.d.	Kirst & Bisson 1979
Gracilaria foliifera	(w)	66 (e)	680 (i)	462 (i)	−81(v)	Gutknecht 1965
Grateloupia filicina	(w)	670	80	380	n.d.	Kirst & Bisson 1979
Griffithsia monilis	(s)	30–90 (e)	500–600 (i)	600–500 (i)	−80(c) −54(v)	Findlay et al. 1969, 1970
Griffithsia monilis	(s)	116	566	627	n.d.	Bisson & Kirst 1979
Griffithsia pacifica	(—)	n.d.	n.d.	n.d.	−60(c) −82(v)	Baumann & Jones 1986
Hypnea valentiae	(w)	360	90	390	n.d.	Kirst & Bisson 1979
Lomentaria umbellata	(w)	400	25	370	n.d.	Kirst & Bisson 1979
Palmaria palmata	(w)	25 (e)	560 (i)	420 (i)	−65(v)	MacRobbie & Dainty 1958
Polysiphonia lanosa (mar)	(w)	110 (e)	765 (i)	484 (i)	−89(v)	Reed 1983a
Polysiphonia lanosa (est)	(w)	102 (e)	617 (i)	422 (i)	−81(v)	Reed 1983a
Polysiphonia urceolata	(w)	96	664	583	n.d.	Kesseler 1964
Porphyra perforata	(w)	51 (e)	482 (i)	81 (—)	−42(c)	Eppley 1958a
Porphyra purpurea†	(w)	24 (e)	435 (i)	341 (i)	−58(c)	Reed & Collins 1980b
Porphyra umbilicalis†	(w)	110	1060	190	n.d.	Wiencke & Läuchli 1981
Rhodymenia foliifera	(w)	210	310	340	n.d.	Kirst & Bisson 1979

mar, marine; est, estuarine; †, ion concentrations calculated with respect to cell osmotic volume; w, whole cell; s, vacuolar sap; i, active influx; e, active efflux; c, potential difference between medium and cytosol; v, potential difference between medium and vacuole.

phates are likely to be linked to vacuolar Na^+/K^+ or to K^+/Ca^{2+} in polyphosphate granules (Adamec et al. 1979; Sicko-Goad et al. 1975).

III. NONMETABOLIZED INORGANIC IONS

Measurements of the intracellular concentrations of the major inorganic ionic solutes Na^+, K^+, and Cl^- in several marine red algae are shown in Table 7-3, together with estimates of the electrical potential difference. In most cases, the ion concentrations shown are whole-cell averages, with appropriate correction for extracellular ions, while the values for Griffithsia monilis represent the vacuolar sap (Bisson & Kirst 1979; Findlay et al. 1970). The electrical potential measurements were obtained by inserting a microelectrode into the cell interior and comparing the electrical potential with a reference electrode in the bathing medium (Findlay & Hope 1976). In some cases (e.g., Porphyra spp.) the plasmalemma potential (between the cytosol and bathing medium) has been measured, whereas other values refer to the potential difference between the vacuole and the bathing medium. In Griffithsia sp. the tonoplast potential was determined by inserting microelectrodes into the cytoplasm and vacuole of a single cell (Findlay et al. 1969). In all cases an inside-negative potential difference was recorded (Table 7-3).

Accumulation of the lipophilic cation triphenylmethylphosphonium ($TPMP^+$) has been used to estimate the cell membrane potential of Porphyra purpurea (Reed & Collins 1980a), giving similar values to those obtained using glass microelectrodes (−61 mV, cf. Table 7-3). However, the use of lipophilic cations as probes for the membrane potential has several practical and theoretical limitations and is not recommended as a substitute for microelectrode techniques (Ritchie 1984).

A. Sodium

Sodium is not an essential element for the growth of all algae (Healey 1973), and the growth of marine red algae is unlikely to be limited by Na^+ availability. However, Na^+ is the major cation in seawater, and the intracellular Na^+ concentration can be considered in terms of the electrochemical potential, calculated from equation 4. In all cases where E_m has been measured (Table 7-3), the intracellular Na^+ concentration was lower than that of the medium. The highest value was reported for Grateloupia filicina, based on chemical assay techniques (Kirst & Bisson 1979), and this is the only instance where internal Na$^+$ exceeds the external concentration. The lowest recorded intracellular Na$^+$ concentration (<5% of the external level: *Porphyra purpurea*) was obtained using radioisotope equilibration techniques (Reed & Collins 1980b).

Some of the variation in intracellular Na$^+$ concentration shown in Table 7-3 may be a result of differences in experimental procedures and the problems of contamination due to Na$^+$ present in extracellular (cell wall) water and associated with cation-exchange groups on the cell wall (Ritchie & Larkum 1982). Some workers (e.g., Eppley 1958a,b, 1959; Gutknecht 1965; MacRobbie & Dainty 1958) used a washing procedure based on an isotonic sucrose solution to remove "extracellular" ions. However, ions bound to cation-exchange groups within the cell wall matrix (Eppley & Blinks 1957) would not be removed using this procedure, leading to an overestimation of internal Na$^+$. Rinsing with isotonic, ice-cold $CaSO_4$ or $Ca(NO_3)_2$ to remove extracellular Na$^+$ (Reed 1983a; Wiencke & Läuchli 1981) may underestimate intracellular Na$^+$ if loss of intracellular ions occurs during the rinsing period. Kirst and Bisson (1979) corrected for external Na$^+$ on the basis of the volume of extracellular (cell wall) water and the Na$^+$ concentration of the bathing medium. This takes no account of the cation exchange capacity of the cell wall and is likely to overestimate intracellular Na$^+$ (Davison & Reed 1985). Radioisotope tracer techniques (using ^{24}Na$^+$ or ^{22}Na$^+$) have been used to quantify the intracellular and extracellular components of tissue Na$^+$, according to their rates of exchange (MacRobbie & Dainty 1958). In *Porphyra purpurea*, tracer exchange was biphasic, and the initial, rapid phase of uptake/loss corresponded to the cell wall component, whereas the second, slower phase was due to an intracellular component (Reed & Collins 1980b). The cell interior behaved as a single compartment with respect to ^{24}Na$^+$ influx and efflux and the rate of tracer exchange (across the plasmalemma) could be described in terms of a single exponential function (Walker & Pitman 1976), giving linear plots on semilog transformation and enabling flux rates and intracellular Na$^+$ concentrations to be calculated (Reed et al. 1981). Transplasmalemma Na$^+$ flux rates in *P. purpurea* (at 18–20 nmol m^{-2} s^{-1}) were found to be an order of magnitude lower than the corresponding fluxes for K$^+$ and Cl$^-$. In contrast, higher Na$^+$ flux rates (up to 170 nmol m^{-2} s^{-1}) have been reported for *Griffithsia* spp. (Findlay et al. 1970).

In those red algae where estimates of the membrane potential and internal Na$^+$ concentration have been obtained, there is evidence of active Na$^+$ extrusion, since the intracellular Na$^+$ concentration is significantly lower than predicted by the Nernst equation. Low concentrations of Na$^+$ in the cytosol are a universal feature of marine algae (see Raven 1976, 1980), and this may be linked to the toxicity of Na$^+$ to metabolic activity in the cytoplasm (Pollard & Wyn Jones 1979).

Compartmental flux analysis showed that Na$^+$ exchange in *Palmaria* (as *Rhodymenia*) *palmata* cells was biphasic (MacRobbie & Dainty 1958). The dual phase exchange kinetics may have been due to variation in the exchange rates of small and large cells (MacRobbie & Dainty 1958) or to a slower rate of exchange at the tonoplast than at the plasmalemma (Walker & Pitman 1976), as in the green seaweed *Ulva lactuca* (West & Pitman 1967a).

In *Porphyra purpurea* the plasmalemma acts as the principal barrier to ^{24}Na$^+$, and no details of the intracellular distribution of Na$^+$ have been obtained from tracer flux analysis (Reed & Collins 1980b). However, the intracellular compartmentation of Na$^+$ in *Porphyra umbilicalis* grown for 14 d in a hypersaline medium (350% sea water) was investigated using electron-probe x-ray microanalysis (Wiencke et al. 1983), showing that intracellular Na$^+$ is preferentially localized within the vacuolar fraction of the cell (at a concentration of 405 mol m^{-3}), whereas the concentration of Na$^+$ in the cytosol and the chloroplast is approximately ten times lower (at 57 mol m^{-3} and 36 mol m^{-3}, respectively). This is in agreement with the hypothesis that the vacuole may act as a compartment where (toxic) Na$^+$ can be contained without detrimental effects on (cytoplasmic) metabolism, and it demonstrates the pitfalls of interpretations based on whole-cell measurements (Table 7-3).

B. Potassium

With an extracellular potassium concentration in seawater of 10 mol m^{-3}, K$^+$ is the ion that is nearest to electrochemical potential equilibrium in most marine algal cells (Gutknecht & Dainty 1968). The low extracellular K$^+$ concentration also reduces the errors due to carry-over of external K$^+$ in estimates of intracellular K$^+$. The values for intracellular K$^+$ shown in Table 7-3 are somewhat higher than predicted from the Nernst equation assuming passive equilibrium in response to E_m, suggesting that an active K$^+$ influx mechanism is operative in *Gracilaria foliifera*, *Griffithsia monilis*, *Palmaria palmata*, *Polysiphonia lanosa*, *Porphyra perforata*, and *Porphyra purpurea*. However, an active transport hypothesis may be weakened if the intracellular K$^+$ activity is significantly lower than the (measured) K^+ concentration (Vorobiev 1967) and by intracellular inhomogeneity of K^+ (Raven 1980). If intracellular inhomogeneity involves a cytoplasmic concentration greater than the mean intracellular concentration, the suggestion of active plasmalemma K^+ influx will be strengthened. The data shown in Table 7-3 suggest that potassium is the principal intracellular cation in most red algae: internal Na^+ was higher than K^+ in only three cases (Table 7-3), and these may be overestimates, as discussed above. There is a growing amount of evidence to support the hypothesis that the cytoplasmic concentration of K^+ in plants is maintained within a critical concentration range (Leigh & Wyn Jones 1984), to ensure optimal metabolic (enzyme, ribosome) activity/stability (Brady et al. 1984), and it is clear that regulation of this kind requires cellular control of K^+ influx and efflux (i.e., active transport). The high transplasmalemma flux rates for K^+ [approximately 180 nmol m^{-2} s^{-1} in *P. palmata* and *P. purpurea* (MacRobbie & Dainty 1958; Reed & Collins 1980b), 500 nmol m^{-2} s^{-1} in *G. foliifera*, (Gutknecht 1965), and 500–3800 nmol m^{-2} s^{-1} in *G. monilis* (Findlay et al. 1970)] and the inhibitory effects of darkness and metabolic inhibitors on transplasmalemma K^+ fluxes (MacRobbie & Dainty 1958; Reed & Collins 1980b) are also consistent with active K^+ transport, although in isolation they do not rule out facilitated diffusion acting to concentrate K^+ in response to a metabolically generated, inside-negative membrane potential.

The efflux of K^+ from cells of *Porphyra purpurea* has been shown to be sensitive to changes in the external K^+ concentration, decreasing on removal of K^+ from the bathing medium (Reed & Collins 1980b). This dependence of unidirectional K^+ flux on the trans concentration of K^+ is often described as "exchange diffusion" and is associated with active transport processes operating near thermodynamic equilibrium (Fletcher 1980), adding further support to the proposal that transplasmalemma K^+ movement is an active, energy-dependent process.

The kinetics of K^+ exchange in marine algae have been studied using the radioisotope $^{42}K^+$ (e.g., West & Pitman 1967a). However, the short half-time for decay of $^{42}K^+$ (12.4 h) precludes long-term experimentation. Radioisotopic $^{86}Rb^+$ has a slower decay rate and is often assumed to act as a "reasonable" tracer for K^+ in plant cells (Läuchli & Epstein 1970). However, several marine algae show qualitative and quantitative differences between $^{86}Rb^+$ and $^{42}K^+$ influx/efflux, since cellular exchange of the former is biphasic, whereas the latter shows monophasic kinetics (West & Pitman 1967b). In the green algae *Ulva lactuca* and *Chaetomorpha darwinii* a specific model was proposed to account for these observations, based on intracellular discrimination against $^{86}Rb^+$ at the tonoplast (Dodd et al. 1966; West & Pitman 1967b). This model was used to study the internal compartmentation of K^+ in *Porphyra purpurea*, showing that more than two-thirds of the intracellular K^+ is localized within a noncytoplasmic compartment (probably the vacuole and/or chloroplast: Reed & Collins 1981). These observations suggest that the concentration of K^+ in the cytosol may be lower than the value shown in Table 7-3 and that the principal energy-requiring step for K^+ uptake is not located at the plasmalemma but may be at the tonoplast or chloroplast membrane(s). This may account for the lack of discrimination between $^{42}K^+$ and $^{86}Rb^+$ at the plasmalemma and the greater specificity for K^+ (and against Rb^+) at intracellular membranes (Reed 1981).

Wiencke et al. (1983) have studied the intracellular distribution of K^+ in *Porphyra umbilicalis* using electron-probe x-ray microanalysis, showing that vacuolar K^+, at 41 mol m^{-3}, was an order of magnitude lower than Na^+ in plants grown for 14 d in a hypersaline medium (350% sea water), whereas the cytoplasm and chloroplast contained K^+ at 751 mol m^{-3} and 564 mol m^{-3}, respectively. With an external K^+ concentration of 32.9 mol m^{-3}, the transplasmalemma K^+ equilibrium potential would be approximately -79 mV, which is higher than the values shown in Table 7-3 for other marine red algae. Although these observations are consistent with the hypothesis that the concentration of K^+ in the cytosol may be due to active K^+ influx, rather than passive uniport in response to E_m, there are indications that an active K^+ efflux mechanism at the tonoplast is required to account for the low vacuolar concentration of potassium in *P. umbilicalis*. These observations contrast with the proposed active influx of K^+ at the tonoplast of *P. purpurea* (Reed 1981) and *Griffithsia monilis* (Findlay et al. 1970), and with arguments against gross variation in the K^+ concentration of different compartments within algal cells (Raven 1976, 1980).

C. Chloride

The values for intracellular Cl^- and E_m shown in Table 7-3 indicate that Cl^- influx is almost certain to be an active process in *Gracilaria foliifera*, *Griffithsia monilis*, *Palmaria palmata*, *Polysiphonia lanosa*, and *Porphyra purpurea*, since the intracellular Cl^- concentration is higher than predicted for passive electrochemical equilibrium (equation 4). This is in agreement with results obtained with other seaweeds, where there is good evidence for active Cl^- influx at the tonoplast (Gutknecht & Dainty 1968). The active uptake of Cl^- in marine algae may be linked to the regulation of vacuolar volume and internal osmotic pressure in walled plant cells (Dainty 1963). Furthermore, Bisson and Kirst (1979) suggested that active uptake of Cl^- is a primary means of generating cell turgor pressure in many marine algae and that the importance of active K^+ transport as a turgor-generating process may have been over-emphasized. Whereas the Cl^- fluxes in G. foliifera and G. monilis are somewhat lower than their corresponding K^+ fluxes (Findlay et al. 1970; Gutknecht 1965), the reverse is true for P. purpurea (Reed & Collins 1980b). The effects of darkness and metabolic inhibitors, together with the sensitivity of Cl^- fluxes to the trans concentration of Cl^- also support the hypothesis of active transport.

Porphyra perforata appears to be an exception, since the value for Cl^- reported by Eppley (1958a) is close to that predicted on the basis of passive electrochemical equilibrium ($E_{Cl} \approx -47$ mV). However, this represents a whole-cell average, and it is possible that the low value reflects the lack of a large central vacuole in this alga. Although tracer flux analysis yielded no information on the intracellular localization of Cl^- in P. purpurea, due to the monophasic nature of cellular $^{36}Cl^-$ fluxes (Reed & Collins 1980b), it is noteworthy that P. purpurea and P. umbilicalis also contain rather less Cl^- than the other species (Table 7-3). In addition, Wiencke et al. (1983) demonstrated that Cl^- is preferentially localized in small vacuoles and in the chloroplast of P. umbilicalis, and it is possible that the Cl^- concentration in the intracellular vacuoles of P. perforata is higher than the value shown in Table 7-3 (due to active Cl^- influx). A similar proposal might be advanced for the large central chloroplast, since Cl^- is known to be a major osmolyte in the chloroplasts of other algal groups (MacRobbie 1970).

IV. ORGANIC SOLUTES

The major low molecular weight organic metabolite in most red algae is the heteroside floridoside (O-α-D galactopyranosyl-(1→2)-glycerol), first identified in Palmaria palmata (Colin & Gueguen 1930). The structure of floridoside (from Iridaea laminarioides) was determined by Putman and Hassid (1954). In addition to floridoside, an isomeric form (isofloridoside), O-α-D galactopyranosyl-(1→1)-glycerol, was characterized (Craigie et al. 1968; Lindberg 1955). Kremer (1980) and Nagashima (1976) showed that floridoside is the principal low molecular weight carbohydrate in all orders of the Rhodophyta except the Ceramiales, isofloridoside occurring in quantity only in members of the Bangiales, where it may exceed the amount of floridoside under certain circumstances.

Radiocarbon tracer studies showed that exogenous inorganic ^{14}C is rapidly assimilated into floridoside, acting as a major photoassimilatory product (Kremer 1981). In contrast, isofloridoside is generally rather weakly ^{14}C-labeled (Craigie et al. 1968) and pulse-chase experiments suggest that ^{14}C is transferred very slowly from floridoside to isofloridoside under light and dark conditions. These observations are consistent with the formation of isofloridoside as a result of a slow isomerization reaction from floridoside rather than as a direct product of photosynthesis.

Floridoside synthesis in Catenella caespitosa (as C. repens), Corallina officinalis and Lomentaria umbellata was studied by Kremer and Kirst (1981). Glycerol and floridoside phosphate were rapidly ^{14}C-labeled in addition to floridoside, and the kinetics of labeling indicated that these two compounds are precursors of floridoside. Two biosynthetic enzymes were identified in cell-free crude extracts: first, glycerol-3-phosphate dehydrogenase, which may catalyze the formation of glycerol-3-phosphate from dihydroxyacetone phosphate, and second, a galactosyltransferase (floridoside phosphate synthase), which may be responsible for the condensation of UDP-galactose and glycerol-3-phosphate to produce floridoside phosphate (Kremer & Kirst 1981). The dephosphorylation of floridoside phosphate would thus result in the formation of floridoside, by analogy with galactosyl-glycerol metabolism in the chrysophyte Poterioochromonas malhamensis (See Kauss 1977, 1979).

The taxonomic significance of galactosyl-glycerol accumulation in algae has been considered by Kremer (1980). The biosynthesis of floridoside is a reliable marker for the Rhodophyta and facilitated the identification of endosymbiotic unicellular red algae (Kremer et al. 1980). The discovery of floridoside and isofloridoside in strains of acidothermophilic algae assigned to Cyanidium caldarium (De Luca & Morelli 1983; Nagashima & Fukuda 1981; Reed 1983b) confirms the proposal that C. caldarium is a red alga (Brock 1978), rather than a symbiotic association of a blue-green alga and an apochlorotic eukaryotic alga (Kremer 1983). De Luca and Morelli (1983) also give details of the taxonomic revision of Cyanidium.

In most of the members of the Ceramiales, digeneaside (O-α-D mannopyranosyl-(1→2)-glyceric acid) is the major low molecular weight carbohydrate (Kremer 1978). This anionic solute is likely to contribute to intracellular charge balance (Raven 1984). In contrast, *Bostrychia* spp. accumulate the hexitols galactitol (dulcitol) and glucitol (sorbitol: Kremer 1976a); early reports of the occurrence of mannitol have not been confirmed (Kremer 1976b). Trehalose is the only disaccharide identified in significant quantities, occurring in floridoside-synthesizing and digeneaside-synthesizing species (see Craigie 1974) but not in *Bostrychia scorpioides*. Kremer (1980) has suggested that accumulation of alditols and that of disaccharides are mutually exclusive phenomena. The cyclitol 1,4/2,5 cyclohexanetetrol occurs in significant amounts in *Porphyridium* sp., as the second most important photoassimilate after floridoside (Craigie 1974).

The tertiary sulphonium compound β-dimethylsulphoniopropionate (dimethylpropiothetin, or DMSP), originally characterized from *Polysiphonia lanosa* (Challenger & Simpson 1948) is present in a range of green and red seaweeds (White 1982; Reed 1983c) and phytoplankton (Ackman et al. 1966). The position of DMSP in the metabolism of methylated compounds is unclear. However, an enzyme catalyzing the breakdown of DMSP to dimethyl sulphide and acrylic acid was identified in crude extracts of *P. lanosa*, and this enzyme may be involved in DMSP metabolism (Cantoni & Anderson 1956).

Amino acids may function as nitrogen reserves in red algae (Section II.A). The major pool amino acid in *P. purpurea* and *P. lanosa* is glutamate (Reed 1983a; Reed et al. 1981). Alanine, glutamate, and taurine predominate in *Porphyra* sp. (Inoue et al. 1973) while proline occurs in significant amounts in seawater-grown *Bangia atropurpurea* (Sheath 1984). Radiocarbon-labeling studies showed that alanine and glutamate are major photoassimilates in several marine red algae (Majak et al. 1966). In contrast, *Gelidium coulteri* exhibits preferential ^{14}C-labeling of asparatate and asparagine, together with glutamate and glutamine (Macler 1986), while di-N-methyl taurine is a major component in *Corallina officinalis* (Craigie et al. 1968).

V. OSMOACCLIMATION

Intertidal marine algae exist in an environment with an unstable salinity regime. Fluctuations in salinity will be greatest during periods of emersion as a result of several climatic factors, including precipitation, temperature, wind velocity, and humidity. Freshwater run-off also may lower the ambient salinity. In contrast, estuarine algae will be exposed to less irregular changes in salinity due to the mixing of seawater and fresh water and the effects of tidal water movements. Den Hartog (1967, 1971) has stressed the fundamental significance of salinity in determining (1) the distribution of marine algae and (2) the community structure of brackish waters (i.e., of unstable salinity).

A plant actively growing in an intertidal or estuarine habitat must have a measurable degree of tolerance to variations in (1) external osmotic pressure and (2) external salt (ion) concentration. If the salinity of the bathing medium is altered, water will flow into or out of the plant until a new water potential equilibrium is reached (equation 2). This process will be rapid and will not be subject to regulation by cellular control systems since the water permeability of algal plasma membranes is large (Dainty 1976).

Cell turgor and volume will vary when ψ_e is altered, in accordance with the relationship:

$$dP_i = \varepsilon(dV/V) \quad \text{(equation 6)}$$

where ε is a coefficient of elasticity (Pa), the volumetric elastic modulus, that governs the fractional change in volume for a given change in P_i (Pa). Intracellular osmotic pressure, turgor, and volume are interrelated via the volumetric elastic modulus as follows:

$$d\psi_e = dP_i - d\pi_i = (\varepsilon + \pi_i)(dV/V) \quad \text{(equation 7)}$$

If ε is large, the volume adjustment associated with a change in ψ_e will be small. In contrast, if it is small, then significant change in cell volume will occur as a result of a change in external water status. The former strategy leads to maximum variation in P_i, rather than V, in response to alterations in external salinity. Thus, in theory, turgor pressure could be used as a passive "buffer" to prevent (potentially damaging) reductions in cell volume in response to salinity increase (Dainty 1976). However, equation 3 shows that P_i provides the driving force for cell expansion and is a fundamental requirement for growth. Conversely, increases in P_i in response to salinity decrease (downshock) may lead to cell wall failure and osmotic lysis (Guggino & Gutknecht 1982). These two aspects of turgor regulation provide effective upper and lower limits to the extent of change in turgor that may be tolerated by a growing algal cell.

In cells with rigid walls (high ε), increases in salinity (upshock) will eventually reduce the turgor pressure to zero. At higher salinities the cells will be plasmolysed and the protoplasts will behave

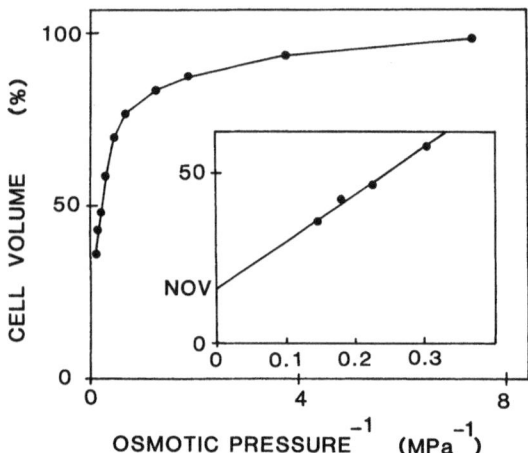

Fig. 7-2. The relationship between the cell volume of *Porphyra purpurea* and the reciprocal of the external osmotic pressure (Boyle–van't Hoff plot: 15-mm incubation in each medium). The insert shows the linear response of cells in hypersaline medium, and the nonosmotic volume (NOV). (From Reed et al. 1981)

as "osmometers," shrinking in proportion to the external osmotic pressure in accordance with the Boyle–van't Hoff relationship (Nobel 1983). Höfler (1930, 1931) used plasmolysis to assess the effects of salinity increase on several red algae, including *Callithamnion tetragonum*, *Ceramium ciliatum*, *Griffithsia* sp., *Heterosiphonia plumosa*, and *Polysiphonia urceolata*. He found that plasmolysis was often lethal to red algae (e.g., *H. plumosa*, *P. urceolata*), although certain species survived at salinities higher than those required to cause incipient plasmolysis (e.g., *C. ciliatum*). These observations suggest that a separation of the plasma membrane and cell wall, rather than a reduction in cell volume, may be the most damaging aspect of plasmolysis in many red algae. Biebl (1962) summarized the results of subsequent studies on the effects of hypersalinity on seaweeds, suggesting that stenohaline red algae may be intolerant of upshock as a result of plasmolysis-induced cell damage, whereas euryhaline species may avoid plasmolysis by increasing π_i, or they may possess the capacity to survive periods of plasmolysis.

In plants with nonrigid cell walls and a low value for ε (e.g., *Porphyra* spp.), cell volume will be reduced as an immediate consequence of salinity increase. Changes in the thallus length of *P. tenera* (Ogata & Takada 1955) and tissue water content of *P. perforata* (Eppley & Cyrus 1960) on transfer to hypersaline media can be interpreted in terms of osmotically induced decreases in cell volume. Decreases in cell volume on transfer to 350% seawater were reported for *P. umbilicalis* (Knoth & Wiencke 1984). Reed et al. (1980a) studied the relationship between cell volume and external salinity in *P. purpurea*. At high external osmotic pressures (3.87–7.74 MPa, equivalent to 150–300% seawater) the cells behaved as osmometers (i.e., $P_i = 0$), with a linear relationship between cell volume and the reciprocal of the external osmotic pressure (Fig. 7-2). However, plasmolysis was not observed, and the nonrigid cell walls accommodated the volume reductions without separating from the protoplasts. *P. purpurea* may thus avoid the damaging effects of plasmolysis-induced separation of the cell wall and plasma membrane. In dilute seawater media (0.161–1.935 MPa), the cells adopted a polygonal, rather than a rounded, appearance in surface view due to an increase in cell turgor, and the cell walls served to limit the increase in cell volume, with consequent increases in P_i and ε (Reed et al. 1980a). These observations are in accord with the effect of P_i on ε in higher plant cells (Zimmermann 1978).

Transfer to dilute or concentrated seawaters may have adverse effects on metabolism due to (1) changes in turgor and volume or (2) disruption of transport processes and transmembrane ion gradients, and it is often impossible to separate these effects on the basis of available experimental evidence. The photosynthetic activity of marine red algae under salinity stress has been studied by several workers. However, in many cases the changes in salinity have been obtained by the dilution of natural seawater with distilled, or deionized water, which reduces the concentration of inorganic carbon and thus affects any interpretation of the results. This was demonstrated by Ogata and Matsui (1965) for *Porphyra tenera* and *Gelidium amansii*. Photosynthesis in plants transferred to fresh water bubbled with 5% CO_2 was greater than in plants in an unsupplemented seawater medium. In an artificial medium with constant $NaHCO_3$ (5 mol m^{-3}) no significant differences between the two species were observed. In contrast, using natural seawater diluted with distilled water, *P. tenera* (intertidal) showed a photosynthetic optimum at 150% seawater, with a somewhat reduced rate in hyposaline media, whereas *G. amansii* (subtidal) gave optimal activity at 100% seawater, with reduced rates in hypersaline and hyposaline media. In other red algae irreversible loss of photosynthetic activity occurs rapidly on exposure to extreme hypoosmotic conditions. Gessner (1971) showed that a 2-min incubation in distilled water results in a total loss of photosynthetic activity in *Halymenia floresia*, due to the damaging effects of water uptake (and high turgor?) rather than ion loss (Gessner 1969), since the response was not observed on transfer to an isotonic solution of mannose.

Short-term effects of salinity change on photosynthesis in *Delesseria sanguinea* (subtidal) have been reported by Nellen (1966); moderate increases or decreases in salinity stimulated photosynthesis transiently by up to 60%, whereas more extreme alternations in salinity (e.g., 0%, 150% seawater) reduced the photosynthetic rate. Longer-term incubation (24 h) restored the original rates of photosynthesis (Nellen 1966). These results are in agreement with a study of the effects of salinity on *Porphyra purpurea*; extreme salinities (e.g., 6% and 300% seawater) caused an immediate decrease in photosynthetic activity that was not observed at less extreme salinities (Reed 1979). In long-term experiments (48 h), *P. purpurea* showed maximum rates of photosynthetic O_2 evolution in 150% seawater, with decreasing rates at higher and lower salinities (Reed et al. 1979).

Whereas the effects of hypersaline conditions on photosynthesis are likely to be osmotic or ionic (Rees 1984) in character, the importance of inorganic carbon supply in hyposaline media was illustrated by several studies where spring waters with a high HCO_3^- content have been used as a diluent. Photosynthetic activity was decreased in hyposaline media prepared using distilled water but increased in media formulated with spring water in *Gracilaria armata* (Hammer 1968), *Porphyra leucosticta* (Zavodnik 1975), and *Bostrychia binderi* (Dawes & McIntosh 1981). The latter authors attribute the growth of *B. binderi* in estuarine waters to a combination of high HCO_3^- and Ca^{2+} salts.

Moderate changes in salinity may also lead to decreases or increases in respiratory activity, and these are often due to alterations in O_2 supply. Extreme salinity stress invariably reduces the respiration rate, presumably due to the inhibition of metabolic activity (Gessner & Schramm 1971).

Although the first phase of osmotic response, involving a change in cell turgor and/or volume, is outside the control of intracellular processes, subsequent changes in internal osmotic pressure may occur as a direct result of metabolic activity (osmoacclimation). These effects are due to adjustments in the total number of osmotically active intracellular solute molecules and may serve to move P_i and/or V towards their original values. In cells with a high volumetric elastic modulus, this process operates as a turgor-regulatory system, whereas wall-less cells exhibit volume regulation (Cram 1976). Although it is not yet known whether P_i or V is the regulated parameter (Zimmermann 1978), it is clear that "osmoregulation" is a misleading term since it implies a constancy of π_i in response to changes in ψ_e (Reed 1984a). The concept of turgor/volume regulation is based on the existence of "ideal" or reference values for P_i and V, maintained by negative feedback control systems that regulate intracellular solute levels (Cram 1976). Osmotic adjustment of internal solutes in response to salinity variation is a fundamental aspect of osmoacclimation in algae.

Eppley and Cyrus (1960) established the osmotic significance of inorganic ionic solutes in *Porphyra perforata*. The internal K^+ content, expressed in terms of the apparent cell volume, was found to increase on incubation (for 24 h) in hypersaline media and decrease under hyposaline conditions. As the major intracellular cation, K^+ was accumulated at >60% of the external Na^+ concentration at all salinities. The ratio of internal:external K^+ was highest in extreme hyposaline media, and the relationship between intracellular K^+ and external salt concentration was linear, with a nonzero intercept on the ordinate. This is in agreement with Fig. 7-3, which shows the dual role of intracellular solutes as agents of (1) turgor generation and (2) osmotic adjustment. The results of Eppley and Cyrus (1960) can be used to estimate the contribution of K^+ to P_i in *P. perforata* under hyposaline conditions at ≈ 0.25 MPa. Intracellular Na^+ was an order of magnitude lower than K^+ at all times, reducing the osmotic

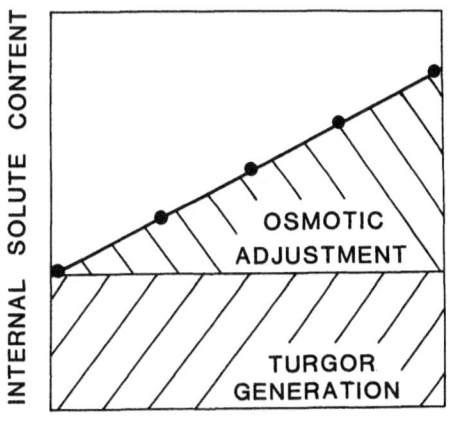

Fig. 7-3. A diagrammatic representation of the relationship between intracellular solute content and external osmotic pressure in a walled algal cell with complete regulation of turgor, showing the contribution of solute to (1) turgor generation and (2) osmotic adjustment.

significance of the observed increases in internal Na$^+$ in hypersaline media.

It should be noted that the changes in intracellular ion concentrations observed by Eppley and Cyrus (1960) in *Porphyra perforata* were, in part at least, due to variation in cell volume as a function of salinity, since no volume recovery phase was observed and the apparent osmotic volume remained constant in dilute and concentrated sea waters between 2 min and 24 h. This phenomenon has been investigated in detail in *P. purpurea* (Reed et al. 1981). This alga also showed minimal volume recovery over a 48-h incubation period in hyposaline and hypersaline media. The major inorganic ions were K$^+$ and Cl$^-$, with lower intracellular concentrations of other cations (e.g., Na$^+$, Ca^{2+}, Mg^{2+}) and anions (phosphate, NO$_3^-$, SO$_4^{2-}$) and all intracellular ionic solutes increased in response to salinity. However, the decrease in cell volume with increasing salinity was identified as the principal cause of intracellular solutes increase. When the data for K$^+$ and Cl$^-$ were expressed in terms of the amount of solute per cell, there were limited changes in the amounts of the ions in plants kept in hypersaline media. In contrast, significant decreases in the levels of K$^+$ and Cl$^-$ per cell were observed in hyposaline media: in 6% seawater internal K$^+$ fell to 25% of the value in 100% seawater, and Cl$^-$ was lowered to 10%. Transplasmalemma flux rates also showed the same dependency on external salinity, decreasing dramatically in hyposaline media. Time courses for K$^+$ and Cl$^-$ showed that new steady-state ion levels were achieved within 24 h, with maximum changes up to 6 h after upshock or downshock. Reed et al. (1981) suggested that the observed effects of salinity on intracellular K$^+$ and Cl$^-$ levels were likely to be due to changes in external K$^+$ concentration, rather than osmotic pressure, since the addition of mannitol had little effect on (1) intracellular K$^+$ content or (2) K$^+$ influx.

The responses of *Porphyra umbilicalis* in dilute and concentrated seawaters appear to be more complex. Wiencke and Läuchli (1981) reported that maximum intracellular K$^+$ accumulation is observed in 100% seawater (Table 7-3), decreasing in hyposaline and hypersaline media (incubation period, 14 d), whereas Na$^+$ is maintained at $\simeq$100 mol m^{-3} up to 300% seawater, rising to >600 mol m^{-3} at 500% seawater. Intracellular Cl$^-$ showed a direct relationship with external salinity up to 300% seawater, decreasing rapidly above this value. The authors suggest that the intracellular osmotic pressure may be lower than that of the external medium at high salinity (above 300% sea water). However, a water potential imbalance of this kind is impossible on a thermodynamic basis in the absence of an external energy input (equation 2) and is inconsistent with the volume recovery observed in 350% seawater (Knoth & Wiencke 1984). The remaining possibility is that other (unidentified) solutes are involved in osmotic balance under hypersaline conditions.

Long-term studies with *Porphyra purpurea* (Reed et al. 1980a) and *P. umbilicalis* (Ogata & Schramm 1971) showed that significant increases in cell volume are caused by prolonged exposure to concentrated sea waters, due to the preferential inhibition of cell division. Whereas such volume increases demonstrate that turgor must be sufficiently greater than the yield threshold to cause cell expansion (equation 3), they also suggest that the concept of turgor/volume regulation and the notion of "ideal" values for P_i and V may not be applicable to all experimental observations.

Turgor regulation has been observed in the giant-celled red alga *Griffithsia monilis*, with a restoration of turgor within 24–36 h in response to salinity perturbation (Bisson & Kirst 1979). Turgor was regulated at approximately 0.4 MPa over the external osmotic pressure range 2.2–3.2 MPa, due to changes in intracellular K$^+$, Na$^+$ and Cl$^-$. The regulatory response always involved Cl$^-$, with preferential loss of K$^+$ in response to downshock, whereas upshock resulted in Na$^+$ accumulation. Turgor regulation was observed in darkness and in response to changes in external osmotic pressure using sorbitol.

The role of inorganic ions as osmotically active solutes in several Australian red algae was studied by Kirst and Bisson (1979). These workers identified three groups of red algae, based on intracellular ion levels: (1) species with high K$^+$ (and Cl$^-$) and low Na$^+$, (2) species with high Na$^+$ (and Cl$^-$) and low K$^+$, and (3) species with approximately equal amounts of K$^+$ and Na$^+$ (and Cl$^-$). In all cases, the major cellular ions (i.e., Cl$^-$ plus Na$^+$ and/or K$^+$) showed some increase with increasing salinity, Cl$^-$ increasing in hypersaline media in all cases and decreasing (four species) or remaining constant (three species) in hyposaline conditions. A characteristic decrease in K$^+$ and increase in Na$^+$ was observed at the highest salinities (equivalent to 3.4–4.5 MPa) in algae assigned to group 1 and may reflect the breakdown of active ion transport processes under hypersaline stress. In contrast, no such effects have been observed in response to upshock in *Porphyra purpurea* (Reed et al. 1981) or *Polysiphonia lanosa* (Reed 1983a).

The effects of Ca^{2+} on the survival of marine red

algae in fresh water and hyposaline media are well established. Hoffmann (1959) showed that *Delesseria sanguinea* survived in a medium containing 1% seawater only in the presence of added Ca^{2+}. Eppley and Cyrus (1960) demonstrated that Ca^{2+} is involved in the maintenance of intracellular ion distribution of *Porphyra perforata*, with rapid loss of K^+ in 30% seawater in the absence of Ca^{2+}. Loss of internal K^+ and accumulation of (external) Na^+ also occurred in Ca^{2+}-free seawater. Marine *Polysiphonia lanosa* survives in low-salinity media for 24 h only if Ca^{2+} is high ($\simeq 10$ mol m^{-3}), and this is associated with the retention of intracellular ions (Reed 1984b). These observations support the hypothesis that the penetration of marine red algae into fresh waters and hyposaline estuarine waters may be governed by the availability of Ca^{2+} (Dawes & McIntosh 1981; Yarish et al. 1980).

The involvement of organic solutes in the osmotic responses of salt-stressed red algae was proposed by Kauss (1968, 1969), who showed that ^{14}C-labeling of floridoside and isofloridoside in *Iridaea splendens* (as *Iridophycus flaccidum*) and *Porphyra perforata* was sensitive to the ambient salinity, with increasing amounts of radioactive tracer incorporated into galactosyl-glycerols as a direct function of the external salt concentration. Subsequent research on *P. umbilicalis* showed that the intracellular concentration of galactosyl-glycerols was increased threefold between 50% and 300% seawater (Wiencke & Läuchli 1981), accounting for 20–40% of the external osmotic pressure after 14 d. Reed et al. (1980b) showed that the cellular content of floridoside, rather than isofloridoside, varies in response to salinity in *P. purpurea*, with the greatest change in hypersaline media, in agreement with ^{14}C-labeling studies showing floridoside to be the major photoassimilate (Section IV). Taken together, floridoside and isofloridoside vary with salinity in accordance with Fig. 7-3, with a minimum intracellular concentration of approximately 100 mol m^{-3} in 6% seawater rising to >800 mol m^{-3} in 300% seawater and thus accounting for a significant fraction of π_i. The responses of plants maintained in darkness were similar to those of illuminated plants, although changes in floridoside concentration occurred at a slower rate. In addition, floridoside synthesis could be initiated by adding nonionic solutes (e.g., mannitol).

In several Australian marine red algae the intracellular galactosyl-glycerol level increases in concentrated seawater and decreases in dilute seawater (Kirst & Bisson 1979). However, the intracellular concentrations were low, accounting for less than 10% of π_e. Kremer (1979, 1981) suggested that the intracellular concentration of floridoside and other low molecular weight carbohydrates passively follows external osmotic fluctuations, e.g., as a result of changes in photosynthetic and respiratory activity (Munda & Kremer 1977), rather than participating in turgor/volume regulation. These suggestions are inconsistent with the changes in floridoside described above and with the 10-fold increase in floridoside concentration in freshwater-grown plants of *Bangia atropurpurea* transferred to seawater (Reed 1985); in this example maximum increases in intracellular floridoside concentrations coincide with reduction in photosynthetic carbon fixation, suggesting that floridean starch may provide fixed carbon for organic osmolyte synthesis in this alga. The enhanced synthesis of floridoside in salt-stressed *Cyanidium caldarium* (Reed 1983b) is also consistent with the proposal that floridoside has an osmotic role and is not solely a short-term (nonosmotic) source of respiratory carbon.

Other organic solutes may have an osmotic role in certain red algae. Bisson and Kirst (1979) showed that the intracellular concentration of digeneaside in *Griffithsia monilis* decreases on downshock and returns to a level close to the original value if the plants are then returned to sea water. The initial decrease in digeneaside on upshock is followed by a sustained increase, reaching new steady-state maxima in approximately 12 h. However, the intracellular concentration is low (<2 mol m^{-3} digeneaside, on a total cell volume basis). Somewhat higher intracellular concentrations of β-dimethylsulphoniopropionate (DMSP) were reported for *Polysiphonia lanosa* (Reed 1983c). Changes in intracellular DMSP concentration in marine and estuarine *P. lanosa* are consistent with a role in turgor regulation, although the measured changes in response to salinity are relatively small. Thus, an increase in external salt concentration of 100-fold produced an increase of less than three-fold in cellular DMSP. These observations are consistent with results for several other algae (e.g., Dickson et al. 1980; Dickson & Kirst 1986; Vairavamurthy et al. 1985). However, recent studies on the effects of hypersaline stress on organic solute accumulation in the DMSP-accumulating green seaweed *Enteromorpha intestinalis* (Edwards et al. 1987) showed that changes in amino acids and saccharides may be more important in osmotic adjustment.

Amino acids have received less attention as intracellular solutes, although their role in other algae is well-documented (e.g., Ahmad & Hellebust 1984; Ben-Amotz & Avron 1983). Changes in the total amino acid level of *Porphyridium purpureum* (Gilles & Pequeux 1977) and *Porphyra purpurea* (Reed et al. 1981) were reported, and Sheath (1984) has shown that a fourfold change in salinity leads to a threefold increase in intracellular proline in *Bangia atropurpurea*, indicating an osmotic function for this metabolite. In several cases the osmotic significance of organic solutes accumulation is unknown due to a lack of quantitative data for (1) the amounts of metabolites and (2) the relative proportion of intracellular water. Natural abundance ^{13}C nuclear magnetic resonance (NMR) spectrometry allows unequivocal identification of the principal organic solutes in living cells (Borowitzka et al. 1980), and this technique has been used to assess the osmotic significance of mannitol accumulation in marine brown macroalgae (Reed et al. 1985). Fig. 7-4 shows a natural abundance ^{13}C NMR spectrum of an extract of *Mastocarpus stellatus*. Eight major resonances are visible, corresponding to floridoside (Meng et al. 1987) and confirming that this is the major organic osmoticum in this alga. The smaller, unidentified peaks must correspond to other, minor organic osmotica. The use of NMR techniques with marine red algae can thus identify organic solutes present in osmotically significant quantities, eliminating the complex procedures of chemical analysis.

Most investigations of the effects of salinity on algae involve abrupt transfer from seawater to a hyposaline or hypersaline medium, with no attempt to reflect field conditions (e.g., in nonstratified estuaries, where salinity may change in a sinusoidal cycle over a tidal period ($\simeq 12$ h). A study of the effects of simulated estuarine fluctuations in salinity on osmotic adjustment in *Porphyra purpurea* (Reed 1980a) showed that the extent of change varied according to the parameter studied. Thus, cell volume and photosynthetic O_2 evolution were as predicted from steady-state experiments (abrupt salinity change), whereas the intracellular solutes K^+, Cl^-, galactosyl-glycerols, and amino acids showed a damped response, decreasing to a lesser extent than predicted from steady-state experiments.

There is evidence that intracellular osmotica are not present in equal concentrations throughout the cell but are preferentially accumulated within specific organelles, or "compartments". The toxic effects of high Na^+ to cytoplasmic enzymes serves to restrict the total salt concentration of the cytoplasm (Pollard & Wyn Jones 1979; Wyn Jones et al. 1977), whereas cytoplasmic K^+ may be maintained at a higher level to give optimum enzyme activation (Wiencke 1984). It is generally accepted that organic osmotica may be localized within the cytosol, where they will function as "compatible solutes" (Brown & Simpson 1972), having no toxic effects on metabolic activity at physiologically relevant concentrations. Schobert (1977) proposed a further role for cytoplasmic organic osmolytes, suggesting that enzyme activity may be protected by the accumulation of organic solutes with amphiphilic properties during periods of osmotic stress, when water availability may be reduced. The changes in galactosyl-glycerol content of hyperosmotically stressed *Porphyra purpurea* may have a similar metabolic function and may be linked to the protection of intracellular macromolecules and the maintenance of protein structure (Low 1985), rather than to turgor regulation (Reed et al. 1980b). In contrast, high intracellular inorganic ion levels may be associated with vacuolar storage. This was shown for *P. umbilicalis*, with preferential accumulation of NaCl in the vacuoles of hyperosmotically stressed cells while the cytoplasm and chloroplast contain the bulk of the cellular K^+ (Wiencke et al. 1983). These results are consistent with measurements of the relative proportions of organic and inorganic osmolytes in a range of red algae. Species with a large cytoplasmic component and without a central vacuole (e.g., *Porphyra* spp.) often contain substantial amounts of organic solutes, although intracellular Na^+ levels are low (Table 7-3). In contrast, in species where the large central vacuole occupies most of the cell interior (e.g., *Griffithsia monilis*) the intracellular solute complement is dominated by inorganic ions, and the contribution of organic solutes to internal osmotic pressure may be small (Bisson & Kirst 1979). Dual-labeling studies using $^{86}Rb^+$ and $^{42}K^+$ (Section III) suggested that intracellular K^+ is mostly located with a noncytoplasmic compartment (composed of chloroplast and vacuoles) in *P. purpurea* and that changes in the distribution of K^+ in response to downshock are consistent with the localization of galactosyl-glycerols in the cytoplasm (Reed 1981). The reduced level of DMSP in estuarine *Polysiphonia lanosa* is accompanied by a decrease in the relative size of the cytoplasm (Reed 1983a), in accordance with the hypothesis that this tertiary sulphonium compound may act as a cytoplasmic osmoticum (Reed 1983c).

At present, the mechanisms by which variations in external water status are sensed and transformed into turgor regulatory processes are not known in detail for any red alga. However, in the green alga *Valonia*, the influx of K^+ appears to be directly affected by changes in turgor pressure (Gutknecht 1968), whereas Cl^- transport is controlled in *Codium* (Bisson & Gutknecht 1980). (In contrasting the responses of *Valonia* with *Codium*, it should be noted that the vacuole is electropositive with respect to the medium in the former, whereas the latter has an electronegative vacuole; no red algae are known to fit the *Valonia* pattern.) These effects on ion transport are comparable with the influence of external osmotic pressure on floridoside accumulation in *Porphyra purpurea* (Reed et al. 1980b) and P_i in *Griffithsia monilis* (Bisson & Kirst 1979). In contrast, the giant-celled green alga *Chaetomorpha* shows changes in P_i and intracellular K^+ as a function of external K^+ concentration, rather than ψ_e (Zimmermann & Steudle 1971), in agreement with the observation that intracellular ion levels of *P. purpurea* are not affected by changes in ψ_e at constant extracellular K^+ concentration (Reed et al. 1981). However, in marine environments, where changes in ψ_e will be linked most exclusively to changes in K^+, these distinctions may not be so important. Gutknecht and co-workers (Bisson & Gutknecht 1980; Gutknecht & Bisson 1977) proposed that an interaction between the plasmalemma and the cell wall may lead to deformation of the plasmalemma and, consequently, to changes in the activity of membrane-associated enzymes and transport systems. Zimmermann and co-workers formulated a more quantitative model, based on electromechanical principles (Zimmermann 1978), which postulates that pressure-induced changes in the dimensions of the plasmalemma are transformed into alterations in the electrical field distribution within the membrane and that osmotic adjustment is linked to (1) changes in plasmalemma thickness due to compression/stretching and/or (2) the plasmalemma electrical potential. The electromechanical model is able to account for the observed effects of external K^+ on intracellular solute accumulation due to changes in E_m. It is not clear how these models could account for the observed changes in organic osmotica in response to salinity variation. However, the regulation of galactosyl-glycerol (isofloridoside) in the chrysophyte *Poterioochromonas malhamensis* provides a model that may be applicable to floridoside synthesis in the Rhodophyta. The studies of Kauss and co-workers (see Kauss 1977, 1979) demonstrated that galactosyl-glycerol accumulation in response to upshock may involve the activation of preexisting enzymes of isofloridoside synthesis, particularly isofloridoside phosphate synthase. Recent studies found that Ca^{2+}/calmodulin (Kauss & Thomson 1982) and/or polar lipids (Kauss 1982) may be important regulators of galactosyl-glycerol synthesis. Although similarities in the biosynthetic pathways of isofloridoside in *P. malhamensis* and floridoside in the Rhodophyta have been noted (Kremer & Kirst 1981), the effects of osmotic stress on enzyme activity in red algae remains unknown.

VI. OSMOADAPTATION

The existence of habitats with dissimilar salinity regimes acts as a mechanism of directional selection and genetically determined variation in salinity tolerances (osmoadaptation) has been reported between different species of red algae. Biebl (1962) suggested that many subtidal red algae are less tolerant of salinity extremes than intertidal species, and Almodovar and Biebl (1961) reported the extreme halotolerance of mangrove red algae. In general, red algae are regarded as less euryhaline than other seaweeds, although this is not always the case (e.g., Bird & McLachlan 1985; Chapter 14).

Several studies indicated that the salinity responses of populations of red algae from habitats subjected to hyposaline stress differ from those of their marine counterparts. Examples include *Bangia atropurpurea* (Geesink 1973; Reed 1980b; Sheath & Cole 1980), *Caloglossa leprieurii*, *Bostrychia radicans* (Yarish & Edwards 1982; Yarish et al. 1979, 1980) and *Polysiphonia lanosa* (Reed 1983a). Differences in the rates of adjustment to hyposaline stress and the relative importance of organic and inorganic osmotica in marine and estuarine *P. lanosa* were interpreted in terms of their adaptive value to the estuarine plants. Thus, estuarine *P. lanosa* are smaller celled, with an increased dependence on inorganic ions, rather than organic solutes, for osmotic adjustment. These features may facilitate rapid changes in π_i in response to salinity variation. An additional aspect is the reduced energy requirement involved in ion uptake, rather than organic solute synthesis (see Bisson & Kirst 1979), although this takes no account of the long-term maintenance costs of ionic osmotica (Yeo 1983). The lower turgor of estuarine *P. lanosa* also is likely to be an adaptive response and may prevent problems associated with high turgor (Guggino & Gutknecht 1982) in response to hyposaline stress (Reed 1983a). The rate of loss of intracellular solutes in response to extreme hyposaline stress in Ca^{2+}-deficient media also is lower in estuarine plants (Reed 1984b).

In other red algae there is less evidence of phenotypic or genetic variation. Nellen (1966) demonstrated that *Delesseria sanguinea* from the Baltic shows evidence of damage in a hypersaline medium (150% seawater), in contrast to plants from the North Sea. However, Lehnberg (1978), in a similar study, found no differences between populations of *D. sanguinea*. In contrast, a survey of the salinity responses of several algae from the Baltic, including *Ceramium tenuicorne* and *Rhodomela confervoides*, has confirmed that Baltic algae show a hyposaline shift in halotolerance when compared with their North Atlantic counterparts (Russell 1985).

VII. SUMMARY

The osmotic pressure of red algal cells is generated principally by inorganic ions and low molecular weight organic solutes. The major inorganic ions include metabolized (e.g., N and P compounds) and nonmetabolized ions (e.g., K^+, Cl^-, and Na^+); their uptake is affected by the combined physical driving forces of the transplasmalemma concentration gradient and the transmembrane electrical potential.

Ammonia uptake from sea water (where NH_4^+ accounts for over 90% of the total ammonia concentration) shows single-phase kinetics in certain (N-sufficient) red algae. In contrast, ammonia uptake exhibits dual-phase kinetics in other red algae, with a low-affinity system due to passive diffusions of NH_3 and a high-affinity system that may represent active transport. Nitrate uptake generally exhibits single-phase saturation kinetics and is likely to be an active process, owing to the electronegativity of the cell interior. Intracellular N-reserves include NO_3^- storage (in vacuoles), the dipeptide citrullinyl-arginine, amino acids, and phycobiliproteins.

Phosphate uptake in marine red algae is likely to be an active process, possibly via a secondary active transport system (Na^+ cotransport). Excess intracellular phosphate may be stored as vacuolar phosphate or as cytoplasmic polyphosphate granules.

Intracellular Na^+ levels generally are lower than the external Na^+ concentration, indicating an efficient active Na^+ extrusion system. The most accurate techniques, based on radioisotope equilibration and compartmental flux analysis, give the lowest values for intracellular Na^+. Low cytosolic Na^+ concentrations are likely to be linked to the toxicity of Na^+ to metabolic activity, and there is evidence that the vacuole may act as a compartment where (toxic) Na^+ can be contained, without detrimental effects on (cytoplasmic) metabolism.

Potassium is the ion nearest to electrochemical potential equilibrium in most marine red algal cells, with a high intracellular level relative to the external K^+ concentration. In some algae (e.g., *Porphyra umbilicalis*), the available data suggest that an active K^+ influx system may operate at the plasmalemma, whereas in other algae (e.g., *P. purpurea*) the active K^+ influx system may be located at the tonoplast.

Chloride influx is almost certain to be an active process in most marine red algae and may be a primary means of generating the cell turgor pressure.

The principal low molecular weight organic solute in most red algae is the heteroside floridoside (O-α-D galactopyranosyl-(1→2)-glycerol). Floridoside is a major photoassimilatory product, showing rapid and substantial incorporation of exogenous inorganic ^{14}C during radiocarbon tracer studies. Isofloridoride (O-α-D galactopyranosyl-(1→1)-glycerol) also occurs in quantity in the Bangiales, whereas members of the Ceramiales may contain digeneaside (O-α-D) mannopyranosyl-(1→2)-glyceric acid). The tertiary sulphonium compound β-dimethylsulphoniopropionate (DMSP) is also present in certain red algae (e.g., *Polysiphonia lanosa*).

Variations in ambient salinity (due to climatic effects during periods of emersion for intertidal algae or to tidal water movements and the mixing of sea water and fresh water for estuarine algae) will lead to changes in cell turgor pressure and volume; the interrelationship of these two parameters depends upon the volumetric elastic modulus. Following a change in salinity, the number of osmotically active solute molecules within the cell may be adjusted to bring turgor and volume toward their original values (i.e., turgor regulation or volume regulation). The available evidence for *Porphyra* spp. suggests that K^+, Cl^- (Na^+), and floridoside may be involved in this osmotic adjustment, whereas in other red algae digeneaside, DMSP, and/or amino acids may vary in response to salinity. There is limited evidence in support of the intracellular compartmentalization of solutes, with organic osmolytes being preferentially contained within the cytoplasm and inorganic ions localized in the vacuole. Decreases in photosynthetic activity in response to salinity variation have been linked to changes in the supply of inorganic C, although extreme salinity stress invariably reduces the photosynthetic rate.

Salinity may act as an agent of directional selection; genetically determined variation in the salinity tolerances of different species of red algae can be correlated with the salinity regimes of their habitats. Additionally, the salinity response of estuarine and marine populations of several red algae (e.g., *Polysiphonia lanosa*, *Bangia atropurpurea*) can be interpreted in terms of intraspecific variation.

A better understanding of the processes responsible for solute accumulation and osmotic adjustment has been achieved as a result of research efforts over the past decade, with a growing literature on the responses of selected red algae at the

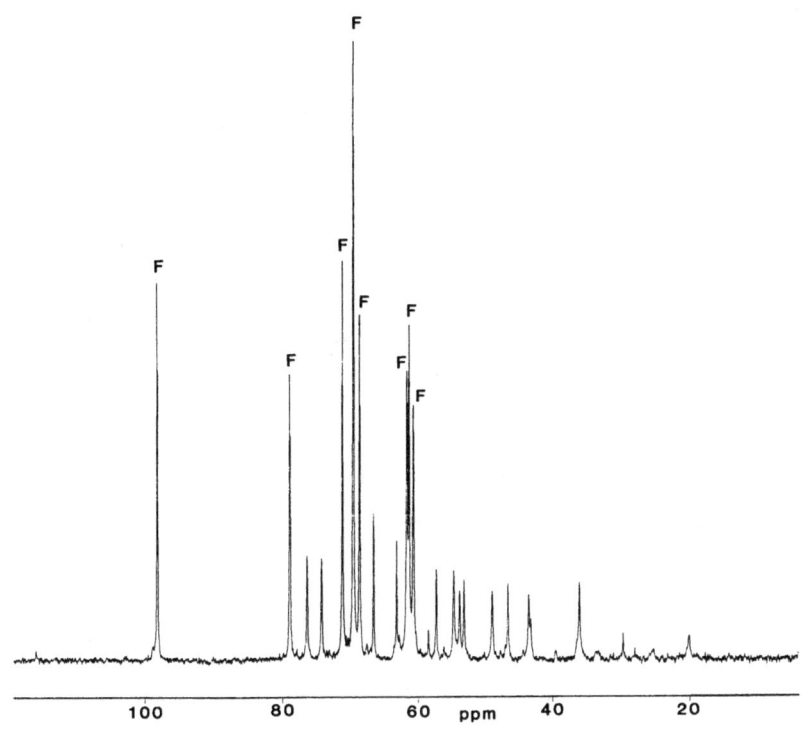

Fig. 7-4. A natural abundance ^{13}C NMR spectrum of an aqueous ethanol (20:80 v/v) extract of *Mastocarpus stellatus* (120 g fresh weight), prepared according to Reed et al. (1985). The principal ^{13}C resonances are due to floridoside (F).

cell, plant, and population level. There is increasing research emphasis on phycocolloid-producing algae, due to the rising demand for such products and the recent expansion of algal mariculture systems (Gellenbeck & Chapman 1983). However, certain topics require further work to resolve conflicting experimental observations, e.g., on the role of low molecular weight organic solutes in osmotic adjustment, the intracellular compartmentation of solutes, the ecological significance of ecotype formation. The application of ^{13}C-labeling and nuclear magnetic resonance techniques (Fig. 7-4) should provide important information on the partitioning of photosynthetically fixed carbon between osmotically active (soluble) and osmotically inactive (storage and structural) metabolites in response to salinity.

The implications of laboratory-based experiments conducted under optimal, defined conditions for plants growing in natural habitats have yet to be fully established, and there is scope for field-based research to evaluate the role of osmotic adjustment and solute accumulation. To date, the emerging techniques of molecular biology and genetics have yet to be applied to the processes of nutrient acquisition and osmoacclimation in red algae. Future research on nutrient uptake in red algae should aim to extend beyond the whole-cell approach, to identify and characterize the specific transport proteins (porters) involved in solute accumulation using subcellular systems and lipid vesicle reconstitution studies. Similar subcellular research is required to characterize the enzymes involved in the biosynthesis and degradation of organic osmotica and the regulatory effects of salinity.

VIII. ACKNOWLEDGMENTS

Dr. J. A. Chudek kindly provided the ^{13}C NMR spectrum of *Mastocarpus stellatus* shown in Fig. 7-4.

IX. REFERENCES

Ackman, R. G., Tocher, C. S. & McLachlan, J. 1966. Dimethyl-β-propiothetin: determination by reactor gas–liquid chromatography; occurrence in algae and implication in fisheries. *Proc. Int. Seaweed Symp.* 5: 235–42.

Adamec, J., Perverly, J. H. & Pathasarathy, M. V. 1979. Potassium in polyphosphate bodies of *Chlorella pyrenoidosa* (Chlorophyceae), as determined by x-ray microanalysis. *J. Phycol.* 15: 466–8.

Ahmad, I. & Hellebust, J. A. 1984. Osmoregulation in the extremely euryhaline marine micro-alga *Chlorella autotrophica*. *Plant Physiol.* 74: 1010–5.

Almodovar, L. R. & Biebl, R. 1961. Osmotic resistance of mangrove algae around La Paraguera, Puerto Rico. *Rev.*

Algol. 6: 203–8.
Asare, S. O. & Harlin, M. M. 1983. Seasonal fluctuations in tissue nitrogen for five species of perennial macroalgae in Rhode Island Sound. *J. Phycol.* 19: 254–7.
Baumann, R. W. & Jones, B. R. 1986. Electrophysiological investigations of the red alga *Griffithsia pacifica* Kyl. *J. Phycol.* 22: 49–56.
Ben-Amotz, A. & Avron, M. 1983. Accumulation of metabolites by halotolerant algae and its industrial potential. *Annu. Rev. Microbiol.* 37: 95–119.
Biebl, R. 1962. Seaweeds. In *Physiology and Biochemistry of Algae*, ed. R. A. Lewin, pp. 799–815. New York: Academic.
Bird, K. T., Habig, C. & DeBusk, T. 1982. Nitrogen allocation and storage patterns in *Gracilaria tikvahiae*. *J. Phycol.* 18: 344–8.
Bird, K. T. & McLachlan, J. A. 1985. The effect of salinity on the distribution of species of *Gracilaria* Grev. (Rhodophyta, Gigartinales): an experimental assessment. *Bot. Mar.* 29: 231–8.
Bisson, M. A. & Gutknecht, J. 1980. Osmotic regulation in algae. In *Membrane Transport Phenomena: Current Conceptual Issues*, eds. W. J. Lucas, R. M. Spanswick & J. Dainty, pp. 131–42. Amsterdam: Elsevier.
Bisson, M. A. & Kirst, G. O. 1979. Osmotic adaptation in the marine alga *Griffithsia monilis* (Rhodophyceae): the role of ions and organic compounds. *Aust. J. Plant Physiol.* 6: 523–38.
Borowitzka, L. J., Demmerle, S., MacKay, M. A. & Norton, R. S. 1980. Carbon-13 nuclear magnetic resonance study of osmoregulation in a blue-green alga. *Science* 210: 650–1.
Brady, C. J., Gibson, T. S., Barlow, E. W. R., Speirs, J. & Wyn Jones, R. G. 1984. Salt-tolerance in plants. I. Ions, compatible organic solutes and the stability of plant ribosomes. *Plant Cell Envir.* 7: 571–8.
Brock, T. D. 1978. *Thermophilic Microorganisms and Life at High Temperatures.* New York: Springer.
Brown, A. D. & Simpson, J. R. 1972. Water relations of sugar-tolerant yeasts: the role of intracellular polyols. *J. Gen. Microbiol.* 72: 589–91.
Butler, M. R. 1936. Seasonal variations in *Chondrus crispus*. *Biochem. J.* 30: 1338–44.
Cantoni, G. L. & Anderson, D. G. 1956. Enzymatic cleavage of dimethylpropiothetin by *Polysiphonia lanosa*. *J. Biol. Chem.* 222: 171–7.
Cembella, A. D., Antia, N. J. & Harrison, P. J. 1983. The utilization of inorganic and organic phosphorous compounds as nutrients by eukaryotic microalgae: a multidisciplinary perspective: part 1. *Crit. Rev. Microbiol.* 10: 317–91.
Challenger, F. & Simpson, M. I. 1948. Studies on biological methylation XII. A precursor of the dimethylsulphide evolved by *Polysiphonia fastigiata*. Dimethyl-2 carboxyethylsulphonium hydroxide and its salts. *J. Chem. Soc.* 1948: 1591–7.
Colin, H. & Gueguen, E. 1930. La constitution du principe sucre de *Rhodymenia palmata*. *C. R. Acad. Sci.* (Paris) 191: 163–4.
Cosgrove, D. 1986. Biophysical control of plant cell growth. *Annu. Rev. Plant Physiol.* 37: 377–405.
Craigie, J. S. 1974. Storage products. In *Algal Physiology and Biochemistry*, ed. W. D. P. Stewart, pp. 206–35. Oxford: Blackwell.
Craigie, J. S., McLachlan, J. & Tocher, R. D. 1968. Some neutral constituents of the Rhodophyceae with special reference to the occurrence of the floridosides. *Can. J. Bot.* 46: 605–11.
Cram, W. J. 1976. Negative feedback regulation of transport in cells. The maintenance of turgor, volume and nutrient supply. In *Encyclopedia of Plant Physiology. Transport in Plants.* IIA. *Cells*, eds. U. Lüttge & M. G. Pitman, pp. 284–316. Berlin: Springer.
Dainty, J. 1963. Water relations of plant cells. *Adv. Bot. Res.* 1: 279–326.
Dainty, J. 1976. Water relations of plant cells. In *Encyclopedia of Plant Physiology. Transport in Plants* IIA. *Cells*, eds. U. Lüttge & M. G. Pitman, pp. 12–35. Berlin: Springer.
Davison, I. R. & Reed, R. H. 1985. Osmotic adjustment in *Laminaria digitata* (Phaeophyta) with particular reference to seasonal changes in internal solute concentrations *J. Phycol.* 21: 41–50.
Dawes, C. J. & McIntosh, R. P. 1981. The effect of organic material and inorganic ions on the photosynthetic rate of the red alga *Bostrychia binderi* from a Florida estuary. *Mar. Biol.* 64: 213–18.
DeBoer, J. A. 1981. Nutrients. In *Biology of Seaweeds*, eds. C. S. Lobban & M. J. Wynne, pp. 356–92. Oxford: Blackwell.
DeBoer, J. A. & Ryther, J. H. 1977. Potential yields from a waste-recycling algal mariculture system. In *The Marine Plant Biomass of the Pacific Northwest Coast*, ed. R. Krauss, pp. 231–49. Corvallis: Oregon State University Press.
DeBoer, J. A. & Whoriskey, F. G. 1983. Production and role of hyaline hairs in *Cermium rubrum*. *Mar. Biol.* 77: 229–34.
D'Elia, C. F. & DeBoer, J. A. 1978. Nutritional studies of two red algae. II. Kinetics of ammonium and nitrate uptake. *J. Phycol.* 14: 266–72.
De Luca, P. & Morelli, A. 1983. Floridosides in *Cyanidium caldarium, Cyanidoschyzon merolae* and *Galdieria sulphuraria* (Rhodophyta, Cyanidiophyceae). *J. Phycol.* 19: 368–9.
Den Hartog, C. 1967. Brackish water as an environment for algae. *Blumea* 15: 31–43.
Den Hartog, C. 1971. The border environment between the sea and the fresh water with special reference to the estuary. *Vie et Milieu* (Suppl.) 22: 739–51.
Dickson, D. M. J. & Kirst, G. O. 1986. The role of β-dimethylsulphoniopropionate, glycine betaine and homarine in the osmoacclimation of *Platymonas subcordiformis*. *Planta* (Berl.) 167: 536–43.
Dickson, D. M. J., Wyn Jones, R. G. & Davenport, J. 1980. Steady state osmotic adaptation in *Ulva lactuca*. *Planta* (Berl.) 150: 158–65.
Dodd, W. A., Pitman, M. G. & West, K. R. 1966. Sodium and potassium transport in the marine alga *Chaetomorpha darwinii*. *Aust. J. Biol. Sci.* 19: 341–54.
Edwards, D. M., Reed, R. H., Chudek, J. A., Foster, R. & Stewart, W. D. P. 1987. Organic solute accumulation in osmotically-stressed *Enteromorpha intestinalis*. *Mar. Biol.* 95: 583–92.
Eppley, R. W., 1958a. Sodium exchange and potassium retention by the red marine algae *Porphyra perforata*. *J. Gen. Physiol.* 41: 901–11.
Eppley, R. W. 1958b. Potassium-dependent sodium extrusion by cells of *Porphyra perforata*, a red marine alga. *J. Gen. Physiol.* 42: 281–8.
Eppley, R. W. 1959. Potassium accumulation and sodium efflux by *Porphyra perforata* tissues in lithium and magnesium sea water. *J. Gen. Physiol.* 43: 29–38.
Eppley, R. W. & Blinks, L. R. 1957. Cell space and apparent free space in the red alga *Porphyra perforata*. *Plant*

Physiol. 32: 63–4.

Eppley, R. W. & Cyrus, C. C. 1960. Cation regulation and survival of the red alga *Porphyra perforata* in diluted and concentrated sea water. *Biol. Bull.* 118: 55–65.

Falkowski, P. G. 1983. Enzymology of nitrogen assimilation. In *Nitrogen in the Marine Environment*, eds. E. J. Carpenter & D. G. Capone, pp. 839–68. New York: Academic.

Findlay, G. P. & Hope, A. B. 1976. Electrical properties of plant cells: methods and findings. In *Encyclopedia of Plant Physiology. Transport in Plants.* IIA. Cells, eds. U. Lüttge & M. G. Pitman, pp. 53–92. Berlin: Springer.

Findlay, G. P., Hope, A. B. & Williams, E. J. 1969. Ionic relations of marine algae. I. *Griffithsia*: membrane electrical properties. *Aust. J. Biol. Sci.* 22: 1163–8.

Findlay, G. P., Hope, A. B. & Williams, E. J. 1970. Ionic relations of marine algae. II. *Griffithsia*: ionic fluxes. *Aust. J. Biol. Sci.* 23: 323–8.

Fletcher, C. R. 1980. The relationship between active transport and the exchange diffusion effect. *J. Theor. Biol.* 82: 643–61.

Friedlander, M. & Dawes, C. J. 1985. In situ uptake kinetics of ammonium and phosphate and chemical composition of the red seaweed *Gracilaria tikvahiae*. *J. Phycol.* 21: 448–53.

Fries, L. 1963. On the cultivation of axenic red algae. *Physiol. Plant.* 16: 695–708.

Fuggi, A. 1985. Mechanism of proton-linked nitrate uptake in *Cyanidium caldarium*, an acidophilic non-vacuolated alga. *Biochim. Biophys. Acta* 815: 392–8.

Fuggi, A., Vona, V., Rigano, V. D. M., Di Martino, C., Martello, A. & Rigano, C. 1984. Evidence for two transport systems for nitrate in the acidophilic thermophilic alga *Cyanidium caldarium*. *Arch. Microbiol.* 137: 281–5.

Geesink, R. 1973. Experimental investigations on marine and freshwater *Bangia* (Rhodophyta) from the Netherlands. *J. Exp. Mar. Biol. Ecol.* 11: 239–47.

Gellenbeck, K. W. & Chapman, D. J. 1983. Seaweed uses: the outlook for mariculture. *Endeavour* 7: 31–7.

Gessner, F. 1969. Photosynthesis and ion loss in the brown algae *Dictyopteris membranacea* and *Fucus virsoides*. *Mar. Biol.* 4: 349–51.

Gessner, F. 1971. Wasserpermeabilität und Photosynthese bei marinen Algen. *Bot. Mar.* 14: 1–3.

Gessner, F. & Schramm, W. 1971. Salinity: Plants. In *Marine Ecology*, I, ed. O. Kinne, pp. 705–820. London: Wiley.

Gilles, R. & Pequeux, A. 1977. Effect of salinity on the free amino aid pool of the red alga *Porphyridium purpureum*. *Comp. Biochem. Physiol. A. Comp. Physiol.* 57: 183–5.

Guarino, L. A. & Cohen, S. S. 1979. Uptake and accumulation of putrescine and its lethality in *Anacystis nidulans*. *Proc. Natl. Acad. Sci. USA* 76: 3184–8.

Guggino, S. & Gutknecht, J. 1982. Turgor regulation in *Valonia macrophysa* following acute osmotic shock. *J. Memb. Biol.* 67: 155–64.

Gutknecht, J. 1965. Ion distribution and transport in the red marine alga *Gracilaria foliifera*. *Biol. Bull.* 129: 495–510.

Gutknecht, J. 1968. Salt transport in *Valonia*: inhibition of potassium uptake by small hydrostatic pressures. *Science* 160: 68–70.

Gutknecht, J. & Bisson, M. A. 1977. Ion transport and osmotic regulation in giant algal cells. In *Water Relations in Membrane Transport in Plants & Animals*, eds. A. M. Jungreis, T. K. Hodges, A. Kleinzeller & S. G. Schulz, pp. 3–14. New York: Academic.

Gutknecht, J. & Dainty, J. 1968. Ionic relations of marine algae. *Ocean. Mar. Biol. Annu. Rev.* 6: 163–200.

Haines, K. C. & Wheeler, P. A. 1978. Ammonium and nitrate uptake by the marine macrophytes *Hypnea musciformis* (Rhodophyta) and *Macrocystis pyrifera* (Phaeophyta). *J. Phycol.* 14: 319–24.

Hammer, K. 1968. Salzgehalt und Photosynthese bei marinen Pflanzen. *Mar. Biol.* 1: 185–90.

Hanisak, M. D. 1983. The nitrogen relationships of marine macroalgae. In *Nitrogen in the Marine Environment*, eds. E. J. Carpenter & D. G. Capone, pp. 699–730. New York: Academic.

Harlin, M. M. & Craigie, J. S. 1978. Nitrate uptake by *Laminaria longicruris* (Phaeophyceae). *J. Phycol.* 14: 464–7.

Healey, F. P. 1973. Inorganic nutrient uptake and deficiency in algae. *Crit. Rev. Microbiol.* 3: 69–113.

Höfler, K. 1930. Das Plasmolyse-Verhalten der Rotalgen. *Z. Bot.* 23: 570–88.

Höfler, K. 1931. Hypotonietod und osmotische Resistenz einiger Rotalgen. *Osterr. Bot. Z.* 80: 51–71.

Hoffmann, C. 1959. Etudes écologiques et physiologiques de quelques algues de la mer Baltique. *Colloque sur l'Écologie des Algues Marines, Dinard*, 1957, 205–18.

Inoue, A., Asakawa, S., Saito, Y. & Yoshikawa, K. 1973. On the changes of extractive amino acid content in the infected purple laver (red alga *Porphyra*). *J. Fac. Fish. Anim. Husb. Hiroshima Univ.* 12: 61–72.

Iwasaki, H. & Matsudaira, C. 1954. Studies of cultural grounds of a laver *Porphyra tenera* Kjellman in Matsukawa-Ura Inlet. I. Environmental characteristics affecting upon nitrogen and phosphorus contents of laver. *Bull. Jap. Soc. Sci. Fish.* 20: 112–19.

Kauss, H. 1968. Galaktosylglyzeride und Osmoregulation in Rotalgen. *Z. Pflanzenphysiol.* 58: 428–33.

Kauss, H. 1969. Osmoregulation mit α-Galactosyglyzeriden bei *Ochromonas* und Rotalgen. *Ber. Deut. Botan. Gesell.* 82: 115–25.

Kauss, H. 1977. Biochemistry of osmotic regulation. In *International Review of Biochemistry: Plant Biochemistry* II, Vol. 13, ed. D. H. Northcote, pp. 119–40. Baltimore: University Park Press.

Kauss, H. 1979. Osmotic regulation in algae. *Prog. Phytochem.* 5: 1–27.

Kauss, H. 1982. Volume regulation: activation of a membrane-associated cryptic enzyme system by detergent-like action of phenothiazine drugs. *Plant Sci. Lett.* 26: 103–9.

Kauss, H. & Thomson, K-S. 1982. Biochemistry of volume control in *Poterioochromonas*. In *Plasmalemma and Tonoplast: their Functions in the Plant Cell*, eds. D. Marme, E. Marré & R. Hertel, pp. 255–62. Amsterdam: Elsevier.

Kesseler, H. 1964. Zellsaftgewinnung, AFS (apparent free space) und Vakuolenkonzentration der osmotisch wichtigsten mineralischen Bestandteile einiger Helgolander Meeresalgen. *Helgol. Meeresunters.* 11: 258–69.

Kirst, G. O. & Bisson, M. A. 1979. Regulation of turgor pressure in marine algae: ions and low molecular weight organic compounds. *Aust. J. Plant Physiol.* 6: 539–56.

Knoth, A. & Wiencke, C. 1984. Dynamic changes of protoplast volume and of fine structure during osmotic adaptation in the intertidal red alga *Porphyra umbilicalis*. *Plant Cell Envir.* 7: 113–19.

Kremer, B. P. 1976a. ^{14}C-assimilate patterns and kinetics of photosynthetic $^{14}CO_2$-assimilation of the marine red alga *Bostrychia scorpioides*. *Planta* (Berl.) 129: 63–7.

Kremer, B. P. 1976b. Mannitol in the Rhodophyceae – a reappraisal. *Phytochemistry* 15: 1135–8.

Kremer, B. P. 1978. Patterns of photoassimilatory products in Pacific Rhodophyceae. *Can. J. Bot.* 56: 1655–9.

Kremer, B. P. 1979. Photoassimilatory products and osmoregulation in marine Rhodophyceae. *Z. Pflanzenphysiol.* 93: 139–47.

Kremer, B. P. 1980. Taxonomic implications of algal photoassimilate patterns. *Br. Phycol. J.* 15: 399–409.

Kremer, B. P. 1981. Carbon metabolism. In *The Biology of Seaweeds*, eds. C. S. Lobban & M. J. Wynne, pp. 493–533. Oxford: Blackwell.

Kremer, B. P. 1983. *Cyanidium caldarium*: rhodophyte, cyanome or transitional species? In *Endocytobiology*. II, eds. H. E. A. Schenk & W. Schwemmler, pp. 963–70. Berlin: W. de Gruyter.

Kremer, B. P. & Kirst, G. O. 1981. Biosynthesis of 2-O-D-glycerol-α-D-galactopyranoside (floridoside) in marine Rhodophyceae. *Plant Sci. Lett.* 23: 349–57.

Kremer, B. P., Schmaljohann, R. & Rottger, R. 1980. Features and nutritional significance of photosynthates produced by unicellular algae symbiotic with larger Foraminifera. *Mar. Ecol. Prog. Ser.* 2: 225–8.

Kuhl, A. 1962. Inorganic phosphorus uptake and metabolism. In *Physiology and Biochemistry of Algae*, ed. R. A. Lewin, pp. 211–29. New York: Academic.

Lapointe, B. E. 1981. The effects of light and nitrogen on growth, pigment content, and biochemical composition of *Gracilaria foliifera* v. *angustissima* (Gigartinales, Rhodophyta). *J. Phycol.* 17: 90–5.

Lapointe, B. D. & Duke, C. S. 1984. Biochemical strategies for growth of *Gracilaria tikvahiae* (Rhodophyta) in relation to light intensity and nitrogen availability. *J. Phycol.* 20: 488–95.

Läuchli, A. & Epstein, E. 1970. Transport of potassium and rubidium in plant roots. *Plant Physiol.* 45: 639–41.

Laycock, M. V. & Craigie, J. S. 1977. The occurrence and seasonal variation of gigartinine and L-citrullinyl-L-arginine in *Chondrus crispus* Stackh. *Can. J. Biochem.* 55: 27–30.

Laycock, M. V., Morgan, K. C. & Craigie, J. S. 1981. Physiological factors affecting the accumulation of L-citrullinyl-L-arginine in *Chondrus crispus*. *Can. J. Bot.* 59: 522–7.

Lehnberg, W. 1978. Die Wirkung eines Licht-Temperatur-Salzgehalt Komplexes auf den Gaswechsel von *Delesseria sanguinea* (Rhodophyta) aus der westlichen Ostee. *Bot. Mar.* 21:485–98.

Leigh, R. A. & Wyn Jones, R. G. 1984. A hypothesis relating critical potassium concentrations for growth to the distribution and functions of this ion in the plant cell. *New Phytol.* 97: 1–13.

Lindberg, B. 1955. Low molecular weight carbohydrates in algae. XI. Investigation of *Porphyra umbilicalis*. *Acta Chem. Scand.* 9: 1097–9.

Lockhart, J. A. 1965. An analysis of irreversible plant cell elongation. *J. Theor. Biol.* 8: 264–75.

Low, P. S. 1985. Molecular basis of the biological compatibility of nature's osmolytes. In *Transport Processes, Iono- and Osmoregulation*, eds. R. Gilles & M. Gilles-Baillien, pp. 469–77. Berlin: Springer.

Macler, B. A. 1986. Regulation of carbon flow by nitrogen and light in the red alga *Gelidum coulteri*. *Plant Physiol.* 82: 136–41.

MacRobbie, E. A. C. 1970. Fluxes and compartmentation in plant cells. *Annu. Rev. Plant Physiol.* 22: 75–96.

MacRobbie, E. A. C. & Dainty, J. 1958. Sodium and potassium distribution in the seaweed *Rhodymenia palmata* (L.) Grev. *Physiol. Plant.* 11: 782–801.

Majak, W., Craigie, J. S. & McLachlan, J. 1966. Photosynthesis in algae. I. Accumulation products in the Rhodophyceae. *Can. J. Bot.* 44: 541–9.

Meng, J., Rosell,. K-G. & Srivastava, L. M. 1987. Chemical characterization of floridosides from *Porphyra perforata*. *Carbohydr. Res.* 61: 171–80.

Morgan, K. C. & Simpson, F. J. 1981. Cultivation of *Palmaria* (*Rhodymenia*) *palmata*: effect of high concentrations of nitrate, and ammonium on growth and nitrogen uptake. *Aquat. Bot.* 11: 167–71.

Munda, I. M. & Kremer, B. P. 1977. Chemical composition and physiological properties of fucoids under conditions of reduced salinity. *Mar. Biol.* 42: 9–15.

Nagashima, H. 1976. Distribution of low molecular weight carbohydrates in marine red algae. *Bull. Jap. Soc. Phycol.* 24: 103–10.

Nagashima, H. & Fukuda, I. 1981. Low molecular weight carbohydrates in *Cyanidium caldarium* and some related algae. *Phytochemistry* 20: 439–42.

Neish, A. C., Shacklock, P. F., Fox, C. H. & Simpson, F. J. 1977. The cultivation of *Chondrus crispus*. Factors affecting growth under greenhouse conditions. *Can. J. Bot.* 55: 2263–71.

Nellen, U. R. 1966. Über die einfluss des Salzgehaltes auf die photosynthetische Leistung verschiedener Standortformen von *Delesseria sanguinea* und *Fucus serratus*. *Helgol. Meeresunters.* 13: 288–313.

Nobel, P. S. 1983. *Biophysical Plant Physiology and Ecology*. San Francisco: Freeman.

Ogata, E. & Matsui, T. 1965. Photosynthesis in several marine plants of Japan as affected by salinity, drying and pH with attention to their growth habits. *Bot. Mar.* 8: 199–217.

Ogata, E. & Schramm, W. 1971. Some observations on the influence of salinity on growth and photosynthesis in *Porphyra umbilicalis*. *Mar. Biol.* 10: 70–6.

Ogata, E. & Takada, H. 1955. Elongation and shrinkage of *Porphyra tenera* and *Ulva pertusa* caused by osmotic changes. *J. Inst. Polytech. Osaka City Univ.* D. 6: 29–41.

Oohusa, T., Araki, S., Sakurai, T. & Saitoh, M. 1977. Physiological studies on diurnal biological rhythms of *Porphyra*. II. The growth and the contents of free and total nitrogen and carbohydrate in the thallus cultured in the laboratory. *Bull. Jap. Soc. Sci. Fish.* 43: 250–4.

Parsons, T. R. & Harrison, P. J. 1983. Nutrient cycling in marine ecosystems. In *Encyclopedia of Plant Physiology Physiological* XIID *Plant Ecology* IV, eds. A. Pirson & M. H. Zimmermann, pp. 77–105. Berlin, Springer.

Passioura, J. B. 1980. The meaning of matric potential. *J. Exp. Bot.* 31: 1161–9.

Pollard, A. & Wyn Jones, R. G. 1979. Enzyme activities in concentrated solutions of glycine betaine and other solutes. *Planta* (Berl.) 144: 291–8.

Putman, E. W. & Hassid, W. Z. 1954. Structure of galactosylglycerol from *Iridaea laminarioides*. *J. Am. Chem. Soc.* 76: 2221–3.

Raven, J. A. 1976. Transport in algal cells. In *Encyclopedia of Plant Physiology. Transport in Plants*. IIA. *Cells*, eds. U. Lüttge & M. G. Pitman, pp. 129–88. Berlin: Springer.

Raven, J. A. 1980. Nutrient transport in microalgae. *Adv. Microb. Physiol.* 21: 47–226.

Raven, J. A. 1984. *Energetics and Transport in Aquatic Plants*. New York: A. R. Liss.

Reed, R. H. 1979. The Osmotic Responses of *Porphyra purpurea* (Roth) C. Ag. Ph.D. thesis, University of Liverpool.

Reed, R. H. 1980a. Physiological responses of *Porphyra* under steady state and fluctuating salinity regimes. *Proc. Int. Seaweed Symp.* 10: 509–13.

Reed, R. H. 1980b. On the conspecificity of marine and freshwater *Bangia* in Britain. *Br. Phycol. J.* 15: 411–16.

Reed, R. H. 1981. Osmotic adaptation in *Porphyra purpurea*: use of $^{86}Rb^+$ and $^{42}K^+$ exchange kinetics to investigate intracellular solute compartmentation. *Plant Sci. Lett.* 23: 215–21.

Reed, R. H. 1983a. The osmotic responses of *Polysiphonia lanosa* (L.) Tandy from marine and estuarine sites: evidence for incomplete recovery of turgor. *J. Exp. Mar. Biol. Ecol.* 68: 169–93.

Reed, R. H. 1983b. Taxonomic implications of osmoacclimation in *Cyanidium caldarium* (Tilden) Geitler. *Phycologia* 22: 351–4.

Reed, R. H. 1983c. Measurement and osmotic significance of β-dimethylsulphoniopropionate in marine macroalgae. *Mar. Biol. Lett.* 4: 173–81.

Reed, R. H. 1984a. Use and abuse of osmo-terminology. *Plant Cell Envir.* 7: 165–70.

Reed, R. H. 1984b. The effects of extreme hyposaline stress upon *Polysiphonia lanosa* (L.) Tandy from marine and estuarine sites. *J. Exp. Mar. Biol. Ecol.* 76: 131–44.

Reed, R. H. 1985. Osmoacclimation in *Bangia atropurpurea* (Rhodophyta, Bangiales): the osmotic role of floridoside. *Br. Phycol. J.* 20: 211–18.

Reed, R. H. & Collins, J. C. 1980a. Membrane potential measurements of marine macroalgae: *Porphyra purpurea* and *Ulva lactuca*. *Plant Cell Envir.* 4: 257–60.

Reed, R. H. & Collins, J. C. 1980b. The ionic relations of *Porphyra purpurea* (Roth) C. Ag. (Rhodophyta, Bangiales). *Plant Cell Envir.* 3: 399–407.

Reed, R. H. & Collins, J. C. 1981. The kinetics of Rb^+ and K^+ exchange in *Porphyra purpurea*. *Plant Sci. Lett.* 20: 281–9.

Reed, R. H., Collins, J. C. & Russell, G. 1979. The influence of variations in salinity upon photosynthesis in the marine alga *Porphyra purpurea* (Roth) C. Ag. (Rhodophyta, Bangiales). *Z. Pflanzenphysiol.* 98: 183–7.

Reed, R. H., Collins, J. C. & Russell, G. 1980a. The effects of salinity upon cellular volume of the marine red alga *Porphyra purpurea* (Roth) C. Ag. *J. Exp. Bot.* 31: 1521–37.

Reed, R. H., Collins, J. C. & Russell, G. 1980b. The effects of salinity upon galactosyl-glycerol content and concentration of the marine red alga *Porphyra purpurea* (Roth) C. Ag. *J. Exp. Bot.* 31: 1539–54.

Reed, R. H., Collins, J. C. & Russell, G. 1981. The effects of salinity upon ion content and ion transport of the marine red alga *Porphyra purpurea* (Roth) C. Ag. *J. Exp. Bot.* 32: 347–67.

Reed, R. H., Davison, I. R., Chudek, J. A. & Foster, R. 1985. The osmotic role of mannitol in the Phaeophyta: an appraisal. *Phycologia* 24: 35–47.

Rees, T. A. V. 1984. Sodium dependent photosynthetic oxygen evolution in a marine diatom. *J. Exp. Bot.* 35: 332–7.

Rigano, C., Di Martin Rigano, V., Vona, V. & Fuggi, A. 1979. Glutamine synthetase activity, ammonia assimilation and control of nitrate reduction the unicellular red alga *Cyanidium caldarium*. *Arch. Microbiol.* 121: 117–20.

Rigby, C. H., Craig, S. R. & Budd, K. 1980. Phosphate uptake by *Synechococcus leopoliensis* (Cyanophyceae): enhancement by calcium ions. *J. Phycol.* 16: 389–94.

Ritchie, R. J. 1984. Permeability of lipophilic cations. *J. Exp. Bot.* 35: 699–706.

Ritchie, R. J. & Larkum, A. W. D. 1982. Cation exchange properties of the cell walls of *Enteromorpha intestinalis* (L.) Link (Ulvales, Chlorophyta). *J. Exp. Bot.* 33: 125–39.

Rosenberg, G. & Ramus, J. 1982. Ecological growth strategies in the seaweeds *Gracilaria foliifera* (Rhodophyceae) and *Ulva* sp. (Chlorophyceae): soluble nitrogen and reserve carbohydrates. *Mar. Biol.* 66: 251–59.

Russell, G. 1985. Recent evolutionary changes in the algae of the Baltic Sea. *Br. Phycol. J.* 20: 87–104.

Ryther, J. H., Corwin, N., DeBusk, T. A. & Williams, L. D. 1981. Nitrogen uptake and storage by the red alga *Gracilaria tikvahiae* (McLachlan, 1979). *Aquaculture* 26: 107–15.

Schobert, B. 1977. Is there an osmotic regulatory mechanism in algae and higher plants? *J. Theor. Biol.* 68: 17–26.

Sheath, R. G. 1984. The biology of freshwater red algae. In *Progress in Phycological Research*, Vol. 3, eds. F. E. Round & D. J. Chapman, pp. 89–157. Bristol: Biopress.

Sheath, R. G. & Cole, K. M. 1980. Distribution and salinity adaptation of *Bangia atropurpurea* (Rhodophyta), a putative migrant into the Laurentian Great Lakes. *J. Phycol.* 16: 412–20.

Sicko-Goad, L. M., Crang, R. E. & Jensen, T. E. 1975. Phosphate metabolism in blue-green algae. IV. In situ analysis of polyphosphate bodies by x-ray energy dispersive analysis. *Cytobiologie* 11: 430–7.

Smith, F. A. & Walker, N. A. 1978. Entry of methylammonium and ammonium ions into *Chara* internodal cells. *J. Exp. Bot.* 29: 107–12.

Stewart, W. D. P. 1980. Transport and utilization of nitrogen sources by algae. In *Microorganisms and Nitrogen Sources*, ed. J. W. Payne, pp. 577–607. Chichester: Wiley.

Syrett, P. J. 1981. Nitrogen metabolism of microalgae. *Can. Bull. Fish. Aquat. Sci.* 210: 182–210.

Thomas, T. E. & Harrison, P. J. 1987. Rapid ammonium uptake and nitrogen interactions in five intertidal seaweeds grown under field conditions. *J. Exp. Mar. Biol. Ecol.* 107: 1–8.

Thomas, T. E., Turpin, D. H. & Harrison, P. J. 1987. Desiccation enhanced nitrogen uptake rates in intertidal seaweeds. *Mar. Biol.* 94: 293–18.

Ullrich-Eberius, C. I. & Yingchol, Y. 1974. Phosphate uptake and its pH-dependence in halophytic and glycophytic algae and higher plants. *Oecologia* (Berl.) 17: 17–26.

Vairavamurthy, A., Andreae, M. O. & Iverson, R. L. 1985. Biosynthesis of dimethlsulphide and dimethylpropiothetin by *Hymenomonas carterae* in relation to sulphur source and salinity variation. *J. Oceanogr.* 30: 59–75.

Vorobiev, L. N. 1967. Potassium ion activity in the cytoplasm and the vacuole of cells of *Chara australis*. *Nature* (Lond.) 216: 1325–7.

Walker, D. A. & Sivak, M. N. 1986. Photosynthesis and phosphate: a cellular affair? *Trends Biochem. Sci.* 11: 176–9.

Walker, N. A. & Pitman, M. G. 1976. Measurement of fluxes across membranes. In *Encyclopedia of Plant Physiology. Transport in Plants* IIA. *Cells*, eds. U. Lüttge & M. G. Pitman, pp. 93–126. Berlin: Springer.

West, K. R. & Pitman, M. G. 1967a. Ionic relations and ultrastructure in *Ulva lactuca*. *Aust. J. Biol. Sci.* 20: 901–14.

West, K. R. & Pitman, M. G. 1967b. Rubidium as a tracer for potassium in the marine algae *Ulva lactuca* L. and *Chaetomorpha darwinii* (Hooker) Kuetzing. *Nature* (Lond.) 214: 1262–3.

Wheeler, P. A. 1979. Uptake of methylamine (an ammonium analogue) by *Macrocystis pyrifera* (Phaeophyta). *J. Phycol.* 15: 12–7.

White, R. H. 1982. Analysis of dimethyl sulphonium compounds in marine algae. *J. Mar. Res.* 40: 529–35.

Wiencke, C. 1984. The response of pyruvate kinase from the intertidal red alga *Porphyra umbilicalis* to sodium and potassium ions. *Z. Pflanzenphysiol.* 116: 447–53.

Wiencke, C. & Läuchli, A. 1981. Inorganic ions and floridoside as osmotic solutes in *Porphyra umbilicalis*. *Z. Pflanzenphysiol.* 103: 247–58.

Wiencke, C., Stelzer, R. & Läuchli, A. 1983. Ion compartmentation in *Porphyra umbilicalis* determinated by electron probe x-ray micro-analysis. *Planta* (Berl.) 159: 336–41.

Wyn Jones, R. G., Storey, R., Leigh, R. A., Ahmad, N. & Pollard, A. 1977. A hypothesis on cytoplasmic osmoregulation. In *Regulation of Cell Membrane Activity in Plants*, eds. E. Marré & O. Ciferri, pp. 121–36. Amsterdam: Elsevier.

Yarish, C. & Edwards, P. 1982. A field and cultural investigation of the horizontal and seasonal distribution of estuarine red algae of New Jersey. *Phycologia* 21: 112–24.

Yarish, C., Edwards, P. & Casey, S. 1979. A culture study of salinity responses in ecotypes of two estuarine red algae. *J. Phycol.* 15: 341–6.

Yarish, C., Edwards, P. & Casey, S. 1980. The effects of salinity, and calcium and potassium variations on the growth of two estuarine red algae. *J. Exp. Mar. Biol. Ecol.* 47: 235–49.

Yeo, A. R. 1983. Salinity resistance: physiologies and prices. *Physiol. Plant.* 58: 213–22.

Zavodnik, N. 1975. Effects of temperature and salinity variations on photosynthesis of some littoral seaweeds of the North Adriatic Sea. *Bot. Mar.* 18: 245–50.

Zimmermann, U. 1978. Physics of turgor- and osmoregulation. *Annu. Rev. Plant Physiol.* 29: 121–48.

Zimmermann, U. & Steudle, E. 1971. Effects of potassium concentration and osmotic pressure of sea water on the cell-turgor pressure of *Chaetomorpha linum*. *Mar. Biol.* 11: 132–7.

Chapter 8

Carbon metabolism

JOHN A. RAVEN
ANDREW M. JOHNSTON
JEFFREY J. MACFARLANE

CONTENTS

I. Introduction / 172
II. Phototrophy and organotrophy: carbon and energy sources / 172
    A. Introduction: some definitions / 172
    B. Economics of rhodophyte phototrophy: the thylakoid reactions / 173
III. Mechanisms of inorganic carbon uptake / 175
    A. Inorganic C availability / 175
    B. Inorganic C interconversions at the plant surface and transport to the plastid stroma / 176
    C. The kinetics of inorganic C and O_2 exchange: C_3-like or C_4-like physiology? / 177
    D. Natural abundance of carbon isotopes and its implications for C transport / 180
    E. Submersed and emersed photosynthesis in benthic Rhodophyta subjected to periodic emersion / 180
    F. Does inorganic C supply limit the growth rate of red algae? / 181
IV. CO_2 fixation and photorespiration / 181
    A. Biochemistry of photosynthetic CO_2 fixation / 181
    B. Biochemistry of anaplerotic CO_2 fixation / 184
    C. Biochemistry of photorespiration / 185
V. Organic C storage and transport / 187
    A. Nature of storage / 187
    B. Type of transport / 187
    C. The forms of stored organic C / 187
    D. Intracellular transport of organic C / 189
    E. Intercellular transport of organic C / 189
    F. Organic C exchange with the bathing medium / 189
VI. Classical respiration / 189
    A. Classical respiration and its role in photolithotrophs / 189
    B. Occurrence, location, and capacity of the pathways of classical respiration / 190
    C. Potential and achieved respiratory rates: the role of respiration in photolithotrophic growth / 191

D. Anaerobic metabolism / 194
E. Controls of the rate of classical respiration / 194
VII. Summary / 196
VIII. Acknowledgments / 196
IX. Endnotes / 196
X. References / 196

I. INTRODUCTION

The aim of this chapter is to outline the way in which the red algae convert inorganic carbon into the substrates for the synthesis of new plant material. The limits to such a discussion are arbitrary. The close quantitative connection between carbon metabolism and energy transformation means that ATP and reductant metabolism are conveniently considered together with carbon metabolism.

Carbon and energy sources for the growth of rhodophytes are discussed with some aspects of the involvement of thylakoid reactions in inorganic carbon assimilation. Inorganic carbon uptake is then considered, followed by an analysis of the mechanism of CO_2 fixation in light and darkness and the extent of photorespiration. The storage and transport of the reduced organic C produced in photosynthesis is then considered. Finally, consumption of the stored or transported photosynthate in respiratory reactions, providing C skeletons, ATP, and reductant for growth and synthesis, are presented.

II. PHOTOTROPHY AND ORGANOTROPHY: CARBON AND ENERGY SOURCES

A. Introduction: some definitions

The Rhodophyta typically grow as photolithotrophs, using light as their energy source and inorganic compounds as their sources of chemical elements (see Table 8-1 for definitions). Many are obligate photolithotrophs; they cannot grow with organic compounds as their sole C and/or energy source (Table 8-1). However, some red algae are able to grow chemo-organotrophically, either facultatively (e.g., *Galderia sulphuraria*) or, apparently, obligately (e.g., many red algae parasitic on other red algae) (Table 8-1). Two points need to be made. One is

Table 8-1. Nutritional types in the Rhodophyta

Term	Definition	Red algal examples
Phototroph	Able to grow with light as energy source	Most
Photolithotroph	Able to grow with light as energy source and with inorganic sources of elements	Probably very common, but frequent auxotrophy (e.g., for vitamin B_{12}), e.g., all Cyanidiophyceae
Photo-organotroph	Able to grow with light as energy source, organic C source	No well authenticated examples from red algae
Obligate photolithotroph	Unable to grow except by photolithotrophy (± vitamin auxotrophy)	Probably very common, e.g., *Cyanidium caldarium*, *Cyanidioschyzon merolae*
Facultative photolithotroph	Able to grow by photolithotrophy or by chemo- or photo-organotrophy	Some, e.g., *Galderia sulphuraria*
Chemo-organotroph	Able to grow with organic energy and carbon sources	Some rhodophytes parasitic on rhodophytes; *Galderia sulphuraria*
Obligate chemo-organotroph	Unable to grow except by chemo-organotrophy	Some parasitic rhodophytes that may well also be obligate biotrophs

For further details see Droop (1974), Merola et al. (1981), and Raven (1984a).

Carbon metabolism

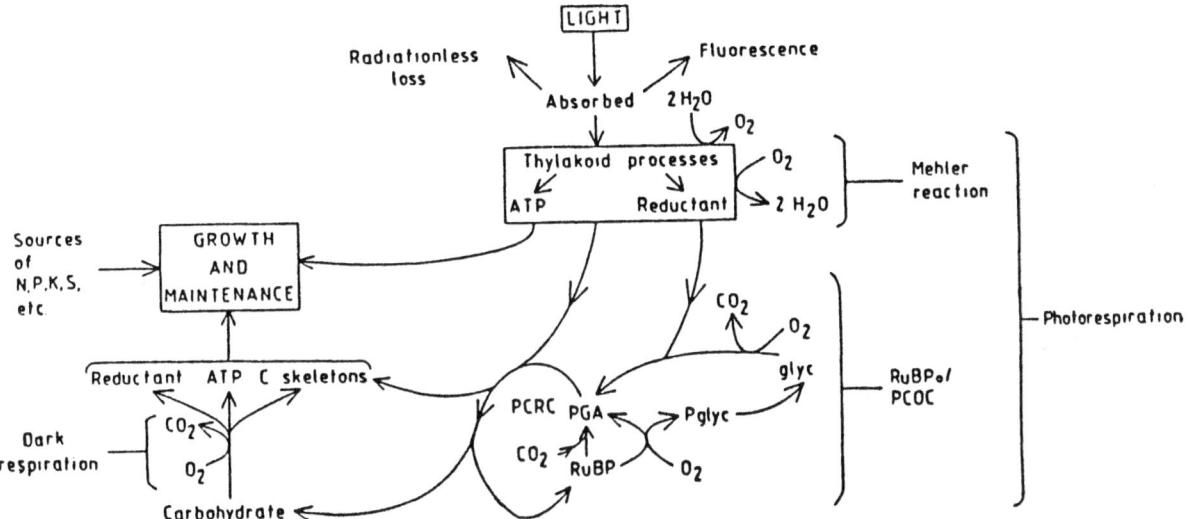

Fig. 8-1. Outline of the pathways of energy, carbon, and oxygen in photosynthesis, photorespiration, dark respiration, and growth of a rhodophyte. No attempt is made to represent stoichiometries. *Glyc*-glycolate, *PCOC*-photorespiratory carbon oxidation cycle, *PCRC*-photosynthetic carbon reduction cycle, *PGA*-3-phosphoglycerate, *Pglyc*-phosphoglycolate, *RuBP*-ribulose bisphosphate, *RuBPo*-ribulose bisphosphate oxygenase. (Modified from Figure 1.1 of Raven 1984a)

that there are nonphotosynthetic (i.e., chemo-organotrophic) cells in many obligately photolithotrophic multicellular red algae. It is possible that the nonphotosynthetic cells receive organic C via pit connections from photosynthetic cells (Chapter 9 of Raven 1984a). This may also occur in the case of nonphotosynthetic rhodophyte parasites, which are living on other, often closely related, photosynthetic red algae (but not, apparently, algae from other divisions: see Kremer 1983) or in carposporophytes, which live physically and largely nutritionally on their female gametophyte (cf. Raven 1984a). In addition, otherwise photolithotrophic red algae are auxotrophic for certain vitamins. The biochemical basis of obligate photolithotrophy is still unclear, as are its benefits in terms of natural selection; this seems to be a case of "generalists" versus "specialists" (Whittenbury & Kelly 1977).

An outline of the reactions involved in photolithotrophy and in chemo-organotrophy is given in Fig. 8-1.

B. Economics of rhodophyte phototrophy: the thylakoid reactions

The thylakoid reactions of red algae are dealt with by Gantt in Chapter 9. Although most of the catalytic proteins involved in these reactions are homologous with those in all other O_2-evolving phototrophs, some red algal catalysts are analagous with those of other organisms (see Raven 1984a). This section analyzes some of the resource use consequences of the presence of a particular catalyst in rhodophytes rather than the occurrence of an analogous catalyst found elsewhere. The three examples are the catalysts of light-harvesting pigment-protein complexes and of redox reactions between the cytochrome b_6–f complex and the oxidizing end of photoreaction one and the reducing end of photoreaction one and $NADP^+$ reductase.

Red algae have phycobilins as their light-harvesting pigment-proteins (Chapter 9; Larkum & Barrett 1983). Raven (1984b) points out that the energy, C, and N input for the synthesis of unit light-absorbing capacity (the ability to absorb photons of a certain wavelength at a given rate from a given light field) is higher than for many of the chlorophyll a + b protein or chlorophyll a + c + carotenoids protein complexes of "chlorophytes" and "chromophytes." This, in turn, may have implications for the N use efficiency and photon use efficiency of overall plant growth in that it may, for example, cost more N and photons to make a given amount of viable spores in a given time. Red algae always have a cytochrome c rather than the analogous plastocyanin as the catalyst of electron transfer from the cytochrome

b_6–f complex to photoreaction one, a trait that they share with chromophyte algae (Raven 1984a, 1987a). The two catalysts have closely similar specific reaction rates, and similar molecular masses; they thus have similar energy, carbon, and nitrogen costs of synthesis. However, cytochrome c uses Fe, whereas plastocyanin uses Cu, with implications for the Cu and Fe use efficiencies of growth (see Raven 1984a).

Ferredoxin is the most common redox agent linking photoreaction one and $NADP^+$ reductase in O_2 evolvers. Some rhodophytes (like some chlorophytes and cyanobacteria) can replace ferredoxin with flavodoxin; in *Chondrus crispus* flavodoxin is constitutive and ferredoxin is not apparently found (Raven 1984a). The catalytically less active flavodoxin lacks Fe, which is essential for ferredoxin. The use of flavodoxin may thus improve Fe use efficiency of growth but at the expense of a decreased nitrogen, carbon, and photon use efficiency due to a requirement for more of the more slowly reacting catalyst (cf. Raven 1984a).

The ecological implications of the above patterns of occurrence are not clear. Whereas the use of phycobilins might be expected to restrict growth at very low photon flux densities, in that the synthesis of a given increment of photon absorption capacity "costs" more absorbed photons in the case of phycobilins than that of many other chromophore–protein complexes (Raven 1984b), the rhodophytes are very adept at growing at very low photon flux densities (e.g., Leclerc 1985). Indeed, a coralline red alga can grow, probably by photolithotrophy, at a depth of 268 m off the Bahamas. In this habitat the incident photon flux density of ~10 nmol photon m^{-2} s^{-1} (400–700 nm) at midday (Littler et al. 1985) scarcely exceeds that at the sea surface on a night with a full moon.

Amplifying somewhat on the cost of producing unit photon absorption capacity in the Rhodophyta relative to other algae, there is no indication that the photon (quantum) efficiency of photosynthetic O_2 evolution (mol O_2 evolved per mol photon absorbed) is different in the red algae from that of other algal divisions, provided the occurrence of the Emerson enhancement effect in natural light fields is taken into account (Brody & Emerson 1959; Dring & Lüning 1985; Lüning & Dring, 1985; Yocum & Blinks 1954). What evidence is available is consistent with the occurrence of similar mechanisms and stoichiometries of processes generating ATP and reductant to photons absorbed and of processes consuming ATP and reductant to growth and maintenance processes in red algae as in other algal divisions (see Raven 1974a, 1976a, 1984a). This means that the assimilation of 1 mol of carbon dioxide or of nitrate requires the absorption of the same number of mol photons by the plant in the Rhodophyta as in other algae. Together with the requirement for more carbon and nitrogen to produce a given increment of photon absorption capacity in phycobiliphytes relative to chromophytes or chlorophytes, we see that the argument that there are higher photon costs of generating photon-harvesting capacity in Rhodophyta than in many other algae is upheld.

In quantitative terms (see Raven 1984a) we can ask how the specific growth rate is altered by the catalyst distribution under conditions of restricted supply of the resources that are involved differentially in the synthesis of the alternative, analogous proteins. For phycobilins (relative to integral light-harvesting complexes of chlorophytes or chromophytes), the diversion of energy into the synthesis of phycobilins rather than "cheaper" pigment complexes could, in limiting oceanic light of 400–600 nm, cause a greater than two-fold difference in specific growth rate between a (faster-growing) chlorophyte or chromophyte and a phycocyanin-dominated phycobiliphyte (Raven 1984b, p. 616). The assumptions made in arriving at this conclusion were those that would maximize the difference. They do not appear to be too extreme to be in accord with some natural situations (small organisms with minimal self-shading, half of their dry weight in light-absorbing apparatus, highest reported protein: chromophore for phycobilins, lowest for chlorophyll complexes). This is potentially a severe restriction on growth of rhodophytes at low light; however, there is a coralline red alga occurring at 268 m (Littler et al. 1985). A similar, three-fold decrement in nitrogen productivity [mol C fixed (mol cell N)$^{-1}$ s^{-1}] can be computed for the phycobiliphyte, making assumptions similar to those used for energy costs (see Raven 1984b).

By comparison, the changes in Fe productivity resulting from the "choice" of redox catalysts is much smaller. The data of Hewitt (1983) on stoichiometries of Fe-containing catalysts of photosynthesis and respiration were used by Raven (1988; see Raven 1987b) to compute a potential Fe productivity [(mol C fixed (mol cell Fe)$^{-1}$ s^{-1}] of 1.29 for an alga with cytochrome c (Fe) rather than plastocyanin (Cu), and ferredoxin (Fe) rather than flavodoxin (no metal). This highest Fe requirement case could refer to a typical rhodophyte; for *Chondrus crispus* with flavodoxin rather than ferredoxin, the potential Fe productivity is only increased to 1.36 mol C fixed (mol cell Fe)$^{-1}$ s^{-1}. For comparison, a chlorophyte with plastocyanin and ferredoxin (Raven 1984a, 1987a) would have a potential Fe productivity of 1.32 mol C fixed (mol cell Fe)$^{-1}$ s^{-1}, whereas a chlorophyte with plastocyanin and flavodoxin (the lowest Fe requirement case; Raven 1984a, 1987a) would potentially fix 1.4 mol C (mol cell Fe)$^{-1}$ s^{-1}. These values refer to NH_4^+-grown organisms at 20°C; lower values, and smaller proportional differences, would occur if Fe in other cell components (e.g., NO_3^- assimilation enzymes, catalase, phytoferritin) were taken into account. For Cu, rather larger differences are predicted. Even if half of the Cu accounted for in known components is in catalysts other than plastocyanin (cytochrome oxidase, CuZn superoxide dismutase in some "higher" photolithotrophs) in organisms with plastocyanin, the replacement of plastocyanin by soluble cytochrome c would double the potential Cu productivity, from 20.5 to 41.0 mol C fixed (mol cell Cu)$^{-1}$ s^{-1} at 20°C (see Hewitt 1983; Raven 1988; Sandmann 1985). This would give a potential advantage to plastocyanin-less phototrophs (Rhodophyta; chromophytes; facultatively cytochrome c-producing Chlorophyta) over obligately plastocyanin-producing Chlorophyta in a Cu-limiting habitat, if such exist. However, Cu toxicity seems more likely than Cu limitation in oxygenated waters (see Morel & Morel-Laurens 1981).

III. MECHANISMS OF INORGANIC CARBON UPTAKE

A. Inorganic C availability

The Rhodophyta cover many of the extremes of carbon availability to the surface of photolithotrophic plants. The flux (mol m^{-2} s^{-1}) of a solute to the surface of an organism that is removing the solute from the medium depends on four factors. These are the bulk phase concentration of the solute, the boundary layer thickness, the diffusion coefficient of the solute in this boundary layer in which the solute transport can occur only by diffusion, and the capacity for solute uptake at the plant surface. In the limiting case the "potential flux" (Raven 1984a, Chapter 1) is a function of bulk phase concentration and boundary layer thickness for a given solute with its (temperature-dependent) diffusion coefficient.

The situation is further complicated for photosynthesis by the occurrence, in a pH-, temperature-, and salinity-dependent manner, of speciation of dissolved inorganic C into CO_2 and HCO_3^- (Raven 1984a, Chapter 5). Dealing, first, with the bulk phase concentration of inorganic C and its speciation, the Rhodophyta include the Cyanidiophyceae (Merola et al. 1981), which can grow at lower pH values (1.0) than other eukaryotic photolithotrophs, as well as at higher CO_2 concentrations (up to at least 0.1 MPa CO_2 in the gas phase: Seckbach et al. 1971). Natural habitats of the Cyanidiophyceae do not, of course, have these very high CO_2 levels, but they are invariably so acidic that all of the inorganic C is CO_2. Habitats of higher pH, including the sea, have significant and frequently dominant HCO_3^- concentrations. Freshwater habitats often have higher CO_2, and hence total inorganic C, concentrations than pertain to air equilibrium at the measured pH. This is due to input of CO_2 from chemo-organotrophic activity, which relies ultimately on terrestrial photosynthesis. Such a CO_2 enrichment is not generally true of marine habitats.

The other great variable is the thickness of the boundary layer. This is a function of organism size and shape and of the velocity of water flow over the organism, smaller characteristic size and more rapid water movement both contributing to a smaller boundary layer thickness. Some benthic red macroalgae, e.g., the freshwater *Lemanea*, live in very rapidly flowing water (1 m s^{-1}) and also generate turbulent flow (MacFarlane & Raven 1985). The flow regimes to which marine bethic red algae are subjected are more spatially variable than those around *Lemanea*; however, Anderson and Charters (1982) have shown that the marine *Gelidium* "manipulates" water flow and generates turbulence.

Red algal habitats are such that the potential flux of inorganic C to the plant surface exceeds that expected for a plant in stagnant water containing no HCO_3^- and air-equilibrated CO_2 levels. This offsets in part the low diffusion coefficient of CO_2 in water ($\sim 10^{-9}$ m^2 s^{-1}) relative to air ($\sim 10^{-5}$ m^2 s^{-1}). For *Lemanea*, with rapid water flow limiting unstirred layer thicknesses to ~ 10–20 μm and a free CO_2 concentration of up to 200 mmol m^{-3} (more than 10 times air equilibrium) the potential flux is ~ 10 μmol CO_2 m^{-2} s^{-1}. This value is five times the maximum rate of C fixation observed at C and light saturation in the laboratory, or from field growth, at 10°C (MacFarlane & Raven 1985, unpubl.). While such an over-supply of free CO_2 may be at the high end of the range found for red algae, these data show that habitat choice, combined with an appropriate plant morphology, can substantially overcome the constraints on photosynthesis imposed by low diffusion coefficients for C species in water (Raven et al. 1987a,b). It must be emphasised that C supply to the plant surface may not be the most limiting process for growth; N or P supply may play this role, so that the morphological "adaptations" that aid C supply may be even more crucial for N and P supply (Raven 1981, 1984a). Indeed, some morphological adaptations, such as colorless hairs, seem to be specifically related to the supply of N, P, or Fe, rather than of C (Raven 1981, 1984a).

B. Inorganic C interconversions at the plant surface and transport to the plastid stroma

The catalysis of HCO_3^- conversion to CO_2 at the plant surface by extracellular carbonic anhydrase (e.g., Aizawa & Miyachi 1986; Price et al. 1985) or by reduction in pH over part of the surface of the plant (e.g., Price & Badger 1985; Walker et al. 1980) is an important part of inorganic C acquisition by many chlorophyll b–containing aquatic plants. However, in the red algae that have been tested – *Palmaria* and *Rhodogramma* (Colman & Cook 1985; Cook et al. 1986), *Porphyridium* (Aizawa & Miyachi 1986; Colman & Gehl 1983; Yagawa et al. 1987) – there seems to be no extracellular carbonic anhydrase (but see Burns & Beardall 1987). The occurrence of acid zones on the surface of macroscopic red algae has been little investigated; no such zones have thus far been detected (Raven & Beardall 1981a; Raven unpubl.). Available evidence suggests that the $HCO_3^- - CO_2$ interconversion is not catalyzed at the surface of red algae. This simplifies the analysis of the C species that is entering the cell; HCO_3^- entry can be demonstrated if the rate of assimilation of exogenous inorganic C exceeds the uncatalyzed rate at which CO_2 can be generated from HCO_3^- in the medium. If the assimilation rate is less than the uncatalyzed conversion rate of HCO_3^- to CO_2, no conclusions can be drawn as to whether HCO_3^- is entering. Based on this rationale, it may be concluded that HCO_3^- entry is necessary to explain the rate of photosynthesis by *Palmaria* (as *Rhodymenia*) *palmata* in seawater (Colman & Cook 1985; Cook & Colman 1987; Cook et al. 1986) and by *Porphyridium purpureum* (as *P. cruentum*) in media at pH 7.9–9.0 (Colman & Gehl, 1983), but not that of *Lemanea mamillosa* at pH 7.0–8.0 (Raven et al. 1987a,b, 1990).

Another method of investigating the inorganic C species entering the cells also involves the relatively slow uncatalyzed equilibration of CO_2 and HCO_3^- outside the plant. Bréchignac and Andre (1985b) and Bréchignac et al. (1986) continously monitored the dissolved free CO_2 in seawater surrounding illuminated *Chondrus crispus*, and found that free CO_2 during steady state photosynthesis was higher than the concentration found when CO_2 and HCO_3^- were equilibrated by addition of carbonic anhydrase to the medium. This is consistent with HCO_3^- entering the cells faster (relative to its concentration in seawater) than does CO_2. It also agrees with the interpretation by Bidwell and McLachlan (1985) of their data on photosynthesis by emersed (air-exposed) *Chondrus crispus* in the presence and absence of added carbonic anhydrase in the surface film and of a carbonic anhydrase inhibitor (see Johnston & Raven 1986; Kerby & Raven 1985). The data of Bidwell and McLachlan (1985) would, in any case, beg the question of which membrane was involved, plasmalemma or plastid inner envelope. Indirect evidence for active influx of CO_2 and/or HCO_3^- and of a steady-state CO_2 concentration available to ribulose bisphosphate carboxylase-oxygenase (RUBISCO) that exceeds that possible by nonconcentrative entry of CO_2 or HCO_3^- comes from gas exchange and ^{14}C labeling studies, which suggest a higher $[CO_2]$, and $[CO_2]/[O_2]$, at the site of RUBISCO action than would be possible for nonconcentrative inorganic C entry (see Sections IV and V). Measurements of this type, which are consistent with operation of an inorganic carbon-concentrating mechanism, are available for a number of red algae (Table 8-2); however, not all red algae show these signs of oxygenase suppression in vivo (Table 8-2). The CO_2 level used for growth (air equilibrium versus higher concentrations) can influence the results on both HCO_3^- use and suppression of RUBISCO oxygenase activity in many microalgae and cyanobacteria. In the case of the red algae, apparently, this has only been tried with *Porphyridium purpureum* (as *P. cruentum* strain R1), where a relatively small effect on the apparent K_m for dissolved inorganic carbon (DIC) was found (Aizawa & Miyachi 1986; Aizawa et al. 1985). In the red algae, apart from *Cyanidioschyzon merolae*, there does seem to be a correlation between the capacity to "use" HCO_3^- and in vivo suppression of oxygenase activity. However, physiological and biochemical evidence for suppression of RuBPo activity also is found for many emersed red algae, in which it is unlikely that HCO_3^- is the C species crossing the plasmalemma (Bidwell & McLachlan 1985).

As a consequence of the lack of certainty about inorganic C entry mechanisms, the relationship between the form of inorganic C entering red algal cells and their capacity to deposit $CaCO_3$ remains unclear (Borowitzka 1984; Pentecost 1985).

The only red alga for which direct evidence for the presence of an inorganic C–concentrating mechanisms has been presented is *Porphyridium purpureum* (Burns & Beardall 1987).

Table 8-2. Occurrence of HCO_3^- use and C_4-like physiology in Rhodophyta

Organism	HCO_3^- use	C_4-like physiology when emersed	C_4-like physiology when submersed	Reference
Cyanidioschyzon merolae	(−)	n.d.	(+) (pH < 5.5)	Zenvirth et al. 1985
Porphyra umbilicalis	n.d.	+	n.d.	Coughlan & Tattersfield 1977
Porphyridium purpureum (sometimes as P. cruentum)	+	+	+	Aizawa & Miyachi 1986; Aizawa et al. 1985; Burns & Beardall 1987; Colman & Gehl 1983; Lloyd et al. 1977
Batrachospermum sp.	−	n.d.	−(pH 6.5–9.0)	Raven & Beardall 1981b
Ceramium rubrum	+	n.d.	n.d.	Sand-Jensen & Gordon 1984
Chondrus crispus	+	+	+(pH ≈ 8.0)	Bidwell & McLachlan 1985; Bréchignac & Andre 1984, 1985a,b; Bréchnignac et al. 1987; Lüning & Dring 1985
Furcellaria foliifera	n.d.	+	n.d.	Bidwell & McLachlan 1985
Gigartina latissima	n.d.	n.d.	+(pH ≈ 5.0)	Brown & Tregunna 1967
Gracilaria foliifera	n.d.	+	n.d.	Bidwell & McLachlan 1985
Gracilaria tikvahiae	n.d.	+	n.d.	Bidwell & McLachlan 1985
Iridaea splendens (as I. cordata)	n.d.	n.d.	±(pH ≈ 5.0)	Brown & Tregunna 1967
Lemanea mamillosa	−	n.d.	−(pH 6.5–9.0)	MacFarlane & Raven 1985; Raven & Beardall 1981b; Raven et al. 1987a,b; Raven et al. 1990
Nemalion helminthoides	n.d.	+	n.d.	Bidwell & McLachlan 1985
Palmaria (sometimes as Rhodymenia) palmata	+	+	+(pH 8.0)	Bidwell & McLachlan 1985; Colman & Cook 1985; Cook & Colman 1987; Cook et al. 1986; Coughlan & Tattersfield 1977
Polyneura latissima	n.d.	n.d.	(pH ≈ 5.0)	Brown & Tregunna 1967
Polysiphonia flexicaulis (as P. violacea)	+	n.d.	n.d.	Sand-Jensen & Gordon 1984

+ = present, − = absent, ± = intermediate, n.d. = no data.

HCO_3^- use by *Porphyridium cruentum* and *Palmaria palmata* was demonstrated as a photosynthetic rate in excess of the rate at which CO_2 could be generated from HCO_3^- outside the plant. The absence of HCO_3^- use in *Lemanea mamillosa* was shown by the absence of a photosynthetic rate in excess of the rate at which HCO_3^- produced CO_2 outside the plant. HCO_3^- use by *Chondrus crispus* was shown by HCO_3^-/CO_2 disequilibrium measurements. Data on other plants involve rather fewer "hard" criteria.

C_4-like physiology when emersed involves macrophytes in air and the microphyte *Porphyridium purpureum* in an artifical leaf.

C_4-like physiology when submersed was examined at the specified pH values: the work at pH 5.0 for marine algae (*Gigartina latissima; Iridaea splendens; Polyneura latissima*) is substantially displaced from their normal pH of ~8.0 (for good reason, in the context of the investigation undertaken) and may explain the equivocal data on C_4 or C_3 physiology.

C_4-like physiology versus C_3-like physiology is decided by examining one or more of the following properties (see Table 8-3): CO_2 compensation concentration at various O_2 concentrations; O_2 effect on net or gross CO_2 fixation rate: apparent CO_2 affinity of whole cells; short-term fluxes (tens of seconds) of $^{12}CO_2$ and $^{14}CO_2$; CO_2 loss to inorganic C–free medium. Discussion of apparently contradictory data on individual species may be found in the text.

The data in this table refer to organisms grown at air-equilibrium levels of CO_2 with the exception of *Lemanea mamillosa*, which was wild material growing in water supersaturated with CO_2; however, no "adaptation" to low CO_2 concentrations (HCO_3^- use or C_4-like physiology) were seen for material held for a week in low CO_2 (Raven unpubl.).

C. The kinetics of inorganic C and O_2 exchange: C_3-like or C_4-like physiology?

The kinetics of gas exchange that bear on the C_3-like or C_4-like physiology of red algae are those that pertain to the CO_2 compensation concentration, the effect of O_2 concentration on inorganic C assimilation, the kinetics of CO_2 and O_2 exchange at dark–light and light–dark transients, and the kinetics of tracer O_2 and CO_2 exchange. Table 8-3 shows

Table 8-3. Some methods by which C_3-like and C_4-like photosynthetic physiology may be distinguished

Attribute	Characteristic of C_3-like physiology	Characteristic of C_4-like physiology	References to work on Rhodophytes
CO_2 compensation concentration in ~250 mmol m^{-3} O_2	"High," > 1 mmol m^{-3}	"Low," < 1 mmol m^{-3}	Bidwell & McLachlan 1985; Brown & Tregunna 1967; Colman & Gehl 1983; Coughlan & Tattersfield 1977; Raven & Beardall 1981a
O_2 inhibition of net photosynthesis in ~10 mmol m^{-3} CO_2	Substantially less net CO_2 fixation at ~250 mmol m^{-3} O_2 than at ~10 mmol m^{-3} O_2	Little difference between net CO_2 fixation at ~250 mmol m^{-3} O_2 and ~10 mmol m^{-3} O_2	Bidwell & McLachlan 1985; Black et al. 1976; Downton et al. 1976; Lüning & Dring 1985
$K_{1/2}$ for inorganic C fixation, expressed as $\{CO_2\}$	"High"	"Low"	Colman & Cook 1985; Cook & Colman 1987; Robbins 1979
"Postillumination burst" of CO_2	Present, if $\{O_2\}$ high in light and dark; result of continued PCOC	If present, $\{O_2\}$ insensitive; a result of CO_2 leaking from internal pool	Burris 1977
Light-stimulated $^{18}O_2$ uptake in ~250 mmol O_2 m^{-3}, ~10 mmol O_2 m^{-3}	Present; result of RUBISCO oxygenase, Mehler reaction, and "dark respiration"	Present; result of Mehler reaction and "dark respiration"; less RUBISCO oxygenase	Bréchignac & Andre 1984, 1985a; Bréchignac et al. 1983, 1987
$^{14}CO_2$ fixation rates and $^{12}CO_2$ fixation rates in first few seconds of exposure of ^{12}C-grown plant to $^{14}CO_2$ in air-equilibrium O_2 concentrations	Large excess of $^{14}CO_2$ uptake betokening substantial $^{12}CO_2$ release of respiratory CO_2 from unlabeled sources	Little or no excess of $^{14}CO_2$ uptake; essentially no $^{12}CO_2$ loss	Bidwell & McLachlan 1985; Lloyd et al. 1977

how the values obtained from these various measurements differ from C_3-like and C_4-like plants, together with some of the provisos pertaining to the measurements.

Dealing first with CO_2 compensation concentrations at air-equilibrium O_2 concentrations, all the freshwater red macroalgae tested have C_3-like compensation concentrations for CO_2 (Raven & Beardall 1981a). The marine red macroalgae usually have values that are C_4-like (Bidwell & McLachlan 1985; Brown & Tregunna 1967; Coughlan & Tattersfield 1977), but *Polyneura latissima* and *Iridaea latissima* measured in seawater at pH 4.5–5.0 have higher values intermediate between C_3 and C_4 compensation concentrations (Brown & Tregunna 1967). The higher values for *Polyneura* and *Iridaea* are reduced at low (~10 mmol m^{-3}) O_2 concentrations (Brown & Tregunna 1967). The microphyte *Porphyridium purpureum* (as *P. cruentum*), grown at air-equilibrium CO_2 concentrations, also has a low CO_2 compensation concentration (Burns & Beardall 1987; Colman & Gehl 1983).

Oxygen inhibition of net photosynthesis has been investigated in a number of rhodophytes. Here, we restrict our discussion mainly to values obtained at air-equilibrium CO_2 concentrations when photosynthetic rates were measured at ~250 mmol O_2 m^{-3} and ~10 mmol O_2 m^{-3}. Lloyd et al. (1977) observed no effect of O_2 concentration in this range on net CO_2 exchange in illuminated *Porphyridium* sp. The most comprehensive set of data on marine rhodophytes is that of Bidwell and McLachlan (1985, Table 4), who found essentially no effect of O_2 concentration on net CO_2 fixation in six species from five genera emersed at approximately normal air CO_2 concentration. However, using one of the same species, *Chondrus crispus*, Bréchignac and Andre (1985a) determined that net CO_2 fixation was increased 45–70% by ~20 mmol O_2 m^{-3} relative to air-equilibrium O_2 in experiments on submersed specimens. This effect was independent of inorganic C concentration and was not due to effects on dark respiration. However, Lüning and Dring (1985) interpreted a 40% increase in photosynthetic O_2

evolution in submersed *Chondrus crispus* with an O_2 decrease from ~240 mmol m^{-3} to ~40 mmol m^{-3} under light-limiting conditions as resulting from constant gross photosynthesis and the observed stimulation of dark respiratory O_2 uptake with increasing O_2 concentration. Lüning and Dring (1985) found no effect of [O_2] changes in this range on net photosynthesis or on dark respiration in the other rhodophyte they tested, *Porphyra umbilicalis*. Cook and Colman (1987) used another of the species (*Palmaria palmata*) investigated by Bidwell and McLachlan (1985) and confirmed the absence of a significant effect of O_2 between 240 mmol m^{-3} and ~0 mmol m^{-3} on photosynthesis at around air-equilibrium CO_2 concentrations. However, Colman and Cook (1985) found a tendency for inorganic C-saturated photosynthesis to be O_2 inhibited over the same O_2 concentration range. Black et al. (1976) report the effects of O_2 concentration on ^{14}C-inorganic C fixation in 10–30 min experiments on *Halymenia durvillaei* (substantial inhibition by 250 mmol O_2 m^{-3} relative to "zero O_2") and *Laurencia* sp. (variable effect). The extent to which these tracer measurements are of net rather than gross photosynthesis is not clear. In similar but shorter-term experiments, Kirst (1981) observed stimulation of C fixation by ~250 mmol O_2 m^{-3} relative to "zero O_2" in *Griffithsia monilis*.

Tracer C (^{14}C) was used in a less equivocal way by Lloyd et al. (1977) and by Bidwell and McLachlan (1985). In these experiments the rationale and procedures of Ludwig and Canvin (1971) were followed; i.e., the uptake of $^{14}CO_2$ and that of $^{12}CO_2$ from a labeled gas stream were measured over a period of 30 to 180 s after $^{14}CO_2$ addition so as to minimize re-evolution of $^{14}CO_2$ after fixation and metabolism to substrates of (photo-) respiration. "True" (gross) photosynthesis is then deduced from $^{14}CO_2$ influx, and apparent (net) photosynthesis from $^{12}CO_2$ influx (taking into account the fractions of the two gases). Using this procedure, Bidwell and McLachlan (1985) investigated light respiration (i.e., the difference between true and apparent photosynthesis) in the same six species of marine red macrophytes. For four of them no CO_2 evolution could be detected, whereas in *Gracilaria foliifera* and in half of the experiments with *Palmaria* (as *Rhodymenia*) *palmata*, no O_2-insensitive "light respiration" was observed. By contrast, Lloyd et al. (1977) found that *Porphyridium* sp. in ~250 mmol O_2 m^{-3} lacked light respiration, whereas in ~10 mmol O_2 m^{-3} it had a very substantial light respiration (e.g., showed the reverse of the O_2-dependence of light respiration in terrestrial C_3 plants) (Table 8-3). We note that the rates of apparent photosynthesis at the two O_2 concentrations were identical, in agreement with other data using no $^{14}CO_2$ (Lloyd et al. 1977). Clearly, on basis of this evidence, the red algae either show no light respiration or have a light respiration whose O_2 response is not that of a typical C_3 plant.

Another important use of tracers in studying gas exchange relates to the O_2 side of light respiration. The temporal constraints due to problems with recycling are less severe than with $^{14}CO_2$ (Jackson & Volk 1970). The only data on $^{18}O_2$ uptake by a red algae that relate to light respiration (as opposed to phycobilin synthesis: Brown et al. 1980) is on *Chondrus crispus* (Bréchignac et al. 1987). Attempts were made to apportion the measured tracer O_2 uptake in the light (substantially in excess of the dark O_2 uptake rate) among photorespiration [i.e., the oxygenase activity of RUBISCO plus any O_2 uptake related to glycolate oxidation and the conversion of glycine to serine in the photorespiratory carbon oxidation cycle (PCOC)], the Mehler reaction, and cytochrome oxidase and the alternate oxidase of mitochondria using varying O_2 and CO_2 concentrations and photon flux densities and a number of metabolic inhibitors. It is important to note that the categories above are not mutually exclusive, in that the O_2 uptake in the PCOC could be mitochondrial if a glycolate dehydrogenase is operating in the mitochondria and the glycine to serine step occurs there. The conclusions reached by Bréchignac et al. (1987) were that, at saturating inorganic C levels (~30 mmol CO_2 m^{-3} in solution) and 250 mmol O_2 m^{-3}, about half of the O_2 uptake in the light was via the Mehler reaction and the rest was via mitochondrial activity. At limiting inorganic C concentrations (~2.5 mmol CO_2 m^{-3}), one-third of the O_2 uptake was attributed to mitochondrial activity, and up to two-thirds to the Mehler reaction, with a rather ill-defined contribution from the oxygenase activity of RUBISCO – see work on a chlorophyte by Glidewell and Raven (1975, 1976). This lack of definition precludes estimation of the achieved ratio of carboxylase-to-oxygenase activity in vivo in *Chondrus*.

Other data that are useful in distinguishing between C_3 and C_4 terrestrial plants include the CO_2 affinity for photosynthesis in terms of CO_2 concentration at the cell surface, and the occurrence and characteristics of the postillumination burst of CO_2 (Table 8-3). However, the postillumination burst has been little studied in red algae, and the estimation of CO_2 affinities, corrected for boundary layers, has not been attempted frequently.

Overall, gas exchange data suggest that the marine red macroalgae and *Porphyridium* have essentially C_4-like gas exchange characteristics, whereas the more limited data on the freshwater red macroalgae suggest that they are C_3-like. However, it is likely that the C_4-like gas exchange and the CO_2 concentrating mechanism it implies do not involve complete suppression of RuBPo activity (Bréchignac et al. 1987), a view supported qualitatively from work on ^{14}C labeling of glycine and serine and on glycolate excretions (Section IV).

D. Natural abundance of carbon isotopes and its implications for C transport

Studies of natural abundance of carbon isotopes can yield useful data on the mechanisms of inorganic carbon transport and metabolism and/or the extent to which transport limits the overall rate of photosynthesis, in both terrestrial and aquatic habitats (see Raven 1987c). In the context of the Rhodophyta our metabolic datum is the carboxylase activity of RUBISCO, the predominant carboxylase in these C_3 plants (Section IV). This carboxylase uses free CO_2 and, provided the CO_2 supply is not diffusion limited, exhibits a pronounced discrimination in favor of the lighter (and ~100-fold commoner) $^{12}CO_2$ relative to the heavier (and rarer) $^{13}CO_2$. The ratio of rate constants for the reaction with $^{12}CO_2$ to that for reaction with $^{13}CO_2$ is 1.03 (see Table 1 of O'Leary 1981 for the relationship between this and the isotope effect in parts per thousand). Liquid-phase diffusion of CO_2 (or HCO_3^-) shows essentially negligible discrimination between ^{12}C and ^{13}C. Thus, if a plant is only using CO_2, the CO_2 enters by diffusion, and CO_2 diffusion is not limiting the rate of photosynthesis, the organic C in the plant will be as enriched in ^{12}C relative to the exogenous CO_2 as would be the carboxyl C derived from CO_2 in the phosphoglycerate produced in the in vitro RUBISCO carboxylase reaction. This is essentially the case in *Lemanea mamillosa*, where a high external CO_2 concentration, thin boundary layers due to rapid water flow and turbulence generation by the plant, and a high activity of RUBISCO with a relatively high CO_2 affinity give plant $^{12}C/^{13}C$ ratios relative to those of source CO_2, which are frequently indistinguishable from those expected of limitation of C fixation by carboxylase activity alone (MacFarlane & Raven unpubl.; Raven 1987a; Raven et al. 1982). It must be emphasized that the interpretation of the $^{12}C/^{13}C$ ratio of even this simple system requires a lot of information other than a "spot" measurement of the $^{12}C/^{13}C$ ratio in the plant material. Such "spot" measurements alone can, however, help to rule out certain possibilities, especially in the sea, where variations in source inorganic C $^{12}C/^{13}C$ ratios are much smaller than in fresh waters. A complication in any organism that can "use" HCO_3^- is that the equilibrium $^{12}C/^{13}C$ of HCO_3^- is lower than is that of free CO_2 in solution; reference of plant values to the correct source $^{12}C/^{13}C$ values is essential. Even if the organism cannot use HCO_3^-, the occurrence of any mediated transport across a membrane of an inorganic C species poses problems of interpretation, since the assumed negligible $^{12}C/^{13}C$ discrimination of such processes has not been rigorously tested. Having said that, the observed $^{12}C/^{13}C$ values of marine red algae (Table III of Kerby & Raven, 1985) are generally not compatible with the $^{12}C/^{13}C$ being determined by RUBISCO carboxylase activity with relatively free CO_2 diffusion between the medium and the enzyme, as is the case for *Lemanea*. The $^{12}C/^{13}C$ is too low relative to seawater CO_2 (or HCO_3^-). If free CO_2 is entering the cell by diffusion, then boundary layers and/or some internal barrier must limit the fixation rate. If mediated transport of inorganic C into the organism occurs, then diffusive leakage out of the accumulated pool must be small (see Raven 1984a; Raven et al. 1987a). Since much of the evidence for marine red algae favors the occurrence of a CO_2-concentrating mechanism, the $^{12}C/^{13}C$ ratio data can be interpreted in these terms. The data on $^{12}C/^{13}C$, then, are compatible with the relatively small detectable CO_2 efflux in the light, despite evidence (Sections III and IV) for significant CO_2 production within the illuminated marine rhodophyte. These data, which severally indicate a relatively nonleaky compartment in which CO_2 is accumulated, are also in accord with the high photon (quantum) efficiency of photosynthesis in marine macroalgae (Dring & Lüning 1985; Lüning & Dring 1985) that possess the CO_2-concentrating mechanism (see Section II; Farquhar 1983; Raven 1987c; Raven & Lucas 1985).

E. Submersed and emersed photosynthesis in benthic Rhodophyta subjected to periodic emersion

Like other algae, the Rhodophyta are poikilohydric; they do not have a pronounced capacity to regulate water status in the face of large changes in environmental water status when emersed in the way that homoiohydric plants do (Jones & Norton 1979; Oates 1986; Raven 1984c, 1986). Intertidal species are subjected to periodic emersion, on a scale of hours to days. Emersion occurs during the summer decrease in water level for many specimens of the freshwater perennial *Lemanea*; this alga must be desiccation tolerant in at least part of its vegetative cycle. The important distinction between poikilohydry and desiccation tolerance is discussed by Raven (1984c, 1986).

Experiments that compared the photosynthetic performance of intertidal rhodophytes (and other algae) under submersed and emersed conditions showed substantial differences between species in the ratio of the photosynthetic rates in air and in seawater, and in the change of emersed photosynthetic rates as the thallus desiccates (see Bidwell & McLachlan 1985; Johnston & Raven 1986; Oates 1986 and references therein; Raven et al. 1987b, 1988). The rate of emersed photosynthesis in ~35 Pa CO_2 and at optimum water content can be greater than, or less than, that in seawater with ~2 mol inorganic C m^{-3} at pH ~8.0. While emersed photosynthesis can be enhanced initially by slight thallus dehydration, both emersed photosynthesis and the initial phase of subsequent submersed photosynthesis can be inhibited by further dehydration.

With this range of responses of different red algae to emersion, it is clearly difficult to draw general conclusions as to the significance of emersed photosynthesis for net photosynthetic carbon acquisition and growth. We accordingly choose two contrasting rhodophytes. The flat thalli of *Porphyra tenera* (Imada et al. 1970) and *Porphyra perforata* (Johnson et al. 1974) show higher rates of emersed photosynthesis (at optimal hydration) than of submersed photosynthesis. However, the growth rate of *Porphyra yezoensis* is decreased by periodic emersion, relative to the rate when permanently submerged (Tajiri & Aruga 1984). Although deprivation of resources other than C during emersion must not be ignored, the rapidity of dehydration inhibiting emersed, and later on submersed, photosynthesis is likely to contribute to the lowered growth rate resulting from periodic emersion.

By contrast, in the saccate *Halosaccion glandiforme* (as *H. americanum*) the rate of CO_2 uptake from air in emersed photosynthesis at optimal hydration is less than the submersed photosynthetic rate at the expense of exogenous inorganic C (Oates, 1986). However, the substantial inorganic C uptake from the water retained in the saccate thallus is at least as high in emersed as in submersed plants, and the retained water greatly slows the dehydration of the thallus. These data suggest that the total photosynthesis in one day shows relatively little change despite variations in the fraction of daylight hours spent emersed (Oates 1986). While growth was not measured directly, it is likely that, in view of the supply of N, P, K, etc., as well as C from the retained water, growth would not be greatly inhibited by increased periods of emersion (Oates 1986). However, we must not forget the effects of water movement during submersion on the uptake of N, P, etc. (this section).

F. Does inorganic C supply limit the growth rate of red algae?

Photosynthesis in some marine red macroalgae can be stimulated by increasing the inorganic C concentration in seawater above its normal level (original data and references in Colman & Cook 1985; Cook & Colman 1987; Holbrook et al. 1988; Robbins 1979; Sand-Jensen & Gordon 1984). Although this clearly means that the plants have the capacity to metabolize inorganic C faster than they do at normal inorganic C levels, it does not necessarily mean that growth in situ is C limited, since light, N or P supply, or even some catalysts of growth process(es) downstream of photosynthesis might be limiting growth rate. Although there is some evidence for physiological and biochemical adaptation to varying inorganic C supplies in *Porphyridium purpureum* (as *P. cruentum*) (Aizawa & Miyachi 1986; Aizawa et al. 1985), much more needs to be done in this area (see work on other major taxa: Aizawa & Miyachi 1986; Bowes & Reiskind 1987).

IV. CO_2 FIXATION AND PHOTORESPIRATION

A. Biochemistry of photosynthetic CO_2 fixation

Light-stimulated CO_2 fixation in the Rhodophyta involves the carboxylase function of RUBISCO and the photosynthetic carbon reduction cycle (PCRC) (see Fig. 8-2). Evidence for the occurrence of the reactions in Fig. 8-2 in red algae comes from the analysis of ^{14}C–inorganic C fixation products and measurements of enzyme activities.

The ^{14}C-labeling data involve supplying ^{14}C inorganic C for short periods (from a few seconds up to tens or hundreds of seconds) and extracting, separating, and quantifying the ^{14}C-labeled products. Such experiments indicate that steady-state photosynthesis has phosphoglycerate as the first fixation product (Bean & Hassid 1955; Döhler et al.

1976; Feige 1973, 1975; Kerby & Raven 1985; Kremer, 1981a,b; Kremer & Kuppers, 1977; Raven & Beardall 1981b). Evidence from pulse-chase experiments on *Palmaria palmata* (Bidwell & McLachlan 1985) shows that the ^{14}C label that does appear in C_4 dicarboxylates (malate and aspartate) in a 5-min labeling period does not show the rapid transfer into sugar phosphates and soluble carbohydrates that occurred in a similar experiment with the C_4 terrestrial vascular plant *Zea mays*. We note that the ^{14}C data apply to both submersed and to emersed algae.

The enzymic evidence shows that members of the Rhodophyta generally have activities of the enzymes of the PCRC for which they have been tested and that the activities are probably adequate to account for the in vivo rate of inorganic C fixation via the cycle (see Akazawa & Osmond 1976; Fewson et al. 1962; Ford 1979, 1986; Kremer 1978a,b; Kremer & Kuppers 1977; MacFarlane & Raven unpubl.; Nisizawa 1978; Ziegler & Ziegler 1967). The qualifications show that, as with other groups of plants, demonstration of adequate activities of all the carbon reduction cycle enzymes in many members of the Rhodophyta requires more experimental evidence than is available. Many of the RUBISCO carboxylase assays, for example, were conducted before recent improvements in technique (Lorimer et al. 1977) became available.

Rhodophyte RUBISCO has been little studied by comparison with the enzyme from many chlorophyll *b*–containing plants, cyanobacteria, and even chromophyte algae. It clearly conforms to the paradigm of having eight large and eight small subunits with a total M_r of $\sim 5.5 \cdot 10^5$ (Ford 1979; Steinmuller et al. 1983). The red algae resemble the

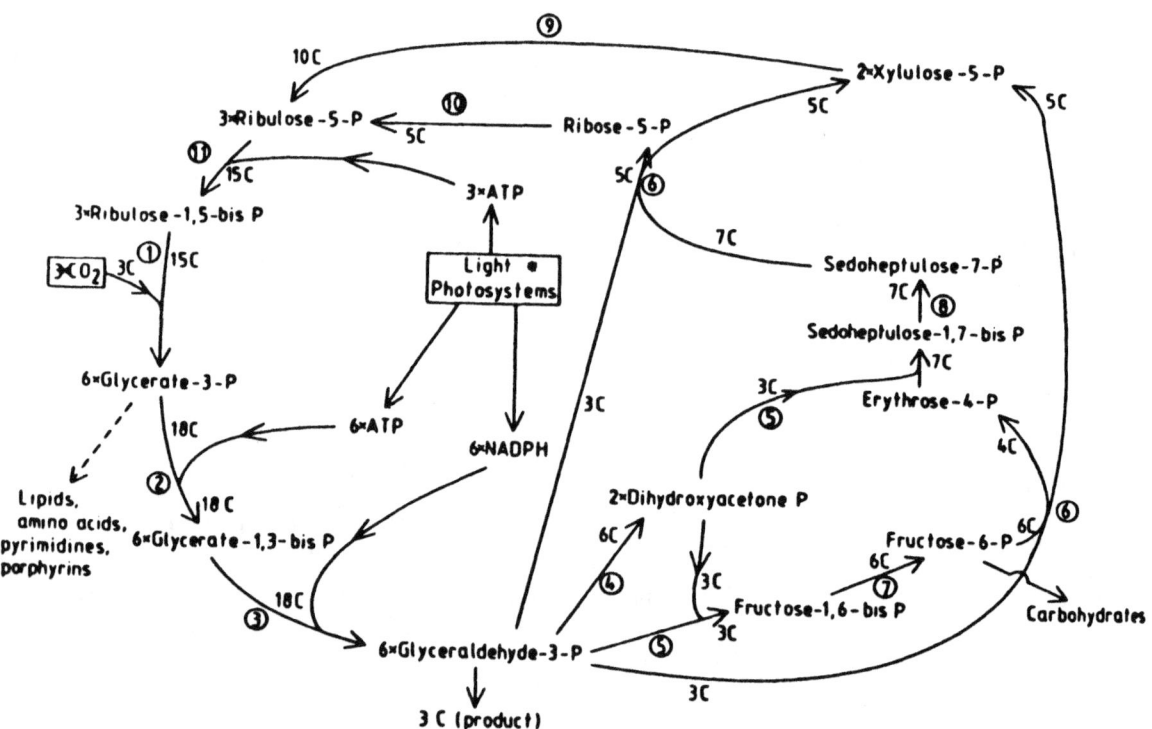

Fig. 8-2. The photosynthetic carbon reduction cycle. This representation does not indicate any oxygenase activity of RUBISCO, i.e., assumes a very high ratio of carbon dioxide to oxygen, which favors carboxylase and suppresses oxygenase. The ratio of carbon dioxide fixed to ATP and NADPH used is $CO_2:3ATP:2NADPH$. Number of C atoms involved in each reaction (per 3C fixed) given in uncircled numbers beside the reaction arrows. The enzymes of the cycle (circled numbers) are ① carboxylase activity of ribulose bisphosphate carboxylase–oxygenase (RUBISCO; regulated enzyme) (E.C.4.1.1.39); ② 3-phosphoglycerate kinase (E.C.2.7.2.3); ③ 3-phosphoglyceraldehyde dehydrogenase, $NADP^+$-linked (regulated enzyme) (E.C.1.2.1.13); ④ phosphotriose isomerase (E.C.5.3.1.1); ⑤ aldolase (E.C.4.1.2.13); ⑥ transketolase (E.C.2.2.1.1); ⑦ fructose-1,6-bisphosphate-1-phosphatase (regulated enzyme) (E.C.3.1.3.11); ⑧ sedoheptulose-1, 7-bisphosphate-1-phosphatase (regulated enzyme) (E.C.3.1.3.37); ⑨ phosphopentose epimerase (E.C.5.1.3.1); ⑩ phosphoribulose isomerase (E.C.5.3.1.6); ⑪ phosphoribulokinase (regulated enzyme) (E.C.2.7.1.19). Asterisk indicates catalytic role of CO_2. (From Figure 5.2 of Raven 1984a)

chromophytes but differ from the Chlorophyta and higher plants in that both the small subunit and the large subunit of the enzyme are coded by chloroplast DNA and synthesized on chloroplast ribosomes (Reith & Cattolico 1985, 1986; Steinmuller et al. 1983; Chapter 5).

The inorganic C–saturated specific reaction rate of carboxylase activity of RUBISCO purified from a cyanidiophycean is relatively low, only reaching 0.4 mol CO_2 fixed (mol enzyme)$^{-1}$s^{-1} at the temperature optimum of 42.5°C and 0.055 mol CO_2 fixed (mol enzyme)$^{-1}$s^{-1} at 20°C (Ford 1979). Although these values are low relative to values for chlorophytan microphytes at 25°C (~15 mol CO_2 (mol enzyme)$^{-1}$s^{-1}; Table 5.2 of Raven 1984a; Seeman et al. 1984), Ford (1979) only found a rate of 0.32 mol CO_2 (mol enzyme)$^{-1}$s^{-1} for the *Chlorella* enzyme at 25°C. Thus, the rhodophytan enzyme may be as catalytically active as that from other eukaryotic sources.

Little appears to be known of the CO_2 affinity of rhodophyte RUBISCO carboxylase activity. Colman and Cook (1985) and Cook and Colman (1987) quote a $K_{1/2}(CO_2)$ value of partially purified RUBISCO of 30 mmol CO_2 m^{-3}, presumably at air levels of O_2 for *Palmaria* (sometimes as *Rhodymenia*) *palmata*. This value may be an overestimate, since the pK_a' used by Colman and Cook (1985) was 6.32 (from the pH of the assay and the quoted $K_{1/2}$ values in terms of CO_2 and of DIC), and although the temperature of the assay was not given, a pK_a' of 6.06–6.10 is likely at the ionic strength of the assay medium (e.g., Yokota & Kitasato 1985). Thus, a $K_{1/2}$ value of 16–18 mmol CO_2 m^{-3} is likely. This is closer to the in vivo $K_{1/2}(CO_2)$ of photosynthesis in *Palmaria* (sometimes as *Rhodymenia*) *palmata* (Colman & Cook 1985; Cook & Colman 1987; Cook et al. 1986; Robbins 1979), and the in vitro RUBISCO kinetics could account for the in vivo CO_2 affinity (but not other gas exchange characteristics), granted limited CO_2 diffusion barriers and an excess of RUBISCO activity in vivo at saturating CO_2 (see Chapter 5 of Raven 1984a; MacFarlane & Raven 1985). However, it is clear that *P. palmata* has many characteristics in vivo that suggest that a CO_2-concentrating mechanism is operative (see Table 8-2), and this alga is in accord with the generally higher $K_{1/2}CO_2$ in aquatic plants than in terrestrial C_3 plants and the common occurrence of a CO_2-concentrating mechanism in aquatics (Badger & Andrews 1987; Raven 1984a; Raven et al. 1985).

Data that may be significant for the CO_2 affinity of rhodophyte RUBISCO are provided by Suzuki and Ikawa (1985), who studied the cryptophyte *Chroomonas* sp. The plastids of the Cryptophyta may have originated from endosymbiotic rhodophytes (Wilcox & Wedermayer 1985; but see Rothschild & Heywood 1987), albeit with a substantial repositioning of the phycobilins (see section II) and with the addition of chlorophyll c_2 to the light-harvesting pigments. At any event, Suzuki and Ikawa (1985) present data in terms of HCO_3^- and of CO_2 at a stated temperature and ionic strength. Recalculating the data, using the pK_a' value appropriate to the temperature and ionic strength (Yokota & Kitasato 1985), yields values of 10–16 mmol CO_2 m^{-3} in 0–21 kPa O_2. These values are still in excess of those found for in vivo photosynthesis in *Chroomonas*, and other data suggest that this algae also has a CO_2-concentrating mechanism (Suzuki & Ikawa 1985).

Returning to the (bona fide) red algae, MacFarlane and Raven (unpubl.; see Raven et al. 1987b, 1988) found a value of ~5 mmol CO_2 m^{-3} for the $K_{1/2}$ of *Lemanea mamillosa* RUBISCO (using the appropriate pK_a' values for the temperature and ionic strength) at low (~40 mmol m^{-3}) O_2 levels. This value, together with the measured CO_2-saturated carboxylation capacity on a unit tissue basis, agrees well with the predictions based on in vivo CO_2 exchange kinetics (MacFarlane & Raven, 1985). Since *Lemanea* apparently lacks a CO_2-concentrating mechanism under natural conditions (which provide a relatively high CO_2 availability; see Section II), perhaps such a low $K_{1/2}(CO_2)$ is not unexpected; indeed, the CO_2 affinity resembles that of terrestrial C_3 plants (Badger & Andrews 1987; Raven 1984a; Raven et al. 1985).

At the other extreme of $K_{1/2}(CO_2)$ values is the value of 41 mmol m^{-3}: Newman and Cattolico (1987) for the marine *Griffithsia pacifica*. On other evidence in the paper it is likely that this value may not have been corrected for the ionic strength effect on pK_a' values and may be an overestimate of the $K_{1/2}(CO_2)$, which might be as low as ~25 mmol m^{-3}. It is likely that *Griffithsia* spp. have a CO_2-concentrating mechanism (see Kirst 1981).

It would thus seem that the $K_{1/2}(CO_2)$ values for RUBISCO from red algae are not exceptional for eukaryotic RUBISCO, although the lowest reported value is rather low for an aquatic eukaryote (Badger & Andrews 1987; Raven 1984a; Raven et al. 1985).

All the evidence suggests that the Rhodophyta are biochemically C_3 plants, with no preparatory C_3–C_4 cycle such as occurs in terrestrial (and possibly aquatic: Bowes & Reiskind 1987) C_4 plants. However, much remains to be achieved in elucidating light-dependent CO_2 fixation in the Rhodophyta. An example is the light–dark regulation of

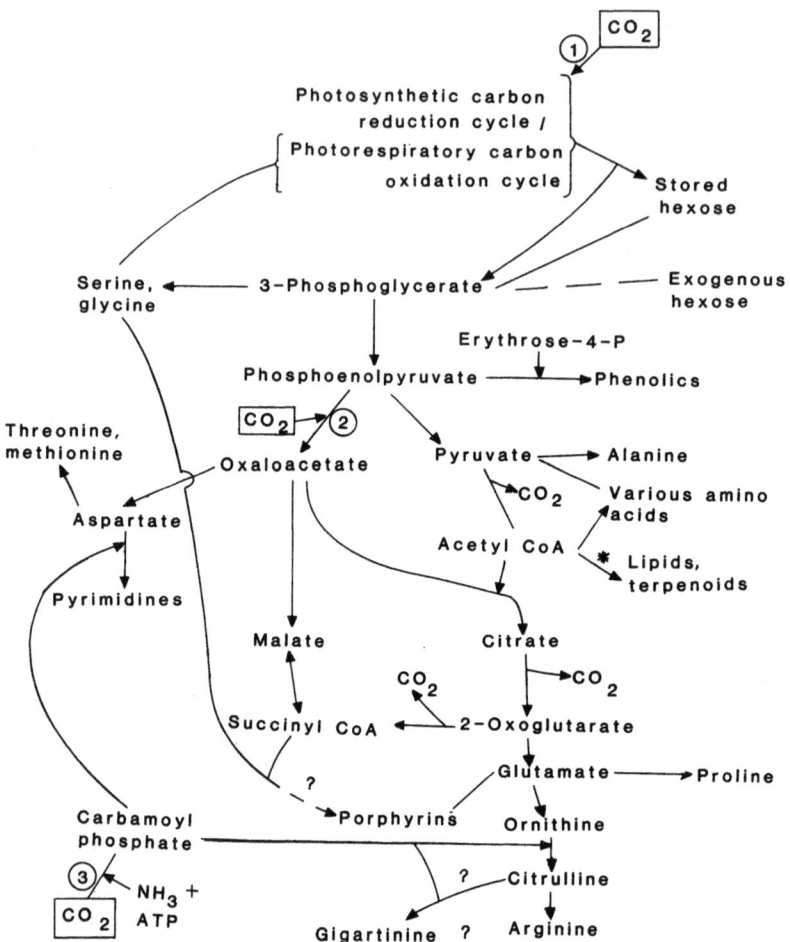

Fig. 8-3. The role of "dark" respiration and of three carboxylation reactions, in synthesis of C skeletons for biosynthesis in Rhodophyta. Porphyrins are indicated as all originating via the C_5 pathway (found for Cyanidiophyceae) rather than from C_5 for plastid components and succinate + glycine for mitochondrial components (found for Euglenophyceae). $\boxed{CO_2}$ = CO_2 fixation reaction. ① RUBISCO carboxylase activity, ② PEP carboxylase (? PEP carboxykinase?), ③ Carbamoyl P synthetase. Asterisk indicates catalytic role for CO_2 in lipid synthesis (From Weinstein et al. 1987a,b). Not shown are the role of the nonoxidative reactions of the oxidative pentose phosphate pathway in producing erythrose-4-P for the shikimate pathway and pentoses, since these can also be produced by the nonreductive reactions of the PCRC. The phosphoenolpyruvate, which is also used in the shikimate pathway, is (as shown) a respiratory product. Also not shown are the sites of NADPH and ATP generation in the oxidative pentose phosphate cycle and in glycolysis/tricarboxylic acid cycle, respectively.

RUBISCO, so far investigated only in eukaryotes with chlorophyll *b* (see Portis et al. 1986, 1987; Salvucci et al 1986; Servaites et al. 1986). Such regulation is important during changes in photon flux density, e.g., in "sunflecks" (cf. Chapter 4 of Raven 1984a) for understory red algae (see Sharkey et al. 1986, for cognate work on vascular land plants).

B. Biochemistry of anaplerotic CO_2 fixation

The Rhodophyta, like all organisms, also have anaplerotic CO_2 fixation mechanisms (Fig. 8-3) that can occur in the dark but also operate in the light (see Raven 1974b). Anaplerotic carboxylation reactions thus differ from RUBISCO, which is essentially completely turned off in the dark in Rhodophyta (see above). The occurrence of label in carbohydrates as a result of dark $^{14}CO_2$ fixation (Macler 1986) cannot necessarily be taken as evidence for RUBISCO activity since gluconeogenesis might not necessarily involve ^{14}C-phosphoglycerate generated via RUBISCO (see Raven 1984a, Chapter 5). Anaplerotic $^{14}CO_2$ fixation in red algae does not normally exceed 1% of the rate of photosynthetic CO_2 fixation or 10% of dark respiratory CO_2 evolution (Kerby & Raven 1985; Kremer 1978a,b, 1979, 1981a,b; see Oates, 1986). Anaplerotic CO_2 fixation rates in Rhodophyta resemble those of the Chlorophyta, but at a lower rate than is found in many members of the Phaeophyta.

The main enzyme of anaplerotic CO_2 fixation in the red algae is a catalyst of a "$C_3 + C_1$" carboxylation. The evidence consistent with this comes from $^{14}CO_2$ fixation studies and from enzyme measurements. The $^{14}CO_2$ fixation measurements very generally indicate a rapid, large labeling of compounds that can be readily derived from the (presumed) primary product of $C_3 + C_1$ carboxylation, oxaloacetate, with some variation in the ratio of amino acids to organic acids (Craigie 1963; Kremer 1979; Macler 1986). The carboxylase(s) involved is still a matter of some dispute. Thus, although Aizawa et al. (1985), Appleby et al. (1980), and Leclerc et al. (1982) report activities of phosphoenolpyruvate carboxylase of a magnitude adequate to account for the dark fixation rate, Bird et al. (1980), Kremer (1978a, 1981a,b), and Kremer and Kuppers (1977) could not detect this carboxylase. Kremer and co-workers (Kremer 1978a, 1981a,b; Kremer & Kuppers 1977) find activity of phosphoenolpyruvate carboxykinase that, faute de mieux, is assumed to be the carboxylase responsible for $C_3 + C_1$ carboxylation (see Kerby & Raven 1985).

The low rate of dark $^{14}CO_2$ fixation in most red algae is quantitatively consistent with the requirement for $C_3 + C_1$ carboxylation in an anaplerotic role necessary for the synthesis of pyrimidines, porphyrins, and several amino acids (see Fig. 8-3). It is unlikely that the glyoxylate pathway to malate relieves the need for anaplerotic $C_3 + C_1$ carboxylation in red algae. A role for $C_3 + C_1$ carboxylation in the entry of the product of glycolysis into mitochondria, where it is consumed by the tricarboxylic acid cycle, is possible (Section VI; Ap Rees et al. 1983). The low rate of dark $^{14}CO_2$ fixation in the Rhodophyta is consistent with the absence – or very low amplitude – of Crassulacean Acid Metabolism–like behavior in those that have been tested (see Table 3 of Raven et al. 1985; Holbrook et al. 1988).

Other dark fixation reactions are even less well characterized than the $C_3 + C_1$ carboxylation(s) in red algae. One such carboxylation, carbamoyl phosphate synthetase, is required for the synthesis of citrulline and hence arginine. The quantitative requirement for the operation of carbamoyl phosphate synthetase per plant mass produced is increased in those perennial red algae that synthesize large quantities of citrullinyl arginine in the autumn when external N becomes more readily available. This N-store can account for half of the organic N in the plant in late winter (Laycock & Craigie 1977; Laycock et al. 1981) in *Chondrus crispus* and several other red algae (Laycock & Craigie 1977). Each mol N stored in citrullinyl arginine requires the fixation, by carbamoyl phosphate synthetase, of $\frac{2}{7}$ mol (≈ 0.29 mol) of HCO_3^- and (in the production of the C_5 skeletons of citrulline and arginine) the fixation of another 0.29 mol HCO_3^- by phosphoenol pyruvate carboxylase. Red algae possessing citrullinyl arginine also usually have gigartinine, $H_2N-C(=NH)-NH-CO-NH-(CH_2)_3-CH(NH_2)-COOH$ (Laycock & Craigie 1977). Synthesis of 1 mol gigartinine presumably requires the fixation of HCO_3^- by phosphoenol pyruvate carboxylase and 2 mol HCO_3^- by carbamoyl phosphate synthetase, i.e., 0.4 mol HCO_3^- per mol N via carbamoyl phosphate synthetase and 0.2 mol HCO_3^- per mol N via phosphoenol pyruvate carboxylase.

By contrast, N storage as a protein – with the composition of a phycocyanin from the Cyanidiophyceae (Troxler et al. 1975), assuming lysine synthesis via the diaminopimelate pathway (Chapman & Ragan 1980) – required only 0.046 mol HCO_3^- assimilated by carbamoyl phosphate synthetase and 0.35 mol HCO_3^- assimilated by phosphoenol pyruvate carboxylase, per mol N assimilated. The high incorporation of $^{14}CO_2$ in the dark into citrulline in cyanobacteria (see Raven 1974b) may relate to their use of the copolymer cyanophycin (equimolar arginine and aspartate) as an insoluble N reserve. High fractional $^{14}CO_2$ incorporation into citrulline has been reported in *Gracilaria verrucosa* in the dark (Bird et al. 1980) and in low photon flux densities of blue, green, and red light (Bird et al. 1981), with highest rates in thalli resupplied with nitrogen after nitrogen depletion.

C. Biochemistry of photorespiration

Photorespiration is defined here in the restricted sense; i.e., it includes the uptake of O_2 and synthesis of phosphoglycolate via the oxygenase activity of RUBISCO and the metabolism of phosphoglycolate to phosphoglycerate via the photorespiratory carbon oxidation cycle (PCOC) (Fig. 8-4).

There are no published attempts to measure O_2 consumption or phosphoglycolate production by rhodophytic RUBISCO in vitro, although O_2 does competitively inhibit CO_2 fixation by RUBISCO from *Griffithsia pacifica* (Newman & Cattolico 1987). Furthermore, RUBISCO from all sources thus far investigated has been shown to have oxygenase activity (Akazawa & Osmond 1976; Badger & Andrews 1987; Keys 1986; Lorimer 1981), and red algal RUBISCO is structurally similar to other eukaryotic RUBISCOs. Thus, we can safely assume that the rhodophyte enzyme does have this activity. However, since data on the in vitro kinetics of carboxylase and oxygenase are not available, their contribution to in vivo metabolism as a function of steady-state CO_2 and O_2 concentrations in the plastid stroma cannot be predicted. Furthermore, attempts to determine the activity of glycolate-oxidizing enzymes (glycolate oxidase or glycolate dehydrogenase) yielded no activity in marine red algae (Tolbert 1976), although there is activity of phosphoglycolate phosphatase (Randall 1976).

That some flux through the RUBISCO oxygenase activity and PCOC occurs in vivo under natural exogenous O_2 and inorganic C concentrations is suggested by work in which ^{14}C-labeled inorganic C incorporation was followed over time courses of seconds to minutes (e.g., Burris 1977, 1980; Burris et al. 1976; Döhler et al. 1976; Feige 1973; Kremer 1978a,b). Here, the percentage of label in glycine and serine (and, in some freshwater examples, glycerate) in a variety of red algae increased with time after addition of ^{14}C-labeled inorganic C to the medium. Evidence consistent with the origin, at least in marine rhodophytes, of the glycine and serine via the oxygenase activity of RUBISCO and the PCOC comes from treatments with O_2 at higher and lower concentrations than pertain to air equilibrium. Lower O_2 concentration gives a smaller fraction of ^{14}C in glycine and serine at a given time after ^{14}C–inorganic C addition, and vice versa for higher O_2 concentration, against a background of seawater inorganic C concentrations (Burris 1977, 1980; Burris et al. 1976; Kirst 1981).

However, we must be aware of the limitations of these data. They do not directly bear on the metabolic source of the glycine and serine, (e.g., from RuBPo and the PCOC, rather than from phosphoglycerate, see Raven & Glidewell 1981). Furthermore, the data cannot be used readily to quantify the fraction of the fixed C that passes through the RUBISCO oxygenase activity and the PCOC (see Kerby & Raven 1985; Raven & Glidewell, 1981), especially if natural inorganic C supply and O_2 loss situations are to be mimicked. Even when "natural" bulk phase concentrations are used, the water flow regime is often not well represented (e.g., the otherwise good work on *Lemanea fluviatilis* by Feige 1973). No data on $^{18}O_2$ incorporation (via RUBISCO oxygenase activity) into PCOC intermediates, a procedure that can lead to better quantitation of PCOC activity, seem to be available for red algae

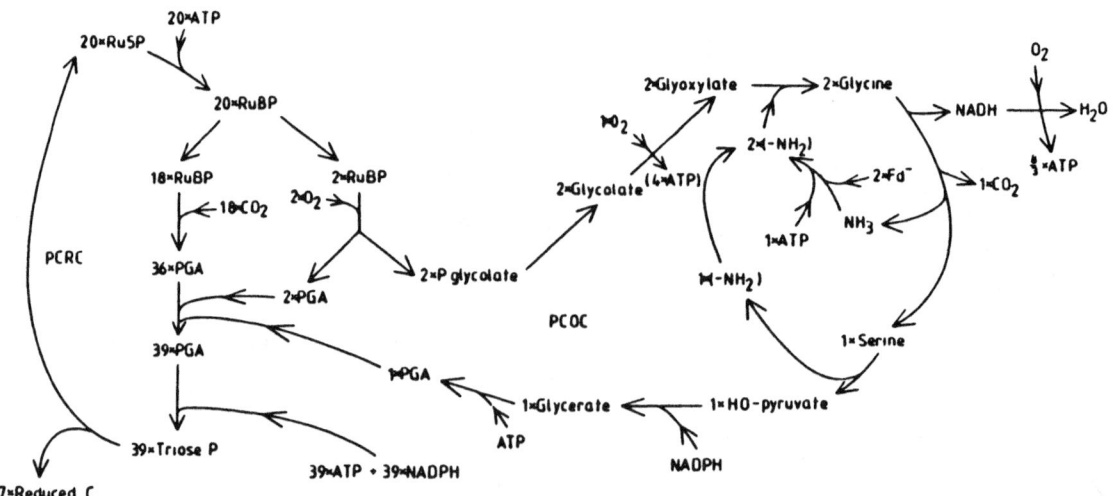

Fig. 8-4. The PCRC + PCOC for a carboxylase/oxygenase ratio in vivo of 9:1. This requires operation of a CO_2-concentrating mechanism for air-equilibrium media granted in vitro RUBISCO kinetics. (From Figure 5.4 of Raven 1984a)

(Kerby & Raven, 1985). Despite the limitations of the data, we can conclude that the inorganic C-concentrating mechanism in marine red algae does not completely suppress the RUBISCO oxygenase reaction, nor does the high free CO_2 concentration around freshwater red algae. The latter taxa appear to lack the concentrating mechanism, although the occurrence of RUBISCO oxygenase activity may be overestimated if water movement in the ^{14}C experiments is less than that found in situ.

Glycolate excretion by red algae can also be used as an indication of RUBISCO oxygenase activity (and phosphoglycolate phosphatase activity), if it is accepted that the RUBISCO oxygenase activity is essentially the only path to glycolate in photosynthetic cells. Such excretion also shows an excess of glycolate synthesis rate over the achieved capacity for glycolate use by the PCOC. The work of Fogg (1976; see Hackney & Hensley 1987) on marine red macroalgae [*Laurencia obtusa* and *Galaxaura oblongata* (as *G. fastigiata*)] showed a glycolate efflux in the light that was inhibited at zero O_2 and stimulated by ~ 1 mol O_2 m^{-3} relative to the rates in ~ 250 mmol O_2 m^{-3}. However, C excretion as glycolate did not exceed 0.5% of the rate of C fixation in photosynthesis. Döhler (1971) reported on glycolate efflux under transient (induction) conditions in *Porphyridium*.

This work on the biochemistry of photorespiration in the red algae clearly needs much extension before we can claim to have any clear understanding of the pathways concerned and the rates at which they function in nature. However, it does seem likely that some RUBISCO oxygenase – and PCOC – activity occur in near-natural conditions in all phototrophic red algae, including those that have a CO_2-concentrating mechanism. This is also the case for green microalgae grown at low CO_2 concentrations (e.g., Moroney et al. 1986) and for C_4 terrestrial plants (e.g., Jolivet-Tournier & Gerster 1984).

V. ORGANIC C STORAGE AND TRANSPORT

A. Nature of storage

The term "storage" is often used rather loosely to encompass any accumulation of metabolites with no obvious immediate role; it is assumed that these metabolite pools will be used later in some event, predictable (night) or unpredictable (sedimentation of planktophytes). Raven (1984a, 1987d) considers the definitions of "storage" and, following Cram (1976), points out that the use of a soluble organic carbon compound in an osmotic role (e.g., as a compatible solute) means that it cannot (unless replaced by a solute of equal utility) simultaneously act as a carbon and energy source at constant, or decreasing, external water potential (Davison & Reed 1985; Raven 1984a, 1987d).

B. Type of transport

Photolithotrophy in eukaryotes involves synthesis of organic compounds from inorganic C in the plastids. The plastids contain less than half of the C in most eukaryotes (see Kost et al. 1984; Raven 1984a). Furthermore, not all of the monomers needed for plastid synthesis can be synthesized from photosynthate in the plastids, so that some of the C produced in photosynthesis and used in plastid synthesis has been cycled through nonplastid cell compartments. This means that well over half of the photosynthate must move out of the plastids.

Photosynthate arriving in the cytosol of Rhodophyta can be converted to insoluble storage polysaccharide ($\alpha 1,4$ glucan) or to comparable solute/soluble storage compounds, or it can be metabolized to compounds used in synthesis of cell material. Some of the low M_r solutes can be transferred to the vacuole (in highly vacuolate cells); another fraction can be used in other organelles (Raven 1984a). In multicellular Rhodophyta with unpigmented cells there is also a requirement for organic C to move from photosynthetic cells to nonphotosynthetic cells, be they of the same ploidy level as the photosynthetic cells (e.g., in the female or male gametophyte, or tetrasporophyte, of Floridiophyceae) or of a higher ploidy level (carposporophytes on female gametophytes of Floridiophyceae). This need also applies to essentially nonphotosynthetic red algae parasitic on photolithotrophic red algae. These intercellular fluxes can be symplastic via intercellular connections or apoplastic with fluxes across plasmalemmas and cell walls (Raven 1984a).

C. The forms of stored organic C

The discussion here is limited to compounds containing only C, H, and O. Table 8-4 shows the occurrence of carbohydrate-based C storage compounds in the Rhodophyta. Floridean starch is probably universal as the polymeric storage compound for organic C and for energy. This compound is an $\alpha 1,4$ glucan that has many $\alpha 1,6$ branch points;

Table 8-4. Organic storage compounds and/or compatible solutes in members of the Rhodophyta

Organism	Floridoside	Isofloridoside	Trehalose	Digeneaside	Galactitol and glucitol	Reference
Cyanidioschzyon merolae	+	+				De Luca & Moretti 1983;
Cyanidium caldarium	+	+				cf. Nagashimi & Fukuda
Galderia sulphuraria	+	+				1981, 1983; Reed 1983
Porphyridium purpureum (as P. cruentum)	+					Nagashimi & Fukuda 1981, 1983
Porphyra yezoensis	+	+				
Chondrus crispus	+	+				Buggeln & Craigie 1973;
Dumontia incrassata	+	+				Craigie 1974; Table II of
Lomentaria umbellata	+					Kremer & Kirst 1982
Batrachospermum moniliforme	+		+			Kremer & Kirst 1982
Compsopogon hookeri	+		+			Kremer 1978b
Lemanea fluviatilis	+		+			
Membranoptera alata				+		Table II of Kremer &
Polysiphonia lanosa				+		Kirst 1982
Bostrychia montagnei					+	Table II of Kremer &
Bostrychia scorpioides					+	Kirst 1982

Floridean starch is probably universal (Craigie 1974; Fredrick & Seckbach 1987). The taxonomic and ecological significance of the occurrence of amylopectin (branched) and amylose (unbranched) in the starch of some Rhodophytes, but of amylopectin alone in other red algae, is not clear (McCracken & Cain, 1981).

Some of the reports of the soluble organic C compounds in the Rhodophyta, e.g., those in Table II of Kremer & Kirst (1981), are based on the pattern of labeling from ^{14}C–inorganic C in the light and thus discriminate against compounds that are labeled more slowly. Kremer and Kirst (1982) thus quote *Chondrus crispus* as just having floridoside, while Buggeln and Craigie (1973), using gas chromatography, found isofloridoside as well as floridoside.

it occurs as particles in the cytosol (see Chapter 2). Its storage role is attested by its accumulation in the light phase, and consumption in the dark phase of partially synchronized cultures of *Porphyridium purpureum* (Sheath et al. 1979b) and from other data (Nagashima et al. 1969; Sheath et al. 1979a, 1981).

The soluble organic C compounds are fairly diverse in this division (see Table 8-4); floridoside (with or without isofloridoside) is by no means universal in the Rhodophyta. The intracellular distribution of the soluble reserves is not well understood. Only in the giant-celled *Griffithsia monilis* has the presence of such a compound (digeneaside) in a specific cell compartment (the vacuole) been determined by microsurgery (Bisson & Kirst 1979). In fact, Bisson and Kirst (1979) used the correlation of digeneaside content of a cell and the fraction of cytoplasm in that cell to argue for a mainly extravacuolar location. At the other extreme of cell size, unicellular Rhodophytes have a very small fraction of the total volume occupied by vacuoles (~3%: Kost et al. 1984), so any compound that contributes more than 3% of the total intracellular osmolarity must be, at least in part, extravacuolar (see Raven 1984a). If diageneaside (and the other solutes) can function as compatible solutes, it would be expected that they are found in osmotically useful concentrations in such enzyme-rich (N) phases as cytosol, stroma, and matrix. It is only recently that the first direct demonstration of any compatible solute (glycine betaine) in protectively useful concentrations in any nonvacuolar compartment (stroma) has been made in any plant (the flowering plant *Spinacia oleracea*: Robinson & Jones 1986).

A separate question from that of steady-state distribution of the solutes is their site of synthesis. Again, few data are available; however, synthesis of digeneaside in *Griffithsia monilis* is nonplastidic. The evidence here is that plastids isolated from the plant by incorporation (naturally) into *Hermea* spp. (Kremer & Schmitz 1976) or (experimentally) into a buffer solution (Lilley & Larkum 1981) do not incorporate ^{14}C–inorganic C into digeneaside, although the plastids can photosynthesize at near–in vivo rates. These data suggest that synthesis of digeneaside is extrachloroplastic, probably in the cytosol, and that the digeneaside found in the vacuole and any other compartment with digeneaside was transported there. Kremer and Kirst (1982)

Carbon metabolism

point out that extrachloroplastic synthesis of di- and oligosaccharides, heterosides, and alditols seems to be the rule in other divisions of phototrophic eukaryotes, so that it is likely that the other soluble storage compounds in the Rhodophyta are produced outside the plastids, as is the case with digeneaside.

The synthesis of floridean starch appears to occur, starting from glucose-1-phosphate, using adenosine diphosphate glucose (ADPG), or uridine diphosphate glycose (UDPG), pyrophosphatase; ADPG, or UDPG, glucosyl transferase; and branching enzyme(s) (Craigie 1974; Fredrick & Seckbach 1987; Sheath et al. 1979a,b, 1981). The ADPG glucosyl transferase is associated with the floridean starch granules (Nagashima et al. 1971; see Craigie 1974), rather than with the plastids as suggested in a preliminary report (Nagashima et al. 1968) and reiterated by Kremer (1981b).

The synthesis of floridoside involves UDP-galactose and glycerol-3-phosphate and hence floridoside phosphate (Kremer & Kirst 1981, 1982). Little direct evidence seems to be available as to the pathway of synthesis of trehalose, digeneaside, galactitol, or glucitol (Kremer & Kirst 1982) in the Rhodophyta.

D. Intracellular transport of organic C

Little is known of the intracellular (transmembrane) fluxes of organic C; the reader is directed to Chapter 7 by Reed in this volume.

E. Intercellular transport of organic C

Raven (1984a, Chapter 9) discusses the transport of organic C between cells in the Rhodophyta in relation to what is known about such transport in other aquatic macrophytes. The occurrence of non-photosynthetic cells in an organism that is (overall) photolithotrophic (or a host–parasite combination, including a carposporophyte growing on a female gametophyte) is presumptive evidence for intercellular organic C transport. Raven (1984a) and Turner and Evans (1986) discuss ^{14}C-labeling data consistent with photosynthate movement from host to parasite, from female gametophyte to carposporophyte and (in *Delesseria*) down midribs and between fronds. The quantitative aspects of the transport [flux: mol C (m^2 plant transectional area)$^{-1}$ s^{-1}; velocity: m s^{-1}; concentration of moving solution from mass flow, or diffusion gradient for diffusion: mol C m^{-3}] are not well understood. The mechanism and the pathway are both still matters of dispute (Raven 1984a). Long-distance transport (e.g., between fronds of *Delesseria*) occurs over distances (substantial fractions of a meter) and at velocities ($\sim 175\,\mu$m s^{-1}) that seem to demand mass flow of solution. This is probably symplastic and thus more similar to phloem than to xylem transport in vascular plants. Short-distance transport of photosynthate (a few millimeters or less) can be by diffusion in an aqueous phase, either symplastically or apoplastically, or with apoplast–symplast alternation. The authors' predilection is to regard diffusion through the symplast, involving pit connection or derivatives thereof, as an important means of transport of photosynthate (or other solutes) over short distances in multicellular Rhodophyta, although apoplastic transport also may be significant (see Chapter 9 of Raven 1984a and the electrical data of Bauman & Jones 1986).

F. Organic C exchange with the bathing medium

Organic C exchange with the external milieu is considered by Reed (Chapter 7). As with other plants, the extent of so-called organic excretion in Rhodophytes is rather poorly characterized as far as near-natural conditions are concerned (Kremer 1981a,b; Raven 1984a).

IV. CLASSICAL RESPIRATION

A. Classical respiration and its role in photolithotrophs

"Classical respiration" is used here to encompass processes generating ATP, reductant, and carbon skeletons, involving O$_2$ uptake and/or CO$_2$ evolution and only dependent on light for their substrates via the photosynthetic formation of such compounds as those mentioned in Section V. It thus excludes photorespiration as defined in Section IV and the Mehler reaction, which depend on light-generated substrates with smaller pool sizes than those that supply dark respiration and are consequently much more tightly coupled, temporally, to energy input from photosynthesis. This definition does not exclude the possibility that light influences the rate of classical respiration both during and after illumination, and much evidence supports this possibility in a range of plants (Raven, 1984a, Chapter 6). Effects of pre-illumination on the rate of dark respiration in *Chondrus crispus* are illustrated by Enright and Craigie (1981).

The core of classical respiration in eukaryotes involves glycolysis, the oxidative pentose phosphate pathway, the tricarboxylic acid cycle, and oxidative phosphorylation. The first three of these processes generate important carbon skeletons, although it is worth remembering that the oxidative operation of the oxidative pentose phosphate pathway is not needed to generate these C skeletons (Raven 1984a). The oxidative pentose phosphate pathway is the major respiratory source of reductant, and oxidative phosphorylation generates most respiratory ATP. All three of these products are needed for growth, whereas only ATP is needed for maintenance. Maintenance must clearly occur in both dark and light, so that the "maintenance," ATP-generating function of respiration (or less commonly of fermentation), must occur in the dark. Growth can be, and in some phototrophs is, restricted to the photoperiod, thus minimizing the need for classical respiration in the dark period. Under these conditions, with growth restricted to the time when photosynthesis can occur, there is the possibility that the reductant and ATP used for growth can be supplied directly from the photoreactions without involvement of respiration. Two constraints apply to replacement of this photosynthetic growth requirement for classical respiration. One is (biochemically) mechanistic; photosynthetic metabolism (PCRC and PCOC) per se does not yield the specific C skeletons needed for certain biosyntheses (e.g., the glutamate family of amino acids; porphyrins), so tricarboxylic acid cycle activity is needed for biosynthesis in the light as well as in the dark (see Fig. 8-3). The other constraint is spatial, in that differentiation in most multicellular plants (including Rhodophyta) involves the occurrence of most of the photosynthetic apparatus in mature cells; the majority of the biosyntheses involved in growth occur in essentially chemo-organotrophic cells. A part of our discussion of the role of classical respiration will relate to the evidence for replacement of ATP and reductant generation from "dark" respiration by the production of ATP and reductant from photosynthetic partial reactions, along with the costs and benefits of so doing.

B. Occurrence, location, and capacity of the pathways of classical respiration

At least some of the enzymes or redox catalysts of all of the major pathways of classical respiration (Fig. 8-3) have been demonstrated in one or more members of the Rhodophyta (e.g., Fewson et al. 1962; Meatyard et al. 1975; Speer & Jones 1964; Sun et al. 1968; Van der Velde et al. 1975; Ziegler & Ziegler 1967). The functioning of the pathways in vivo, using techniques like the feeding of radiolabeled tracers of putative intermediates (preferably with labeling of specific C atoms) and their onward conversion, inhibition of respiration by selective inhibitors of respiration, and the study of spectroscopically detectable changes in the redox state of electron transfer catalysts, has not received very much attention in the Rhodophyta (Furbank & Rebeille 1986; Moyse 1961; Speer & Jones 1964). Difficulties abound; many red algae are obligate photolithotrophs, with little capacity to take up putative substrates and with low respiratory (as opposed to photosynthetic) redox catalyst concentration.

We note that none of the above work has been directed to establishing if the TCAC is complete, i.e., if 2-oxoglutarate dehydrogenase is present. Such an absence may be relevant to obligate photolithotrophy (Raven 1984a).

Evidence as to the location of the respiratory pathways in the Rhodophyta is also meager. In the Chlorophyta and higher plants the tricarboxylic acid cycle and the oxidative phosphosphorylation pathways are (as in all eukaryotes) mitochondrial; the oxidative pentose phosphate pathway is present in plastid stroma and in cytosol; glycolysis is present in the cytosol, at least part of the glycolytic pathway (in addition to enzymes common to the PCRC and glycolysis) is also in the plastids (see Douce 1986; Raven 1984a, 1987a). No comparable fractionation seems to have been done on Rhodophyta using, for example, isolated plastids of *Griffithsia monilis* (Lilley & Larkum 1981). It would be unprecedented if the tricarboxylic acid cycle and oxidative phosphorylation were not mitochondrial. Stewart and Mattox (1980, 1984; see Section IV) have argued that the mitochondria of Rhodophyta more closely resemble those of the chlorophyll *b*–containing organisms than those of chromophytes (excluding the Cryptophyta). Certainly, the amino acid sequence of the respiratory, soluble cytochrome *c* of *Porphyra* and *Palmaria* is not very dissimilar from that of chlorophytan cytochrome *c* (Meatyard et al. 1975). However, the correlations suggested by Stewart and Mattox (1980, 1984) have been upset by data on *Chlorarachnion* (Hibberd & Norris 1984).

We note that the fraction of the nonvacuolar, nonplastid cell volume taken up by mitochondria ($\sim 2\%$) in *Porphyridium purpureum* (as *P. cruentum*: Kost et al. 1984) is less than in other unicellular phototrophic eukaryotes (Table 6.4 of Raven 1984a). Concerning the intracellular distribution of catalysts specific to the oxidative pentose phosphate pathway and of the catalysts specific to glycolysis in Rhodophyta, the arguments of Raven (1976a, 1984a,

1987a) on the location of polysaccharide food reserves predict that the Rhodophyta would not have these catalysts in the plastids. The argument here is that *Euglena*, which, like the Rhodophyta, differs from the Chlorophyta and higher plants in having extraplastidial polyglucan, does not have catalysts specific to the oxidative pentose phosphate pathway or to glycolysis in the plastids (at least as far as studies with non-aqueously extracted chloroplasts bear on this question). It is not clear to what extent the occurrence of three isoenzymes of glucose-6-phosphate dehydrogenase and two isoenzymes of 6-phosphogluconate dehydrogenase in the rhodophyte *Audouinella* (as *Acrochaetium*) *daviesii* can be used to suggest multiple intracellular locations (Van der Velde et al. 1975).

A final point worth considering is the fraction of the total respiratory flux that can be attributed to parallel pathways. For CO_2 production, the competing processes are the oxidative pentose phosphate pathway and the tricarboxylic acid cycle. Other decarboxylations do not account for a large fraction of the total CO_2 production in the dark. O_2 uptake in the dark is largely via mitochondrial reactions, either the coupled pathway via cytochrome oxidase or the uncoupled (at least beyond UQ) pathway via the alternate oxidase. No data seem to be available on red algae with respect to the fractional contribution of the oxidative pentose phosphate pathway to dark CO_2 production using the fractional $^{14}CO_2$ production from specifically (C_1; C_6) labeled glucose molecules. For the occurrence of the cytochrome oxidase relative to the alternate oxidase in dark O_2 uptake, there is the inhibition and O_2 affinity work of Furbank and Rebeille (1986) on *Chondrus crispus*. Their work showed that the maximum inhibition of the two "mitochondrial" oxidases (cytochrome oxidase and alternate oxidase) left a "rump" of O_2 uptake with a lower O_2 affinity (see also Lüning & Dring 1985).

The advances made in the decade and more since Lloyd (1974a,b) reviewed the occurrence of respiratory pathways in algae and other microorganisms have been significant for the red algae, but much remains to be done.

C. Potential and achieved respiratory rates: the role of respiration in photolithotrophic growth

Raven (1976a,b, 1984a) and Raven and Beardall (1981b) compared the capacity for "dark" respiratory production of reductant and ATP with the requirement for reductant and ATP to support the measured (light-saturated) growth rate in terms of conversion of photosynthate (carbohydrate) plus exogenous N, P, K, etc. into cell material in a number of nonrhodophyte unicells. These capacity estimates came from literature values of uncoupled or organic C–saturated O_2 uptake rates in vivo for ATP synthesis and of enzyme (glucose-6-phosphate dehydrogenase, 6-phosphogluconate dehydrogenase) activity for $NADP^+$ reduction. The utility of such comparisons lies in the setting of an upper limit on the rate at which dark respiration can supply reductant and ATP in light, granted the difficulty of estimating the achieved rate of these processes in the light. For an algal culture growing in continuous light, this capacity estimate applies to the whole of growth; for growth in a light–dark cycle, the achieved rate in the dark can be used in conjunction with capacity in the light phase. These estimates include a shortfall of reductant and ATP supply by "dark" respiration, compared to what is needed for growth, and imply an input of ATP and reductant from partial reactions of photosynthesis that makes up the deficit. The shortfall is, of course, greater if it is assumed that "dark" respiratory ATP and reductant generation occurs less rapidly in the light than the achieved rate in the dark. A subsidy from thylakoid processes requires an excess capacity of reductant generation (noncyclic photophosphorylation) and ATP production (cyclic and pseudocyclic, or Mehler reaction, photophosphorylation, possibly with a contribution for noncyclic photophosphorylation). These requirements are met in at least some cases, again using data on nonrhodophyte unicells (Raven 1976a,b, 1984a; Raven & Beardall 1981b).

Evidence relating to the argument in the last paragraph is not readily available for the Rhodophyta. We shall attempt to relate the capacity for the dark respiratory production of ATP or reductant to the requirement for these substrates in growth, although very few data are available. However, Van der Velde et al. (1975) have measured the activity of the two enzymes specific to the oxidative pentose phosphate pathway (glucose-6-phosphate dehydrogenase; 6-phosphogluconate dehydrogenase) in *Audouinella* (as *Acrochaetium*) *daviesii*. The activities reported would be enough to reduce NO_3^- to NH_4^+ at a rate sufficient to yield an exponential (specific) rate of protein synthesis in the alga of 3.5×10^{-3} s^{-1}; although the plant growth rate is not quoted, the highest exponential growth rate in photolithotrophic eukaryotes at 20°C is 3×10^{-5} s^{-1} (Raven 1987a), so there would appear to be substantial overprovision of enzyme activity, even allowing for suboptimal conditions in vivo and other calls on reductant. Thus, reductant could be generated at a rate adequate for conversion of photosynthate into cell material via a combination of overt activity in the dark and cryptic activity in the light.

In terms of ATP, Furbank and Rebeille (1986) demonstrated a capacity for O_2 uptake by cytochrome oxidase immediately after cessation of illumination of 2.2 nmol O_2 $(gfw)^{-1} s^{-1}$, although the steady-state rates was only 1.1 nmol O_2 $(gfw)^{-1} s^{-1}$ in prolonged darkness in Chondrus crispus. These values are rather lower than those found earlier (Enright & Craigie 1981) at 17–20°C in the same species, i.e., 2.45 nmol O_2 $(gfw)^{-1} s^{-1}$ just after illumination and 1.45 nmol $(gfw)^{-1} s^{-1}$ after 8 hours darkness. Taking 2.45 nmol O_2 $(gwf)^{-1} s^{-1}$ as the capacity for electron transport in oxidative phosphorylation, we can estimate the capacity for ATP production in the dark at 13.1 nmol ATP $(gfw)^{-1} s^{-1}$, assuming 5.33 ATP per O_2 (Raven, 1984a). Chondrus crispus has an exponential (specific) growth rate of up to $\sim 10^{-6} s^{-1}$ under conditions of photon flux density, photoperiod, nutrition, and temperature similar to those used by Bréchignac and collaborators (Enright & Craigie 1981; see Neish et al. 1977). With this growth rate, 0.05 g C per g fresh weight (Enright & Craigie 1981) and 2 mol ATP used to convert 1 mol C in photosynthate into cell material (Raven 1984a), the required ATP production rate for growth on photosynthate is 8.3 nmol ATP $(gfw.s)^{-1}$. Thus, the capacity for oxidative phosphorylation is adequate to supply the ATP used in photosynthate processing, even allowing for ATP use for maintenance. However, the achieved rate of oxidative phosphorylation in the dark is only some 7.7 nmol ATP $(gfw)^{-1} s^{-1}$, assuming an O_2 uptake rate of 1.45 nmol O_2 $(gfw)^{-1} s^{-1}$, so in a 12L:12D cycle, the mean rate of oxidative phosphorylation would be 10.4 nmol ATP $(gfw.s)^{-1}$, assuming that oxidative phosphorylation occurs at its maximum rate in the light. The rate of oxidative phosphorylation in the light in Chondrus crispus is probably less than the maximal rate (Bréchignac et al. 1987), so the possibility still exists that oxidative phosphorylation is inadequate to provide all of the ATP needed to process photosynthate.

A less detailed data set, especially on the capacity for oxidative phosphorylation, is that of Lapointe et al. (1984a,b) on growth of Gracilaria tikvahiae. Taking their data for growth at 25°C and 267 ly d^{-1} (970 µmol photon $m^{-2} s^{-1}$ averaged over a 16-h photoperiod), the dark O_2 uptake rate would, if constant over the light and dark periods with an ATP/O_2 ratio of 5.33 (Raven 1984a), produce 4.48 µmol ATP (mol cell C)$^{-1} s^{-1}$. Growth at the observed 2.87 µmol C assimilated (mol cell C)$^{-1} s^{-1}$ (average over the light and dark periods) would, if 2 mol ATP are required to assimilate 1 mol photosynthate into cell material (Raven, 1976a,b, 1984a), need 5.75 µmol ATP (mol cell C)$^{-1} s^{-1}$. With a low estimate of the maintenance ATP requirement of 80 nmol ATP (mol cell C)$^{-1} s^{-1}$, the (growth plus maintenance) ATP requirement would be increased only to 5.83 µmol ATP (mol cell C)$^{-1} s^{-1}$, i.e., still not much more than the computed achieved ATP synthesis by "dark" respiration. However, if light inhibited ATP synthesis by dark respiration relative to the rate achieved in the dark (Raven 1972a,b, 1976c, 1984a), the data of Lapointe et al. (1984a,b) would indicate a need for an "ATP subsidy". We note that the data of Lapointe et al. (1984a,b) do not permit conclusions to be drawn as to NADPH generation, since no data on CO_2 production rate are given (Raven 1976a,b,c, 1984a). Of course, this discussion must be put in the context of the differentiated multicellular nature of Gracilaria tikvahiae and that of Chondrus crispus, considered earlier, in which there is likely to be considerable spatial separation of photosynthetic capacity and growth (Sections V and VI), so photosynthetic partial processes are substantially disadvantaged as suppliers of reductant and ATP for growth. Temporal separation of growth and photosynthesis (i.e., use of photosynthate (organic C) for growth in the dark period) can lower the overall efficacy of photosynthetic partial reactions in subsidizing "dark" respiration in unicells as well as in differentiated multicellular algae.

How could a shortfall of ATP and NADPH production by "dark" respiration be supplemented in rhodophytes? In other words, what qualitative and quantitative evidence is available on the occurrence of cyclic and pseudocyclic photophosphorylation and an excess of noncyclic electron flow over what is needed to account for the observed net CO_2 assimilation in growth? The last question is relatively easily answered: Short-term photosynthesis measurements yield substantially higher rates than are needed to account for C assimilated in growth. Again, using data of Lapointe et al. (1984a,b) for Gracilaria tikvahiae growing at 25°C and 970 µmol photon $m^{-2} s^{-1}$ averaged over a 16-h photoperiod, an assimilation rate of 2.87 µmol C (mol cell C)$^{-1} s^{-1}$ averaged over 24 h needs, when corrected for C respired in the dark period (assuming a respiratory quotient of 1.0), a net photosynthetic rate of 4.68 µmol CO_2 (mol cell C)$^{-1} s^{-1}$ in the light. In terms of net O_2 evolved, we must add the O_2 evolution equivalent to NO_3^- reduction (little different for respiratory or direct photoreduction of NO_3^-: Raven 1976a,b, 1984a, 1985), i.e., 0.74 μmol O_2 (mol cell C)$^{-1}$s^{-1} in the light period (C:N in cells = 11.7; 2 mol O_2 per mol NO_3^-). Thus, total net O_2 production would need to be 5.42 μmol O_2 (mol cell C)$^{-1}$s^{-1}. The measured rate is 16.6 μmol O_2 (mol cell C)$^{-1}$s^{-1}. The difference presumably reflects non-steady-state storage and/or excretion of organic C (see Section V).

For pseudocyclic photophosphorylation (coupled to electron flow measurable as the Mehler reaction), the work of Bréchignac and Andre (1984, 1985a) and Bréchignac et al. (1987) on *Chondrus crispus* suggests a substantial Mehler reaction under both high and low CO_2 supply conditions, amounting to up to 0.2 of the gross rate of O_2 evolution measured with isotopic oxygen. No data are available as to coupling of this Mehler reaction to ATP synthesis (see Raven 1976a,b, 1984a), but if it were coupled with an H$^+$/e$^-$ (at light saturation, with no Q cycle) of 2 and an H$^+$/ATP of 3 (Raven 1984a), it could provide 2.67 mol ATP per mol O_2 taken up. From Figure 2 of Bréchignac et al. (1987), at 17.6°C and at air-equilibrium CO_2 and O_2 concentrations, it would appear that "dark" respiration occurs (in the dark or the light: Table 3 of Bréchignac et al. 1987) at ~4 μmol (gfw)$^{-1}$h^{-1}, whereas O_2 uptake (isotopic) in the light occurs at ~10 μmol (gfw)$^{-1}$h^{-1}. Since, at saturating CO_2 concentration, Mehler reaction and "dark" respiration seem to be the only O_2 uptake processes in the light (Bréchignac et al. 1987), the Mehler reaction would consume ~6 μmol O_2 (gfw)$^{-1}$h^{-1}. With the ATP/O_2 values quoted above (5.33 ATP/O_2 for "dark" respiration; 2.67 ATP/O_2 for Mehler reaction), the Mehler reaction would generate 16 μmol ATP (gfw)$^{-1}$h^{-1} in the light, whereas dark respiration would generate 21.3 μmol ATP (gfw)$^{-1}$ in the light or the dark. With the 18 h light: 6 h dark regime in which *Chondrus crispus* was grown, the Mehler reaction only produced 288 μmol ATP (gfw)$^{-1}$d^{-1}, whereas "dark" respiration produced 512 μmol ATP (gfw)$^{-1}$d^{-1}. In this instance pseudocyclic photophosphorylation produces only just over half as much ATP as does "dark" respiration in the overall light-dark cycle. We note that *Chondrus crispus*, like *Gracilaria tikvahiae*, considered above, is a differentiated multicellar organism that is thus prone to both temporal and spatial constraints on a subsidy from photosynthetic partial processes to dark respiratory ATP and reductant supplies to growth processes. In this context it is pertinent that Bréchignac and Andre (1985a) and Bréchignac et al. (1987) suggest that the Mehler reaction in *Chondrus crispus* contributes to energizing the (putative) CO_2-concentrating mechanism. This is a spatially and temporally compatible sink for ATP generated only in the light in photosynthetically competent cells (see Raven & Lucas 1985).

Turning to the occurrence of cyclic photophosphorylation as a means of supplementing respiratory ATP, there is excellent evidence that, as in other O_2-evolving phototrophs, members of the Rhodophyta can conduct cyclic photophosphorylation. Thus, Biggins (1973, 1974) and Maxwell and Biggins (1976) showed that 3' (3,4 dichlorophenyl) 1, 1 dimethyl urea (DCMU), which inhibits intersystem electron flow, increases approximately 10-fold the half-time of dark reduction of cytochrome *f* oxidized by a light flash in *Porphyridium purpureum* (as *P. cruentum*) and *Porphyra umbilicalis*. In both rhodophytes the reoxidation rate is about doubled by the uncoupler carbonyl cyanide m-chlorophenyl hydrazone (CCCP), suggesting that normally the reduction rate is limited by coupling to ADP phosphorylation. These data suggest that cyclic electron transport has a light-saturated rate about one-tenth that of noncyclic (including pseudocyclic) electron transport (but see Myers 1986 for an alternative interpretation). Assuming that, at light saturation, the noncyclic and the cyclic processes have H$^+$/e$^-$ ratios of two (i.e., a Q cycle in cyclic, but not in noncyclic, electron flow), the ratio of ATP generation capacity in the noncyclic and cyclic processes is also about 10. This means that, if CO_2 accumulation and conversion to carbohydrate needs 4 ATP per C, all from open-chain phosphorylation, and carbohydrate assimilation into cell C needs 2 ATP per C, cyclic photophosphorylation could supply one-fifth of the ATP needed for cell growth, provided that cyclic and noncyclic photophosphorylation could proceed simultaneously. If, however, the full capacity of open-chain phosphorylation is not used for light-saturated growth, cyclic photophosphorylation could contribute a large fraction to the growth ATP requirement.

Evidence that cyclic photophosphorylation can proceed at a substantial fraction of the rate of open-chain photophosphorylation comes from work on the acidophilic Cyanidiophyceae (Enami & Kura-Hotta 1984; Kura-Hotta & Enami 1981, 1984). Active H$^+$ efflux occurs when an aerobic cell suspension held in the dark at pH 3.0 is illuminated. This efflux is not sensitive to DCMU and has the action spectrum of photoreaction one; furthermore, the efflux is inhibited by diethylstilbesterol (DES) and dicyclohexylcarbodi-imide (DCCD) under conditions in which they do not reduce the intracellular ATP concentration. The H$^+$ efflux is thus powered by

ATP produced in cyclic photophosphorylation. This ATP-powered H$^+$ efflux occurs at 0.263 the rate of photosynthesis, i.e., 0.263 mol H$^+$ actively transported out of the cells per mol O$_2$ evolved in net photosynthesis. The H$^+$ free energy difference across the plasmalemma of the acidophilic Cyanidiophyceae (Beardall & Entwisle 1984; Enami et al. 1986; Zenvirth et al. 1985), i.e., up to 30 kJ mol^{-1}, requires that only 1 H$^+$ is extruded per ATP converted to ADP and P$_i$, since this hydrolysis yields not more than 55 kJ per mol ATP used. Thus, with 1 ATP used per H$^+$ extruded, and 2.67 ATP produced per O$_2$ produced in noncyclic electron flow, the rate of cyclic photophosphorylation is 0.1 that of noncyclic photophosphorylation; this is about the same as that deduced by Maxwell and Biggins (1976) for *Porphyridium* and *Porphyra*. The H$^+$ efflux supportable by respiration (after prior treatment with N$_2$ in the dark) only occurs at about half the rate supported by cyclic photophosphorylation (Kura-Hotta & Enami 1981, 1984). Cyclic photophosphorylation that is able to synthesize ATP at least as rapidly as oxidative phosphorylation is compatible with the characteristics of $^{35}SO_4^{2-}$ influx in the bangiophycean *Rhodella maculata* (Millard & Evans 1982).

This survey suggests that the total capacity for photophosphorylation in Rhodophyta, like that in many other algae, is about an order of magnitude greater than that for oxidative phosphorylation, and it is likely that oxidative phosphorylation is supplemented by photophosphorylation in supplying ATP to growth and maintenance processes. Such a subsidy does not necessarily mean that one cannot identify a growth component of classical respiration (corrected for the maintenance requirement, and for any uncoupled respiration) that is proportional to growth rate when the latter is varied within a genotype by changed light or temperature conditions. This sort of correlation has been shown for *Gracilaria tikvahiae* by Lapointe et al. (1984a,b); in this organism it appears that oxidative phosphorylation may be quantitatively inadequate to power "growth" processes converting organic C photosynthate plus inorganic nutrients into new plant (see above).

A final point about the role of classical respiration in providing reductant and ATP for biosynthesis relates to light-limited growth. Few data seem to be available for Rhodophyta (see Lapointe et al. 1984a,b). In general, however, two points can be made. One is that, even with photophosphorylation running with its maximal efficiency (i.e., with a Q cycle in cyclic and noncyclic photophosphorylation), it is not easy to produce a scheme for the direct photoproduction of ATP that has a substantially higher photon yield than does the fixation of CO$_2$ to produce carbohydrate that is subsequently used to power oxidative phosphorylation (see Raven 1984a,b). Thus, in energetic terms, the use of oxidative phosphorylation to energize processes involved in the consumption of organic C photosynthate, as opposed to its synthesis, may be advantageous at low photon flux densities. A second point relates to the lower capacities for photophosphorylation and oxidative phosphorylation in organisms genetically or phenotypically adjusted to growth at lower photon flux densities (Raven 1976a,b). If the plant is growing at limiting photon flux densities, even the lower sum of capacities for ATP synthesis (on a biomass basis) in the low light plant may still be greater, relative to the requirement for light-limited growth, than in a high light plant growing at or near light saturation. Thus, growth at low light is likely to be more constrained by the efficiency (ATP per photon absorbed) than by the capacity (ATP per unit biomass per second) of the various mechanisms of ATP synthesis (Raven 1976a,b, 1984a,b).

D. Anaerobic metabolism

Although anaerobic conditions may not be a common occurrence for most of the phototrophic red algae, it is clear that several rhodophytes are capable of withstanding prolonged (days of) dark anoxia (e.g., Frenkel & Rieger 1951; Ben-Amotz et al. 1975). Anoxia in the absence of exogenous H$_2$ leads to H$_2$ efflux (Ben-Amotz et al. 1975), probably as a means of disposing of some of the reductant generated in fermentative ATP synthesis; little is known of the products of rhodophyte fermentation. Exogenous H$_2$ can be taken up by anoxically pretreated and maintained cells in light and in the dark; light stimulation of H$_2$ uptake, with coupling to reductive CO$_2$ fixation, is not universal in those red algae that can take up H$_2$ (Ben-Amotz et al. 1975; Frenkel & Rieger 1951). Light-dependent H$_2$ evolution appears to be absent from red algae (Ben-Amotz et al. 1975).

E. Controls of the rate of classical respiration

Classical concepts of "respiratory control" are based on observations consistent with a tight coupling of the flux through respiratory pathways and the demand for the products of these pathways in growth (using C skeletons, ATP, and NADPH) and in maintenance (using ATP). The view that respiratory rate is determined by the demand for the products of respiration in essential processes requires, inter alia, that the capacity of the catalysis of respiration – and for the supply of organic C substrate and of O_2 – exceeds the maximum demand for C skeletons, ATP, and NADPH in growth and maintenance in the dark. The correlation observed between the rate of respiration in the dark and growth rate, using temperature or light as the factors altering growth rate, in *Gracilaria tikvahiae* (Lapointe et al. 1984a,b) suggests that respiratory control is operating here. However, data on respiratory capacity would be useful (see Furbank & Rebeille 1986, who showed uncoupler stimulation of respiration in *Chondrus crispus*).

Respiratory control may not account for all phenotypic changes in respiratory rate in rhodophytes (see Lapointe et al. 1984a,b). At the extremes of temperature and light tolerance for a given organism, substrates and energy may be needed to repair damage (e.g., photoinhibition: Raven 1984a; Raven & Samuelsson 1986). Although it is possible that coupled respiration provides these, resulting in a respiration rate higher than is predicted from growth and maintenance costs, this has not been tested rigorously. The phenomenon of a protein synthesis–dependent repair of photoinhibitory damage certainly occurs in at least one rhodophyte (Setlik et al. 1984; Setlikova et al. 1984). However, some of the increased respiration could reflect damage to the respiratory mechanism itself, and might be manifest as "uncoupled" respiration (Beevers 1970). The decline in the rate of respiration in prolonged (3-week) darkness in the conchocelis phase of *Porphyra leucosticta* (Sheath et al. 1979c) may reflect reduced substrate supply to (maintenance) respiration, or it may relate to a gradual decline in growth (synthesis) activities or to decreased capacity for respiration. At the other end of the respiratory pathway, and occurring over much shorter timescales, the respiratory rate of some red macrophytes responds to changes in external O_2 concentration in the range of hundreds of mmol m^{-3}, i.e., much higher than the 100 mmol m^{-3} at which cytochrome oxidase and the alternate oxidase are O_2 limited (Downton et al. 1976; Furbank & Rebeille, 1986; Lüning & Dring 1985; Newell & Pye 1968; Raven 1984a). The extent to which this O_2 dependence at concentrations that are high but ecologically meaningful (e.g., in rock pools) reflects diffusive limitation of O_2 supply to some of the plants complement of "normal," high-affinity oxidases, rather than to the presence of some additional oxidase(s) or oxygenases(s), is still unclear (Furbank & Rebeille 1986; Chapter 6 of Raven 1984a). At all events, the opposing effects of changes in external O_2 concentration on rates of net photosynthesis and of dark respiration in the hundreds of mmol O_2 m^{-3} range in *Chondrus crispus* (see Bidwell & McLachlan 1985; Bréchignac & Andre 1985a; Furbank & Rebeille 1986; Lüning & Dring 1985; Newell & Pye 1968) are likely to upset the stoichiometry between growth and dark gas exchange in this alga. A final example of possible non–respiratory control regulation of classical respiration relates to the control of ATP and NADPH production by classical respiratory processes in illuminated cells. Raven (1976a,b, 1984a) argues for a partial suppression of ATP and NADPH synthesis via classical respiration in the light in photosynthesizing cells despite the large increase in demand for ATP and NADPH in the light; the demand is met by photosynthetic partial processes. Raven (1976c, 1984a) further relates the transient postillumination increase in classical respiration observed in many plants to, at least in part, a postillumination "respiratory control" related to slow inactivation of light-activated consumption of ATP and NADPH. For a number of plants, including *Chondrus crispus* (Enright & Craigie 1981; Furbank & Rebeille 1986), it has been suggested that this stimulation represents a greater respiratory flux caused by increased availability of organic C after a period of photosynthesis. However, it is often difficult to distinguish this possibility from an action of photosynthesis (or, in facultative chemoorganotrophs, exogenous organic C) in stimulating biosynthesis. In this case a slow decline in the rate of biosynthesis in the dark would account for the effect on classical respiration in terms of respiratory control (Raven 1976a,b,c). In such a situation the rate of "dark" respiration would be limited by the supply of ADP, P_i, and $NADP^+$, rather than an organic C substrate. Either phenomenon could account for an absence of uncoupler stimulation of "dark" respiration immediately after illumination, since they could both stimulate respiration to the limit of catalytic capacity. However, the data of Enright and Craigie (1981) and Furbank and Rebeille (1986) on *Chondrus crispus* show that the rate of "dark" respiration immediately after cessation of photosynthesis is substantially greater than the rate found in the presence of an uncoupler during the steady-state dark respiration found more than 30 min after illumination ceased. In this case, the steady-state rate of respiration in the dark is probably limited by supply of photosynthate.

VII. SUMMARY

What can be concluded about the state of knowledge of carbon metabolism in the red algae? C acquisition is generally photolithotrophic, via the C_3 pathway, i.e., using RUBISCO and the PCRC. It appears that, under natural conditions of inorganic C availability, the oxygenase activity of RUBISCO, and hence C flux through the PCOC, is generally restricted relative to that expected for diffusive CO_2 entry and O_2 exit in air-equilibirum solution. These observations can be explained as due either to diffusive entry of CO_2 from a high bulk phase CO_2 concentration via a minimal diffusion boundary layer or to the occurrence of active influx of CO_2 and/or HCO_3^- from media with a smaller potential diffusive flux of CO_2 from the bulk phase to the plastids.

Photosynthate can be stored as floridean starch and, probably, as soluble carbohydrates, although changes in the latter have osmoregulatory repercussions. Intercellular fluxes of soluble carbohydrates, at least in part, are symplastic and involve mass flow of solution, at least for transport over long distances.

Classical respiratory reactions in red algae appear to be qualitatively conventional, although it is not certain that their maximum catalytic capacity can always supply all of the reductant and ATP needed for growth at the expense of photosynthate. In illuminated photosynthetic cells, photosynthetic reductant and ATP supply can supplement classical respiratory supply; pseudocyclic electron flow (and coupled phosphorylation?), cyclic photophosphorylation, and noncyclic photophosphorylation, in excess of the rate of CO_2 fixation, occur in red algae at rates commensurate with a role in photosynthate use for growth.

One role of classical respiration that cannot be replaced by photosynthetic apartial reactions is that of C skeleton synthesis (e.g., aspartate and glutamate and their derivatives). This C skeleton synthesis requires a $C_3 + C_1$ carboxylation, as well as carbamoyl phosphate synthetase.

Overall, there is little evidence that the red algae function in a radically different way with respect to carbon metabolism from those of other algae from similar habitats, granted differences like the chemistry of the storage products. However, much more work is needed to test the outline of C metabolism presented as the conclusions to this chapter.

VIII. ACKNOWLEDGMENTS

Work from the authors' laboratory that is discussed herein was supported by Research Grants and Studentships from S.E.R.C. and N.E.R.C. This support is gratefully acknowledged.

IX. ENDNOTES

1. Recent work (Smith & Bidwell 1987, 1989) on the mechanism of inorganic C uptake by *Chondrus crispus* agrees with neither the previously reported ability of this organism to take up HCO_3^- (Section III.B) nor the interpretation of C_4-like physiology in terms of an inorganic C-concentrating mechanism (Section III.C) but without providing a convincing alternative explanation.

2. The work of Bréchignac and co-workers (Sections III.C and VI.C) on the partitioning of $^{18}O_2$ uptake in illuminated *Chondrus crispus* between various oxygenases and oxidases has been extended by Bréchignac and Furbank (1987). Myers (1986) has questioned the interpretation of data concerning the occurrence of cyclic electron transport in cyanobacteria; this could have implications for the red algae.

3. The occurrence of pyrophosphate-dependent 6-phosphofructokinase (an enzyme that is very widespread in green algae and higher plants and that catalyzes a reaction analogous to the "normal" ATP-dependent enzyme, but with a greater potential for reversibility Carnal & Black 1983) has been shown to occur in all six species of marine red algae examined (Dancer 1987; see Section VI).

X. REFERENCES

Aizawa, K. & Miyachi, S. 1986. Carbonic anhydrase and CO_2 concentrating mechanisms in microalgae and cyanobacteria. *FEMS Microbiol. Rev.* 39: 215–33.

Aizawa, K., Nakamura, Y. & Miyachi, S. 1985. Variation of PEPc activity in *Dunaliella* associated with changes in atmospheric CO_2 concentration. *Plant Cell Physiol.* 26: 1199–203.

Akazawa, T. & Osmond, C. B. 1976. Structural properties and ribulose bisphosphate carboxylase and oxygenase activity of fraction-1 protein from the marine alga *Halimeda cylindracea* (Chlorophyta). *Aust. J. Plant Physiol.* 3: 93–103.

Anderson, S. M. & Charters, A. C. 1982. A fluid dynamic study of seawater flow through *Gelidium nudifrons*. *Limnol. Oceanogr.* 27: 399–412.

Appleby, G., Colbeck, J. & Holdsworth, E. S. 1980. β-carboxylation enzymes in marine phytoplankton and isolation and purification of pyruvate carboxylase from *Amphidinium carterae* (Dinophyceae). *J. Phycol.* 16: 260–95.

Ap Rees, T., Bryce, J. H., Wilson, P. M. & Green, J. H. 1983. Role and location of NAD malic enzyme in thermogenic tissues of Araceae. *Arch. Biochem. Biophys.* 227: 511–21.

Badger, M. R. & Andrews, T. J. 1987. Co-evolution of RUBISCO and CO_2 concentrating mechanisms. In *Progress in Photosynthesis Research, III*, ed. J. Biggins, pp. 501–609. Dordrecht: Nijhoff.

Bauman, R. W., Jr. & Jones, B. R. 1986. Electrophysiological investigations of the red alga *Griffithsia pacifica* Kyl. *J. Phycol.* 52: 49–56.

Bean, R. C. & Hassid, Z. 1955. Assimilation of $^{14}CO_2$ by a photosynthesising red alga, *Iridophycus flaccidum*. *J. Biol. Chem.* 212: 411–26.

Beardall, J. & Entwisle, L. 1984. Internal pH of the obligate acidophile *Cyanidium caldarium* Geitler (Rhodophyta?). *Phycologia* 23: 397–9.

Beevers, H. 1970. Respiration in plants and its regulation. In *Prediction and Measurement of Photosynthetic Productivity*, pp. 209–14. Pudoc: Wageningen.

Ben-Amotz, A., Erbes, D. L., Reiderer-Henderson, M. A., Pearey, D. G. & Gibbs, M. 1975. H_2 metabolism in photosynthetic organisms. I. Dark H_2 evolution and uptake by algae and mosses. *Plant Physiol.* 56: 72–7.

Bidwell, R. G. S. & McLachlan, J. 1985. Carbon nutrition of seaweeds: photosynthesis, photorespiration and respiration. *J. Exp. Mar. Biol. Ecol.* 86: 15–46.

Biggins, J. 1973. Kinetic behaviour of cytochrome *f* in cyclic and non-cyclic electron transport in *Porphyridium cruentum*. *Biochemistry* 12: 1165–9.

Biggins, J. 1974. The role of plastoquinone on the *in vivo* photosynthetic cyclic electron transport pathway in algae. *FEBS Letters* 38: 311–14.

Bird, K. T., Dawes, C. J. & Romeo, T. 1980. Patterns of non-photosynthetic CO_2 fixation in dark held, respiring thalli of *Gracilaria verrucosa*. *Z. Pflanzenphysiol.* 98: 359–64.

Bird, K. T., Dawes, C. J. & Romeo, T. 1981. Light quality effects on carbon metabolism and allocation in *Gracilaria verrucosa*. *Mar. Biol.* 64: 219–23.

Bisson, M. A. & Kirst, G. O. 1979. Osmotic adaptation in the marine algae *Griffithsia monilis* (Rhodophyceae): The role of ions and organic compounds. *Aust. J. Plant Physiol.* 6: 523–38.

Black, C. C., Jr., Burris, J. E. & Everson, R. G. 1976. Influence of O_2 concentration on photosynthesis in marine plants. *Aust. J. Plant Physiol.* 3: 81–6.

Borowitzka, M. A. 1984. Calcification in aquatic plants. *Plant Cell Environ.* 7: 457–66.

Bowes, G. & Reiskind, J. B. 1987. Inorganic carbon concentrating mechanisms in an aquatic environment. In *Progress in Photosynthesis Research IV*, ed. J. Biggins, pp. 345–52. Dordrecht: Nijhoff.

Bréchignac, F. & Andre, M. 1984. Oxygen uptake and photosynthesis of the red macroalga, *Chondrus crispus*, in seawater. Effects of light and CO_2 concentration. *Plant Physiol.* 75: 919–23.

Bréchignac, F. & Andre, M. 1985a. Oxygen uptake and photosynthesis of the red macroalga, *Chondrus crispus*, in seawater. Effects of oxygen concentration. *Plant Physiol.* 78: 545–50.

Bréchignac, F. & Andre, M. 1985b. Continuous measurements of the free dissolved CO_2 concentration during photosynthesis of marine plants. Evidence for HCO_3^- use in *Chondrus crispus*. *Plant Physiol.* 78: 551–4.

Bréchignac, F., Andre, M. & Gerbaud, A. 1986. Preferential photosynthetic uptake of exogenous HCO_3^- in the marine macroalgae *Chondrus crispus*. *Plant Physiol.* 80: 1059–62.

Bréchignac, F., Andre, A., Daguenet, A. & Massimino, D. 1983. Mesur en continu des échanges de O_2 et de CO_2 d'un végétal aquatique. *Physiol. Vég.* 21: 665–76.

Bréchignac, F. & Furbank, R. T. 1987. On the nature of the oxygen uptake in the light by *Chondrus crispus*. Effects of inhibitors, temperatures and light intensity. *Photosynth. Res.* 11: 45–59.

Bréchignac, F., Ranger, C., Andre, M., Daguenet, A. & Massimino, D. 1987. Oxygen exchanges in marine algae. In *Progress in Photosynthesis Research, III*, ed. J. Biggins, pp. 657–60. Dordrecht: Nijhoff.

Brody, M. & Emerson, R. 1959. The quantum yield of photosynthesis in *Porphyridium cruentum*, and the role of chlorophyll *a* in the photosynthesis of red algae. *J. Gen. Physiol.* 43: 251–64.

Brown, D. L. & Tregunna, E. B. 1967. Inhibition of respiration during photosynthesis by some algae. *Can. J. Bot.* 45: 1135–43.

Brown, S. B., Holroyd, A. J. & Troxler, R. F. 1980. Mechanism of bile-pigment synthesis in algae. ^{18}O incorporation into phycocyanobilin in the unicellular rhodophyte, *Cyanidium caldarium*. *Biochem. J.* 190: 445–9.

Buggeln, R. G. & Craigie, J. S. 1973. The physiology and biochemistry of *Chondrus crispus* Stackhouse. *Proc. N. S. Inst. Sci.* 27 (Suppl): 81–102.

Burns, B. S. & Beardall, J. 1987. Utilization of inorganic carbon by marine microalgae. *J. Exp. Mar. Biol. Ecol.* 107: 75–86.

Burris, J. E. 1977. Photosynthesis, photorespiration and dark respiration in eight species of algae. *Mar. Biol.* 39: 371–91.

Burris, J. E. 1980. Respiration and photorespiration in marine algae. In *Primary Productivity in the Sea*, ed. P. G. Falkowski, pp. 411–32. New York: Plenum.

Burris, J. E., Holm-Hansen, O. & Black, C., Jr. 1976. Glycine and serine production in marine plants as a measure of photorespiration. *Aust. J. Plant Physiol.* 3: 87–92.

Carnal, N. W. & Black, C. C. 1983. Phosphofructokinase activities in photosynthetic organisms. The occurrence of pyrophosphate-dependent 6-phosphofructokinase in plants and algae. *Plant Physiol.* 71: 150–5.

Chapman, A. R. O. & Ragan, M. A. 1980. Evolution of biochemical pathways: evidence from comparative biochemistry. *Annu. Rev. Plant Physiol.* 31: 639–78.

Colman, B. & Cook, C. M. 1985. Photosynthetic characteristics of the marine macrophytic red alga *Rhodymenia palmata*: evidence for bicarbonate transport. In *Inorganic Carbon Uptake by Aquatic Photosynthetic Organisms*, eds. W. J. Lucas & J. A. Berry, pp. 97–110. Rockwell: American Society of Plant Physiologists.

Colman, B. & Gehl, K. A. 1983. Physiological characteristics of photosynthesis in *Porphyridium cruentum*: evidence for bicarbonate transport in a unicellular red algae. *J. Phycol.* 19: 216–9.

Cook, C. M. & Colman, B. 1987. Some characteristics of photosynthetic inorganic carbon uptake of a marine macrophytic red algae. *Plant Cell Environ.* 10: 275–8.

Cook, C. M., Lanaras, T. & Colman, B. 1986. Evidence for bicarbonate transport in species of red and brown macrophytic marine algae. *J. Exp. Bot.* 37: 977–84.

Coughlan, S. & Tattersfield, D. 1977. Photorespiration in larger littoral algae. *Bot. Mar.* 20: 265–6.

Craigie, J. S. 1963. Dark fixation of ^{14}C-bicarbonate by marine algae. *Can. J. Bot.* 41: 317–25.

Craigie, J. S. 1974. Storage products. In *Algal Physiology and Biochemistry*, ed. W. D. P. Stewart, pp. 206–35. Oxford: Blackwell.

Cram, W. J. 1976. Negative feedback regulation of transport in cells. The maintenance of turgor, volume and nutrient supply. In *Encyclopedia of Plant Physiology, New Series, Vol 2A*, eds. U. Lüttge & M. G. Pitman, pp. 284–316. Berlin: Springer.

Dancer, J. E. 1987. The role of pyrophosphate: fructose-6-phosphate-1-phosphotransferase in plants. Ph.D. thesis, University of Cambridge.

Davison, I. R. & Reed, R. H. 1985. Osmotic adjustment in *Laminaria digitata* (Phaeophyta) with particular reference to seasonal changes in internal solute concentration. *J. Phycol.* 21: 41–50.

De Luca, P. & Moretti, A. 1983. Floridosides in *Cyanidium caldarium*, *Cyanidioschyzon merolae* and *Galderia sulphuraria*. *J. Phycol.* 19: 319–20.

Döhler, G. 1971. Induction phenomena in CO_2 exchange and glycolate metabolism of the blue-green alga *Anacystis* and the red alga *Porphyridium*. In *Proceedings of the Second International Congress on Photosynthetic Research*, eds. G. Forte, M. Avron & A. Melandri, Vol. III, pp. 2071–6. The Hague: W. Junk.

Döhler, G., Burstell, H. & Jilg-Winter, G. 1976. Pigmentzusammensetzung und photosynthetische CO_2-fixierung von *Cyanidium caldarium* und *Porphyridium aerugineum*. *Biochem. Physiol. Plant.* 170: 103–10.

Douce, R. 1986. *Mitochondria in Higher Plants*. Orlando: Academic.

Downton, W. J. S., Bishop, D. G., Larkum, A. W. D. & Osmond, C. B. 1976. Oxygen inhibition of photosynthetic oxygen evolution in marine plants. *Aust. J. Plant Physiol.* 3: 73–9.

Dring, M. J. & Lüning, K. 1985. Emerson enhancement effect and quantum yield of photosynthesis for marine macroalgae in simulated underwater light fields. *Mar. Biol.* 87: 109–17.

Droop, M. R. 1974. Heterotrophy of carbon. In *Algal Physiology and Biochemistry*, ed. W. D. P. Stewart, pp. 530–59. Oxford: Blackwell.

Enami, I. & Kura-Hotta, M. 1984. Effect of intracellular ATP levels on the light induced H^+ efflux from intact cells of *Cyanidium caldarium*. *Plant Cell Physiol.* 25: 1107–13.

Enami, I., Akutsu, H. & Kyagaku, Y. 1986. Intracellular pH regulation in an acidophilic unicellular alga, *Cyanidium caldarium*: ^{31}P nmr determination of intracellular pH. *Plant Cell Physiol.* 27: 1351–9.

Enright, C. T. & Craigie, J. S. 1981. Effects of temperature and irradiance on growth and respiration of *Chondrus crispus* Stackh. *Proc. Int. Seaweed Symp.* 10: 271–6.

Farquhar, G. D. 1983. On the nature of carbon isotope discrimination in C_4 species. *Aust. J. Plant Physiol.* 10: 205–36.

Feige, G. B. 1973. Beitrage zur Physiologie einheimischer Algen. 2. Untersuchungen zur Kinetik der $^{14}CO_2$-assimilation der Süsswasser-rotalge *Lemanea fluviatilis* C. Ag. *Z. Pflanzenphysiol.* 69: 290–2.

Feige, G. B. 1975. Beitrage zur Physiologie einheimischer Algen. 5. Einige Aspekte des photosynthetischer C-metabolismus der Süsswasserrotalge *Audouinella violacea* (Kütz) Hamel. *Z. Pflanzenphysiol.* 75: 339–45.

Fewson, C. A., Al-Hafidh, H. & Gibbs, M. 1962. Role of aldolase in photosynthesis. I. Enzyme studies with photosynthetic organisms with special reference to blue-green algae. *Plant Physiol.* 37: 402–6.

Fogg, G. E. 1976. Release of glycollate from tropical marine plants. *Aust. J. Plant. Physiol.* 3: 57–61.

Ford, T. W. 1979. Ribulose 1,5-bisphosphate carboxylase from the thermophilic, acidophilic algae *Cyanidium caldarium* (Geitler). Purification, characterisation and thermostability of the enzyme. *Biochim. Biophys. Acta* 569: 239–48.

Ford, T. W. 1986. Thermostability of the photosynthetic system of the thermoacidophilic alga *Cyanidium caldarium* in continuous culture. *J. Exp. Bot.* 37: 1698–707.

Fredrick, J. F. & Seckbach, J. 1987. Storage glucan and glucosyltransferase isozymes of *Cyanidioschyzon merolae*: a primitive eukaryote. *Phytochemistry* 25: 363–5.

Frenkel, A. W. & Rieger, C. 1951. Photoreduction in algae. *Nature* (Lond.) 167: 1030.

Furbank, R. T. & Rebeille, F. 1986. Dark respiration in the marine macroalga *Chondrus crispus* (Rhodophyceae). *Planta* (Berl.) 168: 267–72.

Glidewell, S. M. & Raven, J. A. 1975. Measurements of simultaneous oxygen uptake and evolution in *Hydrodictyon africanum*. *J. Exp. Bot.* 26: 479–88.

Glidewell, S. M. & Raven, J. A. 1976. Photorespiration: RuBP oxygenase or hydrogen peroxide? *J. Exp. Bot.* 27: 200–4.

Hackney, J. M. & Hensley, P. 1987. Use of an enzyme assay to detect glycolate in aquatic systems. *Aquat. Bot.* 27: 395–402.

Hewitt, E. J. 1983. A perspective of mineral nutrition: essential and functional minerals in plants. In *Metals and Micronutrients. Uptake and Utilization by Plants*, eds. D. A. Robb & W. S. Pierpoint, pp. 277–323. London: Academic.

Hibberd, D. J. & Norris, R. E. 1984. Cytology and ultrastructure of *Chlorarachnion reptans* (Chlorarachniophyta divisio nova, Chlorarachniophyceae classis nova). *J. Phycol.* 20: 310–30.

Holbrook, G. P., Beer, S., Spencer, W. E., Reiskind, J. B., Davis, J. S. & Bowes, G. 1988. Photosynthesis in marine macroalgae: evidence for carbon limitation. *Can. J. Bot.* 66, 577–82.

Imada, O., Saito, Y. & Maeki, S. 1970. Relationships between the growth of *Porphyra tenera* and its culturing conditions in the sea. II. Influence of atmospheric exposure on photosynthesis, growth and others on *Porphyra* fronds. *Bull. Jap. Soc. Sci. Fish.* 36: 369–76.

Jackson, W. A. & Volk, R. J. 1970. Photorespiration. *Annu. Rev. Plant Physiol.* 21: 385–432.

Johnson, W. S., Gigon, A., Gulman, S. L. & Mooney, H. A. 1974. Comparative photosynthetic capacities of intertidal algae under exposed and submerged conditions. *Ecology* 55: 450–3.

Johnston, A. M. & Raven, J. A. 1986. The analysis of photosynthesis in air and water by *Ascophyllum nodosum* (L.) Le Jol. *Oecologia* (Berl.) 69: 288–95.

Jolivet-Tournier, P. & Gerster, R. 1984. Incorporation of oxygen into glycolate, glycine and serine during photorespiration in maize leaves. *Plant Physiol.* 74: 108–11.

Jones, H. G. & Norton, T. A. 1979. Internal factors controlling the rate of evaporation from fronds of some intertidal algae. *New Phytol.* 83: 771–82.

Kerby, N. W. & Raven, J. A. 1985. Transport and fixation of inorganic carbon by marine algae. *Adv. Bot. Res.* 11: 71–123.

Keys, A. J. 1986. Rubisco: its role in photorespiration. *Phil. Trans. R. Soc. Lond. B* 313: 325–36.

Kirst, G-O. 1981. Photosynthesis and respiration of *Griffithsia monilis* (Rhodophyceae): effect of light, salinity, and oxygen. *Planta* (Berl.) 151: 281–8.

Kost, H. P., Senser, M. & Wanner, G. 1984. Effect of nitrate and sulphate starvation on *Porphyridium cruentum* cells. *Z. Pflanzenphysiol.* 113: 231–49.

Kremer, B. P. 1978a. Studies on $^{14}CO_2$-dissimilation in marine Rhodophyceae. *Mar. Biol.* 48: 47–54.

Kremer, B. P. 1978b. Aspects of CO_2-fixation in some freshwater Rhodophyceae. *Phycologia* 17: 430–4.

Kremer, B. P. 1979. Light independent carbon fixation in macroalgae. *J. Phycol.* 15: 244–7.

Kremer, B. P. 1981a. Aspects of carbon metabolism in marine macroalgae. *Oceanogr. Mar. Biol. Annu. Rev.* 19: 41–94.

Kremer, B. P. 1981b. Carbon metabolism. In *The Biology of Seaweeds*, eds. C. S. Lobban & M. J. Wynne, pp. 493–533. Oxford: Blackwell.

Kremer, B. P. 1983. Carbon economy and nutrition of the alloparasitic red alga *Harveyella mirabilis*. *Mar. Biol.* 76: 321–39.

Kremer, B. P. & Kirst, G. O. 1981. Biosynthesis of 2-O-D-glycerol-α-D-galactopyranoside (floridoside) in marine Rhodophyceae. *Plant Sci. Lett.* 23: 349–57.

Kremer, B. P. & Kirst, G. O. 1982. Biosynthesis of photosynthates and taxonomy of algae. *Z. Naturforsch.* 37C: 761–71.

Kremer, B. P. & Kuppers, U. 1977. Carboxylating enzymes and the pathway of photosynthetic carbon assimilation in different marine algae – evidence for the C_4 pathway? *Planta* (Berl.) 133: 191–6.

Kremer, B. P. & Schmitz, K. 1976. Aspects of $^{14}CO_2$-fixation by endosymbiotic rhodoplasts in the marine opisthobranchiate *Hermaea bifida*. *Mar. Biol.* 34: 313–6.

Kura-Hotta, M. & Enami, I. 1981. Light-induced H^+ efflux from intact cells of *Cyanidium caldarium*. *Plant Cell Physiol.* 22: 1175–84.

Kura-Hotta, M. & Enami, I. 1984. Respiration-dependent H^+ efflux from intact cells of *Cyanidium caldarium*. *Plant Cell Physiol.* 25: 1115–22.

Lapointe, B. E., Dawes, C. J. & Tenore, K. R. 1984a. Interactions between light and temperature on the physiological ecology of *Gracilaria tikvahiae* (Gigartinales, Rhodophyta). II. Nitrate uptake and levels of pigments and chemical constitutents. *Mar. Biol.* 80: 171–8.

Lapointe, B. E., Tenore, K. R. & Dawes, C. J. 1984b. Interactions between light and temperature on the physiological ecology of *Gracilaria tikvahiae* I. Growth, photosynthesis and respiration. *Mar. Biol.* 80: 161–70.

Larkum, A. W. D. & Barrett, J. 1983. Light-harvesting processes in algae. *Adv. Bot. Res.* 10: 1–219.

Laycock, M. V. & Craigie, J. S. 1977. The occurrence and seasonal variation of gigartinine and L-citrullinyl-L-arginine in *Chondrus crispus* Stackh. *Can. J. Biochem.* 55: 27–30.

Laycock, M. V., Morgan, K. C. & Craigie, J. S. 1981. Physiological factors affecting the accumulation of L-citrullinyl-L-arginine in *Chondrus crispus*. *Can. J. Bot.* 59: 522–7.

Leclerc, J. C. 1985. Premières données sur l'activité photosynthetique de quelques algues subaériennes vivant en milieu très peu éclaires. *Can. J. Bot.* 63: 1893–9.

Leclerc, J. C., Döhler, G. & Rosslenbroich, H. J. 1982. $^{14}CO_2$-fixation under various limited light conditions in *Porphyridium cruentum*. *Plant Sci. Lett.* 24: 225–9.

Lilley, R. M. & Larkum, A. W. D. 1981. Isolation of functionally intact rhodoplasts from *Griffithsia monilis* (Ceramiaceae, Rhodophyta). *Plant Physiol.* 67: 5–8.

Littler, M. M., Littler, D. S., Blair, S. M. & Norris, J. N. 1985. Deepest known plant life discovered on an uncharted seamount. *Science* 227: 57–9.

Lloyd, D. 1974a. Dark respiration. In *Algal Physiology and Biochemistry*, ed. W. D. P. Stewart, pp. 505–29. Oxford: Blackwell.

Lloyd, D. 1974b. *The Mitochondria of Microorganisms*. London: Academic.

Lloyd, N. D. H., Canvin, D. T. & Culver, D. A. 1977. Photosynthesis and photorespiration in algae. *Plant Physiol.* 59: 936–40.

Lorimer, G. H. 1981. The carboxylation and oxygenation of ribulose bisphosphate: the primary events in photosynthesis and photorespiration. *Annu. Rev. Plant Physiol.* 32: 349–83.

Lorimer, G. H., Badger, M. R. & Andrews, T. J. 1977. D-ribulose-1,5-bisphosphate carboxylase-oxygenase. Improved methods for the activation and assay of catalytic activites. *Analyt. Biochem.* 78: 66–75.

Ludwig, L. J. & Canvin, D. T. 1971. An open gas exchange system for the simultaneous measurement of the CO_2 and $^{14}CO_2$ fluxes from leaves. *Can. J. Bot.* 49: 1299–313.

Lüning, K. & Dring, M. J. 1985. Action spectra and spectral quantum yields of photosynthesis in marine macroalgae with thin and thick thalli. *Mar. Biol.* 87: 119–29.

MacFarlane, J. J. & Raven, J. A. 1985. External and internal CO_2 transport in *Lemanea*: interactions with the kinetics of ribulose bisphosphate carboxylase. *J. Exp. Bot.* 36: 610–22.

McCracken, D. A. & Cain, J. R. 1981. Amylose in floridean starch. *New Phytol.* 88: 67–71.

Macler, B. A. 1986. Regulation of carbon flow by nitrogen and light in the red alga, *Gelidium coulteri*. *Plant Physiol.* 82: 136–41.

Maxwell, P. C. & Biggins, J. 1976. Role of cyclic electron transport in photosynthesis as measured by the photoinduced turnover of P_{700} in vivo. *Biochemistry* 15: 3975–81.

Meatyard, B. T., Scawen, M. D., Ramshaw, J. A. M. & Boulter, D. 1975. Cytochrome cs from *Rhodymenia palmata* and *Porphyra umbilicalis* and the amino acid sequence of their N-terminal regions. *Phytochemistry* 14: 1493–7.

Merola, A., Astaldo, R., Deluca, P., Gombardella, R., Musacchio, A. & Taddei, R. 1981. Revision of *Cyanidium caldarium*. Three species of acidophilic algae. *Giorn. Bot. Ital.* 115: 189–95.

Millard, P. & Evans, L. T. 1982. Sulphate uptake in the unicellular marine red alga *Rhodella maculata*. *Arch. Microbiol.* 131: 165–9.

Morel, F. M. M. & Morel-Laurens, N. M. L. 1981. Trace metals and plankton in the oceans: facts and speculations. In *Trace Elements in Sea Water*, eds. C. S. Wong, E. Boyle, K. W. Bruland, J. D. Brunton & E. D. Goldberg, pp. 841–69. New York, Plenum.

Moroney, J. V., Wilson, B. J. & Tolbert, W. E. 1986. Glycolate metabolism and excretion by *Chlamydomonas reinhardtii*. *Plant Physiol.* 82: 821–6.

Moyse, A. 1961. The products of photosynthesis in *Rhodosorus marinus*. Effect of quantity of energy converted with different wavelengths of radiation. *Coll. Int. Centr. Natl. Rech. Sci.* 103: 69–82.

Myers, J. 1986. Photosynthetic and respiratory electron transport in a cyanobacterium. *Photosynth. Res.* 9: 135–147.

Nagashima, H., Nakamura, S. & Nisizawa, K. 1968. Biosynthesis of floridean starch by chloroplast preparations from a marine red alga, *Serraticardia maxima*. *Bot. Mag.* 81: 411–13.

Nagashima, H., Ozaki, H., Nakamira, S. & Nisizawa, K.

1969. Physiological studies on floridean starch, floridoside and trehalose in a red alga, *Serraticardia maxima*. Bot. Mag. 82: 462–73.

Nagashima, H., Nakamyra, S., Nisizawa, K. & Hori, T. 1971. Enzymatic synthesis of floridean starch in a red alga, *Serraticardia maxima*. Plant Cell. Physiol. 12: 243–53.

Nagashimi, H. & Fukuda, I. 1981. Low molecular weight carbohydrates in *Cyanidium caldarium* and some related algae. Phytochemistry 20: 439–42.

Nagashimi, H. & Fukuda, I. 1983. Floridosides in unicellular hot spring algae. Phytochemistry 22: 1949–57.

Neish, A. C., Shacklock, P. F., Fox, C. H. & Simpson, F. J. 1977. The cultivation of *Chondrus crispus*. Factors affecting growth under greenhouse conditions. Can. J. Bot. 55: 2263–71.

Newell, R. C. & Pye, V. J. 1968. Seasonal variations in the effect of temperature on the respiration of certain intertidal algae. J. Mar. Biol. Ass. U.K. 48: 341–8.

Newman, S. M. & Cattolico, R. A. 1987. Structural and functional relatedness of chromophytic and rhodophytic RuBP carboxylase enzymes. In *Progress in Photosynthesis Research, IV*, ed. J. Biggins, pp. 671–4. Dordrecht: Nijhoff.

Nisizawa, K. 1978. Aldolases in multicellular marine algae. In *Handbook of Phycological Methods: Physiological and Biochemical Methods*, eds. J. A. Hellebust & J. S. Craigie, pp. 239–44. Cambridge: Cambridge University Press.

Oates, B. R. 1986. Components of photosynthesis in the intertidal saccate algae *Halosaccion americanum* (Rhodophyta, Palmariales). J. Phycol. 22: 217–23.

O'Leary, M. H. 1981. Carbon isotope fractionation in plants. Phytochemistry 20: 553–68.

Pentecost, A. 1985. Photosynthetic plants as intermediary agents between environmental HCO_3^- and carbonate assimilation. In *Inorganic Carbon Uptake by Aquatic Photosynthetic Organisms*, eds. W. J. Lucas & J. A. Berry, pp. 459–80. Rockwell: American Society of Plant Physiologists.

Portis, A. E., Jr., Salvucci, M. E. & Ogren, W. L. 1986. Activation of ribulose bisphosphate carboxylase/oxygenase at physiological CO_2 and ribulose bisphosphate concentrations by rubisco activase. Plant Physiol. 82: 967–71.

Portis, A. E., Jr., Salvucci, M. E., Ogren, W. L. & Wernecke, J. A. 1987. Rubisco activase: a new enzyme in the regulation of photosynthesis. In *Progress in Photosynthesis Research. III*, ed. J. Biggins, pp. 371–9. Dordrecht: Nijhoff.

Price, G. D. & Badger, M. R. 1985. Inhibition by proton buffers of photosynthetic use of bicarbonate in *Chara corallina*. Austr. J. Plant Physiol. 12: 257–67.

Price, G. D., Badger, M. R., Bassett, M. E. & Whitecross, M. I. 1985. Involvement of plasmalemmasomes and carbonic anydrase in photosynthetic ulitization of bicarbonate in *Chara corallina*. Aust. J. Plant Physiol. 12: 241–56.

Randall, D. D. 1976. Phosphoglycollate phosphatase in marine algae: Isolation and characterisation from *Halimeda cylindracea*. Aust. J. Plant Physiol. 3: 105–11.

Raven, J. A. 1972a. Endogenous inorganic carbon sources in plant photosynthesis. I. Occurrence of the dark respiratory pathways in illuminated green cells. New Phytol. 71: 227–47.

Raven, J. A. 1972b. Endogenous inorganic carbon sources in plant photosynthesis. II. Comparison of total CO_2 production with measured CO_2 evolution in the light. New Phytol. 71: 995–1014.

Raven, J. A. 1974a. Photosynthetic electron flow and photophosphorylation. In *Algal Physiology and Biochemistry*, ed. W. D. P. Stewart, pp. 391–423. Oxford: Blackwell.

Raven, J. A. 1974b. Carbon dioxide fixation. In *Algal Physiology and Biochemistry*, ed. W. D. P. Stewart, pp. 434–55. Oxford: Blackwell.

Raven, J. A. 1976a. Division of labour between chloroplasts and cytoplasm. In *The Intact Chloroplast*, ed. J. Barber, pp. 403–43. Amsterdam: Elsevier.

Raven, J. A. 1976b. The rate of cyclic and non-cyclic photophosphorylation and oxidative phosphorylation, and regulation of the rate of ATP consumption in *Hydrodictyon africanum*. New Phytol. 76: 205–12.

Raven, J. A. 1976c. The quantitative role of "dark" respiratory processes in heterotrophic and photolithotrophic plant growth. Ann. Bot. 40: 587–602.

Raven, J. A. 1981. Nutritional strategies of submer ed benthic plants: the acquisition of C, N and P by rnizophytes and haptophytes. New Phytol. 88: 1–30.

Raven, J. A. 1984a. *Energetics and Transport in Aquatic Plants*. New York: A. R. Liss.

Raven, J. A. 1984b. A cost–benefit analysis of photon absorption by photosynthetic unicells. New Phytol. 98: 593–625.

Raven, J. A. 1984c. Physiological correlates of the morphology of early vascular plants. Bot. J. Linn. Soc. 88: 105–26.

Raven, J. A. 1985. Regulation of pH and generation of osmolarity in vascular land plants: costs and benefits in relation to efficiency of use of water, energy and nitrogen. New Phytol. 101: 25–77.

Raven, J. A. 1986. Evolution of plant life forms. In *On the Economy of Plant Form and Function*, ed. T. Givnish, pp. 421–92. New York: Cambridge University Press.

Raven, J. A. 1987a. Biochemistry, biophysics and physiology of chlorophyll *b*–containing algae: implications for taxonomy and phylogeny. Prog. Phycol. Res. 5: 1–121.

Raven, J. A. 1987b. Limits to growth. In *Microalgal Biotechnology*, eds. M. A. Borowitzka & L. J. Borowitzka, pp. 311–56. Cambridge: Cambridge University Press.

Raven, J. A. 1987c. The application of mass spectrometry to biochemical and physiological studies. In *The Biochemistry of Plants*, Vol. 13, ed. D. D. Davies, pp. 127–79. New York: Academic.

Raven, J. A. 1987d. The role of vacuoles. New Phytol. 106: 357–422.

Raven, J. A. 1988. The iron and molybdenum use efficiencies of plant growth with different energy, carbon and nitrogen sources. New Phytol. 109: 279–87.

Raven, J. A. & Beardall, J. 1981a. Carbon dioxide as the exogenous inorganic carbon source for *Batrachospermum* and *Lemanea*. Br. Phycol. J. 16: 165–75.

Raven, J. A. & Beardall, J. 1981b. Respiration and photorespiration. In *Physiological Bases of Phytoplankton Ecology*, ed. T. Platt, pp. 52–82. Can. Fish. Aq. Sci. Bull. No. 210.

Raven, J. A., Beardall, J. & Griffiths, H. 1982. Inorganic C-sources for *Lemanea*, *Cladophora* and *Ranunculus* in a fast-flowing stream: measurements of gas exchange and of carbon isotope ratio and their ecological implications. Oecologia (Berl.) 53: 68–78.

Raven, J. A. & Glidewell, S. M. 1981. Processes limiting photosynthetic conductance. In *Physiological Processes Limiting Plant Productivity*. ed. C. B. Johnson, pp. 109–36. London: Butterworth.

Raven, J. A. & Lucas, W. J. 1985. The energetics of carbon acquisition. In *Inorganic Carbon Uptake by Aquatic Photosynthetic Organisms*, eds. W. J. Lucas & J. A. Berry, pp. 305–24. Rockwell: American Society of Plant Physiologists.

Raven, J. A., MacFarlane, J. J. & Griffiths, H. 1987a. The application of carbon isotope discrimination techniques. In *Plant Life in Aquatic and Amphibious Habitats*, ed. R. M. M. Crawford, pp. 129–49. Oxford: Blackwell.

Raven, J. A., Johnston, A. M., MacFarlane, J. J., bin Surif, M. & McInroy, S. 1987b. Diffusion and active transport of inorganic carbon species in freshwater and marine macroalgae. In *Progress in Photosynthesis Research, IV*, ed. J. Biggins, pp. 333–40. Dordrecht: Nijhoff.

Raven, J. A., MacFarlane, J. J., Johnston, A. M., bin Surif, M. & McInroy, S. 1990. Inorganic carbon transport in relation to habitat, and resource use efficiency, in the macro-algae *Lemanea mamillosa* (Rhodophyta) and *Ascophyllum nodosum* (Phaeophyta). In *Transport Across Membranes in Plants and Fungi*, eds. M. J. Beilly, N. A. Walker & J. R. Smith, pp. 338–42. Sydney: Sydney University Press.

Raven, J. A., Osborne, B. A. & Johnston, A. M. 1985. Uptake of CO_2 by aquatic vegetation. *Plant Cell Environ.* 8: 417–25.

Raven, J. A. & Samuelsson, G. 1986. Repair of photoinhibitory damage in *Anacystis nidulans* 625 (*Synechococcus* 6301); relation to catalytic capacity for, and energy supply to, protein synthesis, and implications for μ_{max} and the efficiency of light-limited growth. *New Phytol.* 103: 625–43.

Reed, R. H. 1983. Taxonomic implications of osmoacclimation in *Cyanidium caldarium* (Tilden) Geitler. *Phycologia* 22: 351–4.

Reith, M. E. & Cattolico, R. A. 1985. *In vivo* chloroplast protein synthesis by the chromophytic alga, *Olisthodiscus luteus*. *Biochemistry* 24: 2556–61.

Reith, M. E. & Cattolico, R. A. 1986. Inverted repeat of *Olisthodiscus luteus* chloroplast DNA contains genes for both subunits of ribulose-1,5-bisposphate carboxylase and the 32,000-dalton Q_B protein: phylogenetic implications. *Proc. Natl. Acad. Sci. USA* 83: 8599–603.

Robbins, J. V. 1979. Effects of physical and chemical factors on photosynthesis and respiration rates of *Palmaria palmata* (Florideophyceae). *Proc. Int. Seaweed Symp.* 9: 273–83.

Robinson, S. P. & Jones, G. P. 1986. Accumulation of glycinebetaine in chloroplasts provides osmotic adjustment during salt stress. *Aust. J. Plant Physiol.* 13: 659–68.

Rothschild, L. J. & Heywood, P. 1987. Protistan phylogeny and chloroplast evolution: conflicts and congruence. *Prog. Protistol.* 2: 1–68.

Salvucci, M. E., Wernecke, J. A., Ogren, W. C. & Portis, A. R., Jr. 1986. Rubisco activase: purification, subunit composition and species distribution. In *Progress in Photosynthesis Research, III*, ed. J. Biggins, pp. 379–82. Dordrecht: Nijhoff.

Sand-Jensen, K. & Gordon, D. M. 1984. Differential ability of marine and freshwater macrophytes to utilize HCO_3^- and CO_2. *Mar. Biol.* 80: 247–53.

Sandmann, G. 1985. Photosynthetic and respiratory electron transport in Ca^{2+}-deficient *Dunaliella*. *Physiol. Plant.* 65: 481–6.

Seckbach, J., Gross, H. & Nathan, M. B. 1971. Growth and photosynthesis of *Cyanidium caldarium* cultured under pure CO_2. *Israeli J. Bot.* 20: 84–90.

Seeman, J. R., Badger, M. R. & Berry, J. A. 1984. Variations in the specific activity of ribulose-1,5-bisphosphate carboxylase between species using different photosynthetic pathways. *Plant Physiol.* 74: 791–4.

Servaites, J. C., Parry, M. A. J., Gutteridge, S. & Keys, A. J. 1986. Species variation in the predawn inhibition of ribulose-1,5-biphosphate carboxylase/oxygenase. *Plant Physiol.* 82: 1161–3.

Setlik, I., Nedbal, L., Masojidek, J. & Setlikova, E. 1984. Irradiance dependence changes in photosystem 2 caused by chloramphenicol and uncouplers in photosynthesising cells. In *Advances in Photosynthetic Research* Vol. III, ed. C. Sybesma, pp. 259–62. The Hague: Nijhoff/Junk.

Setlikova, E., Masojidek, J., Nedbal, L. & Setlik, I. 1984. The irradiance dependent control of the Q_B-polypeptide turnover is a widespread phenomenon in oxygenic photosynthesis. In *Advances in Photosynthetic Research*, Vol. III, ed. C. Sybesma, pp. 255–8. The Hague: Nijhoff/Junk.

Sharkey, T. D., Seeman, J. R. & Pearcy, R. W. 1986. Contribution of metabolites of photosynthesis to postillumination CO_2 assimilation in response to light-flecks. *Plant Physiol.* 82: 1063–9.

Sheath, R. G., Hellebust, J. A. & Sawa, T. 1979a. Floridean starch metabolism of *Porphyridium purpureum* (Rhodophyta). I. Changes during ageing of batch culture *Phycologia* 18: 149–63.

Sheath, R. G., Hellebust, J. A. and Sawa, T. 1979b. Floridean starch metabolism of *Porphyridium purpureum* (Rhodophyta). II. Changes during the cell cycle. *Phycologia* 18: 185–90.

Sheath, R. G., Hellebust, J. A. & Sawa, T. 1979c. Effects of low light and darkness on structural transformations in plastids of the Rhodophyta. *Phycologia* 18: 1–12.

Sheath, R. G., Hellebust, J. A. & Sawa, T. 1981. Floridean starch metabolism of *Porphyridium purpureum* (Rhodophyta). III. Effects of darkness and metabolic inhibitors. *Phycologia.* 20: 22–31.

Smith, R. G. & Bidwell, R. G. S. 1987. Carbonic anhydrase-dependent inorganic carbon uptake by the red macroalga, *Chondrus crispus*. *Plant Physiol.* 83: 735–8.

Smith, R. G. & Bidwell, R. G. S. 1989. Mechanism of photosynthetic carbon dioxide uptake by the red macroalga, *Chondrus crispus*. *Plant Physiol.* 89: 93–8.

Speer, H. L. & Jones, R. F. 1964. Studies on the respiration of whole cells and cell-free preparations of the red alga *Porphyridium cruentum*. *Physiol. Plant.* 17: 287–98.

Steinmuller, K., Kaling, M. & Zetsche, K. 1983. In vitro synthesis of phycobiliproteins and ribulose-1,5-bisphosphate carboxylase by non-polyadenylated-RNA of *Cyanidium caldarium* and *Porphyridium aerugineum*. *Planta* (Berl.) 159: 308–13.

Stewart, K. D. & Mattox, K. R. 1980. Phylogeny of phytoflagellates. In *Phytoflagellates* ed. E. R. Cox, pp. 433–62. New York: Elsevier/North Holland.

Stewart, K. D. & Mattox, K. R. 1984. The case for a polyphyletic origin of mitochondria: morphological and molecular comparisons. *J. Mol. Evol.* 21: 54–7.

Sun, E., Barr, R. & Carne, F. L. 1968. Comparative studies of plastoquinones. IV. Plastoquinones in algae. *Plant Physiol.* 43: 1935–40.

Suzuki, K. & Ikawa, T. 1985. Effect of oxygen on photosynthetic $^{14}CO_2$ fixation in *Chroomonas* sp. (Cryptophyta). III. Effect of oxygen on photosynthetic carbon metabolism. *Plant Cell Physiol.* 26: 1003–10.

Tajiri, S. & Aruga, Y. 1984. Effect of emersion on the growth and photosynthesis of the *Porphyra yezoensis* thallus. *Jap. J. Phycol.* 32: 134–46.

Tolbert, N. E. 1976. Glycollate oxidase and glycollate dehydrogenase in marine plants and algae. *Aust. J. Plant Physiol.* 3: 129–32.

Troxler, R. F., Foster, J. A., Brown, A. S. & Fraxblau, C. 1975. The α and β subunits of *Cyanidium caldarium* phycocyanin: properties and amino acid sequences of the amino terminus. *Biochemistry* 14: 268–94.

Turner, R. D. & Evans, L. T. 1986. Structural and physiological aspects of translocation in the red alga *Delesseria*. *Br. Phycol. J.* 21: 338.

Van der Velde, H. H., Guiking, P. & Van der Wulp, D. 1975. Glucose-6-phosphate dehydrogenase and 6-phosphogluconate dehydrogenase in *Acrochaetium deviesii* cultured under red, white and blue light. *Z. Pflanzenphysiol.* 76: 95–108.

Walker, N. A., Smith, F. A. & Cathers, I. R. 1980. Bicarbonate assimilation by freshwater charophytes and higher plants. I. Membrane transport of bicarbonate ions is not proven. *J. Memb. Biol.* 57: 51–8.

Weinstein, J. D., Mayer, S. M. & Beale, S. I. 1987a. Formation of γ-aminolaevulinic acid from glutamic acid in algal extracts: fractionation of activities and biological constraints on the RNA requirement. In *Progress in Photosynthesis Research*, IV, ed. J. Biggins, pp. 435–8. Dordrecht: Nijhoff.

Weinstein, J. D., Schneegurt, M. A. & Beale, S. I. 1987b. Biosynthetic precursors of γ-amino laevulinic acid in plants and algae. In *Progress in Photosynthesis Research*, IV, ed. J. Biggins, pp. 431–4. Dordrecht: Nijhoff.

Whittenbury, R. & Kelly, D. R. 1977. Autotrophy: a conceptual phoenix. *Symp. Soc. Gen. Microbiol.* 27: 121–49.

Wilcox, L. W. & Wedermayer, G. J. 1985. Dinoflagellate with blue-green chloroplasts derived from an endosymbiotic eukaryote. *Science* 227: 192–4.

Yagawa, Y., Muto, S. & Miyachi, S. 1987. Carbonic anhydrase of a unicellular red alga *Porphyridium cruentum* R-1. I. Purification and properties of the enzyme. *Plant Cell Physiol.* 28: 1253–62.

Yocum, C. S. & Blinks, L. R. 1954. Photosynthetic efficiency of marine plants. *J. Gen. Physiol.* 38: 1–16.

Yokota, A. & Kitasato, S. 1985. Correct pK values for dissociation constant of carbonic acid lower the reported K_m values of ribulose bisphosphate carboxylase to half. Presentation of a nomogram and an equation for determining the pK values. *Biochem. Biophys. Res. Communs.* 131: 1075–9.

Zenvirth, D., Volokita, M. & Kaplan, A. 1985. Photosynthesis and inorganic carbon accumulation in the acidophilic alga *Cyanidioschyzon merolae*. *Plant Physiol.* 77: 237–9.

Ziegler, H. & Ziegler, I. 1967. Die Lichtinduzierte aktivitatssteigerung der $NADP^+$-abhangigen glycerinaldehyde-3-phosphate dehydrogenase. V. Das verhaltern von Meersesalgen. *Planta* (Berl.) 72: 162–9.

Note added in proof: Work mentioned as MacFarlane and Raven (unpubl.) is now found in the following two references:

MacFarlane, J. J. & Raven, J. A. 1989. Quantitative determination of the unstirred layer permeability in *Lemanea mamillosa*. *J. Exp. Bot.* 40: 321–7.

MacFarlane, J. J. & Raven, J. A. 1990. C, N and P nutrition of *Lemanea mamillosa* Kütz. (Batrachospermales, Rhodophyta) in the Dighty Burn, Angus, Scotland. *Plant Cell Envir.* 13: 1–13.

Chapter 9

Pigmentation and photoacclimation

ELISABETH GANTT

CONTENTS

I. Introduction / 203
II. Irradiance levels / 204
III. Photoautotrophy and heterotrophy / 205
IV. Photosystems / 206
V. Photosynthetic pigments / 207
VI. Phycobilisomes / 210
VII. Energy distribution / 211
VIII. Photoinhibition / 212
IX. Photoregulation / 213
    A. Light quality / 213
    B. Light intensity acclimation / 214
X. Chloroplast genes, translation, and transcription / 215
XI. Summary / 216
XII. Acknowledgments / 217
XIII. References / 217

I. INTRODUCTION

Absorbing light and converting it to chemical energy through photosynthesis constitute a basic function of photosynthetic organisms. This chapter will review some of the background on the light-gathering systems in red algae and is intended to provide a framework for studies leading to a better understanding of light acclimation processes in this algal group. "Acclimation", rather than "adaptation", is the preferred term. As used in this chapter, acclimation is an expression of the adjustments that an organism can make to its environment within the limits of its genome, whereas alteration in the genome is more accurately designated as adaptation.

Most studies concerned with acclimation processes have been at the whole plant level, but presently we are in a transition where the holistic approach is being integrated with studies at the cellular and biochemical level. Following the exhaustive and lucid exposition of holistic aspects of light capture and utilization by Ramus (1981) and the molecular bases for light absorption in phytoplankton with implications for oceanography by Prezelin and Boczar (1986), this chapter in part serves as a continuation but will be concerned mostly with the latest studies on red algae. The major emphasis will be on the biochemical and physiological aspects, with particular emphasis given to studies on the unicellular red alga *Porphyridium purpureum* (*P. cruentum*) (see Gantt 1989), one of the most suitable red algae for laboratory studies.

Results from controlled experiments, which are more easily carried out in the laboratory, can enhance our understanding of acclimation processes in the natural environment.

Red algae, along with all other oxygen-evolving organisms, have two photosystems. However, the photosynthetic apparati of red algae and cyanophytes are unique in that they contain phycobilisomes as light-harvesting antennae and do not have stacked or fused thylakoids. Cryptophytes, although possessing phycobiliproteins, do not contain phycobilisomes (Gantt 1980; MacColl & Guard-Friar 1987). The red algae and cyanophytes do not appear to have a sequestration of photosystem I and II into certain regions of the thylakoids as is found in plants, where photosystem I is primarily in the stroma lamellae and photosystem II in the stacked grana regions (Anderson & Barrett 1986; Gantt 1986; Staehelin 1986).

Under field or laboratory conditions, most algal classes respond to low irradiance levels by increasing their photosynthetic pigment level. However, relatively little is known about the level at which this occurs; i.e., are there changes in the number and/or size of the photosystems or of the antennae complexes? Information on red algae is particularly scarce. Characteristic responses of all algal classes to light environments have been effectively summarized by Larkum and Barrett (1983), Prezelin and Boczar (1986), and Ramus (1981). Recent reviews by Glazer (1985), Moerschel and Rhiel (1987), and Zilinskas and Greenwald (1986) on detailed phycobilisome structure and composition are available. The biochemistry and synthesis of phycobilins is covered by Troxler (1986), and the molecular genetics by Bryant et al. (1987, 1988) and Grossman et al. (1986). A book with a historical perspective by MacColl and Guard-Friar (1987) encompasses all aspects of phycobiliprotein characteristics, including a comprehensible treatment on energy transfer mechanisms. A comparative approach of the light-harvesting systems, including higher plants and algae, has been followed by Anderson and Barrett (1986) and Bryant (1986).

II. IRRADIANCE LEVELS

Red algae can exist at a variable range of irradiance, but many species grow at relatively low light levels (Table 9-1). Exceptions to this are some of the intertidal algae, which can survive at full sunlight intensities (2000 $\mu E\, m^{-2}\, s^{-1}$) (Hodgson 1981). It is of interest to compare some of the physiological expressions by which photosynthesis is defined, especially since it has sometimes been assumed that red algae have lower photosynthetic activity than green plants and some other algae. A comparison of the compensation points, which is the irradiance level at which a net gain of photosynthesis occurs, shows that the range is relatively narrow (Table

Table 9-1. Compensation points of red algae grown under various growth conditions[a]

Species	Compensation level ($\mu E\, m^{-2}\, s^{-1}$)	Reference
Antarctic collection		
Gigartina apoda	14	Drew 1977
Iridaea obovata	28	Drew 1977
Leptosarca simplex	24	Drew 1977
Porphyra umbilicalis	9	Drew 1977
Laboratory-acclimated cultures		
Gracilaria foliifera	30	Rosenberg & Ramus 1982
Gracilaria tikvahiae		
Winter plants at 15°C	10	Penniman & Mathieson 1985
Winter plants at 25°C	8	Penniman & Mathieson 1985
Summer plants at 15°C	8	Penniman & Mathieson 1985
Summer plants at 25°C	8	Penniman & Mathieson 1985
Porphyra leucosticta	15	Sheath et al. 1977
Porphyridium purpureum		
Grown at 180 $\mu E\, m^{-2}\, s^{-1}$	20	Levy & Gantt 1988
Grown at 35 $\mu E\, m^{-2}\, s^{-1}$	6	Levy & Gantt 1988
Grown at 10 $\mu E\, m^{-2}\, s^{-1}$	4	Levy & Gantt 1988

[a] See also Table 15-1.

9-1), whether measured by oxygen evolved or carbon dioxide fixed. Determinations made on algae in the Antarctic (Drew 1977) are very similar to those made in laboratory cultures. Species differences are no doubt responsible for some of the variation, as are temperature and nutrient conditions under which the alga is grown. However, the irradiance level to which the organisms are subjected is certainly a very significant factor. This is clearly illustrated in *Porphyridium purpureum*, where the compensation point of cultures grown for many generations at high irradiance (180 $\mu E\, m^{-2} s^{-1}$) was five times greater than that of cells grown at low irradiance (10 $\mu E\, m^{-2} s^{-1}$) for the same time period. The compensation points and the light saturation points are comparable to many other algae (Richardson et al. 1983), indicating that reds have similar physiologically limiting mechanisms and are not better suited for growing at lower irradiance levels than are any other algal class. The photon flux density required for saturation of photosynthesis, like the compensation point, also depends on the growth history of the organism and, again, is most convincingly illustrated with results from laboratory cultures of *P. purpureum* (Table 9-2). The results clearly indicate acclimation to irradiance alone.

III. PHOTOAUTOTROPHY AND HETEROTROPHY

Nutrient metabolism has the potential to alter photosynthetic characteristics, especially if the organism is not a photoautotroph. Heterotrophic growth in red algae is not commonly reported, yet the capacity to grow heterotrophically would be clearly advantageous to algae, especially during the winter months. In fact, survival and growth during periods of prolonged darkness and very low light have been reported for several red algae. Growth of *Porphyridium purpureum* in 0.5 M glycerol was noted in darkness with a significant increase in cell number (Cheng & Antia 1970). Interestingly, the photosynthetic competency does not appear to be diminished after long dark periods. The same cultures resumed growth within an hour after illumination following 8 weeks of darkness in a seawater medium (Bisalputra & Antia 1980). Even pigment synthesis is possible in darkness, although on a limited level, as noted by Sheath et al. (1977). Thylakoid structural rearrangement was noted by Sheath et al. (1977) and by Lüning and Schmitz (1988) in algae that had been in darkness for extended periods. In *Delessaria*, phycobiliproteins persisted in dark-grown blades, but they were not present as phycobilisomes (Lüning & Schmitz 1988). Chlorophyll was not detected in such blades, nor was photosynthetic activity. In cyanobacteria, on the other hand, retention of full photosynthetic capacity in darkness is not uncommon. Certain cyanobacteria that grow best under photoautotrophic conditions can form a fully competent photosynthetic apparatus including phycobilisomes in complete darkness (Ohki & Gantt 1983).

The most unusual condition thus far reported is of an unclassified crustose red alga that was collected on a sea mount in the Bahamas at 268 m (Littler et al. 1985). The alga existed at calculated irradiance level of approximately 0.008 $\mu E\, m^{-2} s^{-1}$,

Table 9-2. Saturating irradiance of red algae grown at various light intensities[a]

Species	Saturating irradiance level ($\mu E\, m^{-2}\, s^{-1}$)	Reference
Gymnogongrus flabelliformis		
Sunny high intertidal	350[b]	Tseng et al. 1981
Shaded high intertidal	130[b]	
Gracilaria foliifera	200	Rosenberg & Ramus 1982
Gastroclonium subarticulatum	120	Hodgson 1981
Grown at 2000 $\mu E\, m^{-2} s^{-1}$		
Porphyra yezoensis	130[b]	Kato & Aruga 1984
Grown at 180 $\mu E\, m^{-2} s^{-1}$		
Porphyridium purpureum		
Grown at 180 $\mu E\, m^{-2} s^{-1}$	240	Levy & Gantt 1988
Grown at 35 $\mu E\, m^{-2} s^{-1}$	175	Levy & Gantt 1988
Grown at 10 $\mu E\, m^{-2} s^{-1}$	110	Levy & Gantt 1988

[a] See also Table 15-1.
[b] Calculated from published curves.

many orders of magnitude lower than any other red algal species, and at a level where photosynthetic activity would not provide sufficient energy for growth. The alga was photosynthetically fully competent when measured at 20 $\mu E\,m^{-2}\,s^{-1}$, suggesting that it had a fully functional photosynthetic apparatus. Pigment studies on this alga have not been made, and it is not known if there are modifications of photosynthetic size or number or if the alga has heterotrophic capability. It would be extremely interesting to test the heterotrophic capacity of this red alga, especially since sponges, obvious nonphotoautotrophs, grew adjacent to the red alga on the same sea mount and extended to even greater depths.

If *Cyanidium caldarium* is indeed a red alga, then it provides the best example of heterotrophic growth in the red algae. It is unusual in that it normally grows at high temperatures (up to 55°C) and acidic conditions (pH 2–4). It can grow in darkness with glucose as the carbon source. Under these conditions it produces only traces of chlorophyll and does not produce phycocyanin and allophycocyanin (cf. Bogorad 1975). The heterotrophic capability and the existence of a number of experimentally useful pigment mutants have made it the model system for studies on phycobilin biosynthesis in plants (Beale & Chen 1983; Belford et al. 1983; Troxler 1986).

IV. PHOTOSYSTEMS

The functional relationship of the two photosystems (I and II) is better understood than is their structural relationship. Typically, the two photosystems are envisioned as being interconnected via the electron transport chain. Each photosystem has its own reaction center and its own set of antennae pigments. For the proper interaction of the photosystems in a specific time-frame, their spatial arrangement within the thylakoid is important, but it is as yet not known. Photosystem I (PSI) has P700 as its reaction center pigment, and it also has many chlorophyll *a* molecules that serve as antennae pigments. Photosystem II (PSII) has P680 as its reaction center pigment and only a few chlorophyll *a* molecules. Red algae typically have considerably fewer chlorophyll *a* molecules associated with PSII, unlike green plants, where most of the chlorophyll (*a* and *b*) is associated with this photosystem (Anderson & Barrett 1986; Larkum & Barrett 1983; Staehelin 1986). Phycobilisomes in red algae and in cyanophytes serve the major light-harvesting function for PSII. Determination of P680 requires specialized equipment that is not routinely available, and thus is usually not directly determined. In many laboratories P700 is routinely ascertained by spectrophotometric methods and is frequently used as a biochemical expression of the photosynthetic unit size (PSU) (Larkum & Barrett 1983). The PSU is simply the ratio of total chlorophyll molecules to P700. There is also a functional definition of PSU, which is expressed as the ratio of chlorophyll molecules to O_2 molecules evolved during photosynthesis with a fully saturating light flash. The definitions are only comparable if it is assumed that all chlorophyll molecules in the photosystems participate fully and cooperatively in the reaction. The values reported, though variable, fall within a relatively narrow range. For example, in *Porphyridium purpureum* the photosynthetic unit size is 100–120 chlorophylls/P700 by oxidized vs. reduced absorbance change measurements (Kawamura et al. 1979; Levy & Gantt 1988), and 140 chlorophyll/P700 for *Agardhiella subulata* (as *Neoagardhiella baileyei*) (Kursar & Alberte 1983). Determinations of P700 made by measuring flash-induced absorbance changes yielded values in *P. purpureum* of about 200 chlorophylls/P700 (Clement-Metral et al. 1985). The values for red algae are relatively low compared to higher plants and some other algae, which may contain as many as 500 to 1000 chlorophylls/P700 (Anderson & Barrett 1986; Larkum & Barrett 1983). Since the phycobiliprotein contribution is not included in the standard definition of PSU size, and since accessory chlorophylls like *b* or *c* are not present in red algae, a straight comparison with other algae, using this criterion, has limited meaning.

Photosynthetic action spectra, such as first made by Duysens (1952) with *Porphyridium purpureum*, clearly show that the photoactive pigments are chlorophyll and phycobiliproteins, especially phycoerythrin, and to a lesser extent the carotenoids. In the production of action spectra it is important that both photosystems interact and that the reaction centers of the photosystems remain in a state such that electron flow can continue. Computed values for the two photosystems have been derived from measurements of the relative absorptance of photosystems I and II (Ley & Butler 1977; Ried et al. 1977; Wang et al. 1977). As illustrated in Fig. 9-1, the major contributor to PSI is clearly chlorophyll *a* and to a lesser extent carotenoids. The direct contribution of the phycobiliproteins to photosystem I is relatively low. However, the contribution of phycobiliproteins to PSII is large and that of chlorophyll *a* relatively small.

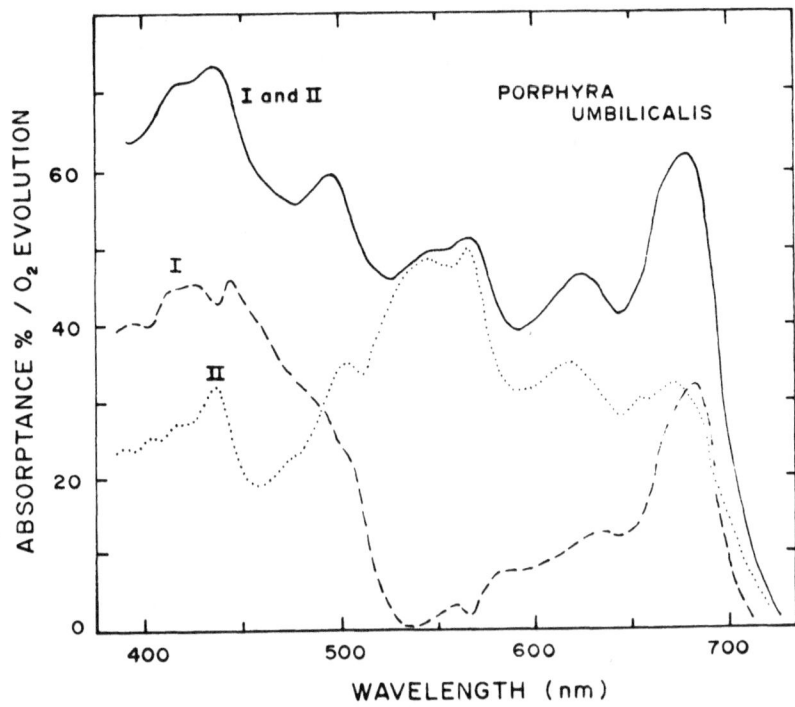

Fig. 9-1. Action spectrum of photosynthesis of *Porphyra umbilicalis*. The combined action of photosystems I and II represents the combined cell absorbance. Chlorophyll and carotenoids predominate in photosystem I, whereas phycobiliproteins (plus chlorophyll) are involved with photosystem II. (According to Ried et al. 1977)

V. PHOTOSYNTHETIC PIGMENTS

All of the chlorophyll *a* and probably most carotenoids are attached to a few specific proteins. The only other chlorophylls are the specialized chlorophylls of PSI (P700) and PSII (P680), which account for less then 0.5% of the total chlorophyll. The presence of chlorophyll *d* has been reported in extracts of *Gigartina*, but its existence in vivo has not been confirmed, and its role as a functional photosynthetic pigment remains to be established (Holt 1966). Even if it is not a degradation product, which is quite possible, chlorophyll *d* does not occur in sufficient quantity to have been designated as making a significant contribution to the action spectrum of any red alga.

The contribution of specific carotenoids to photosynthesis is not clear, except for β-carotene, which is active in PSI and probably also in PSII. This pigment has also been identified in isolated PSI and PSII pigment–protein complexes (Redlinger & Gantt, 1983). A number of other carotenoids have been found in red algae (Table 9-3), but only β-carotene, zeaxanthin, and β-cryptoxanthin have been identified in *Porphyridium purpureum* (Stransky & Hager 1970). The latter two have not been shown to perform a light-harvesting function, and it is doubtful that they are photosynthetically active.

The thylakoids of *Porphyridium purpureum*, when fractionated under mild denaturing conditions, resolve into three Chl–protein complexes (Redlinger & Gantt 1983; Fig. 9-2). The PSI complex (CPI) typically has a 720 nm fluoresence emission, contains P700, and appears to have 82% of the chlorophyll. Two other complexes, CPIII and CPIV, belong to PSII by their 685 nm fluoresence emission. They respectively account for 6 and 8% of the chlorophyll. On full denaturation with sodium dodecyl

Table 9-3. Chlorophyll and carotenoid pigments found in red algae

Chlorophyll *a*	Jeffrey 1980
β-Carotene	Bjornland et al. 1984; Stransky & Hager 1970
α-Cryptoxanthin	Bjornland et al. 1984; Stransky & Hager 1970
β-Cryptoxanthin	Bjornland et al. 1984; Stransky & Hager 1970
Lutein	Bjornland et al. 1984; Stransky & Hager 1970
Zeaxanthin	Bjornland et al. 1984; Stransky & Hager 1970
Antheraxanthin	Bjornland et al. 1984; Stransky & Hager 1970

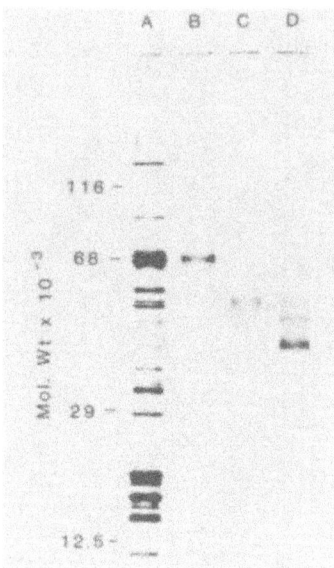

Fig. 9-2. Polypeptides of the major chlorophyll–protein complexes from Porphyridium purpureum. The complexes had been separated on preparative gels, isolated, denatured, and electrophoresed. The gel was stained with Coomassie blue. A: unfractioned thylakoids, B: CPI, C: CP III, D: CP IV. (From Redlinger & Gantt 1983)

sulfate and separation on gel electrophoresis the protein band of CPI is at ca. 62–68 kd, CPIII at 52 kd, and CPIV at 40 and 48 kd. The 52 kd is closely associated with phycobilisomes and is probably the PSII core antenna giving rise to the 695 nm fluorescence emission that is characteristic of PSII (Chereskin et al. 1985). A major photosystem I fraction obtained from Griffithsia (Hiller & Larkum 1981) is similar to that of Porphyridium. In fact, the apoprotein pattern of both is similar to that in green plants and cyanobacteria, suggesting that PSI is evolutionarily highly conserved; this is also born out by high conservation noted in gene sequence analyses (Bryant et al. 1987). The lower chlorophyll content in the PSII complexes is consistent with action spectra determinations, which show that much of the light absorbed by chlorophyll a is involved with PSI.

The phycobiliproteins, located in the phycobilisomes, are major photosynthetic antennae of PSII, as already noted above. They greatly extend the light absorbance capacity by filling in where chlorophyll absorption is low. This is illustrated in Fig. 9-3, which also shows the absorption characteristics of the purified phycobiliproteins from Porphyridium purpureum. Phycoerythrin has a double absorption maximum at 545 and 563 nm, typical of most red algal phycoerythrins, and a shoulder at 500 nm

(Table 9-4). Phycocyanin, with a double peak at 550 nm and 617 nm, is of the R-type, with two types of chromophores, unlike the C-type in many other red algae. Allophycocyanin has a typical absorption maximum at 650 nm. Considerable spectral variation exists among the phycobiliproteins, with the exception of allophycocyanins, as seen by selected examples in Table 9-4. In solution phycobiliproteins are highly fluorescent, with phycoerythrins ranging in emission from 575 to 585 nm, phycocyanins from 637 to 655 nm, and allophycocyanin at 660 nm. The fluorescence emission peaks are characteristic for each pigment type, and much of the variation within each class is caused by the protein environment rather than differences in the chromophore composition. Several recent publications on phycobiliproteins can be consulted for highlights of the biochemical characteristics (Glazer 1977, 1985; Moerschel & Rhiel 1987; Zilinskas & Greenwald 1986), excitation energy transfer in phycobiliproteins (Scheer 1986), and a comprehensive monograph dealing with all of these, including their role in photosynthesis (MacColl & Guard-Friar 1987). The principal chromophore types are phycocyanobilins, mainly present in phycocyanins and allophycocyanins, and phycoerythrobilins and phycourobilins in phycoerythrins. The chromophores are linear tetrapyrroles that are covalently bound to the apoprotein by a thioether linkage through ring A and/or through A and D. Amino acid sequence analysis has shown that the chromophores are always attached to cysteine residues and that the sequences surrounding the chromophore attachment are highly conserved (Bryant 1986; Troxler 1986; Zuber 1986).

The common phycobiliproteins phycoerythrin, phycocyanin, and allophycocyanin consist of α- and β-polypeptides with a size range of 15–22 kd; some phycoerythrins contain γ-subunits in addition. Allophycocyanin is the simplest phycobiliprotein containing one phycocyanobilin each per α- and β-polypeptide. The assumed in vivo state of allophycocyanin is a heteromere comprised of six α and six β subunits, with a total size of about 180 kd. Phycocyanins generally have one phycocyanobilin chromophore per α-polypeptide and two per β-polypeptide, with a combined heteromere size of about 220 kd (α_6, β_6). In R-phycocyanin a phycoerythrobilin chromophore is present together with a phycocyanobilin on the β-polypeptide, and one phycocyanobilin is present on the α-polypeptide. Although the R-phycocyanin possesses a mixture of chromophores, the fluoresence emission is typical of phycocyanins (Table 9-4), since phycocyanobilin

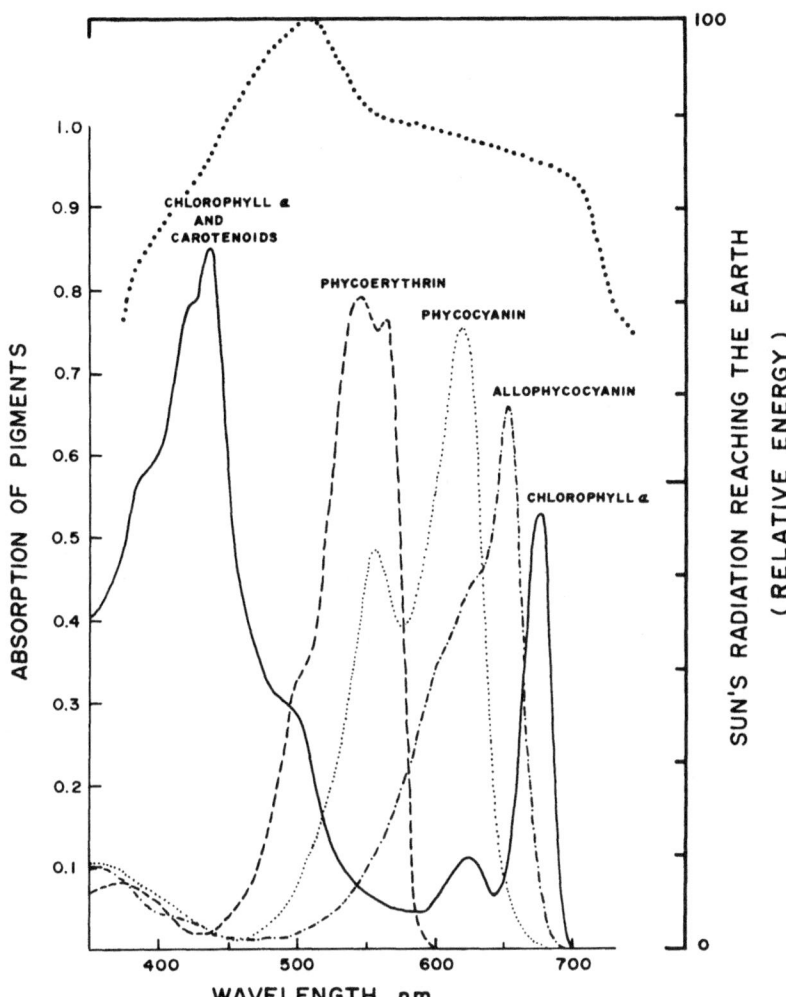

Fig. 9-3. Spectrum of solar energy output (upper dotted curve) and the absorption spectra of photosynthetically active pigments from *Porphyridium purpureum*. (From Gantt 1975)

Table 9-4. Spectral properties of phycobiliproteins purified from phycobilisomes of representative red algal species[a]

Alga	Phycobiliprotein	Absorption maxima (nm)	Fluorescence emission (nm)
Compsopogon coeruleus	B-Phycoerythrin	(500), 545, 563	578
	C-Phycocyanin-1	618	648
	C-Phycocyanin-2	630	652
	Allophycocyanin	653	664
Griffithsia pacifica	R-Phycoerythrin	498, 542, 565	580
	r-Phycoerythrin	498, 542, 560	575
	R-Phycocyanin	553, 615	640
	Allophycocyanin	650	660
Porphyridium aerugineum	C-Phycocyanin	625	650
	Allophycocyanin	650	660
Porphyridium purpureum	B-Phycoerythrin	(500), 545, 563	578
	b-Phycoerythrin	545, 563	575
	R-Phycocyanin	555, 617	637
	Allophycocyanin	650	660

[a] From studies in the author's laboratory (Gantt et al. 1979, 1986).

is the accepter (or fluorescent) chromophore (cf. MacColl & Guard-Friar 1987). In *Porphyridium purpureum* two types of phycoerythrin, B and b, are present in about equal concentrations. Of these, b is somewhat less stable than B-phycoerythrin when released from phycobilisomes and probably is comprised of a heteromere (α_6, β_6). Both b-phycoerythrin subunits contain several phycoerythrobilins. B-phycoerythrin is very stable and is relatively large, with a size of about 260 kd ($\alpha_6, \beta_6, \gamma_1$). In addition to the phycoerythrobilin chromophores, there are also some phycourobilin chromophores on all the γ-subunits (31 kd); the latter are responsible for the 500 nm absorption. In R-phycoerythrins a distinct peak is present instead of just a shoulder. Phycoerythrins have larger numbers of chromophores than the other phycobiliproteins. For example, a B-phycoerythrin heteromer is estimated to contain 32 phycoerythrobilin and 2 phycourobilin chromophores, while an allophycocyanin heteromere only has 12 phycocyanobilin chromophores (Glazer 1977, 1985; MacColl & Guard-Friar 1987). Additional phycobiliprotein types are found in cores of phycobilisomes and, although they comprise only a small percentage of the phycobilisome total content, they are very important in the energy transfer pathway. Of special interest are two long-wavelength pigments with fluorescence emissions of ca. 670–678 nm, because they are terminal pigments in the phycobilisomes (Gantt 1986; Glazer 1985; Zilinskas & Greenwald 1986). These are the high molecular weight L_{CM}, also referred to as the anchor polypeptide, and allophycocyanin-B. The size of the anchor polypeptide in red algae (92–94 kd) is fairly uniform and on SDS-gels retains its blue phycocyanobilin color. Allophycocyanin-B is now known to be a subunit (ca. 16 kd) that can associate with the more common allophycocyanin subunits. Although the anchor polypeptide in *Porphyridium purpureum* is only a minor component (ca. 1–2%), there is at least one per phycobilisome. The protein structure appears to be highly conserved in red algae and cyanobacteria, according to immunological reactivity tests and from partial amino acid sequence analysis (Gantt et al. 1985, 1988). It has been shown that antibodies to the purified anchor polypeptide from a cyanophyte (*Nostoc*) react with high molecular weight polypeptides of reds and blue-greens, but not with the allophycocyanin-B subunit. Further, a comparison of the partial N-terminal sequence of the anchor polypeptide of *P. purpureum* and *Nostoc* showed considerable homology, with 11 of the first 20 residues being identical and five others differing by only one base pair change in the genetic code.

However, there was virtually no homology when compared with any other phycobiliprotein or phycobilisome-linker polypeptide (Fueglistaller et al. 1986; Troxler 1986; Zuber 1986). Bryant (1988), in sequencing the gene from *Cyanophora*, showed that this protein is indeed unique and unlike other phycobiliproteins.

VI. PHYCOBILISOMES

The phycobilisome structure has been established for several algae, although not to the same degree of detail in red algae as has been attained in several cyanobacteria (Glazer 1985; Moerschel & Rhiel 1987; Zilinskas & Greenwald 1986). The phycobilisome is a macromolecular complex consisting of phycobiliproteins bound together by linker polypeptides that ensure the proper orientation of the phycobiliprotein chromophores. The linker polypeptides are basic (ca. 6.0–8.5 pI), whereas the phycobiliproteins are more acidic (ca. 6.3–4.3 pI) in nature. In *Porphyridium purpureum* 18 polypeptides are routinely resolved on SDS–PAGE (Fig. 9-4). Chromophores are associated with half of the polypeptides. The

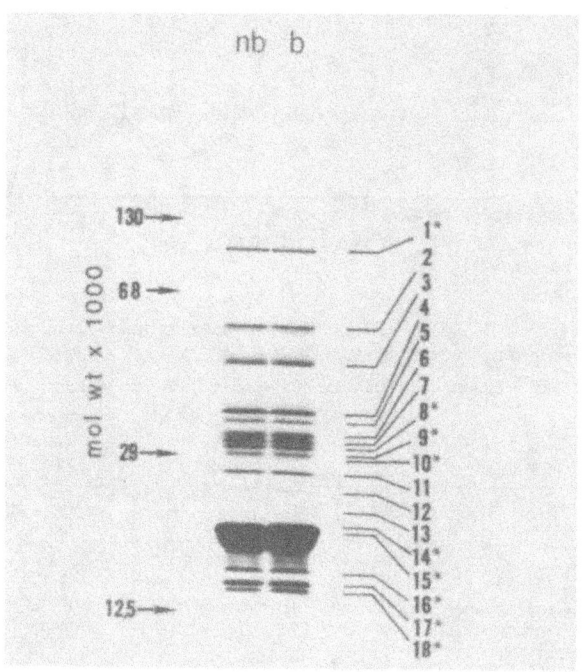

Fig. 9-4. Polypeptides of *Porphyridium purpureum* phycobilisomes separated by SDS polyacrylamide gel electrophoresis. Bands were sequentially numbered from the largest to the smallest. Asterisked numbers indicate colored bands with phycobiliprotein chromophores. Uncolored bands are presumed to be linkers. Heated (left) and unheated (right). (From Redlinger & Gantt 1981)

Pigmentation and photoacclimation

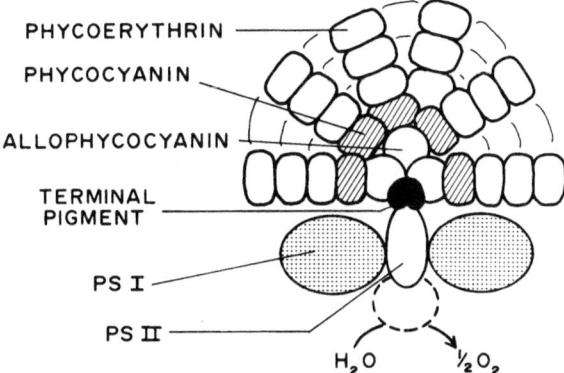

Fig. 9-5. Schematic arrangment of phycobiliprotein organization in a phycobilisome of *Porphyridium purpureum* and the presumed arrangement of the phycobilisome with the photosystems. The terminal pigment may mediate excitation energy transfer between the phycobilisome and the photosystem II–antennae chlorophyll.

colored polypeptides have been identified as being derived from specific phycobiliproteins. The other polypeptides (uncolored) are assumed to be linker polypeptides (Redlinger & Gantt 1981).

The size and shape of phycobilisomes varies with the species (Gantt 1980, 1986). Most are hemidiscoidally shaped. Others, such as in *Griffithsia pacifica* and *Porphyridium purpureum* are hemiellipsoidally or hemispherically shaped. The hemispherically shaped phycobilisomes are generally larger and more complex. For example, a phycobilisome of *P. purpureum* has been estimated to contain between 65 and 75 heteromeres (Dilworth & Gantt 1981). Isolation of phycobilisomes is readily accomplished. Required are high ionic conditions, such as 0.5–0.75 M phosphate buffer (pH 7), and a nonionic detergent like Triton-X 100, which is used to release the phycobilisome from the thylakoid membranes.

On the periphery of the phycobilisomes are rods, which are attached to a central core, which in turn is located nearest to the thylakoid (Fig. 9-5). In *Porphyridium purpureum* the peripheral arms consist of B- and b-phycoerythrin and R-phycocyanin held together by rod-linker polypeptides (Redlinger & Gantt 1981). Analysis of the hemidiscoidally shaped phycobilisomes of *Rhodella violacea* provided the first conclusive evidence for a definitive phycobilisome core structure (Moerschel et al. 1977; Moerschel & Rhiel 1987). The core was found to contain all of the allophycocyanin, whereas the rods contained phycocyanin and phycoerythrin. By electron microscopy it was shown that three pairs of disks (each a presumed α_6, β_6 heteromere) were oriented perpendicularly to the peripheral rods. Most cyanobacteria of those thus far examined have the same overall phycobilisome structure (Glazer 1985).

VII. ENERGY DISTRIBUTION

The arrangement in the phycobilisome of phycoerythrin on the periphery, followed by phycocyanin, and then by allophycocyanin, corresponds energetically to an excitation chain, where the progression is from a higher to a lower energy level. The phycobiliproteins form an energy transfer chain in which the fluorescence emission of the preceding member overlaps the absorption of the next one (Gantt 1975; Scheer 1986). Such a sequence is also consistent with the fluorescence decay of the individual phycobiliproteins within a phycobilisome. Energy transfer between phycobiliproteins from higher to lower energy states probably occurs by the Foerster-type mechanism. Transfer times are in the nanosecond range, with the requirement that the primary accepter must be within 10 nm of the secondary accepter (Scheer 1986). The progress of energy migration has been followed in living cells by picosecond time-resolved fluorescence in *Porphyridium* and in several cyanobacteria (Yamazaki et al. 1984) and occurs as follows:

PE ⇒ PC ⇒ APC ⇒ anchor polypeptides ⇒ PSII
                    ↓
                  APC-B

According to evidence from experiments using the cyanophyte *Nostoc*, the anchor polypeptide (L_{CM}) is the terminal pigment in the primary pathway and is the main conduit between the phycobilisome and PSII (Mimuro et al. 1986). It has long been known that phycobiliproteins transfer most of the energy initially to the chlorophyll core antennae of PSII (Ley & Butler 1977; Wang et al. 1977). Thus, a very close physical association between the phycobilisomes of PSII is required. Such an association has indeed been shown to exist in *P. purpureum* from which PSII–phycobilisome particles were isolated as one supramolecular complex (Chereskin et al. 1985; Clement-Metral & Gantt 1983; Clement-Metral et al. 1985).

Excitation energy within the phycobilisomes is first transferred to the chlorophyll antennae of photosystem II, from which it can then follow one of two paths. From the PSII antennae it may go to the reaction center of II, leading to the electron transport reactions to PSI. Alternatively, excitation energy may go directly from the antennae of PSII to the antennae of PSI, which can occur when cells

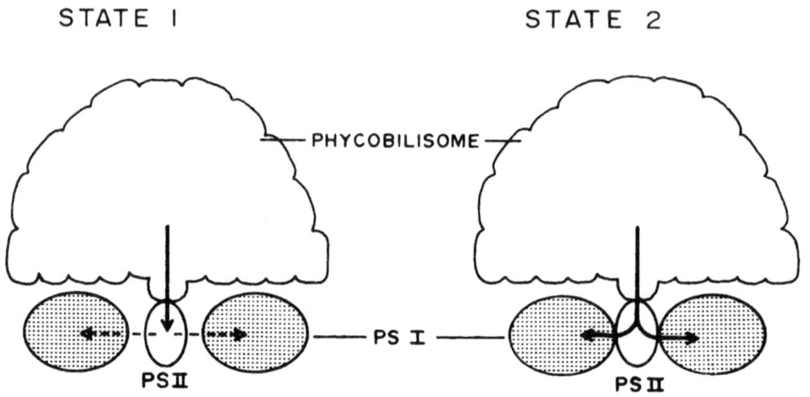

Fig. 9-6. Proposed model of the mechanism by which distribution of excitation energy can occur. Irradiating cells with red light results in state 1, and irradiating them with green light results in state 2. In state 2 "spillover" to photosytem I from the antennae of photosystem II can occur if the antennae complexes of the photosystems are in close proximity. (Modified from Biggins et al. 1984)

undergo a "change in state." If the reaction centers of PSII are closed, the excitation energy can be transferred directly to PSI. The state-change phenomenon by which distribution of excitation energy occurs was anticipated by Myers (1963), who initially termed it spillover; i.e., excitation energy absorbed by chlorophyll in photosystem II could "spill over" to photosystem I. Murata (1969) demonstrated the first evidence for this phenomenon in *Porphyridium purpureum*. The amount of spillover is faster, and appears to a greater extent in red algae and cyanobacteria than it is in green plants (Biggins et al. 1984; Fork & Satoh 1986; Ley & Butler 1977; Wang et al. 1977).

Experimentally, state 1 is brought about by irradiation with blue or red light (absorbed by PSI, which contains most of the chlorophyll) after cells have been in darkness (Biggins et al. 1984). Under such conditions the transfer from PSII to PSI is reduced, and this is manifested by a higher fluorescence from chlorophyll. State 2 can be brought about by exposure to green-orange light (absorbed by phycobiliproteins). In this state the transfer from PSII to PSI is enhanced and the degree of chlorophyll fluorescence is lower. To follow the transitions, monitored by chlorophyll fluorescence, special instrumentation is required. Using fast-resolution instrumentation on *Porphyridium purpureum*, Bruce et al. (1986) showed by picosecond fluorescence that in state 2 PSII chlorophyll emission decayed more rapidly than in state 1. This is consistent with an increase in the rate of spillover. The mechanism by which state changes may occur is illustrated in Fig. 9-6.

There has been considerable confusion on the state of cells kept in darkness. This now seems to be resolved, and according to Fork and Satoh (1986) the dark state now appears to be equivalent to state 2, providing photosynthetically active cells are kept in darkness for a specified time (ca. 30 min). These same authors also noted a state 3, where state 2 light decreased PSII activity without affecting PSI. This phenomenon has thus far only been noted in *Porphyra* and will require further study.

State changes in red and blue-green algae are orders of magnitude faster than those in green plants – microseconds vs. minutes, respectively. What proteins are involved in the state transitions is still an open question. In red algae and cyanobacteria there is no direct evidence that phosphorylation occurs either in chlorophyll-binding polypeptides, or in phycobilisome components with state changes (Biggins & Bruce 1989; Biggins et al. 1984), as appears to be the case in green plants and *Chlamydomonas* (see Fork & Satoh 1986; Staehelin 1986). Phosphorylation of a 32 kd thylakoid polypeptide has been reported to occur in *Porphyridium* in light but not in darkness (Kirschner & Senger 1986). However, it is not known whether the phosphorylation event correlates with a state change transition.

VIII. PHOTOINHIBITION

Photoinhibition usually occurs at light intensities somewhat higher than light saturation. It may be manifested by a decrease in the photosynthetic activity and an increase of fluorescence, and it may be followed by photooxidation of chlorophyll (bleaching). It has been suggested that photoinhibition may result from photooxidation of components of the photosynthetic apparatus, backing up the energy flow from photosystems II and I and causing a reduction in the concentration of NADPH required for CO_2 fixation (see Kyle & Ohad 1986). Photoinhibition is not clearly defined in any system and probably involves the interaction of several factors. In land plants, where most studies have been carried out, there is the additional complication of stomatal closing, thus decreasing CO_2 exchange and at the same time decreasing the water loss from transpiration. Although this complexity is not present in algae, photoinhibition is still difficult to elucidate.

In *Chlamydomonas*, PSII is the photosystem affected during photoinhibition, with little or no effect occurring on PSI, according to Kyle and Ohad (1986). The secondary accepter, also known as the Q_B-protein or D1, which mediates electron flow between the primary accepter and plastoquinone, is thought to be the primary site of the photoinhibition. This protein is quickly turned over, allowing rapid recovery from the photoinhibitory state, if photooxidation of other components has not yet occurred. Photorespiration might be a mechanism to counteract photoinhibition by maintaining electron transport between the photosystems and allowing for the generation of energy for CO_2 fixation (Kyle & Ohad 1986). At increasing photon flux densities photorespiration is known to increase, but in red algae there may be other mechanisms for preventing excessive photoinhibition. The most important of these is probably due to the state transition mechanism (Fork & Satoh 1986). Another may be an elaboration of the types or number of Q_B-proteins.

Photoinhibition is well documented in green plants, diatoms and dinoflagellates (see Richardson et al. 1983), but in red algae it appears to be less obvious. It is most readily observed when plants grown at low irradiance levels are switched to higher levels. In *Gracilaria foliifera* Rosenberg and Ramus (1982) recorded photoinhibition at an irradiance level (ca. 600 $\mu E\,m^{-2}s^{-1}$) very close to P_{max} (ca. 400 $\mu E\,m^{-2}s^{-1}$). Usually much higher light intensities are tolerated by red algae. In *Gastroclonium subarticulatum* (as *G. coulteri*), Hodgson (1981) found that net photosynthesis increased with increasing light intensity but that, as in *Gracilaria tikvahiae* (Penniman & Mathieson 1985) and *Porphyra yezoensis* (Kato & Aruga 1984), no inhibition was evident even up to 1600 $\mu E\,m^{-2}s^{-1}$. However, upon exposure to desiccation conditions photodamage was observed (Hodgson 1981). Furthermore, photoinhibition was not observed in any of 20 red algal species exposed to photon flux densities up to 700 $\mu E\,m^{-2}s^{-1}$, whether they were sun or shade adapted (Tseng et al. 1981). Interestingly, *Porphyridium purpureum* can tolerate very high photon flux densities without photoinhibition. According to Lee and Vonshak (1988) this alga did not show photoinhibition (in 45 min) at 2300 $\mu E\,m^{-2}s^{-1}$ at 15°C, and although photoinhibition occurred at 35°C, it was reversible within minutes in dim light (see also Table 15-1).

IX. PHOTOREGULATION

A. Light quality

The theory of complementary chromatic adaptation suggests that algae produce the pigments that maximally absorb the prevailing light (Bogorad 1975). It has been an attractive theory and has been invoked to explain vertical zonation patterns of differentially pigmented algae. Actually, there is considerable variation in the wavelengths penetrating to certain depths, although green light often penetrates more deeply through the water column relative to most other wavelengths (see Fig. 4, Larkum & Barrett 1983). Consequently, it has been assumed that phycoerythrin-containing red algae can grow more successfully at greater depths. There is little, if any, experimental evidence to support this hypothesis (Larkum & Barrett 1983; Ramus 1981, 1983). Under growth conditions, where the photosynthetically active light was equalized, there was little or no effect of light quality in growth studies with *Gracilaria tikvahiae* (Ramus 1983; Ramus & van der Meer 1983). Definitive data that chromatic complementary adaptation can occur in red algae are as yet lacking.

Light quality has been shown to affect the phycobiliprotein-to-chlorophyll ratio in *Porphyridium purpureum* (Ley & Butler 1980), but in whole cell spectra there was no detectable variation in the ratio of the phycobiliprotein types. In material collected from the field the pigmentation can be easily misinterpreted if pigment extractions are not performed. Furthermore, unless actively growing cultures are examined, the apparent pigment variations may be a result of photooxidation of some phycobiliproteins and the lack of new synthesis of pigments. When *P. purpureum* was grown in continuous red light (660 nm and longer) and compared with cells grown in white light, we found that the pigment composition was highly similar in cells under the two conditions but that the red-light-grown cells were less efficient in light utilization (Fig. 9-7) and the quantum yield

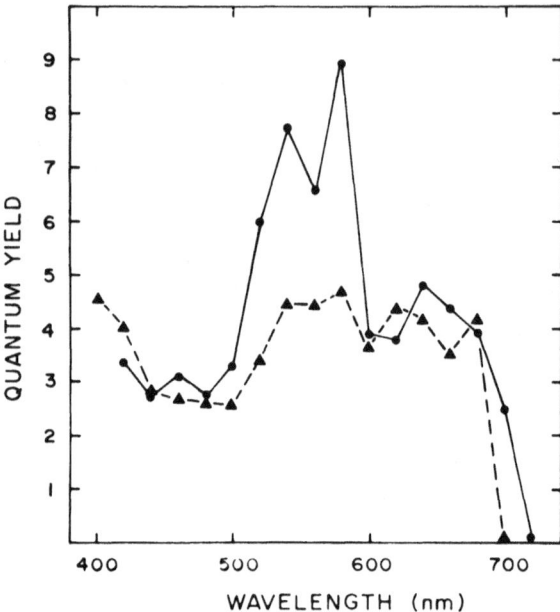

Fig. 9-7. *Porphyridium purpureum* cells grown under continuous red light (>660 nm) (------) had a lower quantum yield, as determined by oxygen evolution, than did those grown under continuous white light (———). (Courtesy of O. Canaani, pers. comm.)

was much lower (Canaani et al. unpubl.). Since the overall pigment composition was virtually the same, other factors must be responsible. Some of these may be an imbalance of the photosystems, a difference in the rate at which the photosystems turn over, a deficiency in the electron transport chain, or a major change in the spatial relationship of the photosystems to one another. In cyanophytes a light quality effect on the stoichiometry of photosystem I:II has been reported (Fujita et al. 1985; Manodori & Melis 1986), and it can be expected that red algae would be similarly affected. Another consideration is the topographic distribution of the photosystems with regard to the phycobilisomes. If they are spatially separated, energy transfer might be impaired. In fact, in *Cyanidium caldarium* an impairment in energy transfer from the phycobilisomes to photosystem II was reported by Diner (1979). He showed that about half of the reaction centers of PSII were not functionally associated with phycobilisomes.

Interestingly, changes in color from pink to red were observed with changing light conditions in whole plants of *Callithamnion* by Yu et al. (1981). However, when the phycobiliproteins were extracted, they found that the color changes were due to differences in the phycoerythrin composition, rather than to a change in the ratio of phycoerythrin: phycocyanin: allophycocyanin. The phycoerythrins had variable ratios of phycourobilin:phycoerythrobilin chromophores. Furthermore, it was determined that this was attributable to changes in light intensity, not in light quality, which again is consistent with the fact that complementary chromatic adaptation does not appear to occur in red algae.

B. Light intensity acclimation

Generally, red algae grow in a relatively low irradiance level. Laboratory studies have shown that at lower irradiances (around 10 $\mu E\, m^{-2}\, s^{-1}$) the cell pigment content is greater than at higher irradiances (around 150–200 $\mu E\, m^{-2}\, s^{-1}$) (Levy & Gantt 1988; Waaland et al. 1974). A similar trend is reflected in observations from field-collected species (Larkum & Barrett 1983; Rosenberg & Ramus 1982). At what point does the acclimation occur? Is there a change in the ratio of PSI:PSII:phycobilisomes? A change in the antennae size of photosystem II is evident, since Waaland et al. (1974) reported that *Griffithsia pacifica* had a lower phycoerythrin content relative to chlorophyll at higher light intensity, along with a reduction in phycobilisomes per thylakoid area.

Studies on *Porphyridium purpureum* grown over a range of from 10 to 180 $\mu E\, m^{-2}\, s^{-1}$ showed that acclimation to the irradiance levels require several days or weeks and active cell division (Levy & Gantt 1988). Cells grown at the highest irradiance had the highest photosynthetic capacity (Table 9-2) and the greatest growth rate. Since the photosynthetic unit size during light acclimation changed little, the number of photosynthetic units per cell was definitely reduced (Cunningham et al. 1988, 1989). Furthermore, the stoichiometry of PSI:PSII:phycobilisome showed no significant change under growth conditions ranging from 6 to 280 $\mu E\, m^{-2}\, s^{-1}$. Jahn et al. (1984) have already shown in this alga that ribulose bisphosphate carboxylase activity is not affected by growth irradiance. It is possible that the PSII reaction centers in these cells have different turnover rates, although the PSII turnover rate (about 0.5 msec, O_2 flash yield) appears to be uniform (Mishkind & Mauzerall 1980). It is worth noting, however, that Stevens and Myers (1976) found a faster turnover rate (1.5×) in a pigment mutant of the cyanophyte *Anacystis* than that found in the wild type.

Staehelin et al. (1978), in an electron microscopic study of *Griffithsia pacifica*, including freeze fracturing, also found a decrease in phycobilisome number in high-irradiance cells; in addition, they attempted to obtain some measure of the PSII number per thylakoid area. For this they counted the 10 nm particles on cleaved EF surfaces, which represent the PSII core particles. They noted only small differences in high- and low-light cells. However, their results suggest that about three to four "photosystem II particles" could be associated with each phycobilisome. From photophysical measurements, Ley (1984) concluded that three to four PSII reaction centers can be functionally linked to each phycobilisome. Similarly, a ratio of four PSII reaction centers per phycobilisome have been found in *Agardhiella subulata* (as *Neoagardhiella bailyei*) by Kursar and Alberte (1983). What functional advantage there would be for an association of multiple PSII reactions per phycobilisome is unknown. Furthermore, whereas direct association of PSI and phycobilisomes has not been found, it cannot be ruled out.

X. CHLOROPLAST GENES, TRANSLATION, AND TRANSCRIPTION

The first characterization of a red algal chloroplast genome was recently published by Li and Cattolico (1987). They found a genome size of 178 and 200 kilobases for *Griffithsia pacifica* and *Porphyra yezoensis*, respectively, which is larger than that for land plants. At least two genes, *psaB* and *psbA*, that code for two thylakoid proteins have been mapped on the chloroplast DNA of *Griffithsia*. The genes for phycobiliproteins are also expected to be part of the chloroplast genome. However, the genes coding for, or controlling the expression of, the phycobiliprotein linker polypeptides will probably be found in the nucleus. This is expected from the results with selective antibiotic inhibitor studies (Egelhoff & Grossman 1983; Gantt et al. 1985). Of 18 or more *Porphyridium purpureum* phycobilisome polypeptides, about half are synthesized in the chloroplast on 70S ribosomes and the other half may be synthesized in the cytosol on 80S ribosomes. Cycloheximide inhibited synthesis of all uncolored (probable linker) polypeptides in the 25–60 kd region, plus the colored τ-polypeptides of B-phycoerythrin (29–30 kd). Chloramphenicol inhibited production of the high-molecular-weight anchor polypeptide, the α and β subunits of phycoerythrin, phycocyanin, and allophycocyanin and a 24 kd uncolored polypeptide. Furthermore, in vitro synthesis studies on *Cyanidium caldarium* and *Porphyridium aerugineum* showed that the apoproteins of phycocyanin and allophycocyanin were translated on nonpolyadenylated mRNA, typical of chloroplast rather than cytoplasmic translation events (Belford et al. 1983; Steinmueller et al. 1983; Troxler 1986). In *C. caldarium* the apoproteins of both the α and β subunits of phycocyanin and the α of allophycocyanin were translated in the rabbit reticulocyte system used by Belford et al. (1983). It is of interest to note that the phycocyanin subunits determined by microsequencing were similar to the mature proteins and the NH_2-terminal sequences were the same. Furthermore, there was a lack of a transit peptide, also suggesting that the polypeptides were unlike chloroplast polypeptides synthesized in the cytosol. In an independent study on *C. caldarium*, Steinmueller et al. (1983) obtained essentially the same results, except that under their conditions only the β subunits of phycocyanin and allophycocyanin were translated, whereas in *P. aerugineum* the α subunit of phycocyanin and the β subunit of allophycocyanin were translated. It is not clear why the α and β subunits are not always cotranslated. Attachment of the covalently bound chromophores does not seem to occur under in vitro conditions.

In cyanobacteria, many genes coding for phycobiliproteins and phycobilisome linker polypeptides have been cloned and sequenced, many of which are listed in Table 9-5. It has been established that the allophycocyanin and phycocyanin genes for the α and β subunits occur in contiguous sequence and are transcribed on dicistronic mRNAs, thus probably ensuring the required 1:1 stoichiometry of the subunits (Bryant 1986; Grossman et al. 1986). The gene sequence characterization accomplished with the eukaryotic *Cyanophora paradoxa* is of special interest because as a cyanelle it may represent a transition stage from free living cyanophyte to chloroplast (see Bryant 1986; Grossman et al. 1986). Because of high similarities in the amino acid sequences of phycobiliproteins in red algae and cyanobacteria, especially the highly conserved regions around the chromophore attachment sites (cf. MacColl & Guard-Friar 1987; Troxler 1986; Zuber 1986), the genes for these proteins are probably highly similar in these two groups. There is general agreement that the phycobiliproteins and the associated linkers originate from one ancestral gene. Whereas support for this idea first came from immunochemical cross-reactivity studies and initial amino acid sequences, future molecular genetic analysis can bring additional support.

Development of the chloroplast is unquestionably regulated by light quality and light quantity. However, it is not known whether this occurs at

Table 9-5. Genes for phycobilisome and photosystem components in cyanophytes

Locus	Product encoded	Species[a]
Phycobilisome		
cpcA	α subunit PC	Agmenellum, Anacystis, Cyanophora
cpcB	β subunit PC	Agmenellum, Anacystis, Cyanophora
cpcC	33 kd rod linker PC	Agmenellum, Fremyella
cpcD	9 kd rod linker	Agmenellum
apcA	α subunit APC	Agmenellum, Anacystis, Fremyella, Cyanophora
apcB	β subunit APC	Agmenellum, Anacystis, Fremyella, Cyanophora
apcC	core linker APC	Agmenellum, Anacystis, Fremyella
cpeA	α subunit PE	Pseudanabaena, Fremyella
cpeB	β subunit PE	Pseudanabaena, Fremyella
gpcA	α subunit constitutive PC	Pseudanabaena, Fremyella
gpcB	β subunit constitutive PC	Pseudanabaena, Fremyella
rpcA	α subunit red inducible PC	Pseudanabaena, Fremyella
rpcB	β subunit red inducible PC	Pseudanabaena, Fremyella
Thylakoid		
psaA	photosystem I apoprotein	Anacystis, Synechocystis
psaB	photosystem I apoprotein	Anacystis, Synechocystis
psbA	Q_B-binding protein (D1)	Synechocystis
psbB	47 kd chl a-binding PII core	Synechocystis
psbC	43 kd chl a-binding PSII core	Synechocystis
psbD	34 kd polypeptide PS II core	Synechocystis
psbE	9 kd subunit cyt. b-559	Synechocystis
psbF	4 kd subunit cyt. b-559	Synechocystis

[a] Composite from data in Bryant (1986) and Grossman et al. (1986).

transcriptional, translational, or post-translational level. The nature of the photoreceptors is a completely open area, although Steinmueller and Zetsche (1984) noted that a blue light photoreceptor is involved in light-stimulated synthesis of phycobiliproteins and chlorophyll in *Cyanidium caldarium*.

XI. SUMMARY

The photosynthetic apparatus of red algae and the cyanophytes is unique in that only chlorophyll *a* occurs in these organisms and because they contain phycobiliproteins as major light-harvesting pigments. Phycobilisomes, which consist of phycobiliproteins and linker polypeptides, are functionally equivalent to the chlorophyll a/b antennae complexes in green algae and higher plants. In red algae there does not seem to be a sequestration of the photosystems, since thylakoid appressions do not occur. It has been shown that PSII is directly associated with phycobilisomes, which is functionally important, since the energy absorbed by phycobilisomes is transferred to the PSII chlorophyll antenna and excess may be delivered to PSI. Energy transfer within phycobilisomes and to photosystem II is by inductive resonance, which requires a close spatial relationship (less than 10 nm) between the interacting pigments.

Red algae generally exist in a relatively narrow range of irradiance, but many can tolerate full sunlight intensity, and they seem surprisingly resistant to photoinhibition. A possible means of defusing damaging photoinhibition lies in their ability to undergo state transitions. By this means, excess energy coming in through phycobilisomes could be transferred directly to the antennae of PSI before photooxidation could occur in the PSII reaction center.

The unicellular red algae and some of the rapidly growing filamentous species are highly suitable for studies on light acclimation. Fractionation of the thylakoid pigment complexes is possible and, from characterization of the three chlorophyll–protein complexes, it appears that they have significant similarities with those of green plants and cyanophytes. Studies on the molecular genetic level are lacking, except for a few preliminary studies. Although some extrapolation on molecular events of the development of the photosynthetic apparatus can be made from cyanobacteria and even from green plants, it is necessary to clone and characterize the genes of red algae and to study their expression.

XII. ACKNOWLEDGMENTS

Work cited from the author's laboratory was supported by the U.S. Department of Energy Contract AS05-76ER-04310 and Grant No. DE-FG05-ER13652. Appreciation for the drawings is expressed to C. A. Lipschultz, as is appreciation to S. VanValkenburg and K. R. Gantt for reading the manuscript.

XIII. REFERENCES

Anderson, J. M. & Barrett, J. 1986. Light-harvesting pigment–protein complexes of algae. In *Encyclopedia of Plant Physiology*. Vol. 19. *Photosynthetic Membranes and Light-Harvesting Systems*, eds. L. A. Staehelin & C. J. Arntzen. pp. 269–85. Berlin: Springer-Verlag.

Beale, S. I. & Chen, N. C. 1983. N-methyl mesoporphyrin IX inhibits phycocyanin, but not chlorophyll synthesis in *Cyanidium caldarium*. *Plant Physiol.* 71: 263–8.

Belford, H. S., Offner, G. D. & Troxler, R. F. 1983. Phycobiliprotein synthesis in the unicellular rhodophyte. *Cyanidium caldarium*: cell-free translation of the mRNAs for the α and β subunit polypeptides of phycocyanin. *J. Biol. Chem.* 258: 4503–10.

Biggins, J. & Bruce, D. 1989. Regulation of excitation energy transfer in organisms containing phycobilins. *Photosynth. Res.* 20: 1–34.

Biggins, J., Campbell, C. L. & Bruce, D. 1984. Mechanism of the light state transition in photosynthesis II. Analysis of phosphorylated polypeptides in the red alga *Porphyridium cruentum*. *Biochim. Biophys. Acta.* 767: 138–44.

Bisalputra, T. & Antia, N. J. 1980. Cytological mechanisms underlying darkness-survival of the unicellular red alga *Porphyridium cruentum*. *Bot. Mar.* 23: 719–30.

Bjornland, T., Borch, G. & Liaaen-Jensen, S. 1984. Configurational studies on red algae carotenoids. *Phytochemistry* 23: 1711–15.

Bogorad, L. 1975. Phycobiliproteins and complementary adaptation. *Annu. Rev. Plant Physiol.* 26: 369–401.

Bruce, D., Hanzlick, C. A., Hancock, L. E., Biggins, J. & Knox, R. S. 1986. Energy distribution in the photochemical apparatus of *Porphyridium cruentum*: picosecond fluorescence spectroscopy of cells in state 1 and state 2 at 77 K. *Photosynth. Res.* 10: 283–90.

Bryant, D. A. 1986. The cyanobacterial photosynthetic apparatus: comparisons to those of higher plants and photosynthetic bacteria. In *Physiological Ecology of Picoplankton*, eds. T. Platt & W. K. W. Li, pp. 423–500. Canadian Bulletin of Fisheries and Aquatic Sciences # 214. Ottawa: Department of Fisheries and Oceans.

Bryant, D. A. 1988. Genetic analysis of phycobilisome biosynthesis, assembly, structure, and function in the cyanobacterium *Synechococcus* sp. PCC 7002. In *Light Energy Transduction in Photosynthesis: Higher Plants and Bacterial Models*. eds. S. E. Stevens & D. A. Bryant, pp. 62–91. American Society of Plant Physiologists.

Bryant, D. A., DeLorimier, R., Guglielmi, G., Stirewalt, V. L., Cantrell, A. & Stevens, S. E., Jr. 1987. The cyanobacterial photosynthetic apparatus: structural and functional analysis employing molecular genetics. In *Progress in Photosynthesis Research*, Vol. IV., ed. J. Biggins, pp. 749–55, Dordrecht: Nijhoff.

Cheng, J. Y. & Antia, N. J. 1970. Enhancement by glycerol of photoautotrophic growth of marine planktonic algae and its significance to ecology of glycerol pollution. *J. Fish. Res. Bd. Can.* 27: 335–54.

Chereskin, B., Clement-Metral, J. D. & Gantt, E. 1985. Characterization of a purified photosystem II–phycobilisome particle preparation from *Porphyridium cruentum*. *Plant Physiol.* 77: 626–9.

Clement-Metral, J. D. & Gantt, E. 1983. Isolation of oxygen-evolving phycobilisome–photosystem II particles from *Porphyridium cruentum*. *FEBS Lett.* 156: 185–8.

Clement-Metral, J. D., Gantt, E. & Redlinger, T. 1985. A photosystem II–phycobilisome preparation from the red alga *Porphyridium cruentum*: oxygen evolution, ultrastructure, and polypeptide resolution. *Arch. Biochem. Biophys.* 238: 10–17.

Cunningham, F. X., Jursinic, P. A., Dennenberg, R., Canaani, O., & Gantt, E. 1988. Photoacclimation in *Porphyridium*: antennae, photosystems I & II. *Plant Physiol.* 86 suppl.: 46.

Cunningham, F. X., Dennenberg, R. J., Mustardy, L., Jursinic, P. & Gantt, E. 1989. Stoichiometry of photosystem I, photosystem II, and phycobilisomes in the red alga *Porphyridium cruentum* as a function of growth irradiance. *Plant Physiol.* 91: 1179–87.

Dilworth, M. F. & Gantt, E. 1981. Phycobilisome–thylakoid topography on photosynthetically active vesicles of *Porphyridium cruentum*. *Plant Physiol.* 67: 608–12.

Diner, B. A. 1979. Energy transfer from the phycobilisomes to photosystem II reaction centers in wild type *Cyanidium caldarium*. *Plant Physiol.* 63: 30–4.

Drew, E. A. 1977. The physiology of photosynthesis and respiration in some Antarctic marine algae. *Br. Antarct. Surv. Bull.* 46: 59–76.

Duysens L. N. M. 1952. Transfer of Excitation Energy in Photosynthesis. Ph.D. Thesis, University of Utrecht, Holland.

Egelhoff, T. & Grossman, A. 1983. Cytoplasmic and chloroplast synthesis of phycobilisome polypeptides. *Proc. Natl. Acad. Sci. USA* 80: 3339–43.

Fork, D. C. & Satoh, K. 1986. The control by state transitions of the distribution of excitation energy transfer in photosynthesis. *Annu. Rev. Plant Physiol.* 37: 335–61.

Fueglistaller, P., Suter, F. & Zuber, H. 1986. Linker polypeptides of the phycobilisomes from the cyanobacterium *Mastigocladus laminosus*: amino acid sequences and function. *Biol. Chem. Hopes-Seylers* 373: 615–26.

Fujita, Y., Ohki, K. & Murakami A., 1985. Chromatic regulation of photosystem composition in the photosynthetic system of red and blue-green algae. *Plant Cell Physiol.* 26: 1541–8.

Gantt, E. 1975. Phycobilisomes: light harvesting pigment complexes. *Bioscience* 25: 781–8.

Gantt, E. 1980. Structure and function of phycobilisomes: light harvesting pigment complexes in red and blue-green algae. *Int. Rev. Cytol.* 66: 45–80.

Gantt, E. 1986. Phycobilisomes. In *Encyclopedia of Plant Physiology*. Vol. 19. *Photosynthetic Membranes and Light-Harvesting Systems*, eds. L. A. Staehelin & C. J. Arnzten, pp. 260–8. Berlin: Springer-Verlag.

Gantt, E. 1989. *Porphyridium* as a red algal model for photosynthesis studies. In *Algae as Experimental*

Systems, eds. A. W. Coleman, L. J. Goff, & J. R. Stein-Taylor, pp. 249–68. New York: Alan R. Liss.

Gantt, E., Lipschultz, C. A., Cunningham F. X. & Mimuro, M. 1988. N-terminus conservation in the terminal pigment of phycobilisomes from a prokaryotic and eukaryotic alga. *Plant Physiol.* 83: 61.

Gantt, E., Lipschultz, C. A., Grabowski, J. & Zimmerman, B. K. 1979. Phycobilisomes from blue-green and red algae. Isolation criteria and dissociation characteristics. *Plant Physiol.* 63: 615–20.

Gantt, E., Lipschultz, C. A. & Redlinger, T. 1985. Phycobilisomes: a terminal acceptor pigment in cyanobacteria and red algae. In *Molecular Biology of the Photosynthetic Apparatus*, eds. K. Steinback, S. Bonitz, C. J. Arntzen, & L. Bogorad, pp. 223–9. Cold Spring Harbor: Cold Spring Harbor Laboratory.

Gantt, E., Scott, J. & Lipschultz, C. 1986. Phycobiliprotein composition and chloroplast structure in the freshwater red alga *Compsopogon coeruleus* (Rhodophyta). *J. Phycol.* 22: 480–4.

Glazer, A. N. 1977. Structure and molecular organization of the photosynthetic accessory pigments of cyanobacteria and red algae. *Mol. Cell. Biochem.* 18: 125–40.

Glazer, A. N. 1985. Light harvesting by phycobilisomes. *Annu. Rev. Biophys.* 14: 47–77.

Grossman, A. R., Lemaux, P. G. & Conley, P. B. 1986. Regulated synthesis of phycobilisome components. *Photochem. Photobiol.* 44: 827–37.

Hiller, R. G. & Larkum, A. W. D. 1981. Chlorophyll-proteins of the red alga *Griffithsia monilis*. In *Photosynthesis III. Structure and Molecular Organization of the Photosynthetic Apparatus*, ed. G. Akoyunoglou, pp. 387–96. Philadelphia: Balaban International Science Services.

Hodgson, L. M. 1981. Photosynthesis of the red alga *Gastroclonium coulteri* (Rhodophyta) in response to changes in temperature, light intensity, and desiccation. *J. Phycol.* 17: 37–42.

Holt, A. S. 1966. Recently characterized chlorophyll. In *The Chlorophylls*, eds. L. O. Vernon & G. R. Seely, pp. 111–18. New York: Academic.

Jahn, W., Steinbiss, J. & Zetsche, K. 1984. Light intensity adaptation of the phycobiliprotein content of the red alga *Porphyridium*. *Planta* (Berl.) 161: 536–9.

Jeffrey, S. W. 1980. Algal pigments. In *Primary Productivity of the Sea*, ed. P. G. Falkowski, pp. 33–58. New York: Plenum.

Kato, M. & Aruga, Y. 1984. Comparative studies on the growth and photosynthesis of the pigmentation mutant of *Porphyra yezoensis* in laboratory culture. *Jap. J. Phycol.* 32: 333–47.

Kawamura, M., Mimuro, M. & Fujita, Y. 1979. Qualitative relationship between two reaction centers in the photosynthetic systems of blue-green algae. *Plant Cell Physiol.* 20: 697–705.

Kirschner, J. & Senger, H. 1986. Thylakoid protein phosphorylation in the red alga *Porphyridium cruentum*. In *Regulation of Chloroplast Differentiation*. Vol. 2, eds. G. Akoyunoglou and H. Senger, pp. 339–44. New York: Alan R. Liss.

Kursar, T. & Alberte, R. 1983. Photosynthetic unit organization in a red alga. Relationships between light-harvesting pigments and reaction centers. *Plant Physiol.* 72: 409–14.

Kyle, D. J. & Ohad, I. 1986. The mechanism of photoinhibition in higher plants and green algae. In *Encyclopedia of Plant Physiology*. Vol. 19. *Photosynthetic Membranes and Light-Harvesting Systems*, eds. L. A. Staehelin & C. J. Arntzen, pp. 468–75. Berlin: Springer-Verlag.

Larkum, A. W. D. & Barrett, J. 1983. Light-harvesting processes in algae. In *Advances in Botanical Research*. Vol. 10, ed. H. W. Woolhouse, pp. 3–219. New York: Academic.

Lee, Y. K. & Vonshak, A. 1988. The kinetics of photoinhibition and recovery of photosynthetic activity in the red alga *Porphyridium cruentum*. *Arch. Microbiol.* 150: 1–5.

Levy, I. & Gantt, E. 1988. Light acclimation of *Porphyridium purpureum* (Rhodophyta): growth, photosynthesis, and phycobilisomes. *J. Phycol.* 24: 452–8.

Ley, A. C. 1984. Effective absorption cross-sections in *Porphyridium cruentum*. Implications for energy transfer between phycobilisomes and photosystem II reaction centers. *Plant Physiol.* 74: 451–4.

Ley, A. C. & Butler, W. L. 1977. The distribution of excitation energy transfer between photosystem I and photosystem II in *Porphyridium cruentum*. In *Photosynthetic Organelles*. Special edn. *Plant Cell Physiol.*, eds. S. Miyachi, K. Katoh, Y. Fujita, & K. Shibata, pp. 33–46. Tokyo: Japanese Society of Plant Physiologists.

Ley, A. C. & Butler, W. L. 1980. Effects of chromatic adaptation on the photochemical apparatus of photosynthesis in *Porphyridium cruentum*. *Plant Physiol.* 65: 714–22.

Li, N. & Cattolico, R. A. 1987. Chloroplast genome characterization in the red alga *Griffithsia pacifica*. *Mol. Gen. Genet.* 209: 343–51.

Littler, M. M., Littler, D. S., Blair, S. M. & Norris, J. 1985. Deepest known plant life discovered on an uncharted seamount. *Science* 227: 57–9.

Lüning, K. & Schmitz, K. 1988. Dark growth of the red alga *Delessaria sanguinea* (Ceramiales): lack of chlorophyll, photosynthetic capability and phycobilisomes. *Phycologia* 27: 72–7.

MacColl, R. & Guard-Friar, D. 1987. *Phycobiliproteins*. Boca Raton: CRC Press.

Manodori, A. & Melis, A. 1986. Light quality regulates photosystem stoichiometry in cyanobacteria. In *Regulation of Chloroplast Development*, eds. G. Akoyunoglou & H. Senger, pp. 653–62. New York: Alan R. Liss.

Mimuro, M., Lipschultz, C. A. & Gantt, E. 1986. Energy flow in the phycobilisome core of *Nostoc* sp. (Mac): two independent terminal pigments. *Biochim. Biophys. Acta.* 852: 126–32.

Mishkind, M. & Mauzerall, D. 1980. Kinetic evidence for a common photosynthetic step in diverse seaweeds. *Mar. Biol.* 56: 261–5.

Moerschel, E., Koller, K.-P., Wehrmeyer, W. & Schneider, H. 1977. Biliprotein assembly in the disc-shaped phycobilisomes of *Rhodella violacea*. I. Electron microscopy of phycobilisomes in situ and analysis of their architecture after isolation and negative staining. *Cytobiologie* 16: 118–29.

Moerschel, E. & Rhiel, E. 1987. Phycobilisomes and thylakoids: the light harvesting system of cyanobacteria and red algae. In *Electron Microscopy of Proteins*. Vol 6. *Membranous Structures*, eds. J. R. Harris & R. W. Horne, pp. 209–54 London: Academic.

Murata, N. 1969. Control of excitation energy transfer in photosynthesis. I. Light induced change of chlorophyll fluorescence in *Porphyridium cruentum*. *Biochim. Biophys. Acta.* 172: 242–51.

Myers, J. 1963. Enhancement. In *Photosynthetic Mechanisms*

of Green Plants, eds. B. Kok & A. Jagendorf, pp. 301–17. Washington, D.C., National Res. Counc. Publ. No. 1145.

Ohki, K. & Gantt, E. 1983. Functional phycobilisomes from *Tolypothrix tenuis.* (Cyanophyta) grown heterotrophically in the dark. *J. Phycol.* 19: 359–64.

Penniman, C. A. & Mathieson, A. C. 1985. Photosynthesis of *Gracilaria tikvahiae* (Gigartinales, Rhodophyta) from the Great Bay Estuary, New Hampshire. *Bot. Mar.* 28: 427–35.

Prezelin, B. B. & Boczar, B. A. 1986. Molecular bases of cell absorption and fluorescence in phytoplankton: potential applications to studies in optical oceanography. *Prog. Phycol. Res.* 4: 349–464.

Ramus, J. 1981. The capture and transduction of light energy. In *The Biology of Seaweeds*, eds. C. Lobban & M. Wynne, pp. 458–92. Oxford: Blackwell Scientific.

Ramus, J. 1983. A physiological test of the theory of complementary chromatic adaptation. II. Brown, green, and red seaweeds. *J. Phycol.* 19: 173–8.

Ramus, J. & van der Meer, J. P. 1983. A physiological test of the theory of complementary chromatic adaptation. I. Color mutants of a red seaweed. *J. Phycol.* 19: 86–91.

Redlinger, T. & Gantt E. 1981. Phycobilisome structure of *Porphyridium cruentum*: polypeptide composition. *Plant Physiol.* 68: 1375–9.

Redlinger, T. & Gantt, E. 1983. Photosynthetic membranes of *Porphyridium cruentum*: an analysis of chlorophyll–protein complexes and heme-binding proteins. *Plant Physiol.* 73: 36–40.

Richardson, K., Beardall, J. & Raven, J. A. 1983. Adaptation of unicellular algae to irradiance: an analysis of strategies. *New Phytol.* 93: 157–91.

Ried, A., Hessenberg, B., Metzler, H. & Ziegler, R. 1977. Distribution of excitation energy among photosystem I and photosystem II in red algae. I. Action spectra of light reactions I and II. *Biochim. Biophys. Acta* 459: 175–86.

Rosenberg, G. & Ramus, J. 1982. Ecological growth strategies in the seaweeds *Gracilaria foliifera* (Rhodophyceae) and *Ulva* sp. (Chlorophyceae): photosynthesis and antenna composition. *Mar. Ecol. Prog. Ser.* 8: 233–41.

Scheer, H. 1986. Excitation transfer in phycobiliproteins. In *Encyclopodia of Plant Physiology. Vol. 19. Photosynthetic Membranes and Light-Harvesting Systems*, eds. L. A. Staehelin & C. J. Arntzen, pp. 325–37. Berlin: Springer-Verlag.

Sheath, R. G., Hellebust, J. A. & Sawa, T. 1977. Changes in plastid structure, pigmentation and photosynthesis of the conchocelis stage of *Porphyra leucosticta* (Rhodophyta, Bangiophyceae) in response to low light and darkness. *Phycologia* 16: 265–76.

Staehelin, L. A. 1986. Chloroplast structure and supramolecular organization of photosynthetic membranes. In *Encyclopedia of Plant Physiology. Vol. 19. Photosynthetic Membranes and Light-Harvesting Systems*, eds. L. A. Staehelin & C. J. Arntzen, pp. 1–84. Berlin: Springer-Verlag.

Staehelin, L. A., Giddings, T. H., Badami, P. & Krsymowski, W. W. 1978. A comparison of the supramolecular architecture of photosynthetic membranes of blue-green, red and green algae and of higher plants. In *Light Transducing Membranes*, ed. D. W. Deamer, pp. 335–55. New York: Academic.

Steinmueller, K., Kaling M. & Zetsche K. 1983. In-vitro synthesis of phycobiliproteins and ribolose-1,5-bisphosphate carboxylase by non-poly-adenylated-RNA of *Cyanidium caldarium* and *Porphyridium aerugineum. Planta* (Berl.) 159: 308–13.

Steinmueller, K. & Zetsche, K. 1984. Photo- and metabolic regulation of the synthesis of ribulose bisphosphate carboxylase/oxygenase and the phycobiliproteins in the alga *Cyanidium caldarium. Plant Physiol.* 76: 935–9.

Stevens, C. L. R. & Myers, J. 1976. Characterization of pigment mutants in a blue-green alga, *Anacystis nidulans. J. Phycol.* 12: 99–105.

Stransky, H. & Hager, A. 1970. Das Carotinoidmuster und die Verbreitung des lichtinduzierten Xanthophyllcyclus in verschiedenen Algenklassen IV. Cyanophyceae und Rhodophyceae. *Arch. Microbiol.* 72: 84–96.

Troxler, R. F. 1986. Bile pigments in plants. In *Bile Pigments and Jaundice: Molecular, Metabolic, and Medical Aspects*, ed J. D. Ostrow, pp. 649–88. New York: Marcel Dekker.

Tseng, C. K., Zhou, B. & Pan, Z. 1981. Comparative photosynthetic studies on benthic seaweeds. III. Effect of light intensity on photosynthesis of intertidal benthic red algae. *Proc. Int. Seaweed Symp.* 10: 515–20.

Waaland, J. R., Waaland, S. D. & Bates, G. 1974. Chloroplast structure and pigment composition in the red alga *Griffithsia pacifica. J. Phycol.* 10: 193–9.

Wang, R. T., Stevens, C. L. & Myers, J. 1977. Action spectra for photoreactions I and II of photosynthesis. *Photochem. Photobiol.* 25: 103–8.

Yamazaki, I., Mimuro, M., Murao, T., Yamazaki, T., Yoshihara, K. & Fujita, Y. 1984. Excitation energy transfer in the light-harvesting antennae system of the red alga *Porphyridium cruentum* and the blue-green alga *Anacystis nidulans*. Analysis of time resolved fluorescence spectra. *Photochem. Photobiol.* 39: 233–40.

Yu, M. H., Glazer, A. N., Spencer, K. G. & West, J. 1981. Phycoerythrins of the red alga *Callithamnion*, variation in phycoerythrobilin and phycourobilin content. *Plant Physiol.* 68: 482–8.

Zilinskas, B. A. & Greenwald, L. S. 1986. Phycobilisome structure and function. *Photosynth. Res.* 10: 7–35.

Zuber, H. 1986. Primary structure and function of the light-harvesting polypeptides from cyanobacteria, red algae, and purple photosynthetic bacteria. In *Encylopedia of Plant Physiology. Vol. 19. Photosynthetic Membranes and Light-Harvesting Systems*, eds. L. A. Staehelin & C. J. Arntzen, pp. 238–251. Berlin. Springer-Verlag.

Chapter 10

Cell walls

JAMES S. CRAIGIE

CONTENTS

- I. Introduction / 221
- II. Cellulose / 222
- III. Mannans / 223
- IV. Xylans / 223
- V. Complex mucilages / 224
- VI. Sulfated galactans / 226
  - A. Agars and carrageenans / 226
  - B. Models for gel formation / 227
  - C. The carrageenan family / 227
  - D. Distribution of carrageenans / 232
  - E. The agarocolloid family / 236
  - F. Distribution of agarocolloids / 237
- VII. Proteins and glycoproteins / 241
- VIII. Other cell wall structures / 242
  - A. Cuticles and pit plugs / 242
  - B. Carbonate minerals / 243
- IX. Seasonal and environmental factors / 244
  - A. Natural populations / 244
  - B. Cultivated algae / 245
- X. Genetic factors / 246
- XI. Biosynthesis / 247
- XII. Summary / 247
- XIII. Acknowledgments / 248
- XIV. Endnote / 249
- XV. References / 249

I. INTRODUCTION

The basic metabolic pathways of red algae are shared in common with those of other algae and higher plants; however, the cell wall organization and composition of the Rhodophyta differ significantly from those of all other plants. The cell wall, except perhaps in some of the unicellular species, is a rather ordered structure as observed optically (Gordon-Mills & McCandless 1977; Gordon-Mills et al. 1978) and in the electron microscope (Fig. 10-1; Chapter 2; Gordon & McCandless 1973). It follows that the synthesis and assembly of the cell wall components are unlikely to be random or haphazard, but rather they are the culmination of many metabolic (enzymatic) processes, each of which is the result of the transcription of information from the algal genome. Environmental variables, physiological state, and reproductive stage can be important in modifying the final cell wall structure. While current taxonomy

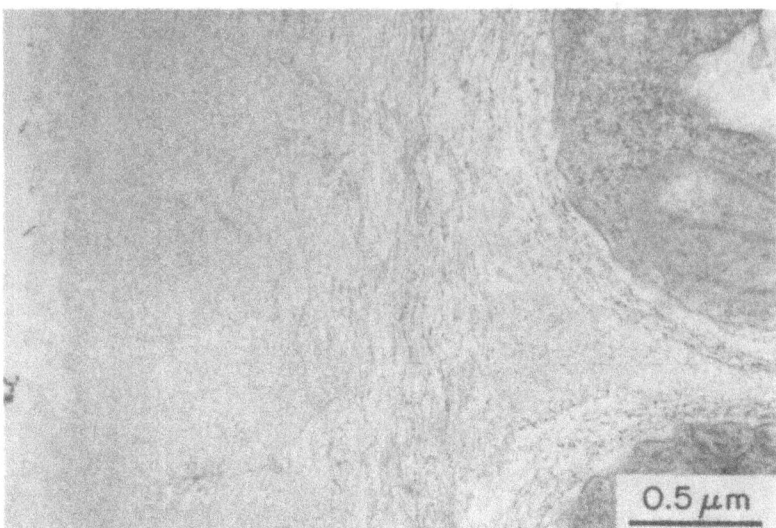

Fig. 10-1. TEM of cell wall of *Lomentaria baileyana*. The inner fibrous matrix surrounding each cell becomes oriented parallel to the thallus surface and is replaced by a fine granular outer matrix beneath the cuticle. The cell wall is swollen by the glutaraldehyde fixative. (Photograph courtesy of J. L. Scott)

of red algae is based upon morphological features and especially reproductive structures, biochemical characters (e.g., pigmentation, reserve metabolic products), including cell wall chemistry, are important determinants to be considered in classification (Gabrielson & Garbary 1986). As more red algal taxa are grouped to reflect their evolutionary affinities, it is to be expected that chemical and biochemical evidence will assume even greater importance (Gabrielson et al. 1985; McCandless 1978; Percival 1978; Stoloff & Silva 1957; Yaphe 1959). In the following discussion attention will be directed to more recent developments in the chemistry and biology of rhodophycean cell walls. The many excellent reviews now available will be cited to provide background information as required.

The cell wall will be considered to include those polymeric materials originating through the metabolic activities of the alga and lying exterior to the plasmalemmal membrane. Most investigators recognize polysaccharides, proteoglycans, peptides, proteins, lipids, and associated inorganic elements as constituents of the native cell wall. The polysaccharides may be grouped according to their presumed biological function as the more rigid structural (β-linked) glycans such as cellulose, mannans and xylans, and the more flexible, frequently sulfated glycans that comprise the matrix in which the skeletal fibers are embedded (Rees 1969; Rees et al. 1982). Relatively little is known of the proteins, peptidoglycans, and lipid constituents of algal cell walls, although these have been extensively studied in other groups.

II. CELLULOSE

Cellulose is undoubtedly the most intensively studied polysaccharide known today. At the molecular level the polymer consists solely of $\beta(1\rightarrow4)$-linked glucopyranosyl units in an extended ribbonlike structure (Marchessault & Sundararajan 1983). These chains are arranged in sheets and further stabilized by hydrogen bonds in two directions along the chain (e.g., OH–3----O–5, and OH–2----OH–6') and intermolecularly (e.g., OH–6----OH–3) to form a highly rigid structure. There are no free hydroxyl groups in crystalline cellulose, and the polymer is insoluble in physiological fluids. Cellulose from the green alga *Valonia ventricosa* is the most crystalline native form yet observed and is the reference material on which accurate physical measurements have been made (Marchessault & Sundararajan 1983).

Four polymorphic forms of cellulose are recognized from x-ray spectroscopy; only the first two are of interest in cell wall studies. Cellulose I is highly crystalline ($\geq70\%$) and is believed to be the native form present in sites in the cell wall. In it the glucan ribbons are arranged in a parallel packing mode unlike that in the less crystalline cellulose II (mercerized cellulose) in which the arrangement is antiparallel. As crystalline forms other than cellulose I are not found in nature, the highly ordered physical state of native cellulose must be imparted at the time of polymer biosynthesis and deposition.

Microscopists had long suspected the presence of cellulose in rhodophycean cell walls on the basis of simple cytochemical staining reactions (e.g., I_2-conc.H_2SO_4; chlor-zinc iodide). Quantitative analyses of 26 red algal species showed that cellulose, as defined by solubility properties, constituted only 1.0–9.0% of the cell wall (Ross 1953). Definitive chemical evidence for the presence of cellulose in *Rhodymenia pertusa* was provided by ^{1}H NMR studies of the permethylated cell wall material (Whyte & Englar 1971). Preston and his collaborators showed that x-ray reflections typical of cellulose I were present in cell walls of six species of Florideophycidae, but were absent in the Bangiophycidae (Frei & Preston 1961; Preston 1974). Other polysaccharides containing D-xylose, D-galactose, and D-mannose were present. X-ray diffraction and methylation analyses have established the presence of cellulose in the freshwater species *Lemanea annulata* (Nemaliales) (Gretz pers. comm.). It may be concluded that red algal cellulose is much lower in quantity and is less crystalline than that from other sources (e.g., *Valonia* and higher plants).

The modern concept that all red algae should be grouped in a single class (see Chapter 16) is supported by the demonstration of cellulose in the conchocelis phases of both *Bangia atropurpurea* and *Porphyra tenera*. In the former, Gretz et al. (1980) proved that the cuprammonium extract of the cell walls could be regenerated into cellulose II; in the latter, Mukai et al. (1981) isolated the cellulose microfibrils and characterized them by chemical methods including methylation analysis. Cellulose was later demonstrated in the conchocelis phases of *P. leucosticta* and *P. umbilicalis* (Gretz et al. 1984). The quantities reported were 3.8%, 5.5%, and 11.8% of the purified cell wall fractions for *B. atropurpurea*, *P. leucosticta*, and *P. tenera*, respectively (Gretz et al. 1986; Mukai et al. 1981). Careful examination revealed no cellulose in the generic phases of either *P. tenera* or *B. atropurpurea*.

III. MANNANS

The existence of an insoluble, structural mannan in members of the order Bangiales is incontrovertible. The chemical linkage of the *Porphyra umbilicalis* mannan was established as β-(1→4) long ago (Jones 1950). The localization of a partially crystalline mannan in the outer layers (cuticle) of the thallus of the same species was reported by Frei and Preston (1964). A similar mannan is known from *Bangia atropurpurea* (as *B. fuscopurpurea*) (Frei & Preston 1964; Usov et al. 1978b). All of these observations were made on the generic phases of the species in question and, as will be seen below, cannot be extrapolated to the conchocelis phases of the corresponding organisms.

Mechanically isolated, purified cell walls of the generic phase of *Porphyra tenera* yielded, on hydrolysis, 61.1% (w/w) of mannose that was shown by methylation analysis to be entirely (1→4)-linked (Mukai et al. 1981). More than 70% of the mannose occurred in an insoluble, granular mannan containing only traces of xylose. The observed chemical properties were consistent with those of a β-(1→4)-mannan. The granular rather than fibrillar appearance of isolated mannan has been attributed to its lability during the isolation procedure (Mackie & Preston 1974). Parallel preparations from the conchocelis phase of the same alga yielded only 2.1% of mannose that was β-(1-4)-linked (Mukai et al. 1981). Cell walls of the generic phase of *Bangia atropurpurea* are rich in mannose (30.8% w/w) (Gretz et al. 1982). The chromatographic properties of partial acid hydrolysis products of this cuprammonium soluble mannan were consistent with the presence of a β-(1→4)-linked polymer. In contrast, hydrolysis of cell walls from the conchocelis phases of *B. atropurpurea* and *P. leucosticta* yielded only 4.0% and 1.3% of mannose, respectively (Gretz et al. 1986). Clearly, mannan is a major structural component of the generic phases of bangialean algae. The conchocelis phases nevertheless appear to biosynthesize the same polymer, but in much smaller amounts.

IV. XYLANS

D-xylose is observed as a common, perhaps universal acid hydrolysis product of red algal cell walls. It is a constituent of the water and alkali soluble, ill-defined mucilage fractions, where it frequently occurs as branches on a complex heteroglycan. It can function as an important bridging molecule in peptide-oligosaccharide conjugates. Where relatively pure xylans have been studied, both β-(1→3) and β-(1→4) glycosidic linkages have been reported (Mackie & Preston 1974; Percival & McDowell 1967). Mild conditions (e.g., hot water or dilute acid) extract xylans with 17–38% of β-(1→3) links. These polymers frequently possess one or two branch points per xylose residue, e.g., rhodymenan from *Palmaria* (as *Rhodymenia*) *palmata*. However, β(1→3) and β(1→4)-linked xylans from *P. stenogona* are reported to be linear, with both linkage types occurring in the same molecules (Usov et al. 1978c). Strong alkali extracts essentially pure β-(1→4) xylan from *P. palmata* and β-(1→3) xylan from both *Bangia atropurpurea* (as *B. fuscopurpurea*) and *Porphyra umbili-*

calis (Painter 1983). Small quantities of xylose are associated with water soluble and insoluble fractions of mechanically purified cell walls from the generic phase of *B. atropurpurea* (Gretz et al. 1982).

Xylans are common in the Nemaliales, where they have been studied in *Nothogenia fastigiata* (as *Chaetangium fastigiatum*), *Galaxaura rugosa* (as *G. squalida*), *Nemalion vermiculare*, and *Audouinella floridula* (as *Rhodochorton floridulum*) (Usov et al. 1973, 1974, 1981, and references therein). In these cases, both 3- and 4-linked β-D-xylose residues occur in the same xylan chain, the separated xylans varying only in the proportions of β-(1→4) to β-(1→3) bonds present (Usov et al. 1981). Gretz (pers. comm.) isolated a β-4-linked xylan as a major alkali soluble constituent of *Lemanea annulata*.

The physical nature of the xylan microfibrils has been investigated (Mackie & Preston 1974), and it is believed that the β-(1→3) xylan molecules are arranged in triads to form macromolecules. The triple helix so formed is stabilized by interchain hydrogen bonds between 0–2 of each xylose unit, and by hydrogen bonds between 0–4 of one chain and the cyclic oxygen in the pyranose ring of an adjacent chain (Atkins & Parker 1969). The helical xylan structure lacks birefringence and is physically quite different from the extended, ribbonlike macromolecules of cellulose and the twofold helical structure of β-(1→4)-linked mannans, both of which are birefringent in polarized light (Mackie & Preston 1974).

The β-(1→3)-linked xylans have been reported only from algae that do not contain cellulose in their cell walls (e.g., certain forms in the Bangiales and members of the chlorophycean order Siphonales) (Iriki et al. 1960). It is interesting to speculate with Painter (1983) that algae capable of synthesizing β-(1→3) xylan microfibrils are "primitive" and may represent evolutionary dead ends as no higher plant is known to produce this type of xylan. The argument becomes less attractive, considering the study of Mukai et al. (1981). These workers were able to prepare essentially pure β-(1→3)-linked xylan microfibrils from the generic phase of *Porphyra tenera*, where the xylan replaced cellulose in the cell walls. The conchocelis phase of the same species did not contain β-(1→3) xylan, but instead formed cellulose microfibrils that were isolated and characterized. Both terminal and 4-linked xylose residues, however, were demonstrated in the conchocelis cell walls. Consistent with these observations is the report that xylose-containing polysaccharides occur together with galactans in the hot water extracts of cell walls from conchocelis phases of *Bangia atropurpurea* and *P. leucosticta* (Gretz et al. 1986). The controlling mechanisms that direct the synthesis of such radically different structural polymers as cellulose or β-(1→3)-linked xylan remain to be discovered.

V. COMPLEX MUCILAGES

Among the more primitive rhodophytes are the unicellular *Porphyridium* spp. and *Rhodella* (Porphyridiales), and the freshwater, multicellular *Batrachospermum* (Batrachospermales). *P. aerugineum* also is a freshwater species, although it grows well in one-tenth seawater medium (pers. obs.), while *P. purpureum* (as *P. cruentum*) thrives in full-strength seawater. Each affords a high molecular weight, viscous mucilage when extracted with hot water or dilute alkali. The unicellular forms release copious amounts of the same polymers into the culture medium (Anderson & Eakin 1986; Arad et al. 1985; Gudin & Thomas 1981; Jones 1962; Ramus 1972, 1973; Witsch et al. 1983). Significant exocellular polysaccharide production occurs only after cell division stops and individual cell size decreases (surface to volume maximized) in the stationary phase; polysaccharide synthesized during photoperiods appears to be solubilized during the dark periods (Thepenier & Gudin 1985).

The data assembled in Table 10-1 illustrate the similarities and differences in the composition of these complex mucilages. While it is obvious that the major monomeric units are D-xylose, D-glucuronic acid, D-glucose, and galactose, there is little similarity in their proportion, even within the same species. Attempts to fractionate the mucilage have not been successful, although a *P. purpureum* (as *P. cruentum*) preparation separated into two components on the analytical centrifuge (Heaney-Kieras & Chapman 1976). Conspicuous also is the moderate degree of sulfation (7.2–10%) of the polysaccharides from the unicells. By contrast, the *Batrachospermum* mucilage contained no sulfate and was rich in mannose residues (Turvey & Griffiths 1973). A variety of methylated carbohydrates have been characterized with the rare 2-O-methyl-D-glucuronic acid appearing in *P. purpureum*, but not in *P. aerugineum*, where mono- and dimethyl galactose derivatives are prominent. *Rhodella maculata* mucilage contains 3-O-methyl-Dk-xylose (Fareed & Percival 1977), in common with both *Porphyridium* spp. The discovery of 3-O-methyl-L-rhamnose (L-acofriose) in *Batrachospermum* mucilage is the first report in red algae.

Structural investigations have revealed that all of

Table 10-1. Composition (mole ratio relative to D-glucuronic acid, or weight percent) of water soluble mucilages of *Porphyridium* spp., *Rhodella*, and *Batrachospermum*

	Porphyridium aerugineum		Porphyridium purpureum (as P. cruentum)			Rhodella maculata	Batrachospermum sp.
Constituent	Ramus[a] (1972, 1973)	Percival & Foyle (1979)	Percival & Foyle (1979)	Heaney-Kieras & Chapman (1976)	Medcalf et al. (1975)	Evans et al. (1974)	Turvey & Griffiths (1973)
D-Gal	11%	1.48	2.38	0.17	0.82	S	1.42
L-Gal	—	0.72	0.75	1.57	7.21	—	0.88
D-Glc	11%	2.00	1.25	0.82	3.57	—	0.59
D-GlcA	—	1.00	1.00	1.00	1.00	12%	1.00
D-Man	—	—	—	—	tr	—	1.55
Ara	—	—	—	—	—	—	0.38
L-Rha	—	—	—	—	tr	+	0.08
D-Xyl	26%	3.40	3.75	1.98	8.93	49%	1.32
3-O-Me-D-Gal	—	1.2	0.16	—	—	—	0.013
3&4-O-MeGal	—	1.0	—	—	—	—	—
2,4-Di-O-MeGal	—	—	0.16	—	—	—	—
2-O-Me-hexose	—	—	0.25	—	—	—	—
2-O-Me-D-GlcA	—	—	—	2.14	0.79	—	—
2-O-Me-L-Rha	—	—	—	—	—	—	0.04
3-O-Me-D-Xyl	—	0.60	0.16	—	—	+	—
Protein	0	~5%	~5%	~1.5%	~5.6%	~16%[b]	~1.75%
Sulfate	~7.6%	~10%[c]	~10%	9.0%	7.2%	~10%	0
Mol. wt	≥2 × 10⁵	~5 × 10⁶	~4 × 10⁶	0.5 × 10⁶ – 10 × 10⁶	High	High	High

[a] Two additional unidentified carbohydrates A and B were reported comprising 35 and 16 percent of the polysaccharide, respectively.
[b] Protein and polysaccharide could be separated by PAGE (Callow & Evans 1981).
[c] Sixty percent of the sulfate is alkali-labile without forming anhydrosugars.

the D-glucuronic acid is internal and forms part of the polysaccharide backbone. It is (1→3)-linked in the *Porphyridium* spp. to D-glucose (Medcalf et al. 1975), solely to D-galactose (Percival & Foyle 1979), or to D-glucose, and D- and L-galactose (Heaney-Kieras & Chapman 1976). However, the 2-O-methyl-D-glucuronic acid (Kieras et al. 1976) is linked to 0-4 of L-galactose (Percival & Foyle 1979) or to D-galactose and D-glucose (Heaney-Kieras & Chapman 1976). Xylose, glucose, and galactose occur both internally in the polysaccharide and as end groups. The location of the sulfate hemiesters remains undefined, although infrared, periodate, and methylation analyses suggest that both xylose and the hexoses may be sulfated; both primary and secondary hydroxyl groups are involved.

The cell wall polysaccharides of other members of the Porphyridiales differ significantly from those of *Porphyridium* and *Rhodella*. Besides having less uronic acid and sulfate and an insignificant methoxyl content, their basic compositions vary as follows: *Chroodactylon* (as *Asterocytis*) *ornata* produces a (1→3)-linked galactan that lacks anhydrogalactose; *Stylonema alsidii* (as *Goniotrichium elegans*) yields an alkali soluble galactan having (1→3), (1→4), and (1→6) linkages, and containing anhydrogalactose but no sulfate; *Rhodosorus marinus* contains an unsulfated (1→4)-linked xylan with terminal galactose groups (Medcalf et al. 1981).

Unlike the situation in *Batrachospermum* (Batrachospermales), in which the backbone of the mucilage appears to be alternating D-glucuronic acid and D-galactose residues (Turvey & Griffiths 1973), members of the Nemaliales are reported to produce sulfated mannans of varying complexity. More than 30% of the dry weight of *Nemalion vermiculare* (Helminthocladiaceae) is a mannan with 15.5% of SO_3Na, which was shown to be composed of α-(1→3)-linked D-mannopyranose, its 6-sulfate, and 4-sulfate in a molar ratio of 6.2:1.3:1 (Usov et al. 1973, 1974, 1975). A small quantity (~2%) of D-xylose is glycosidically bonded to C-2 of the D-mannopyranose residues of the main chain (Usov & Yarotskii 1975). A novel sulfated xylogalactomannan from *Galaxaura rugosa* (as *G. squalida*) (Chaetangiaceae) was shown to have a linear backbone of β-(1-3)-linked D-mannopyranose residues (Usov et al. 1981). Carbon-13 nuclear magnetic resonance (^{13}C NMR), Smith degradation, and methylation analyses revealed that three of every five mannose residues were substituted at position 4 either with α-D-xylopyranosyl or L-galactopyranosyl residues. The sulfate hemiester was carried on C-4 of mannopyranose. More species need to be examined to determine whether the presence of D-glucuronic acid in the backbone structure of mucilages from the Porphyridiales and the freshwater *Batrachospermum* and its replacement by sulfated mannans in the marine macrophytes *Nemalion* and *Galaxaura* have particular significance.

VI. SULFATED GALACTANS

A. Agars and carrageenans

The commonest and most abundant cell wall constituents yet encountered in the Rhodophyta are families of galactans bearing the trivial names agars and carrageenans. The pioneering studies conducted in Japan (reviewed in Araki 1966) and Canada (O'Neill 1955a,b) established that the backbone structure of both agars and carrageenans was based on repeating galactose and 3,6-anhydrogalactose residues linked β-(1→4) and α-(1→3), respectively. The principal feature distinguishing the highly sulfated carrageenans from the less sulfated agars was the presence of D-galactose and anhydro-D-galactose in the former and D-galactose, L-galactose, or anhydro-L-galactose in the latter. Classification of these polysaccharides based on their solubilities and gelling properties proved unsatisfactory, so attention was focused on the common underlying structural patterns (Anderson et al. 1965, 1968a,b). The seminal concept of the masked repeating structure first reported for the agarlike porphyran (Anderson & Rees 1965) is now widely accepted for both agars and carrageenans. The evidence leading to these conclusions has been summarized in several excellent reviews (Haug 1974; Mackie & Preston 1974; O'Colla 1962; Peat & Turvey 1965; Percival 1970; Percival & McDowell 1967).

The repeating units may be substituted or modified in a number of ways to mask the underlying pattern. The repeating carrabiose structure of carrageenans (Fig. 10-2) may be altered by replacing the 3,6-anhydro-D-galactose with D-galactopyranose residues. The glycosyl units may be substituted with sulfate hemiesters in various patterns (Table 10-2) or more rarely with methyl ethers, and occasionally with pyruvic acid as the 4,6-O-carboxyethylidene group. Similarly, the agarobiose repeating structure of agars (Fig. 10-2) may be masked by replacing 3,6-anhydro-L-galactose by L-galactose, and/or adding methyl ethers and sulfate hemiesters at specific sites on either glycosyl unit. A pyruvic acid ketal group also may be present. Single residues of xylose or glucose may be attached to C-6

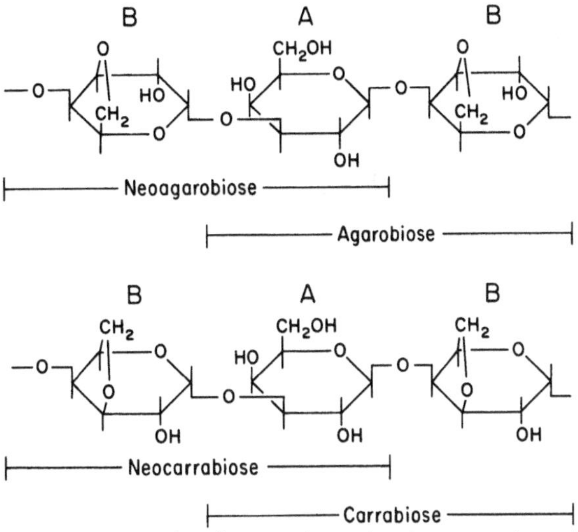

Fig. 10-2. Disaccharide repeating units of agaroses (upper) and carrageenans (lower).

of the 3-linked galactose in some agars (Hirase et al. 1983). Chemical, enzymatic, and NMR evidence showed that 4-O-methyl-α-L-galactopyranose is similarly linked in agar from *Gracilaria tikvahiae* (Craigie & Jurgens 1989; Craigie & Wen 1984). In the carrageenan-related polysaccharides of *Aeodes ulvoidea*, 4-O-methyl-L-galactose is attached through C-6 of the 4-linked residue (Allsobrook et al. 1974).

B. Models for gel formation

The elegant studies of Rees and his collaborators provide some details of the physical interactions of purified phycocolloid polymers (Dea 1981; Rees 1981; Rees et al. 1982). X-ray analysis of oriented dry films of iota carrageenan indicate that the individual molecules are arranged in a right-handed double helix with the strands parallel and threefold with a pitch of 2.66 nm (Anderson et al. 1969). The helix is fully stabilized by interchain hydrogen bonds through the only unsubstituted positions, O-6 and O-2, of the complementary D-galactose units. The sulfate hemiesters project outward from the main axis of the helix. The interchain packing of kappa carrageenan is more disordered than that of iota, but otherwise it forms an analogous double helix with a shorter (2.46 nm) pitch (Rees et al. 1982). Agarose is also visualized as having parallel threefold chains; however, the helix in this case is left-handed and less extended (1.90 nm pitch) than that in either iota or kappa carrageenans (Arnott et al. 1974a,b). According to Rees et al. (1982) the three polysaccharides can exist as double helices in solutions and in gels, but this concept has been challenged (see Letherby & Young 1981). On heating, gels become increasingly disordered and the strands disperse into a random coil form, whereas on cooling, they reaggregate to form gels. In the domain model of gelation, the primary association of molecules on cooling a carrageenan solution is through the formation of double helices. Gelation may be promoted by specific ions (e.g. K^+, Rb^+, Cs^+, or NH_4^+, or Ca^{2+}) that facilitate side-by-side aggregation of double helices to form a three-dimensional gel network (Fig. 10-3). This thermoreversible property and the effects of specific cations on carrageenans are of great importance in the commercial applications of the phycocolloids (Brant & Buliga 1985; Rees et al. 1982; Sanford 1985; Smidsrød & Grasdalen 1984a; Witt 1985). Lambda carrageenan lacks the 3,6-anhydride residues, does not form helices, and will not gel. Its molecules are perceived as having a rather flat, highly extended, ribbonlike form (Brant & Buliga 1985; Rees 1969; Rees et al. 1982).

New experimental evidence conflicts with the domain model for double helix formation during the disorder-to-order transformation of kappa carrageenan (Rinaudo & Rochas 1981; Smidsrød et al. 1980). First, a correlation has been established between ion binding and the strength of a carrageenan gel, and second, no increase in molecular weight could be confirmed during the development of the ordered conformation. Because these observations are incompatible with the formation of a double helix, a cation-specific, nested, single-helix model (the Smidsrød-Grasdalen model) was developed in which the cation promotes intermolecular aggregation of single kappa carrageenan helices (Fig. 10-3). An extensive interpretation of the evidence favoring this model has been put forward (Paoletti et al. 1984; Smidsrød & Grasdalen, 1984a,b). The molecular basis for gelation of kappa and iota carrageenans may not be identical as the disorder–order transition in kappa carrageenan is significantly influenced by anions whereas helix growth and nucleation in iota carrageenan are unaffected (Austen et al. 1985).

C. The carrageenan family

Because the carrageenans observed in the red algal cell walls are end products of the several enzymatic steps leading to their production, a rational grouping of the carrageenan molecules necessitates a

Table 10-2. Variations in patterns of sulfation and in distribution of 3,6-anhydro-D-galactose residues encountered in polysaccharides from selected carrageenophytes[a]

Taxon	Kappa family[b] (a) / (b)	Beta family[b]	Lambda family[b]	References
Cryptonemiales				
HALYMENIACEAE				
Pachymenia hymantophora	Wb?		Ma	Penman & Rees 1973
TICHOCARPACEAE				
Tichocarpus crinitis	Sb	Mb		Usov & Arkhipova 1981
Gigartinales				
PETROCELIDACEAE				
Mastocarpus papillatus (as *Gigartina papillata*)	Sb			Matsuhiro 1983
Mastocarpus stellatus (as *Gigartina stellata*)	Sb; Wa Wa			Bellion et al. 1983
Petrocelis middendorfii			Sa Sb	McCandless et al. 1983; DiNinno et al. 1979
GIGARTINACEAE				
Chondrus crispus ♂, ♀	Sb; Wa Wb		Sa	Bellion et al. 1983
Chondrus crispus ⊕	Sb Wb			McCandless et al. 1973
Chondrus ocellatus gam.	Sb Wb		Sa	McCandless et al. 1983
Chondrus ocellatus ⊕	Sb Wb			McCandless et al. 1983
Chondrus verrucosa gam.			Sa	McCandless et al. 1983
Chondrus verrucosa ⊕			Sa	McCandless et al. 1983
Gigartina atropurpurea	Wa; Wb Wa?		Wa? Sa	Penman & Rees 1973
Gigartina canaliculata	Sb; Wa Wb		Wa? Sa	Penman & Rees 1973
Gigartina canaliculata gam.	Sb Wb			McCandless et al. 1983
Gigartina canaliculata ⊕			Sa; Wb	McCandless et al. 1983
Gigartina chamissoi	Sb Wb; Wa		Wa? Sa	Penman & Rees 1973
Iridaea splendens (as *I. cordata*) ♂, ♀	Sb; Wa Wb			McCandless et al. 1975; 1983
Iridaea splendens ⊕			Sa	McCandless et al. 1975

Species					Reference
Iridaea cornucopiae gam.	Sb				McCandless et al. 1983
Iridaea cornucopiae ⊕	Wb		Sa		McCandless et al. 1983
Rhodoglossum californicum gam.	Wb		Sa		McCandless et al. 1983
Rhodoglossum californicum ⊕	Sb				McCandless et al. 1983
Rhodoglossum hemisphaericum	Sb				Yarotskii et al. 1978
PHYLLOPHORACEAE					
Ahnfeltia concinna	Sb	Sb; Wa			Bellion et al. 1981; 1983
Ahnfeltia durvillaei	Mb	Wb; Wa	Wa		Penman & Rees 1973
Ahnfeltia gigartinoides	Sb	Sb; Wa		Sa	Bellion et al. 1981; 1983
Ahnfeltia gigartinoides ⊕					McCandless et al. 1982
Ahnfeltia torulosa	Wa; Mb	Ma; Sb			Furneaux & Miller 1985
Ahnfeltia sp. B. (Japan) ⊕	Wa	Sb		Sa	McCandless et al. 1982
Gymnogongrus crustiforme ♂, ♀					McCandless et al. 1982
Gymnogongrus crustiforme ⊕	Wb	Sb		Sa	McCandless et al. 1982
Gymnogongrus flabelliformis gam.					McCandless et al. 1982
Gymnogongrus flabelliformis ⊕	Sb	Mb	Wa	Sa	McCandless et al. 1982
Gymnogongrus furcellatus					Penman & Rees 1973
Gymnogongrus furcellatus ⊕	Mb	Sb		Sa	McCandless et al. 1982
Gymnogongrus humilis		Sb			Furneaux & Miller 1985
Gymnogongrus leptophyllus ♀		Sb		Sa	McCandless et al. 1982
Gymnogongrus leptophyllus ⊕		Sb			McCandless et al. 1982
Gymnogongrus nodiferus		Sb			Furneaux & Miller 1985
Phyllophora truncata (as *P. brodiaei*)			Sb		Usov & Shashkov 1985
Phyllophora nervosa (*crispa*)	Mb				Usov & Arkhipova 1981
Phyllophora pseudoceranoides ♀		Sb			McCandless et al. 1982
Phyllophora pseudoceranoides ⊕		Sb			McCandless et al. 1982
Phyllophora truncata		Sb			McCandless et al. 1981
Stenogramma interrupta ♀	Wb	Sb		Sa	McCandless et al. 1982
Stenogramma interrupta ⊕		Sb			McCandless et al. 1982
SOLIERIACEAE					
Agardhiella subulata (as *A. tenera*)	Wa	Sb	Wa		Penman & Rees 1973; Cheney et al. 1987
Eucheuma gelatinae	Wb				Greer & Yaphe 1984a
Eucheuma isiforme	Wb	Sb; Wa	Wa		Penman & Rees 1973
Eucheuma isiforme (as *E. nudum*)	Sb				Shi et al. 1987
Eucheuma okamuri	Wb	Sb; Wa	Wa		Greer & Yaphe 1984b
Eucheuma spinosum		Sb; Wa			Usov & Shashkov 1985
Eucheuma uncinatum	Wa?	Sb; Wa		Wa?	Penman & Rees 1973
Kappaphycus (as *Eucheuma*) *cottonii*	Sb	Wb; Wa			Bellion et al. 1983
Opuntiella californica	Sb				Whyte et al. 1984

Table 10-2. (cont.)

Taxon	Kappa family[b]	Beta family[b]	Lambda family[b]	References
Sarcomena filiforme	Sb			Semesi & Mshigeni 1977b
Sarcodiotheca furcata	Sb			Whyte et al. 1985
Sarcodiotheca gaudichaudii (as Neoagardhiella baileyi)	Sb			Whyte et al. 1984
Turnerella mertensiana	Sb			Whyte et al. 1984
CAULACANTHACEAE				
Catenella nipae			Sb; Wa	Zablackis & Santos 1986
Caulacanthus ustulatus	Sb			Whyte et al. 1984
CYSTOCLONIACEAE				
Calliblepharis ciliata	Sb			Deslandes et al. 1985
Calliblepharis jubata	Sb			Deslandes et al. 1985
Cystoclonium purpureum	Mb	Wb		Deslandes et al. 1985
FURCELLARIACEAE				
Furcellaria lumbricalis	Wb?	Mb Wb	Wb?	Usov & Arkhipova 1981
RISSOELLACEAE				
Rissoella verruculosa	Wb	Sb		Mollion et al. 1986
HYPNEACEAE				
Hypnea musciformis	Sb			Greer et al. 1984
Hypneocolax stellaris	Sb; Wb?			Apt 1984

[a] Row (a) suggests a putative biosynthetic precursor of the disaccharide illustrated in row (b); the Greek symbols indicate the carrageenan type represented by a hypothetical polymer composed of these repeating disaccharide units. The analysis of separated gametophytes and sporophytes is indicated. Concentrations (W = weak, M = moderate; S = strong) of the disaccharide units in the isolated polymers are estimates from the references shown.
[b] The chemical structures illustrated show the positions of the glycosidic bonds and the sulfate hemiesters in the repeating disaccharide units of the carrageenans (e.g., κ-carrageenan = carrabiose-4-sulfate, Fig. 10-2; π-carrageenan bears pyruvic acid as the 4,6-O-carboxyethylidene group). λ-Carrageenan sensu Anderson et al. (1968a; Dolan & Rees 1965) carries two sulfate groups distributed among the three positions of the disaccharide repeating unit shown.

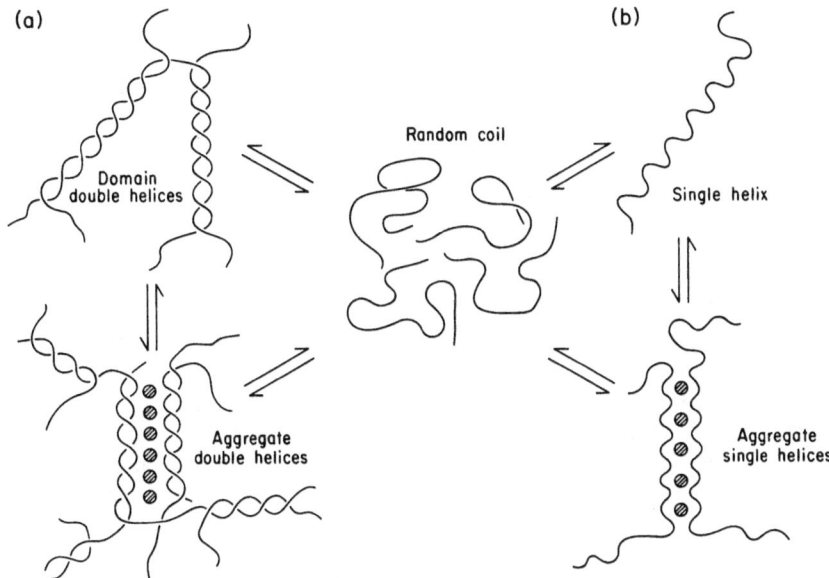

Fig. 10-3. Models for thermally reversible carrageenan gelation. a. The domain model with double stranded helices. The random coil to domain transition is believed to be the principal mode of association of iota carrageenan molecules; cations promote the aggregation of double helices. The random coil to aggregate of double helices is thought to be the sole mechanism of gelation in kappa carrageenan. (From Morris et al. 1980; Rees et al. 1982; with permission.) b. The single chain model. On cooling, individual, random carrageenan molecules form single helices that form side-by-side aggregates with suitable cations. (From Smidsrød & Grasdalen 1984a, with permission)

consideration of their biosynthetic history (Craigie & Wong 1979; McCandless & Craigie 1979).

Although some carrageenans may be methylated or pyruvated (Hirase & Watanabe 1972a), the overriding structural features are the degree and patterns of sulfation (Table 10-2). It seems reasonable to assume that the repeating disaccharide backbone unit, β-(1→4)-D-galactopyranosyl-α-(1→3)-D-galactopyranosyl, is assembled from appropriate nucleotides in an entirely conventional manner. Sometime during the elongation of the galactan chain or immediately after its formation, before it is released from the Golgi vesicles (Fig. 10-4), the polymer is enzymatically sulfated (Millard & Evans 1982a). Specificity of the various sulfotransferase enzymes provides a variety of substitution patterns for the sulfate hemiesters. The further metabolism of carrageenan through the action of a specific sulfoeliminase (sulfohydrolase) produces the commonly encountered 3,6-anhydrogalactose residue (Rees 1961a; Wong & Craigie 1978). Our knowledge of the details of the biosynthetic steps is insufficient to permit a satisfactory grouping of presumably interrelated carrageenans. Reflections along these lines led to the suggestion that the kappa and lambda carrageenans (see Dolan & Rees 1965) form natural biosynthetic groupings (see McCandless & Craigie 1979). The validity of this distinction is supported by the observation that gametophytes of members of the Gigartinaceae produce 4-sulfated carrageenans (the kappa family) whereas the tetrasporophytes do not (see Table 10-2 for references). It follows that the sulfotransferases must be rather specific; those capable of sulfating at C-4 of the 3-linked D-galactose do not act at C-2 of either hexose residue. In an analogous manner, sulfation solely at C-6 may occur. The monosulfated galactan postulated in Craigie and Wong (1979) was isolated from *Eucheuma gelatinae* and is now known as gamma carrageenan (Table 10-2); its desulfated end product is beta carrageenan (Greer & Yaphe 1984a). The recent discovery in *Rissoella verruculosa* (Rissoellaceae) of a galactan consisting of regularly repeating 3-linked-β-D-galactose-6-sulfate and 3,6-anhydro-α-D-galactose (Mollion et al. 1986) is a further example of variations on a theme. This omega carrageenan and its presumed biosynthetic precursor (suggested name psi, Greek symbol ψ) may be considered as more complex members of the beta carrageenan family (Table 10-2). The mono-

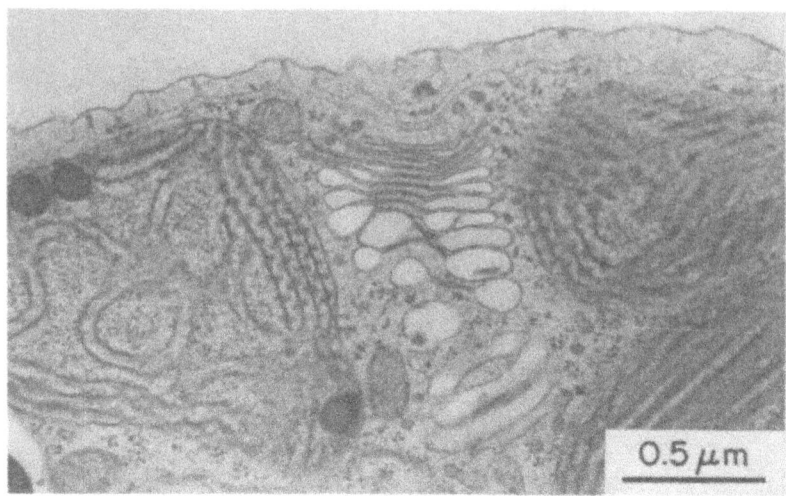

Fig. 10-4. TEM of a Golgi apparatus in *Rhodella violacea*. Note mitochondria adjacent to Golgi; vesicles stacked perpendicular to the plasmalemma. Phycobilisomes are seen on the thylakoid membranes. (Photograph courtesy of J. L. Scott)

sulfated repeating unit of omega carrageenan was previously identified in hybrid kappa carrageenans from both *Furcellaria lumbricalis* (as *F. fastigiata*) and *Phyllophora nervosa* (Usov & Arkhipova 1981).

In addition to the sulfoeliminase that converts the 4-linked galactose-6-sulfate to the 3,6-anhydride, there may exist sulfohydrolases that specifically remove sulfate hemiesters from other positions (e.g., C-2 of the 4-linked residue). For example, Bellion et al. (1983) detected nu but not mu carrageenan in *Kappaphycus* (as *Eucheuma*) *cottonii* and suggested that kappa carrageenan may be biosynthesized differently in this species than in *Chondrus crispus*, where the biological precursor appears to be mu carrageenan (Bodeau-Bellion 1983). It is perhaps significant that small segments of iota carrageenan occur in the kappa carrageenans of both *C. crispus* (Bhattacharjee et al. 1978; Craigie & Wong 1979) and *K. cottonii* (Bellion et al. 1983; Rochas & Heyraud 1981). In both *Ahnfeltia concinna* and *A. gigartinoides*, only nu carrageenan was reported in the "precursor" carrageenan fraction (Bellion et al. 1983); however, *A. torulosa*, which contains kappa and iotacarrageenan in a 30:50 ratio, synthesizes both mu and nu carrageenans as precursors (Furneaux & Miller 1985). In the absence of direct enzymatic evidence and without knowing the kinetics of carbon flow into the carrageenans, it is not possible to conclude whether nu carrageenan is formed from mu, or conversely, and whether or not kappa can be formed from iota.

Enzymatic, infrared, and NMR studies suggest that carrageenan structural hybrids (the main repeating disaccharide units are interrupted by segments of another carrageenan type) are common (see Bellion et al. 1982). The interrupting segments may be closely spaced or widely (randomly?) distributed in the macromolecule. Nothing is known of mechanisms that may control the insertion or number of such segments in a polysaccharide.

D. Distribution of carrageenans

The cell walls of comparatively few red algae have been carefully examined to date. Investigations tend to center on species that contain large quantities of polysaccharide of commercial importance (e.g., carrageenans and agars) and are available in quantity in nature. Therefore, members of the Gigartinales have received the most study. This order, including almost half of the florideophycidaean families, consists of some 100 genera and 700 species (Kraft 1981). Examples of the range of carrageenan structures presently known are provided in Table 10-2. They are grouped according to their sulfation patterns into the kappa, beta, and lambda families of carrageenans. Most of the gigartinalean species listed contain two or more recognizable carrageenan types; it is unusual for an alga to be reported as containing a single carrageenan. It is often difficult to determine whether or not the various carrageenans occur as separate chemical entities in a mixture of cell wall polysaccharides or whether they are glycosidically linked in a single hybrid macromolecule (see Haug 1974; Painter 1983).

The examples given in Table 10-2, together with the surveys of Abbott (1980), Lawson et al. (1973), McCandless et al. (1983), Parsons et al. (1977), Peats (1981), Pickmere et al. (1973, 1975), Stancioff and

Stanley (1969), and Whyte et al. (1984), establish that the Petrocelidaceae and the Gigartinaceae are rather homogeneous with respect to their carrageenan chemistries. Representatives of all genera in the families have now been examined. Carrageenans of the kappa family (kappa-iota hybrids) are produced by gametophytes, whether male or female, whereas carrageenans of the lambda family (lambda, xi, pi) characterize tetrasporophytes whether the alga is foliaceous or crustose in form.

Only a few exceptions to this general pattern of carrageenan distribution have been reported. In *Mastocarpus stellatus* (as *Gigartina stellata*), the flattened basal disk of an apomictic (direct type of life history) strain grown in culture contained lambda carrageenan, whereas erect fronds arising from the same disk contained the expected kappa-iota carrageenan; the corresponding structures from *Chondrus crispus* gametophytes contained kappa carrageenan (Chen & Craigie 1981). The basal disks and erect fronds of *Mastocarpus pacificus* (as *M. ochotensis*) follow the carrageenan distribution pattern observed for *M. stellatus*, whether or not they exhibit direct or heteromorphic life histories (Masuda et al. 1984, p. 121). Carrageenan preparations from field-collected carposporic and tetrasporic *C. canaliculatus* showed no marked differences in sensitivity to K^+ or in their chemical compositions (Ayal & Matsuhiro 1986). However, the present author has extracted independently carposporic and tetrasporic fronds of Chilean *C. canaliculatus* and found them to follow the normal pattern of carrageenan separation characteristic of the Gigartinaceae (Craigie unpubl.).

Although fronds from different phases of the life cycle of members of the Gigartinaceae differ markedly in their carrageenan chemistries (Chen et al. 1973; Hosford & McCandless 1975; McCandless et al. 1973), it must be reemphasized that the carrageenan extract from individual fronds is a complex mixture of related molecular species differing in average composition, and perhaps in molecular weight as well. Further, molecules exist of the hybrid type in which the repeating units of different carrageenans exist in the same molecule (i.e., kappa-iota hybrids, see Usov et al. 1980). Examples of such heterogeneity are well documented by fractional precipitation of *Chondrus crispus* extracts with KCl (Pernas et al. 1967). Similarly, extracts of carposporophytic *C. crispus* were shown to contain at least three types of carrageenan molecules (McCandless et al. 1973). Well-fructified *C. crispus* (female) gametophytes are reported to contain 4.8% (defatted dry wt) of lambda carrageenan when extracted after pretreatment of the alga with alcoholic HCl (Bremond et al. 1987). The carposporophyte-carposporangium complex, borne within ripe female gametophytes of *Chondrus*, is known to synthesize lambda carrageenan (Gordon-Mills & McCandless 1975).

Recent evidence establishes that the extract from *Iridaea membranacea* is heterogenous based on KCl fractionation and polyacrylamide gel electrophoresis (Ibañez & Matsuhiro 1986). Cystocarpic plants of both *I. ciliata* and *I. membranacea* make kappa carrageenans, but the tetrasporophyte carrageenans, although KCl soluble, do not conform to the classical lambda carrageenan structure (Ayal & Matsuhiro 1987). An iota carrageenan alternates with a nongelling lambda type in gametophyte and tetrasporophyte, respectively, of *I. boryana* (as *I. laminarioides*) (Goddard & Fernandez 1986). The carrageenans from *I. undulosa* (reproductive stages undefined) are predominantly soluble in KCl solutions until they are alkali modified (Matulewicz & Cerezo 1980). This extract was further separated into mu-like and lambda-like carrageenans, each containing segments with unusual sulfation patterns, especially on the 3-linked galactose; an undersulfated iota carrageenan was isolated after alkali modification, and the presence of unusual glycosidic linkages was suggested (Stortz & Cerezo 1986, 1987).

A similar, although less uniform, pattern of carrageenan distribution appears in the Phyllophoraceae (Table 10-2; McCandless et al. 1982), where gametophytes again produce kappa, iota, or more commonly, iota-kappa hybrid carrageenans. The tetrasporophytes, where unequivocal data have been obtained, produce the lambda family of carrageenans. A notable exception is *Phyllophora pseudoceranoides* in which iota carrageenan is reported to occur in both phases (McCandless et al. 1982). The gametophytes of *Gymnogongrus* spp. appear to be principally iota-kappa producers (Table 10-2); however, Whyte et al. (1984) showed that *G. norvegicus*, *G. chiton* (as *G. platyphyllus*), and *G. linearis* all produce kappa carrageenan with little or no iota in the first species and slightly more iota in the second and third, respectively. Cell walls of the tetrasporangium of *G. chiton* react with anti-lambda carrageenan antibodies, whereas walls of the tetraspores contained inside react with anti-kappa antibodies (McCandless & Vollmer 1984).

Phyllophora nervosa, formerly believed to be an agarophyte, contains an unusual carrageenan of the beta family as omega or an omega-kappa hybrid (Table 10-2; Usov & Arkhipova 1981). Unfortunately, it is not known in this case whether gametophytes, tetrasporophytes, or a mixture were examined, but a male plant of *P. crispa* contained an unidentifiable carrageenan, as did *Ozophora norrisii* (McCandless et al. 1982). The latter appears to be an iota carrageenan (McCandless & Gretz 1984). The carrageenan chemistry of *Stenogramma interrupta* shown in Table 10-2 may require reinvestigation, as the ^{13}C NMR spectrum of a polysaccharide extract from a carposporic plant is not characteristic of that expected for a carrageenan of the kappa family (Furneaux & Miller 1985). *Besa stipitata* (as *B. papillaeformis*) makes an iota-like carrageenan of rather low 3,6-anhydrogalactose and sulfate contents similar to that of several other members of the Phyllophoraceae (McCandless et al. 1982). An interesting exception to this pattern of carrageenan chemistry is found in *Ahnfeltia plicata*, the type species for the genus, which produces agar, not carrageenan (Bhattacharjee et al. 1978), and would appear on the basis of its cell wall composition to be misclassified.

The polysaccharide chemistries of approximately half of the genera in the Solieriaceae have been examined. The carrageenans thus far encountered belong to the kappa family with the exception of that in *Eucheuma gelatinae* which has a beta-kappa-gamma hybrid (Table 10-2; Greer & Yaphe 1984a). The carrageenan of *Kappaphycus* (as *Eucheuma*) *cottonii* approaches the idealized structure more closely than kappa carrageenan from any other source; nevertheless, it contains minor segments of iota carrageenan and small amounts of 6-O-methylgalactose (Bellion et al. 1983). Cystocarpic and tetrasporic thalli from *Kappaphycus* and *Eucheuma* spp. were shown to contain the same carrageenan type in each: kappa carrageenan in *K. cottonii*, *E. odontophorum*, *E. platycladum*, and *K.* (as *E.*) *procrusteanum* and iota carrageenan in *E. arnoldii* and *Eucheuma* sp. (Doty & Santos 1978). Kappa carrageenan, similar to that in *K. cottonii*, occurs also in *Opuntiella californica* and *Turnerella mertensiana*. It is perhaps significant that in each of these three species the carrageenan contains low levels of 6-O-methylgalactose (Bellion et al. 1983; Whyte et al. 1984), a sugar now reported in carrageenans of the Gigartinaceae (Ayal & Matsuhiro 1986; Ibañez & Matsuhiro 1986). *Solieria chordalis* is reported to contain iota carrageenan (Deslandes et al. 1985), as does *Tenaciphyllum lobatum* (Solieriaceae) (Semesi 1979).

A "deviant" iota carrageenan, in which the regular iota structure is kinked when a 4-O-α-D-galactopyranose-2-sulfate residue replaces the 3,6-anhydride, has been reported in certain of the Phyllophoraceae and Solieriaceae (Table 10-2; Dawes et al. 1977; DiNinno & McCandless 1978; Penman & Rees 1973). In a recent reinvestigation of *Eucheuma isiforme* (as *E. nudum*), the Greek letter delta (δ) was used parenthetically to indicate the repeating deviant unit (Greer & Yaphe 1984b, p. 483); however, the O-β-D-galactopyranosyl-4-sulfate (1→4)-O-α-D-galactopyranosyl-2-sulfate (1→3) residue was not found. It is suggested that the letter delta be reserved to designate the presumed precursor of alpha carrageenan (Table 10-2; Zablackis & Santos 1986). Segments with the deviant iota repeating unit may be termed omicron carrageenan (Table 10-2), following Rees' system for naming carrageenans (Rees 1969).

The gametophyte of *Anatheca montagnei* appears to be quite normal as it is reported to contain an undersulfated iota carrageenan (Mollion 1980). The only other species examined in this genus, *A. dentata*, differs completely from all other members of the Solieriaceae described to date. Its highly sulfated galactan shares the common alternating α-(1→3), β-(1→4) glycosydic linkages, but has a D:L-galactose ratio of 1.57:1 (Nunn et al. 1971, 1973, 1981). All of the 3-linked units are D-galactose; all of the L-galactose is 4-linked. However, about 20% of the 4-linked units are D-galactose, and this residue may be distributed widely in the polysaccharide molecule. The sulfate is alkali stable and appears to be present as the uncommon L-galactopyranosyl 2,3,6-trisulfate. Some of the 3-linked β-D-galactopyranosyl residues bear 4,6-O-(1-carboxyethylidene) groups, whereas others carry xylopyranosyl branches. D-glucopyranosyluronic acid residues were present and shown to occupy internal positions in the polysaccharide chain. In addition, traces of 3-O-methylgalactose were detected. Thus, the *A. dentata* polysaccharide is a complex agaroid-carrageenan hybrid with a previously unknown sulfation pattern.

Representatives of five other families in the Gigartinales have been reported as carrageenophytes (Table 10-2). *Catenella nipae* produces the novel alpha carrageenan (lambda family) as a major cell wall polysaccharide, together with its presumed precursor delta carrageenan (Zablackis & Santos 1986). The logic for the Greek lettering of carrageenans (Rees 1969) broke down with the naming of pi carrageenan (DiNinno et al. 1979), and various authors have named new carrageenan structures (Table 10-2) virtually at random (see deviant iota carrageenan above). The *Furcellaria lumbricalis* (as *F. fastigiata*) polysaccharide, furcellaran, was shown to be a complex kappa-beta hybrid containing some segments of omega carrageenan. The latter was identified as the principal galactan of *Rissoella verruculosa* on the basis of the complete assignment of its ^{13}C NMR spectrum (Mollion et al. 1986). The carrageenan of *R. verruculosa* was previously described as a kappa type somewhat similar to furcellaran (Combaut et al. 1985); however, the ^{13}C chemical shifts reported for the anomeric carbons (δ 105.3, 97.8, 97.2) are rather far downfield, being closer to those of precursor carrageenans (nu + mu) at δ 104.5, 98.0, and 97.5 (Bellion et al. 1983) than to those of kappa, iota (Usov 1984), or omega carrageenans. Kappa carrageenan characterizes both *Hypnea musciformis* and its parasitic red alga, *Hypneocolax stellaris* (Table 10-2). Similarly, *Hypnea ceramioides*, *H. cervicornis*, *H. japonica*, *H. nidifica*, and *H. spicifera* all have been shown to contain kappa carrageenan (see Santos & Doty 1979). The infrared spectra of several of these carrageenan preparations show a weak absorption near 805 cm^{-1}, which suggests that traces of iota carrageenan may be present. There is no change in the carrageenan chemistry with alteration of the life cycle in members of either the Furcellariaceae or the Hypneaceae (McCandless 1978, 1981).

A major departure from the regular repeating α-(1→3)-, β-(1→4)-linked galactans so common in the Gigartinales was reported in two families (Whyte et al. 1984). On the basis of infrared, methylation, and gas chromatographic analyses, the galactans from *Schizymenia pacifica* (family Gymnophloeceae, formerly Nemastomaceae) and from both *Plocamium cartilagineum* and *P. violaceum* (Plocamiaceae) were shown to be composed predominantly of (1→3)-linked D-galactopyranose units. Sulfate hemiesters were present on C-4 and to a lesser extent on C-2, together with a small amount of methoxyl at C-6. Some (1→4)-linked units were present, as was a low level of 3,6-anhydrogalactose. It is obvious that such galactans belong neither to the carrageenans nor to the agars. Further characterization of these polymers is required.

The polysaccharide of *Porteria* (as *Chondroccus*) *hornemannii* (Rhizophyllidaceae, Gigartinales) is reported to be a carrageenan, perhaps of the lambda family (Semesi & Mshigeni 1977b). The carrageenan of *Polyides rotundus* (Polyideaceae) also requires further investigation.

Rather few members of the Cryptonemiales have been examined in detail for wall structure. Evidence for a kappa-beta hybrid carrageenan (Table 10-2) in *Tichocarpus crinitis* (Tichocarpaceae) is derived from ^{13}C NMR and methylation analyses (Usov & Arkhipova 1981). The water soluble polysaccharides extracted from several members of the Halymeniaceae (formerly Grateloupiaceae or Cryptonemiaceae) are highly sulfated galactans containing low levels of 3,6-anhydrogalactose. They exhibit a positive specific optical rotation and do not gel in the presence of K$^+$. The principal glycosidic linkages in these galactans are generally agreed as being alternating α-(1→3) and β-(1→4) with all of the 3-linked residue being in the D-configuration (Usov & Barbakadze 1978). Some of the 4-linked D-galactose may be replaced by the L-enantiomer as in *Grateloupia divaricata* (Barbakadze & Usov 1978; Usov & Barbakadze 1978; Usov et al. 1978a) and *G. elliptica* (Hirase et al. 1967); however, this may not always occur (e.g., *G. lanceola*, Baeza & Matsuhiro 1977). Sulfate hemiesters are located predominantly on the 3-linked residues at the C-4 or C-2 positions (Table 10-2), with occasional sulfation of the 4-linked residue at C-6 (or C-3) and perhaps at the C-2 positions. Partial substitution by methoxyl at C-6 of the 3-linked unit and C-2 of the 4-linked unit is commonly reported. 4-O-methyl-L-galactose, although of infrequent occurrence, can be a major constituent as in aeodan from *Aeodes ulvoides* (Allsobrook et al. 1971), where it is glycosidically α-linked to C-6 of the 4-linked D-galactose as a single branch unit (Allsobrook et al. 1974, 1975). The enantiomer, 4-O-methyl-D-galactose, co-occurs in *Pachymenia carnosa* (Farrant et al. 1971; Parolis 1978). In addition, a small percentage of D-xylose is often found in acid hydrolyzates of the polysaccharides and appears to be attached as individual branches to the main galactan chain (Barbakadze & Usov 1978).

The precise location of sulfate groups in the *Pachymenia carnosa* galactan was achieved only following Hakomori methylation as other methods of methylation caused partial desulfation and artifactual results (Parolis 1978). All the sulfate was located on D-galactose at positions 2,4 and 2,6 of the 3-linked residue. A maximum of only 60% of the polymer could have a regular repeating α-(1→3), β-(1→4) structure. The term "aeodan," implying a D-galactan consisting of α-(1→3) and β-(1→4) linkages, sulfated only on the 3-linked residue and lacking anhydrogalactose, is useful to distinguish such polysaccharides from the carrageenans.

Further examples of structural variations in the sulfated galactans of the Halymeniaceae are the aeodans from *Pachymenia hieroglyphica* and *P. hymantophora*, in which the ratio of 3-linked to 4-linked residues is 4:1 in the former and about 1:1 in the latter (Parolis 1981). Approximately half of the 3-linked residues of the *P. hieroglyphica* aeodan carry all of the sulfate hemiester as β-D-galactopyranose-2-sulfate and β-D-galactopyranose-4-sulfate (in 3:1

proportions), whereas the 4-linked residue is solely α-D-galactopyranose. The xylose present and perhaps some galactose are believed to be glycosidically bound to C-6 of the 4-linked residues. In *P. hymantophora*, virtually all of the 3-linked residues are esterified principally as β-D-galactopyranose-2-sulfate and the 2,6-disulfate; the 4-linked residue is predominantly α-D-galactopyranose, with small amounts of the 6-sulfate and the 3,6-anhydride. Parolis (1981) has cautioned correctly not to classify the aeodans with the carrageenans before it is conclusively established that the 3-linked and 4-linked residues alternate regularly. Indeed, if the aeodan of *P. hieroglyphica* is a true homogenous polymer, only 40% of it could have the classical alternating structure of a carrageenan. In several respects, aeodans seem closely allied to the sulfated galactans reported in the Nemastomaceae and Plocamiaceae (see above). A KCl soluble, sulfated polysaccharide (lambda carrageenan?) has been isolated from *Halymenia venusta* (Semesi & Mshigeni 1977a).

More extreme is the contrast in cell wall galactan composition between the two genera within the Endocladiaceae (Cryptonemiales). *Gloiopeltis furcata* produces funoran, a highly sulfated agarose (Hirase & Watanabe 1972b; Whyte et al. 1985) in which the 3-linked galactose residue is largely sulfated (or sometimes O-methylated) at C-6, while some sulfate hemiester or methoxyl also occurs at C-2 of the 4-linked residue. A similarly sulfated agar was extracted from *G. cervicornis* (Penman & Rees 1973). However, *Endocladia muricata* is reported to synthesize a partly desulfated carrageenan containing about 6% of agarobiose units (Whyte et al. 1985). Therefore, the *E. muricata* polysaccharide appears to be a carrageenan-agar hybrid similar to those found in some *Grateloupia* spp. (Halymeniaceae).

Hot water extracts of *Corallina officinalis* (Corallinaceae, Cryptonemiales) afford a galactan sulfate containing both D- and L-galactose in a 1.3:1 molar ratio (Turvey & Simpson 1966). Although its structure was not fully elucidated, the galactan was composed of equal quantities of (1→3)- and (1→4)-linked units, lacked 3,6-anhydrogalactose, but contained some sulfate hemiester at C-6 of the L-galactose position, as well as at the C-4 position. In addition, the D-xylose content of the polysaccharide was unusually high at 22.7%. More recently, the uniqueness of the cell wall structure of coralline algae was shown by the isolation and characterization of alginate from *Serraticardia maxima* (Okazaki et al. 1982a). The alginate accounted for approximately 2% of the organic matter of the alga and was relatively rich in guluronic acid (mannuronic: guluronic acid ratio = 0.8). Similar alginate-like polymers were observed in other coralline algae, *Marginisporum* (as *Amphiroa*) *aberrans* and *Lithothamnion japonicum* (Okazaki et al. 1982a). While glucuronic acid has been reported as a fairly common minor constituent of rhodophycean polysaccharides, this is the first report of the polyuronide alginate as a constituent of red algae.

Early workers investigating the mucilages of *Dilsea edulis* and *Dumontia contorta* (as *D. incrassata*) (both Dumontiaceae, Cryptonemiales) found them to be sulfated and very complex (see Percival & McDowell 1967). The *D. edulis* galactan is predominantly (1→3)-linked, but residues of glucuronic acid and xylose are also present. Both mucilages should be reexamined to ascertain the nature of the branching and the location of the sulfate hemiesters.

E. The agarocolloid family

The agarocolloid family of galactans may be regarded as analogs of the carrageenan family on both chemical and biological grounds. A high polymer of strictly alternating 3-O-linked β-D-galactopyranose and 4-O-linked 3,6-anhydro-α-L-galactopyranose (Fig. 10-2) constitutes the strongly gelling polysaccharide agarose (see Araki 1966; Percival & McDowell 1967; Rees 1969). As in the case of the carrageenans, the basic structure of agarose may be masked or altered in a number of ways by the substitution of hydroxyl groups with methoxyl or sulfate in various combinations (see McCandless & Craigie 1979; Rees 1969; Turvey 1978). Less frequently, pyruvate residues are present as the 4,6-O-(1-carboxyethylidene) groups (Araki 1966; Duckworth et al. 1971). In general, agarocolloids are more heavily methoxylated than carrageenans, whereas the converse is true for the occurrence of sulfate hemiesters. When the 3,6-anhydride residue in the polymer is replaced by α-L-galactose-6-sulfate, we have the chemical precursor of agarose (Rees 1961b). The early concept of agar as a mixture of neutral agarose and charged "agaropectin" (see Araki 1966) must be abandoned as it is clear that a number of charged polymers can be separated from crude agars by anion exchange chromatography (Duckworth & Yaphe 1971a,b; Izumi 1970, 1971, 1972; Ji et al. 1985). The agarose macromolecule itself can be heterogeneous; the repeating disaccharide structures may be interrupted by sequences of masked repeating units to create block structures within the polymer as seen in the carrageenans. Further, the basic agarose and carrageenan structures can coexist in the same

Cell walls

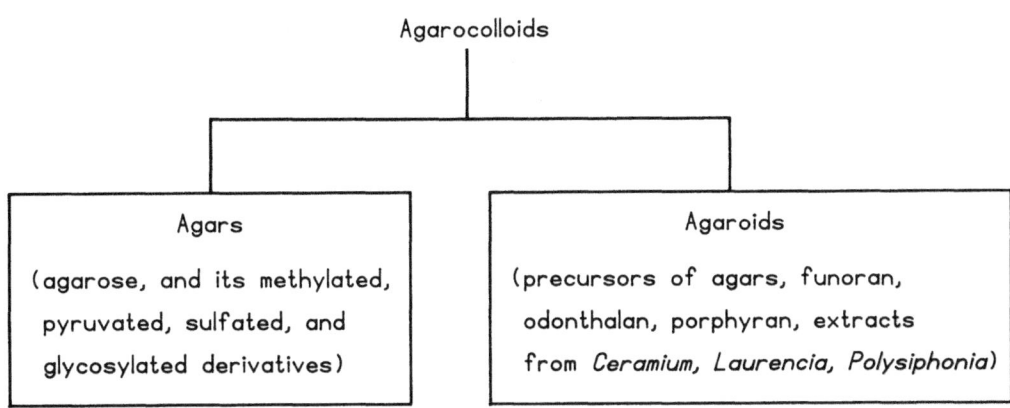

Fig. 10-5. Water soluble galactans (agarocolloids) from agarophytes arranged in gelling (agars) and weakly gelling or nongelling (agaroids) groupings.

polysaccharide to create hybrid polymers in some species.

On biological grounds, the analogy also holds as these polymers constitute much of the intercellular matrix in the species in which they occur (Gordon & McCandless 1973; Verdus et al. 1986). Both the agarocolloids and carrageenans thus appear to serve similar structural and physiological functions in the living algal thallus. The removal of sulfate from galactose-6-sulfate and the concomitant cyclization to form 3,6-anhydrogalactose, although catalyzed by different enzymes in the agars and the carrageenans (Rees 1961a; Wong & Craigie 1978), indicate that at least one step in their biogenesis is analogous.

Caution must be exercised in interpreting reports of the occurrence, yields, and properties of "agar" extracts. Agarocolloids with high charge densities often have low 3,6-anhydrogalactose levels and, because they do not gel, can be discarded during preparation of the agar (see Durairatnam & Santos 1981; Miller & Furneaux 1987). In *Gracilaria sjoestedtii*, the unmodified agar forms a strong gel, but if the nongelling agarlike polymers are alkali treated, the total yield of gel can be increased by ≈50%. However, the resulting "agar" is chemically more heterogeneous and has poorer gelling properties (Wen & Craigie 1984). Similar conclusions have been reached for other *Gracilaria* spp. (Friedlander et al. 1981; Lahaye et al. 1986; Miller & Furneaux 1987). Polysaccharides with highly masked structures (see Rees 1969) such as funoran, odonthalan, porphyran, and sulfated galactans from species of *Ceramium*, *Laurencia*, and *Polysiphonia*, therefore, belong to the agarocolloid family and undoubtedly share several steps of a common biosynthetic pathway.

To encourage the use of more precise yet practical terms to describe galactans from agarophytes, it is suggested that the definition of agar proposed by Tseng (1945) be revised to reflect the present knowledge of the chemical structures involved. Thus agar becomes a subset of the agar family of colloids, which collectively may be termed agarocolloids (Fig. 10-5). In this way the longstanding practical concept of agar as a thermoreversible gel is retained. The masked agarlike galactans, including the extreme case in which all of the 4-linked residues consist of L-galactose and/or its 6-sulfate (and their methyl ethers), may be termed agaroid(s). Much of the original meaning of the term "agaroid" (see Tseng 1945) would be preserved.

F. Distribution of agarocolloids

The agaroid fractions (Fig. 10-5), which are of relatively little commercial interest, are more difficult to purify and examine structurally and so are less well known than agarose. From the biological viewpoint, however, these molecules, being more complex, have involved more biosynthetic steps in their production and thereby provide more points of comparison among potentially related algal species. The body of Table 10-3 provides examples of the range of masking structures reported in agaroids to date.

Agarocolloids are found in at least eight families distributed over five rhodophytan orders (Table 10-3). The Phyllophoraceae (not listed) contains an anomalous member, *Ahnfeltia plicata*, known to synthesize an agarose (Bhattacharjee et al. 1978), but its agaroid fraction has not been investigated.

The fewest variations in masking structure appear in the agaroids (porphyrans) from the

Table 10-3. Examples of masking residues reported in agarocolloids from various species

Taxon	3-Linked residue[a]	4-Linked residue[b]	Reference
Bangiales			
BANGIACEAE			
Bangia atropurpurea (as *B. fuscopurpurea*)	(tr) 2-O-Me and 4-O-Me? (W) 6-O-Me	(tr) 2-O-Me? (S) 6-SO$_4$	Shashkov et al. 1978; Usov et al. 1978b
Bangia atropurpurea (conchocelis)	O	(S) 6-SO$_4$	Gretz et al. 1983
Porphyra capensis	(S) 6-O-Me	(M) L-Gal & 6-SO$_4$	Nunn & von Holdt 1957
Porphyra columbina (Chile & New Zealand)	6-O-Me	6-SO$_4$	Brasch et al. 1981a,b; Villarroel & Zanlungo 1981
Porphyra haitanensis	(W) 6-O-Me	(S) 6-SO$_4$	Lahaye et al. 1985
Porphyra leucosticta	(M) 6-O-Me	(S) 6-SO$_4$	Rees & Conway 1962a
Porphyra leucosticta (conchocelis)	O	(S) 6-SO$_4$	Gretz et al. 1983
Porphyra linearis	(W) 6-O-Me	(S) 6-SO$_4$ L-Gal	Rees & Conway 1962a
Porphyra perforata	(S) 6-O-Me	(S) 6-SO$_4$	Morrice et al. 1983; Rees & Conway 1962a; Turvey & Williams 1964
Porphyra umbilicalis	(S) 6-O-Me	(S) 6-SO$_4$	Rees & Conway 1962a
Nemaliales			
GELIDIACEAE[c]			
Gelidium lingulatum	(M) 6-O-Me	(W) 2-O-Me-L-Gal	Zanlungo 1980
Gelidium purpurascens	(W) 6-O-Me	(M) 6-SO$_4$	Whyte & Englar 1981a
Pterocladia lucida/pinnata	(tr) 6-O-Me	—	Brasch et al. 1981b
Cryptonemiales			
ENDOCLADIACEAE			
Gloiopeltis cervicornis	(S) 6-SO$_4$ (W) 6-O-Me	(W) L-Gal-2-SO$_4$	Penman & Rees 1973
Gloiopeltis furcata	(S) 6-SO$_4$ (W) 6-O-Me	(M) 3,6-An-2-O-Me-L-Gal (W) 3,6-An-L-Gal-2-SO$_4$	Hirase & Watanabe 1972b; Whyte et al. 1985
Gigartinales			
GRACILARIACEAE			
Gracilaria blodgettii	(tr) 6-O-Me	(W) 3,6-An-2-O-Me-L-Gal (W) 6-SO$_4$	Ji et al. 1985; Lahaye et al. 1986
Gracilaria cervicornis (as *G. ferox*)	(M) 6-O-Me	(W) 6-SO$_4$	Duckworth et al. 1971
Gracilaria bursa-pastoris	(W) 6-O-Me	(M) 6-SO$_4$	Duckworth et al. 1971
Gracilaria damaecornis	(M) 6-O-Me	(tr) 6-SO$_4$	Duckworth et al. 1971
Gracilaria debilis	(M) 6-O-Me	(M) 6-SO$_4$	Duckworth et al. 1971
Gracilaria domingensis	(M) 6-O-Me	(W) 6-SO$_4$	Duckworth et al. 1971
Gracilaria eucheumoides	(W) 6-O-Me	(S) 3,6-An-2-O-Me-L-Gal (S) 2-O-Me-L-Gal-6-SO$_4$	Ji et al. 1985; Lahaye et al. 1986
Gracilaria secundata	(S) 6-O-Me	—	Brasch et al. 1981b, 1983
Gracilaria "sjoestedtii"	(M) 6-O-Me	(M) 6-SO$_4$	Craigie et al. 1984
Gracilaria tenuistipitata	(S) 6-O-Me	(W) 3,6-An-2-O-Me-L-Gal (W) 6-SO$_4$	Ji et al. 1985; Lahaye et al. 1986
Gracilaria textorii	(M) 6-O-Me	(W) 6-SO$_4$	Craigie et al. 1984
Gracilaria tikvahiae (as *G. foliifera*)	(S) 6-O-Me	(W) 6-SO$_4$	Craigie et al. 1984; Duckworth et al. 1971
Gracilaria verrucosa	(W) 6-O-Me	(W) 6-SO$_4$	Craigie et al. 1984; Ji et al. 1985
Ceramiales			
CERAMIACEAE[d]			
Ceramium boydenii	(S) 6-O-Me	—	Hirase & Araki 1961
Ceramium rubrum	6-O-Me; 6-SO$_4$ 2-SO$_4$ and 4-SO$_4$?	2-O-Me-L-Gal 3,6-An-2-O-Me-L-Gal 6-SO$_4$; L-Gal	Turvey & Williams 1976

Table 10-3. (cont.)

Taxon	3-Linked residue[a]	4-Linked residue[b]	Reference
DELESSERIACEAE			
Delesseria crassifolia (as *Laingia pacifica*)	6-SO$_4$; 4-SO$_4$	3,6-An-2-O-Me-L-Gal	Kochetkov et al. 1967, 1970a,b
RHODOMELACEAE			
Bryothamnion triquetrum	(S) 6-O-Me	(W) 2-O-Me-Gal	Fernandez et al. 1987
Digenea simplex	(W) 6-O-Me	—	El-Sayed 1983
Laurencia pinnatifida	(M) 6-O-Me 2-SO$_4$	L-Gal; 6-SO$_4$; (W) 2-O-Me-L-Gal; 2-O-Me-L-Gal-6-SO$_4$; 3,6-An-2-O-Me-L-Gal	Bowker & Turvey 1968a,b
Neorhodomela (as *Rhodomela*) *larix*	2-SO$_4$; 4-SO$_4$; 6-SO$_4$	6-SO$_4$; L-Gal (S) 3-6-An-2-O-Me-L-Gal	Shashkov et al. 1978; Usov 1984; Usov & Ivanova 1981
Odonthalia corymbifera	(S) 6-O-Me (M) 4-SO$_4$	—	Shashkov et al. 1978;
Polysiphonia lanosa	6-O-Me; 6-SO$_4$ 6-O-Me-D-Gal-4-SO$_4$	6-SO$_4$ 2-O-Me-L-Gal-6-SO$_4$	Batey & Turvey 1975

[a] The substituent groups are shown for β-D-galactopyranose.
[b] The substituent groups (or residues) of α-L-galactopyranose are shown. These replace 3,6-anhydro-α-L-galactopyranose (3,6-An-L-Gal).
[c] The 6-O-methyl-D-galactose content of agarose is reported as *Acanthopeltis japonica* 3.2%; *Gelidium amansii* 1.4%; *G. japonicum* 1.6%; *G. subcostatum* 7.3%; *Pterocladia tenuis* 0.8% (Araki 1966).
[d] The 6-O-methyl-D-galactose content of agarose is reported as: *Campylaephora hypnaeoides* 0.8%; *Ceramium boydenii* 20.8% (Araki 1966).
tr = trace; W = weak; M = moderate; S = strong.

Bangiales where 6-O-methyl-D-galactopyranose and L-galactopyranose-6-sulfate are commonly encountered. The relative occurrence of these residues can vary widely even within the same species. In some 40 collections involving four *Porphyra* spp., the 6-O-methylgalactose ranged from 3 to 28% of the porphyran, whereas ester sulfate (mostly L-galactose-6-sulfate) ranged from 6 to 11%, the remaining mass being the normal agarose constituents, D-galactose and 3,6-anhydro-L-galactose residues (Rees & Conway 1962a). In contrast, the conchocelis phases of both *Bangia atropurpurea* and *Porphyra leucosticta* are devoid of 6-O-methylgalactose (Gretz et al. 1983). 6-O-methylgalactose was not reported as a constituent of purified cell walls from either the conchocelis or generic phases of *Porphyra tenera* (Mukai et al. 1981). The possible occurrence of other methylated hexoses (Table 10-3) in the generic phase of *B. atropurpurea* (Shashkov et al. 1978; Usov et al. 1978b) deserves further investigation. The cell wall polysaccharides of *Smithora* (as *Porphyra*) *naiadum* (Erythropeltidaceae) are distinguished from those of *Porphyra* spp. by having more ester sulfate (16–20%), little 3,6-anhydrogalactose, and undetectable levels (≤2%) of 6-O-methylgalactose (Rees & Conway 1962b).

Studies of porphyran from *P. umbilicalis* using highly purified β-agarase I (*Pseudomonas atlantica*) and ^{13}C NMR show that only 23% of the porphyran is released as oligosaccharides (≈92% as 6^3-O-methylneoagarotetraose) (Morrice et al. 1983, 1984). An earlier investigation of *P. umbilicalis* porphyran with β-agarase (from *Cytophaga* sp.) showed that neoagarotetraose and methylneoagarotetraose were liberated in approximately equal proportions (Duckworth & Turvey 1969). Such dissimilar results are a further indication that the composition of porphyran is not constant even for the same species. The ^{13}C NMR analysis of Morrice et al. (1983, 1984) established that the sulfate residues occur in blocks averaging 2.0 to 2.5 contiguous units in the charged oligosaccharides of D.P. 4 to 40 released by the action of β-agarase on porphyran.

The distribution and composition of the agarocolloids from the Gelidiaceae were reviewed by Araki (1966). Again, 6-O-methyl-D-galactose is always present, but the levels are low. It is evident that some 4-linked residues also can be substituted with methoxyl at the C-2 position. Similar analyses are given for agar from *Gelidium filicinum* (Zanlungo 1979). As in the Bangiaceae, sulfation of the 3-linked-D-galactose residue does not appear to be a prominent feature (Table 10-3). The gel strengths of agars from *Gelidium* spp. and *Pterocladia* spp. are considerably higher than those of *Gracilaria* spp. (Santos & Doty 1983).

In sharp contrast, however, is the nongelling agaroid funoran, from *Gloiopeltis* spp. (Endocladiaceae, Cryptonemiales), in which many of the D-galactose residues are sulfated at C-6, and others are methylated instead. The 4-linked L-galactose residues of *Gloiopeltis* spp. may be either methylated or sulfated at the C-2 position (Table 10-3). It is noteworthy that a member of the only other genus in the Endocladiaceae, *Endocladia muricata*, synthesizes a carrageenan that contains about 6% of agarobiose residues (Whyte et al. 1985). Other examples in which the 4-linked galactose residue is partially replaced by its enantiomer are found in *Gloiopeltis* spp. (Halymeniaceae) and in *Anatheca dentata* (Solieriaceae).

The complexity of the agarocolloids from the Gracilariaceae is quite variable. Methylation of the 3-linked residue is usual and can be extensive; the 6-O-methyl-D-galactose content ranges from 2.3% in *Gracilaria compressa* to approximately 37% in *G. tikvahiae* (Craigie et al. 1984; Duckworth et al. 1971), with high levels also reported in *G. cunefolia* (Usov et al. 1983). Methylation of the 4-linked residue is reported for several species (Table 10-3) and is virtually complete in *G. eucheumoides* (Ji et al. 1985). Pyruvic acid, bound as 4,6-O-(1-carboxyethylidene)-D-galactose, was discovered first in *Gelidium amansii* and subsequently in *G. subcostatum*, where it is associated solely with the agaroid fraction (Araki 1966). A survey of eight agarophytes, including *Gelidiella acerosa*, showed that pyruvate was detectable in all, ranging from 0.04% in *Gelidium sesquipedale* to 2.92% in *Gracilaria compressa*. Ion exchange chromatography indicates that the pyruvated polysaccharides can be partially separated from the more heavily sulfated agaroids (Duckworth & Yaphe 1971a,b; Young et al. 1971).

Sulfate hemiester quantity in the agarocolloids of the Gracilariaceae ranges from 2.5% in *Gracilaria tikvahiae* to 7.0% in *G. cerricornis* (*G. ferox*) (Duckworth et al. 1971), 13.9% in *G. textorii* (Craigie et al. 1984), and 16.2% in *G. abbottiana* (Santos & Doty 1983). Except for the L-galactose-6-sulfate residues, the position of the sulfate groups is not known with certainty. In those cases where most of the sulfate ester is alkali stable (e.g., *G. damaecornis*, *G. domingensis*, *G. ferox*, *G. textorii*) it may be expected that some of the 3-linked residues will be sulfated.

The low quantity of 4-O-methyl-L-galactose, originally discovered in hydrolyzates of *Gelidium amansii* agars, led Araki (1966) to conclude that this residue was of no structural significance. Recently it was shown that this compound could comprise up to 8.8% of the mass of agars from mature *Gracilaria tikvahiae*; it is glycosidically linked to C-6 of the main chain D-galactose residues (Craigie & Jurgens 1989; Craigie & Wen 1984; Craigie et al. 1984). The distribution of 4-O-methyl-L-galactose among taxa is poorly known, but it is a significant constituent of agars of *G. damaecornis*, *G. domingensis*, *G. ferox* (Duckworth et al. 1971), *G. blodgettii*, and *G. verrucosa* (Bird et al. 1987). Although its quantity can increase as the thallus ages (Craigie & Wen 1984), the traces detected in *G. cervicornis* and *G. cylindrica* appear to be maxima for these species, suggesting that the sugar may have value as a taxonomic marker (Bird et al. 1987).

The most complex agarocolloids are to be found in the Ceramiales, although *Ceramium boydenii* and possibly *Digenea simplex* appear as exceptions (Table 10-3). The agarocolloids of the former species contain 82% of agarose which, in turn, is composed of 20.8% of 6-O-methyl-D-galactose residues (Araki 1966). In contrast, no 6-O-methylgalactose was detected in the agaroid from *C. pacificum*, which instead was moderately sulfated (12.7% w/w); the sulfate apparently did not exist as L-galactose-6-sulfate (Matsuhiro 1982). Besides the usual 6-O-methyl ethers, the 3-linked residues may be sulfated at C-6 or even at C-2 or C-4 (Table 10-3). The novel natural sugar derivative, 6-O-methyl-D-galactopyranose-4-sulfate, was discovered in hydrolyzates of agaroids from *Polysiphonia lanosa* (Batey & Turvey 1975). Methylation at C-2 of the 4-linked residue is a common feature of agaroids from the Ceramiales (see Usov 1974; Usov et al. 1983). The sulfated galactan of *Odonthalia kamtschatica* was recently shown to be an agaroid on the basis of its high yield of agarobiose dimethyl acetal following mild methanolysis (Whyte et al. 1985). Agarocolloids have been confirmed in *Campylaephora hypnaeoides* (Ceramiaceae) and *P. morrowii*, whereas more complex galactans occur in *Chondria decipiens* and *Delesseria serrulata* (Usov et al. 1983).

Evidence based on an unexpectedly high differential rate of release of products (ratio of 3,6-anhydro-2-O-methylgalactose:2,4,6-tri-O-methylgalactose) of permethylated agarocolloids subjected to formolysis suggests that at least some 3,6-anhydro-L-galactose residues may be contiguous in species of *Gracilaria*, *Porphyra*, and *Pterocladia* (Brasch et al. 1984).

The extent to which agaroids are distributed in the Rhodophyta is uncertain. A galactan from *Palmaria* (as *Rhodymenia*) *stenogona* was shown to contain 10% sulfate and approximately equal amounts of D- and L-galactose although no 3,6-anhydride was detected (Usov et al. 1978c). The sulfated galactan from *Corallina officinalis* also lacks 3,6-anhydrogalactose, but has a D- to L-galactose ratio of 1.3:1, with the sulfate hemiesters occurring as L-galactose-6-sulfate and to a lesser extent as galactose-4-sulfate (Turvey & Simpson 1966). Further examination of these galactans is required to establish whether they are true agaroids.

VII. PROTEINS AND GLYCOPROTEINS

Relative to the polysaccharides, proteins are quantitatively minor components of red algal cell walls. Some are extracted readily with the water soluble mucilages, where they can account for several percent of the mass (Table 10-1). Others, such as the cuticular proteins (see below), are highly insoluble in most solvents. The few studies reported have not addressed the function or diversity of the cell wall proteins. It can be surmised that the proteins may be associated with ion transport, movement of organic molecules, and the biosynthesis of cell wall polymers and their modification and degradation, as well as acting as template and structural materials. They also may play important roles in cell adhesion (Chamberlain & Evans 1973; Peyrière 1970; Pueschel 1979), detoxification (Pedersén et al. 1979), and physical protection of the thallus surface and as antifouling mechanisms through sloughing off of the outermost cuticular layers (Sieburth 1975).

Protein-polysaccharide conjugates have been reported from red algae only rarely. Centrifugation in CsCl gradients was used to isolate a glycoprotein from the extracellular mucilage of *Porphyridium purpureum* (as *P. cruentum*) (Heaney-Kieras et al. 1977). The products of β-elimination in alkali showed that both serine and threonine participated in a peptide-O-glycosidic linkage and that xylose was the only sugar involved. No hydroxyproline was reported. The protein moiety had a molecular weight of ≤10,000. It appears to be linked to the reducing end of a polysaccharide to form linear glycoproteins at $2.5-6.3 \times 10^5$ daltons, with some molecules reaching $0.5-1.0 \times 10^6$ daltons (Heaney-Kieras & Swift 1979). The polysaccharide consisted of galactose, xylose, glucose, uronic acid, and sulfate in the molar proportions 2.12:2.42:1.00:1.22:2.61, respectively (Heaney-Kieras et al. 1977). Serine was also shown to form the O-glycoside bonds in a glycoprotein from *Furcellaria lumbricalis* (as *F. fastigiata*) (Krasil'nikova & Medvedeva 1975). However, the β-carboxyl of aspartic acid was reported to form an ester with a branched chain carbohydrate polymer from *Phyllophora* (Medvedeva et al. 1973).

A proteoglycan of quite different composition was extracted in low yields from *Laurencia spectabilis* and purified to electrophoretic homogeneity (Court & Taylor 1979). It was shown to be a galactan containing 9% uronic acid, 8% protein, 5–6% each of glucose and xylose, and no sulfate. Low levels of arabinose together with variable quantities of hexosamines were reported. The protein moiety was rich in the acidic residues aspartate and glutamate, and contained about 0.3% of hydroxyproline.

The fascinating investigations of Waaland (see Chapter 11) and her colleagues on rhodomorphin, a species-specific wound repair hormone discovered in *Griffithsia pacifica* (Waaland 1975; Waaland & Cleland 1974) reveal that it is a glycoprotein (Watson & Waaland 1983). Purification of this morphogenetic substance was effected on Con A, a lectin that specifically binds α-mannosyl and/or α-glucosyl residues. The molecular weight of the biologically active fraction was ≈14,000. Disulfide cleaving agents and proteases destroyed hormonal activity; the α-mannosyl residues are not required for biological activity (Watson & Waaland 1983, 1986). The concentration of hormone necessary to maintain shoot growth in 50% of the cells exposed was estimated at $\approx 6.7 \times 10^{-14}$ M (Watson & Waaland 1986).

Purified cell walls from the conchocelis and generic phases of *Porphyra tenera* were prepared following cell disruption in a French press (Mukai et al. 1981). The proteins of the conchocelis cell walls were distinguished readily by their high levels of arginine, histidine (basic amino acids), and aspartic acid relative to proteins of the generic phase, which were richer in alanine and glycine. Serine and threonine were important constituents of the cell walls of both phases. Closely parallel trends in amino acid composition may be seen when the data for purified cell walls of *Bangia atropurpurea* generic phase (Gretz et al. 1982) are compared with those of its conchocelis phase (Gretz et al. 1986). However, the *Bangia* cell walls apparently lack proline (although the generic phase is reported to have hydroxyproline), and they contain little or no valine. Proline and valine are significant constituents of the cell walls of both phases of *P. tenera* (Mukai et al. 1981). The reasons for the changes in bulk protein composition between the two phases are not known.

The proteins are not uniformly distributed throughout the cell wall because protein-enriched fragments (e.g., cuticles, pit plugs) can be isolated as physically intact entities. Covalent bonds be-

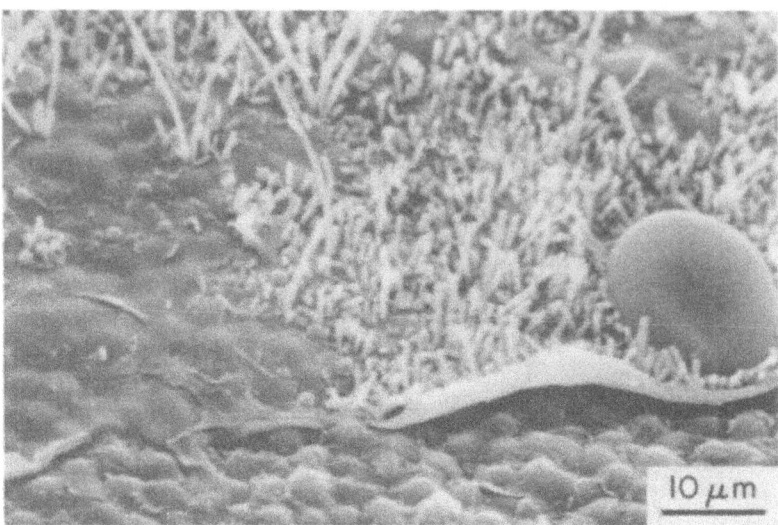

Fig. 10-6. SEM of *Chondrus crispus* surface showing epiphytic microflora and layered cuticle. Note clean areas where the outer layers have been peeled or sloughed off. Where the entire cuticle has been removed, the outline of the cortical cells may be seen in the carrageenan matrix. (Photograph courtesy of J. McN. Sieburth; *Journal of Phycology*, with permission)

tween proteins and polysaccharides involve the hydroxylated amino acids hydroxyproline, serine, and threonine and various sugars, most notably the pentoses arabinose and xylose. The arabinose-hydroxyproline linkage is of wide occurrence in higher plants in the cell wall proteoglycan extensin. A similar linkage is known for green algae but not in the Rhodophyta (see Siegel & Siegel 1973). It may be no coincidence that both arabinose and hydroxyproline are so rarely reported from red algae. Court and Taylor (1979) have pointed out that the role of hydroxyproline in red algal cell walls must differ at least quantitatively from that in green algae and higher plants. Xylose-containing polymers, however, are widely reported in red algae, and in the case of *Porphyridium* it is xylose that is covalently bonded to serine and threonine (Heaney-Kieras et al. 1977).

VIII. OTHER CELL WALL STRUCTURES

A. Cuticles and pit plugs

Many investigators have conducted histochemical tests to ascertain the nature of the heterogenous components of cell walls. Few have physically isolated and attempted to characterize the purified structures chemically. The outer surfaces of many algae appear to be layered when examined micro-scopically (Fig. 10-6). In some cases these layers can be removed by physical and/or chemical means. The analytical data obtained for the preparations depend on the amount of subjacent cell wall material that is removed along with the outer layers. Frei and Preston (1964) showed that the outer layers of *Porphyra umbilicalis* contained a mannan. Cuticle from the same species prepared using more stringent conditions proved to be largely an insoluble protein (Hanic & Craigie 1969). The cuticle isolated from *Iridaea* spp. by mild techniques was composed of protein ($\approx 50\%$) and carbohydrate (40%) along with small amounts of inorganic salts and fatty acids (Gerwick & Lang 1977). Up to 17 electron-dense layers (35–65 nm thick) alternate with translucent layers (75–250 nm thick) to provide the physical basis for the iridescence so characteristic of *Iridaea* spp. Similarly laminated cuticles were demonstrated in *Chondrus crispus*, but only in fronds that were iridescent (Pedersén et al. 1980). X-ray analyses of the cuticles showed bromine in the *C. crispus* but not in the *Iridaea* spp. cuticles. Sulfur was found in cuticles from both sources and is probably mainly associated with polysaccharides. Bromine was also reported in the cuticles of *Polysiphonia nigrescens* (Pedersén et al. 1981) and in middle lamella, pit plugs, and intercellular matrix of *Lenormandia prolifera*, but not in the inner cell walls, radial cell walls, or subepidermal cell walls (Pedersén et al. 1979). Concentrations of bromine of ≈1% have been detected in the medulla of *Callophycus* (as *Thysanocladia*) *densa*, apparently associated with isolated granular deposits embedded in the mucilage of the cell wall (Pallaghy et al. 1983). The significance of the occurrence and distribution of bromine in these cell walls is not known, although deposition in the cuticle may be a method for disposing of toxic organobromine compounds (Pedersén et al. 1979).

The cores of pit plugs from several red algae have been found to contain only proteins, whereas the outer cap layer, when present, stains for polysaccharides (Pueschel 1980; Ramus 1971; Chapter 2). No chemical analysis of pit plugs has been reported.

B. Carbonate minerals

Frequently, carbonate deposition is initiated in such a highly controlled manner that the resulting pattern of calcification is useful in taxonomic identifications. Valuable background information (Darley 1974; Lewin 1962; Siegel & Siegel 1973), together with the treatises of Borowitzka (1977, 1982) and of Pentecost (1980, 1985), outline current thinking and evaluate various models proposed to explain calcification in algae.

Of the five known crystalline forms of $CaCO_3$, only calcite and aragonite are deposited by algae. The birefringent, rhombohedric isomorph calcite is the most stable structure, but the widely occurring aragonite (orthorhombic form) is stable at physiological temperatures and pressures. The presence of calcite in the cell walls of coralline algae was selected as a key character in defining the new order Corallinales (Silva & Johansen 1986). Other rhodophytes deposit aragonite as in calcified members of the Peyssoneliaceae and the Nemaliales (e.g., *Galaxaura* and *Liagora*) (see Pentecost 1980).

The physical and chemical properties of these two $CaCO_3$ isomorphs differ sufficiently that their contaminant cations are distinctive. The high content of $MgCO_3$ in coralline algae (Ca:Mg molar ratio ≈3.5; Furuya 1960) may be explained by the relative abundance of Mg in sea water and the fact that Ca^{2+} in calcite can be replaced by cations of smaller radii, such as Mg, Fe, Zn, and Cd (Siegel & Siegel 1973). Aragonite, on the other hand, accepts cations with radii larger than Ca^{2+} (e.g., Sr, Pb, and Ba) and can be enriched in $SrCO_3$.

Calcite deposition occurs as early as the first cell division in cultured tetraspores of *Amphiroa ephedraea* (Furuya 1960) and is most intense in meristematic regions (Miyata et al. 1980; Pentecost 1980).

$CaCO_3$ deposition in calcifying algae is some 10-fold faster in light than in darkness, where the rate is low and similar to that of killed cells (Borowitzka 1982). Photosynthetic inhibitors CCCP, DCMU, and DNP also inhibit calcification, as does exogenously applied orthophosphate (reviewed in Kremer 1981). Additional evidence that calcification is energy dependent, at least in the Corallinaceae, was provided by the demonstration of a highly active Ca-dependent ATPase in three coralline species (Okazaki 1977). No similar activity could be detected in several aragonite depositing algae.

Photosynthesis may favor carbonate crystallization indirectly by raising the pH near the surface of the alga if CO_2 and/or HCO_3^- is being taken up. Structural modification of the algal surface may favor the rise in pH if the modifications restrict the free movement of sea water at the site of carbonate deposition (see Borowitzka 1982; Miyata et al. 1980; Okazaki et al. 1986; Pentecost 1985). The stable isotopes ^{13}C and ^{16}O are enriched in coralline calcite relative to their abundance in seawater, indicating that a portion of the deposited $CaCO_3$ derives from photosynthetically fixed carbon via cellular respiration. Evidence has been obtained for translocation of both organic carbon and calcium from pigmented parts of the articulated species *Calliarthron tuberculosum* to the rapidly calcifying colorless tips of its fronds. A significant portion of the translocated organic carbon was respired and appeared within 2 h in the calcite of the growing tips (see Borowitzka 1982).

The orderly deposition of carbonates is thought to be initiated and controlled on or within a special organic matrix (Pentecost 1980, 1985). Networks of organic fibers have been demonstrated following decalcification of coralline algae (Giraud & Cabioch 1979). The secretion of fibrous materials into the walls of apical cells of *Corallina* involves the Golgi system, but it has not been demonstrated that the fibers themselves facilitate calcification. An electron microscopic examination of *Galaxaura oblongata* (as *G. fastigiata*) revealed that a fibrous material filled the intercellular spaces. Electron-dense particles (20–400 nm diam.) appeared on this fiber matrix, followed by the deposition of minute $CaCO_3$ crystals. Larger granular crystals from older tissue were identified as aragonite (Okazaki et al. 1982b).

The involvement of acidic polysaccharides in algal calcification has been suggested by analogy with bone mineralization in animals. The widespread distribution of polysaccharide ester sulfates in the cell walls of algae, whether or not they are calcified, makes this analogy rather difficult to defend. Calcium-binding substances have been isolated from the cell walls of calcareous algae; however, no direct relationship between Ca^{2+} binding and the content of acidic groups could be demonstrated (Misonou et al. 1980). The discovery of alginate in the cell walls of coralline algae (Okazaki et al. 1982a) is especially intriguing as alginates containing blocks of guluronic acid residues are known to have an affinity for calcium (see Painter 1983; Rees et al. 1982 for reviews). According to the "egg box" junction model for alginates, Ca^{2+} is bound between adjacent polyguluronate chains, which have a pleated shape, thus forming the egg box structure. It is not clear how the carbonate anion could participate in such a structure to form calcite. Further, guluronate-containing alginates are widespread in cell walls of the Phaeophyta, but calcification (as aragonite) occurs only in the genus *Padina*. The anionic polysaccharides investigated to date do not appear to be the sole effectors of calcification in red algae. The mechanism that initiates $CaCO_3$ deposition remains unknown.

IX. SEASONAL AND ENVIRONMENTAL FACTORS

A. Natural populations

Studies relating seasonal changes in phycolloid yields and properties are difficult to interpret since major variables affecting growth rates and cell wall deposition (e.g., irradiance, temperature, salinity, nutrient cycles) are often not known or are inadequately documented. The influence of short-term local or microenvironmental factors affecting the algae are virtually ignored, as are age of the tissues, growth rates, and reproductive status at the time of harvest. Variations in chemical protocols used for preparing samples add a further level of complexity (see Miller & Furneaux 1987).

Changes in the chemical and physical properties of agar from *Gracilaria verrucosa* (Whyte et al. 1981) nicely illustrate the complex of variables that must be considered in surveys of natural populations. The waxing and waning of the algal biomass followed the solar irradiance curve during the growing season (May to September). Agar content was not related to algal dry weight or to seasonal change in any obvious manner. The chemical composition of the agar changed as the season advanced, but differences were noted among the frond samples depending on their vegetative or reproductive status. In general, the total sulfate content of the agar declined, whereas 3,6-anhydrogalactose content and gel strength increased in the period of rapid biomass accumulation (May to July). The 6-O-methylgalactose content increased in agar from vegetative but not from tetrasporic fronds, whereas cystocarpic and male fronds showed no clear trends. Temperature change appeared not to be an important factor as the range was small (9–12°C). Although data on seawater nutrients and nitrogen content of the *G. verrucosa* were not presented, the observed differences in agar quality probably cannot be attributed to nutritional status in this case as all fronds in the study were exposed to a common water mass.

Seasonal and site-specific changes were noted for the agars from 10 species of Chinese *Gracilaria*. Positive correlations were observed between gel strength, gelation and melting temperatures, and 3,6-anhydrogalactose content; negative correlations occurred between total sulfate and 3,6-anhydrogalactose in the agars (Shi et al. 1983, 1986). However, no close relationship was found between the total sulfate and 3,6-anhydrogalactose contents of agar from vegetative *G. tikvahiae* fronds growing in Rhode Island; the sulfate content was highest between July and October (Asare 1980).

Agars from both *Gracilaria coronopifolia* and *G. bursapastoris* collected in Hawaii showed seasonal changes in gel strength, but these changes did not occur at the same time in both species. The gel strength of *G. coronopifolia* agar declined in late summer, a time of maximum water temperature (Hoyle 1978b). Agar from *Gracilaria* cultivated in Taiwan also had a lower quality in summer, when temperatures and salinities were high; however, *G. lichenoides* from Saipan produced a high-quality agar at relatively high temperatures (26–30°C) (Nelson et al. 1983). The yield of unmodified agar from *G. corticata* collected near Verval, India, was lowest during late summer and coincided with the shedding of branches after tetraspore liberation; the gel strength of the agar, while low, varied little throughout the year (Oza 1978). A similar pattern was observed for *G. edulis* from Brazil (Durairatnam 1987). The marked variations observed in yield, sulfate, and 3,6-anhydrogalactose contents and gel strengths of agar from *G. cylindrica* collected in Malaysia over a 15-month period were attributed to "unknown factors relating to age of crop harvested, environment, or post harvest treatment" (Doty & Santos 1983). The content of agar in South African *Gelidium pristoides* was highest (48% w/w) in January (midsummer) and lowest (32% w/w) in April, being roughly inversely related to the thallus nitrogen content (Carter & Anderson 1986). Variations in gelling ability of agars may relate to the extent of concurrent unsubstituted agarose blocks rather than molecular size (Whyte & Englar 1981b).

Low carrageenan levels characterize winter populations of both tetrasporic and gametophytic *Chondrus crispus* (Fuller & Mathieson 1972; Kopp & Perez 1978; Mathieson & Tveter 1975; McCandless & Craigie 1974). Higher carrageenan levels with improved viscosities/gel strengths were found in late summer to early winter; changes in thallus protein content mirrored those for carrageenans (Mathieson & Tveter 1975). Little seasonal change in the ratios of KCl soluble to insoluble carrageenans was noted for female *C. crispus*, but the gel strength of kappa carrageenan was maximum during winter; viscosities of lambda carrageenan from sporophytes were widely variable, being low in the later summer (Kopp & Perez 1978). A similar pattern of seasonal responses in total carbohydrate and total nitrogen were observed by Butler for the same species (reviewed in Buggeln & Craigie 1973; Mathieson & Prince 1973). Parallel trends with winter minimum and summer maximum were noted for the carrageenans in *Gymnogongrus crenulatus*, *Phyllophora pseudoceranoides*, and *P. truncata* (Mathieson et al. 1984). No pronounced seasonal changes were detected in kappa-iota and mu carrageenans of *C. crispus* female gametophytes taken from the same site, even though thallus morphologies were quite different (Chopin et al. 1987).

The iota carrageenan content of *Eucheuma* and *Meristella* increased in spring and decreased in fall in Florida populations. There was little change in protein content throughout the seasons. The molar ratios of galactose to 3,6-anhydrogalactose varied little and were similar for all populations except for *M.* (as *E.*) *gelidium* and *E. isiforme* (as *E. nudum*) from the Atlantic coast (Dawes et al. 1974, 1977). Both sulfate and 3,6-anhydrogalactose contents changed in parallel seasonally in iota carrageenan from *Agardhiella subulata* (as *Neoagardhiella baileyi*), being lowest in summer and highest in winter (Asare 1980). The kappa carrageenan content of *Hypnea musciformis* varied considerably with season and local environmental conditions (Friedlander & Zelikovitch 1984; John & Asare 1975; Mollion 1973). In Florida the carrageenan content of *H. musciformis* was highest in spring and lowest in the fall, with Atlantic coast populations having generally higher levels than Gulf coast specimens. The inverse relationship between protein and carbohydrate content was observed for both populations (Durako & Dawes 1980).

No significant seasonal changes could be observed in the carrageenan contents of several New Zealand *Gigartina* species (Pickmere et al. 1975) or when plants were taken from geographically separated but ecologically similar localities (Parsons et al. 1977). Finely branched *Gigartina* spp. may contain more kappa carrageenan than broad-bladed forms (Abbott 1980; Parsons et al. 1977). The relative proportion of mu to kappa carrageenan may prove useful in conjunction with morphological features in distinguishing related species of *Gigartina* (Parsons et al. 1977). In another member of the Gigartinales, *Mastocarpus stellatus* (as *G. stellata*) also exhibited little seasonal change in carrageenan content, although protein levels followed the usual cycle (Mathieson & Tveter 1976).

B. Cultivated algae

Photosynthesis in algae under conditions of nitrogen deprivation can result in the accumulation of nonnitrogenous organic compounds (e.g., carbohydrates, lipids), depending on the species (Fogg 1964). A clear reciprocal relationship between the carrageenan and nitrogen contents was established for greenhouse-grown *Chondrus crispus* strain T4 (see Mathieson & Prince 1973; Neish & Shacklock 1971; Simpson & Shacklock 1979). This relationship, known as the 'Neish effect' (Macler 1986; Neish et al. 1977), has been confirmed for a variety of cultivated agarophytes and carrageenophytes (Bird 1984; Bird et al. 1981; DeBoer 1979; Guist et al. 1982; Lapointe 1981; Lapointe & Ryther 1979; Patwary & van der Meer 1983; Rotem et al. 1986). Nitrate starvation may be a useful strategy for enhancing polysaccharide production in *Porphyridium* (Adda et al. 1986).

The nitrogen status of the thallus is important in regulating the quality and composition of agar. Nitrogen-deficient fronds of *Gracilaria tikvahiae* contained more agar than those supplied with nitrogen, but the gel strength was poorer (Bird et al. 1981; Cote & Hanisak 1986; Patwary & van der Meer 1983). Agar from the nitrogen-deficient specimens contained more galactose and 4-O-methyl-L-galactose than did that from plants supplied nitrogen (Craigie et al. 1984). These changes would result in a less regular tertiary helical structure for the agar and weaken its ability to form gels. Exogenously supplied nitrogen directs the flow of carbon from agar synthesis into other carbohydrates and nitrogenous constituents in *Gelidium coulteri* (Macler 1986). It has also been reported that agar from *Gracilaria tikvahiae* grown on NH_4^+ has a higher melting temperature, a lower gelling temperature, and a greater gel strength than does agar from NO_3^--grown plants (Bird et al. 1981). Chlorophyll content of *Gracilaria* was negatively correlated with agar yield and 3,6-anhydrogalactose content (Liu et al. 1981). No obvious relationships between agar content and gel strengths or protein content could be demonstrated with *G.* cf. *conferta* in tank cultures (Friedlander et al. 1987).

In several reports, the growth temperature and age of the agarophyte at harvest have been implicated as possible variables affecting agar quantity and/or quality. Hoyle (1978b) noted the coincidence of maximum water temperature and low gel strength of agar from *Gracilaria coronopifolia*, although he discounted the significance of the relationship. *Gracilaria* cultivated in Taiwan also gave agar with low gel strength when water temperatures and salinities were highest (see Nelson et al. 1983; Yang et al. 1981). Agars from *G. tikvahiae* had the highest 3,6-anhydrogalactose content in spring and early summer, whereas in late summer and fall the sulfate levels were maximum (Asare 1980). These changes in agar composition are in accord with maximum gel strengths being achieved in the spring months.

Agar gel strength rose in *Gracilaria corticata* at times of maximum growth (Oza 1978). The best yields of agar from *G. edulis* were obtained after one month's regrowth; gel strength became maximum after three months (Thomas & Krishnamurthy 1976). More agar could be extracted from the distal halves (younger tissue) of *Gelidium pristoides* than from the older, proximal halves (Carter & Anderson 1986). Juvenile *Gracilaria verrucosa* contained the same quantity of agarocolloid as mature plants, but a much larger proportion of it was sulfated at C-6 of the 4-linked residue (Usov et al. 1979). This "biosynthetically unfinished" agar had a high gel strength after alkali modification.

When *Gracilaria tikvahiae* was cultivated under controlled conditions, agar quality and composition varied in relation to tissue age and growth temperature (Craigie & Wen 1984; Helleur et al. 1985). Young portions of the frond contained less agar than the older mature parts, but the quality was much higher, as judged by the gel strength of the alkali-modified agar. This improvement in quality became especially marked when the alga was grown at lower temperatures. The changes in agar quality were reflected in the chemical composition inasmuch as agar from young tissue, especially that grown at low temperatures, contained very low levels of 4-O-methyl-L-galactose and high levels of alkali-labile sulfate. Agar from old tissue was rich in this methylated sugar and contained more alkali-stable sulfate. Low temperatures favored agarose production in *G. verrucosa* (Christiaen et al. 1987).

Recent histochemical observations on *Bangia atropurpurea* further emphasize that the polysaccharides must be regarded as dynamic components of the cell wall structure. The staining intensity for sulfated polysaccharides decreased in both male and female filaments during sexual reproduction, as compared with those from vegetative filaments (Cole et al. 1985). In female filaments only there appeared to be an increase in carboxylated polysaccharides. The water soluble, acidic polysaccharides from vegetative filaments and asexual reproductive regions gave a single band on electrophoresis, whereas two bands were resolved when tips of sexual filaments were examined (Cole et al. 1986).

X. GENETIC FACTORS

Life history and phenotypic variations in cell wall composition in the Bangiaceae, Gigartinaceae, and Phyllophoraceae cannot be attributed to environmental circumstances when the algae examined are grown under quite similar conditions. Identification of the controlling mechanism for gene expression in these cases remains a challenge. It would be of interest to construct gametophyte and sporophyte generations with increased ploidy levels to ascertain whether their cell wall chemistry parallels that of haploid and diploid thalli in these species.

Agars from cystocarpic, male, and tetrasporic agarophytes have been examined with respect to gel strength and composition, but no generalizations can be made on the basis of existing data. Cystocarpic plants of *Gracilaria verrucosa* from Chile contained more agar than tetrasporic plants, but it had a lower gel strength (Kim & Henriquez 1979). Cystocarpic *G. verrucosa* type in British Columbia generally contained more agar than did male or tetrasporic plants, but the gel strength differed little among them until late in the growing season (Whyte et al. 1981). *G. tikvahiae* showed no differences in agar content among vegetative, cystocarpic, and tetrasporic fronds (Penniman 1977) or in agar composition as tetrasporophytes and female gametophytes gave similar pyrograms (Bird et al. 1987). Female, male, and tetrasporic stages could not be distinguished on the basis of agar yields or gel strengths in either *G. bursapastoris* or *G. coronopifolia* from Hawaii (Hoyle 1978a). In six Chinese collections over the course of a year, carposporic plants of *G. verrucosa* yielded slightly more agar than did tetrasporic plants. No significant differences could be established for these samples in total sulfate content, gelation and melting temperatures, and gel strengths (Shi et al. 1984).

More sophisticated examination of agars from *Gracilaria* species collected in the Sea of Japan revealed that subtle structural differences occur among the samples prepared from the different reproductive phases. In *G. verrucosa*, female, male, and tetrasporic phases yielded similar quantities of agar, whereas juvenile forms contained much less. Virtually the same molar ratios of the principal agar components occurred in the mature reproductive plants, all of which contained more 3,6-anhydrogalactose than did the immature vegetative fronds. However, the agar from male plants contained proportionately more agarose than agaroids and produced stronger gels than the other samples (Usov et al. 1979). Data from an unidentified *Gracilaria* sp. showed that differences can occur in agars from carposporic and tetrasporic plants. In this species, tetrasporic plants yielded much less agar than did cystocarpic plants, but the gel strength was superior. It was shown that more sulfate was esterified to C-6 of the 4-linked residue in agar of tetrasporic plants than in agar of carposporophytes (Usov et al. 1985).

Few attempts have been made to alter the cell wall composition of algae by genetic manipulation. Patwary and van der Meer isolated mutants of *Gracilaria tikvahiae* with widely differing gross morphologies. Agar extracted from a highly branched mutant, MP-40, was much superior in gel strength to the wild type, whereas that from a "puffy" mutant, MP-61, was poorer (Patwary & van der Meer 1983). Analyses showed that the MP-40 agar contained less sulfate and 4-O-methyl-L-galactose than did the wild type agar (Craigie et al. 1984).

Strains with superior agar qualities also have been selected from natural populations. Two of seven strains of *Gracilaria tikvahiae* and two of three strains of *G. verrucosa* cultured in Florida produced native agars with relatively high gel strengths (Cote & Hanisak 1986). More effort is required to understand the basis for these differences so that strains with more valuable properties can be created deliberately.

XI. BIOSYNTHESIS

Several aspects of rhodophytan cell wall biosynthesis have been reviewed, with disappointingly little in the way of definitive conclusions (see Bodeau-Bellion 1983; James et al. 1985; McCandless & Craigie 1979; Painter 1983; Percival & McDowell 1981). The basic principles of polysaccharide formation (Karr 1976; Northcote 1985) appear to be followed, but direct evidence for the activated intermediates is rare (Barber 1971; Su & Hassid 1962); enzymes involved in the polysaccharide synthesis remain to be characterized. It is clear that the partitioning of carbon flow between cell wall and storage polysaccharides can depend on nitrogen nutrient availability (Macler 1986). Difficulties in isolating and purifying small pools of the polysaccharide intermediates complicate the interpretation of $H^{14}CO_3$ incorporation studies (Jackson & McCandless 1979; Macler 1986). Conclusions about the mass flow of carbon or even about the sequence of formation of related polymers in either the carrageenans or the agarocolloids remain speculative.

The facile uptake of SO_4^{2-} into cell wall polysaccharides has been observed by several groups (see McCandless & Craigie 1979). Present evidence indicates that sulfate esterification occurs in the Golgi dictyosomes (Millard & Evans 1982a,b; Ramus & Robins 1975; Tveter-Gallagher et al. 1981) and not in the cell wall (La Claire & Dawes 1976). Both ATP-sulfurylase (ATP: sulfate adenylyl-transferase, E.C.2.7.74) and APS-kinase (ATP: adenylyl-sulfate 3'-phosphotransferase, E.C.2.7.1.25) activities were demonstrated in *Rhodella maculata* (Møller & Evans 1976). The sulfate-activating system is similar in *Porphyridium* (Ramus 1974; Ramus & Groves 1972, 1974).

A rapid exchange (turnover) of SO_4^{2-} has been postulated to occur between the cell walls of *Catenella daespitosa* (as *C. opuntia*) and the adjacent water (de Lestang-Brémond & Quillet 1981). In this case the sulfate is believed to be esterified to a minor 0.5 M KCl soluble carrageenan fraction; the active form of sulfate involved is reported as 5-methylcytosine-3'-phospho-5'-phosphosulfate (MeCDPS) (Quillet & de Lestang-Brémond 1981). Whether MeCDPS is involved in sulfation of carrageenan in the Golgi has not been established. The occurrence of a rapid turnover of sulfate in algal cell walls has been challenged (de Reviers et al. 1986; Millard & Evans 1982b). The mechanisms for introducing other groups (e.g., glycosyl, pyruvate, methoxyl) into rhodophytan polysaccharides await elucidation.

XII. SUMMARY

During the past 15 years, advances in knowledge of rhodophytan cell walls have focused mainly on hydrocolloids with commercial application (e.g., agarocolloids, carrageenans, and mucilages of the

Porphyridiales). Some observations have been made on structural components, especially in the Bangiaceae, where cellulose is synthesized by the conchocelis, but not by the generic phase, where it is replaced by a β-(1→4)-mannan. Xylans of varying chemical structures are conspicuous in cell walls of the Bangiaceae, Chaetangiaceae, Nemaliaceae, and Palmariaceae. Complex sulfated polysaccharides based on glucuronic acid, xylose, galactose, and glucose characterize members of the Porphyridiales, whereas sulfated mannans and xylogalactomannans occur in the Nemaliales.

The cell wall hydrocolloids from members of the Gigartinales, Ceramiales, and Cryptonemiales have richly varied structures, often, but not exclusively, of the agarocolloid or carrageenan families. Some 17 basic carrageenan structures have been identified. It is recognized that polymers conforming in their entirety to these chemical entities are likely to be rare in nature, but hybrid polysaccharides (e.g., kappa-iota carrageenan) are commonly encountered. Indeed, recent evidence indicates that carrageenan-agarocolloid hybrid molecules comprise the cell walls of some species (e.g., *Endocladia muricata* and *Gloiopeltis* spp.). The fact that some sulfated galactans (e.g., aeodans) do not conform to either the carrageenan or the agarocolloid structures and the report of alginate in certain coralline species suggest that generalizations should be made with caution.

The cell wall components of species within present taxonomic groupings often exhibit family characteristics, with some exceptions (e.g., *Endocladia* spp. and *Gloiopeltis* spp.). In the case of *Ahnfeltia plicata*, taxonomic reassignment is underway (C. A. Maggs pers. comm.), thus removing this agarophyte from the Phyllophoraceae. Within the Gigartinaceae and the Phyllophoraceae, it appears that the major cell wall constituents of the gametophytes are kappa-iota carrageenans, whereas sporophytes, except *Phyllophora pseudoceranoides*, produce carrageenans belonging to the lambda family. A beta-kappa carrageenan was reported in *P. crispa*.

Nonpolysaccharide structures in cell walls have received comparatively little attention. Cuticles and pit plugs are known to contain proteins. The composition of the total protein in the cell wall varies among species and stage of the life history, but the function of these proteins is not known. The enzymes involved in secondary modification of the polysaccharide matrix (Cole et al. 1985, 1986) need to be identified. The mechanisms of cuticle deposition and the functions of the pit plugs require examination. An unidentified organic network of fibers has been implicated in $CaCO_3$ mineralization in coralline algae, but whether it is proteinaceous remains to be determined.

Seasonal- and site-specific variations in the composition of cell wall polysaccharides have been demonstrated. The availability of fixed nitrogen is a key determinant in this context, as mild nitrogen starvation increases the flow of carbon into cell wall polysaccharides. Factors like growth temperatures and age of the tissue at harvest also can influence the composition and yields of polysaccharides.

Detailed chemical structures of the minor cell wall components are required as many phycocolloids appear to be hybrid structures. Attention should be directed to the glycoprotein linkages involved in maintaining the integrity of the cell wall. The enzymes involved in methylation, sulfation, pyruvation, and glycosylation of the polysaccharides remain to be characterized. The interrelations between chemical and physical structures and biological function (see Williamson 1981) of red algal cell walls have scarcely been addressed. The cell wall structure influences a wide range of biological processes, ranging from nutrient uptake to colonization by endophytes, epiphytes, and the susceptibility of the thallus to microbial attack.

The development of routine methods for preparing viable red algal protoplasts capable of regeneration will allow the biogenesis of cell wall constituents to be dissected experimentally. Ultimately, genetic probes can be constructed to identify the genes associated with the synthesis of a specific component. The isolated genes may then be cloned and expressed in an organism more suitable than the original alga for large-scale production. Such an approach, if successful, would both revolutionize the phycocolloid industry and provide a means of producing new polymers with novel properties.

XIII. ACKNOWLEDGMENTS

This chapter is dedicated to the memory of E. L. McCandless and W. Yaphe, esteemed colleagues who contributed substantially to our current knowledge of agarocolloids and carrageenans. I am indebted to C. J. Bird, L. C-M. Chen, M. R. Gretz, C. A. Maggs, P. Odense, E. Percival, and J. P. van der Meer for valuable advice and suggestions. Figures 10-1, 10-4, and 10-6 were generously provided by J. McN. Sieburth (University of Rhode Island) and J. Scott (The College of William and Mary), to whom I am grateful.

XIV. ENDNOTE

Issued as NRCC No. 29081.

XV. REFERENCES

Abbott, I. A. 1980. Some field and laboratory studies on colloid-producing red algae in central California. *Aquat. Bot.* 8: 255–66.

Adda, M., Merchuk, J. C. & Arad, S. M. 1986. Effect of nitrate on growth and production of cell-wall polysaccharide by the unicellular red alga *Porphyridium*. *Biomass* 10: 131–40.

Allsobrook, A. J. R., Nunn, J. R. & Parolis, H. 1971. Sulphated polysaccharides of the Grateloupiaceae family. Part V. A polysaccharide from *Aeodes ulvoidea*. *Carbohydr. Res.* 16: 71–8.

Allsobrook, A. J. R., Nunn, J. R. & Parolis, H. 1974. The linkage of 4-O-methyl-L-galactose in the sulphated polysaccharide of *Aeodes ulvoidea*. *Carbohydr. Res.* 36: 139–45.

Allsobrook, A. J. R., Nunn, J. R. & Parolis, H. 1975. Investigation of the acetolysis products of the sulphated polysaccharide of *Aeodes ulvoidea*. *Carbohydr. Res.* 40: 337–44.

Anderson, D. B. and Eakin, D. E. 1986. A process for the production of polysaccharides from microalgae. *Biotechnol. Bioengn. Symp.* 15: 533–47.

Anderson, N. S., Campbell, J. W., Harding, M. M., Rees, D. A. & Samuel, J. W. B. 1969. X-ray diffraction studies of polysaccharide sulphates: Double helix models for $\varkappa$- and ι-carrageenans. *J. Mol. Biol.* 45: 85–99.

Anderson, N. S., Dolan, T. C. S., Lawson, C. J., Penman, A. & Rees, D. A. 1968a. Carrageenans. Part V. The masked repeating structure of λ- and μ-carrageenans. *Carbohydr. Res.* 7: 468–73.

Anderson, N. S., Dolan, T. C. S. & Rees, D. A. 1965. Evidence for a common structural pattern in the polysaccharide sulphates of the Rhodophyceae. *Nature* (Lond.) 205: 1060–2.

Anderson, N. S., Dolan, T. C. S., Penman, A., Rees, D. A., Mueller, G. P., Stancioff, D. J. & Stanley, N. F. 1968b. Carrageenans. Part IV. Variations in the structure and gel properties of $\varkappa$-carrageenan, and the characteristics of sulphate esters by infrared spectroscopy. *J. Chem. Soc. C.* 602–6.

Anderson, N. S. & Rees, D. A. 1965. Porphyran: a polysaccharide with a masked repeating structure. *J. Chem. Soc.* 5880–7.

Apt, K. E. 1984. The polysaccharide of the parasitic red alga *Hypneocolax stellaris* Boergesen. *Bot. Mar.* 27: 489–90.

Arad, S., Adda, M. & Cohen, E. 1985. The potential of production of sulfated polysaccharides from *Porphyridium*. *Plant Soil* 89: 117–27.

Araki, C. 1966. Some recent studies on polysaccharides of agarophytes. *Proc. Int. Seaweed Symp.* 5: 3–17.

Arnott, S., Fulmer, A., Scott, W. E., Dea, I. C. M., Moorehouse, R. & Rees, D. A. 1974a. The agarose double helix and its function in agarose gel structure. *J. Mol. Biol.* 90: 269–84.

Arnott, S., Scott, W. E., Rees, D. A. & McNab, C. G. A. 1974b. ι-Carrageenan: molecular structure and packing of polysaccharide double helices in oriented fibres of divalent cation salts. *J. Mol. Biol.* 90: 253–67.

Asare, S. D. 1980. Seasonal changes in sulphate and 3,6-anhydrogalactose content in phycocolloids from two red algae. *Bot. Mar.* 23: 595–8.

Atkins, E. D. T. & Parker, K. D. 1969. The helical structure of a β-1,3 xylan. *J. Polymer Sci.* C28: 69–81.

Austen, K. R. J., Goodall, D. M. & Norton, I. T. 1985. Anion-independent conformational ordering in iota-carrageenan: Disorder-order equilibria and dynamics. *Carbohydr. Res.* 140: 251–62.

Ayal, H. A. & Matsuhiro, B. 1986. Carrageenans from tetrasporic and cystocarpic *Chondrus canaliculatus*. *Phytochemistry* 25: 1895–7.

Ayal, H. & Matsuhiro, B. 1987. Polysaccharides from nuclear phases of *Iridaea ciliata* and *I. membranacea*. *Hydrobiologia* 151/152: 531–4.

Baeza, P. & Matsuhiro, B. 1977. Polysaccharides from Chilean seaweeds. IV. A sulphated galactan from *Grateloupia lanceola*. *Bot. Mar.* 20: 355–7.

Barbakadze, V. V. & Usov, A. I. 1978. Polysaccharides of algae. 26. Methylation and periodate oxidation of the polysaccharide from the red alga *Grateloupia divaricata* Okam. *Bioorg. Khim.* 4: 1100–6 (in Russian).

Barber, G. A. 1971. The synthesis of L-glucose by plant enzyme systems. *Arch. Biochem. Biophys.* 147: 619–23.

Batey, J. F. & Turvey, J. R. 1975. The galactan sulfate of the red alga *Polysiphonia lanosa*. *Carbohydr. Res.* 43: 133–43.

Bellion, C., Brigand, G., Prome, J.-C., Welti, D. & Bociek, S. 1983. Identification et caractérisation des précurseurs biologiques des carraghénanes par spectroscopie de R.M.N.-^{13}C. *Carbohydr. Res.* 119: 31–48.

Bellion, C., Hamer, G. K. & Yaphe, W. 1981. Analysis of kappa-iota hybrid carrageenans with kappa-carrageenase, iota-carrageenase and ^{13}C NMR. *Proc. Int. Seaweed Symp.* 10: 379–84.

Bellion, C., Hamer, G. K. & Yaphe, W. 1982. The degradation of *Eucheuma spinosum* and *Eucheuma cottonii* carrageenans by ι-carrageenases and $\varkappa$-carrageenases from marine bacteria. *Can. J. Microbiol.* 28: 874–80.

Bhattacharjee, S. S., Yaphe, W. & Hamer, G. K. 1978. ^{13}C-N.m.r. spectroscopic analysis of agar, $\varkappa$-carrageenan and ι-carrageenan. *Carbohydr. Res.* 60: C1–C3.

Bird, C. J., Helleur, R. J., Hayes, E. R. & McLachlan, J. 1987. Analytical pyrolysis as a taxonomic tool in *Gracilaria* (Rhodophyta: Gigartinales). *Hydrobiologia* 151/152: 207–11.

Bird, K. T. 1984. Seasonal variation in protein: carbohydrate ratios in a subtropical estuarine alga, *Gracilaria verrucosa*, and the determination of nitrogen limitation status using these ratios. *Bot. Mar.* 27: 111–15.

Bird, K. T., Hanisak, M. D. & Ryther, J. 1981. Chemical quality and production of agars extracted from *Gracilaria tikvahiae* grown in different nitrogen enrichment conditions. *Bot. Mar.* 24: 441–4.

Bodeau-Bellion, C. 1983. Analysis of carrageenan structure. *Physiol. Vég.* 21: 785–93.

Borowitzka, M. A. 1977. Algal calcification. *Oceanogr. Mar. Biol. Annu. Rev.* 15: 189–223.

Borowitzka, M. A. 1982. Mechanisms in algal calcification. *Prog. Phycol. Res.* 1: 137–77.

Bowker, D. M. & Turvey, J. R. 1968a. Water-soluble polysaccharides of the red alga *Laurencia pinnatifida*. Part I. Constituent units. *J. Chem. Soc.* 983–8.

Bowker, D. M. & Turvey, J. R. 1968b. Water-soluble polysaccharides of the red alga *Laurencia pinnatifida*. Part I. Methylation analysis of the galactan sulphate. *J. Chem. Soc.* 989–92.

Brant, D. A. & Buliga, G. S. 1985. Chemical structure and macromolecular conformation of polysaccharides in the aqueous environment. In *Biotechnology of Marine Polysaccharides*, eds. R. R. Colwell, E. R. Pariser & A. J. Sinskey, pp. 29–73. Washington: Hemisphere.

Brasch, D. J., Chang, H. M., Chuah, C-T. & Melton, L. D. 1981a. The galactan sulfate from the edible, red alga *Porphyra columbina*. *Carbohydr. Res.* 97: 113–25.

Brasch, D. J., Chuah, C-T. & Melton, L. D. 1981b. A ^{13}C N.M.R. study of some agar-related polysaccharides from New Zealand seaweeds. *Aust. J. Chem.* 34: 1095–105.

Brasch, D. J., Chuah, C-T. & Melton, L. D. 1983. The agar-type polysaccharide from the red alga *Gracilaria secundata*. *Carbohydr. Res.* 115: 191–8.

Brasch, D. J., Chuah, C-T. & Melton, L. D. 1984. Formolysis studies on some polysaccharides from red seaweeds. *Aust. J. Chem.* 37: 1539–44.

Bremond, G. de L., Quillet, M. & Bremond, M. 1987. λ-Carrageenan in the gametophytes of *Chondrus crispus*. *Phytochemistry* 26: 1705–7.

Buggeln, R. G. & Craigie, J. S. 1973. The physiology and biochemistry of *Chondrus crispus* Stackhouse. *Proc. N. S. Inst. Sci.* Suppl. 27: 81–102.

Callow, M. E. & Evans, L. V. 1981. Studies on the synthesis and secretion of mucilage in the unicellular red alga *Rhodella*. *Proc. Int. Seaweed Symp.* 8: 154–8.

Carter, A. R. & Anderson, R. J. 1986. Seasonal growth and agar contents in *Gelidium pristoides* (Gelidiales, Rhodophyta) from Port Alfred, South Africa. *Bot. Mar.* 29: 117–23.

Chamberlain, A. H. L. & Evans, L. V. 1973. Aspects of spore production in the red alga *Ceramium*. *Protoplasma* 76: 139–59.

Chen, L. C-M. & Craigie, J. S. 1981. Carrageenan analysis in apomictic *Gigartina stellata* – new puzzles. *Proc. Int. Seaweed Symp.* 10: 391–6.

Chen, L. C-M., McLachlan, J., Neish, A. C. & Shacklock, P. F. 1973. The ratio of kappa- to lambda-carrageenan in nuclear phases of the rhodophycean algae, *Chondrus crispus* and *Gigartina stellata*. *J. Mar. Biol. Assoc. U.K.* 53: 11–16.

Cheney, D. P., Luistro, A. H. & Bradley, P. M. 1987. Carrageenan analysis of tissue cultures and whole plants of *Agardhiella subulata*. *Hydrobiologia* 151/152: 161–6.

Chopin, T., Bodeau-Bellion, C., Floc'h, J.-Y., Guittet, E. & Lallemand, J.-Y. 1987. Seasonal study of carrageenan structures from female gametophytes of *Chondrus crispus* Stackhouse (Rhodophyta). *Hydrobiologia* 151/152: 535–9.

Christiaen, D., Stadler, T., Ondarza, M. & Verdus, M. C. 1987. Structures and functions of the polysaccharides from the cell wall of *Gracilaria verrucosa* (Rhodophyta, Gigartinales). *Hydrobiologia* 151/152: 139–46.

Cole, K. M., Park, C. M., Reid, P. E. & Sheath, R. G. 1985. Comparative studies in the cell walls of sexual and asexual *Bangia atropurpurea* (Rhodophyta). I. Histochemistry of polysaccharides. *J. Phycol.* 21: 585–92.

Cole, K. M., Park, C. M., Reid, P. E. & Sheath, R. G. 1986. Comparative studies on the cell walls of sexual and asexual *Bangia atropurpurea* (Rhodophyta). II. Electrophoretic patterns of polysaccharides. *J. Phycol.* 22: 406–9.

Combaut, G., Piovetti, L., Canal, G. & Sancho, A. 1985. Polysaccharide of the red alga *Rissoella verruculosa*. *Phytochemistry* 24: 1597–9.

Cote, G. L. & Hanisak, M. D. 1986. Production and properties of native agars from *Gracilaria tikvahiae* and other red algae. *Bot. Mar.* 29: 359–66.

Court, G. J. & Taylor, I. E. P. 1979. Isolation and partial characterization of a proteoglycan from the red alga *Laurencia spectabilis*. *Phytochemistry* 18: 411–14.

Craigie, J. S. & Jurgens, A. 1989. Structure of agars from *Gracilaria tikvahiae*: Location of 4-O-methyl-L-galactose and sulfate. *Carbohydr. Polym.* 11: 265–78.

Craigie, J. S. & Wen, Z. C. 1984. Effects of temperature and tissue age on gel strength and composition of agar from *Gracilaria tikvahiae* (Rhodophyceae). *Can. J. Bot.* 62: 1665–70.

Craigie, J. S., Wen, Z. C. & van der Meer, J. 1984. Interspecific, intraspecific and nutritionally determined variations in the composition of agars from *Gracilaria* spp. *Bot. Mar.* 27: 55–61.

Craigie, J. S. & Wong, K. F. 1979. Carrageenan biosynthesis. *Proc. Int. Seaweed Symp.* 9: 369–77.

Darley, W. M. 1974. Silicification and calcification. In *Algal Physiology and Biochemistry*, ed. W. D. P. Stewart, pp. 655–75. Oxford: Blackwell Scientific.

Dawes, C. J., Lawrence, J. M., Cheney, D. P. & Mathieson, A. C. 1974. Ecological studies of Floridean *Eucheuma* (Rhodophyta, Gigartinales). III. Seasonal variation of carrageenan, total carbohydrate, protein, and lipid. *Bull. Mar. Sci.* 24: 286–99.

Dawes, C. J., Stanley, N. F. & Stancioff, D. J. 1977. Seasonal and reproductive aspects of plant chemistry and ι-carrageenan from Floridean *Eucheuma* (Rhodophyta, Gigartinales). *Bot. Mar.* 20: 137–47.

Dea, I. C. M. 1981. Specificity of interactions between polysaccharide helices and β-1,4-linked polysaccharides. In *Solution Properties of Polysaccharides*, ed. D. A. Brant, pp. 439–54. Washington, D.C.: Americal Chemical Society.

DeBoer, J. A. 1979. Effects of nitrogen enrichment on growth rate and phycocolloid content in *Gracilaria foliifera* and *Neogardhiella baileyi* (Florideophyceae). *Proc. Int. Seaweed Symp.* 9: 263–71.

de Lestang-Brémond, G. & Quillet, M. 1981. The turn-over of sulphates on the lambda-carrageenan of the cell-walls of the red seaweed Gigartinale: *Catenella opuntia* (Grev.). *Proc. Int. Seaweed Symp.* 10: 449–54.

de Reviers, B., Mabeau, S., Kloareg, B. & Demarty, M. 1986. Recherche de l'existence d'un turn-over des sulfates des polyosides pariétaux de *Pelvetia canaliculata* (Pheophycées) et de *Cladophora rupestris* (Chlorophycées). *Physiol. Vég.* 24: 347–53.

Deslandes, E., Floc'h, J. Y., Bodeau-Bellion, C., Brault, D. & Braud, J. P. 1985. Evidence for ι-carrageenans in *Solieria chordalis* (Solieriaceae) and *Callibepharis jubata*, *Callibepharis ciliata*, *Cystoclonium purpureum* (Rhodophyllidaceae). *Bot. Mar.* 28: 317–18.

DiNinno, V. & McCandless, E. L. 1978. The chemistry and immunochemistry of carrageenans from *Eucheuma* and related algal species. *Carbohydr. Res.* 66: 85–93.

DiNinno, V. L., McCandless, E. L. & Bell, R. A. 1979. Pyruvic acid derivative of carrageenan from a marine red alga (*Petrocelis* species). *Carbohydr. Res.* 71: C1–C4.

Dolan, T. C. S. & Rees, D. A. 1965. The carrageenans. Part II. The positions of the glycosidic linkages and sulfate esters in λ-carrageenan. *J. Chem. Soc.* 3534–9.

Doty, M. S. & Santos, G. A. 1978. Carrageenans from tetrasporic and cystocarpic *Eucheuma* species. *Aquat. Bot.* 4: 143–9.

Doty, M. S. & Santos, G. A. 1983. Agar from *Gracilaria*

cylindrica. *Aquat. Bot.* 15: 299–306.

Duckworth, M., Hong, K. C. & Yaphe, W. 1971. The agar polysaccharides of *Gracilaria* species. *Carbohydr. Res.* 18: 1–9.

Duckworth, M. & Turvey, J. R. 1969. The action of a bacterial agarase on agarose, porphyran and alkali-treated porphyran. *Biochem. J.* 113: 687–92.

Duckworth, M. & Yaphe, W. 1971a. Preparation of agarose by fractionation from the spectrum of polysaccharides in agar. *Anal. Biochem.* 44: 636–41.

Duckworth, M. & Yaphe, W. 1971b. The structure of agar. Part I. Fractionation of a complex mixture of polysaccharides. *Carbohydr. Res.* 16: 189–97.

Durairatnam, M. 1987. Studies of the yield of agar, gel strength and quality of agar of *Gracilaria edulis* (Gmel.) Silva from Brazil. *Hydrobiologia* 151/152: 509–12.

Durairatnam, M. & Santos, N. Q. 1981. Agar from *Gracilaria verrucosa* (Hudson) Papenfuss and *Gracilaria sjoestedtii* Kylin from northeast Brasil. *Proc. Int. Seaweed Symp.* 10: 669–74.

Durako, M. J. & Dawes, C. J. 1980. A comparative seasonal study of two populations of *Hypnea musciformis* from the east and west coasts of Florida, U.S.A. I. Growth and chemistry. *Mar. Biol.* 59: 151–6.

El-Sayed, M. M. 1983. Purification and characterization of agar from *Digenea simplex*. *Carbohydr. Res.* 118: 119–26.

Evans, L. V., Callow, M. E., Percival, E. & Fareed, V. S. 1974. Synthesis and composition of extracellular mucilage in the unicellular red alga *Rhodella*. *J. Cell Sci.* 16: 1–16.

Fareed, V. S. & Percival, E. 1977. The presence of rhamnose and 3-O-methylxylose in the extracellular mucilage from the red alga *Rhodella maculata*. *Carbohydr. Res.* 53: 276–7.

Farrant, A. J., Nunn, J. R. & Parolis, H. 1971. Sulphated polysaccharides of the Grateloupiaceae family. Part VI. A polysaccharide from *Pachymenia carnosa*. *Carbohydr. Res.* 19: 161–8.

Fernandez, L. E., Valiente, O. G., Garcia, R., Castro, H. V., Machytka, D., Zsoldos-Mady, V. & Neszmélyi, A. 1987. A ^{1}H- and ^{13}C-n.m.r. study of an agar polysaccharide from *Bryothamnion triquetrum*. *Carbohydr. Res.* 163: 143–7.

Fogg, G. E. 1964. Environmental conditions and the pattern of metabolism in algae. In *Algae and Man*, ed. D. Jackson, pp. 77–85. New York: Plenum.

Frei, E. & Preston, R. D. 1961. Variants in the structural polysaccharides of algal cell walls. *Nature* (Lond.) 192: 939–43.

Frei, E. & Preston, R. D. 1964. Non-cellulosic structural polysaccharides in algal cell walls. II. Association of xylan and mannan in *Porphyra umbilicalis*. *Proc. R. Soc. Lond. Ser. B. Biol. Sci.* 160: 314–17.

Friedlander, M., Lipkin, Y. & Yaphe, W. 1981. Composition of agars from *Gracilaria* cf. *verrucosa* and *Pterocladia capillaceae*. *Bot. Mar.* 24: 595–8.

Friedlander, M., Shalev, R., Ganor, T., Strimling, S., Ben-Amotz, A., Klar, H. & Wax, Y. 1987. Seasonal fluctuations of growth rate and chemical composition of *Gracilaria* cf. *conferta* in outdoor culture in Israel. *Hydrobiologia* 151/152: 501–7.

Friedlander, M. & Zelikovitch, N. 1984. Growth rates, phycocolloid yield and quality of the red seaweeds, *Gracilaria* sp., *Pterocladia capillaceae*, *Hypnea musciformis*, and *Hypnea cornuta*, in field studies in Israel. *Aquaculture* 40: 57–66.

Fuller, S. W. & Mathieson, A. C. 1972. Ecological studies of economic red algae. IV. Variations of carrageenan concentration and properties in *Chondrus crispus* Stackhouse. *J. Exp. Mar. Biol. Ecol.* 10: 49–58.

Furneaux, R. H. & Miller, I. J. 1985. Water soluble polysaccharides from the New Zealand red algae in the family Phyllophoraceae. *Bot. Mar.* 28: 419–25.

Furuya, K. 1960. Biochemical studies on calcareous algae. 1. Major inorganic constituents of some calcareous red algae. *Bot. Mag. Tokyo* 73: 355–9.

Gabrielson, P. W. & Garbary, D. 1986. Systematics of red algae (Rhodophyta). *CRC Crit. Rev. Plant Sci.* 3: 325–66.

Gabrielson, P. W., Garbary, D. J. & Scagel, R. F. 1985. The nature of the ancestral red alga: Inferences from a cladistic analysis. *BioSystems* 18: 335–46.

Gerwick, W. H. & Lang, N. J. 1977. Structural, chemical and ecological studies on iridescence in *Iridaea* (Rhodophyta). *J. Phycol.* 13: 121–7.

Giraud, G. & Cabioch, J. 1979. Ultrastructure and elaboration of calcified cell-walls in the coralline algae (Rhodophyta, Cryptonemiales). *Biol. Cell.* 36: 81–6.

Goddard, M. R. & Fernandez, E. B. 1986. Un estudio del carragenano de las fases cistocarpica tetrasporica e inmadura de *Iridaea boryana* (Setch. et Gardn.) Skottsb. *Rev. Latinoamer. Quim.* 17: 35–6.

Gordon, E. M. & McCandless, E. L. 1973. Ultrastructure and histochemistry of *Chondrus crispus* Stackhouse. In *Chondrus crispus*. *Proc. N. S. Inst. Sci. Suppl.* 27: 111–33.

Gordon-Mills, E. M. & McCandless, E. L. 1975. Carrageenans in the cell walls of *Chondrus crispus* Stack. (Rhodophyceae, Gigartinales). I. Localization with fluorescent antibody. *Phycologia* 14: 275–81.

Gordon-Mills, E. M. & McCandless, E. L. 1977. Carrageenans in the cell walls of *Chondrus crispus* Stack. (Rhodophyceae, Gigartinales). II. Birefringence. *Phycologia* 16: 169–76.

Gordon-Mills, E. M., Tas, J. & McCandless, E. L. 1978. Carrageenans in the cell walls of *Chondrus crispus* Stack. (Rhodophyceae, Gigartinales). III. Metachromasia and the topooptical reaction. *Phycologia* 17: 95–104.

Greer, C. W., Shomer, I., Goldstein, M. E. & Yaphe, W. 1984. Analysis of carrageenan from *Hypnea musciformis* by using κ- and ι-carrageenases and ^{13}C-n.m.r. spectroscopy. *Carbohydr. Res.* 129: 189–96.

Greer, C. W. & Yaphe, W. 1984a. Characterization of hybrid (beta-kappa-gamma) carrageenan from *Eucheuma gelatinae* J. Agardh (Rhodophyta, Solieriaceae) using carrageenases, infrared and ^{13}C–nuclear magnetic resonance spectroscopy. *Bot. Mar.* 27: 473–8.

Greer, C. W. & Yaphe, W. 1984b. Hybrid (iota-nu-kappa) carrageenan from *Eucheuma nudum* (Rhodophyta, Solieriaceae), identified using iota- and kappa-carrageenases and ^{13}C–nuclear magnetic resonance spectroscopy. *Bot. Mar.* 27: 479–84.

Gretz, M. R., Aronson, J. M. & Sommerfeld, M. R. 1980. Cellulose in the cell walls of the Bangiophyceae (Rhodophyta). *Science* 207: 779–81.

Gretz, M. R., Aronson, J. M. & Sommerfeld, M. R. 1984. Taxonomic significance of cellulosic cell walls in the Bangiales (Rhodophyta). *Phytochemistry* 23: 2513–14.

Gretz, M. R., Aronson, J. M. & Sommerfeld, M. R. 1986. Cell wall composition of the conchocelis phases of *Bangia atropurpurea* and *Porphyra leucosticta* (Rhodophyta). *Bot. Mar.* 29: 91–6.

Gretz, M. R., McCandless, E. L., Aronson, J. M. & Sommerfeld, M. R. 1983. The galactan sulphates of the conchocelis phases of *Porphyra leucosticta* and *Bangia atropurpurea* (Rhodophyta). *J. Exp. Bot.* 34: 705–11.

Gretz, M. R., Sommerfeld, M. R. & Aronson, J. M. 1982. Cell wall composition of the generic phase of *Bangia atropurpurea* (Rhodophyta). *Bot. Mar.* 25: 529–35.

Gudin, C. & Thomas, D. 1981. Sulfated polysaccharide production in a biophotoreactor by immobilized cells of *Porphyridium cruentum*. *C. R. Acad. Sci.* (Paris) 293: 35–7.

Guist, G. G., Dawes, C. J. & Castle, J. R. 1982. Mariculture of the red seaweed, *Hypnea musciformis*. *Aquaculture* 28: 375–84.

Hanic, L. A. & Craigie, J. S. 1969. Studies on the algal cuticle. *J. Phycol.* 5: 89–102.

Haug, A. 1974. Chemistry and biochemistry of algal cell-wall polysaccharides. In *MTP Int. Rev. Sci. Biochem. Ser.* 1, Vol. 11, *Plant Biochemistry*, ed. D. H. Northcote, pp. 51–88. London: Butterworths.

Heaney-Kieras, J. & Chapman, D. J. 1976. Structural studies on the extracellular polysaccharide of the red alga, *Porphyridium cruentum*. *Carbohydr. Res.* 52: 169–77.

Heaney-Kieras, J., Rodén, L. & Chapman, D. J. 1977. The covalent linkage of protein to carbohydrate in the extracellular protein-polysaccharide from the red alga *Porphyridium cruentum*. *Biochem. J.* 165: 1–9.

Heaney-Kieras, J. & Swift, H. 1979. Electron microscopy of the extracellular protein-polysaccharide from the red alga, *Porphyridium cruentum*. In *Glycoconjugate Research* vol. 1, eds. J. D. Gregory & R. W. Jeanloz, pp. 165–6. New York: Academic.

Helleur, R. J., Hayes, E. R., Craigie, J. S. & McLachlan, J. L. 1985. Characterization of polysaccharides of red algae by pyrolysis–capillary gas chromatography. *J. Anal. Appl. Pyrolysis* 8: 349–57.

Hirase, S. & Araki, C. 1961. Isolation of 6-O-methyl-D-galactose from the agar of *Ceramium boydenii*. *Bull. Chem. Soc. Jap.* 34: 1048.

Hirase, S., Araki, C. & Watanabe, K. 1967. Component sugars of the polysaccharide of the red seaweed *Grateloupia elliptica*. *Bull. Chem. Soc. Jap.* 40: 1445–8.

Hirase, S. & Watanabe, K. 1972a. The presence of pyruvate residues in λ-carrageenan and a similar polysaccharide. *Bull. Inst. Chem. Res., Kyoto Univ.* 50: 332–6.

Hirase, S. & Watanabe, K. 1972b. Fractionation and structural investigation of funoran. *Proc. Int. Seaweed Symp.* 7: 451–4.

Hirase, S., Watanabe, K., Takano, R. & Tamura, J. 1983. Structural features of the sulfated polysaccharide isolated from the red seaweed *Laurencia undulata*. In *Abstracts of the XIth International Seaweed Symposium*. Inst. Oceanology, Academia Sinica, Qingdao, People's Republic of China. p. 93.

Hosford, S. P. C. & McCandless, E. L. 1975. Immunochemistry of carrageenan from gametophytes and sporophytes of certain red algae. *Can. J. Bot.* 53: 2835–41.

Hoyle, M. D. 1978a. Agar studies in two *Gracilaria* species (*G. bursapastoris* (Gmelin) Silva and *G. coronopifolia* J. Ag.) from Hawaii. I. Yield and gel strength in the gametophyte and tetrasporophyte generations. *Bot. Mar.* 21: 343–5.

Hoyle, M. D. 1978b. Agar studies in two *Gracilaria* species (*G. bursapastoris* (Gmelin) Silva and *G. coronopifolia* J. Ag.) from Hawaii. II. Seasonal aspects. *Bot. Mar.* 21: 347–52.

Ibañez, C. M. & Matsuhiro, B. 1986. Structural studies on the soluble polysaccharide from *Iridaea membranacea*. *Carbohydr. Res.* 146: 327–34.

Iriki, Y., Suzuki, T., Nisizawa, K. & Miwa, T. 1960. Xylan of siphonaceous green algae. *Nature* (Lond.) 187: 82–3.

Izumi, K. 1970. A new method for fractionation of agar. *J. Agr. Biol. Chem.* 34: 1739–40.

Izumi, K. 1971. Chemical heterogeneity of the agar from *Gelidium amansii*. *Carbohydr. Res.* 17: 227–30.

Izumi, K. 1972. Chemical heterogeneity of the agar from *Gracilaria verrucosa*. *J. Biochem.* 72: 135–40.

Jackson, S. J. & McCandless, E. L. 1979. Incorporation of [^{35}S] sulfate and [^{14}C] bicarbonate into karyotype-specific polysaccharides of *Chondrus crispus*. *Plant Physiol.* 64: 585–9.

James, D. W., Jr., Preiss, J. & Elbein, A. D. 1985. Biosynthesis of polysaccharides. In *The Polysaccharides*, vol. 3, ed. G. O. Aspinall, pp. 107–207. Orlando: Academic.

Ji, M., Lahaye, M. & Yaphe, W. 1985. Structure of agar from *Gracilaria* spp. (Rhodophyta) collected in the People's Republic of China. *Bot. Mar.* 28: 521–8.

John, D. M. & Asare, S. O. 1975. A preliminary study of the variations in yield and properties of phycocolloids from Ghanaian seaweeds. *Mar. Biol.* 30: 325–30.

Jones, J. K. N. 1950. The structure of the mannan present in *Porphyra umbilicalis*. *J. Chem. Soc.* 3292–5.

Jones, R. F. 1962. Extracellular mucilage of the red alga *Porphyridium cruentum*. *J. Cell Compar. Physiol.* 60: 61–4.

Karr, A. L. 1976. Cell wall biogenesis. In *Plant Biochemistry*, 3rd edition, eds. J. Bonner & J. E. Varner, pp. 405–26. New York: Academic.

Kieras, J. H., Kieras, F. J. & Bowen, D. V. 1976. 2-O-Methyl-D-glucuronic acid, a new hexuronic acid of biological origin. *Biochem. J.* 155: 181–5.

Kim, D. H. & Henriquez, N. P. 1979. Yields and gel strengths of agar from cystocarpic and tetrasporic plants of *Gracilaria verrucosa* (Florideophyceae). *Proc. Int. Seaweed Symp.* 9: 257–62.

Kochetkov, N. K., Usov, A. I. & Miroshnikova, L. I. 1967. Polysaccharides of algae. I. Water-soluble polysaccharides of a red alga, *Laingia pacifica*. *Zh. Obshch. Khim.* 37: 792–6.

Kochetkov, N. K., Usov, A. I. & Miroshnikova, L. I. 1970a. Polysaccharides of algae. IV. Fractionation and methanolysis of a sulfonated polysaccharide from *Laingia pacifica*. *Zh. Obshch. Khim.* 40: 2469–73.

Kochetkov, N. K., Usov, A. I. & Miroshnikova, L. I. 1970b. Polysaccharides of algae. V. Composition and structure of a sulfonated polysaccharide from *Laingia pacifica*. *Zh. Obshch. Khim.* 40: 2473–8.

Kopp, J. & Perez, R. 1978. Contribution à l'étude de l'algue rouge *Chondrus crispus* Stack. Relation entre la croissance, la potentialite sexuelle, la quantité et la composition des carraghénanes. *Rev. Trav. Inst. Pêches Marit.* 42: 291–324.

Kraft, G. T. 1981. Rhodophyta: Morphology and classification. In *The Biology of Seaweeds*, eds. C. S. Lobban & M. J. Wynne, pp. 6–51. Berkeley: University of California Press.

Krasil'nikova, S. V. & Medvedeva, E. I. 1975. Alkali-soluble protein of the red alga *Furcellaria fastigiata*. *Khim. Prir. Soedin.* 11: 400–4 (in Russian).

Kremer, B. P. 1981. Carbon metabolism. In *The Biology of Seaweeds*, eds. C. S. Lobban & M. J. Wynne, pp. 493–533. Berkeley: University of California Press.

La Claire II, J. W. & Dawes, C. J. 1976. An autoradiographic and histochemical localization of sulfated polysaccharides in *Eucheuma nudum* (Rhodophyta). *J. Phycol.* 12: 368–75.

Lahaye, M., Rochas, C. & Yaphe, W. 1986. A new procedure for determining the heterogeneity of agar polymers in the cell walls of *Gracilaria* spp. (Gracilariaceae, Rhodophyta). *Can. J. Bot.* 64: 579–85.

Lahaye, M., Yaphe, W. & Rochas, C. 1985. ^{13}C-N.m.r.-spectral analysis of sulfated and desulfated polysaccharides of the agar type. *Carbohydr. Res.* 143: 240–5.

Lapointe, B. E. 1981. The effects of light and nitrogen on growth, pigment content and biochemical composition of *Gracilaria foliifera* v. *angustissima* (Gigartinales, Rhodophyta). *J. Phycol.* 17: 90–5.

Lapointe, B. E. & Ryther, J. H. 1979. The effects of nitrogen and seawater flow rate on the growth and biochemical composition of *Gracilaria foliifera* v. *angustissima* in mass outdoor cultures. *Bot. Mar.* 22: 529–37.

Lawson, C. J., Rees, D. A., Stancioff, D. J. & Stanley, N. F. 1973. Carrageenans. Part VIII. Repeating structures of galactose sulphates from *Furcellaria fastigiata, Gigartina canaliculata, Gigartina chamissoi, Gigartina atropurpurea, Ahnfeltia durvillaei, Gymnogongrus furcellatus, Eucheuma cottonii, Eucheuma spinosum, Eucheuma isiforme, Eucheuma uncinatum, Aghardhiella tenera, Pachymenia hymantophora,* and *Gloiopeltis cervicornis. J. Chem. Soc. (Perkin 1):* 2177–82.

Letherby, M. & Young, D. A. 1981. The gelation of agarose. *J. Chem. Soc., Faraday Trans.* 1 (77): 1953–66.

Lewin, J. C. 1962. Calcification. In *Physiology and Biochemistry of Algae*, ed. R. A. Lewin, pp. 457–65. London: Academic.

Liu, C-Y., Wang, C-Y. & Yang, S-S. 1981. Seasonal variation of the chlorophyll contents of *Gracilaria* cultivated in Taiwan. *Proc. Int. Seaweed Symp.* 10: 455–60.

Mackie, W. & Preston, R. D. 1974. Cell wall and intercellular region polysaccharides. In *Algal Physiology and Biochemistry*, ed. W. D. P. Stewart, pp. 40–85. Oxford: Blackwell Scientific.

Macler, B. A. 1986. Regulation of carbon flow by nitrogen and light in the red alga, *Gelidium coulteri. Plant Physiol.* 82: 136–41.

Marchessault, R. H. & Sundararajan, P. R. 1983. Cellulose. In *The Polysaccharides*, vol. 2, ed. G. O. Aspinall, pp. 11–95. New York: Academic.

Masuda, M., West, J. A., Ohno, Y. & Kurogi, M. 1984. Comparative reproductive patterns in culture of different *Gigartina* subgenus *Mastocarpus* and *Petrocelis* populations from northern Japan. *Bot. Mag. Tokyo* 97: 107–25.

Mathieson, A. C., Penniman, C. E. & Tveter-Gallagher, E. 1984. Phycocolloid ecology of underutilized economic red algae. *Hydrobiologia* 116/117: 542–6.

Mathieson, A. C. & Prince, J. S. 1973. Ecology of *Chondrus crispus* Stackhouse. In *Chondrus crispus. Proc. N. S. Inst. Sci.* 27 (Suppl.): 53–77.

Mathieson, A. C. & Tveter, E. 1975. Carrageenan ecology of *Chondrus crispus* Stackhouse. *Aquat. Bot.* 1: 25–43.

Mathieson, A. C. & Tveter, E. 1976. Carrageenan ecology of *Gigartina stellata* (Stackhouse) Batters. *Aquat. Bot.* 2: 353–61.

Matsuhiro, B. 1982. Polysaccharides from Chilean seaweeds. Part XII. Studies on the soluble polysaccharide from *Ceramium pacificum. Bot. Mar.* 25: 139–41.

Matsuhiro, B. 1983. Polysaccharides from Chilean seaweeds. Part XIV. Methylation analysis of the fraction precipitated with 0.125 N potassium chloride of *Gigartina papillata. Rev. Latinoam. Quim.* 14: 77–9.

Matulewicz, M. C. & Cerezo, A. S. 1980. The carrageenan from *Iridaea undulosa* B.; analysis, fractionation and alkaline treatment. *J. Sci. Food Agric.* 31: 203–13.

McCandless, E. L. 1978. The importance of cell wall constituents in algal taxonomy. In *Modern Approaches to the Taxonomy of Red and Brown Algae*, eds. D. E. G. Irvine & J. H. Price, pp. 63–85. London: Academic.

McCandless, E. L. 1981. Biological control of carrageenan structure: effects conferred by the phase of life cycle of the carrageenophyte. *Proc. Int. Seaweed Symp.* 8: 1–18.

McCandless, E. L. & Craigie, J. S. 1974. Reevaluation of seasonal factors involved in carrageenan production by *Chondrus crispus*: carrageenans of carposporic plants. *Bot. Mar.* 17: 125–9.

McCandless, E. L. & Craigie, J. S. 1979. Sulfated polysaccharides in red and brown algae. *Annu. Rev. Plant Physiol.* 30: 41–53.

McCandless, E. L., Craigie, J. S. & Hansen, J. E. 1975. Carrageenans of gametangial and tetrasporangial stages of *Iridaea cordata* (Gigartinaceae). *Can. J. Bot.* 53: 2315–18.

McCandless, E. L., Craigie, J. S. & Walter, J. A. 1973. Carrageenans in the gametophytic and sporophytic stages of *Chondrus crispus. Planta* (Berl.) 112: 201–12.

McCandless, E. L. & Gretz, M. R. 1984. Biochemical and immunochemical analysis of carrageenans of the Gigartinaceae and Phyllophoraceae. *Hydrobiologia* 116/117: 175–8.

McCandless, E. L. & Vollmer, C. M. 1984. The nemathecium of *Gymnogongrus chiton* (Rhodophyceae, Gigartinales): immunochemical evidence of meiosis. *Phycologia* 23: 119–23.

McCandless, E. L., West, J. A. & Guiry, M. D. 1982. Carrageenan patterns in the Phyllophoraceae. *Biochem. Syst. Ecol.* 10: 275–84.

McCandless, E. L., West, J. A. & Guiry, M. D. 1983. Carrageenan patterns in the Gigartinaceae. *Biochem. Syst. Ecol.* 11: 175–82.

McCandless, E. L., West, J. A. & Vollmer, C. M. 1981. Carrageenans of species of the genus *Phyllophora. Proc. Int. Seaweed Symp.* 10: 473–8.

Medcalf, D. G., Brannon, J. H., Scott, J. R., Allen, G. G., Lewis, J. & Norris, R. E. 1981. Polysaccharides from microscopic red algae and diatoms. *Proc. Int. Seaweed Symp.* 8: 582–8.

Medcalf, D. G., Scott, J. R., Brannon, J. H., Hemerick, G. A., Cunningham, R. L., Chessen, J. H. & Shah, J. 1975. Some structural features and viscometric properties of the extracellular polysaccharide from *Porphyridium cruentum. Carbohydr. Res.* 44: 87–96.

Medvedeva, E. I., Lukina, G. D., Selich, E. F. & Bozhko, I. G. 1973. Structure and some characteristics of glycoproteins of the alga *Phyllophora. Biokhimiya* 38: 1181–5.

Millard, P. & Evans, L. V. 1982a. Sulphate uptake in the unicellular marine red alga *Rhodella maculata. Arch. Microbiol.* 131: 165–9.

Millard, P. & Evans, L. V. 1982b. The fate of sulphate within the capsular polysaccharide of the unicellular red alga *Rhodella maculata. J. Exp. Bot.* 136: 854–64.

Miller, I. J., & Furneaux, R. H. 1987. The chemical substitution of the agar-type polysaccharide from *Gracilaria secundata* f. *pseudoflagellifera* (Rhodophyta). *Hydrobiologia* 151/152: 523–9.

Misonou, T., Okazaki, M. & Furuya, K. 1980. Particular Ca-binding substances in marine algae. II. Incorporation of ^{45}Ca into acid-insoluble residues from various algae and the solubilization of Ca-binding substances from the residues. *Jap. J. Phycol.* 28: 105–12 (in Japanese).

Miyata, M., Okazaki, M. & Furuya, K. 1980. Initial calcification site of the calcareous red alga *Serraticardia maxima* (Yendo) Silva (Studies on the calcium carbonate deposition of algae III). In *The Mechanisms of Biomineralization in Animals and Plants*, eds. M. Omori & N. Watabe, pp. 205–10. Tokyo: Tokai University Press.

Møller, M. E. & Evans, L. V. 1976. Sulphate activation in the unicellular red alga *Rhodella*. *Phytochemistry* 15: 1623–6.

Mollion, M. J. 1973. Étude préliminaire des *Hypnea* au Sénégal comme source de phycocolloides. *Bot. Mar.* 16: 221–5.

Mollion, J. 1980. Infrared and chemical studies of the carrageenan from *Anatheca montagnei* Schmitz, (Solieriaceae) from Senegal, West Africa. *Bot. Mar.* 23: 197–9.

Mollion, J., Moreau, S. & Christiaen, D. 1986. Isolation of a new type of carrageenan from *Rissoella verruculosa* (Bert.) J. Ag. (Rhodophyta, Gigartinales). *Bot. Mar.* 29: 549–52.

Morrice, L. M., McLean, M. W., Long, W. F. & Williamson, F. B. 1983. Porphyran primary structure. An investigation using β-agarase I from *Pseudomonas atlantica* and ^{13}C-NMR spectroscopy. *Eur. J. Biochem.* 133: 673–84.

Morrice, L. M., McLean, M. W., Long, W. F. & Williamson, F. B. 1984. Porphyran primary structure. *Hydrobiologia* 116/117: 572–5.

Morris, E. R., Rees, D. A. & Robinson, G. 1980. Cation-specific aggregation of carrageenan helices; domain model of polymer gel structure. *J. Mol. Biol.* 138: 349–62.

Mukai, L. S., Craigie, J. S. & Brown, R. G. 1981. Chemical composition and structure of the cell walls of the conchocelis and thallus phases of *Porphyra tenera* (Rhodophyceae). *J. Phycol.* 17: 192–8.

Neish, A. C. & Shacklock, P. F. 1971. Greenhouse Experiments (1971) on the Propagation of Strain T4 of Irish Moss. National Research Council of Canada, Atlantic Regional Laboratory, Tech. Rep. No. 14.

Neish, A. C., Shacklock, P. F., Fox, C. H. & Simpson, F. J. 1977. The cultivation of *Chondrus crispus*. Factors affecting growth under greenhouse conditions. *Can. J. Bot.* 55: 2263–71.

Nelson, S. G., Yang, S. S., Yang, C. Y. & Chiang, Y. M. 1983. Yield and quality of agar from species of *Gracilaria* (Rhodophyta) collected from Taiwan and Micronesia. *Bot. Mar.* 26: 361–6.

Northcote, D. H. 1985. Control of cell wall formation during growth. In *Biochemistry of Plant Cell Walls*, eds. C. T. Brett & J. R. Hillman, pp. 177–97. Cambridge: Cambridge University Press.

Nunn, J. R., Parolis, H. & Russell, I. 1971. Sulphated polysaccharides of the Solieriaceae family. Part I. A polysaccharide from *Anatheca dentata*. *Carbohydr. Res.* 20: 205–15.

Nunn, J. R., Parolis, H. & Russell, I. 1973. Sulphated polysaccharides of the Solieriaceae family. Part II. The acidic components of the polysaccharide from the red alga *Anatheca dentata*. *Carbohydr. Res.* 29: 281–9.

Nunn, J. R., Parolis, H. & Russell, I. 1981. The desulphated polysaccharide of *Anatheca dentata*. *Carbohydr. Res.* 95: 219–26.

Nunn, J. R. & von Holdt, (Mrs.) M. M. 1957. Red seaweed polysaccharides. Part II. *Porphyra capensis* and the separation of D- and L-galactose by crystallization. *J. Chem. Soc.* 1094–7.

O'Colla, P. S. 1962. Mucilages. In *Physiology and Biochemistry of Algae*, ed. R. A. Lewin, pp. 337–56. New York: Academic.

Okazaki, M. 1977. Some enzymatic properties of Ca^{2+} dependent adenosine triphosphatase from a calcareous red alga, *Serraticardia maxima* and its distribution in marine algae. *Bot. Mar.* 20: 347–54.

Okazaki, M., Furuya, K., Tsukayama, K. & Nisizawa, K. 1982a. Isolation and identification of alginic acid from a calcareous red alga *Serraticardia maxima*. *Bot. Mar.* 25: 123–31.

Okazaki, M., Ichikawa, I. & Furuya, K. 1982b. Studies on the calcium carbonate deposition of algae. IV. Initial calcification site of calcareous red alga *Galaxaura fastigiata* Decaisne. *Bot. Mar.* 25: 511–17.

Okazaki, M., Pentecost, A., Tanaka, Y. & Miyata, M. 1986. A study of calcium carbonate deposition in the genus *Padina* (Phaeophyceae, Dictyotales). *Br. Phycol. J.* 21: 217–24.

O'Neill, A. N. 1955a. 3,6-Anhydrogalactose as a constituent of ϰ-carrageenan. *J. Am. Chem. Soc.* 77: 2837–9.

O'Neill, A. N. 1955b. Derivatives of 4-O-β-D-galactopyranosyl-3,6-anhydro-D-galactose from ϰ-carrageenan. *J. Am. Chem. Soc.* 77: 6324–6.

Oza, R. M. 1978. Studies on Indian *Gracilaria*. IV. Seasonal variation in agar and gel strength of *Gracilaria corticata* J. Ag. occurring on the coast of Verval. *Bot. Mar.* 21: 165–7.

Painter, T. J. 1983. Algal polysaccharides. In *The Polysaccharides*, vol. 2, ed. G. O. Aspinall, pp. 195–285. New York: Academic.

Pallaghy, C. K., Minchinton, J., Kraft, G. T. & Wetherbee, R. 1983. Presence and distribution of bromine in *Thysanocladia densa* (Solieriaceae, Gigartinales), a marine red alga from the Great Barrier Reef. *J. Phycol.* 19: 204–8.

Paoletti, S., Smidsrød, O. & Grasdalen, H. 1984. Thermodynamic stability of the ordered conformation of carrageenan polyelectrolytes. *Biopolymers* 23: 1771–94.

Parolis, H. 1978. The structure of the polysaccharide of *Pachymenia carnosa*. *Carbohydr. Res.* 62: 313–20.

Parolis, H. 1981. The polysaccharides of *Phyllymenia hieroglyphica* (= *P. belangeri*) and *Pachymenia hymantophora*. *Carbohydr. Res.* 93: 261–7.

Parsons, M. J., Pickmere, S. E. & Bailey, R. W. 1977. Carrageenan composition in New Zealand species of *Gigartina* (Rhodophyta): geographic variation and interspecific differences. *N. Z. J. Bot.* 15: 589–95.

Patway, M. U. & van der Meer, J. P.1983. Genetics of *Gracilaria tikvahiae* (Rhodophyceae). IX. Some properties of agars extracted from morphological mutants. *Bot. Mar.* 26: 295–9.

Peat, S. & Turvey, J. R. 1965. Polysaccharides of marine algae. In *Progress in the Chemistry of Organic Natural Products*, Vol. 23, ed. L. Zechmeister, pp. 1–45. Vienna: Springer-Verlag.

Peats, S. 1981. The infrared spectra of carrageenans extracted from various algae. *Proc. Int. Seaweed Symp.* 10: 495–502.

Pedersén, M., Roomans, G. M. & Hofsten, A. v. 1980. Blue iridescence and bromine in the cuticle of the red alga *Chondrus crispus* Stackh. *Bot. Mar.* 23: 193–6.

Pedersén, M. E. E., Roomans, G. M. & Hofsten, A. v. 1981. Bromine in the cuticle of *Polysiphonia nigrescens*: localization and content. *J. Phycol.* 17: 105–8.

Pedersén, M., Saenger, P., Rowan, K. S. & Hofsten, A. v. 1979. Bromine, bromophenols and floridorubin in the red alga *Lenormandia prolifera*. *Physiol. Plant.* 46: 121–6.

Penman, A. & Rees, D. A. 1973. Carrageenans. Part IX.

Methylation analysis of galactan sulphates from *Furcellaria fastigiata*, *Gigartina canaliculata*, *Gigartina chamissoi*, *Gigartina atropurpurea*, *Ahnfeltia durvillaei*, *Gymnogongrus furcellatus*, *Eucheuma isiforme*, *Eucheuma uncinatum*, *Aghardhiella tenera*, *Pachymenia hymantophora*, and *Gloiopeltis cervicornis*. Structure of ξ-carrageenan. *J. Chem. Soc. (Perkin Trans. 1)*: 2182-7.

Penniman, C. A. 1977. Seasonal chemical and reproductive changes in *Gracilaria foliifera* (Forssk.) Borg. from Great Bay, New Hampshire (U.S.A.). *J. Phycol.* 13 (Suppl.): 53.

Pentecost, A. 1980. Calcification in plants. *Int. Rev. Cytol.* 62: 1-27.

Pentecost, A. 1985. Photosynthetic plants as intermediary agents between environmental bicarbonate and carbonate deposition. In *Inorganic Carbon Uptake by Aquatic Photosynthetic Organisms*, eds. W. J. Lucas & J. Berry, pp. 459-76. Davis: Am. Soc. Plant Physiol.

Percival, E. 1970. Algal polysaccharides. In *The Carbohydrates: Chemistry and Biochemistry*, 2nd ed., vol. 2B, eds. W. Pigman, D. Horton & A. Herp, pp. 537-68. New York: Academic.

Percival, E. 1978. Do the polysaccharides of brown and red seaweeds ignore taxonomy? In *Modern Approaches to the Taxonomy of Red and Brown Algae*, eds. D. E. G. Irvine & J. H. Price, pp. 47-62. London: Academic.

Percival, E. & Foyle, R. A. J. 1979. The extracellular polysaccharides of *Porphyridium cruentum* and *Porphyridium aerugineum*. *Carbohydr. Res.* 72: 165-76.

Percival, E. & McDowell, R. H. 1967. *Chemistry and Enzymology of Marine Algal Polysaccharides*. London: Academic.

Percival, E. & McDowell, R. H. 1981. Algal walls - composition and biosynthesis. *Encyclopedia Plant Physiol.* N. S. 13B: 277-316.

Pernas, A. J., Smidsrød, O., Larsen, B. & Haug, A. 1967. Chemical heterogeneity of carrageenans as shown by fractional precipitation with potassium chloride. *Acta Chem. Scand.* 21: 98-110.

Peyrière, M. 1970. Évolution de l'appareil de Golgi de la tétrasporogenèse de *Griffithsia flosculosa* (Rhodophycée, Ceramiacée). *C. R. Acad. Sci.* (Paris) sér. D, 270: 2071-4.

Pickmere, S. E., Parsons, M. J. & Bailey, R. W. 1973. Composition of *Gigartina* carrageenan in relation to sporophyte and gametophyte stages of the life cycle. *Phytochemistry* 12: 2441-4.

Pickmere, S. E., Parsons, M. J. & Bailey, R. W. 1975. Variations in carrageenan levels and composition in three New Zealand species of *Gigartina*. *N. Z. J. Sci.* 18: 585-90.

Preston, R. D. 1974. *Physical Biology of Plant Cell Walls*. London: Chapman & Hall.

Pueschel, C. M. 1979. Ultrastructure of tetrasporogenesis in *Palmaria palmata* (Rhodophyta). *J. Phycol.* 15: 409-24.

Pueschel, C. M. 1980. A reappraisal of the cytochemical properties of rhodophycean pit plugs. *Phycologia* 19: 210-17.

Quillet, M. & de Lestang-Brémond, G. 1981. The MeCDPS, a carrying sulphate's nucleotide of the red seaweed *Catenella opuntia* (Grev.). *Proc. Int. Seaweed Symp.* 10: 503-7.

Ramus, J. 1971. Properties of septal plugs from the red alga *Griffithsia pacifica*. *Phycologia* 10: 99-103.

Ramus, J. 1972. The production of extracellular polysaccharide by the unicellular red alga *Porphyridium aerugineum*. *J. Phycol.* 8: 97-111.

Ramus, J. 1973. Cell surface polysaccharides of the red alga *Porphyridium*. In *Biogenesis of Plant Cell Wall Polysaccharides*, ed. F. Loewus, pp. 333-59. New York: Academic.

Ramus, J. 1974. *In vivo* molybdate inhibition of sulfate transfer to *Porphyridium* capsular polysaccharide. *Plant Physiol.* 54: 945-9.

Ramus, J. & Groves, S. T. 1972. Incorporation of sulfate into the capsular polysaccharide of the red alga *Porphyridium*. *J. Cell. Biol.* 54: 399-407.

Ramus, J. & Groves, S. T. 1974. Precursor-product relationships during sulfate incorporation into *Porphyridium* capsular polysaccharide. *Plant Physiol.* 53: 434-9.

Ramus, J. & Robins, D. M. 1975. The correlation of Golgi activity and polysaccharide secretion in *Porphyridium*. *J. Phycol.* 11: 70-4.

Rees, D. A. 1961a. Enzymic synthesis of the 3:6-anhydro-L-galactose within porphyran from L-galactose-6-sulphate units. *Biochem. J.* 81: 347-52.

Rees, D. A. 1961b. Estimation of the relative amounts of isomeric sulphate esters in some sulphated polysaccharides. *J. Chem. Soc.* 5168-71.

Rees, D. A. 1969. Structure, conformation and mechanisms in the formation of polysaccharide gels and networks. *Adv. Carbohydr. Chem. Biochem.* 24: 267-332.

Rees, D. A. 1981. Polysaccharide shapes and their interactions - some recent advances. *Pure Appl. Chem.* 53: 1-14.

Rees, D. A. & Conway, E. 1962a. The structure and biosynthesis of porphyran: a comparison of some samples. *Biochem. J.* 84: 411-16.

Rees, D. A. & Conway, E. 1962b. Water-soluble polysaccharides of *Porphyra* species: a note on the classification of *P. naiadum*. *Nature* (Lond.) 195: 398-9.

Rees, D. A., Morris, E. R., Thom, D. & Madden, J. K. 1982. Shapes and interactions of carbohydrate chains. In *The Polysaccharides*, vol. 1, ed. G. O. Aspinal, pp. 195-290. New York: Academic.

Rinaudo, M. & Rochas, C. 1981. Investigations on aqueous solution properties of ϰ-carrageenans. In *Solution Properties of Polysaccharides*, ed. D. A. Brant, pp. 367-78. Washington, D. C.: American Chemical Society.

Rochas, C. & Heyraud, A. 1981. Acid and enzymic hydrolysis of kappa carrageenan. *Polymer Bull.* 5: 81-6.

Ross, A. G. 1953. Some typical analyses of red seaweeds. *J. Sci. Food Agric.* 4: 333-5.

Rotem, A., Roth-Bejerand, N. & Arad, S. 1986. Effect of controlled environmental conditions on starch and agar contents of *Gracilaria* sp. (Rhodophyceae). *J. Phycol.* 22: 117-21.

Sanford, P. A. 1985. Applications of marine polysaccharides in the chemical industries. In *Biotechnology of Marine Polysaccharides*, eds. R. R. Colwell, E. R. Pariser & A. J. Sinskey, pp. 453-516. Washington: Hemisphere.

Santos, G. A. & Doty, M. S. 1979. Carrageenans from some Hawaiian red algae. *Proc. Int. Seaweed Symp.* 9: 361-7.

Santos, G. A. & Doty, M. S. 1983. Agar from some Hawaiian red algae. *Aquat. Bot.* 16: 385-9.

Semesi, A. K. 1979. Contribution on the content and nature of the phycocolloid from *Tenaciphyllum lobatum* Borgesen. *Inf. Ser. N. Z. Rep. Sci. Ind. Res.* 137: 473-84. (*Proc. Int. Symp. Mar. Biogeogr. Evol. Southern Hemisphere*, Vol. 2, 1978.)

Semesi, A. K. & Mshigeni, K. E. 1977a. Contributions on the content and nature of the phycocolloid from *Halymenia venusta* Boergesen (Rhodophyta, Cryptonemiales).

Bot. Mar. 20: 233–7.
Semesi, A. K. & Mshigeni, K. E. 1977b. Studies on the yield and infrared spectra of phycocolloids from *Chondrococcus hornemannii* (Lyngbye) Schmitz and *Sarconema filiforme* (Sonder) Kylin from Tanzania. Bot. Mar. 20: 271–5.
Shashkov, A. S., Usov, A. I. & Yarotskii, S. V. 1978. Polysaccharides of algae. 24. The application of ^{13}C nmr spectroscopy to the analysis of the structures of polysaccharides of the agar group. Bioorg. Khim. 4: 74–81 (in Russian).
Shi, S. Y., Chang, Y. X., Liu, W. Q. & Li, Z. E. 1983. The seasonal variation in yield, physical properties and chemical composition of agar from *Gracilaria verrucosa*. Oceanol. Limnol. Sinica 14: 272–8. (in Chinese).
Shi, S. Y., Liu, W. & Li, Z. 1987. ^{13}-C nmr spectroscopic analysis of carrageenans from Chinese *Eucheuma* species. Oceanol. Limnol. Sinica 18: 265–72 (in Chinese).
Shi, S. Y., Zhang, Y. X., Li, Z. E. & Liu, W. Q. 1984. The yield and properties of agar extracted from different life stages of *Gracilaria verrucosa*. Hydrobiologia 116/117: 551–3.
Shi, S. Y., Zhang, Z. X., Liu, W. Q. & Li, Z. E. 1986. Comparative studies on the yield and properties of agar from Chinese species of *Gracilaria*. Stud. Mar. Sinica 26: 57–64.
Sieburth, J. McN. 1975. *Microbial Seascapes. A Pictorial Essay on Marine Microorganisms and Their Environments*. Baltimore: University Park.
Siegel, B. Z. & Siegel, S. M. 1973. The chemical composition of algal cell walls. Crit. Rev. Microbiol. 3: 1–26.
Silva, P. C. & Johansen, H. W. 1986. A reappraisal of the order Corallinales (Rhodophyceae). Br. Phycol. J. 21: 245–54.
Simpson, F. J. & Shacklock, P. F. 1979. The cultivation of *Chondrus crispus*. Effect of temperature on growth and carrageenan production. Bot. Mar. 22: 295–8.
Smidsrød, O., Andresen, I-L., Grasdalen, H., Larsen, B. & Painter, T. 1980. Evidence for a salt-promoted "freeze-out" of linkage conformations in carrageenans as a prerequisite for gel-formation. Carbohydr. Res. 80: C11–C16.
Smidsrød, O. & Grasdalen, H. 1984a. Polyelectrolytes from seaweeds. Hydrobiologia 116/117: 19–28.
Smidsrød, O. & Grasdalen, H. 1984b. Conformations of ϰ-carrageenan in solution. Hydrobiologia 116/117: 178–86.
Stancioff, D. J. & Stanley, N. F. 1969. Infrared and chemical studies on algal polysaccharides. Proc. Int. Seaweed Symp. 6: 595–609.
Stoloff, L. & Silva, P. 1957. An attempt to determine possible taxonomic significance of the properties of water extractable polysaccharides in red algae. Econ. Bot. 11: 327–30.
Stortz, C. A. & Cerezo, A. S. 1986. The potassium chloride-soluble carrageenans of the red seaweed *Iridaea undulosa* B. Carbohydr. Res. 145: 219–35.
Stortz, C. A. & Cerezo, A. S. 1987. Specific fragmentation of carrageenans. Carbohydr. Res. 166: 317–23.
Su, J.-C. & Hassid, W. Z. 1962. Carbohydrates and nucleotides in the red alga *Porphyra perforata*. I. Isolation and identification of carbohydrates. Biochemistry 1: 468–74.
Thepenier, C. & Gudin, C. 1985. Studies on optimal conditions for polysaccharide production by *Porphyridium cruentum*. Mircen J. Appl. Microbiol. Biotechnol. 1: 257–68.
Thomas, P. C. & Krishnamurthy, V. 1976. Agar from cultured *Gracilaria edulis* (Gmel.) Silva. Bot. Mar. 19: 115–17.
Tseng, C. K. 1945. The terminology of seaweed colloids. Science 101: 597–602.
Turvey, J. R. 1978. Biochemistry of algal polysaccharides. In *International Review of Biochemistry. Biochemistry of Carbohydrates II*, vol. 16, ed. D. J. Manners, pp. 151–77. Baltimore: University Park.
Turvey, J. R. & Griffiths, L. M. 1973. Mucilage from a fresh-water red alga of the genus *Batrachospermum*. Phytochemistry 12: 2901–7.
Turvey, J. R. & Simpson, P. R. 1966. Polysaccharides from *Corallina officinalis*. Proc. Int. Seaweed Symp. 5: 323–7.
Turvey, J. R. & Williams, T. P. 1964. Sugar sulphates from the mucilage of *Porphyra umbilicalis*. Proc. Int. Seaweed Symp. 4: 370–3.
Turvey, J. R. & Williams, E. L. 1976. The agar-type polysaccharide from the red alga *Ceramium rubrum*. Carbohydr. Res. 49: 419–25.
Tveter-Gallagher, E., Cheney, D. & Mathieson, A. C. 1981. Uptake and incorporation of ^{35}S into carrageenan among different strains of *Chondrus crispus*. Proc. Int. Seaweed Symp. 10: 521–30.
Usov, A. I. 1974. Polysaccharides of algae. 13. Monosaccharide compositions of polysaccharides of some red algae from the Sea of Japan. Zh. Obshch. Khim. 44: 191–6. (in Russian).
Usov, A. I. 1984. NMR spectroscopy of red seaweed polysaccharides: agars, carrageenans, and xylans. Bot. Mar. 27: 189–202.
Usov, A. I., Adamyants, K. S., Yarotskii, S. V. & Anoshina, A. A. 1975. Polysaccharides of algae. 16. Structure of sulfated mannan from the red alga *Nemalion vermiculare* studied by methylation. Zh. Obshch. Khim. 45: 916–21 (in Russian).
Usov, A. I., Adamyants, K. S., Yarotskii, S. V., Anoshina, A. A. & Kochetkov, N.K. 1973. The isolation of a sulphated mannan and a neutral xylan from the red seaweed *Nemalion vermiculare* Sur. Carbohydr. Res. 26: 282–3.
Usov, A. I., Adamyants, K. S., Yarotskii, S. V., Anoshina, A. A. & Kochetkov, N. K. 1974. Polysaccharides of algae. 14. Separation of sulfated mannan and neutral xylan from the red alga *Nemalion vermiculare*. Zh. Obshch. Khim. 44: 416–20 (in Russian).
Usov, A. I. & Arkhipova, V. S. 1981. Polysaccharides of algae. 30. Methylation of ϰ-carrageenan-type polysaccharides of the red algae *Tichocarpus crinitus* (Gmel.) Rupr., *Furcellaria fastigiata* (Huds.) Lam. and *Phyllophora nervosa* (de Cand.) Grev. Bioorg. Khim. 7: 385–90 (in Russian).
Usov, A. I. & Barbakadze, V. V. 1978. Polysaccharides of algae. 27. Partial acetolysis of the sulfated galactans from the red alga *Grateloupia divaricata* Okam. Bioorg. Khim. 4: 1107–15 (in Russian).
Usov, A. I., Barbakadze, V. V., Yarotskii, S. V. & Shashkov, A. S. 1978a. Polysaccharides of algae. 28. Use of carbon-13-nmr spectroscopy for structural studies of galactan from *Grateloupia divaricata* Okam. Bioorg. Khim. 4: 1507–12 (in Russian).
Usov, A. I. & Ivanova, E. G. 1981. Polysaccharides of algae. 31. Enzymatic cleavage of the agar-like polysaccharide from the red alga *Rhodomela larix* (Turn.) C. Ag. Bioorg. Khim. 7: 1060–8 (in Russian).
Usov, A. I., Ivanova, E. G. & Makienko, V. F. 1979. Polysaccharides of algae. 29. Comparison of samples of agar from different generations of *Gracilaria verrucosa* (Huds.) Papenf. Bioorg. Khim. 5: 1647–53 (in Russian).
Usov, A. I., Ivanova, E. G. & Przhemenetskaya, V. F.

1985. Polysaccharides of algae. 36. Composition and properties of the agar from Far Eastern *Gracilaria* sp. *Bioorg. Khim.* 11: 1119–24 (in Russian).

Usov, A. I., Ivanova, E. G. & Shashkov, A. S. 1983. Polysaccharides of algae. 33. Isolation and ^{13}C-nmr spectral study of some new gel-forming polysaccharides from Japan Sea red seaweeds. *Bot. Mar.* 26: 285–94.

Usov, A. I. & Shashkov, A. S. 1985. Polysaccharides of algae. 34. Detection of iota-carrageenan in *Phyllophora brodiaei* (Turn.) J. Ag. (Rhodophyta) using ^{13}C-nmr spectroscopy. *Bot. Mar.* 28: 367–73.

Usov, A. I. & Yarotskii, S. V. 1975. Polysaccharides of algae. 21. Alkaline degradation of the sulfated mannan from the red alga *Nemalion vermiculare* Sur. *Bioorg. Khim.* 1: 919–22 (in Russian).

Usov, A. I., Yarotskii, S. V. & Esteves, M. L. 1978b. Polysaccharides of algae. 23. Polysaccharides of the red alga *Bangia fuscopurpurea* (Dillw.) Lyngb. *Bioorg. Khim.* 4: 66–73 (in Russian).

Usov, A. I., Yarotskii, S. V. & Esteves, M. L. 1981. Polysaccharides of algae. 32. Polysaccharides of the red alga *Galaxaura squalida* Kjelm. *Bioorg. Khim.* 7: 1261–70 (in Russian).

Usov, A. I., Yarotsky, S. V. & Shashkov, A. S. 1980. ^{13}C-NMR spectroscopy of red algal galactans. *Biopolymers* 19: 977–90.

Usov, A. I., Yarotskii, S. V., Shashkov, A. S. & Tishchenko, V. P. 1978c. Polysaccharides of algae. 22. Polysaccharide composition of *Rhodymenia stenogona* Perest. and use of carbon-13 nmr spectroscopy for determining the structure of xylans. *Bioorg. Khim.* 4: 57–65 (in Russian).

Verdus, M. C., Christiaen, D., Stadler, T. & Morvan, H. 1986. Étude ultrastructurale et cytochimique de la paroi cellulaire chez *Gracilaria verrucosa* (Rhodophyceae). *Can. J. Bot.* 64: 96–101.

Villarroel, L. H. & Zanlungo, A. B. 1981. Structural studies on the porphyran from *Porphyra columbina* (Montagne). *Carbohydr. Res.* 88: 139–45.

Waaland, S. D. 1975. Evidence for a species-specific cell fusion hormone in red algae. *Protoplasma* 86: 253–61.

Waaland, S. D. & Cleland, R. D. 1974. Cell repair through cell fusion in the red alga *Griffithsia pacifica*. *Protoplasma* 79: 185–96.

Watson, B. A. & Waaland, S. D. 1983. Partial purification and characterization of a glycoprotein cell fusion hormone from *Griffithsia pacifica*, a red alga. *Plant Physiol.* 71: 327–32.

Watson, B. A. & Waaland, S. D. 1986. Further biochemical characterization of a cell fusion hormone from the red alga, *Griffithsia pacifica*. *Plant Cell Physiol.* 27: 1043–50.

Wen, Zongcun (Z. C. Wen) & Craigie, J. S. 1984. Composition and properties of agar-type polysaccharides from *Gracilaria sjoestedtii* Kylin. *Chin. J. Oceanol. Limnol.* 2: 88–91.

Whyte, J. N. C. & Englar, J. R. 1971. Polysaccharides of the red alga *Rhodymenia pertusa*. II. Cell-wall glucan; proton magnetic resonance studies on permethylated polysaccharides. *Can. J. Chem.* 49: 1302–5.

Whyte, J. N. C. & Englar, J. R. 1981a. The agar component of the red seaweed *Gelidium purpurescens*. *Phytochemistry* 20: 237–40.

Whyte, J. N. C. & Englar, J. R. 1981b. Agar from an intertidal population of *Gracilaria* sp. *Proc. Int. Seaweed Symp.* 10: 537–42.

Whyte, J. N. C., Englar, J. R., Saunders, R. G. & Lindsay, J. C. 1981. Seasonal variations in the biomass, quantity and quality of agar, from the reproductive and vegetative stages of *Gracilaria* (*verrucosa* type). *Bot. Mar.* 24: 493–501.

Whyte, J. N. C., Foreman, R. E. & DeWreede, R. E. 1984. Phycocolloid screening of British Columbia red algae. *Hydrobiologia* 116/117: 537–41.

Whyte, J. N. C., Hosford, S. P. C. & Englar, J. R. 1985. Assignment of agar or carrageenan structures to red algal polysaccharides. *Carbohydr. Res.* 140: 336–41.

Williamson, F. B. 1981. Structure, conformations, and biological functions of ionic polysaccharides of red seaweeds. *Proc. Int. Seaweed Symp.* 8: 650–4.

Witsch, H. von, Bolze, A. & Hornung, U. 1983. Growth and production of biomass and extracellular polysaccharides in batch cultures of *Porphyridium aerugineum*, Rhodophyceae. *Ber. Dtsch. Bot. Ges.* 96: 469–81.

Witt, H. J. 1985. Carrageenan. Nature's most versatile hydrocolloid. In *Biotechnology of Marine Polysaccharides*, eds. R. R. Colwell, E. R. Pariser & A. J. Sinskey, pp. 345–60. Washington, D.C.: Hemisphere.

Wong, K. F. & Craigie, J. S. 1978. Sulfohydrolase activity and carrageenan biosynthesis in *Chondrus crispus* (Rhodophyceae). *Plant Physiol.* 61: 663–6.

Yang, S-S., Wang, C-Y. & Wang, H-H. 1981. Seasonal variation of agar-agar produced in Taiwan area. *Proc. Int. Seaweed Symp.* 10: 737–42.

Yaphe, W. 1959. The determination of ϰ-carrageenan as a factor in the classification of the Rhodophyceae. *Can. J. Bot.* 37: 751–7.

Yarotskii, S. V., Shashkov, A. S. & Usov, A. I. 1978. Polysaccharides of algae. 25. The use of carbon-13 nmr spectroscopy for the structural analysis of ϰ-carrageenan group of polysaccharides. *Bioorg. Khim.* 4: 745–51 (in Russian).

Young, K., Duckworth, M. & Yaphe, W. 1971. The structure of agar. Part III. Pyruvic acid, a common feature of agars from different agarophytes. *Carbohydr. Res.* 16: 446–8.

Zablackis, E. & Santos, G. A. 1986. The carrageenan of *Catenella nipae* Zanard., a marine red alga. *Bot. Mar.* 29: 319–22.

Zanlungo, A. B. 1979. Polisacaridos de algas Chilenas. V. Composicion del agar de *Gelidium filicinum*. *Rev. Latinoam. Quim.* 10: 149–51.

Zanlungo, A. B. 1980. Polysaccharides from Chilean seaweeds. Part IX. Composition of the agar from *Gelidium lingulatum*. *Bot. Mar.* 23: 741–3.

Chapter 11

Development

SUSAN D. WAALAND

CONTENTS

I. Introduction / 259
II. Cell division / 261
  A. Spore germination / 261
  B. Apical cell division / 261
  C. Branch initiation / 262
III. Cell enlargement / 263
IV. Cell differentiation / 265
  A. Reproductive cell differentiation / 266
  B. Polarity and and evidence for intercellular communication / 266
V. Cell repair by cell fusion: a model system for studying development / 268
VI. Summary / 269
VII. Acknowledgments / 270
VIII. References / 270

I. INTRODUCTION

Development is a coordinated series of morphological, physiological, and biochemical events by which an organism progresses from a single cell (zygote or spore) to a multicellular organism. This progression leads eventually to a reproductively mature adult and finally to senescence and death of the individual. Development of adult form usually involves (1) an increase in cell number, (2) a change in cell size and shape, and (3) production of differentiated cells and tissues with distinct biochemical and physiological characteristics. Developmental events may be controlled at the level of individual cells, at the levels of tissues and organs through interactions between cells, and at the level of the whole individual by interactions between the individual and its biotic and abiotic environment. To understand how development is accomplished and controlled, one must first describe developmental events at the cellular, tissue, and organismal level. Then, by studying changes in the normal pattern of development that are caused by environmental stimuli, one can investigate internal controls that coordinate development throughout an organism.

Classical developmental biology, grounded in animal embryology, has concentrated on early morphological, physiological, and biochemical changes as the zygote develops into a multicellular embryo. In red algae and other plants, studies of developmental processes are not limited to analysis of early spore or zygote development because these organisms have totipotent cells; that is, individual cells and pieces of tissue can be induced to divide and produce whole new organisms. Therefore, developmental processes in red algae can be analyzed using isolated vegetative cells (Duffield et al. 1972; Lewis 1909; L'Hardy-Halos 1971a,b), excised filaments (Aghajanian & Hommersand 1980; Garbary 1979; Konrad-Hawkins 1964a,b), and tissue fragments

Fig. 11-1. Indeterminate filament of *Antithamnion kylinii* stained with Feulgen stain for DNA. Note the regular branching pattern, the increasing cell size basipetally down the filament, and the concomitant increase in nuclear size. Scale = 0.04 mm. (Photograph courtesy of Jay Charleston)

(Kling & Bodard 1974; Perrone & Felicini 1972, 1974; Sylvester & Waaland 1983), as well as spores and zygotes.

Although red algae have often been overlooked as subjects for developmental studies, they have several advantages for these investigations:

1. The thallus construction of many genera is relatively simple, consisting of single cells, simple filaments, and sheets of cells one to two layers thick (Fig. 11-1) (see also Chapter 12; illustrations in Bold & Wynne 1985; Fritsch 1945). This allows for the observation and manipulation of individual cells. Many larger thalli are constructed from packed filaments (pseudoparenchyma).

2. Even with their simple morphology, red algae do show cellular and thallus differentiation. For example, holdfast regions differ from upright blades, adhesive rhizoid cells differ from highly pigmented, shoot cells, and vegetative cells are morphologically and physiologically distinct from reproductive organs (i.e., carpogonia, spermatangia, and sporangia).

3. Many species of red algae grow rapidly under controlled laboratory conditions.

4. Both vegetative and reproductive material can be used in developmental studies; consequently, starting material is available at all times. Of particular interest in this respect is the growing list of species in which reproduction and spore

release can be controlled by photoperiod (Breeman & ten Hoopen 1987; Brodie & Guiry 1987; Dickson & Waaland 1985; Dring 1984). In these species, not only can photoperiodic control be used to ensure production of spores for experimental studies, but also reproductive differentiation can be timed precisely for biochemical studies of cell differentiation. In addition, a new source of single cells from more complex tissues is now being exploited: protoplasts can be isolated both from filamentous and bladelike thalli, e.g., several species of *Porphyra* and *Griffithsia* (Chen 1986; Cheney et al. 1986; Fujita & Migata 1985; Fujita & Waaland unpubl.; Liu et al. 1984; Polne-Fuller & Gibor 1984). Protoplasts offer three advantages: (a) they can be produced in large numbers at any time as long as a source is available, (b) they offer great potential for genetic manipulation both by creation of hybrid plants from fused protoplasts and by introduction of foreign DNA, and (c) they represent a source of single cells from complex thalli.

The final morphology of an organism depends on a coordination of patterns of cell division, cell enlargement, and cell differentiation. In red algae, as in other organisms, each of these processes may be influenced by both genetics and environment (Norton et al. 1981). In reviewing what is known about development and its control in red algae, each of these processes will be discussed separately and then one particular developmental sequence will be presented in which control of cell division, enlargement, and differentiation can be probed using an endogenous glycoprotein hormone.

II. CELL DIVISION

Cell division plays several roles in development. First, it adds to the size of an organism by increasing the number of cells. In addition, cytokinesis may effectively separate cells with different biochemical and physiological functions, thus facilitating cell and tissue differentiation.

A. Spore germination

To understand morphogenesis, one would like to know the factors that control where, when, and how division occurs within a thallus. Patterns of cell division during germination of carpospores and tetraspores of many red algal species have been described (Chemin 1937; Chihara 1974; Dixon 1973; Chapters 12 and 14). In many cases, such as in the Corallinaceae, these patterns are rigid enough to be used to establish taxonomic and phylogenetic relationships (Cabioch 1972; Chamberlain 1984; Chemin 1937; Chihara 1974). In the genus *Porphyra*, carpospores and monospores that produce the filamentous sporophyte (conchocelis) stage show a monopolar germination pattern, whereas conchospores and monospores that produce the haploid gametophyte blade germinate in a bipolar pattern (Cole & Conway 1980). However, the timing of cell division during the process of carpospore germination in *Porphyra variegata* can be quite variable (Pueschel & Cole 1985). In certain members of the Gigartinaceae, such as *Gigartina exasperata* and *Chondrus crispus*, spores may germinate to form either discoid or filamentous sporelings; the factors that determine which pattern is followed are not known (Chen & Taylor 1976; Sylvester & Waaland 1984). To date, few studies have been reported that deal with factors that can alter or direct germination patterns. Weber (1960) found that the site of germination of carpospores of several species of red algae is set during sporogenesis and is not altered by light. L'Hardy-Halos (1971b) also noted that the polarity of spores in the Ceramiaceae does not appear to be altered by environmental gradients.

B. Apical cell division

Classical studies of thallus development within the Rhodophyta have documented patterns of cell division within thalli of many genera (Dixon 1973; Fritsch 1945; Kylin 1956; Chapter 12). Although division is not localized within thalli of the lower red algae (Bangiophycidae), in most of the higher red algae (Florideophycidae) division to increase the number of cells in a filament occurs only in the apical cell. As a consequence, there is an apico-basal gradient in cell age along any axis. Subapical cells may divide, however, to produce lateral branches and adventitious rhizoids. Structural aspects of the processes of mitosis and cytokinesis in a number of red algae have been investigated, with the goal of finding clues to phylogenetic relationships (Chapter 6).

The rate of apical cell division and the frequency and position of branch initiation is important in determining the final size and morphology of a plant. Control of the rate and timing of apical cell division appears to vary from species to species; in some species total daily illuminance influences the rate of apical cell division (i.e., number of divisions per 24 h). Garbary (1979) studied the effect of varied photoperiod on development in three species of

Table 11-1. Effect of light intensity and photoperiod on apical cell division in Griffithsia pacifica[a]

Illuminance and photoperiod	Number of cells/primary filament/week
$8\mu M\ m^{-2}\ s^{-1}$	
16L:8D	7.0
$60\mu M\ m^{-2}\ s^{-1}$	
8L:16D	6.6
12L:12D	8.2
16L:8D	8.2
24L:0D	8.4

[a] After Waaland & Cleland (1972).

Ceramium and one species of *Antithamnionella*. He found that the rate of apical cell division on a given daylength varied from 2.1 divisions per day in *A. spirographidis* to 0.4 divisions per day in *C. echionotum*. Within each species, the rate of apical cell division increased with increasing numbers of hours of light per cycle. Edwards (1977) noted that the rate of shoot apical cell division in *Callithamnion hookeri* sporelings was directly related to photon flux density. In *Plumaria elegans*, division has been found to be slower at higher photon flux densities (Boney & Corner 1962). In *Griffithsia pacifica*, although light is required for cell division (cell division stops in continuous dark), the rate of cell division is relatively insensitive to total daily illuminance, averaging about one division per day over a range of daylengths and photon flux densities (Table 11-1; Waaland & Cleland 1972). In this species, the timing of both apical cell division and branch initiation has been shown to be controlled by an endogenous rhythm (Waaland & Cleland 1972). On a 16L:8D photoperiod, cell division occurs early in the dark period. The rhythm in cell division continues on a 24L:0D photoperiod and can be altered by shifting photoperiod. Although endogenous rhythms in cell division have not been documented for other species, it is likely that they do exist. Since an endogenous rhythm in cell division allows one to control the time at which division will occur, these rhythms provide an important tool for studying the mechanics of mitosis and cytokinesis.

C. Branch initiation

The timing and patterns of branch initiation, both on the main axis and on lateral determinate and indeterminate branches, are important factors in establishing the final morphology of a plant. The timing and position of branch formation may be genetically predetermined or may be subject to environmental control. For example, in *Callithamnion hookeri* the length of time between formation of a subapical cell and the division of that cell to form a branch is genetically determined (Rueness & Rueness 1982). In many species, the position of branch initiation also appears to be determined genetically. For example, in *Antithamnion kylinii* each cell of the main axis bears two branches (Fig. 11-1), whereas in *Callithamnion roseum* each cell of the main filament bears a single branch. In species with polysiphonous construction, the number of pericentral cells per central cell (equivalent to the number of branches per axial cell) is so regular that it can be used as a taxonomic characteristic (Abbott & Hollenberg 1976; Kylin 1956). The position of branches on individual axial cells is also significant. In some species, branches are produced unilaterally (e.g., *Microcladia borealis*), giving the axis the appearance of a comb; in other cases, branches may be produced alternately (e.g., *C. roseum*) or branch position may be irregular (e.g., *Griffithsia*) (see illustrations in Abbott & Hollenberg 1976). Whereas in many species branching patterns may be set, in some species branch initiation is under environmental control. In *Griffithsia pacifica*, for example, each axial cell may or may not have a branch. In this species, the number of branches per plant is correlated with the total daily illuminance (Waaland & Cleland 1972). At low photon flux density or short photoperiod, plants may have no branches, whereas under continuous light and/or high photon flux density, nearly every axial cell bears a branch. Increased branching in response to increased photon flux density has been reported in a number of other species (D'Antonio & Gibor 1985; Knaggs 1966; West 1972).

An important question about branch initiation is how the position of a branch is determined at the cellular level. In *Callithamnion roseum*, which has alternate branches, a slanted plane of cytokinesis in the apical cell of a filament produces subapical cells in which one side is longer than the other. Branches are always initiated on the longer of the two sides (Konrad-Hawkins 1964a,b). Since attempts to disrupt this pattern have not been made, it is not known if the relationship is coincidental or causal.

In many organisms, including green algae, land plants, and animals, there is a causal relationship between the position of the mitotic spindle and the position of the cytokinetic plane; i.e., the plane and position of cytokinesis are strongly correlated with the plane and position of the metaphase plate (Clayton 1985; Gunning et al. 1978; Mattox & Stewart 1984; Pickett-Heaps & Northcote 1966). In red algae this relationship is often lacking. The timing of mitosis and the plane of the metaphase plate of the dividing nucleus appear to be closely correlated with the timing and plane of branch formation in uninucleate cells; however, the position of the metaphase plate does not necessarily coincide with the position of new crosswall formation (L'Hardy-Halos 1971a). Even coincidence of the planes of mitosis and cytokinesis may not be causal. For example, in *Antithamnion kylinii*, when an excised filament regenerates a rhizoid cell at its base, the plane and position of the metaphase plate of the dividing nucleus often do not coincide with that of the cytokinetic plane; in fact, the position of new crosswall formation corresponds most closely to the site of the original crosswall of the cell (J. Charleston unpubl.).

In multinucleate cells, the relationship between mitosis and cytokinesis is even less precise. For example, in the multinucleate apical cells of several species (e.g., *Griffithsia* sp., *Anotrichium tenue*) a nearly synchronous division of all nuclei precedes cytokinesis, but the nuclei do not position themselves with respect to the plane or position of the new crosswall (Lewis 1909; L'Hardy-Halos 1971a; Sylvester 1987). This is very different from the case of certain multinucleate green algae (e.g., *Acrosiphonia*), in which a group of nuclei line up along the potential site of the cytokinesis; these nuclei divide synchronously, with their mitotic spindles aligned perpendicular to the plane of cytokinesis (Hudson & Waaland 1976). The question of what intracellular factors control the site and plane of cytokinesis in red algae is very intriguing; it could be investigated using a number of uninucleate and multinucleate species (e.g., *Antithamnion* spp., *Antithamnionella* spp., *Scagelia*, *Griffithsia* spp., *Tiffaniella snyderae*, *Callithamnion cordatum*, *Anotrichium tenue*).

III. CELL ENLARGEMENT

Cell enlargement plays an important role in red algal morphogenesis. Whereas division may be limited to the apical cell of filaments, dramatic enlargement occurs in subapical cells. The increase in cell size moving basipetally along an axis may be seen easily in the drawings of thallus morphology of many genera (e.g., Fig. 11-1; Abbott & Hollenberg 1976; Dixon 1973; Fritsch 1945; Kylin 1956; Chapter 12). Dixon (1970, 1971) measured increase in cell size along axes in a number of filamentous, red algal species; some of these algae undergo spectacular cell elongation. For example, intercalary cells of *Ceramium echionotum* show a 100-fold increase in length (from 4 μm to 420 μm) and a 14,000-fold increase in volume; those in *Lemanea fluviatilis* elongate 1000-fold (from 8 μm to 8000 μm), with a 44,000-fold increase in volume. Similar increases in cell size also occur in thallose species; however, it is much more of difficult to follow cell enlargement in these species. Changes in patterns of enlargement are important in the acquisition of form by specialized cells like tetrasporangia, which acquire a spherical shape (in contrast to elongate vegetative cells), and carpogonia, which produce elongate trichogynes. How this enlargement is accomplished and how it is regulated at the cellular and organismal levels are not understood.

To evaluate the role of cell enlargement in development and to understand the effects that experimental manipulations have on enlargement, it is important to know which cells in a thallus are enlarging and how that enlargement is taking place. Since there is an apicobasal gradient in cell age within a filament, one can estimate which cells are elongating by measuring the lengths of all the cells along a filament and plotting cell length vs. position (Dixon 1970, 1971). Another way to determine which cells are elongating is to use markers or vital stains and follow elongation over time (Waaland & Waaland 1975; Waaland et al. 1972). This method also allows a direct comparison of rates of cell elongation in different cells.

The pattern of elongation within individual cells has been analyzed in several red algae (Aghajanian & Hommersand 1980; Dixon 1970, 1971; Garbary 1979; Hymes & Cole 1983; Waaland & Waaland 1975; Waaland et al. 1972). In rhizoids and some shoot axes (e.g., *Audouinella*), elongation is confined to the apical cell of a filament; subapical cells do not elongate. However, in most shoot axes, intercalary cells, as well as apical cells, enlarge (Dixon 1973). In *Batrachospermum* and *Callithamnion* the pattern of cell elongation has been shown to be different in apical and subapical cells (Fig. 11-2; Aghajanian & Hommersand 1980; Waaland & Waaland 1975).

Elongating apical cells of shoot filaments and rhizoids extend by localized tip growth (Hymes & Cole 1983; Waaland & Waaland 1975; Waaland et al. 1972). This type of elongation is found in apical cells and rhizoids of other algae, in fungal hyphae, filamentous stages of bryophytes and ferns, and in angiosperm pollen tubes and root hairs (Gooday & Trinci 1980; Sievers & Schnepf 1981). In this pattern, elongation is confined to the apical dome of an apical cell; the side walls of the cell are rigid (Green 1969). Elongation is maintained by localized deposition of wall-loosening enzymes and new cell wall material; the shape of the cell is regulated by a

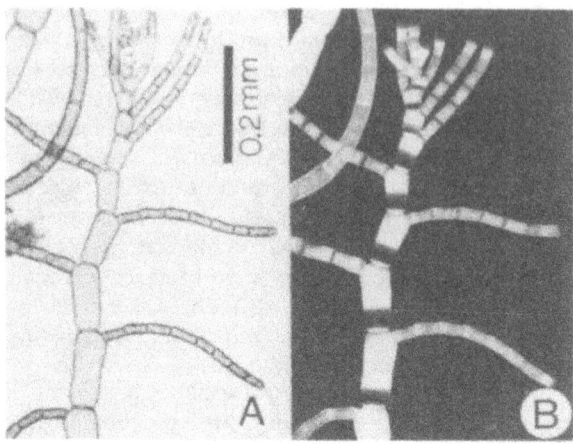

Fig. 11-2. Cell elongation in *Callithamnion* sp. Plant stained with Calcofluor White ST; then removed and allowed to grow in the absence of stain for 24 h. A. Plant photographed in white light. B. Plant photographed in UV light. New cell wall material added during the 24-h chase period is unstained (i.e., dark). Note new cell wall material added apically in apical cells of branches (arrow) (tip growth) and in bands at the bases of intercalary cells (band growth). Scale = 0.2 mm. (Modified from Waaland & Waaland 1975)

stiffening of the lateral walls of the cell so that only the apical dome yields to turgor pressure (Bartnicki-Garcia 1973; Green 1969).

When intercalary cells elongate, a pattern of growth different from tip growth is found. In this case, the lateral walls of the cell must extend; this growth might be diffuse, occurring all along the side walls, or it might be localized to only a portion of the wall. For several genera of red algae (e.g., *Griffithsia*, *Batrachospermum*, *Anotrichium*, *Antithamnion kylinii*, *Callithamnion*), it has been shown that elongation in intercalary cells is localized, taking place in narrow bands at the bottom and/or top of each cell (Aghajanian & Hommersand 1980; Waaland & Waaland 1975; Waaland et al. 1972). In *Griffithsia pacifica*, the zones of extension are only 20 μm wide. Therefore, the fraction of an individual cell that is actually elongating varies, depending on cell age, ranging from about 40% of the length of a cell near the apex of filament to only about 5% of the length of an older cell. It is important to consider this fact in evaluating elongation of entire axes.

In order to determine how elongation is regulated at the cellular level, the effects of environmental parameters or experimental manipulation on growth have been investigated. One approach to such studies has been to measure the rate of elongation of entire filaments under different conditions. This method can be misleading because several parameters will contribute to the total growth rate of the filament. These include not only the rate of elongation of individual cells, but also the length of time over which an individual cell continues to elongate, the proportion of a given cell that is elongating, and any developmental variation in rates of elongation; in addition, the rate at which new cells are added to a filament will contribute to the rate of filament growth. Each of these parameters can be affected by the experimental conditions. First, the rate of cell elongation may change with cell age. For example, in species with both apical and basal zones of elongation, the relative rate of extension in apical and basal growth bands does appear to change with cell age (Garbary 1979; Murray & Dixon 1975; Waaland & Waaland 1975). Second, the rate of cell elongation may not be constant even over a 24-h period. For example, when *Griffithsia pacifica* is grown on a 16L:8D photoperiod and elongation of specific cells is followed, the rate of elongation is most rapid in the first 8 h of light, slows in the second 8 h of light, and is slowest at night (Waaland & Cleland 1972). In contrast, on the same light cycle, intercalary cells of *Anotrichium tenue* elongate most rapidly during the 8-h night (Sylvester 1987). Thus, in 24 h on a 16L:8D cycle, an individual cell of *G. pacifica* would have twice as many hours of elongation as a cell of *A. tenue*.

Third, the time over which an individual cell elongates may vary for different cells in a filament and under different environmental conditions. In *Griffithsia pacifica*, cells with branches elongate at the same rate as cells without branches, but those with branches stop elongating sooner than unbranched cells; thus, cells that bear branches are shorter than those without branches (Duffield et al. 1972). This difference in final length of cells with and without branches has been observed in other species. Dixon (1970, 1971), measuring individual cell lengths along filaments of several species of red algae, found that cells that bear branches are significantly different in size from those with no branches. He did not determine if these differences in size were due to differences in rate of elongation or to a change in the length of time over which elongation occurs. In *Ceramium shuttleworthianum*, filaments grown under long and short days elongate at the same rate, but elongation stops sooner on short days (Garbary 1979).

Finally, only a few cells in a filament may be elongating; in this case, the apparent rate of elongation will decrease with time as a larger number of cells finish enlarging. Thus, when analyzing the effect of various treatments on the rate of growth of entire filaments, it is important to establish which parameters are being affected.

Light affects cell enlargement in several different ways; its effects may vary depending on the species or type of cell studied. First, light is required for elongation in some species. In *Griffithsia pacifica*, elongation of both shoot and rhizoid cells stops in continuous darkness (Waaland 1978b; Waaland & Cleland 1972); however, in *Constantinea subulifera*, production of a complete new blade will take place in continuous darkness (Powell 1964, 1986). Total daily illuminance has been shown to affect elongation rate in a number of species (Garbary 1979; Garbary et al. 1978; Murray & Dixon 1975; Norton et al. 1981). Garbary (1979), measuring elongation of entire filaments in species of *Ceramium* and *Antithamnionella*, and correcting for rates of apical cell division, found that increasing illuminance enhanced the rate of elongation in some species (*C. rubrum* and *A. spirographidis*), but had no effect in others (*C. echionotum* and *C. shuttleworthianum*). Elongation of specific cells in *G. pacifica* has been followed over time; individual intercalary cells elongate for 6–8 days (Duffield et al. 1972). In this species a fivefold increase in illuminance does not change the rate of cell elongation (Waaland & Cleland 1972).

Unidirectional light affects elongation in both upright filaments and rhizoids of red algae (Jones 1959; L'Hardy-Halos 1971b; Neushul et al. 1967; Waaland et al. 1977). In general, filaments and blades are positively phototropic and rhizoids are negatively phototropic. In *Porphyra* conchocelis, vegetative filaments grow away from light and penetrate into shells, but sporangium-bearing branches grow to the surface of the shell toward light (Melvin et al. 1986). The phototropic response of rhizoids has been studied in *Griffithsia pacifica* (Waaland et al. 1977). In this species, the rhizoid elongates by tip growth; its response to unidirectional light begins after a lag of 45–60 min, at which time the rhizoid apex grows slowly away from light. If the unidirectional light is turned off, the rhizoid tip, after a lag time, does not continue to turn away from the light and in fact begins to return to its original direction of elongation. The action spectrum of the response suggests a blue-light photoreceptor. It would be particularly interesting to investigate how directional growth is achieved during the phototropic response of shoot cells and blades. Such studies should yield insight into the control of band growth in individual cells and coordination of growth in blades.

Cell enlargement may also be linked to DNA synthesis and/or mitosis. In some species there appears to be a correlation between the size of a cell and its DNA content. In uninucleate species, nuclear size, and presumably DNA content, may increase with cell size (Fig. 11-1; L'Hardy-Halos 1971a). In multinucleate species, the number of nuclei per cell increases as cell size increases (Goff & Coleman 1987; Lewis 1909; L'Hardy-Halos 1971a; Sylvester 1987). Whether increases in DNA content are required for increases in cell size, or vice versa, is not known. Goff and Coleman (1987) measured nuclear DNA content and nucleus/cytoplasm ratios in haploid and diploid stages of *Griffithsia pacifica*; they found that the total amount of DNA per volume of cytoplasm remains relatively constant over large increases in cell size. Sylvester (1987) found in *Anotrichium tenue* that, when nuclear division was inhibited with griseofulvin, cell enlargement continued. Thus, increase in cell size could be uncoupled from increase in nuclear number; however, since nuclei in griseofulvin-treated cells were larger than those in control cells, it is likely that DNA synthesis continued even though mitosis was inhibited (see Chapter 3).

IV. CELL DIFFERENTIATION

During the process of cell differentiation, cells acquire distinctive morphological, physiological, and biochemical characteristics. For example, rhizoid cells, specialized for anchoring, are usually elongate, adhesive, and pale in color, whereas shoot (axial) cells, whose function is primarily photosynthetic, are broader and more deeply pigmented. As mentioned above (see Section III), these two types of cells also differ in their response to unilateral light and in their mechanism of cell elongation.

Another specialized type of vegetative cell is the spherical, refractile gland cells found in a number of genera, including *Antithamnion*, *Bonnemaisonia*, and *Schizymenia* (Abbott & Hollenberg 1976). In blades of genera like *Iridaea* and *Gigartina*, cortical cells on the outside of the thallus are small and deeply pigmented, whereas medullary cells, in the interior of the thallus, are much larger and pale in color (Bold & Wynne 1985; personal observations). Finally, there are a number of specialized cells and structures associated with reproduction, including the reproductive organs themselves (e.g., sporangia, carpogonia, spermatangia) and auxiliary reproductive structures (e.g., involucres, pericarps, carposporangial branches, and postfertilization structures) (Dixon 1973). To understand how cell differentiation is accomplished, one needs to describe the structural changes that take place and then investigate the physiological and biochemical changes that accompany these morphological events.

A. Reproductive cell differentiation

In many cases the morphological consequences of cell differentiation are well documented. For example, a number of ultrastructural investigations have examined the stages in development and release of spores (Chamberlain & Evans 1973; Delivopoulos & Kugrens 1984; Kugrens & Delivopoulos 1986; Kugrens & West 1973, 1974; Pueschel 1979; Tripodi 1974; Tsekos 1981, 1982, 1983). In addition, the pattern of spore formation within sporangia is so regular that it is used as a taxonomic character (Guiry 1978; Chapter 14). The next step is to explore one of these developmental sequences from an experimental standpoint in an attempt to elucidate the physiological and biochemical basis for that differentiation. Of particular interest in this regard might be strains that produce unusual combinations of reproductive structures (e.g., gametangia and tetrasporangia) on the same plant (Rueness & Rueness 1973, 1985; van der Meer & Todd 1977; West & Norris 1966; Whittick & West 1979). The elegant genetic studies of van der Meer and his colleagues on *Gracilaria tikvahiae* (van der Meer 1986; van der Meer & Todd 1977; van der Meer et al. 1984) have begun to elucidate the genetic regulation of reproductive cell differentiation (see Chapter 5).

The developmental events that take place during sexual reproduction in red algae have been described extensively from preserved material using the light microscope (see Chapter 13); the sequence of pre- and postfertilization events are the basis used to classify red algae at the ordinal level (Bold & Wynne 1985; Dixon 1973; Kraft 1981; Kylin 1956). More recently, ultrastructural studies have confirmed and clarified early postfertilization events (Broadwater & Scott 1982; Delivopoulos & Kugrens 1985; Kugrens & Delivopoulos 1986; Ramm-Anderson & Wetherbee 1982). An exciting technical advance for studies of reproductive development is the use of the quantitative DNA stain, DAPI. Using this stain, Goff and Coleman (1984) were able to trace nuclear movements and evaluate changes in nuclear DNA content during spermatogenesis, carpogonium formation, and postfertilization cell fusions, and also in tetrasporogenesis in *Choreocolax polysiphoniae*, a parasitic species.

In order to study these events at a physiological or biochemical level, it is important to know the kinetics of the developmental sequence. Even though a number of red algae complete their life cycles in culture, there are few reports of the actual time course of the events of sexual reproduction (DeCew & West 1981; Edwards 1971; Lindstrom 1981; O'Kelly & Baca 1984). O'Kelly and Baca (1984) have described a controlled system for studying several aspects of reproduction in *Callithamnion cordatum*. By adding spermatia to virgin female cultures, they were able to establish a chronology of development of the carpogonial branch and of postfertilization events. Using this species, it should be possible to investigate the intercellular interactions that take place at the biochemical level after fertilization.

B. Polarity and evidence for intercellular communication

1. Regenerating single cells and individual filaments

One aspect of the control of cell differentiation that has been studied experimentally is regeneration of cells from excised cells and filaments. Several workers have shown that isolated vegetative cells and intercalary filaments regenerate in a polar fashion, producing shoot cells from their exposed apices and rhizoid cells from their newly exposed bases (Duffield et al. 1972; Konrad-Hawkins 1964a, b, 1972; Lewis 1909; L'Hardy-Halos 1971b; L'Hardy-Halos & Larpent 1983; L'Hardy-Halos et al. 1984; Waaland & Cleland 1974). The regeneration of rhizoids from shoot cells is of particular interest because it involves not only cell division but also cell differentiation. No obvious polarity in distribution of cytoplasmic components has been observed in shoot cells prior to regeneration. The cellular polarity shown during regeneration is quite strong and is not changed by gravity or unidirectional light. However, certain externally applied factors may affect polar regeneration. For example, in *Griffithsia globulifera* (as *G. bornetiana*) Schechter (1934) found that electric currents and centrifugation could induce the formation of either shoot cells from cell bases or rhizoids from cell apices. However, he did not find a treatment that could completely reverse polarity. There is some evidence that cells in one part of a filament may influence the polarity of other cells at some distance from them. For example, L'Hardy-Halos (1971b,c) observed in several species (e.g., *Aglaothamnion tripinnatum*, *A. feldmanniae*, *Pleonosporium borreri*) that apical cells of shoot filaments may occasionally produce rhizoids instead of normal shoot apical cells. In *Antithamnion plumula*, she isolated apical filaments from determinate branches in which apical cells had become rhizoidal; in this case, the isolated filaments regenerated shoot apical cells instead of rhizoids from their bases. This observation suggests that the presence of a rhizoid at the apical end of a filament can control polarity of more basipetal cells of that filament. When apical cells of lateral shoot branches of *Aglaothamnion* occasionally became rhizoidal, subapical cells formed branches from their bases instead of from their apices (L'Hardy-Halos 1971b). Therefore, it appears that rhizoid cells can exert an influence on the polarity of adjacent shoot cells; the mechanism for such intercellular interaction has not been elucidated.

2. Regenerating blades

Polar regeneration has been observed in tissues as well as individual cells or filaments. For example, Perrone and Felicini (1972, 1974) and Perrone – Pesola and Felicini (1981) excised pieces of blades of *Schottera nicaeensis* (as *Petroglossum nicaeense*). These fragments regenerated in a polar fashion, producing blades from their apices and creeping axes (or stolons) from their bases. Basal regeneration was initiated only if apical regeneration had occurred. If newly regenerated apical blades were removed just after basal primordia were established, the basal primordia developed into blades rather than creeping axes. In this latter case, the polarity of the segment was reversed and the apical end now produced creeping axes. Thus, the presence of developing blades, at the apical end of a fragment, appears to be required for the formation of creeping axes and also affects the developmental fate of basally regenerating primordia. These observations strongly suggest that intercellular communication takes place along a blade. The mechanism of communication has not been determined.

3. Branch initiation and cell elongation

Developmental events in one cell can be influenced by cells at a distance. In *Antithamnion*, decapitation of a main axis appears to induce the formation of new indeterminate axes from the basal cells of determinate branches (L'Hardy-Halos 1971b, 1975). Thus, the apical cell of an indeterminate axis seems to exert an inhibitory effect on initiation of new axes of indeterminate growth. Other aspects of development in filamentous red algae also suggest intercellular communication. For example, L'Hardy-Halos (1971b) found that the rate of elongation of whole filaments of *Antithamnion* could be altered either by excising these filaments from the plant or by removing the apex of the main axis. In addition, elongation and the developmental fate of determinate branches was influenced by the presence or absence of indeterminate filaments. She suggested that both growth-promoting and -inhibiting substances move along filaments, controlling elongation in lateral branches.

4. Sexual reproduction

Intercellular communication is probably involved in coordinating sexual reproduction as well. In the Ceramiales, the supporting cell upon which the female branch (carpogonial branch) is borne divides only after fertilization to produce an auxiliary cell; O'Kelly and Baca (1984), using synchronized cultures of *Callithamnion cordatum*, have confirmed this precise timing. The fact that fertilization has occurred in the carpogonium at the tip of the female branch appears to be communicated to the supporting cell at its base. Subsequently, both in the Ceramiales and in certain other orders, the fertilized carpogonium sends out a connecting filament to the auxiliary cell and transfers the diploid nucleus to the latter cell; in some orders (e.g., Gigartinales) the connecting filament must contact auxiliary cells at some distance from the female branch. Again, there must be some communication to guide the connecting filament to the auxiliary cell. In addition to putative communication associated with sexual fusions in *Callithamnion*, O'Kelly and Baca (1984) observed morphogenesis of the vegetative apex influenced by the presence of a developing carposporophyte.

Thus, during both vegetative and reproductive morphogenesis in red algae, there are a number of cases in which one can hypothesize the presence of differentiation-controlling substances that are transported between cells in a plant and coordinate the growth and development of the whole plant. Intercellular transport of substances between adjacent cells in red algae has been the subject of much speculation (Goff 1979; Koslowsky & Waaland 1984, 1987; Turner & Evans 1978; Wetherbee 1979). How developmental messages are transmitted between cells is not known. These messages may be chemical in nature; to date, only one endogenous, development-regulating substance, rhodomorphin, has been isolated and purified from red algae (see Section V). It is also possible that communication between cells is electrical as well as chemical. Bauman and Jones (1986) have presented evidence for electrical coupling between adjacent shoot cells of *Griffithsia pacifica*.

V. CELL REPAIR BY CELL FUSION: A MODEL SYSTEM FOR STUDYING DEVELOPMENT

One developmental process that has been investigated in detail is cell repair by cell fusion (Waaland 1984, 1986). In at least three genera of red algae, Griffithsia, Anotrichium, and Antithamnion, dead intercalary cells may be replaced by the process of cell repair (Fig. 11-3; L'Hardy-Halos 1971a; Waaland 1986; Waaland & Cleland 1974). When a cell in the middle of a filament is killed, the cells on either side of the dead cell divide to form new cells. The cell superjacent to the dead cell produces a rhizoid cell at its base and the subjacent cell produces a long, thin repair shoot cell that elongates by rapid tip growth. The rhizoid and repair shoot cell grow toward each other through the empty wall of the dead cell; when they meet, they fuse to form a single new cell that replaces the dead shoot cell. Cell repair, in which the dead cell wall serves to connect two halves of a filament, differs in several ways from regeneration following complete severing of a cell (Waaland 1986; Waaland & Cleland 1974) in at least three ways. First, the appearance and growth of a repair shoot cell are quite different from that of the dome-shaped shoot apical cell formed during apical regeneration. Secondly, when the kinetics of repair and regeneration processes are compared, it is found that a repair shoot cell is produced in half the time that is required for regeneration of a shoot apical cell (8 h vs. 17 h in G. pacifica). Thirdly, the repair shoot cell and rhizoid cell are attracted to one another and fuse.

In several species of Griffithsia, it has been demonstrated that cell repair is coordinated by a substance or substances that are produced by the rhizoid from the superjacent cell (Waaland 1975). This substance induces the precocious division of the subjacent cell, is required to maintain the rhizoidlike morphology and growth pattern of the repair shoot cell, and may be involved in the attraction of the repair shoot cell to the rhizoid for fusion (Waaland 1984, 1986; Waaland & Watson 1980). This rhizoid-specific substance has been called 'rhodomorphin' (Waaland 1975). Experimental evidence suggests that rhodomorphins are involved in cell repair in all species studied to date. Rhodomorphins appear to be species specific, that is, rhodomorphin from one species of Griffithsia is ineffective in inducing repair shoot cell formation in other species (Waaland 1975, unpubl.). The rhodomorphin from G. pacifica has been isolated, partially purified, and characterized as a glycoprotein with a molecular weight of about 15,000 (14–17.5 kd) (Waaland & Watson 1980; Watson & Waaland 1983, 1986). It is stable at room temperature for several days and at 0°C for at least 5 yr; however, its activity is destroyed by heating to 50°C for 8 min (Watson & Waaland 1983). The carbohydrate portion of the molecule appears to contain terminal mannosyl linkages that can be removed without affecting the activity of the molecule. Rhodomorphin from G. pacifica is active at 10^{-13} to 10^{-14} M.

Partially purified rhodomorphin can be used to probe several developmental processes. Rhodomorphin has been used to study the cellular factors involved in tip growth in the repair shoot cell. In the presence of rhodomorphin, a repair shoot cell, which has a refractile zone at its tip, elongates by tip growth. If a repair shoot cell is removed from rhodomorphin, its refractile zone slowly disappears and, within 1.5–2 h, elongation ceases. Over the next 12 h the repair shoot widens laterally and eventually produces a dome-shaped shoot apical cell at its apex (Waaland & Cleland 1974; Waaland & Lucas 1984). Self-generated inflowing currents, found in some cells to be associated with the induction and maintenance of tip growth (Nuccitelli 1986), have been shown to be insufficient to maintain tip growth in Griffithsia (Waaland & Lucas 1984). Rhodomorphin can also be used to study cell differentiation. Continuous rhodomorphin treatment is necessary to maintain repair shoot differen-

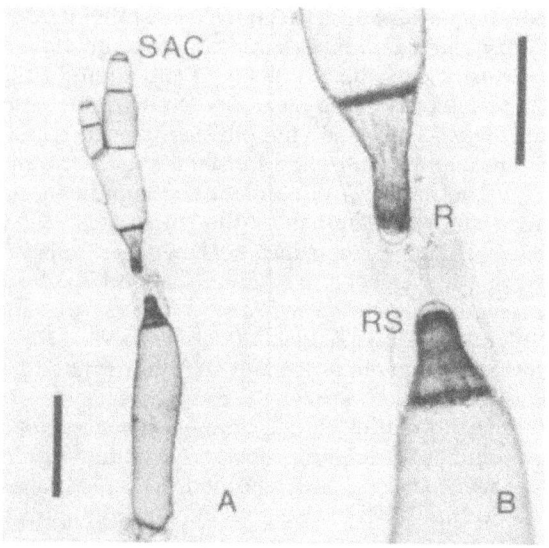

Fig. 11-3. Cell repair by cell fusion in Griffithsia pacifica. A. Whole plant with a dead intercalary cell being replaced by the process of cell repair. Scale = 0.5 mm. B. Close-up view of repair process. SAC = normal shoot apical cell, R = rhizoid, RS = repair shoot cell. Scale = 0.2 mm.

tiation. When repair shoot cells are removed from rhodomorphin, they revert to the normal shoot cell differentiation pattern (Waaland 1986; Waaland & Cleland 1974). This system could be used to assess the biochemical events associated with the conversion from repair shoot cell to normal shoot cell and with the conversion from repair shoot cell to rhizoidal cell differentiation.

Other cell fusion events that take place during red algal development, such as secondary pit connection formation, may involve rhodomorphins. Secondary connections between cells are common in complex thalli. In addition, in a number of filamentous red algae, secondary connections between filaments may be formed when lateral rhizoids from one filament meet and fuse with lateral outgrowths from adjacent plants, establishing a secondary connection between filaments (e.g., L'Hardy-Halos 1969, 1971a; Chapter 3). The kinetics of this process suggest that the presence of a lateral rhizoid close to a shoot cell induces the formation of a small outgrowth from an adjacent cell. In *Griffithsia pacifica*, prolonged treatment with partially purified rhodomorphin induces shoot cells to form multiple lateral outgrowths whose elongation is rhodomorphin dependent (Watson & Waaland unpubl.). Secondary pit connection formation may be a modified cell repair process in which a lateral rhizoid produces rhodomorphin and induces the formation of a lateral repair shoot cell with which it fuses.

By inducing cell repair in an artificial system, it is possible to obtain somatic cell fusion between cells from different plants, such as between cells from haploid and diploid plants, from male and female plants, and from plants from different sites (Waaland 1975), yielding intriguing results. For example, in *Anotrichium tenue* (as *Griffithsia tenuis*), somatic fusion between male and female cells produces a hybrid cell that, when isolated, regenerates a plant that produces tetrasporangia; the tetraspores produced by these plants are not viable (Waaland 1978a). When cell repair is induced between filaments of *G. pacifica* from geographically distinct isolates (e.g., Mexico and British Columbia, Canada), an incompatibility reaction occurs in the hybrid cell (Koslowsky & Waaland 1984, 1987). The chloroplasts from only one parent clump, fuse and degenerate; the incompatibility reaction occurs not only in the hybrid cell but also moves into the attached cells of the sensitive strain. Thus the cell repair process is a good model system for studying the control of cell division, regulation of cell elongation, process of cell differentiation, and organelle–organelle interaction.

VI. SUMMARY

Red algae are excellent organisms for studies of developmental control at the cellular level. A variety of types of tissue (e.g., spores, zygotes, individual vegetative cells, fragments of thalli, and protoplasts) can be used as inoculum for developmental studies. In filamentous genera like *Antithamnion*, *Callithamnion*, and *Griffithsia* individual cells are easily accessible for observation and manipulation. Specific cells can be tagged and their developmental fate followed.

Cell division can be studied during spore germination, apical growth, and branch initiation. In addition, division can be induced by excision of cells or tissue fragments. The effect of environmental parameters on cell division has been studied primarily in filamentous members of the Ceramiaceae. In these species, the rate of cell division is strongly influenced by environmental conditions, such as light. This effect may be seen as a change in the rate of apical cell division in some species and as an increase in branch initiation in others.

One aspect of division in red algae that should be investigated is how the position and plane of cytokinesis are controlled in both uninucleate and multinucleate cells. In both cases there does not appear to be a tight correspondence between the position of dividing nuclei and the location of the cytokinetic plane. Another aspect of interest is why intercalary cells of higher red algae do not divide. That these cells are capable of division is shown by the fact that they do divide during regeneration when adjacent cells are removed. It would appear that the presence of adjacent cells, in some way, prevents cell division.

Cell elongation may occur by tip growth in apical cells of both shoot axes and rhizoids. Tip growth in red algal cells, as in those of fungi and land plants, is associated with organelle localization. In some species, such as *Griffithsia pacifica*, tip growth can be altered by unidirectional light and, in repair shoot cells, by an endogenous hormone. Thus, one can use such cells to study mechanisms of localized growth. For example, in rhizoids and repair shoot cells of *G. pacifica*, the role of endogenous currents in directing tip growth has been evaluated; in these cells, transcellular currents do not appear necessary or sufficient to maintain elongation (Waaland & Lucas 1984).

In many species, intercalary cells elongate also (Dixon 1973). In *Batrachospermum*, *Griffithsia*, and *Antithamnion*, for example, intercalary cells elongate in a unique pattern, termed 'band growth' (Aghajanian & Hommersand 1980; Waaland & Waaland 1975). The rate and direction of intercalary cell elongation can be altered by illuminance, light direction, and correlative effects of adjacent cells. By using these parameters to shift growth, it should be possible to determine the cellular and biochemical factors that regulate such growth.

In the red algae, cell differentiation can be studied both in vegetative and reproductive thalli. A strong foundation of careful morphological observation of reproductive development has been laid at both the light and electron microscope level. In a large number of species, reproductive differentiation can be induced by environmental factors like photoperiod (Dring 1984). Species like *Callithamnion cordatum*, in which reproductive organs are borne without encircling tissue (O'Kelly & Baca 1984), should be excellent subjects for studies of the physiological basis of reproductive differentiation.

Red algal cells, filaments, and thalli show a strong polarity of development, regenerating upright thalli from apical portions and rhizoids from basal portions. Evidence from a number of different species suggests that there is intercellular communication involved in the expression of polarity. Direct communication between cells in a filament also appears to be involved in the transmission of the causative factor for a chloroplast-destroying incompatibility reaction in *Griffithsia pacifica*. In addition, there is some evidence that one part of a thallus may exert correlative effects on the rate of elongation and the polarity of regeneration of cells in other parts of the thallus. The means by which such aspects of thallus development are coordinated has not yet been determined.

One substance that is involved in the control of cell division, elongation, and differentiation has been isolated and characterized from *Griffithsia pacifica*. This hormone, rhodomorphin, is a glycoprotein that has been shown to coordinate the process of cell repair by cell fusion. Similar substances may be involved in coordinating other events in red algal development, such as secondary pit connection formation and fusion between the fertilized carpogonium and the auxiliary cell.

Thus, experimental studies of development have shown that red algae have a number of unusual development features, such as bipolar band growth, postfertilization transfer of the zygote nucleus, a glycoprotein that controls morphogenesis, and cytoplasmic incompatibility following somatic cell fusion. Analysis of these features at the biochemical and physiological levels should enhance our knowledge about cellular controls of development in general and in red algae in particular.

VII. ACKNOWLEDGMENTS

This work was supported in part by National Science Foundation grant DCB 8541028.

VIII. REFERENCES

Abbott, I. A. & Hollenberg G. J. 1976. *Marine Algae of California*. Stanford: Stanford University Press.

Aghajanian, J. G. & Hommersand, M. H. 1980. Growth and differentiation of axial and lateral filaments in *Batrachospermum sirodotii* (Rhodophyta). *J. Phycol.* 16: 15–28.

Bartnicki-Garcia, S. 1973. Fundamental aspects of hyphal morphogenesis. *Symp. Soc. Gen. Microbiol.* 23: 245–67.

Bauman, R. W. & Jones, B. R. 1986. Electrophysiological investigations of the red alga *Griffithsia pacifica* Kyl. *J. Phycol.* 22: 49–56.

Boney, A. D. & Corner, E. D. S. 1962. The effect of light on the growth of sporelings of the intertidal red alga *Plumaria elegans* (Bonnem.) Schm. *J. Mar. Biol. Ass. U.K.* 42: 65–92.

Bold, H. C. & Wynne, M. J. 1985 *Introduction to the Algae*, 2nd ed. Englewood Cliffs, N. J.: Prentice-Hall, pp. 513–633.

Breeman, A. M. & ten Hoopen, A. 1987. The mechanism of daylength perception in the red alga *Acrosymphyton purpuriferum*. *J. Phycol.* 23: 36–42.

Broadwater, S. T. & Scott, J. 1982. Ultrastructure of early development in the female reproductive system of *Polysiphonia harveyi* Bailey (Ceramiales, Rhodophyta). *J. Phycol.* 18: 427–41.

Brodie, J. & Guiry, M. D. 1987. Life history and photoperiodic responses in *Cordylecladia erecta* (Rhodophyta). *Br. Phycol. J.* 22: 300–1.

Cabioch, J. 1972. Étude sur les Corallinaces, 2. La morphogénèse: conséquences systématiques et phylogènétiques. *Cah. Biol. Mar.* 13: 137–288.

Chamberlain, Y. M. 1984. Spore size and germination in *Fosliella*, *Pneophyllum* and *Melobesia* (Rhodophyta, Corallinaceae). *Phycologia* 23: 433–42.

Chamberlain, A. H. L. & Evans, L. V. 1973. Aspects of spore production in the red alga *Ceramium*. *Protoplasma* 76: 139–59.

Chemin, E. 1937. Le développement des spores chez les Rhodophycées. *Rev. Gén. Bot.* 49: 205–34, 300–27, 353–74, 424–48, 478–536.

Chen, L. C.-M. 1986. Cell development of *Porphyra miniata* (Rhodophyceae) under axenic culture. *Bot. Mar.* 29: 435–9.

Chen, L. C.-M. & Taylor, A. R. A. 1976. Scanning electron microscopy of early sporeling ontogeny of *Chondrus crispus*. *Can. J. Bot* 54: 672–8.

Cheney, D. P., Mar, E., Saga, N. & van der Meer, J. 1986. Protoplast isolation and cell division in the agar-producing seaweed *Gracilaria* (Rhodophyta). *J. Phycol.* 22: 238–43.

Chihara, M. 1974. The significance of reproductive and spore germination characteristics to the systematics of the Corallinaceae: nonarticulated coralline algae. *J. Phycol.* 10: 266–74.

Clayton, L. 1985. The cytoskeleton and the plant cell cycle. In *The Cell Division Cycle in Plants*, eds. J. A. Bryant & D. Francis, pp. 114–31. Cambridge: Cambridge University Press.

Cole, K. & Conway, E. 1980. Studies in the Bangiaceae:

reproductive modes. *Bot. Mar.* 23: 545–53.
D'Antonio, C. M. & Gibor, A. 1985. A note on some influences of photon flux density on the morphology of germlings of *Gelidium robustum* (Gelidiales, Rhodophyta) in culture. *Bot. Mar.* 28: 313–16.
DeCew, T. C. & West, J. A. 1981. Life histories in the Phyllophoraceae (Rhodophyceae, Gigartinales) from the Pacific coast of North America. I. *Gymnogongrus linearis* and *G. leptophyllus*. *J. Phycol* 17: 240–50.
Delivopoulos, S. G. & Kugrens, P. 1984. Ultrastructure of carposporogenesis in the parasitic red alga *Faucheocolax attenuata* Setch. (Rhodymeniales, Rhodymeniaceae) *Am. J. Bot.* 71: 1245–59.
Delivopoulos, S. G. & Kugrens, P. 1985. Ultrastructure of carposporophyte development in the red alga *Gloiosiphonia verticillaris* (Cryptonemiales, Gloiosiphoniaceae). *Am. J. Bot.* 72: 1926–38.
Dickson, L. G. & Waaland, J. R. 1985. *Porphyra nereocystis*: a dual-daylength seaweed. *Planta* (Berl.) 165: 548–53.
Dixon, P. S. 1970. The Rhodophyta: some aspects of their biology. II. *Oceanogr. Mar. Biol. Annu. Rev.* 8: 307–52.
Dixon, P. S. 1971. Cell enlargement in relation to the development of thallus form in Florideophyceae. *Br. Phycol. J.* 6: 195–205.
Dixon, P. S. 1973. *Biology of the Rhodophyta*. New York: Hafner.
Dring, M. J. 1984. Photoperiodism and phycology. *Prog. Phycol. Res.* 3: 159–92.
Duffield, E. C. S., Waaland, S. D. & Cleland R. E. 1972. Morphogenesis in the red alga, *Griffithsia pacifica*: regeneration from single cells. *Planta* (Berl.) 105: 185–95.
Edwards, P. 1971. Effects of light intensity, daylength, and temperature on growth and reproduction of *Callithamnion byssoides*. In *Contributions in Phycology*, eds. B. C. Parker & R. M. Brown, pp. 163–74. Lawrence, Kan.: Allen.
Edwards, P. 1977. An analysis of the pattern and rate of cell division and morphogenesis of sporelings of *Callithamnion hookeri* (Dillw.) S. F. Gray (Rhodophyta, Ceramiales). *Phycologia* 16: 189–96.
Fritsch, F. E. 1945. *The Structure and Reproduction of the Algae*, Vol. 2. Cambridge: Cambridge University Press, pp. 397–767.
Fujita, Y. & Migata, S. 1985. Isolation and culture of protoplasts from some seaweeds. *Bull. Fac. Fish. Nagasaki Univ.* 57: 39–45.
Garbary, D. 1979. Daylength and development in four species of Ceramiaceae (Rhodophyta). *Helgol. Meeresunters.* 32: 213–27.
Garbary, D. J., Grund, D. & McLachlan, J. 1978. The taxonomic status of *Ceramium rubrum* (Huds.) C. Ag. (Ceramiales, Rhodophyceae) based on culture experiments. *Phycologia* 17: 85–94.
Goff, L. J. 1979. The biology of *Harveyella mirabilis* (Cryptonemiales, Rhodophyceae) VI. Translocation of photoassimiliated C-14. *J. Phycol.* 15: 82–7.
Goff, L. J. & Coleman, A. W. 1984. Elucidation of fertilization and development in a red alga by quantitative DNA microspectrofluorometry. *Devel. Biol.* 102: 173–94.
Goff, L. J. & Coleman, A. W. 1987. The solution to the cytological paradox of isomorphy. *J. Cell Biol.* 104: 739–48.
Gooday, G. W. & Trinci, A. P. J. 1980. Wall structure and biosynthesis in fungi. *Symp. Soc. Gen. Microbiol.* 30: 207–51.
Green, P. B. 1969. Cell morphogenesis. *Annu. Rev. Plant Physiol.* 20: 365–94.
Guiry, M. D. 1978. The importance of sporangia in the classification of the Florideophyceae. In *Modern Approaches to the Taxonomy of the Red and Brown Algae*, eds. D. E. G. Irvine & J. H. Price, pp. 111–44. London: Academic.
Gunning, B. E. S., Hardham, A. R. & Hughes, J. E. 1978. Pre-prophase bands of microtubules in all categories of formative and proliferative cell division in *Azolla* roots. *Planta* (Berl.) 143: 161–79.
Hudson, P. R. & Waaland, J. R. 1976. Ultrastructure of mitosis and cytokinesis in the multinucleate green alga *Acrosiphonia*. *J. Cell Biol.* 62: 274–94.
Hymes, B. J. & Cole, K. M. 1983. The cytology of *Audouinella hermannii* (Rhodophyta, Florideophyceae). I. Vegetative and hair cells. *Can. J. Bot.* 61: 3366–76.
Jones, W. E. 1959. Experiments on some effects of certain environmental factors on *Gracilaria verrucosa* (Hudson) Pappenfuss. *J. Mar. Biol. Ass. U.K.* 38: 153–67.
Kling, R. & Bodard, M. 1974. Les néoformations chez les algues rouges. *Soc. Phycol. France Bull.* 19: 31–5.
Knaggs, F. W. 1966. *Rhodochorton purpureum* (Lighf.) Rosenvinge. Observations on the relationship between morphology and environment II. *Nova Hedwigia* 11: 337–49.
Konrad-Hawkins, E. 1964a. Developmental studies on regenerates of *Callithamnion roseum* Harvey. I. The development of a typical regenerate. *Protoplasma* 58: 42–59.
Konrad-Hawkins, E. 1964b. Developmental studies on regenerates of *Callithamnion roseum* Harvey. II. Analysis of apical growth. Regulation of growth and form. *Protoplasma* 58: 60–74.
Konrad-Hawkins, E. 1972. Cell differentiation in tetrasporophytes of *Callithamnion roseum* Harvey (Rhodophyceae, Ceramiales). *Phycologia* 11: 37–41.
Koslowsky, D. J. & Waaland, S. D. 1984. Cytoplasmic incompatibility following somatic cell fusion in *Griffithsia pacifica*, a red alga. *Protoplasma* 123: 8–17.
Koslowsky, D. J. & Waaland, S. D. 1987. Ultrastructure of selective chloroplast destruction after somatic cell fusion in *Griffithsia pacifica* Kylin (Rhodophyta). *J. Phycol.* 23: 638–48.
Kraft, G. T. 1981. Rhodophyta: morphology and classification. In *The Biology of Seaweeds*, eds. C. S. Lobban & M. J. Wynne, pp. 6–51. Oxford: Blackwell Scientific.
Kugrens, P. & Delivopoulos, S. G. 1986. Ultrastructure of the carposporophyte and carposporogenesis in the parasitic red alga *Plocamiocolax pulvinata* Setch. (Gigartinales, Plocamiaceae). *J. Phycol.* 22: 8–21.
Kugrens, P. & West, J. A. 1973. The ultrastructure of carpospore differentiation in the parasitic red alga *Levringiella gardneri* (Setch.) Kylin. *Phycologia* 12: 163–73.
Kugrens, P. & West, J. A. 1974. The ultrastructure of carposporogenesis in the marine hemiparasitic red alga *Erythrocystis saccata*. *J. Phycol.* 10: 139–47.
Kylin, H. 1956. *Die Gattungen der Rhodophyceen*. Lund: Gleerup.
Lewis, I. F. 1909. The life history of *Griffithsia bornetiana*. *Ann. Bot.* 23: 639–90.
L'Hardy-Halos, M.-Th. 1969. La formation des anastomoses chez *Pleonosporium borreri* (Smith) Naegeli ex Hauck et *Bornetia secundiflora* (J. Ag.) Thuret (Rhodophyceae-Ceramiaceae). *C. R. Acad. Sci.* (Paris) sér. D, 268: 276–8.
L'Hardy-Halos, M.-Th. 1971a. Recherches sur les Céramiacées (Rhodophycées, Ceramiales) et leur morphogénèse. II. Les modalités de la croissance et les remaniements cellulaires. *Rev. Gén. Bot.* 78: 201–56.

L'Hardy-Halos, M.-Th. 1971b. Recherches sur les Céramiacées (Rhodophycées, Céramiales) et leur morphogénèse. III. Observations et recherches expérimentales sur la polarité cellulaire et la hiérarchisation des éléments de la fronde. *Rev. Gén. Bot.* 78: 407–91.

L'Hardy-Halos, M.-Th. 1971c. Recherches sur les Céramiacées (Rhodophycées, Céramiales) et sur quelques aspects de leur morphogénèse. *Bull. Soc. Sci. Bretagne* 46: 99–112.

L'Hardy-Halos, M.-Th. 1975. A propos des corrélations morphogénes controlant l'initiation des ramifications latérales chez les algues à structure cladomienne. *Soc. Phycol. France Bull.* 20: 1–6.

L'Hardy-Halos, M.-Th. & Larpent, J.-P. 1983. Régénération chez les algues. *Rev. Gén. Bot.* 90: 81–116.

L'Hardy-Halos, M.-Th., Larpent, J.-P. Gaillard, J., Pellegrini, L. & Pellegrini, M. 1984. Morphogénès expérimentale chez les algues. *Rev. Cytol. Biol. Végét. Bot.* 7: 311–62.

Lindstrom, S. 1981. Female reproductive structures and strategy in a red alga, *Constantinea rosa-marina* (Gmelin) Postels et Ruprecht (Dumontiaceae, Cryptonemiales). *Jap. J. Phycol* 29: 251–7.

Liu, W. S., Rang, Y. L., Liu, X. W. & Fang, T. C. 1984. Studies on the preparation and on the properties of sea snail enzymes. *Proc. Int. Seaweed Symp.* 11: 319–20.

Mattox, K. R. & Stewart, K. D. 1984. Classification of the green algae: a concept based on comparative cytology. In *Systematics of the Green Algae*, eds. D. E. G. Irvine & D. M. John, pp. 29–72. London: Academic.

Melvin, D. J., Mumford, T. F., Byce, W. J., Inayoshi, M. & Bryant, V. M. 1986. *Conchocelis Culture*. Vol. 1: *Equipment and Techniques for Nori Farming in Washington State*. Olympia: Wash. St. Dept. of Nat. Resources.

Murray, S. N. & Dixon, P. S. 1975. The effects of light intensity and light period on the development of thallus form in the marine red alga *Pleonosporium squarrulosum* (Harvey) Abbott (Rhodophyta, Ceramiales). II. Cell enlargement. *J. Exp. Mar. Biol. Ecol.* 19: 165–76.

Neushul, M., Scott, J., Dahl, A. L. & Olsen, D. 1967. Growth and development of *Sciadophycus stellatus* Dawson. *Bull. S. Calif. Acad. Sci.* 66: 195–200.

Norton, T. A., Mathieson, A. C. & Neushul, M. 1981. Morphology and environment. In *The Biology of Seaweeds*, eds. C. S. Lobban & M. J. Wynne, pp. 421–51. Berkeley: University of California Press.

Nuccitelli, R., ed. 1986. *Ionic Currents in Development*. New York: Alan R. Liss.

O'Kelly, C. J. & Baca, B. J. 1984. The time course of carpogonial branch and carposporophyte development in *Callithamnion cordatum* (Rhodophyta, Ceramiales). *Phycologia* 23: 407–17.

Perrone, C. & Felicini, G. P. 1972. Sur les bourgeons adventifs de *Petroglossum nicaeense* (Duby) Schotter (Rhodophycées, Gigartinales) en culture. *Phycologia* 11: 87–95.

Perrone, C. & Felicini, G. P. 1974. Dominance apicale et morphogénèse chez *Petroglossum nicaeense*. *Phycologia* 13: 187–94.

Perrone-Pesola, C. & Felicini, G. P. 1981. Polarité dans la fronde de *Schottera nicaeensis* (Phyllophoracées). *Phycologia* 20: 142–6.

Pickett-Heaps, J. D. & Northcote, D. H. 1966. Organisation of microtubules and endoplasmic reticulum during mitosis and cytokinesis in wheat meristems. *J. Cell Sci* 1: 109–20.

Polne-Fuller, M. & Gibor, A. 1984. Developmental studies in *Porphyra* I. Blade differentiation in *Porphyra perforata* as expressed by morphology, enzymatic digestion and protoplast regeneration. *J. Phycol.* 20: 607–18.

Powell, J. 1964. The life history of the red alga *Constantinea*. Ph.D. Dissertation, Seattle, University of Washington.

Powell, J. 1986. A short day photoperiodic response in *Constantinea subulifera*. *Am. Zool.* 26: 479–87.

Pueschel, C. M. 1979. Ultrastructure of tetrasporogenesis in *Palmaria palmata* (Rhodophyta). *J. Phycol* 15: 409–24.

Pueschel, C. M. & Cole, K. M. 1985. Ultrastructure of germinating carpospores of *Porphyra variegata* (Kjellm.) Hus (Bangiales, Rhodophyta). *J. Phycol.* 21: 146–54.

Ramm-Anderson S. M. & Wetherbee R. 1982. Structure and development of the carposporophyte of *Nemalion helminthoides* (Nemalionales, Rhodophyta). *J. Phycol.* 18: 133–41.

Rueness, J. & Rueness, M. 1973. Life history and nuclear phases of *Antithamnion tenuissimum* with special reference to plants bearing both tetrasporangia and spermatia. *Norw. J. Bot.* 20: 205–10.

Rueness M. & Rueness, J. 1982. Hybridization and morphogenesis in *Callithamnion hookeri* (Dillw.) S. F. Gray (Rhodophyceae, Ceramiales) from disjunct northeastern Atlantic populations. *Phycologia* 21: 137–44.

Rueness, J. & Rueness, M. 1985. Regular and irregular sequences in the life history of *Callithamnion tetragonum* (Rhodophyta, Ceramiales). *Br. Phycol. J.* 20: 329–33.

Schechter, V. 1934. Electrical control of rhizoid formation in the red alga, *Griffithsia bornetiana*. *J. Gen. Physiol.* 18: 1–21.

Sievers, A. & Schnepf, E. 1981. Morphogenesis and polarity of tubular cells with tip growth. In *Cytomorphogenesis in Plants*, ed. O. Kiermayer, pp. 265–99. Vienna–New York: Springer-Verlag.

Sylvester, A. 1987. Organization of the cytoplasm during cell growth in *Anotrichium tenue*. Ph.D. Dissertation, Seattle, University of Washington.

Sylvester, A. W. & Waaland, J. R. 1983. Cloning the red alga *Gigartina exasperata* for culture on artificial substrates. *Aquaculture* 31: 305–18.

Sylvester, A. W. & Waaland, J. R. 1984. Sporeling dimorphism in the red alga *Gigartina exasperata* Harvey & Bailey. *Phycologia* 23: 427–32.

Tripodi, G. 1974. Ultrastructural changes during carpospore formation in the red alga *Polysiphonia*. *J. Submicr. Cytol.* 6: 275–86.

Tsekos, I. 1981. Growth and differentiation of the Golgi apparatus and wall formation during carposporogenesis in the red alga *Gigartina teedii* (Roth) Lamour. *J. Cell. Sci.* 52: 71–84.

Tsekos, I. 1982. Plastid development and floridean starch grain formation during carposporogenesis in the red alga *Gigartina teedii*. *Cryptogam. Algol.* 2: 91–103.

Tsekos, I. 1983. The ultrastructure of carposporogenesis in *Gigartina teedii* (Roth) Lamour. (Gigartinales, Rhodophyceae): gonimoblast cells and carpospores. *Flora* 174: 191–211.

Turner, C. H. C. & Evans, L. V. 1978. Translocation of photoassimilated C-14 in the red alga *Polysiphonia lanosa*. *Br. Phycol. J.* 13: 51–5.

van der Meer, J. P. 1986. Genetics of *Gracilaria tikvahiae* (Rhodophyceae). XI. Further characterization of a bisexual mutant. *J. Phycol.* 22: 151–8.

van der Meer, J. P. & Todd, E. R. 1977. Genetics of *Gracil-*

aria sp. (Rhodophyceae, Gigartinales). IV. Mitotic recombination and its relationship to mixed phases in the life history. *Can. J. Bot.* 55: 2810–17.

van der Meer, J. P., Patwary, M. U. & Bird, C. J. 1984. Genetics of *Gracilaria tikvahiae* (Rhodophyceae). X. Studies on a bi-sexual clone. *J. Phycol.* 20: 42–6.

Waaland, S. D. 1975. Evidence for a species-specific cell fusion hormone in red algae. *Protoplasma* 86: 253–61.

Waaland, S. D. 1978a. Parasexually produced hybrids between male and female plants of *Griffithsia tenuis* C. Agardh, a red alga. *Planta* (Berl.) 138: 65–8.

Waaland, S. D. 1978b. Production of the cell fusion hormone, rhodomorphin, by *Griffithsia tenuis* (Rhodophyta). *J. Phycol.* 14 (Suppl.): 35.

Waaland, S. D. 1984. Positional control of development in algae. In *Positional Controls in Plant Development.* eds. P. W. Barlow & D. J. Carr, pp. 137–56. Cambridge: Cambridge University Press.

Waaland, S. D. 1986. Hormonal coordination of the processes leading to cell fusion in algae: a glycoprotein hormone from red algae. In *Plant Growth Substances 1985*, ed. M. Bopp, pp. 257–62. Heidelberg: Springer-Verlag.

Waaland, S. D. & Cleland, R. E. 1972. Development in the red alga, *Griffithsia pacifica*: control by internal and external factors. *Planta* (Berl.) 105: 196–204.

Waaland, S. D. & Cleland, R. E. 1974. Cell repair through cell fusion in the red alga *Griffithsia pacifica*. *Protoplasma* 79: 185–96.

Waaland, S. D. & Lucas, W. J. 1984. An investigation of the role of transcellular ion currents in morphogenesis of *Griffithsia pacifica* Kylin. *Protoplasma* 123: 184–91.

Waaland, S. D., Nehlsen, W. & Waaland, J. R. 1977. Phototropism in the red alga, *Griffithsia pacifica*. *Plant & Cell Physiol.* 18: 603–12.

Waaland, S. D. & Waaland, J. R. 1975. Analysis of cell elongation in red algae by fluorescent labelling. *Planta* (Berl.) 126: 127–38.

Waaland, S. D., Waaland, J. R. & Cleland, R. 1972. A new pattern of plant cell elongation: bipolar band growth. *J. Cell Biol.* 54: 184–90.

Waaland, S. D. & Watson, B. A. 1980. Isolation of a cell-fusion hormone from *Griffithsia pacifica* Kylin, a red alga. *Planta* (Berl.) 149: 493–7.

Watson, B. A. & Waaland, S. D. 1983. Partial purification and characterization of a glycoprotein cell fusion hormone from *Griffithsia pacifica*, a red alga. *Plant Physiol.* 71: 327–32.

Watson, B. A. & Waaland, S. D. 1986. Further biochemical characterization of a cell fusion hormone from the red alga, *Griffithsia pacifica*. *Plant Cell Physiol.* 27: 1043–50.

Weber, W. 1960. Polarity of spore germination. *Bot. Mar.* 2: 182–8.

West, J. 1972. Environmental control of hair and sporangial formation in the marine red alga *Acrochaetium proskaueri* sp. nov. *Proc. Int. Seaweed Symp.* 7: 377–84.

West, J. A. & Norris, R. E. 1966. Unusual phenomena in the life histories of Florideae in culture. *J. Phycol.* 2: 54–7.

Wetherbee, R. 1979. "Transfer connections": specialized pathways for nutrient translocation in a red alga? *Science* 204: 858–9.

Whittick, A. & West, J. A. 1979. The life history of a monoecious species of *Callithamnion* (Rhodophyta, Ceramiaceae) in culture. *Phycologia* 18: 30–7.

Chapter 12

Vegetative growth and organization

ROY J. COOMANS
MAX H. HOMMERSAND

CONTENTS

I. Introduction / 275
II. Bangiophycidae / 276
III. Lower Florideophycidae: Acrochaetiales, Nemaliales, and Batrachospermales / 280
   A. Acrochaetiales / 280
   B. Nemaliales / 280
   C. Batrachospermales / 285
IV. Higher Florideophycidae: Gelidiales, Bonnemaisoniales, Gigartinales, Rhodymeniales, and Ceramiales / 288
   A. Mitosis and cytokinesis / 288
   B. Development of prostrate systems / 290
   C. Initiation of upright axes / 290
   D. The erect thallus: simple uniaxial types / 291
   E. Organization of determinate lateral filaments / 294
   F. Facultative branching of the thallus / 294
   G. Patterned branching of the thallus / 296
   H. Secondary pit connections, wall growth, and increasing nuclear volume / 297
V. Summary / 299
VI. References / 301

I. INTRODUCTION

The red algae have traditionally been divided into two subclasses, the Bangiophycidae and the Florideophycidae, based on differences in vegetative morphology, reproduction, and life histories. Thalli of Florideophycidae are essentially pseudoparenchymatous, consisting of aggregations of filaments that grow almost exclusively through the division of apical cells. As we will show, filamentous growth also prevails in early stages of development in most Bangiophycidae, although the morphology of the adult thallus is highly variable, ranging from unicellular or colonial to filamentous or parenchymatous. The vegetative morphology of red algae has been described and illustrated by Oltmanns (1922), Fritsch (1945), Kylin (1956), and Dixon (1973). A recent review by Gabrielson and Garbary (1986) brings the literature up to date, with coverage of more recent observations on heterotrichy, the initiation of upright axes, and thallus differentiation.

Several markers of morphogenetic events can be

observed with light microscopy, including wall growth, changes in cell size and shape, changes in the density and distribution of the cytoplasm, internal rearrangements of organelles, nuclear division, and cell division. In this chapter we will describe the major patterns of vegetative growth in Rhodophyta, emphasizing the critical role played by mitosis and cytokinesis. We will focus on primary growth. The formation of secondary tissues is not well documented and will not be covered. The majority of the observations reported here are based on research conducted by the senior author as a part of his Ph.D. dissertation at the University of North Carolina at Chapel Hill (Coomans 1986). Material was preserved in 5% formalin-seawater and stained according to the aceto-iron-hematoxylin-chloral hydrate method of Wittmann (1965), as modified by Coomans (1986), or with aniline blue (Gabrielson & Hommersand 1982a).

Three major patterns of mitosis and cytokinesis were seen that relate to the mode of vegetative growth and level of thallus organization. One of these characterizes the Bangiophycidae. A second type is found in the Acrochaetiales, Nemaliales, and Batrachospermales, orders regarded as primitive among the Florideophycidae. The third type predominates in the Bonnemaisoniales, Gelidiales, Gigartinales, Rhodymeniales, and Ceramiales, orders usually considered to be advanced within the Florideophycidae. The Palmariales, Hildenbrandiales, and Corallinales were not studied and will not be discussed. Distinguishing characters include the location and orientation of the mitotic apparatus within the dividing cell, the arrangement of the chromosomes during metaphase and anaphase, the path of migration of the two sets of daughter chromosomes during anaphase, and the timing of septum initiation.

II. BANGIOPHYCIDAE

Light microscope observations of cell division in *Porphyridium* sp. (Porphyridiales) (Figs. 12-1 to 12-4), *Erythrotrichia carnea* (Erythropeltidales) (Figs. 12-5 to 12-8), and *Porphyra carolinensis* (Bangiales) (Figs. 12-9 to 12-12) suggest that a common pattern of mitosis and cytokinesis prevails throughout the Bangiophycidae. Our observations are in good agreement with published descriptions and figures of *Porphyra* (Krishnamurthy 1959; Pringle & Austin 1970; Yabu 1963, 1970), *Bangia* (Yabu 1966), and *Boldia* (Nichols 1964a). The nucleus is elongated immediately prior to and during prophase (Figs. 12-1, 12-5, 12-9). As division proceeds, the chromosomes aggregate to form a plate or ring at metaphase within an extended, spindle-shaped nucleus (Fig. 12-6). During anaphase the interzonal spindle elongates greatly, often becoming flexed over the surface of the pyrenoid, when present (Fig. 12-2), and each set of daughter chromosomes is arranged in a partial or complete ring (Figs. 12-7, 12-10). At late anaphase the dividing nucleus can span the entire length or breadth of the cell (Figs. 12-3, 12-11). Cytokinesis takes place after daughter nuclei have formed through annular ingrowth of a broad cleavage furrow (Figs. 12-4, 12-8, 12-12).

Vegetative growth is apical, intercalary, or diffuse (Garbary et al. 1980). Early stages of development following spore germination often show apical growth followed by a shift to intercalary divisions and diffuse growth. *Rhodochaete parvula*, the sole member of the Rhodochaetales, is strictly filamentous. According to Schmitz (1897), vegetative growth takes place exclusively by means of transversely dividing apical cells. The wall of apical cells is thinner and less dense at the tip (Pueschel & Magne 1987), presumably reflecting tip growth. The presence of structurally simple pit plugs has been confirmed with electron microscopy (Pueschel & Magne 1987). Boillot (1975) demonstrated the presence of a sexual alternation of generations in which monospores of the gametophyte and presumed meiospores of the sporophyte undergo bipolar germination, whereas the diploid, sexually produced carpospores and diploid monospores of the sporophyte undergo unipolar germination. The mature thallus consists of pinnately to irregularly branched, uniseriate filaments in both generations.

The Porphyridiales contains unicellular and colonial genera; colonial taxa are either amorphous or pseudofilamentous in construction (Garbary et al. 1980). The thallus of *Stylonema alsidii* (formerly *Goniotrichum alsidii*; see Wynne 1985) is composed of uniseriate filaments that branch in a manner resembling the false branching of certain cyanobacteria (Fig. 12-13), but the absence of longitudinal divisions during branch initiation has never been clearly demonstrated. *Stylonema cornu-cervi* and *Goniotrichopsis sublittoralis* are pleuriseriate (Garbary et al. 1980). Isolates of *Chroodactylon* investigated in culture by Lewin and Robertson (1971) (as *Asterocytis*) produced a unicellular form resembling *Chroothece* when grown in reduced salinity. Gargiulo et al. (1987) found that the thallus of *Erythrocladia irregularis* (Erythropeltidales) converts to an irregular aggregate of unicells when grown under nutrient-enriched conditions and cautioned that some unicellular entities may be incorrectly placed

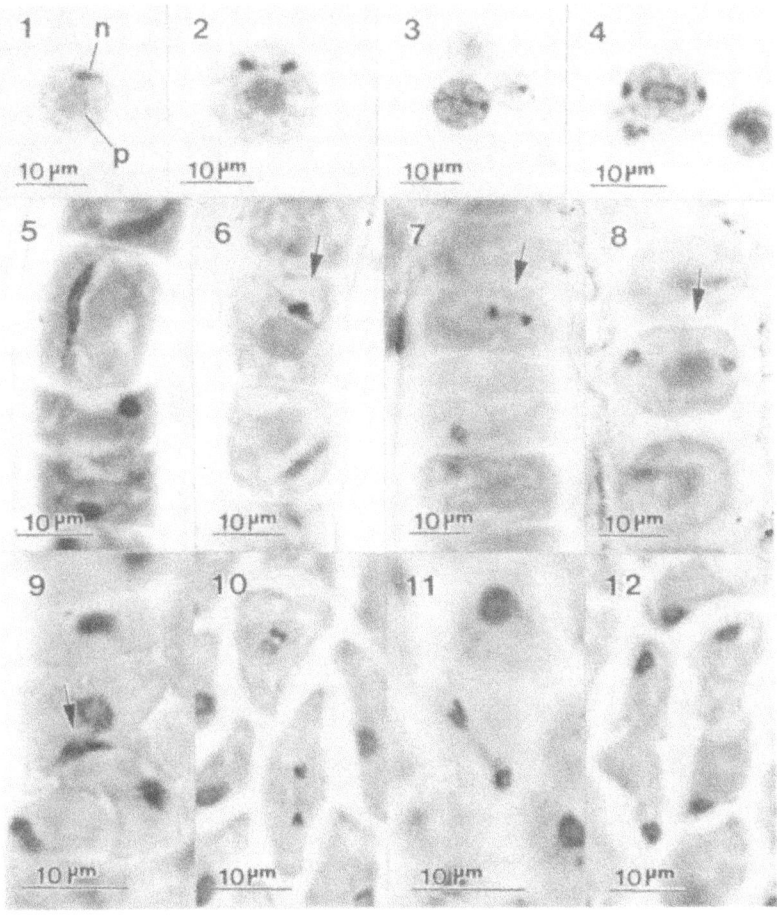

Figs. 12-1 to 12-4. *Porphyridium* sp. (cultured material, Carolina Biological Supply Co.). *12-1*: Prophase; nucleus (n), pyrenoid (p). *12-2*: Early anaphase. *12-3*: Late anaphase. *12-4*: Cytokinesis.
Figs. 12-5 to 12-8. *Erythrotrichia carnea* (North Carolina). *12-5*: Extended nucleus at early prophase. *12-6*: Metaphase (arrow). *12-7*: Anaphase (arrow). *12-8*: Cytokinesis; broad cleavage furrow (arrow).
Figs. 12-9 to 12-12. *Porphyra carolinensis* (North Carolina). *12-9*: Prophase (arrow). *12-10*: Two cells in anaphase; chromosomes form partial or complete rings. *12-11*: Late anaphase. *12-12*: Two pairs of daughter cells of recent divisions.

in the Porphyridiales (see Chapter 18 for further discussion).

The Erythropeltidales contains species that are crustose, filamentous, saccate, or foliose in construction (Garbary et al. 1980). Germinating spores characteristically produce a prostrate filamentous holdfast or a monostromatic or polystromatic disk. All members appear to go through a filamentous stage following spore germination (Heerebout 1968; Howard & Parker 1980; Hus 1902). Growth is apical in the earliest stages of development, but intercalary divisions soon begin to contribute to the growth of the thallus. The thallus of *Erythrocladia subintegra* is discoid. Apical cells at the thallus margin become retuse to Y-shaped and divide obliquely in alternating sequence to generate a pseudodichotomous branching pattern (Fig. 12-14). Repeated divisions of intercalary cells generate a raised, saccate region in the central part of the disk (Gargiulo et al. 1987).

In *Porphyropsis coccinea* monospores germinate into monostromatic disks like those of *Erythrocladia*, but with continued growth vesicles produced from the central area rupture, forming monostromatic

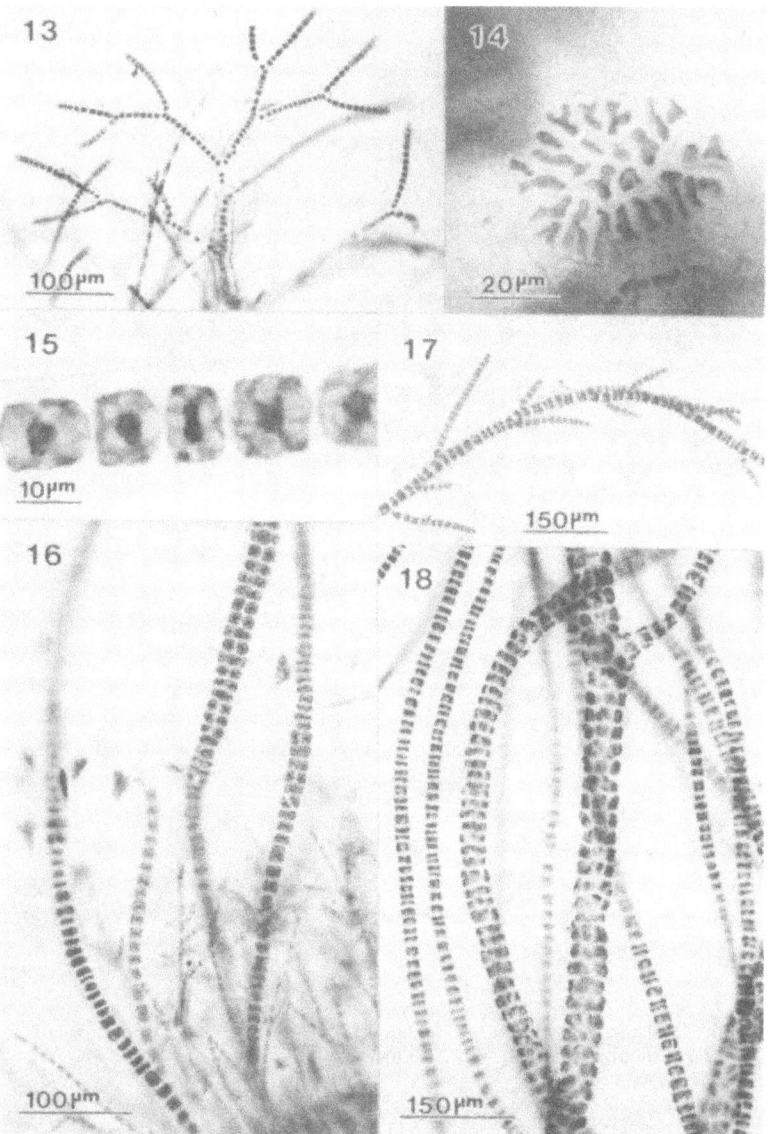

Figs. 12-13 to 12-18. *12-13*: *Stylonema alsidii* (North Carolina). False branching. *12-14*: *Erythrocladia subintegra* epiphytic on *Polysiphonia* sp. (North Carolina). *12-15*, *12-16*: *Erythrotrichia carnea* (North Carolina). *12-15*: Uniseriate portion showing thick, common wall. *12-16*: Filaments becoming multiseriate. *12-17*, *12-18*: *Compsopogon* sp. (cultured material). *12-17*: Branch initiation near the apex. *12-18*: Uncorticated and corticated filaments.

blades (Garbary et al. 1980; Murray et al. 1972). *Smithora naiadum* first produces a polystromatic basal cushion (Garbary et al. 1980) that gives rise to ovate, monostromatic blades. A filamentous conchocelis-like stage has been reported for both *Porphyropsis* (Murray et al. 1972) and *Smithora* (Richardson & Dixon 1969).

Erythrotrichia carnea possesses an erect, filamentous thallus that may be supported by a small, discoid holdfast. Thalli of *E. carnea* are uniseriate when young but may become pleuriseriate through repeated longitudinal divisions (Figs. 12-15, 12-16). In *E. boryana* longitudinal divisions of cells of the primary filament are restricted to a single plane,

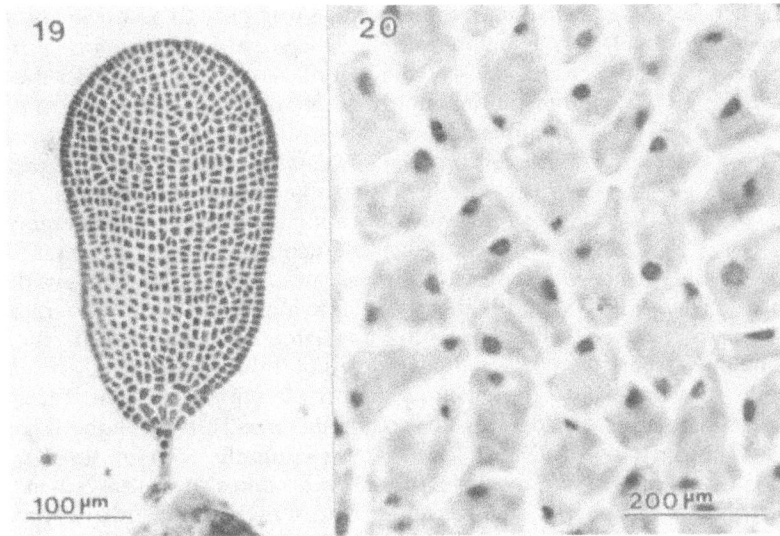

Figs. 12-19, 12-20. *Porphyra carolinensis* (North Carolina). *12-19*: Young blade with margin of large, dividing cells. *12-20*: Interior of thallus, cells dividing in various planes.

producing a monostromatic blade (Heerebout 1968). *Porphyropsis vexillaris* develops similarly (Heerebout 1968). Kornmann (1984, 1987) has described a heteromorphic sexual life cycle in two species of *Porphyrostromium* that includes a haploid, filamentous phase resembling *Erythrotrichia* and a presumably diploid crustose phase resembling *Erythrocladia*. Thus, both morphological types that characterize the Erythropeltidales occur in the life history of a single species.

The genus *Boldia* (Boldiaceae) possesses a complex, pleiomorphic life history that includes stages composed of a monostromatic disk, apically growing uniseriate filaments, erect and prostrate filaments with intercalary growth, and an erect saccate thallus with diffuse growth (Howard & Parker 1980).

The Compsopogonales is treated as an order distinct from the Erythropeltidales by Gabrielson & Garbary (1986). In *Compsopogon* germination of monospores is bipolar. Rhizoidal and erect filaments may arise directly (Nichols 1964b), or a well-developed rhizoidal holdfast may be produced prior to the initiation of one or more uprights (Krishnamurthy 1962). Both apical and intercalary divisions increase the length of the primary filament and any lateral branches present (Fig. 12-17). Behind the apex, cells undergo oblique longitudinal divisions to produce an irregular cortex (Fig. 12-18; Krishnamurthy 1962). Axial cells that have ceased dividing enlarge, reaching several times their original dimensions at maturity. In *Compsopogonopsis* rhizoids are produced by axial cells of the primary filament (*C. leptocladis*, Seto 1982; *C. fruticosa*, Seto 1987) or by the primary cortical cells (*C. japonica*, Seto 1982). A primitive type of pit plug has been described in *Compsopogon* (Chapter 2).

The Bangiales has a heteromorphic life history. The dominant gametophyte is filamentous or foliose and exhibits diffuse growth. The thallus of *Bangia atropurpurea* (reported as *B. fuscopurpurea*) is initially an upright, uniseriate filament with a rhizoidal holdfast. Apical and intercalary divisions increase the length of the filament (Sommerfeld & Nichols 1970). The thallus becomes pleuriseriate through oblique and longitudinal divisions (Drew 1952; Sommerfeld & Nichols 1970), and normally remains unbranched. Germlings of *Porphyra* are also initially filamentous but become foliose through longitudinal divisions of both apical and intercalary cells, beginning at an early stage of thallus development (Rosenvinge 1909). In very young blades there may be an obvious marginal meristem (Fig. 12-19). Subsequent expansion of the blade occurs through diffuse growth, the orientation of cell divisions being variable (Fig. 12-20). Species are either monostromatic or distromatic (Garbary et al. 1980). The sporophytic (conchocelis) stage in Bangiales is filamentous, and cells are joined by pit connections with simple plugs (Cole & Conway 1975; Chapter 2). Spore germination is unipolar and is accompanied by a change in plastid morphology (Pueschel & Cole 1985) and a shift from the production of xylan to cellulose in the cell walls (Gretz et al. 1980; Mukai et al. 1981). Early growth appears to result from apical cell divisions (Cole & Conway 1980), but cytological confirmation is needed.

III. LOWER FLORIDEOPHYCIDAE: ACROCHAETIALES, NEMALIALES, AND BATRACHOSPERMALES

As Fritsch (1945, p. 450) pointed out, the most primitive vegetative structures among Florideophycidae are found in members of the *Acrochaetium* complex, now placed in the Acrochaetiales (Garbary & Gabrielson 1987). Branching patterns are simple and there is little differentiation between vegetative cells, even in heterotrichous species. Though the Nemaliales and Batrachospermales possess more complex morphologies, a survey of their life histories reveals the presence of acrochaetioid thalli in both the tetrasporophyte generation and the protonemal stage of the gametophyte generation (West & Hommersand 1981). These three orders also possess pit plugs with two cap layers, a feature otherwise known only in the Palmariales (Chapter 2).

Acrochaetiales and Nemaliales exhibit a common mode of mitosis and cytokinesis, in which nuclear division does not ordinarily take place at the site of future septum formation, but rather basal to it, and the orientation of the mitotic apparatus is variable. The same nuclear behavior is seen in the whorled determinate lateral filaments of *Batrachospermum boryanum* and the pseudochantransia stage of *Lemanea australis* (Batrachospermales).

A. Acrochaetiales

Whereas most recent authors have followed Garbary (1979a) and Woelkerling (1983) in recognizing a single genus, *Audouinella*, in the Acrochaetiales, Stegenga (1979, 1985) distinguishes some five to seven genera based on combinations of chloroplast type, spore germination pattern, and life history characters (see Chapter 18). Spore germination may be unipolar, bipolar, or septate, and thalli may have a unicellular base, a multicellular base, or a heterotrichous habit consisting of filamentous prostrate and erect systems. Gametophyte and sporophyte generations may be heteromorphic and often have distinctly different germination patterns and basal systems (Stegenga 1979, 1985).

Strictly apical deposition of new wall material has been demonstrated in *Audouinella hermannii* by Hymes and Cole (1983) using calcofluor white. Garbary (1979b), working with five species of *Audouinella*, found that most wall growth occurs in apical cells. Subterminal cells were at or near their mature dimensions when cut off, and all cells reached their maximum length when only a few cells behind the apex.

In *Acrochaetium secundatum* from South Africa (Stegenga 1985) spore germination is septate, the prostrate system forming a pseudoparenchymatous disk (Fig. 12-21). Branching of erect filaments is characteristically secund (Fig. 12-22). Nuclear division is initiated following extension of the apical cell or formation of a lateral bud. Nuclear division occurs in the basal half of the dividing cell, proximal to the ensuing septum. Orientation of the mitotic apparatus is variable but most often oblique (Figs. 12-23, 12-24). As is frequently seen in the Bangiophycidae, the chromosomes are arranged in a partial or complete ring at metaphase and anaphase (Figs. 12-25 to 12-27). There are two distinct stages to anaphase migration. The first involves the separation of daughter chromosomes. At its completion one set has come to lie against the side wall; the other set has moved to the opposite side wall or come to lie against the basal crosswall. During migration the two sets of daughter chromosomes frequently twist with respect to one another, rather than moving perpendicular to the plane of the preceding metaphase plate (Fig. 12-28). In the second stage of anaphase the more distal set of daughter chromosomes migrates into the apex of the dividing cell (Fig. 12-29). Nuclear divisions associated with the formation of a lateral initial take place within the parent cell, not in close association with the lateral bud (Fig. 12-30). In *Audouinella violacea*, as in many Acrochaetiales, there is a precocious initiation of the septum prior to the onset of mitosis or during its early stages (Figs. 12-31 to 12-33).

Vegetative cells of most species remain uninucleate due to an absence of cell fusions, secondary pit connections, and nuclear divisions without ensuing cell divisions. Exceptions are the lateral fusions between cells of the pseudoparenchymatous basal system in *Audouinella spetsbergensis* (Dixon & Irvine 1977; Garbary et al. 1982) and *A. concrescens* (West 1970).

B. Nemaliales

Excluding *Galaxaura*, which has an isomorphic alternation of generations, the Nemaliales possesses a heteromorphic life history in which the dominant gametophyte is multiaxial and the tetrasporophyte resembles *Acrochaetium* (West & Hommersand 1981). Germinating tetraspores give rise to a mono-

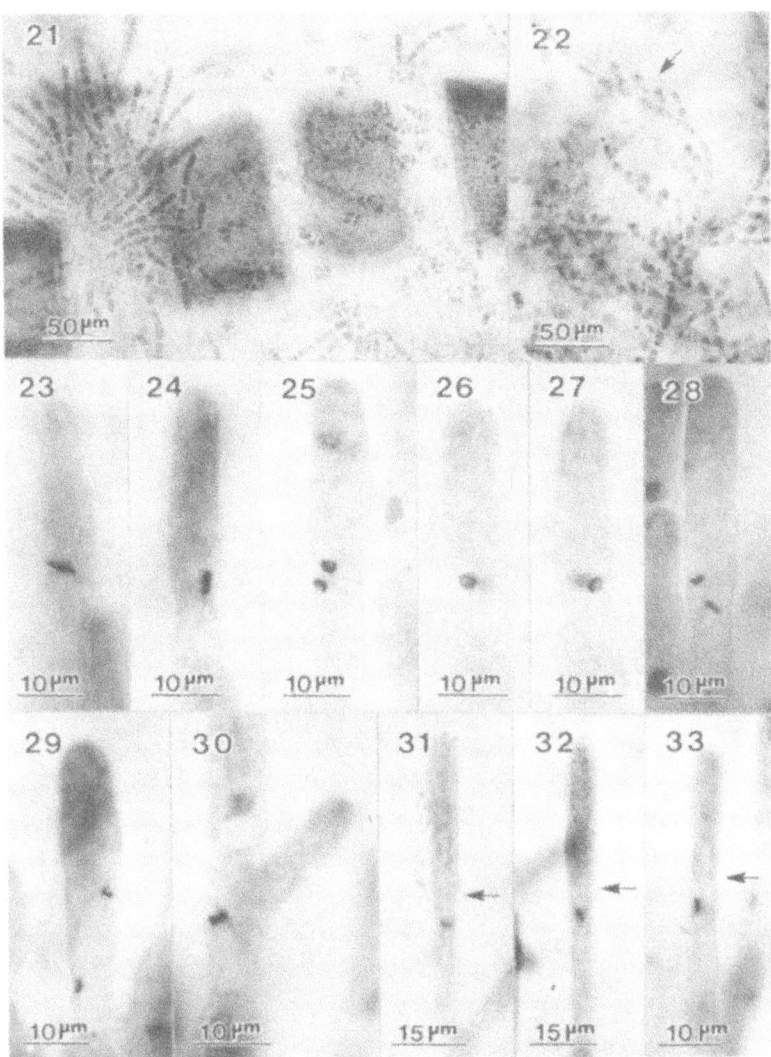

Figs. 12-21 to 12-30. *Acrochaetium secundatum* (South Africa). *12-21*: Numerous sporelings epiphytic on *Chaetomorpha linum*. *12-22*: Secund branching. *12-23, 12-24*: Variation in metaphase orientation. *12-25*: Early anaphase. Oblique division; daughter chromosomes form partial rings. *12-26, 12-27*: Two focal planes of a cell at the end of the first stage of anaphase migration. Chromosomes form rings and have migrated to opposite sidewalls. *12-28*: Anaphase; daughter chromosomes twisting with respect to one another. *12-29*: Late anaphase; distal set of daughter chromosomes migrating apically. *12-30*: Metaphase; division to form a lateral initial.
Figs. 12-31 to 12-33. *Audouinella violacea* (North Carolina). Variation in orientation of metaphase plate. Division located just proximal to precocious septum (arrows).

siphonous protonemal stage. Uprights produced by the protonemal stage may initially be uniaxial or multiaxial.

Little has been added to our knowledge of apical organization in Nemaliales since Oltmanns (1904) introduced the term *Springbrunnentypus* for taxa in which growth is maintained by a cluster of apical initials. *Liagora mucosa* is representative of the multiaxial habit in Nemaliales (Fig. 12-34). In *L. mucosa* growth results from the activity of five to ten indeterminate filaments, each of which develops sympodially. Apical cells of indeterminate

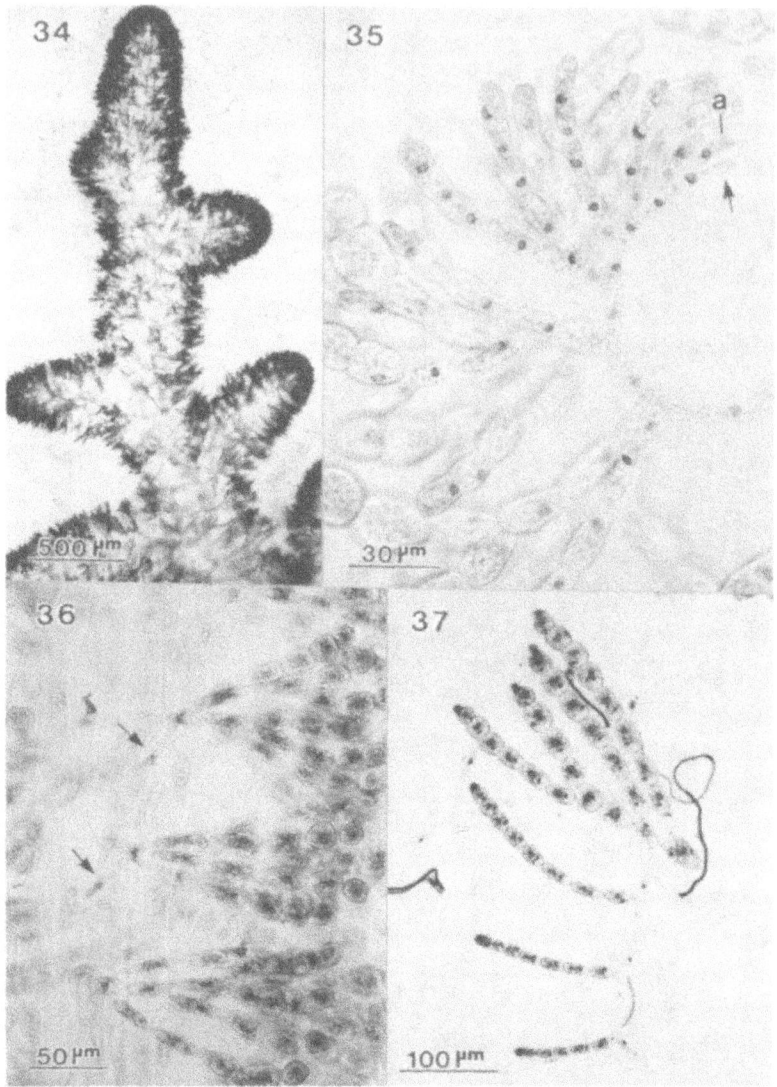

Figs. 12-34 to 12-37. *Liagora mucosa* (Florida). *12-34*: Alternately branched axis. *12-35*: Cellulosympodial development of indeterminate filament. Subapical cell with bud (arrow) to thallus interior, will continue axis; current apical cell (a) will initiate a determinate filament. *12-36*: Initials of rhizoidal filaments (arrows) on basal cells of determinate filaments. *12-37*: Determinate lateral filaments initiated by rhizoidal filament.

filaments elongate and divide transversely. The resulting subapical cell initiates a lateral bud toward the interior of the thallus (Fig. 12-35) that, when cut off, becomes the new apical cell. The former apical cell is displaced to the outside and gives rise to a determinate lateral filament. Determinate filaments branch dichotomously or trichotomously, often with the primary filament curved apically and bearing abaxial derivatives (Fig. 12-35). Repetition of this pattern produces indeterminate axial filaments, each segment of which bears a single determinate lateral filament that contributes to the cortex. Rhizoidal filaments initiated from the basal, abaxial surface of cells within the determinate lateral filaments add to the medulla (Fig. 12-36). These can initiate additional determinate filaments that add to the cortex (Fig. 12-37).

The type of sympodial growth exhibited by

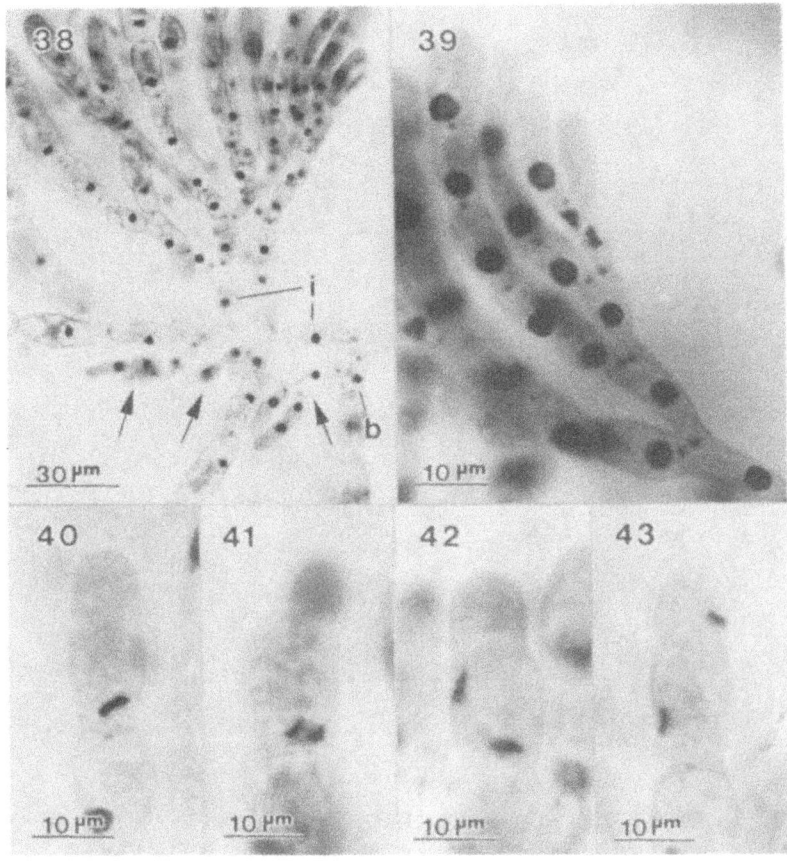

Figs. 12-38 to 12-43. *Liagora mucosa* (Florida). *12-38*: Young indeterminate lateral filaments (arrows) arising from two adjacent indeterminate axial filaments (*i*); each is borne on the basal cell (*b*) of a determinate lateral filament. *12-39*: Mitosis in subapical cell of indeterminate filament, end of first stage of anaphase migration. Lateral initial will assume the apex. *12-40*: Metaphase; determinate filament. *12-41*: Early anaphase, determinate filament. *12-42*: End of first stage of anaphase migration, determinate filament. *12-43*: Second stage of anaphase, distal set of daughter chromosomes migrating into apex.

Liagora has been termed cellulosympodial by Norris et al. (1984). Cellulosympodial growth of indeterminate filaments is not unique to *Liagora mucosa*. Our preliminary observations have verified its presence in *L. farinosa* and *Helminthora australis*. Careful scrutiny of the literature supports the likelihood that cellulosympodial development of axial filaments is a common feature among Nemaliales. Cleland's illustration (1919, text Fig. 1) of the apex of *Nemalion helminthoides* (as *N. multifidum*), Desikachary's drawings of axial filaments of *Helminthocladia australis* (1957, Fig. 15), *Cumagloia andersonii* (1962, Fig. 1), and *Dermonema frappieri* (1962, Fig. 31), Svedelius' illustration of the apex of *Galaxaura diesingiana* (1942, Fig. 6b), and Kjellman's depiction of an indeterminate axial filament of *Galaxaura falcata* (1900, Pl. 11, Fig. 13) could all be interpreted as illustrating cellulosympodial development.

It is commonly assumed that branching of the thallus in Nemaliales takes place by "periodical separation of the apical threads into two groups" (Fritsch 1945, p. 469). This is not true of *Liagora mucosa*, where indeterminate initials cut off from the basal cell of determinate lateral filaments are responsible for branching. One to two indeterminate filaments are initiated from the basal cell of a determinate lateral filament following the production of its normal complement of two to three determinate filaments (Fig. 12-38). Indeterminate initials are cut off from the determinate filaments of two to four

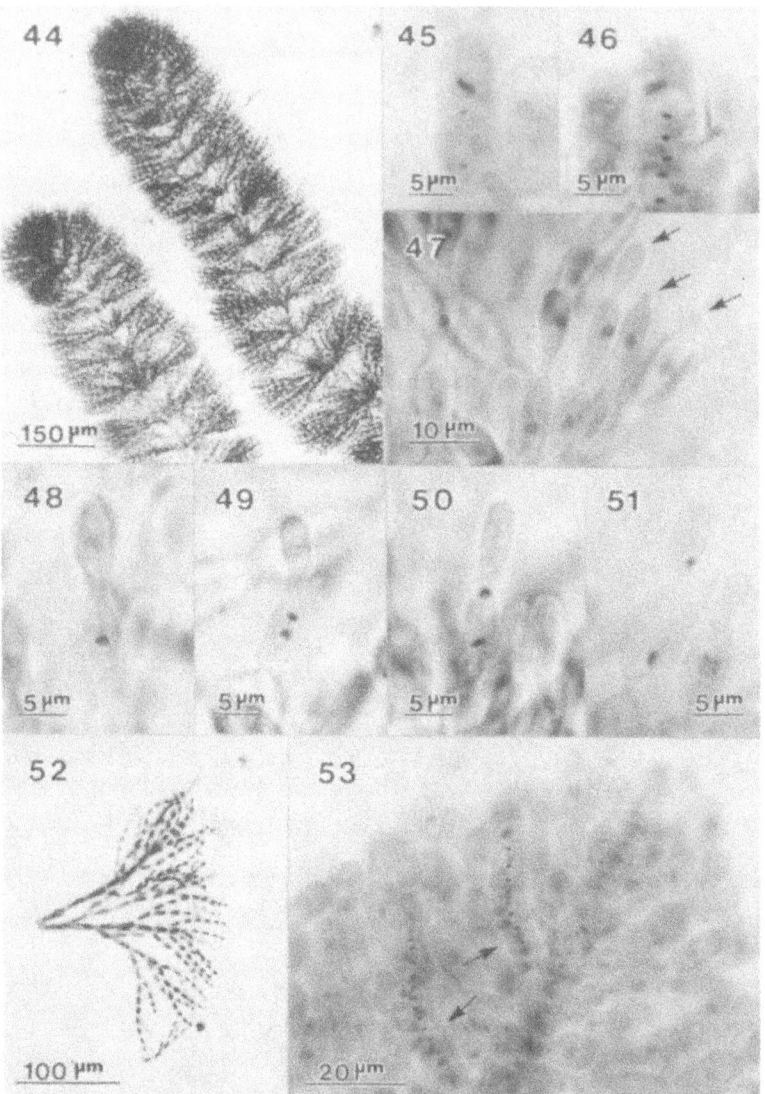

Figs. 12-44 to 12-53. *Batrachospermum boryanum* (North Carolina). *12-44*: Axes with whorled determinate lateral filaments. *12-45*: Tilted metaphase plate in apical cell of indeterminate filament. *12-46*: Early anaphase in apical cell of indeterminate filament. *12-47*: Buds (arrows) forming on apical cells of determinate filaments. *12-48*: Early anaphase in apical cell of a determinate filament. *12-49*: Midanaphase, determinate filament. One set of daughter chromosomes forms a partial ring. *12-50*: Anaphase, determinate filament. Distal set of daughter chromosomes moving into bud through its constricted base. *12-51*: Late anaphase, determinate filament. *12-52*: Mature determinate lateral filament (basal cell not present). *12-53*: Axial filament bearing two abaxial indeterminate lateral filaments (arrows).

adjacent axial filaments over a length of two to four axial segments. The filaments they initiate coalesce to form the axis of a lateral branch. At times the thallus of *L. mucosa* appears to branch at the apex through a splitting of the indeterminate filaments that compose an axis, but this occurs infrequently.

Nuclear divisions in *Liagora mucosa* take place near the base of a dividing cell in both indeter-

Vegetative growth and organization

minate and determinate filaments (Figs. 12-39 to 12-43). The pattern of chromosome migration during anaphase is the same as was described for *Arochaetium secundatum*.

C. Batrachospermales

In the characteristic batrachospermalean life history a macroscopic gametophyte alternates with a diminutive acrochaetioid sporophyte, the pseudochantransia stage (Sheath 1984). The gametophyte of *Batrachospermum* is uniaxial. It consists of branched indeterminate axes, each segment of which bears a whorl of determinate lateral filaments (Fig. 12-44). Indeterminate axes and determinate filaments are distinct in their morphology, cytology, and developmental potentials (Aghajanian & Hommersand 1980).

The apical cell of an indeterminate filament is a domed cylinder that cuts off short, discoid segments basally. In *Batrachospermum sirodotii* apical cells extend by means of tip growth (Aghajanian & Hommersand 1980). In *B. boryanum* nuclear division takes place at or near the site of subsequent septum formation and the metaphase plate is tilted with respect to the cell's axis (Figs. 12-45, 12-46). The basal set of daughter chromosomes appears to migrate to the side of the incipient subapical segment from which the first periaxial cell will be cut off; however, a regular pattern in the arrangement of first periaxial cells on successive axial segments has not been found. With the completion of anaphase migration, daughter nuclei form, a vacuole enlarges between them, and a septum cuts off the new segment.

Periaxial cells are initiated several segments behind the apex as protrusions that extend laterally and curve apically before being cut off by a transverse septum (Aghajanian & Hommersand 1980). The number of periaxial cells per segment varies among species but is most often between four and six (Mori 1975). We have been unable to find a pattern in the sequence of initiation of periaxial cells on an axial segment in *Batrachospermum boryanum*. Each periaxial cell gives rise to a system of branched, determinate filaments. Growth of determinate filaments takes place by budding of apical and subapical cells (Fig. 12-47). Nuclear divisions in determinate filaments of *B. boryanum* resemble divisions in the Acrochaetiales and Nemaliales. Mitosis takes place within the body of the parent cell. The exact location and orientation of division is variable; most often, the metaphase plate is oriented obliquely with respect to the long axis of the cell (Fig. 12-48). As daughter chromosomes separate at anaphase, each set is frequently seen to be arranged in a partial or complete ring (Fig. 12-49). After the two sets of chromosomes are well separated, the more apical set moves into the bud through its constricted base (Figs. 12-50, 12-51). This is contrary to the report that in *B. sirodotii* the nucleus migrates from the parent cell into the bud, where it divides, followed by migration of one of the daughter nuclei back into the parent cell (Aghajanian & Hommersand 1980). Branching of determinate filaments is either trichotomous or dichotomous, with the formation of up to six or more orders of branches in *B. boryanum* (Fig. 12-52).

Periaxial cells initiate descending rhizoidal filaments proximally, from their lower side, after the determinate lateral filament is well developed. Descending rhizoidal filaments of successive segments may overlap, enveloping the axis (Sirodot 1884), and can give rise to secondary determinate lateral filaments (Entwistle & Kraft 1984; Israelson 1942).

Indeterminate lateral branches are produced at irregular intervals along the length of an axis. In *Batrachospermum boryanum* an initial is cut off from the distal end of a periaxial cell and rapidly grows to a length of 15–30 segments (Fig. 12-53). Many indeterminate laterals will cease growth at this stage or at a slightly later stage, when periaxial cells and immature determinate lateral filaments have been produced along the base of the indeterminate branch. Others continue to grow, becoming identical to the parent axis. No more than one indeterminate lateral is produced per axial segment, and many segments have none. Though there is no discernible pattern to the arrangement of indeterminate laterals along the length of an axis in *B. boryanum*, axes tend to have a greater number of more fully developed branches on their abaxial surface (Fig. 12-53).

Wall growth has been investigated in *Batrachospermum sirodotii* (Aghajanian & Hommersand 1980). Apical cells of indeterminate filaments, periaxial cell initials, and buds associated with the growth of determinate lateral filaments all show tip growth. Maturing axial cells elongate through the deposition of new wall material in a basal band, whereas intercalary cells of determinate filaments show bipolar band growth.

Apical development and morphology of *Tuomeya americana* is similar to that of *Batrachospermum* (Fig. 12-54), but the determinate lateral filaments are

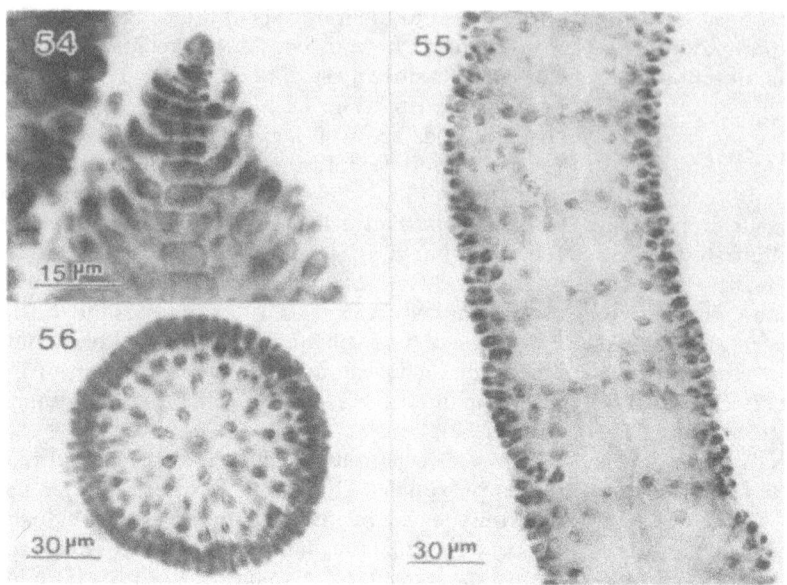

Figs. 12-54 to 12-56. *Tuomeya americana* (North Carolina). *12-54*: Apex. *12-55*: Median longitudinal section. *12-56*: Cross section; axial cell bearing six periaxial cells.

more compactly arranged. *Tuomeya* possesses a confluent cortex at the surface of the thallus that is separated from the axial filament by a space resulting from the elongation of periaxial cells and the loose arrangement of inner cells of the determinate filaments (Figs. 12-55, 12-56).

The gametophyte of *Lemanea australis* is cylindrical and tapers gently toward the apex. Growth is uniaxial. The apical cell of an indeterminate filament cuts off short discoid segments basally (Fig. 12-57). Behind the apex, axial cells expand centrifugally. Each undergoes successive periclinal divisions to cut off four periaxial cells in an ordered sequence, with the second opposite the first and the fourth opposite the third. The first periaxial cells on successive axial segments are arranged in a 60-degree spiral.

Development of the cortex from periaxial cells occurs through a similar process of cell expansion and cleavage, and follows a regular pattern of cell divisions. First and second periaxial cells undergo a slightly different series of divisions than the third and fourth (Atkinson 1890; Sirodot 1872). Cortical development in *Lemanea australis* has been described in detail by Mullahy (1952). First and second periaxial cells cut off two initials of ascending filaments, one of which divides to give a third ascending filament, two initials of descending filaments, and a ray cell. Third and fourth periaxial cells each initiate a single ascending filament and two descending filaments. Initials of ascending and descending filaments and ray cells undergo anticlinal and periclinal divisions (Fig. 12-58) to produce a pseudoparenchymatous cortex. Growth of a segment of the cortex occurs in concert with the elongation of the axial cell that bears it. Regions of the cortex derived from adjacent axial segments abut, forming a confluent surface (Fig. 12-59). As a segment of the axis matures, periaxial cells and inner cortical cells elongate, pushing the cortex out and creating a hollow space between the axial filament and the surface of the thallus (Fig. 12-60). Species in which the gametophyte is branched cut off indeterminate lateral initials from an axial cell near the apex or from the surface of the developing cortex (Atkinson 1890).

The gametophyte of *Lemanea australis* is initiated directly from the pseudochantransia stage following a presumed meiosis (Fig. 12-61). Nuclear divisions in cells of the pseudochantransia stage resemble mitosis in Acrochaetiales (Figs. 12-62 to 12-65). Division takes place within the basal half of the dividing cell, orientation of the metaphase plate is variable but most often oblique, the chromosomes are often arranged in a ring at metaphase and anaphase, and there is a tendency toward precocious initiation of the septum as a septal ring. Nuclear divisions throughout the gametophyte resemble those in indeterminate filaments of *Batrachospermum boryanum* (Figs. 12-66 to 12-68). Division takes place at the site of subsequent septum formation, and the metaphase plate is oriented parallel

Vegetative growth and organization

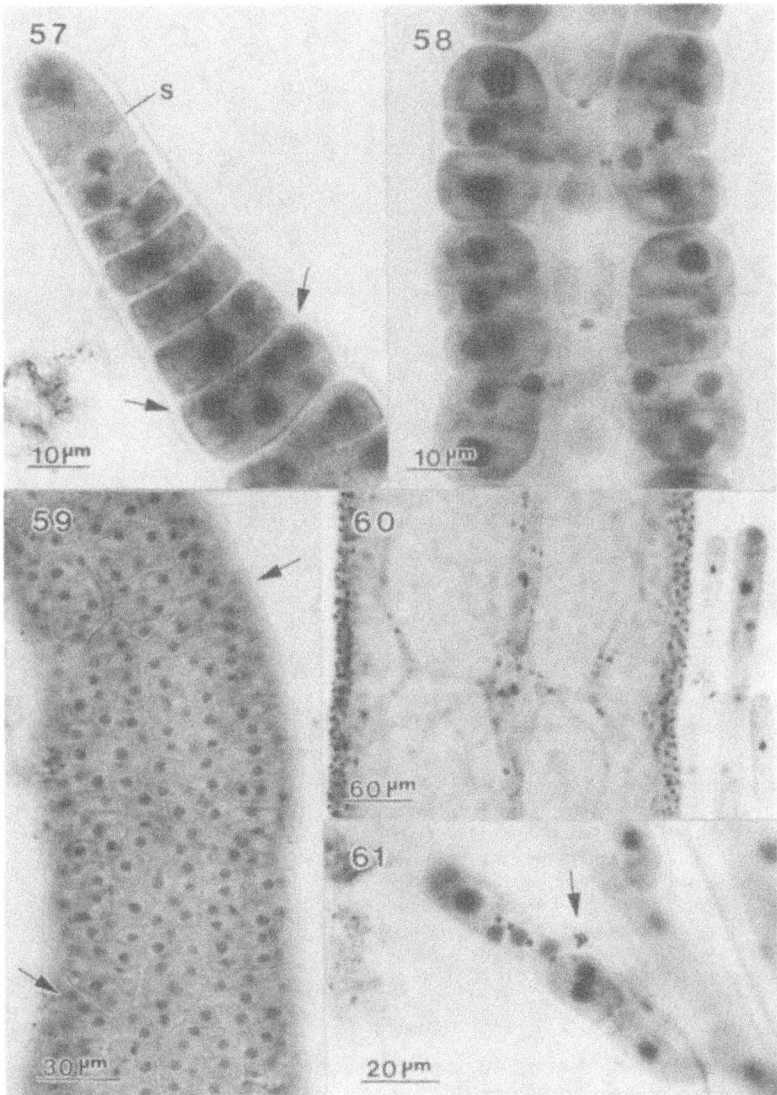

Figs. 12-57 to 12-61. *Lemanea australis* (North Carolina). *12-57*: Apex. Cytokinesis in the apical cell; septum (s) grows centripetally. Behind the apex a periaxial cell has cut off initials of ascending files (arrows). *12-58*: Optical longitudinal section. Ascending and descending files have been initiated. *12-59*: Surface of thallus; ascending and descending files abut (arrows). *12-60*: Optical longitudinal section; periaxial cells and their lower order derivatives have elongated. *12-61*: Gametophyte axis has been initiated on a short lateral branch of pseudochantransia stage following a presumed meiosis, leaving two polar bodies (arrow).

to the ensuing crosswall (Fig. 12-68). This pattern is slightly altered in apical cells of indeterminate filaments, where the orientation of division is slightly oblique but the subsequent septum is normally transverse (Figs. 12-66, 12-67). The tilt of the metaphase plate is such that during anaphase migration the basal set of daughter chromosomes moves to the side of the incipient axial segment from which the first periaxial cell will be cut off. The chromosomes are at times arranged in a ring at metaphase and anaphase (Fig. 12-69), a feature that was noted by Mullahy (1952).

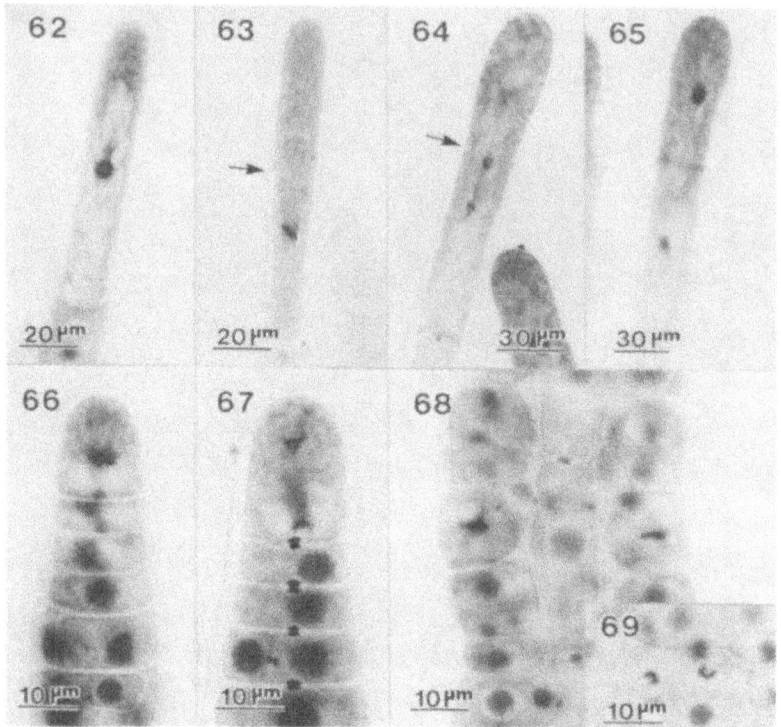

Figs. 12-62 to 12-69. *Lemanea australis* (North Carolina). *12-62*: Pseudochantransia, prophase. *12-63*: Pseudochantransia, tilted metaphase proximal to the precocious septum (arrow). *12-64*: Pseudochantransia, midanaphase. Division is just proximal to the precocious septum (arrow); apical set of daughter chromosomes in polar view forms a ring. *12-65*: Pseudochantransia, cytokinesis. Highly concavo-convex septum cuts into basal daughter cell. *12-66*: Slightly tilted metaphase plate; apex of gametophyte. *12-67*: Late anaphase; apex of gametophyte. *12-68*: Metaphase in initials of descending files. *12-69*: Late anaphase; thallus surface. Daughter chromosomes form partial rings.

The Thoreaceae contains the multiaxial genera *Thorea* and *Nemalionopsis*. In *Thorea ramosissima*, growth of indeterminate filaments is reported to be cellulosympodial (Swale 1962).

IV. HIGHER FLORIDEOPHYCIDAE: GELIDIALES, BONNEMAISONIALES, GIGARTINALES, RHODYMENIALES, AND CERAMIALES

A. Mitosis and cytokinesis

A common mode of mitosis and cytokinesis occurs in representatives of the Gelidiales, Bonnemaisoniales, Gigartinales, Rhodymeniales, and Ceramiales. Nuclear division takes place at the future site of septum formation, and the metaphase plate is oriented parallel to the ensuing crosswall (Figs. 12-70 to 12-72, 12-74, 12-76, 12-78, 12-81). L'Hardy-Halos (1971) has commented on this behavior in the Ceramiaceae. This pattern of development occurs in both determinate and indeterminate filaments. At metaphase and anaphase the chromosomes are normally arranged in a linear band. During anaphase the two sets of daughter chromosomes migrate perpendicular to the plane of the preceding metaphase plate, moving to opposite ends of the dividing cell (Figs. 12-71, 12-73, 12-75, 12-77, 12-79, 12-82). A vacuole appears late in anaphase that separates daughter nuclei at telophase and maintains separation during cytokinesis. A septum is initiated after the completion of nuclear division that grows by annular furrowing.

The orientation of nuclear division is oblique

Vegetative growth and organization

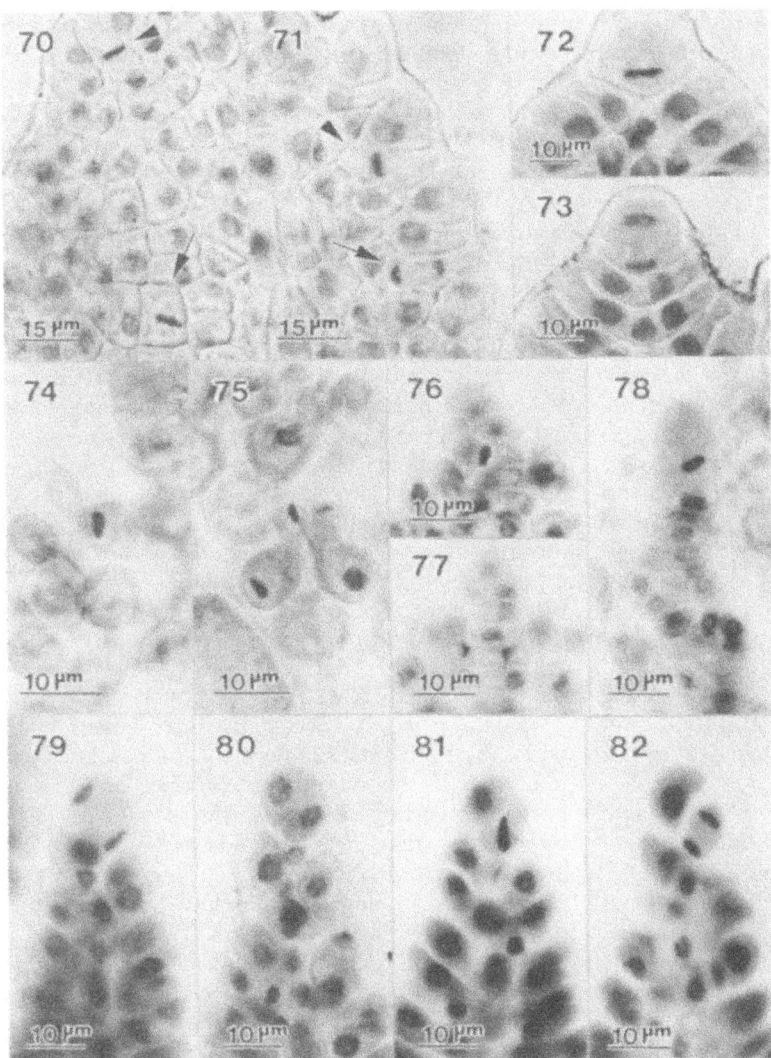

Figs. 12-70 to 12-73. *Membranoptera alata* (Nova Scotia). *12-70*: Metaphase figures in apical (arrowhead) and subapical (arrow) cells of determinate filaments. Subapical cell forming an abaxial derivative. *12-71*: Metaphase (arrowhead) and anaphase (arrow) in subapical cells of determinate filaments. *12-72, 12-73*: Metaphase and anaphase in apical cell of indeterminate axial filament.

Figs. 12-74 to 12-78. *Crouania attenuata* (Florida). *12-74*: Metaphase, determinate filament. *12-75*: Anaphase, determinate filament. *12-76*: Metaphase, periaxial cell formation. *12-77*: Anaphase, periaxial cell formation. *12-78*: Tilted metaphase plate, apical cell of indeterminate filament.

Figs. 12-79 to 12-82. *Rhabdonia coccinea* (Tasmania). *12-79*: Late anaphase; apical cell of indeterminate filament. *12-80*: Recently divided apical cell of indeterminate filament. *12-81*: Metaphase; subapical cell of indeterminate filament initiating a single periaxial cell. *12-82*: Anaphase; subapical cell of indeterminate filament.

in apical cells of indeterminate filaments of many genera (Fig. 12-78). When this occurs, the tilt of the metaphase plate is such that at anaphase the basal set of daughter chromosomes migrates to the side of the incipient subapical segment from which the first periaxial cell will be cut off (Figs. 12-79 to 12-82). The septum that forms following nuclear division is tilted correspondingly. The wedge-shaped segmental cell that results cuts off the first periaxial cell from its high or long side.

B. Development of prostrate systems

Guiry (Chapter 14) recognizes five types of spore germination, most of which occur in several orders of red algae. Spore germination is bipolar, leading to the early differentiation of an erect thallus and basal rhizoidal system in Ceramiales (Dixon 1973; Fritsch 1945) and a few species of Bonnemaisoniales (Chihara 1961). By contrast, most red algae produce an initial prostrate basal system from which one or more erect thalli may later develop. In many genera the prostrate system forms an extensive crust of varying thickness before initiating one or more upright axes; in others either the tetrasporophyte or both the gametophyte and tetrasporophyte are entirely crustose (West & Hommersand 1981).

Crustose prostrate systems arise in one of two ways: Either the attached spore divides perpendicular to the substrate one or more times and the resultant cells form lateral protuberances that are cut off as apical initials of prostrate filaments (*Naccaria*-type) or, following one or more divisions perpendicular to the substrate, horizontal divisions take place to produce a hemispherical mass of cells within the original cell wall (*Dumontia*-type). In the first instance the crust is initially monostromatic; in the second it is polystromatic from its inception (Chapter 14). The *Naccaria*-type germination pattern predominates in Bonnemaisoniales and in many of the more primitive members of the Gigartinales; the *Dumontia*-type is more common in advanced members of the Gigartinales and in Rhodymeniales.

The prostrate phase of *Bonnemaisonia asparagoides* is pinnately branched when growing attached to the substrate but is sparsely and irregularly branched, and bears numerous gland cells when growing free (Rueness & Åsen 1982). Uniaxial prostrate filaments bearing paired opposite laterals are reported in *Delisea* (Chihara 1961), *Hummbrella hydra* (Hawkes 1983), *Calosiphonia vermicularis* (Mayhoub 1973, 1975), and *Schmitzia hiscockiana* (Maggs & Guiry 1985). *Dudresnaya japonica* (Umezaki 1968) has a filamentous prostrate system that is predominantly unilaterlly branched. In *Acrosymphyton purpuriferum* primary filaments of the prostrate system cut off pairs of lenticular cells from their apical pole that develop into lateral filaments. Later, additional lateral filaments may be formed from initials cut off at the antapical pole (Cortel-Breeman 1975). Very often an early filamentous stage develops into a fan-shaped polystromatic crust that has a monostromatic marginal meristem of creeping filaments. This is seen in *Acrosymphyton purpuriferum* (Cortel-Breeman 1975), *Gloiosiphonia verticillaris* (DeCew et al. 1981), *Farlowia mollis* (DeCew & West 1981), *Pikea californica* and *Schimmelmannia plumosa* (Chihara 1972). A pinnately branched, initially monostromatic crust has also been described for *Meredithia microphylla* (Guiry & Maggs 1985), *Cirrulicarpus carolinensis* (Hansen 1977), and several species of *Kallymenia* (Codomier 1972a, 1973a). The crusts of these three genera of Kallymeniaceae are unusual in that cells of neighboring filaments frequently fuse. Spores of *Sebdenia dichotoma* produce irregularly branched, monostromatic crusts and apical cells of filaments derived from different spores may fuse directly (Codomier 1973b).

Crusts of most species become polystromatic when the horizontal filaments of the basal system, referred to as the hypothallus, produce closely compacted ascending filaments, the perithallus. The hypothallus may bear rhizoids or even produce an inferior perithallus of descending filaments. A range of morphological types known from nature have been described by Denizot (1968).

Rietema and Klein (1981) have described the production of discoid microthalli that become polystromatic within a common sheath in *Dumontia contorta*. Radial growth of these disks takes place by a marginal meristem. Chen and Taylor (1976) followed the development of the *Dumontia*-type germling in *Chondrus crispus* with scanning electron microscopy. They observed that an extracellular sheath is produced prior to internal cell differentiation and suggest that this sheath is responsible for the discoid pattern of sporeling development.

C. Initiation of upright axes

The initiation of erect thalli from prostrate filaments or crusts has been reviewed by Gabrielson and Garbary (1986). In *Bonnemaisonia asparagoides*, which has a freely branched, filamentous prostrate system, erect axes originate as lateral initials that quickly become budlike and assume the form of the adult axis (Rueness & Åsen 1982). In *Gloiosiphonia* erect axes are initiated from the crust by rapid elongation and transverse divisions of apical cells of individual filaments. Subapical cells soon cut off periaxial cells that initiate cortical filaments (DeCew et al. 1981; Mohoroshi & Masuda 1980). Uniaxial uprights develop in a similar fashion in *Farlowia* (DeCew & West 1981). In *Hummbrella hydra* the filamentous prostrate system gives rise to enlarged, thin-walled cells before initiating uniaxial uprights (Hawkes 1983).

Kuckuck (1912) illustrated the initiation of the multiaxial upright thallus in *Platoma bairdii*. A group of perithallial cells simultaneously produce apical initials that develop in concert to generate a bundle of erect axes bearing cortical filaments. A similar pattern of development may occur in *Neurocaulon* and other Furcellariaceae (Codomier 1972b). The *Platoma*-type of multiaxial development has also been documented for *Dumontia contorta* (Rietema 1984). Here, the erect filaments can diverge to produce uniaxial side branches (Rietma 1984; Wilce & Davis 1984). *Schizymenia* is interesting in that, although cells of the crust interconnect by means of secondary pit connections, the erect filaments are free and separate (Ardré 1980).

Secondary pit connections are produced regularly in both prostrate and erect systems in more advanced families. The erect thallus often begins as a mound of tissue on the surface of the crust. In uniaxial species an apical initial is soon cut out. A tetrahedral initial with three cutting faces has been identified in *Gracilaria verrucosa* that produces a primary axis in a seemingly multiaxial thallus (Kling & Bodard 1986). The initiation of multiaxial erect thalli from crusts has been investigated in culture in many species, including *Sebdenia dichotoma* (Codomier 1973b), *Meredithia microphylla* (Guiry & Maggs 1985), *Gymnogongrus crenulatus* and *G. devoniensis* (Ardré 1978), and *Mastocarpus* sp. (Masuda et al. 1987). In these particular examples the erect thallus begins as a bulge or mound of tissue that is scarcely differentiated from the prostrate system.

D. The erect thallus: simple uniaxial types

Several genera belonging to the more advanced orders of Florideophycidae have a simple, uniaxial construction superficially similar to that of *Batrachospermum*, in which there is a high level of differentiation between indeterminate axes and the whorled determinate lateral filaments. Examples include *Atractophora* (Bonnemaisoniales), *Schmitzia*, *Calosiphonia*, *Acrosymphyton*, *Dudresnaya* and *Thuretella* (Gigartinales), and *Crouania* and *Gulsonia* (Ceramiales) (Kylin 1956). In *Dudresnaya* cortical filaments are initiated two ways: some arise from the axis a few cells behind the apex, whereas others are produced secondarily from preexisting determinate filaments or from rhizoids (Kraft & Robins 1985).

Although *Crouania attenuata* is similar to *Batrachospermum* in its morphology (Fig. 12-83), mitosis and cytokinesis take place at the site of future septum formation in both indeterminate axes and determinate lateral filaments. In addition, the initiation of determinate lateral filaments and lateral indeterminate axes follows a regular pattern. Apical cells of indeterminate filaments divide transversely to cut off axial segments, each of which initiates three periaxial cells. The position of the first periaxial cells in successive segments describes a 40-degree spiral (Fig. 12-84). The second and third periaxial cells are cut off 120 degrees to the right and left of the first. During division the mitotic apparatus of an apical cell is tilted such that the basal set of daughter chromosomes will migrate to the side of the incipient subapical cell from which the first periaxial cell will be cut off (Figs. 12-85 to 12-88).

Nuclear divisions associated with periaxial cell formation take place at the juncture of the bud and the axial cell. The metaphase plate is oriented vertically, parallel to the lateral wall of the axial cell (Fig. 12-76). During anaphase one set of chromosomes moves out into the apical portion of the bud, the other back into the axial segment (Fig. 12-77), followed by formation of daughter nuclei and a septum.

During growth of determinate lateral filaments buds are formed as bulbous outgrowths from the distal end or shoulder of a cell. Division takes place just within the parent cell or at the juncture of the bud and the parent cell, and the metaphase plate is oriented perpendicular to the long axis of the initial (Fig. 12-74). During anaphase daughter chromosomes move perpendicular to the plane of the preceding metaphase plate (Fig. 12-75). When migration has ceased, daughter nuclei form and a septum develops by annular ingrowth.

Branching of the thallus is largely abaxial and follows a regular pattern. Most indeterminate axes bear an indeterminate lateral abaxially eight to fifteen segments from their base. A second indeterminate lateral is initiated two to six (most often four or five) segments beyond the first, also on the abaxial surface (Fig. 12-83). Indeterminate laterals tend to overgrow the parent axis (Fig. 12-83), as in *Crouania mucosa* (Wollaston 1968). This type of thallus development has been termed ramisympodial

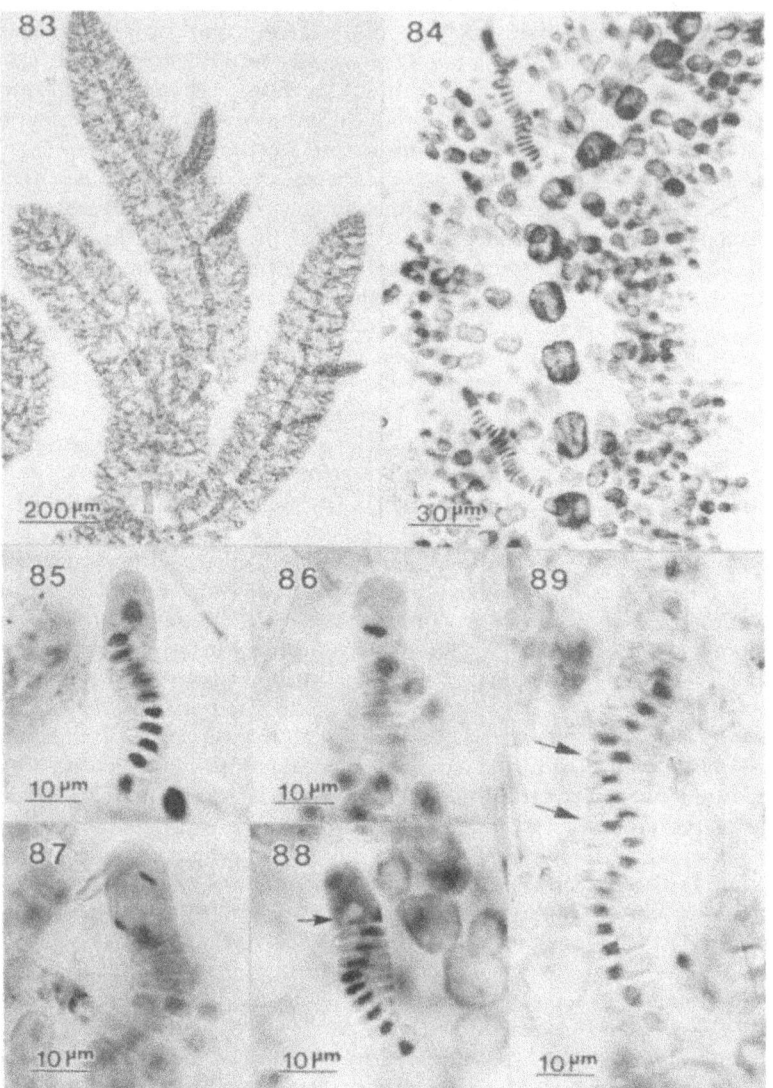

Figs. 12-83 to 12-89. *Crouania attenuata* (Florida). *12-83*: Ramisympodial branching. *12-84*: Two indeterminate lateral filaments with spirally arranged first periaxial cells. *12-85*: Spiral arrangement of nuclei in cells of a young indeterminate lateral filament. *12-86*: Slightly tilted metaphase plate; apical cell of indeterminate filament. *12-87*: Late anaphase; apical cell of indeterminate filament. *12-88*: Telophase; apical cell of indeterminate filament. Vacuole (arrow) separates daughter nuclei; basal daughter nucleus in regular 40-degree spiral. *12-89*: Indeterminate filament. Nuclei of axial segments that will initiate indeterminate lateral filaments (arrows) from their abaxial surface fall out of the regular 40-degree spiral.

branching (Norris et al. 1984). The initial of an indeterminate lateral is cut off prior to the three periaxial cells that give rise to determinate filaments and slightly distal to them. The regular 40-degree rotation of the mitotic apparatus is interrupted when a segment that will bear an indeterminate lateral is cut off. Again, the mitotic apparatus is oriented such that the basal set of daughter chromosomes comes to lie at the site from which the first derivative will be produced, normally the abaxial surface of the incipient axial cell (Fig. 12-89). It is evident from these observations that the fate of the

Vegetative growth and organization

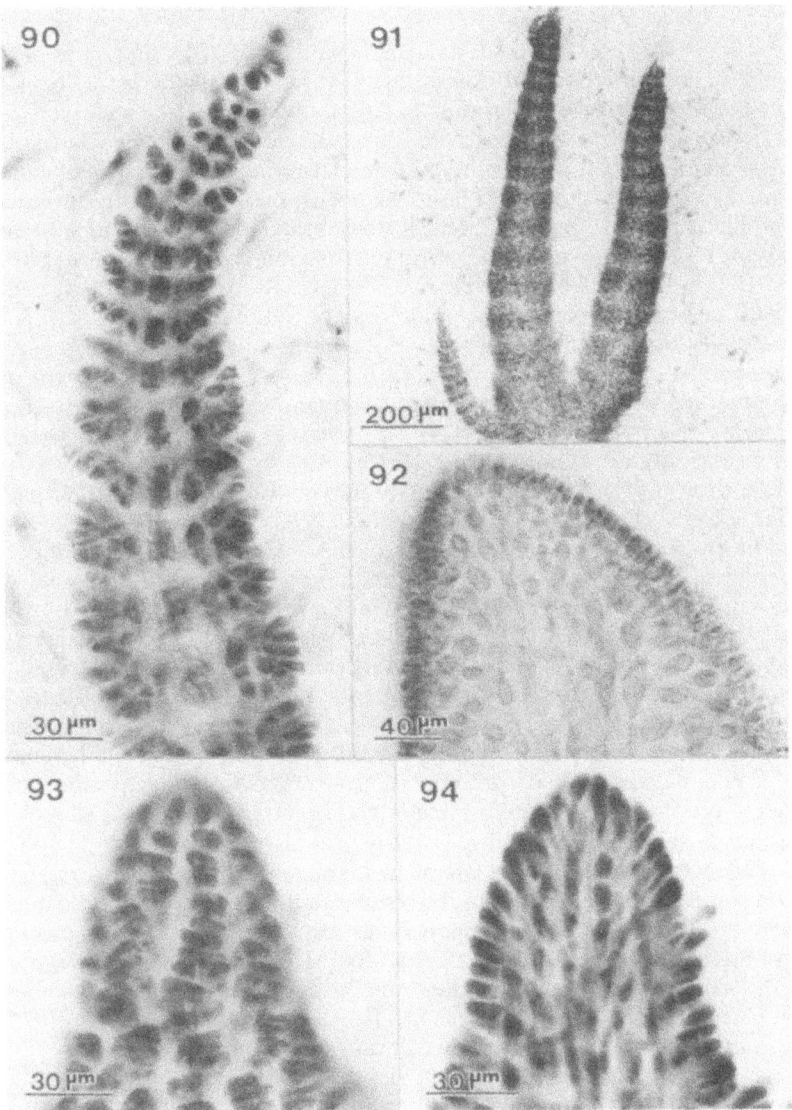

Figs. 12-90, 12-91. *Gloiosiphonia capillaris* (Massachusetts, Newfoundland). *12-90*: Optical longitudinal section. Determinate filaments form ascending, abaxially branched files (lateral derivatives out of focal plane). *12-91*: Surface of thallus; each band is formed of the determinate filaments derived from a single axial cell.
Figs. 12-92 to 12-94. *Agardhiella subulata* (North Carolina, New Jersey). *12-92*: Longitudinal section. Determinate filaments form ascending, abaxially branched files. *12-93*: Cortex formed of vertical sectors, each sector composed of the determinate lateral filaments of a single axial filament. *12-94*: Optical longitudinal section of Fig. 12-93, showing indeterminate filaments.

first derivative of an axial cell is determined at the time the apical cell divides to cut off that axial cell. In many taxa of the Crouanieae, including some species of *Crouania*, indeterminate laterals either replace one of the whorled determinate laterals on a segment or are borne on the basal or suprabasal cell of a determinate lateral filament (Hommersand 1963; Wollaston 1968).

E. Organization of determinate lateral filaments

In many higher Florideophycidae branching of determinate lateral filaments is organized around a leading filament, and the filaments coalesce to form a well-defined cortex. Chadefaud (1952) referred to this tendency as "axialisation des pleuridies," "pleuridie" being the term given to the whorled determinate lateral filaments of algae like *Batrachospermum*. *Gloiosiphonia capillaris* exemplifies this type of thallus organization (Figs. 12-90, 12-91). Each axial cell initiates four periaxial cells; each periaxial cell undergoes repeated transverse divisions to produce a file of cells that ascends at a 30- to 50-degree angle. When three to four cells long, a determinate filament initiates laterals from its basal segments. Each segment of a primary determinate lateral filament gives rise to four derivatives: two laterally, one abaxially, and one basally. The basal derivative initiates a descending rhizoidal filament. The same pattern of branching is repeated by cells of higher order filaments, though initiation of a basal derivative ceases four to six cells from the axis. A mature determinate lateral filament normally contains eight to twelve orders of branches.

As a determinate filament matures, the inner two to four cells elongate, becoming columnar, the middle one to two enlarge to an intermediate size, and the outer four to six remain small and ovoid. The space created by the extension of the basal cells of the determinate lateral filaments becomes filled by a mass of thick-walled descending rhizoidal filaments (Fritsch 1945).

The basic pattern of development of determinate lateral filaments exhibited by *Gloiosiphonia capillaris* occurs in uniaxial and multiaxial genera throughout the advanced Florideophycidae. It is common in the Gelidiales, Gigartinales, and Rhodymeniales (see illustrations in Kylin 1956). In the Ceramiales it is found with modification in the Delesseriaceae. Here, the two lateral periaxial cells initiate determinate filaments that branch abaxially to form a monostromatic blade (Figs. 12-70, 12-71).

In multiaxial plants each segment of an indeterminate axial filament cuts off a limited number of periaxial cells (usually one or two) toward the thallus surface. Each initiates a determinate lateral filament that develops in the manner described (Fig. 12-92). In *Agardhiella subulata* only one periaxial cell is produced by each axial cell (Gabrielson & Hommersand 1982a) and the determinate lateral filaments derived from successive cells of a given indeterminate filament comprise a vertical sector of the cortex (Figs. 12-93, 12-94).

F. Facultative branching of the thallus

Many Rhodophyta initiate lateral indeterminate axes through a transformation of the apical cell of a determinate lateral filament. *Callithamnion byssoides* exhibits this behavior (Fig. 12-95). Divisions in apical cells of indeterminate filaments are oblique (Figs. 12-96 to 12-98). In material attributed to this species from North Carolina the tilt of the metaphase plate and ensuing crosswall shifts 180 degrees with each division, the high side of the resultant segments lying alternately to the left and right. There is no indication of spiral branching as in European *C. byssoides* (Dixon & Price 1981). Each segment of an indeterminate axis cuts off a single derivative from its high side following formation of a lateral bud (Fig. 12-99).

All laterals are initially determinate in their pattern of growth. Apical cells extend, then divide by means of a transverse septum. Some continue to grow in this manner, producing a file of transversely septate, unbranched segments (Fig. 12-95). In others the apical cell begins to divide obliquely in an alternate-distichous fashion after a variable number of segments have been formed. Lateral initials are cut off from the high side of each segment two to five cells behind the apex. Filaments that continue to grow in this manner produce an indeterminate lateral branch. A variety of irregular branching patterns are also encountered. In some cases a lateral filament will shift from an irregular pattern to the regular pattern characteristic of either indeterminate axes or determinate, unbranched filaments (Fig. 12-100). The lack of a clear distinction between determinate and indeterminate filaments in *Callithamnion* has been noted by Dixon (1973) and Hommersand (1963).

Konrad-Hawkins (1964a,b, 1968) has described a similar situation in *Callithamnion roseum*. During growth of some lateral filaments there is a shift from transversely septate, unbranched segments to obliquely septate segments that branch from their high side. Associated with this shift in the orientation of nuclear and cell division is a change in the length of the apical cell at the time of division, a change in the relative lengths of the daughter cells, and a decrease in the rate of elongation of the filament. This transformation appears to be involved in the regulation of thallus form (Konrad-Hawkins 1964b). A similar mechanism seems to operate in *C. byssoides*. Those lateral filaments whose rate of growth is greatest will tend to extend past their neighbors. It is these filaments that are transformed into indeterminate axes, slowing their growth rate

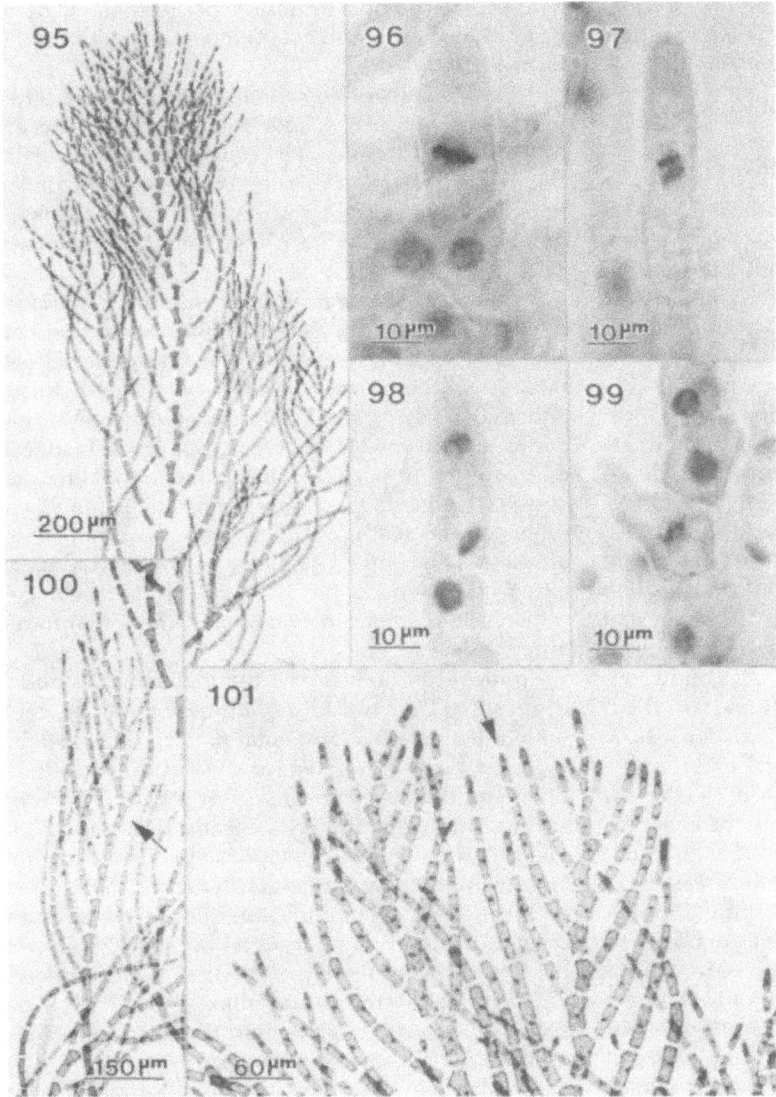

Figs. 12-95 to 12-101. *Callithamnion byssoides* (North Carolina). *12-95*: Alternate-distichous branching of main axis, laterals showing variation in branching patterns. *12-96*: Metaphase, apical cell of indeterminate filament. *12-97*: Early anaphase, apical cell of indeterminate filament. *12-98*: Telophase, apical cell of indeterminate filament. *12-99*: Metaphase, formation of lateral initial. *12-100*: Filament that has shifted (arrow) from unilateral to alternate-distichous branching. *12-101*: Apical cells of lateral derivatives of two indeterminate filaments lie in a line; one (arrow) has shifted to oblique divisions.

and preserving a monopodial habit (Fig. 12-101).

Facultative branching has been described in two species of *Gelidium* (Dixon 1958) and appears to be the mechanism of branch initiation in many thalloid taxa, but individual cases are not well documented. In *Agardhiella subulata* indeterminate branches are initiated at the thallus surface, and inner cortical cells directly below a branch primordium elongate, taking on the appearance of medullary cells (Gabrielson & Hommersand 1982a). Many taxa that produce indeterminate lateral branches in a regular pattern at the thallus apex also have the ability to initiate indeterminate branches facultatively, a process often referred to as adventitious branching (Dixon 1960; Hommersand 1963).

G. Patterned branching of the thallus

In the complex thalli of some uniaxial genera there is a regular pattern to the initiation of lateral branches. Branches may develop similiarly, as in *Microcladia* (Hommersand 1963), or there may be differentiation between axes of various orders (Cramer 1864; Dixon 1973). In the Dasyaceae the regular arrangement of laterals is a result of cellulosympodial development (L'Hardy-Halos 1968, 1971). Some advanced Florideophycidae initiate axes of limited growth, here referred to as determinate branches, in addition to determinate lateral filaments and indeterminate lateral branches. Determinate branches are produced in a prescribed arrangement on indeterminate axes and exhibit a regular pattern of development distinct from that of the indeterminate axes that bear them. This is commonly seen in the Bonnemaisoniales and Ceramiales and also occurs in certain uniaxial Gigartinales (Chihara & Yoshizaki 1972; Dixon 1973; Hommersand 1963; Kylin 1956; Scagel 1953).

In *Bonnemaisonia hamifera* the apical cell of an indeterminate filament cuts off segment cells basally by means of oblique crosswalls (Fig. 12-102). The plane of division and high side of the resultant segment cells describe a three-eighths spiral (Chihara 1961; Kylin 1928). Each axial segment cuts off two periaxial cells: the first from the high side shortly after formation of the segment and prior to the next division of the apical cell (Fig. 12-103), the second much later, opposite the first. The first periaxial cell initiates a determinate branch or short shoot. The apical cell of a determinate branch undergoes slightly oblique divisions (Fig. 12-104). The tilt is such that the abaxial side of the newly formed subapical segment is longer than its adaxial side. Each segment of a determinate lateral branch cuts off three periaxial cells, the first abaxial (Figs. 12-105, 12-106), the second and third adaxial. Each undergoes a series of divisions to form cortex.

The fate of the second periaxial cell of an indeterminate axial segment is variable (Chihara 1961). In female plants it may produce a fertile branch (Fig. 12-102) that bears a carpogonial branch on its fifth or sixth segment, arresting development. Alternatively, it can produce an indeterminate lateral branch that may remain dwarfed or may develop into a major branch of the thallus. When neither a fertile branch nor an indeterminate lateral is produced, the second periaxial cell divides to form cortex.

Development of the thallus is similar in other Bonnemaisoniaceae. There is variation in the arrangement of first periaxial cells on successive axial segments, and thus in the arrangement of determinate branches, and in the mode of initiation of indeterminate lateral branches (Chihara & Yoshizaki 1972).

A similar situation is encountered in *Phacelocarpus* (Gigartinales) (Searles 1968). Each axial cell cuts off four periaxial cells. The first periaxial cell initiates a determinate branch, the second, third, and fourth give rise to determinate filaments that corticate the axis. In most species there is a 180-degree rotation between first periaxial cells in successive segments, and the determinate branches, often termed teeth, are distichous. In others they are spirally arranged. When the thallus branches either the second, third, or fourth periaxial cell initiates an indeterminate lateral filament that forms a new axis.

In most Rhodomelaceae the thallus is composed of axes that bear two or more types of laterals, each with a distinct developmental pattern (Fig. 12-107). In *Polysiphonia harveyi* apical cells of indeterminate filaments divide by means of an oblique septum, cutting off segmental cells basally (Figs. 12-108 to 12-110). The tilt of the metaphase plate and ensuing crosswall describes a one-quarter spiral. When two to three cells behind the apex, an axial segment cuts off a lateral initial from its high side, followed by four pericentral cells. The first pericentral cell is cut off below the lateral initial (Fig. 12-111), the second to the right of the first, the third to the left, and the fourth opposite the first.

The lateral initial gives rise to either an indeterminate lateral branch or a trichoblast. Development of indeterminate lateral branches is identical to the parent axis after the initial three to five segments. Trichoblasts are uniseriate and deciduous. They develop monopodially but appear dichotomously branched at maturity (Fritsch 1945, p. 545). In some species of *Polysiphonia* indeterminate laterals are borne on the basal cell of a trichoblast. Most Rhodomelaceae possess trichoblasts, though they are absent in some genera (Hommersand 1963). In some Rhodomelaceae determinate branches analogous to the short shoots of Bonnemaisoniales are produced in addition to trichoblasts, indeterminate laterals, and pericentral cells (Falkenberg 1901; Kylin 1956).

In each of the examples discussed above, the

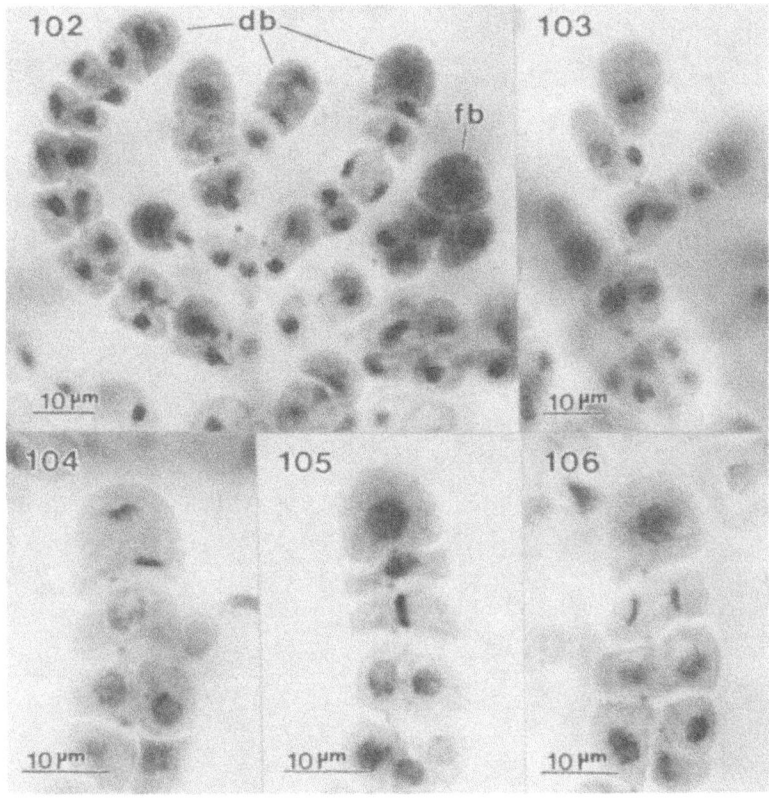

Figs. 12-102 to 12-106. *Bonnemaisonia hamifera* (Maine). *12-102*: Apex of indeterminate axis bearing determinate branches (*db*) and a fertile female branch (*fb*). *12-103*: Metaphase, apical cell of indeterminate filament. *12-104*: Anaphase, apical cell of determinate branch. *12-105*: Metaphase, axial cell of determinate branch. Initiation of abaxial derivative. *12-106*: Anaphase, axial cell of determinate branch. Initiation of abaxial derivative.

site of initiation of determinate filaments, determinate branches, and sometimes indeterminate branches is controlled by a precise pattern of cell divisions at the apex of an indeterminate filament. Development of each type of filament and branch is in turn controlled at its apex by the pattern of division. Maturation is largely a result of precise patterns of cell division and cell enlargement behind the apex.

H. Secondary pit connections, wall growth, and increasing nuclear volume

Secondary pit connections are formed as a result of the fusion of conjunctor cells with adjacent vegetative cells in most higher Florideophycidae. Vegetative fusions and secondary pit connections are absent in genera having the simplest thallus construction, found in the Naccariaceae, Dumontiaceae (Robins & Kraft 1985), Nemastomataceae (Kraft 1975; Kraft & John 1976), Calosiphoniaceae (Feldmann 1954), and Ceramiaceae (Fritsch 1945).

In uniaxial genera and the lateral margins of multiaxial genera secondary pit connections are most often formed basipetally and laterally between intercalary cells of adjacent determinate lateral filaments (Figs. 12-112, 12-113). In multiaxial genera adjacent indeterminate axial filaments become linked, either directly by means of secondary pit connections or through interconnecting filaments (Chiang 1970; Gabrielson & Hommersand 1982a,b; Kraft 1977; Lee 1978). The axial cells of *Agardhiella* contain large numbers of nuclei (Fig. 12-114), most likely as a result of fusion with multinucleate cells analogous to conjunctor cells. Although it has been thought that secondary pit connections are absent

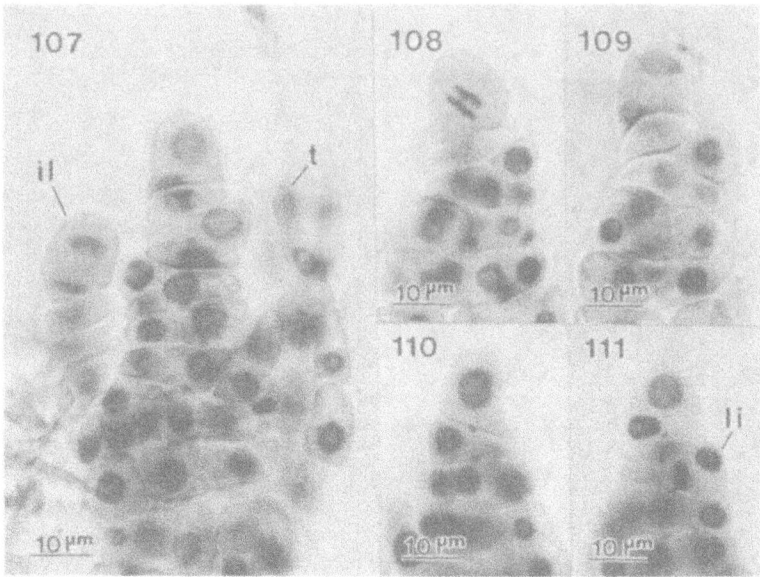

Figs. 12-107 to 12-111. *Polysiphonia harveyi* (North Carolina). *12-107*: Apex bearing indeterminate lateral branch (*il*; apical cell in telophase) and trichoblast (*t*). *12-108*: Anaphase, apical cell of indeterminate filament. *12-109*: Telophase, apical cell of indeterminate filament. *12-110*: Oblique septum, apex of indeterminate filament. *12-111*: Axial segment has cut off a lateral initial (*li*), in metaphase of division that will produce the first periaxial cell.

in Bonnemaisoniales (Fritsch 1945), in *Bonnemaisonia hamifera* they are formed between the cortical cells of both indeterminate axes and determinate branches (Fig. 12-115; Coomans 1986). The benefits of vegetative fusions resulting in secondary pit connections include increased structural integrity and new avenues of vegetative modification.

In many higher Florideophycidae cells of both axial and determinate lateral filaments increase greatly in size during maturation. Cell enlargement continues well behind the apex, and cells can reach many times their original volume (Dixon 1971, 1973). Wall growth associated with the elongation of intercalary cells has been investigated in several genera of Ceramiaceae (Chapter 11) and was found to be localized in unipolar or bipolar bands.

As Magne (1964) has noted, Florideophycidae tend to maintain a fairly constant ratio of nuclear volume to cell volume. Increasing the number of nuclei in a cell as it enlarges through the formation of secondary pit connections is one strategy employed to accomplish this objective. In other taxa nuclear divisions take place within intercalary cells without ensuing cell divisions, leading to the production of multinucleate cells (Goff & Coleman 1986; L'Hardy-Halos 1971). An increase in cell volume is also achieved by increasing the size of a single nucleus. This is seen in axial cells of *Bonnemaisonia hamifera* (Fig. 12-116) and in numerous other taxa that lack secondary pit connections or in which the formation of secondary pit connections is restricted to the cortex. Single nuclei that enlarge in conjunction with cell enlargement are common in Ceramiaceae (L'Hardy-Halos 1971). Axial cells of *Portieria hornemannii* (formerly *Chondrococcus hornemannii*; see Silva et al. 1987) combine these two behaviors: subapical cells become multinucleate through nuclear divisions in the absence of cell division, and the nuclei produced enlarge in conjunction with growth of the axial cell (Fig. 12-117).

That there is an actual increase in the amount of DNA associated with the observed increase in nuclear volume has been demonstrated through quantitative microspectrofluorometry (Goff & Coleman 1984, 1986; Chapter 3). Increase in nuclear DNA content in association with an increase in nuclear volume may prove to be the norm in those algae in which there is extensive cell enlargement in the absence of internal nuclear divisions or the formation of secondary pit connections.

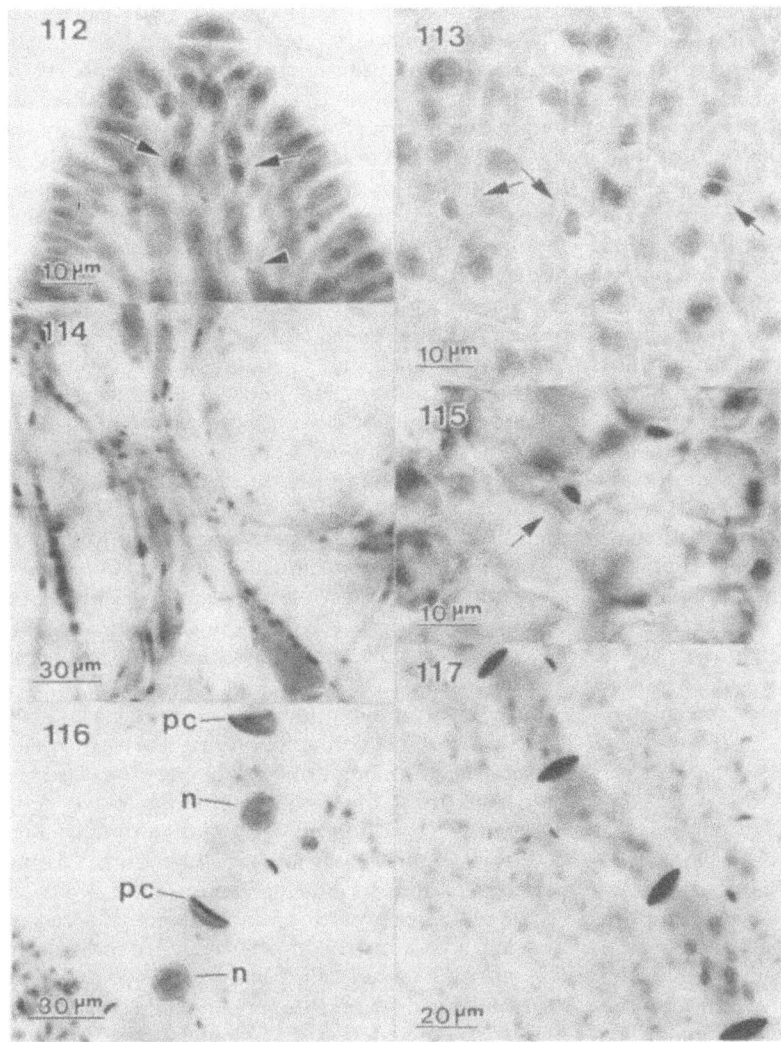

Fig. 12-112. *Gelidium crinale* (North Carolina). Conjunctor cell (arrows) cut off from periaxial cell will fuse with subtending periaxial cell, forming a secondary pit connection (arrowhead).
Fig. 12-113. *Membranoptera alata* (Nova Scotia). Conjunctor cell (arrows) fusing with subtending cell, depositing nucleus and forming secondary pit connection.
Fig. 12-114. *Agardhiella subulata* (North Carolina). Multinucleate axial cells.
Figs. 12-115, 12-116. *Bonnemaisonia hamifera* (Maine). *12-115*: Conjunctor cell fusing with adjacent cortical cell (arrow). *12-116*: Axial cells with enlarged nuclei (n) and pit connections (pc).
Fig. 12-117. *Portieria hornemannii* (South Africa). Axial cells, each with many enlarged nuclei.

V. SUMMARY

We recognize three major groups in the Rhodophyta based on features of vegetative morphology and development: the Bangiophycidae; the lower Florideophycidae, consisting of the Acrochaetiales, Nemaliales, and Batrachospermales; and the higher Florideophycidae, containing the Gelidiales, Bonnemaisoniales, Gigartinales, Rhodymeniales, and Ceramiales.

The Bangiophycidae form a closely knit group based on features of mitosis and vegetative development. Filamentous growth characterizes the earliest stages of thallus development in most Bangiophycidae. Growth takes place through strictly apical or both apical and intercalary cell devisions. In many taxa diffuse growth, with the formation of true parenchyma, characterizes the mature thallus. The thallus is filamentous and growth is apical in *Rhodochaete* (Rhodochaetales) and probably in the conchocelis phase of *Bangia* and *Porphyra* (Bangiales). These also possess simple pit plugs, and the latter have cellulose as the principal fibrillar wall component. It may be that the filament is the ancestral unit of thallus construction in extant Bangiophycidae, and a unicellular or colonial habit, parenchyma, and even intercalary growth represent derived conditions associated with changes in cell wall composition and the absence of pit connections.

Patterns of mitosis and cytokinesis are similar in all orders of Bangiophycidae. Distinguishing features include an extended nucleus at prophase, the aggregation of the chromosomes into a ring during metaphase and anaphase, a greatly extended interzonal spindle during late anaphase, the involvement of the chloroplast and pyrenoid in maintaining separation of daughter nuclei during telophase and cytokinesis, and septation by means of a broad cleavage furrow.

The Florideophycidae are filamentous in construction, and growth is apical with few exceptions. Evolutionary advances in vegetative organization have taken place within the bounds of these developmental constraints. The lower Florideophycidae is a distinct assemblage characterized by the presence of one or more acrochaetioid stages in the life history, filaments that exhibit a primitive type of mitosis, cytokinesis and growth, and similarities in pit plug ultrastructure (Chapter 2) and post-fertilization development (Chapter 13). The relationships between the orders grouped as the higher Florideophycidae are less certain. They may represent several evolutionary lines with parallel modifications of thallus construction (Chapter 18).

In the Acrochaetiales, Nemaliales, determinate lateral filaments of *Batrachospermum* and pseudochantransia stage of *Lemanea* (Batrachospermales) growth takes place through a process best described as budding. Meristematic cells initiate apical or lateral buds. Buds may be conspicuous, as in the determinate filaments of *Batrachospermum* and *Liagora*, or indistinguishable, as in *Audouinella*. Nuclear division takes place in the body of the parent cell. There is no relationship between the location of nuclear division and the site of cytokinesis. Mitosis here resembles that in the Bangiophycidae. The chromosomes are arranged in a ring at metaphase and anaphase, and the interzonal spindle is greatly extended during late anaphase and telophase. In contrast to mitosis in the Bangiophycidae, there are two distinct stages to anaphase, the second involving migration of one set of daughter chromosomes into the apical or lateral bud. In addition, division occurs around an enlarging basal vacuole rather than a central chloroplast and pyrenoid. New wall deposition is strictly apical in *Audouinella* (Acrochaetiales) and cell elongation is completed a few cells behind the apex. The same appears to be true of *Rhodochaete*.

The freely branched, loosely organized determinate filaments of *Batrachospermum* are replaced in *Lemanea* by a confluent cortex composed of filaments that follow a prescribed pattern of development. Associated with this change is a shift from the budding process of *Batrachospermum* to growth by means of cell expansion and cleavage in *Lemanea*, and mitosis becomes located at and oriented perpendicular to the plane of the ensuing crosswall.

The advent of morphologically and developmentally distinct axes was a major step in the evolution of thallus form. A central axis characterizes thallus construction in *Batrachospermum* and *Lemanea* among lower Florideophycidae and predominates in the higher Florideophycidae. Axial filaments are indeterminate and normally bear determinate lateral assimilatory filaments. Distinct patterns of mitosis and cell maturation are associated with the growth of axial filaments. Nuclear division is located at and oriented perpendicular to the ensuing crosswall plane. Cell enlargement continues well behind the apex, axial cells increasing greatly in length and diameter during maturation. Where known, growth of axial cells occurs through localized deposition of new wall material. Unipolar or bipolar band growth has been documented in *Batrachospermum* and several genera of Ceramiaceae.

In the higher Florideophycidae nuclear division takes place at the future site of cytokinesis, and the metaphase plate is oriented parallel to the plane of the ensuing septum. The chromosomes are arranged in a linear band during metaphase and anaphase, and anaphase migration is perpendicular to the plane of the preceding metaphase plate. This is true of divisions in determinate lateral filaments as well as indeterminate axial filaments. Axial filaments exhibit patterned development, established at the apex by the plane of division of the apical cell. In many taxa, especially those with radial symmetry, the metaphase plate is tilted such that the basal set of daughter chromosomes migrates to the side of the incipient axial segment from which the first periaxial cell will be cut off. Often, the first periaxial cell on successive segments describes a spiral, and subsequent periaxial cells on a given axial segment are cut off in a regular sequence.

In many Ceramiales, Bonnemaisoniales, and uniaxial Gigartinales the position of lateral branches, as well as determinate filaments, is determined by the pattern of division at the apex. Lateral branch initials are cut off prior to ordinary periaxial cells in many Ceramiaceae, as are trichoblast and branch initials in many Rhodomelaceae. A similar, but more complex behavior is seen in the Bonnemaisoniaceae and several families of the Gigartinales. In all of these cases the location of lateral initials is determined by the orientation of the mitotic apparatus at the time the axial segment that bears them is cut off. In other higher Florideophycidae branching of the thallus is facultative, involving a transformation of the apical cell of a determinate filament into the initial of an indeterminate axis.

The most primitive genera of the higher Florideophycidae have whorled determinate lateral filaments that develop equally, as in *Batrachospermum*. Evolution of the cortex has led to the establishment of a primary, ascending filament. Cells behind the apex of a primary filament initiate lateral, abaxial, and basal derivatives. This pattern of cortical development is common in the Gigartinales. There is also a shift from budding and filamentous growth to cell expansion and cleavage. Further modification involves the reduction or loss of certain of these filaments and changes in patterns of cell maturation.

A variety of mechanisms that effect an increase in nuclear volume, and presumably ploidy level, are associated with the extensive cell enlargement that takes place in the axial and determinate lateral filaments of higher Florideophycidae. This is accomplished through an increase in the size of nuclei, the number of nuclei, or both. Increased numbers of nuclei result from nuclear divisions without accompanying cell divisions or from the formation of secondary pit connections. The ability to form secondary pit connections by means of vegetative cell fusions, a behavior restricted to the higher Florideophycidae, expands the potential for thallus modification.

VI. REFERENCES

Aghajanian, J. G. & Hommersand, M. H. 1980. Growth and differentiation of axial and lateral filaments in *Batrachospermum sirodotii*. *J. Phycol.* 16: 15–28.

André, F. 1978. Sur les cycles morphologiques du *Gymnogongrus crenulatus* (Turn.) J. Ag. et du *Gymnogongrus devoniensis* (Grev.) Schott. (Gigartinales, Phyllophoracées) en culture. *Rev. Algol.* N.S. 8: 151–76.

André, F. 1980. Observations sur le cycle de développement du *Schizymenia dubyi* (Rhodophycée, Gigartinale) en culture. *Cryptogam. Algol.* 1: 111–40.

Atkinson, G. F. 1890. Monograph of the Lemaneaceae of the United States. *Ann. Bot.* 4: 177–229.

Boillot, A. 1975. Cycle biologique de *Rhodochaete parvula* (Thuret) (Rhodophycées, Bangiophycidées). *Pubbl. Staz. Zool. Napoli* 36 (Suppl.): 67–83.

Chadefaud, M. 1952. La leçon des Algues. Colloque C. N. R. S. – Évolution et phylogénie chez les végétaux. *Ann. Bot.*, sér. 3, 28: 9–25.

Chen, L. C.-M. & Taylor, A. R. A. 1976. Scanning electron microscopy of early sporeling ontogeny of *Chondrus crispus*. *Can. J. Bot.* 54: 672–8.

Chiang, Y.-M. 1970. Morphological studies of red algae of the family Cryptonemiaceae. *Univ. Calif. Pub. Bot.* 58: 1–95.

Chihara, M. 1961. Life cycle of the Bonnemaisoniaceous algae in Japan (1). *Sci. Rep. Tokyo Kyoiku Daigaku* Sect. B, 10: 121–53 + 6 plates.

Chihara, M. 1972. Germination of carpospores of *Pikea californica* and *Schimmelmannia plumosa* as found in Japan, with special reference to their life history. *Bull. Soc. Bot. Fr. Mém.* 1972: 313–22.

Chihara, M. & Yoshizaki, M. 1972. Bonnemaisoniaceae: their gonimoblast development, life history and systematics. In *Contributions to the Systematics of Benthic Marine Algae of the North Pacific*, eds. I. A. Abbott & M. Kurogi, pp. 243–64. Kobe, Japan: Japanese Society of Phycology.

Cleland, R. E. 1919. The cytology and life-history of *Nemalion multifidum* Ag. *Ann. Bot.* 33: 323–51.

Codomier, L. 1972a. Sur le développement des spores et sur l'origine des cellules étoilées médullaires des *Kallymenia* (Rhodophycées, Cryptonemiales). *C. R. Acad. Sci.* (Paris), sér. D 274: 369–71.

Codomier, L. 1972b. Le cycle du *Neurocaulon reniforme* (P. et R.) Zanardini (Rhodophycée, Gigartinale). *Bull Soc. Bot. Fr. Mém.* 1972: 293–310.

Codomier, L. 1973a. Sur le développement des spores et la formation du thalle rampant de *Kallymenia microphylla* J. Ag. (Rhodophyceae, Cryptonemiales). *Giorn. Bot. Ital.* 107: 269–80.

Codomier, L. 1973b. Caractères généraux et développement des spores de *Sebdenia dichotoma* (J. Ag.) Berthold (Rhodophycées, Gigartinales). *Phycologia* 12: 97–105.

Cole, K. & Conway, E. 1975. Phenetic implications of structural features of the perennating phase in the life history of *Porphyra* and *Bangia* (Bangiophycidae, Rhodophyta). *Phycologia* 14: 239–45.

Cole, K. & Conway, E. 1980. Studies in the Bangiaceae: Reproductive modes. *Bot. Mar.* 23: 545–53.

Coomans, R. J. 1986. A light microscope study of mitosis and vegetative development in Rhodophyta. Ph.D. dissertation, University of North Carolina, Chapel Hill, N.C.

Cortel-Breeman, A. M. 1975. The life history of *Acrosymphyton purpuriferum* (J. Ag.) Sjost. (Rhodophyceae, Cryptonemiales). Isolation of tetrasporophytes. *Acta Bot. Neerl.* 24: 111–27.

Cramer, C. 1864. Physiologisch-systematische untersuchungen über die Ceramiaceen. *Neue Denkschr. Allg.*

Schweiz. Ges. Naturwiss. 20: iv + 131 pp., 13 pls.
DeCew, T. C. & West, J. A. 1981. Investigations on the life histories of three *Farlowia* species (Rhodophyta: Cryptonemiales, Dumontiaceae) from Pacific North America. *Phycologia* 20: 342–51.
DeCew, T. C., West, J. A. & Ganesan, E. K. 1981. The life histories and developmental morphology of two species of *Gloiosiphonia* (Rhodophyta: Cryptonemiales, Gloiosiphoniaceae) from the Pacific coast of North America. *Phycologia* 20: 415–23.
Denizot, M. 1968. *Les Algues Floridées Encroutantes (à l'exclusion des Corallinacées)*. Paris: Laboratoire de Cryptogamie Muséum national d'Histoire naturelle.
Desikachary, T. V. 1957. *Helminthocladia* from India and New Zealand. *J. Indian Bot. Soc.* 36: 441–56.
Desikachary, T. V. 1962. *Cumagloia* Setchell et Gardner and *Dermonema* (Grev.) Harvey. *J. Indian Bot. Soc.* 41: 132–46.
Dixon, P. S. 1958. The structure and development of the thallus in the British species of *Gelidium* and *Pterocladia*. *Ann. Bot.* N.S. 22: 353–68.
Dixon, P. S. 1960. Studies on marine algae of the British Isles: the genus *Ceramium*. *J. Mar. Biol. Ass. U.K.* 39: 331–74.
Dixon, P. S. 1971. Cell enlargement in relation to the development of thallus form in Florideophycidae. *Br. Phycol. J.* 6: 195–205.
Dixon, P. S. 1973. *Biology of the Rhodophyta*. New York: Hafner.
Dixon, P. S. & Irvine, L. M. 1977. *Seaweeds of the British Isles*, Vol. 1, Rhodophyta, Part I, Introduction, Nemaliales, Gigartinales. London: British Museum (Nat. Hist.).
Dixon, P. S. & Price, J. H. 1981. The genus *Callithamnion* (Rhodophyta: Ceramiaceae) in the British Isles. *Bull. Br. Mus. Nat. Hist. (Bot.)* 9: 99–141.
Drew, K. M. 1952. Studies on the Bangioideae 1. Observations on *Bangia fuscopurpurea* (Dillw.) Lyngb. in culture. *Phytomorphology* 2: 38–51.
Entwistle, T. J. & Kraft, G. T. 1984. Survey of freshwater red algae (Rhodophyta) of southeastern Australia. *Aust. J. Mar. Freshw. Res.* 35: 213–59.
Falkenberg, P. 1901. Die Rhodomelaceen des Golfes von Neapel und der angrenzenden Meeresabschnitte. *Fauna Flora Golfes Neapel* 26: xvi + 754 pp., 24 plates.
Feldmann, J. 1954. Recherches sur la structure et la développement des Calosiphoniacées. *Rev. Gén. Bot.* 61: 453–99.
Fritsch, F. E. 1945. *The Structure and Reproduction of the Algae*. Vol. 2, Cambridge: Cambridge University Press.
Gabrielson, P. W. & Garbary, D. 1986. Systematics of red algae (Rhodophyta). *CRC Crit. Rev. Plant Sci.* 3: 325–66.
Gabrielson, P. W. & Hommersand, M. H. 1982a. The morphology of *Agardhiella subulata* representing the tribe Agardhielleae, a new tribe in the Solieriaceae (Gigartinales, Rhodophyta). *J. Phycol.* 18: 46–58.
Gabrielson, P. W. & Hommersand, M. H. 1982b. The Atlantic species of *Solieria* (Gigartinales, Rhodophyta): their morphology, distribution and affinities. *J. Phycol.* 18: 31–45.
Garbary, D. 1979a. Numerical taxonomy and generic circumscription in the Acrochaetiaceae (Rhodophyta). *Bot. Mar.* 22: 477–92.
Garbary, D. 1979b. Patterns of cell elongation in some *Audouinella* spp. (Acrochaetiaceae: Rhodophyta). *J. Mar. Biol. Ass. U.K.* 59: 951–60.
Garbary, D. J. & Gabrielson, P. W. 1987. Acrochaetiales (Rhodophyta): Taxonomy and evolution. *Cryptogam. Algol.* 8: 241–52.
Garbary, D. J., Hansen, G. I. & Scagel, R. F. 1980. A revised classification of the Bangiophycidae (Rhodophyta). *Nova Hedwigia* 33: 145–66.
Garbary, D. J., Hansen, G. I. & Scagel, R. F. 1982. The marine algae of British Columbia and northern Washington: Division Rhodophyta (red algae), Class Florideophyceae, Orders Acrochaetiales and Nemaliales. *Syesis* 15 (Suppl. 1): 1–102.
Gargiulo, G. M., De Masi, F. & Tripodi, G. 1987. Thallus morphology and spore formation in cultured *Erythrocladia irregularis* Rosenvinge (Rhodophyta, Bangiophycidae). *J. Phycol.* 23: 351–9.
Goff, L. J. & Coleman, A. W. 1984. Elucidation of fertilization and development in a red alga by quantitative DNA microspectrofluorometry. *Devel. Biol.* 102: 173–94.
Goff, L. J. & Coleman, A. W. 1986. A novel pattern of apical cell polyploidy, sequential polyploidy reduction and intercellular nuclear transfer in the red alga *Polysiphonia*. *Am. J. Bot.* 73: 1109–30.
Gretz, M. R., Aronson, J. M. & Sommerfeld, M. R. 1980. Cellulose in the cell walls of the Bangiophyceae (Rhodophyta). *Science* 207: 779–81.
Guiry, M. D. & Maggs, C. A. 1985. Reproduction and life history of *Meredithia microphylla* (J. Ag.) J. Ag. (Kallymeniaceae, Rhodophyta) from Ireland. *Giorn. Bot. Ital.* 118: 105–25.
Hansen, G. I. 1977. *Cirrulicarpus carolinensis*, a new species in the Kallymeniaceae (Rhodophyta). *Occas. Pap. Farlow Herb.* No. 12: 1–22.
Hawkes, M. W. 1983. *Hummbrella hydra* Earle (Rhodophyta, Gigartinales): seasonality, distribution, and development in laboratory culture. *Phycologia* 22: 403–13.
Heerebout, G. R. 1968. Studies on the Erythropeltidaceae (Rhodophyceae-Bangiophycidae). *Blumea* 16: 139–57.
Hommersand, M. H. 1963. The morphology and classification of some Ceramiaceae and Rhodomelaceae. *Univ. Calif. Pub. Bot.* 35: 165–366.
Howard, R. V. & Parker, B. P. 1980. Revision of *Boldia erythrosiphon* Herndon (Rhodophyta, Bangiales). *Am. J. Bot.* 67: 413–22.
Hus, H. T. A. 1902. An account of the species of *Porphyra* found on the Pacific coast of North America. *Calif. Acad. Sci., Proc., 3rd ser., Bot.* 2: 173–240.
Hymes, B. J. & Cole, K. M. 1983. The cytology of *Audouinella hermannii* (Rhodophyta, Florideophyceae). I. Vegetative and hair cells. *Can. J. Bot.* 61: 3366–76.
Israelson, G. 1942. The freshwater Florideae of Sweden. *Symb. Bot. Upsal.* 6: 1–135 + 2 plates.
Kjellman, F. R. 1900. Om Floridé-slägtet *Galaxaura* dess organografi och systematik. *Vet. Akad. Handl.* Bd. 33: 1–109 + 19 plates.
Kling, R. & Bodard, M. 1986. La construction du thalle de *Gracilaria verrucosa* (Rhodophyceae, Gigartinales): édification de la fronde; essai d'interprétation phylogénétique. *Cryptogam. Algol.* 7: 231–46.
Konrad-Hawkins, E. 1964a. Developmental studies on regenerates of *Callithamnion roseum* Harvey, Part I. The development of a typical regenerate. *Protoplasma* 58: 42–59.
Konrad-Hawkins, E. 1964b. Developmental studies on regenerates of *Callithamnion roseum* Harvey, Part II. Analysis of apical growth. Regulation of growth and form. *Protoplasma* 58: 60–74.
Konrad-Hawkins, E. 1968. Induction of cell differentiation

in dissociated cells and fragments of *Callithamnion roseum*. *Am. J. Bot.* 55: 255–64.

Kornmann, P. 1984. *Erythrotrichopeltis*, eine neue Gattung der Erythropeltidaceae (Bangiophyceae, Rhodophyta). *Helgol. Meeresunters.* 38: 207–24.

Kornmann, P. 1987. Der Lebenszyklus von *Porphyrostromium obscurum* (Bangiophyceae, Rhodophyta). *Helgol. Meeresunters.* 41: 127–37.

Kraft, G. T. 1975. Consideration of the order Cryptonemiales and the families Nemastomaceae and Furcellariaceae (Gigartinales, Rhodophyta) in light of the morphology of *Adelophyton corneum* (J. Agardh) gen. et comb. nov. from southern Australia. *Br. Phycol. J.* 10: 279–90.

Kraft, G. T. 1977. Studies of marine algae in the lesser-known families of the Gigartinales (Rhodophyta). II. Dicranemaceae. *Aust. J. Bot.* 25: 219–67.

Kraft, G. T. & John, D. M. 1976. The morphology and ecology of *Nemastoma* and *Predaea* (Nemastomaceae, Rhodophyta) from Ghana. *Br. Phycol. J.* 11: 331–44.

Kraft, G. T. & Robins, P. A. 1985. Is the order Cryptonemiales (Rhodophyta) defensible? *Phycologia* 24: 67–77.

Krishnamurthy, V. 1959. Cytological investigations on *Porphyra umbilicalis* (L.) Kütz. var. *laciniata* (Lightf.) J. Ag. *Ann. Bot.* N.S. 23: 147–76 + 2 plates.

Krishnamurthy, B. V. 1962. The morphology and taxonomy of the genus *Compsopogon* Montagne. *J. Linn. Soc. (Bot.)* 58: 207–22 + 4 plates.

Kuckuck, P. 1912. Ueber *Platoma bairdii* (Farl.) Kuck. Bieträge zur Kenntnis der Meeresalgen 12. Wiss. Meeresunters., Abt. Helgoland, N.F. 5: 187–208 + 3 plates.

Kylin, H. 1928. Entwicklungsgeschichtliche Florideenstudien. *Lunds Univ. Årsskr.*, N.F., Avd. 2(4), 26: 1–127.

Kylin, H. 1956. *Die Gattungen der Rhodophyceen*. Lund: Gleerups.

L'Hardy-Halos, M.-Th. 1968. Sur la structure de la fronde chez les Dasyacées (Rhodophycées-Céramiales). *C. R. Acad. Sc.* (Paris), sér. D 266: 1833–5.

L'Hardy-Halos, M.-Th. 1971. Recherches sur les Céramiacées (Rhodophycées-Céramiales) et leur morphogénèse. II. Les modalités de la croissance et les remaniements cellulaires. *Rev. Gén. Bot.* 78: 201–56.

Lee, I. K. 1978. Studies on Rhodymeniales from Hokkaido. *J. Fac. Sci., Hokkaido Univ., Ser. V (Botany)* 11: 1–194 + 5 plates.

Lewin, R. A. & Robertson, J. A. 1971. Influence of salinity on the form of *Asterocytis* in pure culture. *J. Phycol.* 7: 236–8.

Maggs, C. A. & Guiry, M. 1985. Life history and reproduction of *Schmitzia hiscockiana* sp. nov. (Rhodophyta, Gigartinales) from the British Isles. *Phycologia* 24: 297–310.

Magne, F. 1964. Recherches caryologiques chez les Floridées (Rhodophycées). *Cah. Biol. Mar.* 5: 461–671.

Masuda, M., West, J. A. & Kurogi, M. 1987. Life history studies in culture of a *Mastocarpus* species (Rhodophyta) from central Japan. *J. Fac. Sci., Hokkaido Univ. Ser. V (Bot.)* 14: 11–38.

Mayhoub, H. 1973. Cycle du développement de *Calosiphonia vermicularis* (J. Agardh) Schmitz (Rhodophycée, Gigartinale). *C. R. Acad. Sci.* (Paris), sér. D 277: 1137–40.

Mayhoub, H. 1975. Nouvelles observations sur le cycle de développement du *Calosiphonia vermicularis* (J. Ag.) Sch. (Rhodophycée, Gigartinale). *C. R. Acad. Sci.* (Paris), sér. D 280: 2441–3.

Mohoroshi, H. & Masuda, M. 1980. The life history of *Gloiosiphonia capillaris* (Hudson) Carmichael (Rhodophyceae, Cryptonemiales). *Jap. J. Phycol.* 28: 81–91.

Mori, M. 1975. Studies on the genus *Batrachospermum* in Japan. *Jap. J. Bot.* 20: 461–85.

Mukai, L. S., Craigie, J. S. & Brown, R. G. 1981. Chemical composition and structure of the cell walls of the conchocelis and thallus phases of *Porphyra tenera* (Rhodophyceae). *J. Phycol.* 17: 192–8.

Mullahy, J. H. 1952. The morphology and cytology of *Lemanea australis* Atk. var. *Chantransia-uniformis* n. var. *Bull. Torrey Bot. Club* 79: 393–406.

Murray, S. N., Dixon, P. S. & Scott, J. L. 1972. The life history of *Porphyropsis coccinea* var. *dawsonii* in culture. *Br. Phycol. J.* 7: 323–33.

Nichols, H. W. 1964a. Developmental morphology and cytology of *Boldia erythrosiphon*. *Am. J. Bot.* 51: 653–9.

Nichols, H. W. 1964b. Culture and developmental morphology of *Compsopogon coeruleus*. *Am. J. Bot.* 51: 180–8.

Norris, R. E., Wollaston, E. M. & Parsons, M. J. 1984. New terminology for sympodial growth in Ceramiales (Rhodophyta). *Phycologia* 23: 233–7.

Oltmanns, F. 1904. *Morphologie und Biologie der Algen*. Vol. I. Jena: Gustav Fischer.

Oltmanns, F. 1922. *Morphologie und Biologie der Algen*. Vol. 2. Jena: Gustav Fischer.

Pringle, J. D. & Austin, A. P. 1970. The mitotic index in selected red algae *in situ*. II. A supralittoral species, *Porphyra lanceolata* (Setchell & Hus) G. M. Smith. *J. Exp. Mar. Biol. Ecol.* 5: 113–37.

Pueschel, C. M. & Cole, K. M. 1985. Ultrastructure of germinating carpospores of *Porphyra variegata* (Kjellm.) Hus (Bangiales, Rhodophyta). *J. Phycol.* 21: 146–54.

Pueschel, C. M. & Magne, F. 1987. Pit plugs and other ultrastructural features of systematic value in *Rhodochaete parvula* (Rhodophyta, Rhodochaetales). *Cryptogam. Algol.* 8: 201–9.

Richardson, N. & Dixon, P. S. 1969. The Conchocelis phase of *Smithora naiadum* (Anders.) Hollenb. *Br. Phycol. J.* 4: 181–3.

Rietema, H. 1984. Development of erect thalli from basal crusts in *Dumontia contora* (Gmel.) Rupr. (Rhodophyta, Cryptonemiales). *Bot. Mar.* 27: 29–36.

Rietema, H. & Klein, A. W. O. 1981. Environmental control of the life cycle of *Dumontia contorta* (Rhodophyta) kept in culture. *Mar. Ecol. Prog. Ser.* 4: 23–9.

Robins, P. A. & Kraft, G. T. 1985. Morphology of the type and Australian species of *Dudresnaya* (Dumontiaceae, Rhodophyta). *Phycologia* 24: 1–34.

Rosenvinge, L. K. 1909. The marine algae of Denmark. Contributions to their natural history. Part I. Introduction. Rhodophyceae I. (Bangiales and Nemalionales). *K. Danske Vidensk. Selsk. Skr.*, ser. 7, Nat. Mat. Afd. 7: 1–151.

Rueness, J. & Åsen, P. A. 1982. Field and culture observations on the life history of *Bonnemaisonia asparagoides* (Woodw.) C. Ag. (Rhodophyta) from Norway. *Bot. Mar.* 25: 577–87.

Scagel, R. F. 1953. A morphological study of some dorsiventral Rhodomelaceae. *Univ. Calif. Pub. Bot.* 27: 1–108.

Schmitz, F. 1897. Rhodochaetaceae. In *Die Natürlichen Pflanzenfamilien*, eds. A. Engler & K. Prantl, pp. 317–18.

Searles, R. B. 1968. Morphological studies of red algae of the order Gigartinales. *Univ. Calif. Pub. Bot.* 43: 1–100.

Seto, R. 1982. Notes on the family Compsopogonaceae (Rhodophyta, Bangiales) in Okinawa Prefecture, Japan.

Jap. J. Phycol. 30: 57–62.

Seto, R. 1987. Study of a freshwater red alga, *Compsopogonopsis fruticosa* (Jao) Seto comb. nov. (Compsopogonales, Rhodophyta) from China. *Jap. J. Phycol.* 35: 265–7.

Sheath, R. G. 1984. The biology of freshwater algae. *Prog. Phycol. Res.* 3: 89–157.

Silva, P. C., Menez, E. G. & Moe, R. L. 1987. Catalog of the benthic marine algae of the Philippines. *Smithson. Contrib. Mar. Sci.* 27: iv + 179 pp.

Sirodot, S. 1872. Étude anatomique, organogénique et physiologique sur les algues d'eau douce de la famille des Lémanéacées. *Ann. Sci. Nat., Bot.*, Paris. ser. 5, T. 16: 5–95.

Sirodot, S. 1884. *Les Batrachospermes: Organisation Functions, Développement, Classification.* Paris: G. Masson.

Sommerfeld, M. R. & Nichols, H. W. 1970. Developmental and cytological studies of *Bangia fuscopurpurea* in culture. *Am. J. Bot.* 57: 640–8.

Stegenga, H. 1979. *Life Histories and Systematics of the Acrochaetiaceae.* Amsterdam: Total Photo/Total Print.

Stegenga, H. 1985. The marine Acrochaetiaceae (Rhodophyta) of southern Africa. *S. Afr. J. Bot.* 51: 291–330.

Svedelius, N. 1942. Zytologisch-entwicklungsgeschichtliche studien über *Galaxaura*. *Nova Acta Soc. Sci. Ups.* ser. 4, vol. 13, no. 4, Uppsala.

Swale, E. M. F. 1962. The development and growth of *Thorea ramosissima* Bory. *Ann. Bot.* N. S., 26: 105–16.

Umezaki, I. 1968. A study on the germination of carpospores of *Dudresnaya japonica* Okamura (Rhodophyta). *Pub. Seto Mar. Biol. Lab.* 16: 263–72 + 2 plates.

West, J. A. 1970. The life history of *Rhodochorton concrescens* in culture. *Br. Phycol. J.* 5: 179–86.

West, J. A. & Hommersand, M. H. 1981. Rhodophyta: Life histories. In *The Biology of Seaweeds*, eds. C. S. Lobban & M. J. Wynne, pp. 133–93. Oxford: Blackwell Scientific.

Wilce, R. T. & Davis, A. N. 1984. Development of *Dumontia contorta* (Dumontiaceae, Cryptonemiales) compared with that of other higher red algae. *J. Phycol.* 20: 336–51.

Wittmann, W. 1965. Aceto-iron-haematoxylin–chloral hydrate for chromosome staining. *Stain Tech.* 40: 161–4.

Woelkerling, W. J. 1983. The *Audouinella* (*Acrochaetium-Rhodochorton*) complex (Rhodophyta): present perspectives. *Phycologia* 22: 59–92.

Wollaston, E. M. 1968. Morphology and taxonomy of southern Australian genera of Crouanieae Schmitz (Ceramiaceae, Rhodophyta). *Aust. J. Bot.* 16: 217–417.

Wynne, M. J. 1985. Nomenclatural assessment of *Goniotrichum* Kützing, *Erythrotrichia* Areschoug, *Diconia* Harvey, and *Stylonema* Reinsch (Rhodophyta). *Taxon* 34: 502–5.

Yabu, H. 1963. Mitosis in *Porphyra*. *Bull. Fac. Fish., Hokkaido Univ.* 14: 131–6 + 6 plates.

Yabu, H. 1966. Nuclear division in *Bangia fuscopurpurea* (Dillwyn) Lyngbye. *Bull. Fac. Fish., Hokkaido Univ.* 17: 163–4 + 5 plates.

Yabu, H. 1970. Observation on chromosomes in some species of *Porphyra* II. *Bull. Fac. Fish., Hokkaido Univ.* 21: 253–8 + 9 plates.

Chapter 13

Sexual reproduction and cystocarp development

MAX H. HOMMERSAND
SUZANNE FREDERICQ

CONTENTS

- I. The sexual system / 306
- II. Developmental homologies of reproductive cells / 307
- III. Life histories and the origin of the carposporophyte / 307
- IV. Auxiliary cell systems / 308
- V. Morphology of the cystocarp / 309
- VI. Spermatangia and spermatia / 310
- VII. The carpogonium and carpogonial branch / 312
- VIII. Fertilization / 312
- IX. Products of the fertilization nucleus / 314
   - A. Direct formation of gonimoblasts: *Nemalion* type / 314
   - B. Connecting filaments and auxiliary cells: *Dudresnaya* type / 315
   - C. Connecting cells and auxiliary cells: *Polysiphonia* type / 316
- X. Functional adaptations of cystocarp morphology / 319
   - A. Absence of special nutritive tissues: Acrochaetiales and Palmariales / 319
   - B. Modified carpogonial branches: Batrachospermales and Nemaliales / 319
   - C. The female conceptacle and fusion cell: Corallinales / 320
   - D. Preformed nutritive tissues: Gelidiales / 320
   - E. Preformed nutritive tissues: Gigartinales / 322
   - F. Transformed cortical cells: Plocamiaceae / 328
   - G. Sterile nutritive filaments: Ceramiales and Bonnemaisoniales / 328
   - H. Fusion cells: Gigartinales / 332
   - I. Special nutritive tissues: Ahnfeltiales and Gracilariales / 332
   - J. Nutritive tissues and fusion cells: Rhodymeniales / 334
   - K. Placentae: Gigartinales / 338
- XI. Summary / 339
- XII. Acknowledgments / 343
- XIII. References / 343

I. THE SEXUAL SYSTEM

Sexual reproduction is oogamous in red algae and involves the union of a nonflagellated male gamete, the spermatium, with a receptive process, the trichogyne or protrichogyne, at the distal end of a special female cell, the carpogonium. Life histories of sexually reproducing members of the Bangiophycidae are variously interpreted as biphasic or triphasic at the present time (e.g., Cole & Conway 1980; Gabrielson & Garbary 1986; Guiry 1987). The cytological studies of Yamanouchi (1906) on *Polysiphonia* and of Magne (1964, 1972) on genera having a heteromorphic life history established a fundamental linkage between the sexual system of the Florideophycidae and a life cycle consisting of three phases: a haploid sexual phase, the gametophyte, a diploid phase that develops directly on the female thallus, the carposporophyte, and a free-living diploid phase bearing meiosporangia, the tetrasporophyte. Most modified life histories encountered in field and laboratory studies during the past 20 years are thought to be derived secondarily from the triphasic life cycle (See Dixon 1973; Guiry 1987; Magne 1972; West & Hommersand 1981).

Starting with the premise that a successful life history tends to maximize the potential for genetic recombination and genetic diversity from the union of a single pair of gametes, Searles (1980) concluded that selection has favored the evolution of a triphasic life history in red algae as compensation for an inefficient fertilization in the absence of motile gametes. The retention of the carposporophyte and its nourishment by the gametophyte are essential components of this adaptation, which balances the ability of the gametophyte to supply nutrients to the carposporophyte and the ability of the carposporophyte to utilize these nutrients effectively for reproduction.

The morphological manifestation of this nutrient-driven interaction between gametophyte and carposporophyte generations is the structure known as the cystocarp. The term "cystocarp" was first proposed by Kützing (1843, p. 100) as a substitute for "capsule", a name given to the larger of the two fruiting bodies found in red algae, and was adopted by J. Agardh (1844, p. 12), with some debate as to its interpretation. In the final analysis J. Agardh (1880, p. 168) came to regard the cystocarp as a complex structure consisting of an interior part (nucleus) and an exterior part (pericarp). He distinguished between external pericarp modified from pigmented cortical tissue and internal pericarp modified from colorless subcortical or medullary tissue. Where a cellular pericarp was absent, as in *Ceramium* or *Halymenia*, the cystocarp was said to consist of the naked nucleus alone.

Schmitz (1883, pp. 243–6) was the first to distinguish clearly between the carpospore-bearing filaments that develop either directly from the carpogonium following fertilization or from an "auxiliary cell" after being contacted by an "ooblast cell" (connecting cell) or "ooblast filament" (connecting filament) and the vegetative tissue of the female plant. Later, Schmitz (1892, p. 17) and Schmitz and Hauptfleisch (1896, p. 303) called these filaments gonimoblasts.

In 1954 Drew adopted the term "carposporophyte" for the phase of development initiated by fertilization that leads to the production of filaments bearing sporangia. The carposporophyte is thus equivalent to all connecting cells, connecting filaments, and gonimoblasts whose development, barring apomixis, is correlated with the event of fertilization. In keeping with Drew's concept, the cystocarp consists of a diploid generation, the carposporophyte, produced from a fertilized carpogonium, together with any gametophytic tissues that may become specially modified for a supporting or protective role.

Classical studies of sexual reproduction in red algae are documented in the standard references of Fritsch (1945), Kylin (1956), and Oltmanns (1922). Recently, Silva and Johansen (1986) reviewed and analyzed the conceptual basis for the sexual systems evolved by Schmitz and Kylin for classifying the red algae. Information published since Kylin can be found in Dixon (1973), Drew (1954), Kraft (1981), Papenfuss (1966), and numerous papers on special taxa.

In the present chapter we will trace the stages of cystocarp development from the time of gamete production and fertilization to the formation and release of the carpospores. We will consider the role of the fertilized carpogonium and of auxiliary cells in gonimoblast formation and document the major evolutionary tendencies in cystocarp morphology and development, with examples taken from representative families. Finally, we will propose a model to account for the functional relationship between carposporophyte and gametophyte generations, based on mechanisms that regulate the production and processing of nutrients by the gametophyte and their transfer to and utilization by the carposporophyte.

II. DEVELOPMENTAL HOMOLOGIES OF REPRODUCTIVE CELLS

The morphogenesis of male gametes (spermatia), female gametes (carpogonia), and sexually or asexually produced spores (monospores, carpospores, tetraspores, etc.) each observe patterns of differentiation that are highly conserved throughout the Rhodophyta. Moreover, spores of all types exhibit similar ultrastructural features at comparable stages of differentiation and may also share characters in common with gamete morphogenesis (Chapters 2 and 14).

The position of reproductive cells on the thallus during early stages of their transformation is another highly conserved character. Three different developmental patterns may be distinguished in red algae. Each type is diagnostic for a distinct evolutionary line. Two of these lines are presently placed in the Bangiophycidae, and the third composes the Florideophycidae. In the orders Rhodochaetales and Compsopogonales (including Erythropeltidales) reproductive cells are intercalary. Spermatia, carpospores, and monospores are cut out one at a time by a curved wall. The mother cell remains, and the process may be repeated (Kornmann 1984, 1987; Magne 1960). In the Bangiales spermatia and carpospores are formed in packets by a series of successive divisions of the mother cell that are perpendicular to one another. The process is the same whether the spores are produced sexually or asexually (Cole & Conway 1980). Additionally, conchospores are generated in fertile cell rows and monospores are formed terminally in the filamentous conchocelis stage. In Florideophycidae, in sharp contrast to the Bangiophycidae, spermatangia, carpogonia, monosporangia, carposporangia, and tetrasporangia are all produced through transformation of apical initials. Only rarely are reproductive bodies intercalary in primitive genera belonging to this subclass (Gabrielson & Garbary 1986). Timing elements that determine the position of cells to be metamorphosed into spores or gametes are evidently conserved separately and expressed independently of elements that determine the type and function of the reproductive body produced.

III. LIFE HISTORIES AND THE ORIGIN OF THE CARPOSPOROPHYTE

The life history of *Bangia* and *Porphyra* (Bangiales) is biphasic according to Cole & Conway (1980). *Rhodochaete* (Rhodochaetales) is said to have a triphasic life history in which the zygote divides into two cells, one of which is released as a single carpospore (Guiry 1987). In our opinion, the life histories of all Bangiophycidae are biphasic and their fruiting bodies are neither equivalent to, nor prototypes of, the floridean carposporophyte.

Traditionally, the carposporophyte of Florideophycidae has been regarded as a somatic phase that develops directly on the female thallus from a zygote retained within the carpogonium (Dixon 1973; Drew 1955). Feldmann (1952) proposed that the original life cycle contained three morphologically identical, free-living generations: gametophyte, carposporophyte, and tetrasporophyte. This was succeeded by one in which the zygote, instead of being released immediately, divides within the carpogonium, giving rise to a carposporophyte that lives parasitically upon the gametophyte. The observation in a cultured strain of *Acrochaetium pectinatum* that the zygote either is released from the carpogonium as a "sporozygote" that can give rise to a free-living tetrasporophyte or develops in situ to produce an attached carposporophyte has been taken as evidence for Feldmann's original hypothesis (Abdel-Rahman & Magne 1983). Carrying the argument a step further, Magne (1972) suggested that evolution is regressive in red algae in such a way that any generation can potentially grow parasitically on the preceding one. Recently, Magne (1987) supported his proposal with a new evaluation of the life cycle of *Palmaria*, in which he interprets the erect diploid thallus as a free-living carposporophyte and the stalk cell and tetrasporangium as a reduced, parasitic tetrasporophyte.

Guiry (1987) rejected the hypothesis of a free-living carposporophyte generation and proposed that zygote amplification came about in two ways: (1) by formation of a mitosporangial generation and a meiosporangial generation (triphasic), or (2) by formation of a meiosporangial generation directly from the zygote (biphasic). In Guiry's model the in situ production of a carposporophyte (most Florideophycidae) or a tetrasporophyte (Palmariaceae and some Acrochaetiaceae) are alternative means of zygote amplification that are derived separately from the ancestral condition.

It is difficult in practice to distinguish between an ancestral biphasic life history and one that has arisen through secondary replacement of a carposporophyte with a tetrasporophyte. For example, in *Rhodochorton purpureum* the tetrasporophyte is said to develop from a clavate gonimoblast cell

(Stegenga 1978; West 1969), whereas in *Rhodothamniella floridula* the tetrasporophyte appears to develop directly from the fertilized carpogonium with only a single erect filament and one rhizoid (Stegenga 1978). The former could represent a reversal from a triphasic to a biphasic condition and the latter an ancestral biphasic life history.

In many Acrochaetiaceae clusters of branches bearing monosporangia, spermatangia, carposporangia, or tetrasporangia are similar in appearance. We suggest that the same or related genetic programs control branch formation and branching pattern in the development of all of these reproductive structures. Presumably, regulatory genes for the production of asexual filaments are expressed in the diploid zygote after the first one or two divisions. If so, the ancestral carposporophyte in Florideophycidae may be a somatic phase, as envisioned by Drew (1954), that is related fundamentally to an asexual fruiting structure bearing monosporangia, and it may be incorrect from a developmental standpoint to speak of in situ germination of a "retained zygote," "sporozygote," or "carpospore."

The ancestral carposporophyte must have been essentially autotrophic in its nutrition. This condition comes closest to being realized at the present time in the order Acrochaetiales. The evolution of the carposporophyte has been one of increasing dependency on the gametophyte. In primitive Florideophycidae it is the earliest stages of carposporophyte development that exhibit the greatest dependency. Nutritional dependency extended progressively to later stages in more advanced taxonomic groups. In the comparatively primitive genus *Nemalion* meristematic gonimoblast cells and carpospores both contain mature chloroplasts with pyrenoids that are substantially identical to plastids found in vegetative cortical filaments (Ramm-Anderson & Wetherbee 1982). In *Scinaia*, a more advanced member of the Nemaliales, developing carposporophytes contain proplastids that do not differentiate into functional chloroplasts until the carposporangia are cut off, and in *Bonnemaisonia* (Bonnemaisoniales) not even the differentiated chloroplasts of the mature carposporangia contain the number of thylakoids seen in vegetative or pericarpic cells (Ramm-Anderson 1983). The conversion of proplastids into mature, functional chloroplasts during carposporogenesis is a normal event in many advanced genera of red algae (Delivopoulos & Kugrens 1984; Delivopoulos & Tsekos 1986; Tsekos & Schnepf 1982; Wetherbee 1980).

IV. AUXILIARY CELL SYSTEMS

Whereas gonimoblasts are produced directly from the carpogonium in primitive Florideophycidae, carposporophyte development is mediated by one or more auxiliary cells in the advanced orders Gigartinales (including Cryptonemiales), Rhodymeniales, and Ceramiales. The carpogonium either produces connecting filaments or connecting cells or fuses directly with one or more auxiliary cells; either gonimoblasts are produced directly from the connecting filament in proximity to an auxiliary cell, or a diploid nucleus is transferred to the auxiliary cell, which in turn gives rise to the gonimoblasts.

Papenfuss (1951) and Drew (1954) distinguished between "nutritive auxiliary cells," which were said to have only a nutritive function, and "generative auxiliary cells," which were said to have both a nutritive and a generative function in that they produce the gonimoblasts. Our observations indicate that the nutrition of the carposporophyte and the regulation of its morphogenesis are separate functions carried out by different cell types that are distinguishable cytologically from one another. Accordingly, we will restrict the term "auxiliary cell" to cells that play a morphogenetic role in gonimoblast formation. All specialized structures that are strictly nutritive will be referred to as nutritive cells, nutritive filaments, or nutritive tissues.

In our opinion, auxiliary cells have evolved to perform two functions in support of zygote amplification: (1) as a site for the introduction of morphogenetic factors that either initiate gonimoblasts or transform their mode of development and (2) as an isolating mechanism operating at a second level that rejects incompatible or disharmonious fertilizations while allowing the rest to proceed.

In some respects an auxiliary cell resembles a carpogonium. Indeed, Schmitz (1883, p. 246) originally proposed that double fertilization occurs in Florideophycidae that possess auxiliary cells, a notion disproved by Oltmanns (1898), who showed that the nucleus derived from the fertilized carpogonium does not unite with the auxiliary cell nucleus. We have observed that, like the carpogonium, the auxiliary cell is granular in appearance with only small vacuoles present and that the nucleus possesses a conspicuous central nucleolus surrounded by a clear region, the hyaloplasm, and a faint network of chromatin (see Figs. 13-24, 13-94, 13-146). Walls of both are secondarily thickened with gelatinous material. The adherence of a connecting filament or connecting cell with an auxiliary cell, like the adhesion of spermatia to a trichogyne, appears to involve cell-specific interactions. Moreover, gonimoblasts produced in the vicinity of an auxiliary cell are often compact and branched like those formed directly from the carpogonium (see Section X).

In the past a distinction has been made between cases in which gonimoblasts develop from connecting filaments and those in which the diploid nucleus enters the auxiliary cell before gonimoblast initials are cut off. Recent studies have tended to downgrade the importance of this difference. For example, gonimoblasts develop directly from the auxiliary cell in *Predaea weldii*, whereas in most species of this genus they are produced from the connecting filament alongside the auxiliary cell (Kraft 1984).

Connecting cells are minute, uninucleate cells cut off by the carpogonium along with a minimal amount of cytoplasm. Typical connecting cells occur in only two orders: Ceramiales and Rhodymeniales. The nuclei are highly condensed and usually surrounded by a hyaline region just inside the cell membrane. A connecting cell expands only when contacted by an auxiliary cell. The behavior of a typical connecting cell is described and illustrated for *Polysiphonia* in Section IX.C.

Schmitz (1883, p. 245) regarded the "ooblast filament" (connecting filament) and the "ooblast cell" (connecting cell) as de novo structures. Drew (1954) interpreted the connecting filament as "primary gonimoblast" that gives rise to "secondary gonimoblast" once fusion with an auxiliary cell has taken place. In our opinion, connecting filaments, connecting cells, and auxiliary cells are structures that originated more than once and are polyphyletic.

The second role we attribute to auxiliary cells is that they function in rejecting disharmonious fertilizations. The experimental observations of Boo and Lee (1983) on interspecific crosses between *Antithamnion sparsum* from Korea and *A. defectum* from California are instructive in this regard. Boo and Lee found that gonimoblast development and release of carpospores occurred in the cross of *A. sparsum* (male) × *A. defectum* (female), but not in the reciprocal cross of *A. sparsum* (female) × *A. defectum* (male). Fertilization occurred in the second cross; the supporting cell cut off an auxiliary cell that enlarged normally, and the carpogonium cut off a connecting cell. The connecting cell, however, failed to fuse with the auxiliary cell, and further development ceased at this point.

In the Ceramiales an early fertilization leading to diploidization of the auxiliary cell and normal carposporophyte development is often followed by fertilizations of nearby carpogonia that fail to produce carposporophytes (Hommersand 1963). Again, a connecting cell or both an auxiliary cell and a connecting cell may have been cut off, but fusion between the two fails to take place. If it should take place, the carposporophyte probably would abort.

Red algae are referred to as either procarpial or nonprocarpial, depending on the position of the auxiliary cell in relation to the carpogonial branch. The term "procarp" has a checkered history (Silva & Johansen 1986) but, in general, procarps are said to be present if auxiliary cells are borne on the same branch system in close proximity to the carpogonial branch, and absent if they are borne in separate, more remote branch systems. The procarpial condition has probably arisen de novo in conjunction with the evolution of connecting cells and auxiliary cells in the Ceramiales and Rhodymeniales. Clusters of filaments that are produced secondarily and bear both carpogonia and auxiliary cells are inherently procarpial in the families Gloiosiphoniaceae (in part), Endocladiaceae, and Tichocarpaceae (Kylin 1956), and perhaps, also in the Sphaerococcaceae and Phacelocarpaceae (Searles 1968; Sjöstedt 1926). *Callophyllis* and a few other genera belonging to the Kallymeniaceae have become procarpial secondarily through the loss of connecting filaments with the transfer of gonimoblast development to a fusion cell formed from part of the carpogonial branch apparatus (Norris 1957). In the Cystocloniaceae and Hypneaceae, families thought to be related to the Solieriaceae, the auxiliary cell is an inner cortical cell that is suprajacent to the supporting cell of the carpogonial branch (Min-Thein & Womersley 1976). The affinities of the procarpial families Mychodeaceae, Acrotylaceae, Dicranemaceae, and Sarcodiaceae are thought to lie ultimately with the nonprocarpial family Solieriaceae or the procarpial family Cystocloniaceae (Kraft 1977b,c, 1978); the origin of procarps in the Plocamiaceae (Kylin 1928) and in the Gigartinaceae–Phyllophoraceae assemblage (Mikami 1965) are entirely unknown.

V. MORPHOLOGY OF THE CYSTOCARP

Structurally the cystocarp may be divided into three compartments: (1) the outer photosynthetic tissues, (2) the modified, nonphotosynthetic inner gametophytic tissues, and (3) the developing carposporophyte. The three compartments are readily differentiated using classical fixation and hematoxylin staining techniques as applied by Oltmanns (1898) and Kylin (1914 and subsequent papers). We have obtained comparable results with material fixed in 8% to 10% formalin-seawater and stored in 5% formalin-seawater using a modification of Wittmann's (1965) aceto–iron–hematoxylin–chloral hydrate technique (Coomans 1986; Hommersand & Fredericq 1988).

The photosynthetic compartment consists of assimilatory filaments or unmodified cortical tissues of the vegetative system, plus any secondary photosynthetic tissues generated before or after fertilization. These latter may include secondary assimilatory filaments, involucre, nemathecia, or pericarps that contain functional chloroplasts. In most red algae a complex multilayered structure called a cuticle overlies the external tissues. Kugrens (1980) has suggested that the cuticle functions as a differentially permeable membrane that allows small molecules to pass through while retaining larger molecules produced by photosynthesis in the spermatangial branches of *Polysiphonia*. The same would hold for the female reproductive system. It is not required that the cuticle be a true membrane, only that it delay or restrain the outward diffusion of metabolites required for carposporophyte nutrition. Most cystocarps we have observed are provided with a cuticle, and ostiolate cystocarps generally have the ostiolar region sealed with a plug prior to carposporophyte maturation, or the ostiole is not formed until after the cystocarp matures.

The second compartment consists of specially modified gametophytic cells or tissues that lie in close proximity to the carpogonia or auxiliary cells, or are so situated that they are readily contacted by developing carposporophytes. Such cells and tissues usually stain deeply with hematoxylin, aniline blue, or other stains that exhibit a degree of specificity for proteins. Some have increased numbers of nuclei, or their nuclei and nucleoli may be enlarged, and they may contain amplified levels of DNA seen with DNA-specific fluorochromes, such as DAPI. Deeply staining tissues have been referred to as nutritive tissues in the past (Drew 1954; Kylin 1956). We distinguish between ordinary nutritive tissues, in which the cell contents are consumed in a single cycle of generative activity, and special nutritive tissues that appear to persist for longer periods during carposporophyte development providing renewable resources. We call this latter type a nutrient-processing center. It is always identifiable by the presence of large numbers of small nuclei, moderate numbers of enlarged nuclei, or single enlarged, often giant nuclei per cell. Examples are given in Section X.

The third compartment is the carposporophyte itself which may consist of simple, unmodified filamentous gonimoblasts or be highly differentiated, sometimes with the production of auxiliary cells and connecting cells or connecting filaments. Morphological modifications of carposporophytic and closely associated gametophytic tissues may involve (1) enlargement and structural modification of existing pit connections, (2) fusions between contiguous cells either through or around the pit connections, sometimes with the formation of a central fusion cell, or (3) fusions or the production of secondary pit connections between gonimoblast cells and noncontiguous gametophytic cells, sometimes with formation of a well-defined placenta (see Sections X, XI).

VI. SPERMATANGIA AND SPERMATIA

Spermatogenesis follows a similar pattern in Bangiophycidae (Hawkes 1978), primitive Florideophycidae (Duckett & Peel 1978), and advanced Florideophycidae (Kugrens 1980; Scott & Dixon 1973). The condition of the nucleus is variable, with the chromatin ranging from dispersed to highly condensed. In the Corallinaceae the chromatin forms discrete cylindrical bodies aligned in two parallel plates (Fig. 13-1), suggesting that the nucleus is in arrested anaphase at the time of spermatial release (Duckett & Peel 1978). (The ultrastructure of spermatogenesis is described in Chapter 2.)

Spermatangia are borne separately in clusters of two to five on a spermatangial mother cell in the more primitive Florideophycidae (Figs. 13-2, 13-3). A single spermatium differentiates within each spermatangium. The spermatangial wall often remains after spermatial release, and new spermatia may proliferate within the older spermatangial walls (Scott & Dixon 1973). Spermatangia are generally initiated successively as subapical protrusions that are cut off by oblique septa in Florideophycidae (Dixon 1973). Initiation by transverse division is reported in *Gelidium* (Renfrew 1988) and occurs in some members of the Gracilariaceae (Fig. 13-4). In a few genera, such as *Holmsella* (Fig. 13-6) and *Endocladia* (Kylin 1956), the spermatangia are produced in linear rows.

The spermatangia of thalloid Florideophycidae commonly occur in superficial sori, as in *Gracilariopsis lemaneiformis* (Fig. 13-4). In *Gracilaria verrucosa* they are generated on branched filaments derived from a single initial and line the surface of a cavity

Sexual reproduction and cystocarp development

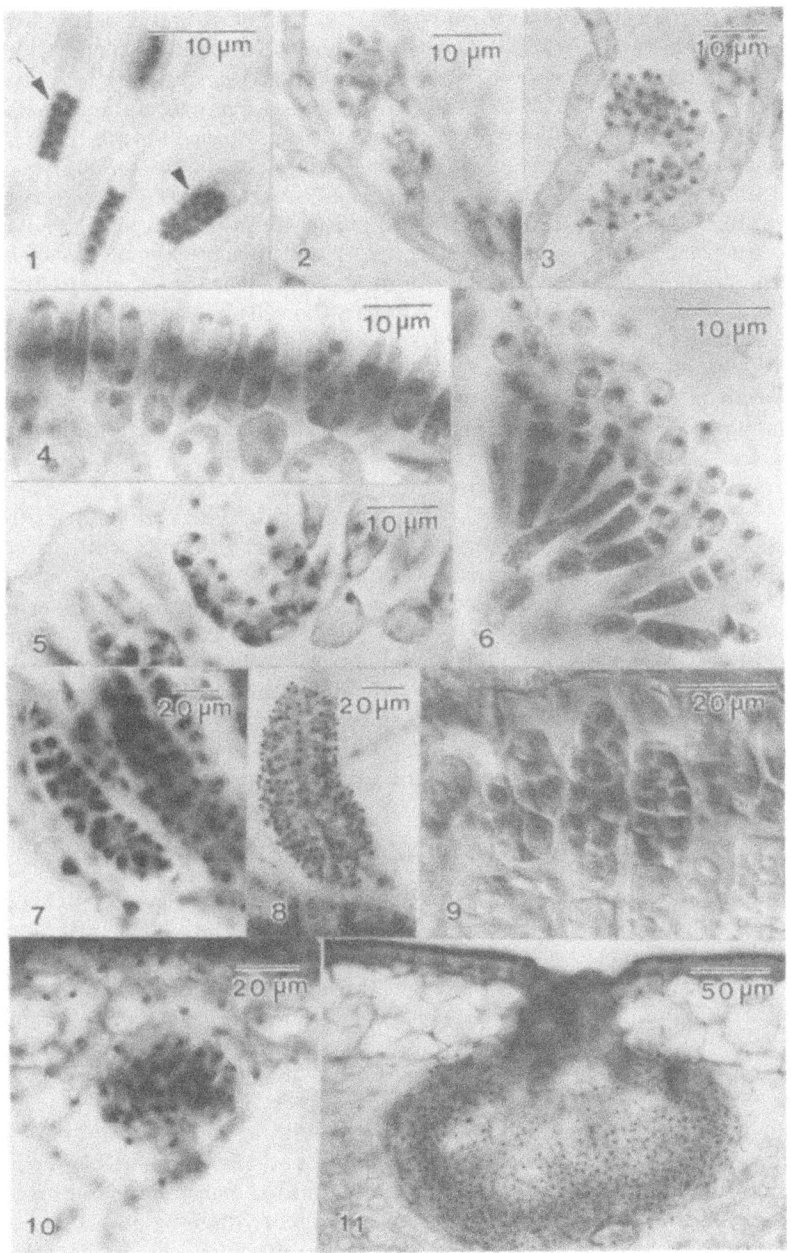

Figs. 13-1 to 13-11. Spermatangia and spermatangial branches [Unless otherwise indicated, all material was fixed in 5–8% Formalin-seawater and stained with aceto–iron–hematoxylin–chloral hydrate according to the method of Wittmann (1965) as modified by Coomans (1986) and Hommersand and Fredericq (1988).] 13-1: *Metamastophora flabellata* (West Australia, material provided by W. Woelkerling), spermatangia with nuclei in anaphase in side view (arrow) and face view (arrowhead). 13-2: *Audouinella violacea* (North Carolina, material provided by R. Coomans). 13-3: *Dudresnaya crassa* (Florida). 13-4: *Gracilariopsis lemaneiformis* (California). 13-5: *Gracilaria verrucosa* (Ireland). 13-6: *Holmsella pachyderma* (Wales, material provided by E. Jones). 13-7, 13-8: *Polysiphonia harveyi* (North Carolina). 13-9: *Peyssonnelia dubyi* (Ireland, material provided by C. Maggs). 13-10, 13-11: *Galaxaura diesingiana* (South Africa). (See text for descriptions.)

(Fig. 13-5). Spermatangia are produced on special spermatangial branches borne on trichoblasts in *Polysiphonia* (Fig. 13-7), with the spermatangia embedded in a confluent matrix beneath a common outer layer (Fig. 13-8). Spermatangial mother cells repeatedly proliferate spermatia within the common matrix, and older spermatia are released as young spermatia expand to take their place (Kugrens 1980).

In several families of red algae the male and female reproductive systems utilize many of the same developmental pathways. The similar male and female conceptacles of the Corallinaceae are a striking example (Johansen 1981; Woelkerling 1988). Spermatangia and carpogonia are produced in superficial pustules in the nemathecial families Rhizophyllidaceae, Peyssonneliaceae, and Polyideaceae. The spermatangia are borne laterally on spermatangial filaments formed during nemathecial development, as in *Peyssonnelia* (Fig. 13-9). In *Galaxaura* (Svedelius 1942) the male structure is produced from a modified filament bearing whorled laterals that resembles the carpogonial filament and forms an ostiolate conceptacle that is very similar to the female conceptacle (compare Figs. 13-10, 13-11 with Figs. 13-58, 13-59).

VII. THE CARPOGONIUM AND CARPOGONIAL BRANCH

Cell organelles do not appear to be modified extensively during carpogonial differentiation (See Chapter 2). In most Florideophycidae the carpogonium is transformed from the apical cell of a special lateral or terminal filament that usually contains a specified number of cells. Such a filament is called a carpogonial filament or carpogonial branch (Dixon 1973). Exceptions are the orders Acrochaetiales, Palmariales, and Gelidiales, in which the carpogonium is borne terminally or laterally on a vegetative filament or is intercalary (Gabrielson & Garbary 1986; Hommersand & Fredericq 1988).

A carpogonial branch can arise in one of two ways: (1) laterally, as a secondary filament produced by an adventitious initial, and (2) terminally, by transformation of the apical cell of an ordinary vegetative filament. In Ceramiales the carpogonial branch is almost always a four-celled filament developed from a lateral initial cut off from a periaxial cell (Hommersand 1963; Kylin 1923). In Liagoraceae it is either a lateral branch, as in most species of *Liagora*, or is terminal on a vegetative filament, as in *Nemalion* (Abbott 1976).

Some of the difficulties encountered in evaluating the ontogeny of carpogonial branches are illustrated by *Dudresnaya*. As Kraft and Robins (1985) showed, both carpogonial branches and auxiliary cell branches are modified from secondary cortical filaments produced in addition to the primary cortical filaments generated at the thallus apex. Early development follows the same pattern in both types of branches. The apical initial of a secondary filament shifts from producing elongated vegetative cells to forming cells that are broader than long. Vacuoles disappear, the cells become densely filled with cytoplasm, and the nuclei enlarge (Siotas & Wetherbee 1982).

Carpogonial branch (and auxiliary cell branch) ontogeny is probably triggered by genetically controlled processes that may or may not have an immediate morphological expression. The transforming event may coincide with formation of the carpogonial branch initial at the time that it is cut off from the supporting cell. Alternatively, transformation to a reproductive state may involve the apical cell of an already formed cortical filament, rhizoidal filament, or modified indeterminate axis. In this latter case delimitation of the carpogonial branch is determined by cytological changes, and filament length and position of the supporting cell are poorly defined.

VIII. FERTILIZATION

Fertilization in Rhodophyta is effected through spermatial adhesion to the surface layer of the carpogonial wall. Spermatia are thought to be transferred passively through the water column (Dixon 1973). In at least one instance, *Tiffaniella* (Ceramiaceae), the spermatia are released from a number of spermatangial heads into spermatial strands that coalesce, extend, contract, and rotate in the water until contact is made with a female plant (Fetter & Neushul 1981). Sheath and Hambrook (Chapter 16) have proposed that fluid dynamics are important in increasing the probability of gamete fusion.

Attachment of a spermatium to a carpogonium appears to require the presence of binding substances secreted by exocytosis from the tip of the trichogyne (Broadwater & Scott 1982). The binding between gametes of *Callithamnion* are not species specific, but occur only between closely related species (Magruder 1984). Trichogynes persist after fertilization in many Florideophycidae, and several spermatia may attach either near the tip of the trichogyne, as in *Batrachospermum* (Fig. 13-42), *Nemalion* (Fig. 13-12) or *Polysiphonia*, (Fig. 13-30), or along

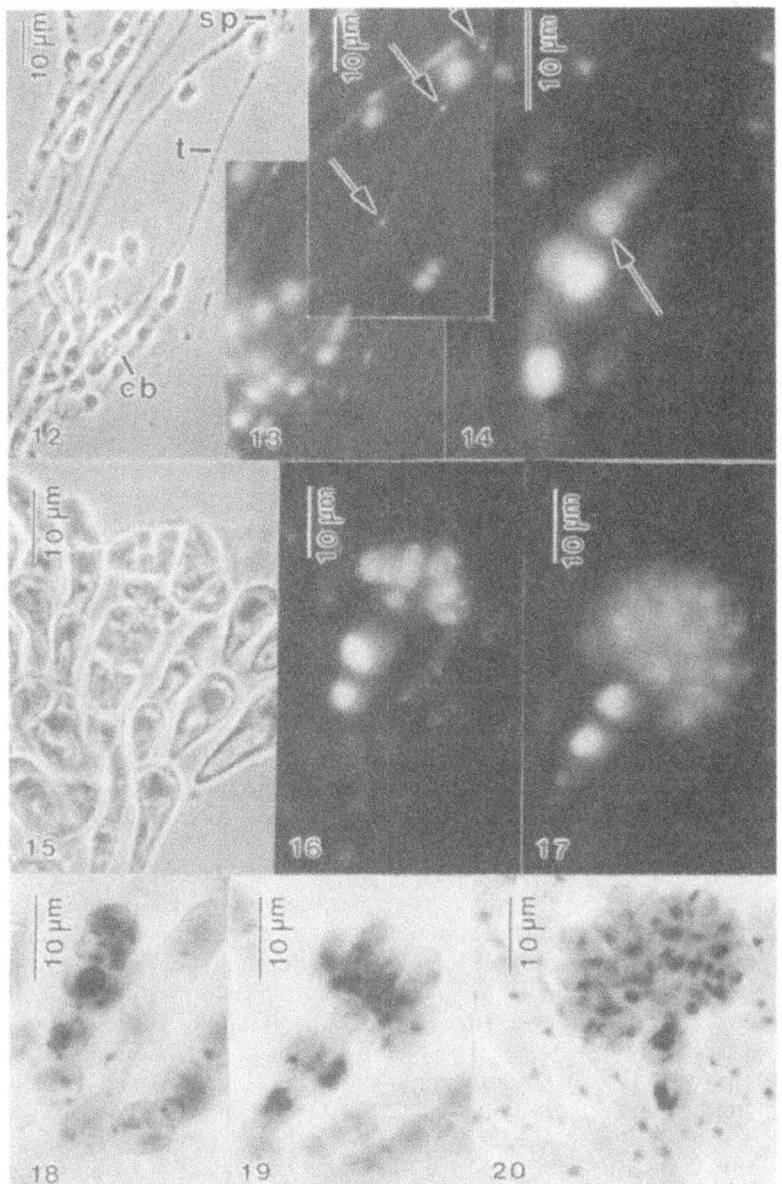

Figs. 13-12 to 13-17. *Nemalion helminthoides* (California, phase contrast and ultraviolet photographs by M. Rosczyk from material stained with DAPI). *13-12*: Carpogonial branch (*cb*) and trichogyne (*t*) with attached spermatia (*sp*) in phase contrast. *13-13*: Sperm nuclei (arrows) within trichogyne (UV). *13-14*: Sperm nucleus (arrow) fused to egg nucleus (UV). (Note enlarged nuclei in hypogynous and subhypogynous cells). *13-15*: Carpogonial branch with gonimoblasts (phase). *13-16*: Same as Fig. 13-15 (UV). *13-17*: Older gonimoblasts (UV).
Figs. 13-18 to 13-20. *Nemalion helminthoides* (Nova Scotia), stages of carpogonial branch fusion and gonimoblast development.

its sides, as in *Acrosymphyton* (Figs. 13-80, 13-81). We have observed that in some genera, such as *Gelidium* and *Gracilaria*, the trichogyne collapses soon after fertilization, and attached spermatia are rarely seen.

Spermatia may first attach to vegetative cells. In *Callithamnion*, the spermatia have fimbriate, cone-shaped appendages projecting from both ends that are visible with scanning electron microscopy (Magruder 1984). These may attach initially to vegetative hair cells and later bind to a nearby trichogyne. In *Hymenena* (Delesseriaceae) spermatia appear to attach to the thallus surface. Trichogynes of receptive carpogonia, which just break the thallus surface, appear to extend toward the spermatium and fuse with it (Hommersand, unpubl.). It is possible that the spermatia of some species secrete a chemotropic substance that promotes the directed growth of the trichogyne.

Breakdown of the carpogonial wall and entry of the spermatium nucleus and cytoplasm appear to involve the enzymatic digestion of the wall. In *Porphyra* a narrow channel, the fertilization canal, is formed, through which the spermatial nucleus passes (Hawkes 1978). The fusion area is broader in other red algae and leads to an open connection between the cytoplasm of the spermatium and that of the carpogonium. Several nuclei from different spermatia may enter the carpogonium in *Porphyra*, although only one appears to fuse with the carpogonial nucleus (Hawkes 1978). With the exception of the Batrachospermales, the trichogyne of Florideophycidae has a narrow neck at its base that admits the passage of a single nucleus. In *Polysiphonia* (and probably most other Florideophycidae) the vacuole contracts immediately after the sperm nucleus has entered the base of the carpogonium and draws the sperm nucleus into contact with the egg nucleus, at the same time pinching off the cytoplasm (Broadwater & Scott 1982). The latter event separates the trichogyne from the inflated base of the carpogonium, effectively preventing supernumerary fertilizations.

IX. PRODUCTS OF THE FERTILIZATION NUCLEUS

We recognize three major patterns of carposporophyte development based on the fate of the first cell derivatives cut off from the carpogonium that carry the division products of the fertilization nucleus: (1) they give rise directly to gonimoblasts (*Nemalion* type), (2) they initiate connecting processes or filaments that unite with one or more auxiliary cells, and either the gonimoblasts develop directly from the connecting filament or a diploid nucleus is transferred to the auxiliary cell, which produces the gonimoblasts (*Dudresnaya* type), or (3) they cut off one, two, or three connecting cells, each of which may fuse with an adjacent auxiliary cell, depositing its nucleus, and the auxiliary cell produces the gonimoblasts (*Polysiphonia* type).

A. Direct formation of gonimoblasts: *Nemalion* type

The following report of carposporophyte development in *Nemalion* is based primarily on the research of Rosczyk (1984). The carpogonium of *Nemalion* is borne terminally on a lateral filament that is not at first distinguishable from a vegetative lateral. At about the same time that the carpogonium initiates a trichogyne, the hypogynous cell and the cell immediately below it expand and their nuclei enlarge. Staining with DAPI reveals that the DNA levels of these nuclei become greatly amplified during differentiation of the carpogonial branch prior to fertilization (Rosczyk 1984). Several spermatia may become attached near the tip of the trichogyne (Fig. 13-12) and sperm nuclei can be seen migrating through the trichogyne towards the carpogonium (Fig. 13-13). Usually, only one sperm nucleus unites with the carpogonial nucleus (Fig. 13-14). The fertilized carpogonium first divides transversely into two cells (Fig. 13-18). Only the terminal cell produces gonimoblasts, while the lower cell fuses with the hypogynous and subhypogynous cells of the carpogonial branch containing the enlarged nuclei (Fig. 13-19). The modified nuclei of these cells persist within the fusion cell throughout the course of gonimoblast development and are conspicuous when stained either with DAPI (Figs. 13-15 to 13-17) or hematoxylin (Figs. 13-19, 13-20). Branching of the carposporophyte is initially monopodial, with the terminal gonimoblast cells differentiating into carposporangia (Fig. 13-19). Once carposporangia have formed, lateral filaments are initiated from subterminal cells, and there is a shift to a pattern of sympodial growth as terminal cells differentiate successively into carposporangia (Ramm-Anderson & Wetherbee 1982).

Direct development of gonimoblasts from the carpogonium is seen in the orders Acrochaetiales, Batrachospermales, Nemaliales, Gelidiales, and Bonnemaisoniales, and in *Ahnfeltia* and the family Gracilariaceae, as described in Section X.

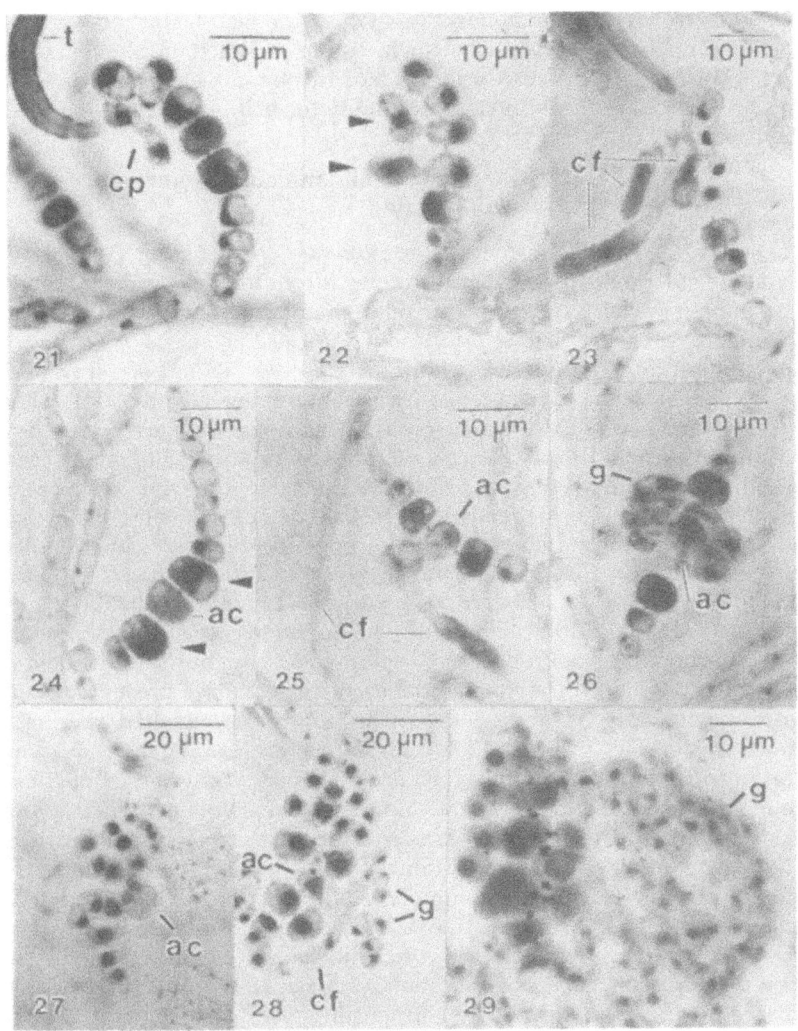

Figs. 13-21 to 13-26. *Dudresnaya crassa* (Florida). *13-21*: Division of fertilized carpogonium (cp) with trichogyne (t). *13-22*: Fusion of carpogonium and its derivative cell (arrowheads). *13-23*: Fusion cells issuing connecting filaments (cf). *13-24*: Auxiliary cell filament, auxiliary cell (ac) and adjoining nutritive cells (arrowheads). *13-25*: Connecting filament (cf) fused to auxiliary cell (ac). *13-26*: Auxiliary cell (ac) bearing gonimoblasts (g).
Figs. 13-27 to 13-29. *Kraftia dichotoma* (South Australia, photographs by M. Knauss). *13-27*: Auxiliary cell branch and auxiliary cell (ac). *13-28*: Auxiliary cell (ac), connecting filament (cf) and gonimoblasts (g). *13-29*: Auxiliary cell branch with modified pit connections and gonimoblasts (g).

B. Connecting filaments and auxiliary cells: *Dudresnaya* type

Dudresnaya is a comparatively primitive member of the family Dumontiaceae (Gigartinales). The details of postfertilization development have been particularly well documented by Robins and Kraft (1985).

Cells of the carpogonial filament become differentiated before fertilization and contain enlarged nuclei, especially the fourth and fifth and sometimes the sixth cells behind the carpogonium. Typically, the carpogonium elongates alongside the carpogonial filament after fertilization (Fig. 13-21) and cuts off a derivative cell. The carpogonium and its derivative fuse with the fourth and fifth cells, respectively (Fig. 13-22). The resulting fusion cells initiate several connecting filaments (Fig. 13-23).

The auxiliary cell in *Dudresnaya* is an intercalary cell in a separate filament that is homologous with the carpogonial branch. It is easily distinguished in hematoxylin-stained material by the presence of an unmodified nucleus containing a nucleolus surrounded by a clear hyaline region (Fig. 13-24). The cells on either side of the auxiliary cell are enlarged and contain highly modified, deeply staining nuclei (Fig. 13-24). When a connecting filament reaches the vicinity of an auxiliary cell, it cuts off an intercalary segment by two successive divisions. The auxiliary cell and the intercalary segment fuse, while the terminal cell remains a connecting filament (Fig. 13-25). An auxiliary cell that is not in close proximity to the intercalary segment will form a process that extends toward and fuses with the segment of the connecting filament. The intercalary segment may branch, cutting off one to two initials that develop into secondary connecting filaments (Robins & Kraft 1985). Ultimately, two to three gonimoblasts are produced from the segment of the connecting filament that has fused to the auxiliary cell, and these coalesce to form a single cluster (Fig. 13-26). Pit connections between the nutritive cells of the auxiliary cell branch increase in surface area, often becoming convoluted (Siotas & Wetherbee 1982).

Kraftia is similar to *Dudresnaya* except that the auxiliary cell filament is branched and contains a greater number of modified cells with enlarged nuclei (Fig. 13-27). The nuclei in the cells adjacent to the auxiliary cell continue to enlarge during gonimoblast development, and the pit plugs broaden asymmetrically, becoming broader on the side facing toward the auxiliary cell (Figs. 13-28, 13-29). Connecting filament behavior and gonimoblast development follow a common pattern in most genera of Dumontiaceae (Lindstrom & Scagel 1987).

Connecting filaments occur only in the orders Cryptonemiales and Gigartinales, as illustrated in Section X. The connecting filaments are septate and frequently branched in the families Calosiphoniaceae, Nemastomataceae, Furcellariaceae, Acrosymphytaceae, Polyideaceae, Rhizophyllidaceae, and Peyssonneliaceae. They are essentially nonseptate and form secondary connecting filaments in the vicinity of auxiliary cells in the Dumontiaceae, Gloiosiphoniaceae, Halymeniaceae, and most Kallymeniaceae. Alternatively, connecting filaments may be unbranched and terminate with the diploidization of a single auxiliary cell in the Caulacanthaceae and Solieriaceae. Other families of Gigartinales have procarps in which the connecting filaments are short or absent and carpogonia and auxiliary cells are in close proximity to each other.

C. Connecting cells and auxiliary cells: *Polysiphonia* type

The female reproductive apparatus of *Polysiphonia* is regularly produced from the fifth periaxial cell of the second-basal segment of a modified trichoblast. Before fertilization it normally consists of a supporting cell bearing a two-celled lateral sterile group, a four-celled carpogonial branch, and a one-celled basal sterile group. It is surrounded by a prefertilization pericarp composed of cortical filaments that develop primarily from the third and fourth periaxial cells of the fertile segment (Kylin 1956). Occasionally, the carpogonial branch is three-celled (Figs. 13-30 to 13-37).

The trichogyne persists after fertilization, and one commonly sees spermatia attached near the tip and sperm nuclei migrating within the trichogyne. The total number of male nuclei present usually corresponds to the number of attached spermatia (Figs. 13-30, 13-31). Before karyogamy, the carpogonium is flask shaped with the carpogonial nucleus and associated vacuole suspended below the base of the trichogyne (Fig. 13-30). Immediately after karyogamy, the trichogyne separates and the carpogonium broadens and forms a process that contacts the surface of the supporting cell (Fig. 13-31). Pit connections between the cells of the carpogonial branch break down, and the nucleus inside the supporting cell divides (Fig. 13-32). Broadwater and Scott (1982) documented this stage with electron microscopy and suggested that a hormone released from the fertilized carpogonium is transported through the carpogonial branch to the supporting cell, where it triggers formation of the auxiliary cell.

The carpogonium next divides (Fig. 13-33), cutting off the first connecting cell toward the outside, which is nonfunctional. The second division of the zygote nucleus produces a connecting cell along the inner side of the auxiliary cell (Fig. 13-34). The auxiliary cell then expands and fuses with the connecting cell (Fig. 13-35). Contrary to reports in the literature (Broadwater & Scott 1982; Kylin 1923), we have never seen evidence of direct fusion between the auxiliary cell and the carpogonium that resulted in the transfer of a derivative of the fertilization nucleus. Concomitant with these events, terminal cells of the pericarpic filaments are converted into

Sexual reproduction and cystocarp development

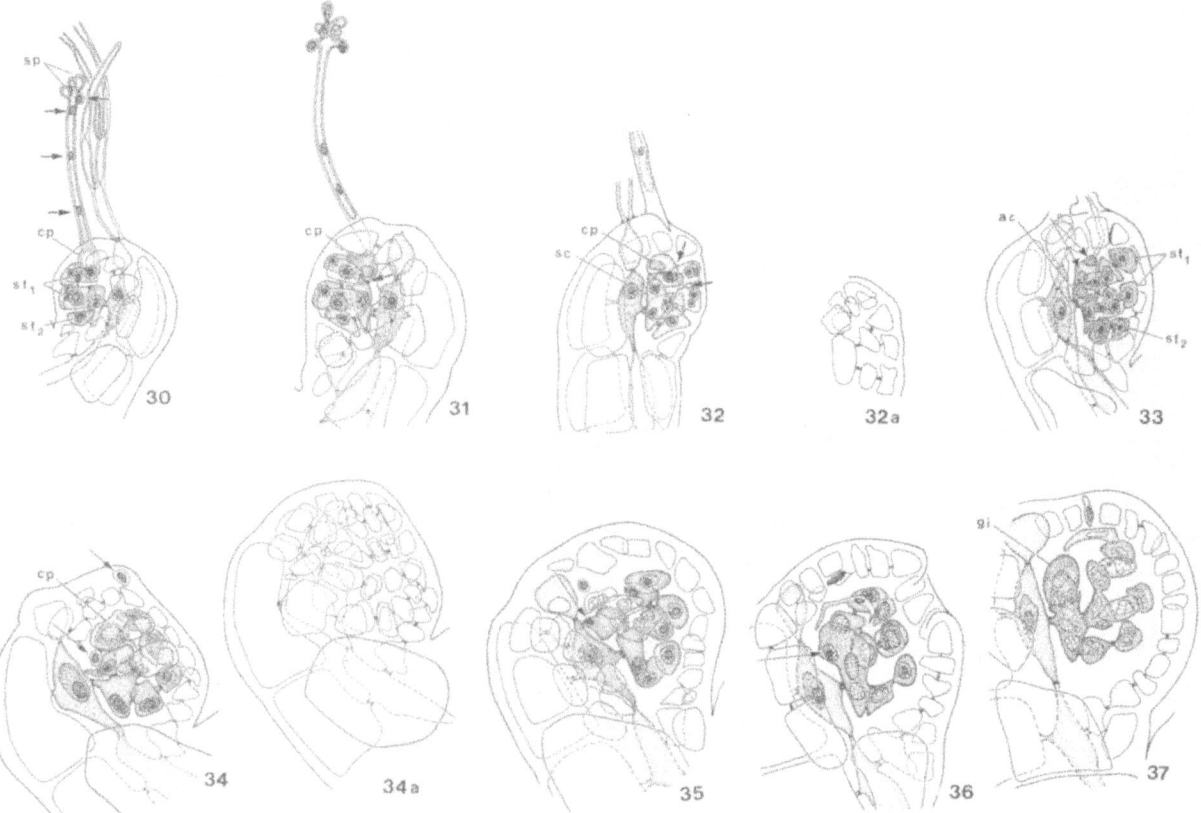

Figs. 13-30 to 13-37. *Polysiphonia harveyi* (North Carolina). *13-30*: Fertilized carpogonium (cp) before karyogamy showing spermatia (sp) and sperm nuclei (small arrows), with two-celled sterile group 1 (st_1) and one-celled sterile group 2 (st_2). *13-31*: Carpogonium (cp) after karyogamy contacting supporting cell (small arrow). *13-32*: Supporting cell (sc) after nuclear division; pit connections between cells of carpogonial branch have degenerated (small arrows). *13-32a*: Pericarp of Fig. 13-32. *13-33*: Fertilization nucleus has divided (arrows), auxiliary cell (ac) cut off, sterile groups 1 and 2 (st_1, st_2) have divided. *Fig. 13-34*: Carpogonium (cp) has cut off two connecting cells (arrows). *13-34a*: Pericarp of Fig. 13-34. *13-35*: Auxiliary cell fused with inner connecting cell (arrow). *13-36*: Diploid nucleus (arrow) expanded inside auxiliary cell. *13-37*: Gonimoblast initial (gi).

apical initials of axial files that produce the post-fertilization pericarp (compare Figs. 13-32a and 13-34a), and the two sterile groups divide once, so that sterile group 1 becomes four-celled and sterile group 2 becomes two-celled (Fig. 13-33). The nuclei within cells of each of the two sterile groups enlarge and stain deeply with hematoxylin. Pit connections break down between the sterile groups and the supporting cell (Figs. 13-35, 13-36), followed by dissolution of the pit connections between the outer and inner cells of the sterile groups themselves (Fig. 13-37). The diploid nucleus inside the auxiliary cell enlarges (Fig. 13-36) and divides, cutting off the gonimoblast initial (Fig. 13-37), which remains quiescent while the photosynthetic pericarp expands to its full size. Branching of the gonimoblast is initially monopodial but becomes sympodial as terminal cells mature into carposporangia and new gonimoblast filaments issue from subterminal cells. The gonimoblast initial and auxiliary cell fuse, and fusions extend progressively to incorporate the inner cells of the gonimoblast. Nuclei and proplastids break down inside the expanding fusion cell, and pit connections between unfused gonimoblast cells lose their cap membranes (Wetherbee 1980).

A derivative of the zygote nucleus is transferred to the auxiliary cell either by a carpogonial process or through direct fusion, or transfer is mediated by a connecting cell in Ceramiales and Rhodymeniales (Dixon 1973). Connecting cells are well documented in the family Dasyaceae (Parsons 1975). Our own

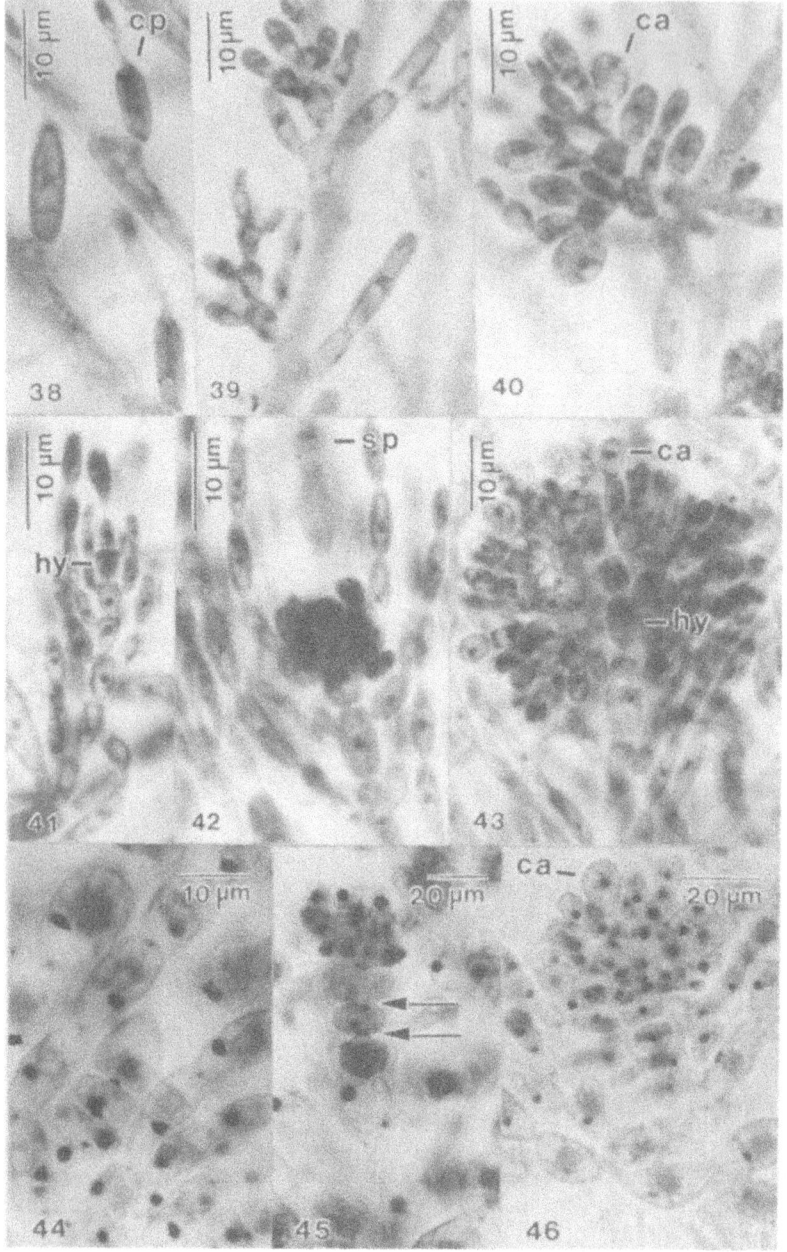

Figs. 13-38 to 13-40. *Audouinella violacea* (North Carolina, material provided by R. Coomans). *13-38*: Unfertilized carpogonia (cp). *13-39, 13-40*: Developing gonimoblasts with young carposporangia (ca).

Figs. 13-41 to 13-43. *Batrachospermum boryanum* (North Carolina, material provided by R. Coomans). *13-41*: Carpogonial branch with deeply staining hypogynous cell (hy). *13-42*: Fertilized carpogonium with attached spermatium (sp) and gonimoblasts. *13-43*: Gonimoblasts with terminal carposporangia (ca) and deeply staining hypogynous cell (hy).

Figs. 13-44 to 13-46. *Liagora mucosa* (Florida). *13-44*: Lateral carpogonial branch. *13-45, 13-46*: Carpogonial branch with broadened pit connections (arrows), developing gonimoblasts, and terminal carposporangia (ca).

survey indicates that direct transfer occurs in the Antithamnieae, Ptiloteae, and Ceramieae, but that diploidization of the auxiliary cell is effected by means of a connecting cell in most other tribes of Ceramiaceae and in the families Delesseriaceae, Dasyaceae, and Rhodomelaceae. Connecting cells have been illustrated for the Rhodymeniaceae (Sparling 1957) and the Champiaceae (Bliding 1928) of the Rhodymeniales, and direct fusions have also been reported between the fused carpogonial branch and the auxiliary cell (Lee 1978).

X. FUNCTIONAL ADAPTATIONS OF CYSTOCARP MORPHOLOGY

A. Absence of special nutritive tissues: Acrochaetiales and Palmariales

Among the Florideophycidae only the Acrochaetiales and Palmariales appear to lack supplementary vegetative nutritive filaments or specially modified vegetative cells. In most Acrochaetiaceae the carpogonium divides transversely after fertilization, and both the upper and lower cells produce gonimoblasts (Woelkerling 1970). A comparison of female thalli of *Audouinella violacea* before and after fertilization (Figs. 13-38 to 13-40) shows little evidence of the production of secondary assimilatory filaments and no obvious cytological differentiation of vegetative cells. Moreover, pit connections between adjoining cells do not seem to enlarge or become modified. In *Palmaria* the fertilized carpogonium gives rise directly to a free-living sporophyte that develops autotrophically (van der Meer & Todd 1980).

B. Modified carpogonial branches: Batrachospermales and Nemaliales

The Batrachospermales comprises an independent line in which the nutritive system is relatively unspecialized in most species of *Batrachospermum* (Batrachospermaceae) but is highly modified in *Lemanea* (Lemaneaceae). Carpogonia are terminal on cortical filaments or on modified indeterminate branches in *Batrachospermum boryanum*. The hypogynous cell is deeply staining, and inner cells of the carpogonial branch bear juvenile filaments before fertilization (Fig. 13-41) that grow into functional assimilatory filaments as the carposporophyte develops (Figs. 13-42 to 13-43). A spermatium often persists, attached to the fertilized carpogonium (Fig. 13-42), and the carpogonium gives rise directly to gonimoblasts bearing terminal carposporangia (Fig. 13-43). Although pit connections between cells of the carpogonial branch and inner gonimoblast cells enlarge somewhat, they are only slightly modified (Kylin 1917). An extensive fusion cell is formed in *Tuomeya* (Webster 1958).

The female reproductive system of the Lemaneaceae is highly specialized. Carpogonial branches of *Lemanea* are restricted to internodal regions and are four to ten cells long. As they mature, the plastids degenerate, and spherical-celled laterals develop profusely from all cells of the carpogonial branch except the carpogonium (Mullahy 1952a). After fertilization, the sterile laterals increase in number and become deeply staining with hematoxylin. Gonimoblasts originate directly from the carpogonium and overgrow and penetrate among the sterile laterals. Subsequently, inner gonimoblast cells fuse, rupturing their pit connections, while the sterile laterals become vacuolate (Mullahy 1952b). Outer gonimoblast filaments are unpigmented and profusely branched, with most of the cells maturing into chains of carposporangia.

The multiaxial family Thoreaceae is presently placed in the Batrachospermales (Yoshizaki 1986). Recent studies have demonstrated that the carposporophyte develops directly from the carpogonium in *Thorea* and ramifies among the surrounding gametophytic filaments (Necchi 1987; Yoshizaki 1986). No specialized vegetative cells or filaments have been described, and cell fusions appear to be absent.

The multiaxial order Nemaliales contains at least two families, the Liagoraceae and Galaxauraceae. Secondary assimilatory filaments are produced directly on the carpogonial branch or are formed by cortical filaments in the immediate vicinity of a fertilized carpogonium. Only cells of the carpogonial branch or laterals borne on it may contain modified nuclei. Basic features of sexual reproduction in Nemaliales were described for *Nemalion helminthoides* in Section IX.A.

Carpogonial branches are lateral in *Liagora mucosa* (Fig. 13-44). The carpogonium divides transversely after fertilization, with only the distal cell producing gonimoblasts (Fig. 13-45). Cortical cells adjacent to the supporting cell characteristically produce secondary assimilatory filaments that coalesce to form an involucre surrounding the carposporophyte. The cells of the carpogonial branch are little modified during carposporophyte development. Nuclei remain small, and pit connections between adjacent cells broaden slightly (Figs. 13-45, 13-46). Cell fusions appear to be absent.

Gonimoblast filaments ramify among the surrounding assimilatory filaments in *Dermonema frappieri* without fusing with them (Svedelius 1939). The supporting cell and cells of the carpogonial branch contain enlarged nuclei and stain heavily for protein before fertilization (Fig. 13-47). They retain their modified nuclei and staining properties during gonimoblast development (Fig. 13-48) and eventually unite into a fusion cell (Fig. 13-49).

An undescribed species of *Trichogloea* possesses several unusual characters. Assimilatory filaments are absent from the carpogonial filament before fertilization (Fig. 13-50) and are only produced after fertilization (Figs. 13-51, 13-52). The unfertilized carpogonium stains deeply and contains two nuclei: an egg nucleus with a typical nucleolus and an enlarged, modified nucleus (Fig. 13-50). Cells of the carpogonial branch fuse around the primary pit connections after fertilization, and the fusion cell becomes multinucleate and rich in protein content (Fig. 13-51). Later, it becomes vacuolate, and the remnant primary pit connections can be seen (Fig. 13-52). Ultimately, fusions extend to include the basal cells of the gonimoblast (Fig. 13-53) and, as was described for *Nemalion*, branching is initially monopodial, becoming sympodial as the terminal cells mature into carposporangia (Fig. 13-54).

Some of the trends seen in the Liagoraceae are carried further in the Galaxauraceae. In both *Scinaia* and *Galaxaura* the carpogonium is the terminal cell of a three-celled fertile branch in which both the hypogynous cell and the basal cell bear lateral filaments (Svedelius 1915, 1942). In *Scinaia* the hypogynous cell and its laterals are enlarged, stain deeply, and contain greatly enlarged nuclei, and the basal cell bears a whorl of juvenile involucral filaments (Fig. 13-55). An extensive involucre develops after fertilization while the nuclei in the hypogynous cell and its laterals persist (Fig. 13-56). The carpogonium produces four gonimoblast initials that fuse inwardly with the carpogonium, the hypogynous cell, and its laterals and outwardly with the inner sterile cells of the gonimoblast (Fig. 13-57).

The lateral filaments borne on the hypogynous cell are extensively branched in *Galaxaura*, and each cell contains an enlarged nucleus, whereas the involucral filaments are only weakly developed on the basal cell prior to fertilization (Fig. 13-58). Fusions become even more extensive after fertilization than in *Scinaia* and incorporate a much larger portion of the enveloping involucral system (Figs. 13-59, 13-60). Filaments of the carposporophyte branch in close association with the developing involucre to form a conceptacle with an ostiole.

C. The female conceptacle and fusion cell: Corallinales

Several theories have been proposed to account for the structure of the female reproductive system in the Corallinales (Lebednik 1977; Woelkerling 1980), but postfertilization events are not fully understood. Woelkerling (1980) observed in *Metamastophora* that a channel forms between the fertilized carpogonium and the supporting cell as a result of direct fusion through the hypogynous cell. Multiple fertilizations may take place, resulting in the formation of several initial fusion cells in *Synarthrophyton* (Woelkerling & Foster 1989). The initial fusion appears to extend laterally through the floor of the conceptacle at the level of the supporting cells of the carpogonial branches to form a plate-like fusion cell characteristic of many Corallinales (Johansen 1981; Woelkerling 1988). Gonimoblast filaments bearing terminal carposporangia are produced at the margin of the fusion cell in *Metamastophora flabellata* (Figs. 13-61, 13-62). In *Metamastophora* the haploid nucleus of each supporting cell remains in a fixed position within the fusion cell, where it enlarges (Figs. 13-62, 13-63). Clavate paraphyses situated adjacent to carpogonial branches persist on each supporting cell after fertilization and develop enlarged nuclei and conspicuous globules (Fig. 13-62). Pit plugs between the fusion cell and the gonimoblast filaments broaden and thicken, and dome-shaped plug caps form over the pit connections between the paraphyses and the fusion cell (Fig. 13-63) and between the gonimoblast filaments and fusion cell (Fig. 13-64). The conceptacle of Corallinales appears to be similar functionally to the cystocarp of other red algae, with the outer cortex serving as the photosynthetic compartment and the fusion cell acting as a nutrient-processing center.

D. Preformed nutritive tissues: Gelidiales

The nutritive system is largely formed before fertilization in the Gelidiales (Fan 1961; Hommersand & Fredericq 1988). Deeply staining, branched nutritive filaments are produced secondarily from inner cortical cells at the tips of fertile female pinnules prior to carpogonium maturation. Nuclei in the cortical cells adjacent to the carpogonium enlarge after fertilization, and a channel develops through the pit connections leading to the carpogonium in *Gelidium pteridifolium* (Hommersand & Fredericq 1988). At the same time that the carpogonium fuses with the adjoining cortical cells it becomes multinucleate and initiates uninucleate gonimoblast fila-

Sexual reproduction and cystocarp development

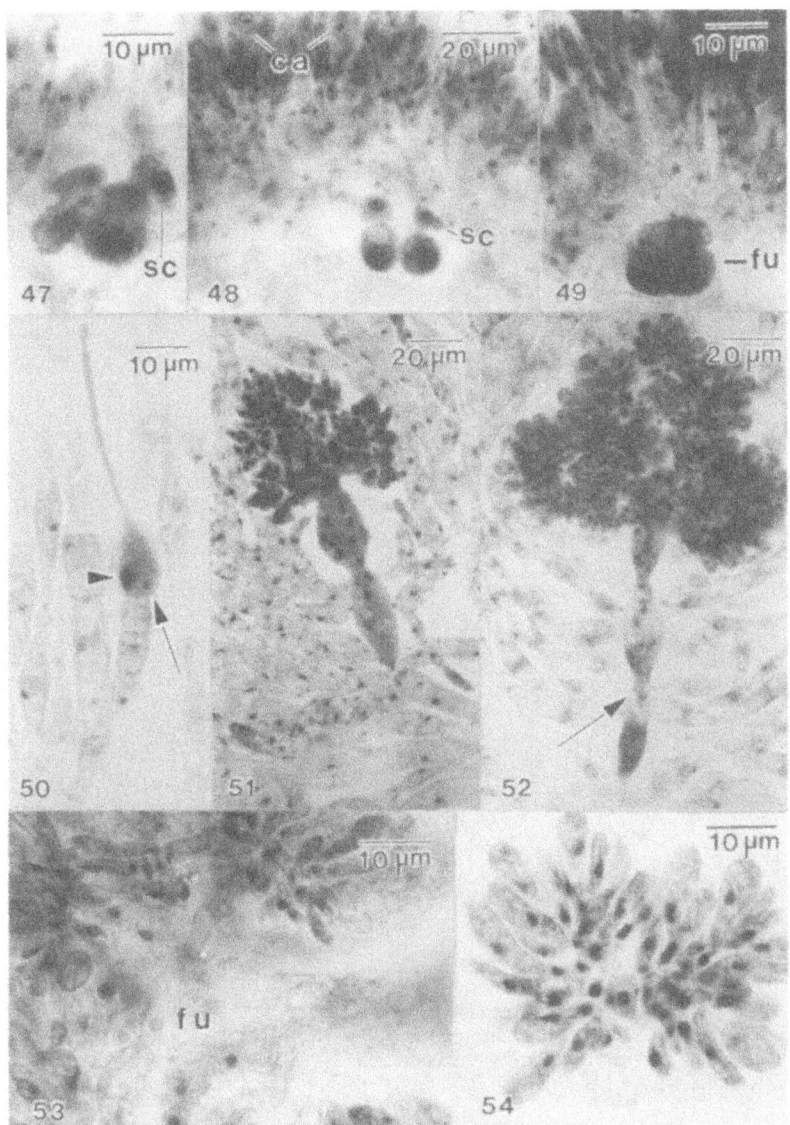

Figs. 13-47 to 13-49. *Dermonema frappieri* (Taiwan, material collected by S. Martinique). *13-47*: Supporting cell (sc) and carpogonial branch. *13-48, 13-49*: Supporting cell (sc), developing fusion cell (fu) and gonimoblasts with terminal carposporangia (ca).

Figs. 13-50 to 13-54. *Trichogloea* sp. (Virgin Islands, material provided by J. Sears). *13-50*: Carpogonium with egg nucleus (arrow) and modified nucleus (arrowhead). *13-51*: Fused, multinucleate carpogonial branch with developing gonimoblasts and assimilatory filaments. *13-52*: Fusions around pit connection (arrow). *13-53*: Gonimoblast fusion cell (fu). *13-54*: Sympodially branched gonimoblast.

ments that ramify among the clusters of nutritive filaments in both *Gelidium* (Fig. 13-65) and *Suhria* (Fig. 13-66). Terminal gonimoblast cells either fuse specifically with the terminal cells of nutritive filaments, as in *G. pteridifolium* (Fig. 13-67), or randomly with terminal and intercalary cells, as in *S. vittata* (Figs. 13-68, 13-69), before cutting off carposporangial initials. At maturity the cystocarp is biconvex in *Gelidium* and *Suhria* and consists of a central partition supporting a plexus of intercon-

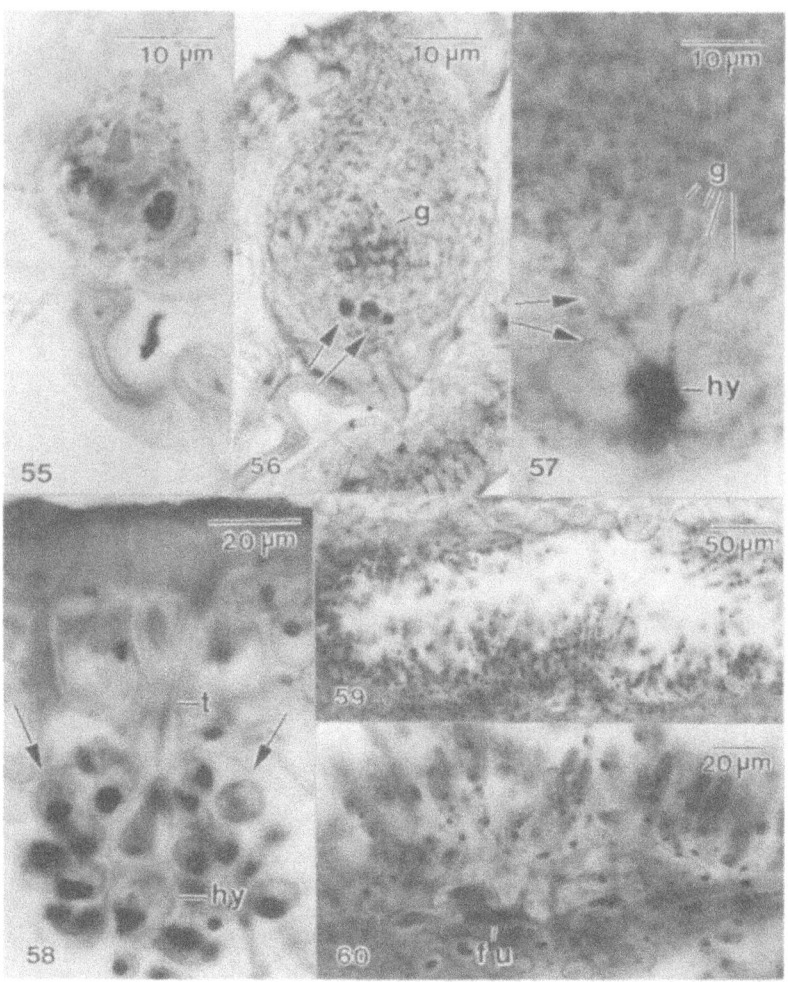

Figs. 13-55 to 13-57. *Scinaia complanata* (North Carolina, material collected by G. Hansen). *13-55*: Mature unfertilized carpogonial branch. *13-56*: Developing gonimoblasts (g), hypogynous cell, and its laterals (arrows) surrounded by involucral filaments. *13-57*: Gonimoblasts (g) fused to hypogynous cell (hy) and its laterals (arrows).

Figs. 13-58 to 13-60. *Galaxaura diesingiana* (South Africa). *13-58*: Carpogonial branch with carpogonium, trichogyne (t) and hypogynous cell (hy) bearing modified lateral filaments with enlarged nuclei. *13-59*: Cross section of cystocarp. *13-60*: Fusion cell (fu) at base of cystocarp.

nected nutritive and gonimoblast filaments bearing carposporangia, surmounted by pericarps transformed from slightly modified cortex on either side (Fig. 13-70).

E. Preformed nutritive tissues: Gigartinales

The primitive type of sexual reproduction in the Gigartinales is thought to be one in which the carpogonia and auxiliary cells are borne in separate branch systems (nonprocarpial) and many connecting filaments issue from a single fertilized carpogonium and give rise directly to gonimoblasts after uniting with auxiliary cells (Drew 1954; Kraft 1981). This type is represented by the genus *Dudresnaya* (Dumontiaceae), described in Section IX.B, and by representatives of the Calosiphoniaceae, Nemastomataceae, and several advanced families.

In *Schmitzia hiscockiana* (Calosiphoniaceae) the carpogonial branch is a three-celled lateral filament

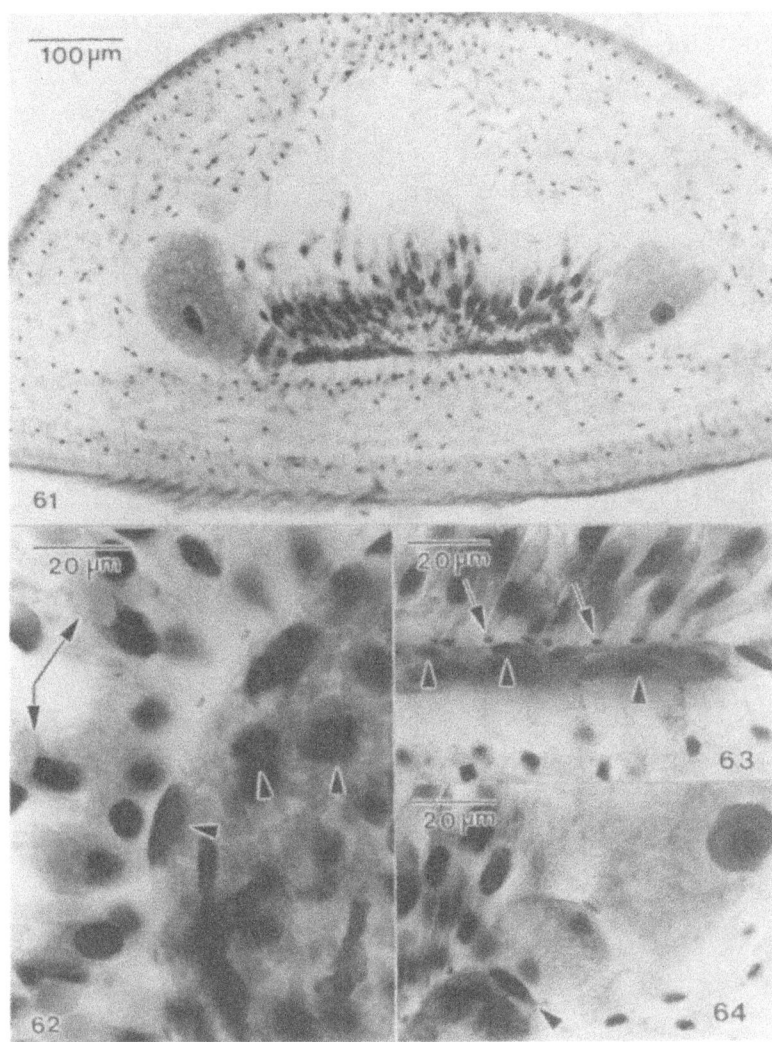

Figs. 13-61 to 13-64. *Metamastophora flabellata* (Western Australia, alcohol preserved material provided by W. Woelkerling). *13-61*: Cross section of mature cystocarpic conceptacle. *13-62*: Fusion cell with enlarged nuclei (arrowheads) and paraphyses with enlarged nuclei and globular bodies (arrows) in surface view. *13-63*: Fusion cell with enlarged nuclei (arrowheads) and paraphyses with modified pit connections (arrows) in side view. *13-64*: Enlarged pit connection (arrowhead) between gonimoblast filament and fusion cell.

(Maggs & Guiry 1985). Following fertilization (Fig. 13-71) the carpogonium and its first derivative fuse with nearby vegetative cells in the same branch system. The resulting fusion cell is multinucleate and issues uninucleate connecting filaments that are septate and branched (Fig. 13-72). Auxiliary cells are ordinary vegetative inner cortical cells. When a connecting filament approaches an auxiliary cell, the auxiliary cell forms a lateral process that fuses with the connecting filament. A cluster of gonimoblast filaments issues directly from the connecting filament near the point of fusion. In Fig. 13-73 two adjacent cortical cells function as auxiliary cells that have formed lateral processes. Only one has fused with the connecting filament.

The nemathecial families Rhizophyllidaceae (Wiseman 1977) and Peyssonneliaceae (Schneider & Reading 1987), Polyideaceae (Rao 1956) and the genus *Rhodopeltis* (Nozawa 1970) have separate carpogonial filaments and auxiliary cells interspersed

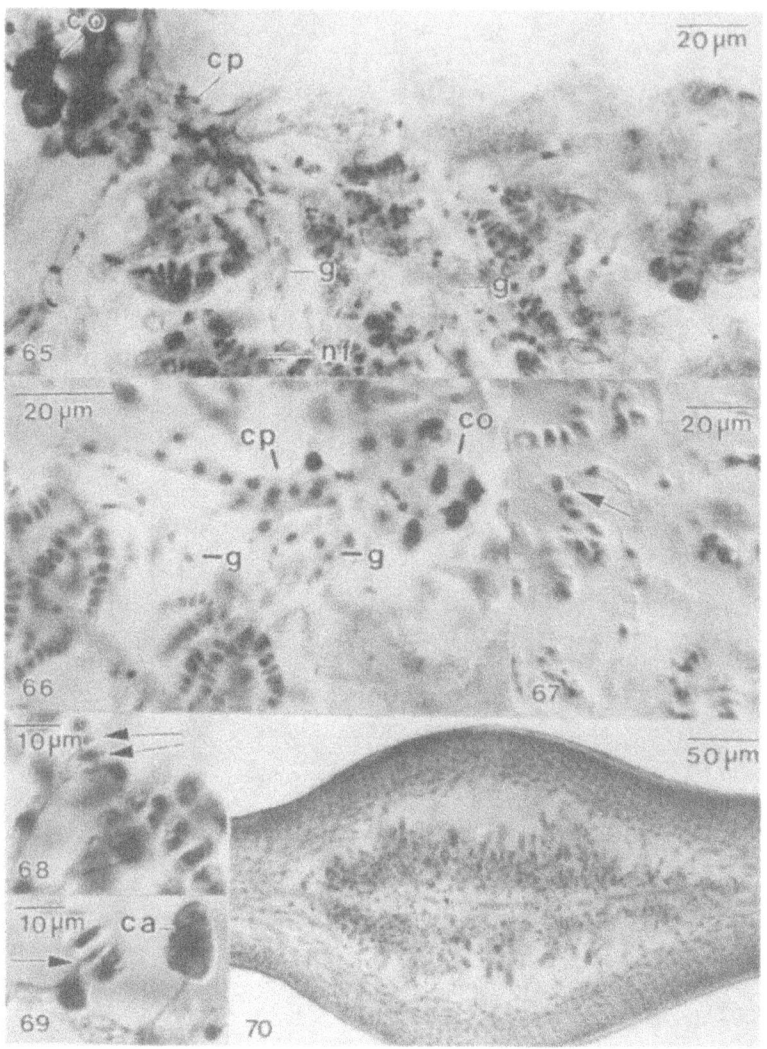

Figs. 13-65, 13-67. *Gelidium pteridifolium* (South Africa). *13-65*: Carpogonium (cp), fused cortical cells (co), gonimoblasts (g) and nutritive filaments. *13-67*: Fusion between terminal gonimoblast and terminal nutritive cells (arrow).

Figs 13-66, 13-68 to 13-70. *Suhria vittata* (South Africa). *13-66*: Carpogonium (cp), fused cortical cells (co), gonimoblasts (g), and nutritive filaments. *13-68*, *13-69*: Fusions between gonimoblast cells and nutritive cells (arrows). *13-70*: Cross section of cystocarp.

among assimilatory filaments within a common nemathecial pustule. *Portieria* (formerly *Chondrococcus*; see Silva et al. 1987) is representative of the Rhizophyllidaceae. The nuclei are enlarged in the basal cell of the carpogonial branch before fertilization (Fig. 13-74), and the cells on either side of the auxiliary cell contain enlarged nuclei in *Portieria* (Figs. 13-75, 13-76) and in *Peyssonnelia* of the Peyssonneliaceae (Fig. 13-77). Gonimoblasts develop directly from the connecting filaments in the vicinity of the auxiliary cells in both genera (Figs. 13-76, 13-78).

Acrosymphyton has recently been removed from the Dumontiaceae to the Acrosymphytaceae (Lindstrom 1986). Carpogonial branches and auxiliary cell branches are initially unbranched filaments that arise from inner cortical cells near the thallus apex (Fig. 13-79). Whereas auxiliary cell branches tend to

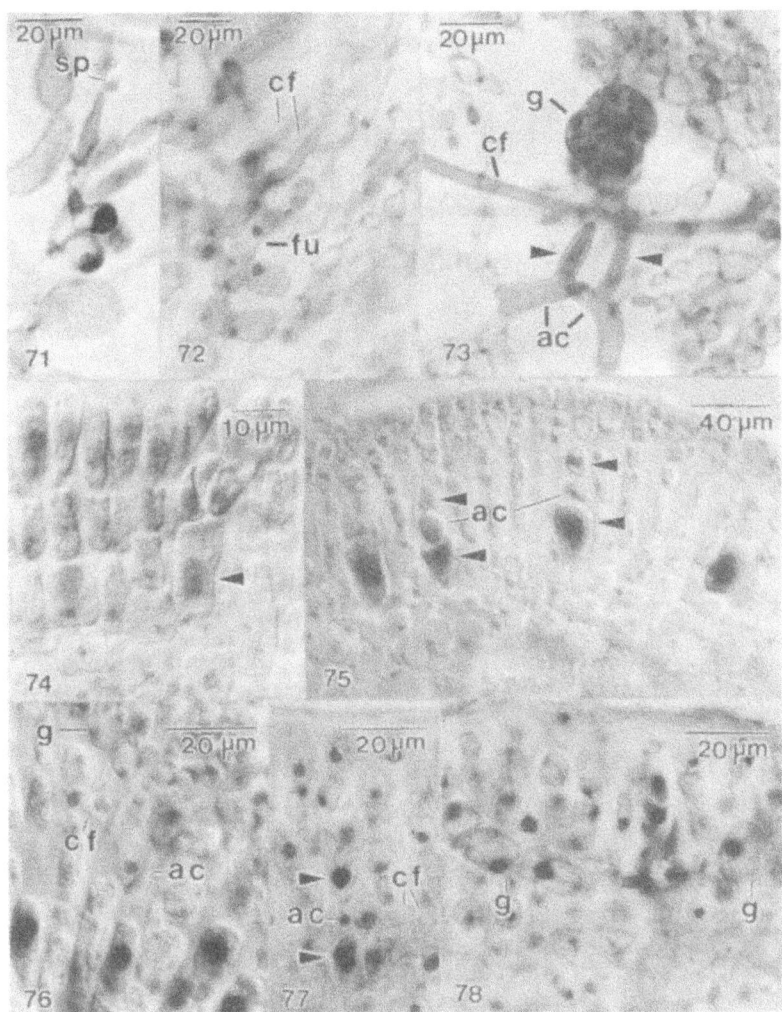

Figs. 13-71 to 13-73. *Schmitzia hiscockiana* (Bardsey, Wales, material provided by C. Maggs, photographs by M. Knauss). *13-71*: Carpogonial branch and carpogonium with attached spermatia (sp). *13-72*: Carpogonial fusion cell (fu) with connecting filaments (cf). *13-73*: Auxiliary cells (ac) with processes (arrowheads) and connecting filament (cf) with gonimoblast (g).

Figs. 13-74 to 13-76. *Portieria (Chondrococcus) hornemannii* (South Africa). *13-74*: Nemathecium with carpogonial branch. Note enlarged nucleus in basal cell (arrowhead). *13-75*: Nemathecium with auxiliary cell branches and auxiliary cells (ac) with adjoining nutritive cells (arrowheads). *13-76*: Connecting filament (cf), auxiliary cell (ac) and gonimoblasts (g).

Figs. 13-77, 13-78. *Peyssonnelia dubyi* (Northern Ireland, material provided by C. Maggs). *13-77*: Connecting filament (cf) and auxiliary cell (ac) with adjoining nutritive cells (arrowheads). *13-78*: Developing gonimoblasts (g).

remain unbranched, the carpogonial branches produce opposite laterals except for the carpogonium and its two subtending cells (Fig. 13-80). The trichogyne is long and spirally coiled, and numerous spermatia can attach to its sides (Fig. 13-80). Individual spermatia are binucleate and appear to be in arrested telophase (Fig. 13-81). After fertilization the carpogonium and the more distal cells of the carpogonial branch are deeply staining and contain enlarged nuclei, whereas the proximal cells

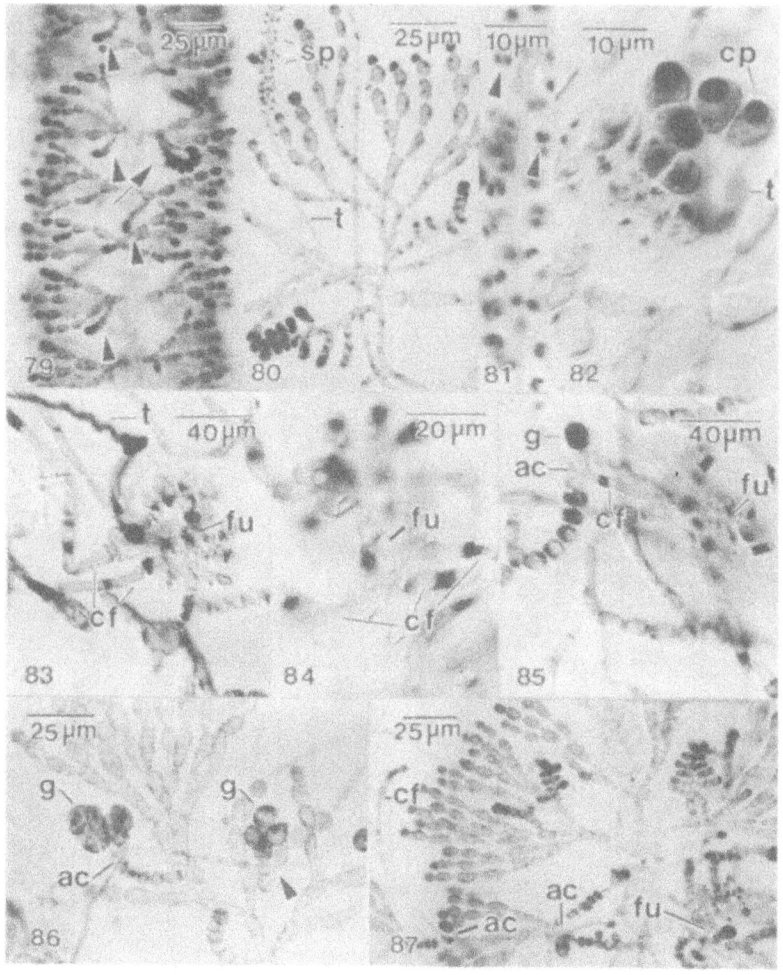

Figs. 13-79 to 13-87. *Acrosymphyton caribaeum* (Florida). *13-79*: Young carpogonial branch (arrow) and auxiliary cell branches (arrowheads). *13-80*: Spermatia (sp) attached to trichogyne (t). *13-81*: Enlarged view of trichogyne in Fig. 13-80, showing binucleate spermatia (arrowheads). *13-82*: Differentiated carpogonial branch and carpogonium (cp) with trichogyne (t). *13-83 to 13-85*: Carpogonial fusion cell (fu) and connecting filaments (cf). *13-86, 13-87*: Auxiliary cells (ac), connecting filaments (cf), and developing gonimoblasts (g). (Note gonimoblast remote from auxiliary cell in Fig. 13-86, arrowhead).

are vacuolate with small nuclei (Fig. 13-82). Processes emerging from the fertilized carpogonium fuse only with vacuolate cells, forming a fusion cell that, in turn, gives rise to numerous connecting filaments (Figs. 13-83 to 13-85). The auxiliary cell is a terminal cell of an auxiliary cell branch in which the subtending cells contain proteinaceous granules (Fig. 13-85). When a connecting filament passes close to an auxiliary cell, the auxiliary cell sends out a process that fuses with it. A gonimoblast initial is usually cut off in close proximity to the auxiliary cell (Figs. 13-85, 13-86) but may be produced at a distance from it (Fig. 13-86). A connecting filament can branch independently of the auxiliary cell, unite with several auxiliary cells, and bear many gonimoblasts (Fig. 13-87).

The genus *Predaea* (Nemastomataceae) is unusual in having clusters of small-celled nutritive filaments produced on cortical cells above and below the auxiliary cell before fertilization (Fig. 13-88) that expand after contact by a connecting filament (Figs. 13-89, 13-90). Ultrastructural studies

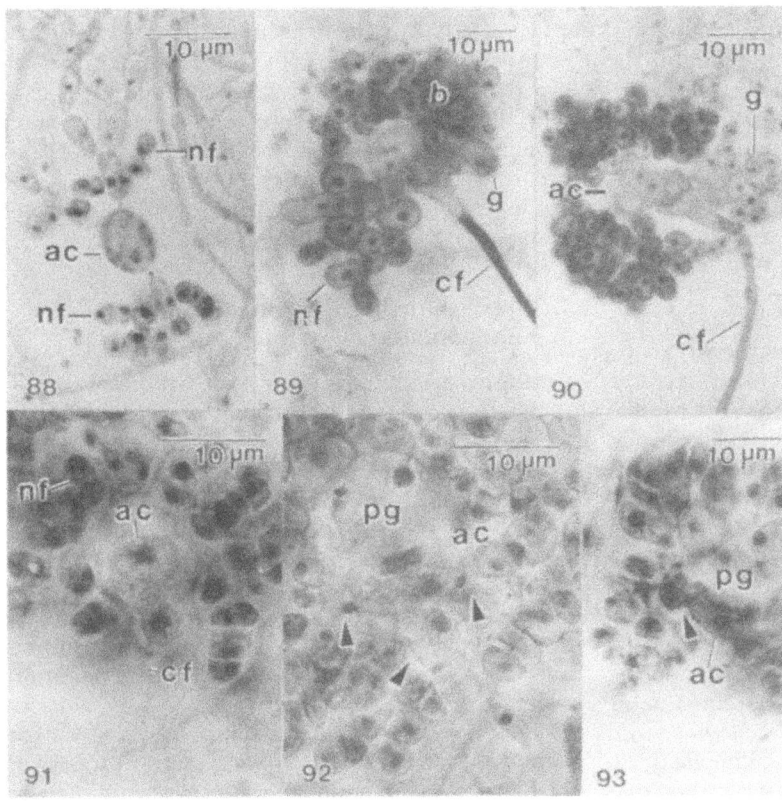

Figs. 13-88 to 13-90. *Predaea feldmannii* (North Carolina, material provided by R. Searles), *13-88*: Auxiliary cell (ac) and nutritive filaments (nf). *13-89, 13-90*: Auxiliary cell (ac), nutritive filaments (nf) and connecting filament (cf) with gonimoblasts (g).

Figs. 13-91 to 13-93. *Grateloupia filicina* (South Africa). *13-91*: Connecting filament (cf), auxiliary cell (ac), and ampullary nutritive filaments (nf). *13-92, 13-93*: Fusions formed around pit connections (arrowheads) near auxiliary cell (ac) and primary gonimoblast cell (pg).

(Siotas & Wetherbee 1982) established that the pit connections between cells of the nutritive filaments break down and disappear at the time of gonimoblast formation, whereas the nuclei and cytoplasm remain intact. As in most species of *Predaea* (Kraft 1984), the connecting filament cuts off a single gonimoblast initial at its point of contact with the auxiliary cell in *P. feldmannii* (Fig. 13-89) and produces a compact cluster of gonimoblast filaments (Fig. 13-90).

Carpogonia and auxiliary cells are borne in separate clusters of filaments called ampullae in Halymeniaceae (Chiang 1970). After fertilization the carpogonium enlarges and fuses with its hypogynous cell in *Grateloupia*, and the resulting fusion cell emits numerous connecting filaments (Kraft 1977a). The auxiliary cell is a large subbasal cell in an auxiliary cell ampulla that otherwise consists of small cells containing prominent nuclei (Fig. 13-91). A connecting filament cuts off a cell that fuses with the inner side of the auxiliary cell (Fig. 13-91). The gonimoblast initial is produced toward the outside and forms a compact cluster of gonimoblast filaments (Fig. 13-92). Channels develop progressively between the nutritive cells of the ampullary filaments commencing next to the auxiliary cell; however, unlike the development in *Predaea*, the fusions in *Grateloupia* take place alongside the primary pit connections (Figs. 13-92, 13-93).

Perhaps the most specialized system of preformed carpogonial and auxiliary cell branches is found in the Kallymeniaceae, a family that contains both procarpial and nonprocarpial genera (Norris 1957). In *Kallymenia reniformis* the auxiliary cell apparatus consists of an auxiliary cell surrounded by a cluster of specially modified cells containing enlarged nuclei (Fig. 13-94). Correspondingly, the carpogonial branch system is composed of a supporting cell surrounded by a cluster of three-celled carpogonial branches in which the basal cell of each branch contains enlarged nuclei (Fig. 13-95). After fertilization, the basal cells of the carpogonial branches fuse with the supporting cell (Fig. 13-96) alongside the primary pit connections (Fig. 13-97) before producing connecting filaments. Primary connecting filaments are nonseptate in *Kallymenia*, with the nucleus and cytoplasm restricted to the filament tip (Fig. 13-98). They unite with remote auxiliary cells and produce the gonimoblast filaments. Womersley and Norris (1971) have illustrated the production of connecting filaments from the carpogonial fusion cell, their union with an auxiliary cell, and the direct development of gonimoblast filaments from the connecting filament in *Kallymenia cribrogloea*.

Gonimoblasts are produced directly from vegetative cells after the connecting filament has united with an auxiliary cell and contacted neighboring gametophytic cells in a *Kallymenia* identified as *K. reniformis* by Hommersand and Ott (1970), and auxiliary cells are absent and the gonimoblasts develop directly from vegetative cells in *Cirrulicarpus carolinensis* (Hansen 1977). In *Hommersandia maximicarpa* connecting filaments branch abundantly, become septate, and fuse with cells of special moniliform branches that, in turn, bear the gonimoblasts (Hansen & Lindstrom 1984). It appears that a nutritive system has evolved secondarily in these species, involving fusions with vegetative cells, either as a supplement to or as a replacement for the nutritive function of the auxiliary cell apparatus.

F. Transformed cortical cells: Plocamiaceae

A nutrient-processing center is generated soon after fertilization through the transformation of ordinary cortical cells adjoining the auxiliary cell in the Plocamiaceae, a procarpial family that is traditionally placed in the Gigartinales. The auxiliary cell is the supporting cell of a three-celled carpogonial branch in *Plocamium* (Kylin 1923). After diploidization, the auxiliary cell cuts off a uninucleate gonimoblast initial and itself becomes multinucleate (Fig. 13-99). At the same time the nucleus in each of the adjoining cortical cells enlarges (Figs. 13-99, 13-100). The cells with their modified nuclei persist throughout gonimoblast development (Fig. 13-101) and are prominent at the base of the gonimoblast during carpospore differentiation (Figs. 13-102, 13-103). The pit connection between the primary gonimoblast cell and the auxiliary cell broadens substantially, but no fusions take place (Fig. 13-103). Fertilized procarps commonly abort in branches in which an earlier fertilization has occurred. The formation of modified cortical cells and enlarged nuclei is triggered even in the event of an early abortion (Fig. 13-104).

G. Sterile nutritive filaments: Ceramiales and Bonnemaisoniales

Cystocarp structure is correlated with thallus morphology in members of the Ceramiales. In the Delesseriaceae the thallus is a thin, membranous blade with ostiolate cystocarps protruding from one side or the other. The procarp consists of a supporting cell, lateral and basal sterile groups, and a four-celled carpogonial branch (Kylin 1956). In *Myriogramme* – and other genera of the Delesseriaceae that we have examined – the nucleus in each sterile group enlarges greatly after fertilization, followed by enlargement of the nuclei in the multinucleate central cells (Fig. 13-105). Fusions take place around the auxiliary cell as the gonimoblast ramifies in the plane of the blade (Fig. 13-106). Later, an ostiolate pericarp is formed from cortical cells and the central cells in the floor of the cystocarp, and their nuclei continue to enlarge (Fig. 13-107). Finally, many of these cells are incorporated into a prominent, multinucleate fusion cell as carposporangia form in chains (Fig. 13-108). The nutrient-processing function, which begins with the sterile cells, extends to the central cells in the floor of the cystocarp and finally to the large central fusion cell in successive stages of gonimoblast development.

Cystocarp formation was described earlier for the filamentous genus *Polysiphonia* (Section IX.C). Similarly, in *Heterosiphonia* (Dasyaceae) a pericarp composed of axial filaments is produced after fertilization. In contrast to *Polysiphonia*, the sterile groups of *Heterosiphonia* become extensively branched after fertilization, forming a tuft that fills the cavity of the young cystocarp (Fig. 13-109). The sterile filaments are still evident long after gonimoblasts have formed within the pericarp and fusions have taken place around the auxiliary cell (Fig. 13-110).

The Bonnemaisoniales have evolved a nutritive system similar to the one found in the Ceramiales. In *Naccaria* (Naccariaceae) the procarp consists of a supporting cell and a two-celled carpogonial

Sexual reproduction and cystocarp development

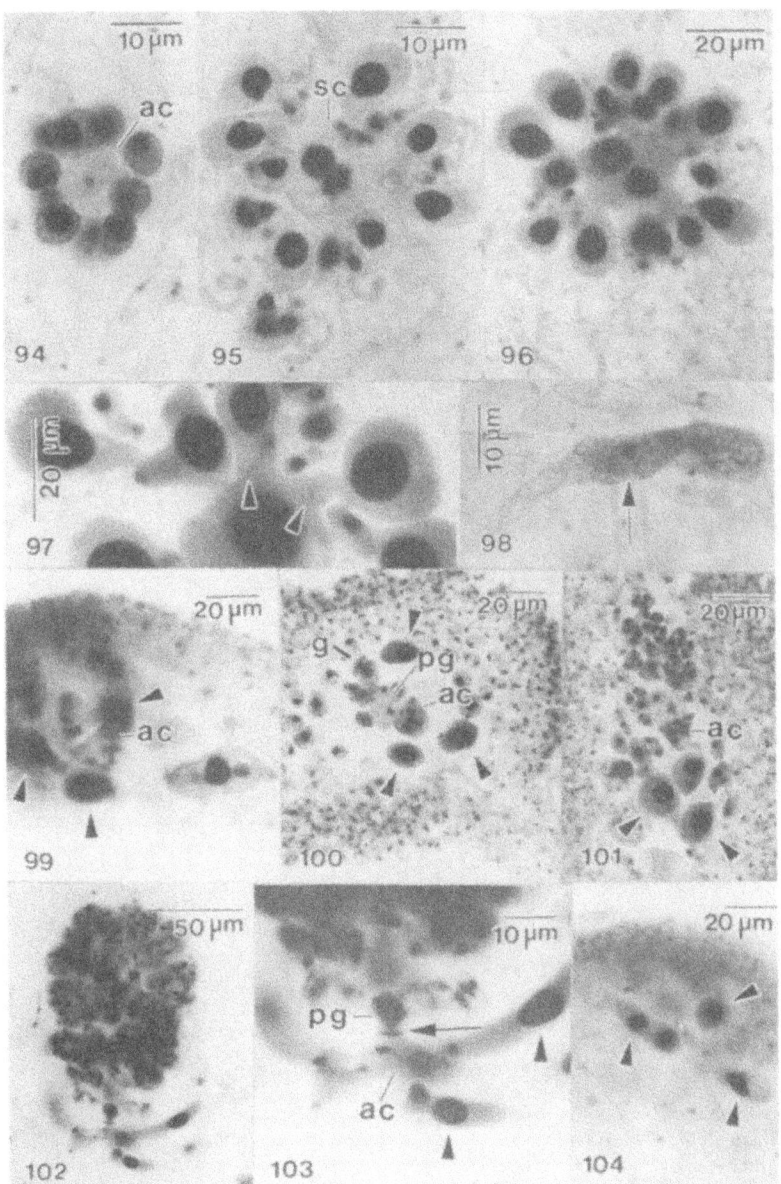

Figs. 13-94 to 13-98. *Kallymenia reniformis* (Ireland). *13-94*: Auxiliary cell apparatus and auxiliary cell (ac). *13-95*: Supporting cell (sc) and carpogonial branches. *13-96*: Carpogonial fusion cell. *13-97*: Fusions around pit connections (arrowheads). *13-98*: Tip of connecting filament containing nucleus (arrow).

Figs. 13-99 to 13-101. *Plocamium cartilagineum* (Ireland). Auxiliary cell (ac), primary gonimoblast cells (pg), and developing gonimoblasts (g) surrounded by cortical cells with enlarged nuclei (arrowheads).

Figs. 13-102 to 13-104. *Plocamium* sp. (New Caledonia). *13-102*: Gonimoblasts removed from cystocarp. *13-103*: Close-up of basal region in Fig. 13-102, showing auxiliary cell (ac), primary gonimoblast cell (pg) attached to auxiliary cell by a broad pit connection (arrow), and vegetative cells with enlarged nuclei (arrowheads). *13-104*: Aborted fertilized procarp with modified cortical cells (arrowheads).

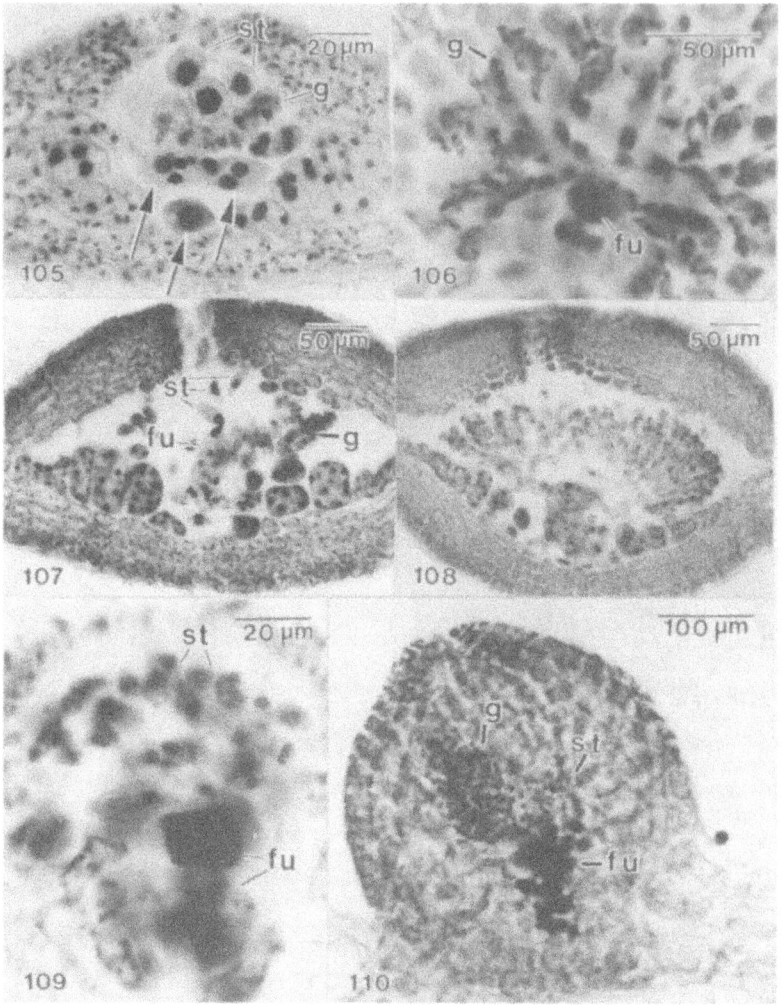

Figs. 13-105 to 13-108. *Myriogramme livida* (southern Chile). *13-105*: Cross section of young cystocarp with gonimoblast (g). Note enlarged nuclei in sterile groups (st) and in floor of cystocarp (arrows). *13-106*: Surface view of fusion cell (fu) bearing gonimoblasts (g). *13-107*, *13-108*: Cross section of cystocarp showing fusion cell (fu), developing gonimoblasts (g), and remnant sterile groups (st).

Figs. 13-109, 13-110. *Heterosiphonia berkeleyi* (southern Chile). *13-109*: Young cystocarp with fusion cell (fu) and sterile groups (st). *13-110*: Older cystocarp with fusion cell (fu), sterile groups (st), and gonimoblasts (g) surrounded by pericarp.

branch. A small tuft of nutritive filaments is produced from the hypogynous cell before fertilization (Kylin 1928). After fertilization, the nuclei of all of the cells of the carpogonial branch apparatus enlarge, the carpogonium fuses with the hypogynous cell, and the nutritive cells are progressively incorporated into the fusion cell (Fig. 13-111). The supporting cell produces extensive secondary assimilatory filaments, and the nuclei in the basal cells of these filaments also enlarge (Fig. 13-112). Pit connections broaden between vegetative cells during gonimoblast development, followed by fusions that extend into the vegetative axis as the carposporangia mature.

Cystocarp development in *Bonnemaisonia asparagoides* (Bonnemaisoniaceae) is fundamentally the

Sexual reproduction and cystocarp development

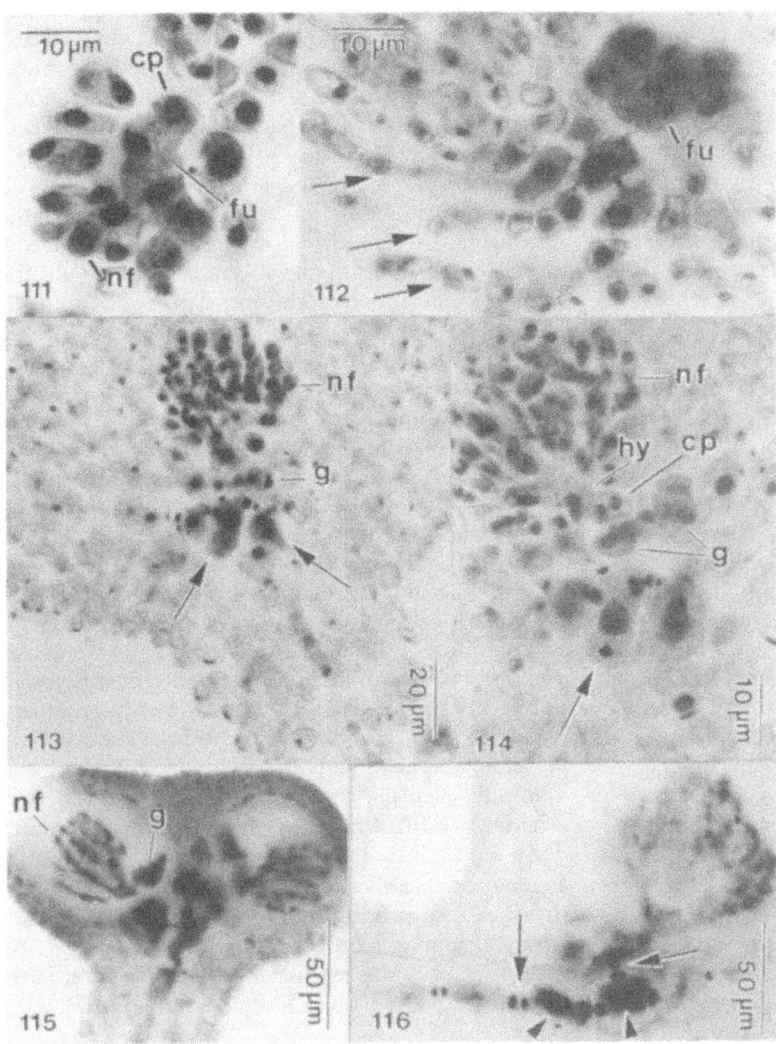

Figs. 13-111, 13-112. *Naccaria wiggii* (Ireland, material provided by C. Maggs). *13-111*: Carpogonium (cp) fused to carpogonial branch (fu) and nutritive filaments (nf). *13-112*: Carpogonial fusion cell (fu) and secondary assimilatory filaments (arrows).

Figs. 13-113, 13-114. *Bonnemaisonia asparagoides* (Sweden). *13-113*: Young cystocarps showing fusing nutritive filaments (nf), gonimoblast (g), and modified vegetative cells (arrows) within pericarp. *13-114*: Later stage, after nutritive filaments (nf) have fused with hypogynous cell (hy) and carpogonium (cp).

Figs. 13-115, 13-116. *Bonnemaisonia nootkana* (California). *13-115*: Young cystocarps with gonimoblast (g) and nutritive filaments (nf) seen in optical section. *13-116*: Axial cells with modified pit connections (arrows) and enlarged nuclei (arrowheads).

same as in *Naccaria*. The hypogynous cell produces a tuft of nutritive filaments, and the cells below it initiate a pericarp before fertilization (Kylin 1916). At the same time that the carpogonium cuts off the gonimoblast initial the nuclei in cells of the nutritive tuft enlarge, nearly filling each cell, and pit connections broaden between neighboring vegetative cells in the axis and basal cells of the developing pericarp (Fig. 13-113). Pit connections between cells of the nutritive filaments break down, forming continuous channels at about the time that the carpogonium fuses with the hypogynous cell (Fig. 13-114). Also, the pit connections between the axial cells expand, becoming dome-shaped, and the nuclei enlarge (Fig. 13-114). Young gonimoblast and candelabra-like fused nutritive filaments surrounded by a pericarp are seen in optical section in *Bonnemaisonia hamifera* (Fig. 13-115). The nuclei and pit connections of the axial cells enlarge while the nuclei in the tuft of nutritive filaments degenerate (Fig. 13-116). Later, inner gonimoblast cells and axial cells will fuse to produce a large central fusion cell as the terminal carposporangia mature (Kylin 1916).

H. Fusion cells: Gigartinales

Fusions between nutritive cells like the ones described earlier in *Predaea* or *Grateloupia* (Section X.E) leave the bounding outline of the filamentous system intact. More often, the initial fusion broadens or the fusions are extended to produce a fusion cell with a continuous outline. Progressive fusions involving the auxiliary cell, the inner gonimoblast cells, and adjoining gametophytic cells may give rise to a prominent central fusion cell.

In *Gloiosiphonia capillaris* (Gloiosiphoniaceae) the carpogonium and auxiliary cell are borne in the same filament cluster (Figs. 13-117, 13-119). The nucleus in the hypogynous cell is enlarged (Figs. 13-118, 13-119) and the carpogonium would normally support the production of connecting filaments (Edelstein 1972); however, the material from British Columbia illustrated here is apparently apomictic, like one of the strains described by DeCew et al. (1981), and lacks connecting filaments. Gonimoblasts are initiated before fusions commence (Fig. 13-120). The primary gonimoblast cell first fuses with the auxiliary cell, followed by progressive fusion between cells of the auxiliary cell branch and the inner gonimoblast cells (Figs. 13-121, 13-122). All fusions take place alongside the primary pit connections (Fig. 13-121). Later, fusions extend to include the inner gonimoblast cells (Fig. 13-123). Modified pit connections separate the outer gonimoblast cells, which mature into carposporangia, from the inner gonimoblast cells, which are incorporated into the fusion cell, leaving remnant pit plugs within the gonimoblast fusion cell (Delivopoulos & Kugrens 1985).

A much larger fusion cell is formed in *Gloiopeltis* (Endocladiaceae), incorporating a substantial portion of the inner gonimoblast filaments in addition to cells of the auxiliary cell apparatus. Pit connections between adjoining cells do not degenerate; they remain well defined within the fusion cell, even in mature cystocarps (Figs. 13-124, 13-125).

Cystocarps are stalked and have thick multicellular pericarps with transverse ostioles and a large central fusion cell in *Heringia* (Caulacanthaceae) (Fig. 13-126). Inner gonimoblast cells unite with the auxiliary cell and neighboring gametophytic cells, forming a central fusion cell (Searles 1968); the outer gonimoblast filaments ramify among the surrounding vegetative filaments and produce carposporangia terminally in short chains (Figs. 13-127, 13-128). The cortical filaments and central fusion cell appear to function as a conduit between the photosynthetic pericarp and the outwardly radiating gonimoblast filaments.

Most genera of the Solieriaceae (including the Rhabdoniaceae) are characterized by a large central fusion cell (Min-Thein & Womersley 1976). In *Solieria* inner gonimoblast cells fuse with the auxiliary cell, which in turn fuses both inwardly and outwardly with cells of the bearing cortical filament (Gabrielson & Hommersand 1982a). The primary pit connections between the auxiliary cell, the inner gonimoblast cells, and the adjoining cortical cells broaden, and fusion takes place through the center of the pit plug (Fig. 13-129). Cells become confluent as fusion proceeds, leading to formation of a central fusion cell supported by a basal stalk (Fig. 13-130). Later, terminal sterile gonimoblast filaments are formed that unite with cells of the surrounding involucre (Gabrielson & Hommersand 1982a).

I. Special nutritive tissues: Ahnfeltiales and Gracilariales

Maggs and Pueschel (1989) demonstrated that the life history of *Ahnfeltia plicata* involves an alternation of generations between erect male (50%) and female (50%) gametophytes, a nemathecial carposporophyte and a crustose tetrasporophyte. Gonimoblast filaments develop directly from the carpogonium and grow over the outermost cortical cells, fusing with them and with each other to produce a complex fusion tissue. This tissue gives rise to outwardly growing filaments that terminate in diploid carposporangia (Maggs & Pueschel 1989). The fusion tissue corresponds to the "primary nemathecium," the outwardly growing filaments to the "secondary nemathecium," and the car-

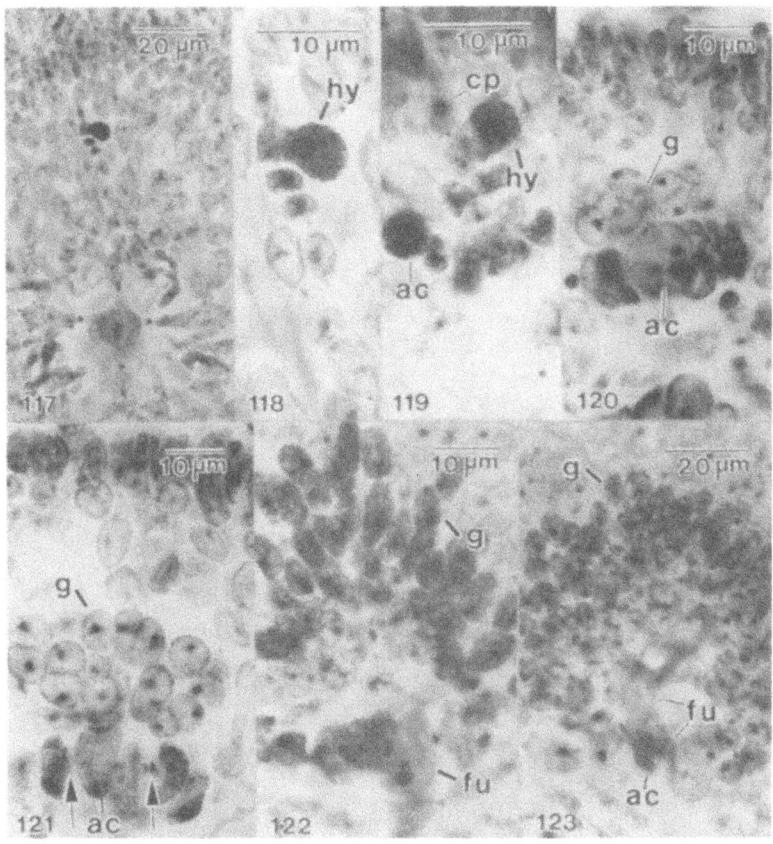

Figs. 13-117 to 13-123. *Gloiosiphonia capillaris* (British Columbia). *13-117, 13-118*: Cross section of fertile area with enlarged hypogynous cell (*hy*). *13-119*: Filament cluster containing carpogonium (*cp*), hypogynous cell (*hy*) and auxiliary cell (*ac*). *13-120 to 13-123*: Stages in fusion cell (*fu*) formation around the auxiliary cell (*ac*) and gonimoblast (*g*) development. (Note fusions around pit connections in Fig. 13-121, arrows.)

posporangia to the "monosporangia" of Rosenvinge (1931).

The Gracilariaceae, traditionally placed in the Gigartinales (Kylin 1956), is now placed in the Gracilariales (Fredericq & Hommersand 1989). In *Gracilaria verrucosa* the carpogonial branch is a two-celled filament borne on a supporting cell that also bears a pair of sterile filaments surrounding the carpogonium and typical auxiliary cells are absent (Fredericq & Hommersand 1989; Sjöstedt 1926). After fertilization, the carpogonium enlarges and initially fuses with cells of the adjacent sterile filaments while the zygote nucleus is displayed prominently in the center of the fusion cell (Fig. 13-131). Additional fusions with neighboring gametophytic cells lead to the formation of a large, lobed fusion cell containing both diploid and haploid nuclei in *G. verrucosa* (Fig. 13-132). Such fusions take place around existing pit connections, which may persist and are readily seen inside the fusion cell (Fig. 13-133). Several uninucleate gonimoblast initials are cut off from lobes of the multinucleate fusion cell (Fig. 13-134), and these produce compact gonimoblasts initially composed of uninucleate cells (Fig. 13-135). A prominent pericarp is formed to the outside by transverse division of the apical cells of cortical files (Figs. 13-131, 13-132, 13-135). After the gonimoblasts have expanded, filling the cystocarp cavity, they initiate chains of carposporangia and also produce tubular nutritive cells that fuse with pericarp cells (Fig. 13-136) or with cells in the floor of the cystocarp (Fig. 13-137).

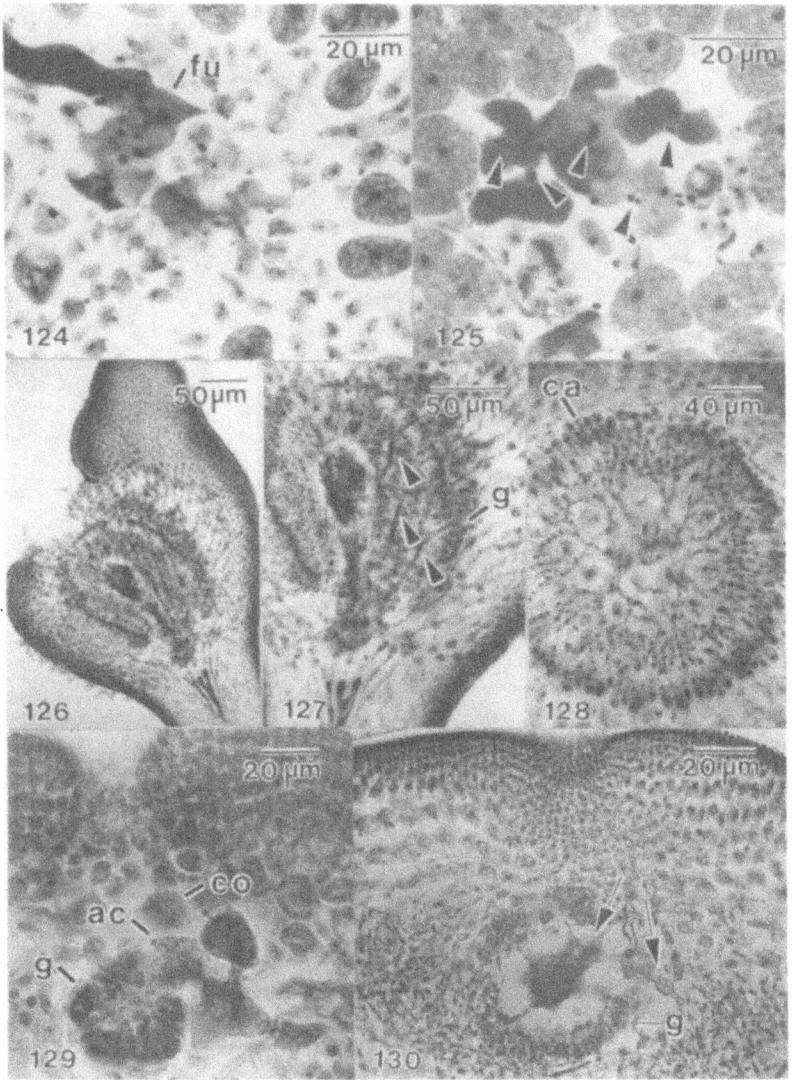

Figs. 13-124, 13-125. *Gloiopeltis furcata* (California). Fusion cell (*fu*) formed around persistent pit connections (arrowheads).
Figs. 13-126 to 13-128. *Heringia mirabilis* (South Africa). *13-126, 13-127*: Longitudinal section of cystocarp with gonimoblasts (*g*), central fusion cell, and elongated cortical filaments (arrowheads). *13-128*: Cross section of cystocarp showing ramifying gonimoblasts bearing carposporangia (*ca*).
Figs. 13-129, 13-130. *Solieria australis* (Natal, South Africa, material provided by G. Lambert). *13-129*: Early stage showing fusions through pit connections between auxiliary cell (*ac*), young gonimoblast (*g*), and vegetative cortical cells (*co*). *13-130*: Older fusion cell (arrows) and gonimoblasts (*g*).

J. Nutritive tissues and fusion cells: Rhodymeniales

The auxiliary cell in Rhodymeniales is the terminal cell of a two-celled filament borne on the supporting cell of a three- to four-celled carpogonial branch (Kylin 1956). A supporting cell bearing a three-celled carpogonial branch is illustrated in *Rhodymenia* (Fig. 13-138). Transfer of the diploid nucleus from the fertilized carpogonium to the auxiliary cell is mediated by a connecting cell (Sparling 1957). The supporting cell, auxiliary mother cell, and

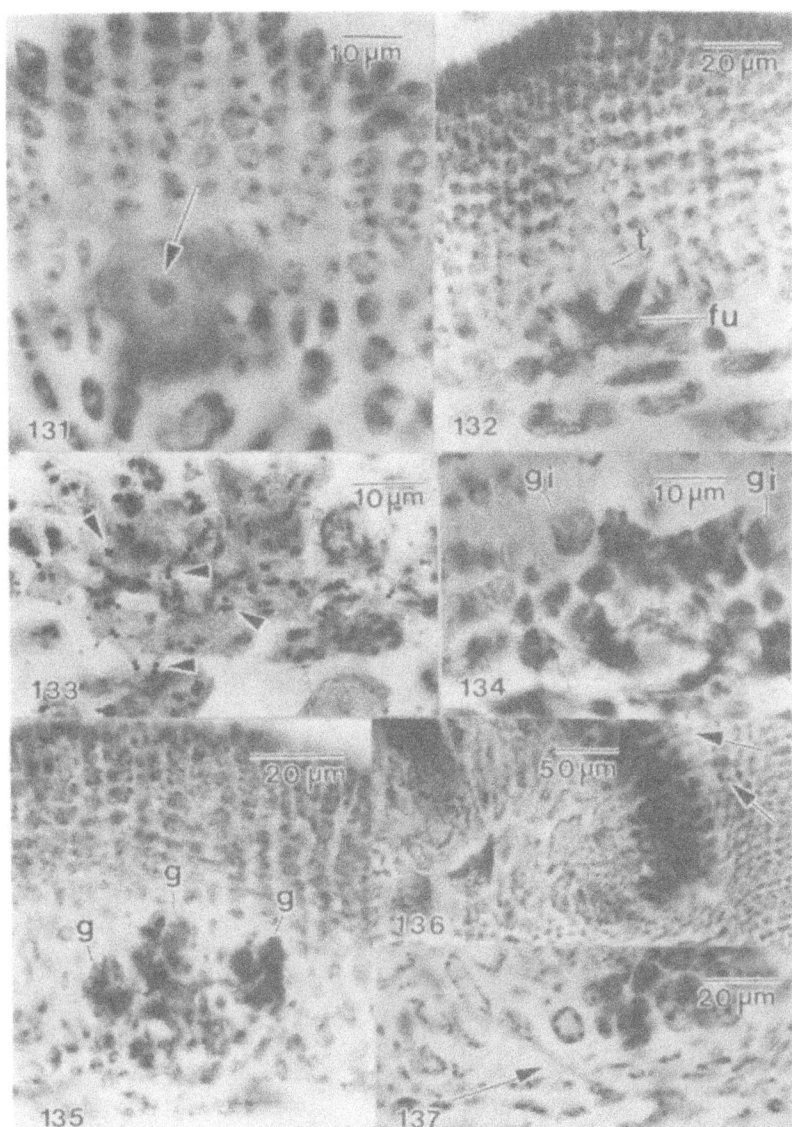

Figs. 13-131 to 13-137. *Gracilaria verrucosa* (Ireland and Wales, material provided by J. Brodie and E. Jones). *13-131*: Fertilized carpogonium containing zygote nucleus (arrow). *13-132*: Young fusion cell (*fu*) and trichogyne remnant (*t*). *13-133*: Fusion cell formed around pit connections (arrowheads). *13-134, 13-135*: Gonimoblast initials (*gi*) and young gonimoblasts (*g*). *13-136, 13-137*: Gonimoblasts with tubular nutritive cells (arrows).

auxiliary cell enlarge, a protein body forms in the auxiliary cell, and a primary gonimoblast cell is cut off that retains the protein body (Fig. 13-139). At the same time an external pericarp is produced and a nutritive tissue forms in the floor of the cystocarp, composed of deeply staining, multinucleate vegetative cells (Fig. 13-139). As the gonimoblasts develop, the auxiliary cell, auxiliary mother cell, and supporting cell fuse, and the nutritive cells commence to fuse with one another and with the supporting cell (Figs. 13-140, 13-141). The protein body persists inside the primary gonimoblast cell, and the

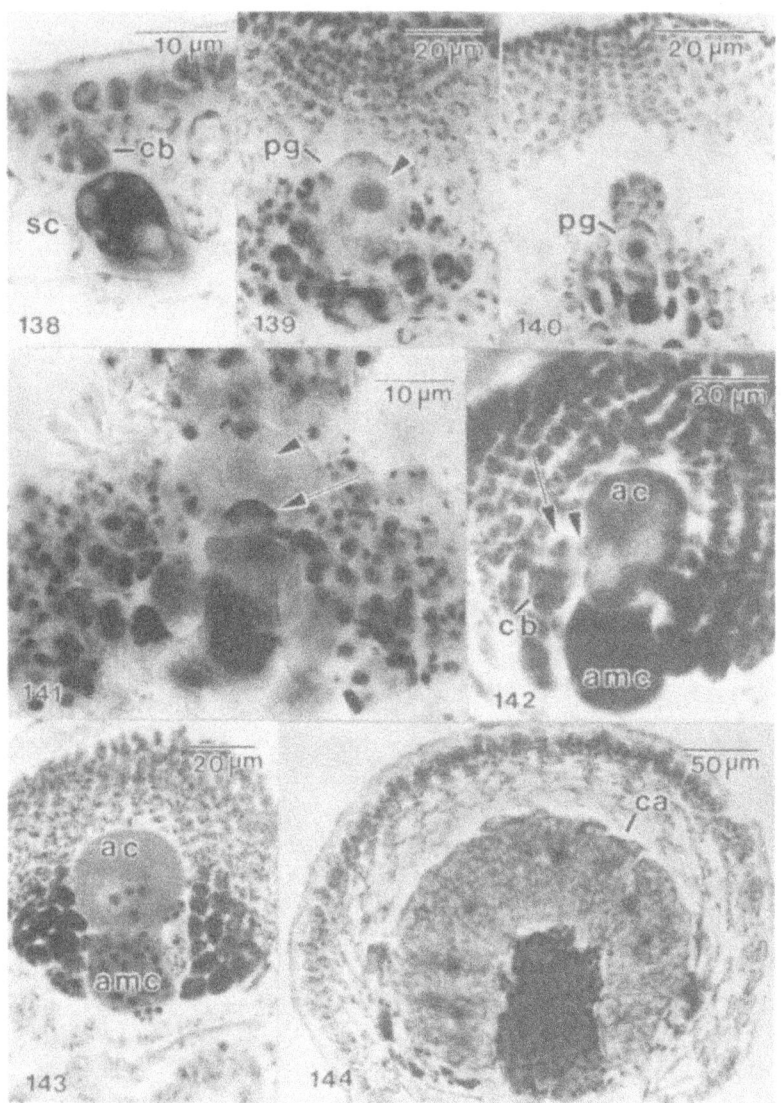

Fig. 13-138. *Rhodymenia pseudopalmata* ssp. *caroliniana* (North Carolina); supporting cell (*sc*), carpogonial branch (*cb*).
Figs. 13-139 to 13-141. *Rhodymenia howeana* (Chile), *13-139*: Cross section of young cystocarp showing primary gonimoblast cell (*pg*) with protein body (arrowhead) surrounded by nutritive tissue. *13-140*: Primary gonimoblast cell (*pg*) bearing gonimoblasts. *13-141*: Close-up of multinucleate nutritive cells and primary gonimoblast cell with protein body (arrowhead) and modified pit connection (arrow).
Figs. 13-142 to 13-144. *Chylocladia verticillata* (Ireland). *13-142*: Fused carpogonial branch with connecting cell (arrow), auxiliary mother cell (*amc*), and auxiliary cell (*ac*) with process (arrowhead). *13-143*: Young cystocarp showing auxiliary mother cell (*amc*) and auxiliary cell (*ac*) surrounded by nutritive tissue. *13-144*: Mature cystocarp with fusion cell and carposporangia (*ca*).

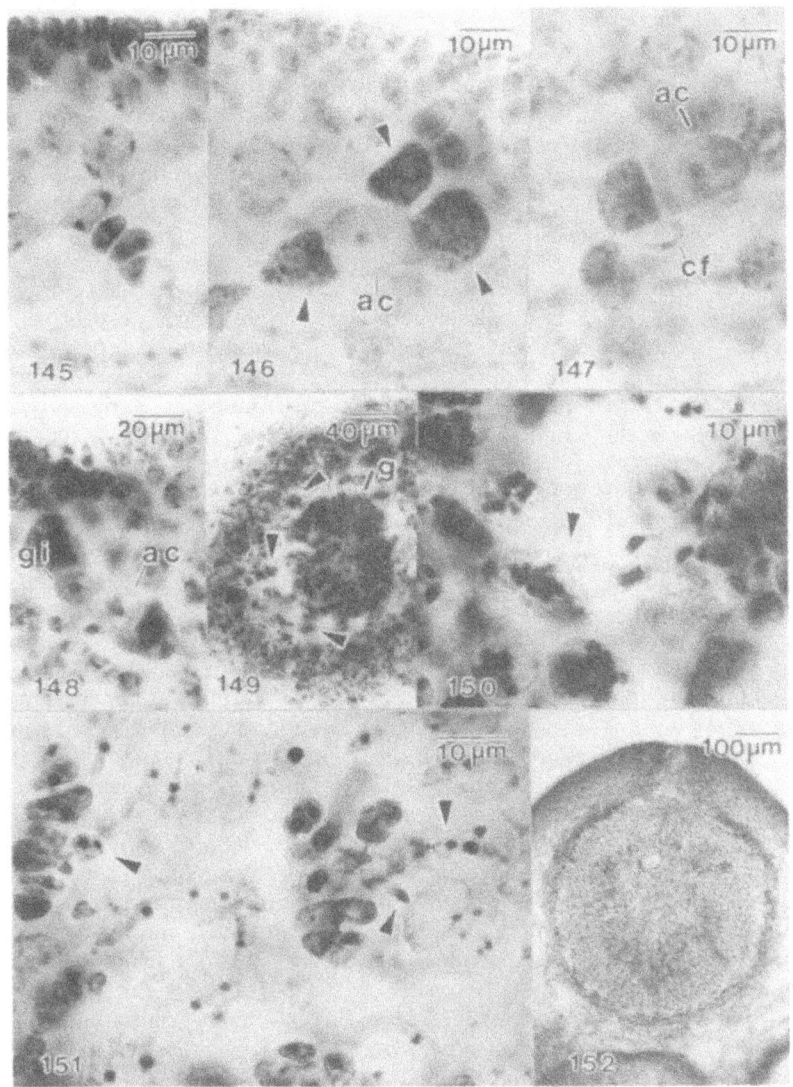

Figs. 13-145 to 13-152. *Agardhiella subulata* (North Carolina, collected by P. Gabrielson). *13-145*: Carpogonial branch. *13-146*: Auxiliary cell (ac) and adjacent multinucleate cells (arrowheads). *13-147*: Connecting filament (cf) approaching auxiliary cell (ac). *13-148*: Auxiliary cell (ac) and gonimoblast initial (gi). *13-149*: Gonimoblasts (g) and secondary nutritive filaments (arrowheads). *13-150*: Gonimoblast cell fusing with vegetative nutritive cell (arrowhead). *13-151*: Later stage, gonimoblast filaments cutting off conjunctor cells that fuse with vegetative cells (arrowheads). *13-152*: Cross section of mature cystocarp.

pit plug between it and the auxiliary cell broadens dramatically (Fig. 13-141). In *Rhodymenia* the fused nutritive cells with their enlarged nuclei persist up to the final stages of carposporangial maturation and evidently function as a major nutrient-processing center.

We have seen a stage in *Chylocladia* (Champiaceae) with an expanded auxiliary cell and fused carpogonial branch with a connecting cell (Fig. 13-142). A process extends from the auxiliary cell toward the connecting cell, perhaps to initiate fusion. The auxiliary mother cell and the surrounding nutritive cells become multinucleate and deeply staining prior to gonimoblast formation (Figs. 13-142, 13-143) and appear to function as a primary nutrient-processing center. Unlike the case in *Rhodymenia*, the cystocarp subsequently forms a large, multinucleate central fusion cell bearing sessile carposporangia at maturity (Fig. 13-144). This fusion cell becomes deeply staining and filled with nuclei as the contents of the surrounding nutritive cells are depleted. It appears to function as a secondary nutrient-processing center.

K. Placentae: Gigartinales

In several families belonging to the order Gigartinales individual gonimoblast cells either fuse directly with noncontiguous gametophytic cells or become linked to them by means of secondary pit connections to form a network in the center or at the base of the cystocarp. Kraft (1977b) referred to such a structure, composed of intermixed, fused carposporophyte and gametophyte tissues having a nutritive function, as a placenta.

A good example of placental development is provided by the genus *Agardhiella*, one of the few genera of the Solieriaceae to have a central placenta instead of a fusion cell. The carpogonial branch is directed inward and is typically three-celled with a reflexed trichogyne (Fig. 13-145), and auxiliary cells are borne in separate primary cortical filaments (Fig. 13-146). Cells proximal and distal to the auxiliary cell become filled with nuclei as a result of synchronous nuclear divisions and are deeply staining prior to diploidization (Fig. 13-146). In *Agardhiella* the fertilized carpogonium produces two connecting filaments that are unbranched and nonseptate (Gabrielson & Hommersand 1982b). A connecting filament contacts an auxiliary cell at its proximal end (Fig. 13-147), and a gonimoblast initial is cut off from the distal end (Fig. 13-148). Inner cortical cells generate files of secondary filaments that grow toward the developing gonimoblast (Fig. 13-149). Terminal, uninucleate gonimoblast cells contact and fuse with the files of multinucleate gametophytic cells (Figs. 13-149, 13-150). Later, as the cystocarp expands, small conjunctor cells are cut off from gonimoblast filaments that fuse with neighboring vegetative cells that are now enlarged and vacuolate (Fig. 13-151). The mature cystocarp is ostiolate and has a cellular center composed of a placenta of gonimoblast and vegetative cells bearing carposporangia in short chains surrounded by a stretched outer involucre (Fig. 13-152). Sterile gonimoblast filaments extend and fuse with the involucral cells during the final stages of carposporangial maturation (Gabrielson & Hommersand 1982b). The nutritive system described above consists of four compartments, each of which is produced and functions at a particular stage of carposporophyte development.

The families Cystocloniaceae and Hypneaceae have procarps in which the auxiliary cell is situated distal to the supporting cell of a three-celled carpogonial branch (Kylin 1956). In *Hypnea* the fertilized carpogonium fuses directly with the auxiliary cell at its proximal end (Fig. 13-153). Inner cortical cells generate a special nutritive tissue composed of multinucleate cells beneath the developing gonimoblast (Fig. 13-154). Uninucleate, terminal gonimoblast cells fuse with the multinucleate cells of the nutritive tissue (Figs. 13-155, 13-156). Later, as the cystocarp expands, clusters of gonimoblast filaments differentiate into carposporangia, and a few form sterile nutritive filaments that will fuse with cells of the pericarp (Figs. 13-157, 13-158).

The Australasian families Acrotylaceae (Kraft 1977b), Dicranemaceae (Kraft 1977c), and Mychodeaceae (Kraft 1978) have placental cystocarps in which fusions between gonimoblast filaments and vegetative cells are abundant. In the polycarpogonial genus *Mychodea* (Mychodeaceae) the supporting cell is an enlarged, multinucleate inner cortical cell that bears a cluster of three-celled carpogonial branches (Figs. 13-159, 13-160). The supporting cell functions as the auxiliary cell (Kraft 1978), which generates gonimoblast filaments that ramify inwardly among the surrounding gametophyte tissues (Figs. 13-161, 13-162) and unite with vegetative cells by means of secondary pit connections (Fig. 13-163). Gonimoblast cells are initially uninucleate, whereas the vegetative cells are all multinucleate (Figs. 13-162, 13-163). Carposporangia are produced in clusters associated with the gametophytic cells (Kraft 1978). Some of the gonimoblast filaments elongate and unite with vegetative cells toward the periphery of the cystocarp at the stage of carpospore differentiation (Fig. 13-164).

Some members of the Phyllophoraceae produce ordinary cystocarps, whereas others produce an external pustule bearing tetrasporangia (André 1978). *Gymnogongrus patens* is representative of members of the Phyllophoraceae in which the gonimoblast develops inwardly and produces masses of carposporangia (Fig. 13-165). As in *Mychodea*, the gonimoblast cells are initially uninucleate and the gametophytic cells are multinucleate (Fig. 13-166). Many gonimoblast cells fuse onto each vegetative cell, forming small fusion cells within the cystocarp

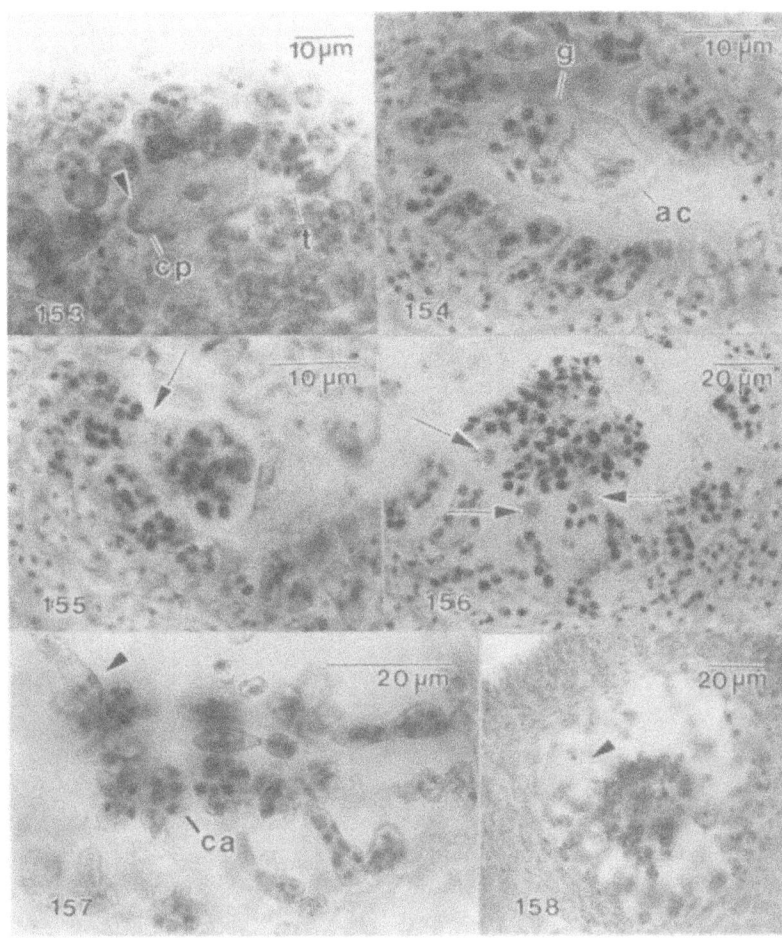

Figs. 13-153 to 13-158. *Hypnea musciformis* (North Carolina). *13-153*: Carpogonium (cp) with trichogyne (t) fused to auxiliary cell (arrowhead). *13-154*: Auxiliary cell (ac) and gonimoblast (g) surrounded by nutritive tissue. *13-155, 13-156*: Fusions between nutritive cells and gonimoblast cells (arrows). *13-157, 13-158*: Clusters of carposporangia (ca) and sterile gonimoblast filaments (arrowheads).

(Fig. 13-167) that serve as initiation sites for the production of carpospore-bearing filaments.

In species of *Chondrus* (Gigartinaceae) very fine gonimoblast filaments ramify through the medullary tissue, making secondary pit connections with medullary cells (Mikami 1965). The cystocarp is more compact in *Gigartina*. In *G. teedii* an extensive secondary tissue derived from subcortical and medullary cells forms around the diploidized auxiliary cell (Tsekos & Schnepf 1983). This tissue consists of clusters of deeply staining, multinucleate cells (Figs. 13-168, 13-169). Gonimoblast filaments composed of uninucleate cells develop directly from the auxiliary cell and grow in a compact formation between the clusters of vegetative cells (Figs. 13-168, 13-169). Later, the gonimoblast filaments cut off small conjunctor cells that fuse with enlarged, vacuolate vegetative cells, forming secondary pit connections at the same time that other files of gonimoblast cells are being converted into carposporangia (Figs. 13-170, 13-171). The mature cystocarp is surrounded by a prominent involucre (Fig. 13-170). Secondary tissue formed around the fertilized procarp functions as the initial nutrient-processing center in *G. teedii*, with the auxiliary cell acting as a conduit to the young gonimoblast filaments. Later, secondary pit connections provide direct linkage between gonimoblast and gametophyte tissues. Finally, contact is made with cells of the involucre during the final stages of carpospore maturation.

XI. SUMMARY

Evolution of the sexual system in Rhodophyta has proceeded subject to two constraints. First, construction of the vegetative thallus is fundamentally filamentous, with growth initiated by apical cells in

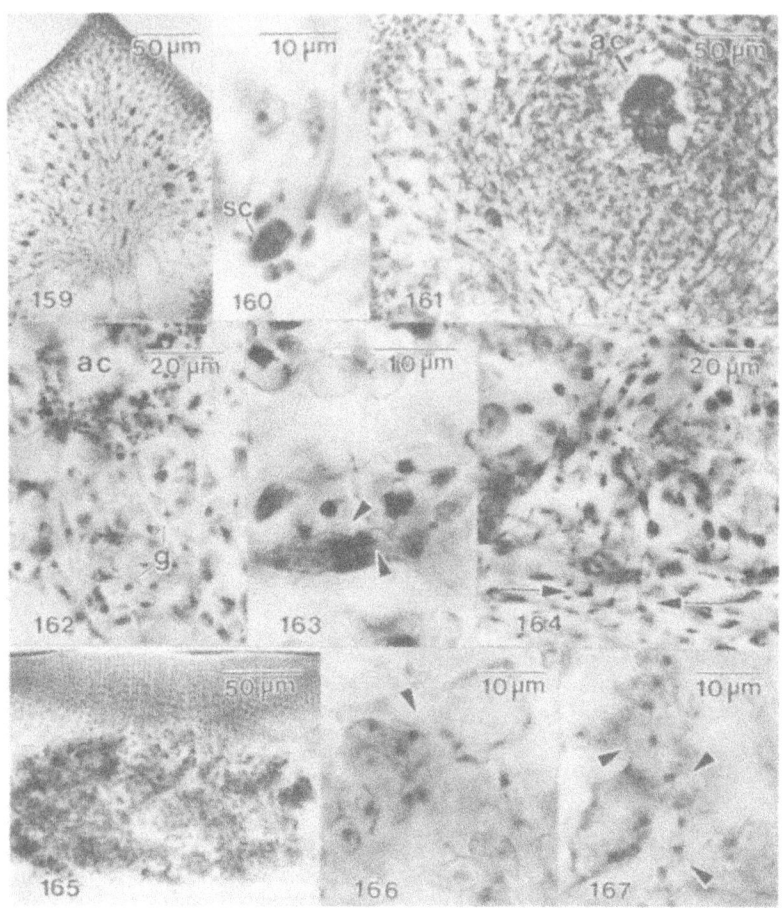

Figs. 13-159 to 13-164. *Mychodea carnosa* (South Australia). *13-159*: Fertile female tip. *13-160*: Supporting cell (sc) bearing carpogonial branches. *13-161*, *13-162*: Multinucleate auxiliary cell (ac) and uninucleate gonimoblast filaments (g) ramifying through medullary tissue. *13-163*: Gonimoblast cells fused to vegetative cells (arrowheads). *13-164*: Sterile gonimoblast cells fused to involucral cells (arrows).
Figs. 13-165 to 13-167. *Gymnogongrus patens* (Morocco). *13-165*: Cross section of cystocarp. *13-166*, *13-167*: Uninucleate gonimoblast cells fused with multinucleate vegetative cells (arrowheads).

the principal subclass, Florideophycidae (Chapter 12). Second, fertilization is inherently inefficient in the absence of gamete motility, a limitation that has been partly compensated for by enhanced spore production associated with biphasic and triphasic life cycles (Searles 1980; Section I; see Chapter 17).

The principal reproductive features that unite the Rhodophyta are homologies in the ultrastructure and developmental pathways of spermatia, carpogonia, and spores of all types. In contrast, the differences in the position of reproductive cells on the thallus separate three major phylogenetic lines: (1) intercalary, with spores and male gametes cut out one at a time by a curved wall (Rhodochaetales, Compsopogonales); (2) intercalary, in packets produced by successive perpendicular divisions (Bangiales); (3) terminal, transformed from apical initials (Florideophycidae). The localization of reproductive functions in apical initials in Florideophycidae is the key event in the evolution of this group that probably stabilized filamentous growth and led to carposporophyte formation and the triphasic life cycle (Section II).

We have argued that the primitive carposporophyte in Florideophycidae was an autotrophic, somatic phase corresponding to an asexual, branched

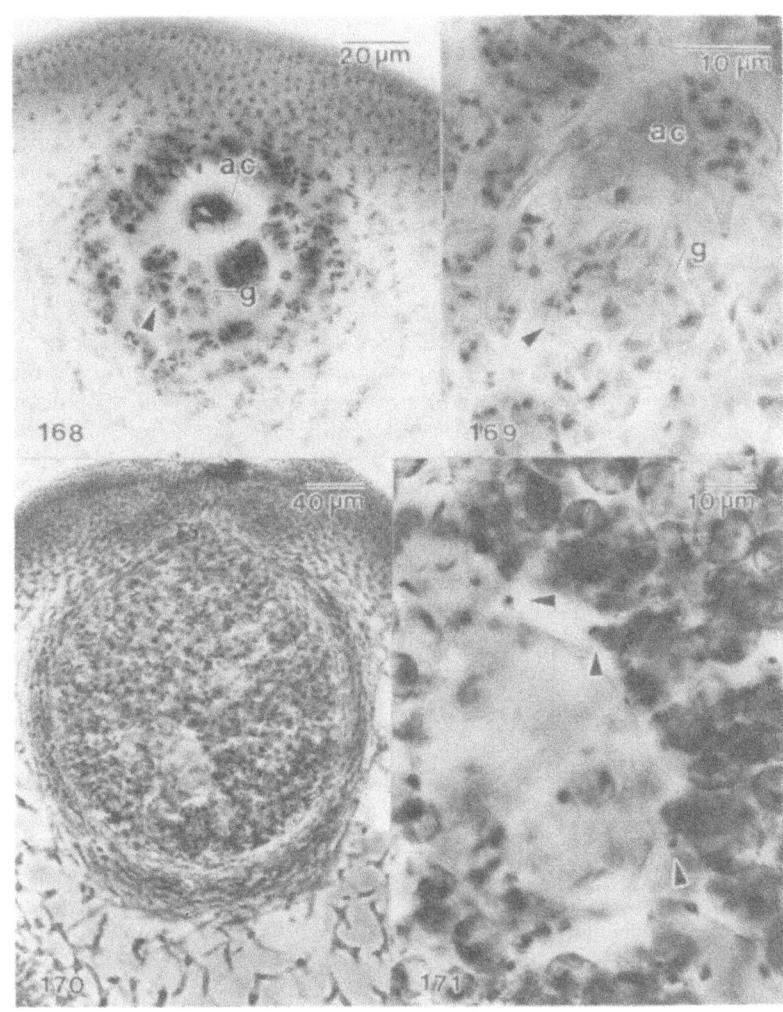

Figs. 13-168 to 13-171. *Gigartina teedii* (Morocco). *13-168, 13-169*: Auxiliary cell (ac), uninucleate gonimoblasts (g), and multinucleate secondary vegetative cells (arrowheads). *13-170*: Cross section of cystocarp. *13-171*: Conjunctor cells (arrowheads) cut off from gonimoblast cells fusing onto large gametophytic cell.

filament bearing monosporangia (Section III). Its subsequent evolution involved increased nutritional dependency on the gametophyte. Some evolutionary lines retained the primitive condition in which gonimoblasts develop directly on the fertilized carpogonium; others evolved auxiliary cells and connecting cells or connecting filaments as a means of amplifying the products of a single fertilization (Section IV). Further evolution has involved the progressive modification of gametophytic tissues and the structural and functional compartmentalization of the cystocarp. We recognize three compartments: (1) the outer photosynthetic tissues, (2) modified inner gametophytic tissues that process and store the metabolites of photosynthesis, and (3) the developing carposporophyte (Section V).

In the first compartment light energy is used to absorb mineral nutrients from the environment and fix carbon dioxide, water, and mineral salts into organic molecules. Although it is possible that complex substances are transported to the carposporophyte through pit connections that link the tissues, we favor a model in which simple molecules are excreted from their "source" in the photosynthetic layers and diffuse along gradients to "sinks" composed of gametophytic tissues in the interior of the cystocarp. In reproductive thalli the sinks consist of specially modified cells or tissues that lie in close proximity to carpogonia or auxiliary cells or are so situated that they are readily contacted by developing gonimoblasts. The efficiency of the proposed model depends on the creation of sinks that are strategically located to support carposporophyte development in competition with sinks that sustain vegetative growth or with the growing vegetative tip itself.

We have observed that carposporophyte growth is not a continuous process in most families of Florideophycidae. Rather, it proceeds in stages, with periods of rapid growth followed by intervals of slow growth or apparent inactivity during which the pattern of development may change. Each stage is preceded by the transformation of existing gametophytic cells into special protein-rich tissues, or new secondary gametophytic tissues are formed that are rich in protein content and often contain modified nuclei (Sections IX, X).

Nutritional tissues fall into two categories: those that are consumed in a single stage of carposporophyte development and those that persist and appear to function through several stages. Both usually contain cells that are rich in proteins, but the latter invariably have one or more enlarged nuclei containing amplified levels of DNA. We have called such a tissue a nutrient-processing center (Section V). The transformed tissues may be ordinary carpogonial branches; secondarily produced carpogonial or auxiliary cell branches; nemathecial filaments; modified uninucleate vegetative cells, sterile "groups," or nutritive "tufts"; special multinucleate vegetative cells, filaments or tissues, and fusion cells or placentae (Section X).

Morphological evidence indicates that the transport of nutriment from nutritive tissues to the carposporophyte may involve (1) the enlargement and structural modification of existing pit connections, (2) fusions between contiguous cells either through or around primary pit connections, sometimes with formation of a central fusion cell, or (3) direct fusions or the formation of secondary pit connections between gonimoblast cells and noncontiguous gametophytic cells, sometimes with the formation of a well-defined placenta.

Fusion cells commonly undergo autolysis of their nuclei and other cell organelles, becoming vacuolate at a late stage of gonimoblast ontogeny, as in *Polysiphonia* (Wetherbee 1980) or *Gloiosiphonia* (Delivopoulos & Kugrens 1985). Delivopoulos & Kugrens (1985) have expressed the opinion that a nutritive function cannot be ascribed to such fusion cells. We take an opposite view and interpret the ultrastructural evidence as indicating that fusion cell formation and organelle autolysis is a process of senescence designed to cannibalize compounds, particularly nitrogen-containing substances, for use during the final stages of carposporophyte development.

In some families of the Gigartinales, such as the

Table 13-1. Evolutionary trends in the nutrition of the carposporophyte

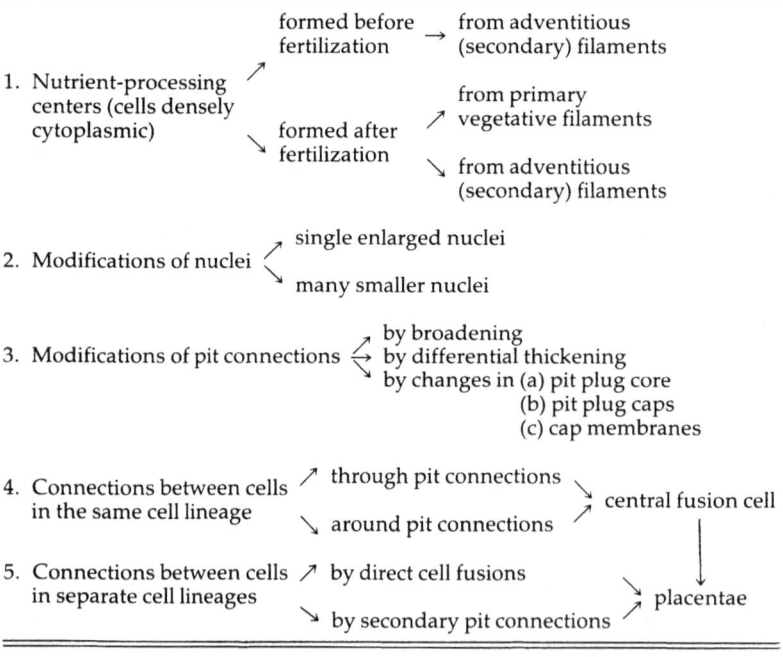

Solieriaceae, fusion cells have been replaced by placentae in the cystocarps of more advanced genera (Section X.K). Individual cystocarps are invariably larger and produce greater numbers of carpospores in families or genera with fusion cells or placentae than in related taxa that lack them.

Similar structures and nutritional strategies have evolved at different times in different groups of red algae. Indeed, convergent evolutionary themes appear to be the norm rather than the exception. Table 13-1 presents some alternative ontogenetic pathways that have produced morphologically similar cystocarps. It is clear in retrospect that Kylin possessed a highly developed nutritional concept of cystocarp morphology, which he used primarily in characterizing families. Many of the developmental characters that we have identified based on a functional model of cystocarp morphology apply to families and infrafamilial taxa recognized by Kylin (1956). A few call for taxonomic revisions.

XII. ACKNOWLEDGMENTS

We thank our many colleagues who have provided material for this study. Their names are listed in the legends under the individual species. This study was supported in part by a grant from the University Research Council, the University of North Carolina at Chapel Hill.

XIII. REFERENCES

Abbott, I. A. 1976. *Dotyophycus pacificum* gen. et sp. nov., with a discussion of some families of Nemaliales (Rhodophyta). *Phycologia* 15: 125–32.

Abdel-Rahman, M. H. & Magne, F. 1983. Existence d'un nouveau type de cycle de développement chez les Rhodophycées. *C. R. Acad. Sci.* (Paris), sér. III. 296: 641–4.

Agardh, J. G. 1844. *In Systemata Algarum Hodierna Adversaria*, pp. 1–56, Lund.

Agardh, J. G. 1880. *Species Genera et Ordines Algarum.* Morphologia Floridearum. vol. 3(2) pp. [1]–199, plates 1–33. Leipzig.

André, F. 1978. Sur les cycles morphologiques du *Gymnogongrus crenulatus* (Turn.) J. Ag. et du *Gymnogongrus devoniensis* (Grev.) Schott. (Gigartinales, Phyllophoracées) en culture. *Rev. Algol.* N.S.13: 151–76.

Bliding C. 1928. Studien über die Florideenordnung Rhodymeniales. *Lunds Univ. Årsskr.* N.F. Avd. 2, 24(3): 1–74.

Boo, S.M. & Lee, I. K. 1983. A life history and hybridization of *Antithamnion sparsum* Tokida (Rhodophyta, Ceramiaceae) in culture. *Korean J. Bot.* 26: 141–50.

Broadwater, S. T. & Scott, J. 1982. Ultrastructure of early development in the female reproductive system of *Polysiphonia harveyi* Bailey (Ceramiales, Rhodophyta). *J. Phycol.* 18: 427–41.

Chiang, Y.-M. 1970. Morphological studies of red algae of the family Cryptonemiaceae. *Univ. Calif. Pub. Bot.* 58: 1–83.

Cole, K. & Conway, E. 1980. Studies in the Bangiaceae: Reproductive modes. *Bot. Mar.* 23: 545–53.

Coomans, R. J. 1986. A light microscope study of mitosis and vegetative development in Rhodophyta. Ph.D. dissertation, University of North Carolina at Chapel Hill.

DeCew, T., West, J. A. & Ganesan, E. K. 1981. The life histories and developmental morphology of two species of *Gloiosiphonia* (Rhodophyta: Cryptonemiales, Gloiosiphoniaceae) from the Pacific Coast of North America. *Phycologia* 20: 415–23.

Delivopoulos, S. G. & Kugrens, P. 1984. Ultrastructure of carposporogenesis in the parasitic red alga *Faucheocolax attenuata* Setch. (Rhodymeniales, Rhodymeniaceae). *Am. J. Bot.* 71: 1245–59.

Delivopoulos, S. G. & Kugrens, P. 1985. Ultrastructure of carposporophyte development in the red alga *Gloiosiphonia verticillaris* (Cryptonemiales, Gloiosiphoniaceae). *Am. J. Bot.* 72: 1926–38.

Delivopoulos, S. G. & Tsekos, I. 1986. Ultrastructure of carposporogenesis in the red alga *Gracilaria verrucosa* (Gigartinales, Gracilariaceae). *Bot. Mar.* 29: 27–35.

Dixon, P. S. 1973. *Biology of the Rhodophyta.* New York: Hafner.

Drew, K. M. 1954. The organization and inter-relationships of the carposporophytes of living Florideae. *Phytomorphology* 4: 55–69.

Drew, K. M. 1955. Life histories in the algae with special reference to the Chlorophyta, Phaeophyta and Rhodophyta. *Biol. Rev.* 30: 343–90.

Duckett, J. G. & Peel, M. C. 1978. The role of transmission electron microscopy in elucidating the taxonomy and phylogeny of the Rhodophyta. In *Modern Approaches to the Taxonomy of Red and Brown Algae*, eds. D. E. G. Irvine & J. H. Price, pp. 157–204. London: Academic.

Edelstein, T. 1972. On the taxonomic status of *Gloiosiphonia californica* (Farlow) J. Agardh (Cryptonemiales, Gloiosiphoniaceae). *Syesis* 5: 227–34.

Fan, K. C. 1961. Morphological studies of the Gelidiales. *Univ. Calif. Pub. Bot.* 32: 315–68.

Feldmann, J. 1952. Les cycles de reproduction des algues et leurs rapports avec la phylogénie. *Rev. Cytol. Biol. Vég.* 13: 1–39.

Fetter, R. & Neushul, M. 1981. Studies on developing and released spermatia in the red alga *Tiffaniella snyderae* (Rhodophyta). *J. Phycol.* 17: 141–59.

Fredericq, S. & Hommersand, M. H. 1989. Proposal of the Gracilariales ord. nov. (Rhodophyta) based on an analysis of the reproductive development of *Gracilaria verrucosa*. *J. Phycol.* 25: 213–27.

Fritsch, F. E. 1945. *The Structure and Reproduction of the Algae.* vol. 2. Cambridge: Cambridge University Press.

Gabrielson, P. W. & Garbary, D. 1986. Systematics of red algae (Rhodophyta). *CRC Critic. Rev. Plant Sci.* 3: 325–66.

Gabrielson, P. W. & Hommersand, M. H. 1982a. The Atlantic species of *Solieria* (Gigartinales, Rhodophyta). II. Their morphology, distribution and affinities. *J. Phycol.* 18: 31–45.

Gabrielson, P. W. & Hommersand, M. H. 1982b. The morphology of *Agardhiella subulata* representing the Agardhielleae, a new tribe in the Solieriaceae (Gigartinales, Rhodophyta). *J. Phycol.* 18: 45–58.

Guiry, M. D. 1987. The evolution of life history types in

the Rhodophyta: an appraisal. *Cryptogam. Algol.* 8: 1–12.

Hansen, G. I. 1977. *Cirrulicarpus carolinensis*, a new species in the Kallymeniaceae (Rhodophyta). *Occas. Pap. Farlow Herb.* No. 12: 1–22.

Hansen, G. I. & Lindstrom, S. C. 1984. A morphological study of *Hommersandia maximicarpa* gen. et sp. nov. (Kallymeniaceae, Rhodophyta) from the North Pacific. *J. Phycol.* 20: 476–88.

Hawkes, M. W. 1978. Sexual reproduction in *Porphyra gardneri* (Smith et Hollenberg) Hawkes (Bangiales, Rhodophyta). *Phycologia* 17: 329–53.

Hommersand, M. H. 1963. The morphology and classification of some Ceramiaceae and Rhodomelaceae. *Univ. Calif. Pub. Bot.* 35: 165–366.

Hommersand, M. H. & Fredericq, S. 1988. An investigation of cystocarp development in *Gelidium pteridifolium* with a revised description of the Gelidiales (Rhodophyta). *Phycologia* 27: 254–72.

Hommersand, M. H. & Ott, D. W. 1970. Development of the carposporophyte of *Kallymenia reniformis* (Turner). J. Agardh. *J. Phycol.* 6: 322–31.

Johansen, H. W. 1981. *Coralline Algae, a First Synthesis.* Boca Raton: CRC Press.

Kornmann, P. 1984. *Erythrotrichopeltis*, eine neue Gattung der Erythropeltidaceae (Bangiophyceae, Rhodophyta). *Helgol. Meeresunters.* 38: 207–24.

Kornmann, P. 1987. Der Lebenszyklus von *Porphyrostromium obscurum* (Bangiophyceae, Rhodophyta). *Helgol. Meeresunters.* 41: 127–37.

Kraft, G. T. 1977a. The morphology of *Grateloupia intestinalis* from New Zealand, with some thoughts on generic criteria within the family Cryptonemiaceae (Rhodophyta). *Phycologia* 16: 43–51.

Kraft, G. T. 1977b. Studies of marine algae in the lesser-known families of the Gigartinales (Rhodophyta). I. The Acrotylaceae. *Aust. J. Bot.* 25: 97–140.

Kraft, G. T. 1977c. Studies of marine algae in the lesser known families of the Gigartinales (Rhodophyta). II. The family Dicranemaceae. *Aust. J. Bot.* 25: 219–67.

Kraft, G. T. 1978. Studies of marine algae in the lesser known families of the Gigartinales (Rhodophyta). III. The Mychodeaceae and Mychodeophyllaceae. *Aust. J. Bot.* 26: 515–610.

Kraft, G. T. 1981. Rhodophyta: morphology and classification. In *The Biology of Seaweeds*, eds. C. S. Lobban & M. J. Wynne, pp. 6–51. Oxford: Blackwell Scientific.

Kraft, G. T. 1984. The red algal genus *Predaea* (Nemastomataceae, Gigartinales) in Australia. *Phycologia* 23: 2–20.

Kraft, G. T. & Robins, P. 1985. Is the order Cryptonemiales defensible? *Phycologia* 24: 67–77.

Kützing, F. T. 1843. *Phycologia Generalis*, pp. xvi + 1–144 + xvii–xxxii + 145–458 + [1], 80 plates. Leipzig: Brockhaus.

Kugrens, P. 1980. Electron microscopic observations on the differentiation and release of spermatia in the marine red alga *Polysiphonia hendryi* (Ceramiales, Rhodomelaceae). *Am. J. Bot.* 67: 519–28.

Kylin, H. 1914. Studien über die Entwicklungsgeschichte von *Rhodomela virgata* Kjellm. *Svensk. Bot. Tidsskr.* 8: 33–69.

Kylin, H. 1916. Die Entwicklungsgeschichte und die systematische Stellung von *Bonnemaisonia asparagoides* (Woodw.) Ag. *Zeitschr. f. Bot.* 8: 545–86.

Kylin, H. 1917. Über die Entwicklungsgeschichte von *Batrachospermum moniliforme*. *Ber. deutsch. bot. Ges.* 35: 155–64.

Kylin, H. 1923. Studien über die Entwicklungsgeschichte der Florideen. *K. Svenska Vetensk. Akad. Handl.* 63: 1–139.

Kylin, H. 1928. Entwicklungsgeschichtliche Florideenstudien. *Lunds Univ. Årsskr.*, N.F., Adv. 2, 26(4): 1–127, 64 figs.

Kylin, H. 1956. *Die Gattungen der Rhodophyceen.* Lund: Gleerups.

Lebednik, P. A. 1977. Post fertilization development in *Clathromorphum*, *Melobesia* and *Mesophyllum* with comments on the evolution of the Corallinaceae and the Cryptonemiales (Rhodophyta). *Phycologia* 16: 379–406.

Lee. I. K. 1978. Studies on Rhodymeniales from Hokkaido. *J. Fac. Sci. Hokkaido Univ.*, ser. V. (Botany). 11(1): 1–194, 5 plates.

Lindstrom, S. C. 1986. Acrosymphytaceae, a new family in the order Gigartinales *sensu lato* (Rhodophyta). *Taxon* 36: 50–3.

Lindstrom, S. C. & Scagel, R. F. 1987. The marine algae of British Columbia, northern Washington, and southeast Alaska: division Rhodophyta (red algae), class Rhodophyceae, order Gigartinales, family Dumontiaceae, with an introduction to the order Gigartinales. *Can. J. Bot.* 65: 2202–32.

Maggs, C. A. & Guiry, M. D. 1985. Life history and reproduction of *Schmitzia hiscockiana* sp. nov. (Rhodophyta, Gigartinales) from the British Isles. *Phycologia* 24: 297–310.

Maggs, C. A. & Pueschel, C. M. 1989. Morphology and development of *Ahnfeltia plicata* (Rhodophyta): Proposal of the Ahnfeltiales ord. nov. *J. Phycol.* 25: 333–51.

Magne, F. 1960. Le *Rhodochaete pc vula* Thuret (Bangioidée) et sa reproduction sexuée. *Cah. Biol. Mar.* 5: 461–671.

Magne, F. 1964. Les Rhodophycées à cycle haplophasique éxistent-elles? *C. R. IV Congrès Int. des Algues Marines*, Biarritz 1961, pp. 112–16.

Magne, F. 1972. Le cycle de développement des Rhodophycées et son évolution. *Bull. Soc. Bot. France*: 247–67.

Magne, F. 1987. La tétrasporogènese et le cycle de développement des Palmariales (Rhodophyta): une nouvelle interprétation. *Cryptogam. Algol.* 8: 273–80.

Magruder, W. H. 1984. Specialized appendages on spermatia from the red alga *Aglaothamnion neglectum* (Ceramiales, Ceramiaceae) specifically bind with trichogynes. *J. Phycol.* 20: 436–40.

Mikami, H. 1965. A systematic study of the Phyllophoraceae and Gigartinaceae from Japan and its vicinity. *Sci. Papers Inst. Algol. Res. Fac. Sci., Hokkaido Univ.* 5: 181–285.

Min-Thein, U. & Womersley, H. B. S. 1976. Studies on Southern Australian taxa of Solieriaceae, Rhabdoniaceae and Rhodophyllidaceae (Rhodophyta). *Aust. J. Bot.* 24: 1–166.

Mullahy, J. H. 1952a. The morphology and cytology of *Lemanea australis* Atk. *Bull. Torrey Bot. Club* 79: 393–406.

Mullahy, J. H. 1952b. The morphology and cytology of *Lemanea australis* Atk. (concluded). *Bull. Torrey Bot. Club* 79: 471–84.

Necchi, O. 1987. Sexual reproduction in *Thorea* Bory (Rhodophyta, Thoreaceae). *Jap. J. Phycol.* 35: 106–12.

Norris, R. E. 1957. Morphological studies on the Kallymeniaceae. *Univ. Calif. Pub. Bot.* 28: 251–334.

Nozawa, Y. 1970. Systematic anatomy of the red algal genus *Rhodopeltis*. *Pacific Sci.* 24: 99–133.

Oltmanns, F. 1898. Zur Entwicklungsgeschichte der Florideen. *Bot. Ztg.* 56: 99–140, plates 4–7.

Oltmanns, F. 1922. *Morphologie und Biologie der Algen.* Vol. 2. Jena: Gustav Fischer.

Papenfuss, G. F. 1951. Problems in the classification of the marine algae. *Sven. Bot. Tidskr.* 45: 4–11.

Papenfuss, G. F. 1966. A review of the present system of classification of the Florideophycidae. *Phycologia* 5: 247–55.

Parsons, M. J. 1975. Morphology and taxonomy of the Dasyaceae and the Lophothaliae (Rhodomelaceae) of the Rhodophyta. *Aust. J. Bot.* 25: 549–713.

Ramm-Anderson, S. M. 1983. The fine structure of postfertilization development in selected members of the Nemalionales. Ph.D. dissertation, University of Melbourne.

Ramm-Anderson, S. M. & Wetherbee, R. 1982. Structure and development of the carposporophyte of *Nemalion helminthoides* (Nemalionales, Rhodophyta). *J. Phycol.* 18: 133–41.

Rao, C. S. P. 1956. The life history and reproduction of *Polyides caprinus*. *Ann. Bot.* 20: 211–30.

Renfrew, D. E. 1988. Spermatogenesis in *Gelidium*. *J. Phycol* 24 (Suppl.): 9.

Robins, P. A. & Kraft, G. T. 1985. Morphology of the type and Australian species of *Dudresnaya* (Dumontiaceae, Rhodophyta). *Phycologia* 24: 1–34.

Rosczyk, M. L. 1984. DNA amplification in *Nemalion helminthoides*. M.A. thesis, California State University, Long Beach.

Rosenvinge, G. L. 1931. The marine algae of Denmark. Contributions to their natural history. Vol. 1. Rhodophyceae, Pt. IV (Gigartinales, Rhodymeniales, Nemastomatales). *K. Danske Vidensk. Selsk. Skrift;, ser. 7, Nat. Mat. Afd.* 7: 497–627.

Schmitz, F. 1883. Untersuchungen über die Befruchtung der Florideen. *K. Preuss. Akad. Wiss.* (Berlin) 10: 215–58, 5 plates.

Schmitz, F. 1892. [6. Klasse Rhodophyceae] 2. Unterklasse Florideae. In *Syllabus . . . Grosse Ausgabe* ed. A. Engler, pp. 16–23. Berlin: G. Borntraeger.

Schmitz, F. & Hauptfleisch, P. 1896. Rhodophyceae. In *Die Natürlichen Pflanzenfamilien*, vol. 1(2), eds. A. Engler & K. Prantl, pp. 298–306.

Schneider, C. W. & Reading, R. P. 1987. A revision of the genus *Peyssonnelia* (Rhodophyta, Cryptonemiales) from North Carolina, including *P. atlantica* sp. nov. *Bull. Mar. Sci.* 40: 175–92.

Scott, J. L. & Dixon, P. S. 1973. Ultrastructure of spermatium liberation in the marine red alga *Ptilota densa*. *J. Phycol.* 9: 85–91.

Searles, R. B. 1968. Morphological studies of red algae of the order Gigartinales. *Univ. Calif. Pub. Bot.* 43: 1–86.

Searles, R. B. 1980. The strategy of the red algal life history. *Am. Nat.* 115: 113–20.

Silva, P. C. & Johansen, H. W. 1986. A reappraisal of the order Corallinales (Rhodophyceae). *Br. Phycol. J.* 21: 245–54.

Silva, P. C., Meñez, E. G. & Moe, R. L. 1987. Catalog of the benthic marine algae of the Philippines.*Smithson. Contrib. Mar. Sci.* 27: iv–179.

Siotas, A. S. & Wetherbee, R. 1982. The structure and apparent role of gametophytic "nutritive cells" during post fertilization development in two genera of red algae. *Micron* 13: 333–4.

Sjöstedt, L. G. 1926. Floridean studies. *Lunds Univ. Årsskr.*, N.F., Avd. 2, 22(4): 1–95.

Sparling, S. R. 1957. The structure and reproduction of some members of the Rhodymeniaceae. *Univ. Calif. Pub. Bot.* 29: 319–96.

Stegenga, H. 1978. The life histories of *Rhodochorton purpureum* and *Rhodochorton floridulum* (Rhodophyta, Nemaliales) in culture. *Br. Phycol. J.* 13: 279–89.

Svedelius, N. 1915. Zytologisch-entwicklungsgeschichtliche Studien über *Scinaia furcellata*. *Nova Acta R. Soc. Sci. Upsal.* ser. 4, 4(4): 1–55.

Svedelius, N. 1939. Anatomisch-entwicklungsgeschichtliche Studien über die Florideengattung *Dermonema* (Grev.) Harv. *Bot. Notiser* 1939: 21–39.

Svedelius, N. 1942. Zytologisch-entwicklungsgeschichtliche Studien über *Galaxaura* eine diplobiontische Nemalionales-Gattung. *Nova Acta Regiae Soc. Sci. Upsal.,* ser. 4, 13(4): 1–154.

Tsekos, I. & Schnepf, E. 1982. Plastid development and floridean starch grain formation during carposporogenesis in the red alga *Gigartina teedii*. *Cryptogam. Algol.* 2: 91–103.

Tsekos, I. & Schnepf, E. 1983. The ultrastructure of carposporogenesis in *Gigartina teedii* (Roth) Lamour. (Gigartinales, Rhodophyceae): auxiliary cell, cystocarpic plant. *Flora* 173: 81–96.

van der Meer, J. P. & Todd, E. R. 1980. The life history of *Palmaria palmata* in culture. A new type for the Rhodophyta. *Can. J. Bot.* 58: 1250–6.

Webster, R. N. 1958. The life history of the freshwater alga *Tuomeya fluviatilis* Harv. *Butler Univ. Bot. Stud.* 13: 141–59.

West, J. A. 1969. The life histories of *Rhodochorton purpureum* and *R. tenue* in culture. *J. Phycol.* 5: 12–21.

West, J. A. & Hommersand, M. H. 1981. Rhodophyta: life histories. In *The Biology of Seaweeds*, eds. C. S. Lobban & M. J. Wynne, pp. 133–93. Oxford: Blackwell Scientific.

Wetherbee, R. 1980. Postfertilization development in the red alga *Polysiphonia*. 1. Proliferation of the carposporophyte. *J. Ultrastr. Res.* 58: 119–33.

Wiseman, D. R. 1977. Observations on the reproductive morphology of the red algal genus *Ochtodes* J. Agardh (Rhizophyllidaceae, Gigartinales). *Phycologia* 16: 1–8.

Wittmann, W. 1965. Aceto–iron–haematoxylin–chloral hydrate for chromosome staining. *Stain Tech.* 40: 161–4.

Woelkerling, W. J. 1970. *Acrochaetium botryocarpum* (Harv.) J. Ag. (Rhodophyta) in Southern Australia. *Br. Phycol. J.* 5: 159–71.

Woelkerling, W. J. 1980. Studies on *Metamastophora* (Corallinaceae, Rhodophyta). 1. *M. flabellata*: morphology and anatomy. *Br. Phycol. J.* 15: 201–5.

Woelkerling, W. J. 1988. *The Coralline Red Algae: An Analysis of the Genera and Subfamilies of Nongeniculate Corallinaceae.* British Museum (Natural History), London & Oxford: Oxford University Press.

Woelkerling, W. J. & Foster, M. S. 1989. A systematic and ecographic account of *Synarthrophyton schielianum* sp. nov. (Corallinaceae, Rhodophyta) from the Chatham Islands. *Phycologia* 28: 39–60.

Womersley, H. B. S. & Norris, R. E. 1971. The morphology and taxonomy of Australian Kallymeniaceae (Rhodophyta). *Austr. J. Bot.*, Suppl. Series No. 2: 1–62.

Yamanouchi, S. 1906. The life history of *Polysiphonia violacea*. *Bot. Gaz.* 42: 401–49.

Yoshizaki, M. 1986. The morphology and reproduction of *Thorea okadai* (Rhodophyta). *Phycologia* 25: 476–81.

Chapter 14

Sporangia and spores

MICHAEL D. GUIRY

CONTENTS

I. Introduction / 347
II. Sporangia and spores / 348
    A. Endosporangia / 351
    B. Monosporangia / 351
    C. Monosporangia-like bodies / 355
    D. Bisporangia / 357
    E. Tetrasporangia / 359
    F. Conchosporangia / 362
    G. Zygotosporangia / 363
    H. Carposporangia / 365
    I. Carpotetrasporangia / 366
    J. Polysporangia / 366
    K. Parasporangia / 368
III. Sporangial regeneration / 368
IV. Spore germination and sporeling development / 369
V. Summary / 371
VI. Acknowledgments / 372
VII. Endnotes / 372
VIII. References / 372

I. INTRODUCTION

Members of the Rhodophyta completely lack flagellated vegetative and reproductive cells, a feature shared elsewhere in the algae only by the prokaryotic Cyanophyta and Prochlorophyta. Perhaps as a result, complex sexual processes and sporangial structures have evolved. Drew (1956) and Dixon (1973) reviewed the form and function of sporangia in the subclass Bangiophycidae, and Dixon (1973), Guiry (1978), and Guiry and Irvine (1989) have undertaken similar analyses for the Florideophycidae. Recent advances in our knowledge of the life histories and genetics of these algae have been so rapid that an overall review of sporangial nomenclature is again necessary, particularly as some terms still remain confused. In this chapter, the naming of red algal sporangial and spore types will be reviewed, with particular regard to their origin, mode of formation, and numbers of spores formed. Emphasis will be placed on taxonomic trends and areas in which further work is needed.

II. SPORANGIA AND SPORES

A spore is a specialized cell that can reproduce a new plant independently, and a sporangium is a structure within which the spore arises (Guiry 1978; Guiry & Irvine 1989).

Hypothetically, red algal sporangia may be divided into two groups, based on the presence (meiosporangia) or absence (mitosporangia) of meiosis (Guiry 1978). Although very important from the point of view of the sequence of life history phases, this distinction is of little practical value in classifying sporangia. For example, diploid conchosporangia and tetrasporangia both produce spores that develop into the haploid gametophytic generation, and it would be reasonable to conclude that they are the site of meiosis. However, mitosis appears to occur in the tetrasporangia of some species, giving rise to diploid spores that either germinate to form diploid plants (West & Hommersand 1981) or undergo meiosis at the sporeling stage (see Chapter 3) to form haploid plants. Nevertheless, the majority of tetrasporangia probably undergo reduction division in the formation of tetraspores. Accumulating evidence also indicates that meiosis probably occurs in germinating conchospores rather than during the formation of conchospores (Burzycki & Waaland 1987; Tseng & Sun 1989). The generally accepted system of naming sporangia on the basis of the origin, mode of formation, and number of their spores is thus more satisfactory, in that it does not require the determination of DNA content or life history studies to identify sporangial types. The greatest difficulties in naming sporangial and spore types are evident in the Bangiophycidae. The sporangial types recognized in this review and their distribution in the two subclasses of the Rhodophyta, the Bangiophycidae, and the Florideophycidae, are shown in Table 14-1, and their occurrence in the orders and families of the Florideophycidae is given in Table 14-2. Distinguishing these types is often difficult in practice, but as more information becomes available, particularly from cell biology, life history, and fine structural studies, the limits of each should become clearer.

Drew (1956) attempted to rationalize the earlier confused nomenclature of spores in the Bangiophycidae by describing three types: (1) monospores produced by differentiated sporangia; (2) monospores produced from undifferentiated vegetative cells; (3) spores formed by the successive divisions of mother cells to give packets of 4–128+. As she took into account only form and developmental

Table 14-1. Types and occurrence of sporangia in the subclasses of the Rhodophyta

Type of sporangium	Subclass	
	Bangiophycidae	Florideophycidae
Endosporangia	Present	Absent
Monosporangia	Present	Present
Bisporangia	Absent	Present
Tetrasporangia	Absent	Present
Conchosporangia	Present	Absent
Zygotosporangia	Present	Rare
Carposporangia	Absent	Present
Carpotetrasporangia	Absent	Present
Polysporangia	Absent	Present
Parasporangia	Absent	Present

See text for further details.

patterns of sporangia and excluded origin, two quite different types of sporangia are included in the third type: endosporangia, in which the spores arise internally by successive mitotic divisions in a sporangium but are not the result of sexual fusion, and zygotosporangia, in which the spores arise internally in the carpogonium by mitosis after fertilization. At present, the only putative representative of the Bangiophycidae that has carposporangia is *Rhodochaete parvula*, a species of uncertain subclass affinities (Kraft 1981) placed in its own order, the Rhodochaetales.

Although the distinctions between sporangial types are becoming better known in the Bangiales, the picture in the Porphyridiales and Compsopogonales[1] (=Erythropeltidales sensu Garbary et al. 1980) is less clear. In the Porphyridiaceae (Porphyridiales), a family of unicellular or pseudofilamentous organisms, liberation of reproductive cells is either referred to as release of vegetative cells (e.g., Garbary et al. 1980) or monospores (e.g., Sheath 1984, p. 96). Distinguishing between these is difficult, and fine structural studies are required to establish whether a sporangial wall and sporangial vesicles (intracellular mucilage-forming organelles that are derived from dictyosomes) are differentiated. Monospores produced by differentiated sporangia were considered by Drew (1956) to be absent in the Bangiales, but Conway and Cole (1977) reported well-defined single spores arising in differentiated monosporangia in the conchocelis phases of both *Porphyra* and *Bangia*.

A much wider range of sporangia occurs in the Florideophycidae than in the Bangiophycidae (Table 14-1), and the individual types are also more

Sporangia and spores

Table 14-2. Occurrence of various types of sporangia in the orders and families[a] of the Florideophycidae

| Taxon | Tetrasporangia | | | | | | Polysporangia | Parasporangia | Seirosporangia | Monosporangia | Bisporangia |
	Cruciate	Zonate	Tetrahedral	Unknown	Irregularly cruciate	Irregularly zonate					
Acrochaetiales											
Acrochaetiaceae[b]	+	−	−	−	+	−	−	−	+?	+	+
Palmariales											
Palmariaceae[c]	+	−	−	−	−	−	−	−	−	−	+
Nemaliales											
Liagoraceae	+	−	−	−	+	−	−	−	−	+	−
Galaxauraceae	+	−	−	−	+	+	−	−	−	+	−
Gymnocodiaceae[d]	?	−	−	−	−	−	−	−	−	−	−
Batrachospermales											
Lemaneaceae	−	−	−	+	−	−	−	−	−	+?	−
Batrachospermaceae	−	−	−	+	−	−	−	−	−	+	−
Thoreaceae	−	−	−	+	−	−	−	−	−	+	−
Bonnemaisoniales											
Bonnemaisoniaceae	+	−	−	−	+	−	−	−	−	−	−
Naccariaceae[e]	+	−	−	−	−	−	−	−	−	−	−
Gelidiales											
Gelidiaceae[f]	+	−	+?	−	+	−	−	−	−	−	+
Corallinales											
Corallinaceae	+?	+	−	−	−	−	−	−	−	−	+
Solenoporaceae[d]	?	−	−	−	−	−	−	−	−	−	−
Hildenbrandiales											
Hildenbrandiaceae	+	+	−	−	+	+	−	−	−	−	−
Gigartinales											
Choreocolacaceae	+	−	−	−	−	−	−	−	−	−	−
Gracilariaceae	+	−	−	−	−	−	−	−	−	−	−
Sebdeniaceae	+	−	−	−	−	−	−	−	−	−	−
Calosiphoniaceae	+	−	−	−	−	−	−	−	−	−	−
Phyllophoraceae	+	−	−	−	−	−	−	−	−	+?	−
Gigartinaceae	+	−	−	−	−	−	−	−	−	−	−
Petrocelidaceae[g]	+	−	−	−	−	−	−	−	−	−	−
Chondriellaceae	+	−	−	−	−	−	−	−	−	−	−
Polyideaceae	+	−	−	−	−	−	−	−	−	−	−
Peyssonneliaceae	+	−	−	−	−	−	−	−	−	−	−
Tichocarpaceae	+?	−	−	−	−	−	−	−	−	−	−
Acrosymphytaceae[h]	+	−	−	−	−	−	−	−	−	−	−
Nemastomataceae	+	+	−	−	+	+	−	−	−	−	−
Dumontiaceae	+	+	−	−	+	+	−	−	+?	+	+
Halymeniaceae	+	−	−	−	+	−	−	−	−	+	−
Corynomorphaceae	+	−	−	−	+	+	−	−	−	−	−
Kallymeniaceae	+	−	−	−	+	+	−	−	−	−	−
Endocladiaceae	+	−	−	−	+	+	−	−	−	−	−
Gloiosiphoniaceae	+	+	−	−	+	+	−	+?	−	−	+
Nizymeniaceae	+	−	−	−	+	−	−	−	−	−	−
Rhizophyllidaceae	+	−	−	−	+	+	−	−	−	−	−
Gainiaceae[i]	+	−	−	−	+	+	−	−	−	−	−

Table 14-2. (cont.)

Taxon	Tetrasporangia										
	Cruciate	Zonate	Tetrahedral	Unknown	Irregularly cruciate	Irregularly zonate	Polysporangia	Parasporangia	Seirosporangia	Monosporangia	Bisporangia
Solieriaceae	−	+	−	−	+	−	−	−	−	−	+
Acrotylaceae	−	+	−	−	−	−	−	−	−	−	−
Pterocladiophilaceae	−	+	−	−	−	−	−	−	−	−	−
Sphaerococcaceae	−	+	−	−	−	−	−	−	−	−	−
Phacelocarpaceae	−	+	−	−	−	−	−	−	−	−	−
Sarcodiaceae	−	+	−	−	−	−	−	−	−	−	−
Furcellariaceae	−	+	−	−	−	−	−	−	−	−	−
Plocamiaceae	−	+	−	−	−	−	−	−	−	−	−
Hypneaceae	−	+	−	−	−	−	−	−	−	−	−
Rissoellaceae	−	+	−	−	−	−	−	−	−	−	−
Caulacanthaceae	−	+	−	−	−	−	−	−	−	−	−
Cubiculosporaceae	−	+	−	−	−	−	−	−	−	−	−
Cystocloniaceae	−	+	−	−	−	−	−	−	−	−	−
Mychodeaceae	−	+	−	−	−	−	−	−	−	−	−
Mychodeophyllaceae	−	+	−	−	−	−	−	−	−	−	−
Dicranemaceae	−	+	−	−	−	−	−	−	−	−	−
Cruoriaceae[j]	−	+	−	−	−	−	−	−	−	−	−
Blinksiaceae	−	+	−	−	−	−	−	−	−	−	−
Wurdemanniaceae	−	+	−	−	−	−	−	−	−	−	−
Pseudoanemoniaceae	−	−	−	+	−	−	−	−	−	−	−
Rhodymeniales											
Rhodymeniaceae	+	−	+	−	−	−	−	−	−	−	−
Champiaceae	−	−	+	−	−	−	+	−	−	−	−
Lomentariaceae[k]	−	−	+	−	−	−	−	−	−	−	−
Ceramiales											
Ceramiaceae	+	−	+	−	−	−	+	+	+	+?	+
Delesseriaceae	−	−	+	−	−	−	+?	−	−	+?	−
Dasyaceae	−	−	+	−	−	−	−	−	−	−	−
Rhodomelaceae	−	−	+	−	−	−	−	−	−	+?	−

[a] For a recent treatment of the nomenclature of the families of the Rhodophyta, see Silva (1980). Since this chapter went to press, the new orders Ahnfeltiales and Gracilariales have been proposed (see Chapters 1, 2, 13, and 18 in this book).
[b] Zygotosporangia are found in only one species of *Acrochaetium* (Acrochaetiaceae) and are not included here.
[c] *Rhodophysema* is included in the Palmariaceae but should probably form the basis for a new family.
[d] Fossil families for which sporangia are known, but the mode of division is not.
[e] The ordinal position of the Naccariaceae is doubtful.
[f] The Gelidiellaceae Fan is included in the Gelidiaceae (see Maggs & Guiry 1988).
[g] See Guiry et al. (1984).
[h] See Lindstrom (1987).
[i] See Moe (1985).
[j] The Cruoriaceae is regarded as including only the type species of the genus *Cruoria* (*C. purpurea*) and *C. cruoriaeformis* but not the cruoria-phase tetrasporophytes of various algae.
[k] See Lee (1978).
Adapted from Guiry (1978).

Sporangia and spores

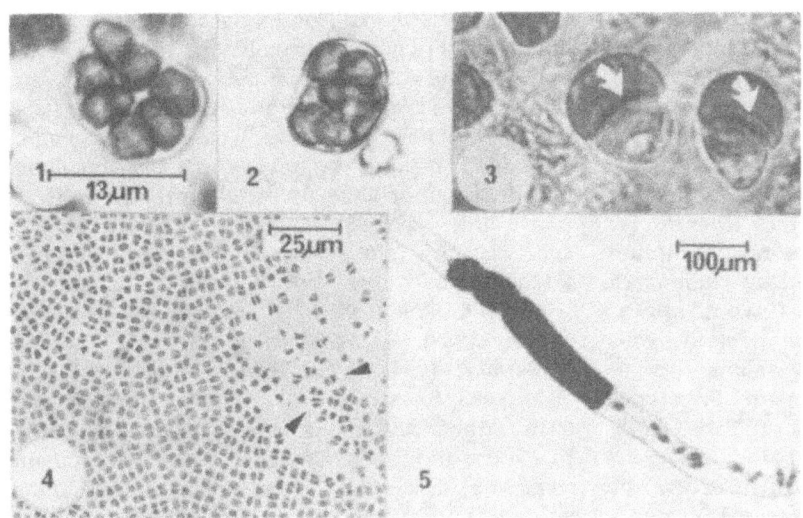

Figs. 14-1 to 14-5. Sporangia in the Bangiophycidae. *14-1, 14-2*: Endosporangia in *Erythrocladia irregularis* (Erythropeltidaceae). (Scales in Figs. 14-2 and 14-3 as in Fig. 14-1.) *14-3*: Monosporangia formed by acropetal division (arrows) of vegetative cells in *E. irregularis*. *14-4*: Monosporangial sorus of *Porphyropsis coccinea* var. *coccinea* (Erythropeltidaceae). Note oblique formation of sporangia (arrowheads). *14-5*: Formation of monosporangia in cultured plant of *Bangia atropurpurea* (Bangiales). (Figs. 14-1 to 14-3 after Gargiulo et al. 1987, with permission; remainder original)

clearly delineated. Guiry (1978) considered that meiotic and mitotic homologues of all sporangial types exist in the Florideophycidae.

A. Endosporangia

Endosporangia form irregularly arranged, uninucleate spores of indefinite number and are mitotic. They were previously thought to occur only in the Phragmonemataceae (Porphyridiales), where they are found in at least half of the included genera and are likely to occur in the remainder (Garbary et al. 1980). Recently, however, Gargiulo et al. (1987) found endosporangia in *Erythrocladia irregularis* (Compsopogonales) in culture (Figs. 14-1, 14-2), and other members of this order may eventually be found to possess such sporangia. A possibility that should not be overlooked is that some of these sporangia are the product of gametic union and are thus actually zygotosporangia.

B. Monosporangia

Monosporangia form a single uninucleate spore on either haploid or diploid plants but are not derived from gonimoblast tissue and are nearly always mitotic. They are generally formed singly but may be formed in packets occasionally.

A range of monosporangial types is found in the Bangiophycidae. Garbary et al. (1980) characterized the Compsopogonales[1] as having monosporangial formation in which undifferentiated cells divide by means of a curved wall, one of the resulting cells becoming a monosporangium (Figs. 14-3, 14-4). The congruity of this character – among other features – led the authors to include the Erythropeltidaceae, Compsopogonaceae, and Boldiaceae, families formerly referred to the Bangiales, in the Compsopogonales.

Kornmann and Sahling (1985) found that all species of Erythropeltidaceae cultured from Helgoland (North Sea) formed monospores from differentiated sporangia. In *Porphyrostromium* (=*Erythrotricopeltis*; see Wynne 1986b) *ciliaris* and *P. obscurum*, monosporangia recycle each of the morphological phases found in culture (Kornmann 1984, 1987). Two types of monosporangia are found in *Smithora naiadum*, a bladelike representative of the Compsopogonales. Knox (1926) found that some monosporangia were formed by direct transformation of vegetative cells in the terminal portion of the thallus; these released monospores singly or in an aggregated mass. This latter phenomenon led Hollenberg (1959) to refer to them as deciduous sori. McBride and Cole (1971) found that monospore production of this type is one of the main methods of propagation in *S. naiadum*. In a fine structural study they showed that there is extensive differentiation in the monosporangia, with dictyosome activity giving rise to two types of vesicles, and that released monospores lack a cell wall (McBride & Cole 1972). A further type of monosporangium found in *S. naiadum* was described by Hollenberg (1959) as "neutral spores" [sic]. These would appear to be single sporangia cut off from vegetative cells in an acropetal manner by curved walls and occur in marginal, rather than terminal, sori. Two types of monosporangia, corresponding to those of *S. naiadum*, were found on cultured plants of *Porphyropsis coccinea* var. *dawsonii* (Erythropeltidaceae) by Murray et al. (1972); further generations of monosporangial plants were formed from both types of monospore, and no sexual reproduction was found.

In *Rhodochaete parvula* (Rhodochaetales), monosporangia are formed on both diploid and haploid plants by a lateral curved wall, rather than an acropetally curved wall as found in the Compsopogonales. Some of the monospores formed on diploid plants develop into gametangial plants, but others give rise to further diploid plants. However, the exact site of meiosis is unknown (Boillot 1975; Magne 1960). The pit connections of this species are plugged and not open as described by Boillot (1978), but they lack plug cap or cap membranes (Pueschel 1987; Pueschel & Magne 1987). These and other ultrastructural features led Pueschel and Magne (1987) to conclude that *R. parvula* "is perhaps the most phylogenetically important red alga known...." In some species of *Porphyra* (Bangiales) monospores are differentiated by direct transformation of vegetative cells of the leafy *Porphyra* phase. Increased dictyosome activity in the monosporangia of *P. gardneri* results in the formation of large and small vesicles. The monosporangia of *Porphyra* species are generally produced on young thalli but may also be formed on larger fronds (Hawkes 1980). Monospore formation seems to be more common in Japanese species of the genus than in those from the northeastern Pacific (Hawkes 1980; Hymes & Cole 1983b), but it has not been found in distromatic entities with two chloroplasts per cell. Monospore production leads to an increased "set" of leafy thalli in the course of cultivation and results in a greater harvest (Miura 1975). Not all populations of *Porphyra* species show accessory reproduction by monospores; Bird (1973) reported that she could find no means of vegetative increase in *P. linearis* from Nova Scotia and that the life history appeared to involve an obligate alternation with the conchocelis phase.

The term "aplanosporangia" was used (Cole & Conway 1980; Hymes & Cole 1983b) for small "packets" of spores (up to 16) formed as a result of a few divisions of vegetative cells of *Porphyra* (leafy phase only) and *Bangia* species. Records of such sporangia are relatively rare: they were found in *P. subtumens* (Conway & Wylie 1972), *P. maculosa* (Hymes & Cole 1983b), *P. sanjuanensis* (Krishnamurthy 1969), *P. argentinensis* (Piriz 1981) and *B. atropurpurea* (Cole 1972a; Cole & Conway 1980; Cole et al. 1983; Sheath & Cole 1980). In these species, each of the individual cells in a "packet" gives rise to a spore and, after release, a distinct wall remains (Fig. 14-5). Essentially, these are packets of monosporangia and appear to be exclusively mitotic; their spores give rise by bipolar germination to individuals similar to the parent thallus. The use of the term "aplanosporangia" for such sporangia is not desirable since, strictly speaking, all red algal spores are aplanospores in that they lack flagella; the term is widely used in this sense in the Chlorophyta.

Monosporangia are also formed on the conchocelis phases of some, but not all, species of *Porphyra* and *Bangia* (Conway & Cole 1977) and are rounded, in contrast to the rectangular cells of conchosporangia. They may be stalked or sessile and are formed singly or occasionally in chains. Monospores released from the parent cell wall grow immediately into new conchocelis-phase filaments. In still culture, monosporangia often germinate in situ and become part of the conchocelis-phase branching system (Conway & Cole 1977).

The range of monosporangial types found in the Bangiophycidae was considered useful in separating the orders of the subclass by Garbary et al. (1980). However, further fine structural and culturing studies are needed to elucidate the form and function of many of the types, and the status of the orders of the Bangiophycidae must still be considered provisional (Gargiulo et al. 1987).

Monosporangia in the Florideophycidae (Figs. 14-6, 14-12 to 14-15) are essentially oblong sporangia enclosing a single, uninucleate spore (Guiry et al. 1988) and formed from terminal cells of branched filaments. They are generally found on loosely filamentous plants or on the creeping phases or attachment organs of more organized entities (Guiry & Irvine 1989). Monosporangia occur most commonly in the Acrochaetiales, Nemaliales, and Batrachospermales.

In the Acrochaetiales, a large number of species

Sporangia and spores 353

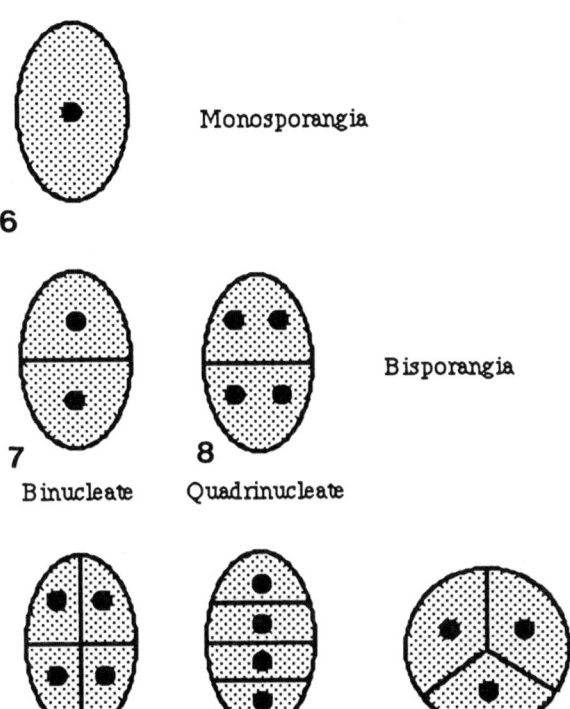

Figs. 14-6 to 14-11. Cleavages and arrangement of nuclei (dark spheres) of some sporangial types in the Florideophycidae. (Redrawn after Guiry 1978, with permission)

reproduce exclusively by monospores (Stegenga 1985; Woelkerling 1983). In some instances, tetrasporangia were observed on plants bearing monosporangia, and it was suggested that such tetrasporangia represent divided monosporangia (Fritsch 1942) or monospores germinating in situ (Boney 1967; Dixon 1973). However, tetrasporangia and monosporangia occur together on wild and cultured plants of *Acrochaetium*[2] *botryocarpum* (Guiry et al. 1988; Woelkerling 1970). Many species of *Acrochaetium* and *Colaconema*[2] have strains that seem to reproduce exclusively by monospores and strains in which sexual reproduction and tetrasporangial formation are confined to certain seasons. In south-eastern Australia, *A. botryocarpum* forms monosporangia in the wild more or less all the year round, whereas sexual reproduction occurs mainly in spring and early summer and tetrasporangial reproduction occurs principally in winter (Woelkering 1970). In Ireland, monosporangia are also formed continuously, but tetrasporangia are present only from December to May, and sexual reproduction is absent (Guiry et al. 1988).

In the Acrochaetiaceae, monosporangia are generally produced singly on lateral branches or in branched clusters or may even be sessile on the filaments. However, Klavestad (1957) has reported terminal unbranched chains of sporangia in *Rhodochorton*[2] *purpureum*, the origin, cytology, and spore release of which are unknown.

The fine structure of the development of monosporangia in the freshwater alga *Acrochaetium hermannii* (as *Audouinella*) has been described by Hymes and Cole (1983a). They concluded that monosporogenesis and carposporogenesis were rather similar, lending some support to the hypothesis (Guiry 1978) that the carposporophyte essentially represents a monosporangial generation. However, the ultrastructural aspects of all types of spore development seem rather similar to date, and only time will tell if there are phylogenetic gains to be made from comparative fine structural studies of sporogenesis.

In the Nemaliales, monosporangia have been reported in all families. In the Liagoraceae, monosporangia were found on the loosely filamentous tetrasporophytes of *Nemalion* (Chen et al. 1978; Cunningham & Guiry unpubl.; Fries 1967), *Liagora* (von Stosch 1965), *Helminthocladia* (Boillot 1972; Guiry unpubl.), *Helminthora* (Cunningham & Guiry unpubl.), and *Cumagloia* (Umezaki 1972) species. Additionally, a report of monosporangia occurring on

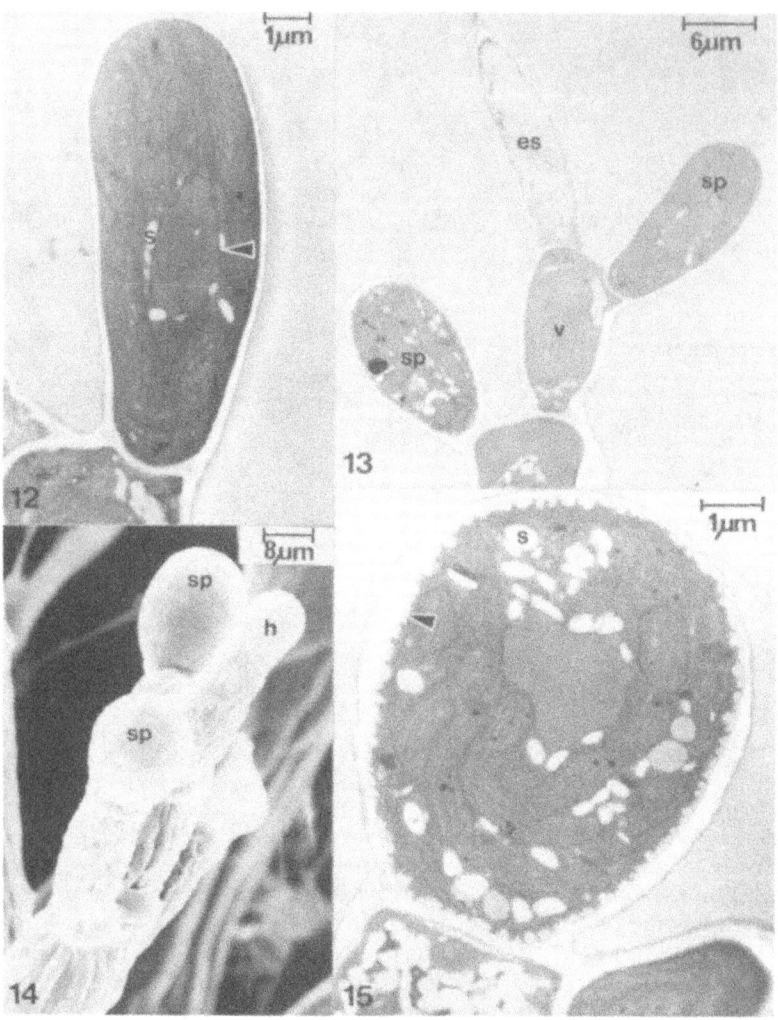

Figs. 14-12 to 14-15. Transmission (TEM) and scanning (SEM) electron microscopy of the development of monosporangia in *Acrochaetium hermannii* (Acrochaetiaceae). *14-12*: TEM of terminal vegetative cell with few starch grains (s), smooth walls and a single nucleus (arrowhead). *14-13*: TEM of terminal portion of plant showing empty sporangium (es), vegetative cells (v) and sporangia (sp) at several stages of development. *14-14*: SEM showing two sporangia (sp) and a developing hair cell (h). *14-15*: Maturing monosporangium with plasmalemma retracting from wall (arrowhead) and numerous starch grains (s). (Figs. 14-12 to 14-15 after Hymes & Cole 1983a, with permission.

erect gametangial plants of *Helminthora divaricata* from the Adriatic (Svedelius 1917) has been confirmed in culture for Irish plants (Cunningham & Guiry unpubl.); these monospores give rise to further gametangial plants. Larger than normal monospores formed by the filamentous tetrasporophytes of *Nemalion vermiculare* from Japan gave rise to erect, supposedly gametangial plants in culture (Masuda & Umezaki 1977); these probably represent undivided tetrasporangia in which, nevertheless, meiosis had taken place. In the Galaxauraceae, monosporangia are also described for *Scinaia* (now considered to include *Pseudogloiophloea* and *Pseudoscinaia*; Huisman 1985) species. These are formed singly or in groups on the outer cortical cells of the gametophytes (Dixon 1973; Huisman 1985;

Maggs & Guiry 1982a), but no spore release has been reported. Ramus (1969) found monosporangia on filamentous tetrasporophytes grown from carpospores of *Scinaia* (as *Pseudogloiophloea*) *confusa* from California; these monospores recycle the tetrasporangial plants. However, Boillot (1968, 1969, 1972) did not encounter monosporangia on the filamentous carpospore germlings of *Scinaia forcellata* and *S. turgida*, and they were not found on the filamentous tetrasporophyte of *Gloiophloea scinaioides* (Huisman 1987).

In the Batrachospermales, species of *Thorea*, *Nemalionopsis*, and *Batrachospermum* frequently form monosporangia (Sheath 1984), and it may be that these are an important dispersal mechanism, as these plants do not have the facility of forming diploid tetrasporangial phases (Guiry & Irvine 1989). Some *Batrachospermum* species form monosporangia in both the erect phase and in the chantransia phase, but in others they are confined to the latter (Dixon 1973).

In the Gigartinales, reports of monosporangia are very rare and are confined to a few genera of Dumontiaceae and Halymeniaceae. In the Dumontiaceae, Richardson and Dixon (1970), Scott and Dixon (1971), and Dixon et al. (1972) observed monosporangia in culture that repeated the parent phase in *Thuretellopsis peggiana* and *Pikea californica*. In the former species they occurred on both the crustose phase and the erect axes, but in the latter they were found only at the margin of the crustose phase. In both of these species the sporangia appear to be formed terminally. In the Halymeniaceae, monosporangia were reported for *Halymenia floresia* on both the basal filamentous growths and on erect plants formed from carpospores in culture (van den Hoek & Cortel-Breeman 1970). Monospores from the filamentous growths formed further filamentous plants, but sporangia that formed over the entire surface of erect thalli released single spores producing gametophytes. This suggests that the latter represent tetrasporangia that failed to cleave after meiosis, although no karyological data are available. In cultures of *H. latifolia* from Ireland and France, however, neither the filamentous basal growths nor the erect axes formed monosporangia, although vegetative reproduction was very efficient in culture by fragmentation of the filamentous growths (Maggs & Guiry 1982b, 1987). The degree to which monosporangia occur in the Gigartinales is in need of assessment. Monosporangia are absent in the Palmariales, Bonnemaisoniales, Gelidiales, Corallinales, Hildenbrandiales, and probably the Ceramiales.

C. Monosporangia-like bodies

Bodies superficially resembling monosporangia have been reported in the Nemaliales, Gigartinales, and Ceramiales. In the Nemaliales, prostrate filamentous growths derived from carpospores of *Helminthocladia calvadosii* in culture were found to produce short chains of rounded cells that, on occasion, superficially resembled tetrasporangia; these detached easily and formed new filaments, but no release of spores was observed (Guiry & Irvine 1989). In cultures of *H. pacifica* from Japan, similar rounded cells were described as forming in long chains and, although no spores were released, the cells broke off and regenerated (Umezaki 1972). In the Gigartinales, structures very similar to those found in *H. calvadosii* and *H. pacifica* were found in a carpospore culture of *Platoma marginifera* (Nemastomataceae; Guiry & Maggs unpubl.); short chains of rounded cells were formed on filamentous plants that otherwise did not form reproductive structures (Figs. 14-16, 14-17). These cells appeared to detach and grow directly into new filamentous plants, suggesting that they are propagules rather than sporangia.

In culture, monoecious gametophytes of *Gloeophycus koreanum* (Gloiosiphoniaceae), as well as producing carposporangia at the bases of lateral fertile branches, formed berrylike masses of sporangia at the branch apices (Notoya 1984). Apparently, these release spores that form tetrasporangial plants in a similar manner to carpospores from the same plants. However, they do not appear to be the result of fertilization. Referring to these sporangia as parasporangia (Notoya 1984) seems incorrect, as the true parasporangia found in the Ceramiales (Section II. K) are quite different structures.

Monosporangia formed in chains on phases other than carposporophytes are generally referred to as seirosporangia. In *Seirospora seirosperma* (Ceramiaceae), the chains are branched and terminal (Figs 14-18 to 14-20), whereas in *Dohrniella neapolitana* (Ceramiaceae) they are terminal but unbranched (Feldmann-Mazoyer 1941; L'Hardy-Halos 1966). Seirosporangia are formed on thalli that may also bear other reproductive structures (tetrasporangia, gametangia, carposporangia). *Gibsmithia hawaiiensis* (Dumontiaceae) forms unbranched chains of cells terminally on the filaments of erect, pseudoparenchymatous plants. These are referred to as seirosporangia by Doty (1963), but Kraft (1986) calls them "putative seirosporangia." Karam-Keriman (1976) reported the release and germination of spores from such cells, but the resulting plants were not grown

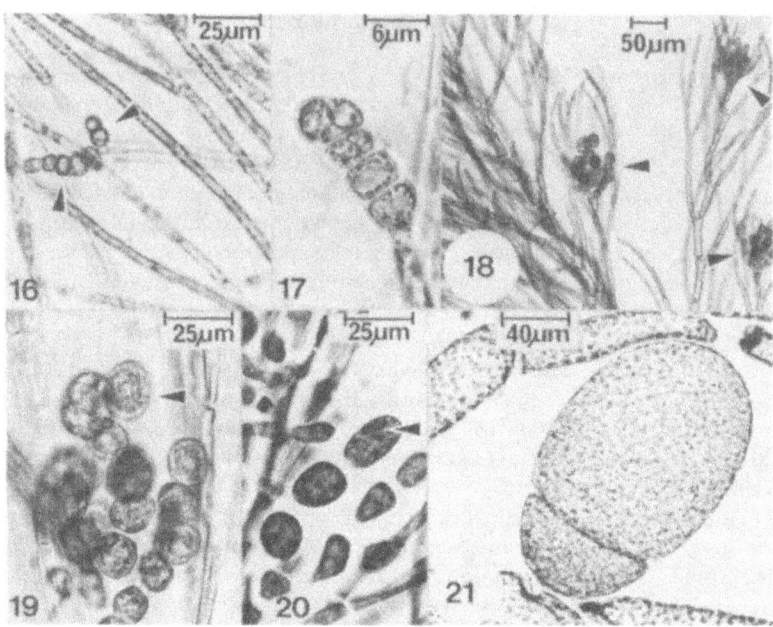

Figs. 14-16 to 14-21. Monosporangia-like structures and propagules in the Florideophycidae. *14-16*: Filamentous growths in culture from carpospores of *Platoma marginifera* (Nemastomataceae) showing reproductive propagules (arrowheads). *14-17*: Another propagule of *P. marginifera*. *14-18 to 14-20*: Seirosporangia (arrowheads) of *Seirospora seirosperma* (Ceramiaceae). *14-21*: Two-celled propagule of *Aniscoschizus propaguli* (Ceramiaceae). (All figures original; Fig. 14-21 courtesy of J. M. Huisman)

to reproductive maturity. These sporangia are absent in a further three species of *Gibsmithia* from Australia (Kraft 1986).

Dixon (1973, p. 136) was of the opinion that the seirosporangia of *Seirospora*, *Dohrniella*, and *Gibsmithia* are best treated as a form of monosporangium. Plattner and Nichols (1978) described the release and germination of seirospores from *S. seirosperma*, but it is not clear whether the sporangia or the spores are being released. Bird and Johnson (1984) refer to these reproductive structures both as seirosporangia and as propagules. In each of these latter investigations only the recycling of seirosporangial plants was reported in culture. Further studies of seirosporangia in the Ceramiaceae and Dumontiaceae are needed, particularly of their fine structure.

The solitary, multinucleate "monosporangia" of *Monosporus pedicellatus* (Ceramiaceae) are considered by Bornet and Thuret (1876), Feldmann-Mazoyer (1941), and L'Hardy-Halos (1970) to represent vegetative reproductive structures, as the whole sporangium is shed and the cell contents germinate without being released (L'Hardy-Halos 1970). They are large and dark in color and are packed with starch grains. They may be sessile or stalked, with the stalk cell losing most of its cell contents at maturity, possibly to aid dehiscence. Guiry (1978) referred to such structures as pseudosporangia, but the term "propagula" seems just as adequate (Huisman & Kraft 1982). Sexual reproduction has not been reported in any species of *Monosporus*, so that the inclusion of the genus in the tribe Griffithsieae (Baldock 1976) is largely based on supposed similarities in vegetative morphology to other members of the tribe. Huisman and Kraft (1982) considered that reports of tetrasporangia in this species, although doubted by Baldock (1976), indicated the possibility of a sexual phase. Huisman and Kraft (1982) found that the propagula of *M. australis* (Fig. 14-38) germinate as a unit in a manner similar to that found in *M. pedicellatus* (L'Hardy-Halos 1970), except that up to three apical filaments may be produced instead of one. Huisman and Kraft (1982) recommended retention of *Monosporus* as a form genus for species reproducing wholly by single-celled propagules. However, they considered that as species of the genus probably represent end products of parallel development in more than one tribe, it should not be specifically assigned to the Griffithsieae or any other tribe of the Ceramiaceae. This opinion is supported by the recent discovery of polysporangia on plants of M. australis (Fig. 14-38), the polyspores of which gave rise in culture to gametangial plants lacking propagula (Huisman pers. comm.).

In *Mazoyerella arachnoidea* (Ceramiaceae, Spermothamnieae), single-celled propagula are released into culture with the sporangial wall intact and germinate without releasing their contents as a spore (Huisman & Kraft 1982). *Anisoschizus propaguli* (Fig. 14-21), also referred to the Spermothamnieae, forms two-celled propagula on plants that also bear sessile polysporangia. Huisman and Kraft (1982) found that these propagula develop in culture without shedding spores. *Deucalion levringii* (Compsothamnieae) forms three-celled, multinucleate, ovoid propagules on plants that may also bear polysporangia with 24–36 spores. The propagules germinate as a unit to form further propagule-forming plants, whereas the polyspores formed male and female gametophytes that did not form propagules. The "monosporangia" of *Tanakaella* (Compsothamnieae; Itono 1977) also appear to be propagules.

Monosporangia-like structures were reported in *Antithamnionella floccosa* by Jassund (1965), but nothing is known of the number of nuclei or of the fate of the sporangia or spores. Whittick and Hooper (1977) found what appear to be zonately divided "tetrasporangia" in *Antithamnion cruciatum* from Newfoundland; again, nothing is known of the cytology of these structures, nor was spore release obtained. Multinucleate, monosporangia-like structures were reported in close association with the procarps of *Nitophyllum punctatum* (Delesseriaceae) by Svedelius (1914). These were thought by Fritsch (1945, p. 723) to be a consequence of arrested development of the procarps. Also in the Ceramiales, there are several questionable reports of small, undivided sporangia on plants of *Polysiphonia urceolata* (Rosenvinge 1923–4), *P. flexicaulis* (Yamanouchi 1906), and *Griffithsia globulifera* (Lewis 1909).

D. Bisporangia

Bisporangia (Figs. 14-7, 14-8, 14-22) are found either on tetrasporangial plants mixed in with cruciately or zonately divided tetrasporangia or on exclusively bisporangial plants. The consensus (Dixon 1973; Fritsch 1945) is that bisporangia are homologous with tetrasporangia principally because, until recently, bisporangia were not found on carpogonial or spermatangial plants. However, gametangial plants of *Acrochaetium*[2] *dictyota* from Australia occasionally form bisporangia (Woelkerling 1971), although nothing is known of the fate of these in culture.

Bauch (1937) concluded that bisporangia in the Corallinaceae are of two types: binucleate with uninucleate bispores and quadrinucleate with binucleate bispores, the former being the most common type. Suneson (1950) later showed that in the crustose coralline *Dermatolithon litorale* from Sweden – plants of which are known to form only bisporangia – two nuclei are formed by mitosis and these pass into the two spores, each of which is uninucleate at maturity. Suneson (1982) cultured bispores from such plants and found that further bisporangial plants were formed. However, *D. corallinae* forms three types of sporangia: quadrinucleate tetrasporangia, quadrinucleate bisporangia, and binucleate bisporangia (Suneson 1950). Meiosis occurs in the first two types but not in the last. In northern Europe tetraspores and binucleate bispores from *D. corallinae* probably form gametangial plants, whereas uninucleate bispores would be expected to form further bisporangial plants (Johansen 1981), but this has yet to be confirmed in culture. Chamberlain (1977, 1983, 1984) established that uninucleate bispores give rise in culture to more bisporangial plants in the crustose species *Fosliella farinosa*, *Pneophyllum zonale*, *P. lobescens*, and *P. plurivalidum* from the British Isles. A wide range of crustose species of Corallinaceae appear to form uninucleate bispores, and it was suggested that this may be the only means of reproduction in plants of some species at the northern limit of their geographical distribution (Foslie 1905; Johansen 1981; Suneson 1950, 1982). Although typically a cold-water phenomenon in crustose Corallinaceae, bisporangia occur with tetrasporangia in *F. farinosa* from India (Balakrishnan 1947).

Bisporangia also are found in articulated Corallinaceae. Suneson (1937) and Johansen (1976) reported bisporangia in *Amphiroa rigida* and *A. zonata*, respectively. About 75% of specimens of *Bossiella orbigniana* collected from the eastern Pacific were bisporangial (Johansen 1971, 1981). Only bisporangial plants of *Lithothrix aspergillum* were found north of Point Conception in California, but bisporangial, tetrasporangial, and gametangial plants occurred further south (Gittins & Dixon 1976). Quadrinucleate bisporangia are usually found intermingled with normal tetrasporangia in the articulated and

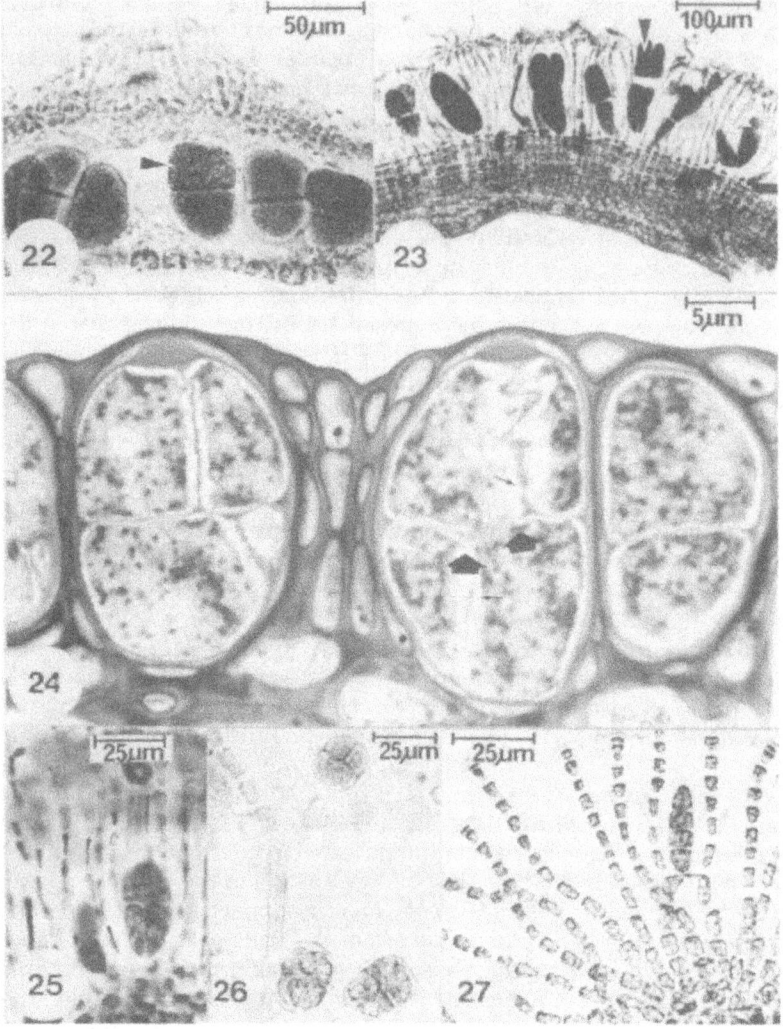

Figs. 14-22 to 14-27. Bisporangia and tetrasporangia in the Florideophycidae. *14-22*: Section of bisporangial conceptacle of *Pneophyllum limitatum* (Corallinaceae), showing bisporangia (arrowhead). *14-23*: Mature tetrasporangial sorus of *Peyssonnelia harveyana* (Peyssonneliaceae), showing cruciately divided tetrasporangia (arrowhead). *14-24*: Large, regularly cruciately divided tetrasporangia of *Palmaria palmata* (Palmariaceae), with sterile filaments between tetrasporangia. Periclinal cleavage is arrested (large arrows), whereas the anticlinal cleavage progresses (small arrows). *14-25*: Cruciately divided tetrasporangia of *Peyssonnelia dubyi* (Peyssonneliaceae) among elongated sterile filaments. *14-26*: Cruciately divided tetrasporangia of *Antithamnionella spirographidis* (Ceramiaceae). *14-27*: Zonately divided tetrasporangia formed laterally (arrow) on the erect filaments of *Cruoria cruoriaeformis* (Cruoriaceae). (Fig. 14-24 after Pueschel 1979, with permission; remainder original: Fig. 14-22 courtesy of Y. M. Chamberlain; Figs. 14-23, 14-25, 14-27 courtesy of C. A. Maggs)

crustose Corallinaceae (Guiry 1978) and are likely to be quite common.

In the Gelidiales, quadrinucleate bisporangia were found in *Suhria vittata* and *Gelidium pristoides* from South Africa (Carter 1985; Fan 1961). Tetrasporangia are unknown in both species, and meiosis may take place in the bisporangial initials.

Shimizu and Masuda (1983) found that in

Masudophycus (as *Farlowia*) *irregularis* (Gigartinales, Dumontiaceae) from Japan bispores from erect plants give rise in culture to erect gametangial plants. This contrasts with the life history of three species of *Farlowia* from Pacific North America, all of which formed crustose tetrasporophytes in culture (DeCew & West 1981). It seems likely that the Japanese species has meiotic quadrinucleate bisporangia (Guiry & Irvine 1989). In *Gardneriella tubifera* (Solieriaceae), a parasite on *Sarcodiotheca gaudichaudii* in California, both uninucleate and binucleate bispores are produced in addition to gametes, carpospores, and tetraspores. In culture, carpospores formed pustules on the host, which produced binucleate bispores and, occasionally, tetraspores (Goff 1981). Strangely, infections by uninucleate bispores resulted in pustules forming binucleate bispores, which in turn developed into plants bearing gametangia. Tetraspores infected additional host tissue, resulting directly in the formation of gametangia and ultimately, carpospores. In no instance did uninucleate bispores give rise to pustules forming uninucleate bispores. Goff's study showed that recycling via uninucleate spores from bisporangia should not be assumed.

Two crustose red algae from the British Isles and northern France provisionally referred to the Gigartinales (Guiry & Irvine 1989) formed only bisporangia in the wild, the bispores giving rise to further bisporangial plants in culture. These plants were not readily identifiable with any previously described species. Maggs (pers. comm.) found an apomictic recycling of bisporangial crustose plants of *Gloiosiphonia capillaris* (Gloiosiphoniaceae), which otherwise has a life history involving tetrasporangial crustose plants. In this instance, however, the bisporangial plants were clearly related to the crustose tetrasporangial phase. *Audouinella occulta* was described as forming only bisporangial plants, spores of which recycled further bisporangial plants in culture (Stegenga 1985); this species has an unusual morphology and may represent a bisporangial strain of gigartinalean alga.

In the Nemastomataceae, Kajimura (1987) described *Predaea bisporifera* [as *Predaeopsis japonica* nom. illeg. in Kajimura (1981), Latin diagnosis not provided] from Japan as having terminal "bisporangia"; however, little information is given as to the mode of formation and release of the "bispores", and Kraft (1984) has argued that these might represent propagules.

In the Ceramiaceae, bisporangial plants have been reported in several genera (*Crouania, Spermothamnion, Seirospora, Dohrniella,* and *Callithamnion*;

see Guiry 1978 for sources). The life histories of such plants are unknown and are in need of further investigation. In most cases it is not known if the spores are uninucleate and whether or not meiosis takes place. In *Callithamnion byssoides*, bisporangia are frequently reported (Dixon & Price 1981; Kylin 1917) and generally occur on plants bearing only bisporangia, although Harris (1962) has described them on plants bearing tetrasporangia, and Rosenvinge (1923-4) found them on plants bearing gametangia. These sporangia are probably similar in function to the mitotic binucleate bisporangia described earlier for the Corallinaceae. In *Crouania attenuata*, bisporangial plants may be commoner in certain areas in the Mediterranean than the tetrasporangial plants (Feldmann-Mazoyer 1941). In the north Atlantic and north Pacific, there is evidence that bisporangia in *C. attenuata* are formed during the winter months, with tetrasporangia being produced on the same plants during the summer (Dixon 1973, p. 134). Nothing is known of the cytology of these two types of sporangia.

E. Tetrasporangia

Tetrasporangia typically give rise to four uninucleate, equal-sized spores (Figs. 14-9 to 14-11). They are usually symmetrically cleaved and are characteristic of the Florideophycidae, being found in all families of this subclass, except in the exclusively freshwater order Batrachospermales and a single family of the Gigartinales, the Pseudoanemoniaceae. The latter family includes only a single species, *Hummbrella hydra*, which may have a *Lemanea*-type life history (Hawkes 1983). In the Florideophycidae, tetrasporangia are usually found on free-living diploid thalli. In a restricted number of cases, however, they may be formed from cells derived directly from gonimoblast tissue, in which case they are known as carpotetrasporangia. It must be strongly stressed at this point that the division of a sporangium into four spores should not be taken as evidence of meiosis. Sporangia with four spores generally form these by means of two or three cleavages, the orientation of which gives rise to three basic types (Figs. 14-9 to 14-11): cruciate, zonate, and tetrahedral.

1. Cruciately divided tetrasporangia

These are ovate or, occasionally, cubical in shape and ordinarily have a median cleavage, normally along the short axis (Figs. 14-9, 14-23 to 14-26). The orientation of the other two cleavages, which typically take place along the long axis, is much

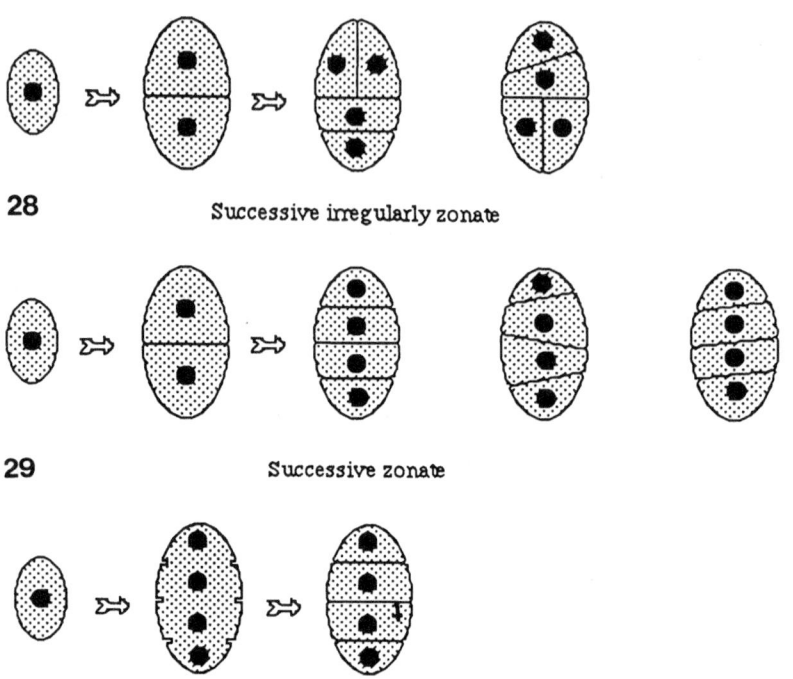

Figs. 14-28 to 14-30. Formation sequences in various types of zonately divided tetrasporangia. (Redrawn after Guiry 1978, with permission)

more irregular (Fig. 14-24). In some cruciately divided tetrasporangia these longitudinal cleavages are in a single plane, but this may or may not be in the plane of view; if it is, the sporangium seems to have only two spores. As Guiry (1978) pointed out, this may partly account for the frequent reports of "bisporangia" being mixed with tetrasporangia in the same sorus. In the vast majority of cruciately divided tetrasporangia, the furrowing of the median cleavage appears to be complete or near complete before the longitudinal furrowing is initiated. Guiry (1978) named this successive cleavage. In the Ceramiaceae there are indications that in some cruciately divided tetrasporangia all the cleavages are initiated at the same time, a condition that Guiry (1978) referred to as simultaneous cleavage. In *Palmaria palmata*, however, the median cleavage is arrested just short of completion, and the other cleavages are initiated then (Fig. 14-24); this could be regarded as being intermediate between simultaneous and successive (Pueschel 1979), but little is known of the timing of cleavage events in other algae with cruciately divided tetrasporangia.

A distinction is made frequently between "cruciate tetrasporangia" and "decussate tetrasporangia," the first referring to tetrasporangia in which the longitudinal cleavages are in the same plane (Fig. 14-9), whereas the latter denotes longitudinal cleavages at an angle of 90 degrees to one another. This terminology is useful but not absolute. In a considerable number of red algae with cruciately divided tetrasporangia, the longitudinal cleavages may not be in the same plane (Fig. 14-24), the degree of divergence varying through 180 degrees (Guiry 1976). Also, sporangia have been found in which one of the longitudinal cleavages does not meet the median cleavage (Fig. 14-28), giving a sporangium with the appearance of being intermediate between cruciate and zonate. A further variation, which seems rather rare, is where two cleavages take place along the long axis. In several species, mainly crustose Gigartinales and Hildenbrandiales, sporangia have been described in which the median division is oblique, giving the appearance of an irregularly zonate tetrasporangium (Fig. 14-32).

Cruciately divided tetrasporangia are known in all orders of the Florideophycidae and seem to be the most primitive of all sporangial types (Guiry & Irvine 1989). Cruciately divided or irregularly cruciately divided tetrasporangia are the only type known in the Nemaliales, whereas all the other orders of the subclass have cruciately or irregularly cruciately divided tetrasporangia and either zonate or tetrahedral types. In the Corallinales, however,

Sporangia and spores

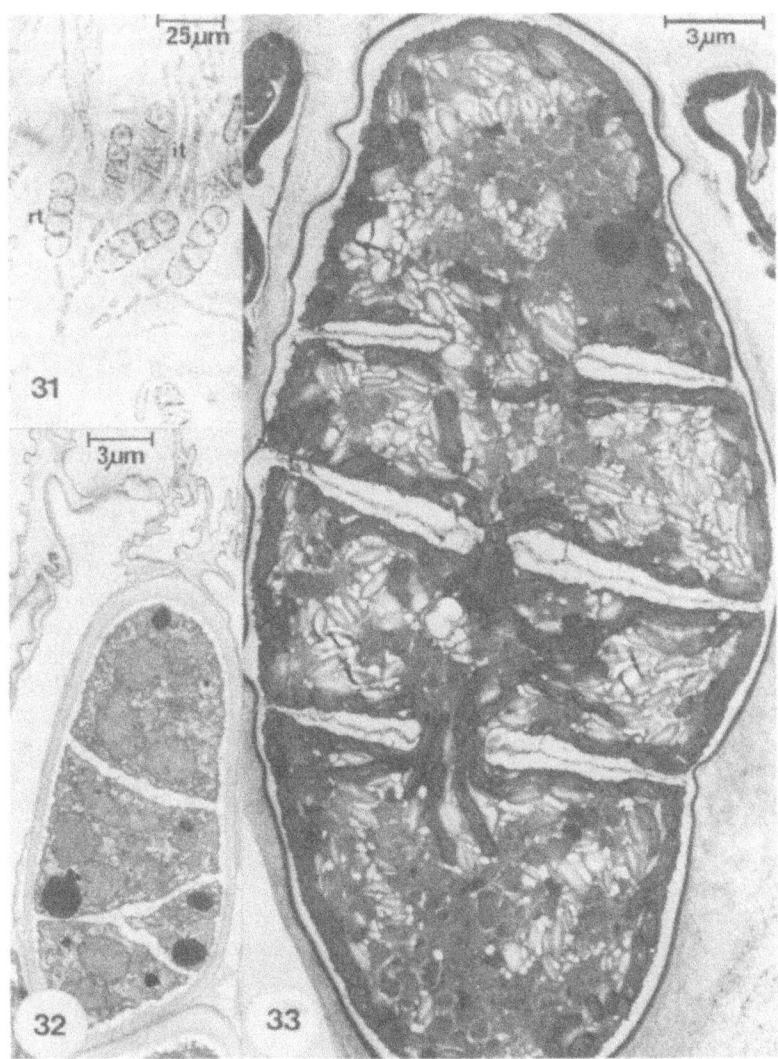

Figs. 14-31 to 14-33. Regularly and irregularly zonately divided tetrasporangia. 14-31: Regularly (rt) and irregularly divided (it) tetrasporangia in *Dudresnaya verticillata* (Dumontiaceae). 14-32: TEM of an irregularly divided tetrasporangium in *Hildenbrandia rubra* (Hildenbrandiaceae). 14-33: TEM of a regularly zonately divided tetrasporangium of *Agardhiella subulata* (Solieriaceae). Note that median cleavage is not complete. (Fig. 14-31 original; Fig. 14-32 after Pueschel 1982, with permission; Fig. 14-33, courtesy of J. L. Scott)

tetrasporangia are almost exclusively zonate, cruciate divisions only occuring occasionally in the primitive genus *Sporolithon* (Cabioch 1972).

In the Gelidiales and Bonnemaisoniales, the tetrasporangia, although appearing to be successively regularly or irregularly cruciately divided, are spherical or near-spherical in shape, often giving rise to apparently erroneous reports that they are tetrahedrally divided (see Maggs & Guiry 1988).

Many of the families of the Gigartinales[3] have irregularly cruciate or irregularly zonately divided tetrasporangia, but within the order there is a strong evolutionary trend toward producing families with either regularly zonate or regularly cruciate tetrasporangia. The reasons for these trends are poorly understood. In the Phyllophoraceae and Gigartinaceae (see Guiry et al. 1984) and *Petrocelis hennedyi*[4] (see Fletcher & Irvine 1982), tetrasporangia are produced in unbranched or branched, generally intercalary chains, but in the closely related Petrocelidaceae and Chondriellaceae, tetrasporangia are formed singly in an intercalary position (Guiry et al. 1984; Levring 1941).

In the Rhodymeniales, cruciately divided tetrasporangia are characteristic of the subfamily Rhodymeniodeae (Rhodymeniaceae), where they are regularly and successively divided. The subfamily Hymenocladioideae (Rhodymeniaceae) and the families Champiaceae and Lomentariaceae are characterized by simultaneous, tetrahedrally divided tetrasporangia. Recent criticisms of the distinctness of the Rhodymeniales from the Gigartinales (Gabrielson & Garbary 1986; Kraft & Robins 1985) do not take adequate account of the occurrence of such sporangia in the Rhodymeniales and their total absence in the Gigartinales. In the Ceramiales, regularly and irregularly cruciately divided tetrasporangia occur in some tribes of the Ceramiaceae (Fig. 14-26; see Guiry 1978, Table III), but most of the tribes have only tetrahedrally divided tetrasporangia.

2. Zonately divided tetrasporangia

These tend to be much narrower (Figs. 14-27, 14-31 to 14-33) than cruciately divided tetrasporangia, except in the Corallinales. It has been suggested that zonately divided tetrasporangia were derived from cruciately divided types, largely because various intermediate forms occur (Fig. 14-28; Guiry 1978; Searles 1968). Two types of zonately divided tetrasporangia occur in the Florideophycidae. Division in the first and most common type is successive zonate, a median division occurring initially, followed by two other roughly parallel divisions (Fig. 14-29) at equal distances on both sides. In the second type, which is found only in the Corallinales, the cleavages are all initiated at the same time (Fig. 14-30), and Guiry (1978) has referred to this as simultaneous zonate. The exclusive occurrence of such tetrasporangia in the Corallinales and the formation of a unique type of roofed conceptacle, among other characters, supports the recognition of the Corallinales as an order (Silva & Johansen 1986). However, Scott (pers. comm.) found that in *Agardhiella subulata* (Solieriaceae, Gigartinales), after the first meiotic division, the median cleavage furrows are almost complete but stop just short of merging, as if room were being left for a pit plug (Fig. 14-33). Following the second meiotic division, the other two furrows invaginate, and once they are at the same depth as the first cleavage furrow, all three merge simultaneously. This would appear to be intermediate between successive and simultaneous zonate division. Further studies of various gigartinalian families may show this furrowing sequence to be of systematic value.

Regularly and irregularly successively divided zonate tetrasporangia are characteristic and exclusive to the Gigartinales[3] and Hildenbrandiales[5] (see Pueschel 1982; Pueschel & Cole 1982). Irregularly cruciate and irregularly zonate tetrasporangia need further study to establish how the cleavages are initiated and the subsequent furrowing events.

3. Tetrahedrally divided tetrasporangia

These are usually spherical or subspherical (Figs. 14-11, 14-34 to 14-37). There is strong evidence that they were derived from cruciately divided tetrasporangia, because many intermediate forms are found in the Ceramiaceae (Guiry 1978). Tetrahedrally divided tetrasporangia are generally regarded as undergoing simultaneous cleavages (Fig. 14-34), although there are exceptions, most frequently in the Ceramiaceae (see Guiry 1978, Table III). Broadwater et al. (1986a,b; Chapter 6) described the ultrastructure of meiotic events in *Dasya baillouviana* (Dasyaceae). During meiosis II, simultaneous division of two separate nuclei was demonstrated, the cleavage furrows detectable as early as telophase I. At the end of meiosis II, the nuclei advance half the distance from the cell periphery to its center, at which stage they appear to migrate simultaneously to the center. Although close juxtaposition of the nuclei was observed, actual fusion was not seen; however, this stage could be mistaken for intranuclear meiosis (Scott & Thomas 1975). The nuclei subsequently migrate centrifugally, and the cleavage furrows are completed. The purpose of this postmeiotic movement is unknown.

Tetrahedrally divided tetrasporangia occur in all families of the Rhodymeniales and Ceramiales but cruciately divided types are known only in a few tribes of the Ceramiaceae that generally show other primitive features of structure and reproduction (Moe & Silva 1979, 1980). Although entities with tetrasporangia that are clearly intermediate between cruciately and tetrahedrally divided occur in the Ceramiaceae, no such intermediate sporangia occur in the Rhodymeniales, the members of which order also have both types of cleavage. Tetrahedrally divided tetrasporangia with simultaneous cleavages are definitely known only in the Rhodymeniales and Ceramiales, where they occur in conjunction with other features that are widely regarded as being advanced.

F. Conchosporangia

Conchosporangia are formed on the conchocelis phases of *Bangia* and *Porphyra* species in irregularly

Sporangia and spores

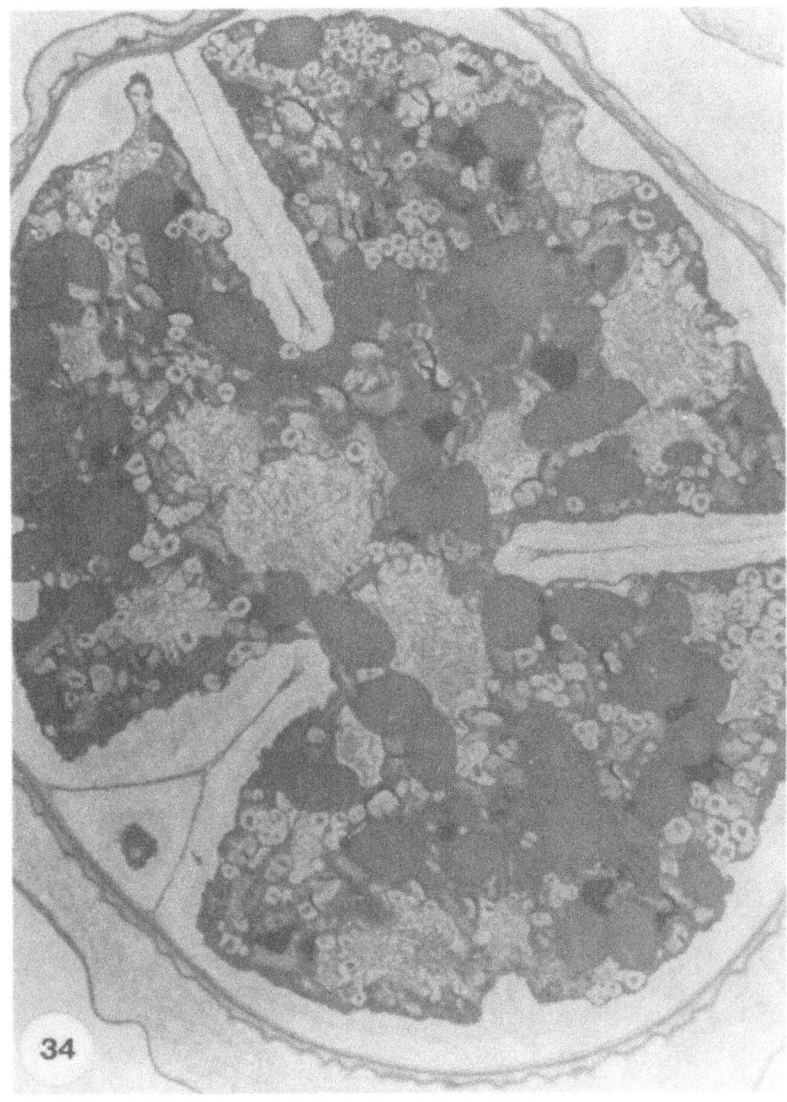

Fig. 14-34. TEM of longitudinal section through a partially cleaved, tetrahedrally divided tetrasporangium of *Ptilota hypnoides* (Ceramiaceae) (After Scott & Dixon 1973, with permission.)

shaped conchosporangial branches (Figs. 14-39, 14-40) and are generally easy to distinguish from vegetative cells because of their rectangular outlines, strong pigmentation, dense cell contents, and sizes and thicknesses of their cell walls. Conchosporangia form a single uninucleate spore and are grouped together in structures that must be regarded as sporangial branches. Pit connections are formed between individual conchosporangia (Cole 1972b), and it seems that each conchosporangium is formed by cell division rather than by furrowing. Migita (1967) and Giraud and Magne (1968) reported stages of reduction division in the conchosporangia of *P. umbilicalis* var. *laciniata* and *P. yezoensis*, respectively. However, Ma and Miura (1984), Burzycki and Waaland (1987), and Tseng and Sun (1989) provided strong evidence that meiosis in *P. yezoensis*, *P. torta*, and *P. tenera* occurred in the released conchospores. Guiry (1987) suggested that it was possible that the site of reduction division is variable depending on the species or environmental factors.

G. Zygotosporangia

The new term "zygotosporangia" is proposed for sporangia formed within a fertilized carpogonium from the zygote or by direct division of the zygote. These were referred to previously as carposporangia (forming α-spores; spermatia were referred

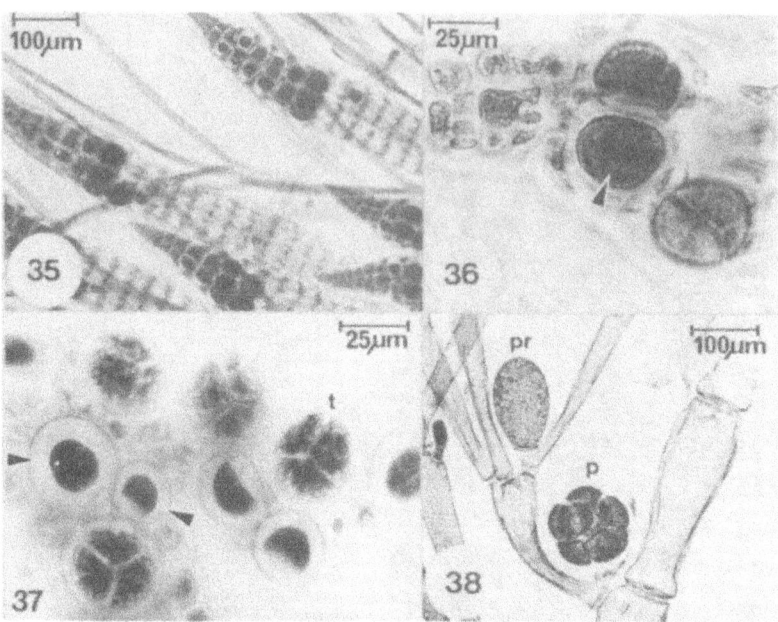

Figs. 14-35 to 14-38. Tetrahedrally divided tetrasporangia and polysporangia in the Ceramiales. *14-35*: Tetrasporangial stichidia of *Dasya corymbifera* (Dasyaceae). *14-36*: Tetrasporocyte (arrowhead) and tetrahedrally divided sporangium of *D. corymbifera*. *14-37*: Tetrahedrally divided tetrasporangia (t) and regenerating tetrasporocytes (arrowheads) of *Ptilota plumosa* (Ceramiaceae). *14-38*: Polysporangia (p) and propagules (pr) on same plant of *Monosporus australis* (Ceramiaceae). (All original; Fig. 14-38 courtesy of J. M. Huisman)

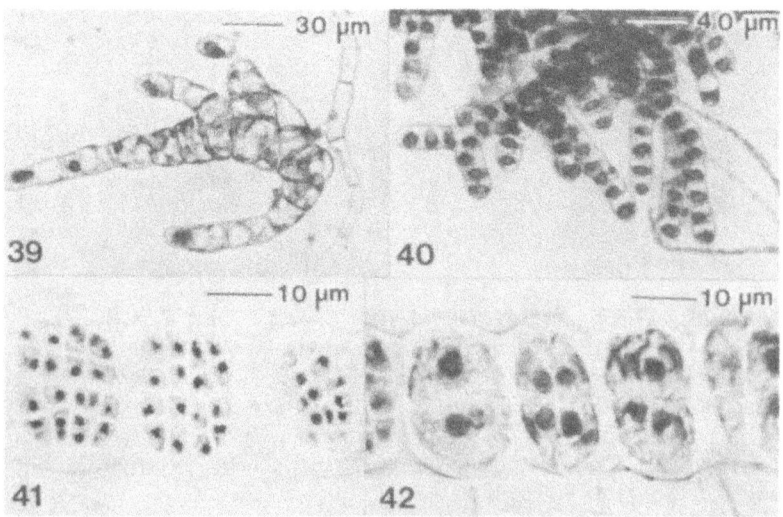

Figs. 14-39 to 14-42. Sporangia and spermatangia in Bangiophycidae. *14-39, 14-40*: Conchosporangial branches of *Porphyra linearis* in culture. *14-39*: Uniform branch with sporangia; note centrally placed plastids. *14-40*: Branch with elongated, vacuolate cells. *14-41*: Spermatangia of *P. gardneri*, stained by Feulgen technique to show nuclei. *14-42*: Zygotosporangia of *P. gardneri* (Bangiales), also stained by Feulgen technique. (Figs. 14-39, 14-40, after Bird et al. 1972, with permission; Figs. 14-41, 14-42 after Hawkes 1978, with permission)

to as β-spores: Conway 1964; Hawkes 1978 and included references). However, carposporangia of the vast majority of Florideophycidae are formed externally from gonimoblast filaments radiating from the site of zygote formation. In the zygotosporangia of *Bangia* and *Porphyra* (Bangiales) formation of zygotospores takes place by parenchymatous (i.e., three-dimensional) cell divisions of the zygote (Fig. 14-42), a radically different ontogeny to that found in carposporangia. For many years there was considerable doubt as to the occurrence of sexual fusion in *Porphyra* species (see Dixon 1973 for an historical review). However, Hawkes (1978) observed gametic union in *P. gardneri*, and Kornmann (1984, 1987) reported similar events in species of the Erythropeltidaceae. Consequently, the fundamental differences in ontogeny between carposporangia and zygotosporangia should now be recognized. Zygotosporangia in the Erythropeltidaceae appear to produce only a single zygotospore, whereas those in the Bangiales always seem to produce more than one. The mode of spermatangial formation in the Bangiales – also by internal, parenchymatous divisions (Fig. 14-41) – seems radically different from that found in the majority of florideophytes, in which spermatangia are formed singly in cells developing outwards from a spermatangial mother cell, although spermatangial formation in the Erythropeltidaceae seems more similar to that found in the Florideophycidae (Berthold 1882; Kornmann 1984, 1987).

Zygotosporangia with internal parenchymatous divisions similar to those described for species of *Bangia* and *Porphyra* have not been described in the Florideophycidae. However, Abdel-Rahman and Magne (1983) found that the contents of the fertilized carpogonium in *Acrochaetium pectinatum* may be released as a single spore in a manner similar to that found by Kornmann (1984, 1987) in the Erythropeltidaceae. Abdel-Rahman and Magne refer to this spore as a sporozygote, but this term is not suitable, as it makes independent reference to spore and sporangium difficult. The fertilized carpogonium may also divide to form an external carposporophyte.

DeCew and West (1982) found that the fertilized carpogonium of *Rhodophysema elegans* (Palmariaceae) divides to form a stalk cell and a tetrasporangial initial that forms four tetraspores, presumably by meiosis as the spores give rise to gametangial plants. Although the sporangia are formed internally in the carpogonium, the zygote divides initially to form a stalk cell and a tetrasporangial initial, and as such should be regarded as a tetrasporophyte rather than zygotosporangia. Guiry (1987) supported the hypothesis of Searles (1980) that two modes of zygote amplification are possible in the Rhodophyta: in the first a mitosporangial generation is produced from the fertilized zygote, the spores of which form diploid plants (carposporangia and zygotosporangia); in the second a meiosporangial generation is formed, the spores of which undergo meiosis and, on release, give rise to haploid, gametangial plants. The latter type is found in species of the Palmariaceae (Palmariales) and in *Rhodophysema* species.

H. Carposporangia

Single-spored carposporangia (Figs. 14-43 to 14-46) are derived from gonimoblast filaments radiating outward from zygotes and are similar to monosporangia. Typically, they give rise to spores that form tetrasporangial plants rather than repeating the parent phase. There is enormous variation in the mode of formation and derivation of gonimoblast filaments (see Chapter 12). The fine structure of carposporogenesis in various Florideophycidae has been described by Ramm-Anderson and Wetherbee (1982), Triemer and Vasconcelos (1977), Tripodi (1971), and Tsekos (1981).

Male plants of the normally dioecious species *Chondrus crispus* from a number of widely separated geographic locations in the North Atlantic formed spores masses in culture which, although they resembled cystocarps, did not appear to be the result of sexual fusion. Spores released from those "sporogenous bodies" (van der Meer et al. 1983) gave rise to plants that closely resembled normal tetrasporophytes but formed tetrasporangia that did not release spores. Van der Meer et al. (1983) concluded that these tetrasporophytes were probably haploid, with an abortive haploid meiosis in the tetrasporangia; such haploid sporophytes were also reported in cultured plants of *Palmaria palmata* (van der Meer & Chen 1979). These phenomena are poorly understood.

The only putative species of the Bangiophycidae for which carposporangia-like sporangia are known is *Rhodochaete parvula* (Rhodochaetales), in which they are virtually indistinguishable from monosporangia, in that they are formed by a lateral curved wall after fertilization (Boillot 1975; Magne 1960). Nevertheless, as they are formed external to the carpogonium, they should be regarded as carposporangia. This is one of a number of good reasons for referring the Rhodochaetales to the Florideophycidae rather than to the Bangiophycidae.[2]

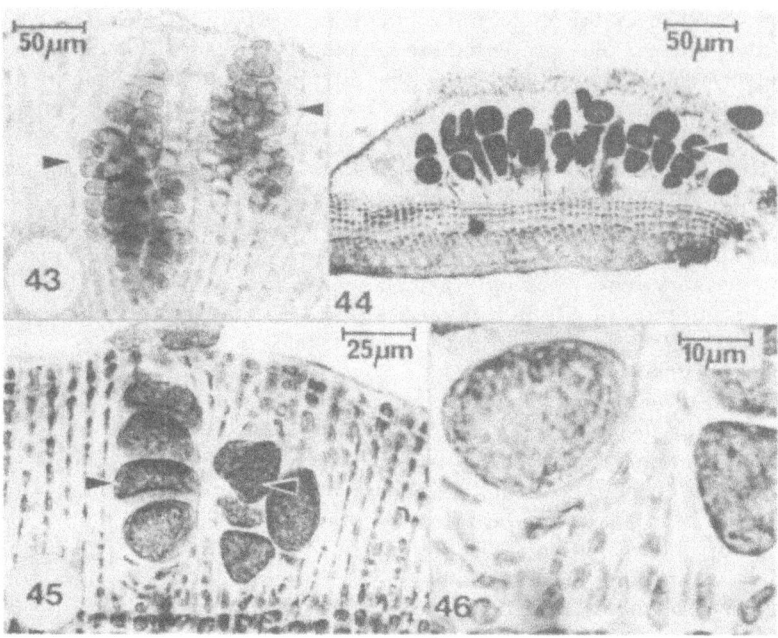

Figs. 14-43 to 14-46. Carposporangia in crustose Florideophycidae. *14-43*: Transverse section of mature carposporangia in *Cruoria cruoriaeformis* (Cruoriaceae). *14-44*: Radial transverse section (RTS) of the carposporangial sorus of *Peyssonnelia harveyana* (Peyssonneliaceae), showing chains of carposporangia (arrowheads). *14-45*: RTS of carposporangial sorus of *P. atropurpurea*, showing chains of carposporangia (arrowheads) among sterile filaments. *14-46*: Detail of Fig. 14-45. (All original, courtesy of C. A. Maggs)

I. Carpotetrasporangia

In a very few instances, quadripartite carposporangia are formed by the cleavage of initials derived from gonimoblast filaments; these are referred to as carpotetrasporangia (Feldmann 1939). In the Florideophycidae there are two basic types of carpotetrasporangia (Guiry 1987; Guiry & Irvine 1989). In the first type, carpotetraspores are formed in sporangia that resemble the carposporangia of related species of the same genus; such carpotetrasporangia are found in the Nemaliales (Liagoraceae and Dermonemataceae). In the second type, gonimoblast filaments spread over the surface of the thallus and form sporangia in chains (see Guiry in McCandless & Vollmer 1984); such carpotetrasporangia are so far known only in the Phyllophoraceae, and this carpotetrasporangial structure (often referred to as a tetrasporoblast) does not resemble the internal carposporophytes found in homologous species (see Schotter 1968). The ontogeny of the zonately cleaved carpotetrasporangial structures described for *Schizymenia epiphytica* from California (Abbott 1967; Smith & Hollenberg 1943) needs further study (Guiry & Irvine 1989).

Ngan and Price (1979) reported that four spores were formed in the "carposporangia" of species of *Chondrococcus*, *Gracilaria*, *Sarconema*, *Solieria* (Gigartinales), and *Scinaia* (Nemaliales) that they examined in the course of a study of spore size, but no details were given. Carpospores or carposporangia germinating in situ in cystocarps – a common occurrence – frequently resemble multispored sporangia.

In a culture study of *Liagora harveyana* (Liagoraceae) from southern Australia, Guiry (unpubl.) found that spores from quadripartite carposporangia gave rise to prostrate plants that formed tetrasporangia, the tetraspores of which produced plants that gave rise to erect gametophytes. It thus appears that the widely held assumption that all quadripartite carpospores are the result of meiosis is erroneous.

J. Polysporangia

Polysporangia (Fig. 14-38) are found only in the Rhodymeniales and Ceramiales. Drew (1937) restricted the term "polysporangia" to sporangia with more than four meiotic spores formed in place

Sporangia and spores

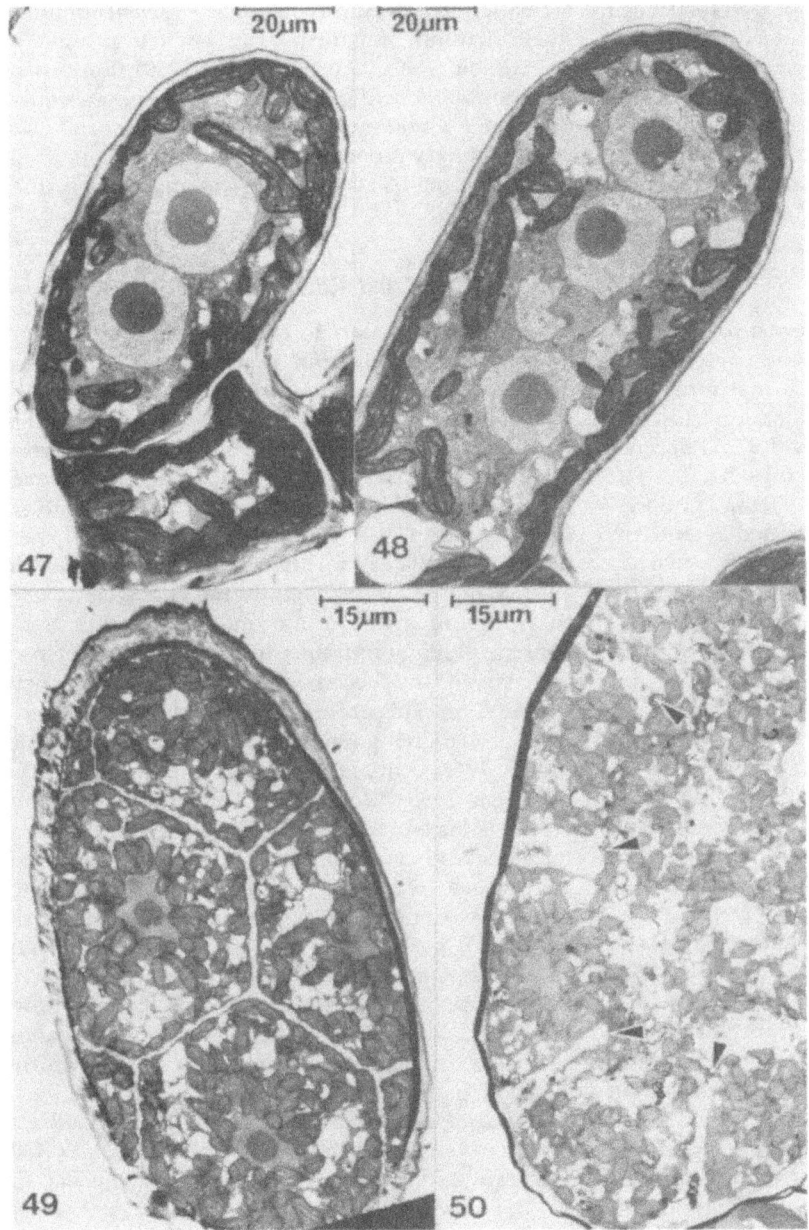

Figs. 14-47 to 14-50. TEMs of the polysporangia of *Pleonosporium vancouverianum* (Ceramiaceae). *14-47, 14-48*: Polysporocytes with two (14-47) and four (14-48) nuclei. *14-49*: Mature polysporangium showing regular cleavages. *14-50*: Initiation of cleavages (arrowheads) in polysporocyte. (After Sheath et al. 1987, with permission.)

of tetrasporangia. Polysporangia with spores in multiples of four, suggesting a meiotic origin, have been reported in about 13 genera of Ceramiaceae and in a genus of Champiaceae (*Gastroclonium*). In the parasite *Gonimophyllum skottsbergii* (Delesseriaceae) from the North Pacific, 22–50 spores are formed around a central body (Wagner 1954), a feature not found in other polysporangia. Polysporangia are simultaneously cleaved (Fig. 14-50) and occur only in families where other species have simultaneously cleaved, tetrahedrally divided tetrasporangia. Westbrook (1930) and Guiry and Irvine (1989) observed both tetrahedrally divided tetrasporangia and polysporangia with eight spores occasionally occurring on the same thallus in *Compsothamnion thuyoides* (Compsothamnieae). Westbrook (1930) saw some binucleate sporangial initial cells in this species and suggested that the polysporangia may have arisen from these; Dixon (1973) considered that if polysporangia arose from such cells, they should be considered multiple tetrasporangia. Kylin (1924) and Drew (1937) found multinucleate initials in various species of Delesseriaceae and in *Tiffaniella snyderae* (Ceramiaceae). Young polysporangial initials of *Pleonosporium vancouverianum* (Ceramiaceae) are formed as lateral protrusions near the multinucleate apical cells (Figs. 14-47, 14-48; Sheath et al. 1987). These contain two, four, or five nuclei, each of which undergoes meiosis, giving rise to polysporangia with 8, 16, or 20 uninucleate spores (Fig. 14-49). Cleavages take place by simultaneous, centripetal invaginations of the plasmalemma (Fig. 14-50). The formation of the polysporocyte as a lateral protrusion accompanied by a basally directed furrowing is unique compared to the typical differentiated apical cell that forms the initial in species producing tetrasporangia (Broadwater et al. 1986a; Kugrens & West 1972; Pueschel 1979; Scott & Dixon 1973; Vesk & Borowitzka 1984). Also, the variable number of nuclei that undergo meiosis differs from the single nucleus found in premeiotic tetrasporocytes (see Vesk & Borowitzka 1984). The occurrence of simultaneous cleavages in *P. vancouverianum* lends some support to the hypothesis (Guiry 1978) that polysporangia are derived from simultaneously tetrahedrally divided, compound tetrasporangia.

K. Parasporangia

Sporangia with more than four irregularly arranged spores that reproduce the ploidy level of the parent thallus and are not formed in place of tetrasporangia were referred to as parasporangia by Drew (1939). She concluded from a cytological investigation of *Plumaria elegans* (Ceramiaceae) that parasporangia were formed mitotically on triploid plants; Rueness (1968) later found that paraspores from this species formed further parasporangial plants in culture. Rueness (1973) also discovered that a triploid strain of *Ceramium strictum* from Norway reproduced solely by means of parasporangia. Paraspores of *Callithamnion hookeri* (Ceramiaceae) from Sweden, a species that reproduces solely by parasporangia confined to the adaxial branches, formed parasporangial plants through four successive generations in culture (Rueness & Rueness 1978). *C. decompositum*, which is often difficult to distinguish from *C. hookeri* (Dixon & Price 1981), also forms parasporangia, sometimes on plants that also bear tetrasporangia. Northern populations may have only parasporangial plants.

Culturing and cytological evidence suggests that parasporangia are uniquely associated with $3n$ plants in the Ceramiaceae and strongly resemble the carposporophytes of the family. It may be that the alleles that control sexuality and tetrasporangial formation in red algae (van der Meer 1985) stimulate the formation of parasporangia when present in the triploid state (Guiry & Irvine 1989). Reports of parasporangia in the Gigartinales discussed earlier (Section II. C) refer to quite different structures. The parasporangia reported on *Tiffaniella apiculata* from Japan (Itono 1977) are not typical and need further study.

III. SPORANGIAL REGENERATION

Regeneration appears to occur in all types of sporangia except perhaps for polysporangia and parasporangia. It is particularly common in species of the Florideophycidae regarded as having primitive characters. As a general rule, species forming relatively small spores that display amoeboid movement also tend to have sporangial regeneration, particularly in the Acrochaetiales, Batrachospermales, Nemaliales, and Gelidiales (Guiry & Irvine 1989; Maggs & Guiry 1988). In the Acrochaetiaceae, the subterminal cell of a monosporangial or tetrasporangial filament frequently grows into the sporangial wall and becomes cut off, forming a new sporangium. This process may take place repeatedly (Fig. 14-51), giving rise to a nest of sporangial walls (Boney 1967; Guiry et al. 1988; Hymes & Cole 1983a; Stegenga 1985; Woelkerling 1971). In species of *Scinaia* (Galaxauraceae), the mode of regeneration of tetrasporangia is similar to that found in the Acrochaetiaceae, with the subterminal cell repeatedly giving rise to new tetrasporocytes and a nest of old sporangial walls (Boillot 1972; Ramus 1969). In *Galaxaura* species, tetrasporangial regeneration is very complex, with several different types (Svedelius 1942), whereas monosporangial regeneration seems to be similar to the one type found in the Acrochaetiaceae.

An unusual mode of sporangial regeneration occurs in the Palmariales, where a specialized stalk cell is enclosed within the tetrasporangial wall (Fig. 14-53; Guiry 1974; Pueschel 1979; Westbrook 1928). On release of tetraspores, the stalk cell divides unequally, the outermost and larger cell giving rise to further tetraspores and the innermost and smaller cell forming another stalk cell, a process that may take place repeatedly. This mode of sporangial regeneration is probably derived from that found in the Acrochaetiales and Nemaliales, suggesting that these orders and the Palmariales arose from a common ancestor (see Gabrielson & Garbary 1986). Ultrastructural studies of the articulated coralline *Haliptilon cuvieri* (Vesk & Borowitska 1984) have shown that its "stalk cell" is not included in the sporangial wall.

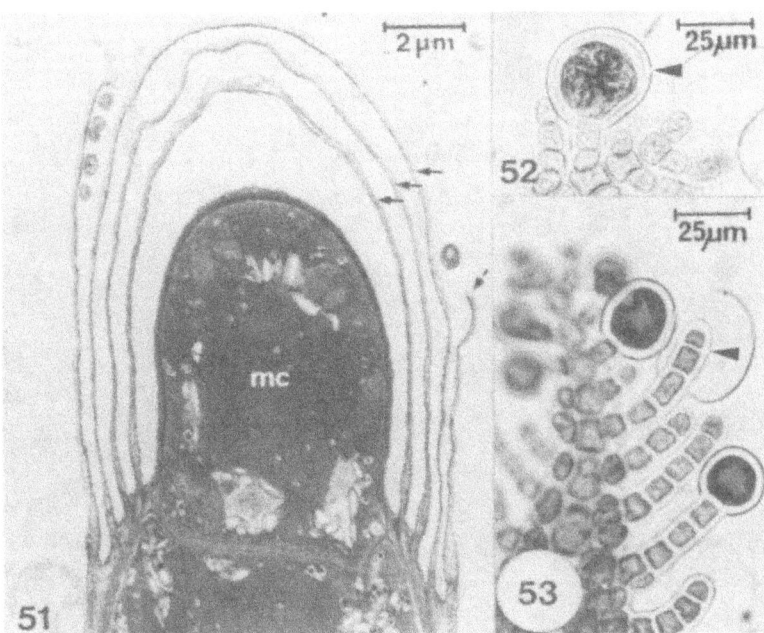

Figs. 14-51 to 14-53. Regeneration of sporangia in Florideophycidae. *14-51*: TEM of regenerating monosporangia in *Acrochaetium secundata* showing monosporocyte (*mc*) and old sporangial walls (arrowheads). *14-52*: Secondary tetrasporangium forming inside old sporangial wall (arrowhead) of *Ptilota plumosa* (Ceramiaceae). *14-53*: Vegetative regeneration of lateral branch (arrowhead) through old sporangial wall of *P. plumosa*. (All original; Fig. 14-51 courtesy of W. R. Kee)

Tetrasporangia with several concentric walls were observed in *Gelidiella calcicola* from the British Isles (Maggs & Guiry 1988), suggesting that sporangial regeneration may be an occasional occurrence in some species of the Gelidiales. Tetrasporangial formation in the Hildenbrandiales is by means of a unique erosive sequence (Pueschel 1982), the tetrasporangia eventually becoming sunken in cavities that, despite their superficial resemblance to the conceptacles of the Corallinales, are quite different in ontogeny. Sporangial regeneration may occur in mature tetrasporangial cavities of *Hildenbrandia* species, but further studies are needed. In the Gigartinales, sporangial regeneration was found in *Peyssonnelia immersa* (Maggs & Irvine 1983). Tetrasporangia are repeatedly regenerated, giving rise to a nest of sporangial walls. It is likely that some other members of the Gigartinales regenerate sporangia in this manner.

The carposporangia of *Nemalion helminthoides* may proliferate from generative mother cells (Ramm-Anderson & Wetherbee 1982), as has been observed in a number of species of Acrochaetiales and Nemaliales. Gabrielson and Garbary (1986) concluded that this feature may prove diagnostic for the Acrochaetiales, Nemaliales, and Bonnemaisoniales, in which the formation of each carposporophyte requires a separate fertilization event, carposporophyte morphology is simple, and carposporangial regeneration may serve to increase spore output. However, Kraft (pers. comm.) found that carposporangial regeneration may be present or absent in various genera and species of the Liagoraceae.

Sporangial regeneration is rare in the Rhodymeniales and Ceramiales and may be secondarily derived. Some evidence of tetrasporangial regeneration was observed in *Botryocladia ardreana* (Rhodymeniaceae) by Brodie and Guiry (1988, Fig. 54). In *Ptilota plumosa* (Ceramiaceae) the subterminal cell of a tetrasporangial branch may divide to form a further initial (Fig. 14-52) or may resume vegetative growth through the old sporangial wall (Fig. 14-53).

IV. SPORE GERMINATION AND SPORELING DEVELOPMENT

The adult thallus of a red alga is only rarely formed directly from a spore, and there are usually intermediate stages in development (Dixon 1973). Spore germination and sporeling development patterns were described for a large number of algae (e.g.,

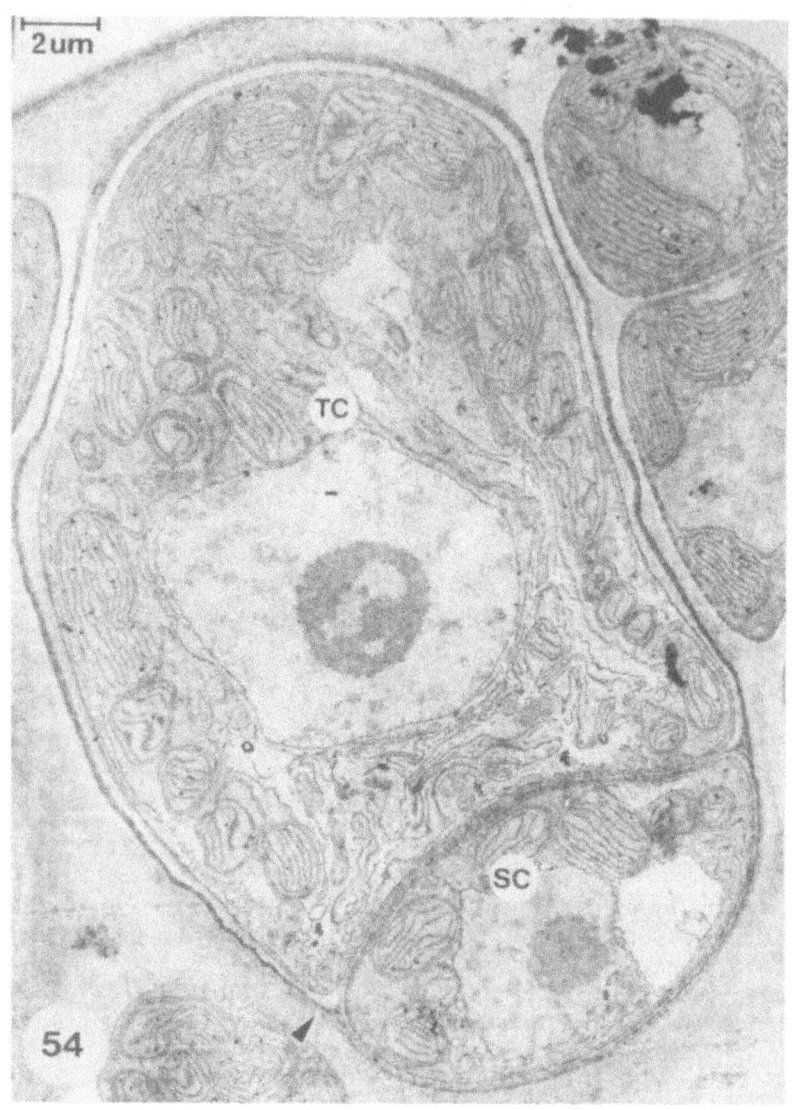

Fig. 14-54. TEM of tetrasporangial initial of *Palmaria palmata* (Palmariaceae), showing inclusion of generative stalk cell (sc) and tetrasporocyte (tc) in tetrasporangial wall (arrowhead). (After Pueschel 1979, with permission)

Chemin 1937; Inoh 1947; Killian 1914; Kylin 1917). Five types of spore germination are usually recognized, four of which were named by Chemin (1937), plus the *Gelidium* type subsequently described by Feldmann (1938). Unfortunately, it seems that there are many intermediate patterns of sporeling development, and the categories outlined here must be regarded as tentative and not comprehensive.

1. *Nemalion* type. After attachment the spore does not divide but puts out a protuberance into which some or all of the spore contents pass. The protuberance is cut off by a cell wall and then functions as an apical cell. A filament is produced by the apical cell, which remains attached to the spore wall. This filament may have a rhizoidal (attachment) function, or a rhizoid may be produced subsequently from a lower cell of the filament adjacent to the spore wall.

2. *Naccaria* type. After attachment the spore usually divides to form a number of subisodiametric cells but may also remain as a single cell. The undivided spore or each of the derivative cells forms one or more protuberances that are then cut off by transverse walls. The spore, or the cells formed by division of the spore, retain their contents. The protuberance(s) function as apical cells and divide to form prostrate filaments that subsequently

branch. The extent to which the spore does or does not divide and the subsequent development of the filaments and branches is such that a prostrate mass of cells of variable outline is produced.

3. *Gelidium* type. After attachment, all the spore contents pass into a protuberance. One end of the protuberance immediately forms a rhizoid, and the cells between the rhizoid and the original spore wall segment to form a mass of cells. This mass of cells then forms one or more apical cells at the opposite end to the rhizoid. This type is superficially similar to the *Nemalion* type.

4. *Dumontia* type. After attachment, the spore divides into two more or less equal halves by a wall perpendicular to the substratum. Further perpendicular and, usually, several horizontal divisions take place very rapidly to form a hemispherical mass of small cells, all of which are within the original spore wall. The marginal cells of this mass then become meristematic, forming branching filaments that grow centrifugally to form a coherent disk. In species with erect fronds, the first upright initial is generally formed from the central mass of cells that constituted the cells of the original spore. After further expansion, the disk or rhizoidal growths from the disk may form further apical cells. The extent of expansion of the disk or rhizoids varies from species to species, and probably also with environmental conditions. A degree of polarity is present in that centrifugally dividing cells form attachment structures and cells of the original spore mass or their derivatives give rise to the upright fronds.

5. *Ceramium* type. After attachment, the spore develops two opposite protuberances, one of which forms a rhizoid while the other develops into an apical cell from which the erect frond is formed. This bipolar germination type is considered to be very different from the previous types (Dixon 1973), but it should be emphasized that strong polarity is displayed by all the spore development patterns.

The types of germination are not absolute, and various intermediate patterns of development are known (Chemin 1937). For example, the germinating polyspores of *Deucalion levringii* (Ceramiaceae) (Huisman & Kraft 1982) form only a single protuberance, the rhizoid being absent. Dixon (1973) concluded that the earlier classifications of these patterns were largely a matter of convenience, with little or no relevance to systematics.

Spore germination patterns appear to be least fixed in the Bangiophycidae (Gargiulo et al. 1987; Heerebout 1968; Murray et al. 1972). Gabrielson and Garbary (1986) stated that in the Bangiales and Rhodochaetales certain patterns are restricted to specific kinds of spores in the life history. Germinating zygotospores of *Porphyra* species always form a single protuberance, but septation of the original spores may or may not occur before cell division takes place (Pueschel & Cole 1985). Monospores of *Bangia* and *Porphyra* species germinate in a manner that is regarded as bipolar (Cole & Conway 1980). Protuberances similar to those of *Porphyra* species were not formed by germinating monospores of *Smithora naiadum* (McBride & Cole 1972). Further clarification of spore germination patterns in the Bangiophycidae is necessary.

In the Florideophycidae, although the patterns are more fixed, there is considerable interspecific and intraspecific variation, and caution should be exercised in the use of spore-germination categories, particularly with regard to cultured material. Spore development patterns are frequently affected by the ages of spores (West pers. comm.).

V. SUMMARY

Of the sporangial types found in the Rhodophyta, only endosporangia and conchosporangia are exclusive to the subclass Bangiophycidae. Several sporangial characters may be used to distinguish the Bangiophycidae from the Florideophycidae: the presence of endosporangia and conchosporangia and the absence of tetrasporangia and carposporangia. Conchosporangia are not tetrasporangial analogues since they are delimited by primary cell walls with pit connections in the conchosporangial branches and, as such, are compound sporangia. In most phylogenetic schemes, insufficient emphasis is placed on the almost universal presence of tetrasporangia in the Florideophycidae and their total absence in the Bangiophycidae. The occurrence of zygotosporangia in the Bangiophycidae should also be emphasized, as well as the absence of carposporangia.[6]

Cruciately divided tetrasporangia are found in all orders of Florideophycidae, except for the Batrachospermales, in which meiotic sporangia are unknown, and it seems reasonable to conclude that they gave rise to all other types (Guiry 1978; Searles 1968). Successive zonate tetrasporangia are unique to the Gigartinales[3] and the Hildenbrandiales[5]; simultaneous zonate tetrasporangia are only found in the Corallinales. Tetrahedrally divided tetrasporangia occur only in a minority of Rhodymeniales and in most Ceramiales. Guiry and Irvine (1989) have suggested that in cruciately divided tetrasporangia, cleavages along the long axis probably close from one side only, i.e., in an asymmetrical manner, whereas in the zonate mode cleavages are probably initiated from both sides in a symmetrical manner. Thus, irregularly divided sporangia could be categorized as fundamentally cruciate or zonate, depending on whether the cleavages are initiated asymmetrically or symmetrically. Polysporangia cleave in a manner similar to simultaneous tetrahedral tetrasporangia and are unique to the Rhodymeniales and Ceramiales. Parasporangia are found only in the Ceramiaceae and may represent parthenogenic development.

The occurrence of monosporangia, sporangial regeneration, small spores, and amoeboid movement are characters generally found in families regarded as having primitive characters, such as a low level of vegetative organization, nonprocarpic female reproductive structures, and cruciately or near-cruciately divided tetrasporangia (Guiry & Irvine 1989) and may thus be regarded as primitive features. Spore germination is a promising area from the systematic point of view. Gabrielson and Garbary (1986) considered that the spore-evacuation sporeling development types (*Nemalion* and *Gelidium* types) probably evolved from the types that do not evacuate, and that they were more specialized and less widely distributed among the red algae. Why this conclusion was reached is not clear, because virtually all types of sporeling development are found in the Acrochaetiaceae, a family considered to be of pivotal phylogenetic importance by these authors. Although the bipolar pattern found in the Ceramiales is said to be widespread in the Bangiophycidae (Gabrielson & Garbary 1986), this seems to be a case of parallel evolution.

Fine structural studies of spores and sporangia, although very interesting at the cellular level, have shown that most spores and sporangia are quite similar. This is not too surprising, as there is certain to be a form/function relationship, and as most spores have comparable functional requirements, the subcellular events leading to their formation are likely to be similar.

Sporangia and spores provide many potentially useful characters for the ordinal classification of the Rhodophyta, but they are often rejected in favor of the more alluring female sexual characteristics. The wide range of sporangial types could be of particular use in phylogenetic systematics (cladistics), but more stringent definitions are needed in many instances.

VI. ACKNOWLEDGMENTS

I wish to thank the Development Fund, University College, Galway, for photographic equipment, and the School of Botany, University of Melbourne, for a Visiting Research Fellowship and facilities in 1987–8 that greatly helped in the completion of the manuscript. I am grateful to those who provided me with comment, unpublished information, and photomicrographs: C. J. Bird, J. Brodie, K. M. Cole, Y. M. Chamberlain, E. M. Cunnigham, M. J. Hawkes, J. M. Huisman, B. J. Hymes, W. R. Kee, C. M. Pueschel, J. L. Scott, and R. G. Sheath. I am particularly indebted to L. M. Irvine, G. T. Kraft, C. A. Maggs, and J. A. West for sharing their insights with me, to P. Cooke for preparing photographs, and to W. Guiry and D. Murphy for typing the manuscript.

VII. ENDNOTES

1. The Compsopogonales Skuja (1939) has nomenclatural priority over the Erythropeltidales sensu Garbary et al. (1980); see Wynne (1986a).
2. The generic concepts of Stegenga (1985) are adopted here.
3. The Gigartinales is considered, provisionally, to include the Cryptonemiales (Kraft & Robins 1985).
4. Features of tetrasporangial formation in *Petrocelis hennedyi*, among other things, suggest that it does not belong in the Petrocelidaceae (Fletcher & Irvine 1982).
5. For this reason, it might be reasonable to retain the Hildenbrandiales in the Gigartinales sensu Kraft & Robins (1985).
6. Despite some recent advances in knowledge of the fine structure of *Rhodochaete parvula* (Pueschel & Magne 1987), the sole species referred to the Rhodochaetales, the placement of this order in either of the currently recognized subclasses still presents problems.

VIII. REFERENCES

Abbott, I. A. 1967. Studies in some foliose red algae of the Pacific coast II. *Schizymenia. Bull. South. Calif. Acad. Sci.* 66: 161–74.

Abdel-Rahman, M. H. & Magne, F. 1983. Existence d'un nouveau type de cycle de développement chez les Rhodophycées. *C. R. Acad. Sci.* (Paris), sér. III, 296: 641–4.

Balakrishnan, M. S. 1947. The morphology and cytology of *Melobesia farinosa* Lamour. *J. Ind. Bot. Soc.* 1946: 305–19.

Baldock, R. N. 1976. The Griffithsieae group of the Ceramiaceae (Rhodophyta) and its southern Australian representatives. *Aust. J. Bot.* 24: 509–93.

Bauch, R. 1937. Die Entwicklung der Bisporen der Corallinaceen. *Planta* (Berl.) 26: 365–90.

Berthold, G. 1882. Die Bangiaceen des Golfes von Neapel und der angrenzenden Meeresabshnitte. *Fauna Flora Golf. Neapel* 8: 1–28.

Bird, C. J. 1973. Aspects of the life history and ecology of *Porphyra linearis* (Bangiales, Rhodophyceae) in nature. *Can. J. Bot.* 51: 2371–9.

Bird, C. J., Chen, L. C.-M. & McLachlan, J. 1972. The culture of *Porphyra linearis* (Bangiales, Rhodophyceae). *Can. J. Bot.* 50: 1859–63.

Bird, C. J. & Johnson, C. R. 1984. *Seirospora seirosperma* (Harvey) Dixon (Rhodophyta, Ceramiaceae) – a first record for Canada. *Proc. N. S. Inst. Sci.* 34: 173–5.

Boillot, A. 1968. Sur l'existance d'un tetrasporophyte dans le cycle de *Scinaia furcellata* (Turner) Bivona, Némalionales. *C. R. Acad. Sci.* (Paris) sér. D, 266: 1831–2.

Boillot A. 1969. Sur le développement des tetraspores et l'édification du gametophyte chez *Scinaia furcellata*. *C. R. Acad. Sci.* (Paris) sér. D, 267: 273–5.

Boillot, A. 1972. Cycle biologique de quelques Némalionales. *Botaniste* 55: 207–50.

Boillot, A. 1975. Cycle biologique de *Rhodochaete parvula* (Thuret) (Rhodophycées, Bangiophycidées). *Pub. Staz. Zool. Napoli* 39 (Suppl.): 67–83.

Boillot, A. 1978. Les ponctations intercellulaires du *Rhodochaete parvula* Thuret (Rhodophycées, Bangiophycées). *Rev. Algol.* N.S. 13: 251–8.

Boney, A. D. 1967. Spore emission, sporangium proliferation and spore germination *in situ* in monosporangia of *Acrochaetium virgatulum*. *Br. Phycol. Bull.* 3: 317–26.

Bornet, E. & Thuret, G. 1876. *Notes Algologiques*. Vol. 1. Paris: Masson.

Broadwater, S., Scott, J. & Pobiner, B. 1986a. Ultrastructure of meiosis in *Dasya baillouviana* (Rhodophyta). I. Prophase I. *J. Phycol.* 22: 490–500.

Broadwater, S., Scott, J. & Pobiner, B. 1986b. Ultrastructure of meiosis in *Dasya baillouviana* (Rhodophyta). II. Prometaphase I–Telophase II and post-division nuclear behavior. *J. Phycol.* 22: 501–12.

Brodie, J. & Guiry, M. D. 1988. Life history and reproduction of *Botryocladia ardreana* sp. nov. (Rhodymeniales, Rhodophyta) from Portugal. *Phycologia* 27: 109–30.

Burzycki, G. M. & Waaland, J. R. 1987. On the position of meiosis in the life history of *Porphyra torta*. *Bot. Mar.* 30: 5–10.

Cabioch, J. 1972. Étude sur les Corallinacées. II. La morphogenèse: conséquences systématiques et phylogénétiques. *Cah. Biol. Mar.* 13: 137–288.

Carter, A. R. 1985. Reproductive morphology and phenology, and culture studies of *Gelidium pristoides* (Rhodophyta) from Port Alfred in South Africa. *Bot. Mar.* 28: 303–11.

Chamberlain, Y. M. 1977. Observations on *Fosliella farinosa* (Lamour.) Howe (Rhodophyta, Corallinaceae) in the British Isles. *Br. Phycol. J.* 12: 343–58.

Chamberlain, Y. M. 1983. Studies in the Corallinaceae with special reference to *Fosiella* and *Pneophyllum* in the British Isles. *Bull. Br. Mus. Nat. Hist.* (Bot.) 11: 291–463.

Chamberlain, Y. M. 1984. Spore size and germination in *Fosliella*, *Pneophyllum* and *Melobesia* (Rhodophyta, Corallinaceae). *Br. Phycol. J.* 23: 433–42.

Chemin, E. 1937. Le développement des spores chez les Rhodophycées. *Rev. Gén. Bot.* 49: 205–34, 300–27, 353–74, 424–48, 478–536.

Chen, L. C.-M., Edelstein, T., Bird, C. & Yabu, H. 1978. A culture and cytological study of the life history of *Nemalion helminthoides* :Rhodophyta, Nemaliales). *Proc. N. S. Inst. Sci.* 28: 191–9.

Cole, K. 1972a. Observations on the life history of *Bangia fuscopurpurea*. *Mém. Soc. Bot. Fr.* 1972: 231–5.

Cole, K. 1972b. Some electron microscopic observations on the cultured conchocelis phase of *Porphyra* species. In *Contributions to the Systematics of Benthic Marine Algae of the North Pacific*, eds. I. A. Abbott & M. Kurogi, pp. 157–66. Kobe: Japanese Society of Phycology.

Cole, K. & Conway, E. 1980. Studies in the Bangiaceae: reproductive modes. *Bot. Mar.* 23: 545–53.

Cole, K., Hymes, B. J. & Sheath, R. G. 1983. Karyotypes and reproductive seasonality of the genus *Bangia* (Rhodophyta) in British Columbia, Canada. *J. Phycol.* 19: 136–45.

Conway, E. 1964. Autecological studies of the genus *Porphyra*: I. The species found in Britain. *Br. Phycol. Bull.* 2: 342–8.

Conway, E. & Cole, K. 1977. Studies in the Bangiaceae: structure and reproduction of the conchocelis of *Porphyra* and *Bangia* in culture (Bangiales, Rhodophyceae). *Phycologia* 16: 205–16.

Conway, E. & Wylie, A. P. 1972. Spore organization and reproductive modes in two species of *Porphyra* from New Zealand. *Proc. Int. Seaweed Symp.* 7: 105–7.

DeCew, T. C. & West, J. A. 1981. Investigations on the life histories of three *Farlowia* species (Rhodophyta: Cryptonemiales, Dumontiaceae) from Pacific North America. *Phycologia* 20: 342–51.

DeCew, T. C. & West, J. A. 1982. A sexual life history in *Rhodophysema* (Rhodophyceae): a re-investigation. *Phycologia* 21: 67–74.

Dixon, P. S. 1973. *Biology of the Rhodophyta*. Edinburgh: Oliver and Boyd.

Dixon, P. S., Murray, S. N., Richardson, W. N. & Scott, J. L. 1972. Life history studies in genera of the Cryptonemiales. *Mém. Soc. Bot. Fr.* 1972: 323–32.

Dixon, P. S. & Price, J. H. 1981. The genus *Callithamnion* (Rhodophyta: Ceramiaceae) in the British Isles. *Bull. Br. Mus. Nat. Hist.* (Bot.) 9: 99–141.

Doty, M. S. 1963. *Gibsmithia hawaiiensis* gen. n. et sp. n. *Pacif. Sci.* 17: 458–65.

Drew, K. M. 1937. *Spermothamnion snyderae* Farlow, a Floridean alga bearing polysporangia. *Ann. Bot.* N. S., 1: 463–76.

Drew, K. M. 1939. An investigation of *Plumaria elegans* (Bonnem.) Schmitz with special reference to triploid plants bearing parasporangia. *Ann. Bot.* N. S. 3: 347–67.

Drew, K. M. 1956. Reproduction in the Bangiophycidae. *Bot. Rev.* (Lond.) 22: 553–611.

Fan, K.-C. 1961. Morphological studies of the Gelidiales. *Univ. Calif. Pub. Bot.* 32: 315–68.

Feldmann, J. 1938. Sur le développement des tétraspores du *Caulacanthus ustulatus* (Mertens) Kützing (Rhodophyceae). *Bull. Soc. Hist. Nat. Afr. N.* 29: 298–303.

Feldmann, J. 1939. Une Némalionale à carpotétraspores: *Helminthocladia Hudsoni* (C. Ag.) J. Ag. *Bull. Soc. Hist. Nat. Afr. N.* 30: 87–97.

Feldmann-Mazoyer, G. 1941. *Recherches sur les Céramiacées de la Méditerranée occidentale*. Alger: Minerva.

Fletcher, R. L. & Irvine, L. M. 1982. Some preliminary observations on the ecology, structure, culture and taxonomic position of *Petrocelis hennedyi* (Harvey) Batters (Rhodophyta) in Britain. *Bot. Mar.* 25: 601–9.

Foslie, M. 1905. Remarks on northern Lithothamnia. *K. Norske Vidensk. Selsk. Skr.* 1905 (3): 1–138.

Fries, L. 1967. The sporophyte of *Nemalion multifidum*

(Weber et Mohr.) J. Ag. *Svensk Bot. Tidskr.* 61: 457–62.
Fritsch, F. E. 1942. Studies in the comparative morphology of the algae I. Heterotrichy and juvenile stages. *Ann. Bot. N. S.*, 6: 397–412.
Fritsch, F. E. 1945. *The Structure and Reproduction of the Algae. Volume 2. Foreword, Phaeophyceae, Rhodophyceae, Myxophyceae.* Cambridge: Cambridge University Press.
Gabrielson, P. W. & Garbary, D. J. 1986. Systematics of red algae (Rhodophyta). *CRC Press Crit. Rev. Plant Sci.* 3(4): 329–66.
Garbary, D. J., Hansen, G. I. & Scagel, R. F. 1980. A revised classification of the Bangiophyceae (Rhodophyta). *Nova Hedwigia* 33: 145–66.
Gargiulo, G. M., DeMasi, F. & Tripodi, G. 1987. Thallus morphology and spore formation in cultured *Erythrocladia irregularis* Rosenvinge (Rhodophyta, Bangiophycidae). *J. Phycol.* 23: 351–9.
Giraud, A. & Magne, F. 1968. La place de la méiose dans le cycle de développement de *Porphyra umbilicalis*. *C. R. Acad. Sci.* (Paris) sér. D, 267: 586–8.
Gittins, B. T. & Dixon, P. S. 1976. Biology of *Lithothrix aspergillum*. *Br. Phycol. J.* 11: 194 (abstract).
Goff, L. J. 1981. The role of bispores in the life history of the parasitic red alga, *Gardneriella tubifera* (Solieriaceae, Gigartinales). *Phycologia* 20: 397–406.
Guiry, M. D. 1974. A preliminary consideration of the taxonomic position of *Palmaria palmata* (Linnaeus) Stackhouse = *Rhodymenia palmata* (Linnaeus) Greville. *J. Mar. Biol. Ass. U.K.* 54: 509–28.
Guiry, M. D. 1976. An assessment of *Palmaria palmata* forma *mollis* (S. et G.) comb. nov. (= *Rhodymenia palmata* forma *mollis* S. et G.) in the eastern North Pacific. *Syesis* 8: 245–61.
Guiry, M. D. 1978. The importance of sporangia in the classification of the Florideophyceae. In *Modern Approaches to the Taxonomy of Red and Brown Algae.* Systematics Association Special Vol. 10, eds. D. E. G. Irvine & J. H. Price, pp. 111–44. London: Academic.
Guiry, M. D. 1987. The evolution of life history types in the Rhodophyta: an appraisal. *Cryptogam. Algol.* 8: 1–12.
Guiry, M. D. & Irvine, L. M. 1989. Sporangial form and function in the Nemaliophycidae (Rhodophyta). In *Phykotalk* Vol. I, ed. H. D. Kumar, pp. 155–84. Meerut: Rastogi Publ. (India).
Guiry, M. D., Kee, W. R. & Garbary, D. J. 1988. Morphology, temperature and photoperiodic responses in *Audouinella botryocarpa* (Harvey) Woelkerling (Acrochaetiaceae, Rhodophyta) from Ireland. *Giorn. Bot. Ital.* 121: 229–46.
Guiry, M. D., West, J. A., Kim, D.-H. & Masuda, M. 1984. Reinstatement of the genus *Mastocarpus* Kützing (Rhodophyta). *Taxon* 33: 53–63.
Harris, R. E. 1962. Contribution to the taxonomy of *Callithamion* Lyngbye emend. Naegeli. *Bot. Not.* 115: 18–28.
Hawkes, M. W. 1978. Sexual reproduction in *Porphyra gardneri* (Smith et Hollenberg) Hawkes Bangiales, Rhodophyta). *Phycologia* 17: 329–53.
Hawkes, M. W. 1980. Ultrastructure characteristics of monospore formation in *Porphyra gardneri* (Rhodophyta). *J. Phycol.* 16: 192–6.
Hawkes, M. W. 1983. *Hummbrella hydra* Earle (Rhodophyta, Gigartinales): seasonality, distribution and development in laboratory culture. *Phycologia* 22: 401–13.
Heerebout, G. R. 1968. Studies in the Erythropeltidaceae (Rhodophyceae–Bangiophycidae). *Blumea* 16: 139–57.
Hollenberg, G. J. 1959. *Smithora*, an interesting new algal genus in the Erythropeltidaceae. *Pacif. Nat.* 1(8): 3–11.

Huisman, J. M. 1985. The *Scinaia* assemblage (Galaxauraceae, Rhodophyta): a re-appraisal. *Phycologia* 24: 403–18.
Huisman, J. M. 1987. The taxonomy and life history of *Gloiophloea* (Galaxauraceae, Rhodophyta). *Phycologia* 26: 167–74.
Huisman, J. M. & Kraft, G. T. 1982. *Deucalion* gen. nov. and *Anisoschizus* gen. nov. (Ceramiaceae, Ceramiales), two new propagule-forming red algae from southern Australia. *J. Phycol.* 18: 177–92.
Hymes, B. J. & Cole, K. M. 1983a. The cytology of *Audouinella hermannii* (Rhodophyta, Florideophyceae). II. Monosporogenesis. *Can. J. Bot.* 61: 3377–85.
Hymes, B. J. & Cole, K. M. 1983b. Aplanospore production in *Porphyra maculosa* (Rhodophyta). *Jap. J. Phycol.* 31: 225–8.
Inoh, S. 1947. *Kaiso no hassei [Development of marine algae].* Tokyo: Hokuryukan (in Japanese; not seen, cited by Dixon 1973).
Itono, H. 1977. Studies on the ceramiaceous algae (Rhodophyta) from southern parts of Japan. *Bibl. Phycol.* 35: 1–499.
Jaasund, E. 1965. Aspects of the marine algal vegetation of north Norway. *Bot. Goth.* 4: 1–174.
Johansen, H. W. 1971. *Bossiella*, a new genus of articulated corallines (Rhodophyceae, Cryptonemiales) in the eastern Pacific. *Phycologia* 10: 381–96.
Johansen, H. W. 1976. Family Corallinaceae. In *Marine Algae of California*, eds. I. A. Abbott & G. J. Hollenberg, pp. 379–419. Stanford: Stanford University Press.
Johansen, H. W. 1981. *Coralline Algae. A First Synthesis.* Boca Raton: CRC Press.
Kajimura, M. 1981. On deep-sea red algal vegetations in the Japan Sea, and a new nemastomataceous alga *Predaeopsis japonica* n.gen. n. sp. *Proc. Int. Seaweed Symp.* 10: 181–6.
Kajimura, M. 1987. Two new species of *Predaea* (Nemastomataceae, Rhodophyta) from the Sea of Japan. *Phycologia* 26: 419–28.
Karam-Keriman, T. B. 1976. Structure, reproduction et discussion sur la position systématique du genre *Gibsmithia*. *Bull. Mus. Natl. Hist. Nat. Paris.*, sér. 3, no. 365: 21–32.
Killian, C. 1914. Über die Entwicklung einiger Florideen. *Z. Bot.* 6: 209–78.
Klavestad, N. 1957. Paraspores in *Rhodochorton rothii*. *Nytt Mag. Bot.* 5: 61–2.
Knox, E. 1926. Some steps in the development of *Porphyra naiadum*. *Pub. Puget Sound Mar. Biol. Stn.* 5: 125–35.
Kornmann, P. 1984. *Erythrotichopeltis*, eine neue Gattung der Erythropeltidaceae (Bangiophyceae, Rhodophyta). *Helgol. Meeresunters.* 38: 207–24.
Kornmann, P. 1987. Der Lebenzyklus von *Porphyrostromium obscurum* (Bangiophyceae, Rhodophyta). *Helgol. Meeresunters.* 41: 127–37.
Kornmann, P. & Sahling, P.-H. 1985. Erythropeltidaceen (Bangiophyceae, Rhodophyta) von Helgoland. *Helgol. Meeresunters.* 39: 213–36.
Kraft, G.T. 1981. Rhodophyta: morphology and classification. In *The Biology of Seaweeds*, eds. C. S. Lobban & M. J. Wynne, pp. 6–51. Oxford: Blackwell.
Kraft, G. T. 1984. The red algal genus *Predaea* (Nematomataceae, Gigartinales) in Australia. *Phycologia* 23: 3–20.
Kraft, G. T. 1986. The genus *Gibsmithia* (Dumontiaceae, Rhodophyta) in Australia. *Phycologia* 25: 423–47.
Kraft, G. T. & Robins, P. A. 1985. Is the order Cryptonemiales defensible? *Phycologia* 24: 67–77.
Krishnamurthy, V. 1969. On two species of *Porphyra* from

San Juan Island, Washington. *Proc. Int. Seaweed Symp.* 6: 225–34.

Kugrens, P. & West, J. A. 1972. Ultrastructure of tetrasporogenesis in the parastic red alga *Levringiella gardneri* (Setchell) Kylin *J. Phycol.* 8: 370–83.

Kylin, H. 1917. Über die Keimung der Florideensporen. *Ark. Bot.* 14(22): 1–25.

Kylin, H. 1924. Studien über die Delesseriaceen. *Acta Univ. Lund. N. F. avd. 2,* 20(6): 1–111.

Lee, I. K. 1978. Studies on Rhodymeniales from Hokkaido. *J. Fac. Sci. Hokkaido Univ.* ser. 5, 11(1): 1–194.

Levring, T. 1941. Die Meeresalgen der Juan Fernandez-Inseln. In *Natural History of Juan Fernandez and Easter Island*, Vol. 2, ed. C. Skottsberg, pp. 601–70. Uppsala: Almquist & Wiksells.

Lewis, I. F. 1909. The life history of *Griffithsia bornetiana. Ann. Bot.* 23: 639–90.

L'Hardy-Halos, M. T. 1966. Observations sur la morphologie et sur la position systématique de *Dohrniella neapolitana* Funk (Floridées-Céramiacées). *Bull. Soc. Bot. Fr.* 113: 295–304.

L'Hardy-Halos, M. T. 1970. Recherches sur les Céramiacées (Rhodophycées, Céramiales) et leur morphogénèse. I. Structure de l'appareil végétatif et des organes reproducteurs. *Rev. Gén. Bot.* 77: 211–87.

Lindstrom, S. C. 1987. Acrosymphytaceae, a new family in the order Gigartinales sensu lato. *Taxon* 36: 50–3.

Ma, J. H. & Miura, A. 1984. Observations of the nuclear division in the conchospores and their germlings in *Porphyra yezoensis* Ueda. *Jap. J. Phycol.* 32: 373–8.

Maggs, C. A. & Guiry, M. D. 1982a. The taxonomy, morphology and distribution of species of *Scinaia* Biv.-Bern. (Nemaliales, Rhodophyta) in north-western Europe. *Nord. J. Bot.* 2: 517–23.

Maggs, C. A. & Guiry, M. D. 1982b. Morphology, phenology and photoperiodism in *Halymenia latifolia* Kütz. (Rhodophyta) from Ireland. *Bot. Mar.* 25: 589–99.

Maggs, C. A. & Guiry, M. D. 1987. Environmental control of macroalgal phenology. In *Plant life in Aquatic and Amphibious Habitats. Special Publication No. 5. of The British Ecological Society*, ed. R. M. M. Crawford, pp. 359–73. Oxford: Blackwell.

Maggs, C. A. & Guiry, M. D. 1988 ["1987"]. *Gelidiella calcicola* sp. nov. (Rhodophyta) from the British Isles and northern France. *Br. Phycol. J.* 22: 417–34.

Maggs, C. A. & Irvine, L. M. 1983. *Peyssonnelia immersa* sp. nov. (Cryptonemiales, Rhodophyta) from the British Isles and France, with a survey of infrageneric classification. *Br. Phycol. J.* 18: 219–38.

Magne, F. 1960. Le *Rhodochaete parvula* Thuret (Bangioideae) et sa réproduction sexuée. *Cah. Biol. Mar.* 1: 407–20.

Masuda, M. & Umezaki, I. 1977. On the life history of *Nemalion vermiculare* Suringar (Rhodophyta) in culture. *Bull. Jap. Soc. Phycol.* 25 (Suppl.): 129–36.

McBride, D. L. & Cole, K. 1971. Electron microscopic observations on the differentiation and release of monospores in the marine red alga *Smithora naiadum*. *Phycologia* 10: 49–61.

McBride, D. L. & Cole, K. 1972. Ultrastructural observations on germinating monospores in *Smithora naiadum* (Rhodophyceae, Bangiophycidae). *Phycologia* 11: 181–91.

McCandless, E. L. & Vollmer, C. M. 1984. The nemathecium of *Gymnogongrus chiton* (Rhodophyceae, Gigartinales): immunochemical evidence of meiosis. *Phycologia* 23: 119–23.

Migita, S. 1967. Cytological studies on *Porphyra yezoensis* Ueda. *Bull. Fac. Fish. Nagasaki Univ.* 24: 55–64.

Miura, A. 1975. *Porphyra* cultivation in Japan. In *Advance of Phycology in Japan*, eds. J. Tokida & H. Hiroyuki, pp. 273–304. The Hague: Junk.

Moe, R. L. 1985. *Gainia* and Ganiaceae, a new genus and family of crustose marine Rhodophyceae from Antarctica. *Phycologia* 24: 419–28.

Moe, R. L. & Silva, P. C. 1979. Morphological and taxonomic studies on Antarctic Ceramiaceae (Rhodophyceae). I. *Antarcticothamnion polysporum* gen. et sp. nov. *Br. Phycol. J.* 14: 385–405.

Moe, R. L. & Silva, P. C. 1980. Morphological and taxonomic studies on Antarctic Ceramiaceae (Rhodophyceae). II. *Pterothamnion antarcticum* (Kylin) comb. nov. (*Antithamnion antarcticum* Kylin). *Br. Phycol. J.* 15: 1–17.

Murray, S. N., Dixon, P. S. & Scott, J. L. 1972. The life history of *Porphyropsis coccinea* var. *dawsonii* in culture. *Br. Phycol. J.* 7: 323–33.

Ngan, Y. & Price, I. R. 1979. Systematic significance of spore size in the Florideophyceae (Rhodophyta). *Br. Phycol. J.* 14: 285–303.

Notoya, M. 1984. The life history of *Gloeophycus koreanum* I. K. Lee and Yoo (Rhodophyta, Gloiosiphoniaceae) in culture. *Hydrobiologia* 116/117: 233–6.

Piriz, M. L. 1981. A new species and a new record of *Porphyra* (Bangiales, Rhodophyta) from Argentina. *Bot. Mar.* 24: 599–602.

Plattner, S. B. & Nichols. H. W. 1978. Asexual development in *Seirospora seirosperma*. *Phytomorphology* 27: 371–7.

Pueschel, C. M. 1979. Ultrastructure of tetrasporogenesis in *Palmaria palmata*. *J. Phycol.* 15: 409–24.

Pueschel, C. M. 1982. Ultrastructural observations of tetrasporangia and conceptacles in *Hildenbrandia* (Rhodophyta: Hildenbrandiales). *Br. Phycol. J.* 17: 333–41.

Pueschel, C. M. 1987. Absence of cap membranes as a characteristic of pit plugs of some red algal orders. *J. Phycol.* 23: 150–6.

Pueschel, C. M. & Cole, K. M. 1982. Rhodophycean pit plugs: an ultrastructural survey with taxonomic implications. *Am. J. Bot.* 69: 703–20.

Pueschel, C. M. & Cole, K. M. 1985. Ultrastructure of germinating carpospores of *Porphyra variegata* (Kjellm.) Hus (Bangiales, Rhodophyta). *J. Phycol.* 21: 146–54.

Pueschel, C. M. & Magne, F. 1987. Pit plugs and other ultrastructural features of systematic value in *Rhodochaete parvula* (Rhodophyta, Rhodochaetales). *Cryptogam. Algol.* 8: 201–9.

Ramm-Anderson, S. M. & Wetherbee, R. 1982. Structure and development of the carposporphyte of *Nemalion helminthoides* (Nemaliales, Rhodophyta). *J. Phycol.* 18: 133–41.

Ramus, J. 1969. The developmental sequence of the marine red alga *Pseudogloiophloea confusa* in culture. *Univ. Calif. Pub. Bot.* 52: 1–28.

Richardson, N. & Dixon, P. S. 1970. Culture studies on *Thuretellopsis peggiana* Kylin. *J. Phycol.* 6: 154–9.

Rosenvinge, L. K. 1923–1924. The marine algae of Denmark. Contributions to their natural history Part III, Rhodophyceae III. (Ceramiales). *K. Dansk. Vidensk. Selsk. Skr. 7 Raekke Nat. Math. Afh.* 7: 287–486.

Rueness, J. 1968. Paraspores from *Plumaria elegans* (Bonnem.) Schmitz in culture. *Nytt Mag. Bot.* 15: 220–4.

Rueness, J. 1973. Culture and field observations of growth and reproduction of *Ceramium strictum* Harv. from the Oslofjord, Norway. *Norw. J. Bot.* 20: 61–5.

Rueness, J. & Rueness, M. 1978. A parasporangia-bearing

strain of *Callithamnion hookeri* (Rhodophyceae, Ceramiales) in culture. *Norw. J. Bot.* 25: 201–5.

Schotter, G. 1968. Recherches sur les Phyllophoracées. Notes postumés publiées par Jean Feldmann et Marie-France Magne. *Bull. Inst. Océanogr. Monaco* 67 (1383): 1–99.

Scott, J. L. & Dixon, P. S. 1971. The life history of *Pikea californica* Harv. *J. Phycol.* 7: 295–300.

Scott, J. L. & Dixon, P. S. 1973. Ultrastructure of tetrasporogenesis in the marine red alga *Ptilota hypnoides*. *J. Phycol.* 9: 29–46.

Scott, J. L. & Thomas, J. P. 1975. Electron microscopy observation of telophase II in the Florideophyceae. *J. Phycol.* 11: 474–6.

Searles, R. B. 1968. Morphological studies of red algae of the order Gigartinales. *Univ. Calif. Pub. Bot.* 43: 1–86.

Searles, R. B. 1980. The strategy of the red algal life history. *Am. Nat.* 115: 113–20.

Sheath, R. G. 1984. The biology of freshwater red algae. *Prog. Phycol. Res.* 3: 89–157.

Sheath, R. G. & Cole, K. M. 1980. Distribution and salinity adaptations of *Bangia atropurpurea* (Rhodophyta), a putative migrant into the Laurentian Great Lakes. *J. Phycol.* 16: 412–20.

Sheath, R. G., Cole, K. M. & Hymes, B. J. 1987. Ultrastructure of polysporogenesis in *Pleonosporium vancouverianum* (Ceramiaceae, Rhodophyta). *Phycologia* 26: 1–8.

Shimizu, T. & Masuda, M. 1983. The life history of *Farlowia irregularis* Yamada (Rhodophyta, Cryptonemiales). *Jap. J. Bot.* 31: 202–7.

Silva, P. C. 1980. Names of classes and families of living algae with special reference to their use in the *Index Nominum Genericorum (Plantarum)*. *Regnum Veg.* 103: 1–156.

Silva, P. C. & Johansen, H. W. 1986. A reappraisal of the order Corallinales (Rhodophyceae). *Br. Phycol. J.* 21: 245–54.

Skuja, H. 1939. Versuch einer systematischen Einteilung der Bangioideen oder Protoflorideen *Acta Horti Bot. Univ. Latv.* 11/12: 23–40.

Smith, G. M. & Hollenberg, G. J. 1943. On some Rhodophyceae from the Monterey Peninsula, California. *Am. J. Bot.* 30: 211–22.

Stegenga, H. 1985. The marine Acrochaetiaceae (Rhodophyta) of southern Africa. *S. Afr. J. Bot.* 51: 291–330.

Suneson, S. 1937. Studien über die entwicklungsgeschichte der Corallinaceen. *Acta Univ. Lund. N. F. adv.* 2, 33(2): 1–102.

Suneson, S. 1950. The cytology of the bispore formation in two species of *Lithophyllum* and the significance of the bispores in the Corallinaceae. *Bot. Notiser* 103: 429–50.

Suneson, S. 1982. The culture of bisporangial plants of *Dermatolithon litorale* (Suneson) Hamel et Lemoine (Rhodophyta, Corallinaceae). *Br. Phycol. J.* 17: 107–16.

Svedelius, N. 1914. Über Sporen an Geschlechtspflanzen von *Nitophyllum punctatum*: ein Beitrag zur Frage des Generationswechsels der Florideen. *Ber. Dtsch. Bot. Ges.* 32: 106–16.

Svedelius, N. 1917. Die monosporen bei *Helminthora divaricata* nebst Notiz über die Zweikernigkeit ihres Karpogons. *Ber. Dtsch. Bot. Ges.* 35: 212–24.

Svedelius, N. 1942. Zytologisch-entwicklungsgeschichtliche studien über *Galaxaura* eine diplobiontische Nemalionales-gattung. *Nova Acta R. Soc. Sci Upsal.*, ser. IV, 13(4): 1–154.

Triemer, R. E. & Vasconcelos, A. C. 1977. The ultrastructure of carposporogenesis in *Caloglossa liprieurii* (Delesseriaceae, Ceramiales). *Am. J. Bot.* 64: 825–34.

Tripodi, G. 1971. The fine structure of the cystocarp in the red alga *Polysiphonia sertularioides* (Grat.) J. Ag. *Submicrosc. Cytol.* 3: 71–9.

Tsekos, I. 1981. Growth and differentiation of the Golgi apparatus and wall formation during carposporogenesis in the red alga, *Gigartina teedii* (Roth) Lamour. *J. Cell Sci.* 52: 71–84.

Tseng, C. K. & Sun, A. 1989. Studies on the alternation of the nuclear phases and chromosome numbers in the life history of seven species of *Porphyra* from China. *Bot. Mar.* 32: 1–8.

Umezaki, I. 1972. The life histories of some Nemaliales whose tetrasporophytes were unknown. In *Contributions to the Systematics of Benthic Marine Algae of the North Pacific*, eds. I. A. Abbott & M. Kurogi, pp. 231–42. Kobe: Japanese Society of Phycology.

van den Hoek, C. & Cortel-Breeman, A. M. 1970. Life history studies on Rhodophyceae. II. *Halymenia floresia* (Clem.) Ag. *Acta Bot. Neerl.* 19: 341–62.

van der Meer, J. P. 1985. Genetic contributions to research on seaweeds. *Prog. Phycol. Res.* 4: 1–38.

van der Meer, J. P. & Chen, L. C.-M. 1979. Evidence for sexual reproduction in the red algae *Palmaria palmata* and *Halosaccion ramentaceum*. *Can. J. Bot.* 57: 2452–9.

van der Meer, J. P., Guiry, M. D. & Bird, C. J. 1983. Sporogenesis in male plants of *Chondrus crispus* (Rhodophyta, Gigartinales). *Can. J. Bot.* 61: 2261–8.

Vesk, M. & Borowitzka, M. A. 1984. Ultrastructure of tetrasporogenesis in the coralline alga *Haliptylon cuvieri* (Rhodophyta). *J. Phycol.* 20: 501–15.

von Stosch, H. A. 1965. The sporophyte of *Liagora farinosa* Lamour. *Br. Phycol. Bull.* 2: 286–96.

Wagner, F. S. 1954. Contributions to the morphology of the Delesseriaceae. *Univ. Calif. Pub. Bot.* 27: 279–346.

West, J. A. & Hommersand, M. H. 1981. Rhodophyta: life histories. In *The Biology of Seaweeds*, eds. C. S. Lobban & M. J. Wynne, pp. 133–93. Oxford: Blackwell.

Westbrook, M. A. 1928. Contributions to the cytology of tetrasporic plants of *Rhodymenia palmata* (L.) Grev., and some other Florideae. *Ann. Bot.* 42: 149–72.

Westbrook, M. A. 1930. *Compsothamnion thuyoides* (Smith) Schmitz. *J. Bot.* (Lond.) 68: 353–64.

Whittick, A. & Hooper, R. G. 1977. The reproduction and phenology of *Antithamnion cruciatum* (Rhodophyta: Ceramiaceae) in insular Newfoundland. *Can. J. Bot.* 55: 520–4.

Woelkerling, W. J. 1970. *Acrochaetium botryocarpum* (Harv.) J. Ag. (Rhodophyta) in southern Australia. *Br. Phycol. J.* 5: 159–71.

Woelkerling, W. J. 1971. Morphology and taxonomy of the *Audouinella* complex (Rhodophyta) in southern Australia. *Aust. J. Bot.*, Suppl. Ser. No. 1: 1–91.

Woelkerling, W. J. 1983. The *Audouinella* (*Acrochaetium–Rhodochorton*) complex (Rhodophyta): present perspectives. *Phycologia* 22: 59–92.

Wynne, M. J. 1986a. A checklist of benthic marine algae of the tropical and subtropical western Atlantic. *Can. J. Bot.* 64: 2239–81.

Wynne, M. J. 1986b. *Prophyrostromium* Trevisan (1848) vs. *Erythrotrichopeltis* Kornmann (1984) (Rhodophyta). *Taxon* 35: 328–9.

Yamanouchi, S. 1906. The life-history of *Polysiphonia violacea*. *Bot. Gaz.* 41: 425–33; 42: 401–49.

Chapter 15

Marine ecology

JOANNA M. KAIN
TREVOR A. NORTON

CONTENTS

I. Introduction / 377
II. Environmental factors / 377
    A. Substratum / 377
    B. Light / 379
    C. Daylength / 380
    D. Temperature / 385
    E. Salinity / 387
    F. Desiccation / 390
    G. Water movement / 391
    H. Nutrients / 392
    I. Grazing / 393
    J. Competition / 395
III. Distribution / 396
    A. Geographical / 396
    B. Intertidal / 398
    C. Subtidal / 398
IV. Communities / 400
    A. Production and growth / 400
    B. Temporal change / 403
    C. Demography / 407
V. Summary / 408
VI. Acknowledgments / 408
VII. References / 409

I. INTRODUCTION

Ecology is perhaps the most neglected aspect of the biology of red seaweeds. No complete review of the topic exists: texts on red algae omit ecology, and texts on marine ecology place understandable emphasis on animals and canopy-forming plants. Nonetheless, a great deal of ecological information exists, scattered through the literature. It is hoped that the compilation presented here will draw attention to the great ecological interest of this group and stimulate attempts to solve the many enigmas that remain.

We have confined ourselves to ecological aspects of the Rhodophyta and have thus excluded any general discussions of ecological theory.

II. ENVIRONMENTAL FACTORS

A. Substratum

Seaweeds do not need to be fixed to the substratum, and large populations of unattached plants can thrive in calm-water localities, where the danger of being washed ashore is reduced. Norton and Mathieson (1983) cite over 100 species of red algae that have been found free floating, loose lying, or entangled. Some even adopt characteristic abnormal morphologies when unattached. Such plants are easily harvested, and in the case of *Furcellaria lumbricalis* (as *F. fastigiata*) (Austin 1960a; Levring et al. 1969) and the maerl-forming calcareous Corallinaceae, they constitute commercial crops.

Subtidal calcareous sand can support many species of red algae (Dawes et al. 1967). Species like *Ahnfeltia plicata* and *Gymnogongrus linearis* thrive on sandy shores, where abrasion and intermittent burial exclude most algae (Daly & Mathieson 1977; Markham & Newroth 1972). Their survival is greatly aided by their ability to regenerate following damage (D'Antonio 1986; Markham & Newroth 1972) and perhaps by having incomplete or asexual life histories (Daly & Mathieson 1977). Such psammophytic algae often have tough, wiry thalli, which may enable them to withstand sand abrasion. In contrast, another sand-tolerant species, *Audouinella floridula*, is constructed of delicate silky filaments. It avoids abrasion damage by binding the sand with its thallus and creating immobile hummocks several centimeters thick. Enmeshed sand grains are a common constituent of many red algal turfs (Stewart 1983).

Most red algae are attached to stable rock, but under the right conditions of depth and water movement cobbles are colonized. If these are disturbed seasonally, annual plants predominate (Lieberman et al. 1979; Sousa 1979).

The immotile spores of red algae sink very slowly (Coon et al. 1972; Okuda & Neushul 1981; Sawada et al. 1972). Fortunately, the parent plants being relatively small, the spores do not have far to fall. Maximum spore deposition from both *Porphyra tenera* and *Polysiphonia japonica* occurs at relatively low rates of water flow (Matsumoto 1959; Sawada et al. 1972).

The mucilage-coated spores probably stick to the substratum on contact, but their initial adhesion is usually insecure, and attached spores have been observed to slide downhill (Sawada et al. 1972). For almost all of the red algal spores tested, tenacity of adhesion progressively increases with time (Chamberlain 1976; Chamberlain & Evans 1973; Charters et al. 1973; Suto 1950), but for those of *Cryptopleura violacea* it appears to decline (Charters et al. 1973). Secure attachment occurs within a few hours with *Gracilaria andersonii* (as *Gracilariopsis sjoestedtii*) or *Sarcodiotheca gaudichaudii* (as *Agardhiella tenera*) (Charters et al. 1973) but may take days for *Corallina officinalis* (Moorjani & Jones 1972), *Membranoptera alata*, and *Phycodrys rubens* (D. A. Rugg unpubl.). Such differences bear no obvious relationship to the degree of water motion prevalent at the sites occupied by these species.

Once fixed, however, the spores of red algae can survive shear stresses of 20 dynes cm^{-2}, a dislodging force nearly 100 times greater than their gravitational weight (Charters et al. 1973; D. A. Rugg unpubl.). The thick mucilage around the spore readily deforms (Boney 1975), possibly dissipating some of this tension.

Although red algae may exhibit distinct morphologies and differential survival on different artificial substrata (Fletcher et al. 1985; West & Crump 1974), there is little evidence to indicate that they thrive better in nature if attached to some rock types rather than to others. Nonetheless, rougher substrata certainly enhance the settlement, anchorage, and subsequent survival of the spores (Harlin & Lindbergh 1977; Ogata 1953). They are "caught" in depressions of the microtopography, and the number caught increases if the depth of the depressions exceeds the diameter of the spores (Ogata 1953). The optimum surface relief seems to differ between species, with *Corallina officinalis* thriving on a coarser substratum than that of *Chondrus crispus* (Harlin & Lindbergh 1977). When the tide recedes, spores lodged in depressions are likely to be in a more humid microclimate than those on open rock, and they may also receive some protection from grazers.

Rhodophyta commonly occur as epiphytes growing on the surface of other organisms. Few are obligate epiphytes unable to colonize rock. Most epiphytes use the "host" organism merely for support, rather than as a source of nutriment. Nonetheless, dense infestations of epiphytes may reduce the growth of *Neorhodomela larix* (D'Antonio 1985). Sometimes there is intimate contact between epiphyte and host. Although the rhizoids of *Polysiphonia lanosa* obtain little nutrition from the host plant *Ascophyllum* (Harlin & Craigie 1975), they ramify deeply, locally breaking down much of the host's tissue (Rawlence 1972). Such deep penetration allows *Polysiphonia* to persist even when *Ascophyllum* sheds its entire outer surface (Filion-Myklebust & Norton 1981).

Many red algae colonize the surface of animals. Obligate epizoic algae are rare, but *Audouinella membranacea* is restricted to chitin-containing substrata, such as the coenosarcs of hydroids and bryozoans. Similarly, the conchocelis phase of *Porphyra* bores into mollusk shells, presumably by dissolving the calcareous matter. Substratum penetration is

Marine ecology

vital for endophytes living within the host's tissue. Spores of *Audouinella bonnemaisoniae* exhibit immediate penetrative growth when germinated on their normal host plant, *Bonnemaisonia nootkana*, but fail to do so on the thalli of other red algae (Boney 1980).

Among the species restricted to a single host species are some insignificant-looking, wartlike plants like *Callocolax neglectus* (found on *Callophyllis*) and *Choreocolax polysiphoniae* (found on *Polysiphonia*). Goff (1982) lists over 100 such plants. Most are poorly pigmented and some lack photosynthetic pigments entirely. The parasitic nature of several species has been confirmed by the translocation of ^{14}C-labeled compounds from host to parasite (Goff 1982). Many endophytic and parasitic red algae are highly host specific, although the reasons for this are largely unknown. Not infrequently, they are taxonomically very closely related to their hosts (Boney 1980; Goff 1982).

B. Light

The red algae can tolerate a wider range of light levels than any other group of photosynthetic plants. The deepest known plant is an undescribed rhodophycean coralline found at 268 m off the Bahamas in about 8 nmol $m^{-2} s^{-1}$ (Littler et al. 1985). Other members of the group abound on tropical shores (Lawson 1966) where full sunlight is 2.3 mmol $m^{-2} s^{-1}$.

Between these extremes the amount of light reaching marine habitats is highly variable. The effect of latitude and season on the mean photon irradiance (400–700 nm) reaching the sea surface during the solstice months is shown in Fig. 15-1. In summer, daylength counteracts the effect of sun altitude, creating a surprisingly similar mean irradiance over a wide latitudinal range. In winter the two factors act in concert. The mean summer value of 400 μmol $m^{-2} s^{-1}$ and the winter level for 45°N or S of 100 μmol $m^{-2} s^{-1}$ are used as surface values in Fig. 15-2, showing light penetration into different water types.

The light requirements of the red algae given in Table 15-1 can be compared with the photon irradiances expected at different depths (Fig. 15-2). The compensation point might be expected to coincide with winter irradiance at a lower light limit. Not many compensation points have been determined reliably, but they are probably of the order of one-twentieth of irradiances saturating photosynthesis (Table 15-1). Crustose corallines have very low requirements, the lowest quoted being that of *Leptophytum laeve*, which could be calculated to extend to 60 m in the clearest coastal water (Table 15-1, Fig. 15-2) but is actually abundant at 75 m in the northwest Atlantic (Adey 1966). Such plants grow slowly (Adey 1970), but filaments of *Plumaria* sporelings, also with a low saturation level, show a fast relative growth rate of 0.2 d^{-1} (Boney & Corner 1962). Other temperate subtidal species have photosynthesis saturation levels of 50–100 μmol $m^{-2} s^{-1}$ (Table 15-1), which should allow winter existence to around 20 m in clear coastal

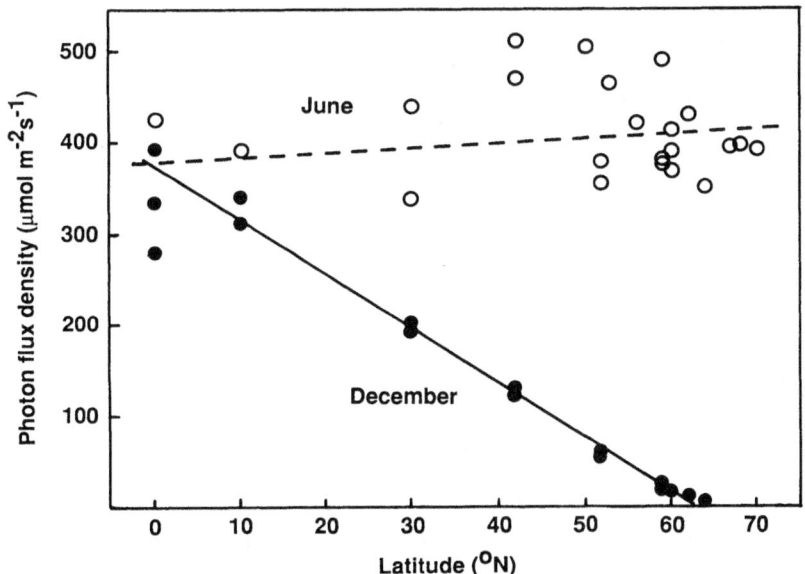

Fig. 15-1. Mean daily photon flux density for the months of June and December in the Northern Hemisphere, plotted against latitude. Data plotted from Kimball, in Sverdrup et al. (1942), and Smayda (1959).

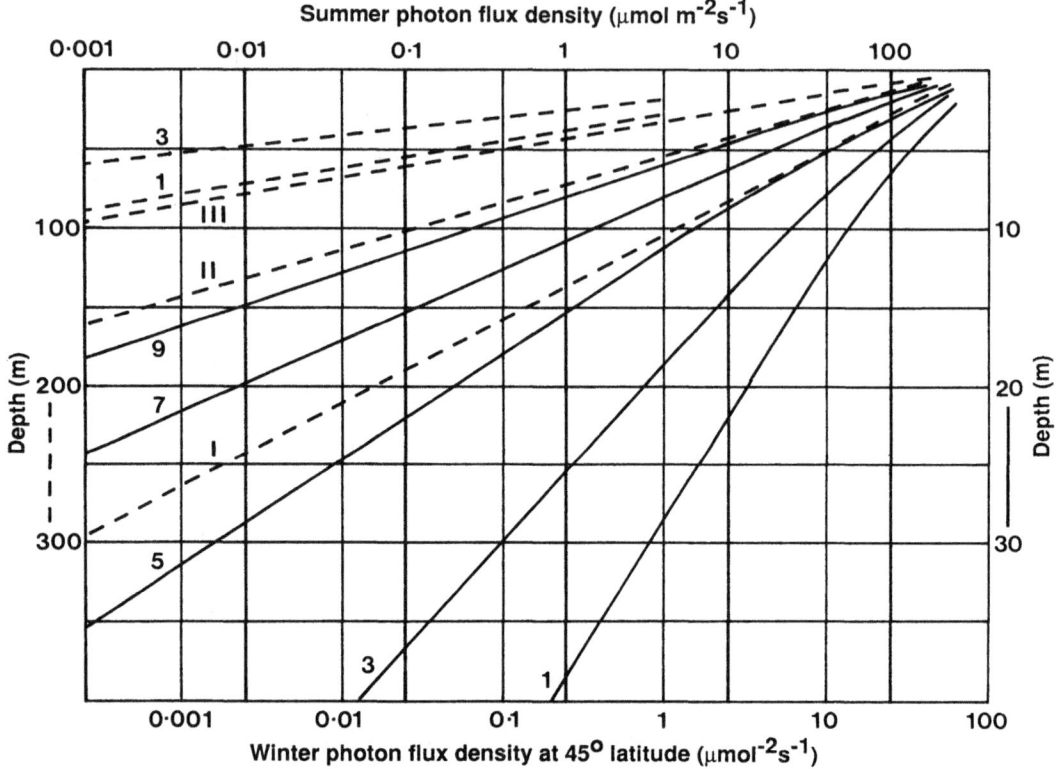

Fig. 15-2. Photon flux density (350–700 nm) at various depths in different Jerlov water types when mean surface photon irradiance is 400 µmol m^{-2} s^{-1} as at the summer solstice (upper scale, heavy vertical lines) and 100 µmol m^{-2} s^{-1} as at 45° latitude at the winter solstice (lower scale, lighter vertical lines). Dashed lines: left-hand scale. Continuous lines: right-hand scale. Replotted from Lüning & Dring (1979) and Jerlov (1976).

water (Fig. 15-2). These can be categorized largely as shade plants, often occurring under a brown algal canopy or in deeper water than the canopy. Intertidal species and shallow subtidal tropical species free of a brown canopy can be regarded as sun plants with high saturation levels, high maximum rates of photosynthesis, and fast growth rates.

Although spores can survive for some weeks in the dark, for example, *Porphyra tenera* and *Gelidium amansii* (Ohno & Arasaki 1969), *Audouinella* spp. (as *Acrochaetium*) (White & Boney 1969), light is necessary for the germination of monospores of *Bangia atropurpurea* (Charnofsky et al. 1982).

Photoinhibition is common, particularly among the subtidal "shade lovers" and at lower light levels when the saturating irradiance is low (Table 15-1). This often involves bleaching, but this in itself may not immediately prevent growth (Jones & Dent 1971). In a shade-loving intertidal species, *Plumaria elegans*, however, growth was enhanced when sporelings were screened by phycoerythrin or other green-absorbing pigments, even at subsaturating irradiances (Boney & Corner 1962), though this did not apply to *Callithamnion hookeri*, also a shade lover and extending to the subtidal (Edwards 1979).

The effects of shading canopies are described in Section II.J.

C. Daylength

Clearly, daylength is important in determining the daily supply of quanta, but the relationship among daylength, irradiance, and growth is not simple. In five members of the Ceramiaceae the rates of apical cell division increase with daylength but the rates of cell elongation do not necessarily keep pace, resulting in various cell sizes at a similar number of segments from the apex (Garbary 1979a; Murray & Dixon 1973, 1975).

Many red algae, however, have become sensitive to daylength as a trigger for seasonal behavior. Table 15-2 summarizes the responses found in laboratory culture. The first group is of species that reproduce in any daylength. Apart from *Porphyra*

Table 15-1. Compensation points (CP), saturating and inhibiting photon irradiances for photosynthesis (PS), and growth in µmol m^{-2} s^{-1}. Where necessary, saturation was calculated from the authors' data as the irradiance allowing 95% of maximum PS or growth using the hyperbolic tangent function model of Jassby & Platt (1976)

Species	CP	Saturation PS	Saturation Growth	Inhibition PS	Inhibition Growth	Reference
Archaeolithothamnion dimotum		100				Vooren 1981
Atractophora hypnoides			15			Maggs & Guiry 1987
Bonnemaisonia asparagoides					50[a]	Jones & Dent 1971
Brongniartella byssoides			6[a]			Boney & Corner 1963
Callithamnion byssoides			40 (14:$\overline{10}$)			Edwards 1971
Callophyllis (as Euthora) cristata		120		500		Mathieson & Norall 1975a
Centroceras clavulatum			70 (14:$\overline{10}$)			Edwards 1969
Ceramium strictum			70 (14:$\overline{10}$)			Edwards 1969
Ceramium tenuicorne		145		>800		Wallentinus 1978
Chondrus crispus		200	80[a]	1000		Mathieson & Burns 1971
						Burns & Mathieson 1972
		120		500		Mathieson & Norall 1975b
			65 (16:$\overline{8}$)			Bird et al. 1979
Clathromorphum circumscriptum	0.6		7 (8:16)			Adey 1970
					13	Adey 1973
Delesseria sanguinea		110				King & Schramm 1976
		50		>200		Lüning 1979
Dumontia contorta		300				King & Schramm 1976
Furcellaria lumbricalis (as F. fastigiata)		150				King & Schramm 1976
Furcellaria lumbricalis			13 (20:$\overline{4}$)			Bird et al. 1979
Gastroclonium coulteri		300		>1400		Hodgson 1981
Gastroclonium subarticulatum (as G. coulteri)		280				Ogata & Matsui 1965a
Gelidium amansii			60a			Ohno 1969
			50 (16:$\overline{8}$)			Correa et al. 1985
Gelidium chilensis			25 (16:8)			Correa et al. 1985
Gelidium lingulatum		400				Merrill & Waaland 1979
Gigartina exasperata		180				Yokohama 1973
Gigartina mamillosa		180				Yokohama 1973
Mastocarpus stellatus (as Gigartina mamillosa)			100 (12:$\overline{12}$)			Bird et al. 1979
Gracilaria textorii		300–470				Lapointe et al. 1984a
Gracilaria tikvahiae		200–600		>1400		Penniman & Mathieson 1985
		140				Yokohama 1973
Grateloupia eliptica		200			50[a]	Jones & Dent 1971
Halarachnion ligulatum		230				Oates 1986
Halosaccion americanum			35[a]			Vooren 1981
Hydrolithon boergesenii			60 (12:$\overline{12}$)[a]			Hannach & Santelices 1985
Iridaea ciliata			0.6 (8:16)			Hannach & Santelices 1985
Iridaea laminarioides			7 (8:$\overline{16}$)			Adey 1970
Leptophytum laeve						
Lithothamnion glaciale						Adey 1970

381

Table 15-1. (cont.)

Species	Saturation			Inhibition		Reference
	CP	PS	Growth	PS	Growth	
Lithophyllum orbiculatum			>7 (8:16)			Adey 1970
Mastocarpus stellatus		200–400		>1000		Mathieson & Burns 1971
			150[a]			Burns & Mathieson 1972
Meredithia microphylla		110			5	Guiry & Maggs 1984
Meristotheca papulosa		90				Yokohama 1973
Palmaria palmata	6.6		100			Robbins 1979
Phycodrys rubens		120				Morgan & Simpson 1981a
		100		>600		Mathieson & Norall 1975a
		50				King & Schramm 1976
		100		>200		Lüning 1979
Phyllophora truncata		120				King & Schramm 1976
				500		Mathieson & Norall 1975a
Phymatolithon polymorphum			12 (14:$\overline{10}$)			Adey 1970
Pleonosporium squarrulosum			20 (16:8)			Murray & Dixon 1975
Plocamium cartilagineum		50			500	Kain 1987a
Plumaria elegans			2.3[a]		9[a]	Yokohama 1973
Polysiphonia boldii			8 (14:$\overline{10}$)			Boney & Corner 1962
Polysiphonia denudata			70 (14:$\overline{10}$)			Edwards 1969
			20 (12:$\overline{12}$)			Edwards 1970
Polysiphonia elongata		37				Kapraun 1978
Polysiphonia ferulacea			20 (12:$\overline{12}$)			Fralick & Mathieson 1975
Polysiphonia lanosa		170				Kapraun 1978
Polysiphonia nigrescens		200				Fralick & Mathieson 1975
		140				Fralick & Mathieson 1975
		200				King & Schramm 1976
Polysiphonia subtilissima			40 (12:$\overline{12}$)			Fralick & Mathieson 1975
Polysiphonia urceolata		150				Kapraun 1978
Porphyra leucosticta	0.6	16				King & Schramm 1976
Porphyra suborbiculata		370				Sheath et al. 1977
Porphyra tenera	6.5	300				Yokohama 1973
			60[a]			Ogata & Matsui 1965a
			34[a]			Ohno 1969
Pterothamnion (as *Antithamnion*) *plumula*						Boney & Corner 1963
Ptilota serrata		120		350		Mathieson & Norall 1975a

[a] Juveniles. (See also Tables 9-1 and 9-2).

Table 15-2. Species of red algae that have been tested for direct effects of daylength

	Hours light		Temperature °C			Reference
	Obs/Opt	Up to	Obs/Opt	Up to	Down to	
No effect on production of						
TETRASPORES						
Rhodochorton concrescens						West 1970
Lomentaria baileyana						Yarish et al. 1984
Halurus equisetifolius						Yarish et al. 1984
Callophyllis laciniata						Yarish et al. 1984
Hypoglossum hypoglossoides						Yarish et al. 1984
Gigartina teedii						Guiry 1984a
Lomentaria articulata						Yarish et al. 1986
Callithamnion tetragonum						Yarish et al. 1986
Lomentaria orcadensis						Yarish et al. 1986
TETRASPORES AND GAMETES						
Callithamnion byssoides						Edwards 1971
Grinnellia americana						Yarish et al. 1984
Agardhiella subulata						Yarish et al. 1984
Polyneura hilliae						Yarish et al. 1986
GAMETES AND MONOSPORES						
Porphyra rosengurttii						Kapraun & Luster 1980
MONOSPORES						
Acrochaetium proskaueri						West 1972a
Long days stimulate production of						
TETRASPORES						
Acrochaetium proskaueri	16		15			West 1972a
Callithamnion hookeri	16		10/15			Whittick 1981
Audouinella alariae	16		10/15			Lee & Kurogi 1983
GAMETES						
Callithamnion hookeri	16		5–15			Whittick 1981
Atractophora hypnoides	11+		15			Maggs 1987
MONOSPORES						
Bangia atropurpurea	16	14				Richardson 1970
UPRIGHT THALLI						
Farlowia mollis (as *F. compressa*)	16		15			DeCew & West 1981
Farlowia conferta	16		15			DeCew & West 1981
Farlowia mollis	12–16		10			DeCew & West 1981
Short days stimulate production of						
TETRASPORES OR CONCHOSPORES						
Gigartina acicularis	8	16+			13	Guiry 1984b
Acrosymphyton purpuriferum	8	16+		16	12	Cortel-Breeman & ten Hoopen 1978

Table 15-2. (cont.)

	Hours light		Temperature °C			
	Obs/Opt	Up to	Obs/Opt	Up to	Down to	Reference
Halymenia latifolia	10	16	10/15			Maggs & Guiry 1982
Audouinella purpurea – California	9	10	10	15+		West 1972b
– Washington	10	12	10	<15		Dring & West 1983
– Alaska	11	14	10	<15		
Bangia atropurpurea	9	12.5				Richardson 1970
Porphyra rosengurttii	8	12	10–25			Kapraun & Luster 1980
Porphyra leucosticta	8	12	23			Edwards 1969
Meredithia microphylla	8	12	15	<18	>10	Guiry & Maggs 1984
Audouinella asparagopsis		10	11			Abdel-Rahman & Magne 1981
Bonnemaisonia hamifera	8	11	15/17	<20	>12	Lüning 1981
Acrochaetium pectinatum		10	5–15			West 1968
Porphyra tenera	8	10				Dring 1967
Thuretellopsis peggiana	9					Richardson & Dixon 1970
Gloiosiphonia capillaris	8					DeCew et al. 1981
Calosiphonia vermicularis	8		20			Mayhoub et al. 1976
Farlowia mollis	8		15			DeCew & West 1981
Farlowia conferta	8		15			DeCew & West 1981
Atractophora hypnoides	8		10	15		Maggs & Guiry 1987
Mastocarpus stellatus	8		10	<15		Guiry & West 1983
Asparagopsis armata	8		15	<20	>10	Oza 1977; Lüning 1981
GAMETES						
Cordylecladia erecta	10	12	15	>17	10	Brodie & Guiry 1987
Audouinella purpurea – California	10	24	10/15			West 1972b
Audouinella alariae	8		5/10			Lee & Kurogi 1983
Delesseria sanguinea	8	12	12			Kain 1987b
Gigartina acicularis	8	12		18	14	Guiry & Cunningham 1984
UPRIGHT THALLI OR BLADES						
Constantinea subulifera	0+	14				Powell 1964
Meredithia microphylla	8	12	15	<18	>10	Guiry & Maggs 1984
Dumontia contorta	4+	13		16	8	Rietema 1982; Rietema & van den Hoek 1984
Bonnemaisonia asparagoides	8	12	15/18		8.5	Rueness & Åsen 1982
Schmitzia hiscockiana	8		10/15			Maggs & Guiry 1985
Constantinea rosa-marina	8		10			Lindstrom 1980

Hours light: in 24 hours. Obs/opt: hours of light per day producing or optimal for observed response. Up to: maximum hours of light per day allowing response. Similarly for temperature.

rosengurttii, to which this insensitivity only applies in the haploid phase, these algae are either not geared to the seasons, even at high latitudes, or are triggered by some other factor, such as temperature. The second group of species becomes fertile in long days. *Porphyra* and *Bangia* diploid monospores reproduce the conchocelis phase throughout the summer months. *Callithamnion hookeri* produces spores in summer at the northern end of the range (Whittick 1981).

At higher latitudes winter light usually limits growth, and wave action results in losses. This is thus the season when there is most bare rock available for colonization. It is not surprising, therefore, that the third group in Table 15-2, consisting of plants reproducing in response to short days, is the largest. These species, mainly from fairly high latitudes, release their propagules, mostly tetraspores, when competition for space is at its minimum. In all taxa reproduction involves a change of ploidy, and in over half the species a change of morphology. In some cases, where photoperiod is combined with temperature constraints, there appears to be only a small window in the seasonal sequence for reproduction to take place, e.g., in *Bonnemaisonia hamifera* (Lüning 1981).

The threshold irradiance detected in daylength response was found in two separate species to be about 0.2 µmol $m^{-2} s^{-1}$ (Breeman & ten Hoopen 1981; Dring & West 1983). The time for which the irradiance in the shallow subtidal is above 0.5 µmol $m^{-2} s^{-1}$ can be as much as 3 h shorter than astronomical daylength in midwinter in Jerlov water type 9 (Dring 1984), but in water type 3 the measured difference is negligible (Kain 1987b). The height of the tide can combine with shading by brown algae to produce artificial short days on the shore (Breeman et al. 1984). Thus, there may be different daylength effects in different environments simultaneously at one latitude.

D. Temperature

As an important determinant of plant performance, temperature is a critical ecological factor. Its effect on respiratory and dark reaction photosynthetic enzyme activity is well known, and their relationship is important in determining compensation points and maximum photosynthetic rates. The former can rise dramatically with temperature even in acclimated plants (*Clathromorphum circumscriptum*: Adey 1973). The classic increase in light-saturated photosynthesis with temperature is apparent in a wide range of red algae, e.g., *Chondrus crispus* and *Mastocarpus stellatus* (Mathieson & Burns 1971), *Phyllophora truncata* (as *P. brodiaei*), *Callophyllis* (as *Euthora*) *cristata* (Mathieson & Norall 1975a), *Gastroclonium subarticulatum* (as *G. coulteri*) (Hodgson 1981), and *Gracilaria tikvahiae* (Penniman & Mathieson 1985). In using laboratory data the importance of adaptation must be realized. An example, for *Phycodrys rubens*, is shown in Fig. 15-3, in which net photosynthesis of winter plants was higher at low temperatures and lower at high. Similar patterns are apparent for *Callophyllis cristata* and *Chondrus crispus* (Mathieson & Norall 1975a,b, respectively). In *Gracilaria tikvahiae*, however, there is little difference between summer and winter photosynthetic rate when expressed per unit of chlorophyll (Penniman & Mathieson 1985).

Growth rates of red algae are influenced greatly by temperature, particularly in saturating irradiance. While the optimum temperature clearly varies geographically, the general pattern is usually of an

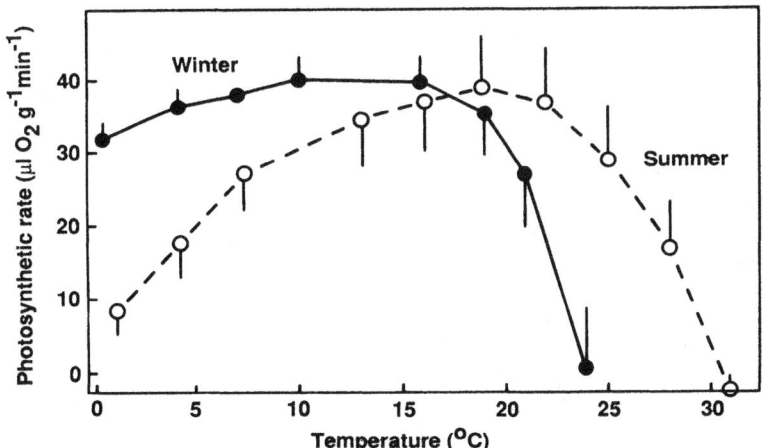

Fig. 15-3. Net photosynthetic rates of *Phycodrys rubens* collected in winter and summer at various temperatures redrawn from Mathieson & Norall (1975a), courtesy of Elsevier Science Publishers B.V.

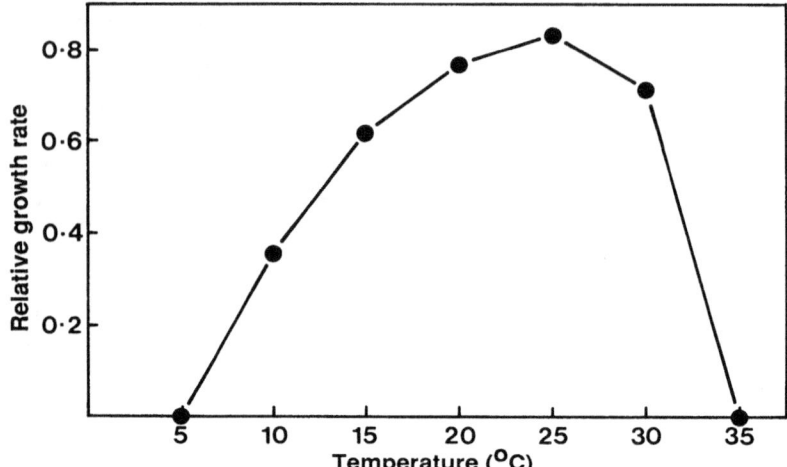

Fig. 15-4. Relative growth rate (day^{-1}) in cell number of tetrasporelings of *Grinnellia americana* in long days at 40 µmol m^{-2} s^{-1} at various temperatures. Recalculated and replotted from Yarish et al. (1984).

increase in growth rate to a maximum that is near the top of the tolerated range, as in *Grinnellia americana*, shown in Fig. 15-4. Similar patterns are shown in *Gelidium amansii* (Ohno 1969), *Callithamnion hookeri* (Edwards 1979), *Chondrus crispus* (Simpson & Shacklock 1979), *Lomentaria baileyana*, and *Hypoglossum hypoglossoides* (as *H. woodwardii*) (Yarish et al. 1984). This may be associated with the occasional observation that the optimum temperature in the laboratory is higher than that encountered in the field, e.g., growth of *Stylonema alsidii* (as *Goniotrichum elegans*), *Nemalion helminthoides* (as *N. multifidum*), and *Polysiphonia urceolata* (Fries 1966a), photosynthesis but not growth of *Phyllophora antarctica* (Ohno 1976). In some species there is a dramatic change in relative growth rate with temperature, e.g., a 20-fold increase between 10° and 20°C in *Gracilaria tikvahiae* (Bird et al. 1979). A widely distributed species like *Palmaria palmata*, however, reacts less strongly to temperature differences (Morgan & Simpson 1981a).

When haploid and diploid phases of red algae have been tested separately for temperature responses, many have shown no difference in growth; examples include *Polysiphonia boldii* (Edwards 1969), *P. denudata* (Edwards 1970), *Callithamnion byssoides* (Edwards 1971), and *Grinnellia americana* and *Lomentaria baileyana* (Yarish et al. 1984). Slight differences between phases have appeared in the growth of *Iridaea laminarioides*, *I. ciliata* (Hannach & Santelices 1985), and *Polyneura hilliae* (Yarish et al. 1986) and in photosynthesis of *Ptilota serrata* (Mathieson & Norall 1975a).

Temperature can affect morphology; filaments of *Audouinella* species branch less and the cells are shorter and wider at higher temperatures (Garbary 1979b).

It is widely accepted that temperature is the principal controlling factor in geographical distribution (see Section III.A). One important criterion in this is the tolerance of extremes. Many temperate intertidal species can withstand freezing (Biebl 1972), which only involves 74% of the water of *Chondrus crispus* even at −15°C (Kanwisher 1957). Survival is aided by desiccation before freezing in *Porphyra tenera*, half the cells of which survive at −60°C (Migita 1966), a hardiness that is harnessed for the storage of cultivated *Porphyra* (Miura 1975). The conchocelis phase of this species is less tolerant, one month at −20°C being lethal (Migita 1967). The greatest frost tolerance is shown by *P. yezoensis*, withstanding −196°C for 24 h (Terumoto 1965). Subtidal species, on the other hand, are unable to withstand freezing, although −2°C is tolerated in liquid seawater (Biebl 1972). Clearly, the lowest temperature tolerated varies with geographical environment, and as much as 14°C may be lethal over 12 hours (Biebl 1962). Within a species there may be widely varied tolerances in different geographical areas [*Asparagopsis armata* (as *Falkenbergia rufolanosa*) and *Laurencia obtusa*; Biebl 1962] or there may be little or no difference (*Dumontia contorta*: Rietema & van den Hoek 1984). Within a geographically widespread genus there can be broad limits (*Gracilaria*: McLachlan & Bird 1984).

The upper temperature limit is less extreme in subtidal than in intertidal species and lower (28–32°C) in eight red algae from deeper water than it is (36°C) in three red algae from shallow water off California (Schölm 1966). Intertidal *Bangia atropurpurea* is more tolerant partly dry than wet (Biebl 1972). *Chondrus crispus* will not grow at 26°C, and spores are killed by a 6-min exposure at 35°C (Prince & Kingsbury 1973). Another type of stress, lowered salinity, reduces the upper temperature tolerance of *Delesseria sanguinea* and *Phycodrys rubens* (Schwenke 1959). Twelve-hour experiments give an upper limit of 40°C for three Puerto Rican species (Biebl 1962), whereas the upper temperature tolerated for a period of a week varies between 18°C and 30°C in 23 species of temperate red algae from both inter- and subtidal habitats (Lüning 1984).

Temperature tolerances or optima are important in regulating local distributions of a number of red algal species (Fralick & Mathieson 1975; Yarish & Edwards 1982). Temperature has been suggested as the most important factor controlling their seasonal variations at fairly low latitudes (Edwards 1969; Kapraun 1979; Yarish & Edwards 1982). However, for subtidal species in general there is no evidence that temperature is an important determinant in seasonality (Kain 1989). At higher latitudes irradiance is likely to assume greater importance in winter (Section II.B, Fig. 15-1).

Temperature can act as a trigger in seasonal behavior. Examples where this is combined with daylength can be seen in Table 15-2. Independently of daylength, however, parasporangia of *Plumaria elegans* develop only at 10°C or above (Whittick 1977), and gametophytes of *Acrosymphyton purpuriferum* mature at 17°C and above (Breeman & ten Hoopen 1981).

E. Salinity

There are several ways in which salinity changes may be important to marine algae. There may be osmotic stress, unfavorable ionic balances, and/or a shortage of essential metabolites. The physiological aspects are dealt with in Chapter 7; of ecological importance is the content of the water with which seawater is diluted. The maintenance of the apparently necessary high level of potassium in algal cells, relative to seawater, depends on the presence of adequate quantitites of calcium ions (Dawes & McIntosh 1981; Eppley & Cyrus 1960; Lehnberg 1978; Reed 1984; Schwenke 1958; Yarish et al. 1980). The calcium content is also critical for calcification in algae like *Phymatolithon calcareum* (King & Schramm 1982). The carbon content of the diluting water, associated with its pH, has a marked effect on photosynthesis at low salinities (Dawes & McIntosh 1981; Hammer 1968; Ogata & Matsui 1965b). Fluctuations in salinity result in changes in cellular ionic levels, though with a 12-h cycle the reaction is damped and incomplete compared with a longer time (Reed 1981).

When red algae are transferred from full to slightly reduced salinity water, the photosynthetic rate may be temporarily lowered in *Porphyra purpurea* (Reed et al. 1980) and *Gracilaria tikvahiae* (Lapointe et al. 1984b) but raised in *Delesseria sanguinea* (Nellen 1966). Slight hypersalinity may stimulate photosynthesis in *Porphyra tenera* (Ogata & Matsui 1965b), *P. purpurea* (Reed et al. 1980), and *Delesseria sanguinea* (Nellen 1966). Higher salinity, of several times normal, results in tissue dehydration, which inhibits at least three processes in the photosynthetic system (Satoh et al. 1983).

The optimum salinities, with some indication of the range tolerated, for photosynthesis and for growth of some red algal species are shown in Table 15-3. Euryhaline species are ranked at the top, with the most stenohaline at the bottom. Clearly, the growth values are of more interest ecologically. In various species of *Gracilaria* peak growth occurs at the lower end of the range tolerated (Bird & McLachlan 1986), contrasting with the photosynthetic reaction to temperature (McLachlan & Bird 1984).

Spores of *Porphyra schizophylla* can survive salt crystal formation but not salinities of less than half seawater (Boney 1978a). The percentage survival of spores of *Callithammion hookeri* (Edwards 1979), *Chondrus crispus*, and *Mastocarpus* (as *Gigartina*) *stellatus* (Burns & Mathieson 1972) follows the same pattern as that of sporeling size after 7 days' growth (e.g., Fig. 15-5).

At high and/or low salinities rates of respiration decrease in *Porphyra tenera* (Ogata 1963) *Ceramium* sp., *Gloiopeltis* spp. and *Gracilaria verrucosa* (Ogata & Takada 1968) but increase in *Odonthalia floccosa* (Kjeldsen & Phinney 1972) and *Eucheuma isiforme* (as *E. nudum*) (Mathieson & Dawes 1974).

Although the foregoing observations indicate that the plants would be under varying degrees of stress when subjected to various salinities in nature, other studies have looked at the actual survival (usually for about a day) of different species. Biebl (1952) claimed that subtidal algae could tolerate salinities of 18 to 52‰, algae from around low water 15 to 70‰, and intertidal algae 3 to 100‰. This is clearly only partly true. The subtidal *Delesseria sanguinea* survived a day at 4‰, and *Phycodrys rubens* a day at 6‰ (Schwenke 1959), and of the euryhaline species in Table 15-3 several, notably *Gracilaria tikvahiae*, are subtidal.

Whether or not laboratory observations fit distributional data has been of interest in the study of estuaries, lochs or fjords, lagoons, and larger water bodies, such as the Baltic Sea. Laboratory tolerances wider than habitat variations have been shown for *Odonthalia floccosa* (Kjeldsen & Phinney 1972), *Chondrus crispus*, and *Mastocarpus stellatus* (Hardwick-Witman & Mathieson 1983), whereas three species

Table 15-3. The salinites (‰ S) preventing (nil), allowing 50% of and optimal for photosynthesis and growth of various red algal species.

Species	Photosynthesis				Growth				Reference
	Nil	50%	Opt	50%	Nil	50%	Opt	50%	Nil	
Bostrychia binderi	—		5–30							Dawes & McIntosh 1981
Caloglossa leprieurii		<5	15–25	>35		0	5–25	35		Yarish & Edwards 1982
Gracilaria tikvahiae		<5	15–35	>40						Yarish et al. 1979
					5	14	17–27	31	60	Penniman & Mathieson 1985
Porphyra tenera					—	9	17–35	61	87	Bird & McLachlan 1986
	—	17	52	>70						Ohno 1969
Gelidium amansii	—	9	35	>70						Ogata & Matsui 1965b
					—	17	35	70		Ogata & Matsui 1965b
										Ohno 1969
Bostrychia radicans		5	25	35	0	5	25	35		Yarish et al. 1979
Porphyra umbilicalis		5	17–35	70			17–35	70		Yarish & Edwards 1982
Polysiphonia subtilissima	0	5	10–35	>40						Ogata & Schramm 1971
Polysiphonia lanosa		5	15	35						Fralick & Mathieson 1975
Porphyra leucosticta		5	30	40						Yarish & Edwards 1982
Polysiphonia elongata		10	25–37	42						Fralick & Mathieson 1975
		15	20–40							Zavodnik 1975
Chondrus crispus					10	15	25–40	45	50	Fralick & Mathieson 1975
Mastocarpus stellatus					10	15	35	45	55	Burns & Mathieson 1972
Polysiphonia nigrescens	0	15	20–35	>40						Burns & Mathieson 1972
Callithamnion hookeri										Fralick & Mathieson 1975
					10	23	34	50	70	Edwards 1979
Dasya baillouviana (as *D. pedicellata*)						<5	20			Nygren 1970
Odonthalia floccosa		7	20	>35						Kjeldsen & Phinney 1972
Meristiella (as *Eucheuma*) *gelidium*	5	20	30	34	>50					Mathieson & Dawes 1974
Eucheuma isiforme	10	20	40	50						Mathieson & Dawes 1974
Wrangelia penicillata	20	25	37							Zavodnik 1975

—: photosynthesis took place at 0‰ S. Approximately ranked with the most euryhaline species first.

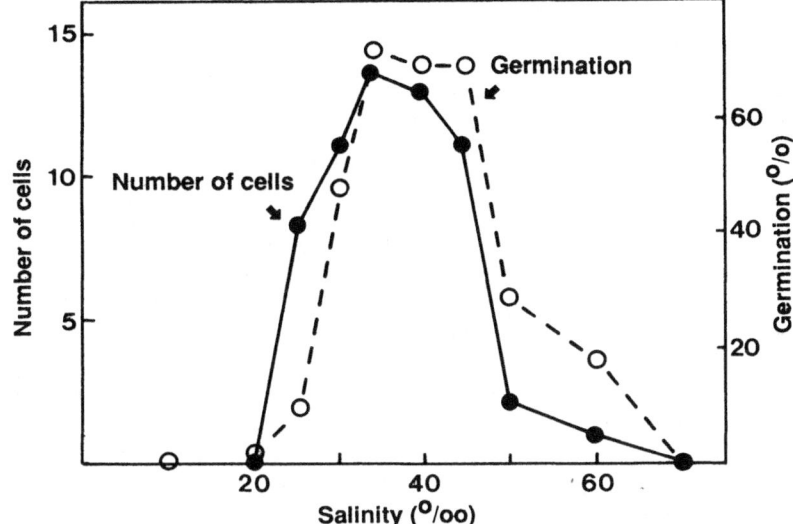

Fig. 15-5. Carpospore percentage germination and cell number of carposporelings after 7 days in *Callithamnion hookeri*, at various salinities. Redrawn from Edwards (1979) in *Phycologia*, courtesy of Blackwell Scientific Publications.

showed abundances broadly in agreement with their salinity optima (Yarish et al. 1979), but their inner limit was probably determined by calcium and potassium levels as well as salinity per se (Yarish et al. 1980).

Passing down a salinity gradient, the number of species of Rhodophyta declines sooner than that of the Phaeophyta (Munda 1978), whereas that of the Chlorophyta may actually increase (Coutinho & Seeliger 1984). This decline in red algal species in reduced salinity is illustrated in Fig. 15-6, which includes plots from areas from the tropics to the subarctic. A salinity of about 20‰ seems to be critical; below it the number of species drops dramatically. Similarly, whereas on the open coast of Norway there are 96 species of red algae (Jorde 1966), at the head of the >100-km-long Hardangerfjord this is reduced to 19 (Jorde & Klavestad 1963) and there is a similar number, 18, at the entrance to the Gulf of Bothnia in the Baltic (Waern 1952).

One might expect evidence of selection in process when a range of salinities is tolerated by a species. This is the case in *Polysiphonia lanosa*, where plants taken from estuarine and marine environments show differential viability at 7‰ (Reed 1983, 1984), but not in *Delesseria sanguinea* (Lehnberg

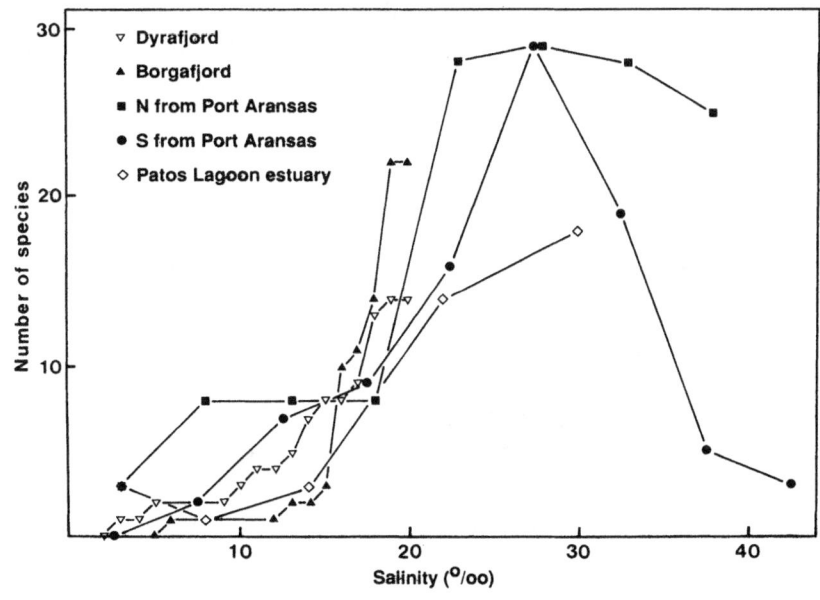

Fig. 15-6. The number of species of red algae recorded along salinity gradients. Dyrafjord and Borgafjord, southeast Iceland, replotted from Munda (1978); northward from Port Aransas, Texas, from Edwards & Kapraun (1973); southwards from Port Aransas, Texas, from Conover (1964); Patos Lagoon estuary, Brazil, from Coutinho & Seeliger (1984).

1978). Marine *Bangia atropurpurea* (as *B. fuscopurpurea*) can be grown in fresh water after passing through several generations (den Hartog 1972), and in a similar way freshwater *B. atropurpurea* can acclimate to seawater (Sheath & Cole 1980).

F. Desiccation

All intertidal red algae that have been tested grow best when permanently submerged, irrespective of how high on the shore they normally grow (Edwards 1977a). Algae stranded by the retreating tide are subjected to aerial conditions until the tide returns. In the case of upper shore dwellers like *Endocladia muricata* periods of aerial exposure may last for many hours, even days.

The severity of damage suffered is related to the duration of exposure. In *Plumaria elegans* drying naturally on the shore in summer, branch tip cells were killed within 4 h and all the outer branches were dead within 6 h; the cells of the main axes were not affected until 8 h had elapsed (Boney 1969, p. 159). Although the vulnerable tips bear most of the growing points, many red algae possess considerable powers of regeneration from the axes or the holdfast. Hence, a single episode of drought may not prove fatal, but a series of exposures may progressively destroy the plants.

Different species of algae seem to be differentially susceptible to desiccation stress. Among the most sensitive are the calcareous lithothamnia, plants growing in the splash zone at Funafuti atoll being severely damaged by a 1-h exposure on a calm day (Finckh 1904). On the coast of California the hot, dry Santa Ana winds can cause heavy mortality of coralline algae (Seapy & Littler 1982). Biebl (1938) demonstrated that a variety of subtidal red algae can survive aerial exposure for 14 h only if the air is close to 100% saturated, whereas intertidal red algae can survive for the same length of time at relative humidities of around 84%. The effects of desiccation are exacerbated at higher air temperatures (Ohno 1969).

Spores and developing germlings are particularly susceptible to drought as they have very large surface-to-volume ratios, although they benefit from the film of water that persists in concavities on the substratum. Germlings of different species of algae may respond differently to desiccation. Boney (1969, p. 154) found that pool dwellers and subtidal species, such as *Ceramium pedicillatum* and *Pterothamnion* (as *Antithamnion*) *plumula*, were less tolerant than intertidal species, such as *Polysiphonia lanosa*, *Ceramium flabelligerum*, and *Plumaria elegans*. The relative tolerance of these species was similar whether they were dried at elevated air temperatures, at subzero temperatures, or in different velocities of air in a wind tunnel.

Thalli of different species undoubtedly lose water at different rates even when dried under identical conditions (Fig. 15-7). Boney (1969) attributed these differences to the habitats from which the plants had been taken. The two fastest-drying species in Fig. 15-7 are pool dwellers, whereas the others inhabit exposed rock, most of them on or under larger brown seaweeds, but *Porphyra* and *Nemalion* often on open rock.

The anatomy and growth form of the plants must also influence the rate of water loss. The fastest-drying plants in Fig. 15-7 are finely divided thin filaments lacking cortication. The slower-drying plants are bulkier, have lower surface area-to-volume ratios

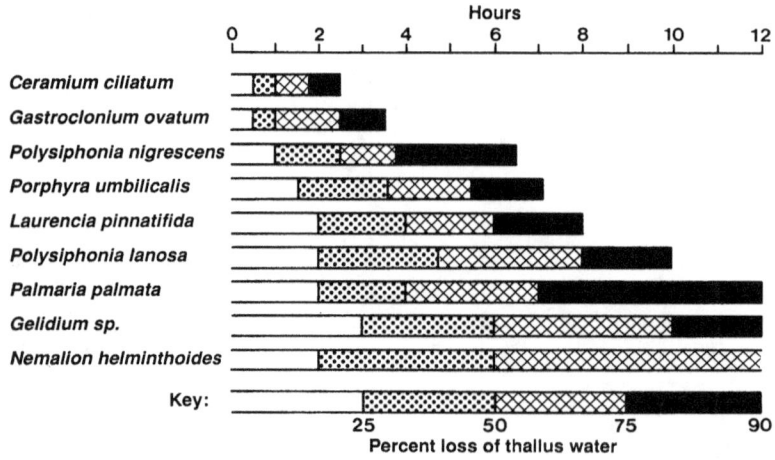

Fig. 15-7. Loss of water from thalli of various intertidal red algae in dry air at 25°C. The times at which 25, 50, 75, and 90% of the total water content of the thallus was lost are indicated. Modified from Boney (1969).

and/or layers of small cortical cells. Perhaps surprisingly, the flux of water vapor from flat laminae, such as those of *Porphyra* and *Palmaria*, is less than from cylinders constructed from similar material (Schonbeck & Norton 1979). Membranous algae also shrink very substantially as they dry, thus reducing the size of the evaporative surface. Drying thalli of *Iridaea splendens* (as *I. flaccida*) can halve their area (Abbott 1972).

The relevance of mucilage content to water loss is unknown, but it is noteworthy that both *Porphyra* and *Nemalion* lose water slowly (Fig. 15-7) and are somewhat mucilaginous. Even if mucilage does not curb moisture loss, it may increase the initial water content of the plant or accelerate hydration on resubmergence.

The degree of tolerance to desiccation seems to determine the upper limits of several species of red algae on the shore (Druehl & Green 1982; Hodgson 1981; Hruby & Norton 1979). Water trickling down the shore enables plants of *Chondrus crispus*, *Corallina officinalis*, and *Mastocarpus stellatus* to range higher upshore than they do on dry rock (Hawkins & Hartnoll 1985). Wave splash has a similar effect (see Section III.B).

Some plants avoid the most severe desiccation by growing in dense humid stands (Hay 1981; Taylor 1985; Taylor & Hay 1984). Plants of *Corallina vancouveriensis* and no doubt many other species may trap large amounts of water between their branches (Padilla 1984). Many intertidal coralline algae are obligate understory species inhabiting the humid spaces between aggregated sea anemones (Taylor 1985) or beneath a canopy of seaweed (Johnson & Mann 1986). Such a canopy also enables subtidal species to venture into the intertidal zone (Dayton 1975a). The vulnerable conchocelis phase of *Porphyra* bores into the shells of intertidal barnacles, which must confer some protection from desiccation, but even so, severe drought can prove fatal (Saito 1956).

For most red algae slight drying causes a small stimulation of photosynthesis, but increasing desiccation causes a dramatic decline (Hodgson 1981, 1984; Oates 1986; Ogata & Matsui 1965a; Quadir et al. 1979). Growth rate decreases as a result of even mild desiccation in the subtidal species *Gracilaria tikvahiae*, and repeated exposures are even more deleterious than a single more severe episode (Hodgson 1984). This species can recover from a 50% water loss (Hodgson 1984), as can *Gastroclonium subarticulatum* (as *G. coulteri*) (Hodgson 1981). *Chondrus crispus* recovers fully after being 65% dehydrated; *Mastocarpus stellatus*, after 88% dehydration (Mathieson & Burns 1971). High shore dwellers like *Bangia atropurpurea* can tolerate being air dried on the shore for 15 consecutive days and revive rapidly on resubmergence (Feldmann 1951).

G. Water movement

Some species of red algae thrive in relatively calm-water sites, others in more turbulent localities (Conover 1968). Moderate water movement is beneficial to seaweeds. It carries a supply of nutrients and gases to the plants, removes waste products, and prevents the settling of silt. Because the coefficient of molecular diffusion in seawater is very low (Sverdrup et al. 1942), seaweeds in still water rapidly deplete the nutrients in their immediate vicinity. In contrast, plants growing in moving water have a steady supply of nutrients. Therefore, it is not surprising that many plants grow faster in agitated or flowing cultures than in stationary media, for example, *Polysiphonia nigrescens* (Whitford & Kim 1966), *Devaleraea* (as *Halosaccion*) *ramentacea* (Munda 1977), *Iridaea* spp. (Hannach & Santelices 1985), *Gracilaria tikvahiae* (Parker 1982), and *Pterocladia capillacea* (Santelices 1978). With *Porphyra tenera*, a given rate of growth can be accelerated by increasing either nutrient concentration in the medium or flow velocity (Matsumoto 1959).

In *Gracilaria verrucosa*, and probably in many other species, the downstream deflection of the thallus caused by flowing water increases the plant's rate of growth by presenting the thallus perpendicular to the incident light (Jones 1959a). As a result of these phenomena, many plants achieve a remarkable stature when growing in steady tidal streams. *Porphyra umbilicalis* can reach 40 cm, and *Palmaria palmata* grows to a meter (Jorde 1966).

Excessive water motion has deleterious effects. In shallow water waves passing overhead or breaking near plants subject algae to a violent oscillating flow. Such stresses are likely to tear or dislodge the plants, and they impose the main restraint on the size and shape of seaweed thalli. In a repeatedly accelerating fluid the stresses suffered by an organism increase with its length (Denny et al. 1985); larger plants are therefore more at risk. Consequently, plants of *Corallina officinalis*, *Chondrus crispus*, and *Callithamnion* spp. appear to be stunted when growing on relatively exposed shores (Dommasnes 1968; Mathieson & Prince 1973; Price 1978). There is also a striking reduction in the length of upright filaments of *Plocamium cartilagineum* in shallower water as a result of wave-induced damage (Kain 1987a). Variations in branching may also be

Table 15-4. Morphological variation in plants of Mastocarpus stellatus (means ± 95% confidence limits) from sites on the Isle of Cumbrae, Scotland[a]

Morphological feature (mm)	Site		
	S. W. Cumbrae, Exposed	Tomont End, Moderately exposed	Marine Laboratory, Relatively Sheltered
Length	34.0 ± 3.60	38.1 ± 2.14	65.0 ± 5.41
Maximum width	15.3 ± 2.20	18.3 ± 2.80	34.6 ± 4.58
Stipe width	2.5 ± 0.34	4.3 ± 3.90	7.0 ± 0.97

[a] Unpublished results.

induced by water motion rather than caused by damage (Munda 1977; Steneck & Adey 1976). Table 15-4 illustrates similar variations.

Red algae tolerate wave action in various ways. Low crusts inhabit the more sluggish water close to the substratum and therefore avoid the swifter flows. Small plants, even if erect, may find shelter behind wave-deflecting features of the topography. Overlying or adjacent seaweeds may also baffle the impact of waves. Even a single plant of *Gelidium nudifrons* sieves the water that passes between its branches, supressing ambient turbulence (Anderson & Charters 1982). *G. nudifrons* has a stiff latticework of angular, wiry branches quite unlike most red algae. Wave-tolerant algae like *Porphyra umbilicalis* give no impression of strength; rather, they are delicate and flimsy. They yield to the wave. Pliable thalli are readily deflected downstream parallel to the direction of flow, vastly reducing the drag. This passive streamlining is aided in flimsy laciniate or much-branched thalli, which can fold and conform to the flow. Deflection also brings the thallus close to the slower-moving water near the substratum. The drag on plants of *Gigartina exasperata* "blown" over close to the substratum is half that of the same plants subjected to the same velocity of flow away from the substratum (Koehl 1986).

The surface of many red algae is mucilaginous, which might reduce skin friction drag. However, skin friction is probably of little significance in comparison with form drag (Koehl 1986). It may be that a slippery surface could reduce abrasion damage by lubricating the passage across the rock of a thallus flailing in the surf.

Among the few upright stiff red algae are calcareous species like *Calliarthron* and *Corallina*. Here, the risk of breakage is minimized by having easily deformed articulations between the calcified segments, thus enabling the plant to deflect (Koehl 1986).

H. Nutrients

As Dring (1982) has pointed out, it is difficult or impossible to determine which of the major elements contained in seawater are essential to marine algae. Most seawater culture media are complicated by the addition of chelating agents that reduce free ions of trace metals; thus, the optimum zinc concentration found for the growth of *Porphyra tenera* in variations of ASP (Iwasaki 1967; Noda & Horiguchi 1972) is four times the mean concentration in seawater, because these media contain chelators. Under these conditions stimulation of growth has been demonstrated in *Porphyra tenera* with copper (Shimo & Nakatani 1967) and *Stylonema alsidii* (as *Goniotrichum elegans*) with selenium (Fries 1982). Remarkably, *Polysiphonia urceolata* requires iodine (Fries 1966b) and bromine (Fries 1975), and the former stimulates growth of *Asparagopsis* (von Stosch 1964). Ion uptake is faster in the light with cobalt (Bunt 1970) and zinc (Gutnecht 1963). A high sulfur requirement might be expected in red algae, with their sulfated polysaccharides. The sulfate uptake rate in *Chondrus crispus* seems to be faster when growth is rapid (Tveter-Gallagher et al. 1981). Iron has not been implicated as a limiting nutrient in the culture of this species (Neish et al. 1977).

When red algae were obtained in bacteria-free culture, many of them were found to require vitamin B_{12} (cobalamin), e.g., *Stylonema alsidii* (as *Goniotrichum elegans*) (Fries 1960), *Nemalion helminthoides* (as *N. multifidum*) (Fries 1961), *Polysiphonia urceolata*, *Erythrotrichia carnea* (Fries 1964), *Bangia atropurpurea* (Provasoli 1964), three species of *Antithamnion* or *Antithamniella* (Tatewaki & Provasoli 1964), *Porphyra tenera* (Iwasaki 1965), *Chroodactylon ramosum* (as *Asterocytis ramosa*) (Fries & Pettersson 1968), and *Rhodella maculata* (Turner 1970). *Nemalion helminthoides* has been shown to be stimulated by hydrolyzed casein (Fries 1970) and by D vitamins (Fries 1984). No requirement has so far appeared for thiamine or biotin.

Of the major nutrients, carbon is the most important. It seems that red algae, along with other seaweeds, use bicarbonate (Bidwell & McLachlan 1985; Jolliffe & Tregunna 1970; Sand-Jensen & Gordon 1984), the concentration of which in full-salinity seawater is rarely likely to limit photosynthesis (Sand-Jensen & Gordon 1984), except in intensive cultivation (*Chondrus crispus*, Bidwell et al. 1985) (see Chapter 8).

Although nitrogen and phosphorus vary in the sea together, it is nitrogen that is usually implicated as being the limiting nutrient (Ryther & Dunstan 1971). Of the inorganic sources, although nitrate is more common, it has to be reduced by the algae before further metabolism can take place. Perhaps for this reason, the ammonium ion is taken up and used in preference by species such as *Agardhiella subulata* (as *Neoagardhiella baileyi*) and *Gracilaria foliifera* (DeBoer et al. 1978; D'Elia & DeBoer 1978; Lapointe & Ryther 1979; Ryther et al. 1981), *Hypnea musciformis* (Haines & Wheeler 1978), *Ceramium tenuicorne*, *Phyllophora truncata*, *Furcellaria lumbricalis* (Wallentinus 1984), and *Porphyra perforata* (Thomas & Harrison 1985). This is not the case in *Gelidium nudifrons* (Bird 1976). The direct use of ammonium may be important close to animal populations from which the ion derives. For the same reason, the use of urea could be advantageous in *Porphyra tenera* (Iwasaki 1967), *Pterocladia capillacea* (Nasr et al. 1968), *Agardhiella subulata*, and *Gracilaria tikvahiae* (DeBoer et al. 1978), but it does not seem to be available to *Chondrus crispus* (Neish et al. 1977).

Recycling of nutrients by animals and/or bacterial colonies must lead to patchiness, both spatial and temporal. In this case the ability of *Gracilaria* to take up nitrogen and phosphorus faster than it can use them for growth must also be advantageous. This has been demonstrated in cultivation experiments where pulse fertilization is applied to plants of *G. verrucosa* (Dawes et al. 1984) and *G. tikvahiae* (Friedlander & Dawes 1985; Lapointe 1985). Cultivated in natural daylight, this species accumulates nitrogen reserves in winter (Rosenberg & Ramus 1982), phycoerythrin pigments even being implicated as effective storage substances (Bird et al. 1982). In *Chondrus crispus*, also, there are marked seasonal variations in tissue nitrate (Asare & Harlin 1983) and L-citrullinyl-L-arginine (Laycock et al. 1981). In *Palmaria palmata* luxury nitrogen accumulates in blades when growth does not keep up with uptake (Morgan et al. 1980; Morgan & Simpson 1981b,c). Such storage could be particularly important at high latitudes, where nutrient availability and light are favorable at different times of year, even though light may be necessary for or stimulatory to nitrate or even ammonium uptake (Bird et al. 1982; D'Elia & DeBoer 1978). Nutrient uptake in situ is intimately connected with water movement (see Section II.G) (see also Chapter 7).

Another important factor is surface/volume ratio. Uptake rates are faster in filamentous species like *Ceramium tenuicorne* than in those with more fleshy thalli (Wallentinus 1984). The development of hairs clearly increases the surface area of an alga. These have been observed to develop under some conditions but not others in *Acrochaetium proskaueri* (West 1972a) and *Gelidium caulacanthum* (Dromgoole & Booth 1985), and in *Ceramium rubrum* their development is clearly a reaction to nutrient deficiency (DeBoer & Whoriskey 1983).

Direct information on the role of nutrients in the field is scarce. Inorganic nitrogen uptake rates of *Gracilaria verrucosa* are stimulated by slight desiccation in high-shore but not in low-shore plants (Thomas 1982). Autumnal decline in *Iridaea splendens* (as *I. cordata*) may be delayed by high ammonium concentrations resulting from a local seal population (Hansen & Doyle 1976).

I. Grazing

Red algae represent a major source of food for a variety of herbivores, including mollusks (e.g., Carefoot 1967; Dayton 1971; Kitting 1980; Shepherd 1973; Steneck 1982), crustaceans (e.g., Brawley & Adey 1981; Carefoot 1973; Nicotri 1980), sea urchins (e.g., Ayling 1981; Larson et al. 1980; Lawrence 1975; Vadas 1977; Verlaque 1984), and fish (e.g., Horn et al. 1982, 1985; Lundberg & Liplin 1979; Montgomery 1980). Grazers may truncate the vertical distribution of some red algae (Hawkins 1981; Lubchenco 1980; Underwood 1980) or eliminate all but the most grazer-resistant species (e.g., Ayling 1981; Bertness et al. 1983; Branch 1981; Leighton 1971; Littler & Doty 1975; Ogden 1976; Paine & Vadas 1969; Shepherd 1973; Vine 1974; Wanders 1977).

Even those plants that survive are often reduced to tiny stunted versions of the tall arborescent individuals that are found in areas relatively free from grazers (Dahl 1973; Hay 1981; Mathieson et al.

1975; Stephenson & Stephenson 1972, p. 9). Grazing Crustacea can consume as much as 20% of the tissue produced by *Chondrus crispus* (Shacklock & Croft 1981) and 38% of that produced by *Gracilaria* (Nicotri 1977). At certain times of year, browsing snails can remove tissue as fast as *Iridaea splendens* (as *I. cordata*) can produce it (Mumford 1978).

A large number of studies has shown that many grazers feed selectively, preferring some algae and shunning others. However, an alga that is apparently inedible to one species of grazer may be consumed by others. For example, *Chondrus crispus*, which is rejected by *Littorina littorea* (Lubchenco 1978), is eaten by other species of snail, as well as by some sea urchins and crustaceans (Larson et al. 1980; Lubchenco & Menge 1978; Shacklock & Croft 1981). Some authors have attempted to relate the relative edibility of some red algae to their calorific content or digestibility, with mixed success (for example, Carefoot 1967; Himmelman & Carefoot 1975; Larson et al. 1980; Lawrence 1975; Littler et al. 1983; Montgomery & Gerking 1980; Nicotri 1980; Vadas 1977).

A major factor influencing edibility is the toughness of the outer part of the thallus, for this barrier must be breached by a successful grazer. Red algae differ considerably with respect to the ease with which their thalli can be punctured or abraded, and this often correlates with their susceptibility to herbivores (Littler et al. 1983; Padilla 1985; Watson & Norton 1985a). The most effective physical defense is seen in the calcareous thalli of Corallinaceae, which are much tougher than those of most algae (Littler et al. 1983; Watson & Norton 1985b; but see Padilla 1985). Not surprisingly, such calcareous forms are among the most grazer resistant of algae, although even these are not immune (Adey & Vassar 1975; Clokie & Norton 1974; Padilla 1984; Shepherd 1973; Steneck 1982). In some calcareous crusts, such as *Clathromorphum circumscriptum*, both the meristem and the reproductive conceptacles are deeply immersed beneath the calcareous epithallium, which must afford considerable protection to these vital organs (Steneck 1982). In addition, calcareous algae are of low nutritive value (Littler 1976; Littler & Littler 1980; Steneck 1982), and some also contain toxins (Paul & Fenical 1980).

Some red algae that are not excessively tough are nonetheless inedible to herbivores (Ogden 1976; Vadas 1977; Watson & Norton 1985a,b); indeed even extracts of them may be unpalatable (D. C. Watson unpubl.). Many red algae contain a variety of potentially distasteful substances, such as an acetylene-containing lipid (Paul & Fenical 1980) and various halogenated secondary metabolites (Fenical 1975; Fenical & Norris 1975; Norris & Fenical 1982), especially brominated phenols (Chevolot-Magueur et al. 1976; Pedersen 1978; Pedersen et al. 1974, 1980; Weinstein et al. 1975).

Algae that lack effective physical or chemical defenses may achieve some respite from grazing by virtue of their growth habit or their habitat. Many red algae simply grow too large to remain vulnerable to some herbivores (Brawley & Adey 1981; Dahl 1964). Turfs of red algae also seem to be less susceptible than scattered individuals (Hay 1981). Undoubtedly many small or encrusting algae escape being eaten because they grow in crevices or other habitats inaccessible to grazers (Adey & Vassar 1975; Underwood 1979). The conchocelis phases of *Bangia* and *Porphyra* acquire protection by living within the calcareous matrix of shells (Lubchenco & Cubit 1980). Even here, however, they are not completely immune from grazers (Farrow & Clokie 1979).

Some red algae seem to encourage the settlement and metamorphosis of the larvae of a variety of invertebrate grazers (Barnes & Gonor 1973; Fretter & Manly 1977; Morse & Morse 1984; Morse et al. 1979; Rumrill & Cameron 1983; Steneck 1982). *Chondrus*, *Laurencia*, and many calcareous crustose species seem in some circumstances to be dependent upon browsing animals to remove epiphytes or competitors that might otherwise swamp the plants (e.g., Ayling 1981; Branch 1981; Brawley & Adey 1981; Hay 1981; Littler & Littler 1980; Lubchenco 1978; Lubchenco & Menge 1978; Paine 1980; Steneck 1982; Vine 1974; Wanders 1977; but see Johnson & Mann 1986). Some territorial limpets and fish maintain "gardens" of red algae that, if the animal is removed, are soon consumed by rival herbivores (Branch 1981; Brawley & Adey 1977; Foster 1972).

Many red seaweeds arise from a basal crust or have a crustose phase in their life history. Such crusts are often more resistant to grazing than the erect plants (Kitting 1980; Littler & Kauker 1984; Slocum 1980; Underwood 1980). Certainly, removing *Collisella*, a herbivorous limpet, allows the erect frondose plants of *Mastocarpus papillatus* to increase at the expense of its crustose alternate phase (Slocum 1980). The same phenomenon has been observed with *Iridaea boryana* when limpets and chitons are removed (Jara & Moreno 1984). Although the erect plants grow faster and therefore are more competitive than the crusts, they are more vulnerable to grazers. The species thus has at its disposal two entirely different forms that are appropriate to different grazing regimes.

J. Competition

The presence of an overlying algal canopy may allow the survival of an undergrowth of red algae by providing protection from excessive insolation (see Section II.B) or desiccation (see Section II.F) or by inhibiting the growth of potential rivals (Kastendiek 1982). The removal of such competitors often enables red algae to extend the limits of their distribution (e.g., Foster 1982a; Hruby 1976) and may even allow characteristically intertidal algae to colonize the subtidal zone (Kain 1975). Conversely, the presence of red algae is said to limit the distribution of other seaweeds (Clokie & Boney 1980a; Lubchenco 1980), corals (Potts 1977) and barnacles (Jernakoff & Fairweather 1985; Lewis 1964, p. 215).

It is rare for the resource for which algae compete to be identified with any degree of certainty. We know of no clear example of red algae competing for nutrients. Nevertheless, opportunistic filamentous plants like *Ceramium tenuicorne* take up nutrients much more rapidly than do more robust plants like *Furcellaria lumbricalis* and *Phyllophora truncata*, which have much lower surface area-to-volume ratios (Wallentinus 1984). An assessment of the nutrient kinetics of these and other red algae indicates that filamentous types have a greater competitive advantage at low nutrient concentrations (Wallentinus 1984), the condition under which competition for nutrients is most likely to occur.

Clearly, light is important for photosynthetic plants, and faster-growing or larger seaweeds may shade rivals. Removal of a shading canopy of *Fucus* permits crusts of *Mastocarpus papillatus* (as *Gigartina papillata*) beneath to produce numerous uprights and form a dominant stand (Dayton 1971). Shading from kelps also limits the distribution of *Iridaea splendens* (as *I. cordata*) (Hruby 1976) and the abundance of red epiphytes on the stipes of *Laminaria* (Harkin 1981). Clearances of *Laminaria* spp. allow dramatic increases in red algal turf (Dayton 1975b) or *Palmaria palmata* (Hawkins & Harkin 1985), whereas clearance of *Macrocystis* canopy in southern Chile had little effect on red algae (Santelices & Ojeda 1984). The development of *Porolithon* on coral reefs is attributed to the increased insolation that results from the removal of frondose algae by wave action (van den Hoek 1969).

Competition for space is likely to be important. Many red algae seem adept at monopolizing the substratum. Filamentous forms creep and reattach at intervals to form a dense mat; others produce congested turfs. Encrusting algae or those with a spreading base may fuse with neighbors to form a continuous sheet, sometimes virtually occluding the substratum (Clokie & Boney 1980a; Johnson & Mann 1986; Littler & Kauker 1984). When red algae of different species are adjacent, one often overgrows the other in a predictable way (e.g., Foster 1975; Hay 1981). When crustose algae come into contact, sometimes both cease growing where they abut, but usually the crust of one species predictably overgrows that of the other (Bertness et al. 1983; Clokie & Norton 1974; Padilla 1984; Paine 1984). The result may be reversed when grazers erode the margins of faster-growing crusts, thus negating their potential advantage (Fig. 15-8). Similar interactions occur between crustose algae and encrusting animals (Breitburg 1984; Sebens 1986).

The occupancy of space can prevent the incursion of potential rivals. Dense turfs can exclude even the most invasive and aggressive competitors (Dayton et al. 1984; De Wreede 1983; Deysher & Norton 1982). Likewise, the spores of some algae seem unable to colonize crusts of *Hildenbrandia* (Lubchenco 1978) and of calcareous Corallinaceae (Breitburg 1984; Masaki et al. 1981). Many calcareous

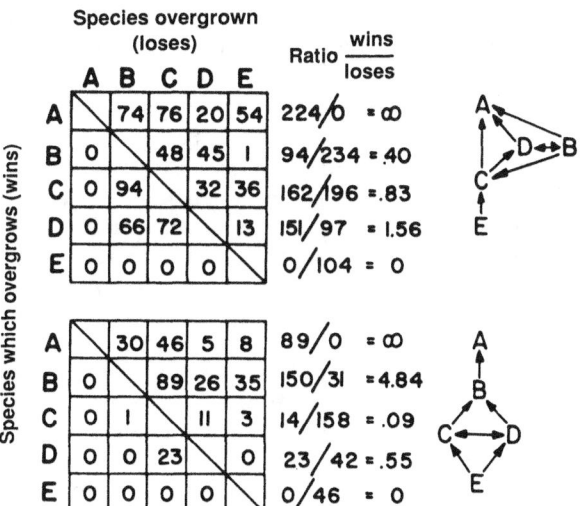

Fig. 15-8. Interspecific competition within a guild of five calcareous crustose species growing together on a natural substratum. Top: with grazers; bottom: without grazers. Letters refer to species: (A) *Pseudolithophyllum lichenare*, (B) *Lithothamnion phymatodeum*, (C) *P. whidbeyense*, (D) *Lithophyllum impressum*, (E) *Bossiella* sp. Numbers in the array are observed overgrowths when two guild members come into contact. The column of wins to losses provides an index of the relative competitive ability of the species in each row. The diagrams to the right of the ratios indicate the hierarchy of competitive interactions that results. Arrows point from losers toward winners. Modified from Paine (1984), courtesy the Ecological Society of America.

algae, both crustose and erect forms, also continually slough their outer layers, thus shedding potential fouling organisms (Bailey & Bisalputra 1970; Johnson & Mann 1986; Steneck 1982).

Faster-growing plants often succeed in the short term, but in all but unstable habitats they are usually replaced eventually by slower-growing but longer-lived types (see Section IV.B). However, the dichotomy between r-strategists and K-strategists sits uncomfortably on some seaweeds. Some species of *Porphyra*, for example, grow rapidly and exhibit a high reproductive output (r-strategists) but are nonetheless extremely stress tolerant (see Section II.D, II.F), as would be expected of K-strategists. The three-strategy model of adaptive specialization proposed by Grime (1977) may be more appropriate to red algae (Littler & Littler 1984a; Russell 1986; Shepherd 1981), but even this scheme may be simplistic.

Losers of competitive interactions are often eliminated altogether (e.g., Russell & Fielding 1974). Allelopathy may be the cause. Certainly, the growth of *Porphyrodiscus simulans* and *Rhodophysema elegans* is inhibited when they lie adjacent to, but not touching, the thallus of *Ralfsia* (Fletcher 1975). Similarly, the growth of germlings of *Chondrus crispus* and especially *Mastocarpus stellatus* can be adversely affected by the presence of some diatoms (Huang & Boney 1984, 1985). *Chondrus* is known to possess chemical secretions capable of eliminating diatoms (Huang & Boney 1984; Khfaji & Boney 1979). Furthermore, many of the substances red algae contain that are claimed to be grazer repellents (see Section II.I) may act against epiphytes or competitors.

Interactions also take place between red algae, competitors, and predators. Clearly, predators that remove grazers are often of enormous benefit to seaweeds, but grazers may remove rival algae, allowing grazing-resistant forms to predominate (see Section II.I). Mussels are major competitors of many intertidal seaweeds, and on some shores colonization by *Chondrus crispus* (Lubchenco & Menge 1978) and *Endocladia muricata* (Paine 1969) may be reliant upon predator removing mussels and making space available.

Little is known of intraspecific competition within the red algae, although plants crowded together in turfs often grow more slowly than do isolated individuals (Hay 1981; Taylor 1985). When *Porphyra* is seeded onto string for cultivation, the plants sown at greater densities achieve a smaller size, but the resultant yield is about the same (Yoshida 1972). Suprisingly, the coalescence of abutting crusts mentioned earlier does not necessarily lead to heightened intraspecific competition. On the contrary, fusion of basal crusts results in enhanced initial production of uprights in *Chondrus crispus*, *Masctocarpus stellatus* (Tveter & Mathieson 1976; Tveter-Gallagher & Mathieson 1980), and *Gracilaria verrucosa* (Jones 1956). There is even an indication that plants of *M. stellatus* will not develop singly on the shore but require mutual protection (Marshall et al. 1949, p. 44). Nevertheless, for most species intraspecific competition is likely to be fierce between densely settled spores. Perhaps the large mucilaginous coat that surrounds most red algal spores (Boney 1975; Ngan & Price 1979) is a spacing mechanism reducing competition at this critical initial period.

III. DISTRIBUTION

A. Geographical

There is probably no species of red alga that is cosmopolitan, although *Digenia simplex* is reputedly found in almost all warm seas. Many genera have representatives in most of the world's oceans, e.g., *Bostrychia*, *Corallina*, *Gracilaria*, *Plocamium*, and *Polysiphonia*. Most species, however, are fairly restricted in their distribution.

The paucity of Rhodophyta in polar seas contrasts with the wide variety of species present in tropical waters (Table 15-5). The proportion of various "life-forms" also seems to vary with latitude (Chapman & Chapman 1976; Garbary 1976). Cluster analyses and ordination techniques have been used to assess the affinities between different floras, for example, for the warmer waters of the eastern Atlantic (Lawson 1978), the cooler north Atlantic (Hooper et al. 1980; van den Hoek 1975), and on a worldwide basis (Joosten & van den Hoek 1986; van den Hoek 1984).

The distribution of many species can only be understood with reference to the distant past (Hommersand 1986; Joosten & van den Hoek 1986). The vegetation of several now separate warm temperate and tropical regions may represent the remnants of a once continuous Tethyan flora, and the discreteness of the temperate water floras between the North Atlantic and the North Pacific is ascribed to their long separation by the Bering land bridge (Joosten & van den Hoek 1986).

For many species, however, their distribution is curbed by present-day conditions. Water temperature often is proposed as a major factor influencing geographical limits. Many authors have attempted to correlate the extreme conditions prevailing near the limits of a species' distribution with the tem-

Table 15-5. The number of species of seaweed recorded and the floral composition at various localities in the northern hemisphere

	Location	Latitude	Total number of species recorded	Ratio Rhodophyta/ Phaeophyta
Eastern Atlantic Ocean	Spitzbergen	76°–80° N	67	0.6
	Troms–Finnmark, Norway	70°–71° N	204	0.7
	British Isles	50°–60° N	679	1.8
Western Atlantic Ocean	Eastern Greenland	71°–73° N	109	0.6
	Labrador–New Brunswick	45°–60° N	354	1.0
	Florida	25°–31° N	414	3.9
Eastern Pacific Ocean	British Columbia and Washington	48°–55° N	478	2.4
	California	32°–41° N	666	2.9
	Mexico	15°–32° N	714	4.6

peratures known to prevent the growth or survival of that species (e.g., Biebl 1972; Bolton 1986; Kapraun 1980; Lüning 1984; McLachlan & Bird 1984; Stewart 1984a; van den Hoek 1982a,b; Yarish et al. 1984, 1986; see also Section II.D).

The temperature limits for reproduction often are narrower than for either growth or survival; thus, close to its geographical limits an alga may lose its power to spread further by means of spores (Dixon 1965; Norton & Parkes 1972; Whittick 1978). Vegetative propagation alone allowed *Pterosiphonia complanata* to persist for a century in the same site at the northern limits of its range (Dixon 1965). As completion of the life history is often not obligatory; the different phases may exhibit somewhat different geographical distributions (Dixon 1965; Feldmann 1957; McLachlan et al. 1969). In several instances the tetrasporophyte seems to have a wider range than the gametophyte (Dixon 1965).

Many widely separated regions of the world that experience similar oceanographic climates do not sustain similar floras. Many species from one region should be able to thrive in the other, but the limitations of their dispersal mechanisms do not give them the opportunity to do so. The fact that it took 3 years for a red alga to colonize Surtsey (Jónsson 1972), a new volcanic island only 5 km from the nearest land, is evidence that dispersal is not easy even over quite short distances.

Many red algae are dependent upon their microscopic reproductive propagules for dispersal. The range of viable spores carried away on currents is largely unknown, but it is thought to be small. A few spores of some red algae have been collected over 2 km from the nearest land, but none was found 8 km from land (Zechman & Mathieson 1985). Nonetheless, the progressive spread in Europe of *Antithamnionella* (as *Antithamnion*) *spirographidis*, *Asparagopsis armata*, and *Bonnemaisonia hamifera* (Farnham 1980; Jones 1974) could be explained by marginal dispersal, each generation of plants establishing itself a little further along the coast.

The movement of detached drifting plants or fragments by wind or currents may extend the dispersal range of a species greatly, providing that the plants are buoyant or are otherwise maintained in the euphotic zone. A drift specimen of *Gelidium versicolor*, for example, has been washed up on the south coast of England, reputedly more than 2700 km from the nearest known attached populations (Dixon & Irvine 1977, p. 134).

Some plants, such as *Asparagopsis armata* and *Bonnemaisonia hamifera*, readily entangle, and thus drifting specimens could easily become anchored. Moreover, drifting red algae are commonly fertile (Neushul et al. 1976), and it is probable that such plants might deposit spores while en route, thus giving rise to new populations. The importance of drifting algae should not be underestimated. For example, ten bottle brushes placed on the shore on the west coast of Scotland entangled 96 species of algae in only 5 days, and in a month they trapped a quarter of the entire algal flora of the region (Clokie & Boney 1980a). No sampling of drifting algae has been carried out on suspected pathways of long-distance dispersal, for example, the west wind drift route, proposed to explain similarities in the floras of subantarctic islands from the Falklands to New Zealand (Hommersand 1986).

Several species of red algae have been deliberately introduced into new regions for the purposes of cultivation, and accidental transplantation is often invoked when a species appears inexplicably far outside its previously known range. Five species of Rhodophyta recently have "invaded" the south coast of England (Farnham 1980), and a European species has appeared in Australian waters (Lewis & Kraft 1979). There is no unequivocal evidence that long-distance transportation by ships has influenced the distribution of red algae. However, several species that have passed from the Red Sea to the Mediterranean would have been unable to survive for long in the Suez Canal, so ships or fishing nets are likely vectors (Lipkin 1972).

In order to recognize with certainty future changes in distribution patterns, we need detailed maps of the present distribution of species, such as the 76 maps of Rhodophyta published recently for the British Isles (Norton 1985).

B. Intertidal

Muddy areas are generally inhospitable to red algae, although a few species, such as *Gracilaria* spp., occasionally occur loose-lying or even embedded in mud (Norton & Mathieson 1983). In such habitats, however, most species of Rhodophyta are epiphytic, perched above the mud surface. Plants of *Bostrychia* and *Catenella* are found characteristically entangled on other plants (Norton & Mathieson 1983) and form a distinctive community worldwide both on temperate salt marshes (Chapman 1960) and, with *Caloglossa*, on the pneumatophores of mangroves (Post 1963). There is even a minizonation of red algae on the pneumatophores (Almodovar & Biebl 1962).

Most red algae inhabit rocky shores. The upper shore is the least favorable habitat, but a few Rhodophyta grow there, such as *Bangia atropurpurea*, *Endocladia muricata*, *Hildenbrandia* spp., *Bostrychia* spp., and *Porphyra* spp. These thrive particularly where spray frequently wets the upper shore; indeed, *Porphyra umbilicalis* may occur up to 15 m above high-tide level on wave-exposed coasts (Børgesen 1908).

On the midshore desiccation remains the major constraint, but some species, such as *Lomentaria articulata* and *Plumaria elegans*, avoid the worst effects by inhabiting shady gullies or caves (Norton et al. 1971). Tide pools are aquaria of species that can inhabit the open rock surface at lower levels on the shore (e.g., Goss-Custard et al. 1979). Often, pools are lined with encrusting corallines that do not extend beyond the water's edge. Some common pool dwellers, such as *Corallina officinalis*, *Chondrus crispus*, and *Dumontia contorta* (as *D. incrassata*), owe part of their success to their relative immunity to the herbivorous snails that often abound in such pools (see Section II.I).

Red algae in the intertidal zone must compete for space with ascidians, oysters, or zoanthids in warm-water regions and with fucoid seaweeds, barnacles, and mussels on temperate shores (Stephenson & Stephenson 1972). Some species, however, are characteristically found growing on top of mussels, e.g., *Callithamnion arbuscula* and *Ceramium shuttleworthianum* (Lewis 1964), or are epiphytes on other plants, e.g., *Polysiphonia lanosa* on *Ascophyllum* and *Smithora naiadum* on seagrasses (see Section II.A).

The most distinctive zone of red algae is usually found close to or below the lower limit of barnacles, where a calcareous carpet of encrusting corallines begins and extends down into the subtidal zone (Stephenson & Stephenson 1972). There is also often a distinct mosslike turf of red algae, especially on tropical or warm temperate shores (Table 15-6). The similarity of zonation on widely separated shores can be quite striking. In cold temperature regions the red algal turf is often a less conspicuous feature of the shore, even though individual plants may be large and robust. The most characteristic species are *Chondrus crispus* and *Mastocarpus stellatus*, together with *Corallina officinalis*, *Laurencia pinnatifida*, and *Palmaria palmata* (e.g., Lewis 1964, p. 83). The similarities and contrasts between the zonation of red algae on rocky shores worldwide are vividly illustrated by Stephenson & Stephenson (1972).

C. Subtidal

In contrast to the intertidal environment, the influences in the subtidal change fairly gradually with depth, so zonation is often not particularly clear-cut unless a change in substratum creates a sharp edge to a zone. The other two important factors, light and wave action, both decrease logarithmically with depth, resulting in greater changes (thus, narrower zones) in shallow than in deep water.

In temperate waters the highly favorable shallow parts are frequently dominated by Phaeophyceae, relegating the red algae to the undergrowth (Bergquist 1960; Edelstein et al. 1969; Funano & Sakai 1967; Kain 1960; McLean 1962). However, the red algae are able to grow in deeper water than are most of these brown dominants, so there is a red algal zone below, consisting of the same species that constitute the undergrowth in Britain (Hiscock & Mitchell 1980; Norton 1968; Norton & Milburn 1972), northeastern Canada (Edelstein et al. 1969), and northwestern United States (Neushul 1967). In contrast, off Australia (Shepherd & Womersley 1970) and parts of the southwest coast of Hokkaido, Japan (Yamada 1980), several red algal species are common only near or below the lower limit of the brown algal

Table 15-6. Similarities in the composition of red algal communities in different regions of the world

San Diego County, U.S.A. (Stewart 1982)	Cabo Frio, Brazil (Yoneshigue 1985)	St. Martin Island, Caribbean (Vroman 1968)	Angola, West Africa (Lawson et al. 1975)	Townsville, Australia (Ngan & Price 1980a)
Splash zone				
—	—	*Bostrychia* spp.	*Bostrychia binderi*	*Bostrychia binderi*
—	—	—	*Bostrychia tenella*	—
—	—	—	*Hildenbrandia rubra* (as *H. prototypus*)	*Hildenbrandia rubra*
—	—	—	*Murrayella periclados*	—
Porphyra sp.	*Porphyra acantophora*	—	*Porphyra* sp.	*Porphyra* sp.
Lower shore turf				
—	*Acanthophora spicifera*	—	*Caulacanthus ustulatus*	*Caulacanthus ustulatus*
Centroceras clavulatum	*Centroceras clavulatum*	*Centroceras clavulatum*	*Centroceras clavulatum*	*Centroceras clavulatum*
—	—	—	*Corallina* spp.	—
Corallina spp.	*Gelidium pusillum*	*Gelidiella acerosa*	*Gelidium* spp.	*Gelidium* spp.
Gelidium pusillum				
Gigartina canaliculata	—	—	*Gigartina acicularis*	*Gigartina* spp.
Hypnea valentiae	*Hypnea musciformis*	*Hypnea musciformis*	*Hypnea musciformis*	—
—	*Hypnea cervicornis*	*Hypnea spinella*	—	—
Jania crassa	*Jania* spp.	*Jania* spp.	*Jania* spp.	—
Laurencia pacifica	*Laurencia* spp.	*Laurencia papillosa*	—	*Laurencia pygmaea*
Polysiphonia simplex	*Polysiphonia scopulorum*	*Polysiphonia ferulacea*	*Polysiphonia subtillissima*	*Polysiphonia subtillissima*
Pterocladia capillacea	*Pterocladia capillacea*	—	—	—

canopy. In many areas there is a greater biomass of red algae at slightly greater depths than this canopy than that under it (Jansson & Kautsky 1977; Lüning 1970; Norton et al. 1977; Shepherd & Womersley 1970, 1976; Sohn et al. 1983). This is illustrated in Fig. 15-9, which also shows how important epiphytism can be (Phillips & Springer 1960; Smith 1967; Whittick 1983), allowing an alternative substratum often considerably exceeding the rock in area (Drach 1949).

In the absence of large brown algae, Rhodophyta may dominate at various levels. Some articulated corallines seem able to withstand breaking waves and may form a surf zone, for example, in California (McLean 1962) and off South Australia (Shepherd & Womersley 1970, 1971, 1976). There may be extensive red algal domination in the tropics, for example, off Ghana (John et al. 1977) and Brazil (Yoneshigue 1985). Coralline red algae are extremely important in helping to cement together coral reefs

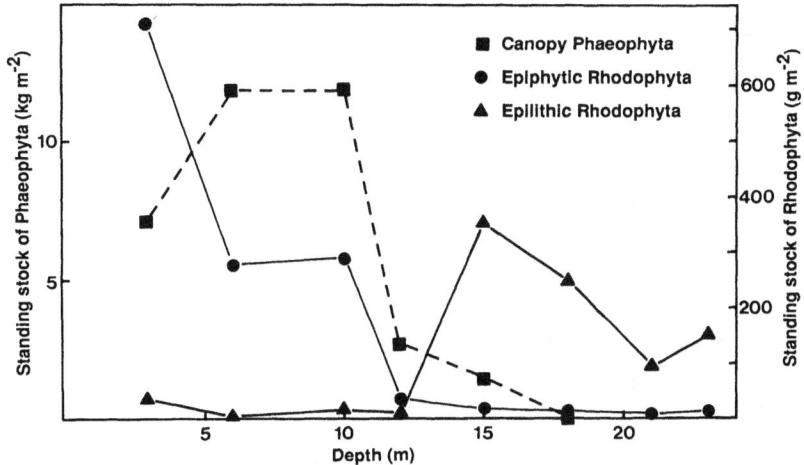

Fig. 15-9. The standing stock (fresh weight) of algae at various depths off Carrigathorna, southwest Ireland. Plotted from Norton et al. (1977).

(van den Hoek et al. 1975; Littler & Littler 1984b; Price 1971). On the other hand, in deep water it is the green algae that are dominant off Eniwetok Atoll in the tropical Pacific (Gilmartin 1960). At higher latitudes red algal domination may be limited to certain zones, for example, below *Fucus* in the Baltic (Kautsky 1974), near low water in the Mediterranean (Lauret 1974), and between brown algal zones in New Zealand (Bergquist 1960) and eastern Canada (Bird et al. 1984). In the Mediterranean codomination with green algae often is followed, in deeper water, by dominance solely by red algae (Boudouresque 1973).

Although red algae frequently fail to dominate in biomass or stature, they usually excel in terms of species numbers (Mathieson 1979; Mathieson et al. 1975; Phillips & Springer 1960; Shepherd & Womersley 1981). Even within large kelp beds 104 and 86 species were found in mid-California (Devinny & Kirkwood 1974) and South California/Mexico (Dawson et al. 1960), respectively. The number of red algal species, as a proportion of the total, increases with depth (Fig. 15-10) in the northwest Atlantic (Bird et al. 1984; Mathieson 1979), the Mediterranean (Boudouresque 1973), and the tropical Atlantic (John et al. 1977). This is not the case, however, in some other tropical areas: the Caribbean Sea (van den Hoek et al. 1975; Phillips et al. 1982) or off Hawaii (Doty et al. 1974).

Lower limits of foliose red algae are frequently determined by substratum, suitable rock giving way to unsuitable sand or mud. At other sites grazing may make colonization in deeper water impossible, either because grazing animals are more abundant there (Jones & Kain 1967) or because with lower growth rates in low light, the plants are less able to compensate for losses (L. M. Vost unpubl.). In sites where neither of these applies, the main factor determining lower limits is the water type (see Section II.B). Some examples are shown in Table 15-7, together with some lower limits for encrusting coralline species. These seem to be able to tolerate lower light levels and therefore can grow deeper than foliose forms (Lüning & Dring 1979). The penetration of algae into caves is probably also influenced by light, with a mainly quantitative rather than qualitative change. In the Mediterranean red species penetrate further than green (Larkum et al. 1967), whereas in Ireland a green alga penetrates as far as the red (Norton et al. 1971). Vertical surfaces may favor red algae relative to the other groups (Castric-Fey et al. 1973; Shepherd & Womersley 1970).

IV. COMMUNITIES

A. Production and growth

In Table 15-8 some net photosynthetic rates are arranged in the categories devised by Littler and Arnold 1982), which should show decreasing rates from tube/sheet to encrusting forms. Although there is some correspondence between rate and form, there is clearly considerable overlap, and the largest group, "coarsely branched," contains the highly productive *Gracilaria* spp., as well as a variety of forms showing varied rates.

The relative productivities of the red algae in the intertidal region are apparent in Table 15-9. The higher value for the exposed compared with the sheltered site, quite close together in central California, is due mainly to coralline algae, already mentioned as sometimes forming a surf zone at

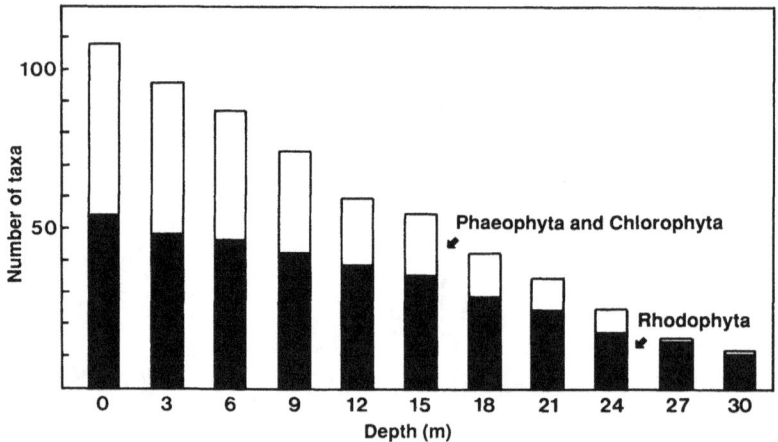

Fig. 15-10. The number of taxa of red algae compared with the total of green, brown, and red algae at various depths: pooled data from 12 coastal sites in New Hampshire and southern Maine, redrawn from Mathieson (1979), courtesy Walter de Gruyter Publishers.

Table 15-7. Some recorded lower limits, as depth in m, of foliose and crustose red algae

Site	Lower Limit (m) Foliose	Lower Limit (m) Crustose	Reference
Menai Straights, British Isles	6		Knight-Jones & Nelson-Smith 1977
Firth of Clyde, British Isles	13	27	Clokie & Boney 1980b
Helgoland, Germany	14		Lüning 1970
Jaffrey Point, New Hampshire, U.S.A.	19	>24	Mathieson et al. 1981
Hood Canal, Washington, U.S.A.	24		Phillips & Fleenor 1970
Prince Edward Island, Nova Scotia, Canada	25		Bird et al. 1984
Southern Ireland	27		Cullinane & Whelan 1983
South of Bergen, Norway	30	55	Jorde 1966
Isle of Shoals, New Hampshire, U.S.A.	32		Mathieson 1979
Glénan Archipelago, Brittany, France	40–47		Castric-Fey et al. 1973
Pearson Island, South Australia	>50		Shepherd & Womersley 1971
Norway		>57	Adey 1971
Gulf of Mexico	>95		Eiseman & Blair 1982
North & South Carolinas, U.S.A.	100		Schneider 1976
Hawaii, U.S.A.	91–128		Doty et al. 1974
Corsica, France	120		Molinier 1960
Jamaica		175	Lang 1974
San Salvadore Island, Bahamas		268	Littler et al. 1985

the top of the subtidal region (Section III.C). In the seasonal study (Table 15-9) red algae were most active in the winter; they were also more important further south on the Pacific coast, in Mexico.

Figures for annual primary production per unit area of substratum have been drawn both from productivity and from seasonal biomass observations (Table 15-10). Some values are surprisingly high, particularly on intertidal rock, where photosynthesis may be dramatically reduced by desiccation (see Section II.F). It seems that red algae forming turfs are better able to withstand desiccation, though through shading, their photosynthetic rate is reduced by up to a half compared with plants of the same species with spatially separated thalli (Hay 1981; Taylor & Hay 1984). An even higher cost is paid by the crustose base of *Corallina officinalis*, which is very much more resistant to grazing than its erect counterpart (Littler & Kauker 1984).

Growth rates of red algae fairly frequently have been measured under artificial or partially artificial conditions. In the laboratory very early stages can grow fast, doubling their cell number in less than a day in *Bonnemaisonia asparagoides* (Boney 1962) and *Callithamnion hookeri* (Edwards 1977b) (see also Fig. 15-4). The maximum measured relative growth rates R (change in $\log_e$ of length or weight per day) of macroscopic plants are given in Table 15-11. There is a 35-fold range in the maximum rate observed for the division. The slow rate of the perennial *Pterocladia capillacea* is inherent and cannot be increased by manipulating the conditions (Stewart 1984b). Growth rates in the field have been measured far less often; some are shown in Table 15-12. Clearly, the "sheet" form of *Iridaea* grows fast and the corallines grow slowly, corresponding with the productivity categories of Littler and Arnold (1982).

When individual plants or the biomass of single species have been measured seasonally in the field, peaks have been observed at various times of year. An early spring peak in growth rate, associated with low water temperature, is shown by *Porphyra umbilicalis* (Munda & Markham 1982), *Dumontia contorta* (Kilar & Mathieson 1978; Munda & Markham 1982), three species of *Eucheuma* (Dawes et al. 1974), and *Furcellaria lumbricalis* (Austin 1960b). In the last case the low temperature requirement was attributed to the northern distribution of the species. Summer peaks in growth rate, associated with the water temperature rising above a minimum, are shown by *Gracilaria verrucosa* (Jones 1959b), *G. tikvahiae* (Penniman et al. 1986), and *Chondrus crispus* (Mathieson & Burns 1975). A summer peak attributed to high ambient light levels is exhibited by *Iridaea splendens* (as *I. cordata*) (Hansen 1977). Almost continuous growth throughout the year, as a series of cohorts, is shown by *Asparagopsis armata* (Aranda et al. 1984).

The onset of reproduction coincides with the end of the growth phase in some species, including *Mastocarpus stellatus* (Dion & Delépine 1983) and *Dumontia contorta* (Kilar & Mathieson 1978). *Iridaea*

Table 15-8. Net primary production rates of red algae in five morphological categories at various sites

Site	Net primary production (mg C g DW^{-1} h^{-1})					Reference
	Tube and sheet	Delicately branched	Coarsely branched	Articulated corallines	Encrusting	
Rocky intertidal						
Arctic Alaska, U.S.A.	5.2					Healey 1972
Kiel, Germany	11	6.3				King & Schramm 1976
Stanley Park, British Columbia, Canada	1.8					Quadir et al. 1979
Pacific Grove, California, U.S.A.	0.5–1.7		0.3–0.6			Johnson et al. 1974
San Clemente Island, California, U.S.A.	0.8		1.6–2.5	1.6–1.7	0.2–0.4	Littler & Murray 1974
Gulf of California, Mexico	3.2		0.4–1.4	0.2		Littler & Littler 1981
Baja, Mexico			0.4–2.8	1.1–1.6		Littler & Littler 1981
Mid-California, U.S.A., to Baja Mexico	2.3	2.8–5.4	0.4–3.2	0.2–0.6	<0.1	Littler & Arnold 1982
Mangrove communities						
Florida, U.S.A.			4.1–9.6			Hoffman & Dawes 1980
Puerto Rico		1.6–1.9	0.3–1.8			Burkholder & Almodovar 1973
Subtidal						
Southeast Sweden		3.4–8.3	0.6–1.3			Wallentinus 1978
Kiel, Germany	1.8–3.3	2.0–6.4	0.7–1.2			King & Schramm 1976
Monterey, California, U.S.A.			0.3–1.1			Heine 1983
Canary Islands			0.2–0.3			Johnston 1969

DW = dry weight.

Table 15-9. The percentage of primary production m^{-2} attributable to red algae in some intertidal sites

Site	Remarks	%	Reference
Central California	Sheltered	29	Seapy & Littler 1978
	Exposed to waves	46	Seapy & Littler 1978
San Clemente Island, California	In May	46	Littler & Murray 1974
	Spring	38	Littler et al. 1979
	Summer	24	Littler et al. 1979
	Fall	29	Littler et al. 1979
	Winter	55	Littler et al. 1979
Baja California, Mexico	Pacific side	91	Littler & Littler 1981
	Gulf of California side	35	Littler & Littler 1981

Table 15-10. Net annual primary production by various red algae at various sites.

Species	Place	Production $m^{-2} y^{-1}$ gC	g DW	kg FW	Reference
Archaeolithothamnion dimotum	Curaçao at 25 m	78			Vooren 1981
Hydrolithon boergesenii	Curaçao at 25 m	77			Vooren 1981
Chondrus crispus	Helgoland, Germany			1.7	Munda & Markham 1982
Corallina officinalis	Helgoland, Germany			2.5	Munda & Markham 1982
Dumontia contorta	Helgoland, Germany			3	Munda & Markham 1982
	New Hampshire		200		Kilar & Mathieson 1978
Chondrus crispus	New Hampshire			8	Mathieson & Burns 1975
Iridaea splendens	California		1500		Hansen 1977
Encrusting corallines	Curaçao	370			Wanders 1976
Intertidal red algae	San Clemente Island, California	112			Littler et al. 1979

C = carbon; DW = dry weight; FW = fresh weight.

Table 15-11. The maximum relative growth rate in length (R^L) or weight (R^M) day^{-1} of species of Florideophyceae in tank, raft, or rope culture.

Species	R^L	R^M	Reference
Gracilaria tikvahiae		0.35	Parker 1982
Devaleraea ramentacea	0.097		Rueness & Tananger 1984
Gracilaria verrucosa	0.096		Rueness & Tananger 1984
Hypnea musciformis		0.19	Humm & Kreuzer 1975
Agardhiella subulata		0.18	DeBoer et al. 1978
Gelidium sp.	0.072		Rueness & Tananger 1984
Gracilaria foliifera		0.14	Guiry & Ottway 1981
Gracilaria andersonii		0.09	Hansen 1983
Iridaea splendens (as I. cordata)		0.090	Waaland 1977
Iridaea cornucopiae		0.083	Waaland 1977
Palmaria palmata		0.083	Morgan et al. 1980
Gigartina exasperata		0.080	Waaland 1977
Iridaea heterocarpa		0.078	Waaland 1977
Eucheuma unciatum		0.077	Polne et al. 1981
Gelidium coulteri		0.076	Hansen 1980
Gracilaria edulis		0.076	Nelson et al. 1980
Chondrus crispus		0.074	Bird et al. 1979
Farlowia mollis		0.065	Waaland, 1977
Gracilaria sp.		0.058	Yoneshigue-Braga & Baeta Neves 1981
Eucheuma spinosum		0.052	Braud & Pérez 1979
Plocamium cartilagineum		0.039	Waaland 1977
Hypnea cervicornis		0.039	Mshigeni 1978
Hypnea nidifica		0.036	Mshigeni 1978
Callophyllis flabellulata		0.035	Waaland 1977
Gracilaria arcuata		0.035	Nelson et al. 1980
Hypnea chordacea		0.032	Mshigeni 1978
Furcellaria lumbricalis		0.030	Bird et al. 1979
Phyllophora antarctica		0.026	Ohno 1984
Pterocladia caerulescens		0.023	Santelices 1976
Schizymenia pacifica		0.017	Waaland 1977
Pterocladia capillacea		0.016	Santelices 1976
Prionitis lanceolata		0.010	Waaland 1977

R calculated from percent increases when necessary. Ranked (using maxima and assuming $R^M \approx 2 \times R^L$). For each species, only the fastest recorded R is presented; see Kain (1987a) for further records.

splendens (as *I. cordata*), however, grows while reproducing (Hansen 1977).

B. Temporal change

Short-term changes in red algae in the field may be diel responses. The mitotic index has been shown to peak in the morning and early afternoon in *Neorhodomela* (as *Rhodomela*) *larix* (Austin & Pringle 1968) and near the end of the daylight period in *Porphyra lanceolata* (Pringle & Austin 1970). Spore discharge of intertidal tropical red algae is related to the tidal cycle, occurring when the plants are immersed (Ngan & Price 1983). On the other hand, *Gelidiella acerosa* appears to liberate tetraspores in response to a light period (Rao 1974).

Seasonal changes in populations are likely to be greater at higher latitudes. One measure of seasonality is the ratio of the number of species observed in winter to the number observed in summer and is plotted against latitude in Fig. 15-11. In the tropics the number of species is greater in winter; at higher latitudes the expected drop of winter species with latitude does not occur. Seasonal changes in numbers of all (not just red) algal species in the subtidal do not seem to be correlated with either latitude or temperature (Kain 1989). There does not seem to be a systematic change in the proportion (about half) that red algae contribute to the flora during different seasons in temperate waters (Coleman & Mathieson 1975; Mathieson et al. 1981; Mathieson & Penniman 1986; Reynolds & Mathieson 1975). More than half the species of red algae in the temperate northwest Atlantic coast are thought to be perennial (Coleman & Mathieson 1975; Mathieson & Penniman 1986; Reynolds & Mathieson 1975). On a deep offshore reef, where seasonal changes in water masses are added to the fact that light levels are very low in winter (Peckol & Searles 1984), the flora consists mainly of seasonal annuals (Peckol, 1982). Seasonal substratum disturbance has a similar effect (see Section II.A). Many

Table 15-12. The maximum elongation rate per day (mm) or relative growth rate in length (R^L), area (R^A), volume (R^V), or weight (R^M) per day of species of red algae in the field — ranked (approximately)

Species	mm	R^L	R^A	R^V	R^M	Reference
Iridaea splendens (as *I. cordata*)					0.061	Hansen 1977
Porphyra linearis	5					Bird 1973
Eucheuma acanthocladum					0.04	Dawes et al. 1974
Constantinea subulifera			0.031			Powell 1964
Plocamium cartilagineum	0.55	0.0085			0.026	Kain 1987a
Gelidium robustum	0.63					Guzmán del Próo & de la Campa de Guzman 1979
	0.35					Barilotti & Silverthorne 1972
Pterocladia capillacea	0.44					Strömgren 1984
Meristiella gelidium					0.020	Dawes et al. 1974
Chondrus crispus			0.012			Taylor 1972
Paragoniolithon solubile	0.29					Adey & Vassar 1975
Neogoniolithon westindianum	0.26					Adey & Vassar 1975
Hydrolithon boergesenii	0.15					Adey & Vassar 1975
Neogoniolithon megacarpum	0.13					Adey & Vassar 1975
Porolithon pachydermum	0.07					Adey & Vassar 1975
Lithophyllum tortuosum				0.009		Boudouresque et al. 1972
Calliarthron tuberculosum	0.10					Johansen & Austin 1970

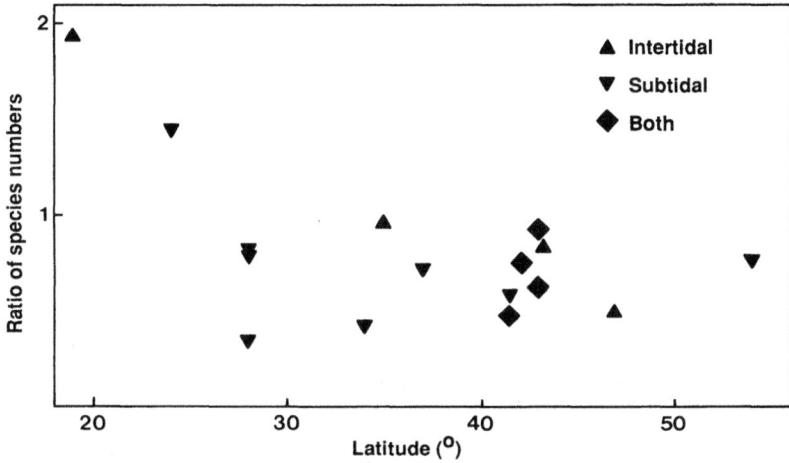

Fig. 15-11. The ratio of the number of red algal species in two winter months (December/January in Northern and July/August in Southern Hemisphere) to the number in two summer months at various latitudes. From (in increasing order of latitude and/or ratio): Ngan & Price (1980b), Mathieson & Dawes (1975) twice, Conover (1964), Cheney & Dyer (1974), Peckol (1982), Matsuura (1958), Breda & Foster (1985), Sears & Wilce (1975), Coleman & Mathieson (1975) twice, Mathieson et al. (1981), Mathieson & Penniman (1986), Femino & Mathieson (1980), Thom (1980), Kain (1960).

species overwinter as "resting" small stages, able to develop when conditions improve (Neushul & Dahl 1967).

A more dramatic seasonal change than that of species number can be that in the quantity of red algae. In Washington the percentage cover of *Neorhodomela* (as *Rhodomela*) *larix* in low tidepools doubles in summer (Dethier 1982). In California the overall percentage cover of a number of species varies little, but their biomass increases several-fold from a winter minimum, both in the intertidal (Horn et al. 1983) and subtidal (Breda & Foster 1985). The numbers of plants of 11 subtidal species of red algae show dramatic seasonal changes in Washington (Hodgson & Waaland 1979), and the number of fronds of *Chondrus crispus* in the Bay of Fundy doubles in the summer (Chen & Taylor 1980). Single plants of *Delesseria sanguinea* and *Odonthalia dentata* show, respectively, a four- and threefold increase in biomass in summer off the Isle of Man (Kain 1984a). In deep water off North Carolina, however, the summer biomass increase of conspicuous species is irregular (Peckol & Searles 1984). In a California kelp forest the annual development of *Polyneura latissima* depends on the disturbance of the kelp canopy by winter storms (Foster 1982b). In South Africa marked seasonal changes in the biomass of several dominant red intertidal algae have been correlated with tidal heights (McQuaid 1985). In the tropics, in Ghana, the seasonal change in the upper limit of *Hypnea musciformis* on the shore follows a seasonal change in the height of the lowest low water occurring during the day (Lawson 1957). Similarly, in California daytime low tides are lowest in the winter and result in the suppression of the zones of several red algal species (Seapy & Littler 1982).

Seasonality in reproductive activity sometimes may be related to a seasonal growth cycle in that the plants become fertile when they are large enough. However, season may influence fertility directly (Sheath et al. 1985) or its effect may be coupled to that of plant size (Kain 1986). When the entire red algal flora is examined at one site, many species appear to reproduce all through the year (Mathieson et al. 1981), and in any one season more than half the species are reproducing (Cormaci et al. 1984; Mathieson et al. 1981). If one takes as a measure of seasonality in perennial species reproduction in two to nine consecutive months with sterility in the remaining months, then there is a marked change with latitude. In the tropics none of the 45 red intertidal species in Queensland qualifies (Ngan & Price 1980b). In the West Atlantic the proportion, around Cape Cod, is 12% (Sears & Wilce 1975) and 10% of the species (Mathieson et al.

1981), but in Newfoundland it is 45% of the species (Hooper et al. 1980). In part this difference could be due to varying definitions of reproductive activity by different workers (R. G. Hooper unpubl.). On the other hand, some species may apparently reproduce all year, or nearly so, but the proportion of reproductive plants may change markedly with season, e.g., *Chondrus crispus* (Tveter-Gallagher et al. 1980), *Gracilaria* spp. (Hay & Norris 1984; Penniman et al. 1986), *Lithophyllum incrustans* (Edyvean & Ford 1986), and several subtidal species (Kain 1982). This seasonal variation in proportion may vary with depth (Kain 1986; Norall et al. 1981). It also may differ between cystocarpic and tetrasporic plants, as in *Hypnea* spp. (Rao 1977), *Gracilaria* spp. (Hoyle 1978), and *Pterocladia heteroplatos* (Kaliaperumal & Rao 1985).

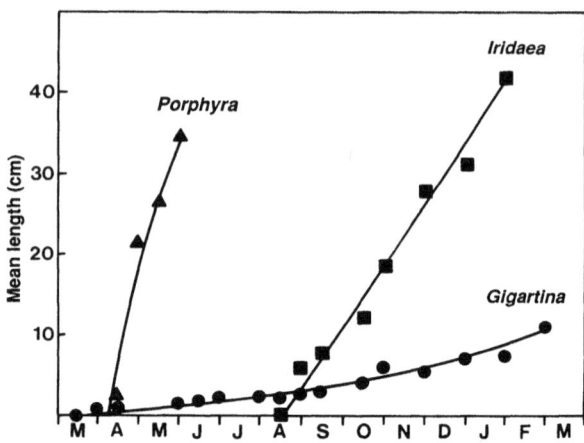

Fig. 15-12. The growth rates of three taxa of red algae — *Porphyra capensis*, *Iridaea capensis*, and *Gigartina radula* with *G. stiriata*. Successive dominants on experimentally denuded rock at Sea Point, Cape Peninsula, South Africa. From Bokenham & Stephenson (1938).

Significant changes in seaweed communities ensue when removal of existing plants exposes the substratum for colonization. The time taken for the community to return to the predisturbance state is clearly influenced by the nature and season of the disturbance and the type of dominant plant. In communities in which red algae are conspicuous, recovery may take less than 2 years (Bokenham & Stephenson 1938; Kitching 1937; Lee 1966) or as much as 3 to 7 years (Dayton 1971; Murray & Littler 1978; Saito et al. 1976, 1977).

Recruitment may result from regeneration from stolons or holdfasts of previous residents left behind on the rock or from an influx of propagules from plants in adjacent areas. The nature of the latter inoculum is obviously dependent on seasonal changes in the fertility of local plants.

In analyses of the recolonization of denuded surfaces many authors have confounded the order in which the macroscopic plants become apparent with the sequence of colonization. Far from being a latecomer, the eventual "climax" dominant is often among the earliest colonists (Bokenham & Stephenson 1938, Foster 1982a; Kennelly 1983; Mathieson & Prince 1973). The succession of different species of algae is not often the result of sequence of colonization but is a reflection of the relative growth rates and longevity of the participating species (e.g., Fig. 15-12). In Japan perennial red algae usually replace annual forms (Saito et al. 1976), and in western Washington *Porphyra* gives way to a variety of longer-lived red algae before *Mastocarpus papillatus* reestablishes its dominance (Dayton 1971). In New England ephemerals, including *Porphyra*, eventually give way to the perennial *Chondrus crispus* (Lubchenco & Menge 1978), and in southern California *Gigartina canaliculata* supplants other perennial reds that earlier displaced *Ulva* that was consumed by crabs (Sousa 1979). *G. canaliculata* replaces red algae that are more susceptible to desiccation and infestation with epiphytes. Once established, *Gigartina* can prevent invasion by other species.

We know of no unequivocal example of the development of a red alga being facilitated by the presence of an earlier colonist. There are, however, many cases of an alga being prevented from recruiting because of the presence of existing plants (see Section II.J). Moreover, established plants also may inhibit the development of juveniles already settled beneath. For example, overgrowing ephemeral green algae often retard the growth of *Chondrus crispus* (Lubchenco & Menge 1978) and coralline algae (Dayton 1975a; Paine 1977), delaying their eventual dominance. Similarly, other red algae can postpone the appearance of established plants of *Gigartina canaliculata* (Sousa 1979), and the demise of *Cryptosiphonia* may allow *Neorhodomela larix* to replace it (Turner 1983).

Longer-term changes in communities of red algae also occur. Sometimes these result from the introduction of new species (see Section III.A), but more often they follow changes in environmental conditions brought about by human activities. The large-scale demise of grazing sea urchins allows a more diverse flora to develop (Novaczek & McLachlan 1986), although presumably it can decline again rapidly, should the grazers return.

Some dramatic fluctuations in algal populations remain unexplained, such as the marked decline in the abundance of *Mastocarpus stellatus* on the southeast coast of Scotland from 1941 to 1961 and the apparent disappearance of *Nemalion helminthoides* from parts of Wales in the 1940s and 1950s and its equally mysterious reappearance in the 1960s at the same sites it had occupied at the turn of the century (Jones 1974).

C. Demography

Little is known about the recruitment and survivorship of populations of red algae. A single plant of *Gelidium robustum* releases between 34,000 and 300,000 carpospores or 11,000 to 27,000 tetraspores per month (Guzmán del Próo et al. 1972). *Botryocladia pseudodichotoma* produces 3.88×10^6 tetraspores per day (Neushul 1981). A single plant of *Rhodymenia pertusa* is the source of 83×10^6 spores (Boney 1978b), and each of the cystocarpic papillae that cover the surface of *Mastocarpus papillatus* can emit 3.6×10^5 spores (West & Crump 1975). The mean release per month from each square meter of a stand of *Chondrus crispus* was estimated to be 961×10^6 carpospores and 204×10^6 tetraspores (Bhattacharya 1985).

The density of microscopic spores settled on the substratum is only rarely known. During spring tides spores of *Porphyra* settle on shells at a density of 200–1000 cm^{-2} (Takeuchi et al. 1956). Some algae, such as species of *Gelidium*, show low recruitment rates (Jernakoff 1985; Mairh & Rao 1978; Montalva & Santelices 1981). This may be because of a low-percentage germination of the spores, as reported for *Mastocarpus stellatus* (Marshall et al. 1949). Alternatively it may result from high mortality. Spores of seaweeds are rarely dormant. They usually germinate immediately, and consequently many perish if conditions are unsuitable (e.g., Bhattacharya 1985). Probably the bulk of mortality occurs very shortly after settlement, perhaps months before any plants become visible to the unaided eye (e.g., Harger et al. 1981; Hruby & Norton 1979). Not all spores may be equally vulnerable. The frequently observed preponderance of tetrasporophytes over sexual plants could result from carpospores exhibiting lower mortality than tetraspores (Kain 1984b), although this does not seem to be the case with *Chondrus crispus* (Bhattacharya 1985).

Neushul et al. (1976) followed the fate of individual algae on fouling plates from the moment the microscopic plantlets could be identified and revealed a pattern of continual recruitment and attrition (Fig. 15-13). For most species more than 50% of the colonists were lost within a month or so of settlement and virtually all the plants might be eliminated within this period. A few plants survived over 5 months in the case of *Chondria*, 7 months for *Antithamnion*, and 9 months for *Platythamnion* (Fig. 15-13). Shepherd (1981) estimated that for various Australian red algae half the plants in a population would be lost within 5 to 28 months, depending on the species, but it is not clear whether the original density was measured at the time of spore settlement or much later. Most estimates of mortality are made on established stands of macroscopic plants; Hansen and Doyle (1976) found that 90% of a population of *Iridaea splendens* (as *I. cordata*) were juveniles. The density of plants remains the same throughout the year, but it is not clear whether individual fronds persist or are continually replaced. Tagging individual uprights of *Chondrus crispus* revealed that they exhibit increasing mortality as

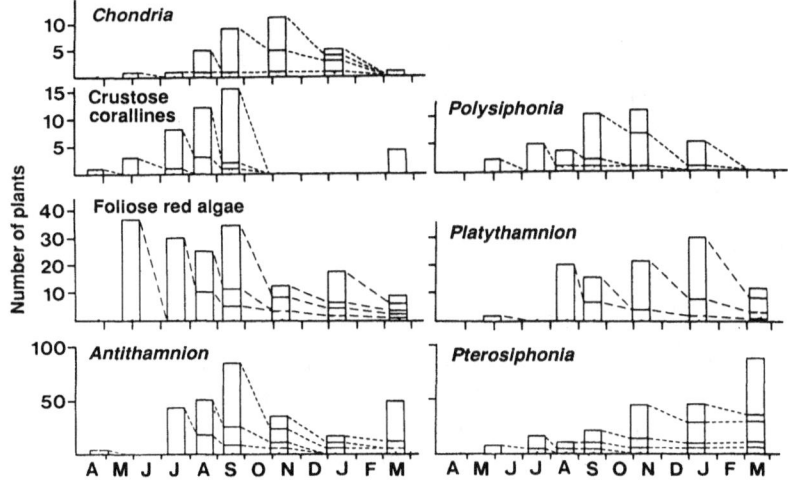

Fig. 15-13. Recruitment and survival rates of red algae on settlement plates. Vertical bars indicate the number of new recruits at each observation time; broken lines indicate the depletion rate of that batch of recruits. Modified from Neushul et al. (1976), courtesy Phycological Society of America.

they grow larger, branch more profusely, or become fertile (Bhattacharya 1985).

Mortality may be greater at higher levels on the shore (Dion & Delépine 1981; Hruby & Norton 1979), and in many species mortality varies seasonally, often being greater in winter as a result of storms (Burns & Mathieson 1972; Kain 1987a; Norall et al. 1981). However, in *Chondrus* mortality of vegetative fronds is greatest in spring and early summer, rather late to be the delayed effects of winter gales, as Bhattacharya (1985) suggests. Survival may be greater for plants attached to barnacles as compared to those on bare rock (Jernakoff 1986), and in members of turfs as opposed to isolated individuals (Hay 1981). In many of these examples mortality refers to the loss of erect fronds rather than to the demise of entire plants. Many red algae can regenerate from creeping stolons or basal crusts after the upright portions have been lost. The holdfasts are thus often considerably older than the uprights they support.

The long-term survival of plants is difficult to assess as most red algae are impossible to age. An exception is *Constantinea* (Lindstrom 1980): A 19-year-old plant of *C. subulifera* has been recorded (Powell 1964). Plants of *Odonthalia dentata* and *Delesseria sanguinea* have been estimated to reach 9 and 16 years of age, respectively (Kain 1984a). By measuring rates of growth, Paine et al. (1979) were able to estimate that larger crusts of *Petrocelis middendorffii* (*Mastocarpus papillatus*) are 25–87 years old. Clearly, the rate of recruitment necessary to maintain a population depends on the longevity of the established plants. In the case of the long-lived *P. middendorffii*, recruitment rates are very low (Paine et al. 1979).

V. SUMMARY

Red algae vary considerably in their responses to environmental factors like temperature and light; for example, some species exist in natural levels of light that are 3×10^5 times that received by others in deep water. Some species are often relegated to the undergrowth beneath larger brown algae and therefore tolerate reduced light (and may be inhibited by high light), but nothing is known about either the nutrient climate or the type of water movement that is prevalent under such conditions. It is known, however, that water movement aids nutrient uptake and that the plants can store nitrogen when it is available in intermittent pulses. Water movement can also be destructive, but few experiments have been made on the direct effects of the turbulent conditions that often prevail in the sea. Most red algae are sensitive to reduced salinity, and few penetrate into waters with a salinity of less than 20‰. Some are highly tolerant of desiccation and occur high on the shore, though others are killed when they lose water. There is usually a distinct belt of red algal turf below the barnacle belt on the shore, especially in the tropics. In deeper water some species may dominate on subtidal rock while others play an important role in the building of coral reefs.

Grazing by animals can have a marked effect on distribution or morphology, with some forms, indeed some phases, being more vulnerable than others. Haploid and diploid phases can show the same temperature tolerances but inexplicably can be very differently distributed. Temperature affects geographical distribution, but it is not known how often other latitudinal differences in the environment determine distribution. Fluctuations may also be important but it is not known how rapidly adaptations to seasonal temperature changes take place. A combination of factors triggers the seasonal development of plants; daylength has been implicated in over 30 species. There is no clear relationship between seasonal variation in species number and latitude, but the number of species showing seasonality in reproduction clearly increases with latitude, whereas the total number of red algal species decreases with latitude.

Only a few red algae have been shown to be long lived; in most, spore production and early mortality are high. Little is known about reproductive ecology, particularly about dispersal distance and the effect of the microtexture of the rock on spore and sporeling survival. One puzzle is how plants without motility in their spores become restricted to very specific habitats.

Although shore ecologists interested in biotic factors have enthusiastically embraced field experimentation as a technique, most students of red algal biology remain firmly laboratory bound. The ecological interpretation of many of their findings awaits the collection of substantive data from the natural environment in situations inhabited by Rhodophyta.

VI. ACKNOWLEDGMENTS

Dr. A. C. Mathieson meticulously reviewed the first draft of this chapter and with remarkable perception made numerous suggestions for its improvement. Barbara Brereton fought with outsize tables and accepted numerous changes in the typescript with great patience. We are extremely grateful to them both.

VII. REFERENCES

Abbott, I. A. 1972. Field studies which evaluate criteria used in separating species of *Iridaea* (Rhodophyta). In *Contributions to the Systematics of Benthic Marine Algae of the North Pacific*, eds. I. A. Abbott & M. Kurogi, pp. 253–64. Kobe, Japan: Japanese Society of Phycology.

Abdel-Rahman, M. H. M. & Magne, F. 1981. Le cycle de développement de l'*Acrochaetium asparagopsis* (Rhodophycées, Acrochaetiales). *Cryptogam. Algol.* 2: 163–70.

Adey, W. H. 1966. Distribution of saxicolous crustose corallines in the northwestern North Atlantic. *J. Phycol.* 2: 49–54.

Adey, W. H. 1970. The effects of light and temperature on growth rates in boreal-subarctic crustose corallines. *J. Phycol.* 6: 269–76.

Adey, W. H. 1971. The sublittoral distribution of crustose corallines on the Norwegian coast. *Sarsia* 46: 41–58.

Adey, W. H. 1973. Temperature control of reproduction and productivity in a subarctic coralline alga. *Phycologia* 12: 111–8.

Adey, W. H. & Vassar J. M. 1975. Colonization, succession and growth rates of tropical crustose coralline algae (Rhodophyta, Cryptonemiales). *Phycologia* 14: 55–69.

Almodovar, L. R. & Biebl, R. 1962. Osmotic resistance of mangrove algae around La Parguera, Puerto Rico. *Rev. Algol.* 6: 203–8.

Anderson, S. M. & Charters, A. C. 1982. A fluid dynamics study of seawater flow through *Gelidium nudifrons*. *Limnol. Oceanogr.* 27: 399–412.

Aranda, J., Niell, F. X. & Fernandez, J. A. 1984. Production of *Asparagopsis armata* (Harvey) in a thermally-stressed intertidal system of low tidal amplitude. *J. Exp. Mar. Biol. Ecol.* 84: 285–95.

Asare, S. O. & Harlin, M. M. 1983. Seasonal fluctuations in tissue nitrogen for five species of perennial macroalgae in Rhode Island Sound. *J. Phycol.* 19: 254–7.

Austin, A. P. 1960a. Observations on *Furcellaria fastigiata* L. Lam. forma *aegagropila* Reinke in Danish waters together with a note on unattached forms. *Hydrobiologia* 14: 255–77.

Austin, A. P. 1960b. Observations on the growth, fruiting and longevity of *Furcellaria fastigiata* (L.) Lam. *Hydrobiologia* 15: 193–207.

Austin, A. P. & Pringle, J. D. 1968. Mitotic index in selected red algae *in situ* I. Preliminary survey. *J. Mar. Biol. Ass. U.K.* 48: 609–35.

Ayling, A. M. 1981. The role of biological disturbance in temperate subtidal encrusting communites. *Ecology* 62: 830–47.

Bailey, A. & Bisalputra, T. 1970. A preliminary account of the application of thin-sectioning, freeze etching, and scanning electron microscopy to the study of coralline algae. *Phycologia* 9: 83–107.

Barilotti, C. D. & Silverthorne, W. 1972. A resource management study of *Gelidium robustum*. *Proc. Int. Seaweed Symp.* 7: 255–61.

Barnes, J. R. & Gonor, J. J. 1973. The larval settling response of the lined chiton *Tonicella lineata*. *Mar. Biol.* 20: 259–64.

Bergquist, P. L. 1960. Notes on the marine algal ecology of some exposed rocky shores of Northland, New Zealand. *Bot. Mar.* 1: 86–94.

Bertness, M. D., Yund, P. O. & Brown, A. F. 1983. Snail grazing and the abundance of algal crusts on a sheltered New England rocky beach. *J. Exp. Mar. Biol. Ecol.* 71: 147–64.

Bhattacharya, D. 1985. The demography of fronds of *Chondrus crispus* Stackhouse. *J. Exp. Mar. Biol. Ecol.* 91: 217–31.

Bidwell, R. G. S. & McLachlan, J. 1985. Carbon nutrition of seaweeds: photosynthesis, photorespiration and respiration. *J. Exp. Mar. Biol. Ecol.* 86: 15–46.

Bidwell, R. G. S., McLachlan, J. & Lloyd, N. D. H. 1985. Tank cultivation of Irish moss, *Chondrus crispus* Stackh. *Bot. Mar.* 28: 87–97.

Biebl, R. 1938. Trochenresistenz und osmotische Empfindlichkeit der Meeresalgen verschieden tiefer Standorte. *Jahrb. Wiss. Botan.* 86: 350–86.

Biebl, R. 1952. Ecological and non-environmental constitutional resistance of the protoplasm of marine algae. *J. Mar. Biol. Ass. U.K.* 31: 307–15.

Biebl, R. 1962. Temperaturresistenz tropischer Meeresalgen. *Bot. Mar.* 4: 241–54.

Biebl, R. 1972. Temperature resistance of marine algae. *Proc. Int. Seaweed Symp.* 7: 23–8.

Bird, C. J. 1973. Aspects of the life history and ecology of *Porphyra linearis* (Bangiales, Rhodophyceae) in nature. *Can. J. Bot.* 51: 2371–9.

Bird, C. J., Greenwell, M. & McLachlan, J. 1984. Benthic marine algal flora of the north shore of Prince Edward Island (Gulf of St Lawrence), Canada. *Aquat. Bot.* 16: 315–35.

Bird, C. J. & McLachlan, J. 1986. The effect of salinity on distribution of species of *Gracilaria* Grev. (Rhodophyta, Gigartinales): an experimental assessment. *Bot. Mar.* 29: 231–8.

Bird, K. T. 1976. Simultaneous assimilation of ammonium and nitrate by *Gelidium nudifrons* (Gelidiales: Rhodophyta). *J. Phycol.* 12: 238–41.

Bird, K. T., Habig, C. & DeBusk, T. 1982. Nitrogen allocation and storage patterns in *Gracilaria tikvahiae* (Rhodophyta). *J. Phycol.* 18: 344–8.

Bird, N. L., Chen, L. C.-M. & McLachlan, J. 1979. Effects of temperature, light and salinity on growth in culture of *Chondrus crispus*, *Furcellaria lumbricalis*, *Gracilaria tikvahiae* (Gigartinales, Rhodophyta), and *Fucus serratus* (Fucales, Phaeophyta). *Bot. Mar.* 22: 521–7.

Bokenham, N. A. H. & Stephenson, T. A. 1938. The colonization of denuded rock surfaces in the intertidal region of the Cape peninsula. *Ann. Natal Mus.* 9: 47–81.

Bolton, J. J. 1986. Marine phytogeography of the Benguela upwelling region on the west coast of southern Africa: a temperature dependent approach. *Bot. Mar.* 29: 251–6.

Boney, A. D. 1962. Observations on the rate of growth of the Hymenoclonium stage of *Bonnemaisonia asparagoides* (Woodw.) Ag. *Br. Phycol. Bull.* 2: 172–3.

Boney, A. D. 1969. *A Biology of Marine Algae*. London: Hutchinson International.

Boney, A. D. 1975. Mucilage sheaths of spores of red algae. *J. Mar. Biol. Ass. U.K.* 44: 511–8.

Boney, A. D. 1978a. Survival and growth of alpha-spores of *Porphyra schizophylla* Hollenberg (Rhodophyta: Bangiophyceae). *J. Exp. Mar. Biol. Ecol.* 35: 7–29.

Boney, A. D. 1978b. The liberation and dispersal of carpospores of the red alga *Rhodymenia pertusa* (Postels et Rupr.) J. Ag. *J. Exp. Mar. Biol. Ecol.* 32: 1–6.

Boney, A. D. 1980. Post attachment responses of monospores of some endophytic *Audouinella* spp. (Nemaliales: Florideophyceae). *Nova Hedwigia* 33: 499–507.

Boney, A. D. & Corner, E. D. S. 1962. The effects of light on the growth of sporelings of the intertidal red algae *Plumaria elegans* (Bonnem.) Schm. *J. Mar. Biol. Ass. U.K.* 42: 65–92.

Boney, A. D. & Corner, E. D. S. 1963. The effect of light on the growth of sporelings of the red algae *Antithamnion plumula* and *Brongniartella byssoides*. *J. Mar. Biol. Ass. U.K.* 43: 319–25.

Børgesen, F. 1908. The algae-vegetation of Faeröese coasts. In *Botany of the Faeroes*, Vol. 3. Copenhagen: Nordiske Forlag.

Boudouresque, C.-F. 1973. Recherches de bionomie analitique, structurale et expérimentale sur les peuplements benthiques sciaphiles de Méditerranée occidentale (fraction algale): Les peuplements sciaphiles de mode relativement calme sur substrats durs. *Bull. Mus. Hist. Nat. Marseille* 33: 147–225.

Boudouresque, C.-F., Augier, A. & Gúenoun, Y.-C. 1972. Végétation marine de L'Il^e de Port-Cros VIII. Premiers résultats de l'étude de la croissance in situ de *Lithophyllum tortuosum* (Rhodophycées, Corallinacées). *Bull. Mus. Hist. Nat. Marseille* 32: 197–215.

Branch, G. M. 1981. The biology of limpets: physical factors, energy flow and ecological interactions. *Oceanogr. Mar. Biol. Annu. Rev.* 19: 235–380.

Braud, J.-P. & Pérez, R. 1979. Farming on pilot scale of *Eucheuma spinosum* (Florideophyceae) in Djibouti water. *Proc. Int. Seaweed Symp.* 9: 533–9.

Brawley, S. H. & Adey, W. H. 1977. Territorial behavior of three-spot damsel-fish (*Eupomacentrus planifrons*) increases reef algal biomass and productivity. *Environ. Biol. Fish.* 2: 45–51.

Brawley, S. H. & Adey, W. H. 1981. The effect of micrograzers on algal community structure in a coral reef microcosm. *Mar. Biol.* 61: 167–77.

Breda, V. A. & Foster, M. S. 1985. Composition, abundance, and phenology of foliose red algae associated with two central California kelp forests. *J. Exp. Mar. Biol. Ecol.* 94: 115–30.

Breeman, A. M., Bos, S., van Essen, S. & van Mulekom, L. L. 1984. Light–dark regimes in the intertidal zone and tetrasporangial periodicity in the red alga *Rhodochorton purpureum*. *Helgol. Meeresunters.* 38: 365–87.

Breeman, A. M. & ten Hoopen, A. 1981. Ecology and distribution of the subtidal red alga *Acrosymphyton purpuriferum* (J. Ag.) Sjöst. (Rhodophyceae, Cryptonemiales). *Aquat. Bot.* 11: 143–66.

Breitburg, D. L. 1984. Residual effects of grazing: inhibition of competitor recruitment by encrusting coralline algae. *Ecology* 65: 1136–43.

Brodie, J. & Guiry, M. D. 1987. Life history and photoperiodic responses in *Cordylecladia erecta* (Rhodophyta) from Ireland. *Br. Phycol. J.* 22: 300–1.

Bunt, J. S. 1970. Uptake of cobalt and vitamin B_{12} by tropical marine macroalgae. *J. Phycol.* 6: 339–43.

Burkholder, P. R. & Almodovar, L. R. 1973. Studies on mangrove algal communities in Puerto Rico. *Florida Sci.* 36: 66–74.

Burns, R. L. & Mathieson, A. C. 1972. Ecological studies of economic red algae. II. Culture studies of *Chondrus crispus* Stackhouse and *Gigartina stellata* (Stackhouse) Batters. *J. Exp. Mar. Biol. Ecol.* 8: 1–6.

Carefoot, T. H. 1967. Growth and nutrition of *Aplysia punctata* feeding on a variety of marine algae. *J. Mar. Biol. Ass. U.K.* 47: 565–90.

Carefoot, T. H. 1973. Feeding, food preference, and the uptake of food energy by the supralittoral isopod *Ligia pallasii*. *Mar. Biol.* 18: 228–36.

Castric-Fey, A., Girard-Descatoire, A., Lafargue, F. & L'Hardy-Halos, M.-T. 1973. Etagement des algues et des invertébrés sessiles dans l'Archipel de Glénan. Définition biologique des horizons bathymétriques. *Helgol. Meeresunters.* 24: 490–509.

Chamberlain, A. H. L. 1976. Algal settlement and secretion of adhesive materials. In *Proceedings of the 3rd International Symposium of Biodegradation*, eds. J. M. Sharpley & A. M. Kaplan, pp. 417–32. London: Applied Science.

Chamberlain, A. H. L. & Evans, L. V. 1973. Aspects of spore production – the red alga *Ceramium*. *Protoplasma* 76: 139–59.

Chapman, V. J. 1960. *Salt Marshes and Salt Deserts of the World*. London: Leonard Hill.

Chapman, V. J. & Chapman, D. J. 1976. Life forms in the algae. *Bot. Mar.* 19: 65–74.

Charnofsky, K., Towill, L. R. & Sommerfeld, M. R. 1982. Light requirements for monospore germination in *Bangia atropurpurea* (Rhodophyta). *J. Phycol.* 18: 417–22.

Charters, A. C., Neushul, M. & Coon, D. 1973. The effect of water motion on algal spore adhesion. *Limnol. Oceanogr.* 18: 884–96.

Chen, L. C.-M. & Taylor, A. R. A. 1980. Investigations of distinct strains of *Chondrus crispus* Stackh. *Bot. Mar.* 23: 435–40.

Cheney, D. P. & Dyer, J. P. III 1974. Deep-water benthic algae of the Florida Middle Ground. *Mar. Biol.* 27: 185–90.

Chevolot-Magueur, A. M., Cave, A., Potier, P., Teste, J., Chiaroni, A. & Riche, C. 1976. Composés bromés de *Rytiphlea tinctoria* (Rhodophyceae). *Phytochemistry* 15: 767–71.

Clokie, J. J. P. & Boney, A. D. 1980a. The assessment of changes in intertidal ecosystems following major reclamation work: framework for interpretation of algal-dominated biota and the use and misuse of data. In *The Shore Environment Vol. 2 Ecosystems*, eds J. H. Price, D. E. G. Irvine & W. F. Farnham, pp. 609–75. London: Academic.

Clokie, J. J. P. & Boney, A. D. 1980b. *Conchocelis* distribution in the Firth of Clyde: estimates of the lower limits of the photic zone. *J. Exp. Mar. Biol. Ecol.* 46: 111–25.

Clokie, J. J. P. & Norton, T. A. 1974. The effects of grazing on the algal vegetation of pebbles from the Firth of Clyde. *Br. Phycol. J.* 9: 216.

Coleman, D. C. & Mathieson, A. C. 1975. Investigations of New England marine algae VII: seasonal occurrence and reproduction of marine algae near Cape Cod, Massachusetts. *Rhodora* 77: 76–104.

Conover, J. T. 1964. The ecology, periodicity and distribution of benthic plants in some Texas lagoons. *Bot. Mar.* 7: 4–41.

Conover, J. T. 1968. The importance of natural diffusion gradients and transport of substances related to benthic marine plant metabolism. *Bot. Mar.* 11: 1–9.

Coon, D. A., Neushul, M. & Charters, A. C. 1972. The settling behaviour of marine algal spores. *Proc. Int.*

Seaweed Symp. 7: 237–42.
Cormaci, M., Duro, A. & Furnani, G. 1984. On reproductive phenology of Ceramiales (Rhodophyta) of east Sicily. Bot. Mar. 27: 95–104.
Correa, J., Avila, M. & Santelices, B. 1985. Effects of some environmental factors on growth of sporelings in two species of Gelidium (Rhodophyta). Aquaculture 44: 221–7.
Cortel-Breeman, A. M. & ten Hoopen, A. 1978. The short day response in Acrosymphyton purpuriferum (J. Ag.) Sjöst. (Rhodophyceae, Cryptonemiales). Phycologia 17: 125–32.
Coutinho, R. & Seeliger, U. 1984. The horizontal distribution of the benthic algal flora in the Patos Lagoon estuary, Brazil, in relation to salinity, substratum and wave exposure. J. Exp. Mar. Biol. Ecol. 80: 247–57.
Cullinane, J. P. & Whelan, P. M. 1983. Subtidal algal communities on the south coast of Ireland. Cryptogam. Algol. 4: 117–25.
Dahl, A. L. 1964. Macroscopic algal foods of Littorina planaxis Philippi and Littorina scutulata Gould. Veliger 7: 139–43.
Dahl, A. L. 1973. Benthic algal ecology in a deep reef and sand habitat off Puerto Rico. Bot. Mar. 16: 171–5.
Daly, M. A. & Mathieson, A. C. 1977. The effects of sand movement on intertidal seaweeds and selected invertebrates at Bound Rock, New Hampshire, U.S.A. Mar. Biol. 43: 45–55.
D'Antonio, C. M. 1985. Epiphytes on the rocky intertidal red alga Rhodomela larix (Turn.) C. Agardh: negative effects on the host and food for herbivores? J. Exp. Mar. Biol. Ecol. 86: 197–218.
D'Antonio, C. M. 1986. Role of sand in the domination of hard substrata by the intertidal alga Rhodomela larix. Mar. Ecol. Prog. Ser. 27: 263–75.
Dawes, C. J., Chen, C.-P., Jewett-Smith, J., Marsh, A. & Watts, S. A. 1984. Effect of phosphate and ammonium levels on photosynthetic and respiratory responses of the red alga Gracilaria verrucosa. Mar. Biol. 78: 325–8.
Dawes, C. J., Earle, S. A. & Croley, F. C. 1967. The offshore benthic flora of the southwest coast of Florida. Bull. Mar. Sci. Gulf Caribb. 17: 211–31.
Dawes, C. J. & McIntosh, R. P. 1981. The effect of organic material and inorganic ions on the photosynthetic rate of the red alga Bostrychia binderi from a Florida estuary. Mar. Biol. 64: 213–8.
Dawes, C. J., Mathieson, A. C. & Cheney, D. P. 1974. Ecological studies of Floridian Eucheuma (Rhodophyta, Gigartinales): I. Seasonal growth and reproduction. Bull. Mar. Sci. Gulf Caribb. 24: 235–73.
Dawson, E. Y., Neushul, M. & Wildman, R. D. 1960. Seaweeds associated with kelp beds along southern California and northwestern Mexico. Pacif. Nat. 1 (14): 1–81.
Dayton, P. K. 1971. Competition, disturbance, and community organization: the provision and subsequent utilization of space in a rocky intertidal community. Ecol. Monogr. 41: 351–89.
Dayton, P. K. 1975a. Experimental evaluation of ecological dominance in a rocky intertidal algal community. Ecol. Monogr. 45: 137–59.
Dayton, P. K. 1975b. Experimental studies of algal canopy interactions in a sea otter-dominated kelp community at Amchitka Island, Alaska. Fish. Bull. Nat. Mar. Fish. Serv. U.S. 73: 230–7.
Dayton, P. K., Currie, V., Gerrodelle, T., Keller, B. D., Rosenthal, R. & ven Tresca, D. 1984. Patch dynamics and stability of some California kelp communities. Ecol. Monogr. 54: 253–89.
DeBoer, J. A., Guigli, H. J., Israel, T. L. & D'Elia, C. F. 1978. Nutritional studies of two red algae. I. Growth rate as a function of nitrogen source and concentration. J. Phycol. 14: 261–6.
DeBoer, J. A. & Whoriskey, F. G. 1983. Production and role of hyaline hairs in Ceramium rubrum. Mar. Biol. 77: 229–34.
DeCew, T. C. & West, J. A. 1981. Investigations on the life histories of three Farlowia species (Rhodophyta: Cryptonemiales, Dumontiaceae) from Pacific North America. Phycologia 20: 342–51.
DeCew, T. C., West, J. A. & Ganesan, E. K. 1981. The life histories and developmental morphology of two species of Gloiosiphonia (Rhodophyta: Cryptonemiales, Gloiosiphoniaceae) from the Pacific coast of North America. Phycologia 20: 415–23.
D'Elia, C. F. & DeBoer, J. A. 1978. Nutritional studies of two red algae. II. Kinetics of ammonium and nitrate uptake. J. Phycol. 14: 266–72.
den Hartog, C. 1972. The effect of the salinity tolerance of algae on their distribution, as exemplified by Bangia. Proc. Int. Seaweed Symp. 7: 274–6.
Denny, M. W., Daniel, T. L. & Koehl, M. A. R. 1985. Mechanical limits to size in wave-swept organisms. Ecol. Monogr. 55: 69–102.
Dethier, M. N. 1982. Pattern and process in tidepool algae: factors influencing seasonality and distribution. Bot. Mar. 25: 55–66.
Devinny, J. S. & Kirkwood, P. D. 1974. Algae associated with kelp beds of the Monterey Peninsula, California. Bot. Mar. 17: 100–6.
De Wreede, R. E. 1983. Sargassum muticum (Fucales, Phaeophyta): regrowth and interaction with Rhodomela larix (Ceramiales, Rhodophyta). Phycologia 22: 153–60.
Deysher, L. & Norton, T. A. 1982. Dispersal and colonization in Sargassum muticum (Yendo) Fensholt. J. Exp. Mar. Biol. Ecol. 56: 179–95.
Dion, P. & Delépine, R. 1981. Studies on the development of Palmaria palmata (Rhodophyceae) using in situ controlled cultures. Proc. Int. Seaweed Symp. 10: 264–70.
Dion, P. & Delépine, R. 1983. Experimental ecology of Gigartina stellata (Rhodophyta) at Roscoff, France, using an in situ culture method. Bot. Mar. 24: 201–11.
Dixon, P. S. 1965. Perennation, vegetative propagation and algal life histories, with special reference to Asparagopsis and other Rhodophyta. Bot. Gothoburg 3: 67–74.
Dixon, P. S. & Irvine, L. M. 1977. Seaweeds of the British Isles Vol. 1. Rhodophyta Pt 1. London: British Museum (Natural History).
Dommasnes, A. 1968. Variations in the meiofauna of Corallina officinalis L. with wave exposure. Sarsia 34: 117–24.
Doty, M. S., Gilbert, W. J. & Abbott, I. A. 1974. Hawaiian marine algae from seaward of the algal ridge. Phycologia 13: 345–57.
Drach, P. 1949. Premières recherches en scaphandre autonome sur les formations de laminaires en zone littorale profonde. C. R. Somm. Séanc. Soc. Biogéog. 227: 46–9.
Dring, M. J. 1967. Effects of daylength on growth and reproduction of the Conchocelis-phase of Porphyra tenera. J. Mar. Biol. Ass. U.K. 47: 501–10.
Dring, M. J. 1982. The Biology of Marine Plants. London: Edward Arnold.

Dring, M. J. 1984. Photoperiodism and phycology. *Prog. Phycol. Res.* 3: 159–92.

Dring, M. J. & West, J. A. 1983. Photoperiodic control of tetrasporangium formation in the red alga *Rhodochorton purpureum*. *Planta* (Berl.) 159: 143–50.

Dromgoole, F. I. & Booth, W. E. 1985. The structure and development of hairs on the thallus of *Gelidium caulacantheum* J. Ag. *N. Z. J. Mar. Freshw. Res.* 19: 43–8.

Druehl, L. D. & Green, J. M. 1982. Vertical distribution of intertidal seaweeds as related to patterns of submersion and emersion. *Mar. Ecol. Prog. Ser.* 9: 163–70.

Edelstein, T., Craigie, J. S. & McLachlan, J. 1969. Preliminary survey of the sublittoral flora of Halifax County. *J. Fish. Res. Bd. Can.* 26: 2703–13.

Edwards, P. 1969. Field and cultural studies on the seasonal periodicity of growth and reproduction of selected Texas benthic marine algae. *Contr. Mar. Sci.* 14: 59–114.

Edwards, P. 1970. Field and cultural observations on the growth and reproduction of *Polysiphonia denudata* from Texas. *Br. Phycol. J.* 5: 145–53.

Edwards, P. 1971. Effects of light intensity, daylength, and temperature on growth and reproduction of *Callithamnion byssoides*. In *Contributions in Phycology*, eds. B. C. Parker & R. M. Brown, pp. 163–73. Lawrence, Kan.: Allen.

Edwards, P. 1977a. An investigation of the vertical distribution of selected benthic marine algae with a tide-simulating apparatus. *J. Phycol.* 13: 62–8.

Edwards, P. 1977b. An analysis of the pattern and rate of cell division, and morphogenesis of the sporelings of *Callithamnion hookeri* (Dillw.) S. F. Gray (Rhodophyta, Ceramiales). *Phycologia* 16: 189–96.

Edwards, P. 1979. A cultural assessment of the distribution of *Callithamnion hookeri* (Dillw.) S. F. Gray (Rhodophyta, Ceramiales) in nature. *Phycologia* 18: 251–63.

Edwards, P. & Kapraun, D. F. 1973. Benthic marine algal ecology in the Port Aransas, Texas area. *Contr. Mar. Sci.* 17: 15–52.

Edyvean, R. G. J. & Ford, H. 1986. Spore production by *Lithophyllum incrustans* (Corallinales, Rhodophyta) in the British Isles. *Br. Phycol. J.* 21: 255–61.

Eiseman, N. J. & Blair, S. M. 1982. New records and range extensions of deepwater algae from East Flower Garden Bank, northwestern Gulf of Mexico. *Contr. Mar. Sci.* 25: 21–6.

Eppley, R. W. & Cyrus, C. C. 1960. Cation regulation and survival of the red alga, *Porphyra perforata*, in diluted and concentrated seawater. *Biol. Bull. Mar. Biol. Lab., Woods Hole* 118: 55–65.

Farnham, W. F. 1980. Studies on aliens in the marine flora of southern England. In *The Shore Environment Vol. 2 Ecosystems*, eds. J. H. Price, D. E. G. Irvine & W. F. Farnham, pp. 875–914. London: Academic.

Farrow, G. E. & Clokie, J. 1979. Molluscan grazing of sublittoral algal-bored shells and the production of carbonate mud in the Firth of Clyde, Scotland. *Trans. Roy. Soc. Edinb.* 70: 139–48.

Feldmann, J. 1951. Ecology of marine algae. In *Manual of Phycology*, ed. G. M. Smith, pp. 313–34. Waltham: Chronica Botanica.

Fledmann, J. 1957. La reproduction des algues marines dans ses rapports avec leur situation géographique. *Coll. Int. Biol. Mar. Sta. Roscoff Ann. Biol.* 33: 49–56.

Femino, R. J. & Mathieson, A. C. 1980. Investigations of New England marine algae IV. The ecology and seasonal succession of tide pool algae at Bald Head Cliff, York, Maine, USA. *Bot. Mar.* 23: 319–32.

Fenical, W. 1975. Halogenation in the Rhodophyta: a review. *J. Phycol.* 11: 245–59.

Fenical, W. & Norris, J. N. 1975. Chemotaxonomy of marine algae: chemical separation of some *Laurencia* species (Rhodophyta) from the Gulf of California. *J. Phycol.* 11: 104–8.

Filion-Myklebust, C. & Norton, T. A. 1981. Epidermis shedding in the brown seaweed *Ascophyllum nodosum* (L.) Le Jolis, and its ecological significance. *Mar. Biol. Lett.* 2: 45–51.

Finckh, A. E. 1904. Biology of the reef-forming organisms at Funafuti Atoll. In *The Atoll of Funafuti. Borings into a Coral Reef and the Results*, pp. 125–50. London: Royal Society.

Fletcher, R. L. 1975. Heteroantagonism observed in mixed algal cultures. *Nature* (Lond.) 253: 534–5.

Fletcher, R. L., Baier, R. E. & Fornalik, M. S. 1985. The effects of surface energy on germling development of some marine macroalgae. *Br. Phycol. J.* 20: 184–5.

Foster, M. S. 1972. The algal turf community in the nest of the Ocean Goldfish, *Hypypops rubicunda*. *Proc. Int. Seaweed Symp.* 7: 55–60.

Foster, M. S. 1975. Regulation of algal community development in a *Macrocystis pyrifera* forest. *Mar. Biol.* 32: 331–42.

Foster, M. S. 1982a. Factors controlling the intertidal zonation of *Iridaea flaccida*. *J. Phycol.* 18: 285–94.

Foster, M. S. 1982b. The regulation of macroalgal associations in kelp forests. In *Synthetic and Degradative Processes in Marine Macrophytes*, ed. L. Srivastava, pp. 185–205. Berlin: DeGruyter.

Fralick, R. A. & Mathieson, A. C. 1975. Physiological ecology of four *Polysiphonia* species (Rhodophyta, Ceramiales). *Mar. Biol.* 29: 29–36.

Fretter, V. & Manly, R. 1977. Algal associations of *Tricolia pullus*, *Lacuna vincta* and *Cerithiopsis tubercularis* (Gastropoda) with special reference to the settlement of their larvae. *J. Mar. Biol. Ass. U.K.* 57: 999–1018.

Friedlander, M. & Dawes, C. J. 1985. In situ uptake kinetics of ammonium and phosphate and chemical composition of the red seaweed *Gracilaria tikvahiae*. *J. Phycol.* 21: 448–53.

Fries, L. 1960. The influence of different B_{12} analogues on the growth of *Goniotrichum elegans* (Chauv.). *Physiol. Plant.* 13: 264–75.

Fries, L. 1961. Vitamin requirements of *Nemalion multifidum*. *Experimentia* 17: 1–4.

Fries, L. 1964. *Polysiphonia urceolata* in axenic culture. *Nature* (Lond.) 202–110.

Fries, L. 1966a. Temperature optima of some red algae in axenic culture. *Bot. Mar.* 9: 12–4.

Fries, L. 1966b. Influence of iodine and bromine on growth of some red algae in axenic culture. *Physiol. Plant.* 19: 800–8.

Fries, L. 1970. The influence of microamounts of organic substances other than vitamins on the growth of some red algae in axenic culture. *Br. Phycol. J.* 5: 39–46.

Fries, L. 1975. Requirement of bromine in a red alga. *Z. Pflanzenphysiol.* 76: 366–8.

Fries, L. 1982. Selenium stimulates growth of marine macroalgae in axenic culture. *J. Phycol.* 18: 328–31.

Fries, L. 1984. D-vitamins and their precursers as growth regulators in axenically cultivated marine macroalgae. *J. Phycol.* 20: 62–6.

Fries, L. & Pettersson, H. 1968. On the physiology of the

red alga *Asterocystis ramosa* in axenic culture. *Br. Phycol. J.* 3: 417–22.

Funano, T. & Sakai, Y. 1967. Ecological studies on 2-year-old *Laminaria religiosa* Prox. growing on the coast of Oshoro Bay, Hokkaido, Japan. *Sci. Rep. Hokkaido Fish. Exp. Stn.* 8: 1–37.

Garbary, D. 1976. Life-forms of algae and their distribution. *Bot. Mar.* 19: 97–106.

Garbary, D. 1979a. Daylength and development in four species of Ceramiaceae (Rhodophyta). *Helgol. Meeresunters.* 32: 213–27.

Garbary, D. 1979b. The effects of temperature on the growth and morphology of some *Audouinella* spp. (Acrochaetiaceae, Rhodophyta). *Bot. Mar.* 22: 493–8.

Gilmartin, M. 1960. The ecological distribution of the deep water algae of Eniwetok Atoll. *Ecology* 41: 210–21.

Goff, L. J. 1982. The biology of parasitic red algae. *Prog. Phycol. Res.* 3: 289–369.

Goss-Custard, S., Jones, J., Kitching, J. A. & Norton, T. A. 1979. The ecology of Lough Ine. XXI. Tide pools of Carrigathorna and Barloge Creek, County Cork. *Phil. Trans. R. Soc. Ser. B* 287: 1–44.

Grime, J. P. 1977. Evidence for the existence of three primary strategies in plants and its relevance to ecological and evolutionary theory. *Am. Nat.* 111: 1169–94.

Guiry, M. D. 1984a. Structure, life history and hybridization of Atlantic *Gigartina teedii* (Rhodophyta) in culture. *Br. Phycol. J.* 19: 37–55.

Guiry, M. D. 1984b. Photoperiodic and temperature responses in the growth and tetrasporogenesis of *Gigartina acicularis* (Rhodophyta) from Ireland. *Helgol. Meeresunters.* 38: 335–47.

Guiry, M. D. & Cunningham, E. M. 1984. Photoperiodic and temperature responses in the reproduction of north-eastern Atlantic *Gigartina acicularis* (Rhodophyta; Gigartinales). *Phycologia* 23: 357–67.

Guiry, M. D. & Ottway, B. 1981. Maricultural studies on *Gracilaria foliifera*, an agar-producing seaweed. *Proc. Econ. Med. Pl. Res. Assoc.* 3: 85–95.

Guiry, M. D. & Maggs, C. A. 1984. Reproduction and life history of *Meredithia microphylla* (J. Ag.) J. Ag. (Kallymeniaceae, Rhodophyta) from Ireland. *Giorn. Bot. Ital.* 118: 105–24.

Guiry, M. D. & West, J. A. 1983. Life history and hybridization studies on *Gigartina stellata* and *Petrocelis cruenta* (Rhodophyta) in the North Atlantic. *J. Phycol.* 19: 474–94.

Gutnecht, J. 1963. Zn^{65} uptake by benthic marine algae. *Limnol. Oceanogr.* 8: 31–8.

Guzmán del Próo, S. A. & de la Campa de Guzman, S. 1979. *Gelidium robustum* (Florideophyceae), an agarophyte of Baja California, Mexico. *Proc. Int. Seaweed Symp.* 9: 303–8.

Guzmán del Próo, S. A., de la Campa de Guzman, S. & Pineda-Barrera, J. 1972. Shedding rhythm and germination of spores in *Gelidium robustum*. *Proc. Int. Seaweed Symp.* 7: 221–30.

Haines, K. C. & Wheeler, P. A. 1978. Ammonium and nitrate uptake by the marine macrophytes *Hypnea musciformis* (Rhodophyta) and *Macrocystis pyrifera* (Phaeophyta). *J. Phycol.* 14: 319–24.

Hammer, L. 1968. Salzgehalt und Photosynthese bei marinen Pflanzen. *Mar. Biol.* 1: 185–90.

Hannach, G. & Santelices, B. 1985. Ecological differences between the isomorphic reproductive phases of two species of *Iridaea* (Rhodophyta: Gigartinales). *Mar. Ecol. Prog. Ser.* 22: 291–303.

Hansen, J. E. 1977. Ecology and natural history of *Iridaea cordata* (Gigartinales, Rhodophyta) growth. *J. Phycol.* 13: 395–402.

Hansen, J. E. 1980. Physiological considerations in the mariculture of red algae. In *Pacific Seaweed Aquaculture*, eds. I. A. Abbott, M. S. Foster & L. F. Eklund, pp. 80–91. La Jolla: Institute of Marine Resources, University of California.

Hansen, J. E. 1983. A physiological approach to mariculture of red algae. *J. Wld. Maricult. Soc.* 14:380–91.

Hansen, J. E. & Doyle, W. T. 1976. Ecology and natural history of *Iridaea cordata* (Rhodopyta: Gigartinaceae): population structure. *J. Phycol.* 12: 273–8.

Hardwick-Witman, M. N. & Mathieson, A. C. 1983. Intertidal macroalgae and macroinvertebrates: seasonal and spatial abundance patterns along an estuarine gradient. *Est. Coast. Shelf Sci.* 16: 113–29.

Harger, B. W. W., Coon, D. A., Foster, M. S., Neushul, M. & Woessner, J. 1981. Settlement and reproduction of benthic marine algae within *Macrocystis* forests in California. *Proc. Int. Seaweed Symp.* 8: 342–52.

Harkin, E. 1981. Fluctuations in epiphyte biomass following *Laminaria hyperborea* canopy removal. *Proc. Int. Seaweed Symp.* 10: 303–8.

Harlin, M. M. & Craigie, J. S. 1975. The distribution of photosynthate in *Ascophyllum nodosum* as it relates to epiphytic *Polysiphonia lanosa*. *J. Phycol.* 11: 109–13.

Harlin, M. M. & Lindbergh, J. M. 1977. Selection of substrata by seaweeds: optimal surface relief. *Mar. Biol.* 40: 33–40.

Hawkins, S. J. 1981. The influence of *Patella* grazing on the fucoid/barnacle mosaic on moderately exposed rocky shores. *Kieler Meeresforsch.* 5: 537–43.

Hawkins, S. J. & Harkin, E. 1985. Preliminary canopy removal experiments in algal dominated communities low on the shore and in the shallow subtidal on the Isle of Man. *Bot. Mar.* 28: 223–30.

Hawkins, S. J. & Hartnoll, R. G. 1985. Factors determining the upper limits of intertidal canopy-forming algae. *Mar. Ecol. Prog. Ser.* 20: 265–71.

Hay, M. E. 1981. The functional morphology of turf-forming seaweeds: persistence in stressful marine habitats. *Ecology* 62: 739–50.

Hay, M. E. & Norris, J. N. 1984. Seasonal reproduction and abundance of six sympatric species of *Gracilaria* Grev. (Gracilariaceae; Rhodophyta) on a Caribbean subtidal sand plain. *Proc. Int. Seaweed Symp.* 11: 63–72.

Healey, F. P. 1972. Photosynthesis and respiration of some Arctic seaweeds. *Phycologia* 11: 267–71.

Heine, J. N. 1983. Seasonal productivity of two red algae in a central California kelp bed. *J. Phycol.* 19: 146–52.

Himmelman, J. H. & Carefoot, T. H. 1975. Seasonal changes in calorific value of three Pacific coast seaweeds and their significance to some marine invertebrate herbivores. *J. Exp. Mar. Biol. Ecol.* 18: 139–51.

Hiscock, K. & Mitchell, R. 1980. The description and classification of sublittoral epibenthic ecosystems. In *The Shore Environment, Vol. 2: Ecosystems*, eds. J. H. Price, D. E. G. Irvine & W. F. Farnham, pp. 323–70. London: Academic.

Hodgson, L. M. 1981. Photosynthesis of the red alga *Gastroclonium coulteri* (Rhodophyta) in response to changes in temperature, light intensity, and desiccation. *J.*

Phycol. 17: 37–42.

Hodgson, L. M. 1984. Desiccation tolerance of *Gracilaria tikvahiae* (Rhodophyta). *J. Phycol.* 20: 444–6.

Hodgson, L. M. & Waaland, J. R. 1979. Seasonal variation in the subtidal macroalgae of Fox Island, Puget Sound, Washington. *Syesis* 12: 107–12.

Hoffman, W. E. & Dawes, C. J. 1980. Photosynthetic rates and primary production by two Florida benthic red algal species from a salt marsh and a mangrove community. *Bull. Mar. Sci.* 30: 358–64.

Hommersand, M. 1986. The biogeography of the South African marine red algae: a model. *Bot. Mar.* 29: 257–70.

Hooper, R. G., South, G. R. & Whittick, A. 1980. Ecological and phenological aspects of the marine phytobenthos of the island of Newfoundland. In *The Shore Environment, Vol. 2: Ecosystems*, eds. J. H. Price, D. E. G. Irvine & W. F. Farnham, pp. 395–423. London: Academic.

Horn, M. H., Murray, S. N. & Edwards, T. W. 1982. Dietary selectivity in the field and food preferences in the laboratory for two herbivorous fishes (*Cebidichthys violaceus* and *Xiphister mucosus*) from temperate intertidal zone. *Mar. Biol.* 67: 237–46.

Horn, M. H., Murray, S. N. & Seapy, R. R. 1983. Seasonal structure of a central California rocky intertidal community in relation to environmental variations. *Bull. S. Calif. Acad. Sci.* 82: 79–94.

Horn, M. H., Neighbors, M. A., Rosenberg, M. J. & Murray, S. N. 1985. Assimilation of carbon from dietary and non-dietary macroalgae by a temperate-zone intertidal fish, *Cebidichthys violaceus* Girard (Teleostei: Stichaediae). *J. Exp. Mar. Biol. Ecol.* 86: 241–53.

Hoyle, M. D. 1978. Reproductive phenology and growth rates in two species of *Gracilaria* from Hawaii. *J. Exp. Mar. Biol. Ecol.* 35: 273–83.

Hruby, T. 1976. Observations of algal zonation resulting from competition. *Est. Coast. Mar. Sci.* 4: 231–3.

Hruby, T. & Norton, T. A. 1979. Algal colonization on rocky shores in the Firth of Clyde. *J. Ecol.* 67: 65–77.

Huang, R. & Boney, A. D. 1984. Growth interactions between littoral diatoms and juvenile marine algae. *J. Exp. Mar. Biol. Ecol.* 81: 21–46.

Huang, R. & Boney, A. D. 1985. Individual and combined interactions between littoral diatoms and sporelings of red algae. *J. Exp. Mar. Biol. Ecol.* 85: 101–12.

Humm, H. J. & Kreuzer, J. 1975. On the growth rate of the red alga, *Hypnea musciformis*, in the Caribbean Sea. *Carib. J. Sci.* 15: 1–4.

Iwasaki, H. 1965. Nutritional studies of the edible seaweed *Porphyra tenera* I. The influence of different B_{12} analogues, plant hormones, purines and pyrimidines on the growth of *Conchocelis*. *Plant Cell Physiol.* 6: 325–36.

Iwasaki, H. 1967. Nutritional studies of the edible seaweed *Porphyra tenera*. II. Nutrition of *Conchocelis*. *J. Phycol.* 3: 30–4.

Jannsson, A.-M. & Kautsky, N. 1977. Quantitative survey of hard bottom communities in a Baltic Archipelago. In *Biology of Benthic Organisms, 11th European Symposium on Marine Biology*, eds. B. F. Keegan, P. O'Céidigh & P. J. S. Boaden, pp. 359–66. Oxford: Pergamon.

Jara, H. F. & Moreno, C. A. 1984. Herbivory and structure in a mid-littoral rocky community: a case in southern Chile. *Ecology* 65: 28–38.

Jassby, A. D. & Platt, T. 1976. Mathematical formulation of the relationship between photosynthesis and light for phytoplankton. *Limnol. Oceanogr.* 21: 540–7.

Jerlov, N. G. 1976. *Marine Optics*. Amsterdam: Elsevier.

Jernakoff, P. 1985. An experimental evaluation of the influence of barnacles, crevices and seasonal patterns of grazing on algal diversity and cover in an intertidal barnacle zone. *J. Exp. Mar. Biol. Ecol.* 88: 287–302.

Jernakoff, P. 1986. Experimental investigation of interactions between the perennial red alga *Gelidium pusillum* and barnacles on a New South Wales rocky shore. *Mar. Ecol. Prog. Ser.* 28: 259–63.

Jernakoff, P. & Fairweather, P. G. 1985. An experimental analysis of interactions among several intertidal organisms. *J. Exp. Mar. Biol. Ecol.* 94: 71–88.

Johansen, H. W. & Austin, L. F. 1970. Growth rates in the articulated coralline *Calliarthron* (Rhodophyta). *Can. J. Bot.* 48: 125–32.

John, D. M., Lieberman, D. & Lieberman, M. 1977. A quantitative study of the structure and dynamics of benthic subtidal algal vegetation in Ghana (tropical West Africa). *J. Ecol.* 65: 497–521.

Johnson, C. R. & Mann, K. H. 1986. The crustose coralline alga *Phymatolithon* Foslie inhibits the overgrowth of seaweeds without relying on herbivores. *J. Exp. Mar. Biol. Ecol.* 96: 127–46.

Johnson, W. S., Gigon, A., Gulmin, S. L. & Mooney, H. A. 1974. Comparative photosynthetic capacities of intertidal algae under exposed and submerged conditions. *Ecology* 55: 450–3.

Johnston, C. S. 1969. Studies on the ecology and primary production of Canary Islands marine algae. *Proc. Int. Seaweed Symp.* 6: 213–22.

Jolliffe, E. A. & Tregunna, E. B. 1970. Studies on HCO_3^- ion uptake during photosynthesis in benthic marine algae. *Phycologia* 9: 293–303.

Jones, N. S. & Kain, J. M. 1967. Subtidal algal colonization following the removal of *Echinus*. *Helgol. Meeresunters.* 15: 460–6.

Jones, W. E. 1956. Effect of spore coalescence on the early development of *Gracilaria verrucosa* (Hudson) Papenfuss. *Nature* (Lond.) 178: 426–7.

Jones, W. E. 1959a. Experiments on some effects of certain environmental factors on *Gracilaria verrucosa* (Hudson) Papenfuss. *J. Mar. Biol. Ass. U.K.* 38: 153–67.

Jones, W. E. 1959b. The growth and fruiting of *Gracilaria verrucosa* (Hudson) Papenfuss. *J. Mar. Biol. Ass. U.K.* 38: 47–56.

Jones, W. E. 1974. Changes in the seaweed flora of the British Isles. In *The Changing Flora and Fauna of Britain*, ed. D. L. Hawksworth, pp. 97–113. London: Academic.

Jones, W. E. & Dent, E. S. 1971. The effect of light on the growth of algal spores. In *Proceedings of the IVth European Marine Biology Symposium*, ed. D. J. Crisp, pp. 363–74. Cambridge: Cambridge University Press.

Jónsson, S. 1972. Marine benthic algae recorded in Surtsey during the field seasons of 1969–1970. *Surtsey Res. Prog. Rep.* 6: 75–6.

Joosten, A. M. T. & van den Hoek, C. 1986. World-wide relationships between red seaweed floras: a multivariate approach. *Bot. Mar.* 29: 195–214.

Jorde, I. 1966. Algal associations of a coastal area south of Bergen, Norway. *Sarsia* 23: 1–52.

Jorde, I. & Klavestad, N. 1963. The natural history of the Hardangerfjord. 4. The benthonic algal vegetation. *Sarsia* 9: 1–99.

Kain, J. M. 1960. Direct observations on some Manx sublittoral algae. *J. Mar. Biol. Ass. U.K.* 39: 609–30.

Kain, J. M. 1975. Algal recolonization of some cleared

subtidal areas. *J. Ecol.* 63: 739–65.
Kain, J. M. 1982. The reproductive phenology of nine species of Rhodophyta in the subtidal region of the Isle of Man. *Br. Phycol. J.* 17: 321–31.
Kain, J. M. 1984a. Seasonal growth of two subtidal species of Rhodophyta off the Isle of Man. *J. Exp. Mar. Biol. Ecol.* 82: 207–20.
Kain, J. M. 1984b. *Plocamium cartilagineum* in the Isle of Man: why are there so many tetrasporophytes? *Br. Phycol. J.* 19: 195.
Kain, J. M. 1986. Plant size and reproductive phenology of six species of Rhodophyta in subtidal Isle of Man. *Br. Phycol. J.* 21: 129–38.
Kain, J. M. 1987a. Seasonal growth and photoinhibition in *Plocamium cartilagineum* (Rhodophyta) off the Isle of Man. *Phycologia* 26: 88–99.
Kain, J. M. 1987b. Photoperiod and temperature as triggers in the seasonality of *Delesseria sanguinea*. *Helgol. Meeresunters.* 41: 355–70.
Kain, J. M. 1989. The seasons in the subtidal. *Br. Phycol. J.* 24: 203–15.
Kaliaperumal, N. & Rao, M. U. 1985. Seasonal growth, reproduction and spore shedding in *Pterocladia heteroplatos*. *Proc. Ind. Acad. Sci (Plant Sci.)* 94: 627–32.
Kanwisher, J. 1957. Freezing and drying in intertidal algae. *Biol. Bull. Mar. Biol. Lab.* (Woods Hole) 113: 275–85.
Kapraun, D. F. 1978. Field and cultural studies on selected North Carolina *Polysiphonia* species. *Bot. Mar.* 21: 143–53.
Kapraun, D. F. 1979. Comparative studies of *Polysiphonia urceolata* from three North Atlantic sites. *Norw. J. Bot.* 26: 269–76.
Kapraun, D. F. 1980. Floristic affinities of North Carolina inshore benthic marine algae. *Phycologia* 19: 245–52.
Kapraun, D. F. & Luster, D. G. 1980. Field and culture studies of *Porphyra rosengurtii* Coll et Cox (Rhodophyta, Bangiales) from North Carolina. *Bot. Mar.* 23: 449–57.
Kastendiek, J. 1982. Competitor-mediated co-existence: interactions among three species of benthic macroalgae. *J. Exp. Mar. Biol. Ecol.* 62: 201–10.
Kautsky, N. 1974. Quantitative investigations of the red algal belt in the Åsko area, Northern Baltic proper. *Contr. Åsko Lab.* 3: 1–29.
Kennelly, S. J. 1983. An experimental approach to the study of factors affecting algal colonization in a sublittoral kelp forest. *J. Exp. Mar. Biol. Ecol.* 68: 257–76.
Khfaji, A. K. & Boney, A. D. 1979. Antibiotic effects of crustose germlings of the red alga *Chondrus crispus* Stackh. on benthic diatoms. *Ann. Bot.* 43: 231–2.
Kilar, J. A. & Mathieson, A. C. 1978. Ecological studies of the annual red alga *Dumontia incrassata* (O. F. Müller) Lamouroux. *Bot. Mar.* 21: 423–37.
King, R. J. & Schramm, W. 1976. Photosynthetic rates of benthic marine algae in relation to light intensity and seasonal variations. *Mar. Biol.* 37: 215–22.
King, R. J. & Schramm, W. 1982. Calcification in the maerl coralline alga *Phymatolithon calcareum*: effects of salinity and temperature. *Mar. Biol.* 70: 197–204.
Kitching, J. A. 1937. Studies in sublittoral ecology. II. Recolonization at the upper margin of the sublittoral region; with a note on the denudation of *Laminaria* forests by storms. *J. Ecol.* 25: 482–95.
Kitting, C. L. 1980. Herbivore/plant interactions of individual limpets maintaining a mixed diet of intertidal marine algae. *Ecol. Monogr.* 50: 527–50.
Kjeldsen, C. K. & Phinney, H. K. 1972. Effects of variations in salinity and temperature on some estuarine macro-algae. *Proc. Int. Seaweed Symp.* 7: 301–8.
Knight-Jones, E. W. & Nelson-Smith, A. 1977. Sublittoral transect in the Menai Straits and Milford Haven. In *Biology of Benthic Organisms, 11th European Symposium on Marine Biology*, eds. B. F. Keegan, P. O'Céidigh & P. J. S. Boaden, pp. 379–98. Oxford: Pergamon.
Koehl, M. A. R. 1986. Seaweeds in moving water: form and mechanical function. In *On the Economy of Plant Form and Function*, ed. T. J. Givnish, pp. 603–34. Cambridge: Cambridge University Press.
Lang, J. C. 1974. Biological zonation at the base of a reef. *Am. Scient.* 62: 272–81.
Lapointe, B. E. 1985. Strategies for pulsed nutrient supply to *Gracilaria* cultures in the Florida Keys: interactions between concentration and frequency of nutrient pulses. *J. Exp. Mar. Biol. Ecol.* 93: 211–22.
Lapointe, B. E., Rice, D. L. & Lawrence, J. M. 1984a. Responses of photosynthesis, respiration, growth and cellular constituents to hypo-osmotic shock in the red alga *Gracilaria tikvahiae*. *Comp. Biochem. Physiol. A* 77: 127–32.
Lapointe, B. E. & Ryther, J. H. 1979. The effects of nitrogen and seawater flow rate on the growth and biochemical composition of *Gracilaria foliifera* var. *angustissima* in mass outdoor cultures. *Bot. Mar.* 22: 529–37.
Lapointe, B. E., Tenore, K. R. & Dawes, C. J. 1984b. Interactions between light and temperature on the physiological ecology of *Gracilaria tikvahiae* (Gigartinales: Rhodophyta). I. Growth, photosynthesis and respiration. *Mar. Biol.* 80: 161–70.
Larkum, A. W. D., Drew, E. A. & Crossett, R. N. 1967. The vertical distribution of attached marine algae in Malta. *J. Ecol.* 55: 361–71.
Larson, B. R., Vadas, R. L. & Keser, M. 1980. Feeding and nutritional ecology in the sea urchin *Strongylocentrotus drobachiensis* in Maine, U.S.A. *Mar. Biol.* 59: 59–62.
Lauret, M. 1974. Étude phytosociologique préliminaire sur les gazons à *Pterosiphonia pennata* (Rhodophycées, Céramiales). *Bull. Soc. Phycol. Fr.* 19: 229–37.
Lawrence, J. M. 1975. Marine plant and sea urchin relationships. *Oceanogr. Mar. Biol. Annu. Rev.* 13: 213–85.
Lawson, G. W. 1957. Seasonal variation of intertidal zonation on the coast of Ghana in relation to tidal factors. *J. Ecol.* 45: 831–60.
Lawson, G. W. 1966. The littoral ecology of West Africa. *Oceanogr. Mar. Biol. Annu. Rev.* 4: 405–48.
Lawson, G. W. 1978. The distribution of seaweed floras in the tropical and subtropical Atlantic Ocean: a quantitative approach. *Bot. J. Linn. Soc.* 76: 177–93.
Lawson, G. W., John, D. M. & Price, J. H. 1975. The marine algal flora of Angola: its distribution and affinities. *Bot. J. Linn. Soc.* 70: 307–24.
Laycock, M. V., Morgan, K. C. & Craigie, J. S. 1981. Physiological factors affecting the accumulation of L-citrullinyl-L-arginine in *Chondrus crispus*. *Can. J. Bot.* 59: 522–7.
Lee, R. K. S. 1966. Development of marine benthic algal communities on Vancouver Island, British Columbia. In *The Evolution of Canada's Flora*. eds. R. L. Taylor & R. A. Ludwig, pp. 100–20. Toronto: University of Toronto Press.
Lee, Y. P. & Kurogi, M. 1983. The life history of *Audouinella alariae* (Jonsson) Woelkerling (Rhodophyta, Acrochaetiaceae) in nature and culture. *J. Fac. Sci. Hokkaido Univ. Ser. 5 (Botany)* 13: 57–76.

Lehnberg, W. 1978. Die Wirkung eines Licht-Temperatur-Salzhalt Komplexes auf den Gaswechsel von *Delesseria sanguinea* (Rhodophyta) aus der westlichen Ostsee. *Bot. Mar.* 21: 485–97.

Leighton, D. L. 1971. Grazing activities of benthic invertebrates in southern California kelp beds. In *The Biology of Giant Kelp Beds (Macrocystis) in California*, ed. W. J. North, pp. 421–53. Lehre, FRG: Cramer.

Levring, T., Hoppe, H. A. & Schmid, O. 1969. *Marine Algae, a Survey of Research and Utilization.* Hamburg: Cram de Gruyter.

Lewis, J. A. & Kraft, G. T. 1979. Occurrence of a European red alga (*Schottera nicaeensis*) in Southern Australian water. *J. Phycol.* 15: 226–30.

Lewis, J. R. 1964. *The Ecology of Rocky Shores.* London: English Universities Press.

Lieberman, M., John, D. M. & Lieberman, D. 1979. Ecology of subtidal algae on seasonally devastated cobble substrates off Ghana. *Ecology* 60: 1151–61.

Lindstrom, S. 1980. New blade initiation in the perennial red alga *Constantinea rosa-marina* (Gmelin) Postels et Ruprecht (Cryptonemiales, Dumontiaceae). *Jap. J. Phycol.* 28: 141–50.

Lipkin, Y. 1972. Marine algal and sea-grass flora of the Suez Canal (the significance of this flora to the understanding of the recent migration through the Canal). *Israel J. Zoo.* 21: 405–46.

Littler, M. M. 1976. Calcification and its role among the macroalgae. *Micronesia* 12: 27–41.

Littler, M. M. & Arnold, K. E. 1982. Primary productivity of marine macroalgal functional-form groups from southwestern North America. *J. Phycol.* 18: 307–11.

Littler, M. M. & Doty, M. S. 1975. Ecological components structuring the seaward edges of tropical Pacific reefs. The distribution of communities and productivity of *Porolithon. J. Ecol.* 63: 117–29.

Littler, M. M. & Kauker, B. J. 1984. Heterotrichy and survival strategies in the red alga *Corallina officinalis* L. *Bot. Mar.* 27: 37–44.

Littler, M. M. & Littler, D. S. 1980. The evolution of thallus form and survival strategies in benthic marine macroalgae: field and laboratory tests of functional form model. *Am. Nat.* 116: 25–44.

Littler, M. M. & Littler, D. S. 1981. Intertidal macrophyte communities from Pacific Baja California: relatively constant vs. environmentally fluctuating systems. *Mar. Ecol. Prog. Ser.* 4: 145–58.

Littler, M. M. & Littler, D. S. 1984a. Relationships between macroalgal functional form groups and substrata stability in a subtropical rocky intertidal system. *J. Exp. Mar. Biol. Ecol.* 74: 13–34.

Littler, M. M. & Littler, D. S. 1984b. Models of tropical reef biogenesis: the contribution of algae. *Prog. Phycol. Res.* 3: 323–64.

Littler, M. M., Littler, D. S., Blair, S. M. & Norris, J. N. 1985. Deepest known plant life discovered on an uncharted seamount. *Repr. Ser. Amer. Assoc. Adv. Sci.* 227: 57–9.

Littler, M. M., Littler, D. S. & Taylor, P. R. 1983. Evolutionary strategies in a tropical barrier reef system: functional-form groups of marine macroalgae. *J. Phycol.* 19: 229–37.

Littler, M. M. & Murray, S. N. 1974. The primary productivity of marine macrophytes from a rocky intertidal community. *Mar. Biol.* 27: 131–6.

Littler, M. M., Murray, S. N. & Arnold, K. E. 1979. Seasonal variations in net photosynthetic performance and cover of intertidal macrophytes. *Aquat. Bot.* 7: 35–46.

Lubchenco, J. 1978. Plant species diversity in a marine intertidal community: importance of herbivore food preference and algal competetive ability. *Am. Nat.* 112: 23–39.

Lubchenco, J. 1980. Algal zonation in the New England rocky intertidal community: an experimental analysis. *Ecology* 61: 333–44.

Lubchenco, J. & Cubit, J. 1980. Heteromorphic life histories of certain marine algae as adaptations to variations in herbivory. *Ecology* 64: 1116–23.

Lubchenco, J. & Menge, B. A. 1978. Community development and persistence in a low rocky intertidal zone. *Ecol. Monogr.* 48: 67–94.

Lundberg, B. & Lipkin, Y. 1979. Natural food of the herbivorous rabbit fish (*Siganus* spp.) in the northern Red Sea. *Bot. Mar.* 22: 173–81.

Lüning, K. 1970. Tauchuntersuchungen zur Vertikalverteilung der sublitoralen Helgoländer Algenvegetation. *Helgol. Meeresunters.* 21: 271–91.

Lüning, K. 1979. Growth strategies of three *Laminaria* species (Phaeophyceae) inhabiting different depth zones in the sublittoral region of Helgoland (North Sea). *Mar. Ecol. Prog. Ser.* 1: 195–207.

Lüning, K. 1981. Photomorphogenesis of reproduction in marine algae. *Ber. Deutsch. Bot. Ges.* 94: 401–17.

Lüning, K. 1984. Temperature tolerance and biogeography of seaweeds: the marine algal flora of Helgoland (North Sea) as an example. *Helgol. Meeresunters.* 38: 305–17.

Lüning, K. & Dring, M. J. 1979. Continuous underwater light measurement near Helgoland (North Sea) and its significance for characteristic light limits in the sublittoral region. *Helgol. Meeresunters.* 32: 403–24.

Maggs, C. A. 1987. A long-day photoperiodic response in *Atractophora* (Rhodophyta). *Br. Phycol. J.* 22: 307–8.

Maggs, C. A. & Guiry, M. D. 1982. Morphology, phenology and photoperiodism in *Halymenia latifolia* Kütz. (Rhodophyta) from Ireland. *Bot. Mar.* 25: 589–99.

Maggs, C. A. & Guiry, M. D. 1985. Life history and reproduction of *Schmitzia hiscockiana* sp. nov. (Rhodophyta, Gigartinales) from the British Isles. *Phycologia* 24: 297–310.

Maggs, C. A. & Guiry, M. D. 1987. Environmental control of macroalgal phenology. In *Plant Life in Aquatic and Amphibious Habitats*, ed. R. M. M. Crawford, pp. 359–73. Special Publication No. 5 of the British Ecological Society.

Mairh, O. P. & Rao, P. S. 1978. Culture studies on *Gelidium pusillum* (Stackh.) Le Jolis. *Bot. Mar.* 28: 169–74.

Markham, J. W. & Newroth, P. R. 1972. Observations on the ecology of *Gymnogongrus linearis* and related species. *Proc. Int. Seaweed Symp.* 7: 126–30.

Marshall, S. M., Newton, L. & Orr, A. P. 1949. *A Study of Certain British Seaweeds and Their Utilisation in the Preparation of Agar.* London: His Majesty's Stationery Office.

Masaki, T., Fujita, D. & Akioka, H. 1981. Observations on the spore germination of *Laminaria japonica* on *Lithophyllum yessoense* (Rhodophyta, Corallinaceae) in culture. *Bull. Fac. Fish. Hokkaido Univ.* 32: 349–56.

Mathieson, A. C. 1979. Vertical distribution and longevity of subtidal seaweeds in northern New England. *Bot. Mar.* 30: 511–20.

Mathieson, A. C. & Burns, R. L. 1971. Ecological studies of economic red algae. I. Photosynthesis and respiration of *Chondrus crispus* Stackhouse and *Gigartina stellata*

(Stackhouse) Batters. *J. Exp. Mar. Biol. Ecol.* 7: 197–206.
Mathieson, A. C. & Burns, R. L. 1975. Ecological studies of economic red algae. V. Growth and reproduction of natural and harvested populations of *Chondrus crispus* Stackhouse in New Hampshire. *J. Exp. Mar. Biol. Ecol.* 17: 137–56.
Mathieson, A. C. & Dawes, C. J. 1974. Ecological studies of Floridian *Eucheuma* (Rhodophyta, Gigartinales). II. Photosynthesis and respiration. *Bull. Mar. Sci.* 24: 274–85.
Mathieson, A. C. & Dawes, C. J. 1975. Seasonal studies of Florida sublittoral marine algae. *Bull. Mar. Sci. Gulf Carib.* 25: 46–65.
Mathieson, A. C., Fralick, R. A., Burns, R. L. & Flashive, W. 1975. Phycological studies during Tektite II, at St. John, U.S.V.I. In *Results of the Tektite Program: Coral Reef Invertebrates and Plants*, eds. S. A. Earle & R. L. Lavenberg, pp. 77–103. Natural History Museum of Los Angeles Country, Science Bulletin 20.
Mathieson, A. C., Hehre, E. J. & Reynolds, N. B. 1981. Investigations of New England marine algae I: a floristic and descriptive ecological study of the marine algae at Jaffrey Point, New Hampshire, U.S.A. *Bot. Mar.* 24: 521–32.
Mathieson, A. C. & Norall, T. L. 1975a. Physiological studies of subtidal red algae. *J. Exp. Mar. Biol. Ecol.* 20: 237–47.
Mathieson, A. C. & Norall, T. L. 1975b. Photosynthetic studies of *Chondrus crispus*. *Mar. Biol.* 33: 207–313.
Mathieson, A. C. & Penniman, C. A. 1986. Species composition and seasonality of New England seaweeds along an open coastal-estuarine gradient. *Bot. Mar.* 29: 161–76.
Mathieson, A. C. & Prince, J. S. 1973. The ecology of *Chondrus crispus* (Stackhouse). In *Chondrus crispus*, eds. M. J. Harvey & J. McLachlan, pp. 53–79. Halifax, Nova Scotia: Nova Scotia Institute of Science.
Matsumoto, F. 1959. Studies on the effect of environmental factors on the growth of "Nori" (*Porphyra tenera* Kjellm.), with special reference to the water current. *J. Fac. Fish. Anim. Husb. Hiroshima Univ.* 2: 249–333.
Matsuura, S. 1958. Observations on the annual growth cycles of marine algae on a reef at Manadzuru on the Pacific coast of Japan. *Bot. Mag. Tokyo* 71: 93–109.
Mayhoub, H., Gayral, P. & Jacques, R. 1976. Action de la composition spectrale de la lumière sur la croissance et la reproduction de *Calosiphonia vermicularis* (J. Agardh) Schmitz (Rhodophycées, Gigartinales). *C. R. Acad. Sci.* (Paris), sér. D, 283. 1041–4.
McLachlan, J. & Bird, C. J. 1984. Geographical and experimental assessment of the distribution of *Gracilaria* species (Rhodophyta: Gigartinales) in relation to temperature. *Helgol. Meeresunters.* 38: 319–34.
McLachlan, J., Chen, L. C.-M. & Edelstein, T. 1969. Distribution and life-history of *Bonnemaisonia hamifera* Haroit. *Proc. Int. Seaweed Symp.* 6: 245–9.
McLean, J. H. 1962. Sublittoral ecology of kelp beds of the open coast area near Carmel, California. *Biol. Bull. Mar. Biol. Lab. Woods Hole* 122: 95–114.
McQuaid, C. D. 1985. Seasonal variation in biomass and zonation of nine intertidal algae in relation to changes in radiation, sea temperature and tidal regime. *Bot. Mar.* 28: 539–44.
Merrill, J. E. & Waaland, J. R. 1979. Photosynthesis and respiration in a fast growing strain of *Gigartina exasperata* (Harvey and Bailey). *J. Exp. Mar. Biol. Ecol.* 39: 281–90.
Migita, S. 1966. Freeze-preservation of *Porphyra* thalli in viable state – II. Effect of cooling velocity and water content of thalli on the frost-resistance. *Bull. Fac. Fish. Nagasaki Univ.* 21: 131–8.
Migita, S. 1967. Viability and spore-liberation of *Conchocelis*-phase, *Porphyra tenera*, freeze-preserved in sea water. *Bull. Fac. Fish. Nagasaki Univ.* 22: 33–43.
Miura, A. 1975. *Porphyra* cultivation in Japan. In *Advances of Phycology in Japan*, eds. J. Tokida & H. Hirose, pp. 273–304. The Hague: W. Junk.
Molinier, R. 1960. Étude des biocenoses marines du Cap Corse. II. *Vegetatio* 9: 217–312.
Montalva, S. & Santelices, B. 1981. Interspecific interference among species of *Gelidium* from central Chile. *J. Exp. Mar. Biol. Ecol.* 53: 77–88.
Montgomery, W. L. 1980. Comparative feeding ecology of two herbivorous damselfish (Panatentridae: Teleosti) from the Gulf of California, Mexico. *J. Exp. Mar. Biol. Ecol.* 47: 9–24.
Montgomery, W. L. & Gerking, S. D. 1980. Marine macroalgae as foods for fishes: an evaluation of potential food quality. *Env. Biol. Fish* 5: 143–53.
Moorjani, S. A. & Jones, W. E. 1972. Spore attachment and development in some coralline algae. *Br. Phycol. J.* 7: 282.
Morgan, K. C. & Simpson, F. J. 1981a. The cultivation of *Palmaria palmata*. Effect of light intensity and temperature on growth and chemical composition. *Bot. Mar.* 24: 547–52.
Morgan, K. C. & Simpson, F. J. 1981b. The cultivation of *Palmaria (Rhodymenia) palmata*: effect of high concentrations of nitrate and ammonium on growth and nitrogen uptake. *Aquat. Bot.* 11: 167–71.
Morgan, K. C. & Simpson, F. J. 1981c. The cultivation of *Palmaria palmata*. Effect of light intensity and nitrate supply on growth and chemical composition. *Bot. Mar.* 24: 273–7.
Morgan, K. C., Shacklock, P. F. & Simpson, F. J. 1980. Some aspects of the culture of *Palmaria palmata* in greenhouse tanks. *Bot. Mar.* 23: 765–70.
Morse, D. E., Hooker, N., Duncan, H. & Jensen, L. 1979. Gamma-aminobutyric acid, a neurotransmitter induces planktonic larvae to settle and begin metamorphosis. *Science* 204: 407–10.
Morse, A. N. C. & Morse, D. E. 1984. Recruitment and metamorphosis of *Haliotis* larvae induced by molecules uniquely available at the surfaces of crustose red algae. *J. Exp. Mar. Biol. Ecol.* 75: 191–215.
Mshigeni, K. E. 1978. Effects of nitrate fertilizer on the growth of the economic seaweed *Hypnea* Lamouroux (Rhodophyta, Gigartinales). *Nova Hedwigia* 19: 231–6.
Mumford, T. F. 1978. Growth of Pacific Northwest marine algae on artificial substrates – potential and practice. In *The Marine Plant Biomass of the Pacific Northwest Coast*, ed. V. R. Krauss, pp. 139–61. Corvallis: Oregon State University Press.
Munda, I. 1977. A note on the growth of *Halosaccion ramentaceum* (L.) J. Ag. under different culturing conditions. *Bot. Mar.* 20: 493–8.
Munda, I. M. 1978. Salinity dependent distribution of benthic algae in estuarine areas of Icelandic fjords. *Bot. Mar.* 21: 451–68.
Munda, I. M. & Markham, J. W. 1982. Seasonal variations of vegetation patterns and biomass constituents in the rocky eulittoral of Helgoland. *Helgol. Meeresunters.* 35: 131–51.
Murray, S. N. & Dixon, P. S. 1973. The effect of light

intensity and light period on the development of thallus form in the marine red algal *Pleonosporium squarrulosum* (Harvey) Abbott (Rhodophyta: Ceramiales). I. Apical cell division – main axis. *J. Exp. Mar. Biol. Ecol.* 13: 15–27.

Murray, S. N. & Dixon, P. S. 1975. The effects of light intensity and light period on the development of thallus form in the marine red algal *Pleonosporium squarrulosum* (Harvey) Abbott (Rhodophyta: Ceramiales). II. Cell enlargement. *J. Exp. Mar. Biol. Ecol.* 19: 165–76.

Murray, S. N. & Littler, M. M. 1978. Patterns of intertidal succession in a perturbated marine intertidal community. *J. Phycol.* 14: 506–12.

Nasr, A. H., Bekheet, I. A. & Ibrahim, R. K. 1968. The effect of different nitrogen and carbon sources on amino acid synthesis in *Ulva*, *Dictyota* and *Pterocladia*. *Hydrobiologia* 31: 7–16.

Neish, A. C., Shacklock, P. F., Fox, C. H. & Simpson, F. J. 1977. The cultivation of *Chondrus crispus*. Factors affecting growth under greenhouse conditions. *Can. J. Bot.* 55: 2263–71.

Nellen, U. R. 1966. Über den Einfluss des Salzgehaltes auf die photosynthetische Leistung verschiedener Standortformen von *Delesseria sanguinea* und *Fucus serratus*. *Helgol. Meeresunters.* 13: 288–313.

Nelson, S. G., Tsutsui, R. N. & Best, B. R. 1980. A preliminary evaluation of the mariculture potential of *Gracilaria* (Rhodophyta) in Micronesia: growth and ammonium uptake. In *Pacific Seaweed Aquaculture*, eds. I. A. Abbott, M. S. Foster & L. F. Eklund, pp. 72–9. La Jolla: Institute of Marine Resources, University of California.

Neushul, M. 1967. Studies of subtidal marine vegetation in Western Washington. *Ecology* 48: 83–94.

Neushul, M. 1981. The domestication and cultivation of Californian macroalgae. *Proc. Int. Seaweed Symp.* 10: 71–96.

Neushul, M. & Dahl, A. L. 1967. Composition and growth of subtidal parvosilvosa from Californian kelp forests. *Helgol. Meeresunters.* 15: 480–8.

Neushul, M., Foster, M. S., Coon, D. A., Woessner, J. W. & Harger, B. W. W. 1976. An in situ study of recruitment, growth and survival of subtidal marine algae: techniques and preliminary results. *J. Phycol.* 12: 397–408.

Ngan, Y. & Price, I. R. 1979. Systematic significance of spore size in the Florideophyceae (Rhodophyta). *Br. Phycol. J.* 14: 285–303.

Ngan, Y. & Price, I. R. 1980a. Distribution of intertidal benthic algae in the vicinity of Townsville, tropical Australia. *Aust. J. Mar. Freshw. Res.* 31: 175–92.

Ngan, Y. & Price, I. R. 1980b. Seasonal growth and reproduction of intertidal algae in the Townsville region (Queensland, Australia). *Aquat. Bot.* 9: 117–34.

Ngan, Y. & Price, I. R. 1983. Periodicity of spore discharge in tropical Florideophyceae (Rhodophyta). *Br. Phycol. J.* 18: 83–95.

Nicotri, M. E. 1977. The impact of crustacean herbivores on cultured seaweed populations. *Aquaculture* 12: 127–36.

Nicotri, M. E. 1980. Factors involved in herbivore food preference. *J. Exp. Mar. Biol. Ecol.* 42: 13–26.

Noda, H. & Horiguchi, Y. 1972. The significance of zinc as a nutrient for the red alga *Porphyra tenera*. *Proc. Int. Seaweed Symp.* 7: 368–72.

Norall, T. L., Mathieson, A. C. & Kilar, J. A. 1981. Reproductive ecology of four subtidal red algae. *J. Exp. Mar. Biol. Ecol.* 54: 119–36.

Norris, J. N. & Fenical, W. 1982. Chemical defense in tropical marine algae. In *The Atlantic Barrier Reef Ecosystem at Carrow Bow Cay, Belize. I. Structure and Communities*, eds. K. Rützler & I. G. MacIntyre, pp. 417–31. Washington D.C.: Smithsonian Contributions to the Marine Sciences 12.

Norton, T. A. 1968. Underwater observations on the vertical distribution of algae at St Mary's, Isle of Scilly. *Br. Phycol. Bull.* 3: 585–8.

Norton, T. A., ed. 1985. *Provisional Atlas of the Marine Algae of Britain and Ireland*. Huntingdon: N.E.R.C. Institute of Terrestrial Ecology.

Norton, T. A., Ebling, F. J. & Kitching, J. A. 1971. Light and the distribution of organisms in a sea cave. In *Proceedings of the IVth European Marine Biology Symposium*, ed. D. J. Crisp, pp. 409–32: Cambridge: Cambridge University Press.

Norton, T. A., Hiscock, K. & Kitching, J. A. 1977. The ecology of Lough Ine. XX. The *Laminaria* forest at Carrigathorna. *J. Ecol.* 65: 919–41.

Norton, T. A. & Mathieson, A. C. 1983. The biology of unattached seaweeds. *Prog. Phycol. Res.* 2: 333–86.

Norton, T. A. & Milburn, J. A. 1972. Direct observations on the sublittoral marine algae of Argyll, Scotland. *Hydrobiologia* 40: 55–68.

Norton, T. A. & Parkes, H. M. 1972. The distribution and reproduction of *Pterosiphonia complanata*. *Br. Phycol. J.* 7: 13–9.

Novaczek, I. & McLachlan, J. 1986. Recolonization by algae of the sublittoral habitat of Halifax County, Nova Scotia, following the demise of sea urchins. *Bot. Mar.* 29: 69–73.

Nygren, S. 1970. Effect of salinity on the growth of *Dasya pedicellata*. *Helgol. Meeresunters.* 20: 126–9.

Oates, B. R. 1986. Components of photosynthesis in the intertidal saccate alga *Halosaccion americanum* (Rhodophyta, Palmariales). *J. Phycol.* 22: 217–23.

Ogata, E. 1953. Some experiments on settling of spores of red algae. *Bull. Soc. Plant Ecol.* 2: 104–7.

Ogata, E. 1963. Manometric studies on the respiration of a marine alga, *Porphyra tenera* – I. Influence of salt concentration, temperature, drying and other factors. *Bull. Jap. Soc. Sci. Fish* 29: 139–45.

Ogata, E. & Matsui, T. 1965a. Photosynthesis in several marine plants of Japan in relation to carbon dioxide supply, light and inhibitors. *Jap. J. Bot.* 19: 83–98.

Ogata, E. & Matsui, T. 1965b. Photosynthesis in several marine plants of Japan as affected by salinity, drying and pH, with attention to their growth habits. *Bot. Mar.* 8: 199–217.

Ogata, E. & Schramm, W. 1971. Some observations on the influence of salinity on growth and photosynthesis in *Porphyra umbilicalis*. *Mar. Biol.* 10: 70–6.

Ogata, E. & Takada, H. 1968. Studies on the relationship between the respiration and the changes in salinity in some marine plants in Japan. *J. Shimonoseki Univ. Fish.* 16: 67–138.

Ogden, J. C. 1976. Some aspects of herbivore plant relationships on Caribbean reefs and seagrass beds. *Aquat. Bot.* 2: 103–16.

Ohno, M. 1969. A physiological ecology of the early stage of some marine algae. *Rep. Usa Mar. Biol. Stn Kochi Univ.* 16 (1): 1–46.

Ohno, M. 1976. A preliminary report on the photosynthetic activity of *Phyllophora antarctica* A. et E. S. Gepp,

Antarctic red alga. *Antarctic Rec., Nat. Inst. Polar Res., Tokyo* 57: 141–5.

Ohno, M. 1984. Culture of an Antarctic seaweed, *Phyllophora antarctica* (Phyllophoraceae, Rhodophyceae). *Mem. Nat. Inst. Polar Res., Tokyo Special Issue* 32: 112–16.

Ohno, M. & Arasaki, S. 1969. Examination of the dark treatment at spore stage of sea weeds. *Bull. Jap. Soc. Phycol.* 17: 37–42.

Okuda, T. & Neushul, M. 1981. Sedimentation studies of red algal spores. *J. Phycol.* 17: 113–8.

Oza, R. M. 1977. Culture studies on induction of tetraspores and their subsequent development in the red alga *Falkenbergia rufolanosa* (Harvey) Schmitz. *Bot. Mar.* 20: 29–32.

Padilla, D. K. 1984. The importance of form: differences in competitive ability, resistance to consumers and environmental stress in an assemblage of coralline algae. *J. Exp. Mar. Biol. Ecol.* 79: 105–27.

Padilla, D. K. 1985. Structural resistance of algae to herbivores. A biomechanical approach. *Mar. Biol.* 90: 103–9.

Paine, R. T. 1969. The *Pisaster – Tegula* interaction: prey patches, predatory food preference and intertidal community structure. *Ecology* 50: 950–61.

Paine, R. T. 1977. Controlled manipulations in the marine intertidal zone and their contributions to ecological theory. In *The Changing Scenes in Natural Sciences 1776–1976*, pp. 245–70. Philadelphia: Academy of Natural Sciences, Special Publication 12.

Paine, R. T. 1980. Food webs: linkage, interaction strength and community infrastructure. *J. Anim. Ecol.* 49: 667–85.

Paine, R. T. 1984. Ecological determinism in the competition for space. *Ecology* 65: 1339–48.

Paine, R. T., Slocum, C. J. & Duggins, D. O. 1979. Growth and longevity in the crustose red alga *Petrocelis middendorffii*. *Mar Biol.* 51: 185–92.

Paine, R. T. & Vadas, R. L. 1969. The effects of grazing by sea urchins, *Strongylocentrotus* spp. on benthic algal populations. *Limnol. Oceanogr.* 14: 710–9.

Parker, H. S. 1982. Effects of simulated current on the growth rate and nitrogen metabolism of *Gracilaria tikvahiae* (Rhodophyta). *Mar. Biol.* 69: 137–45.

Paul, V. & Fenical, W. 1980. Toxic acetylene containing lipids from the red alga *Liagora farinosa* Lamouroux. *Tetrahedron Letters* 21: 3327–30.

Peckol, P. 1982. Seasonal occurrence and reproduction of some marine algae of the continental shelf, North Carolina. *Bot. Mar.* 25: 185–90.

Peckol, P. & Searles, R. B. 1984. Temporal and spatial patterns of growth and survival of invertebrate and algal populations of a North Carolina continental shelf community. *Est. Coast. Shelf Sci.* 18: 133–43.

Pedersen, M. 1978. Bromochlorophenols and brominated diphenylmethane in red algae. *Phytochemistry* 17: 291–3.

Pedersen, M., Roomans, G. M. & Hofsten, A. V. 1980. Cell inclusions containing bromine in *Rhodomela confervoides* (Huds.) Lamour. and *Polysiphonia elongata* Harv. (Rhodophyta, Ceramiales). *Phycologia* 19: 153–8.

Pedersen, M., Saenger, P. & Fries, L. 1974. Simple brominated phenols in red algae. *Phytochemistry* 13: 2273–9.

Penniman, C. A. & Mathieson, A. C. 1985. Photosynthesis of *Gracilaria tikvahiae* McLachlan (Gigartinales, Rhodophyta) from the Great Bay Estuary, New Hampshire. *Bot. Mar.* 28: 427–35.

Penniman, C. A., Mathieson, A. C. & Penniman, C. E. 1986. Reproductive phenology and growth of *Gracilaria tikvahiae* McLachlan (Gigartinales, Rhodophyta) in the Great Bay Estuary, New Hampshire. *Bot. Mar.* 29: 147–54.

Phillips, R. C. & Fleenor, B. 1970. Investigation of the benthic marine flora of Hood Canal, Washington. *Pacif. Sci.* 24: 275–82.

Phillips, R. C. & Springer, V. G. 1960. Observations on the offshore benthic flora in the Gulf of Mexico off Pinellas County, Florida. *Am. Midl. Nat.* 64: 362–81.

Phillips, R. C., Vadas, R. L. & Ogden, N. 1982. The marine algae and seagrasses of the Miskito Bank, Nicaragua. *Aquat. Bot.* 13: 187–95.

Polne, M., Neushul, M. & Gibor, A. 1981. Studies in domestication of *Eucheuma uncinatum*. *Proc. Int. Seaweed Symp.* 10: 619–24.

Post, E. 1963. Zur verbreitung und Okologie der *Bostrychia Caloglossa* Assoziation. *Int. Rev. Ges. Hydrobiol. Hydrogr.* 48: 47–152.

Potts, D. C. 1977. Suppression of coral populations by filamentous algae within damselfish territories. *J. Exp. Mar. Biol. Ecol.* 28: 207–16.

Powell, J. H. 1964. The life-history of a red alga, *Constantinea*. Ph.D. thesis, University of Washington.

Price, J. H. 1971. The shallow sublittoral marine ecology of Aldabra. *Phil. Trans. R. Soc.* Ser. B 260: 123–71.

Price, J. H. 1978. Ecological determinants of adult form in *Callithamnion*: its taxonomic implications. In *Modern Approaches to the Taxonomy of Red and Brown Algae*, eds, D. E. G. Irvine & J. H. Price, pp. 263–80. London: Academic.

Prince, J. S. & Kingsbury, J. M. 1973. The ecology of *Chondrus crispus* at Plymouth, Massachusetts. III. Effect of elevated temperature on growth and survival. *Biol. Bull. Mar. Biol. Lab., Woods Hole* 145: 580–8.

Pringle, J. D. & Austin, A. P. 1970. The mitotic index in selected red algae in situ. II. A supralittoral species, *Porphyra lanceolata* (Setchell & Hus.) G. M. Smith. *J. Exp. Mar. Biol. Ecol.* 5: 113–37.

Provasoli, L. 1964. Growing marine seaweeds. *Proc. Int. Seaweed Symp.* 4: 9–17.

Quadir, A., Harrison, P. J. & De Wreede, R. E. 1979. The effects of emergence and submergence on the photosynthesis and respiration of marine macrophytes. *Phycologia* 18: 83–8.

Rao, K. R. 1977. Studies on Indian Hypneaceae. II. Reproductive capacity in the two species of *Hypnea* over the different seasons. *Bot. Mar.* 20: 33–9.

Rao, M. U. 1974. Observations on fruiting cycle, spore output and germination of tetraspores of *Gelidiella acerosa* in the Gulf of Mannar. *Bot. Mar.* 27: 204–7.

Rawlence, D. J. 1972. An ultrastructural study of the relationship between rhizoids of *Polysiphonia lanosa* (L.) Tandy (Rhodophyta) and tissue of *Ascophyllum nodosum* (L.) Le Jolis (Phaeophyceae). *Phycologia* 11: 279–90.

Reed, R. H. 1981. Physiological responses of *Porphyra* under steady state and fluctuating salinity regimes. *Proc. Int. Seaweed Symp.* 10: 509–14.

Reed, R. H. 1983. The osmotic responses of *Polysiphonia lanosa* (L.) Tandy from marine and estuarine sites: evidence for incomplete recovery of turgor. *J. Exp. Mar. Biol. Ecol.* 68: 169–93.

Reed, R. H. 1984. The effects of extreme hyposaline stress upon *Polysiphonia lanosa* (L.) Tandy from marine and estuarine sites. *J. Exp. Mar. Biol. Ecol.* 76: 131–44.

Reed, R. H., Collins, J. C. & Russell, G. 1980. The influence of variations in salinity upon photosynthesis in the marine alga *Porphyra purpurea* (Roth) C. Ág. (Rhodophyta, Bangiales). *Z. Pflanzenphysiol.* 98: 183–7.

Reynolds, N. B. & Mathieson, A. C. 1975. Seasonal occurrence and ecology of marine algae in a New Hampshire tidal rapid. *Rhodora* 77: 512–33.
Richardson, N. 1970. Studies on the photobiology of *Bangia fuscopurpurea*. *J. Phycol.* 6: 215–9.
Richardson, N. & Dixon, P. S. 1970. Culture studies on *Thuretellopsis peggiana* Kylin. *J. Phycol.* 6: 154–9.
Rietema, H. 1982. Effects of photoperiod and temperature on macrothallus initiation in *Dumontia contorta* (Rhodophyta). *Mar. Ecol. Prog. Ser.* 8: 187–96.
Rietema, H. & van den Hoek, C. 1984. Search for possible latitudinal ecotypes in *Dumontia contorta* (Rhodophyta). *Helgol. Meeresunters.* 38: 389–99.
Robbins, J. V. 1979. Effects of physical and chemical factors on photosynthetic and respiratory rates of *Palmaria palmata* (Florideophyceae). *Proc. Int. Seaweed Symp.* 9: 273–83.
Rosenberg, G. & Ramus, J. 1982. Ecological growth strategies in the seaweeds *Gracilaria foliifera* (Rhodophyceae) and *Ulva* sp. (Chlorophyceae): soluble nitrogen and reserve carbohydrates. *Mar. Biol.* 66: 251–9.
Rueness, J. & Åsen, A. 1982. Field and culture observations on the life history of *Bonnemaisonia asparagoides* (Woodw.) C. Ag. (Rhodophyta) from Norway. *Bot. Mar.* 25: 577–87.
Rueness, J. & Tananger, T. 1984. Growth in culture of four red algae from Norway with potential for mariculture. *Proc. Int. Seaweed Symp.* 11: 303–7.
Rumrill, S. S. & Cameron, R. A. 1983. Effects of gamma-aminobutyric acid on the settlement of larvae of the black chiton *Katherina tunicata*. *Mar. Biol.* 72: 243–7.
Russell, G. 1986. Variation and natural selection in marine macroalgae. *Oceanogr. Mar. Biol. Annu. Rev.* 24: 309–77.
Russell, G. & Fielding, A. H. 1974. The competitive properties of marine algae in culture. *J. Ecol.* 62: 689–98.
Ryther, J. H., Corwin, N., DeBusk, T. A. & Williams, L. D. 1981. Nitrogen uptake and storage by the red alga *Gracilaria tikvahiae* (McLachlan, 1979). *Aquaculture* 26: 107–15.
Ryther, J. H. & Dunstan, W. M. 1971. Nitrogen, phosphorus and eutrophication in the coastal marine environment. *Science* 171: 1008–13.
Saito, Y. 1956. A study of some environmental factors upon the development and maturity of the Conchocelis-phase of the laver, *Porphyra tenera* Kjellman. *Bull. Jap. Soc. Sci. Fish* 22: 21–9.
Saito, Y., Naganawa, S. & Miyasaka, H. 1977. The climax phase and its recognition in intertidal algal vegetation. *Jap. J. Ecol.* 27: 33–43.
Saito, Y., Sasaki, H. & Watanabe, K. 1976. Succession of algal communities on the vertical substratum faces of breakwaters in Japan. *Phycologia* 15: 93–100.
Sand-Jensen, K. & Gordon, D. M. 1984. Differential ability of marine and freshwater macrophytes to utilize HCO_3 and CO_2. *Mar. Biol.* 80: 247–53.
Santelices, B. 1976. Nota sobre cultivo masivo de algunas especies de Gelidiales (Rhodophyta). *Rev. Biol. Mar. Dep. Oceanol. Univ. Chile* 16: 27–33.
Santelices, B. 1978. Multiple interaction of factors in the distribution of some Hawaiian Gelidiales (Rhodophyta). *Pacif. Sci.* 32: 119–47.
Santelices, B. & Ojeda, F. P. 1984. Effects of canopy removal on the understory algal community structure of coastal forests of *Macrocystis pyrifera* from southern South America. *Mar. Ecol. Prog. Ser.* 14: 165–73.
Satoh, K., Smith, C. M. & Fork, D. C. 1983. Effects of salinity on primary processes of photosynthesis in the red alga *Porphyra perforata*. *Plant Physiol.* 73: 643–7.
Sawada, T., Koga, S. & Uchiyama, S. 1972. Some observations on carpospore adherence in *Polysiphonia japonica* Harvey. *Sci. Bull. Fac. Agric., Kyushu Univ.* 26: 223–36.
Schneider, C. W. 1976. Subtidal and temporal distributions of benthic marine algae on the continental shelf of the Carolinas. *Bull. Mar. Sci.* 26: 133–51.
Schölm, H. 1966. Untersuchungen zur Wärmeresistenz-von Tiefenalgen. *Bot. Mar.* 9: 54–61.
Schonbeck, M. W. & Norton, T. A. 1979. An investigation of drought avoidance in fucoid algae. *Bot. Mar.* 22: 133–44.
Schwenke, H. 1958. Über einige zellphysiologische Faktoren der Hypotonieresistenz mariner Rotalgen. *Kieler Meeresforsch.* 14: 130–50.
Schwenke, H. 1959. Untersuchungen zur Temperaturresistenz mariner Algen der westlichen Ostsee. I. Das Resistenzverhalten von Tiefenrotalgen bei ökologischen und nichtökologischen Temperaturen. *Kieler Meeresforsch.* 15: 34–50.
Seapy, R. R. & Littler, M. M. 1978. The distribution, abundance, community structure, and primary productivity of macroorganisms from two central California rocky intertidal habitats. *Pacif. Sci.* 32: 293–314.
Seapy, R. R. & Littler, M. M. 1982. Population and species diversity fluctuations in a rocky intertidal community relative to severe aerial exposure and sediment burial. *Mar. Biol.* 71: 87–96.
Sears, J. R. & Wilce, R. T. 1975. Sublittoral, benthic marine algae of southern Cape Cod and adjacent islands: seasonal periodicity, associations, diversity, and floristic composition. *Ecol. Monogr.* 45: 337–65.
Sebens, K. P. 1986. Spatial relationships among encrusting marine organisms in the New England subtidal zone. *Ecol. Monogr.* 56: 73–96.
Shacklock, P. F. & Croft, G. B. 1981. Effect of grazers on *Chondrus crispus* in culture. *Aquaculture* 22: 331–42.
Sheath, R. G. & Cole, K. M. 1980. Distribution and salinity adaptations of *Bangia atropurpurea* (Rhodophyta), a putative migrant into the Laurentian Great Lakes. *J. Phycol.* 16: 412–20.
Sheath, R. G., Hellebust, J. A. & Sawa, T. 1977. Changes in plastid structure, pigmentation and photosynthesis of the conchocelis phase of *Porphyra leucosticta* (Rhodophyta, Bangiophyceae) in response to low light and darkness. *Phycologia* 16: 265–76.
Sheath, R. G., Van Alstyne, K. L. & Cole, K. M. 1985. Distribution, seasonality and reproductive phenology of *Bangia atropurpurea* (Rhodophyta) in Rhode Island, U.S.A. *J. Phycol.* 21: 297–303.
Shepherd, S. A. 1973. Studies on southern Australian abalone (genus *Haliotis*). 1. Ecology of five sympatric species. *Aust. J. Mar. Freshw. Res.* 24: 215–57.
Shepherd, S. A. 1981. Ecological strategies in a deep water red algal community. *Bot. Mar.* 24: 457–63.
Shepherd, S. A. & Womersley, H. B. S. 1970. The sublittoral ecology of West Island, South Australia. 1. Environmental features and algal ecology. *Trans. R. Soc. S. Aust.* 94: 105–37.
Shepherd, S. A. & Womersley, H. B. S. 1971. Pearson Island expedition 1969. 7. The subtidal ecology of benthic algae. *Trans. R. Soc. S. Aust.* 95: 155–67.
Shepherd, S. A. & Womersley, H. B. S. 1976. The subtidal algal and seagrass ecology of St Francis Island, South Australia. *Trans. R. Soc. S. Aust.* 100: 177–91.

Shepherd, S. A. & Womersley, H. B. S. 1981. The algal and seagrass ecology of Waterloo Bay, South Australia. *Aquat. Bot.* 11: 305–71.

Shimo, S. & Nakatani, S. 1967. The influence of various concentrations of copper and mercury ions on the growth of *Porphyra tenera* and its tolerance to them. *J. Agr. Lab.* (Chiba) 9: 109–25.

Simpson, F. J. & Shacklock, P. F. 1979. The cultivation of *Chondrus crispus*. Effect of temperature on growth and carrageenan production. *Bot. Mar.* 22: 259–98.

Slocum, C. J. 1980. Differential susceptibility to grazers in two phases of an intertidal alga: advantages of heteromorphic generations. *J. Exp. Mar. Biol. Ecol.* 46: 99–110.

Smayda, T. J. 1959. The seasonal incoming radiation in Norwegian and arctic waters and indirect methods of measurements. *J. Cons. Int. Explor. Mer* 24: 215–20.

Smith, R. M. 1967. Sublittoral ecology of marine algae on the North Wales coast. *Helgol. Meeresunters.* 15: 467–79.

Sohn, C. H., Lee, K. & Kang, J. W. 1983. Benthic marine algae of Dolsan-Island in the southern coast of Korea. *Bull. Korean Fish Soc.* 16: 379–88.

Sousa, W. P. 1979. Experimental investigations of disturbance and ecological succession in a rocky intertidal algal community. *Ecol. Monogr.* 49: 227–54.

Steneck, R. S. 1982. A limpet–coralline alga association: adaptations and defenses between a selective herbivore and its prey. *Ecology* 63: 507–22.

Steneck, R. S. & Adey, W. H. 1976. The role of environment in control of morphology in *Lithophyllum congestum*, a caribbean algal ridge builder. *Bot. Mar.* 19: 197–215.

Stephenson, T. A. & Stephenson, A. 1972. *Life Between Tidemarks on Rocky Shores*. San Francisco: W. H. Freeman.

Stewart, J. G. 1982. Anchor species and epiphytes in intertidal algal turf. *Pacif. Sci.* 36: 45–57.

Stewart, J. G. 1983. Fluctuations in the quantity of sediments trapped among algal thalli on intertidal rock platforms in southern California. *J. Exp. Mar. Biol. Ecol.* 73: 205–11.

Stewart, J. G. 1984a. Algal distributions and temperature: test of an hypothesis based on vegetative growth rates. *Bull. S. Calif. Acad. Sci.* 83: 57–68.

Stewart, J. G. 1984b. Vegetative growth rates of *Pterocladia capillacea* (Gelidiaceae, Rhodophyta). *Bot. Mar.* 27: 85–94.

Strömgren, T. 1984. Diurnal variation in the length growthrate of three intertidal algae from the Pacific West Coast. *Aquat. Bot.* 20: 1–10.

Suto, S. 1950. Studies on the shedding, swimming and fixing of spores of seaweeds. *Bull. Jap. Soc. Fish.* 16: 1–9.

Sverdrup, H. U., Johnson, M. W. & Fleming, R. H. 1942. *The Oceans*. New York: Prentice-Hall.

Takeuchi, T., Matsubara, T., Shitanaka, M. & Suto, S. 1956. On the shedding of spores from the cultured *Conchocelis*-phase of *Porphyra tenera* set in the sea. *Bull. Jap. Soc. Sci. Fish.* 20: 487–9.

Tatewaki, M. & Provasoli, L. 1964. Vitamin requirements of three species of *Antithamnion*. *Bot. Mar.* 6: 193–203.

Taylor, A. R. A. 1972. Growth studies of *Chondrus crispus* in Prince Edward Island. *Proc. Mtg. Can. Atl. Seaweeds Industry Ind. Dev. Can., Fish. Serv. Environ. Can.* 1972: 29–36.

Taylor, P. R. 1985. The influence of sea anemones on the morphology and productivity of two intertidal seaweeds. *J. Phycol.* 21: 335–40.

Taylor, P. R. & Hay, M. E. 1984. Functional morphology of intertidal seaweeds: adaptive significance of aggregate vs. solitary forms. *Mar. Ecol. Prog. Ser.* 18: 295–302.

Terumoto, I. 1965. Freezing and drying in a red marine alga, *Porphyra yezoensis* Ueda. *Low Temp. Sci.*, Ser. B 23: 11–20.

Thom, R. M. 1980. Seasonality in low intertidal benthic marine algal communities in central Puget Sound, Washington, USA. *Bot. Mar.* 23: 7–11.

Thomas, T. W. 1982. Nitrogen uptake and assimilation in intertidal *Gracilaria verrucosa*. First Int. Phycol. Congr. Abstr. A49.

Thomas, T. W. & Harrison, P. J. 1985. Effect of nitrogen supply on nitrogen uptake, accumulation and assimilation in *Porphyra perforata* (Rhodophyta). *Mar. Biol.* 85: 269–78.

Turner, M. F. 1970. A note on the nutrition of *Rhodella*. *Br. Phycol. J.* 5: 15–8.

Turner, T. 1983. Complexity of early and middle successional stages in a rocky intertidal surfgrass community. *Oecologia* (Berl.). 60: 56–65.

Tveter, E. & Mathieson, A. C. 1976. Sporeling coalescence in *Chondrus crispus* (Rhodophyceae). *J. Phycol.* 12: 110–8.

Tveter-Gallagher, E., Cheney, D. & Mathieson, A. C. 1981. Uptake and incorporation of ^{35}S into carrageenan among different strains of *Chondrus crispus*. *Proc. Int. Seaweed Symp.* 10: 521–30.

Tveter-Gallagher, E. & Mathieson, A. C. 1980. An electron microscopic study of sporeling coalescence in the red alga *Chondrus crispus*. *Scan. Elect. Microsc.* 3: 571–81.

Tveter-Gallagher, E., Mathieson, A. C. & Cheney, D. P. 1980. Ecology and developmental morphology of male plants of *Chondrus crispus* (Gigartinales, Rhodophyta). *J. Phycol.* 16: 257–64.

Underwood, A. J. 1979. The ecology of intertidal gastropods. *Adv. Mar. Biol.* 16: 111–210.

Underwood, A. J. 1980. The effects of grazing by gastropods and physical factors on the upper limits of intertidal macroalgae. *Oecologia* (Berl.) 46: 210–3.

Vadas, R. L. 1977. Preferential feeding: an optimization strategy in sea urchins. *Ecol. Monogr.* 47: 337–71.

van den Hoek, C. 1969. Algal vegetation types along the open coasts of Curaçao, Netherlands Antilles I and II. *Proc. K. Ned. Akad. Wet. Ser. C.* 72: 537–77.

van den Hoek, C. 1975. Phytogeographic provinces along the coasts of the northern Atlantic Ocean. *Phycologia* 14: 317–30.

van den Hoek, C. 1982a. The distribution of benthic marine algae in relation to the temperature regulation of their life histories. *Biol. J. Linn. Soc.* 18: 81–144.

van den Hoek, C. 1982b. Phytogeographic distribution groups in the North Atlantic Ocean. A review of experimental evidence from life history studies. *Helgol. Meeresunters.* 35: 153–214.

van den Hoek, C. 1984. World-wide latitudinal and longitudinal seaweed distribution patterns and their possible causes, as illustrated by the distribution of Rhodophycean genera. *Helgol. Meeresunters.* 38: 227–57.

van den Hoek, C., Cortal-Breeman, A. M. & Wanders, J. B. W. 1975. Algal zonation in the fringing coral reef of Curaçao, Netherlands Antilles, in relation to zonation of corals and gorgonians. *Aquat. Bot.* 1: 269–308.

Verlaque, M. 1984. Biologie des juvéniles de l'oursin herbivore *Paracentrolus lividus* (Lamark): sélectivité du broutage et impact de l'espèce sur les communantés algales de substrat rocheux en Corse (Méditerranée, France). *Bot. Mar.* 27: 401–24.

Vine, P. J. 1974. Effects of algal grazing and aggressive behaviour of the fishes *Pomacentrus lividus* and *Acanthrus sohol* on coral reef ecology. *Mar. Biol.* 24: 131–6.

von Stosch, H. A. 1964. Wirkungen von jod und arsenit auf meeresalgen in kultur. *Proc. Int. Seaweed Symp.* 4: 142–50.

Vooren, C. M. 1981. Photosynthetic rates of benthic algae from the deep coral reef of Curaçao. *Aquat. Bot.* 10: 143–54.

Vroman, M. 1968. *Studies on the Flora of Curaçao and Other Caribbean Islands. Vol. II. The Marine Algal Vegetation of St. Martin, St. Eustatius and Saba (Netherlands Antilles).* The Hague: Martinus Nijhoff.

Waaland, J. R. 1977. Growth of Pacific Northwest marine algae in semi-closed culture. In *The Marine Plant Biomass of the Pacific Northwest Coast*, ed. R. W. Krauss, pp. 117–37. La Jolla: Institute of Marine Resources, University of California.

Waern, M. 1952. Rocky-shore algae in the Oregrund Archipelago. *Acta Phytogeogr. Suecica* 30: 1–298.

Wallentinus, I. 1978. Productivity studies on Baltic macroalgae. *Bot. Mar.* 21: 365–80.

Wallentinus, I. 1984. Comparisons of nutrient uptake rates for Baltic macroalgae with different thallus morphologies. *Mar. Biol.* 80: 215–25.

Wanders, J. B. W. 1976. The role of benthic algae in the shallow reef of Curaçao (Netherlands Antilles). I. Primary productivity in the coral reef. *Aquat. Bot.* 2: 235–70.

Wanders, J. B. W. 1977. The role of benthic algae in the shallow reef of Curaçao (Netherlands Antilles) III. The significance of grazing. *Aquat. Bot.* 3: 357–90.

Watson, D. C. & Norton, T. A. 1985a. Dietary preferences of the common periwinkle, *Littorina littorea*. *J. Exp. Mar. Biol. Ecol.* 88: 193–211.

Watson, D. C. & Norton, T. A. 1985b. The physical characteristics of seaweed thalli as deterrents to littorine grazers. *Bot. Mar.* 28: 383–7.

Weinstein, B., Rold, T. L., Harrell, C. E., Burns, M. W. & Wealand, J. R. 1975. Re-examination of the bromophenols in the red alga *Rhodomela larix*. *Phyto. Chem.* 14: 2667–70.

West, J. A. 1968. Morphology and reproduction of the red alga *Acrochaetium pectinatum* in culture. *J. Phycol.* 4: 89–99.

West, J. A. 1970. The life history of *Rhodochorton concrescens* in culture. *Br. Phycol. J.* 5: 179–86.

West, J. A. 1972a. Environmental control of hair and sporangial formation in the marine red alga *Acrochaetium proskaueri* sp. nov. *Proc. Int. Seaweed Symp.* 7: 377–84.

West, J. A. 1972b. Environmental regulation of reproduction in *Rhodochorton purpureum*. In *Contributions to the Systematics of Benthic Marine Algae of the North Pacific*, eds. I. A. Abbott & M. Kurogi, pp. 213–30. Kobe: Japanese Society of Phycology.

West, J. A. & Crump, E. 1974. The influence of substrate and spore concentration on spore survival and germination in *Gigartina* and *Petrocelis* (Rhodophyta). *J. Phycol.* 10 (Suppl): 12.

West, J. A. & Crump, E. 1975. Carpospore discharge periodicity in excised cystocarpic papillae of *Gigartina–Petrocelis* (Rhodophyta). *J. Phycol.* 11 (Suppl.): 17.

White, E. B. & Boney, A. D. 1969. Experiments with some endophytic and endozoic *Acrochaetium* species. *J. Exp. Mar. Biol. Ecol.* 3: 246–74.

Whitford, L. A. & Kim, C. S. 1966. The effect of light and water movement on some species of marine algae. *Rev. Algol.* 8: 251–4.

Whittick, A. 1977. The reproductive ecology of *Plumaria elegans* (Bonnem.) Schmitz (Ceramiaceae: Rhodophyta) at its northern limits in the western Atlantic. *J. Exp. Mar. Biol. Ecol.* 29: 223–30.

Whittick, A. 1978. The life history of *Callithamnion corymbosum* (Rhodophyta: Ceramiacea) in Newfoundland. *Can. J. Bot.* 56: 2497–9.

Whittick, A. 1981. Culture and field studies on *Callithamnion hookeri* (Dillw.) S. F. Gray (Rhodophyta: Ceramiaceae) from Newfoundland. *Br. Phycol. J.* 16: 289–95.

Whittick, A. 1983. Spatial and temporal distributions of dominant epiphytes on the stipes of *Laminaria hyperborea* (Gunn.) Fosl. (Phaeophyta: Laminariales) in S.E. Scotland. *J. Exp. Mar. Biol. Ecol.* 73: 1–10.

Yamada, I. 1980. Benthic marine algal vegetation along the coasts of Hokkaido, with special reference to the vertical distribution. *J. Fac. Sci., Hokkaido Univ. Ser. 5 (Botany)* 12: 11–98.

Yarish, C., Breeman, A. M. & van den Hoek, C. 1984. Temperature, light, and photoperid responses of some northeast American and west European endemic Rhodophytes in relation to their geographic distribution. *Helgol. Meeresunters.* 38: 273–304.

Yarish, C., Breeman, A. M. & van den Hoek, C. 1986. Survival strategies and temperature responses of seaweeds belonging to different biogeographic distribution groups. *Bot. Mar.* 24: 215–30.

Yarish, C. & Edwards, P. 1982. A field and cultural investigation of the horizontal and seasonal distribution of estuarine red algae of New Jersey. *Phycologia* 21: 112–24.

Yarish, C., Edwards, P. & Casey, S. 1979. Acclimation responses to salinity of three estuarine red algae from New Jersey. *Mar. Biol.* 51: 289–94.

Yarish, C., Edwards, P. & Casey, S. 1980. The effects of salinity, and calcium and potassium variations on the growth of two estuarine red algae. *J. Exp. Mar. Biol. Ecol.* 47: 235–49.

Yokohama, Y. 1973. Photosynthetic properties of marine benthic red algae from different depths in coastal area. *Bull. Jap. Soc. Phycol.* 21: 119–24.

Yoneshigue, Y. 1985. Taxonomie et ecologie des algues marines dans la région de Cabo Frio (Rio de Janeiro, Bresil). Ph.D. thesis, Univ. d'Aix-Marseille II.

Yoneshigue-Braga, Y. & Baeta Neves, M. H. C. 1981. Preliminary studies on mass culture of *Gracilaria* sp. using different nutrient media. *Proc. Int. Seaweed Symp.* 10: 643–8.

Yoshida, T. 1972. Relations between the density of individuals and the final yield in the cultivated *Porphyra*. *Bull. Tohoku Reg. Fish. Res. Lab.* 32: 89–94.

Zavodnik, N. 1975. Effects of temperature and salinity variations on photosynthesis of some littoral seaweeds of the North Adriatic Sea. *Bot. Mar.* 28: 245–53.

Zechman, F. W. & Mathieson, A. C. 1985. The distribution of seaweed propagules in estuarine, coastal and offshore water of New Hampshire, U.S.A. *Bot. Mar.* 28: 283–94.

Chapter 16

Freshwater ecology

ROBERT G. SHEATH
JULIE A. HAMBROOK

CONTENTS

I. Introduction / 423
II. Physical factors / 425
   A. Water motion / 425
   B. Illumination / 427
   C. Temperature / 430
III. Chemical factors / 431
   A. pH, inorganic carbon, and ions / 431
   B. Oxygen / 433
   C. Nutrients / 435
   D. Pollutants / 436
IV. Biotic factors / 438
   A. Grazing / 438
   B. Species associations / 442
   C. Reproductive ecology / 444
V. Biogeography / 447
VI. Summary / 448
VII. Acknowledgments / 449
VIII. References / 449

I. INTRODUCTION

"Rhodophyceae ... seem to be out of place in inland waters." (Skuja 1938, p. 665)

The above view in part reflects the paucity of detailed studies on freshwater red algae. There is no major synthesis of the ecology of these algae. Aspects of the subject are covered in general reviews, such as those of Friedrich (1980), Ott and Sommerfeld (1982), Sheath (1984), and Skuja (1938). However, certain areas have been greatly neglected, including nutrient uptake mechanisms, population biology, reproductive ecology, and biogeography. This review will attempt to bridge some of the gaps in our knowledge by integrating existing literature with several unpublished works.

Freshwater red algae exhibit a smaller size range than do marine species (Fig. 16-1). The majority (80%) of freshwater rhodophytes have a relatively narrow range of lengths (1 to 10 cm) compared with that of marine red algae (<1 to 30 cm). Thus, it would be expected that most freshwater species would encounter a more limited range of substratum texture, grazer size, and competitor type. However, this is contrasted with the more variable chemical environment in fresh waters, particularly ion level, pH, and form of inorganic carbon (Sheath 1984).

The majority of freshwater Rhodophyta are localized in running waters or waters that are at

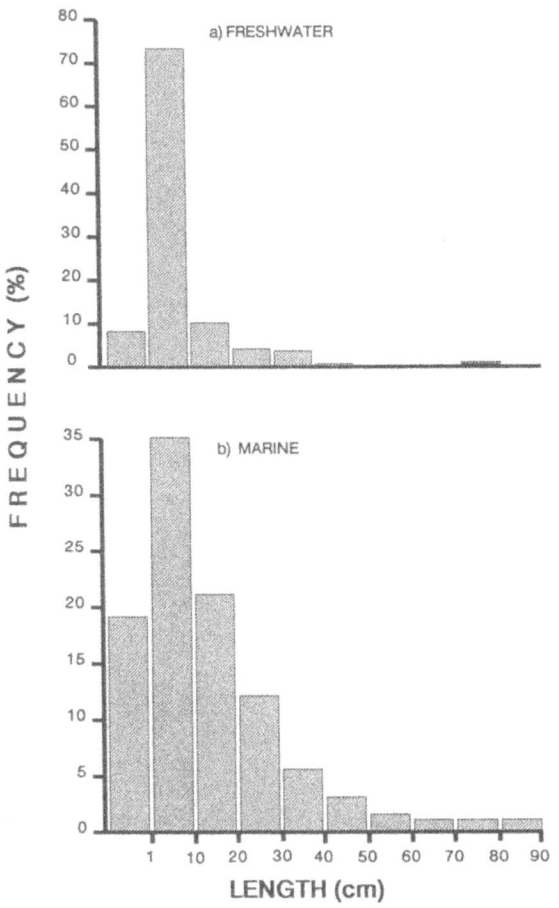

Fig. 16-1. Size distribution of freshwater and marine red algal species; n = 111 and 477, respectively. Data from Abbott & Hollenberg (1976), Starmach (1977), and Taylor (1957).

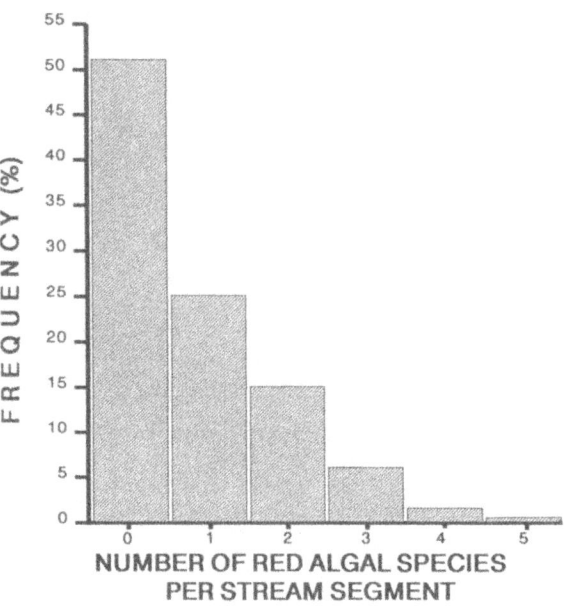

Fig. 16-2. Number of red algal species in 480 streams throughout North America. Data from Sheath & Burkholder (1985), Sheath & Hymes (1980), Sheath et al. (1986b), and Sheath et al. (unpubl.)

least in motion (Kumano 1980; Sheath 1984; Sheath & Hambrook 1988; Sheath & Hymes 1980; Skuja 1938). Exceptions include *Batrachospermum keratophytum*, which occurs widely in bogs in Europe and North America, and *Compsopogon* spp., which are common in stagnant ponds and pools (Sheath 1984). In this presentation, we will concentrate on an analysis of the ecology of macrophytic rhodophytes in rivers and streams.

Red algae are widespread in drainage basins from arctic to tropical areas. Major surveys that have examined many streams in a minimum area of 100,000 km^2 have found the following frequencies of Rhodophyta occurrence (as a percentage of streams studied): 65% in Sweden (Israelson 1942), 51% in Great Britain (N. T. H. Holmes unpubl.), 49% in North America (Fig. 16-2), and 18% in southeastern Australia (Entwisle & Kraft 1984). In North America 24% of the streams surveyed have two or more red algal species. Rhodophyte abundance in stream segments varies considerably. In terms of the amount of stream bottom covered by Rhodophyta, values ranging from <1 to 90% have been found in North America (Fig. 16-3). However, half of the streams have a range of 1–10% and the mean cover is 11%. The amount of biomass contributed by various red algal forms also is quite variable. There is a 10 fold range in dry weight produced by a 1% cover value for three common genera (Table 16-1). This value is 74 mg m^{-2} for *Batrachospermum*, 203 mg m^{-2} for *Tuomeya*, and 618 mg m^{-2} for *Lemanea* in three Rhode Island streams. Similarly, saturating photosynthesis rates vary considerably for Japanese and Malaysian species (3.0–54.3 mg O$_2$ g^{-1} h^{-1}: Ikushima & Kumano 1982; Kumano unpubl.). *Lemanea fucina* makes up two-thirds of the late summer biomass in a Swedish stream (Muller 1978), *Boldia erythrosiphon* accounts for one-third of the seasonal primary production in an Alabama stream (Stock et al. 1987), and *Batrachospermum moniliforme* contributes 12% of the total periphyton primary production in a Quebec stream (Duthie & Hamilton 1983). The latter genus has been found to have among the highest specific primary productivity values measured for freshwater periphyton (Duthie & Hamilton 1983; Westlake

Freshwater ecology

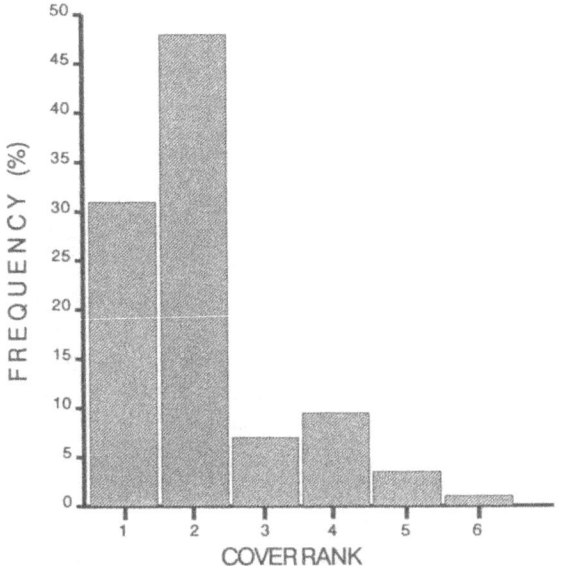

Fig. 16-3. Abundance of freshwater rhodophytes in 480 streams throughout North America. Cover ranks are as follows: 1 as <1%, 2 as 1–10%, 3 as 11–25%, 4 as 26–50%, 5 as 51–75%, and 6 as 76–100%. Data as described in Fig. 16-2.

1980). Therefore, stream-inhabiting red algae are common and periodically contribute significantly to stream community structure and dynamics.

Freshwater rhodophytes can be grouped into a series of forms in much the same manner that marine macroalgae have been categorized (Littler & Littler 1980; Norton et al. 1982). Crusts and tufts occur within the boundary layer or at least in a region of reduced current velocity (Fig. 16-4; Sheath & Hambrook 1988). These forms tend to be avoiders of stress caused by flow, according to the terminology of Levitt (1980). This group accounts for approximately 12% of the total taxa (Sheath 1984). The remaining species can be regarded as semierect, experiencing bending, tensile, and compressive forces and perhaps torsional stresses in flowing waters (Vogel 1984; Fig. 16-4). This group includes mucilaginous and nonmucilaginous filaments and tissue-like forms (Sheath 1984). It would be expected that the semi-erect forms possess adaptive mechanisms to tolerate flow (Sheath & Hambrook 1988). In addition, there should be differences in other aspects of ecology between the various forms of freshwater Rhodophtya, such as productivity, calorific value, susceptibility to predation, nutrient uptake and reproductive capacity (Hambrook & Sheath 1987; Littler & Littler 1980; Norton et al. 1982). Ecological adaptations of the freshwater red algal forms will constitute the central theme of this chapter.

II. PHYSICAL FACTORS

A. Water motion

Positive effects of moderate flow have been observed on various aspects of freshwater rhodophyte occurrence, including distribution in stream segments (Minckley & Tindall 1963; Rider & Wagner 1972; Sheath 1984; Sheath & Hymes 1980; Thirb & Benson-Evans 1985; Traden & Lindstrom 1983), productivity and pigment content (Thirb & Benson-Evans 1982), growth in the laboratory (Whitford 1960), respiration levels (Schumacher & Whitford 1965), and phosphorus uptake rates (Schumacher & Whitford 1965). Most taxa exhibit a wide range of occurrence with respect to flow (Table 16-2). However, the range of mean current velocities at which freshwater red algae occur is relatively narrow, 29–57 cm s^{-1}. The potential benefits of moderate flow include washout of loosely attached competitors (Whitton 1975), constant replenishment of nutrients and gases (Hynes 1970), and reduction of the boundary layers of depletion around the alga (MacFarlane & Raven 1985; Whitford 1960). The latter two effects are important in that nutrients, particularly phosphorus, may be limiting in undisturbed drainage basins. For example, Sheath and Burkholder (1985) reported active growth of six species of red algae in streams with total phosphorus

Table 16-1. Abundance of Rhodophyta in three Rhode Island streams

Stream	Abundance (n = 160)	
	Average cover (%)	Biomass[a] (mg dry wt m^{-2})
Chipuxet		
Batrachospermum virgatum	2.3	170
Batrachospermum moniliforme	0.3	24
Beaver		
Tuomeya americana	2.3	467
Batrachospermum virgatum	0.02	2
Batrachospermum moniliforme	0.01	0.4
Wood		
Lemanea fucina	2.7	1670

[a] Data from mapping 625 cm^2 quadrats in a 10 m^2 area at the end of season, estimating percentage of cover in each quadrat and then removing the algae for dry weight determination (Sheath & Carlson unpubl.).

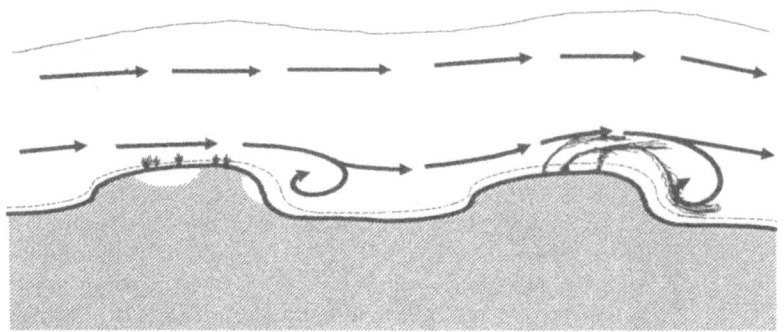

Fig. 16-4. Tufts and crusts (left rock) and semi-erect filaments (right rock) in relation to the flow pattern in a stream and the boundary layer (dotted line).

levels less than 145 nmol and PO_4^{-3} less than 113 nmol.

There appears to be no relationship between distribution of freshwater red algal forms and mean current velocity (Sheath & Hambrook 1988; Table 16-2). Hence, these algae have evolved various mechanisms to grow in moderate flow. This is seen in stress-extension tests. Species can be divided into three groups based on their breaking stress (Table 16-3). There is a significant increase in strength ($p <$ .05) along a gradient from tufts ($\bar{x} = 12 \pm 7$ kN m^{-2}) to mucilaginous filaments ($\bar{x} = 530 \pm 160$ kN m^{-2}) to nonmucilaginous filaments and tissues ($\bar{x} = 1400 \pm 400$ kN m^{-2}). In terms of breaking extension, there is no consistent trend among the morphological forms. The range of breaking extension seen in freshwater rhodophytes tested is 11.3–29.2% beyond the original lengths. MacFarlane and Raven (1985) measured no stretching in *Lemanea mamillosa* at current velocities of 3.0 to 5.4 cm s^{-1}, but at 25.5 cm s^{-1}, plants were stretched 3–5% of their normal length.

Another method to demonstrate differences in biomechanical adaptations to flow is the relationship between drag (force tending to move an object downstream) and current velocity as well as the value E (relative reduction of drag as speed is increased) (Sheath & Hambrook 1988; Table 16-4; Fig.

Table 16-2. Distribution of freshwater Rhodophyta in North America in relation to mean current velocity

From group	Sample size	Current velocity (cm s^{-1})	
		Range	Mean
Boundary layer forms			
Crusts			
Hildenbrandia rivularis	26	8–88	34
Tufts			
Audouinella violacea	37	4–88	46
Semi-erect forms			
Mucilaginous filaments			
Batrachospermum boryanum	19	11–119	46
Batrachospermum moniliforme	48	4–119	48
Batrachospermum virgatum	15	11–80	46
Sirodotia suecica	5	25–60	49
Nonmucilaginous filaments			
Batrachospermum keratophytum	8	25–100	53
Bostrychia radicans	8	8–66	29
Compsopogon coeruleus	12	10–66	38
Tissue-like forms			
Boldia erythrosiphon	4	33–65	47
Lemanea fucina	35	11–96	49
Tuomeya americana	21	16–119	57

Data from Sheath and Hambrook (1988) and unpublished surveys of Belize, California, Grenada, Louisiana, Mississippi, and Nevada.

Table 16-3. Stress and extension of freshwater red algae[a]

Morphological form	Breaking stress (kN m^{-2}) (n = 5)	Breaking extension ratio (%) (n = 5)
Tufts		
Lemanea fucina "chantransia" stage	12 ± 7[b]	27.0 ± 10.1
Mucilaginous filaments		
Batrachospermum boryanum	270 ± 50	11.3 ± 3.6
Batrachospermum moniliforme		
RI 1	610 ± 340	23.8 ± 8.0
WA 110	510 ± 250	29.2 ± 12.5
Batrachospermum virgatum		
RI 1	660 ± 450	11.7 ± 7.7
WA 308	710 ± 380	28.8 ± 9.2
Sirodotia suecica	440 ± 200	22.2 ± 13.0
Nonmucilaginous filaments		
Batrachospermum keratophytum	1730 ± 670	20.6 ± 7.1
Compsopogon coeruleus	1530 ± 640	25.4 ± 8.0
Tissue-like forms		
Boldia erythrosiphon	1040 ± 240	26.0 ± 8.2
Lemanea fucina[c]	910 ± 430	23.5 ± 5.0
Tuomeya americana	1780 ± 850	21.8 ± 9.0

[a] Data from Sheath & Hambrook (1988). Techniques according to Koehl & Wainwright (1985)
[b] One standard deviation.
[c] releasing carpospores.

Table 16-4. Values of E and predicted breaking current velocities for freshwater Rhodophyta

Morphological form	E[a]	Breaking current velocity (cm s^{-1})
Tufts		
Audouinella violacea	−0.92	80 ± 40[b]
Batrachospermum moniliforme ("chantransia" stage)	−0.67	80 ± 40
Mucilaginous filaments		
Batrachospermum boryanum	−0.33	80 ± 20
Batrachospermum moniliforme	−0.65	140 ± 70
Batrachospermum virgatum	−0.45	140 ± 80
Sirodotia suecica	−1.27	290 ± 140
Tissue-like forms		
Lemanea fucina[c]	−0.83	450 ± 205
Tuomeya americana	−0.64	710 ± 340

Data from Sheath and Hambrook (1988). Techniques according to Vogel and LaBarbera (1978) and Vogel (1984).
[a] $E = (\log D/U^2)/\log U$, for 9 speeds ranging from 0.2 to 0.75 ms^{-1} for the mean of five plants, D = drag (newtons), U = speed, R^2 = 0.75 to 0.99.
[b] One standard deviation.
[c] Releasing carpospores.

16-5). There is considerable overlap among the morphological forms in each of these analyses. *Sirodotia* has the least increase in drag force with increasing velocity, thereby giving the lowest E value (−1.27). This value is low enough that there may be adaptive design for branch reconfiguration at high flows so that drag is almost independent of current velocity (Vogel 1984). It is clear that the large mucilaginous forms *Batrachospermum boryanum* and *B. virgatum* have little ability to reconfigure to reduce drag forces at high current speeds.

Predicted current velocities necessary to break apart various freshwater red algae can be determined using the strengths determined by the stress measurements and the values of drag versus flow (Sheath & Hambrook 1988). There is almost a 10-fold range in these values for the species examined (Table 16-4). The predicted velocities at which the various morphological groups would break are as follows: tufts 80 ± 30 cm s^{-1}, mucilaginous filaments 160 ± 90 cm s^{-1}, and tissues 580 ± 150 cm s^{-1}. The tufts and tissue-like forms are significantly different in this regard ($p < .001$). *Lemanea* has been reported to occur at current velocities as high as 200 cm s^{-1} (Sirjola 1969), which is less than half of the breaking flow estimated. It is clear that all red algae examined are more resistant to removal than the periphytic communities reported by Horner and Welch (1981), which are greatly eroded at velocities of 50 cm s^{-1}.

B. Illumination

The light regime, which includes fluctuations in intensity, quality, and photoperiod, is one of the major factors affecting distribution and seasonality of freshwater Rhodophyta (Sheath 1984; Sheath & Burkholder 1985; Whitton 1975). Hellebust (1970) noted that illumination affects algal growth via photosynthesis, by processes indirectly related to photosynthesis and by those processes unrelated to photosynthesis. In the case of freshwater red algae, distribution within a drainage basin is partially determined by the photoregime established by the surrounding canopy. A dense tree canopy can re-

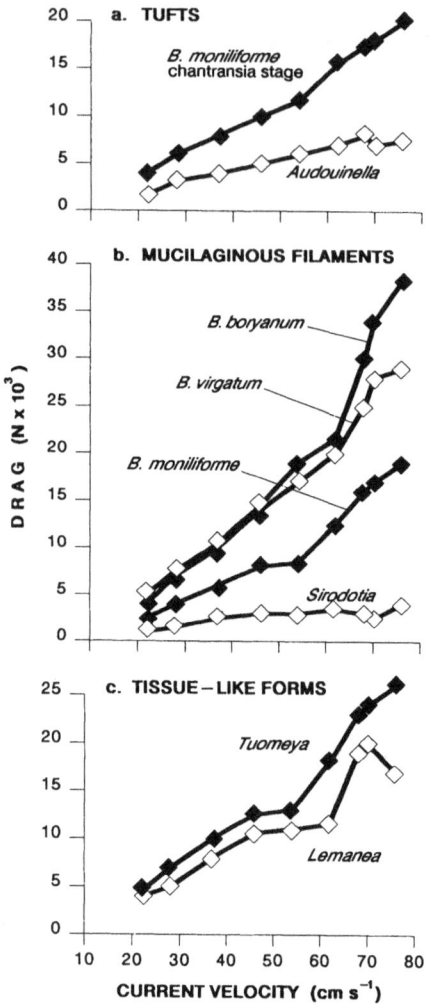

Fig. 16-5. Drag values of different freshwater red algal forms with increasing current velocities. Data from Sheath & Hambrook (1988), with permission.

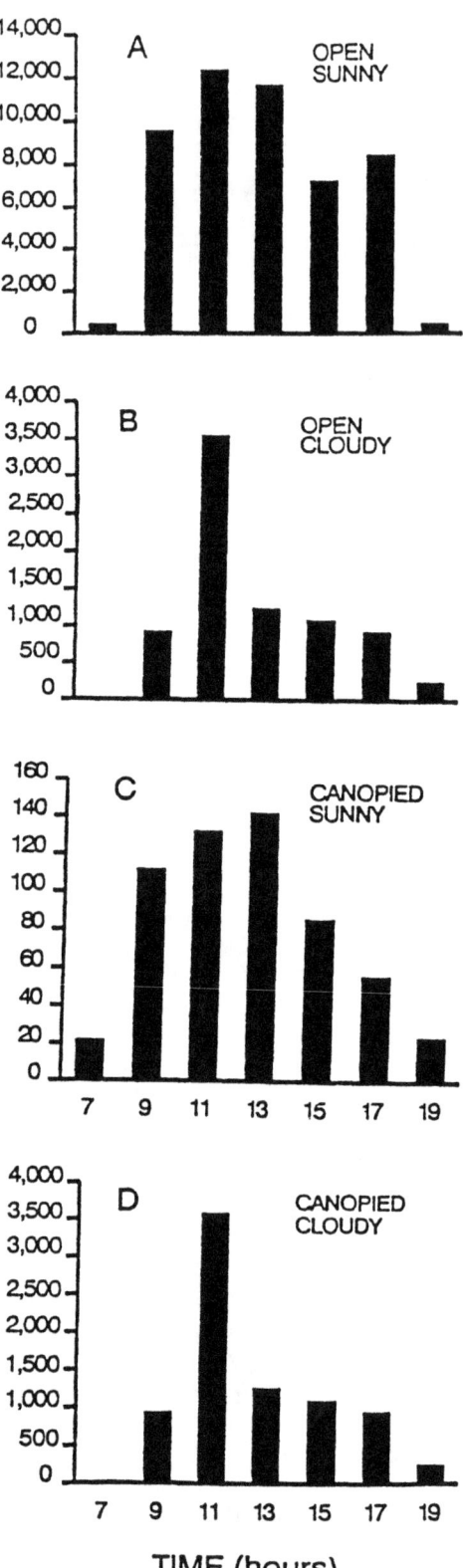

duce the total illumination reaching a small stream by approximately 90% on both sunny and cloudy days (Fig. 16-6). There is also a shift in the quality of the light penetrating a canopy (Fig. 16-7). This trend is particularly evident on a sunny day, when most of illumination is direct. At the beginning and end of the day there is an enhancement of violet light reaching the stream; in the middle of the day there is relative decrease in orange and red light. This spectral shift will vary depending on the canopy

Fig. 16-6. Diurnal changes in total illumination reaching two Rhode Island stream segments in September 1987 on sunny and cloudy days. One site is a headwater stream that is well shaded by surrounding tree canopy and the other is an open farmland stream. Data from Kaczmarczyk & Sheath (unpubl.)

Freshwater ecology

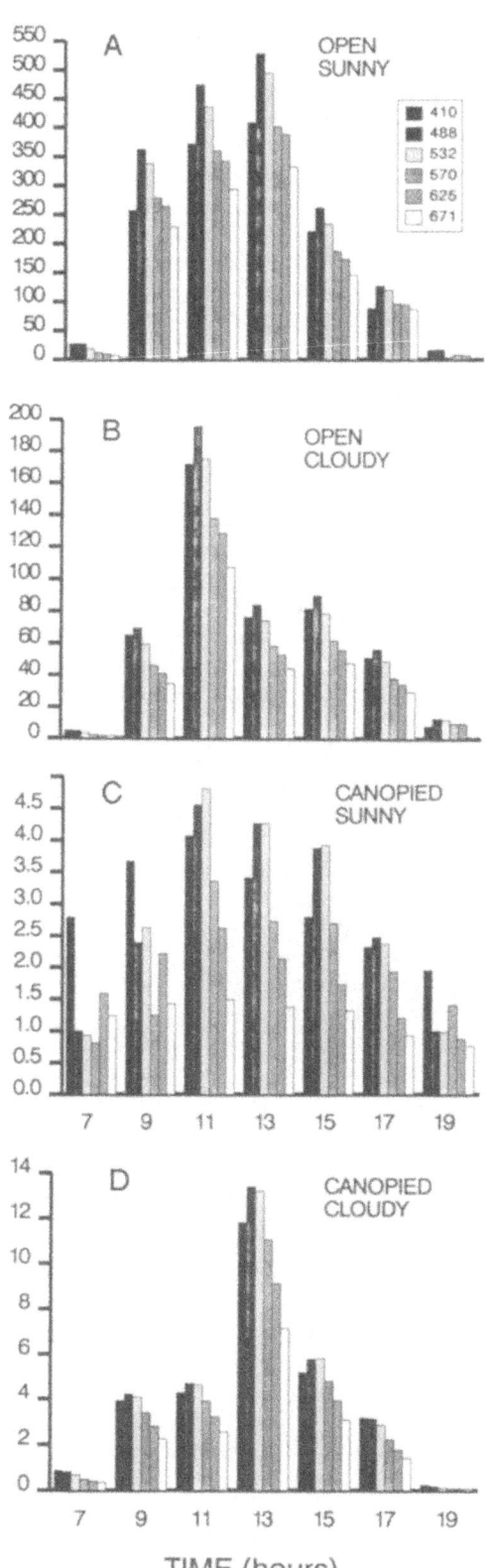

Fig. 16-7. Diurnal changes in violet (410), blue (488), green (532), yellow (570), orange (625), and red (671) illumination reaching two Rhode Island stream segments described in Fig. 16-6.

composition and density, but relatively greater removal of red wavelengths is a common phenomenon (Federer & Tanner 1966). It is predicted that most benthic macroalgae are distributed predominantly in the midregions of large forested drainage basins (Vannote et al. 1980). This is the zone where the tree canopy surrounding the stream opens up but the water is not too deep or turbid to support growth. However, this prediction has not been adequately tested since there are no detailed examinations of algal taxa from such basins. In moderately to greatly disturbed basins in Europe, red algal diversity is quite variable along large rivers, ranging from a headwater distribution to occurrence along the entire river length (Fig. 16-8).

Localized distribution of freshwater Rhodophyta within a stream segment will also be affected by the photoregime. A variety of light intensity requirements of red algae are reported in the literature. *Audouinella violacea*, *Batrachospermum macrosporum*, and *B. vagum* grow well in shaded stream reaches but exhibit rapid deterioration under high light (Dillard 1966; Parker et al. 1973; Rider & Wagner 1972). *B. macrosporum* and *B. vagum* can grow in sunny portions only if streams contain brown waters. In contrast, *Batrachospermum moniliforme* appears to be euryphotic, tolerating a wide range of illumination (Goodwin 1926; Johansson 1982; Rider

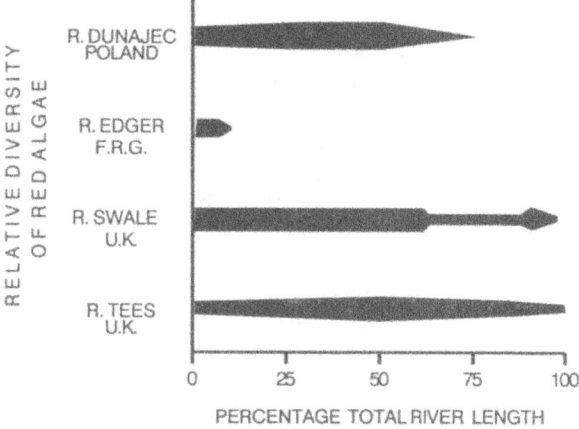

Fig. 16-8. Relative diversity of freshwater red algae along large rivers (100- to 241-km lengths) in Europe. Data from Kawecka & Szczęsny (1984), Steubing et al. (1983), and Holmes & Whitton (1977, 1981), respectively.

& Wagner 1972; Sheath et al. 1979, 1986a). In a Rhode Island headwater stream, *B. moniliforme* exhibited the same periodicity and relative cover in the year prior to and during a major defoliation caused by gypsy moth larvae (Sheath et al. 1986a). Nonetheless, Johansson (1982) found, in a survey of 324 Swedish streams, that this species occurred most frequently in shaded streams with colored waters. The "chantransia" stage of *B. moniliforme* can be maintained for prolonged periods of darkness in a viable state (Sheath et al. 1979). Kremer (1983) determined photosynthesis–light intensity curves for four species of freshwater red algae and found the following saturation points: 100, 150, 250, and 400 µmol Quanta m^{-2} s^{-1} for *Hildenbrandia rivularis*, *Lemanea fluviatilis*, *Batrachospermum* sp., and *Compsopogon hookeri*, respectively. These values are all relatively small, further indicating a preference for low light regimes.

Seasonality of freshwater Rhodophyta can be influenced greatly by a changing light regime. Minckley and Tindall (1963) studied the periodicity of *Batrachospermum* sp. compared to that of the canopy development in a small stream in Kentucky. They observed a strong correlation between production of the canopy leaves and distintegration of the algal thalli. Sheath and Burkholder (1985) examined four streams in Rhode Island on a biweekly basis, two of which had little riparian canopy development and two of which displayed definite seasonality of dense canopy shading. The growth of most red algal species occurred between late fall and early summer, regardless of the light periodicity (Sheath 1984). The one species that exhibited a difference between the two regimes was *B. keratophytum*, which was an aseasonal annual in the open streams but a winter form in the shaded habitats.

The mechanisms by which freshwater red algae acclimate to the photoregime involve modifications in the pigment complement in response to fluctuating intensity and wavelength. Rider and Wagner (1972) found little difference in the chlorophyll content of *Batrachospermum moniliforme* and *B. vagum* plants maintained in a range of intensities from 110 to 11,000 lx for 25 days. In contrast, Thirb and Benson-Evans (1983) measured a four-fold change in chlorophyll *a* and a 10-fold change in phycoerythrin content in *Lemanea* apical tips exposed to a range of illumination from 0.88 to 20.4 W m^{-2} for 21 days. The chlorophyll *a*–to–phycoerythrin ratio varied from 1.7 to 8.6 from the low to high intensity. The chlorophyll content in *B. vagum* is reduced by 2–3x when exposed to green and far red light, compared to white, blue, or red wavelengths, for one month (Rider & Wagner 1972). This indicates that there may be chromatic acclimation in this alga, that is, preferential synthesis of pigments with a high absorption for the incident wavelengths (Hellebust 1970). However, recent studies on marine Rhodophyta indicate that responses to a limited light spectrum represent intensity acclimation (Saffo 1987). In accordance with this conclusion, the freshwater red alga *Lemanea* exhibits no significant differences in chlorophyll *a* or phycoerythrin content when exposed to a variety of wavelengths (Thirb & Benson-Evans 1983).

C. Temperature

Temperature regime influences the latitude, elevation, and drainage basin distribution patterns, as well as seasonality, of freshwater red algae. In North America, the greatest species diversity has been found in temperate and tropical latitudes, with boreal regions having the lowest numbers (Table 16-5). On a global basis, certain taxa appear to have the center of their distribution in lowland tropical and subtropical regions, including most taxa of the section Contorta of the genus *Batrachospermum*, *Bostrychia radicans*, most ceramialean species, *Compsopogon* spp., and *Thorea* spp. (Krishnamurthy 1963; Kumano 1980; Sheath et al. 1987; Skuja 1938; Starmach 1977; Figs. 16-9, 16-10). However, *Compsopogon* and *Thorea* spp. can become abundant in temperate environments during the warm months (Descy & Empain 1974; Entwisle & Kraft 1984; Kremer 1983; Tomas et al. 1980). On the other hand, the genus *Lemanea* has its greatest occurrence in boreal and alpine habitats (Kann 1978a; Kumano 1980; Palmer 1941; Skuja 1938; Fig. 16-11). Kremer (1983) explained these geographic patterns based on photosynthetic response to temperature. *Batrachospermum* sp. and *Lemanea fluviatilis* exhibit maximum photosynthesis at approximately 15°C, whereas *Compsopogon hookeri* has highest rates at 30° to 35°C. These patterns are supported by distribution data in North America (Fig. 16-9).

In large drainage basins, elevation and basin distribution patterns are interrelated. Mean temperatures tend to increase from the source to mouth, although the amplitude of diurnal fluctuations in temperatures tends to become less and less (Whitton 1975). Israelson (1942) found that most rhodophytes were restricted to elevations less than 700–900 m above sea level in Sweden. The exceptions were *Audouinella* (as *Chantransia*) *hermannii* and *Lemanea condensata*, which occurred up to 1200 m. *Batrachospermum vagum* and *B. testale* can be

Table 16-5. Geographic distribution of freshwater macroscopic red algal species numbers in North America

Latitude (°N)	Location	Species numbers	Stream segments examined
Atlantic Basin			
17	Belize	9	14
22	mid-Mexico[a]	5	16
28	mid-FL	5	22
31	MS–LA	11	14
37	Appalachians	7	40
42	RI–MA	11	43
45	s. ON–MI	9	65
46	MT–WY	4	12
49	NF	7	16
53	mid-Labrador	4	11
65	s. Baffin I.	2	11
73	n. Baffin I.	1	6
Pacific Basin			
10	Costa Rica	6	10
33	s. CA[b]	3	—
34	AZ–NM	3	11
40	n. CA	5	10
44	mid OR	5	12
47	n. WA	6	15
55	s. BC	6	15
62	sc. AK	5	40
65	w. AK[c]	1	—
Great Basin			
39	w. CA–e. NV	1	14
Arctic Basin			
52	PQ–James Bay	4	15
56	Belcher I.	1	8
60	PQ–Hudson Bay	1	6
70	n. AK[d]	1	—
75	Devon I.[e]	1	—
Caribbean Islands			
12	Grenada	6	10
15	Guadeloupe–Dominica	5	15
18	Jamaica[f]	5	32
19	Dominican Republic	4	13

[a] From G. Montjano-Zurita (unpubl.).
[b] From Univ. of Calif. Herbarium at Berkeley.
[c] From Lowe (1923).
[d] From B. J. Peterson (pers. comm.).
[e] From H. Croasdale (pers. comm.).
[f] From Boon et al. (1986), Whitford and Robertson (1981) and Sheath and Cole (unpubl.).
Other data from studies as described in Table 16-2.

found at high altitudes in the Rockies, Andes, Papua, and New Guinea (Kumano & Watanabe 1983; Parker et al. 1973).

In temperate regions most freshwater red algae exhibit maximum biomass, growth, and reproduction in the period from late fall to early summer (Dillard 1966; Holmes & Whitton 1981; Israelson 1942; Korch & Sheath 1989; Kremer 1983; Rider & Wagner 1972; Sheath 1984; Sheath & Burkholder 1985; Sheath & Hymes 1980; Table 16-6). Sheath and Burkholder (1985) determined that there is a negative correlation between abundance of six species of red algae in Rhode Island streams and water temperature. However, as noted previously, for some species, such as *Batrachospermum keratophytum*, the negative correlation to temperature results from growth confined to the period when there is no tree canopy shading of streams. This trend is further substantiated in a study of the seasonality of *B. boryanum* in two adjacent Rhode Island headwater stream segments (Hambrook & Sheath unpubl.). There is a distinct periodicity of light penetration to each segment, but one site is spring fed and exhibits little change in water temperature, whereas the second site has a 13°C annual temperature range (Figs. 16-12, 16-13). The seasonality of *B. boryanum* in each segment is the same, paralleling that of light penetration. Hence, in shaded habitats illumination appears to be more important than temperature fluctuations in determining red algal seasonality. Dillard (1966) and Rider and Wagner (1972) also noted similar interactions between light and temperature in determining the appearance of *B. macrosporum*, *B. moniliforme*, and *B. vagum* in North Carolina and Pennsylvania. There have been no detailed studies of seasonality of freshwater red algae in tropical and arctic regions. It does appear that spring occurrence is typical, such that plants are predominant from January to February in the tropics, from March to April in temperate climates, and from June to July in the arctic regions of North America (Sheath & Burkholder 1985; Sheath & Hymes 1980; Sheath et al. 1986b, unpubl.).

III. CHEMICAL FACTORS

A. pH, inorganic carbon, and ions

The interaction between pH and the form of inorganic carbon can greatly influence productivity and distribution of freshwater plants in general. Although widespread freshwater red algae are found in a large range of pH values, they are most frequently found in mildly acidic waters, usually between pH 6.0 and 7.0 (dos Reis 1974; Entwisle & Kraft 1984; Johansson 1982; Kremer 1983; Ratnasabapathy & Seto 1981; Sheath & Burkholder 1985; Sheath & Hymes 1980; Sheath et al. 1986b; Starmach

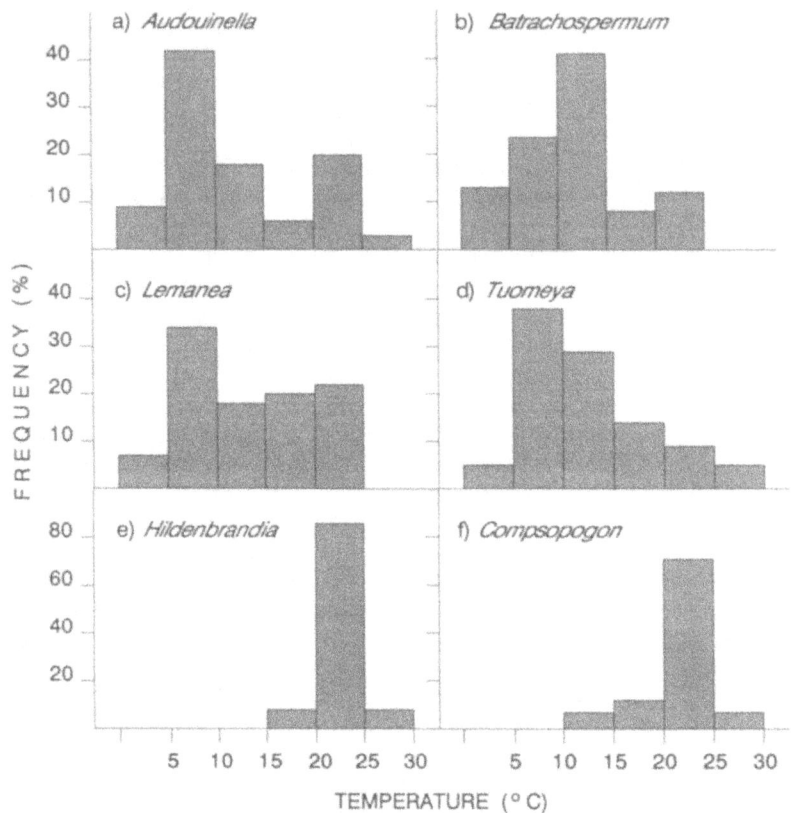

Fig. 16-9. Occurrence of common freshwater red algae in relation to water temperature. The species are A. violacea (n = 36), B. moniliforme (n = 47), L. fucina (n = 32), T. americana (n = 21), H. rivularis (n = 26), and C. coeruleus (n = 12). Data from sites described in Fig. 16-2 and Table 16-5.

1984) (Fig. 16-14). Certain taxa, such as *Audouinella* (as *Rhodochorton*) sp. can become prominent in the periphyton of highly acidified streams (Mulholland et al. 1986). In contrast, a small number of taxa are common in alkaline waters, particularly *Compsopogon* spp. (Sheath et al. 1987; Sinha & Srivastava 1979; Tomas et al. 1980; Zanveld et al. 1976; Fig. 16-14) and *Hildenbrandia rivularis* (Holmes & Whit-

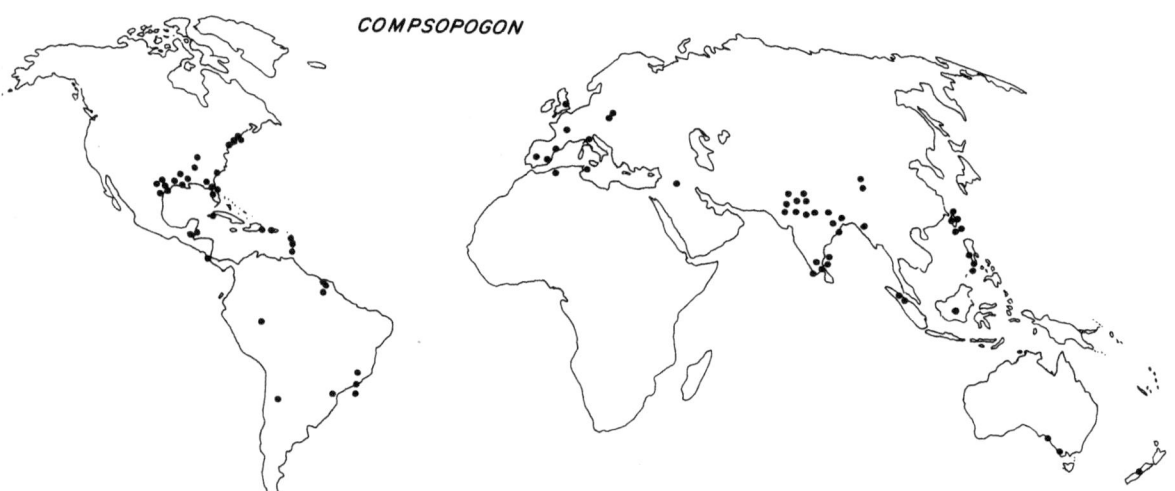

Fig. 16-10 Global distribution of the genus *Compsopogon*. Data based on available published data, including Krishnamurthy (1963) and Starmach (1977).

Fig. 16-11. Global distribution of the genus *Lemanea*. Data based on available published data, including Atkinson (1931), Palmer (1941), dosReis (1962), and Starmach (1977).

ton 1977; Kann 1978a; Khan 1974; Seto 1977; Sheath et al. 1987; Fig. 16-14).

The typical occurrence in mildly acidic waters of many freshwater rhodophytes can be attributed to the form of inorganic carbon. Two species of *Batrachospermum* and *Lemanea mamillosa* have been shown to utilize only free carbon dioxide, and not bicarbonate, as a carbon source (MacFarlane & Raven 1985; Raven & Beardall 1981; Raven et al. 1982; Ruttner 1960). Above pH 8, the proportion of free CO_2 drops below 2–5% of the total inorganic carbon (Stumm & Morgan 1981). Therefore, the aforementioned species would require flow replenishment to exist at pH values greater than 8 (Raven et al. 1982). At moderate current velocities, the external unstirred layer around the plants can be thin (<10 μm for *L. mamillosa*), and the plants are stretched, leading to increased CO_2 transport from the bulk solution (MacFarlane & Raven 1985).

Specific conductance and pH are related in that alkaline waters are commonly high in ions, such as carbonates, and are buffered strongly above pH 8 (Stumm & Morgan 1981). In contrast, waters draining igneous rock catchment areas are less well buffered and are more acidic, with pH values less than 7. Four common freshwater red algal species in North America exhibit a negative distribution pattern in relation to conductance (Fig. 16-15). Nonetheless, some species, such as *Audouinella violacea*, *Batrachospermum moniliforme*, and *Lemanea fucina*, can be found in waters having over a 10-fold range in ions. The greater frequency of occurrence in soft waters rather than hard waters appears to be common to many freshwater red algae (Israelson 1942; Johansson 1982; Woelkerling 1975). This is due, in part, to the form of inorganic carbon at high and low ion levels, as previously described. However, there are certain freshwater rhodophyte species that are more typically found in hard waters, such as *Compsopogon* spp. (Hinton & Maulood 1980; Sheath et al. 1987; Sinha & Srivastava 1979; Tomas et al. 1980; Fig. 16-15) and *Hildenbrandia rivularis* (Israelson 1942; Kann 1978a; Sheath et al. 1987; Fig. 16-15). Putative red algal secondary invaders of fresh waters, such as *Bangia atropurpurea*, *Bostrychia radicans*, and *Caloglossa leprieurii*, are also largely restricted to ion-rich waters (Kumano 1980; Reed 1980; Sheath 1984; Sheath et al. 1987; Skuja 1938). A minimum amount of calcium might be necessary for these species for the functioning of Ca-ATPase (Yarish et al. 1980).

B. Oxygen

Freshwater red algae are found at a wide range of oxygen concentrations, ranging from 0.2 to 21 mg L^{-1} (Howard & Parker 1981; Kremer 1983; Sheath & Hymes 1980; Thirb & Benson-Evans 1985; Woelkerling 1975). There is a slight increase in frequency of occurrence with higher O_2 concentrations (Sheath 1984). To some extent, this relationship results from the fact that most species occur in the period from late fall to late spring, when oxygen solubility is higher than in the summer. Nevertheless, fresh-

Table 16-6. Phenology of freshwater red algae

Taxon	Macroscopic populations	Seasonality Mono-sporangia	Gametangia	Tetra-sporangia	Location	Reference
Acrochaetiales						
Audouinella macrospora	Jan–Dec	Jan–Dec	—	—	R.I.	Sheath & Burkholder 1985
Audouinella violacea	Oct–Jul	Oct–Jul	May–Jun	—	U.K.	Drew 1935
	Sep–May	Sep–May	May	Dec–Jun	R.I.	Korch & Sheath 1989
	May–Jul	May–Jul	—	Feb	Idaho	Reed 1970
Balbiania investiens	Feb–Jul	Feb–Jul	Feb–Jul	—	U.K.	Swale & Belcher 1963
	Oct–Aug	Oct–Aug	Aug	Apr	DDR	Huth 1979
Batrachospermales						
Batrachospermum arcuatum	Jul–Sep	—	Jul–Sep	—	Sweden	Kylin 1912
Batrachospermum atrum	Apr–Aug	—	Apr–Aug	—	Sweden	Kylin 1912
Batrachospermum boryanum	Mar–Sep	—	Mar–Sep	—	Sweden	Kylin 1912
	Sep–Jun	—	Sep–Jun	—	R.I.	Sheath & Burkholder 1985
Batrachospermum crouanianum	Apr–Aug	—	Apr–Aug	—	Sweden	Kylin 1912
Batrachospermum densum	Jul–Aug	—	Jul–Aug	—	Sweden	Kylin 1912
	Sep–Jun	—	Sep–Jun	—	Ont.	Sheath & Hymes 1980
Batrachospermum distensum	Jul–Aug	—	Jul–Aug	—	Sweden	Kylin 1912
Batrachospermum durum	Mar–Oct	—	Mar–Oct	—	Poland	Starmach 1984
Batrachospermum keratophytum	Jun–Aug	Jun–Aug	Aug	—	Sweden	Kylin 1912
	Jan–Dec	Mar	Jun	—	R.I.	Sheath & Burkholder 1985
Batrachospermum moniliforme	Apr–Oct	—	Apr–Oct	—	Sweden	Kylin 1912
	Oct–Aug	—	Oct–Aug	—	R.I.	Sheath & Burkholder 1985
Batrachospermum sporulans	Jun–Jul	—	Jun–Jul	—	Sweden	Kylin 1912
Batrachospermum testale	Jul–Aug	—	Jul–Aug	—	Sweden	Kylin 1912
Batrachospermum virgatum	Jun–Aug	—	Jun–Aug	—	Sweden	Kylin 1912
	Sep–Jun	—	Sep–Jun	—	R.I.	Sheath & Burkholder 1985
Batrachospermum sp.	Sep–Jun	—	Sep–Jun	—	Kentucky	Minckley & Tindall 1963
	Nov–Jan	—	Nov–Jan	—	Japan	Yoshida 1959
Lemanea fluviatilis	Aug–Sep	—	—	—	Sweden	Muller 1978
Lemanea fucina	Oct–Jun	—	Jan–May	—	Idaho	Reed 1970
Sirodotia suecica	Jul–Aug	—	Jul–Aug	—	Sweden	Kylin 1912
	Sep–Jul	—	Sep–Jul	—	R.I.	Sheath & Burkholder 1985

Freshwater ecology

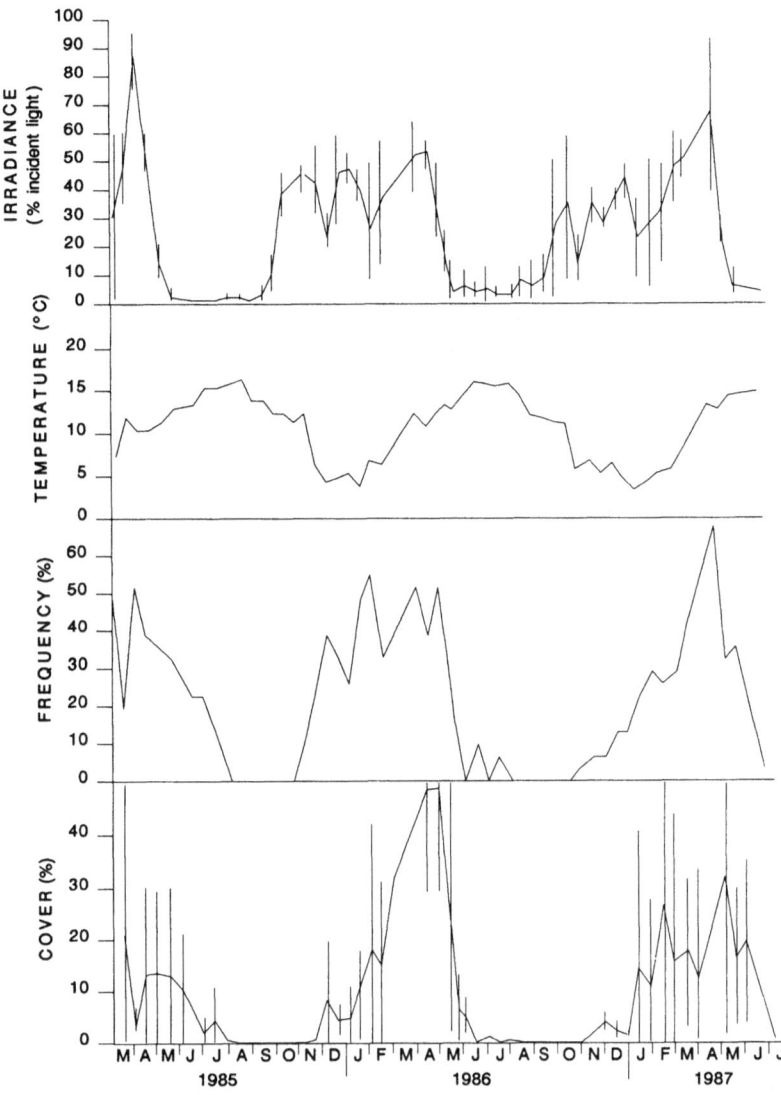

Fig. 16-12. Seasonality of illumination, water temperature, and abundance of *Batrachospermum boryanum* in a Rhode Island headwater stream. Data from Hambrook & Sheath (unpubl.).

water Rhodophyta are not commonly associated with stagnant, organic-rich waters with very low oxygen contents (Flint 1970; Sheath 1984; Sheath & Hymes 1980; Skuja 1938).

C. Nutrients

Nutrients are usually considered to spiral through a lotic ecosystem as organic matter is processed along the river channel (Elwood et al. 1983). This concept is based on the view that the river system is an uninterrupted continuum. However, natural debris dams are common, resulting in an alternating series of lentic and lotic reaches. The serial discontinuity concept accounts for these disruptions in nutrient spiraling (Ward & Stanford 1983a). In general, phosphorus is the least abundant and most commonly limiting nutrient in fresh waters (Wetzel 1983). Therefore, algal biomass can increase greatly with localized inputs of this nutrient in stream systems (Krewer & Holm 1982; Stockner & Shortreed 1978).

There are no detailed studies of the nutrient relationships of stream-inhabiting red algae. However, these algae have been recorded in a wide variety of streams in which nutrients have been measured (Table 16-7). In general, all morphological forms of freshwater rhodophytes can be found in water having low nutrient concentrations. For example, the minimum PO_4^{-3} concentration recorded for the various species examined ranges from levels below detection to 100 μg L^{-1} ($\bar{x} = 13$). Since nutrient concentrations in streams often fluc-

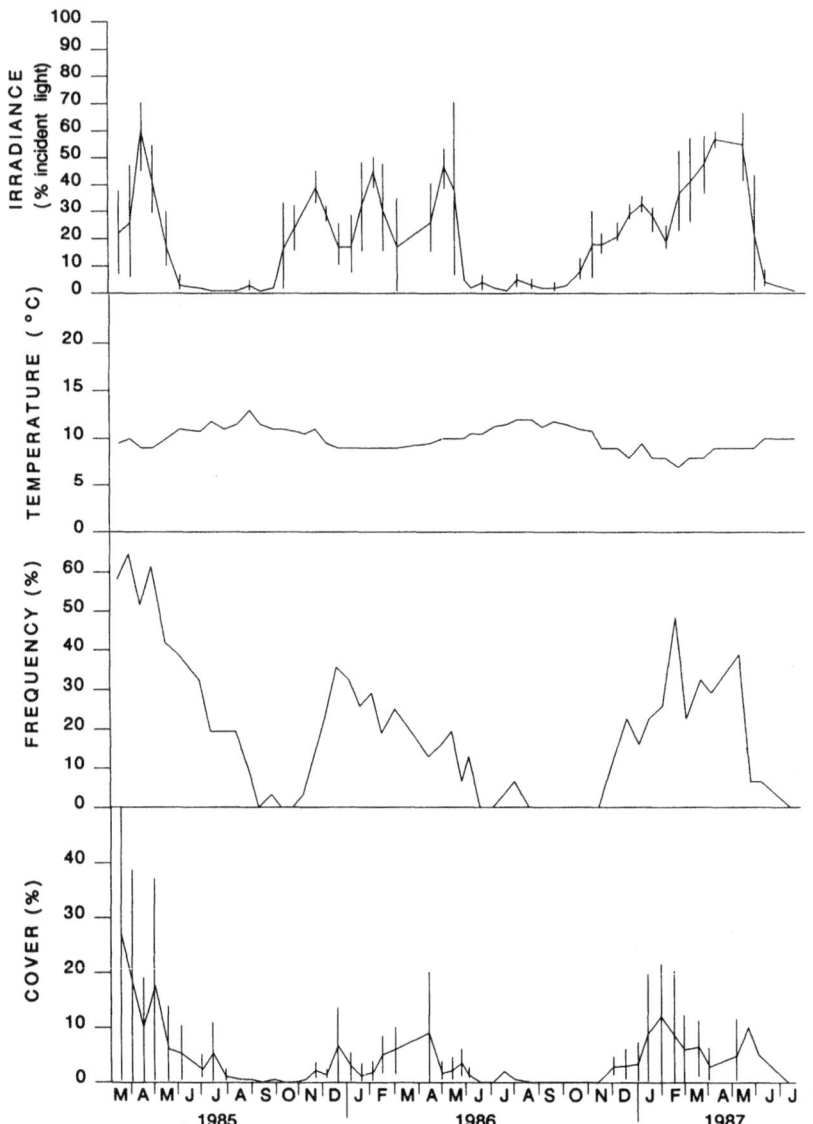

Fig. 16-13. Seasonality of illumination, water temperature, and abundance of *Batrachospermum boryanum* in a Rhode Island spring-fed stream. Data from Hambrook and Sheath (unpubl.).

tuate more rapidly than algal abundance, no significant correlation has been found between these parameters (Sheath & Burkholder 1985; Thirb & Benson-Evans 1985). The phosphorus content of *Lemanea fluviatilis* tissue in a Swedish stream is only 0.48% (Muller 1978). The common occurrence of freshwater red algae at low nutrient levels is partially due to flow replenishment and reduction of the boundary layer of depletion in lotic systems. In addition, most red algae have been found to contain colorless hair cells that may be produced in response to nutrient deficiency, as is the case for some green algal filaments (Gibson & Whitton 1987). In the latter group, phosphatase can be localized on hairs, indicating that they function to increase the surface area of phosphorus uptake.

D. Pollutants

In general, freshwater red algae are localized in reasonably unpolluted waters and are infrequent to absent in streams and rivers that are organically enriched, greatly silted, or high in nutrients (Sheath 1984). However, *Lemanea fluviatilis* and *Bangia atropurpurea* appear to be tolerant of heavy metal pollution (Carpenter 1924; Deb et al. 1974; Harding & Whitton 1981; Lin & Blum 1977). For example, *L. fluviatilis* can occur at aqueous concentrations of

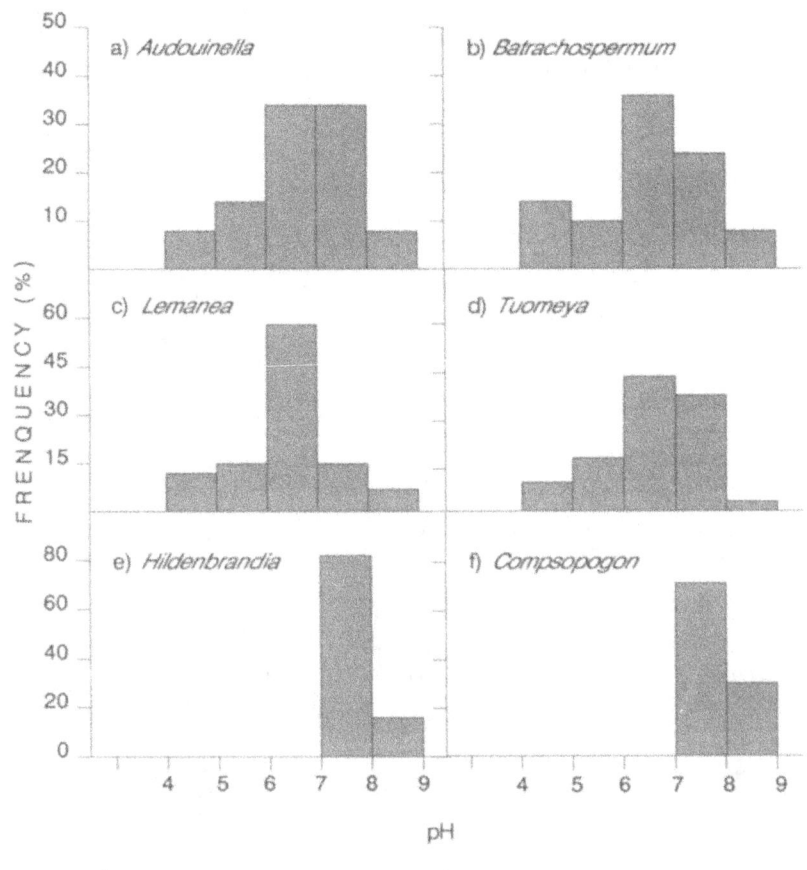

Fig. 16-14. Occurrence of common freshwater red algae in relation to pH. Data and species as described in Fig. 16-9.

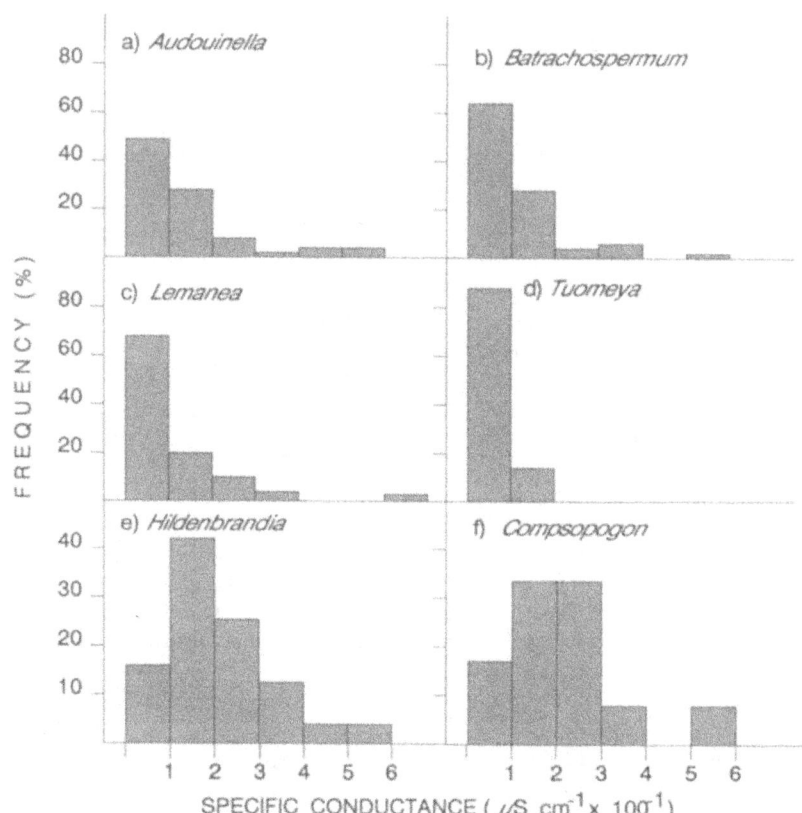

Fig. 16-15. Distribution of common freshwater red algae in relation to specific conductance. Data and species as described in Fig. 16-9.

Table 16-7. Freshwater red algal occurrence in relation to nutrient concentration

Taxon	Nutrient Concentration			Reference
	PO_4^{3-} ($\mu g\,l^{-1}$)	NO_3^- ($mg\,l^{-1}$)	NH_4^+ ($\mu g\,l^{-1}$)	
Audouinella hermannii	12–46	0.2–0.3	ND	Kann 1978b
	4–8	0.2–0.3	20–25	Wehr 1981
Audouinella macrospora	0–11	0–1	0–4	Sheath & Burkholder 1985
Audouinella violacea	20–175	0.7–0.9	ND	Holmes & Whitton 1977
	12–16	0.2–0.3	ND	Kann 1978b
	0–8	0–3	0–1	Sheath & Burkholder 1985
Audouinella sp.	1.3	4	130	Mulholland et al. 1986
Batrachospermum boryanum	0–8	0–8	0–4	Sheath & Burkholder 1985
	7–4000	1–5	ND	Woelkerling 1975
Batrachospermum keratophytum	0–14	0–1	0–4	Sheath & Burkholder 1985
Batrachospermum moniliforme	1.8–16	0.4–1	ND	Kann 1978b
	0–3	0–8	0–4	Sheath & Burkholder 1985
	50	4	ND	Steubing et al. 1983
	4–8	0.2–0.3	20–25	Wehr 1981
	20–4900	0–17	ND	Woelkerling 1975
Batrachospermum virgatum	0–8	0–8	0–3	Sheath & Burkholder 1985
Batrachospermum sp.	100–600	0.7–2	ND	Minckley 1963
	20–8100	2–10	ND	Woelkerling 1975
Compsopogon coeruleus	5–1300	2–6	ND	Tomas et al. 1980
Hildenbrandia rivularis	20–175	0.8–3.2	ND	Holmes & Whitton 1977
	11–14	0.3–1	3–110	Holmes & Whitton 1981
	12	0.6	ND	Kann 1978b
Lemanea fluviatilis	20–175	0.8–1.2	ND	Holmes & Whitton 1977
	26–1272	0.1–1	22–280	Holmes & Whitton 1981
	12–16	0.2–0.3	ND	Kann 1978b
	10–470	ND	10–820	Thirb & Benson-Evans 1985
Lemanea fucina	0–4	ND	ND	Reed 1970
Lemanea sp.	50–600	0.65–2	ND	Minckley 1963
Sirodotia suecica	0–11	1–8	0–4	Sheath & Burkholder 1985
Sirodotia tenuissima	0–11	0–1	0–4	Sheath & Burkholder 1985
Tuomeya americana	0–3	1–8	0–13	Sheath & Burkholder 1985

ND = not determined.

zinc up to 1.16 mg L^{-1}. Whitton et al. (1981) have recommended that this species be used in a "package" of ten species as a bioassay organism to monitor heavy metals in Europe. Kann (1986) concluded that other freshwater red algae, such as *Batrachospermum moniliforme* and *Hildenbrandia rivularis*, cannot be used with certainty as indicators of the state of pollution of a stream. The pollution intolerance of most freshwater Rhodophyta partly accounts for their typical paucity in the lower reaches of large river basins.

IV. BIOTIC FACTORS

A. Grazing

Thirty-seven lotic animals have been found to ingest freshwater red algae, based on gut contents or observed feeding (Table 16-8). These grazers include two amphipods, six mayfly larvae, twelve caddisfly larvae, six stonefly larvae, six chironomids, one beetle larva, two snails, and two cyprinoid fish. Red algal fragments in the guts of grazers are shown in Fig. 16-16. Most animals remove small pieces consisting of 5–20 cells. The contents are then digested, leaving the empty cell walls. As noted by Cummins (1973), the majority of these animals are polyphagous, consuming a wide variety of food matter, including detritus, leaf fragments, and several algal taxa. In studies of gut contents by Hambrook and Sheath (1987), Jones (1949), Mecom and Cummins (1964), and Shih (1941) several algal species were found, though diatoms were the most frequently encountered gut material. Nonetheless, red algae can be a primary food source of herbivorous and omnivorous invertebrates during the autumn, win-

Table 16-8. Lotic grazers with red algal fragments in their guts or observed feeding on rhodophytes

Grazer	A	B	C	L	T	Reference
Amphipoda						
Gammarus fasciatus	+	+	+		+	Hambrook & Sheath 1987
Synurella sp.		+	+			Hambrook & Sheath 1987
Ephemeroptera						
Baetis rhodani		+				Jones 1950
Baetis vagans			+			Minckley 1963
Cloeon sp.	+	+				Hambrook & Sheath 1987
Ephemerella sp.					+	Hambrook & Sheath 1987
Rhithrogena semicolorato		+	+			Jones 1950
Stenonema sp.		+	+			Hambrook & Sheath 1987
Trichoptera						
Anabolia nervosa		+		+		Jones 1950
Brachycentrus americanus		+				Mecom & Cummins 1964
Brachycentrus numerosus			+			Hambrook & Sheath 1987
Brachycentrus sp.	+	+			+	Hambrook & Sheath 1987
Cheumatopysche sp.		+				Hambrook & Sheath 1987
Dibusa angata				+		Resh & Houp 1986
Halesus radiatus		+		+		Jones 1950
Hydropsyche instabilis		+		+		Jones 1949, 1950
Maydenoptila cuneola		+				Wells 1985
Mystacides sepulchralis					+	Hambrook & Sheath 1987
Sericostoma personatum		+				Jones 1950
limnephilid larva		+				Chapman & Demory 1963
Plectoptera						
Amphinemura cinerea		+		+		Jones 1950
Chloroperla tripunctata		+		+		Jones 1950
Isoperla grammatica		+		+		Jones 1950
Leuctra hippopus		+				Jones 1950
Perlodes mortoni				+		Jones 1950
Protonemura meyeri		+				Jones 1950
Diptera						
Cardiocladius sp.		+				Rider & Wagner 1972
Diamesa sp.	+					Hambrook & Sheath 1987
Orthacladius sp.	+	+	+			Hambrook & Sheath 1987
Pentaneura sp.		+				Rider & Wagner 1972
Psectrocladus sp.	+					Hambrook & Sheath 1987
unknown chironomid					+	Hambrook & Sheath 1987
Coleoptera						
Promoresia tardella					+	Hambrook & Sheath 1987
Gastropoda						
Goniobasis sp.			+			Minckley 1963
Physa sp.		+				Hambrook & Sheath 1987
Cyprinoid fishes						
Onychostoma laticeps				+		Shih 1941
Oreinus prenanti				+		Shih 1941

[a] A = *Audouinella*
B = *Batrachospermum*
C = "chantransia" stages
L = *Lemanea*
T = *Tuomeya*

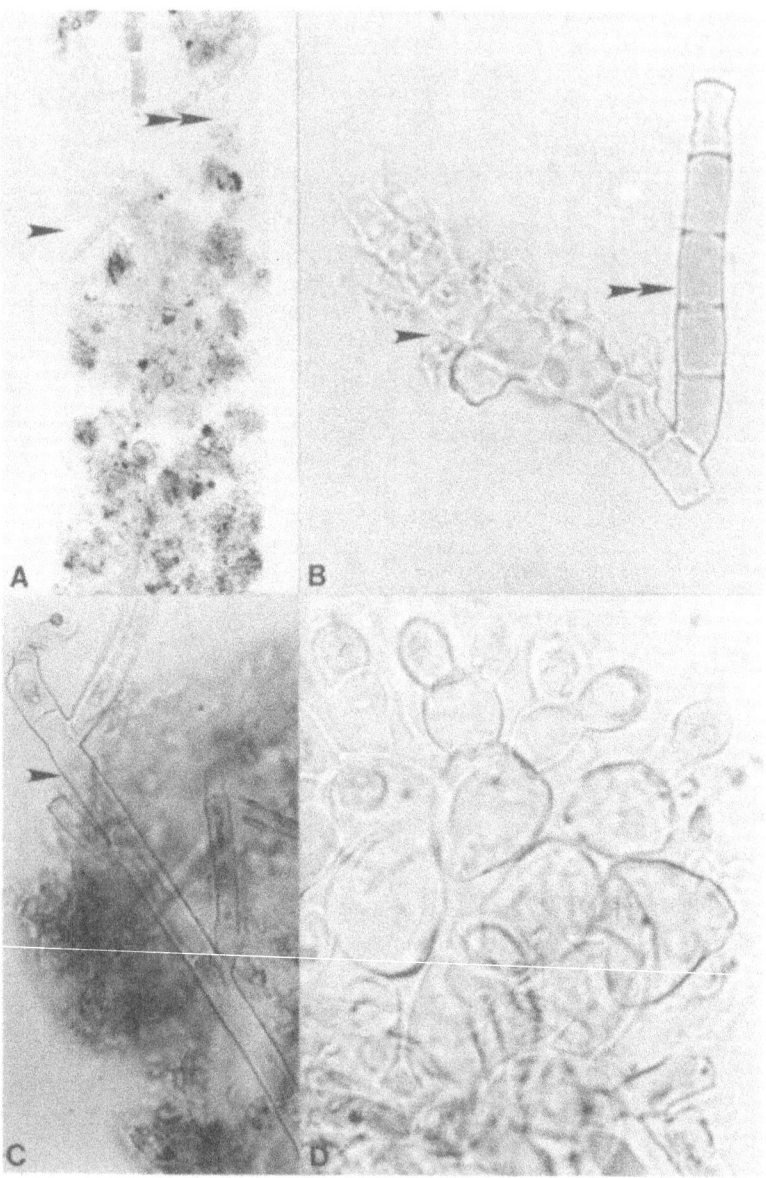

Fig. 16-16. Gut contents of grazers ingesting freshwater red algae. A. Intact chironomid gut showing outline (single arrow) and piece of *Audouinella* (double arrow). Note the predominance of diatoms. ×90. B. Piece of "chantransia" (double arrow) with an attached juvenile gametophyte (single arrow) of *Batrachospermum* in a mayfly gut. Note that most cell contents are completely digested. ×440. C. Piece of *A. violacea* (arrow) in the gut of *Gammarus fasciatus*. All of the cells are empty. ×300; D. Cross section of *Tuomeya americana* in a mayfly gut in which most cells are digested. ×220.

ter, and spring, when diatoms and green algae are relatively unimportant (Jones 1950).

Hambrook and Sheath (1987) examined ingestion rates of three widespread rhodophyte species, *Audouinella violacea*, *Batrachospermum virgatum*, and *Tuomeya americana*, by the amphipod *Gammarus fasciatus*, the mayfly larva *Cloeon* sp., and the caddisfly larva *Brachycentrus* sp. In both individual and choice

Freshwater ecology

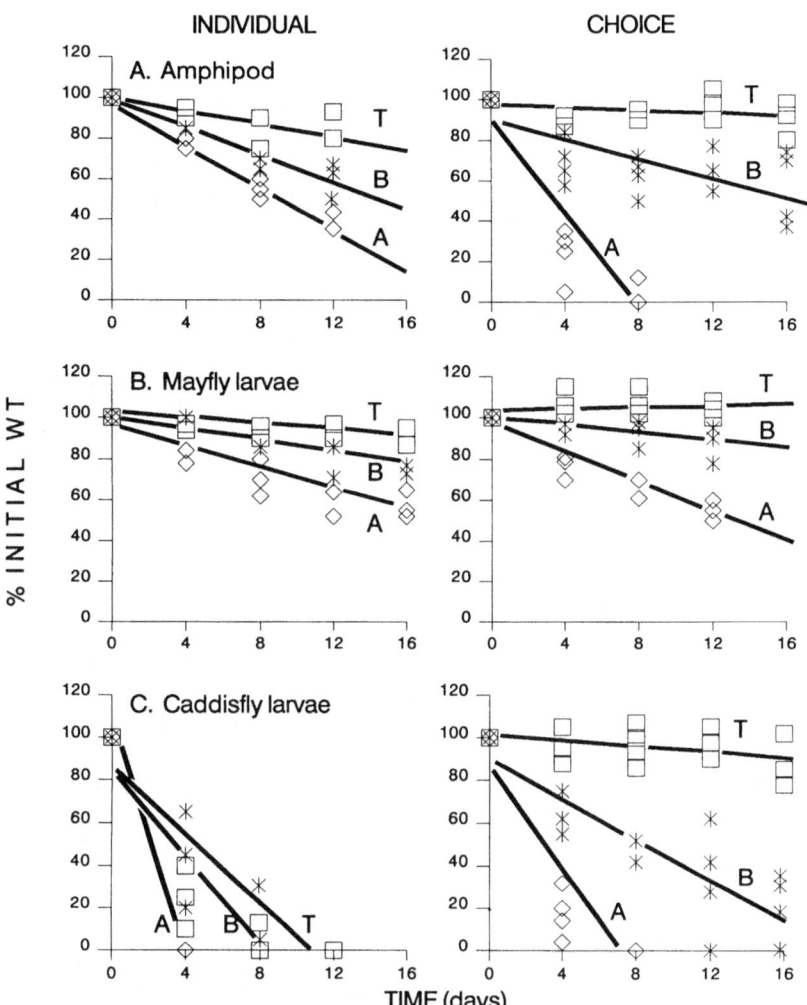

Fig. 16-17. Ingestion rates of three freshwater red algae by *Gammarus fasciatus*, *Cloeon* sp., and *Brachycentrus* sp. in single diet and choice experiments. The red algal sources are *Audouinella violacea* (A), *Batrachospermum virgatum* (B), and *Tuomeya americana* (T). Each experiment was performed in quadruplicate, with each value shown and plotted as relative loss of fresh weight of each alga. The symbols are closed squares for *Tuomeya*, open circles for *Batrachospermum*, and open squares for *Audouinella*. The lines represent regression lines and hence do not necessarily go through an initial value of 100%. Data from Hambrook & Sheath (1987), with permission.

experiments of all grazers, the simple, sparsely branched *Audouinella* is the preferred food source, followed by the mucilaginous and densely branched *Batrachospermum* and then by the cartilaginous pseudoparenchyma *Tuomeya* (Fig. 16-17). The relative ingestion rates are inversely proportional to toughness (Tables 16-3, 16-4), as has been shown for marine macroalgae (Littler & Littler 1980).

Biochemical composition is also important in determining grazing rates. *Audouinella*, *Batrachospermum*, and *Tuomeya* all have large calorific values (5.0–6.8 kcal g^{-1} dry weight; Table 16-9). These values are high compared with the freshwater chlorophyte *Cladophora glomerata*, which has a caloric content of 1.6 kcal g^{-1} dry weight. In general marine red algal species have a range of 0.2–4.52, as summarized by Hambrook and Sheath (1987). It is the low ash content and high lipid content of the freshwater rhodophytes studied that give them their high calorific values (Table 16-9). Hence,

Table 16-9. Biochemical composition of freshwater red algae

Composition (% dry weight)	Taxon		
	Audouinella	Batracho-spermum	Tuomeya
Lipid	22.2 ± 2.1[a]	43.8 ± 9.4	59.2 ± 6.3
Carbohydrate	25.4 ± 5.4	17.8 ± 1.3	15.8 ± 1.3
Protein	32.3 ± 5.5	26.7 ± 1.9	21.2 ± 0.2
Ash	5.6 ± 2.3	13.8 ± 3.7	6.5 ± 1.4
kcal g^{-1} dry wt.	5.0	5.7	6.8

Data from Hambrook and Sheath (1987).
[a] One standard deviation.

discrimination among forms by selective grazers is likely based on morphology and biochemical content rather than on energy needs (Fig. 16-17). Ingestion rates of *Audouinella*, *Batrachospermum*, and *Tuomeya* are positively correlated with protein and negatively correlated with lipid and carbohydrate contents (Table 16-9; Fig. 16-17). The relative lipid contents of the freshwater Rhodophyta are higher than those of marine red algae (1–7.8%) summarized by Hambrook and Sheath (1987). However, this is offset by lower ash contents in freshwater taxa (5.6–13.8%), compared to most marine species (9–81%). The preference for the more protein-rich *Audouinella* by the grazers tested agrees with the conclusion of Cummins (1973) that protein content is a basis for selective grazing in streams. Cargill et al. (1985) observed that shredding caddisflies selected for lipid-rich substrates when triglycerides were necessary for metamorphosis and reproduction. Therefore, the preferential herbivory of freshwater Rhodophyta with high protein content and delicate construction may also vary, depending on developmental stage, season, and relative nutritional and energy needs of the grazers.

Certain caddisfly larvae, such as *Maydenoptila cuneola* and *Dibusa angata*, appear to occur exclusively among plants of *Batrachospermum* and *Lemanea*, respectively, using the red algae for case making and as a food source during final larvae instar stages (Resh & Houp 1986; Wells 1985). Scanning electron micrographs of portions of these cases are shown in Fig. 16-18. The case of *Dibusa* is composed of concentric strips of *Lemanea* intricately intermeshed with silk (Resh & Houp 1986). In contrast, the case of *Maydenoptila* appears to be a loose arrangement of *Batrachospermum* fascicles, epiphytized by other stream algae. The ability of *Dibusa* to cut uniform strips of *Lemanea* is probably related to the shape of the mandibles, which have lateral projections along the inner edge (Resh & Houp 1986). The *Lemanea* pieces incorporated into the case remain alive throughout the duration of the final larval and pupal stages. On occasion, larvae eat portions of their case. Apparently, these host-specific algal feeding relationships are unusual among stream-inhabiting insect larvae (Resh & Houp 1986).

B. Species associations

It is not clear whether lotic communities containing red algae are controlled and structured by biotic interactions or whether they are simply concentrations of opportunistic species existing together because of appropriate environmental conditions (Shiozawa 1983). Streams are often considered to be physically controlled environments in which flooding, droughts, and rapid temperature changes can act as major sources of density-independent mortality (Hart 1983). However, this observation does not negate the importance of biotic interactions in such systems. Since streams are so patchy with regard to substrate, flow, depth, and light penetration, competition for suitable habitats may be intense during certain times of the year (Sheath 1984). For Rhodophyta, this competition can occur at different size levels. Microscopic stages and low-growing forms, such as "chantransia" filaments, *Audouinella* tufts, and *Hildenbrandia* crusts, are components of the stream epilithic periphyton. As such, they compete for substrate with a complex assortment of microalgae, usually dominated by diatoms during early colonization stages (Hoagland et al. 1982; Steinman & McIntire 1986). Fritsch (1929) noted that *Hildenbrandia* thalli are often overgrown by diatoms or blue-green algae in British streams. In later stages of succession, filamentous and stalked species can form an upper story where they have a competitive advantage for light and nutrient replenishment. The semi-erect rhodophyte forms fit into this latter category. As these species grow beyond the boundary layer, they are subjected to space competition with lotic macrophytes. Bryophytes are frequent dominants in upper reaches of rivers, where red algae occur (Holmes & Whitton 1977; Hynes 1970; Sheath et al. 1986a; Westlake 1975). Macrophytic rhodophytes also encounter flow-related forces, as previously discussed. As a result, their long-term distribution will be partially controlled by density-independent disturbances like flooding. These disturbances create openings in the community, allowing new periphyton colonization.

Freshwater ecology

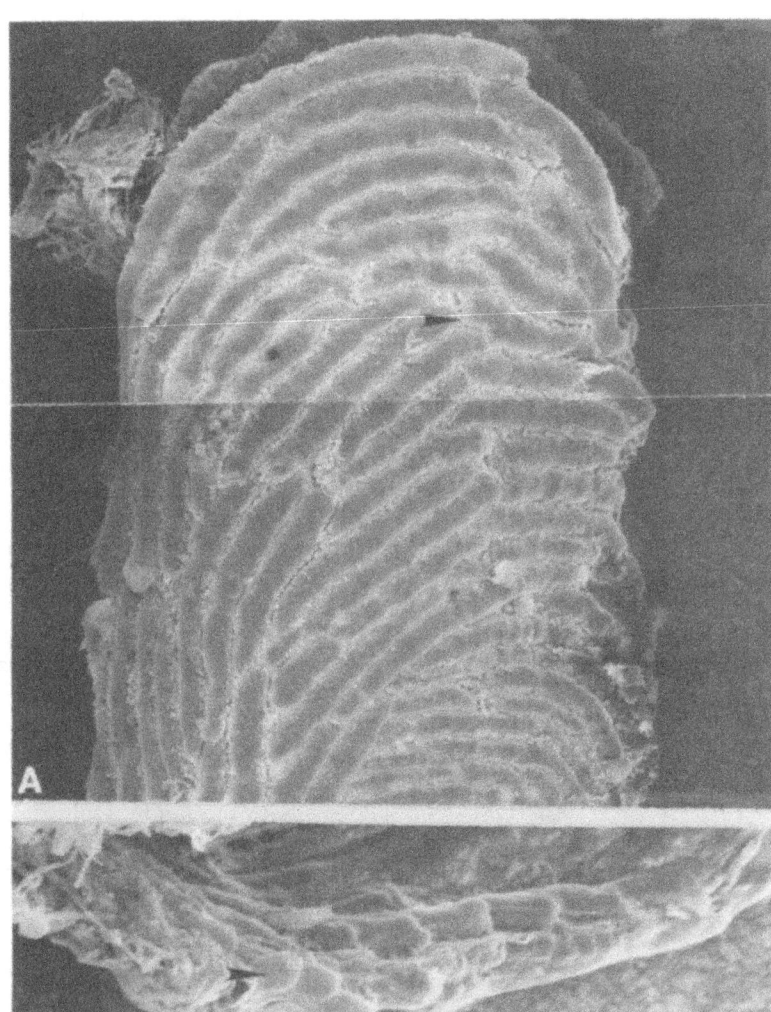
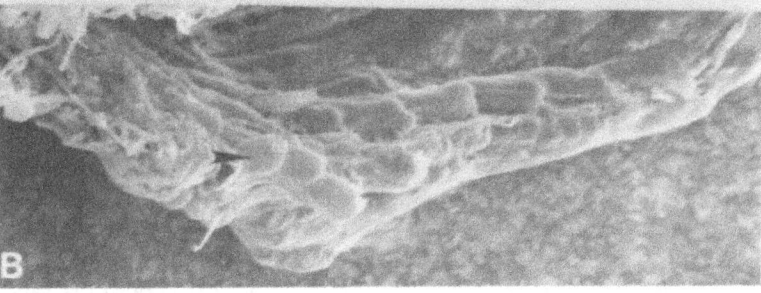

Fig. 16-18. Scanning electron micrographs of cases of the caddisfly larvae. A. *Dibusa angata* (sample compliments of G. B. Wiggins). ×18. B. *Maydenoptila cuneola* (sample compliments of A. Wells). ×360. The pieces of *Lemanea* and *Batrachospermum* composing each case respectively are indicated by the arrows.

Therefore, lotic communities containing red algae are generally in a nonequilibrium state, consisting of most successional stages, which are regulated by both density-dependent and density-independent factors. This spatial–temporal heterogeneity is consistent with the intermediate disturbance hypothesis, which predicts that biotic diversity will be greatest in communities with moderate levels of disturbance (Ward & Stanford 1983b).

On a larger scale, the occurrence of certain freshwater rhodophyte species is correlated with that of other species. A cluster analysis of 11 widespread North American species shows three groupings (Fig. 16-19). The tropical cluster, consisting of *Bostrychia*, *Hildenbrandia*, and *Compsopogon*, is the most distinct, having no co-occurrences with species in the two temperate–boreal clusters. The two closest species associations are between *Audouinella violacea* and *Lemanea fucina* and between *Bostrychia radicans* and *Hildenbrandia rivularis*. The common co-occurrence of *Audouinella* and *Lemanea* is also true in Europe (Holmes & Whitton 1977; Israelson 1942; Steubing et al. 1983). However, the lack of overlap in distribution between *Hildenbrandia* and *Audouinella* and between *Batrachospermum* and *Lemanea* in North America is quite different from observations in Europe (Budde 1926; den Hartog 1956; Holmes & Whitton 1977, 1981; Israelson 1942; Kann 1986;

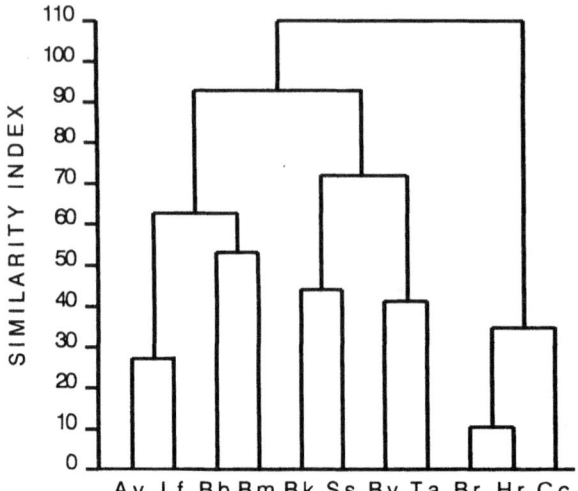

Fig. 16-19. Cluster diagram of 11 widespread freshwater red algae in North America. The analysis is based on Sorenson's similarity index as described by Mueller-Dombois & Ellenberg (1974). Species are as follows: A.v. = *Audouinella violacea*, L.f. = *Lemanea fucina*, B.b. = *Batrachospermum boryanum*, B.m. = *B. moniliforme*, B.k. = *B. keratophytum*, S.s. = *Sirodotia suecica*, B.v. = *B. virgatum*, T.a. = *Tuomeya americana*, B.r. = *Bostrychia radicans*, H.r. = *Hildenbrandia rivularis*, C.c. = *Compsopogon coeruleus*. Data from survey described in Fig. 16-2.

Starmach 1984). The low similarity indices in North America and observations in Europe and Asia indicate that there are probably no associations in which the occurrence of one freshwater red algal species is dependent on that of another.

Lotic rhodophytes can also coexist with various macroalgae from other divisions. In North America the cyanophyte *Phormidium* spp., the green algal genera *Microspora, Spirogyra, Stigeoclonium, Tetraspora*, and *Zygnema*, and the xanthophyte *Vaucheria* spp. commonly occur in streams containing red algae (Minckley 1963; Sheath & Burkholder 1985; Sheath et al. 1986a,b,c, unpubl.; Wehr 1981; Wehr & Stein 1985; Whitford & Schumacher 1963). In central Europe and Great Britain crustose communities containing the rhodophyte *Hildenbrandia rivularis*, the brown alga *Heribauldiella fluviatilis*, and the lichen *Verrucaria* spp. are common (Fritsch 1929; Holmes & Whitton 1977; Israelson 1942; Kann 1986). In contrast, within the North American range of *Heribauldiella*, no *Hildenbrandia* has been found (Wehr & Stein 1985). Other macroalgae associated with lotic Rhodophyta in Europe are diverse, overlapping in part with those reported for North America. Like red algal co-occurrences, there do not appear to be any dependent associations.

Freshwater rhodophytes vary in their degree of epiphytism. As stated earlier, tufts and crusts in the periphyton are commonly covered with diatoms and blue-green algae. The semi-erect forms are not as readily colonized, and much of their epiphytes represent loosely attached microalgae that became entangled during downstream drift. Sheath and Hymes (1980) recorded 63 epiphytes on 10 species of red algae from southern Ontario. The dominant epiphytic groups were the diatoms and green algae. Mucilaginous filaments, such as *Batrachospermum* and *Sirodotia*, were not as readily colonized during active growth. However, at the end of the season, they can become heavily overgrown by microalgae. The cartilaginous forms *Lemanea* and *Tuomeya* support the widest diversity and greatest abundance of epiphytes. Periodically, the mucilage of *Batrachospermum* can support the growth of the chrysophyte *Chrysocapsa epiphytica* (Starmach 1973) or certain blue-green algae (Starmach 1980). Freshwater red algae can be common epiphytes of larger algae, including other rhodophytes. Examples are *Balbiana investiens* (Rieth 1979; Starmach 1977), *Audouinella violacea* (Sheath & Hymes 1980), and *Chroodactylon ramosum* (Sheath 1987).

Occasionally lotic rhodophytes are epizoic. *Boldia erythrosiphon* and *Nemalionopsis shawii* are frequently attached to aquatic snails of the family Pleuroceridae in the southeastern United States (Howard & Parker 1979, 1981). There is a relatively high manganese content in some of these snail shells, which may account for the *Boldia* attachment (Howard & Parker 1981). Camburn and Warren (1983) have found *Compsopogon coeruleus* growing on the parasitic copepod *Lernaea* sp., which is attached to cyprinoid fish in the Mud River, Kentucky. This association appears to be uncommon in that less than 2% of the fish examined had both organisms attached.

Parasitism of freshwater red algae has not been widely observed in situ. Starmach (1936) reported a heavy infection of the bacteria *Siderocystis confervarum* and *Sidercapsa major* on "chantransia" filaments. In addition, viruslike particles have been observed in cells of *Sirodotia tenuissima* (Lee 1971). Water molds can be a problem in culturing these algae in soil-water media (Sheath unpubl.).

C. Reproductive ecology

Most freshwater red algae are not well differentiated into growth, photosynthetic, storage, and reproductive regions. Hence, analysis of resource allocation for reproduction is not straightforward.

This is further complicated by the fact that some taxa can produce both asexual monosporangia and gametangia on the same plant, including *Audouinella* (as *Rhodochorton*) *violacea* (Drew 1935), *Batrachospermum intortum* (Kumano & Bowden-Kerby 1986; Starmach 1977), *B. lochmodes* (Starmach 1977), *B. pseudocarpum* (Kumano & Bowden-Kerby 1986), *B. sporulans* (Starmach 1977), *B. vagum* (Starmach 1977), and *Thorea bachmanii* (Necchi 1987). In 127 macroalgal species where reproduction of the dominant phase has been reported, 29% are strictly asexual, 5% are both sexual and asexual, 53% are monoecious, 10% are dioecious, and 8% are polyoecious (as summarized by Sheath 1984, as well as in recent papers by Kumano & Bowden-Kerby 1986; Kumano & Liao 1987; Kumano & Necchi 1985; Kumano & Ratnasabapathy 1984; Necchi 1986, 1987; Sheath et al. 1986d; Yadava & Kumano 1985). Approximately 99% of the sexually reproducing species have a heteromorphic life history pattern (Sheath 1984). *Audouinella violacea* is the notable exception, with isomorphic alternation of gametophyte and tetrasporophyte (dos Reis 1961; Drew 1935). Littler and Littler (1980) have proposed that heteromorphic life history strategies in marine red algae are "bet hedging," spreading the risk from different sources of mortality among alternate stages. The higher proportion of heteromorphy in stream species compared to those in marine habitats indicates that freshwater red algae may be subjected to greater environmental fluctuations, such as temperature, current velocity, light penetration, and desiccation (Sheath 1984).

It is of ecological significance that almost every sexually reproducing freshwater rhodophyte belongs to the Batrachospermales. This order is characterized by its lack of tetrasporic meiosis (Pueschel & Cole 1982), contrasting with the typical "*Polysiphonia*" life history pattern common to many marine red algae in which there is a sequence of gametangial, carposporangial, and tetrasporangial phases (Dixon 1982). Both the gametophyte and the tetrasporophyte release spores in the "*Polysiphonia*" strategy. The life history pattern of the Batrachospermales, designated the "*Lemanea*" type (Dixon 1982), involves release of only one spore type, the carpospore. Tetraspores are not formed by the "chantransia" filaments. Instead, the macroscopic, semi-erect gametophyte is produced directly attached to the microscopic "chantransia" phase (Sheath 1984). As such, this life history type is atypical of marine rhodophytes. Rather, the "*Lemanea*" life history strategy appears to be an adaptation for growth in the unidirectional flow pattern of streams. The "chantransia" phase is often perennial, seasonally forming the attached gametophyte (Hambrook & Sheath unpubl.; Sheath & Hymes 1980; Yoshida 1959). Therefore, the population can continue to proliferate in the upper part of the drainage basin while still colonizing downstream. If these algae possessed the "*Polysiphonia*" pattern, release of both carpospores and tetraspores would result in a gradual shift of populations downstream until they were in larger trunk rivers, which are too deep, turbid, and sedimented to support the growth of most autotrophs (Vannote et al. 1980).

The net result of the batrachospermalean life history is that for every meiotic event, only one product forms the gametophyte (Sheath 1984). Thus, the maximum attainable gametophyte biomass for each generation is reduced by 75%, compared with that of a life history employing tetrasporic meiosis. In addition, by eliminating three products of meiosis, potential diversity of genotypes is reduced. In marine florideophytes with a typical triphasic life history, each occurrence of syngamy is amplified by the formation of a carposporophyte (Searles 1980). The Batrachospermales have the same adaptive advantage, but this is offset by the elimination of three meiotic products (Sheath 1984). On the other hand, the batrachospermalean life history conveys a much greater probability of success to the one product of meiosis through attachment in a favorable habitat than does the tetrasporic system in releasing all four meiotic products.

Population dynamics of male and female gametophytes is of significance for the 18% of freshwater red algal species that employ dioecious reproduction. In addition, certain species, such as *Audouinella violacea*, produce gametangia for only a brief period of time (Korch & Sheath 1989). Flint (1970) noted that a single "chantransia" crust of *Batrachospermum boryanum* may produce both male and female plants. Thus, spermatia can be released in close proximity to carpogonia enhancing the probability of fertilization. Hambrook and Sheath (unpubl.) have determined the proportion of female-to-male filaments of this species in a small Rhode Island stream during a growing season (Fig. 16-20). In January this ratio is approximately 1, rising to 1.5 in March and 4 in May. This change is largely due to a relative decline in the number of male plants. The question arises as to the success of fertilization in such a dioecious species at an average current velocity of 20 cm s^{-1} (Hambrook & Sheath unpubl.). It would be predicted that at this flow rate the probability is extremely small that a spermatium with an

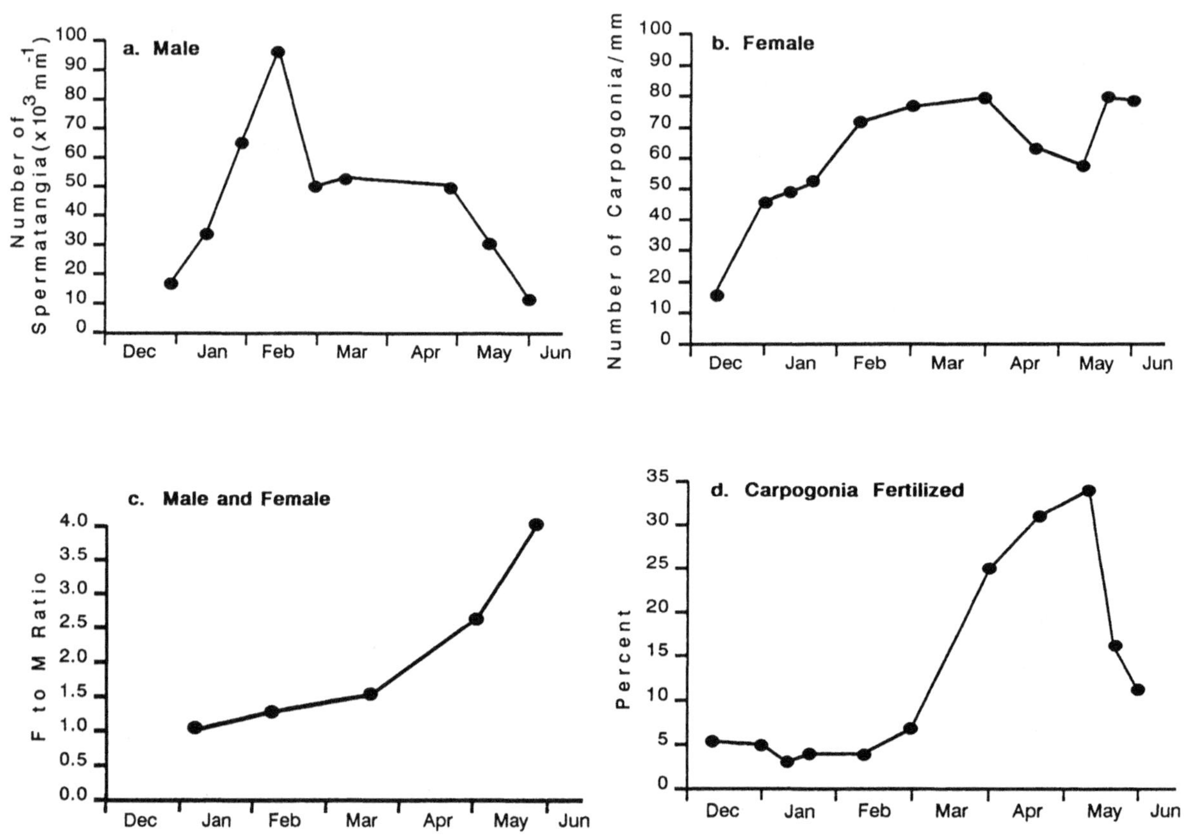

Fig. 16-20. Seasonal fluctuations in spermatangia, carpogonia, ratios of female to male plants, and percentage fertilization of *Batrachospermum boryanum* in a Rhode Island headwater stream in 1986 and 1987. Data from Hambrook & Sheath (unpubl.).

average diameter of 6.5 µm would contact a trichogyne that is 8 µm wide, even when over 1000 spermatia are released for each carpogonium present (Sheath & Burkholder 1983; Fig. 16-20). However, our analysis shows that the percentage of carpogonia that have been fertilized in *B. boryanum* attains a peak of 34%, which is equal to 24 mm^{-1} length (Fig. 16-20). It is our hypothesis that fertilization does not typically result from spermatia carried by laminar flow currents between the branches of a female plant. Rather, most fertilization occurs because of turbulent flow patterns downstream of

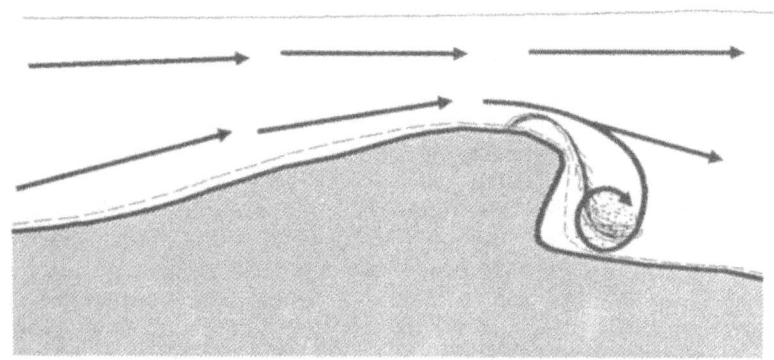

Fig. 16-21. Potential for fertilization of a dioecious freshwater red alga is increased by release of spermatia into turbulent eddies downstream of each rock rather than into the laminar flow currents.

Freshwater ecology

each rock, which create eddies that carry spermatia through the female plant numerous times as it is moving back and forth (Fig. 16-21). Thus, gamete contact is controlled by fluid dynamics in much the same way that wind pollination is effected in angiosperms (Niklas 1985). The possibility of fertilization is further increased by the fact that most freshwater red algae have trichogynes that are wider and more long lived than marine species (Sheath 1984).

V. BIOGEOGRAPHY

On a global basis the geographic distribution of freshwater Rhodophyta will be influenced by the temperature and hydrological regime, the evolutionary divergence of species in relation to continental drift and glaciation as well as vector-assisted transport. Since biomes represent communities with similar amounts and periodicity of temperature and precipitation, it is of interest to compare diversity of freshwater red algae in relation to these systems. In North America deciduous forests contain the greatest number of lotic red algal species, followed by the coniferous forest and highlands and the rain forests (Table 16-10). Biomes with little precipitation, the desert, tundra, and chaparral have the lowest diversities.

Freshwater rhodophytes have no mechanisms of long-distance transport, and hence it would be expected that colonization of glaciated regions is a relatively slow process compared to that of algal species with air-borne cells. This is substantiated in an analysis of diversity patterns in North America (Table 16-10). All recognized species can be found in nonglaciated regions, and there are progressively fewer species in areas where glaciers retreated more recently.

It is of interest that certain freshwater red algal species are relatively cosmopolitan, whereas others are apparently quite rare. This is further compounded by the fact that there are pairs of such species that overlap in part of their distribution and have very similar morphology and reproduction. For example, *Audouinella violacea* and *A. macrospora* occur in close proximity in New Jersey and Rhode Island (Flint 1970; Sheath & Burkholder 1985; Wolle 1887). However, *A. macrospora* is reported from only five locations, whereas *A. violacea* is perhaps the most widely distributed freshwater red alga in North America (Flint 1970) and is common throughout Europe (Starmach 1977). Another example is *Batrachospermum carpocontortum* and *B. moniliforme*, which coexist in one stream in Washington (Sheath et al. 1986b); the former species has been found in only one other stream, but *B. moniliforme* is widespread throughout North America and Europe (Flint 1970; Starmach 1977). Theoretically, these pairs of species would be expected to have similar component niches as described by Grubb (1986), namely habitat, life form, phenological, and regeneration niches. The fact that they apparently do not may be due to several reasons. *A. macrospora* and *B. carpocontortum* may be truly intolerant of wide environmental fluctuations compared with *A. violacea* and *B. moniliforme*. Alternatively, the former species may have limited dispersability, or they may be evolutionarily recent taxa. Grubb (1986) noted that such sparse and patchily distributed species are important in species-rich communities for insuring that most species are not encountering and affecting each other.

Common European, North American, and South American species, such as *Audouinella violacea*, *Batrachospermum moniliforme*, *B. vagum*, and *Lemanea fluviatilis* (Flint 1970; Necchi 1984; Starmach 1977), were probably present prior to the separation of the land masses between 49 and 180 million years B.P. (Pielou 1979). Mechanisms of dispersal of such widespread freshwater rhodophytes between drainage basins have not been studied. In terms of localized vector-assisted transport, it is important to note that freshwater red algae do not form resting spores (Sheath 1984). Hence, vegetative fragments (which may contain sporangia) must be tolerant of being removed from stream water and must be able to initiate growth in another location. Both phenomena are possible since these steps are followed in establishing cultures of these algae (Sheath et al.

Table 16-10. *Geographic distribution of freshwater red algae in North America[a]*

	Number of species
In relation to biome	
Deciduous forest	26
Coniferous forest and highlands	18
Rain forest	11
Grassland	7
Desert scrub	4
Tundra	3
Chaparral	1
In relation to glaciation	
Nonglaciated region	41
Glaciated region	
Total extent	17
Extent at 10,000 years B.P.	5
Extent at 7,000 years B.P.	2

[a] Data from survey described in Fig. 16-2.

1986d, unpubl.). Wading waterfowl may account for some transport, since it has been shown that several types of algal spores and fragments can be viably transported on their feet, among the feathers, on bills, and in gullets and fecal material (Atkinson 1980; Schlichting 1960). In addition, the muskrat (*Ondatra zibethicus*) has been found to contain viable algae in its gut contents (Roscher 1967).

Within a stream channel, downstream colonization of freshwater red algae would seem to be easily facilitated by unidirectional flow. However, transport of spores or fragments downstream may be slow because 60–70% of cells are sedimented within a distance of 20 m in moderate current velocities (Chandler 1937). In addition, most riverine systems are an alternating series of lentic and lotic reaches due to the presence of small and large impoundments (Ward & Stanford 1983a). Few freshwater rhodophytes grow well in large pools (Sheath 1984). Therefore, lentic reaches would be major barriers for further downstream dispersal. Upstream migration resulting from growth of the perennial "chantransia" stage is extremely slow. The vectors described above may contribute to more rapid dispersal within the stream channel. Additional transporters of these algae include herbivorous and omnivorous fish and insect larvae. When the golden shiner (*Notemigonus crysoleucas*) is given a single diet of *Audouinella violacea*, the alga is not digested and dense interwoven mats of viable filaments are present in the feces (Fig. 16-22). *Audouinella* readily regenerates if such filaments are placed in appropriate media. The grazing caddisfly *Dicosmoecus gilvipes* can move more than 25 m upstream in a few days (Hart & Resh 1980). Thus, partially digested fragments could be transported widely by this and similar larvae.

VI. SUMMARY

The majority of freshwater red algae are tufts, filaments, or tissue-like forms, ranging from 1 to 10 cm in length. They occur with an average frequency of 45% and with a mean cover of 11%.

The average current velocity at which freshwater rhodophytes occur is a relatively narrow range, 29–57 cm s^{-1}. The potential benefits of moderate flow include washout of loosely attached competitors, constant replenishment of nutrients and gases, and reduction of the boundary layer of depletion. There is a significant increase in breaking stress and predicted breaking current velocity from tufts to mucilaginous filaments to tissue-like forms. Large mucilaginous filaments create the largest drag values.

Light and temperature interact to determine drainage basin distribution, seasonality, latitude, and altitude. It is predicted that lotic red algae are predominantly distributed in midregions of large forested watersheds. They also tend to occur between late fall and early summer, thereby avoiding dense canopy shading. Most rhodophytes are restricted to elevations less than 700–900 m.

Inland Rhodophyta are most often found in mildly acidic waters, between pH 6 and 7. This is due to an inability of some species to utilize bicar-

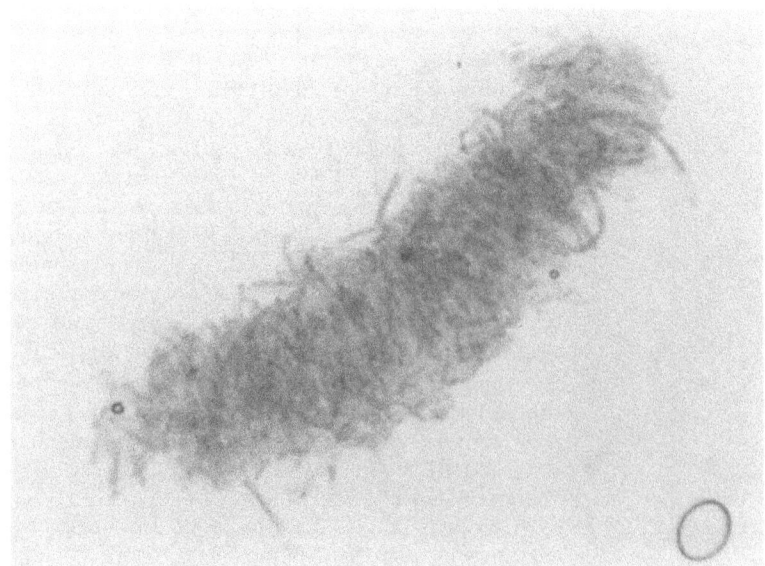

Fig. 16-22. Fecal pellet of the golden shiner, composed of tightly interwoven, viable filaments of *Audouinella violacea*.

bonate, which is the predominant form of inorganic carbon above pH 8. As a result, these species also tend to be more common in soft waters. Freshwater red algae occur at a wide range of oxygen concentrations, but there is a slight increase in frequency of occurrence with higher values. In general, all morphological forms can be found at low PO_4^{-3} concentrations, but there is no detectable correlation between nutrient fluctuations and seasonality. Pollution intolerance of many species partly accounts for their paucity in lower reaches of large rivers.

Thirty-seven lotic animals have been found to ingest freshwater red algae. There appears to be preferential herbivory for delicate forms with high protein contents, such as *Audouinella violacea*. All species examined have large calorific values (5.0–6.8 kcal g^{-1} dry weight). Certain caddisfly larvae, such as *Maydenoptila cuneola* and *Dibusa angata*, appear to occur exclusively among plants of *Batrachospermum* and *Lemanea*, respectively, using the red algae for case making and as a food source during final instar stages.

Lotic communities containing red algae are generally in a nonequilibrium state, consisting of most successional stages, which are regulated by both density-independent and -dependent factors. There do not appear to be any well-defined associations between species. Diatoms, chlorophytes, and cyanophytes are common epiphytes of freshwater rhodophytes, particularly on nonmucilaginous forms. Infrequently, lotic red algae are epizoic in occurrence, or they are parasitized by bacteria, virus-like particles, or aquatic fungi.

In the macroalgal rhodophyte species where reproduction of the dominant phase has been reported, 29% are strictly asexual, 5% are both sexual and asexual, 53% are monoecious, 10% are dioecious, and 8% are polyoecious. Approximately 99% of the sexually reproducing species have a heteromorphic life history; *Audouinella violacea* is the key exception. This strategy may be "bet hedging" in that different sources of mortality are spread over the two stages. The life history of almost every sexually reproducing species lacks tetrasporic meiosis, and hence only carpospores are released. Therefore, the population can continue to proliferate upstream by growth of the perennial "chantransia" stage while still colonizing downstream by carpospore release. However, the maximum attainable gametophyte biomass for each generation is reduced by 75% compared with that of a life history employing tetrasporic meiosis. Fertilization in dioecious species appears to be reasonably successful in spite of the small gamete size. It is hypothesized that most fertilization occurs in the turbulent flow patterns downstream of each rock, where eddies carry spermatia between the branches of the female plant many times. The probability of fertilization is further increased by the relatively large and long-lived trichogyne compared to marine red algal species.

In North America, freshwater rhodophytes are most frequent in nonglaciated, forested biomes. Certain globally distributed species, such as *Audouinella violacea*, *Batrachospermum moniliforme*, *B. vagum*, and *Lemanea fucina*, were probably present prior to the separation of land masses between 49 and 180 million years B.P. Sparse and patchily distributed species are still important in community dynamics. Waterfowl, muskrats, and grazing fish may act as vectors for rapid transport between and within river channels, particularly for species that remain viable in fecal material. Slower colonization rates can take place by release of spores downstream and spread of the perennial "chantransia" stage. Lentic stretches act as major barriers within channels.

VII. ACKNOWLEDGMENTS

The authors gratefully acknowledge the critical review of this chapter by the following people: David Hart, Keith Killingbeck, Shigeru Kumano, and Brian Whitton. Our sincere thanks are extended to Donald Kaczmarczyk, Alice Wells, and Glen Wiggens for making available samples or unpublished data. Assistance in preparation of this manuscript from Marion Jenkins and Glen Thursby is greatly appreciated. Financial support was made available by the University of Rhode Island Botany Department and by N.S.F. Grant No. BSR-8607092 to R.G.S.

VIII. REFERENCES

Abbott, I. A. & Hollenberg, G. J. 1976. *Marine Algae of California*. Stanford: Stanford University Press.

Atkinson, G. F. 1931. Notes on the genus *Lemanea* in North America. *Bot. Gaz.* 92: 225–42.

Atkinson, K. M. 1980. Experiments in dispersal of phytoplankton by ducks. *Br. Phycol. J.* 15: 49–58.

Boon, P. J., Jupp, B. P. & Lee, D. G. 1986. The benthic ecology of rivers in the Blue Mountains (Jamaica) prior to construction of a water regulation scheme. *Arch. Hydrobiol.* Suppl. 74: 315–55.

Budde, H. 1926. Erster Beitrag zur Entwicklungsgeschichte von *Hildenbrandia rivularis* (Liebmann) Bréb. *Ber. Deutsch. Bot. Ges.* 44: 280–9.

Camburn, K. E. & Warren, M. L., Jr. 1983. Epizoic occurrence of *Compsopogon coeruleus* (Rhodophyta) on *Lernaea* (Copepoda) from Kentucky fishes. *Can. J. Bot.* 61: 3545–8.

Cargill A. S. II, Cummins, K. W., Hanson, B. J. & Lowry, P. R. 1985. The role of lipids as feeding stimulants for shredding insects. *Freshwat. Biol.* 15: 455–64.

Carpenter, K. E. 1924. A study of the fauna of rivers polluted by lead mining in the Aberystwyth district of Cardiganshire. *Ann. Appl. Biol.* 9: 1–23.

Chandler, D. C. 1937. Fate of typical lake plankton in streams. *Ecol. Monogr.* 7: 445–79.

Chapman, D. W. & Demory, R. L. 1963. Seasonal changes in the food ingested by aquatic insect larvae and nymphs in two Oregon streams. *Ecology* 44: 140–6.

Cummins, K. W. 1973. Trophic relations of aquatic insects. *Annu. Rev. Entomol.* 18: 183–206.

Deb, D. B., Krishna, B., Mukherjee, K., Bhattacharya, S., Chowdhury, A. N., Das, H. B. & Singh, S. T. 1974. An edible alga of Manipur (*Lemanea australis*): Presence of silver. *Current Sci.* 43: 629.

den Hartog, C. 1956. Deux stations d'*Hildenbrandia rivularis* en Bretagne. *Rev. Algol.* 3: 194–5.

Descy, J.-P. & Empain, A. 1974. *Thorea ramosissima* Bory (Rhodophyceae, Nemalionales) dans le bassin Mosan Belge. *Bull. Soc. Roy. Bot. Belg.* 107: 23–6.

Dillard, G. E. 1966. The seasonal periodicity of *Batrachospermum macrosporum* Mont. and *Audouinella violacea* (Kütz.) Ham. in Turkey Creek, Moore County, North Carolina. *J. Elisha Mitchell Sci. Soc.* 82: 204–7.

Dixon, P. S. 1982. Life-histories in the Florideophyceae with particular reference to the Nemaliales *sensu lato*. *Bot. Mar.* 25: 611–21.

dos Reis, P. M. P. 1961. Sobre a identificação de *Chantransia violacea* Kütz. *Bol. Soc. Brot.* 35: 141–7.

dos Reis, P. M. P. 1962. Subsídos para o conhecimento das Rodofíceas de água doce de Portugal III. *Mem. Soc. Brot.* 15: 57–71.

dos Reis, M. P. 1974. Chaves para o identificação dos espécies Portuguesas de *Batrachospermum* Roth. *Ann. Soc. Brot.* 40: 37–125.

Drew, K. M. 1935. The life-history of *Rhodochorton violaceum* (Kütz.) comb. nov. (*Chantransia violacea* Kütz.). *Ann. Bot.* 49: 439–50.

Duthie, H. C. & Hamilton, P. B. 1983. Studies on periphyton community dynamics of acidic streams using track autoradiography. In *Periphyton of Freshwater Ecosystems*, ed. R. G. Wetzel, pp. 185–9. The Hague: Junk.

Elwood, J. W., Newbold, J. D., O'Neill, R. V. & Van Winkle, W. 1983. Resource spiralling: An operational paradigm for analyzing lotic ecosystems. In *Dynamics of Lotic Ecosystems*, eds. T. D. Fontaine III & S. M. Bartell, pp. 3–27. Ann Arbor: Ann Arbor Science.

Entwisle, T. J. & Kraft, G. T. 1984. Survey of freshwater red algae (Rhodophyta) of south-eastern Australia. *Aust. J. Mar. Freshw. Res.* 35: 213–59.

Federer, C. A. & Tanner, C. B. 1966. Spectral distribution of light in the forest. *Ecology* 47: 555–60.

Flint, L. H. 1970. *Freshwater Red Algae of North America*. New York: Vantage.

Friedrich, G. 1980. Rotalgen in unseren Gewässern, *Niederrhein. Jahrb.* 14: 19–25.

Fritsch, F. E. 1929. The encrusting algal communities of certain fast-flowing streams. *New Phytol.* 28: 165–96.

Gibson, M. T. & Whitton, B. A. 1987. Hairs, phosphatase activity and environmental chemistry in *Stigeoclonium*, *Chaetophora* and *Draparnaldia* (Chaetophorales). *Br. Phycol. J.* 22: 11–22.

Goodwin, K. M. 1926. Some observations on *Batrachospermum moniliforme*. *New Phytol.* 25: 51–4.

Grubb, P. J. 1986. Problems posed by sparse and patchily distributed species in species-rich plant communities. In *Community Ecology*, eds. J. Diamond & T. J. Case, pp. 207–25. New York: Harper & Row.

Hambrook, J. A. & Sheath, R. G. 1987. Grazing of freshwater Rhodophyta. *J. Phycol.* 23: 656–62.

Harding, J. P. C. & Whitton, B. A. 1981. Accumulation of zinc, cadmium and lead by field populations of *Lemanea*. *Water Res.* 15: 301–19.

Hart, D. D. 1983. The importance of competitive interactions within stream populations and communities. In *Stream Ecology. Application and Testing of General Ecological Theory*, eds. J. R. Barnes & G. W. Minshall, pp. 99–136. New York: Plenum.

Hart, D. D. & Resh, V. H. 1980. Movement patterns and foraging ecology of a stream caddisfly larva. *Can. J. Zool.* 58: 1174–85.

Hellebust, J. A. 1970. Light-plants. In *Marine Ecology*, Vol. 1, Part 1, ed. O. Kinne, pp. 125–48. London: Wiley Interscience.

Hinton, G. C. F. & Maulood, B. K. 1980. Freshwater red algae a new addition to the Iraqi flora. *Nova Hedwigia* 33: 487–97.

Hoagland, K. D., Roemer, S. C. & Rosowski, J. R. 1982. Colonization and community structure of two periphyton assemblages, with emphasis on the diatoms (Bacillariophyceae). *Am. J. Bot.* 69: 188–213.

Holmes, N. T. & Whitton, B. A. 1977. Macrophytic vegetation of the Rivers Swale, Yorkshire. *Freshwat. Biol.* 7: 545–58.

Holmes, N. T. & Whitton, B. A. 1981. Phytobenthos of the River Tees and its tributaries. *Freshwat. Biol.* 11: 139–63.

Horner, R. R. & Welch, E. B. 1981. Stream periphyton development in relation to current velocity and nutrients. *Can. J. Fish. Aquat. Sci.* 38: 449–57.

Howard, R. V. & Parker, B. C. 1979. *Nemalionopsis shawii* forma *caroliniana* (forma nov.) (Rhodophyta: Nemaliales) from the southeastern United States. *Phycologia* 18: 330–7.

Howard, R. V. & Parker, B. C. 1981. A non-obligate association between the red alga, *Boldia*, and pleurocerid snails. *Sterkiana* 71: 18–23.

Huth, K. 1979. Einfluß von Tageslänge und Beleuchtungsstärke auf den Generationswechsel bei *Batrachospermum moniliforme*. *Ber. Deutsch. Bot. Ges.* 92: 467–72.

Hynes, H. B. N. 1970. *The Ecology of Running Waters*. Liverpool: Liverpool University Press.

Ikushima, I. & Kumano, S. 1982. Primary production, submerged macrophytes. In *Tasek Bera*, eds. J. I. Furtado & S. Mori, pp. 262–7. The Hague: Junk

Israelson, G. 1942. The freshwater Florideae of Sweden. *Symb. Bot. Upsal.* 6(1): 1–134.

Johansson, C. 1982. Attached algal vegetation in running waters of Jämtland, Sweden. *Acta Phytogeogr. Suec.* 71: 1–84.

Jones, J. R. E. 1949. A further ecological study of calcareous streams in the 'Black Mountain' district of South Wales. *J. Anim. Ecol.* 18: 142–59.

Jones, J. R. E. 1950. A further ecological study of the River Rheidol: The food of the common insects of the mainstream. *J. Anim. Ecol.* 19: 159–74.

Kann, E. 1978a. Typification of Austrian streams concerning algae. *Verh. Internat. Verein. Limnol.* 20: 1523–6.

Kann, E. 1978b. Systematik und Ökologie der Algen österreichischer Bergbäche. *Arch. Hydrobiol. Suppl.* 53: 405–643.

Kann, E. 1986. Können benthische Algen zur Wassergutebestimmung herangezogen werden? *Arch. Hydroiol.* 73 (Suppl.): 405–23.

Kawecka, B. & Szczęsny, B. 1984. Dunajec. In *Ecology of European Rivers*, ed. B. A. Whitton, pp. 499–525. Oxford: Blackwell Scientific.

Khan, M. 1974. On a fresh water *Hildenbrandia* Nardo. from India. *Hydrobiologia* 44: 237–40.

Koehl, M. A. R. & Wainwright, S. A. 1985. Biomechanics. In *Handbook of Phycological Methods, Ecological Field Methods: Macroalgae*, ed. M. M. Littler & D. S. Littler. pp. 292–313. Cambridge: Cambridge University Press.

Korch, J. E. & Sheath, R. G. 1989. The phenology of *Audouinella violacea* (Acrochaetiaceae, Rhodophyta) in a Rhode Island stream (USA). *Phycologia* 28: 228–36.

Kremer, B. 1983. Untersuchungen zur Ökophysiologie einiger Süsswasserrotalgen. *Decheniana* (Bonn) 136: 31–42.

Krewer, J. A. & Holm, H. W. 1982. The phosphorus–chlorophyll *a* relationship in periphytic communities in a controlled ecosystem. *Hydrobiologia* 94: 173–6.

Krishnamurthy, V. 1963. The morphology and taxonomy of the genus *Compsopogon* Montagne. *J. Linn. Soc. (Bot.)* 58: 207–22.

Kumano, S. 1980. On the distribution of some freshwater red algae in Japan and southeast Asia. *Proc. Workshop Prom. Limnol. Dev. Count.* 1: 3–6.

Kumano, S. & Bowden-Kerby, W. A. 1986. Studies on the freshwater Rhodophyta of Micronesia. I. Six new species of *Batrachospermum* Roth. *Jap. J. Phycol.* 34: 107–28.

Kumano, S. & Liao, L. M. 1987. A new species of the section Contorta of the genus *Batrachospermum* (Rhodophyta, Nemalionales) from Nonac Island, the Philippines. *Jap. J. Phycol.* 35: 99–105.

Kumano, S. & Necchi, O., Jr. 1985. Studies on the freshwater Rhodophyta of Brazil II. Two new species of *Batrachospermum* Roth from States of Amazonas and Minas Gerais. *Jap. J. Phycol.* 33: 181–9.

Kumano, S. & Ratnasabapathy, M. 1984. Studies on freshwater red algae of Malaysia IV. *Batrachospermum bakarense*, sp. nov. from Sungai Baker, Kelanton, West Malaysia. *Jap. J. Phycol.* 27: 19–23.

Kumano, S. & Watanabe, M. 1983. Two new varieties of *Batrachospermum* (Rhodophyta) from Mt. Albert Edward, Papua New Guinea. *Bull. Nat. Sci. Mus. Tokyo* Ser. B 9: 85–94.

Kylin, H. 1912. Studien über die schwedischen Arten der Gattungen *Batrachospermum* Roth und *Sirodotia* nov. gen. *N. Acta Reg. Soc. Sci. Upsal.* Ser. 4, Vol. 4(3): 1–40.

Lee, R. E. 1971. Systemic viral material in the cells of the freshwater red alga *Sirodotia tenuissima* (Holden) Skuja. *J. Cell Sci.* 8: 623–31.

Levitt, J. 1980. *Responses of Plants to Environmental Stresses. Vol. 1, Chilling, Freezing, and High Temperatures Stresses.* New York: Academic.

Lin, C. K. & Blum, J. L. 1977. Recent invasion of a red alga *Bangia atropurpurea* in Lake Michigan. *J. Fish. Res. Bd. Can.* 34: 2413–16.

Littler, M. M. & Littler, D. S. 1980. The evolution of thallus form and survival strategies in benthic marine macroalgae: Field and laboratory tests of a functional form model. *Am. Nat.* 116: 25–44.

Lowe, C. W. 1923. *Report of the Canadian Arctic Expedition 1913–18.* Vol. IV. Part A. Ottawa: King's Printer.

MacFarlane, J. J. & Raven, J. A. 1985. External and internal CO_2 transport in *Lemanea*: interactions with the kinetics of ribulose bisphosphate carboxylase. *J. Exp. Bot.* 36: 610–22.

Mecom, J. O. & Cummins, K. W. 1964. A preliminary study of the trophic relationships of the larvae of *Brachycentrus americanus* (Banks) (Trichoptera: Brachycentridae). *Trans. Am. Microsc. Soc.* 83: 233–43.

Minckley, W. L. 1963. The ecology of a spring stream. Doe Run, Meade County, Kentucky. *Widl. Monogr.* 11: 1–124.

Minckley, W. L. & Tindall, D. R. 1963. Ecology of *Batrachospermum* sp. (Rhodophyta) in Doe Run, Meade County, Kentucky. *Bull. Torrey Bot. Club.* 90: 391–400.

Mueller-Dombois, D. & Ellenberg, H. 1974. *Aims and Methods of Vegetation Ecology.* New York: Wiley.

Mulholland, P. J., Elwood, J. W., Palumbo, A. V. & Stevenson, R. J. 1986. Effect of stream acidification on periphyton, chlorophyll, and productivity. *Can. J. Fish. Aquat. Sci.* 43: 1846–58.

Muller, C. 1978. On the productivity and chemical composition of some benthic algae in hard-water streams. *Verh. Internat. Verein. Limnol.* 20: 1457–62.

Necchi, O., Jr. 1984. Catálogo das Rhodophyta de águas continent do Brasil. *Rickia* 11: 99–107.

Necchi, O., Jr. 1986. Studies on the freshwater Rhodophyta of Brazil. 4: Four new species of *Batrachospermum* (Section Contorta) from the southern state of São Paulo. *Rev. Brasil Biol.* 46: 517–25.

Necchi, O., Jr. 1987. Sexual reproduction in *Thorea* Bory (Rhodophyta, Thoraceae). *Jap. J. Phycol.* 35: 106–12.

Niklas, K. J. 1985. The aerodynamics of wind pollination. *Bot. Rev.* 51: 328–86.

Norton, T. A., Mathieson, A. C. & Neushul, M. 1982. A review of some aspects of form and function in seaweeds. *Bot. Mar.* 25: 501–10.

Ott, F. D. & Sommerfeld, M. R. 1982. Freshwater Rhodophyceae: introduction and bibliography. In *Selected Papers in Phycology* II, eds. J. R. Rosowski & B. C. Parker, pp. 671–81. Lawrence, Kan.: Phycological Society of America.

Palmer, C. M. 1941. A study of *Lemanea* in Indiana with notes on its distribution in North America. *Butler Univ. Bot. Stud.* 5: 1–26.

Parker, B. C., Samsel, G. L. & Prescott, G. W. 1973. Comparison of microhabitats of macroscopic subalpine stream algae. *Am. Midl. Nat.* 90: 143–53.

Pielou, E. C. 1979. *Biogeography.* New York: Wiley-Interscience.

Pueschel, C. M. & Cole, K. M. 1982. Rhodophycean pit plugs: an ultrastructural survey with taxonomic implications. *Am. J. Bot.* 69: 703–20.

Ratnasabapathy, M. & Seto, R. 1981. *Thorea prowsei* sp. nov. and *Thorea clavata* sp. nov. (Rhodophyta, Nemaliales) from West Malaysia. *Jap. J. Phycol.* 29: 243–50.

Raven, J. A. & Beardall, J. 1981. Carbon dioxide as the exogenous inorganic source for *Batrachospermum* and *Lemanea*. *Br. Phycol. J.* 16: 165–75.

Raven, J., Beardall, J. & Griffiths, H. 1982. Inorganic C–sources for *Lemanea, Cladophora* and *Ranunculus* in a fast-flowing stream: Measurements of gas exchange and of carbon isotope ratio and their ecological implications. *Oecologia* (Berl.) 53: 68–78.

Reed, E. W. 1970. Studies on the ecology of *Lemanea fucina*

Bory with notes on *Audouinella violacea* Kütz. Ph.D. thesis, Washington State University.

Reed, R. H. 1980. On the conspecificity of marine and freshwater *Bangia* in Britain. *Br. Phycol. J.* 15: 411–16.

Resh, V. H. & Houp, R. E. 1986. Life history of the caddisfly *Dibusa angata* and its association with the red alga *Lemanea australis*. *J. N. Am. Benthol. Soc.* 5: 28–40.

Rider, D. E. & Wagner, R. H. 1972. The relationship of light, temperature, and current to the seasonal distribution of *Batrachospermum* (Rhodophyta). *J. Phycol.* 8: 323–31.

Rieth, A. 1979. Ein Standort der epiphytischen Süsswasser-Rotalge *Balbiania investiens* (Lenormand) Sirodot 1876 in Mitteleuropa. *Arch. Protistenk.* 121: 401–16.

Roscher, J. 1967. Algal dispersal by muskrat intestinal contents. *Trans. Am. Microsc. Soc.* 86: 497–8.

Ruttner, F. 1960. Über die Kohlenstoff aufnahme bei Algen aus Rhodophyceen-Gattung *Batrachospermum*. *Schweiz. Z. Hydrol.* 22: 280–91.

Saffo, M. B. 1987. New light on seaweeds. *Bioscience* 37: 654–64.

Schlichting, H. E., Jr. 1960. The role of waterfowl in the dispersal of algae. *Trans. Am. Microsc. Soc.* 79: 160–6.

Schumacher, G. J. & Whitford, L. A. 1965. Respiration and P^{32} uptake in various species of freshwater algae as affected by a current. *J. Phycol.* 1: 78–80.

Searles, R. B. 1980. The strategy of the red algal life history. *Am. Nat.* 115: 113–20.

Seto, R. 1977. On the vegetative propagation of a freshwater red alga, *Hildenbrandia rivularis* (Liebm.) J. Ag. *Bull. Jap. Soc. Phycol.* 25: 129–36.

Sheath, R. G. 1984. The biology of freshwater algae. *Prog. Phycol. Res.* 3: 89–157.

Sheath, R. G. 1987. Invasions into the Laurentian Great Lakes by marine algae. *Arch. Hydrobiol. Beih.* 25: 165–86.

Sheath, R. G. & Burkholder, J. M. 1983. Morphometry of *Batrachospermum* populations intermediate between *B. boryanum* and *B. ectocarpum* (Rhodophyta). *J. Phycol.* 19: 324–31.

Sheath, R. G. & Burkholder, J. M. 1985. Characteristics of softwater streams in Rhode Island. II. Composition and seasonal dynamics of macroalgal communities. *Hydrobiologia* 128: 109–18.

Sheath, R. G., Burkholder, J. M., Hambrook, J. A., Hogeland, A. M., Hoy, E., Kane, M. E., Morison, M. O., Steinman, A. D. & VanAlstyne, K. L. 1986c. Characteristics of softwater streams in Rhode Island. III. Distribution of macrophytic vegetation in a small drainage basin. *Hydrobiologia* 140: 183–91.

Sheath, R. G., Burkholder, J. M., Morison, M. O., Steinman, A. D. & VanAlstyne, K. L. 1986a. Effect of tree canopy removal by gypsy moth larvae on the macroalgae of a Rhode Island headwater stream. *J. Phycol.* 22: 567–70.

Sheath, R. G. & Hambrook, J. A. 1988. Mechanical adaptations to flow in freshwater red algae. *J. Phycol.* 24: 107–11.

Sheath, R. G., Hellebust, J. A. & Sawa, T. 1979. Effects of low light and darkness on structural transformations in plastids of the Rhodophyta. *Phycologia* 18: 1–12.

Sheath, R. G. & Hymes, B. J. 1980. A preliminary investigation of the freshwater red algae in streams of southern Ontario, Canada. *Can. J. Bot.* 583: 1295–318.

Sheath, R. G., Kullberg, P. G. & Cole, K. M. 1987. Distribution of stream macroalgae in Dominica. *J. Phycol.* 23 (Suppl.): 20 (Abstract).

Sheath, R. G., Morison, M. O., Cole, K. M. & VanAlstyne, K. L. 1986d. A new species of freshwater Rhodophyta, *Batrachospermum carpocontortum*. *Phycologia* 25: 321–30.

Sheath, R. G., Morison, M. O., Korch, J. E., Kaczmarczyk, D. & Cole, K. M. 1986b. Distribution of stream macroalgae in south-central Alaska. *Hydrobiologia* 135: 259–69.

Shih, H. J. 1941. On the foods of some cyprinoid fishes. *Sinesia* 12: 235–7.

Shiozawa, D. K. 1983. Density independence versus dependence in streams. In *Stream Ecology. Application and Testing of General Ecological Theory*, eds. J. R. Barnes & G. W. Minshall, pp. 55–77. New York: Plenum.

Sinha, B. D. & Srivastava, N. K. 1979. Taxonomy and biology of the freshwater red alga *Compsopogon* in Bihar, India. *Nova Hedwigia*. Ex: Beih. 63: 71–6.

Sirjola, E. 1969. Aquatic vegetation of the river Teuronjoki, south Finland, and its relation to water velocity. *Ann. Bot. Fenn.* 6: 68–75.

Skuja, H. 1938. Comments on fresh-water Rhodophyceae. *Bot. Rev.* 4: 665–76.

Starmach, K. 1936. Powtoki wodorotlenku zelaza na galazkach *Chantransia chalybaea* Fries. *Acta Soc. Bot. Pol.* 13: 137–49.

Starmach, K. 1973. *Chrysocapsa epiphytica* spec. nova (Chrysophyceae). *Bull. Acad. Pol. Sci. Ser.* 5, 21: 611–13.

Starmach, K. 1977. *Flora Stodkowodna Polski* Tom. 14, Warsaw and Krakow: Polska Akademia Nauk Instytut Botaniki.

Starmach, K. 1980. *Batrachospermum vagum* (Roth) Ag. und epiphytische Blaualgen im See Silm bei Itawa. *Frag. Flor. Geobot.* 26: 163–74.

Starmach, K. 1984. Red algae in the Kryniczanka stream *Frag. Flor. Geobot. Ann.* 28: 257–93.

Steinman, A. D. & McIntire, C. D. 1986. Effects of current velocity and light energy on the structure of periphyton assemblages in laboratory streams. *J. Phycol.* 22: 352–61.

Steubing, L., Fricke, G. & Jehn, H. 1983. Veranderung des Algenspektrums der Eder im Verlauf von vier Jahrzehnten. *Arch. Hydrobiol.* 96: 205–22.

Stock, M. S., Richardson, T. D. & Ward, A. K. 1987. Distribution and primary productivity of the epizoic macroalga *Boldia erythrosiphon* (Rhodophyta) in a small Alabama stream. *J. N. Am. Benthol. Soc.* 6: 168–74.

Stockner, J. G. & Shortreed, K. R. S. 1978. Enhancement of autotrophic production by nutrient addition in a coastal rainforest stream on Vancouver Island. *J. Fish. Res. Bd. Can.* 35: 28–34.

Stumm, W. & Morgan, J. J. 1981. *Aquatic Chemistry. An Introduction Emphasizing Chemical Equilibria in Natural Waters*, 2nd ed. New York: Wiley.

Swale, C. M. F. & Belcher, J. H. 1963. Morphological observations on wild and cultured material of *Rhodochorton investiens* (Lenormand) nov. comb. (*Balbiana investiens* (Lenorm.) Sirodot). *Ann. Bot. N. S.* 27: 281–90.

Taylor, W. R. 1957. *Marine Algae of the Northeastern Coast of North America*, 2nd ed. Ann Arbor: University of Michigan Press.

Thirb, H. H. & Benson-Evans, K. 1982. The effect of different current velocities on the red alga *Lemanea* in a laboratory stream. *Arch. Hydrobiol.* 96: 65–72.

Thirb, H. H. & Benson-Evans, K. 1983. The effect of different light intensities and wavelengths on carpospore germination and the apical tips of the red alga *Lemanea* Bory (1808). *Nova Hedwigia* 37: 669–82.

Thirb, H. H. & Benson-Evans, K. 1985. The effect of water temperature, current velocity and suspended solids on the distribution growth and seasonality of *Lemanea fluviatilis* (C. Ag.), Rhodophyta in the River Usk and other South Wales rivers. *Hydrobiologia* 127: 63–78.

Tomas, X., Lopez, P., Margalef-Mir, R. & Comin, F. A. 1980. Distribution and ecology of *Compsopogon coeruleus* (Balbis) Montagne (Rhodophyta, Bangiophycidae) in Eastern Spain. *Cryptogam. Algol.* 1: 179–86.

Traden, T. S. & Lindstrom, E.-A. 1983. Influence of current velocity on periphyton distribution. In *Periphyton of Freshwater Ecosystems*, ed. R. G. Wetzel, pp. 97–9. The Hague: Junk.

Vannote, R. L., Minshall, G. W., Cummins, K. W., Sedell, J. R. & Cushing, C. E. 1980. The river continuum concept. *Can. J. Fish. Aquat. Sci.* 37: 130–7.

Vogel, S. 1984. Drag and flexibility in sessile organisms. *Am. Zool.* 24: 37–44.

Vogel, S. & LaBarbera, M. 1978. Simple flow tanks for research and teaching. *Bioscience* 28: 638–43.

Ward, J. V. & Stanford, J. A. 1983a. The serial discontinuity concept of lotic ecosystems. In *Dynamics of Lotic Ecosystems*, eds. T. D. Fontaine III & S. M. Bartell, pp. 29–42. Ann Arbor: Ann Arbor Science.

Ward, J. V. & Stanford, J. A. 1983b. The intermediate-disturbance hypothesis: An explanation for biotic diversity patterns in lotic ecosystems. In *Dynamics of Lotic Ecosystems*, eds. T. D. Fontaine III & S. M. Bartell, pp. 347–56. Ann Arbor: Ann Arbor Science.

Wehr, J. D. 1981. Analysis of seasonal succession of attached algae in a mountain stream, the North Alouette River, British Columbia. *Can. J. Bot.* 59: 1465–74.

Wehr, J. D. & Stein, J. R. 1985. Studies on the biogeography and ecology of the freshwater phaeophycean alga *Heribauldiella fluviatilis*. *J. Phycol.* 21: 81–93.

Wells, A. 1985. Larvae and pupae of Australian Hydroptilidae (Trichoptera) with observations on general biology and relationships. *Aust. J. Zool.* (Suppl.) 113: 1–69.

Westlake, D. F. 1975. Macrophytes. In *River Ecology*, ed. B. A. Whitton, pp. 106–28. Berkeley & Los Angeles: University of California Press.

Westlake, D. F. 1980. Primary production. In *The Functioning of Freshwater Ecosystems*, eds. E. D. Lecren & R. H. Lowe-McConnell, pp. 141–246. Cambridge: Cambridge University Press.

Wetzel, R. G. 1983. *Limnology*, 2nd ed. Philadelphia: Saunders College.

Whitford, L. A. 1960. The current effect and growth of fresh-water algae. *Trans. Am. Microsc. Soc.* 74: 302–9.

Whitford, L. A. & Robertson, E. T. 1981. Some freshwater algae from Jamaica. *Nova Hedwigia* 34: 521–4.

Whitford, L. A. & Schumacher, G. J. 1963. Communities of algae in North Carolina streams and their seasonal relations. *Hydrobiologia* 22: 133–61.

Whitton, B. A. 1975. Algae. In *River Ecology*, ed. B. A. Whitton, pp. 81–105. Berkeley & Los Angeles: University of California Press.

Whitton, B. A., Say, P. J. & Wehr, J. D. 1981. Use of plants to monitor heavy metals in rivers. In *Heavy Metals in Northern England: Environmental and Biological Aspects*, eds. P. J. Say & B. A. Whitton, pp. 135–45. Durham: Durham University Press.

Woelkerling, W. J. 1975. Observations on *Batrachospermum* (Rhodophyta) in southeastern Wisconsin streams. *Rhodora* 77: 467–77.

Wolle, F. 1887. *Fresh-water Algae of the United States*. Bethlehem, Pa.: Comenius.

Yadava, R. N. & Kumano, S. 1985. *Compsopogon prolificus* sp. nov. (Compsopogonaceae, Rhodophyta) from Allahabad, Uttar Pradesh in India. *Jap. J. Phycol.* 33: 13–20.

Yarish, C., Edwards, P. & Casey, S. 1980. The effects of salinity, and calcium and potassium variations on the growth of two estuarine red algae. *J. Exp. Mar. Biol. Ecol.* 47: 235–49.

Yoshida, T. 1959. Life-cycle of a species of *Batrachospermum* found in northern Kyushu, Japan. *Jap. J. Bot.* 17: 29–42.

Zanveld, J. S., Fott, B. & Nováková, M. 1976. *Compsopogon coeruleus*, a red alga newly reported for freshwater aquaria in Prague. *Preslia* (Praha) 48: 17–20.

Note added in proof: Recently, MacFarlane and Raven (1990) have shown that 'knobbles' on *Lemanea mamillosa* generate turbulence, thereby reducing inorganic carbon limitation at high pH. Membrane transporters in this alga have high affinity for ammonia and phosphate ($K_{1/2}$ = 2 and < 2 mmol m^{-3}, respectively). Additional information has been published on nutrient levels (Entwisle 1989) and geographical patterns (Necchi 1989; Sheath et al. 1989).

Entwisle, T. J. 1989. Macroalgae in the Yarra River Basin: flora and distribution. *Proc. Roy. Soc. Victoria* 101: 1–76.

MacFarlane, J. J. & Raven, J. A. 1990. C, N and P nutrition of *Lemanea mamillosa* Kütz. (Batrachospermales, Rhodophyta) in the Dighty Burn, Angus, U.K. *Plant Cell Envir.* 13: 1–13.

Necchi, O., Jr. 1989. Geographical distribution of the genus *Batrachospermum* (Rhodophyta, Batrachospermales) in Brazil. *Rev. Brasil. Biol.* 49: 663–9.

Sheath, R. G., Hamilton, P. B., Hambrook, J. A. & Cole, K. M. 1989. Stream macroalgae of the eastern boreal forest region of North America. *Can. J. Bot.* 67: 3553–62.

Chapter 17

Reproductive strategies

MICHAEL W. HAWKES

CONTENTS

I. Introduction / 455
II. Biosystematics: patterns of variation and species recognition / 456
    A. Life history diversity / 456
    B. Hybridization studies / 457
    C. Breeding systems and population biology / 460
III. Evolution and ecology / 466
    A. Life history evolution / 466
    B. Breeding systems and genotype diversification / 468
    C. Life history theory / 468
IV. Summary / 470
V. Acknowledgments / 470
VI. References / 470

I. INTRODUCTION

"The range of forms, life histories and reproductive intricacies is greater in the Rhodophyta than in any other algal division." (Kraft 1981, p. 6)

The diversity of red algal reproductive strategies, such as life history traits and type of breeding system, presents challenging opportunities for both theoretical and empirical research. Despite a plethora of detail about the sexual process in red algae, there is little information available on breeding systems. Electrophoretic studies of enzymes in higher plants have demonstrated breeding system effects on population genetic structure and morphological variability and correlated life history features with levels of variability (Gottlieb 1981; Richards 1986). Such variation has important systematic, ecological, and evolutionary implications.

In this chapter I will present a synopsis of higher plant literature on breeding systems, then summarize current knowledge of red algal breeding systems. Of special interest in this regard is the frequency of uniparental reproduction, either sexual, in the case of self-compatible, inbreeding species, or asexual, in the case of vegetative reproduction or apomixis. The importance of facultative asexual reproduction as an alternate pathway in sexual life histories has been largely ignored, despite its apparent widespread occurrence (Dixon 1963, 1965; Rueness 1978).

I will also review life history diversity in red algae. To date the main focus of life history studies has been biosystematic, with major contributions being made to species recognition. Red algal life history diversity is largely unexplored in terms of ecological and evolutionary process. Life history theorists, most notably zoologists, are becoming aware of the potential rhodophytes offer for testing theory on the origin and maintenance of complex life histories and on the advantages versus disadvantages of sexual and asexual reproduction. In this chapter I will discuss life history theory, most of which is based on higher plants and animals, from an ecological and evolutionary perspective. I also suggest that revision will be necessary if the complexities and diversity of red algal reproductive strategies are to be incorporated into this body of theory.

II. BIOSYSTEMATICS: PATTERNS OF VARIATION AND SPECIES RECOGNITION

> "To attempt a higher-level classification with no understanding of reproductive processes at the populational level is to construct a 'house of cards' of the most fragile nature.... The real action is at the populational level, the original and central domain of biosystematics." (Stuessy 1985, p. 376)

A. Life history diversity

> "... a review of life history diversity can itself challenge theorists by revealing the complex nature of the phenomena." (Stearns 1977, p. 146)

The main focus of life history investigations to date has been biosystematic and has yielded much valuable information for species recognition; especially in the case of heteromorphic life histories, where the haploid gametophyte and diploid sporophyte generations have frequently been referred to different genera, often in different orders (Table 17-1).

Rhodophyta employ both sexual and asexual reproductive modes. Reproductive biology of the group is unusual in that all spores and gametes are nonflagellated, and there is an evolutionary trend for zygote amplification (Figs. 17-1, 17-3) (Searles 1980; Guiry 1987). Some rhodophytes lack zygote amplification; a single diploid spore is dispersed (e.g., *Rhodochaete*, *Smithora*, *Porphyrostromium*) (Figs. 17-1, 17-3). In *Palmaria* the zygote develops directly into a multicellular meiosporangial generation (Figs. 17-1, 17-3). Other postfertilization variations are indicated in Fig. 17-3. The meiosporangial generation of most red algae is free-living, and meiosis occurs in a sporangium; however, cases of somatic meiosis with no dispersable products are known (e.g., *Porphyra*, *Lemanea*) (Figs. 17-1, 17-3).

Rhodophytes can have one, two, or three free-living, morphological phases in their life history, and these may be morphologically similar (isomorphic) or different (heteromorphic). In the case of a single morphological phase (e.g., *Porphyridium*), only asexual reproduction is involved in the life history. However, in the cases of two or three morphological phases, sexual and/or asexual cycles may be involved (Fig. 17-2). In Fig. 17-3 only free-living morphological phases are recognized, for example, gametangial generation (one individual in the case of monoecious species, two in the case of dioecious species) and a meiosporangial generation. In a few instances the meiosporangial generation is not free-living, i.e., separate from the gametangial generation, for example, *Rhodophysema* (Fig. 17-3).

Complicating the morphological picture are one, two or three karyological phases (haploid, diploid, or rarely polyploid) (Figs. 17-2, 17-3). The most common red algal life history pattern is exemplified by the *Polysiphonia*-type (Fig. 17-3), which is karyologically trigenetic (i.e., one haploid and two diploid phases) and morphologically has two free-living,

Table 17-1. Examples of biosystematic studies that elucidated heteromorphic life histories

Taxon (gametophyte/sporophyte)	Reference
Ahnfeltia/"Porphyrodiscus"	Farnham & Fletcher 1976; Chen 1977
Asparagopsis/"Falkenbergia"	Bonin & Hawkes 1987; Feldmann & Feldmann 1939, 1942
Bonnemaisonia/"Hymenoclonium"	Breeman 1979; Feldmann & Mazoyer 1937; Rueness & Åsen 1982
Farlowia/"Haematocelis-Cruoriopsis"	DeCew & West 1982
Gloiosiphonia/"Cruoriopsis"	DeCew et al. 1982; Edelstein 1970, 1972; Edelstein & McLachlan 1971
Gymnogongrus/"Petrocelis–Erythrodermis"	DeCew & West 1981
Halarachnion/"Cruoria"	Boillot 1965
Mastocarpus/"Petrocelis"	Polanshek & West 1975, 1977; West 1972b; West et al. 1978
Porphyra/"Conchocelis"	Drew 1949, 1954; Kurogi 1953
Schizymenia/"Haematocelis"	André 1977
Turnerella/"Cruoria"	South et al. 1972

Reproductive strategies

Fig. 17-1. Flow chart illustrating main trends of red algal sexual reproduction, emphasizing amplification of dispersable products from single fertilization and site of meiosis and number of dispersable products. *: Most red algae in these categories.

isomorphic generations: gametangial (monoecious or dioecious) and meiosporangial.

For details of reproductive structures and post-fertilization development, see Chapter 13, Bold and Wynne (1985), Dixon (1973), Gabrielson and Garbary (1986), and West and Hommersand (1981). A plethora of life history types is now reported for the Rhodophyta (Fig. 17-3) (for reviews, see Gabrielson & Garbary 1986; West & Hommersand 1981). In most cases we only know the sequence of morphological phases and lack karyological details, especially regarding the site of meiosis.

B. Hybridization studies

Recognizing species and determining their phylogenetic relationships are central goals of systematics. Phycologists employ both morphological and biological species concepts, although the former is most often used. Attempts to apply a biological species concept have led to recent biosystematic studies of rhodophytes employing hybridization to help characterize breeding relationships within and between species (e.g., Guiry 1984; Guiry & Freamhainn 1986; Guiry & West 1984; Guiry et al. 1987; Koch 1986; Masuda et al. 1984; Polanshek & West 1977; Rueness 1978; van der Meer & Bird 1985; West et al. 1983). On the basis of their pioneering hybridization studies with *Petrocelis* species Polanshek and West (1975) demonstrated the conspecificity of *P. franciscana* with *P. middendorffii*. Subsequent investigation employing hybridization (Guiry & West, 1984; Polanshek & West 1977) led to the synonymizing of these *Petrocelis* species under *Mastocarpus papillatus*.

Van der Meer and Bird (1985) showed that Pa-

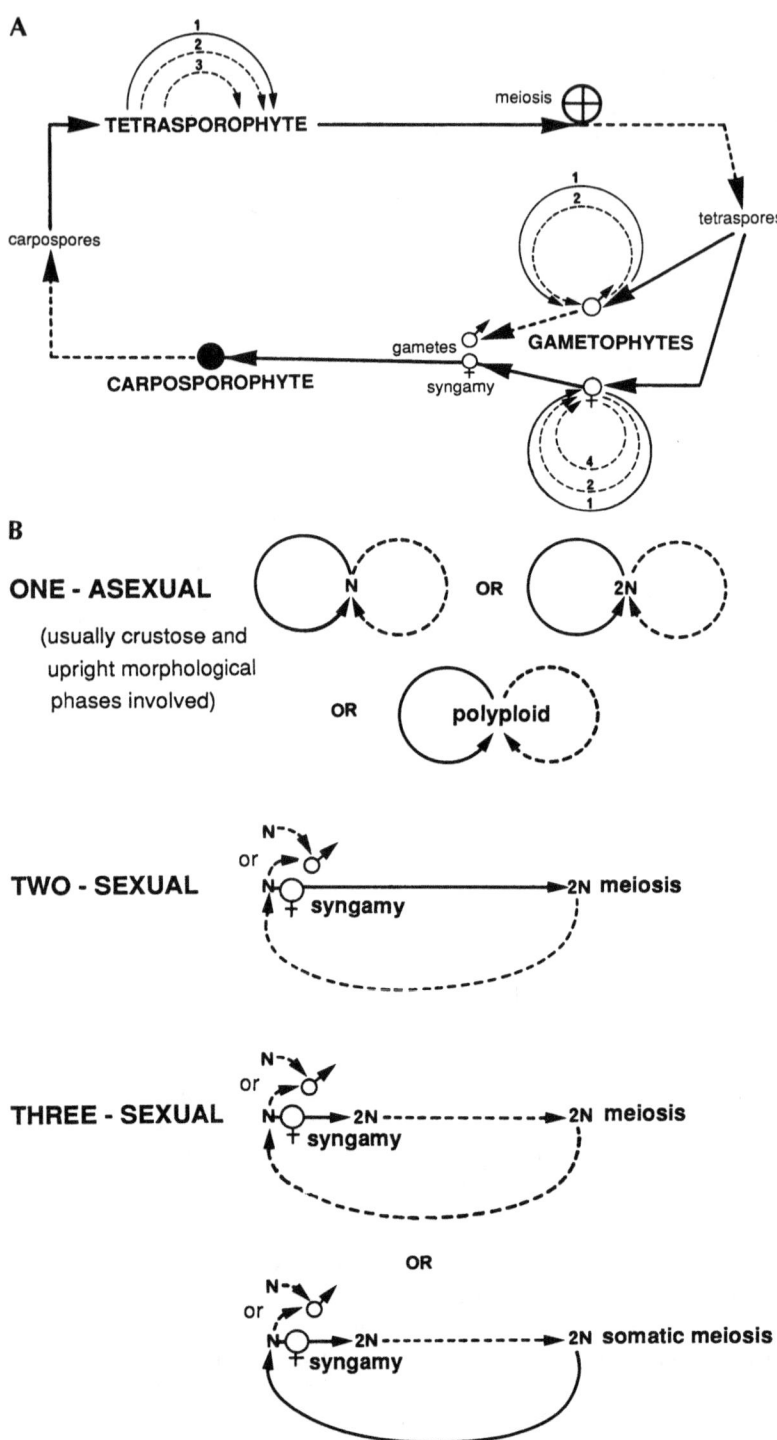

Fig. 17-2. Morphological and karyological phases that have been reported in red algal life histories. A. Diagrammatic summary of a theoretical red algal life history, showing the various potential sexual and asexual pathways of development. A dioecious species is illustrated, but species may be monoecious. Asexual cycles illustrated are (1) vegetative propagation, (2) asexual spores (e.g., monospores) or special propagules, (3) apomeiosis, and (4) apogamy and parthenogenesis. These cycles may be facultative or obligate in some or all individuals in a population and some or all populations of a species. B. Summary diagram of the various karyological phases that have been documented in red algal life histories. ------ = Dispersal event (spore or gamete). ——— = No dispersal event, direct development.

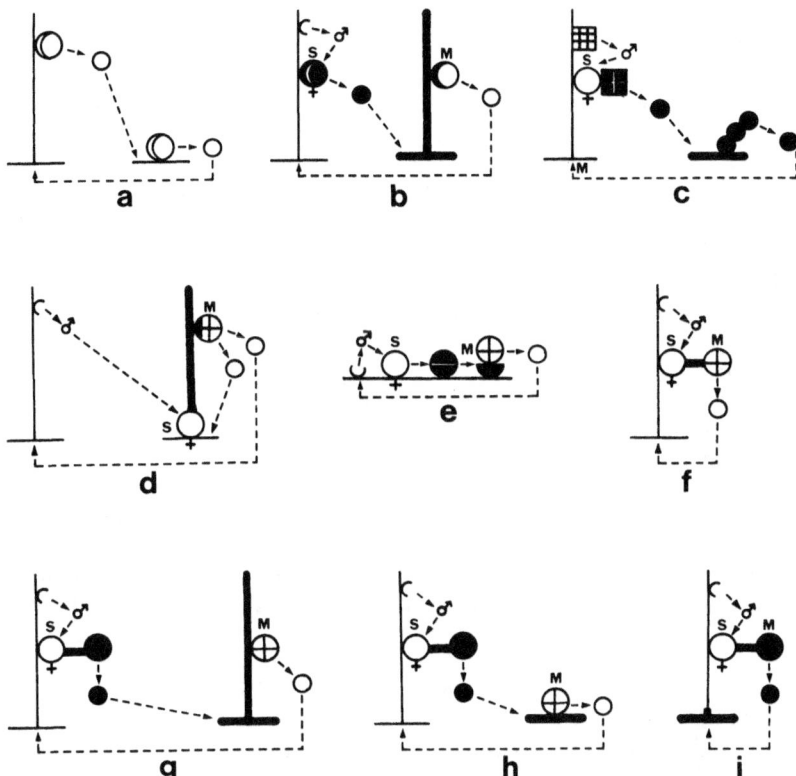

Fig. 17-3. Examples of life history diversity in the Rhodophyta (schematic; types c, f, g, h drawn as monoecious, but species may be dioecious). With the exception of a, which is obligately asexual, all life history types illustrated are sexual. a = Porphyropsis-type; b = Rhodochaete-type; c = Porphyra-type; d = Palmaria-type; e = Rhodophysema-type; f = Liagora tetrasporifera-type; g = Polysiphonia-type; h = Nemalion-type; i = Lemanea-type. Symbols: — haploid thallus; ▬ diploid thallus; ◐ monosporangium; ⊂ or ▦ spermatangium; ♂ spermatium; ♀ carpogonium; ⊸ or ■ carposporangial phase; ● carpospore; ❢ zygote following a single unequal, mitotic cytokinesis; ⊕ tetrasporangium; ⊕ tetrasporocyte and diploid generative stalk cell; ⊕ tetrasporangium and diploid stalk cell; ○ meiosporangium; ⚫ conchosporangia ○ tetraspore; ⊕ carpotetrasporangium; S = syngamy; M = meiosis. (Types b and c after Gabrielson & Garbary 1986; types d, f, g, h and i from Bonin 1981.)

cific coast plants referred to *Palmaria palmata* var. *mollis* were not interfertile with Atlantic *P. palmata* var. *palmata* and therefore elevated the former taxon to specific status. Rueness (1973a) crossed *Polysiphonia hemisphaerica* from Scandinavia with *P. boldii* from Texas. On the basis of their partial fertility, and morphological criteria also suggesting conspecificity, Rueness reduced *P. boldii* to a variety of *P. hemisphaerica*.

Rueness (1973b, 1978) has also done extensive hybridization studies in the *Ceramium diaphanum–strictum* species complex using isolates from Norway and New Jersey. The American isolate, referred to *C. strictum*, is morphologically indistinguishable from the Norwegian isolate, also referred to this taxon, yet the two are intersterile. Rueness (1978) concluded that the numerous taxa described within this species complex can best be interpreted as representing a hybrid zone consisting of the two parent species, *C. tenuicorne* and *C. strictum*, F_1 hybrids, backcrosses, and asexually reproducing clones (via parasporangia). This work is notable for demonstrating the complexity of the problem facing the systematist.

Sundene (1975) attempted hybridizations with several isolates of three varieties of *Antithamnion* (*Pterothamnion*?) *plumula* from Britain, Scandinavia, and the Mediterranean. Variety *crispum* and an undescribed Mediterranean isolate were intersterile with all other strains. Rueness (1978) considered var. *crispum* to be sufficiently distinct to warrant specific status. Two of the varieties of *A. plumula*, var. *plumula* and var. *bebbii*, were interfertile. In some very elegant hybridization work, using F_1 and F_2 hybrids of these two varieties, Rueness and Rueness (1975) demonstrated that the different branching patterns of the two varieties were a single gene character controlled by two alleles.

Where feasible, when making taxonomic decisions, results of crossability tests should always be employed in conjunction with morphological, ecological, karyological, and genetic data. Complete or partial interfertility between otherwise "good" species is widely known for higher plants (e.g., Helenurm & Ganders 1985).

C. Breeding systems and population biology

> "... it is clear that a detailed consideration of the breeding system is essential for a complete understanding of any group." (Stace 1980, p. 137)

In higher plant studies, population biology is often a close associate of biosystematics especially when breeding systems are being investigated (Schaal 1984); it would be valuable for phycologists to adopt a similar approach.

Type of breeding system, especially the frequency of uniparental reproduction (asexual and/or sexual), can greatly complicate the patterns of morphological and genetic variation, thereby frustrating the systematist's attempts to recognize taxa on the basis of discrete breaks in the patterns of variation.

1. Higher plant-breeding systems

Isozyme/allozyme electrophoresis is a technique widely used by higher plant botanists in breeding system studies to obtain information on genetic structure of populations. Useful applications in population genetics, ecology, and biosystematics include (a) determining genetic diversity in population(s), (b) quantifying inbreeding or outcrossing rates, (c) establishing levels of heterozygosity, and (d) detection of apomixis and other forms of asexual reproduction (Briggs & Walters 1984; Buth 1984; Crawford 1983, 1985; Gottlieb 1981, 1984; Richards 1986; Ritland & Ganders 1985). Enzyme electrophoresis can also provide evidence for speciation via polyploidy (Soltis & Rieseberg 1986).

In higher plants three main types of breeding system are recognized:

1. *Obligate outcrossing*. As a result of biparental reproduction due to separate male and female individuals or, where one individual possesses both sexes, to self-incompatibility. For obligate outcrossers, the within-population variation is greater than the between-population variation. Species that present few taxonomic problems are typically sexual, outbreeding, and nonhybridizing. Stace (1980) refers to such taxa as ideal species.
2. *Mixed mating*. Individuals that possess both sexes and are self-compatible can have a mixture of both outcrossing and self-fertilization. Mixed mating, with high rates of inbreeding due to selfing, results in a within-population variation that is less than the between-population variation. Over a number of generations individuals in such a population tend to become homozygous at most loci. This type of breeding system often leads to a large number of morphologically distinguishable microspecies (Briggs & Walters 1984).
3. *Asexual reproduction* via vegetative reproduction or apomixis (Briggs & Walters 1984):

In this chapter I am using the term vegetative reproduction for growth of plant parts that can survive as independent entities, i.e., clonal reproduction or ramet production *sensu* Harper (1977). Vegetative reproduction is frequently facultative, occurring as an alternative asexual reproductive mode to sexual reproduction at certain times and under certain environmental conditions. Ellstrand and Roose (1987) reported intermediate patterns of genotypic diversity in clonal higher plant species.

In higher plants apomixis occurs by agamospermy; seeds are produced asexually as a result of female parthenogenesis. Agamospermy can be obligate, e.g., polyploid *Taraxacum* (Stace 1980; Urbanska 1985), but more typically it is facultative. Apogamy, i.e., seed formation from a synergid or an antipodal cell, is very rare (Urbanska 1985). Details of some other modes of higher plant apomixis are summarized by Asker (1984).

Obligate apomixis can fix high levels of heterozygosity in a population and over time may result in a number of microspecies. Groups that are facultatively apomictic are often taxonomically difficult because they tend to have large morphological variation without discrete breaks in the pattern.

Ecological determinants of genetic structure in higher plant populations have been reported by Loveless and Hamrick (1984). Hamrick et al. (1979) correlated different life history characteristics with electrophoretically detectable genetic variation in a number of angiosperms and gymnosperms. They found that species with wide ranges, high fecundities, outcrossing breeding system, long generation time, and later successional habitats had more genetic variation than did species with other combinations of characteristics. This type of information is unknown for red algae.

2. Red algal breeding systems

In rhodophytes we know little about breeding systems. A priori one would assume that inbreeding due to selfing in monoecious taxa and assortative mating in dioecious taxa may be common phenomena because of presumed limited dispersal of spores and gametes. The importance of facultative asexual reproduction, which is a widespread phenomenon (Dixon 1963, 1965; Rueness 1978), has been largely ignored.

Population genetics of rhodophytes utilizing isozyme/allozyme analysis is in its infancy. Innes (1984) and Cheney (1985) have pointed out the potential of electrophoresis for studying genetic variation within taxa and/or genetic differentiation between taxa of marine algae. Using isozyme data, Innes and Yarish (1984) and Innes (1988) demonstrated that populations of the chlorophyte *Enteromorpha* in Long Island Sound were maintaining themselves by asexual reproduction.

Cheney and Babbel (1978) have done some pioneering isozyme work with *Eucheuma* and were able to demonstrate genetic differentiation among populations of *E. isiforme* (as *E. nudum*) from different depths. They also reported a deficiency of heterozygotes in populations with a high incidence of vegetative reproduction. Japanese workers have examined about a dozen enzyme systems in *Porphyra*, and in general the results have demonstrated high levels of polymorphism and genetic differentiation (e.g., Fujio et al. 1985; Miura et al. 1978, 1979); however, the genetic structure of populations of *P. yezoensis* was taken as being indicative of clonal propagation (Fujio et al. 1987).

a. Sexual and asexual reproduction

"The overall breeding system of a species which is successful over a long period of time is likely to represent a combination of sexual and nonsexual processes." (Grant 1981, p. 5)

The biology and distribution patterns of sexual and asexual reproduction in higher plant and animal species are well documented (Bierzychudek 1985), but there is no consensus on possible causes of the observed patterns. There is a trend for sexual reproduction to be replaced by asexual reproduction at low population densities (Lovlie & Bryhni 1978) or low gamete densities (Gerritsen 1980), in environments that are physically and temporally stochastic, in disturbed environments (Bierzychudek 1985), and at higher latitudes (Michaels & Bazzaz 1986).

Most red algae combine both sexual and asexual reproductive modes in their life history repertoires (Tables 17-2, 17-3), but much more investigation is needed to clarify the various sexual and asexual reproductive pathways that rhodophytes utilize to maintain themselves throughout their range and under different environmental conditions. Field studies have shown variation in gametophyte-to-sporophyte ratios, as well as deviation from a 1:1 gametophyte sex ratio, in the case of dioecious taxa (De Wreede & Klinger 1988). Such variations from

Table 17-2. Examples of asexual reproduction in Rhodophyta via spores, propagules or vegetative propagation

Taxon	Type	Reference
Acrochaetiales		
Audouinella purpurea	Vegetative; f	Breeman & Hoeksema 1987 (as *Rhodochorton purpureum*)
Audouinella simplex	Monospores; f	West 1968 (as *Acrochaetium pectinatum*)
Audouinella spp.	Monospores; f & o	Woelkerling 1983
Bangiales		
Bangia atropurpurea	Monospores; o in all freshwater populations in Great Lakes	Sheath & Cole 1980; Sheath et al. 1985
Bangia vermicularis	Monospores; upper-shore thalli obligately monosporic, thalli lower down on shore are sexual	Cole et al. 1983 (as *B. fuscopurpurea*); Sheath & Cole 1984
Porphyra spp (13 Japanese spp.)	Monospores; f	Hawkes 1980
Porphyra gardneri	Monospores; f	Hawkes 1980
Porphyra sanjuanensis	Aplanospores; o, entire species	Conway et al. 1976

Table 17-2. (cont.)

Taxon	Type	Reference
Porphyra subtumens[a]	Monospores; f	Hawkes, unpubl. observ.
Bonnemaisoniales		
Asparagopsis armata	Vegetative propagation of tetrasporophyte; f	Dixon & Irvine 1977; Farnham 1980; Bonin & Hawkes 1988a & b
Delisea japonica	Vegetative; f	Chihara 1962 (as *D. pulchra*)
Ceramiales[b]		
Anisoschizus propaguli	Propagules; o	Huisman & Kraft 1982
Antithamnionella sarniensis	Vegetative; f	L'Hardy-Halos 1986
Antithamnionella spirographidis	Vegetative; f	L'Hardy-Halos 1986
Callithamnion corymbosum	Vegetative; o?	Whittick 1984
Ceramium strictum	Paraspores; o in one strain	Rueness 1973b
Deucalion levringii	Propagules; f	Huisman & Kraft 1982
Plumaria elegans	Paraspores; o in some populations; all triploid, but sexual diploids also occur	Drew 1939; Rueness 1968
Polysiphonia ferulacea	Propagules on gametophytes and sporophytes; f	Kapraun 1977
Polysiphonia fibrillosa	Propagules; f	Koch 1986
Compsopogonales		
Erythrocladia irregularis	Monospores; o	Gargiulo et al. 1987; Kornmann & Sahling 1985
Sahlingia subintegra	Monospores; o	Kornmann & Sahling 1985 (as *Ethropeltis*)
Erythrotrichia carnea	Monospores; o?	Kornmann & Sahling 1985
Porphyropsis coccinea	Monospores; o	Kornmann & Sahling 1985; Murray et al. 1972
Porphyrostromium ciliare	Monospores; f	Kornmann 1984 (as *Erythrotrichopeltis*)
Porphyrostromium obscurum	Monospores; f	Kornmann 1987
Smithora naiadum	Monospores; f, some populations may be o	Hawkes 1988; McBride & Cole 1971
Gigartinales		
Acrosymphyton purpuriferum	Tetrasporophyte propagation by spores; f	Breeman 1979
Acrosymphyton firmum	Vegetative propagation of putative tetrasporophyte; f	Hawkes 1986a
Calosiphonia vermicularis	Monospores; f	Mayhoub 1973, 1975; Mayhoub et al. 1976
	Vegetative; f	Gayral & Mayhoub 1981
Chondrus crispus	"Spores" on tetrasporophytes (known only in culture)	van der Meer 1987
Halymenia floresia	Monospores; f	van den Hoek & Cortel-Breeman 1970
Schmitzia evanescens	Vegative propagation of putative tetrasporophyte; f	Hawkes 1986a
Thuretellopsis peggiana	Monospores; f	Richardson & Dixon 1970
Nemaliales		
Liagora farinosa	Monospores; f	von Stosch 1965
Nemalion vermiculare	Monospores; f	Masuda & Umezaki 1977
Scinaia confusa	Monospores; f	Ramus 1969 (as *Pseudogloiophloea*)
Rhodymeniales		
Lomentaria hakodatensis	Vegetative propagation; o in most British Columbian populations	Hawkes & Scagel 1986b
Rhodymenia pacifica	Vegetative propagation; o in British Columbian populations	Hawkes & Scagel 1986b

[a] Although Conway et al. (1976) reported this species as obligately monosporic it does have a sexual cycle (Hawkes unpubl.)
[b] Bisporangia, monosporangia, seirosporangia, pseudosporangia, parasporangia, gemmae, propagules, and fragmentation are all asexual modes of reproduction reported in Ceramiales.
o = obligate; f = facultative.

Table 17-3. Examples of asexual reproduction in Rhodophyta via apomixis (apogamy/apomeiosis)

Taxon	Type	Reference
Acrochaetiales		
Acrochaetium proskaueri	Apomeiosis; o and f	West 1972a
Audouinella spp.	Apomeiosis; o and f	Woelkerling 1983
Rhodochorton concrescens	Apomeiosis; o	West 1970a
Bangiales		
Porphyra carolinensis	Apogamy and apomeiosis; o, entire species	Freshwater & Kapraun 1986
Porphyra kanakaensis[a]	Apogamy and apomeiosis; f?	Mumford & Cole 1977
Porphyra papenfussii[a]	Apogamy and apomeiosis; f?	Conway & Cole 1973; Conway et al. 1976;
Porphyra rosengurtii	Apogamy and apomeiosis; o, entire species	Kapraun & Luster 1980
Ceramiales		
Antithamnion cruciatum	Apomeiosis; o in Newfoundland pop[ns], may be polyploid	Whittick & Hooper 1977
Antithamnionella sarniensis	Apomeiosis; freq. varying with strain	Magne 1986a,b, 1987a[c]
Heterosiphonia densiuscula	Apomeiosis; o	West 1970b (as *H. densiuscula* & *H. asymmetrica*)
Ptilota serrata	Apomeiosis; f in 80–85% of Newfoundland pop[ns]	Whittick 1984
Corallinales		
Dermatolithon litorale	Apomeiosis; o	Suneson 1982
Lithophyllum incrustans	Apomeiosis; o in some pop[ns]	Edyvean & Ford 1984, 1986; Ford et al. 1983
Gigartinales		
Atractophora hypnoides	Apomeiosis; f	Maggs 1985
Gloiosiphonia californica	Apomeiosis; o in some pop[ns]	DeCew et al. 1982
Gloiosiphonia capillaris	Apomeiosis; f	Maggs 1985
Gymnogongrus leptophyllus	Apogamy; o in some pop[ns]	DeCew et al. 1982
Iridaea splendens	Apomeiosis; o in some pop[ns]	Kim 1976 (as *I. cordata* var. *splendens*)
Iridaea cornucopiae	Apomeiosis in one pop[n] from Washington	Kim 1976
Mastocarpus jardinii	Apogamy in 25% of isolates sampled, no geographical pattern	West & Hommersand 1981; West et al. 1978 (as *G. agardhii*)
Mastocarpus pacificus	Apogamy in 87% of isolates from 21 pop[ns]	Masuda & Uchida 1976 (as *G. ochotensis*); Masuda et al. 1984 (as *G. pacifica*)
Mastocarpus papillatus	Apogamy; o in some isolates	Polanshek & West 1977 (as *G. papillata*); Zupan & West 1988
Mastocarpus stellatus	Apogamy; o in many northern isolates,[b] southern isolates sexual	West et al. 1977; West & Hommersand 1981
Schmitzia hiscockiana	Apomeiosis; f	Maggs 1985
Hildenbrandiales		
Apophlaea sinclairii	Apomeiosis; o	Hawkes 1983b
Hildenbrandia spp.	Apomeiosis; o	DeCew & West 1977; Umezaki 1969
Palmariales		
Rhodophysema elegans	Apomeiosis; o in some pop[ns]	Fletcher 1975, 1977; Masuda & Ohta 1975; South & Whittick 1976
Rhodymeniales		
Fauchea fryeana	Apomeiosis/apogamy?; o in all pop[ns]?	Hawkes & Scagel 1986b
Fauchea laciniata	Apomeiosis/apogamy? o in most pop[ns]?	Hawkes & Scagel 1986b; West & Norris 1966
Gastroclonium subarticulatum	Apomeiosis in northern pop[ns]?	Hawkes & Scagel 1986b
Lomentaria orcadensis	Apomeiosis; o?	Svedelius 1937; West & Hommersand 1981
Rhodymenia californica	Apomeiosis; o in northern pop[ns]?	Hawkes & Scagel 1986b

Note: The inference of apomixis must be considered tentative because in most cases critical karyological data are lacking.
[a] Same chromosome number in gametophytes and sporophytes is circumstantial evidence for apomixis; probably facultative.
[b] West and Hommersand (1981) caution that some of these isolates thought to be apomictic may actually be sexual, monoecious gametophytes with somatic meiosis. Resorcinol testing has shown lambda carrageenan in basal disk and kappa carrageenan in upright. If the carrageenan type is indeed determined by ploidy level, then this is good evidence of somatic meiosis.
[c] These studies present the first karyological proof of apomeiosis in Rhodophyta.
o = obligate; f = facultative; pop[n], pop[ns] = population (s)

the theoretical life history may be attributable to greater fecundity and survivorship (relative fitness) of one phase or to asexual phenomena, such as vegetative propagation and/or apomixis (apogamy/ parthenogenesis, apomeiosis) (e.g., DeCew et al. 1982, Edyvean & Ford 1984; Feldmann & L'Hardy-Halos 1977; Guiry & West 1984; Maggs 1985, 1988; Masuda et al. 1984; Zupan & West 1988). To further complicate the situation, these asexual phenomena may be obligate or facultative, occurring in some or all individuals in a population, or in some or all populations of a species.

Clonal propagation via ramets has the advantage of allowing continued occupation of space, which is usually in short supply in the marine benthic environment. Vegetative reproduction is probably more widespread and more important in maintaining populations of rhodophytes than currently recognized and represents an exciting area for further investigation; population and karyological studies are required.

Rhodophytan algae are also capable of asexual reproduction via monospores or other dispersable propagules (Tables 17-2, 17-3) (Rueness 1978). Monospore production is common in "lower" red algae like Compsopogonales, Bangiales, Acrochaetiales, and Nemaliales and rare or absent in Gigartinales, Rhodymeniales, and Ceramiales (Table 17-2). Monospore-producing taxa like *Porphyra gardneri* (Hawkes 1978, 1980), *Bangia vermicularis* (Cole et al. 1983), and *Smithora naiadum* (Hawkes 1988) take particular advantage of both sexual and asexual reproductive modes. Presumably, asexual reproduction allows for persistence of the species when environmental conditions are not favorable to sexual reproduction (Templeton 1982). Asexual reproduction via monospores is a seasonal phenomenon in *Porphyra* and *Bangia* (Hawkes, 1977; Sheath et al. 1985). In *Porphyra gardneri* early spring thalli reproduce asexually via monospores but switch to sexual reproduction later in the spring and summer (Hawkes 1978). Sheath et al. (1985) noted that for *Bangia atropurpurea* the ecological and physiological tolerances are quite different for asexual and sexual plants.

> "... apomixis represents the preservation of a fit genotype in a hostile environment when the Sisyphean nature of sexually reproducing genotypes would produce inappropriate and disadvantageous variation." (Edyvean & Ford 1984, p. 363)

In algae apomixis can occur via apogamy/parthenogenesis or apomeiosis. In the algal literature, "apogamy" has been used with a broader meaning than in higher plants. Thus, all cases of gamete (male or female) development without sexual fusion, to produce an adult plant, have been referred to as apogamy.

Apomixis has been reported or implied for several red algal species (Table 17-3), but in all but one case, *Antithamnionella sarniensis* (Magne 1986a,b; 1987a), the data must be regarded as tentative because karyological proof is lacking. Based on the meager data available, species that are obligate apomicts (i.e., all populations completely asexual) would appear to be relatively few; for example, obligate apomeiosis has been reported in all populations of *Dermatolithon litorale* (Suneson 1982), *Hildenbrandia* (DeCew & West 1977), *Rhodochorton concrescens*, and *Acrochaetium proskaueri* (West 1970a, 1972a).

Facultative apomixis is a more common phenomenon. In her study of *Atractophora hypnoides*, *Gloiosiphonia capillaris*, and *Schmitzia hiscockiana*, Maggs (1985) observed that each species had three potential life history patterns: (1) tetraspore germination to give 100% erect, gametophytic thalli, (2) tetraspore germination to give 100% crustose, putatively tetrasporophytic thalli, and (3) any ratio of crusts to erect thalli. She reported that the majority of tetraspores recycled the crustose tetrasporophyte, presumably as a result of apomeiosis.

There are many reports of gametophytes being initiated directly from a carpospore-derived crust without the production of tetrasporangia (Table 17-4; Hawkes 1983a, 1986a). It is tempting to assume that such a pattern of "direct" development is simply apomictic recycling of the gametangial generation. However, somatic meiosis is known to occur in *Lemanea* (Magne 1972) and *Batrachospermum* (von Stosch & Theil 1979) and should not be ruled out. Guiry and West (1984) cautioned that populations of *Mastocarpus* species, reported to be apomictic, may in fact be sexual, monoecious gametophytes, with somatic meiosis occurring at the time of upright initiation. It may be the case that a species can have sexual and asexual direct pathways of development, as well as sexual pathways involving tetraspore production. To complicate the picture further, these three pathways may occur in the same or different populations of a species. Guiry and West's (1984) results indicated just this type of flexible life history scenario for *Mastocarpus stellatus*, where some populations had a sexual, heteromorphic life history, some populations had direct development and cycling of female gametophytes, and some populations had both life history types. Guiry and Coleman (1982) reported that 90% of the population of *Mastocarpus stellatus* (as *Gigartina stellata*) had a direct life history (probably sexual

Table 17-4. Examples of marine rhodophytes showing "direct" development of gametophyte[a]

Taxon	Reference
Bonnemaisoniales	
Bonnemaisonia asparagoides	Breeman 1979; G. Feldmann 1966; Rueness & Åsen 1982
Delisea compressa	Bonin & Hawkes (1988a)
Delisea fimbriata	Chihara 1962
Delisea japonica	Chihara 1962 (as *D. pulchra*)
Ptilonia mooreana	Bonin & Hawkes (1988b)
Ptilonia okadae	Chihara 1962 (as *P. okadai*)
Gigartinales	
Atractophora hypnoides	Boillot 1967
Calosiphonia vermicularis[b]	Mayhoub 1973, 1975, 1976; Gayral & Mayhoub 1981
Dudresnaya japonica	Umezaki 1968
Hummbrella hydra	Hawkes 1983a
Naccaria wiggii[b]	Boillot 1967, Boillot & L'Hardy-Halos 1975; Jones & Smith 1970
Pikea californica[b]	Chihara 1972; Scott & Dixon 1971
Schimmelmania plumosa[b]	Chihara 1972; Scott & Dixon 1971
Schmitzia evanescens	Hawkes 1986a

[a] Life history is either asexual, or sexual, and of the *Lemanea*-type (cytological data lacking).
[b] Both direct development without tetrasporangium formation and development with formation of tetrasporangia have been reported for these taxa.

with somatic meiosis) and 10% had a heteromorphic sexual life history. In *Mastocarpus* sp. (Masuda et al. 1987) plants with a heteromorphic life history occur throughout the species range, but plants with direct development have a narrow range.

There may also be a geographical component to asexual cycles. Preliminary data suggest that, for some species, asexual reproduction is prevalent in populations at the high latitude ends of their distribution (Dixon 1965). For example, many red algae at the entrance to the Baltic Sea reproduce by vegetative means only (Rueness 1978). Edyvean and Ford (1986) reported that the percentage of gametophytes in populations of *Lithophyllum incrustans* increased with decreasing latitude. Similar patterns are reported for *Mastocarpus papillatus* and *M. stellatus* (Guiry & West 1984 as *Gigartina stellata*); however, *M. pacificus* showed an opposite pattern, with the largest percentage of gametophytes at higher latitudes (Masuda et al. 1984).

In British Columbian waters only vegetative populations of *Rhodymenia pacifica* have been reported, and for *R. californica* only tetrasporangial or vegetative populations are known (Hawkes 1986b; Hawkes & Scagel 1986b). However, further south in Californian waters sexual gametophytes have been reported for both these *Rhodymenia* species. *Lithothrix aspergillum* shows similar variation in reproduction with latitude (De Wreede & Vandermeulen pers. comm.). Detailed population studies are needed to verify these suspected trends.

The apparent absence of sexual reproduction in some bangiophyte taxa has led to the suggestion that they are primitively asexual (e.g., Tappan 1980). Demonstration of sexuality in *Porphyra* (Hawkes 1978; Cole & Conway 1980), *Bangia* (Cole et al. 1983), *Porphyrostromium* (Kornmann 1984 as *Erythrotrichopeltis*; Kornmann 1987), and *Smithora* (Hawkes 1988) makes the primitively asexual hypothesis less likely. An alternative interpretation is that the ancestral rhodophyte was sexual and that asexual reproduction is a derived rather than an ancestral reproductive mode. A few rhodophytes are apparently obligately asexual, although the cryptic nature of sexuality in many red algae demands that these taxa be more thoroughly investigated, for example, *Stylonema*, *Erythrocladia*, *Porphyropsis*. These taxa are not clearly derived from sexual forms, but other obligately asexual taxa are more clearly derived from sexual cyles (e.g., *Hildenbrandia*).

Knowledge of breeding systems in red algae, and the patterns of morphological and genetic variation they produce, may provide insights into taxonomically difficult groups. Rueness's (1978) work on *Ceramium* has already been cited as an excellent example. In particular, such data could help us to revise our species concepts and assist in understanding speciation in rhodophytes.

Higher animals and many higher plants are obligately sexual and outbreeding. This type of breeding system yields a particular pattern of morphological and genetic variation. Some speciation theory defines "real" species in these terms (Bernstein et al. 1985) and specifically excludes organisms with other types of reproduction as "gross alterations of the sexually reproducing, cross-fertilizing system" (Carson & Templeton 1984, p. 103). It is clear, even with our meager knowledge of rhodophyte breeding systems, that such definitions and theories will need modification if they are to incorporate rhodophytes.

b. Monoecy, dioecy, and compatibility. The occurrence of monoecy and dioecy is incompletely known for many red algae (Table 17-5), as are the factors favoring one or the other. In higher plants dioecy is correlated with long-lived perennials of woody habit; fleshy, animal-dispersed fruit; and pollination

Table 17-5. Examples of the distribution of monoecy and dioecy in Rhodophyta

Order	Monoecious	Dioecious	Polyoecious
Rhodochaetales	+	−	−
Compsopogonales	+	−	−
Bangiales	mainly		
Porphyra			
33 Japanese spp. (Kurogi 1972)	61%	27% (+ 6% androdioecious)	6%
16 British Columbia spp. (Conway et al. 1976)	87%	6.7%	6.7%
Achrochaetiales		mainly	
25 *Audouinella* spp. (Garbary et al. 1983: only 2% sexual)	−	80%	20%
Nemaliales	+	+	−
Bonnemaisonales		mainly	
Palmariales		mainly	
Gigartinales			
British spp. (Dixon & Irvine 1977)	31%	51%	−
Dumontiaceae (Lindstrom & Scagel 1987)	8%	77%	15%
Rhodymeniales	−	+	−
Ceramiales		mainly	

by small bee species or wind pollination (Charlesworth 1985; Fox 1985). Dioecy is also more common on islands (Baker & Cox 1984) and in the tropics (Bawa 1980) than on mainlands or in temperate floras (see also Chapter 16).

Even less is known about compatibility (the ability to self-fertilize), although most reports indicate that monoecious species are self-compatible (Table 17-6). For these self-compatible species no data are available on outcrossing rates. In the few self-incompatible species reported to date, nothing is known about the genetic basis of the incompatibility, nor whether there are incompatibility systems, as in higher plants (Lewis 1979), that restrict crossing between close relatives, thereby avoiding inbreeding depression. One area of focus for future studies would be the distribution, both systematic and ecological, of dioecy and self-incompatibility in Rhodophyta.

III. EVOLUTION AND ECOLOGY

A. Life history evolution

"... the possibility of a phylogenetic interpretation of red algal life histories seems almost to be receding from us. It is not obvious in any of the more difficult groups whether an isomorphic or a heteromorphic life cycle is the more primitive. It is not even certain whether a biphasic or a triphasic cycle is primitive in extant genera." (West & Hommersand 1981, p. 136)

Understanding of the evolution of red algal life history diversity is far from clear, and fresh approaches are needed. Feldmann (1952, 1972) hypothesized that, because it was the most common, the *Polysiphonia*-type life history was the ancestral florideophyte pattern, all other types being derivations, through reduction, from this gametophyte–carposporophyte/tetrasporophyte, isomorphic pattern. Magne (1972, 1987b) has been a strong advocate of this view, which suggests that the carposporophyte was once a free-living, spore-forming generation that has subsequently become "parasitic" on the female gametophyte. The term "triphasic" (referring to morphological phases) has been used to describe life history patterns having a gametophyte–carposporophyte–tetrasporophyte alternation.

As advocated by Searles (1980), the red algal carposporophyte may have had its evolutionary origin as a means of zygote amplification, to compensate for the presumed infrequency of syngamy

Table 17-6. Compatibility in rhodophytes

Taxon	Compatibility	References
Bangiales		
Porphyra yezoensis	sc	Ohme et al. 1986
Bonnemaisoniales		
Bonnemaisonia asparagoides	sc	Rueness & Åsen 1982
Ceramiales		
Antithamnionella spirographidis	sc	L'Hardy-Halos 1986
Callithamnion sp.	si	Magruder pers. comm.
Callithamnion sp.	sc	Whittick & West 1979
Callithamnion bipinnatum	sc	Rueness & Rueness 1980
Callithamnion tetragonum	sc	Rueness & Rueness 1985
Gelidiales		
Gelidium sp.	sc	van der Meer 1987, (documented using a color marker gene)
Gigartinales		
Gigartina sp.[a]	sc	West & Guiry 1982
Rhodymeniales		
Minium parvum[b]	si	West & Hommersand 1981
Lomentaria sp.	si	as above

[a] Unisexual species became bisexual and self-compatible in culture.
[b] This species is dioecious, but male and female plants derived from the same tetrasporophyte may be incompatible.
sc = self-compatible; si = self-incompatible.

due to the nonflagellated male gamete. Hawkes and Scagel (1986a) pointed out that the success of Palmariales taxa lacking a carposporophyte argues against Searles' (1980) hypothesis. Dispersal, rather than infrequency of fertilization, may also be a significant factor in carposporophyte evolution.

Life histories lacking a carposporophyte have been designated biphasic. Implicit in the biphasic/triphasic terminology has been the assumption of a single evolutionary origin for each of these patterns – Guiry (1987) being a notable exception. If we view the carposporophyte as a means of zygote amplification, then I see no reason to assume a single origin for the carposporophyte.

Available evidence argues against the view that the biphasic and triphasic patterns evolved only once (Guiry 1987). Because of this and the difficulty of applying the terms in some cases, I have avoided using the biphasic/triphasic terminology. If triphasic/biphasic continue to be used, it will be necessary to clarify that both patterns probably originated more than once in different phylogenetic lines. I believe this to be the case and feel that failure to point this out has led to confusion. For example the *Palmaria*-type life history (gametophyte/tetrasporophyte) has been subject to various interpretations. Abdel-Rahman and Magne (1983), Gabrielson and Garbary (1986), and Magne (1982, 1987b) would derive it by reduction (i.e., loss of zygote amplification by a carposporophyte) from a triphasic florideophyte pattern. However, Guiry (1987) and Hawkes and Scagel (1986a) suggest that this is not the case and speculate that the carposporophyte was never present in these algae in the first place. One must not assume that all life history patterns lacking zygote amplification evolved only once (Guiry 1987). For example, the *Liagora tetrasporifera*-type life history, where a diploid tetraspore-forming generation develops directly on the gametangial generation following fertilization, may indeed be derived by reduction from a "gametophyte–carposporophyte/tetrasporophyte" life history type.

Bangialean life histories are another confusing area for application of the biphasic/triphasic terminology. In the *Porphyra* life history zygote amplification does occur. However, we should not assume that this is homologous to zygote amplification (i.e., the carposporophyte) in florideophytes. I believe they are independently evolved analogies, similar solutions to zygote amplification. This is implied by West and Hommersand (1981), but their treatment of the *Porphyra* life history as biphasic is misleading, because there is clearly zygote amplification in this case.

Other bangiophytes pose similar problems of interpretation. The zygote in *Rhodochaete* divides once to form two diploid cells, but only a single carpospore is released. Functionally the zygote is not amplified. Guiry (1987) presumably regards the two postfertilization cleavage products as a carposporophyte, because he refers to the life history as triphasic. Although he does not elaborate on this, it is clear from his placement of *Rhodochaete* in the bangiophytes that he does not regard *Rhodochaete* triphasic as homologous with florideophyte triphasic. One encounters the same definitional problem in *Porphyrostromium* (Kornmann 1984, 1987) and *Smithora* (Hawkes 1988). Zygote development in all these taxa is probably not homologous to the zygote development (=carposporophyte) of florideophytes.

It may indeed be the case that *Rhodochaete* should be grouped with the compsopogonalean taxa (=Compsopogonales and Erythropeltidales). This would give three major rhodophyte evolutionary lines: florideophyte, bangialean, and compsopogonalean. At the base of the florideophyte line are two lineages: One lineage never had a carposporophyte, i.e., zygote amplification (as exemplified by the Palmarialean life history), and the other had a carposporophyte (exemplified by the other florideophytes). This also raises the question: Are the carposporophytes homologous in those florideophytes having zygote amplification, or do some of them represent independently evolved analogies from a biphasic pattern?

The cladistic analysis of Gabrielson and Garbary (1986) suggests that isomorphy and heteromorphy evolved several times in different ordinal lineages, and it is unclear which is the primitive type.

B. Breeding systems and genotype diversification

When looked at in terms of amplification and diversification of genotypes, red algal breeding systems and life histories are especially intriguing. Depending on compatibility and postfertilization development, the genotypes amplified by the carposporophytes borne on a single gametophyte can be the same or different. For example, in procarpic/dioecious species or procarpic/monoecious and self-incompatible species each carposporophyte borne on a single female gametophyte could potentially be genetically different, each being the product of a fertilization by a different male gametophyte. On the other hand, in a self-compatible species some carposporophytes could be genetically identical as a result of selfing, whereas others could be genetically different due to outcrossing. Similar scenarios apply for those species that produce a connecting filament following fertilization. In some species one fertilization event produces one connecting filament that contacts and diploidizes only one auxiliary cell, thereby initiating a carposporophyte. In such species fertilization by different males could result in many genetically different carposporophytes on a single gametophyte (comparable to the procarpic/dioecious or procarpic/monoecious and self-incompatible situation). In taxa where the connecting filament diploidizes several auxiliary cells there is the potential for a single fertilization event to produce many genetically identical carposporophytes on the same female gametophyte.

In addition to syngamy as a means of genotype diversification, meiosis also provides this opportunity. Those taxa that produce many meiosporangia per sporophyte have potentially greater genetic recombination than those with few; for example, sporophytes with somatic meiosis, i.e., *Lemanea*-type life history, presumably have fewer meioses than foliose sporophytes that are covered in meiosporangia (see Chapter 16). A further variation worthy of comment is that taxa with a *Palmaria*-type life history have the ability to produce successive crops of meiospores from a generative stalk cell (see Guiry 1987; Hawkes & Scagel 1986a). Perhaps this increased ability for genotype diversification is a means of compensation for lack of genotype amplification in the *Palmaria*-type life history, due to absence of a carposporophyte.

C. Life history theory

"The algae exhibit a wide range of life history variation that remains mostly unexplored in terms of evolutionary processes." (De Wreede & Klinger 1988, p. 274)

What mechanisms played a role in the origin and maintenance of red algal life histories? Guiry (1987) suggests that development of the various life history types has been mediated by selective pressures on dispersal, ability to capture male gametes, herbivore grazing, and increasing surface area/volume. Is spermatium dispersal analogous to wind pollination (Niklas 1985), so that hydrodynamics account for fertilization (as discussed in Chapter 16), or does the microinvertebrate fauna play a "pollinator" role in carrying spermatia to trichogynes? What are the levels and consequences of inbreeding? Does inbreeding promote parthenogenesis in haplodiploids, as theory predicts (Uyenoyama 1985)?

There is a vast body of literature, based on higher plant and animal systems, that deals with life history theory (Caswell 1982, 1985; Stearns 1977, 1980). Hastings and Caswell (1979) discussed theoretical aspects of the role of environmental variability in the evolution of annual and perennial life history strategies. De Wreede and Klinger (1988) noted that most of our information on algal reproductive strategies consists of correlations between reproductive patterns and environment types.

De Wreede and Klinger (1988) pointed out that resource allocation theory – Charnov's (1982) sex allocation theory being a specific aspect of this – has not been applied to the algae and cautioned that reproductive cost to an autotrophic plant may not be comparable to the costs of reproduction for a heterotrophic animal. Margulis and Sagan (1986) also rejected such socioeconomic analysis, i.e., cost of sex arguments, as applied in biology. Complex cryptogamic life histories need to be incorporated into this theoretical framework.

1. Sexual versus asexual reproduction

There have been many theoretical considerations on the possible advantages of sexual reproduction (Bell 1982; Bremermann 1985; Margulis & Sagan 1986; Maynard Smith 1978; Michod & Levin 1987; Shields 1982; Stearns 1985, 1987; Tooby 1982) despite its apparent costs (Williams 1975). Bell (1982, 1985) and Bierzychudek (1985) present compelling evidence arguing against the hypothesis that sex is a preadaptation to unpredictable environments (Willson 1981). Margulis and Sagan (1986) emphasize that we must be careful, when looking for evolutionary reasons for the significance of sex, to distinguish between the origin of sex and the maintenance of sex, two very different things.

Asexuality in animals and higher plants is correlated with disturbed, ephemeral environments where biotic interactions are less intense (Mladenov & Emson 1988). This lends support to Bell's (1982) hypothesis that sexual reproduction is favored in spatially heterogeneous, stable environments, where intraspecific competition predominates. Prevailing ideas would attribute two advantages to being facultatively asexual: It allows replication of successful genotypes unchanged, and it provides a means of generating genetic diversity via sexual reproduction (syngamy and meiosis) at irregular or regular intervals (Bierzychudek 1985; Edyvean & Ford 1984). According to Bell's (1982) hypothesis, facultatively asexual taxa should reproduce asexually at low population densities and switch to sexual reproduction when density increases.

An experimental approach to the evolutionary significance of sexual versus asexual reproduction is needed, and Antonovics and Ellstrand (1984), Ellstrand and Antonovics (1985), and Schmitt and Antonovics (1986) provide excellent examples of the type of tests that could be conducted using marine red algae. These studies focus on determining the importance of genetic variability to survival under various types of selection in a field situation. Both sexually derived and asexually derived material of the grass *Anthoxanthum odoratum* were used.

Unequal sex ratios and spatial or temporal variation in gametophyte and tetrasporophyte dominance require population studies of reproductive ecology (Dyck et al. 1985; Hannach & Santilices 1985; May 1986) and demography (Bhattacharya 1985; see Chapman 1986 for general overview). Such studies may shed light on the ecological and evolutionary aspects of sexual versus asexual reproduction in marine red algae.

2. Form–function

Marine red algae have played an important role in testing the hypothesis that herbivore grazing (biological disturbance) has been a selective force determining algal form and life histories (especially heteromorphic ones) (Lubchenco & Cubit 1980; Steneck 1985). Abiotic factors, such as seasonal physical disturbance of the habitat, are also often correlated with crustose habits and/or heteromorphic life histories, e.g., cobble flora's (Davis & Wilce 1987; Hawkes 1983a; Lieberman et al. 1984).

Form–function theory attempts to provide ecological explanations for the evolution of various algal life history types and morphologies (Littler & Littler 1980, 1984; Littler et al. 1983). The theory predicts correlations between certain morphologies and grazer resistance (toughness), reproduction, calorific value, productivity, successional status, and growth characteristics. Field and laboratory tests of the theory have been done (Littler & Littler 1980) and revealed several interesting trends. For example, sheet and filamentous algae are the highest primary producers, followed by coarsely branched and thick, leathery groups. The articulate and crustose coralline algae had the lowest productivity but highest grazer resistance. Rubbery or leathery species also showed high resistance to grazing

(Littler et al. 1983). Form–function theory, in synthesizing morphology, physiology, and ecology from a selection perspective, has shown the complexity of the problem and suggests that a suite of factors, rather than a single factor, must be considered. However, Hay (1984), S. M. Lewis (1985), and Padilla (1985) question the general applicability of the form–function model.

3. Haploidy versus diploidy

Persistance of diplohaplontic alternation of generations, typical of most rhodophytes, suggests that hypotheses invoking diploid superiority via genetic buffering of recessive alleles, more accurate control of development (W. M. Lewis 1985), and possibly heterosis are flawed. Hannach and Santelices' (1985) study of *Iridaea*, in which gametophytes were more abundant than sporophytes, casts further doubt on arguments invoking increased adaptivity of diploidy over haploidy (Stebbins & Hill 1980). Ewing (1977) presented a deterministic genetic model for the maintenance of haploid and diploid life history phases. He was able to demonstrate that selection at both the haploid and diploid phase can make a difference to retention of variability in the population.

Work by Goff and Coleman (1984, 1986, Chapter 3) on DNA variation has shown a much more complicated karyological life history than previously envisioned. They reported gametophytes and sporophytes with the same levels of DNA, high levels of endopolyploidy, or polyploidy via a multinucleate condition. These data do not support W. M. Lewis's (1985) hypothesis that nutrient scarcity is an evolutionary cause of haploidy. In higher plants polyploidy is known to buffer the effects of inbreeding. Such novel phenomena challenge our previous concepts of haploidy and diploidy in the algae.

IV. SUMMARY

Life history diversity and variation in reproductive biology are greater in Rhodophyta than in any other algal division. Rhodophytes usually have both sexual and asexual reproductive modes. Following fertilization, zygote amplification occurs in most rhodophytes. This is achieved by a diploid mitosporangial generation (the carposporophyte), which produces many dispersable products from a single fertilization event. It is hypothesized that zygote amplification evolved at least twice in red algal phylogeny: once in the bangialean line and once in the florideophyte line.

In some rhodophytes zygote amplification does not occur. It is hypothesized that this condition also arose more than once in red algal evolution. In the compsopogonalean, rhodochaetalian, and palmarialean lineages absence of zygote amplification is regarded as the primitive condition. In taxa like *Liagora tetrasporifera*, this absence is regarded as being derived from the typical florideophyte life history pattern, as a result of loss of the carposporophyte. The ancestral rhodophyte life history is regarded as being sexual, with no zygote amplification, and consisting of two free-living morphological generations: gametangial and meiosporangial.

Reproductive biology, in particular breeding systems, has an important effect on the patterns of morphological and genetic variation that we observe in populations of rhodophytes. A knowledge of such variation is important to the systematist attempting to delimit species and may help revise our concepts of species and speciation in the Rhodophyta. There are many opportunities for pioneering work on red algal breeding systems. For example, we need data on the frequency of uniparental reproduction, either sexual, in the case of self-compatible taxa, or asexual. Preliminary studies indicate that facultative asexual reproduction, via vegetative propagation or apomixis (apogamy/parthenogenesis and apomeiosis), can be an important component of red algal life histories.

Most life history theory is based on higher plants and animals. There is a need to incorporate the complexities of red algal reproductive strategies into this body of theory. Rhodophytes present opportunities for testing theory on the origin and maintenance of complex life histories and on the advantages versus disadvantages of sexual and asexual reproduction.

V. ACKNOWLEDGMENTS

Research support was provided by Canadian NSERC operating grants U-0318 and 80384. Dr. Robert De Wreede kindly made available a prepublication copy of his and T. Klinger's paper on algal reproductive strategies. Drs. K. Cole, R. De Wreede, and R. Sheath and Ms. Denise Bonin suggested several improvements to the manuscript. Discussions with Dr. P. Gabrielson were useful for clarifying some of the ideas presented in this paper. The University of Washington's Friday Harbor Laboratories provided support and facilities during preparation of the figures.

VI. REFERENCES

Abdel-Rahman, M. H. & Magne, F. 1983. Existence d'un nouveau type de cycle de développement chez les

Rhodophycées. *C. R. Acad. Sci.* (Paris), sér. D. 296: 641–4.
Antonovics, J. & Ellstrand, N. C. 1984. Experimental studies of the evolutionary significance of sexual reproduction. I. A test of the frequency-dependent selection hypothesis. *Evolution* 38: 103–15.
Ardré, F. 1977. Sur le cycle de *Schizymenia dubyi* (Chauvin ex Duby) J. Agardh (Némastomacée, Gigartinale). *Rev. Algol.* N.S. 12: 73–86.
Asker, S. 1984. Apomixis and biosystematics. In *Plant Biosystematics*, ed. W. F. Grant, pp. 237–48. London: Academic.
Baker, H. G. & Cox, P. A. 1984. Further thoughts on dioecism and islands. *Ann. Missouri Bot. Gard.* 71: 244–53.
Bawa, K. S. 1980. Evolution of dioecy in flowering plants. *Annu. Rev. Ecol. Syst.* 11: 15–39.
Bell, G. 1982. *The Masterpiece of Nature: The Evolution and Genetics of Sexuality.* Berkeley: University California Press.
Bell, G. 1985. Two theories of sex and variation. *Experimentia* 41: 1235–45.
Bernstein, H., Byerly, H. C., Hopf, F. A. & Michod, R. E. 1985. Sex and the emergence of species. *J. Theor. Biol.* 117: 665–90.
Bhattacharya, D. 1985. The demography of fronds of *Chondrus crispus* Stackhouse. *J. Exp. Mar. Biol. Ecol.* 91: 217–31.
Bierzychudek, P. 1985. Patterns of plant parthenogenesis. *Experimentia* 41: 1255–64.
Boillot, A. 1965. Sur l'alternance de générations hétéromorphes d'une rhodophycée, *Halarachnion ligulatum* (Woodward) Kützing (Gigartinales, Furcellariacées). *C. R. Acad Sci.* (Paris), sér. D, 261: 4191–3.
Boillot, A. 1967. Sur le développement des carpospores de *Naccaria wiggii* (Turner) Endlicher et d'*Atractophora hypnoides* Crouan (Naccariacées, Bonnemaisoniales). *C. R. Acad. Sci.* (Paris), sér. D, 264: 257–60.
Boillot, A. & L'Hardy-Halos, M.-Th. 1975. Observations en culture d'une Rhodophycée Bonnemaisonale: le *Naccaria wiggii* (Turner) Endlicher. *Bull. Soc. Phycol. France* 20: 30–6.
Bold H. C. & Wynne, M. J. 1985. *Introduction to the Algae,* 2nd ed. Englewood Cliffs, N.J.: Prentice-Hall.
Bonin, D. R. 1981. Systematics and life histories of New Zealand Bonnemaisoniaceae (Rhodophyta). M.Sc. thesis, University of Auckland.
Bonin, D. R. & Hawkes, M. W. 1987. Systematics and life histories of New Zealand Bonnemaisoniaceae (Bonnemaisoniales, Rhodophyta): I. The genus *Asparagopsis*. *N. Z. J. Bot.* 25: 577–90.
Bonin, D. R. & Hawkes, M. W. 1988a. Systematics and life histories of New Zealand Bonnemaisoniaceae (Bonnemaisoniales, Rhodophyta): II. The genus *Delisea*. *N. Z. J. Bot.* 26: 1–14.
Bonin, D. R. & Hawkes, M. W. 1988b. Systematics and life histories of New Zealand Bonnemaisoniaceae (Bonnemaisoniales, Rhodophyta): III. The genus *Ptilonia*. *N. Z. J. Bot.* 26: 15–25.
Breeman, A. M. 1979. The life history and its environmental regulation in the subtidal red alga *Acrosymphyton purpuriferum* (J. Ag.) Sjost. Ph.D. thesis, Rijksuniversiteit te Groningen. Groningen: Stichting Drukkerij C. Regenboog.
Breeman, A. M. & Hoeksema, B. W. 1987. Vegetative propagation of the red alga *Rhodochorton purpureum* by means of fragments that escape digestion by herbivores. *Mar. Ecol. Prog. Ser.* 35: 197–201.
Bremermann, H. J. 1985. The adaptive significance of sexuality. *Experimentia* 41: 1245–54.
Briggs, D. & Walters, S. M. 1984. *Plant Variation and Evolution*, 2nd ed. Cambridge: Cambridge University Press.
Buth, D. G. 1984. The application of electrophoretic data in systematic studies. *Annu. Rev. Ecol. Syst.* 15: 501–22.
Carson, H. L. & Templeton, A. R. 1984. Genetic revolutions in relation to speciation phenomena: the founding of new populations. *Annu. Rev. Ecol. Syst.* 15: 97–131.
Caswell, H. 1982. Life history theory and the equilibrium status of populations. *Am. Nat.* 120: 317–39.
Caswell, H. 1985. The evolutionary demography of clonal reproduction. In *Population Biology and Evolution of Clonal Organisms*, eds. J. B. C. Jackson, L. W. Buss & R. E. Cook, pp. 187–224. New Haven & London: Yale University Press.
Chapman, A. R. O. 1986. Population and community ecology of seaweeds. *Adv. Mar. Biol.* 23: 1–161.
Charlesworth, D. 1985. Distribution of dioecy and self-incompatibility in angiosperms. In *Evolution. Essays in Honour of John Maynard Smith*, eds. P. J. Greenwood, P. H. Harvey & M. Slatkin, pp. 237–68. Cambridge: Cambridge University Press.
Charnov, E. L. 1982. *The Theory of Sex Allocation.* Princeton: Princeton University Press.
Chen, L. C.-M. 1977. The sporophyte of *Ahnfeltia plicata* (Huds.) Fries (Rhodophyceae, Gigartinales) in culture. *Phycologia* 16: 163–8.
Cheney, D. P. 1985. Electrophoresis. In *Handbook of Phycological Methods: Ecological Methods for Macroalgae*, eds. M. Littler & D. Littler, pp. 87–119. Cambridge: Cambridge University Press.
Cheney, D. P. & Babbel, G. R. 1978. Biosystematic studies of the red algal genus *Eucheuma*. I. Electrophoretic variation among Florida populations. *Mar. Biol.* 47: 251–64.
Chihara, M. 1962. Life cycle of the Bonnemaisoniaceous algae in Japan (2). *Sc. Rep. Tokyo Kyoiku Daigaku Sect. B,* 11: 27–53.
Chihara, M. 1972. Germination of carpospores of *Pikea californica* and *Schimmelmania plumosa* as found in Japan, with special reference to their life history. *Mém. Soc. Bot. France* 1972: 313–22.
Cole, K. & Conway, E. 1980. Studies in the Bangiaceae: reproductive modes. *Bot. Mar.* 23: 545–53.
Cole, K. M., Hymes, B. J. & Sheath, R. G. 1983. Karyotypes and reproductive seasonality of the genus *Bangia* (Rhodophyta) in British Columbia, Canada. *J. Phycol.* 19: 136–45.
Conway, E. & Cole, K. 1973. Observations on an unusual form of reproduction in *Porphyra* (Rhodophyceae, Bangiales). *Phycologia* 12: 213–25.
Conway, E., Mumford, T. F. & Scagel, R. F. 1976 [dated 1975]. The genus *Porphyra* in British Columbia and Washington. *Syesis* 8: 185–244.
Crawford, D. J. 1983. Phylogenetic and systematic inferences from electrophoretic studies. In *Isozymes in Plant Genetics and Breeding*, eds. S. O. Tanksley & T. J. Orton, pp. 257–87. Amsterdam: Elsevier.
Crawford, D. J. 1985. Electrophoretic data and plant speciation. *Syst. Bot.* 10: 405–16.
Davis, A. N. & Wilce, R. T. 1987. Algal diversity in relation to physical disturbance: a mosaic of successional stages

in a subtidal cobble habitat. *Mar. Ecol. Prog. Ser.* 37: 229-37.

DeCew, T. C. & West, J. A. 1977. Culture studies on the marine red algae *Hildenbrandia occidentalis* and *H. prototypus* (Cryptonemiales, Hildenbrandiaceae). *Bull. Jap. Soc. Phycol.* 25 (Suppl.): 31-41.

DeCew, T. C. & West, J. A. 1981. Life histories in the Phyllophoraceae (Rhodophyta: Gigartinales) from the Pacific coast of North America. I. *Gymnogongrus linearis* and *G. leptophyllus. J. Phycol.* 17: 240-50.

DeCew, T. C. & West, J. A. 1982 [dated 1981]. Investigations on the life histories of three *Farlowia* species (Rhodophyta: Cryptonemiales, Dumontiaceae) from Pacific North America. *Phycologia* 20: 342-51.

DeCew, T. C., West, J. A. & Ganesan, E. K. 1982 [dated 1981]. The life histories and developmental morphology of two species of *Gloiosiphonia* (Rhodophyta: Cryptonemiales, Gloiosiphoniaceae) from the Pacific coast of North America. *Phycologia* 20: 415-23.

De Wreede, R. E. & Klinger, T. 1988. Reproductive strategies in algae. In *Reproductive Strategies of Plants*, eds. J. Lovett Doust & L. Lovett Doust, pp. 267-84. Oxford: Oxford University Press.

Dixon, P. S. 1963. Variation and speciation in marine Rhodophyta. In *Speciation in the Sea*, eds. J. P. Harding & N. Tebble, pp. 51-62. London & New York: Academic.

Dixon, P. S. 1965. Perennation, vegetative propagation and algal life histories, with special reference to *Asparagopsis* and other Rhodophyta. *Bot. Gothoburg.* 3: 67-74.

Dixon, P. S. 1973. *Biology of the Rhodophyta*. Edinburgh: Oliver & Boyd.

Dixon, P. S. & Irvine, L. M. 1977. *Seaweeds of the British Isles. Vol. 1 Rhodophyta, Part 1 Introduction, Nemaliales, Gigartinales*. London: British Museum (Natural History).

Drew, K. M. 1939. An investigation of *Plumaria elegans* (Bonnem.) Schmitz with special reference to triploid plants bearing parasporangia. *Ann. Bot.* N.S. 7:347-67.

Drew, K. M. 1949. *Conchocelis*-phase in the life history of *Porphyra umbilicalis* (L.) Kuetz. *Nature* (Lond.) 164: 748-9.

Drew, K. M. 1954. Studies in the Bangioideae. III. The life history of *Porphyra umbilicalis* (L.) Kütz. var. *laciniata* (Lightf.) J. Ag. *Ann. Bot.* N.S. 18: 183-211.

Dyck, L., De Wreede, R. E. & Garbary, D. 1985. Life history phases in *Iridaea cordata* (Gigartinaceae): relative abundance and distribution from British Columbia to California. *Jap. J. Phycol.* 33: 225-32.

Edelstein, T. 1970. The life history of *Gloiosiphonia capillaris* (Hudson) Carmichael. *Phycologia* 9: 55-9.

Edelstein, T. 1972. On the taxonomic status of *Gloiosiphonia californica* (Farlow) J. Agardh (Cryptonemiales, Gloiosiphoniaceae). *Syesis* 5: 227-34.

Edelstein, T. & McLachlan, J. 1971. Further observations on *Gloiosiphonia capillaris* (Hudson) Carmichael in culture. *Phycologia* 10: 215-19.

Edyvean, R. G. J. & Ford, H. 1984. Population biology of the crustose red alga *Lithophyllum incrustans* Phil. 2. A comparison of populations from three areas of Britain. *Biol. J. Linn. Soc.* 23: 353-63.

Edyvean, R. G. J. & Ford, H. 1986. Spore production by *Lithophyllum incrustans* (Corallinales, Rhodophyta) in the British Isles. *Br. Phycol. J.* 21: 255-61.

Ellstrand, N. C. & Antonovics, J. 1985. Experimental studies of the evolutionary significance of sexual reproduction. II. A test of the density-dependent selection hypothesis. *Evolution* 39: 657-66.

Ellstrand, N. C. & Roose, M. L. 1987. Patterns of genotypic diversity in clonal plant species. *Am. J. Bot.* 74: 123-31.

Ewing, E. P. 1977. Selection at the haploid and diploid phases: Cyclical variation. *Genetics* 87: 195-208.

Farnham, W. F. 1980. Studies on aliens in the marine flora of southern England. In *The Shore Environment, Vol. 2: Ecosystems*, eds. J. H. Price, D. E. G. Irvine & W. F. Farnham, pp. 875-914. New York: Academic.

Farnham, W. F. & Fletcher, R. L. 1976. The occurrence of *Porphyrodiscus simulans* Batt. in the life-history of *Ahnfeltia plicata* (Huds.) Fries. *Br. Phycol. J.* 11: 183-90.

Feldmann, G. 1966. Sur le cycle haplobiontique du *Bonnemaisonia asparagoides* (Woods.) Ag. *C. R. Acad. Sci.* (Paris), sér. D, 262: 1695-8.

Feldmann, J. 1952. Les cycles de reproduction des algues et leurs rapports avec la phylogénie. *Rev. Cytol. Biol. Vég.* 13: 1-49.

Feldmann, J. 1972. Les problemes actuels de l'alternance de générations chez les algues. *Soc. Bot. Fr. Mém.*, 1972: 7-38.

Feldmann, J. & Feldmann, G. 1939. Sur l'alternance de générations chez les Bonnemaisoniacées. *C. R. Acad. Sci.* (Paris), sér D, 208: 1425-7.

Feldmann, J. & Feldmann, G. 1942. Recherches sur les Bonnemaisoniacées et leur alternance de générations. *Ann. Sc. Nat. Bot. Biol. Vég.*, sér. 2, 3: 75-175.

Feldmann, J. & L'Hardy-Halos, M.-Th. 1977. La multiplication végétative chez les Algues: ses principaux aspects morphologiques. *Soc. Bot. Fr., Coll. Multipl. Végét.* 1977: 13-41.

Feldmann, J. & Mazoyer, G. 1937. Sur l'identité de l'*Hymenoclonium serpens* (Crouan) Batters et du protonéma du *Bonnemaisonia asparagoides* (Woodw.) C. Ag. *C. R. Acad. Sci.* (Paris), sér. D, 205: 1084-5.

Fletcher, R. L. 1975. The life history of *Rhodophysema georgii* in laboratory culture. *Mar. Biol.* 31: 299-304.

Fletcher, R. L. 1977. Studies on the life history of *Rhodophysema elegans* in laboratory culture. *Mar. Biol.* 40: 291-7.

Ford, H., Hardy, F. G. & Edyvean, R. G. J. 1983. Population biology of the crustose red alga *Lithophyllum incrustans* Phil. Three populations on the east coast of Britain. *Biol. J. Linn. Soc.* 23: 353-63.

Fox, J. F. 1985. Incidence of dioecy in relation to growth form, pollination and dispersal. *Oecologia* (Berl.) 67: 244-9.

Freshwater, D. W. & Kapraun, D. F. 1986. Field, culture and cytological studies of *Porphyra carolinensis* Coll et Cox (Bangiales, Rhodophyta) from North Carolina. *Jap. J. Phycol.* 34: 251-62.

Fujio, Y., Kodaka, P. L. G. & Hara, M. 1985. Genetic differentiation and amount of genetic variability in natural populations of the haploid laver *Porphyra yezoensis. Jap. J. Genet.* 60: 347-54.

Fujio, Y., Tanaka, M. Y., Hara, M. & Akiyama, K. 1987. Enzyme polymorphism and population structure of the haploid laver *Porphyra yezoensis. Nippon Suisan Gakkaishi* 53: 357-62.

Gabrielson, P. W. & Garbary, D. G. 1986. Systematics of red algae (Rhodophyta). *CRC Crit. Rev. Plant Sci.* 3: 325-66.

Garbary, D. J., Hansen, G. I. & Scagel, R. F. 1983 [dated 1982]. The marine algae of British Columbia and northern Washington: division Rhodophyta (red algae), class Florideophyceae, orders Acrochaetiales and Nemaliales. *Syesis* 15 (Suppl. 1): 1–102.

Gargiulo, G. M., De Masi, F. & Tripodi, G. 1987. Thallus morphology and spore formation in cultured *Erythrocladia irregularis* Rosenvinge (Rhodophyta, Bangiophycidae). *J. Phycol.* 23: 351–9.

Gayral, P. & Mayhoub, H. 1981. Influence of some physical factors on the development of *Calosiphonia vermicularis* (J. Ag.) Schmitz. *Proc. Int. Seaweed Symp.* 8: 98–105.

Gerritsen, J. 1980. Sex and parthenogenesis in sparse populations. *Am. Nat.* 115: 718–42.

Goff, L. J. & Coleman, A. W. 1984. Transfer of nuclei from a parasite to its host. *Proc. Natl. Acad. Sci. USA* 81: 5420–4.

Goff, L. J. & Coleman, A. W. 1986. A novel pattern of apical cell polyploidy, sequential polyploidy reduction and intercellular nuclear transfer in the red alga *Polysiphonia*. *Am. J. Bot.* 73: 1109–30.

Gottlieb, L. D. 1981. Electrophoretic evidence and plant populations. *Prog. Phytochem.* 7: 1–46.

Gottlieb, L. D. 1984. Isozyme evidence and problem solving in plant systematics. In *Plant Biosystematics*, ed. W. F. Grant, pp. 343–57. London: Academic.

Grant, V. 1981. *Plant Speciation*, 2nd ed. New York: Columbia University Press.

Guiry, M. D. 1984. Structure, life history and hybridization of Atlantic *Gigartina teedii* (Rhodophyta) in culture. *Br. Phycol. J.* 19: 37–55.

Guiry, M. D. 1987. The evolution of life history types in the Rhodophyta: an appraisal. *Cryptogam. Algol.* 8: 1–12.

Guiry, M. & Coleman, M. 1982. Observations on the phenology and life history of a monoecious strain of *Gigartina stellata* (Stackh.) Batters (Rhodophyta) in Galway Bay. *Br. Phycol. J.* 17: 232 (Abstract).

Guiry, M. D. & Freamhainn, M. T. 1986. Biosystematics of *Gracilaria foliifera*(Gigartinales, Rhodophyta). *Nord. J. Bot.* 5: 629–37.

Guiry, M. D. & West, J. A. 1984 [dated 1983]. Life history and hybridization studies on *Gigartina stellata* and *Petrocelis cruenta* (Rhodophyta) in the North Atlantic. *J. Phycol.* 19: 474–94.

Guiry, M. D., Tripodi, G. & Luning, K. 1987. Biosystematics, genetics and upper temperature tolerance of *Gigartina teedii* (Rhodophyta) from the Atlantic and Mediterranean. *Helgol. Meeresunters.* 41: 283–95.

Hamrick, J. L., Linhart, Y. B. & Mitton, J. B. 1979. Relationships between life history characteristics and electrophoretically detectable genetic variation in plants. *Annu. Rev. Syst. Ecol.* 10: 173–200.

Hannach, G. & Santilices, B. 1985. Ecological differences between the isomorphic reproductive phases of two species of *Iridaea* (Rhodophyta: Gigartinales). *Mar. Ecol. Prog. Ser.* 22: 291–303.

Harper, J. L. 1977. *Population Biology of Plants*. London: Academic.

Hastings, A. & Caswell, H. 1979. Environmental variability in the evolution of life history strategies. *Proc. Natl. Acad. Sci. USA* 76: 4700–3.

Hawkes, M. W. 1977. A field, culture and cytological study of *Porphyra gardneri* (Smith & Hollenberg) comb. nov., (= *Porphyrella gardneri* Smith & Hollenberg), (Bangiales, Rhodophyta). *Phycologia* 16: 457–69.

Hawkes, M. W. 1978. Sexual reproduction in *Porphyra gardneri* (Smith et Hollenberg) Hawkes (Bangiales, Rhodophyta). *Phycologia* 17: 329–53.

Hawkes, M. W. 1980. Ultrastructure characteristics of monospore formation in *Porphyra gardneri* (Rhodophyta). *J. Phycol.* 16: 192–6.

Hawkes, M. W. 1983a. *Hummbrella hydra* Earle (Rhodophyta, Gigartinales): seasonality, distribution, and development in laboratory culture. *Phycologia* 22: 403–13.

Hawkes, M. W. 1983b. Anatomy of *Apophlaea sinclairii* – An enigmatic red alga endemic to New Zealand. *Jap. J. Phycol.* 31: 55–64.

Hawkes, M. W. 1986a. Life histories of *Acrosymphyton firmum* and *Schmitzia evanescens* (Rhodophyta, Gigartinales): Carpospore germination and development. *N. Z. J. Bot.* 24: 343–50.

Hawkes, M. W. 1986b. Apomixis in the Rhodymeniales flora of British Columbia. *2nd N. W. Algal Symposium, Bamfield Marine Station*, Abstr., p. 6.

Hawkes, M. W. 1988. Evidence of sexual reproduction in *Smithora naiadum* (Erythropeltidales, Rhodophyta) and its evolutionary significance. *Br. Phycol. J.* 23: 327–36.

Hawkes, M. W. & Scagel, R. F. 1986a. The marine algae of British Columbia and northern Washington: division Rhodophyta (red algae), class Rhodophyceae, order Palmariales. *Can. J. Bot.* 64: 1148–73.

Hawkes, M. W. & Scagel, R. F. 1986b. The marine algae of British Columbia and northern Washington: division Rhodophyta (red algae), class Rhodophyceae, order Rhodymeniales. *Can. J. Bot.* 64: 1549–80.

Hay, M. E. 1984. Predictable spatial escapes from herbivory: how do these affect the evolution of herbivore resistance in tropical marine communities? *Oecologia* (Berl.) 64: 396–407.

Helenurm, K. & Ganders, F. R. 1985. Adaptive radiation and genetic differentiation in Hawaiian *Bidens*. *Evolution* 39: 753–65.

Huisman, J. M. & Kraft, G. T. 1982. *Deucalion* gen. nov. and *Anisoschizus* gen. nov. (Ceramiaceae, Ceramiales), two new propagule-forming red algae from southern Australia. *J. Phycol.* 18: 177–192.

Innes, D. J. 1984. Genetic differentiation among populations of marine algae. *Helgol. Meeresunters.* 38: 401–17.

Innes, D. J. 1988. Genetic differentiation in the intertidal zone in populations of the alga *Enteromorpha linza* (Ulvales: Chlorophyta). *Mar. Biol.* 97: 9–16.

Innes, D. J. & Yarish, C. 1984. Genetic evidence for the occurrence of asexual reproduction in populations of *Enteromorpha linza* (L.) J. Ag. (Chlorophyta, Ulvales) from Long Island Sound. *Phycologia* 23: 311–20.

Jones, W. E. & Smith, R. M. 1970. The occurrence of tetraspores in the life history of *Naccaria wiggii* (Turn.) Endl. *Br. Phycol. J.* 5: 91–5.

Kapraun, D. F. 1977. Asexual propagules in the life history of *Polysiphonia ferulacea* (Rhodophyta, Ceramiales). *Phycologia* 16: 417–26.

Kapraun, D. F. & Luster, D. G. 1980. Field and culture studies of *Porphyra rosengurtii* Coll et Cox (Rhodophyta, Bangiales) from North Carolina. *Bot. Mar.* 23: 449–57.

Kim, D. H. 1976. A study of the development of cystocarps and tetrasporangial sori in Gigartinaceae (Rhodophyta, Gigartinales). *Nova Hedwigia* 27: 1–146.

Koch, C. 1986. Attempted hybridization between *Polysiphonia fibrillosa* and *P. violacea* (Bangiophyceae) [sic] from Denmark; with culture studies primarily on *P. fibrillosa*. *Nord. J. Bot.* 6: 123–8.

Kornmann, P. 1984. *Erythrotrichopeltis*, eine neue gattung

der Erythropeltidaceae (Bangiophyceae, Rhodophyta). *Helgol. Meeresunters.* 38: 207–24.

Kornmann, P. 1987. Der lebenszyklus von *Porphyrostromium obscurum* (Bangiophyceae, Rhodophyta). *Helgol. Meeresunters.* 41: 127–37.

Kornmann, P. & Sahling, P.-H. 1985. Erythropeltidaceen (Bangiophyceae, Rhodophyta) von Helgoland. *Helgol. Meeresunters.* 39: 213–36.

Kraft, G. T. 1981. Rhodophyta: morphology and classification. In *The Biology of Seaweeds*, eds. C. S. Lobban & M. J. Wynne, pp. 6–51. Oxford: Blackwell Scientific.

Kurogi, M. 1953. Study of the life-history of *Porphyra*. I. The germination and development of carpospores. *Bull. Tohoku Reg. Fish. Res. Lab.* 3: 67–103 (in Japanese, English summary).

Kurogi, M. 1972. Systematics of *Porphyra* in Japan. In *Contributions to the Systematics of Benthic Marine Algae of the North Pacific*, eds. I. A. Abbott & M. Kurogi, pp. 167–91. Kobe: Jap. Soc. Phycol.

Lewis, D. 1979. *Sexual Incompatibility in Plants*. London: Arnold.

Lewis, S. M. 1985. Herbivory on coral reefs: algal susceptibility to herbivorous fishes. *Oecologia* (Berl.) 65: 370–5.

Lewis, W. M. Jr. 1985. Nutrient scarcity as an evolutionary cause of haploidy. *Am. Nat.* 125: 692–701.

L'Hardy-Halos, M,-Th. 1986. Observations on two species of *Antithamnionella* from the coast of Brittany. *Bot. Mar.* 29: 37–42.

Lieberman, M., John, D. M. & Lieberman, D. 1984. Factors influencing algal species assemblages on reef and cobble substrata off Ghana. *J. Exp. Mar. Biol. Ecol.* 75: 129–43.

Lindstrom, S. C. & Scagel, R. F. 1987. The marine algae of British Columbia, northern Washington, and southeast Alaska: division Rhodophyta (red algae), class Rhodophyceae, order Gigartinales, family Dumoniaceae, with an introduction to the order Gigartinales. *Can. J. Bot.* 65: 2202–32.

Littler, M. M. & Littler, D. S. 1980. The evolution of thallus form and survival strategies in benthic marine macroalgae: Field and laboratory tests of a functional form model. *Am. Nat.* 116: 25–44.

Littler, M. M. & Littler, D. S. 1984. Relationships between macroalgal functional form groups and substrata stability in a subtropical rocky-intertidal system. *J. Exp. Mar. Biol. Ecol.* 74: 13–34.

Littler, M. M., Littler, D. S. & Taylor, P. R. 1983. Evolutionary strategies in a tropical barrier reef system: Functional-form groups of marine macroalgae. *J. Phycol.* 19: 229–37.

Loveless, M. D. & Hamrick, J. L. 1984. Ecological determinants of genetic structure in plant populations. *Annu. Rev. Ecol. Syst.* 15: 65–95.

Løvlie, A. & Bryhni, E. 1978. On the relation between sexual and parthenogenetic reproduction in haplodiplontic algae. *Bot. Mar.* 21: 155–63.

Lubchenco, J. & Cubit, J. 1980. Heteromorphic life histories of certain marine algae as adaptations to variation in herbivory. *Ecology* 61: 676–87.

Maggs, C. A. 1985. Environmental regulation of tetrasporophyte recycling in some heteromorphic red algae. *2nd Int. Phycol. Congr.*, Copenhagen: 100 (abstr.).

Maggs, C. A. 1988. Intraspecific life history variability in the Florideophycidae. *Bot. Mar.* 31: 465–90.

Magne, F. 1972. Le cycle de développement des Rhodophycées et son évolution. *Soc. Bot. Fr. Mém.* 1972: 247–68.

Magne, F. 1982. On two new types of life history in the Rhodophyta. *Cryptogam. Algol.* 3: 265–71.

Magne, F. 1986a. Anomalies du développement chez *Antithamnionella sarniensis* (Rhodophyceae, Ceramiaceae). I: formation et début du développement des tétraspores. *Cryptogam. Algol.* 7: 135–47.

Magne, F. 1986b. Anomalies du développement chez *Antithamnionella sarniensis* (Rhodophyceae, Ceramiaceae). II: nature des individus issus des tétraspores. *Cryptogam. Algol.* 7: 215–29.

Magne, F. 1987a. Is the frequency of apomeiosis in the Rhodophyta a genetic character? *Hydrobiologia* 151/152: 221–32.

Magne, F. 1987b. La tétrasporogenese et le cycle de développement des Palmariales (Rhodophyta): Une nouvelle interpretation. *Cryptogam. Algol.* 8: 273–80.

Margulis, L. & Sagan, D. 1986. *Origins of Sex. Three Billion Years of Genetic Recombination*. New Haven: Yale University Press.

Masuda, M. & Ohta, M. 1975. The life history of *Rhodophysema georgii* Batters (Rhodophyta, Cryptonemiales). *J. Jap. Bot.* 50: 1–10.

Masuda, M. & Uchida, T. 1976. On the life history of *Gigartina ochotensis* (Rupr.) Rupr. from Muroran, Hokkaido. *Bull. Jap. Soc. Phycol.* 24: 41–7.

Masuda, M. & Umezaki, I. 1977. On the life history of *Nemalion vermiculare* Suringer (Rhodophyta) in culture. *Bull. Jap. Soc. Phycol.* 25 (Suppl.): 129–36.

Masuda, M., West, J. A. & Kurogi, M. 1987. Life history studies in culture of a *Mastocarpus* species (Rhodophyta) from central Japan. *J. Fac. Sci., Hokkaido Univ.*, ser. 5, 14: 11–38.

Masuda, M., West, J. A., Ohno, Y. & Kurogi, M. 1984. Comparative reproductive patterns in culture of different *Gigartina* subgenus *Mastocarpus* and *Petrocelis* populations from northern Japan. *Bot. Mag. Tokyo* 97: 107–25.

May, G. 1986. Life history variations in a predominantly gametophytic population of *Iridaea cordata* (Gigartinaceae, Rhodophyta). *J. Phycol.* 22: 448–55.

Mayhoub, H. 1973. Cycle du développement de *Calosiphonia vermicularis* (J. Agardh) Schmitz (Rhodophycée, Gigartinale). *C. R. Acad. Sci.* (Paris), sér. D, 277: 1137–40.

Mayhoub, H. 1975. Nouvelles observations sur le cycle de développement du *Calosiphonia vermicularis* (J. Ag.) Sch. (Rhodophycée, Gigartinale). *C. R. Acad. Sci.* (Paris), sér. D, 280: 2441–3.

Mayhoub, H. 1976. Cycle de développement du *Calosiphonia vermicularis* (J. Ag.) Sch. (Rhodophycées, Gigartinales). Mise en evidence d'une réponse photopériodique. *Bull. Soc. Phycol. France* 21: 48.

Mayhoub, H., Gayral, P. & Jacques, R. 1976. Action de la composition spectrale de la lumière sur la croissance et la reproduction de *Calosiphonia vermicularis* (J. Agardh) Schmitz (Rhodophycées, Gigartinales). *C. R. Acad. Sci.* (Paris), sér. D, 283: 1041–4.

Maynard Smith, J. 1978. *The Evolution of Sex*. Cambridge: Cambridge University Press.

McBride, D. L. & Cole, K. 1971. Electron microscopic observations on the differentiation and release of monospores in the marine red alga *Smithora naiadum*. *Phycologia* 10: 49–61.

Michaels, H. J. & Bazzaz, F. A. 1986. Resource allocation and demography of sexual and apomictic *Antennaria parlinii*. *Ecology* 67: 27–36.

Michod, R. E. & Levin, B. R. (eds.). 1987. *The Evolution of Sex: An Examination of Current Ideas.* Sunderland, Ma.: Sinauer.

Miura, W., Fujio, Y. & Suto, S. 1978. Isozymes from individual thallus of *Porphyra* species. *Jap. J. Phycol.* 26: 139–43.

Miura, W., Fujio, Y. & Suto, S. 1979. Genetic differentiation between the wild and cultured populations of *Porphyra yezoensis*. *Tohoku J. Agr. Res.* 30: 114–25.

Mladenov, P. V. & Emson, R. H. 1988. Density, size structure and reproductive characteristics of fissiparous brittle stars in algae and sponges: evidence for interpopulational variation in levels of sexual and asexual reproduction. *Mar. Ecol. Prog. Ser.* 42: 181–194.

Mumford, T. F. & Cole, K. 1977. Chromosome numbers for fifteen species in the genus *Porphyra* (Bangiales, Rhodophyta) from the West coast of North America. *Phycologia* 16: 373–7.

Murray, S. N., Dixon, P. S. & Scott, J. L. 1972. The life history of *Porphyropsis coccinea* var. *dawsonii* in culture. *Br. Phycol. J.* 7: 323–33.

Niklas, K. J. 1985. The aerodynamics of wind pollination. *Bot. Rev.* 51: 328–86.

Ohme, M., Kunifuji, Y. & Miura, A. 1986. Cross experiments of the color mutants in *Porphyra yezoensis* Ueda. *Jap. J. Phycol.* 34: 101–6.

Padilla, D. K. 1985. Structural resistance of algae to herbivores. A biomechanical approach. *Mar. Biol.* 90: 103–9.

Polanshek, A. R. & West, J. A. 1975. Culture and hybridization studies on *Petrocelis* (Rhodophyta) from Alaska and California. *J. Phycol.* 11: 434–9.

Polanshek, A. R. & West, J. A. 1977. Culture and hybridization studies on *Gigartina papillata* (Rhodophyta). *J. Phycol.* 13: 141–9.

Ramus, J. 1969. The development sequence of the marine red alga *Pseudogloiophloea* in culture. *Univ. Calif. Pub. Bot.* 52: 1–42.

Richards, A. J. 1986. *Plant Breeding Systems.* London: Allen and Unwin.

Richardson, N. & Dixon, P. 1970. Culture studies on *Thuretellopsis peggiana* Kylin. *J. Phycol.* 6: 154–9.

Ritland, K. & Ganders, F. R. 1985. Variation in the mating system of *Bidens menziesii* (Asteraceae) in relation to population substructure. *Heredity* 55: 235–44.

Rueness, J. 1968. Paraspores from *Plumaria elegans* (Hornem.) Schmitz in culture. *Norw. J. Bot.* 15: 220–4.

Rueness, J. 1973a. Speciation in *Polysiphonia* (Rhodophyceae, Ceramiales) in view of hybridization experiments: *P. hemisphaerica* and *P. boldii*. *Phycologia* 12: 107–9.

Rueness, J. 1973b. Culture and field observations on growth and reproduction of *Ceramium strictum* Harv. from the Oslofjord, Norway. *Norw. J. Bot.* 20: 61–5.

Rueness, J. 1978. Hybridization in red algae. In *Modern Approaches to the Taxonomy of Red and Brown Algae*, eds. D. E. G. Irvine & J. H. Price, pp. 247–62. London: Academic.

Rueness, J. & Åsen, P. A. 1982. Field and culture observations on the life history of *Bonnemaisonia asparagoides* (Woodw.) C. Ag. (Rhodophyta) from Norway. *Bot. Mar.* 25: 577–87.

Rueness, J. & Rueness, M. 1975. Genetic control of morphogenesis in two varieties of *Antithamnion plumula* (Rhodophyceae, Ceramiales). *Phycologia* 14: 81–5.

Rueness, J. & Rueness, M. 1980. Culture and field observations on *Callithamnion bipinnatum* and *C. byssoides* (Rhodophyta, Ceramiales) from Norway. *Sarsia* 65: 29–34.

Rueness, J. & Rueness, M. 1985. Regular and irregular sequences in the life history of *Callithamnion tetragonum* (Rhodophyta, Ceramiales). *Br. Phycol. J.* 20: 329–33.

Schaal, B. A. 1984. Population biology and biosystematics: current experimental approaches. In *Plant Biosystematics*, ed. W. F. Grant, pp. 439–52. New York: Academic.

Schmitt, J. & Antonovics, J. 1986. Experimental studies of the evolutionary significance of sexual reproduction. IV. Effect of neighbor relatedness and aphid infestation on seedling performance. *Evolution* 40: 830–6.

Scott, J. L. & Dixon, P. S. 1971. The life history of *Pikea californica* Harv. *J. Phycol.* 7: 295–300.

Searles, R. B. 1980. The strategy of the red algal life history. *Am. Nat.* 115: 113–20.

Sheath, R. G. & Cole, K. M. 1980. Distribution and salinity adaptations of *Bangia atropurpurea* (Rhodophyta), a putative migrant into the Laurentian Great Lakes. *J. Phycol.* 16: 412–20.

Sheath, R. G. & Cole, K. M. 1984. Systematics of *Bangia* (Rhodophyta) in North America. I. Biogeographic trends in morphology. *Phycologia* 23: 383–96.

Sheath, R. G., Van Alstyne, K. L. & Cole, K. M. 1985. Distribution, seasonality and reproductive phenology of *Bangia atropurpurea* (Rhodophyta) in Rhode Island, U.S.A. *J. Phycol.* 21: 297–303.

Shields, W. M. 1982. *Philopatry, Inbreeding, and the Evolution of Sex.* Albany: State University of New York Press.

Soltis, D. E. & Rieseberg, L. H. 1986. Autopolyploidy in *Tolmiea menziesii* (Saxifragaceae): Genetic insights from enzyme electrophoresis. *Am. J. Bot.* 73: 310–18.

South, G. R. & Whittick, A. 1976. Aspects of the life history of *Rhodophysema elegans* (Rhodophyta, Peyssonneliaceae). *Br. Phycol. J.* 11: 349–54.

South, G. R., Hooper, R. G. & Irvine, L. M. 1972. The life history of *Turnerella pennyi* (Harv.) Schmitz. *Br. Phycol. J.* 7: 221–33.

Stace, C. A. 1980. *Plant Taxonomy and Biosystematics.* London: Edward Arnold.

Stearns, S. C. 1977. The evolution of life history traits: A critique of the theory and a review of the data. *Annu. Rev. Ecol. Syst.* 8: 145–71.

Stearns, S. C. 1980. A new view of life-history evolution. *Oikos* 35: 266–81.

Stearns, S. C. 1985. The evolution of sex and the role of sex in evolution. *Experimentia* 41: 1231–5.

Stearns, S. C. (ed.). 1987. *The Evolution of Sex and Its Consequences.* Basel: Birkhauser.

Stebbins, G. L. & Hill, G. J. C. 1980. Did multicellular plants invade the land? *Am. Nat.* 115: 342–53.

Steneck, R. S. 1985. Adaptations of crustose coralline algae to herbivory: patterns in space and time. In *Paleoalgology*, eds. D. F. Toomey & M. H. Nitecki, pp. 352–66. Berlin: Springer-Verlag.

Stuessy, T. F. 1985. Review [of] *Plant Biosystematics*. Grant, W. F. (ed.). *Syst. Zool.* 34: 375–7.

Sundene, O. 1975. Experimental studies on form variation in *Antithamnion plumula* (Rhodophyceae). *Norw. J. Bot.* 22: 35–42.

Suneson, S. 1982. The culture of bisporangial plants of *Dermatolithon litorale* (Suneson) Hamel et Lemoine (Rhodophyta, Corallinaceae). *Br. Phycol. J.* 17: 107–16.

Svedelius, N. 1937. The apomeiotic tetrad division in *Lomentaria rosea* in comparison with the normal development in *Lomentaria clavellosa*. *Symb. Bot. Upsal.* 2: 1–53.

Tappan, H. 1980. The Rhodophyta. In *The Paleobiology of Plant Protists*, ed. H. Tappan, pp. 107–47. San Francisco: W. H. Freeman.

Templeton, A. R. 1982. The prophecies of parthenogenesis. In *Evolution and Genetics of Life Histories*, eds. H. Dingle & J. P. Hegmann, pp. 75–101. Berlin: Springer-Verlag.

Tooby, J. 1982. Pathogens, polymorphism, and the evolution of sex. *J. Theor. Biol.* 97: 557–76.

Umezaki, I. 1968. A study on the germination of carpospores of *Dudresnaya japonica* Okamura (Rhodophyta). *Pub. Seto Mar. Biol. Lab.* 16: 263–72.

Umezaki, I. 1969. The germination of tetraspores of *Hildenbrandia prototypus* Nardo and its life history. *J. Jap. Bot.* 44: 17–29.

Urbanska, K. M. 1985. Some life history strategies and population structure in asexually reproducing plants. *Bot. Helv.* 95: 81–97.

Uyenoyama, M. C. 1985. On the evolution of parthenogenesis. II. Inbreeding and the cost of meiosis. *Evolution* 39: 1194–1206.

van den Hoek, C. & Cortel-Breeman, A. M. 1970. Life history studies on Rhodophyceae. II. *Halymenia floresia* (Clem.) Ag. *Acta Bot. Neerl.* 19: 341–62.

van der Meer, J. P. 1987. Using genetic markers in phycological research. *Hydrobiologia* 151/152: 49–56.

van der Meer, J. P. & Bird, C. J. 1985. *Palmaria mollis* (Setchell & Gardner) stat. nov.: A newly recognized species of *Palmaria* (Rhodophyceae) from the northeast Pacific Ocean. *Can. J. Bot.* 63: 398–403.

von Stosch, H. A. 1965. The sporophyte of *Liagora farinosa* Lamour. *Br. Phycol. Bull.* 2: 486–96.

von Stosch, H. A. & Theil, G. 1979. New mode of life history in the freshwater red algal genus *Batrachospermum*. *Am. J. Bot.* 66: 105–7.

West, J. A. 1968. Morphology and reproduction of the red alga *Acrochaetium pectinatum* in culture. *J. Phycol.* 4: 89–99.

West, J. A. 1970a. The life history of *Rhodochorton concrescens* in culture. *Br. Phycol. J.* 5: 179–86.

West, J. A. 1970b. The conspecificity of *Heterosiphonia asymmetrica* and *H. densiuscula* and their life histories in culture. *Madroño* 20: 313–19.

West, J. A. 1972a. Environmental control of hair and sporangial formation in the marine red alga *Acrochaetium proskaueri* sp. nov. *Proc. Int. Seaweed Symp.* 7: 377–84.

West, J. A. 1972b. The life history of *Petrocelis franciscana*. *Br. Phycol. J.* 7: 299–308.

West, J. A. & Guiry, M. D. 1982. A life history study of *Gigartina johnstonii* (Rhodophyta) from the Gulf of California. *Bot. Mar.* 25: 205–11.

West, J. A., Guiry, M. D. & Masuda, M. 1983. Further investigations of the genetic affinities and life history patterns of the red alga *Gigartina*. In *Proceedings of the Joint China–U.S. Phycology Symposium*, ed. C. K. Tseng, pp. 137–66. Beijing: Science Press.

West, J. A. & Hommersand, M. H. 1981. Rhodophyta: Life histories. In *The Biology of Seaweeds*, eds. C. S. Lobban & M. J. Wynne, pp. 133–93. Oxford: Blackwell Scientific.

West, J. A. & Norris, R. E. 1966. Unusual phenomena in the life histories of Florideae in culture. *J. Phycol.* 2: 54–7.

West, J. A., Polanshek, A. & Guiry, M. 1977. The life history in culture of *Petrocelis cruenta* J. Agardh (Rhodophyta) from Ireland. *Br. Phycol. J.* 12: 45–53.

West, J. A., Polanshek, A. R. & Shevlin, D. E. 1978. Field and culture studies on *Gigartina agardhii* (Rhodophyta). *J. Phycol.* 14: 416–26.

Whittick, A. 1984. The Newfoundland Ceramiaceae, why are there so many tetrasporophytes? *Br. Phycol. J.* 19: 201.

Whittick, A. & Hooper, R. G. 1977. The reproduction and phenology of *Antithamnion cruciatum* (Rhodophyta: Ceramiaceae) in insular Newfoundland. *Can. J. Bot.* 55: 520–4.

Whittick, A. & West, J. A. 1979. The life history of a monoecious species of *Callithamnion* (Rhodophyta, Ceramiaceae) in culture. *Phycologia* 18: 30–7.

Williams, G. C. 1975. *Sex and Evolution*. Princeton: Princeton University Press.

Willson, M. F. 1981. The evolution of complex life cycles in plants: A review and ecological perspective. *Ann. Missouri Bot. Gard.* 68: 275–300.

Woelkerling, W. J. 1983. The *Audouinella* (*Acrochaetium–Rhodochorton*) complex (Rhodophyta): present perspectives. *Phycologia* 22: 59–92.

Zupan, J. R. & West, J. A. 1988. Geographic variation in the life history of *Mastocarpus papillatus*. *J. Phycol.* 24: 223–9.

Chapter 18

Taxonomy and evolution

DAVID J. GARBARY
PAUL W. GABRIELSON

CONTENTS

- I. Introduction / 478
- II. Historical perspective / 478
- III. Approaches to red algal taxonomy and evolution / 482
  - A. Chemotaxonomy / 482
  - B. Vegetative morphogenesis / 482
  - C. Quantitative analysis / 484
  - D. Life histories / 484
  - E. Monographs / 485
  - F. Biogeography / 486
- IV. Recognition of classes/subclasses / 486
- V. Ordinal classification and relationships / 487
  - A. Porphyridiales / 487
  - B. Compsopogonales / 487
  - C. Rhodochaetales / 488
  - D. Bangiales / 489
  - E. Acrochaetiales / 489
  - F. Nemaliales / 489
  - G. Batrachospermales / 489
  - H. Gelidiales / 490
  - I. Hildenbrandiales / 490
  - J. Corallinales / 490
  - K. Palmariales / 491
  - L. Gigartinales / 491
  - M. Bonnemaisoniales / 492
  - N. Rhodymeniales / 492
  - O. Ceramiales / 492
- VI. Red algae in the eukaryotic domain / 493
- VII. Summary / 493
- VIII. Acknowledgments / 494
- IX. References / 494

I. INTRODUCTION

Our understanding of the phylogeny of red algae, and hence their classification, is presently in a period of major upheaval unprecedented since the end of the nineteenth century, when the first comprehensive classification scheme was established by Schmitz and Hauptfleisch (1897). Current revisions are potentially as far-reaching as those that occurred in the Chlorophyta in the 1970s and early 1980s (Irvine & John 1984). Many new taxa have been recognized at higher taxonomic ranks, and some traditional groups have been discarded or are undergoing major revision. New characters are being applied to classically defined groups, and the circumscriptions of these taxa are being modified.

Several recent reviews by Bold and Wynne (1985), Brawley and Wetherbee (1981), Gabrielson and Garbary (1986), Kraft (1981), and West and Hommersand (1981), as well as the various chapters in this volume, provide much of the background information on morphology, ultrastructure, development, and life histories. As a framework for systematic research, Gabrielson et al. (1985) and Gabrielson and Garbary (1987), using cladistic methodology, analyzed the relationships among all currently recognized red algal orders and presented the resulting phylogenies as cladograms. Herein, we continue to use the terminology of cladistics, since we consider this the most unambiguous way of expressing taxonomic and phylogenetic information (Wiley 1981). As noted by Stevens (1986), evolutionary classifications often are treated as if they were cladistic, even if this was not the intent of the original authors. Therefore, the classification of red algae (or any other group) should reflect the known (or presumed) phylogeny of the group.

In this chapter, we provide a brief history of the major themes in the development of red algal taxonomy. Methodologies and areas of research that we believe will resolve systematic problems from specific through ordinal ranks are discussed. We highlight new techniques that are being applied to the traditional fields of vegetative and reproductive ontogeny and stress their continuing contributions to red algal systematics. Infra- and interordinal relationships that have been proposed are considered in light of recent contributions and advances, and a classification of red algal orders is presented.

II. HISTORICAL PERSPECTIVE

Below we only briefly summarize red algal systematics before 1955, the publication date of G. F. Papenfuss's "Classification of the Algae," an excellent overview of the history of red algal classification. From a historical viewpoint, knowledge of red algae predates Linnaeus, but the concept that these organisms formed a taxonomic group was not recognized until early in the nineteenth century. Lamouroux (1813) first used color to separate certain algal genera from others with a similar morphology, and his "Floridees" was circumscribed in part as a group with a purple or red color. Harvey (1836) first recognized the three major taxa of macroalgae, i.e., Rhodospermae (red algae), Melanospermae (brown algae), and Chlorospermae (green algae), based on his observations that the spores are the same color as the parent thalli. Coralline red algae initially were classified with corals by Linnaeus, but with the observations of Schweigger (1819), Gray (1821), and Philippi (1837), both the articulated and crustose forms began to be recognized as rhodophytes. Harvey did not include any bangiophyte red algae in his Rhodospermae, and it was not until the careful studies of Berthold (1882) on the morphology and reproduction of *Bangia*, *Porphyra*, and *Erythrotrichia* that these genera were classified as red algae.

In the early part of the nineteenth century, the more important classifications were those of Lamouroux (1813), C. A. Agardh (1820, 1822, 1824), Harvey (1836), and Kützing (1843). C. A. Agardh's *Species Algarum* . . . (1820, 1822) and his *Systema Algarum* (1824) are noteworthy for their use of vegetative anatomy and macroscopic reproductive features, as well as the gross morphology of thalli, in characterizing and classifying algae. In the latter half of the century, the aforementioned classifications largely were supplanted by those of J. G. Agardh (1851–63, 1876), who used the vegetative anatomy of thalli, the mode of division and position of tetrasporangia, and especially the structure and position of cystocarps as the basis of his classifications.

Two important advances in the later nineteenth century enabled the establishment of what is considered the first classification system of florideophycidean red algae that reflected relationships among taxa. First were Bornet and Thuret's (1866, 1867) observations on several red algae, including *Ceramium*, *Dudresnaya*, and *Nemalion*, that antheridia (spermatia) fused with trichogynes, thereby effecting fertilization, and that cystocarp formation is a consequence of fertilization. This was followed by Schmitz's (1883) observation that in some red algae sporogenous filaments develop directly from the fertilized carpogonium, whereas in others, sporogenous filaments first fuse with other cells in the female gametophyte before cutting off carposporangia. These cells Schmitz called auxiliary cells, and various attributes of auxiliary cells formed the basis of his classification system.

Schmitz (1892) recognized four orders: (1) Nemaliales (as Nemalionales), characterized by gonimoblast developing directly from the zygote (auxiliary cell absent), (2) Gigartinales, characterized by paired carpogonial branches and auxiliary cells (procarpic) and inwardly developing gonimoblast, (3) Rhodymeniales, characterized by paired carpogonial branches and auxiliary mother cells (procarpic) and outwardly developing gonimoblast, and (4) Cryptonemiales, characterized by carpogonial branches and auxiliary cells scattered in the thallus (nonprocarpic) and either outwardly or inwardly developing gonimoblasts. A major refinement of this system, accepted by all subsequent workers, was the segregation of Ceramiales, in which the auxiliary cell is cut off only after fertilization of the carpogonium, from Rhodymeniales, where the auxiliary cell is present before fertilization (Oltmanns 1904–5).

Kylin (1923) recognized six florideophyte orders, adding Gelidiales, a segregate of Nemaliales (as Nemalionales) to those recognized by Oltmanns (1904–5). Kylin segregated Gelidiales (apparently) because, although auxiliary cells are present, they function only as nutritive cells and do not initiate gonimoblast formation. Kylin believed the carpogonium was the site of gonimoblast initiation in both Nemaliales and Gelidiales. Circumscription of the remaining orders followed Schmitz (1892). Like Oltmanns, Kylin considered Cryptonemiales to be evolutionarily the lowest among the higher Florideae, excluding Nemaliales and Gelidiales, and he regarded Ceramiales as having the most advanced reproductive development. Gigartinales, Rhodymeniales, and Ceramiales he considered independent groups evolved from cryptonemialean ancestors.

A new character, i.e., whether the auxiliary cell is borne in a special (accessory) filament or is a normal vegetative cell, was introduced by Kylin (1925) and used to distinguish his new order Nemastomales (nonprocarpic and nonaccessory auxiliary cells) from Cryptonemiales (nonprocarpic and accessory auxiliary cells). Sjöstedt (1926) emended the family Sphaerococcaceae to include only the genera *Sphaerococcus* and *Plocamium* and elevated the family to ordinal rank. Kylin (1928) maintained his 1925 classification and did not recognize Sjöstedt's Sphaerococcales, considering them as a family in Nemastomales.

In Kylin's (1932) summary of his final classification system, Nemaliales (as Nemalionales) and Gelidiales were characterized as lacking typical auxiliary (=generative) cells, i.e., where the auxiliary cell functions as the starting point of gonimoblast formation, whereas generative auxiliary cells were present in the remaining orders. Kylin also had abandoned the presence or absence of procarps as a distinguishing character at the ordinal rank, thereby negating the feature that previously had segregated Nemastomales and Cryptonemiales from Gigartinales. Kylin chose the older name Gigartinales as a replacement for Nemastomales. Thus, Cryptonemiales contained taxa with auxiliary cells borne in accessory filaments, whereas in Gigartinales normal intercalary vegetative cells served as auxiliary cells. Kylin also abandoned the direction of gonimoblast development as a feature distinguishing Rhodymeniales from Gigartinales and now segregated Rhodymeniales based on a daughter cell of the supporting cell serving as the auxiliary cell.

In Kylin's (1932) classification, Gelidiales were segregated from Nemaliales based on the diplohaplontic nature of the former in contrast to the haplontic nature of the latter. The notion of haplontic Nemaliales was based on the absence of a sporophyte generation isomorphic with gametophytes in field collections of some genera and the supposed occurrence of meiosis in fertilized carpogonia in the genera *Scinaia* (Svedelius 1915), *Nemalion* (Kylin 1916), and *Batrachospermum* (Kylin 1917). Feldmann and Feldmann (1942) demonstrated in culture that carpospores of *Asparagopsis armata* and *Bonnemaisonia asparagoides* gave rise to *Falkenbergia rufolanosa* and *Hymenoclonium serpens*, respectively, thereby inferring that a heteromorphic, diplohaplontic life history was present in the nemalialean family Bonnemaisoniaceae. Based on this life history feature, they elevated Bonnemaisoniaceae to ordinal rank and transferred to it the nemalialean family Naccariaceae due to similarities in gonimoblast development (especially the presence of nutritive cells) and to the fact that carpospores of *Naccaria* and *Atractophora* each germinate to produce a protonemal stage, which might potentially be the tetrasporophyte of the erect gametophyte, as in Bonnemaisoniaceae. Later, Magne reinvestigated all of the critical genera in which zygotic meiosis had been reported, including *Scinaia* (1960a), *Nemalion* (1961), *Bonnemaisonia* (1960b), and *Lemanea* (1967), by counting chromosomes in both gametangia and carposporangia. Zygotic meiosis in carpogonia was repudiated in every case. Magne's conclusions thus supported Feldmann and Feldmann's (1942) culture studies, and his own observations later were substantiated by additional culture studies (e.g., Fries 1967; Huisman 1987; Ramus 1969; Umezaki 1967).

Feldmann (1953) further dismantled Nemaliales, elevating the family Acrochaetiaceae to ordinal rank because of the extreme simplicity of its vegetative thallus and the absence of a carpogonial branch. Desikachary (1958) suggested that if Gelidiaceae, Acrochaetiaceae, and Bonnemaisoniaceae were recognized as segregate orders of Nemaliales, then the remaining taxa could be divided into two orders, Nemaliales and Chaetangiales (=Galaxaurales). Subsequently, Desikachary (1964) formally proposed Galaxaurales (as Chaetangiales), characterized by the fusion of the fertilized carpogonium with the hypogynous cell, transfer of the zygote nucleus or its products to that cell and initiation of gonimoblast from that cell. Recently, Huisman (1985) has shown that in the galaxauracean genera *Gloiophloea* and *Scinaia* the gonimoblast arises from the carpogonium, not the hypogynous cell.

Despite the recognized heterogeneity of the Nemaliales, no consensus was apparent among red algal systematists during the 1960s and 1970s regarding which and how many of the proposed segregate orders of Nemaliales should be recognized. Dixon (1961), Abbott (1962), and Papenfuss (1966) argued against recognizing Acrochaetiales. Gelidiales (Papenfuss 1966) and, among mainland European phycologists, Acrochaetiales gained some measure of acceptance, but only Chihara and Yoshizaki (1972) supported recognition of Bonnemaisoniales.

The essentially Kylinian classification (Table 18-1) began to be reconstructed during the late 1970s as new interpretations of existing characters, as well as new ultrastructural characters, began to be utilized in red algal systematics. The first major departure was, somewhat surprisingly, in Rhodymeniales, an order that had been regarded as very homogeneous. Guiry and Irvine (in Guiry 1978) elevated the family Palmariaceae to ordinal rank based on the presence of a stalk cell within each tetrasporangium and the absence of reports of female gametophytes and carposporophytes in field-collected species in the new order. Concomitantly, Garbary (1978) renewed the argument for recognition of Acrochaetiales based on its phylogenetic position. The cladistic analyses of Gabrielson et al. (1985) and Gabrielson and Garbary (1987) supported recognition of Acrochaetiales based on a suite of vegetative and reproductive characters.

An important contribution to red algal systematics was the report by Pueschel and Cole (1982) that there were differences in the morphology and number of cap layers that overlay red algal pit plugs and that these differences appeared to be useful in distinguishing among taxa at higher ranks. Their findings provided another line of evidence that

Table 18-1. A comparison of various classifications of red algae

Fritsch (1945)	Kylin (1956)	Dixon (1973)	Pueschel & Cole (1982)[a]	This paper[b]
Bangioideae	Bangioideae	Bangiophyceae	Bangiophyceae	Rhodophyceae
	Porphyridiales	Porphyridiales		Porphyridiales
	Goniotrichales			
	Compsopogonales	Comsopogonales		Compsopogonales
	Rhodochaetales	Rhodochaetales		Rhodochaetales
Bangiales	Bangiales	Bangiales		Bangiales
Florideae	Florideae	Florideophyceae	Florideophyceae	
Nemalionales	Nemalionales	Nemalionales	Palmariales	Acrochaetiales
			Corallinales	Palmariales
Gelidiales	Gelidiales		Gelidiales	Corallinales
			Hildenbrandiales	Gelidiales
			Nemaliales	Hildenbrandiales
			Batrachospermales	Nemaliales
			Bonnemaisoniales	Batrachospermales
Gigartinales	Gigartinales	Gigartinales	Gigartinales	Bonnemaisoniales
Cryptonemiales	Cryptonemiales	Cryptonemiales	Cryptonemiales	Gigartinales
Rhodymeniales	Rhodymeniales	Rhodymeniales	Rhodymeniales	Rhodymeniales
Ceramiales	Ceramiales	Ceramiales	Ceramiales	Ceramiales

[a] Did not consider taxa in which pit plugs are absent or lack both caps and cap membranes.
[b] This classification should include the recently described Anfeltiales and Gracilariales (see Chapter 1).

supported segregation of Gelidiales and Bonnemaisoniales from Nemaliales and of Palmariales from Rhodymeniales. In addition, they proposed two new orders, Batrachospermales for the freshwater nemalialean families Batrachospermaceae, Lemaneaceae and Thoreaceae, and Hildenbrandiales for the monotypic cryptonemialean family Hildenbrandiaceae. Pueschel and Cole also provided evidence for elevating the cryptonemialean family Corallinaceae to ordinal status, a proposal formally made by Silva and Johansen (1986). Recently, Pueschel (1987) has indicated that another pit plug character, presence or absence of pit plug membranes, may have taxonomic significance, and, whereas this feature has been used to support recognition of Bangiales and Corallinales, no additional orders have been segregated using this feature. Thus far, pit plug characters have been useful in defining monophyletic groups, but they have not aided in resolving phylogenetic relationships among higher-order taxa (Gabrielson & Garbary 1987).

Just as Kylin found unworkable Schmitz's use of procarpy versus nonprocarpy to segregate Gigartinales from Cryptonemiales [see Silva & Johansen (1986) and Chapter 13 for review of the concept of procarpy], red algal systematists in the latter half of this century likewise have become increasingly dissatisfied with Kylin's use of the accessory versus nonaccessory origin of auxiliary cell filaments as a criterion to distinguish between these two orders. Extensive discussions of how to determine whether a filament is accessory or not and/or the ramifications of the various interpretations on the characterization and classification of the affected taxa are provided by Searles (1968, 1983), Lebednik (1977), Kraft and Robins (1985), Moe (1985), and Gabrielson and Garbary (1987). Searles (1983) concluded that "accessory auxiliary cell branches have probably arisen more than once and from both carpogonial branches and vegetative branches." He suggested that Cryptonemiales is likely polyphyletic, with different families being related to taxa in either Nemaliales or Gigartinales. Kraft and Robins (1985), on the other hand, merged Cryptonemiales with Gigartinales, arguing that in some species of *Dudresnaya*, a cryptonemialean examplar, there are no clearly accessory auxiliary cell filaments, whereas in some species of *Predaea*, a gigartinalean examplar, auxiliary cells appear to be borne in "subsidiary" cortical filaments. They further noted that the heterogeneity of an enlarged Gigartinales (not including Hildenbrandiaceae and Corallinaceae) may be more apparent than real, the remaining cryptonemialean and gigartinalean families sharing a continuum of reproductive morphology as well as ultrastructural and biochemical similarities. Moe (1985) and Silva and Johansen (1986) do not support Kraft and Robins' proposed merger, whereas Gabrielson and Garbary (1986, 1987) see this merger as an interim step toward establishing monophyletic groups among cryptonemialean and gigartinalean genera and families.

Skuja (1939) presented the first comprehensive classification of bangiophyte red algae, Schmitz (in Schmitz & Hauptfleisch 1897) only having assigned them with reservation to Rhodophyceae. Skuja recognized four orders, Porphyridiales, Goniotrichales, Bangiales, and Compsopogonales, mainly based on morphological features. This classification was not followed by Fritsch (1945), who recognized only one order, Bangiales, but it was adopted by Kylin (1956), who also recognized Rhodochaetales, an order that previously had been suggested by Skuja (1939). Feldmann (1955) mentioned that the orders Goniotrichales and Porphyridiales should be merged and later elaborated on his suggestion (Feldmann 1967), pointing out that there was little difference between unicells that form amorphous colonies (classified in Porphyridiales) and taxa that form somewhat filamentous thalli (classified in Goniotrichales). Chapman (1974) followed Feldmann's suggestion and merged the two orders, recognizing four families, Cyanidiaceae, Goniotrichaceae, Phragmonemataceae, and Porphyridiaceae, primarily based on features of asexual reproduction and chloroplast and thallus morphology. Garbary et al. (1980) reduced the number of families to two, Porphyridiaceae and Phragmonemataceae, characterizing them by using chloroplast and pyrenoid features.

The classification of bangiophyte red algae was revised substantially by Garbary et al. (1980). The families Erythropeltidaceae and Boldiaceae, formerly placed in Bangiales, were grouped with the family Compsopogonaceae in a new order, Erythropeltidales (=Compsopogonales), leaving only Bangiaceae in Bangiales. Compsopogonales were distinguished from other bangiophyte orders by monosporangium formation, in which an undifferentiated cell of the thallus cleaves by a curved wall, producing a monosporangium. Rhodochaetales also exhibit this type of monosporangium formation. The realignment of Erythropeltidaceae and Boldiaceae has been supported by the ultrastructural observations of Scott (1984) on organelle associations in the families assigned to Compsopogonales and Bangiales.

III. APPROACHES TO RED ALGAL TAXONOMY AND EVOLUTION

In this section we briefly discuss characters and methodologies that have been used to construct the current classification system for red algae. Topics that historically have made, and currently are making, major taxonomic and evolutionary contributions to red algae, including vegetative and reproductive development, chromosomes, genetics, and ultrastructure, received a detailed treatment in Gabrielson and Garbary (1986) and are discussed extensively elsewhere in this volume (e.g., Chapters 4, 6, 12, 13). We also highlight the importance and problems of some new kinds of systematic data (chemotaxonomy and vegetative morphogenesis) and methods of analysis (quantitative approaches) that are being applied to red algae. Within the topics that are covered below, our comments are selective rather than comprehensive.

A. Chemotaxonomy

The application of chemotaxonomy to red algal systematics has been a slow process, and there are few examples where this approach has been applied to, and actually resolved, particular taxonomic problems. Chemotaxonomic characters rarely have been used as the primary basis for taxonomic conclusions except at the division rank. In general, they are used to support taxonomic conclusions based on morphological evidence. Recently, we reviewed chemotaxonomic data, including information on pigments, polysaccharides, halogenated compounds, and proteins in red algae (Gabrielson & Garbary 1986). Herein, we restrict our comments to some limitations in using data from halogenated compounds as taxonomic markers. A wide variety of halogenated mono-, di-, and sesquiterpenes have been described from red algae, particularly from the genera *Plocamium* (Plocamiaceae, Gigartinales), *Laurencia* (Rhodomelaceae, Ceramiales), and *Bonnemaisonia* and *Asparagopsis* (Bonnemaisoniaceae, Bonnemaisoniales). Naylor et al. (1983) argued that if chemical data are to be applied to taxonomic problems, three criteria must be met: (1) An array of compounds must be present, (2) the composition of the various species should not depend upon habitat (i.e., there should be no ecotypes), and (3) the composition of the mixtures should not vary among individuals of the same species (i.e., there should be no chemotypes).

One study (Crews & Selover 1986) that suggested that there may be major limitations in using halogenated compounds as systematic characters is an analysis of sesquiterpenes from *Laurencia pacifica* and its adelphoparasite, *Erythrocystis saccata*. Crews and Selover found identical sesquiterpene compounds in both host and parasite and proposed three hypotheses to account for this result: (1) The two organisms are closely related and produce the same compounds via the same pathways, (2) there is direct transfer of sesquiterpenes from host to parasite, or (3) the parasite is able to concentrate these compounds from sea water after they have been released by *Laurencia*. The first explanation was not considered, perhaps because the authors mistakenly classified *E. saccata* in the wrong order (Gigartinales instead of Ceramiales) and therefore believed host and parasite to be unrelated. The second explanation also is possible as shown by studies of the movement of labeled compounds from hosts into parasites (Goff 1982). If true, this is a handicap to using terpenoids as taxonomic markers. The final possibility, that algae are able to concentrate such compounds from seawater, is more disquieting, and there is some evidence to support this explanation. Rivera et al. (1987) found the same halogenated monoterpenes in *Schottera* (Phyllophoraceae, Gigartinales) as were found in plants of *Plocamium cartilagineum* collected from the same site in Chile. Similarly, adjacent plants of *Microcladia* (Ceramiaceae, Ceramiales) and *Plocamium* from Washington yielded the same complement of compounds (Naylor et al. 1983). Uptake of compounds from seawater might account, in part, for the observations of Blunt et al. (1984), in which different sesquiterpenes were identified in morphologically identical samples of *Laurencia distichophylla* from different tidal heights, and for the variation found in populations of *L. pacifica* by Fenical and Norris (1975), as well as the variation in populations of *P. cartilagineum* from Chile found by San-Martin and Rovirosa (1986). As was suggested by Fenical and Norris, more extensive studies are needed before such chemotypes are segregated taxonomically. The above studies indicate that much of the terpenoid data do not (or may not) meet the last two criteria of Naylor et al. (1983). This problem places severe restrictions on the taxonomic application of terpenoid data and means that more rigorous experimental controls are necessary to ensure that compounds being identified are being produced by the organism under study.

B. Vegetative morphogenesis

Vegetative morphogenesis was treated in considerable detail by Gabrielson and Garbary (1986), who discussed aspects of spore germination, initiation of upright axes, and thallus differentiation. Historically, red algal systematics at ordinal and infraordinal ranks has been based primarily on carposporophyte development, but since the 1960s there have been several technical advances that have enabled workers to study vegetative morphogenesis in greater detail than previously was possible. These include using Wittmann's (1965) hematoxylin–chloral hydrate stain to study patterns of karyokinesis and cytokinesis in multicellular red algae (Chapter 12), using epifluorescence microscopy to study cell wall elongation by means of Calcofluor stain (Waaland & Waaland 1975; Chapter 11), and using DAPI and related fluorochromes to study the amount of DNA in vegetative cells (Goff & Coleman 1986; Chapter 3).

Within florideophytes there are several different patterns of cell wall elongation. In Acrochaetiales elongation is primarily or exclusively by means of tip growth (Garbary 1979b; Hymes & Cole 1983), and band elongation appears to be absent. In *Batrachospermum* (Batrachospermales) one filament type has only tip growth, whereas another filament type has both tip growth and band growth (Aghajanian & Hommersand 1980), and in Ceramiales only band growth has been reported in intercalary cells (Waaland & Waaland 1975). This sequence of characters would be consistent with the cladistic relationships of these orders (Fig. 18-1), but tip growth without band growth, is present in at least one genus of Ceramiaceae (Garbary, unpubl.). An extensive, comparative study is required to determine if this developmental feature can be used as a taxonomic character at specific, generic, or higher taxonomic ranks.

The use of DAPI and similar fluorochromes to investigate changes in DNA levels during vegetative development (Goff & Coleman 1986, 1987) provides exciting possibilities for taxonomic characters. The pattern of polyploidy and then sequential ploidy reduction that occurs in apices of *Polysiphonia* (Ceramiales) during pericentral cell development is extremely precise and should be examined in other red algae. Such ontogenetic sequences (Goff & Coleman 1986; Chapter 3) likely will be of systematic importance, although the classification ranks at which such information will be applied remain to be determined. This approach to investigating morphology may be critical in elucidating how filamentous and syntagmatic thalli in red algae are assembled. The underlying assumption of the DAPI technique, that increase in relative fluorescence represents an increase in DNA, has been demonstrated in *Griffithsia* (Ceramiales), where there is a correlation between the amount of fluorescence and chromosome numbers (Goff & Coleman 1987). It is important that this assumption be verified in other experimental systems where DAPI is being utilized.

The techniques and advances presented above are descriptive in nature, but more experimental approaches also have been utilized. In culture it is possible to manipulate the environment so that the

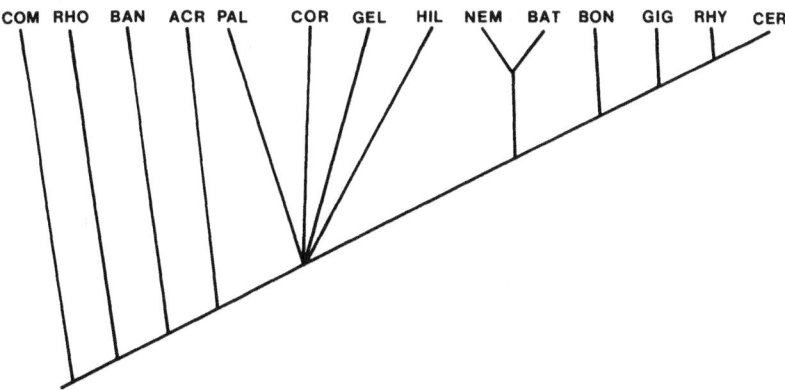

Fig. 18-1. Cladogram showing consensus tree for orders of red algae (except Porphyridiales). Note: Figure redrawn from Gabrielson et al. (1985); see that paper for details of cladistic methods and characters. COM = Compsopogonales, RHO = Rhodochaetales, BAN = Bangiales, ACR = Acrochaetiales, NEM = Nemaliales, BAT = Batrachospermales, COR = Corallinales, HIL = Hildenbrandiales, GEL = Gelidiales, PAL = Palmariales, GIG = Gigartinales, BON = Bonnemaisoniales, RHY = Rhodymeniales, CER = Ceramiales.

effects of light, temperature, and photoperiod on apical cell division and cell elongation can be determined (see review by L'Hardy-Halos et al. 1984). Along this line, Garbary et al. (1988) showed that the location of band deposition in cell walls of *Antithamnion* (Ceramiales) was under apical control. As such studies are extended to other organisms, a greater understanding of thallus construction and of the homology of morphological types will be attained.

C. Quantitative analysis

Taxonomic analysis of red algae historically has dealt with qualitative characters, with taxonomic conclusions based on a straightforward inspection of morphological features. Most current taxonomic work on red algae remains at this level of investigation. This situation is counter to one of the common trends in modern systematics: the increasing use of quantitative analysis. Quantification can take two approaches, the first being a numerical description of morphological and developmental features, and the second the quantitative analysis or interpretation of qualitative and/or quantitative data. Some studies on red algae that utilized these approaches are discussed below.

Quantitative characters often are defined more readily and enable comparisons to be made more easily between individuals and/or taxa. We advocate the use of such characters, not because we believe that they are inherently better than qualitative characters, but because they often enable a more sophisticated level of analysis of taxonomic structure than do qualitative characters alone. This is particularly true for studies involving infraspecific or populational differences. Papers that attempt to quantify aspects of red algal morphology and that provide some statistical comparison among individuals or taxa include Akatsuka (1986) on Gelidiales, Garbary (1979b) on *Audouinella* (Acrochaetiales), King et al. (1988) on *Bostrychia* (Ceramiales), Sheath and Burkholder (1983) and Sheath et al. (1986) on *Batrachospermum* (Batrachospermales), and Sheath and Cole (1984) on *Bangia* (Bangiales).

Three primary methods of quantifying taxonomic data in the search for relative similarity or phylogenetic relationships among operational taxonomic units (OTUs) are cluster analysis, ordination, and cladistics. Two examples of ordination (e.g., principal component analysis) are by Garbary and Scagel (1980) on *Iridaea* (Gigartinales), and King et al. (1988) for *Bostrychia*. Other data sets where this approach is suitable include Sheath and Cole (1984), on variation in *Bangia*, and Sheath and Burkholder (1983), on variation in *Batrachospermum*.

Cluster analysis has been used widely in studies on a variety of organisms, especially higher plants (Baum et al. 1984). One early study of red algae using cluster analysis dealt with similarities among genera and families and Nemaliales (Levin and Rogers 1964). The most elaborate application of cluster analysis in red algae was by Garbary (1979a) in a study of 170 OTUs representing 112 species of Acrochaetiaceae. Although this study was strongly criticized by Woelkerling (1983) for the choice of characters, the resulting dendrogram did support the one genus classification scheme that was advocated by Garbary and others (see Woelkerling 1983 for review). Despite the criticism that cluster analysis has received, particularly by phylogeneticists (Wiley 1981), this approach remains widely used in current taxonomic literature and is the method of choice for many workers (e.g., Abbott et al. 1985).

The third primary method of taxonomic analysis referred to above is cladistics or phylogenetic systematics (Hennig 1966; Wiley 1981). This approach only recently is beginning to be applied to various algal divisions including Rhodophyta, and some recent studies that show the potential of this technique include Prud'Homme van Reine (1982) on Sphacelariales (Phaeophyceae) and Kociolek and Stoermer (1986) on Bacillariophyceae. Within red algae the studies of Gabrielson et al. (1985) and Gabrielson and Garbary (1987) have shown the method to be a useful tool for making and testing phylogenetic hypotheses and character homologies. Cladistic methodology also was used by Lindstrom (1987, 1988) in studies on Dumontiaceae.

D. Life histories

The pairing of genera originally independently named as tetrasporophytes or gametophytes but comprising a single heteromorphic life history and the resolution of other life histories with a "missing" generation have had significant taxonomic consequences at ordinal and infraordinal ranks (see West and Hommersand 1981 for review). Numerous inconspicuous, filamentous genera (e.g., *Conchocelis, Trailliella, Falkenbergia, Hymenoclonium*), as well as numerous inconspicuous, or conspicuous but for the most part ignored, crustose genera (e.g., *Cruoria, Cruoriopsis, Erythrodermis, Haematocelis, Petrocelis, Porphyrodiscus*) reported to bear only tetrasporangia (in some cases female reproductive structures were reported but have not been confirmed), have been shown to be the alternate in the life history phase of a conspicuous macrophyte. Exemplifying this approach is the work of Guiry, Masuda, West, and their colleagues – see Guiry & West (1983) and Guiry et al. (1984) and references therein – on genera in the families Gigartinaceae, Petrocelidaceae and Phyllophoraceae (Gigartinales). This research culminated in the removal of *Mastocarpus* from the Gigartinaceae to its own family, Petrocelidaceae, based on a suite of characters that included a heteromorphic life history. At present only seven families – Galaxauraceae (Nemaliales), Dumontiaceae, Furcellariaceae, Gymnophloeaceae, Phyllophoraceae and Solieriaceae (Gigartinales), and Bonnemaisoniaceae (Bonnemaisoniales) – are heterogeneous with regard to this life history feature. For at least some of these families (e.g., Dumontiaceae, Furcellariaceae, Solieriaceae) West and Hommersand (1981) suggested that genera with heteromorphic life histories be removed to their own families, and for Furcellariaceae and Solieriaceae their suggestion merits further research. In Dumontiaceae, however, such a move would split apart a taxon that, based on features of anatomical and reproductive morphology, appears quite homogeneous, especially with regard to the northeast Pacific genera (Lindstrom 1988).

An order of red algae in which life history work coupled with characters from other disciplines could significantly improve our understanding of generic concepts is Acrochaetiales. Woelkerling (1983) noted that at least 24 different classifications of generic and suprageneric ranks have been proposed for acrochaetioid algae since 1900 and that 14 have been used since 1970. The proposals that the approximately 400 described species (see Garbary 1987b for list) belong to one genus, *Audouinella*, or two genera, *Audouinella* and *Colaconema* (Woelkerling 1971), based on sexual and asexual reproduction, respectively, clearly are unsatisfactory as a final solution, even though both classifications have received wide acceptance. A promising line of investigation was taken up by Stegenga and coworkers (Stegenga & van Wissen 1979 and references therein) in which acrochaetioid algae that were known only as gametophytes or tetrasporophytes were cultured under varying light and temperature conditions in an attempt to complete their life histories. Observations also were made on the effects of culture conditions on various morphological features. Subsequently, between three and six genera have been recognized (Stegenga 1985; Lee 1987) based on a suite of characters including life history features, chloroplast form and position, and spore germination patterns. The above features need to be assessed together with ultrastructural and biochemical characters, e.g., R- versus B-phycoerythrin (Glazer et al. 1982), to further resolve generic relationships and to determine if Acrochaetiales is a monophyletic taxon.

Resolution of problematic life histories at the species and generic ranks, and of relationships among related genera and families in which both isomorphic and hetermorphic life histories occur, will contribute significantly to the ecological question of the adaptive significance of heteromorphic versus isomorphic life histories (Lubchenco & Cubit 1980). Families that are especially amenable to studies on evolutionary ecology are those wherein the classified genera have different life history types and overlapping distributions, and where an evolutionary framework (i.e., a cladogram) exists within which the results of ecological experiments can be interpreted. Studies like that of Littler and Littler (1983) on *Scytosiphon* (Phaeophyta) need to be done on red algae that meet the above criteria.

E. Monographs

Despite recent trends toward experimental and biochemical approaches in taxonomy, it is clear that traditional morphological studies will continue to provide the foundation for red algal systematics, particularly at generic and familial ranks. Monographic studies, as practiced at the familial rank by, for example, Akatsuka (1986 and included references) on Gelidiaceae (Gelidiales), Lindstrom (1988) and Lindstrom and Scagel (1987) on Dumontiaceae (Gigartinales), Woelkerling and colleagues on Corallinaceae (Corallinales) (Penrose & Woelkerling 1988; Woelkerling 1987), and Wynne and colleagues on Delesseriaceae (Ceramiales) (Wynne 1983 and included references) and at the generic rank by, for example, Bird and colleagues on *Gracilaria* (Gigartinales) (Bird et al. 1986 and included references), Kapraun and colleagues on *Polysiphonia* (Ceramiales) (Kapraun et al. 1983 and included references), Kraft and colleagues on *Predaea* (Gigartinales) (Kraft 1984 and included references), and Huisman (1985) on *Scinaia* (Nemaliales), continue to provide the essential taxonomic data that are necessary for more synthetic studies on phylogeny (Lindstrom 1988), biogeography, and ecology (Adey & Steneck 1985; Steneck & Watling 1982). In some cases monographic studies are supplemented by cytological research (Bird et al. 1982) or by field and culture studies (Kapraun 1978; Maggs & Guiry

1987) that contribute additional data to evolutionary and ecological interpretations.

The use of cladistics as an analytical tool in systematic studies should provide new impetus for encouraging monographic approaches to systematic problems. This is especially true with the emphasis placed on outgroup comparison as the best method for determining character polarity, i.e., which character state is derived (apomorphic) and which ancestral (plesiomorphic) (Maddison et al. 1984). Thus, in a cladistic analysis of any taxon of a given rank (the ingroup), character states of the characters being analyzed must also be known in related taxa [the outgroup(s)]. The primary assumption is that the character state for a given character that is present in both the outgroup and the ingroup is the primitive condition; deviations from this represent advanced states. For more detailed treatments of cladistic methodology, see Nelson and Platnick (1981) and Wiley (1981).

F. Biogeography

The application of biogeography to the systematics of red algae is in its infancy, but there is great potential for the use of this tool in conjunction with other kinds of characters. Biogeographic studies, as applied to red algae, fall into two major categories: (1) floristic comparisons, in which the presence or absence of species or genera are tabulated for a given locality, usually along with other macroalgae, and this list is compared to lists from other localities, and (2) studies where the physiological limits, typically temperature, light, and photoperiod, of a given species are tested in culture, and the theoretical laboratory tolerances of a species are compared with actual tolerances based on distribution. Examples of the former include Pielou (1977), Womersley (1981), Parsons (1985), and Joosten and van den Hoek (1986), and of the latter, van den Hoek (1982), McLachlan and Bird (1984), and Yarish et al. (1986). Whereas all of these studies have contributed to our knowledge of the distribution of red algae, none has contributed to the systematics of red algae (see critique of marine algal biogeography by Garbary 1987a).

Hommersand (1986) analyzed the biogeographic affinities of southern Africa based on the distribution of 300 red algal species from that region compared with their presumed sister taxa (i.e., their closest relatives) from other localities. His speculations regarding species migration and distribution patterns and the origins of genera and families were related to paleo-oceanographic events and climatic changes during the Tertiary period. However, the taxonomic evidence on which these interpretations are based was not provided. It should be pointed out that the paper of Hommersand (1986) is among the only studies of marine algae that utilize, at least in part, the methodology of panbiogeography advocated by Croizat (1964).

With the advent of cladistics, phylogenies of taxa (i.e., cladograms) can be generated and the distributions of the taxa plotted on the cladogram(s) to examine the extent of congruence of phylogeny with distribution. Regions of noncongruence may indicate problems with character evolution, distribution patterns, or the taxonomy of the taxa. Garbary (1987a) discussed this approach in a marine algal context, and Lindstrom (1987) gave specific examples using red algal genera and species. Lindstrom (1988) used a combination of cladistics and biogeography in her study of Dumontiaceae and concluded that a new genus, *Masudophycus*, needed to be erected. Three biogeographic patterns also were identified, and Lindstrom suggested paleoecological phenomena that may be reflected in the observed distribution patterns. The synthesis of cladistic and biogeographic studies on red algae promises to shed additional light on the evolutionary relationships of red algal species, genera, and families.

IV. RECOGNITION OF CLASSES/SUBCLASSES

Based on cladistic analyses in which the orders Porphyridiales, Bangiales, and Rhodochaetales were used as hypothetical sister groups (functional outgroups), Gabrielson et al. (1985, Figs. 1–3) recommended that only a single class of red algae, Rhodophyceae, be recognized. No shared, derived characters (synapomorphies) could be found for Bangiophyceae/Bangiophycidae or for the class Cyanidiophyceae proposed by Merola et al. (1981). Lee (1980) also did not recognize bangiophytes as a separate class/subclass of red algae because of a lack of defining characters. Cyanidiophyceae originally was segregated from other red algae based on the supposed presence of a single membrane in the chloroplast envelope (Merola et al. 1981). Recently, Ueda and Chida (1987) examined freeze-substituted chloroplasts of *Cyanidium caldarium* and demonstrated that chloroplast envelopes in *Cyanidium* are like those in other unicellular red algae. *Cyanidium* appears, therefore, to be a unicellular red alga and not an endocyanome or a "bridge alga" between Cyanobacteria/Cyanophyta and Rhodophyta as postulated by Kremer (1982) and

Seckbach (1987), respectively. For further discussion of this controversy see Gabrielson et al. (1985).

Two characters, presence of a filamentous gonimoblast and of tetrasporangia, were synapomorphous for florideophytes (Gabrielson et al. 1985; Gabrielson & Garbary 1987). These features are present in nearly all florideophytes, and exceptions can be interpreted as evolutionary reductions or modifications. Florideophyceae/Florideophycidae appears to be monophyletic, whereas bangiophytes certainly are paraphyletic.

Lim et al. (1986), based on an analysis of 5S rRNA sequences from one bangiophyte taxon, *Porphyra tenera*, and seven species of florideophytes, concluded: "Thus, the 5S rRNA data are consistent with the view that Rhodophyta is divided into two classes, 'primitive' Bangiophyceae and 'more advanced' Florideophyceae." However, no comparisons were made with other bangiophyte taxa to determine whether or not the class is monophyletic, and their results equally support the recognition of only one class of red algae. Based on the limited number of organisms examined, their data cannot be used to support the ancestral position of Bangiales relative to florideophytes as postulated by the cladograms of Gabrielson et al. (1985). Because inter- as well as infraordinal relationships among taxa previously assigned to Bangiophyceae/Bangiophycidae currently are unresolved (see Section V), the recognition of one class, Rhodophyceae, appears to be the best taxonomic alternative at present.

V. ORDINAL CLASSIFICATION AND RELATIONSHIPS

In this section we discuss phylogenetic relationships that have been proposed for the orders of red algae. Where orders have been suggested to be paraphyletic or polyphyletic, we indicate what we believe to be monophyletic subgroups and the relationships of these subgroups to other taxa. These discussions summarize or complement the analyses in Gabrielson et al. (1985), Gabrielson and Garbary (1987), and Garbary and Gabrielson (1987). For descriptions of most of the taxa treated here, see Bold and Wynne (1985). See Table 18-1 for the ordinal classification utilized in this work.

A. Porphyridiales

Porphyridiales remains the most problematic order of red algae in terms of classification and potential relationships. Garbary et al. (1980) and Gabrielson et al. (1985) regarded this order as an assemblage of organisms united solely by their unicellular to palmelloid habit. These and other authors have suggested that the order does not form a natural assemblage and is both polyphyletic and paraphyletic. Thus, not only may Porphyridiales be derived from a variety of different ancestors that would be included in different orders (polyphyletic), but even if Porphyridiales has a single ancestor, not all the descendants of that ancestor are included within the order (paraphyletic). Although it frequently has been assumed that unicells are primitive in Rhodophyta (e.g., Dixon 1973), the absence of sexual reproduction in all unicellular taxa argues against this. If unicellular taxa are primitive within Rhodophyta, then sexuality (including meiosis) would have to have evolved twice, once in red algae and once in the evolutionary line leading to other eukaryotes. The finding of sexuality in some Porphyridiales would provide evidence that the unicellular condition was primitive in red algae.

It has been suggested that some unicells might have evolved through reduction from filamentous or parenchymatous ancestors (Fritsch 1945; Garbary et al. 1980; Lee 1974), but direct evidence for this remains inconclusive. The most intriguing observations are those of Scott (1986), who pointed out that cell division in *Flintiella* was similar to that in *Batrachospermum* (Batrachospermales) and not to that in *Porphyridium* (Porphyridiales). This suggests that *Flintiella* may have evolved through reduction from a *Batrachospermum*-like (syntagmatic) ancestor, but we must await observations on mitosis from taxa in other orders to test this hypothesis (see Chapter 6 for further discussion). At the very least, Scott (1986) has demonstrated that Porphyridiales is paraphyletic, and not a natural group. The observations of Gargiulo et al. (1987) on development of *Erythrocladia irregularis* suggests that the separation of unicellular and multicellular organisms into different orders is highly artificial.

Regarding infraordinal classification, neither the Porphyridiaceae and Phragmonemataceae recognized by Garbary et al. (1980), based on the presence or absence of pyrenoids, nor the Porphyridiaceae and Goniotrichaceae recognized by Bold and Wynne (1985), based on unicellular versus "multicellular" taxa, are supported by other ultrastructural features, i.e., dictyosome associations (Scott 1984; Scott & Gabrielson 1987) and mitosis (Scott 1986).

B. Compsopogonales

Compsopogonales includes three families, Compsopogonaceae, Erythropeltidaceae, and Boldiaceae

(Garbary et al. 1980), and it is possible that the family Rhodochaetaceae should be transferred here. Relationships among the included families have not been resolved. There have been two recent advances in our understanding of this order. Kornmann (1984, 1987) reported a sexual life history in *Porphyrostromium* (as *Erythrotrichopeltis*), in which an erect, gametophytic, trichoid phase alternates with a sporophytic, peltoid phase. These reports are the first well-documented accounts of sexuality in this order since Berthold (1882) and confirm that three of four orders of bangiophytes have sexual life histories (see Magne 1960b on Rhodochaetales and Hawkes 1978 on Bangiales). Another genus in Erythropeltidaceae where a sexual life history seems likely is *Smithora*, where well-differentiated gametangia are reported (Hollenberg 1959; Hawkes 1988). Should sexual reproduction in *Smithora* be similar to that in *Porphyrostromium*, this would provide further confirmation for removing Erythropeltidaceae from Bangiales.

Scott et al. (1988), using electron microscopy, have confirmed the earlier light microscope report of Fan (1959) of pit connections in *Compsopogon*. If *Compsopogon* is the only member of the order with pit connections, then this might be grounds for maintaining Compsopogonaceae distinct from Erythropeltidaceae and Boldiaceae at ordinal rank. This was done for operational reasons in Gabrielson et al. (1985), but we now consider a more detailed study of the distribution of pit connections both within Compsopogonaceae (i.e., additional species of *Compsopogon* and *Compsopogonopsis*) and in other families of the order to be necessary before taxonomic changes are instituted. Potential locations for pit connections are the pseudoconchocelis phases and/or pseudoparenchymatous bases of *Boldia*, *Smithora*, *Porphyropsis*, and *Erythrotrichia*. Even a strict presence or absence of pit connections in Compsopogonaceae versus other families may be insufficient to define two orders unless an evolutionary dichotomy is present.

Observations of sexuality and pit connections in this order are of particular interest in that they establish these features in all multicellular bangiophyte orders (i.e., all orders except Porphyridiales). As a result, such characters cannot be used to distinguish Bangiophycidae from Florideophycidae (e.g., Bold & Wynne 1985).

The recent developmental studies of Gargiulo et al. (1987) on *Erythrocladia irregularis* have shown that thalli in this species vary from well-defined crusts that are apparently multicellular to loose aggregations of unicells. In addition, both monospores and endospores are produced. *Erythrocladia* appears to represent an intermediate evolutionary stage between unicellular and multicellular forms.

C. Rhodochaetales

The Rhodochaetales with its single species, *Rhodochaete parvula*, is possibly central to our understanding of red algal evolution. With its filamentous organization, apical growth, triphasic, slightly heteromorphic life history, and pit connections devoid of cap layers and cap membranes (Boillot 1975, 1978; Magne 1960b; Pueschel & Magne 1987), this taxon may be closest to the the ancestral red alga.

The morphology of both sexual and asexual reproduction in *Rhodochaete* is basically similar to that in *Porphyrostromium*, and pit connections in *Rhodochaete* (Pueschel & Magne 1987) appear similar to those in *Compsopogon* (Scott et al. 1988). One could argue that *Rhodochaete* be placed in Compsopogonales, where, based on its pit connection morphology, it would be related most closely to *Compsopogon*. *Rhodochaete* differs from all other taxa in the order, however, by its apical growth and slightly heteromorphic life history, both of which argue for retention in its own family.

If one assumes that the ancestral red alga, like *Rhodochaete*, had a sexual life history, apical growth, and pit plug cores, then other taxa in Compsopogonales represent a reduction series with a loss of sexuality, apical growth, and pit plugs. This evolutionary progression is more parsimonious than if Rhodochaetaceae is considered advanced in Compsopogonales, in which case apical growth, a sexual life history, and pit plugs all would be homoplasious, (i.e., independently derived in two evolutionary lineages), occurring once in Compsopogonales and again in the lineage leading to other red algae. If Rhodochaetaceae is ancestral in the order, then diffuse growth in red algae would be homoplasious, arising once in the remaining families of Compsopogonales and again in Bangiales. If Rhodochaetaceae is the sister group of Compsopogonales, recognition of one or two orders is consistent with cladistic principles. Regardless of these complications, it would appear that *Rhodochaete* and Compsopogonales likely comprise a monophyletic group that may also include some unicellular forms (Gargiulo et al. 1987). If there are real "bangiophytes," *Rhodochaete*, some taxa of Porphyridiales, and all Compsopogonales would be included, but the Bangiales would be omitted.

D. Bangiales

The suggestion of Garbary et al. (1980) that Bangiales comprises the single family Bangiaceae has been accepted (Bold & Wynne 1985). With Bangiales as a monophyletic group, its relationships with other red algal orders become more apparent. In virtually all cladograms produced by Gabrielson et al. (1985) and Gabrielson and Garbary (1987), Bangiales were resolved as the outgroup for Florideophycidae. The features shared by these two taxa include pit plugs with cap layers, dictyosomes associated with mitochondria, presence of R-phycoerythrin, presence of cellulose, and similar monosporangia (i.e., different from the Compsopogonales type). The absence of cap membranes in Bangiales (Pueschel 1987), as well as in *Compsopogon* and *Rhodochaete*, supports the contention that they are primitive relative to florideophyte taxa.

Bangiaceae, with its two genera, *Bangia* and *Porphyra*, form a well-circumscribed family based on life history and developmental features. Because of the economic importance of *Porphyra*, there are extensive taxonomic studies in this genus, particularly in the north Pacific, where numerous species occur. Taxonomy of *Bangia* has been shown to be more complex than previously thought, as demonstrated by the cytological and morphological studies of Cole et al. (1983) and Sheath and Cole (1984). At least several species appear to be present, but definite conclusions regarding species delimitation still are lacking. Using *Bangia* as the outgroup, it should now be possible to resolve relationships in *Porphyra* using cladistic analyses. Of particular interest in both genera are details of the life history, particularly the site of meiosis, which has been inferred for only a few species. Ma and Miura (1984) and Ohme et al. (1986), both in *P. yezoensis*, and Burzycki and Waaland (1987), in *P. torta*, provided cytological and genetical evidence that meiosis occurs during germination of the conchospores. This is inconsistent with earlier observations on *P. umbilicalis* by Giraud and Magne (1968), who indicated that at least part of meiosis occurs in the conchosporangial branch.

E. Acrochaetiales

Recognition of Acrochaetiales at ordinal rank remains controversial, with many authors (e.g., Bold & Wynne 1985) retaining the family in Nemaliales. Garbary (1978) argued for ordinal recognition based on evolutionary position, i.e., that Acrochaetiaceae were the most primitive extant florideophytes. This was supported by the cladistic analyses of Gabrielson et al. (1985) and Gabrielson and Garbary (1987). Garbary and Gabrielson (1987) extensively discussed relationships with other florideophyte orders and concluded that there are no obvious sister groups among extant florideophytes. Glazer et al. (1982), based on pigment analyses and consideration of life histories and morphology, transferred *Rhodophysema* from Palmariales to Acrochaetiales. Guiry (1987) argued on the basis of life histories, that Acrochaetiales and Palmariales may be related. Saunders (1987) proposed that many species in Acrochaetiales, particularly those with biphasic life histories or with crustose morphologies, should be included in new families and transferred to an expanded Palmariales.

F. Nemaliales

With the removal of Acrochaetiaceae, Gelidiaceae, Bonnemaisoniaceae, and Batrachospermaceae (and related families) to their own orders, Nemaliales becomes better circumscribed with its central core of families (Liagoraceae sensu lato and Galaxauraceae), and is possibly monophyletic. To be monophyletic a taxon should have a single ancestor, and all the descendants of that ancestor must be included in the taxon. This may not be true for Nemaliales if Batrachospermales have evolved from a nemalialean ancestor, a distinct possibility considering the cladograms produced by Gabrielson and Garbary (1987). A further problem with regarding Nemaliales as monophyletic is that there are no apparent autapomorphies (=uniquely derived characters) that define the order. More detailed studies of vegetative and reproductive morphology and of ultrastructure in the Galaxauraceae and Liagoraceae sensu lato are required before relationships among these taxa, and with other red algae, are resolved. The vegetative and reproductive morphology and generally tropical distribution suggest an ancient origin for the order as part of the original diversification of florideophyte red algae.

G. Batrachospermales

Batrachospermales, with their *Lemanea*-type life history, exclusively freshwater distribution, and two pit plug cap layers, the outer layer of which is enlarged, are a monophyletic taxon. Their relationships with other red algal orders that possess two-layered pit plug caps have not been resolved (Gabrielson & Garbary 1987, Fig. 7). Only with Nemaliales, however, do Batrachospermales share significant similarities in reproductive morphology including (1) presence of morphologically differentiated carpogonial branches, (2) initiation of the gonimoblast directly from the carpogonial branch, and (3) carpogonial branches modified for nutrient processing (see Chapter 13). Batrachospermales and Corallinales share the pit plug cap feature of having the outer cap layer enlarged (Pueschel & Cole 1982), but this character obviously has arisen independently in these two taxa. Corallinales also lack a cap membrane, but whether or not this structure is present in Batrachospermales has not been elucidated (Pueschel 1987).

The three families included in Batrachospermales – Batrachospermaceae, Lemaneaceae, and Thoreaceae – are well circumscribed.

H. Gelidiales

By virtue of their spore germination pattern, vegetative and male and female reproductive morphology, and one-layered pit plug caps, Gelidiales appears to be a monophyletic group. A sister taxon relationship was postulated between Gelidiales and Hildenbrandiales (Gabrielson & Garbary 1987) based on spore germination features and a one-layered plug cap. Hommersand and Fredericq (1988) suggested a possible relationship between Gelidiales and Gracilariaceae based on similarities in wall chemistry, patterns of spermatiogenesis, direct development of the carposporophyte from the carpogonium, and the absence of typical auxiliary cells. If this relationship proves correct, then it will be the first close link between taxa with different numbers of pit plug cap layers. We consider it critical for workers to resolve the homology of one-layered pit plug caps, as well as the homology of one-layered pit plug caps with the inner layer of two-layered pit plug caps. If this layer is not homologous among taxa in which it occurs, then other characters, such as those suggested by Hommersand and Fredericq (1988), will have to be employed to resolve relationships. At present, the relationships of Gelidiales with other red algal orders remains unresolved.

I. Hildenbrandiales

Hildenbrandiales is an order whose phylogenetic relationships remain obscure. Cladograms of Gabrielson and Garbary (1987) (e.g., Fig. 18-1) suggested relationships with Gelidiales. This conclusion, however, was based on several assumptions of homology relating to one-layered plug caps and *Gelidium*-type spore germination, in addition to the assumption of an ancestral isomorphic life history in Hildenbrandiales. One or more of these assumptions may be incorrect. *Hildenbrandia* undergoes secondary pit connection formation without the formation of conjunctor cells (Pueschel 1988), which suggests that not all secondary pit connections in red algae are homologous. Finding a similar ontogenetic sequence of secondary pit connection formation in other red algae may help resolve the systematic position of Hildenbrandiales.

J. Corallinales

Establishment of the order Corallinales was implicit in Pueschel and Cole (1982), and Silva and Johansen (1986) provided a formal diagnosis. In the cladograms of Gabrielson and Garbary (1987), Corallinales were highly variable with regard to their position on the trees and in relation to other orders. Characters relating to spore germination and to the morphology of thalli and of cap layers served as synapomorphies, placing corallines in lineages either with Palmariales, Hildenbrandiales, and Gelidiales or with Batrachospermales. The first two features are highly variable among orders of red algae, and problems with the homology of both the number and morphology of pit plug caps were discussed above and by Gabrielson and Garbary (1987). In the consensus tree, Corallinales is grouped in a large polytomy of orders and lineages of red algae but clearly is separated from the order Gigartinales (as Cryptonemiales), in which it previously was classified. Corallinales may represent one lineage of the original diversification of florideophyte red algae as indicated by the polytomy in the consensus cladogram. The apparent absence of cap membranes associated with the pit plugs (Pueschel 1987) supports this conclusion.

Silva and Johansen (1986) discuss only the extant family Corallinaceae and make no mention of the fossil family Solenoporaceae or of its relationship to Corallinaceae. The Solenoporaceae are considered close to the Corallinaceae (Wray 1977), and the two families should be considered in the same order. Steneck (1986) discussed the evolution of Corallinaceae in terms of morphological adaptations to environment and ecological interactions, particularly with herbivores. In a preliminary report Townsend (1982) presented a phylogenetic analysis of Corallinaceae in which *Sporolithon* was used as the outgroup. This is consistent with Cabioch (1970), who considered *Sporolithon* (as *Archaeolithothamnion*) ancestral in Corallinaceae. Cladistic and paleontological analyses (Wray 1977) indicated that corallines evolved early relative to many other florideophytes, and they may be part of the original diversification of florideophytes.

Two conflicting classifications have been proposed for the family, one by Cabioch (1972) and the other by Johansen (1981). The former classification is based on concepts of evolutionary relationships, whereas the latter is more strictly phenetic. Thus, in the Johansen (1981) classification, subfamilies are either geniculate or nongeniculate, whereas Cabioch (1972) allows both morphological types to occur within subfamilies. Woelkerling (1987) has extended the Johansen scheme by describing an additional subfamily, Choreonematoideae, for the parasitic genus *Choreonema*, based on a suite of vegetative and reproductive features. Because of their rigid construction and wide variety of morphological and developmental characters, the Corallinaceae is an ideal group in which to investigate generic relationships (e.g., Garbary & Johansen 1987 on *Lithothrix* and *Amphiroa*). The ongoing nomenclatural and taxonomic studies of Woelkerling (1987 and references therein) provide an excellent basis on which to pursue such investigations.

K. Palmariales

Hawkes and Scagel (1986a) provided a comprehensive review of Palmariales in a floristic treatment from British Columbia. The biphasic life history (carposporophytes absent) with macroscopic males and microscopic females (when unisexual) and the presence of a generative stalk cell in tetrasporogenesis provide a unique suite of characters for this monophyletic order. Magne (1982) and Gabrielson and Garbary (1986) considered this life history pattern an evolutionary reduction from the *Polysiphonia*-type, but Hawkes and Scagel (1986a) argued that "the absence of carposporophytes in Palmariales may not be due to a reduction and elimination of this phase ... because it may never have existed in these groups in the first place," thus inferring derivation from biphasic ancestors. Guiry (1987) also concluded that the ancestor for Palmariales had a biphasic life history. Magne (1987) has reinterpreted the life history of *Palmaria* as being triphasic. Accordingly, the tetrasporangium and its foot cell is considered the entire tetrasporophyte generation, and the morphological phase that develops after fertilization – but before tetrasporangium initiation – represents the carposporophyte. Thus, the *Palmaria* life history is interpreted as having an independent, free-living carposporophyte. With this diversity of opinion, it is clear that further studies are required before a more definite conclusion can be reached regarding the homologies of the different vegetative and reproductive phases involved in the *Palmaria* life history.

Guiry (1987) considered Palmariales to be among the primitive groups of florideophytes and suggested that they be placed with Nemaliales and Acrochaetiales as a primitive assemblage. Saunders (1987) considered there to be a close taxonomic link between Palmariales and Acrochaetiales and suggested that some species of Acrochaetiales with biphasic life histories might be better classified in Palmariales. These included such species as *Audouinella floridula*, *A. purpurea*, and *A. subimmersa*, where a tetrasporophyte developed after fertilization, rather than a carposporophyte [see Lee & Kurogi (1978) and Stegenga (1978) for details on reproduction].

L. Gigartinales

Following Kraft and Robins (1985), we include Cryptonemiales in Gigartinales. The merger of these orders clearly is controversial. Silva and Johansen (1986) provided the historical background for understanding this controversy, and Gabrielson and Garbary (1986) reviewed more recent arguments beginning with Searles (1968, 1983) and including Kraft (1975), Lebednik (1977), and finally Kraft and Robins (1985). Inasmuch as we recognize that Gigartinales sensu lato is paraphyletic, as it was before the inclusion of Cryptonemiales, we believe this is a necessary interim step in sorting out relationships among the included families. Recently, Maggs and Pueschel (1989) have segregated the agarophyte species of the genus *Ahnfeltia*, formerly classified in the Phyllophoraceae, to a new order, the Ahnfeltiales, based on features of reproductive morphology and pit plug ultrastructure. The agarophyte family Gracilariaceae likewise has been raised to ordinal status as Gracilariales (Fredericq & Hommersand 1989), based on the absence of auxiliary cells and of connecting filaments or cells. Fredericq and Hommersand also speculated that the above new orders, together with the Gelidiales, are a related assemblage. We support the removal of these orders from Gigartinales but believe that further evidence is needed to resolve the relationships among Ahnfeltiales, Gelidiales, and Gracilariales.

Recently a wide variety of morphological, ultrastructural, and biochemical studies have been carried out using Gigartinales. Despite this work, additional studies of reproductive morphology, particularly the details of early postfertilization stages, are needed for many genera, especially in the families Gigartinaceae, Gracilariaceae, Nemastomataceae, Hypneaceae, Plocamiaceae, and Sarcodiaceae. Such observations have proved useful in circumscribing families and in indicating relationships among families (see Kraft 1981 for review). These need to be supplemented with electron microscope studies of early stages of carposporophyte development along the lines of Ramm-Anderson and Wetherbee (1982) in Nemaliales and Broadwater and Scott (1982) in Ceramiales. Extensive studies of carposporogenesis have been carried out in several families of Gigartinales (see Kugrens & Delivopoulos 1986 and Tsekos & Schnefp 1983 and included references) and, although these studies did not provide discontinuities useful in segregating taxa, they provided a better understanding of the process of sporogenesis. Biochemical characterization of the cell wall matrix is providing useful taxonomic characters (Doty & Norris 1985; McCandless et al. 1982, 1983), and studies of spore germination and vegetative ontogeny (see review in Gabrielson & Garbary 1986) likewise will aid in elucidating relationships. The criteria of Kylin (1932) for circumscribing families in Gigartinales sensu stricto have proved inadequate, and all families in the order need to be recircumscribed using a suite of features (e.g., Gabrielson & Kraft 1984; Kraft 1978; Lindstrom & Scagel 1987). Not until families classified in Gigartinales have been rendered monophyletic will it be possible to resolve interfamilial, as well as interordinal relationships.

M. Bonnemaisoniales

This order, first proposed by Feldmann and Feldmann (1942) and originally based on the presence of triphasic, heteromorphic life histories, recently was supported by the morphological studies of Chihara and Yoshizaki (1972) and by the pit plug studies of Pueschel and Cole (1982). The two families, Bonnemaisoniaceae and Naccariaceae, currently placed in the order, are very different vegetatively. Gabrielson and Garbary (1987) suggested that Naccariaceae might have closer affinities with Gigartinales than with Bonnemaisoniaceae. The vegetative morphology of Bonnemaisoniaceae suggests that the family is related to Ceramiales. This conclusion previously was reached by Chihara and Yoshizaki (1978), but there are no synapomorphies (other than possibly bipolar spore germination) that support a sister taxon relationship between these orders.

Bonnemaisoniaceae appears to be monophyletic (Chihara & Yoshizaki 1972), but relationships among the included genera and species have yet to be resolved.

N. Rhodymeniales

Implicit in the Kraft and Robins' (1985) suggestion that Cryptonemiales and Gigartinales be merged is the suggestion that Rhodymeniales also be merged with Gigartinales. The sole diagnostic criterion for recognizing Rhodymeniales is the particular location of the auxiliary cell relative to the carpogonial branch. This feature alone is inadequate for distinguishing Rhodymeniales as an order coordinate with other Florideophytes, considering the variation in the position of carpogonial branches and auxiliary cells in the enlarged Gigartinales. We continue to recognize Rhodymeniales, however, because they appear to be a monophyletic group and would probably be one of the taxa segregated from a further enlarged Gigartinales when interfamilial relationships are resolved. In a modern floristic treatment, Hawkes and Scagel (1986b) discuss the morphology and systematics of the order.

O. Ceramiales

With a suite of defining characters (Bold & Wynne 1985; Gabrielson & Garbary 1986), this order is monophyletic. Various proposals for the relationships of the four families (Ceramiaceae, Delesseriaceae, Rhodomelaceae, and Dasyaceae) were made by Papenfuss (1944) and Hommersand (1963). Both authors agreed that Ceramiaceae is the most primitive family and that other families were derived from ceramiacean ancestors. Papenfuss suggested that Delesseriaceae, Rhodomelaceae, and Dasyaceae formed a monophyletic group, whereas Hommersand (1963, p. 323) suggested that these families "have originated independently from different prototypes within the Ceramiaceae, none of which are now living." If Hommersand is correct, then Ceramiaceae becomes paraphyletic with some of the descendants of the common ancestor being relegated to other families. This might account for the lack of resolution in the evolutionary relationships proposed by Moe and Silva (1979) for tribes of Ceramiaceae. With the detailed knowledge of morphology and development currently available in Ceramiales, we consider cladistic analyses to be a useful approach. The evolutionary trees of

Gordon (1972) in which character changes and putative synapomorphies are indicated directly on the trees provide excellent hypotheses for relationships that could be improved by more formal cladistic treatments.

Hommersand (1963, p. 322) commented on the similarity between the vegetative and reproductive organs of *Atractophora* (Naccariaceae) and *Wrangelia* (Ceramiaceae). Chihara and Yoshizaki (1972) suggested that Bonnemaisoniaceae and "probably" Naccariaceae be placed near the Ceramiales. These suggestions are more or less consistent with the cladogram of Gabrielson and Garbary (1987) (Fig. 18-1). It should be noted, however, that Hommersand (1963) considered Ceramiales to be derived from "one of the major independent lines of evolution that has originated from ancestors in the Nemalionales."

VI. RED ALGAE IN THE EUKARYOTIC DOMAIN

The relationship of red algae to other eukaryotic organisms is unknown and problematic. There has been a tendency over the past 20 years to suggest that Rhodophyta are either the most primitive eukaryotes or the sister group of remaining eukaryotes (e.g., Dodge 1980; Taylor 1978). The evidence supporting this is derived from two major features: (1) the absence of flagella, basal bodies, etc. and (2) the presence of phycobiliprotein pigments. This line of reasoning culminated with the analysis of Lipscomb (1985), who derived a phylogeny of eukaryotes in which red algae were the sister group of all other eukaryotes. No other organisms even were hypothesized as outgroups. This conclusion was supported by data from 5S rRNA sequences (Hori et al. 1985). A variety of alternative hypotheses have been expounded by other authors. These include the idea that Rhodophyta are related to (1) Cryptophyta (Lee 1974), (2) the higher fungi, i.e., the Eumycota (Demoulin 1974, 1985), and (3) Chlorophyta and/or Glaucophyta (Cavalier-Smith 1986; Melkonian 1984). Detailed evolutionary models exist to explain the origin of the red algal chloroplast (Whatley & Whatley 1981), but equivalent explanations for the origin of the remainder of the red algal cell are unresolved.

We believe that resolving the phylogenetic position of the red algae among eukaryotes is an important question for red algal systematics because it will help elucidate the nature of the ancestral red alga and the pattern of relationships within Rhodophyta. Relationships among orders closest to the ancestral progenitors of red algae (i.e., Compsopogonales, Rhodochaetales, Bangiales) are variable [see different cladograms in Gabrielson et al. (1985)] because some morphological and ultrastructural characters are polarized in different ways depending upon the outgroup that is used. Recognition of the correct sister group(s) for red algae will aid in resolving this problem by polarizing characters whose ancestral and derived states currently are unresolved.

VII. SUMMARY

The most important advance in red algal taxonomy is seen in the kind of questions that currently are being investigated. The information base has expanded vastly over the past decades through the examination of unstudied or poorly known groups, as has the technological sophistication in investigating taxonomic problems. Both the new information and the new techniques have enabled us to ask questions about the homologies of morphological and developmental features rather than merely describing features. Genetic studies should allow more fundamental understanding of morphological and developmental characters widely used in taxonomy. Sufficient model systems are available so that taxonomic and evolutionary problems can be examined using a variety of ultrastructural, chemotaxonomic, and developmental approaches. Ultrastructural studies on pit plugs (Pueschel & Cole 1982; Pueschel 1987) and mitosis (Scott 1986) and the more recent accounts of the life history of *Choreocolax* (Gigartinales) (Goff & Coleman 1984) and the development and consequences of secondary pit connection formation in *Polysiphonia* (Ceramiales) (Goff & Coleman 1986) and *Hildenbrandia* (Pueschel 1988) permit a sophistication in taxonomic analysis that previously was impossible. These approaches are providing new data to resolve problems at the ordinal and infra-ordinal ranks posed by the cladistic studies of Gabrielson et al. (1985) and Gabrielson and Garbary (1987). We see morphological studies of vegetative and reproductive morphology as the primary approach for resolving relationships among genera and families of red algae. The onus will be on workers carrying out such studies to express their results as explicit evolutionary hypotheses. Without the phylogenetic, biogeographic, and ecological interpretations of taxonomic work that are becoming standard in most groups of organisms, taxonomic analyses of red algae are a relatively sterile endeavor that will not attract future generations of scientists.

VIII. ACKNOWLEDGMENTS

Dr. M. Hommersand and Dr. M. Guiry provided valuable commentary on our initial manuscript. We thank Dan Belliveau and Alka Gautam for assistance in the preparation of the manuscript. This work was supported by NSERC Grant A2931 to D. G. P. W. G. acknowledges financial support from NSERC Grant A4471 to Dr. R. F. Scagel.

IX. REFERENCES

Abbott, I. A. 1962. Some *Liagora*-inhabiting species of *Acrochaetium*. *Occas. Pap. Bernice P. Bishop Mus.* 23: 77–120.

Abbott, L. A., Bisby, F. A. & Rogers, D. J. 1985. *Taxonomic Analysis in Biology*. New York: Columbia University Press.

Adey, W. H. & Steneck, R. S. 1985. Highly productive eastern Caribbean reefs: synergistic effects of biological, chemical, physical and geological factors. *NOAA Undersea Research Program* 3: 163–87.

Agardh, C. A. 1820. *Species Algarum. Vol. 1, part 1*. Lund: Berling.

Agardh, C. A. 1822. *Species Algarum. Vol. 1, part 2*. Lund: Berling.

Agardh, C. A. 1824. *Systema Algarum*. Lund: Berling.

Agardh, J. G. 1851–63. *Species Genera et Ordines Algarum . . .* Lund: Gleerup.

Agardh, J. G. 1876. *Species Genera et Ordines Algarum . . .* Weigel: Leipzig.

Aghajanian, J. G. & Hommersand, M. H. 1980. Growth and differentiation of axial and lateral filaments in *Batrachospermum sirodotii* (Rhodophyta). *J. Phycol.* 16: 15–28.

Akatsuka, I. 1986. Surface cell morphology and its relationship to other generic characters in non-parasitic Gelidiaceae (Rhodophyta). *Bot. Mar.* 29: 59–68.

Baum, B. R., Duncan, T. & Phillips, R. B. 1984. A bibliography on numerical phenetic studies in systematic botany. *Ann. Missouri Bot. Gard.* 71: 1044–60.

Berthold, G. 1882. Fauna und Flora des Golfes von Neapel. VIII. Die Bangiaceen des Golfes von Neapel. Leipzig.

Bird, C. J., van der Meer, J. P. & McLachlan, J. 1982. A comment on *Gracilaria verrucosa* (Huds.) Papenf. (Rhodophyta: Gigartinales). *J. Mar. Biol. Ass. U. K.* 62: 453–9.

Bird, C. J., Oliveira, E. C. de & McLachlan, J. 1986. *Gracilaria cornea*, the correct name for the western Atlantic alga hitherto known as *G. debilis* (Rhodophyta, Gigartinales). *Can. J. Bot.* 64: 2045–51.

Blunt, J. W., Lake, R. J. & Munro, M. H. G. 1984. Sesquiterpenes from the marine red alga *Laurencia distichophylla*. *Phytochemistry* 23: 1951–4.

Boillot, A. 1975. Cycle biologique de *Rhodochaete parvula* Thuret (Rhodophycées, Bangiophycidées). *Pub. Stn. Zool. Napoli* (supplement) 39: 67–83.

Boillot, A. 1978. Les ponctuations intercellulaire du *Rhodochaete parvula* Thuret (Rhodophycées, Bangiophycidées). *Rev. Algol.* N. S. 13: 252–8.

Bold, H. C. & Wynne, M. J. 1985. *Introduction to the Algae*, 2nd ed. Englewood Cliffs, N.J.: Prentice-Hall.

Bornet, E. & Thuret, G. 1866. Note sur la fécondation des Floridées. *Mem. Soc. Imp. Sci. Nat. Cherbourg* 12: 257–62.

Bornet, E. & Thuret, G. 1867. Recherches sur la fécondation des Floridées. *Ann. Sci. Nat. (Bot.)*, sér. 5, 7: 137–66.

Brawley, S. H. & Wetherbee, R. 1981. Cytology and ultrastructure. In *The Biology of Seaweeds*, eds. C. S. Lobban & M. J. Wynne, pp. 248–99. Berkeley: University of California Press.

Broadwater, S. T. & Scott, J. 1982. Ultrastructure of early development in the female reproductive system of *Polysiphonia harveyi* Bailey (Ceramiales, Rhodophyta). *J. Phycol.* 18: 427–41.

Burzycki, G. M. & Waaland, J. R. 1987. On the position of meiosis in the life history of *Porphyra torta* (Rhodophyta). *Bot. Mar.* 30: 5–10.

Cabioch, J. 1970. Sur l'importance de phénomènes cytologiques pour la systématique et la phylogénie des Corallinacées (Rhodophycées, Cryptonémiales). *C. R. Acad. Sci. (Paris)* sér. D 271: 296–9.

Cabioch, J. 1972. Étude sur les Corallinacées. II. La morphogenèse: conséquence systématique et phylogénétiques. *Cah. Biol. Mar.* 13: 137–288.

Cavalier-Smith, T. 1986. The kingdom Chromista: origin and systematics. *Prog. Phycol. Res.* 4: 309–47.

Chapman, D. J. 1974. Taxonomic status of *Cyanidium caldarium*, the Porphyridiales and Goniotrichales. *Nova Hedwigia* 25: 673–82.

Chihara, M. & Yoshizaki, M. 1972. Bonnemaisoniaceae: their gonimoblast development, life history and systematics. In *Contributions to the Systematics of Benthic Marine Algae of the North Pacific*, eds. I. A. Abbott & M. Kurogi, pp. 243–51. Kobe Jap. Soc. Phycology.

Chihara, M. & Yoshizaki, M. 1978. Anatomical and reproductive features of *Ptilonia okadai* (Rhodophyta, Bonnemaisoniaceae). *Phycologia* 17: 382–7.

Cole, K. M., Hymes, B. J. & Sheath, R. G. 1983. Karyotypes and reproductive seasonality of the genus *Bangia* (Rhodophyta) in British Columbia. *J. Phycol.* 19: 136–45.

Crews, P. & Selover, S. J. 1986. Comparison of the sesquiterpenes from the seaweed *Laurencia pacifica* and its epiphyte *Erythrocystis saccata*. *Phytochemistry* 25: 1847–52.

Croizat, L. 1964. *Space, Time, Form: the Biological Synthesis*. Caracas: published by the author.

Demoulin, V. 1974. The origin of Ascomycetes and Basidiomycetes: the case for a red algal origin. *Bot. Rev.* 40: 315–45.

Demoulin, V. 1985. The red algal–higher fungal phylogenetic link: the last ten years. *BioSystems* 18: 347–56.

Desikachary, T. V. 1958. Taxonomy of algae. *Mem. Ind. Bot. Soc.* 1: 52–62.

Desikachary, T. V. 1964. Status of the order Chaetangiales (Rhodophyta). *J. Ind. Bot. Soc.* 42a: 16–26.

Dixon, P. S. 1961. On the classification of the Florideae with particular reference to the position of the Gelidiaceae. *Bot. Mar.* 3: 1–16.

Dixon, P. S. 1973. *Biology of the Rhodophyta*. Edinburgh: Oliver & Boyd.

Dodge, J. D. 1980. Morphology and phylogeny of flagellated protists. In *Endocytobiology*, eds. W. Schwemmler & H. E. A. Schenk, pp. 33–50. Berlin: Walter de Gruyter.

Doty, M. S. & Norris, J. N. 1985. *Eucheuma* species (Solieriaceae, Rhodophyta) that are major sources of carrageenan. In *Taxonomy of Economic Seaweeds*, eds. I. A. Abbott & J. N. Norris, pp. 47–61. La Jolla: California Sea Grant College Program.

Fan, K.-C. 1959. On pit-connections in Bangiophycidae. *Nova Hedwigia* 1: 305–7.
Feldmann, J. 1953. L'évolution des organs femelles chez les Floridées. *Proc. Int. Seaweed Symp.* 1952: 11–12.
Feldmann, J. 1955. Un nouveau genre de Protofloridée: Colacodictyon, nov. gen. *Bull. Soc. Bot. Fr.* 102: 23–8.
Feldmann, J. 1967. Sur une Bangiophycées endozoique (*Neevea repens* Batters) et ses affinités (Rhodophyta). *Blumea* 15: 25–9.
Feldmann, J. & Feldmann, G. 1942. Recherches sur les Bonnemaisoniacées et leur alternance de générations. *Ann. Sci. Natl. Bot.*, sér. 11, 3: 75–175.
Fenical, W. & Norris, J. N. 1975. Chemotaxonomy in marine algae: chemical separation of some *Laurencia* species (Rhodophyta) from the Gulf of California. *J. Phycol.* 11: 104–8.
Fredericq, S. & Hommersand, M. H. 1989. Proposal of the Gracilariales ord. nov. (Rhodophyta) based on an analysis of the reproductive development of *Gracilaria verrucosa*. *J. Phycol.* 25: 213–27.
Fries, L. 1967. The sporophyte of *Nemalion multifidum* (Weber & Mohr) J. Ag. *Svensk. Bot. Tidskr.* 61: 457–62.
Fritsch, F. E. 1945. *The Structure and Reproduction of the Algae*. Vol. 2. Cambridge: Cambridge University Press.
Gabrielson, P. W. & Garbary, D. J. 1986. Systematics of red algae (Rhodophyta). *CRC Crit. Rev. Plant Sci.* 3: 325–66.
Gabrielson, P. W. & Garbary, D. J. 1987. A cladistic analysis of Rhodophyta: florideophycidean orders. *Br. Phycol. J.* 22: 125–38.
Gabrielson, P. W., Garbary, D. J. & Scagel, R. F. 1985. The nature of the ancestral red alga: inferences from a cladistic analysis. *BioSystems* 18: 335–46.
Gabrielson, P. W. & Kraft, G. T. 1984. The marine algae of Lord Howe Island (N.S.W.): the family Solieriaceae (Gigartinales, Rhodophyta). *Brunonia* 7: 217–51.
Garbary, D. 1978. On the phylogenetic relationships of the Acrochaetiaceae (Rhodophyta). *Br. Phycol. J.* 13: 247–54.
Garbary, D. 1979a. Numerical taxonomy and generic circumscription in the Acrochaetiaceae (Rhodophyta). *Bot. Mar.* 22: 477–92.
Garbary, D. 1979b. Patterns of cell elongation in some *Audouinella* spp. (Acrochaetiaceae: Rhodophyta). *J. Mar. Biol. Ass. U.K.* 59: 951–60.
Garbary, D. 1987a. A critique of traditional approaches to seaweed distribution in light of the development of vicariance biogeography. *Helgol. Meeresunters.* 41: 235–44.
Garbary, D. 1987b. *The Acrochaetiaceae (Rhodophyta): An Annotated Bibliography*. Bibliotheca Phycologica, Vol. 77. Berlin/Stuttgart: J. Cramer.
Garbary, D., Belliveau, D. & Irwin, R. 1988. Apical control of band elongation in *Antithamnion defectum* (Ceramiaceae). *Can. J. Bot.* 66: 1308–15.
Garbary, D. J. & Gabrielson, P. W. 1987. Acrochaetiales (Rhodophyta): taxonomy and evolution. *Cryptogam. Algol.* 8: 241–52.
Garbary, D. J., Hansen, G. I. & Scagel, R. F. 1980. A revised classification of the Bangiophyceae (Rhodophyta). *Nova Hedwigia* 33: 145–66.
Garbary, D. & Johansen, H. W. 1987. Morphogenesis and evolution in the Amphiroideae (Rhodophyta, Corallinaceae). *Br. Phycol. J.* 22: 1–10.
Garbary, D. & Scagel, R. F. 1980. Cross sections and species characterization in some *Iridaea* spp. *J. Phycol.* 16 (Suppl.): 12.
Gargiulo, G. M., De Masi, F. & Tripodi, G. 1987. Thallus morphology and spore formation in cultured *Erythrocladia irregularis* Rosenvinge (Rhodophyta, Bangiophycidae). *J. Phycol.* 23: 351–9.
Giraud, A. & Magne, F. 1968. La place de la méoise dans le cycle de développement de *Porphyra umbilicalis*. *C. R. Acad. Sci.* (Paris), sér. D 267: 586–8.
Glazer, A. N., West, J. A. & Chan, C. 1982. Phycoerythrins as chemotaxonomic markers in red algae: a survey. *Biochem. Syst. Ecol.* 10: 203–15.
Goff, L. J. 1982. The biology of parasitic red algae. *Prog. Phycol. Res.* 1: 289–369.
Goff, L. J. & Coleman, A. W. 1984. Elucidation of fertilization and development in a red alga by quantitative microspectrofluorometry. *Devel. Biol.* 102: 173–94.
Goff, L. J. & Coleman, A. W. 1986. A novel pattern of apical cell polyploidy, sequential polyploidy reduction and intercellular nuclear transfer in the red alga *Polysiphonia*. *Am. J. Bot.* 73: 1109–30.
Goff, L. J. & Coleman, A. W. 1987. The solution to the cytological paradox of isomorphy. *J. Cell Biol.* 104: 739–48.
Gordon, E. M. 1972. Comparative morphology and taxonomy of the Wrangelieae, Spondylothamnieae, and Spermothamnieae (Ceramiaceae, Rhodophyta). *Aust. J. Bot.* Suppl. 4: 1–180.
Gray, S. F. 1821. *A Natural Arrangement of British Plants*. 2 Vols. London: Baldwin, Craddock & Joy.
Guiry, M. D. 1978. The importance of sporangia in the classification of the Florideophyceae. In *Modern Approaches to the Taxonomy of Red and Brown Algae*, eds. D. E. G. Irvine & J. H. Price, pp. 111–44. London: Academic.
Guiry, M. D. 1987. The evolution of life history types in the Rhodophyta: an appraisal. *Cryptogam. Algol.* 8: 1–12.
Guiry, M. D. & West, J. A. 1983. Life history and hybridization studies on *Gigartina stellata* and *Petrocelis cruenta* (Rhodophyta) in the North Atlantic. *J. Phycol.* 19: 474–94.
Guiry, M. D., West, J. A., Kim, D.-H. & Masuda, M. 1984. Reinstatement of the genus *Mastocarpus* Kützing (Rhodophyta). *Taxon* 33: 53–63.
Harvey, W. H. 1836. Algae. In *Flora Hibernica*, 2 vols, ed. J. T. Mackay, pp. 157–256. Dublin: N. Curry & Son.
Hawkes, M. W. 1978. Sexual reproduction in *Porphyra gardneri* (Smith et Hollenberg) Hawkes (Bangiales, Rhodophyta). *Phycologia* 17: 329–53.
Hawkes, M. W. 1988. Evidence for sexual reproduction in *Smithora naiadum* (Erythropeltidales, Rhodophyta). *Br. Phycol. J.* 23: 327–36.
Hawkes, M. W. & Scagel, R. F. 1986a. The marine algae of British Columbia and northern Washington: division Rhodophyta (red algae): class Rhodophyceae, order Palmariales. *Can. J. Bot.* 64: 1148–73.
Hawkes, M. W. & Scagel, R. F. 1986b. The marine algae of British Columbia and northern Washington: division Rhodophyta (red algae), class Rhodophyceae, order Rhodymeniales. *Can. J. Bot.* 64: 1549–80.
Hennig, W. 1966. *Phylogenetic Systematics*. Urbana: University of Illinois Press.
Hollenberg, G. J. 1959. *Smithora*, an interesting new genus in the Erythropeltidaceae. *Pac. Nat.* 1: 3–11.
Hommersand, M. H. 1963. The morphology and classification of some Ceramiaceae and Rhodomelaceae. *Univ. Calif. Pub. Bot.* 35: 165–366.

Hommersand, M. H. 1986. The biogeography of South African marine red algae: a model. *Bot. Mar.* 29: 257–70.

Hommersand, M. H. & Fredericq, S. 1988. An investigation of cystocarp development in *Gelidium pteridifolium* with a revised description of the Gelidiales (Rhodophyta). *Phycologia* 27: 254–272.

Hori, H., Lim, B.-L. & Osawa, S. 1985. Evolution of green plants as deduced from 5S rRNA sequences. *Proc. Natl. Acad. Sci. USA* 82: 820–3.

Huisman, J. M. 1985. The *Scinaia* assemblage (Galaxauraceae, Rhodophyta): a re-appraisal. *Phycologia* 24: 403–18.

Huisman, J. M. 1987. The taxonomy and life history of *Gloiophloea* (Galaxauraceae, Rhodophyta). *Phycologia* 26: 167–74.

Hymes, B. J. & Cole, K. M. 1983. The cytology of *Audouinella hermannii* (Rhodophyta, Florideophycidae). I. Vegetative and hair cells. *Can. J. Bot.* 61: 3366–76.

Irvine, D. E. G. & John, D. M., eds. 1984. *Systematics of Green Algae*. London: Academic.

Johansen, H. W. 1981. *Coralline Algae, a First Synthesis*. Boca Raton, Fl.: CRC Press.

Joosten, A. M. T. & van den Hoek, C. 1986. World-wide relationships between red seaweed floras: a multivariate approach. *Bot. Mar.* 29: 195–214.

Kapraun, D. F. 1978. Field and cultural studies on selected North Carolina *Polysiphonia* species. *Bot. Mar.* 21: 143–53.

Kapraun, D. F., Lemus, A. J. & Bula-Meyer, G. 1983. The genus *Polysiphonia* (Rhodophyta, Ceramiales) in the tropical western Atlantic I. Columbia and Venezuela. *Bull. Mar. Sci.* 33: 881–98.

King, R. J., Puttock, C. F. & Vickery, R. S. 1988. A taxonomic study on the *Bostrychia tenella* complex (Rhodomelaceae, Rhodophyta). *Phycologia* 28: 10–19.

Kociolek, J. P. & Stoermer, E. F. 1986. Phylogenetic relationships and classification of monoraphid diatoms based on phenetic and cladistic methodologies. *Phycologia* 25: 297–303.

Kornmann, P. 1984. *Erythrotrichopeltis*, eine neue Gattung der Erythropeltidaceae (Bangiophyceae, Rhodophyta). *Helgol. Meeresunters.* 38: 207–24.

Kornmann, P. 1987. Der Lebenszyklus von *Porphyrostromium obscurum* (Bangiophyceae, Rhodophyta). *Helgol. Meeresunters.* 41: 127–37.

Kraft, G. T. 1975. Consideration of the order Cryptonemiales and of the families Nemastomataceae and Furcellariaceae (Gigartinales, Rhodophyta) in light of the morphology of *Adelophyton corneum* (J. Agardh) gen. et comb. nov. from southern Australia. *Br. Phycol. J.* 10: 279–90.

Kraft, G. T. 1978. Studies on the marine algae in the lesser-known families of the Gigartinales (Rhodophyta). III. The Mychodiaceae and Mychodeophyllaceae. *Aust. J. Bot.* 26: 515–610.

Kraft, G. T. 1981. Rhodophyta: morphology and classification. In *The Biology of Seaweeds*, eds. C. Lobban & M. J. Wynne, pp. 6–51. Berkeley: University of California Press.

Kraft, G. T. 1984. The red algal genus *Predaea* (Nemastomataceae, Gigartinales) in Australia. *Phycologia* 23: 3–20.

Kraft, G. T. & Robins, P. 1985. Is the order Cryptonemiales defensible? *Phycologia* 24: 67–77.

Kremer, B. P. 1982. *Cyanidium caldarium*: a discussion of biochemical features and taxonomic problems. *Br. Phycol. J.* 17: 51–61.

Kugrens, P. & Delivopoulos, S. G. 1986. Ultrastructure of the carposporophyte and carposporogenesis in the parasitic red alga *Plocamiocolax pulvinata* Setch. (Gigartinales, Plocamiaceae). *J. Phycol.* 22: 8–21.

Kützing, F. T. 1843. *Phycologia Generalis*. Leipzig: Brockhaus.

Kylin, H. 1916. Über die Befruchtung und Reduktionsteilung bei *Nemalion multifidum*. *Ber. Deut. Bot. Ges.* 34: 257–71.

Kylin, H. 1917. Über die Entwicklungsgeschichte von *Batrachospermum mcniliforme*. *Ber. Deut. Bot. Ges.* 35: 155–64.

Kylin, H. 1923. Studien über die Entwicklungsgeschichte der Florideen. *K. Svensk. Vet.-Akad. Handl.* 63 (11): 1–139.

Kylin, H. 1925. The marine red algae in the vicinity of the biological station at Friday Harbor, Wash. *Lunds Univ. Arsskr., N.F., Avd. 2* 21 (9): 1–87.

Kylin, H. 1928. Entwicklungsgeschichtliche Florideenstudien. *Lunds Univ. Arsskr., N.F., Avd. 2* 24 (4): 1–127.

Kylin, H. 1932. Die Florideenordnung Gigartinales. *Lunds Univ. Arsskr., N.F., Avd. 2* 28 (8): 1–88.

Kylin, H. 1956. *Die Gattungen der Rhodophyceen*. Lund: Gleerup.

Lamouroux, J. V. F. 1813. Essai sur les genres de la famille de Thalassiophytes, non articulées. *Ann. Mus. Natl. Hist. Nat., Paris* 20: 115–39, 267–93.

Lebednik, P. A. 1977. Postfertilization development in *Clathromorphum*, *Melobesia*, and *Mesophyllum* with comments on the evolution of the Corallinaceae and Cryptonemiales (Rhodophyta). *Phycologia* 16: 379–406.

Lee, R. E. 1974. Chloroplast structure and starch grain production as phylogenetic indicators in the lower Rhodophyceae. *Br. Phycol. J.* 9: 291–5.

Lee, R. E. 1980. *Phycology*. Cambridge: Cambridge University Press.

Lee, Y. P. 1987. Taxonomy of the Rhodochortonaceae (Rhodophyta) in Korea. *Kor. J. Phycol.* 2: 1–50.

Lee, Y.-P. & Kurogi, M. 1978. Sexual reproductive structures and postfertilization in *Rhodochorton subimmersum* Setchell et Gardner. *Jap. J. Phycol.* 26: 115–19.

Levin, M. H. & Rogers, D. J. 1964. A taximetric analysis of genera and families of the Nemalionales (Rhodophyceae, Florideae). *Am. J. Bot.* 51: 689.

L'Hardy-Halos, M.-Th., Larpent, J. P., Gaillard, J., Pellegrini, L. & Pellegrini, M. 1984. Morphogenèse expérimentale chez les algues. *Rev. Cytol. Biol. Végét. Bot.* 7: 311–62.

Lim, B.-L., Kawai, H., Hori, H. & Osawa, S. 1986. Molecular evolution of 5S ribosomal RNA from red and brown algae. *Jap. J. Genet.* 61: 169–76.

Lindstrom, S. C. 1987. Possible sister groups and phylogenetic relationships among North Pacific and North Atlantic red algae. *Helgol. Meeresunters.* 41: 245–60.

Lindstrom, S. C. 1988. The Dumontieae J. Agardh, a resurrected tribe of red algae (Dumontiaceae, Rhodophyta). *Phycologia* 27: 89–102.

Lindstrom, S. C. & Scagel, R. F. 1987. The marine algae of British Columbia, northern Washington and southeast Alaska: division Rhodophyta (red algae), class Rhodophyceae, order Gigartinales, family Dumontiaceae, with an introduction to the order Gigartinales. *Can. J. Bot.* 65: 2202–32.

Lipscomb, D. L. 1985. The eukaryotic kingdoms. *Cladistics*

1: 127–40.
Littler, M. M. & Littler, D. S. 1983. Heteromorphic life-history strategies in the brown alga *Scytosiphon lomentaria* (Lyngb.) Link. *J. Phycol.* 19: 425–31.
Lubchenco, J. & Cubit, J. 1980. Heteromorphic life histories of certain marine algae as adaptations to variations in herbivory. *Ecology* 61: 676–87.
Ma, J. H. & Miura, A. 1984. Observations of the nuclear division in the conchospores and their germlings in *Porphyra yezoensis*. *Jap. J. Phycol.* 32: 373–8.
Maddison, W. P., Donoghue, M. J. & Maddison, D. R. 1984. Outgroup analysis and parsimony. *Syst. Zool.* 33: 83–103.
Maggs, C. A. & Guiry, M. D. 1987. *Gelidiella calcicola* sp. nov. (Rhodophyta) from the British Isles and northern France. *Br. Phycol. J.* 22: 417–34.
Maggs, C. A. & Pueschel, C. M. 1989. Morphology and development of *Ahnfeltia plicata* (Rhodophyta): Proposal of Ahnfeltiales ord. nov. *J. Phycol.* 25: 333–51.
Magne, F. 1960a. Sur le lieu de la méiose chez le *Bonnemaisonia asparagoides* (Wood.) C. Ag. *C. R. Acad. Sci.* (Paris) sér. D 250: 2742–4.
Magne, F. 1960b. Le *Rhodochaete parvula* Thuret (Bangioidée) et sa reproduction sexuée. *Cah. Biol. Mar.* 1: 407–20.
Magne, F. 1961. Sur le cycle cytologique du *Nemalion helminthoides* (Velley) Batters. *C. R. Acad. Sci.* (Paris), sér. D 252: 157–9.
Magne, F. 1967. Sur le deroulement et le lieu de la meiose chez Lemaneacées (Rhodophycées, Nemalionales). *C. R. Acad. Sci.* (Paris), sér. D 265: 670–3.
Magne, F. 1982. On two new types of life history in the Rhodophyta. *Cryptogam. Algol.* 3: 265–71.
Magne, F. 1987. La tétrasporegenèse et le cycle de développement des Palmariales (Rhodophyta): une nouvelle interprétation. *Cryptogam. Algol.* 8: 273–80.
McCandless, E. L., West, J. A. & Guiry, M. D. 1982. Carrageenan patterns in the Phyllophoraceae. *Biochem. Syst. Ecol.* 10: 275–84.
McCandless, E. L., West, J. A. & Guiry, M. D. 1983. Carrageenan patterns in the Gigartinaceae. *Biochem. Syst. Ecol.* 11: 175–82.
McLachlan, J. & Bird, C. J. 1984. Geographical and experimental assessment of the distribution of *Gracilaria* species (Rhodophyta: Gigartinales) in relation to temperature. *Helgol. Meeresunters.* 38: 319–34.
Melkonian, M. 1984. D. Taxonomy I. Systematics and evolution of the algae. *Prog. Bot.* 46: 249–73.
Merola, A., Castaldo, R., De Luca, P., Gambardella, R., Musacchio, A. & Taddei, R. 1981. Revision of *Cyanidium caldarium*. Three species of acidophilic algae. *Giorn. Bot. Ital.* 115: 189–95.
Moe, R. L. 1985. *Gainia* and Gainiaceae, a new genus and family of crustose marine Rhodophyceae from Antarctica. *Phycologia* 24: 419–28.
Moe, R. L. & Silva, P. C. 1979. Morphological and taxonomic studies on Antarctic Ceramiaceae (Rhodophyceae). I. *Antarcticothamnion polysporum* gen. et sp. nov. *Br. Phycol. J.* 14: 385–405.
Naylor, S., Hanke, F. J., Manes, L. V. & Crews, P. 1983. Chemical and biological aspects of marine monoterpenes. In *Progress in the Chemistry of Organic Natural Products 44*, ed. W. Herz, H. Grisebach & G. W. Kirby, pp. 189–241. New York: Springer-Verlag.
Nelson, G. & Platnick, N. 1981. *Systematics and Biogeography: Cladistics and Vicariance*. New York: Columbia University Press.
Ohme, M., Kunifuji, Y. & Miura, A. 1986. Cross experiments of the color mutants in *Porphyra yezoensis* Ueda. *Jap. J. Phycol.* 34: 101–6.
Oltmanns, F. 1904–5. *Morphologie und Biologie der Algen*. 2 Vols. Jena: Fischer.
Papenfuss, G. F. 1944. Structure and taxonomy of *Taenioma*, including a discussion of the phylogeny of the Ceramiales. *Madroño* 7: 193–214.
Papenfuss, G. F. 1955. Classification of the algae. In *A Century of Progress in the Natural Sciences 1853–1953*, ed. E. L. Kessel, pp. 115–224. San Francisco: California Academy of Sciences.
Papenfuss, G. F. 1966. A review of the present system of classification of the Florideophycidae. *Phycologia* 5: 247–55.
Parsons, M. J. 1985. New Zealand seaweed flora and its relationships. *N. Z. J. Mar. Freshw. Res.* 19: 131–9.
Penrose, D. & Woelkerling, W. J. 1988. A taxonomic reassessment of *Hydrolithon* Foslie, *Porolithon* Foslie and *Pseudolithophyllum* Lemoine emend. Adey (Corallinaceae, Rhodophyta) and their relationships to *Spongites* Kützing. *Phycologia* 27: 159–76.
Philippi, R. A. 1837. Beweis, dass die Nulliporen Pflanzen sind. *Arch. Naturgesch.* 3 (1): 387–93.
Pielou, E. C. 1977. The latitudinal spans of seaweed species and their patterns of overlap. *J. Biogeogr.* 4: 299–311.
Prud'Homme van Reine, W. F. 1982. *A Taxonomic Revision of the European Sphacelariaceae (Sphacelariales, Phaeophyceae)*. Leiden: E. J. Brill, Leiden University Press.
Pueschel, C. M. 1987. Absence of cap membranes as a characteristic of pit plugs of some red algal orders. *J. Phycol.* 23: 150–6.
Pueschel, C. M. 1988. Secondary pit connections in *Hildenbrandia* (Rhodophyta, Hildenbrandiales). *Br. Phycol. J.* 23: 25–32.
Pueschel, C. M. & Cole, K. M. 1982. Rhodophycean pit plugs: an ultrastructural survey with taxonomic implications. *Am. J. Bot.* 69: 703–20.
Pueschel, C. M. & Magne, F. 1987. Pit plugs and other ultrastructural features of systematic value in *Rhodochaete parvula* (Rhodophyta, Rhodochaetales). *Cryptogam. Algol.* 8: 201–9.
Ramm-Anderson, S. M. & Wetherbee, R. 1982. Structure and development of the carposporophyte of *Nemalion helminthoides* (Nemaliales, Rhodophyta). *J. Phycol.* 18: 133–41.
Ramus, J. 1969. The developmental sequence of the marine red alga *Pseudogloiophloea* in culture. *Univ. Cal. Pub. Bot.* 52: 1–28.
Rivera, P., Astudillo, L., Rovirosa, J. & San-Martin, A. 1987. Halogenated monoterpenes of the red alga *Shottera nicaensis*. *Biochem. Syst. Ecol.* 15: 3–4.
San-Martin, A. & Rovirosa, J. 1986. Variations in the halogenated monoterpene metabolites of *Plocamium cartilagineum* of the Chilean coast. *Biochem. Syst. Ecol.* 14: 459–61.
Saunders, G. W. 1987. Investigation and taxonomic considerations of the red algal genus *Rhodophysema* Batters and a discussion of the Acrochaetiales–Palmariales complex of the Florideophycidae, M.Sc. thesis, Acadia University, Wolfville, Nova Scotia.
Schmitz, F. 1883. Untersuchungen über die Befruchtung der Florideen. *Sber. Akad. Wiss.* 1883: 215–58.
Schmitz, F. 1892. Florideae. In *Syllabus der Pflanzenfamilien*,

ed. A. Engler, pp. 16–23. Berlin: G. Borntraeger.
Schmitz, F. & Hauptfleisch, P. 1897. Rhodophyceae. In *Die Natürlichen Pflanzenfamilien, Teil 1, Abt. 2*, eds. A. Engler & K. Prantl, pp. 298–544. Leipzig: Engelmann.
Schweigger, A. F. 1819. *Beobachtungen auf naturhistorischen Reisen, anatomisch-physiologische Untersuchungen über Corallen*. Berlin: Reimer.
Scott, J. 1984. Electron microscope contributions to red algal phylogeny. *J. Phycol.* (Suppl.) 20: 6.
Scott, J. 1986. Ultrastructure of cell division in the unicellular red alga *Flintiella sanguinaria*. *Can. J. Bot.* 64: 516–24.
Scott, J. & Gabrielson, P. W. 1987. Bangiophycidae (Rhodophyta), a polyphyletic taxon: new evidence from ultrastructural studies. *J. Phycol.* (Suppl.) 23: 12.
Scott, J., Thomas, J. & Saunders, B. 1988. Primary pit connections in the freshwater red alga *Compsopogon coeruleus* (Balbis) Montagne (Compsopogonales, Rhodophyta). *Phycologia* 27: 327–33.
Searles, R. B. 1968. Morphological studies of red algae of the order Gigartinales. *Univ. Calif. Pub. Bot.* 43: 1–86.
Searles, R. B. 1983. Vegetative and reproductive morphology of *Dudresnaya georgiana* sp. nov. (Rhodophyta, Dumontiaceae). *Phycologia* 22: 309–16.
Seckback, J. 1987. Evolution of eukaryotic cells via bridge algae. *Ann. N.Y. Acad. Sci.* 503: 424–37.
Sheath, R. G. & Burkholder, J. M. 1983. Morphometry of *Batrachospermum* populations intermediate between *B. boryanum* and *B. ectocarpum* (Rhodophyta). *J. Phycol.* 19: 324–31.
Sheath, R. G. & Cole, K. M. 1984. Systematics of *Bangia* (Rhodophyta) in North America. I. Biogeographic trends in morphology. *Phycologia* 23: 383–96.
Sheath, R. G., Morison, M. O., Cole, K. M. & Vanalstyne, K. L. 1986. A new species of freshwater Rhodophyta, *Batrachospermum carpocontortum*. *Phycologia* 25: 321–30.
Silva, P. C. & Johansen, H. W. 1986. A reappraisal of the order Corallinales (Rhodophyceae). *Br. Phycol. J.* 21: 245–54.
Sjöstedt, L. G. 1926. Floridean studies. *Acta Univ. Lund, N.F., Avd. 2* 22 (4): 1–95.
Skuja, H. 1939. Versuch einer systematischen Einteilung der Bangioideen oder Protoflorideen. *Acta Horti Bot. Univ. Latv.* 11/12: 23–40.
Stegenga, H. 1978. The life histories of *Rhodochorton purpureum* and *Rhodochorton floridulum* (Rhodophyta, Nemaliales) in culture. *Br. Phycol. J.* 13: 279–89.
Stegenga, H. 1985. The marine Acrochaetiaceae (Rhodophyta) of southern Africa. *S. Afr. J. Bot.* 51: 291–330.
Stegenga, H. & van Wissen, M. J. 1979. Remarks on the life histories of three acrochaetioid algae (Rhodophyta, Nemaliales). *Acta Bot. Neerl.* 28: 97–115.
Steneck, R. S. 1986. The ecology of coralline algal crusts: convergent patterns and adaptive strategies. *Annu. Rev. Ecol. Syst.* 17: 272–303.
Steneck, R. S. & Watling, L. 1982. Feeding capabilities and limitation of herbivorous molluscs: a functional group approach. *Mar. Biol.* 68: 299–319.

Stevens, P. F. 1986. Evolutionary classification in botany, 1960–1985. *J. Arnold Arbor.* 67: 313–39.
Svedelius, N. 1915. Zytologisch-Entwicklungsgeschichtliche studien über *Scinaia furcellata* ein Beitrag zur Frage der Reductionsteilung der nicht tetrasporenbildenden Florideen. *Nova Acta Reg. Soc. Sci. Upsal., Ser. IV* 4: 1–55.
Taylor, F. J. R. 1978. Problems in the development of an explicit hypothetical phylogeny of the lower eukaryotes. *BioSystems* 10: 67–89.
Townsend, R. 1982. *Sporolithon*, an important genus of Corallinaceae (Rhodophyta). *Scientific Programme and Abstracts, First International Phycological Congress*, St. John's, Newfoundland, August 8–14, p. a50.
Tsekos, I. & Schnefp, E. 1983. The ultrastructure of carposporogenesis in *Gigartina teedii* (Roth) Lamour. (Gigartinales, Rhodophyceae): gonimoblast cells and carpospores. *Flora* 174: 191–211.
Ueda, K. & Chida, Y. 1987. Chloroplast structure of *Cyanidium caldarium* shown by freeze-substitution. *Br. Phycol. J.* 22: 61–5.
Umezaki, I. 1967. The tetrasporophyte of *Nemalion vermiculare* Sur. *Rev. Algol.* 9: 19–24.
van den Hoek, C. 1982. The distribution of benthic marine algae in relation to the temperature regulation of their life histories. *Biol. J. Linn. Soc.* 18: 81–144.
Waaland, S. D. & Waaland, J. R. 1975. Analysis of cell elongation in red algae by fluorescent labelling. *Planta* (Berl.) 126: 127–38.
West, J. A. & Hommersand, M. H. 1981. Rhodophyta: life histories. In *The Biology of Seaweeds*, eds. C. S. Lobban & M. J. Wynne, pp. 133–93. Berkeley: University of California Press.
Whatley, J. M. & Whatley, F. R. 1981. Chloroplast evolution. *New Phytol.* 87: 233–47.
Wiley, E. O. 1981. *Phylogenetics, the Theory and Practice of Phylogenetic Systematics*. New York: Wiley.
Wittmann, W. 1965. Aceto–iron–haematoxylin–chloral hydrate for chromosome staining. *Stain Tech.* 40: 161–4.
Woelkerling, W. J. 1971. Morphology and taxonomy of the *Audouinella* complex (Rhodophyta) in southern Australia. *Aust. J. Bot.* 1 (Suppl.): 1–91.
Woelkerling, W. J. 1983. The *Audouinella* (*Acrochaetium–Rhodochorton*) complex (Rhodophyta): present perspectives. *Phycologia* 2: 59–92.
Woelkerling, W. J. 1987. The genus *Choreonema* in southern Australia and its subfamilial classification within the Corallinaceae (Rhodophyta). *Phycologia* 26: 111–27.
Womersley, H. B. S. 1981. Biogeography of Australasian marine macroalgae. In *Marine Botany, An Australasian Perspective*, eds. M. N. Clayton & R. J. King, pp. 292–307. Melbourne: Longman Cheshire.
Wray, J. L. 1977. *Calcareous Algae*. Amsterdam: Elsevier.
Wynne, M. J. 1983. The current status of genera in the Delesseriaceae (Rhodophyta). *Bot. Mar.* 26: 437–50.
Yarish, C., Breeman, A. M. & van den Hoek, C. 1986. Survival strategies and temperature responses of seaweeds belonging to different biogeographic distribution groups. *Bot. Mar.* 29: 215–30.

Note added in proof: Magne (1989) proposed a classification with three subclasses Archeorhodophycidae, Metarhodophycidae and Eurhodophycidae.

Magne, E. 1989. Classification et phylogénie des Rhodophycées. *Cryptogam. Algol.* 10: 101–15.

Taxonomic index

Acanthopeltis japonica, 82, 239
Acanthophora spicifera, 80, 399
Acrochaetiaceae
  (= Rhodochortaceae), 76, 307–8, 319, 480, 484, 489
Acrochaetiales, 3, 26, 48, 75–6, 87, 276, 280, 308, 312, 314, 319, 434, 463–4, 466, 480, 485, 489
Acrochaetium. See also *Audouinella*, *Rhodochorton*
Acrochaetium botryocarpum, 76, 353. See also *Audouinella botryocarpa*
Acrochaetium daviesii, 191
Acrochaetium dictyota, 357
Acrochaetium hermannii, 353. See also *Audouinella hermannii*, *Audouinella violacea*
Acrochaetium pectinatum, 48, 50–1, 307, 365, 384, 461. See also *Audouinella simplex*
Acrochaetium proskaueri, 383, 393, 463–4
Acrochaetium secundatum, 280–1
Acrosymphytaceae, 316, 324, 350
Acrosymphyton, 314, 324
Acrosymphyton caribaeum, 326
Acrosymphyton firmum, 462
Acrosymphyton purpuriferum, 290, 383, 387, 462
Acrotylaceae, 309, 338
Aeodes ulvoides, 277, 235
Agardhiella, 338
Agardhiella subulata, 117, 127, 132, 134–5, 138, 150–2, 206, 215, 229, 245, 337, 362, 383, 393, 403
Agardhiella tenera. See *Agardhiella subulata*, *Sarcodiotheca gaudichaudii*
Aglaothamnion, 26, 267
Aglaothamnion byssoides, 79
Aglaothamnion feldmanniae, 266
Aglaothamnion tripinnatum, 266
Agmenellum, 216
Agmenellum quadruplicatum, 116
Ahnfeltia, 26, 229, 314, 456, 491
Ahnfeltia concinna, 229
Ahnfeltia durvillaei, 229
Ahnfeltia qiqartinoides, 229, 232
Ahnfeltia plicata, 83, 234, 237, 332, 378
Ahnfeltia torulosa, 229, 232
Ahnfeltiales, 3, 26, 332, 491
Amphineura cinerea, 439
Amphipoda, 438–9, 441

Amphiroa, 491
Amphiroa abberans. See *Marginisporum abberans*
Amphiroa ephedraea, 243
Amphiroa rigida, 357
Amphiroa zonata, 357
Anabolia nervosa, 439
Anacystis, 214, 216
Anatheca dentata, 234, 240
Anatheca montagnei, 83, 234
Anisoschizus propaguli, 357, 462
Anotrichum, 264, 268
Anotrichum tenue, 263–5, 269
Anthoxanthum odoratum, 469
Antithamnieae, 319
Antithamnion, 15, 24, 29, 88, 262, 265, 267–9, 392, 407, 484
Antithamnion cruciatum, 79, 357, 463
Antithamnion defectum, 49, 309
Antithamnion glanduliferum. See *Antithamnionella spirographidis*
Antithamnion kylinii, 130, 137, 260, 263–4
Antithamnion plumula. See *Pterothamnion plumula*
Antithamnion sparsum, 309. See also *Antithamnion defectum*
Antithamnion spirographidis. See *Antithamnionella spirographidis*
Antithamnion tenuissimum, 79
Antithamnionella, 263, 392
Antithamnionella floccosa, 357
Antithamnionella pacifica, 51, 58
Antithamnionella sarniensis, 462–4
Antithamnionella spirographidis, 21, 79, 262, 265, 397, 462, 467
Apoglossum ruscifolium, 80
Apophlaea sinclairii, 463
Archaeolithothamnion, 490
Archaeolithothamnion dimotum, 381, 403
Ascomycota, 126
Ascophyllum, 378, 398
Asparagopsis, 392, 456, 479, 482
Asparagopsis armata, 79, 384, 386, 397, 401, 462, 479
Asterocytis. See *Chroodactylon*
Atractophora hypnoides, 381, 383–4, 463–5
Audouinella, 280, 380, 386, 428, 437–43, 461, 463, 466, 484–5. See also *Rhodochorton*

Audouinella alariae, 383–4
Audouinella asparagopsis, 384
Audouinella bonnemaisoniae, 379
Audouinella botryocarpa, 76
Audouinella concrescens 76, 280. See also *Rhodochorton concrescens*
Audouinella floridula, 76, 224, 378, 491
Audouinella hermannii, 29, 76, 85, 280, 430, 438. See also *Acrochaetium hermannii*, *Audouinella violacea*
Audouinella macrospora, 434, 438, 447
Audouinella membranacea, 378
Audouinella occulta, 359
Audouinella purpurea, 384, 461, 491. See also *Rhodochorton purpureum*
Audouinella simplex, 461
Audouinella subimmersa, 491
Audouinella violacea, 280–1, 311, 318–19, 426–7, 429, 432–4, 438, 440–1, 443–5, 447–9. See also *Audouinella*, *Acrochaetium hermannii*
Asterocolax, 32

Bacillariophyceae (diatoms), 213, 396, 440, 442
Baetis rhodanii, 439
Baetis vagans, 439
Balbiania investiens, 434–4
Bangia, 17–18, 30, 32, 74, 93–4, 276, 307, 352, 365, 371, 464–5, 484, 489
Bangia atropurpurea, 48, 76, 85, 157, 161–4, 223–4, 238–9, 241, 246, 279, 380, 383–4, 386, 390–2, 433, 436, 461, 464
Bangia fuscopurpurea. See *Bangia atropurpurea*, *Bangia vermicularis*
Bangia vermicularis, 74–6, 85, 87, 90, 93, 461, 464
Bangiaceae, 75–6, 481, 489
Bangiales, 3, 11, 15, 26, 48, 75–6, 85, 94, 223, 279–80, 307, 340, 348, 351, 461, 463–4, 466–7, 481, 489
Bangiophycidae, 46, 74, 84, 276, 306–7, 310, 351, 363, 365
Basidiomycota, 126
Batrachospermaceae, 75, 78, 319, 349, 490
Batrachospermales, 3, 26, 75, 78, 87, 276, 280, 314, 319, 349, 352, 359, 368, 371, 434, 481, 489–90
Batrachospermum, 13, 15, 78, 177,

499

Batrachospermales (cont.)
224–6, 263–4, 269, 285, 312, 319,
355, 424, 430, 434, 437–44, 448,
464, 479, 483–4, 487
Batrachospermum arcuatum, 434
Batrachospermum atrum, 434
Batrachospermum boryanum, 284–5,
318–9, 426–8, 431, 434–6, 438,
444–6. See also *Batrachospermum
ectocarpum*
Batrachospermum carpocontortum, 447
Batrachospermum crouanianum, 434
Batrachospermum distensum, 434
Batrachospermum durum, 434
Batrachospermum ectocarpum, 124–7,
130, 132, 134–5, 137–8, 142–3. See
also *Batrachospermum boryanum*
Batrachospermum intortum, 445
Batrachospermum keratophytum, 424,
426–7, 430–1, 434, 438, 444. See
also *Batrachospermum vagum*
Batrachospermum lochmodes, 445
Batrachospermum macrosporum, 429, 431
Batrachospermum mahabaleshwarensis,
78
Batrachospermum moniliforme, 78, 188,
424–34, 438, 444, 447, 449
Batrachospermum pseudocarpum, 445
Batrachospermum sirodotii, 48, 285.
See also *Batrachospermum virgatum*
Batrachospermum testale, 430, 434
Batrachospermum vagum, 429–31, 445,
447, 449. See also *Batrachospermum
keratophytum*
Batrachospermum virgatum, 425–7,
434, 438, 440, 444
Besa papillaeformis. See *Besa stipitata*
Besa stipitata, 234
Blinksiaceae, 350
Boldia, 276, 279, 444, 488
Boldia erythrosiphon, 81, 127, 131, 424,
426–7
Boldiaceae, 81, 351, 481, 487–8
Bonnemaisonia, 308, 457, 479, 482
Bonnemaisonia asparagoides, 79,
330–1, 381, 384, 401, 465, 467, 479
Bonnemaisonia hamifera, 79, 296–9,
384–5, 397
Bonnemaisonia nootkana, 130, 331–2,
379
Bonnemaisoniaceae, 29, 79, 330, 349,
480, 482, 485, 493
Bonnemaisoniales, 24, 26, 28, 79, 87,
276, 290, 296–8, 308, 314, 328, 349,
355, 361, 462, 465–7, 492
Bornetia californica, 48, 51, 62–3, 67
Bossiella, 395
Bossiella orbigniana, 357
Bostrychia, 51, 157, 396, 398–9, 484
Bostrychia binderi, 159, 388–9
Bostrychia montagnei, 188
Bostrychia radicans, 49, 163, 426, 430,
443–4
Bostrychia scorpioides, 157, 188
Bostrychia tenella, 399

Botryocladia, 29
Botryocladia ardreana, 369
Botryocladia pseudodichotoma, 407
Botryoglossum farlowianum, 49
Brachycentrus, 439–41
Brachycentrus americanus, 439
Brachycentrus numerosus, 439
Brongniartella byssoides, 381
Bryophyta (mosses), 442
Bryothamnion triquetrum, 239

Calliarthron, 392
Calliarthron tuberculosum, 243, 404
Calliblepharis ciliata, 230
Calliblepharis jubata, 82, 230
Calliblepharis lanceolata. See
Calliblepharis jubata
Callithamnion, 89, 214, 263–4, 267,
269, 312, 314, 391, 467
Callithamnion arbuscula, 398
Callithamnion baileyi, 49
Callithamnion bipinnatum, 467
Callithamnion brachiatum. See
Callithamnion tetragonum
Callithamnion byssoides, 51, 79, 105,
111–12, 294–5, 359, 381, 383, 386
Callithamnion cordatum, 263, 266–7,
270
Callithamnion corymbosum, 49, 79
Callithamnion decompositum, 368
Callithamnion halliae, 49
Callithamnion hookeri, 79, 262, 368,
380, 383, 385–9, 401
Callithamnion paschale, 51
Callithamnion purpurascens, 79
Callithamnion roseum, 79, 262, 294
Callithamnion tetragonum, 79, 85, 158,
383, 467
Callithamnion tetricum, 79
Callithamnion vermicularis, 290, 462,
465
Callocolax neglectus, 379
Callophycus densa, 243
Callophyllis, 309, 379
Callophyllis cristata (= *Euthora
cristata*), 381, 385
Callophyllis flabellulata, 403
Callophyllis laciniata, 383
Caloglossa, 398
Caloglossa leprieurii, 80, 163, 388, 433
Calosiphonia vermicularis, 384, 462,
465
Calosiphoniaceae, 316, 322
Campylaeophora hypnaeoides, 239–40
Cardiocladus, 439
Catenella, 398
Catenella caespitosa, 156, 247
Catenella nipae, 230
Catenella opuntia. See *Catenella
caespitosa*
Catenella repens. See *Catenella
caespitosa*
Caulacanthaceae, 230, 316, 322, 350
Caulacanthus ustulatus, 229, 399
Centroceras clavulatum, 153, 381, 399

Ceramiaceae, 79, 90, 288, 298, 312,
319, 349, 362, 482, 492–3
Ceramiales, 3, 13, 24, 26, 29, 33, 75,
79, 85, 87–9, 93–4, 156, 276, 290,
294, 296, 308–9, 312, 317, 328, 350,
362, 430, 457, 462–4, 466–7, 492–3
Ceramium, 89, 152, 237, 264, 306, 371,
387, 465, 478
Ceramium boydenii, 238–40
Ceramium ciliatum, 158, 390
Ceramium cruciatum, 79, 93
Ceramium deslongchampii, 79
Ceramium diaphanum, 459
Ceramium echionotum, 262–3
Ceramium fastigiatum, 79, 93
Ceramium fimbriatum, 79, 88, 93
Ceramium flabelligerum, 390
Ceramium gracillimum, 79, 89, 93
Ceramium japonicum, 79
Ceramium pacificum, 240
Ceramium pedicillatum, 390
Ceramium rubrum, 49, 153, 177, 238,
265, 393
Ceramium shuttleworthianum, 264–5,
398
Ceramium strictum, 368, 381, 459, 462
Ceramium tenuicorne, 164, 381, 393,
395
Chaetangium fastigiatum. See
Nothogenia fastigiata
Chaetangium saccatum. See *Nothogenia
erinacea*
Chaetomorpha darwinii, 155
Champia parvula, 84, 104–5, 111–12,
115
Champiaceae, 84, 319, 337, 350, 362
Chantransia. See subject index
Chantransia hermannii. See
Acrochaetium, *Audouinella
hermannii*
Chara, 21–2
Cheumatophysche, 439
Chlamydomonas, 212–13
Chlorella, 183
Chloroperla tripunctata, 439
Chlorophyta (chlorophytes or green
algae), 3, 67, 173, 176, 185, 190,
289, 436, 440–1, 444, 478
Chondria armata, 80
Chondria baileyana, 126, 140
Chondria crassicaulis, 80
Chondria dasyphylla, 80
Chondria decipiens, 240
Chondria nidifica, 80
Chondria tenuissima, 80, 86
Chondriellaceae, 349
Chondrococcus, 324, 366. See also
Portieria
Chondrococcus hornemannii. See
Portieria hornemannii
Chondrus, 48, 51, 63, 112, 179, 339,
394, 408
Chondrus canaliculatus, 233
Chondrus crispus, 80, 85, 89, 104,
111–12, 115, 118, 151–2, 174,

Taxonomic index

176–7, 179, 185, 188, 191–3, 195–6, 228, 232–3, 242, 245, 261, 290, 365, 378, 381, 385–8, 391–4, 396, 398, 401, 403–7, 462
Chondrus ocellatus, 228
Chondrus verrucosa, 228
Choreocolacaceae, 82, 349
Choreocolax, 493
Choreocolax polysiphoniae, 48–9, 66–7, 89, 266, 379
Chroodactylon, 276
Chroodactylon ornata, 226. See also *Chroodactylon ramosum*
Chroodactylon ramosum, 393, 444. See also *Chroodactylon ornata*
Chroomonas, 183
Chroothece, 276
Chrysocapsa epiphytica, 444
Chrysophyceae (chrysophytes or golden algae), 156, 163, 176, 444
Chylocladia, 337
Chylocladia kaliformis. See *Chylocladia verticillata*
Chylocladia verticillata, 84, 336
Cirrulicarpus carolinensis, 290, 328
Cladophora glomerata, 441
Clathromorphum circumscriptum, 381, 385, 394
Cloeon, 439, 441
Codium, 162
Colaconema, 485
Coleoptera (beetles), 438–9
Collisella, 394
Compsopogon, 12, 15, 22, 26, 82, 278–9, 424, 432–3, 437, 443, 488–9
Compsopogon coeruleus, 12, 81, 127, 131, 138, 209, 426–7, 432, 444
Compsopogon hookeri, 188, 430
Compsopogonaceae, 81, 351, 487–8
Compsopogonales, 3, 26, 75, 81, 84, 87, 279, 307, 340, 348, 351, 372, 462, 464, 466, 468, 481, 487–8
Compsopogonopsis, 279, 488
Compsothamnion thuyoides, 367
Constantinea, 19
Constantinea rosa-marina, 384
Constantinea subulifera, 265, 384, 404, 408
Corallina, 21, 243, 392, 396, 399
Corallina officinalis, 27, 48, 60–1, 82, 90, 153, 156–7, 236, 241, 378, 391, 398, 401, 403
Corallina rubens. See *Jania rubens*
Corallina vancouveriensis, 391
Corallinaceae, 13, 82, 310, 312, 349, 395, 481, 485, 490
Corallinales, 3, 26, 29, 32, 82, 320, 349, 362, 371, 463, 481, 490–1
Cordylecladia erecta, 384
Crouania attenuata, 289, 291–3, 359
Cruoria, 484
Cruoria cruoriaeformis, 350
Cruoria purpurea, 350
Cruoriopsis, 484
Crustacea, 393–4

Cryptonemiales, 3, 26, 47, 228, 308, 316, 372
Cryptophyta (cryptophytes), 183, 204, 493
Cryptopleura ramosa, 80
Cryptopleura violacea, 378
Cryptosiphonia, 406
Cubiculosporaceae, 350
Cumagloia, 353
Cumagloia andersonii, 47–8, 283
Cyanidiaceae, 481
Cyanidiophyceae, 172, 184–5, 193–4, 486
Cyanidioschyzon merolae, 172, 176–7, 188
Cyanidium, 12, 15, 17–18, 46, 48, 486
Cyanidium caldarium, 115, 151, 156, 161, 172, 188, 206, 214–16, 486
Cyanophora, 216
Cyanophora paradoxa, 115–16, 215
cyanophyta (blue-green algae or cyanobacteria), 2, 182, 185, 206, 210–12, 214, 216, 347, 441, 444
Cystocloniaceae, 82, 230, 309, 338
Cystoclonium purpurascens. See *Cystoclonium purpureum*
Cystoclonium purpureum, 82, 230

Dasya, 31–2
Dasya arbuscula. See *Dasya hutchinsiae*
Dasya baillouviana, 49, 51, 80, 124–7, 129–30, 136–8, 140–1, 362, 388
Dasya elegans. See *Dasya baillouviana*
Dasya hutchinsiae, 80
Dasya pedicellata. See *Dasya baillouviana*
Dasyaceae, 80, 296, 317, 319, 328, 492
Delisea, 290
Delisea compressa, 465
Delisea fimbriata, 465
Delisea japonica, 462, 465
Delisea pulchra, 462
Delesseria, 189, 205
Delesseria crassifolia, 239
Delesseria sanguinea, 80, 90, 159, 163, 381, 384, 387, 389, 405, 408
Delesseria serrulata, 240
Delesseriaceae, 80, 294, 309, 338, 350, 485, 492
Dermatolithon corallinae, 357
Dermatolithon litorale, 357, 463–4
Dermonema frappieri, 283, 320–1
Deucalion levringii, 357, 462
Devaleraea, 88
Devaleraea ramentacea, 27, 84, 86, 104, 111, 114, 391, 403
Devaleraea yendoi, 84, 87, 90
Diamesa, 439
Dibusa angata, 439, 442, 449
Dicranemaceae, 314, 319, 328
Dicusmoecus gilvipes, 448
Digenia simplex, 239–40, 396
Dilsea carnosa, 82
Diptera (particularly chironomids), 438–40

Dohrniella neapolitana, 355
Drosophila melanogaster, 103
Dudresnaya, 291, 312, 314–16, 322, 478, 481
Dudresnaya crassa, 48–9, 311, 315
Dudresnaya japonica, 290, 465
Dumontia, 371
Dumontia contorta, 82, 188, 236, 290–1, 381, 384, 386, 398, 401, 403
Dumontia filiformis. See *Dumontia contorta*
Dumontia incrassata. See *Dumontia contorta*
Dumontiaceae, 82, 315–16, 322, 324, 349, 355–6, 466, 484–5
Dunaliella bardawil, 117
Dunaliella salina, 117

Endocladia, 310
Endocladia muricata, 47–8, 50, 236, 240, 248, 390, 396, 398
Endocladiaceae, 349
Enteromorpha, 461
Enteromorpha intestinalis, 117, 161
Ephemerella, 439
Ephemeroptera (mayflies), 438–9, 440–1
Erythrocladia, 48, 465, 487–8
Erythrocladia irregularis, 276–7, 462
Erythrocladia subintegra, 277–8. See also *Sahlingia subintegra*
Erythrocystis saccata, 482
Erythrodermis, 484
Erythropeltidaceae, 82, 351, 365, 481, 487–8
Erythropeltidales, 3, 82, 276–7. See also Compsopogonales
Erythropeltis subintegra. See *Sahlingia subintegra*
Erythrotrichia, 488
Erythrotrichia boryana, 278–9
Erythrotrichia carnea, 48, 276–8, 392, 462
Erythrotrichopeltis. See *Porphyrostromium*
Escherichia coli, 103
Eucheuma, 234, 461
Eucheuma acanthocladum. See *Meristiella gelidium*
Eucheuma arnoldii, 234
Eucheuma cottonii, 229. See *Kappaphycus cottonii*
Eucheuma gelatinae, 229, 231, 234
Eucheuma gelidium, 245, 388. See also *Meristiella gelidium*
Eucheuma isiforme, 229, 234, 387–8, 461
Eucheuma nudum. See *Eucheuma isiforme*
Eucheuma odontophorum, 234
Eucheuma okamurai, 229. See also *Kappaphycus cottonii*
Eucheuma platycladum, 234
Eucheuma procrusteanum. See *Kappaphycus procrusteanum*

Eucheuma spinosum, 229, 403
Eucheuma uncinatum, 229, 403
Euglena, 188
Euglenophyceae, 184
Euthora cristata. See *Callophyllis cristata*

Falkenbergia rufolanosa. See *Asparagopsis armata*
Farlowia, 291, 456
Farlowia compressa. See *Farlowia mollis*
Farlowia conferta, 383–4
Farlowia irregularis. See *Masudaphycus irregularis*
Farlowia mollis, 290, 383–4, 403
Fauchea laciniata, 463
Faucheocolax, 20
Flintiella, 11–12, 21, 487
Flintiella sanguinaria, 124–7, 129, 133–4, 136–8, 142–3
Florideophyceae, 187–8
Florideophycidae, 47, 73–4, 84, 86, 90, 275–6, 298, 306–8, 310, 312, 314, 319, 340, 342, 352
Foslielia farinosa, 357
Fremyella, 216
Fucus, 395, 400
Furcellaria fastigiata. See *Furcellaria lumbricalis*
Furcellaria foliifera, 177
Furcellaria lumbricalis, 74, 82, 87, 90, 230, 232, 234, 241, 378, 381, 393, 395, 401, 403
Furcellariaceae, 82, 230, 291, 316, 350, 485

Gainiaceae, 349
Galaxaura, 226, 243, 312, 320, 368
Galaxaura corymbifera, 83
Galaxaura diesingiana, 83, 283, 311, 322
Galaxaura falcata, 283
Galaxaura fastigiata. See *Galaxaura oblongata*
Galaxaura oblongata, 187, 243
Galaxaura rugosa, 224, 226
Galaxaura squalida. See *Galaxaura rugosa*
Galaxaura tenera, 83
Galaxauraceae, 83, 319–20, 349, 485, 489
Galaxaurales, 480
Galdieria sulphuraria, 172, 188
Gammarus fasciatus, 439–41
Gardneriella tuberifera, 48, 60, 359
Gastroclonium, 367
Gastroclonium coulteri. See *Gastroclonium subarticulatum*
Gastroclonium ovatum, 390
Gastroclonium subarticulatum, 205, 213, 381, 385, 391, 463
Gastropoda, 393–5, 398, 438–9, 444
Gelidiaceae, 82, 480, 485
Gelidiales, 3, 26, 75, 82, 87–8, 276, 294, 312, 314, 320, 349, 358, 360, 368, 467, 490–1
Gelidiella acerosa, 82, 399, 404
Gelidiella calcicola, 369
Gelidiellaceae, 82
Gelidium, 20, 105–6, 111–12, 239, 295, 310, 314, 321, 371, 390, 399, 403, 407, 467
Gelidium amansii, 158, 239–40, 380–1, 386, 388
Gelidium caulacanthum, 393
Gelidium chilensis, 381
Gelidium corneum. See *Gelidium sesquipedale*
Gelidium coulteri, 105, 111, 157, 245, 403
Gelidium crinale, 299
Gelidium filicinum, 239
Gelidium japonicum, 239
Gelidium latifolium, 82
Gelidium lingulatum, 238, 381
Gelidium nudifrons, 392–3
Gelidium pristoides, 244, 246, 358
Gelidium pteridifolium, 320–1, 324
Gelidium purpurascens, 238
Gelidium pusillum, 23, 399
Gelidium robustum, 404, 407
Gelidium sesquipedale, 82, 240
Gelidium subcostatum, 239–40
Gelidium vagum, 82
Gelidium versicolor, 397
Gibsmithia, 356
Gibsmithia hawaiiensis, 355
Gigartina, 51, 207, 245, 265, 339, 399, 467
Gigartina acicularis, 383–4, 399
Gigartina agardhii. See *Mastocarpus jardinii*
Gigartina apoda, 204
Gigartina atropurpurea, 228
Gigartina canaliculata, 228, 399, 406
Gigartina chamissoi, 228
Gigartina exasperata, 118, 261, 381, 392, 403
Gigartina latissima, 177
Gigartina mamillosa. See *Mastocarpus stellatus*
Gigartina ochotensis. See *Mastocarpus pacificus*
Gigartina papillata. See *Mastocarpus papillatus*
Gigartina pistillata, 82
Gigartina radula, 406
Gigartina stellata. See *Mastocarpus stellatus*
Gigartina stiriata, 406
Gigartina teedii, 383
Gigartinaceae, 82, 228, 233, 309, 339, 349, 485, 492
Gigartinales, 3, 26, 47, 75, 82, 87, 94, 228, 276, 290, 294, 296, 308, 315, 322, 328, 333, 338, 342, 361, 371–2, 462–7, 481, 491–2
Glaucophyta, 493
Glaucosphaera vacuolata, 132

Gloeophycus koreanum, 355
Gloiopeltis, 240, 248, 332, 387
Gloiopeltis cervicornis, 236, 238
Gloiopeltis furcata, 236, 334
Gloiophloea, 480
Gloiophloea scinaioides, 355
Gloiosiphonia, 290–1, 342, 384
Gloiosiphonia californica, 463
Gloiosiphonia capillaris, 293–4, 332–3, 359, 384, 463–4
Gloiosiphonia verticillaris, 290
Gloiosiphoniaceae, 309, 316, 332, 349
Gonimophyllum skottsbergii, 367
Goniobasis, 439
Goniotrichaceae, 481, 487
Goniotrichales, 3, 481. See also Porphyridiales
Goniotrichopsis sublittoralis, 276
Goniotrichum. See *Stylonema*
Gracilaria, 17, 63, 74, 83, 90, 94, 112, 115–16, 237, 240, 244, 247, 314, 386, 393–4, 396, 398, 400, 403, 406, 485
Gracilaria abbottiana, 240
Gracilaria andersonii, 378, 403
Gracilaria arcuata, 404
Gracilaria bursa–pastoris, 82, 238, 240, 246
Gracilaria cervicornis, 238, 240
Gracilaria chilensis, 82
Gracilaria compressa. See *Gracilaria bursa–pastoris*
Gracilaria conferta, 246
Gracilaria coronopifolia, 244, 246
Gracilaria corticata, 246
Gracilaria cylindrica, 240, 244
Gracilaria damaecornis, 238, 240
Gracilaria edulis, 244, 246, 403
Gracilaria elliptica, 235
Gracilaria eucheumoides, 238, 240
Gracilaria ferox. See *Gracilaria cervicornis*
Gracilaria foliifera, 82, 111, 154–6, 177, 204–5, 213
Gracilaria lichenoides, 244
Gracilaria multipartita, 83
Gracilaria pacifica, 83
Gracilaria pristoides, 246
Gracilaria secundata, 104, 238
Gracilaria sjoestedtii, 83, 111, 117, 237–8. See also *Gracilariopsis lemaneiformis*
Gracilaria tenuistipitata, 238
Gracilaria textorii, 238, 240, 381
Gracilaria tikvahiae, 83, 88, 90–1, 104–18, 149–52, 192–5, 204, 213, 227, 238, 240, 244–7, 266, 381, 385–8, 391, 393, 401, 403
Gracilaria verrucosa, 83, 90, 185, 238, 240, 244–7, 291, 310–11, 333, 335, 387, 391, 393, 396, 401, 403
Gracilariaceae, 82, 240, 310, 314, 333, 349, 491–2
Gracilariales, 3, 26, 332–3, 491
Gracilariophila, 63

Taxonomic index

Gracilariophila oryzoides, 48
Gracilariopsis lemaneiformis, 48, 310–11
Gracilariopsis sjoestedtii. See *Gracilaria andersonii, Gracilariopsis lemaneiformis*
Grateloupia, 236, 327, 332
Grateloupia divaricata, 235
Grateloupia elliptica, 381
Grateloupia filicina, 153, 327
Grateloupia lanceolata, 235
Griffithsia, 22, 28, 153–4, 158, 183, 261, 263–4, 268–9, 483
Griffithsia bornetiana. See *Griffithsia globulifera*
Griffithsia corallina. See *Griffithsia corallinoides*
Griffithsia corallinoides, 79
Griffithsia floculosa, 47, 51, 67, 69, 124, 127
Griffithsia globulifera, 50, 266, 357
Griffithsia monilis, 149, 153–6, 160–3, 179, 188, 190
Griffithsia pacifica, 9, 14, 17, 21, 49, 80, 89, 116–17, 153, 183, 186, 209, 211, 214–15, 241, 261, 264–5, 267–70
Griffithsia tegis, 62–3
Grinnellia americana, 49, 383, 386
Gymnocodiaceae, 349
Gymnogongrus, 233, 291, 456
Gymnogongrus chiton, 233
Gymnogongrus crenulatus, 245
Gymnogongrus crustiforme, 229
Gymnogongrus devoniensis, 83
Gymnogongrus flabelliformis, 205, 229
Gymnogongrus griffithsiae, 83
Gymnogongrus humilus, 229
Gymnogongrus leptophyllus, 229, 463
Gymnogongrus linearis, 83, 233, 378
Gymnogongrus nodiferus, 229
Gumnogongrus norvegicus, 233
Gymnogongrus patens, 338, 340
Gymnogongrus platyphyllus. See *Gymnogongrus chiton*
Gymnophloeaceae. See Nemastomataceae

Haematocelis, 484
Halarachnion, 456
Halarachnion ligulatum, 381
Halesus radiatus, 439
Haliptilon cuvieri, 141, 368
Halopitys pinastroides, 81
Halosaccion, 88
Halosaccion americanum, 381. See also *Halosaccion glandiforme*
Halosaccion glandiforme, 180
Halosaccion ramentacea. See *Devaleraea ramentacea*
Halosaccion saccatum. See *Devaleraea yendoi*
Halosaccion yendoi. See *Devaleraea yendoi*
Halurus equisetifolius, 80, 383

Halymenia, 306
Halymenia durvillaei, 179
Halymenia floresia, 158, 355, 462
Halymenia latifolia, 355, 384
Halymenia venusta, 236
Halymeniaceae, 228, 235, 316, 327, 349, 355
Harveyella, 47
Harveyella mirabilis, 48, 82, 90
Helminthocladia australis, 283
Helminthocladia calvadosii, 83, 355
Helminthocladia pacifica, 355
Helminthocladia papenfussii, 83
Helminthocladia senegalensis, 83
Helminthocladiaceae (= Liagoraceae), 83
Helminthora, 353
Helminthora australis, 283
Helminthora divaricata, 83, 354
Heribaudiella fluviatilis, 444
Heringia, 332
Heringia mirabilis, 334
Hermea, 188
Heterosiphonia, 328
Heterosiphonia asymmetrica. See *Heterosiphonia densiuscula*
Heterosiphonia berkeleyi, 330
Heterosiphonia densiuscula, 463
Heterosiphonia japonica. See *Heterosiphonia densiuscula*
Heterosiphonia plumosa, 80, 158
Heterosiphonia pulchra, 80
Hildenbrandia, 28, 32, 369, 437, 442–3, 463, 490, 493
Hildenbrandia crouanii, 25
Hildenbrandia prototypus. See *Hildenbrandia rubra*
Hildenbrandia rivularis, 426, 430, 432–3, 438, 443–4
Hildenbrandia rubra, 16, 19, 399
Hildenbrandiales, 3, 26, 75, 349, 371, 463, 490
Holmsella, 310
Holmsella pachyderma, 48, 311
Hommersandia maximacarpa, 328
Hummbrella hydra, 290–1, 359, 465
Hydrolithon boergesenii, 381, 403–4
Hydropsyche instabilis, 439
Hymenea, 314
Hymenocladioideae, 362
Hymenoclonium, 479, 484
Hymenoclonium serpens, 479
Hypnea, 48, 338, 406
Hypnea ceramioides, 235
Hypnea cervicornis, 235, 399, 403
Hypnea chordacea, 403
Hypnea japonica, 235
Hypnea musciformis, 150–1, 230, 235, 245, 339, 393, 399, 403, 405
Hypnea nidifica, 235, 403
Hypnea spicifera, 235
Hypnea spinella, 399
Hypnea valentiae, 153, 399
Hypneaceae, 230, 309, 338, 492
Hypneocolax stellaris, 230, 235

Hypoglossum hypoglossoides, 80, 383, 386
Hypoglossum woodwardii. See *Hypoglossum hypoglossoides*

Iridaea, 29, 51, 178, 242, 265, 391, 401, 470, 484
Iridaea boryana, 233, 394
Iridaea ciliata, 233, 381, 386
Iridaea cordata. See *Iridaea splendens*
Iridaea cornucopiae, 228–9, 403, 463
Iridaea flaccida. See *Iridaea splendens*
Iridaea heterocarpa, 82, 403
Iridaea laminarioides, 156, 381, 386
Iridaea latissima, 178
Iridaea membranacea, 233
Iridaea obovata, 204
Iridaea splendens, 48, 82, 161, 177, 228, 391, 393–5, 401, 403–4, 407, 463
Iridaea undulosa, 233
Isoperla grammatica, 439

Jania, 399
Jania crassa, 399
Jania rubens, 82, 90

Kallymenia, 290, 328
Kallymenia cribrogloea, 328
Kallymenia reniformis, 327–9
Kallymeniaceae, 290, 309, 316, 327, 349
Kappaphycus cottonii, 229, 232, 234
Kappaphycus procrusteanum, 234
Kraftia, 316
Kraftia dichotoma, 315

Laminaria, 395
Laminariales, 90
Laurencia, 24, 29, 51, 88, 93, 237, 394, 399, 482
Laurencia arbuscula, 81
Laurencia caraibica, 81, 87
Laurencia catarinensis, 81
Laurencia distichophylla, 482
Laurencia hybrida, 81
Laurencia nana. See *Laurencia caraibica*
Laurencia nipponica, 81, 90
Laurencia obtusa, 81, 187, 386
Laurencia okamurae, 81, 90–1
Laurencia pacifica, 399, 482
Laurencia papillosa, 81, 399
Laurencia parvula, 81
Laurencia pinnata, 81, 90
Laurencia pinnatifida, 81, 239
Laurencia pygmaea, 399
Laurencia spectabilis, 48, 66, 241
Laurencia undulata, 81, 90
Leachiella, 31
Lemanea, 85, 175, 180, 183, 319, 359, 424, 427–8, 430, 433, 437, 439, 443–5, 449, 456–7, 459, 464, 479
Lemanea annulata, 223–4
Lemanea australis, 79, 286–8
Lemanea condensata, 430

Lemanea fluviatilis, 79, 186, 188, 263, 430, 434, 436, 438, 447
Lemanea fucina, 424–7, 432–4, 444, 449
Lemanea mamillosa, 79, 176–7, 180, 183, 427, 433
Lemaneaceae, 79, 319, 349, 490
Lemaneales, 3. See also Batrachospermales
Lenormandia prolifera, 242
Leptophytum laeve, 379, 381
Lernaea, 444
Leuctra hippopus, 439
Leveillea jungermannioides, 81
Liagora, 243, 312
Liagora farinosa, 283, 462
Liagora harveyana, 366
Liagora mucosa, 281–5, 318–19
Liagora tetrasporifera, 457, 459, 470
Liagoraceae, 349, 353, 369, 489
Lithophyllum corallinae, 82
Lithophyllum impressum, 395
Lithophyllum incrustans, 406, 463, 465
Lithophyllum litorale, 82
Lithophyllum orbiculatum, 382
Lithophyllum tortuosum, 404
Lithothamnion glaciale, 381
Lithothamnion japonicum, 236
Lithothamnion phymatodeum, 395
Lithothamnion whidbeyense, 395
Lithothrix, 491
Lithothrix aspergillum, 357, 465
Littorina littorea, 394
Lomentaria, 467
Lomentaria articulata, 383, 398
Lomentaria baileyana, 84, 124–9, 132–5, 137, 142–3, 224, 383, 386
Lomentaria clavellosa, 84, 89
Lomentaria hakodatensis, 462
Lomentaria orcadensis, 84, 93, 383, 463
Lomentaria rosea, 84
Lomentaria umbellata, 156, 188
Lomentariaceae, 84, 357, 362

Macrocystis, 395
Macrocystis pyrifera, 151
Marchantia polymorpha, 115
Marginisporum aberrans, 82, 236
Mastocarpus, 291, 456, 464–5, 485
Mastocarpus jardinii, 463
Mastocarpus ochotensis. See *Mastocarpus pacificus*
Mastocarpus pacificus, 233, 463, 465
Mastocarpus papillatus, 228, 394–5, 406–8, 457, 463, 465
Mastocarpus stellatus, 82, 162, 164, 228, 233, 245, 381–2, 384–5, 387–8, 391–2, 396, 398, 401, 407, 463–5
Masudaphycus, 486
Masudaphycus irregularis, 359
Maydenoptila cuneola, 439, 442, 449
Mazoyerella arachnoidea, 357
Melobesia farinosa, 82. See also *Fosliella farinosa*

Melobesia mediocris, 48
Membranoptera alata, 80, 188, 289, 299, 378
Membranoptera platyphylla, 124, 126–7, 129–30, 137
Meredithia microphylla, 290–1, 382, 384
Meristiella gelidium, 245, 388, 404
Meristotheca papulosa, 382
Metamastophora, 320
Metamastophora flabellata, 311, 320
Microcladia, 51–2, 67–8, 482
Microcladia borealis, 262
Microcladia coulteri, 48
Microspora, 444
Minium, 51
Minium parvum, 49, 63–5, 467
Monosporus australis, 356
Monosporus pedicellatus, 356
Murrayella periclados, 399
Mychodea, 338
Mychodea carnosa, 340
Mychodeaceae, 309, 338
Myriogramme, 328
Myriogramme livida, 330
Mystacides sepulchralis, 439

Naccaria, 328, 331, 370, 479
Naccaria wiggii, 331, 465
Naccariaceae, 328, 493
Nemaliales, 3, 26, 48–9, 75, 83, 243, 276, 280–5, 308, 314, 319, 462, 464–5, 480–1, 489
Nemalion, 226, 308, 312, 314, 320, 370, 459, 489
Nemalion helminthoides, 83, 85, 150, 177, 283, 313, 319, 369, 386, 390, 392–3, 407
Nemalion multifidum. See *Nemalion helminthoides*
Nemalion vermiculare, 224, 226, 354, 462
Nemalionales. See Nemaliales
Nemalionopsis, 355
Nemalionopsis shawii, 444
Nemastomataceae, 236, 316, 322, 326, 349, 369, 492
Nemastomales, 479
Neoagardhiella baileyi. See *Agardhiella subulata*, *Sarcodiotheca gaudichaudii*
Neodilsea borealis, 27
Neogoniolithon megacarpum, 404
Neogoniolithon westindianum, 404
Neorhodomela, 51
Neorhodomela larix, 239, 378, 404–6
Neurocaulon, 291
Nienburgia andersoniana, 80
Nitophyllum hollenbergii, 80
Nitophyllum laceratum. See *Cryptopleura ramosa*
Nitophyllum punctatum, 80, 357
Nostoc, 210
Notemigonus crysoleucas, 448
Nothogenia erinacea, 83
Nothogenia fastigiata, 224

Odonthalia corymbifera, 239
Odonthalia dentata, 405, 408
Odonthalia floccosa, 81, 90, 387–8
Odonthalia kamtschatica, 240
Oenothera, 112
Olisthodiscus luteus, 115
Onychostoma laticeps, 439
Opuntiella, 23, 29
Opuntiella californica, 229, 234
Oreinus prenanti, 439
Orthocladius, 439
Ozophora norrisii, 234

Pachymenia carnosa, 235
Pachymenia hieroglyphica, 235–6
Pachymenia hymantophora, 228, 235
Padina, 244
Palmaria, 17, 20, 24, 26, 32, 115, 176, 307, 319, 491
Palmaria mollis, 84, 90
Palmaria palmata, 15–16, 19, 23, 24, 27, 30, 48, 84–5, 90, 93, 104, 106, 108, 111–12, 114, 116, 118, 130, 141, 153–6, 176–7, 179, 182, 223, 360, 365, 382, 386, 390–1, 393, 395, 398, 403, 459
Palmaria stenogona, 223, 240
Palmariaceae, 84, 94, 307, 349
Palmariales, 3, 26, 47, 84, 312, 319, 463, 466–7, 489, 491
Paragoniolithon solubile, 404
Pelargonium, 112
Pentaneura, 439
Perlodes, 439
Petrocelidaceae, 228, 233, 349, 362, 372, 485
Petrocelis, 484
Petrocelis hennedyi, 361, 372
Petrocelis franciscana, 457. See also *Mastocarpus papillatus*
Petrocelis middendorfii, 408. See also *Mastocarpus papillatus*
Peyssonnelia, 312, 324
Peyssonnelia dubyi, 311, 325
Peyssonnelia immersa, 369
Peyssonnelia obscura, 83
Peyssonnelia squamaria, 83
Peyssonneliaceae, 312, 316, 323–4, 349
Phacelocarpaceae, 309, 359
Phacelocarpus, 296
Phaeophyta (brown algae), 3, 65, 90, 185, 389, 397–9, 444
Phormidium, 444
Phycodrys rubens, 80, 378, 382, 385, 387
Phycodrys setchellii, 49
Phycodrys sinuosa. See *Phycodrys rubens*
Phyllophora, 241
Phyllophora antarctica, 386, 403
Phyllophora brodiaei. See *Phyllophora truncata*
Phyllophora crispa, 248
Phyllophora nervosa, 229, 232–3

Taxonomic index

Phyllophora pseudoceranoides, 83, 229, 233, 245, 248
Phyllophora truncata, 83, 229, 245, 382, 385, 395
Phyllophoraceae, 229, 349, 361, 366, 491
Phymatolithon calcareum, 387
Phymatolithon polymorphum, 382
Physa, 439, 444
Pikea californica, 290, 355, 462
Plantae, 1
Platoma bairdii, 291
Platoma marginifera, 355
Platythamnion, 52, 407
Plectoptera (stoneflies), 438–9
Pleonosporium, 51
Pleonosporium borreri, 266
Pleonosporium squarrulosum, 49
Pleonosporium vancouverianum, 124, 367–8
Pleuroceridae, 444
Plocamiaceae, 309, 328, 350, 482, 492
Plocamium, 328, 396, 482
Plocamium cartilagineum, 47, 49, 51, 235, 329, 382, 391, 403–4, 482
Plocamium telfaire, 382
Plocamium violaceum, 235
Plumaria, 379
Plumaria elegans, 262, 368, 380, 382, 387, 390, 398
Pneophyllum lobescens, 357
Pneophyllum plurivalidum, 357
Pneophyllum zonale, 357
Polyideaceae, 75, 83, 312, 316, 323, 349
Polyides caprinus. See *Polyides rotundus*
Polyides rotundus, 83, 235
Polyneura, 178
Polyneura hillae, 383, 386
Polyneura latissima, 49, 177–8, 405
Polysiphonia, 20, 28, 31, 51, 60, 63, 74–5, 81, 85–9, 93–4, 110, 237, 306, 309–10, 312, 314, 316, 328, 342, 379, 396, 407, 445, 459, 483, 485, 493
Polysiphonia binneyi, 75, 81, 93
Polysiphonia boldii, 382, 386, 459
Polysiphonia breviarticulata, 81
Polysiphonia brodiaei, 81
Polysiphonia confusa, 66, 89
Polysiphonia denudata, 11, 75, 81, 93, 124–8, 130, 132–3, 137, 139, 142–3, 382, 386
Polysiphonia elongata, 81
Polysiphonia ferulacea, 81, 87, 91, 382, 399, 462
Polysiphonia flexicaulis, 73, 81, 177, 357
Polysiphonia fibrillosa, 462
Polysiphonia harveyi, 28, 30–1, 33, 81, 296, 298, 311, 317
Polysiphonia hemisphaerica, 459
Polysiphonia japonica, 81, 378
Polysiphonia lanosa, 81, 152–3, 157, 160–4, 188, 239–40, 378, 382, 388–9, 398
Polysiphonia mollis, 49, 67–8, 81, 85, 89
Polysiphonia morrowii, 240
Polysiphonia nigrescens, 81, 242, 382, 388, 390–1
Polysiphonia pacifica, 25
Polysiphonia platycarpa, 81
Polysiphonia scopulorum, 399
Polysiphonia simplex, 399
Polysiphonia subtilissima, 81, 382, 388–9
Polysiphonia urceolata, 75, 81, 93, 153, 158, 382, 386, 392
Polysiphonia violacea. See *Polysiphonia flexicaulis*
Porolithon, 395
Porolithon pachydermum, 404
Porphyra, 18, 22, 74–5, 78, 85, 87, 90–1, 93–4, 104, 106, 112, 117–18, 153, 158, 162, 164, 194, 212, 239–40, 261, 265, 276, 279–80, 307, 314, 378, 385–6, 390–1, 394, 396, 398–9, 406–7, 456–7, 459, 461, 464–8, 487, 489
Porphyra abbottae, 76, 91
Porphyra acanthophora, 85, 399
Porphyra argentinensis, 352
Porphyra atropurpurea, 76
Porphyra capensis, 238, 406
Porphyra carolinensis, 76, 87, 93, 276–7, 279, 463
Porphyra columbina, 76, 238
Porphyra crispata, 76
Porphyra cuneiformis, 77
Porphyra dentata, 76
Porphyra drewii, 85
Porphyra gardneri, 30, 76, 352, 365, 461, 464
Porphyra haitanensis, 76, 238
Porphyra irregularis, 76
Porphyra kanakaensis, 76, 463
Porphyra katadae, 76–7
Porphyra kinositae, 77
Porphyra lanceolata, 77, 93, 404
Porphyra leucosticta, 77, 86, 93, 159, 195, 223–4, 238–9, 382, 384, 388
Porphyra linearis, 77, 117, 238, 352, 404
Porphyra lucasii, 85
Porphyra maculosa, 77, 87, 352
Porphyra marginata, 77
Porphyra miniata, 14, 77
Porphyra moriensis, 77
Porphyra naiadum. See *Smithora naiadum*
Porphyra nereocystis, 77
Porphyra norrisii, 77, 85
Porphyra okamurae, 75, 77
Porphyra oligospermatangia, 77
Porphyra onoii, 77
Porphyra papenfussii, 77, 87, 90, 92, 463
Porphyra perforata, 77, 91–3, 117, 153–4, 156, 158–61, 180, 238, 393
Porphyra pseudocrassa, 75, 77
Porphyra pseudolanceolata, 77, 87, 93
Porphyra pseudolinearis, 75, 77
Porphyra pujalsii, 85
Porphyra purpurea, 77, 151, 153–60, 162–4, 387
Porphyra rizzinii, 85
Porphyra rosengurttii, 77, 87, 383–5, 463
Porphyra sanjuanensis, 77, 85, 352, 461
Porphyra schizophylla, 77, 85, 90, 92–3, 387
Porphyra segregata, 85
Porphyra seriata, 75, 77
Porphyra smithii, 77
Porphyra spiralis, 77, 87, 93
Porphyra suborbiculata, 382
Porphyra subtumens, 78, 85, 462
Porphyra tasa, 78
Porphyra tenera, 77, 87, 149, 158, 180, 223–4, 239, 241, 363, 378, 380, 382, 384, 387–8, 391–3
Porphyra thuretii, 85
Porphyra torta, 78, 87, 91, 363
Porphyra umbilicalis, 77–8, 86, 93, 104–5, 179, 193, 204, 207, 223, 238–9, 242, 363, 388, 390–2, 398, 401
Porphyra variegata, 14, 34, 78, 87, 261
Porphyra yezoensis, 75, 78, 86–7, 106, 111, 117, 152, 188, 205, 213, 215, 363, 386, 461, 467
Porphyridiaceae, 75, 84, 348, 481–487
Porphyridiales, 3, 15, 23, 26, 75, 84, 276, 348, 481, 487
Porphyridium, 10–12, 18, 20, 22, 24, 157, 176, 178–80, 187, 194, 276–7, 456, 487
Porphyridium aerugineum, 84, 115, 152, 209, 215
Porphyridium cruentum. See *Porphyridium purpureum*
Porphyridium purpureum, 8, 19, 84, 117, 124–30, 132–3, 135, 137–9, 142–3, 152, 162, 176–9, 181, 188, 190, 193, 203–17
Porphyrodiscus, 484
Porphyrodiscus simulans, 396
Porphyropsis, 278, 459, 465, 488
Porphyropsis coccinea, 277–8, 352, 462
Porphyropsis vexillaris, 279
Porphyrostromium, 279, 456–7, 465, 468, 488
Porphyrostromium ciliare (ciliaris), 351, 462
Porphyrostromium obscurum, 351, 462
Portieria, 324
Portieria hornemannii, 235, 298–9, 325
Poterioochromonas malhamensis, 156, 163
Predaea, 326–7, 332, 481, 485
Predaea bisporifera, 359
Predaea feldmannii, 327

Predaea weldii, 309
Prionitis lanceolata, 403
Prochlorophyta, 347
Promoresia tardella, 439
Protoctista, 1
Protonemura meyeri, 439
Psectrocladus, 439
Pseudoanabaena, 216
Pseudogloiophloea. See *Scinaia*
Pseudolithophyllum lichenare, 395
Pseudoscinaia. See *Scinaia*
Pterocladia, 239–40
Pterocladia caerulescens, 403
Pterocladia capillacea, 391, 393, 399, 401, 403
Pterocladia heteroplatos, 406
Pterocladia lucida/pinnata, 238
Pterocladia tenuis, 239
Pterocladiophilaceae, 350
Pterosiphonia, 407
Pterosiphonia complanata, 397
Pterothamnion crispum, 80
Pterothamnion plumula, 80, 111, 382, 390, 459–60
Ptilonia mooreana, 465
Ptilonia okadae, 465
Ptilota plumosa, 369
Ptilota serrata, 80, 90–1, 382, 386, 463
Ptiloteae, 319
Pyrrhophyta (dinoflagellates), 213

Ralfsia, 396
Rhabdonia coccinea, 289
Rhabdoniaceae, 332
Rhithrogenia semicolorata, 439
Rhizophyllidaceae, 235, 312, 316, 323
Rhodella, 10, 12, 18, 46, 48, 224, 226
Rhodella cyanea, 12
Rhodella maculata, 12, 133, 194, 224–5, 247, 393
Rhodella reticulata, 12, 18–19, 127, 131–2
Rhodella violacea, 12, 127, 133–4, 211, 232
Rhodochaetaceae, 84, 488
Rhodochaetales, 3, 26, 75, 84, 87, 276, 307, 340, 348, 372, 466, 481
Rhodochaete, 12, 25, 56, 307, 456–7, 459, 468
Rhodochaete parvula, 84, 276, 348, 352, 365, 372
Rhodochorton, 51. See also *Audouinella*
Rhodochorton concrescens, 383, 463–4. See also *Audouinella concrescens*
Rhodochorton floridulum. See *Audouinella floridula*
Rhodochorton purpurpeum, 307, 365, 461
Rhodoglossum californicum, 229
Rhodoglossum hemisphaericum, 229
Rhodogramma, 176
Rhodomela confervoides, 81, 164

Rhodomela larix. See *Neorhodomela larix*
Rhodomela subfusca. See *Rhodomela confervoides*
Rhodomela virgata, 81
Rhodomelaceae, 49, 56, 80, 296, 319, 482, 492
Rhodopeltis, 323
Rhodophycophyta, 1
Rhodophycota, 1
Rhodophyllis bifida. See *Rhodophyllis divaricata*
Rhodophyllis divaricata, 82
Rhodophysema, 456–7, 459, 489
Rhodophysema elegans, 83, 365, 396, 463
Rhodosorus marinus, 127, 138, 226
Rhodothamniella floridula, 76, 90, 308
Rhodymenia, 334, 337–8
Rhodymenia californica, 463, 465
Rhodymenia foliifera, 153
Rhodymenia howeana, 336
Rhodymenia pacifica, 462, 465
Rhodymenia palmata. See *Palmaria palmata*
Rhodymenia pertusa, 84, 223, 407
Rhodymenia pseudopalmata, 84, 336
Rhodymeniaceae, 84, 319, 362
Rhodymeniales, 3, 26, 49, 84, 276, 290, 294, 308–9, 317, 319, 334, 350, 355, 362, 371, 462–4, 466–7, 492
Rhodymenioideae, 362
Rissoella verruculosa, 230–1, 235
Rissoellaceae, 230, 350

Sahlingia subintegra, 462
Sarcodiaceae, 309, 350, 492
Sarcodiotheca furcata, 230
Sarcodiotheca gaudichaudii, 48, 60, 230, 352, 378
Sarconema, 366
Sarconema filiforme, 230
Scagelia, 51–2, 67–8
Scagelia pylaisaei, 54, 58–60, 80, 89
Schimmelmannia plumosa, 290, 465
Schizymenia, 265, 291, 456
Schizymenia pacifica, 235, 403
Schmitzia, 457
Schmitzia evanescens, 462, 465
Schmitzia hiscockiana, 290, 322, 325, 384, 463–4
Schottera, 482
Schottera nicaeensis, 267
Scinaia, 308, 320, 479–80, 485
Scinaia capensis, 83
Scinaia complanata, 322
Scinaia confusa, 83, 355, 462
Scinaia forcellata, 83, 355
Scinaia furcellata. See *Scinaia forcellata*
Scinaia turgida, 83, 355
Scytosiphon, 485
Sebdenia dichotoma, 290–1
Sebdeniaceae, 349
Seirospora, 51

Seirospora griffithsiana, 49
Seirospora seirosperma, 355–6
Sericostoma personatum, 439
Serraticardia maxima, 236
Siderocystis confervarum, 444
Siercapsa major, 444
Sirodotia, 427–8, 444
Sirodotia huillensis, 78
Sirodotia suecica, 426–7, 434, 438
Sirodotia tenuissima, 438, 444
Smithora, 12, 278, 456, 465, 468, 488
Smithora naiadum, 239, 278, 351–2, 371, 398, 456, 462, 464
Solenoporaceae, 470
Solieria, 332, 366
Solieria australis, 334
Solieria chordalis, 234
Solieriaceae, 83, 229, 309, 316, 332, 338, 343, 350, 362, 485
Sorella repens, 80, 93
Spermothamnion repens, 73, 80, 87–9
Spermothamnion roseolum, 80
Spermothamnion snyderae. See *Tiffaniella snyderae*
Spermothamnion turneri, 73, 80, 87
Sphaerococcaceae, 309, 350
Spinacia, 188
Sporolithon, 490
Stenogramme interrupta, 83, 229, 234
Stenonema, 439
Stigeoclonium, 444
Stylonema, 465
Stylonema alsidii, 48, 150, 226, 276, 278, 386, 392
Stylonema cornu–cervi, 276
Suhria, 321
Suhria vittata, 321, 324, 358
Synarthrophyton, 320
Synechococcus leopoliensis, 152
Synechocystis, 216

Tanakaella, 357
Taraxacum, 460
Tenaciphyllum lobatum, 234
Tetraspora, 444
Thorea, 319, 355, 430
Thorea bachmanii, 445
Thorea ramosissima, 288
Thoroeaceae, 319, 490
Thuretellopsis peggiana, 384, 462
Tichocarpaceae, 228, 309
Tichocarpus crinitis, 228, 235
Tiffaniella, 312
Tiffaniella apiculata, 366
Tiffaniella snyderae, 80, 263, 368
Trichogloea, 320–1
Trichoptera (caddisflies), 438, 441, 448
Tuomeya, 286, 319
Tuomeya americana, 79, 286, 425–9, 440–1, 444
Tuomeya fluviatilis. See *Tuomeya americana*
Turnerella, 456

Turnerella mertensiana, 230, 234
Turnerella pennyi, 83

Ulva, 406
Ulva lactuca, 154–5

Valonia, 162–3, 223
Valonia ventricosa, 222
Vaucheria, 21, 444
Verrucaria, 444

Wrangelia, 24, 51–2, 67–8, 493
Wrangelia argus, 80, 90–1
Wrangelia penicillata, 80, 388
Wrangelia plumosa, 49–50, 54–8

Xanthophyceae (xanthophytes or yellow-green algae), 444

Zea mays, 182
Zygnema, 444

Subject index

Unless otherwise specified, these entries refer to the Rhodopyta. Chapter subjects are indicated in bold face. Parentheses following crossreferences refer to subentries.

abrasion, 378
acclimation
 definition, 203
 desiccation, 181–2
 light, 203–16, 430
 osmotic, 10, 22–3, 157–63, 165
 temperature, 385
acetyl CoA, 184
acid phosphatase, 10–11, 13, 20
acidity (pH). *See also* carbon dioxide (pH effects).
 hot springs, 151, 156, 175, 193–4, 206
 streams, 175, 423, 431–4, 442, 447–9
actin, 12, 21–2
action spectrum, 207, 214
active transport, 148–9, 154. *See also* bicarbonate
 antiport, 149
 carriers (porters), 148
 ion. *See* individual entries
 primary, 149, 152
 secondary, 149, 152
 symport, 149
adaptation
 carbon supply, 181
 definition, 203
 desiccation, 390
 grazing, 394
 heavy metals, 436
 inorganic carbon, 182, 433
 light, 203, 213
 osmotic, 163–4, 387, 389
 temperature, 385
 water motion, 392, 425–7
adelphoparasite, 63, 66
adhesive vesicle, 14, 32, 34. *See also* spore (adhesion)
ADP glucosyl transferase, 22, 189
aeodan, 235
Africa, 486
agarocolloids
 composition, 238–9
 distribution, 237–41
 family, 236–7
 gels, 237
 nomenclature, 237
agaroses, 227, 236
 agarobiose, 227
 composition, 239
 gelation, 227
 helix formation, 227
 neoagarobiose, 227
agars, 226–7
 content, 246
 gelling. *See* ammonium, nitrate
 in mutants, 110, 112
 synthesis, 236–7
 yield and gel strength. *See* carposporophyte, gametophyte, tetrasporophyte
Alabama, 424
Alaska, 431
alginate, 236, 244
alkali-modification, 237, 246
alkalinity, 432–3
alloparasite, 63, 66
allophycocyanin, 2, 208–11
 mutants in *Gracilaria tikvahiae*, 109, 113
alpha and beta spores, 363, 365
alternation of generations. *See* life history
amino acids
 alanine, 157, 184, 241
 arginine, 184–6, 241
 asparagine, 157
 aspartate, 157, 182, 184–5, 196, 241
 glutamate, 157, 184, 190, 196, 241
 glycine, 179, 186, 241
 histidine, 241
 hydroxyproline, 241–2
 lysine, 185
 in metabolism, 182, 184, 241
 methionine, 184
 nitrogen reserve, 24, 152, 157, 164
 osmotic role, 161–2
 proline, 157, 162, 184
 serine, 179, 185–6
 taurine, 157
 threonine, 184, 242
ammonium
 effects on agar gelling, 245–6
 freshwater, 438
 marine, 393
 in metabolism, 175, 184–6
 methylamine analogue, 151
 transport, 150–1, 393
amylase, 22
Andes, 431
aneuploidy. *See* polyploidy
annual, 404, 445. *See also* life history
annulate lamellae, 13–14
Antarctic, 204–5
apical cell. *See also* cell fusion, DNA, growth, mitosis (orientation)
 division, 47, 261–3
 initials, 307, 312, 340
aplanosporangium, 352
aplanospore, 461
apomixis, 306, 332, 458, 460, 463–5
 apogamy, 87, 458, 463
 apomeiosis, 58, 89, 458, 463–4
 apomictic recycling, 359
apoplastic. *See* intercellular transport
Appalachians, 431
arabinose, 225, 242
arctic, 402, 431
Arizona, 431
Arkansas, 431
asexual reproduction, 383, 397, 445, 460–5, 469. *See also* monospore
ash content, 441–2
Asia, 4, 440, 444
Atlantic, 3–4, 93, 164, 245, 359, 365, 379, 396–7, 400, 404–5, 459
ATP, 149, 172–4, 182, 184, 189–96
ATPase, 243, 433
Australia, 2–3, 76, 311, 315, 323, 340, 353, 356–7, 398–9, 401, 405, 407, 424
autopolyploidy. *See* polyploidy
auxillary cell, 3, 323, 328, 334, 338, 479, 490–2. *See also* gonimoblast formation
 ampullae, 327

Subject index

branch, 312, 324, 334, 342
  definition, 308–10
  fusion. *See* connecting cell, connecting filament
  homologies, 316
  modification, 316, 337
  mother cell, 335–7
  nucleus, 308
  ultrastructure, 13, 31
axial cell, 284–9, 292–4, 296, 298
  quantitative DNA, 49–51, 53

Baffin Island, 431
Bahamas, 205, 401
Baltic, 163–4, 387, 389, 400, 465
band growth, 264, 269, 483
Bay of Fundy, 405
Belize, 431
Bermuda, 81, 91
bicarbonate
  active transport, 175–7, 180, 196
  fixation, 184–5. *See also* carbon (C4 physiology)
  freshwater, 433
  marine, 393
biogeography. *See* geographical distribution
biological significance, 1
biphasic life history. *See* life history
bisporangium, 349–50, 357–9, 462
bivalents. *See* chromosome pairing
boundary layer, 175, 180–1, 196, 425–6, 436, 492, 498
branching
  adventitious, 295–6
  correlation to growth, 264
  determinate, 68, 296–7
  facultative, 294–6
  false, 276
  initiation, 262, 267
  patterned, 296–7
  pseudodichotomous, 277
  taxonomic distribution, 283–6, 296–8
Brazil, 78, 244, 389, 399
breeding systems, 460–6, 468. *See also* apomixis, dioecy, monoecy
  biparental, 460
  compatibility, 90, 465–8
  higher plant, 460–1
  mixed mating, 460
  obligate out-crossing, 460
  uniparental, 460
  vegetative, 458, 460–2
British Columbia, 75, 246, 333, 397, 402, 431, 462–3, 465–6
British Isles (Great Britain), 397, 401, 424, 434, 444
bromine (and brominated compounds), 24, 29, 242, 392, 394

calcareous forms, 1, 29, 174, 248, 378, 392, 394–5
calcium

aragonite deposition, 243–4
calcite deposition, 243–4
calmodulin, 163
maerl, 378
marine/estuarine, 378, 387–9, 433
nucleus association, 138
transport, 160–1, 163
calcofluor staining, 264, 280, 483
California, 309, 313, 331, 334, 357, 359, 386, 390, 397, 399, 409, 402–3, 405–6, 431
caloric (calorific) content, 394, 425, 441–2, 469
Canada, 94, 398, 400
Canary Islands, 402
canopy
  marine macroalgal, 380, 391, 395, 398
  tree, 427–31, 448
undergrowth, 184, 395, 398, 408
carbamyl phosphate synthetase, 185
carbohydrates, 13, 34, 182, 245, 442
carbon
  C3 pathway, 177, 179, 183–4. *See also* carbon dioxide (fixation), ribulose bisphosphate carboxylase-oxygenase
  C3 + C1 carboxylation, 185, 196
  C4 physiology, 177–80, 183–4, 187, 196. *See also* bicarbonate fixation, phosphoenolpyruvate carboxylase
  content, 387, 393
  partitioning, 247
  skeletons, 172–3, 190, 195–6
  storage, 187
carbon dioxide
  affinity, 180, 183
  anaplerotic fixation, 184–5
  compensation concentration, 178
  concentrating mechanisms, 176, 180, 183, 187, 196
  diffusion coefficient, 175
  fixation, 161, 173–4, 178, 181–5. *See also* carbon (C3 pathway), ribulose bisphosphate carboxylase-oxygenase
  influx, 173, 175, 178–81, 183, 185, 196
  pH effects, 176–8, 180, 387, 431–3
carbon metabolism, 171–202
carbonic anhydrase, 176
Caribbean, 399–400, 431
Carnoy's fixative, 45
carotenoids, 1, 206–7, 209
carpogonial branch, 13, 30–1, 312, 316, 319, 321, 324, 328, 333–4, 338, 342, 490, 492. *See also* auxiliary cell
  hypogynous cell, 319–20, 327, 330–2
  modification, 315, 319–20, 324, 328, 330, 337
  sterile group, 316–17, 328, 342
  supporting cell, 309, 316, 320, 328, 333–5, 338
  ultrastructure, 30, 316
carpogonium. *See also* gonimoblast formation, trichogyne
  chromosome numbers, 76
  definition, 306
  development, 309, 325
  fertilized, chromosomes, 76–9, 81–4
  freshwater fertilization, 445–6
  fusions, 319–20, 323, 326–8, 330–3, 338
  modification, 315, 320, 325
  position, 312–14, 316, 341
  quantitative DNA, 60, 68
  ultrastructure, 30–2
carposporangium, 307–8, 319–22, 328, 332–3, 338–9, 366
  chromosome numbers, 76–84
  definition, 365
  ultrastructure, 11, 32, 34
carpospore, 106, 276, 307, 328, 338, 343, 407, 445
  chromosome numbers, 76–84, 87
  in situ germination, 366
  quantitative DNA, 57–8
  ultrastructure, 14, 17–18, 34, 308
carposporogenesis, 365, 492
carposporophyte, 445, 466–8, 491–2. *See also* cystocarp, gonimoblast, life history
  agar yield and gel strength, 246–7
  carbon metabolism, 173, 187, 189
  definition, 306
  development, 307–10, 312–39
  evolution, 306–9, 339–42
  nutrient model, 306–7, 310, 319–43
  ultrastructure, 13, 28, 31–2, 308
carpotetrasporangium (tetrasporoblast)
  definition, 359, 366
  types, 366
carrageenan-agarocolloid hybrids, 240, 248
carrageenans, 226–7. *See also* gametophyte, tetrasporophyte
  alpha, 234
  beta, 228–31, 248
  carrabiose, 227
  content, 245
  delta, 234
  distribution, 228
  families, 228–30
  gamma, 231
  gelation, 227
  helix formation, 227
  iota, 227, 233–4, 245, 248
  kappa, 227–30, 233–45, 245, 463
  lambda, 228–30, 233–5, 248, 463
  mu, 232, 245
  neocarrabiose, 227
  nu, 232
  omega, 231, 233–5
  pi, 233–4
  psi, 228

carrageenans (cont.)
  theta, 228
  xi, 228, 233
caves, 398, 400
cell division, 123–45. See also mitosis
  in development, 261–3
  endogenous rhythm. See photoperiod
  mitotic index. See photoperiod
  rate. See light quantity
cell elongation, 147, 263–4, 380, 483. See also growth
cell fusion, 268–9, 280, 290
  chromosome numbers, 85
  gonimoblast formation, 319–20, 323, 332–3, 337, 339–41
  quantitative DNA, 60–5
  ultrastructure, 9, 13, 26
cell membrane. See plasmalemma
cell motility, 18, 22
cell repair, 268–70. See also regeneration
  compatibility, 269
cell structure, 7–41
cell walls, 221–57, 300, 483–4, 492. See also calcium, cuticle, individual components
  deposition, 10, 231–2, 243, 247, 300, 308, 483–3
  digestion, 314
  gel strength, 244–7
  growth, 148, 276, 280, 285, 298–9
  microfibrils, 223–4
  role in osmoacclimation, 157–63
  ultrastructure, 27–9, 30, 32, 222
cellulose, 28, 222–3, 300
Central America, 2
centrioles, absence of, 22, 126
centromere (kinetochore). See chromosome morphology
chantransia stages, 285–8, 355
  chromosome numbers, 76, 78–9
  ecology, 427, 430, 439–45, 448–9
characteristics of Rhodophyta, 1–2
  in common with Cyanophyta and Prochlorophyta, 347
Chile, 76, 233, 246, 330, 336
China, 244, 246
chloride transport, 149, 154–6, 160–4
chlorophyll a, 1, 173, 205, 207, 209, 216, 430
  mutants in Gracilaria tikvahiae, 109
chlorophyll b, 173, 184, 206, 216
chlorophyll c, 173, 206
chlorophyll d, 207
chloroplast (plastid), 1, 2, 62, 123, 487. See also cytokinesis, DNA, phycobilisome, proplastid
  development, 17
  envelope, 14–15, 17–18, 486
  eyespot, 18
  ion content, 154, 156, 162
  plastoglobuli, 14, 17–18

pyrenoid, 8, 14, 18, 22, 133–4, 276, 487
thylakoid (photosynthetic membrane), 2, 14–17, 173
ultrastructure, 8–9, 12, 14–19, 22–3, 29, 31–2, 34, 134, 139
chromatic adaptation. See acclimation (light)
chromatin, 13–14. See also chromosome morphology
  euchromatin, 126
  heterochromatin, 125
chromomere. See chromosome morphology
chromosome morphology, 89–90, 92. See also sex chromosome
  banding patterns, 89–90, 92
  centromere position (primary constriction, kinetochore), 74–5, 89–90, 92, 124
  chromomere, 89
  eu- and heterochromatic regions, 90–2
  nucleolar organizing region (secondary constriction), 90, 92
  size and shape, 74–5, 87, 89–92, 94
  structural changes, 90, 93–4
  taxonomic aspects, 73, 90–4
  ultrastructure (including kinetochore), 124–7, 128–9
chromosome number. See also polyploidy
  in asexual material, 76, 80–4
  basic, 87–8, 93
  basic genome, 87–8
  discrepancies, 73–4, 81, 93
  in freshwater red algae, 75, 85
  in life history studies, 73, 86–7, 94
  meiotic counts, 76–84
  in rhodophytan species, 74, 76–85, 94
  taxonomic aspects, 73, 90–4
  trends, 74–5, 86, 94
  in various cell types and life history stages, 76–85. See also individual entries
chromosome pairing (synapsis). See also synaptonemal complex
  bivalents, 74–5, 86–8, 91
  homologues, 74–5, 85–6, 91
  meiotic, 74, 87
  polyvalents, 87–8
  somatic (mitotic), 84, 87
  univalents, 86–8
chromosomes, 73–101
cladistics, 372, 478, 483–4, 486, 488–90, 493
classification of Rhodophyta, 2, 4, 486–93. See also systematics, individual taxa
  historical perspective, 478–81
cleavage furrow. See cytokinesis
cluster analysis, 443–4, 484

cobalt, 392
cold sensitive mutants. See mutants and mutagenesis
colonization, 406, 445, 449. See also dispersion, vector-assisted transport
commercial utilization, 3–4
communities, 400–8, 425, 442, 447. See also individual factors
compatibility. See breeding systems, cell repair
compensation point, 204–5, 379
competition, 395–6, 423, 425, 442
  intraspecific, 396
conchocelis stage, 106, 117, 223, 241, 261, 265, 278–9, 307, 352, 378, 391, 394, 484, 488
  chromosome numbers, 76–8
  compensation point, 204
  grazing pressure, 394
conchosporangium
  chromosome numbers, 76–8
  definition, 362–3
  meiosis, 363
conchospore, 106, 307, 384, 489
  meiosis, 106, 363, 368, 489
conjunctor cell, 60, 297, 338–9, 490
connecting cell, 306, 309, 341
  fusion with auxiliary cell, 316–19
connecting filament, 306, 309, 316, 323, 326, 328, 338, 491. See also gonimoblast formation
  fusion with auxilliary cell, 316, 323, 326, 328, 338
  occurrence, 316
continental drift, 447–9, 486. See also geographic distribution
copper, 174–5, 392
corticating filament, 279, 285–6, 291
  quantitative DNA (ploidy), 50, 56–8
Costa Rica, 431
cruciate tetrasporangium. See tetrasporangium
crustose (crust), 18, 63–5, 233, 279, 290–1, 332, 355, 357, 359, 366, 379, 390, 394–6, 398, 400–3, 407–8, 425–6, 444–5, 464, 469, 484, 488–9
Curaçao, 403
currents, transcellular, 268–9
cuticle, 29, 223, 241–3, 310
cyclosis, lack of, 49, 69
cystocarp. See also carposporophyte
  compartmentalization, 309–10, 320, 338, 341–2
  definition, 306
  development, 106, 309–10, 316–39
  functional adaptations, 309–10, 319–43
  morphology, 306, 309–10, 320
  nutritional strategies, 309–10, 319–43
cytochromes, 173–5, 190–1, 193

Subject index

cytokinesis. *See also* tetrasporangium
  cleavage furrow, 138–9, 141–2
  multinucleate cells, 263, 269
  postmeiotic nuclear movement, 139–42
  role of chloroplasts and vacuoles, 125, 138–9, 142
  ultrastructure, 13, 21, 32, 124–5, 138–9
  uninucleate cells, 262–3
cytological paradox, 63
cytoplasmic domain, 47, 52–3
cytoplasmic streaming, 20–2, 69
cytoplasmic volume, 47
cytoskeleton, 12, 20–2
cytrulline, 185
  cytrullinyl arginine, 152, 164, 186
C-value
  definition, 54
  paradox, 52–3

DAPI (4′-6-diamidino-2-phenylindole, DNA fluorochrome), 44–69, 73, 310, 314, 483
daylength. *See* photoperiod
decussate tetrasporangium. *See* tetrasporangium
dehiscence of sporangia, 356
depth, 379. *See also* light quality, light quantity
  light effects, 379–80, 398
  lower limits, 174, 400
desiccation (emersion), 151, 180–2, 386, 390–1, 393, 398, 401, 406
development, 259–73
  sporeling. *See* spore
dictyosome. *See* Golgi body
differentiation, cellular
  hormonal effects, 268–70
  reproductive cells, 265–6, 270
  vegetative cells, 265
diffusion 183, 196
  coefficient, 175
  exchange, 155
  facilitated, 148
  passive, 148
digeneaside, 156–7, 161, 164, 188–9
β-dimethylsulphoniopropionate (DMSP), 157, 161–4
dioecy, 365, 445–6, 449, 465–6. *See also* breeding systems
dispersion, 355, 397, 447–8. *See also* monospore, vector-assisted transport
disturbance
  freshwater, 442–3
  marine, 405–6
DNA
  cytoplasmic ratios, 260, 265, 298
  fluorochromes, 44–5
  microspectrofluorometric studies, 43–71
  nuclear (C level), 46–61, 63–69, 266, 298

organelle, 14–16, 18, 45, 47, 69, 113, 115–17
quantitative. *See* cell fusion, life history, *individual cell and tissue types*
drainage basin, 425, 429, 445, 447–8
  lentic, 448–9
  lotic, 442–3, 449

elastic modulus, 157, 161
electrophoresis, 210, 246, 455, 460–1
elevation, 430–1, 438. *See also* drainage basin
endoplasmic reticulum (ER), 11–13, 15, 19–20, 23–4, 26, 32–3, 130, 132, 138, 141
  perinuclear, 125, 127–8, 137–8, 142
endopolyploidy. *See* polyploidy
endosporangium, 348, 351, 488
endosymbiosis, 156, 183
England, 90
epiphytes
  freshwater, 442, 444, 449
  marine, 242, 378, 394–5, 398–9, 406
erythrose-4-phosphate, 182, 184
estuaries, 157, 161–3, 387, 389. *See also* salinity
eukaryotic, 2, 63, 123, 186, 189
Europe, 4, 397–8, 424, 429, 438, 443–4, 447
euryphotic. *See* light quantity
evolution, 2. *See also* carposporophyte, homologies, karyotype, life history, mitosis, taxonomy, tetrasporangium, tetrasporophyte

Falkland Islands, 397
fat bodies, 24
ferredoxin, 174–5
fertilization, 88, 104, 312–14, 342, 457, –60. *See also* breeding systems
  freshwater, 445–7
  ultrastructure, 13, 29–31
Feulgen staining, 85, 90–1, 260, 364
fibrous lamina, 13
fibrous vacuole associated organelles, 13
filamentous, 8, 51, 260–9, 275–99, 340, 395, 425–7, 442, 448, 469, 488
flagella, absence of, 2, 8, 22, 33, 123, 306, 467, 493
flavodoxin, 174–5
Florida, 245, 247, 326, 397, 402, 431
floridean starch
  carbon storage, 172, 179, 187–8, 196
  cellular localization, 8, 11, 14, 18, 20, 22–4, 32

composition, 2, 187–8
granule structure, 22
synthesis, 189
taxonomic significance, 2
floridoside
  carbon storage, 22, 24, 189–9
  distribution, 156
  osmotic regulation, 161–4
  synthesis, 156–7
  taxonomic significance, 156
floristics, 3, 486. *See also* species diversity
flow. *See* water motion
fluorescence, photosynthesis, 173, 207–12
form-function, 393–6, 425–7, 438–42, 469–70, 490
  K-strategist, 396
  r-strategist, 396
fossil record, 2
France, 84, 401
freshwater, 1, 4, 147–8, 175, 390, 489. *See also* chromosome number, streams and rivers
  ecology, 423–53
furcellaran, 234–5
fusion cell, 316, 319–20, 323, 326, 328, 330, 332–5, 337–8, 342–3
  ultrastructure, 13, 20, 31–2, 342

galactose, 225–6, 236, 238–9, 245, 248
  anhydrogalactose, 226–7
  galactitol (dulcitol), 157, 189
  galactosyl-glycerol, 146, 161
  six-sulfate, 238–41
  two-sulfate, 234–8
  UDP-galactose, 189
gametic union, 314, 365
gametophyte, 187, 279–81, 285–6, 306–7, 341, 354, 357, 359, 485. *See also* life history
  agar yield and gel structure, 246–7
  carrageenan type, 233–4
  chromosome numbers, 76–84
  direct development, 464–5
  genetic mutants, 107
  modification of tissues, 306, 309–10, 341
  polyploid, 108, 118
  quantitative DNA, 54, 56–8, 60, 63
gels, gelation. *See* agarocolloids, agaroses, ammonium, carposporophyte, cell walls, gametophyte, nitrate, tetrasporophyte
genetic buffering, 68–9
genetic regulation
  life history phase determination, 114–15
  reproductive cell differentiation, 266
  sex determination, 105, 113–15
genetics, 103–21. *See also* nonnuclear inheritance, nuclear

genetics *(cont.)*
  inheritance
    Gracilaria tikvahiae, 106–14, 118
    molecular, 115–18
    quantitative, 117–18
    somatic cell, 117, 261
genophore. *See* DNA
geographical distribution. *See also* karyotype (comparative studies), vector-assisted transport, *individual continents and oceans*
  freshwater, 423, 430–3, 447–9
  marine, 4, 357, 359, 368, 386, 396–8, 404, 486
  taxonomic analysis, 486
germ lines, 68–9
Germany, 40–3, 434
germination of spores. *See* carpospore, conchocelis, meiosis, spore
Giemsa staining, 90, 92
glaciation, 447, 449
gland cell, 29, 47, 265, 290
glucose, 22, 225–6, 241
  ADPG (adenosine diphosphate glucose), 189
  in cellulose, 222–3
  glucitol (sorbitol), 157, 179, 188–9
  glucose-1-phosphate, 189
  UDPG (uridine diphosphate glucose), 189
glycolate, 173, 179, 186–7
  excretion, 180, 187
  phosphoglycolate, 173, 186
glycoproteins, 241, 261
glyoxylate pathway, 180, 185
Golgi body (apparatus), 8–12, 20, 22, 26, 29, 32–4, 231–2, 243, 247, 352, 487, 489
gonimoblast, 310, 332, 339, 341, 359
  branching, 317, 320
  chromosome numbers, 76, 87
  definition, 306
  fusions, 317, 332
  initial, 317, 327–8, 331, 333, 338
  quantitative DNA, 57–8
  sterile, 338
  ultrastructure, 10, 308
gonimoblast formation. *See also* cell fusion
  from auxiliary cells, 308–9, 315, 321, 323, 325, 328, 338–9
  from carpogonium directly, 308, 314, 319–20, 332–3, 341
  from connecting filament, 308–9, 315–16, 323–4, 326–8, 341
grazing (herbivory)
  freshwater, 423, 438–42, 449
  marine, 393–4, 396, 399, 406, 468–9
  pressure. *See* conchocelis stage
Great Lakes, 461
Greenland, 397

Grenada, 426, 431
growth, patterns
  apical (tip), 47, 263, 269, 276–7, 279–80, 483, 488
  cellulosympodial, 283
  determinate, 68, 282, 285–6, 288–9, 291, 293–4
  diffuse, 276, 279, 488
  indeterminate, 281–6, 288–92, 294–6
  intercalary, 276–7, 279
  multiaxial, 319
  sympodial, 281–3, 288, 291–2, 296
Guana, 405
gut contents, 438–40, 449. *See also* grazing

hair cell
  hairless mutants, 112
  nutrient absorption role, 150–1, 176, 393, 436
  ultrastructure, 29
halogenated compounds, 20, 24, 29, 392, 482
haploid genetic analysis. *See* nuclear inheritance
Hawaii, 244, 246, 400–1
hematoxylin staining, 43, 73, 86, 91–2, 276, 310–11, 483
heterokaryon (heterokaryotic), 60, 63
heteromorphic life history. *See* life history
heterotrophy, 172, 175, 190, 205–6. *See also* parasite
holdfast, 108, 260, 277
homologies, evolutionary
  auxiliary cell and carpogonial branch, 316
  life histories, 468
  pit connections and plug caps, 490
  reproductive cells, 307–8, 340
  sporangia, 351, 355
  taxonomic, 361, 493
homologous. *See* chromosome pairing
hormones. *See also* cell repair, rhodomorphin
  evidence, 267–8
hybridization, 91, 457–60. *See also* nuclear inheritance, chromosome morphology, chromosome number
hybrid vigor, 18
hypogynous cell. *See* carpogonial branch

Iceland, 389
Idaho, 434
ideogram. *See* karyogram
indeterminant growth, 68
India, 87, 244, 351
induced mutation. *See* mutants and mutagenesis

insect larvae in streams, 438–43, 449
intercellular transport
  apoplastic, 187, 189, 196
  symplastic, 28, 187, 189, 196
  transfer cell, 29
  transfer connection, 28. *See also* pit plug
involucre, 310, 320, 332, 338–9
Ireland, 94, 111, 311, 325, 329, 331, 335–6, 354–5, 400–1
iridescence, 29
iron, 174–5, 392
Isle of Man, 405
isofloridoside
  carbon storage, 188
  distribution, 156
  osmotic regulation, 161, 163–4
isomorphic life history. *See* life history

Jamaica, 401, 431
Japan, 74–5, 85, 93, 117, 247, 352, 359, 398, 406, 424, 434, 466

karyogram 74–5. *See also* karyotype
karyokinesis, 13, 32. *See also* mitosis, nuclear envelope, spindle and spindle poles
  comparative ultrastructural features, 125
  taxa, EM studies, 124
karyotype, 74
  comparative studies, 74–5, 91, 93–4
  evolution of, 93–4
  genetic distinctiveness, 91, 93
  meiotic and mitotic, 74–5
  symmetry, 75
Kentucky, 430, 434, 444
kinetochore (centromere). *See* chromosome morphology
Korea, 309

Labrador, 397
latitude, 379, 385, 387, 396, 403, 405, 430–3, 447–8, 465. *See also* geographical distribution
length, 423–4
life history, 276, 279–80, 285, 291–2. *See also* chromosome number, nuclear inheritance, genetic regulation
  biphasic, 306–7, 340, 467–8, 491
  diversity, 456, 458–9
  evolution, 446–8, 485
  history, 479
  genotype diversification, 468
  haploidy vs diploidy, 470
  heteromorphic, 280, 445, 456, 485, 488, 492
  isomorphic, 65–7, 69, 73, 445, 456, 466, 491
  pleomorphic, 279
  quantitative DNA, 65–7, 69

Subject index

taxonomic, 483–5
theory, 468–70
triphasic, 306–7, 340, 445, 456, 466–7, 488, 492
types, 459, 466–8
wall changes, 233
light quality, 185, 213, 380, 400, 427–9. See also acclimation (light)
light quantity (intensity, irradiance), 173–5, 189, 204–5, 212–14, 379–82, 429–30. See also acclimation, compensation point, photoinhibition, saturating irradiance
branch initiation, 262
cell division, 261–2
darkness effects, 17, 155, 185, 194, 205, 212, 380
euryphotic, 429
growth, 181, 193, 265, 379, 401, 408, 429–30
shade plants, 205, 380
sun plants, 205, 380
lipids, 9, 16, 24, 182, 184, 442. See also fat bodies
lomasome, 10
longevity, 185, 408, 445
Louisiana, 431
lysosome, 13, 20

Maine, 400, 430
malate, 82, 184–5
Malaysia, 244, 424
mangroves, 163, 398, 402
mannose, 225–6
mannan, 223, 226, 242
marine ecology, 377–422
marker gene. See mutants and mutagenesis maternal
maternal inheritance. See nonnuclear inheritance
mats, 395, 448
Mediterranean, 74–5, 93, 359, 398, 400, 459
Mehler reaction, 173, 178–9, 189, 191, 193
meiosis
meiosporangium, 348
site of, 54, 57–60, 69, 87, 106, 363, 445, 449, 464, 479, 487, 489
ultrastructure, 138–42
membrane potential, 148, 152–4
methylated carbohydrates, 223–5, 227, 235, 238–40, 244–5
Mexico, 397, 402, 431
Michaelis-Menten kinetics, 150–1
Michigan, 431
microbody 8, 19–20, 131, 133
microfibrils. See cell walls
microfilament, 21–2, 138
microscopy, epifluorescence, 44–5
microspectrofluorometer, 46
microtubule, 20–2, 124–5, 128, 137

in karyokinesis, 124–5, 128, 137
organizing center (MTOC), 137
microwave fixation, 45
Mississippi, 431
mithramycin staining, 44–5
mitochondrion (mainly ultrastructure), 8–9, 11–12, 19–20, 22, 44, 133, 136, 141
respiration, 179, 185, 190–1
mitosis. See also cytokinesis, karyokinesis, nuclear envelope, photoperiod
orientation, 276–81, 283–9, 291–5
taxonomic and evolutionary considerations, 142–4, 275–301
types of, 124–5, 142–4, 276
ultrastructure, 123–38, 142–4
mitosporangium, 348
mitotic recombination. See nuclear inheritance
molecular genetics. See genetics
monoecy, 104, 115, 445, 449, 465–6. See also breeding systems
monosporangia-like bodies, 355–7
monosporangium, 307–8, 341, 462
curved walls, 351, 481
definition, 351
freshwater distribution, 434, 445
taxonomic distribution, 349–55, 357
ultrastructure, 352–4
monospore, 276, 307, 383, 385, 461–2, 464, 488
chromosome numbers, 76
dispersal, 355
Montana, 431
Morocco, 341
mucilage, 8, 10–11, 15, 22, 28–9, 32, 34, 138, 224–6, 378, 391–2, 396, 425–7, 441, 444, 448
sacs, 13, 34
mutants and mutagenesis. See also phycobilisome, hair cell
chemical induction (EMS, MMS, NNG), 104–5
cold sensitive, 111
frequency, 104–5
genetic buffering effects, 68–9
marker gene, 104
morphological, 104–5, 109–12
phenotypes, 104–5, 108–13
pigments, 104, 108–13
sectorial, 69, 104–6
sex, 105, 110–11, 113–14
spontaneous, 104–5
sterile, 110–12
unstable, 108

NADP
NADPH, 173, 192, 195
reductase, 173–4
nemathecium, 310, 312, 323–4, 333, 342

Nernst equation, 148
Nevada, 431
New Brunswick, 397
New Caledonia, 329
New England, 406
Newfoundland, 357, 406, 431
New Guinea, 431
New Hampshire, 400
New Jersey, 447
New Mexico, 431
New Zealand, 245, 397, 400
nitrate
effects on agar gelling, 246
freshwater, 438
marine, 24, 393
in metabolism, 175
transport, 150–1
nitrogen. See also ammonium, nitrate
limitation, 149, 185
marine, 393
metabolism, 173–5, 181, 185, 191
reserves, 24, 149, 151, 185
transport, 149, 151
nonnuclear inheritance, 110–13
maternal transmission, 112–13
phenotype. See mutants and mutagenesis
North America, 4, 93, 359, 424–6, 430–1, 433, 443–4, 447, 449
North Carolina, 81, 91, 311, 317–18, 322, 327, 336–7, 339, 401, 405, 431
North Sea, 163, 351
Norway, 397, 401
Nova Scotia, 111, 313, 352, 401
nuclear division. See cell division, karyokinesis, mitosis
nuclear envelope, 12–13, 15. See also nucleus associated organelle
chromosome relationship, 125
in mitosis and meiosis, 124–5, 127–32, 134–5, 137–8, 141–2
polar gaps and fenestrations, 124, 129–30, 133–7, 142
pores, 13, 131, 133, 136
nuclear fusion, 60–1, 89, 314–18
nuclear inheritance, 106–8, 110–12. See also mutants and mutagenesis, zygote amplification
complementation, 68–9, 107
crossing techniques, 106
dominance, 107
epistasis, 107
haploid genetic analysis, 106
independent assortment, 107
life history considerations, 106
linkage, 107–8
mitotic recombination, 108, 113–14
multiple alleles, 107
parental effects, 108
recombination, 107
segregation, 107

nuclear inheritance *(cont.)*
  sorting-out, 104–12
  tetrad analysis, 106, 114
nuclear magnetic resonance (NMR), 162, 165, 223, 232
nuclear pores. *See* nuclear envelope
nuclear transfer, 47–9, 63, 89, 317, 319
nuclear volume. *See also* DNA
  ratio to cytoplasmic volume, 298, 301
nucleolar organizing region. *See* chromosome morphology
nucleolus, 12–13, 15, 53, 85, 124, 126–7, 140–1, 308
nucleus associated organelle (NAO), 124, 126–35, 142. *See also* nuclear envelope, spindles and spindle poles
  polar ring, 126, 130–1, 142
  rough ER association, 138
  types, 124–5
nucleus, cell number
  multi-, 8, 13, 47–53, 55, 62–3, 67–9, 89, 297–9
  uni-, 8, 46–5, 53, 55–60, 67–9, 85, 89
nucleus, morphology and localization, 8–9, 12–13, 15, 19, 21, 30–2, 47, 50–3, 60–3, 69, 298–9, 309–10, 312, 314–20, 324–6, 330–3, 338, 362. *See also* karyokinesis
nutrients, 391–3, 395, 423, 435–6, 438, 449. *See also* cystocarp, *individual nutrients*
  absorption role. *See* hair cell

ordination, 484
osmoregulation. *See* acclimation (osmotic)
osmotic adjustment, 159, 162, 164–5. *See also* acclimation, salinity
  organic solutes, 148, 156, 161–2
  plasmolysis, 158
  volume regulation, 149–60, 164
osmotic pressure, 147–8, 156–9
osmotic techniques. *See also* nuclear magnetic resonance
  electron-probe x-ray microanalysis, 154–5
  flux analysis, 154, 156
  microelectrode, 153
  radioisotope, 154, 164
ostiole, 310, 312, 320, 328, 332, 338
oxygen supply, 159, 178–9, 187, 192–5, 433, 435

Pacific, 74, 90, 93, 352, 357, 359, 367, 397, 400–1, 485, 489
Papua, 431
parasite, 1, 359, 379, 482. *See also* adelphoparasite, alloparasite
  carbon metabolism, 172, 188–9

ultrastructure, 10, 13, 15, 29, 31
parasporangium, 348–50, 355, 368, 371–2
  chromosome numbers, 76, 80
paraspore, 462
parenchymatous, 8, 275, 365
perennial, 19, 186, 403, 405–6, 445
pericarp, 306, 316–17, 321, 330–3, 335
pericentral cell, quantitative DNA, 52, 67–8
perinuclear endoplasmic reticulum. *See* endoplasmic reticulum
peroxidase, 20
pH in the environment. *See* acidity
phenolics, 16, 18, 24, 184
phenotype, nuclear and nonnuclear. *See* mutants and mutagenesis
Philippines, 118
phosphate (orthophosphate)
  freshwater, 425–6, 435, 438
  marine, 393
  transport, 152
phosphoenolpyruvate (PEP)
  carboxykinase, 184–5
  carboxylase, 184–5
3-phosphoglycerate (PGA), 173, 180–1, 184–6
phosphorylase, 22
photoadaptation. *See* acclimation (light)
photoinhibition, 195, 212–13, 216, 380–2, 429
  Q_B-protein in, 213
photolithotrophy, 172–3, 188–9, 191, 196. *See also* light quality, light quantity, photosynthetic mechanisms/properties
photoperiod (daylength), 379–85, 430, 486
  branch initiation, 262
  carbon metabolism, 187
  cell division, 261–2
  growth, 264
  mitotic index, 404
  reproduction, 261, 270
photorespiration, 172–3, 183. *See also* ribulose bisphosphate carboxylase-oxygenase
photorespiratory carbon oxidation cycle (PCOC), 173, 179, 185–7, 190, 196
photosynthetic carbon reduction cycle (PCRC), 173, 181–2, 184, 190
photosynthetic mechanisms/ properties
  action spectrum 207, 214
  carotenoid participation, 207
  chlorophyll-binding complexes, 208
  energy transfer, 210–14, 216–17
  oxygen inhibition, 178–9
  pathway, 173

photophosphorylation, 192–4, 196
  rates of O_2 evolution, 204–5
  state transitions, 211, 216
  thylakoid reactions, 173
  unit size, 206
photosystems (PSI and PSII), 174, 182, 206–8
  P680/P700 determination, 206
  transfer to, 211–12, 217
phototaxis, 18, 22
phycobiliprotein (phycobilin), 15, 24, 173–4, 183, 186, 493
  absorption, 208–10
  chromophores, 208–10, 215
  energy "cost", 174
  fluorescence, 209
  genes coding for, 215–16
  mutants in *Gracilaria*, 108–9
  as nitrogen reserves, 152
  polypeptides, 208, 210
phycobilisome
  composition, 210
  energy transfer to PSII, 211–12, 216
  isolation, 211
  mutants in *Gracilaria*, 108–9, 113
  regulation of, 214, 216
  thylakoid-linker, 210–11
  terminal pigments, 211
  ultrastructure, 14–15, 17
phycocolloids. *See* seasonality
phycocyanin, 1–2, 185
  mutants in *Gracilaria*, 108–9, 113
  mutants in *Porphyra*, 112
  types, 208–11
phycoerythrin, 1–2, 380, 430, 485, 489
  mutants in *Callithamnion, Chondrus, Palmaria* and *Porphyra*, 112
  mutants in *Gracilaria*, 108–9, 113
  types, 208–11
phylogeny, 2, 116, 309, 353, 372, 467, 470, 487, 489–93
  monophyletic, 487, 491–90, 492
  paraphyletic, 487, 491–2
  polyphyletic, 487
  using ribosomal RNA, 116–17
phytoferritin, 19
pigmentation and photoacclimation, 203–19
pit connection, 189, 279, 310, 327–8, 332–3, 342, 352, 488
  breakdown, 327, 331
  modification, 310, 316, 319–20, 331–2, 335, 342
  primary, 25–6, 28, 32, 60–1
  secondary, 25–6, 28, 47, 51, 60–1, 64–6, 69, 269–70, 291, 297–8, 338–9, 342, 490
pit plug, 10, 26–8, 31, 49, 60–1, 241–3, 276, 279–80, 480–1, 488–92
placenta, 338–9, 342–3

Subject index

plant breeding, 117–18
　selection of *Porphyra* cultivars, 117–18
　strain improvement, 118
plasmalemma (cell membrane, plasma membrane), 9–10, 12–13, 19, 26, 27
　attached tubules, 10
　transport, 148–9, 154–5, 158, 163–4
plasmalemmavilli, 10
plastid. *See* chloroplast
plastocyanin, 173–5
plastoglobuli. *See* chloroplast
pleomorphic life history. *See* life history
ploidy, 53–4, 58, 68
polar gap. *See* nuclear envelope
polar ring. *See* nucleus associated organelle
Poland, 434
polarity
　blade fragments, 267, 270
　intercellular communication, 266–7, 270
　single cells and filaments, 266, 270
polycomplex. *See* synaptonemal complex
polygenomy, 46, 63, 67–9
polyphosphate granules, 44, 152–3, 157
polyploidy. *See also* DNA
　aneuploidy, 75, 85, 89, 93–4
　autopolyploidy, 88–9
　cell size, 47, 67
　chromosome numbers, 80, 83, 87–9, 94, 483
　endopolyploidy, 47, 67–8, 89, 94
　origin of, 88, 94, 118
　polytene, 46–7, 89
　role of hybridization, 87
polysporangium, 60, 89, 348–50, 357, 366–8, 372
　definition, 367
　ultrastructure, 367–8
polytene. *See* polyploidy
polyvalents. *See* chromosome pairing
porphyran, 226, 239
postfertilization development, ultrastructure, 28, 31–2
potassium transport, 149, 154–5, 159–64
principal components analysis, 484
procarp, 309, 316, 327–8, 332, 338–9
　noncarpial, 309, 323, 327, 338
productivity, primary. *See also* individual factors
　freshwater, 424–5, 431
　marine, 400–3
prokaryotic, 2, 63, 347
prolamellar body, 16–17
proplastid, 17–18, 30–1. *See also* chloroplast (development)

prostrate system, 277–8, 280, 290, 445. *See also* life history
protein bodies and crystals, 9, 13, 19, 23–4, 29
proton transport, 149, 151, 193–4
protoplast. *See* genetics (somatic cell)
pseudochantransia. *See* chantransia stage
pseudoparenchymatous, 8, 260
pseudosporangium. *See* propagule
Puerto Rico, 387, 402
purines, 104
pyrenoid. *See* chloroplast
pyrimidines, 184–5
pyruvate, 184–6, 240

quantitative genetics. *See* genetics
Quebec, 424, 431

recruitment, 407–8
Red Sea, 398
reduction division. *See* meiosis
regeneration, 266–7, 390, 406, 408. *See also* sporangium (regeneration)
relative fluorescence value (rfu), 52, 54–60, 67–8
reproductive strategies, 455–76
respiration. *See also* mitochondrion
　anaerobic, 194
　dark, 179, 184, 189–94, 196, 387
　glycolysis, 185, 189–90
　oxidative phosphorylation, 190–2
　pathway, 173
　tricarboxylic acid cycle (TCAC), 185, 190–1
restriction analysis and map. *See* genetics (molecular)
rhizoid, 58–9, 69, 261, 266–70, 279, 282, 285, 290–1, 378
　cell differentiation, polarity, 58, 68–9, 266–7
　ultrastructure, 13, 29
Rhode Island, 244, 424–5, 428–9, 430–1, 434–6, 445–7
rhodomorphin, 241, 267–70
ribosomes, 15
　subunit sequencing, 116–17, 215
ribulose bisphosphate carboxylase-oxygenase (RUBISCO), 18, 115, 173, 176, 179–80, 182–7, 196, 214. *See also* carbon (C3 pathway), carbon dioxide (fixation), DNA (organelles)
RNA, 116–17, 138, 215, 487, 493
Rockies, 431

Saipan, 244
salinity, 157–64, 387–90. *See also* acclimation (osmotic), adaptation (osmotic)
　downshock, 157, 161
　euryhaline, 387
　specific conductance, 433

stenohaline, 387
　upshock, 157
saturating irradiance, 193, 205, 379, 381–2, 430
Scotland, 392, 397
seasonality, 149, 152, 185, 353, 359, 403–6, 408, 424, 427, 430–1, 436, 445–6. *See also* temperature
phycolloids, 244–6
secretory cell, ultrastructure, 29
sectorial mutant. *See* mutants and mutagenesis
seirosporangium, 349–50, 355–6, 462
　definition, 355
senescence, 19
sex chromosome, 90–1
sex determination. *See* genetic regulation, mutants and mutagenesis
sexual reproduction and cystocarpic development, 305–45
sheetlike thallus, 260, 279, 401, 469
sodium transport, 149, 152–4, 160, 164
solute accumulation and osmotic adjustment, 147–70
somatic cell genetics. *See* genetics
somatic pairing. *See* chromosome pairing
South Africa, 244, 311, 322, 324–5, 327, 334, 358, 405–6, 460
South America, 2, 447
species associations, 442–4
species concepts, 2–3, 457, 460, 465
species diversity (numbers), 400, 403, 429, 445–7, 486
　Chlorophyta, 3–4
　Phaeophyta, 3–4
　Rhodophyta, 2–3, 103
species introductions, 397, 406
specific conductance, 433, 437
spermatangium
　initiation and position, 310–12, 363
　quantitative DNA, 57–8, 61
　spermatangial mother cell, 310
　spermatogenesis, 310, 312
spermatium, 266, 306, 309–14, 316, 319, 325, 340, 445–7
　chromosome numbers, 76–84, 86–7
　quantitative DNA, 57, 68–9
　ultrastructure, 10, 13, 20, 28–9, 30–3, 310
spindles and spindle poles, 124–9, 133–9, 142–3. *See also* microbody
　interzonal spindle, 124, 128–9, 138–9, 142–3
　spindle pole body, 13
　types, 137, 142–3
spontaneous mutation. *See* mutants

spontaneous mutation *(cont.)*
  and mutagenesis
sporangia and spores, 347–76
sporangium
  definition, 348
  distinguishing features, 348
  mitosis and meiosis in, 348
  regeneration, 368–9
  types and distribution of, 348–50
  vesicles, 348, 352
spore. *See also individual types*
  adhesion, 32, 378
  amoeboid movement, 22, 368
  definition, 348
  development, 369–71
  ecology of, 378, 380, 385, 390, 403, 407
  germination, 87, 261, 276, 279–80, 290, 352, 366, 369–71, 407, 483, 492
  mortality, 407
  numbers released, 407
  sinking, 378, 407
  ultrastructure, 10, 20, 23, 28, 32–4, 372
sporeling coalescence, 63–5, 69
sporozygote, 307–8, 365
squashing of tissue, 46
staining techniques
  nuclei and chromosomes, 44–6, 73, 86, 90–2. *See also* DAPI, Feulgen, Giemsa staining, hematoxylin staining
  walls. *See* calcofluor staining
stalk cell, 368–9, 468
starch, 2, 22. *See also* floridean starch
streams and rivers. *See also* water motion
  freshwater, 423–449
  tidal, 391
stress, mechanical, 391–2, 425–7, 448. *See also* water motion
substratum
  cobbles, 378, 469
  freshwater, 423, 442
  marine, 377–9, 395
  microtopography, 378
  sand, 378
succession
  freshwater, 442–3, 449
  marine, 406–7, 469
sulfate, 10, 225, 234, 238–9, 392
  active exchange, 247
  esterification, 247
  seasonal changes, 244
  sulfotransferase, 231
  turnover, 247
supporting cell. *See* carpogonial branch
Surtsey Volcanic Island, 397
Sweden, 331, 357, 368, 402, 424, 430, 434, 436
symplastic. *See* intercellular transport
synapomorphies, 487–7, 490, 492–3

synapsis. *See* chromosome pairing
synaptonemal complex, 13, 138, 141
  polycomplex, 13, 141
systematics, 11–12, 15, 18, 24, 26, 482–6. *See also individual methods and taxa*

Taiwan, 244, 246, 321
taxonomy and evolution, 472–98
temperature. *See also* acclimation, adaptation
  cold sensitive mutant. *See* mutants and mutagenesis
  effects on branching, 386
  effects on growth, 384–6
  elevation distribution, 430–1
  freezing, 386
  latitude, 386, 401, 430–1
  photosynthetic rate, 385
  seasonality, 387, 430–1
  thermophilic, 156
temporal change, 403–7, 430–1, 434–6, 445. *See also* photoperiod, seasonality
  diel, 403, 430
terpenoids, 184, 482
tetrad analysis. *See* nuclear inheritance
tetrahedral tetrasporangium. *See* tetrasporangium
tetrasporangium, 106, 307–8, 359–62, 434, 445, 459, 487, 491. *See also* stalk cell
  chains, 361
  chromosome numbers, 76, 79–85
  cruciate, 349–50, 355, 359–62, 371–2
  decussate, 360
  definition, 359
  evolution, 362
  irregular divisions, 360, 372
  ordinal distribution, 371–2
  quantitative DNA, 57–9
  simultaneous cleavage, 360, 362, 371
  successive cleavage, 360, 362
  tetrahedral, 349–50, 362, 371
  ultrastructure, 11, 13, 25, 31–3, 138, 140–1, 363
  zonate, 349–50, 362–3, 371–2
tetraspore, 105, 307, 383–4, 407, 445
  quantitative DNA, 52, 54, 56–8
tetrasporophyte, 281, 355, 359, 459, 485. *See also* life history
  agar yield and gel strength, 246–7
  carrageenan type, 233
  chromosome numbers, 76, 78–84
  evolution, 306–8, 332
  genetic mutants, 107
  haploid, 114–15, 365
  polyploid, 118
  quantitative DNA, 52, 54, 56–8
thylakoid. *See* chloroplast, photosynthetic mechanisms/properties, phycobilisome

tidal distribution, 398–400. *See also* depth
  intertidal, 398
  subtidal, 174, 398–400
tidal pools, 195, 390, 398, 404
tissue and cell culture. *See* genetics (somatic cell)
tonoplast. *See* vacuole
trace metals, 392, 436, 438
transfer cell. *See* intercellular transport
transfer connection. *See* intercellular transport, pit plug
transport. *See* intercellular transport, plasmalemma, *individual elements*
trehalose, 157, 188–9
trichoblast, 24, 29, 296, 316
trichogyne, 29, 30–2, 306, 309, 312–16, 325, 338, 446–7. *See also* carpogonium
triphasic life history. *See* life history
tropical. *See also* geographical distribution
  freshwater, 424, 430–1, 443
  marine, 396–7, 400, 404, 489
tuft, 426–7, 442
turf, 378, 391, 394–5, 398, 401
turgor
  pressure, 147–8, 156–7, 162, 164
  regulation, 148, 157–64

unattached forms, 378, 397
unicells, 138, 221, 275–7, 487
United States, 398, 444
univalents. *See* chromosome pairing
unstable mutants. *See* mutations and mutagenesis
upright axes, 281, 290–1, 445. *See also* life history
uronic acids, 224–5, 236, 241
  glucuronic acid, 225

vacuole, 45, 288, 308, 319–20, 326, 342
  cell transport, 151–2, 155, 162–3, 188
  osmotic role, 20
  tonoplast, 153, 164
  ultrastructure, 20, 153, 164
van't Hoff relationship, 147, 158
vector-assisted transport, 398, 447–9
vegetative growth and organization, 275–304
vegetative propagation, 461–2, 464, 470
Venezuela, 77–8
vesicle. *See also* sporangium
  spermatial, 13, 21–2, 32–3
  striated, 11, 33–4
  striped, 33–4
vesicle cell, 15, 24, 29
Virgin Islands, 321

Wales, 335

Washington, 397, 401, 405, 431
water motion (flow, currents)
    boundary layer effects, 175, 425
    drag, 392, 426–7
    lotic, 175, 425–8, 445–9
    stress and extension, 426–6
    wave action, 391–2, 395, 398–9
water potential, 147–8, 157, 160

West Africa, 399
Wyoming, 431

xylose, 225, 236, 241–2, 248
    xylan, 223–4, 280

zinc, 392, 438

zonate tetrasporangium. *See* tetrasporangium
zygote amplification, 2, 106, 307–8, 341, 365, 445, 456, 466–8, 470
zygotosporangium, 348, 351, 363–5
    definition, 363
    occurrence, 351
    origin, 348

For EU product safety concerns, contact us at Calle de José Abascal, 56–1°,
28003 Madrid, Spain or eugpsr@cambridge.org.

www.ingramcontent.com/pod-product-compliance
Lightning Source LLC
LaVergne TN
LVHW081523060526
838200LV00044B/1978